## Physical Constants

| | |
|---|---|
| Standard acceleration of terrestrial gravity | $g = 9.80665$ m s$^{-2}$ (exactly) |
| Avogadro's number | $N_0 = 6.022137 \times 10^{23}$ |
| Bohr radius | $a_0 = 0.52917725$ Å $= 5.2917725 \times 10^{-11}$ m |
| Boltzmann's constant | $k_B = 1.38066 \times 10^{-23}$ J K$^{-1}$ |
| Electron charge | $e = 1.6021773 \times 10^{-19}$ C |
| Faraday constant | $\mathscr{F} = 96{,}485.31$ C mol$^{-1}$ |
| Masses of fundamental particles: | |
|    Electron | $m_e = 9.109390 \times 10^{-31}$ kg |
|    Proton | $m_p = 1.672623 \times 10^{-27}$ kg |
|    Neutron | $m_n = 1.674929 \times 10^{-27}$ kg |
| Ratio of proton mass to electron mass | $m_p/m_e = 1836.15270$ |
| Permittivity of vacuum | $\epsilon_0 = 8.8541878 \times 10^{-12}$ C$^2$ J$^{-1}$ m$^{-1}$ |
| Planck's constant | $h = 6.626076 \times 10^{-34}$ J s |
| Speed of light in a vacuum | $c = 2.99792458 \times 10^8$ m s$^{-1}$ (exactly) |
| Universal gas constant | $R = 8.31451$ J mol$^{-1}$ K$^{-1}$ |
| | $= 0.0820578$ L atm mol$^{-1}$ K$^{-1}$ |

Values are taken from "Quantities, Units and Symbols in Physical Chemistry," International Union of Pure and Applied Chemistry, Blackwell Scientific Publications, 1988.

## Conversion Factors

| | |
|---|---|
| Standard atmosphere | $1$ atm $= 1.01325 \times 10^5$ Pa $= 1.01325 \times 10^5$ kg m$^{-1}$ s$^{-2}$ (exactly) |
| Atomic mass unit | $1$ u $= 1.660540 \times 10^{-27}$ kg |
| | $1$ u $= 1.492419 \times 10^{-10}$ J $= 931.4943$ MeV (energy equivalent from $E = mc^2$) |
| Calorie | $1$ cal $= 4.184$ J (exactly) |
| Electron volt | $1$ eV $= 1.6021773 \times 10^{-19}$ J $= 96.48531$ kJ mol$^{-1}$ |
| Foot | $1$ ft $= 12$ in $= 0.3048$ m (exactly) |
| Gallon (U.S.) | $1$ gallon $= 4$ quarts $= 3.78541$ L (exactly) |
| Liter-atmosphere | $1$ L atm $= 101.325$ J (exactly) |
| Metric ton | $1$ metric ton $= 1000$ kg (exactly) |
| Pound | $1$ lb $= 16$ oz $= 0.45359237$ kg (exactly) |

# PRINCIPLES OF
# MODERN CHEMISTRY

# DAVID W. OXTOBY
The University of Chicago

# NORMAN H. NACHTRIEB†
The University of Chicago
University of Illinois at Chicago

# THIRD EDITION

*Saunders Golden Sunburst Series*
**SAUNDERS COLLEGE PUBLISHING**
Harcourt Brace College Publishers

Fort Worth   Philadelphia   San Diego   New York   Orlando   Austin
San Antonio   Toronto   Montreal   London   Sydney   Tokyo

# PRINCIPLES OF
# MODERN CHEMISTRY

Requests for permission to make copies of any part of the work should be mailed to: Permissions Department, Harcourt Brace & Company, 6277 Sea Harbor Drive, Orlando, Florida 32887-6777.

Text Typeface: Times Roman
Compositor: York Graphic Services
Publisher: John Vondeling
Developmental Editor: Elizabeth Rosato
Picture Developmental Editor: George Semple
Managing Editor: Carol Field
Project Editors: Margaret Mary Anderson and Bonnie Boehme
Copy Editor: Mary Patton
Manager of Art and Design: Carol Bleistine
Art Directors: Anne Muldrow and Caroline McGowan
Art Assistants: Sue Kinney and Kathleen Flanagan
Text Designer: Kathryn Needle
Cover Designer: Kathryn Needle
Text Artwork: Rolin Graphics
Layout Artist: Dorothy Chattin
Director of EDP: Tim Frelick
Production Manager: Joanne M. Cassetti
Marketing Manager: Marjorie Waldron

Cover Credit: Leonardo Ferrante/Tony Stone Images

Credit for Frontmatter and Backmatter Photos and for All Non-credited Photos Appearing in the Text: C. D. Winters

Title page: The structure of diamond consists of a network of interlocking tetrahedral carbon atoms.

Printed in the United States of America

PRINCIPLES OF MODERN CHEMISTRY
THIRD EDITION

ISBN 0-03-005904-6

Library of Congress Catalog Card Number: 95-067824

 678901234   032   10   9876543

*The reaction of sodium with water.*

# PREFACE

The organization of the third edition of *Principles of Modern Chemistry* is fundamentally the same as that of the first two editions, although a number of changes and additions have been made to improve the text. Chemistry, as an experimental science, is introduced to students in terms of macroscopic concepts and principles that have their origins in the laboratory before the more abstract, theoretical, interpretative aspects of the subject are presented. Of course, the long-range goal is to persuade students that both theory and experiment are indispensable to the progress of science and to help them appreciate that each of these approaches informs and guides the other.

## CHANGES IN THIS EDITION

Several changes in organization have been made in this edition, and several chapters have been extensively rewritten:

A brief introduction to the constituents of the atoms (electrons, protons, neutrons) and their masses and charges has been moved to Chapter 1, so that the subsequent discussion of Lewis structures and the periodic table in Chapter 2 has a more solid foundation. This material also connects closely to the general theme of Chapter 1: the analysis of what makes up matter.

Intermolecular forces are introduced earlier in this edition, first in the context of nonideal gases (Chapter 3) and then in connection with liquids and solids (Chapter 4). This allows a more modern, molecular-level approach to a number of topics in general chemistry.

Chapters 13 through 15 have been extensively revised in response to comments from reviewers. Chapter 13 now includes many-electron as well as single-electron atoms, and Chapter 14 is a comprehensive treatment of chemical bonding. Chapter 15 contains an expanded treatment of modern spectroscopy and two important areas of application: photosynthesis and atmospheric photochemistry.

End-of-chapter problems have been revised, with a new category introduced: Cumulative Problems. These problems relate the new material in the chapter closely to that in earlier chapters, and help to show interconnections between different aspects of chemistry. Chapters 18 through 23 do not include such problems, because most of the problems in these chapters are cumulative in their very nature, incorporating applications of chemical principles to the study of chemical processes.

Many small changes have been made to clarify the presentations and to make the figures more useful. Up-to-date topics such as the fullerenes have been introduced.

## TEACHING OPTIONS

As in the previous editions, an instructor may choose to alter the order in which the chapters are presented for a chemistry course; the text is structured to permit this kind of flexibility. For example, some instructors may wish to cover the material on atomic structure and chemical bonding earlier in their courses. In this case, they can move directly to Chapters 13 through 15 after Chapter 7, or even after Chapter 2. Certain topics may be omitted without loss of continuity. For example, a principles-oriented course might cover the first 17 chapters thoroughly and then select only one or two of the last chapters for close attention. A course with a more descriptive orientation might omit the sections entitled "A Deeper Look," which are more advanced mathematically than the main part of the book, and cover the last six chapters more systematically. Further suggestions for ways of using this book in courses with different plans of organization are given in the Instructor's Manual.

## FEATURES

### Mathematical Level

This book presupposes a solid high school background in algebra and coordinate geometry. We introduce the concepts of slope and area in the physical and chemical contexts in which they arise, and we use differential and integral notation only where necessary. The book is designed to be fully self-contained in its use of mathematical methods. In this context, Appendix C should prove particularly useful to the student and the instructor.

Key equations in the text are indicated by boldface numbers on the right, and practice should be gained in using them for chemical calculations. Other equations, such as intermediate steps in mathematical derivations, are less central to the overall line of reasoning in the book.

## Worked Examples

This textbook includes worked examples, which help to show the methods of reasoning applied in solving chemical problems. The examples are inserted immediately after the corresponding principles have been presented, and references are made to related problems at the end of the chapter. Careful attention is given in the examples to the proper use of units, significant figures, and scientific notation.

## A Deeper Look

Sections titled "A Deeper Look" provide students with the physical origins of observed chemical behavior. They are sometimes more advanced mathematically than the main parts of the book. Their use allows the instructor considerable flexibility in the level of course material taught.

## Key Terms

Key terms appear in **boldface** where they are first introduced. In addition, definitions for most key terms are included in the Index/Glossary for ready reference.

## Concepts and Skills

Each chapter concludes with a list of concepts and skills for review by the student. Included in this list are references back to the section in which the topic was covered, and forward to problems that help to test mastery of the particular skill involved. This feature is helpful for self-testing and review of material.

## Practice Exercises

At the end of each of Chapters 1 through 17 there is a practice exercise that is built around a problem of chemical interest and draws on material from the entire chapter for its solution. Working through these exercises provides a useful review of material from the chapter, helps students to put principles into practice, and prepares them to solve the problems that follow.

## Problems

Problems are grouped into three categories. Answers to odd-numbered "Paired Problems" are collected in Appendix G and enable the student to check the answer to the first problem in a pair before undertaking the second problem, which is parallel to it. The "Additional Problems," which are unpaired, provide further applications of the principles developed in the chapter. The "Cumulative Problems" integrate material from the chapter with topics presented earlier in the book. The more challenging problems are indicated with an asterisk.

## Appendices

Appendices A, B, and C are important pedagogically. Appendix A discusses experimental error and scientific notation, while Appendix B introduces the SI system of units used throughout the book and describes the methods used for converting units. Appendix C provides a review of mathematics for general chemistry. Appendices D, E, and F are compilations of thermodynamic, electrochemical, and physical data, respectively. Students gain experience in using them through many of the problems in the book.

## Index/Glossary

The index/glossary at the back of the book gives a brief definition of key terms and a reference to the pages on which that term appears.

## SUPPLEMENTS

***Study and Problem-Solving Guide***   The best way to master physical and chemical concepts is to solve problems; even students with outstanding mathematical skills can encounter difficulties with chemistry problems. Trouble usually arises in passing from the words of the problem through formal statements of concepts to manipulating mathematical equations. The *Study and Problem-Solving Guide* by Wade A. Freeman of the University of Illinois at Chicago helps students meet such difficulties head-on. It summarizes definitions, concepts, and equations, gives some additional insights into the material presented in the text, provides problem-solving hints, and, as a practical illustration, presents detailed solutions to all the odd-numbered problems in this book. Each student is strongly encouraged to purchase a copy of this book.

***CalTech Chemistry Animation Project (CAP)***   This consists of a set of six video units that cover, with unmatched quality and clarity, the chemical topics of atomic orbitals, valence shell electron pair repulsion theory, crystals and unit cells, molecular orbitals in diatomic molecules, periodic trends, and hybridization and resonance.

***Instructor's Manual***   The Instructor's Manual contains solutions to even-numbered problems as well as suggestions for ways to use this book in courses with different plans of organization.

***Overhead Transparencies***   A set of at least 110 full-color figures and tables from the text is available.

## ACKNOWLEDGMENTS

In preparing this third edition, I have benefited greatly from the comments of students who used the first two editions over the past several years. I would also like to acknowledge the many helpful suggestions of colleagues in Chicago and at other colleges and universities who have taught from this book.

    I would like to extend special thanks to the following reviewers of the third edition:

Curtis B. Anderson, University of California, Santa Barbara
Joseph J. BelBruno, Dartmouth College
Stuart Belli, Vassar College
Ernest R. Davidson, Indiana University
James E. Davis, Harvard University
Robert V. Dilts, Vanderbilt University
Bruce Hudson, University of Oregon
John Lamb, Brigham Young University
Michael Moran, West Chester University
David J. Morrissey, Michigan State University
Keith Peterson, Wesleyan College
Kenneth Poeppelmeier, Northwestern University
Russel J. Wiley, University of Northern Iowa
John Winn, Dartmouth College

Reviewers of the first two editions, whose comments helped to shape the book, include:

Donald Berry, University of Pennsylvania
Francis Bonner, SUNY at Stony Brook
Lewis Brubacher, University of Waterloo
John DeKorte, Northern Arizona University
Thomas Dunn, University of Michigan at Ann Arbor
Leonard Fine, Columbia University
Wade Freeman, University of Illinois at Chicago
Robert Gavin, Macalester College
Graham Glass, Rice University
Alton Hassell, Baylor University
Anthony Haymet, University of California at Berkeley
Michael Heinekey, Yale University
Donald Jicha, University of North Carolina at Chapel Hill
Joseph Kushik, Amherst College
William Masterton, University of Connecticut
Leonard Nash, Harvard University
Malcolm Nicol, UCLA
Ronald Ragsdale, University of Utah
Robert Reeves, Rensselaer Polytechnic Institute
Charles Rowell, United States Naval Academy
Richard Schmitt, Texas A & M University
Maurice Schwartz, Notre Dame University
Robert Sharp, University of Michigan
Conrad Stanitski, Randolph Macon College
Fred Tabbutt, Evergreen State College
J. C. Thompson, University of Toronto
Thomas Weaver, Northwestern University
Donald Wink, New York University
Diane Wolff, University of California at San Diego

The staff at Saunders College Publishing has been extremely helpful in preparing this third edition. In particular I would like to single out the Developmental

Editor, Beth Rosato, who helped with many suggestions for revisions, and the Project Editors, Margaret Mary Anderson and Bonnie Boehme, who helped to keep the revision on an efficient schedule. The continuing support and encouragement of Publisher John Vondeling is also gratefully acknowledged.

Several new photos by C. D. Winters appear in this edition. Fay Yates assisted with printing, copying, and mailing. I am particularly grateful to my wife and children for their encouragement in the revision of this book.

Sadly, my friend and colleague Norman Nachtrieb died in September, 1991, and thus was not able to take part in the preparation of this edition. I have greatly missed his sound advice, and his deep knowledge of and love for chemistry. I trust that he would have been proud of this new edition of our book.

*David W. Oxtoby*

*The University of Chicago*
June 1995

*The search for truth is in one way hard and in another easy, for it is evident that no one can master it fully or miss it completely. But each adds a little to our knowledge of nature, and from all the facts assembled there arises a certain grandeur.*

*(Greek inscription, taken from Aristotle, on the facade of the National Academy of Sciences building in Washington, D.C.)*

*Azurite.*

# CONTENTS OVERVIEW

# CONTENTS

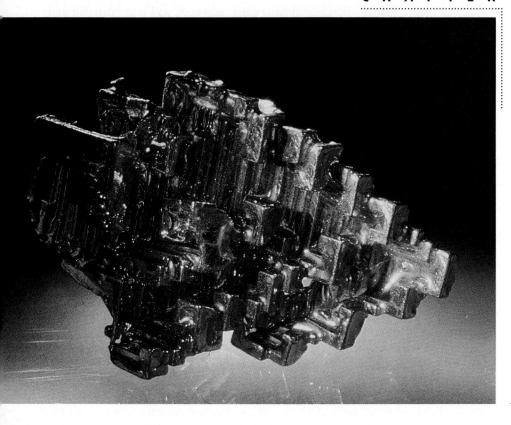

*A crystal of elemental bismuth.*

# Stoichiometry and the Atomic Theory of Matter

## 1-1 CHEMISTRY AND THE NATURAL SCIENCES

What is chemistry? It is the study of the properties of substances and of the reactions that transform them into other substances. But these are just words, which fail to convey the range of problems studied by chemists and the excitement of their quest for knowledge in the field. An understanding of the meaning of chemistry can be gained only by doing it—by carrying out chemical experiments in the laboratory and by applying the principles of chemistry to explain observations and make predictions. The discoveries made and the insights gained open up new and exciting vistas and in turn suggest further questions to be asked.

An 18th-century chemist would recognize many objects in a modern chemical laboratory. Beakers and test tubes are still very much a part of the modern chemist's laboratory apparatus, and chemists continue to mix substances and watch the reactions that result. The sophistication of the methods available for examining sub-

1

**Figure 1-1**

Modern studies of reaction rates employ sophisticated instruments that can select molecules with particular speeds and energies and examine their reactions. This apparatus is designed to bombard a solid surface with molecules and examine their reactions on the surface. (Leon Lewandowski)

stances and reactions has increased dramatically, however, and more detailed and more fundamental questions can now be posed and answered. Many aspects of a modern chemical laboratory would be unfamiliar not only to an 18th-century chemist but even to one from the first part of the 20th century.

One tool used in modern chemical research, for example, is the molecular beam spectrometer, a large piece of equipment in which two beams of atoms or molecules are prepared with well-defined speeds and internal motions (Fig. 1-1). The particles are allowed to collide, and the products are analyzed. Lasers, powerful and versatile sources of light, are used to prepare the molecular beams going in and to probe the product molecules coming out, with the whole experiment controlled by a small computer programmed to analyze the results in terms of the forces between atoms and molecules. Such an experiment certainly fits into the tradition of classical chemistry (mixing chemicals and observing reactions), but it is a long way from the simple beaker-and-test-tube operations of the past.

In another laboratory, a chemist may analyze a product made by a small marine organism for use as a possible anticancer agent. The first step is to extract the product by adding a chemical solvent. The problem is then to separate the substance of interest from others that are present. In the past, a lengthy series of mechanical separations using different solvents, or a series of distillations, might have been necessary. Now the chemist may simply inject the mixture into an instrument called a gas chromatograph and obtain the separated fractions at the other end. The fraction of interest can be analyzed using its light-absorption properties to determine its molecular structure, and the chemist may then develop a procedure to synthesize the same product from commonly occurring substances.

These are just two examples of the types of research carried out by modern chemists; there are many more. A feature of 20th-century natural science is that the traditional ''compartments'' used to identify scientific fields (physics, astronomy, chemistry, geology, biology) are no longer capable of holding separately all the parts of modern scientific research. Instead, nature keeps spilling out of the compartments into which we try to force it, making interdisciplinary research rewarding (see Fig. 1-2). Some of the most exciting results in modern science come from the interface between two fields once considered essentially distinct.

Because chemistry occupies a central role in the natural sciences, its overlap with other fields provides fruitful sources of new discoveries. For example, both chemistry and physics gain from each other: chemistry advances by applying physical laws and techniques, and physics is enriched through the discovery by chemists of new substances with interesting physical properties. Astronomers now use results from chemical laboratories to search for new compounds in distant galaxies and to learn from them about the origins of the universe. Geologists analyze the structures of minerals and the transformations between different forms of matter in the earth's crust by using principles from chemical bonding and chemical thermodynamics. Biologists increasingly probe the chemical base of life processes. These contacts between chemistry and other fields are by no means peripheral. Indeed, they provide the impetus for much modern chemical research.

### The Methods of Science

Progress in chemistry, and indeed in all of the sciences, occurs through a balance between two complementary aspects: theory (predictions and conclusions) and experiment (discovery of facts). During certain periods in the development of science,

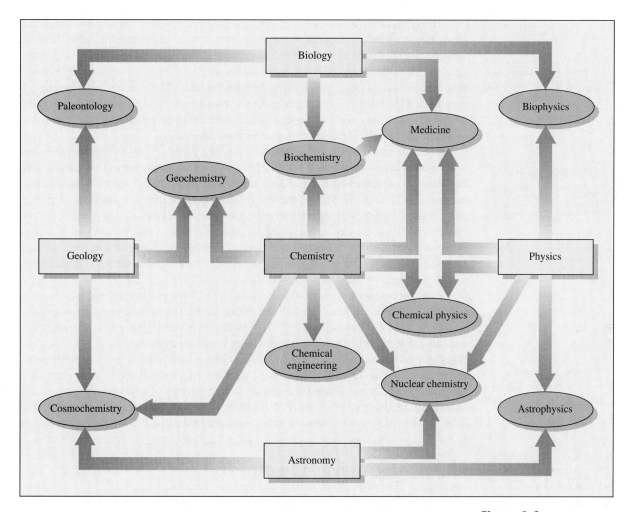

**Figure 1-2**

Interrelations among the natural sciences.

one may lag behind the other. When the gap becomes too great, scientific progress slows down.

Experimentation without the guidance of theory (or hypothesis, or ''hunch'') does not proceed very far. Because innumerable experiments can be performed, the problem for the scientist is to choose which experiments are important and then to carry them out in a carefully controlled fashion. Asking intelligent questions of nature is what experiment is about. Sometimes the answers that come back fit in with other pieces of information and a correlation emerges, suggesting further questions (experiments). At other times an apparent contradiction appears, and additional experiments are needed before a more comprehensive picture emerges.

Just as experiment requires guidance from theory, so theories need to be tested continually against the real world of nature; the ultimate test of every theory is that it must agree with experiment. Predictions from Albert Einstein's general theory of relativity, for example, continue to be verified many years after they were first made. On the other hand, his efforts during the last years of his life to develop a generalized field theory that would encompass all known interactions of matter did not succeed. The physicist Enrico Fermi said of Einstein's effort, ''Beautiful theory, wrong universe.''

Throughout the study of chemistry we should make an effort to distinguish fact from theory. Ideally, facts are enduring. Of course, experimental facts must be verified, and the only acceptable facts are those that can be obtained by anyone who takes sufficient care in repeating an experiment. Sometimes we do not agree on the facts; different experimentalists find different results. This occurs because experimental conditions have not been properly controlled, or errors have been made. Mistakes can occur, but sooner or later experimentalists agree upon the facts, and more careful work improves the accuracy of those facts.

Theories are in a different category. They represent our attempt to explain some body of fact and to predict others that have not yet been discovered by experiment. Sometimes several theories are in contention, and the preferred theory is the one that accounts for the largest body of fact with the fewest arbitrary assumptions. Theories are always subject to modification, improvement, or rejection. They are never the ''last word,'' but they represent our best current effort to understand physical phenomena. Scientists hold theories *provisionally,* believing them, building upon them, criticizing them, trying to disprove them, always seeking to deepen the level of our understanding.

There are many hills and valleys in this search for understanding. Some of the valleys are blind, and we have to retrace our steps. The analogy of a mountainous terrain is, in fact, a rather good one. We climb a number of foothills until eventually we come to a pass or summit from which the view is panoramic. Formerly disjoint facts and incongruous observations fall into place, and we suddenly have a much-enlarged understanding. If we can view the landscape of science from several vantage points (i.e., using several theoretical and experimental approaches), our understanding is deepened. This thrill of understanding is what scientists live for, and what impels them to develop ever more sensitive probes with which to query nature.

## 1-2 THE COMPOSITION OF MATTER

Two traditions intertwine through the history of chemistry: **analysis** (taking things apart) and **synthesis** (putting things together). These two directions can be traced from the earliest periods of Greek science to the present. The early Greek philosophers attempted to analyze the constituents of matter in terms of four elements: air, earth, fire, and water. Early artisans and metallurgists synthesized new materials of practical importance—including bronze, glass, and cement—both by design and by accident. The dual traditions of analysis and synthesis have remained driving forces to the present day. When a medicinally important product is discovered in a natural source, the first step is to analyze it to find its composition and structure, and then a strategy can be designed to synthesize it from easily available starting materials. Penicillin was discovered in 1929 through the accidental contamination of a bacterial culture by mold; its laboratory synthesis was not achieved until 1957.

### Substances and Mixtures

Let us begin our study of chemistry with the process of analysis—the determination of what makes up matter (Fig. 1-3). Suppose we take a sample of a material, which may be gas, liquid, or solid, and examine various properties, or distinguishing characteristics, such as color, odor, or density. A piece of wood, for example, has regions

(a)

(b)

(c)

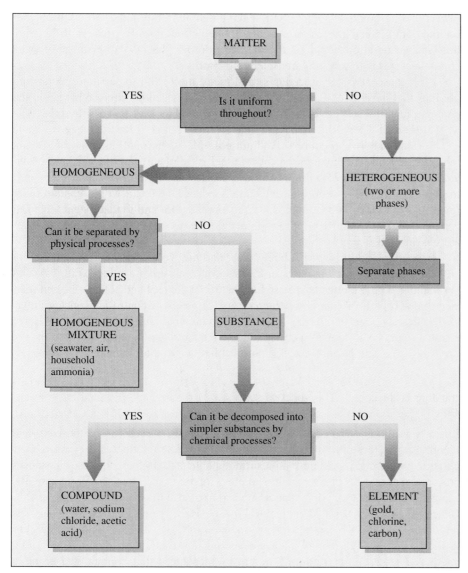

**Figure 1-3**

An outline of the steps in the analysis of matter.

**Figure 1-4**

(a) A solid mixture of blue $Cu(NO_3)_2 \cdot 6H_2O$ and yellow CdS is added to water. (b) Although the $Cu(NO_3)_2 \cdot 6H_2O$ dissolves readily and passes through the filter, the CdS remains largely undissolved and is held on the filter. (c) Evaporation of the solution leaves nearly pure crystals of $Cu(NO_3)_2 \cdot 6H_2O$. (Leon Lewandowski)

that possess different properties (some portions are darker than others) and is said to be **heterogeneous.** Other materials, such as air or a mixture of salt and water, are **homogeneous** because their properties do not vary through the sample. They are still **mixtures,** however, because it is possible to separate them into components by ordinary physical means such as melting, freezing, boiling, or dissolving in solvents (Fig. 1-4). These operations provide ways of separating materials from one another on the basis of their properties (freezing point, boiling point, solubility, and so on). For example, air is a mixture of several components (oxygen, nitrogen, argon, and various other gases). If it is liquefied and then warmed slowly, those gases with the

(a)

(b)

**Figure 1-5**

(a) Nearly pure elemental silicon is produced by pulling a 10-inch-long solid cylinder called a boule out of the melt, leaving most of the impurities behind. (b) The boule can then be sliced into thin wafers for use in integrated circuits. (b, Leon Lewandowski)

lowest boiling points evaporate first, leaving behind in the liquid those with higher boiling points. Such a separation would not be perfect, but the process of liquefaction and evaporation could be repeated to improve the resolution of air into its component gases to any required degree of purity.

If all these physical procedures (and many more) fail to separate matter into portions that have different properties, the material is a **substance.** What about the common material sodium chloride, which we call table salt? Is it a substance? The answer is yes if we use the term ''sodium chloride'' but no if we use the term ''table salt.'' Table salt is a mixture of sodium chloride with small amounts of sodium iodide (to provide for the requirements of the thyroid gland) and magnesium carbonate (to coat each grain of sodium chloride and prevent the salt from caking). Even if these two components were not deliberately added, table salt would contain other impurities that had not been removed in its preparation, and to that extent table salt is a mixture. But when we refer to sodium chloride, we imply that all other materials are absent, and it is a substance.

In practice, nothing is absolutely pure, and the word ''substance'' is therefore an idealization. Among the purest materials ever prepared are silicon (Fig. 1-5) and germanium. These elements are used in electronic devices and solar cells, and their electronic properties require very high purity (or precisely controlled concentrations of deliberately added impurities). Meticulous chemical and physical methods have enabled scientists to prepare germanium and silicon that contain concentrations not exceeding one part per billion of impurities that would alter their electrical properties.

## The Elements

Literally millions of substances have so far been discovered (or synthesized) and identified. Are these the fundamental building blocks of matter? If they were, their classification alone would pose an insurmountable task. In fact, all these substances are combinations of much smaller numbers of building blocks called **elements.** Elements are substances that cannot be decomposed into two or more simpler substances by ordinary physical or chemical means. Here again, the word ''ordinary'' is important, because we exclude the processes of radioactive decay (either natural or artificial) and of high-energy nuclear reactions that *do* transform one element into another. The term ''**compound**'' refers to substances that contain two or more chemical elements. Hydrogen and oxygen are elements because no further chemical separation is possible, but water is a compound because it can be separated into hydrogen and oxygen by passing an electric current through it (Fig. 1-6). *Binary* compounds are substances (such as water) that contain two elements, *ternary* compounds contain three elements, *quaternary* four, and so on.

At present there are some 111 chemical elements. A few have been known since before recorded history, principally because they occur in nature as elements rather than in combination with one another in compounds. Gold, silver, lead, copper, and sulfur are chief among them. Gold was found in streams in the form of little granules (placer gold) or nuggets in loosely consolidated rock. Sulfur was associated with volcanoes, and copper could often be found in its native state in shallow mines. Iron occurred in its elemental state only rarely (in meteorites). Usually it was combined with oxygen or other elements. In the second millennium B.C., ancient metallurgists somehow learned to reduce iron oxide to iron with charcoal in forced-draft fires, and the Iron Age was inaugurated.

**Figure 1-6**
As electric current passes through water containing dissolved sulfuric acid, gaseous hydrogen and oxygen form as bubbles at the electrodes. This method of electrolysis can be used to prepare very pure gases. Note that the volume of hydrogen produced (*left*) is twice that of the oxygen (*right*). This is a consequence of the law of combining volumes, discussed in Section 1-3.

A sample of native gold attached to a small piece of quartz. (Gary Retherford/Photo Researchers)

The names of the chemical elements and the symbols that designate them are fascinating subjects. Many have Latin roots, such as gold (*aurum,* symbol Au), copper (*cuprum,* Cu), iron (*ferrum,* Fe), and mercury (*hydrargyrum,* Hg). Hydrogen (H) means "water former." Potassium (*kalium,* K) takes its common name from potash (potassium carbonate), a useful chemical obtained in early times by leaching the ashes of wood fires with water. Many elements take their names from Greek and Roman mythology: cerium (Ce) from Ceres, goddess of plenty; tantalum (Ta) from Tantalus; niobium (Nb) from Niobe, daughter of Tantalus. Some elements are named for continents: europium (Eu), americium (Am). Others are named after countries: germanium (Ge), francium (Fr), polonium (Po). Cities provide the names of others: holmium (Stockholm, Ho), ytterbium (Ytterby, Yb), berkelium (Berkeley, Bk). Some are named for the planets: uranium (U), plutonium (Pu), neptunium (Np). Others take their names from colors: praseodymium (green, Pr), rubidium (red, Rb), cesium (sky blue, Cs). Still others honor great scientists: curium (Marie Curie, Cm), mendelevium (Dmitri Mendeleyev, Md), fermium (Enrico Fermi, Fm), einsteinium (Albert Einstein, Es).

## 1-3 THE ATOMIC THEORY OF MATTER

The existence of elements does not resolve the question of the nature of matter on a microscopic scale. Two descriptions are possible, both of which can be traced back to ancient philosophy. One is a view of matter in which elements are continuous "qualities" that are mixed to form substances. In this view, matter is divisible without limit, with the elements remaining mixed at every scale. The other view postulates the existence of **atoms,** and holds that there is a lower limit to which compounds can be divided before they are separated into atoms of the elements that constitute them. Democritus (c. 460–c. 370 B.C.) postulated the existence of such unchangeable atoms of the elements, which he imagined to undergo continuous random motion in the vacuum, a remarkably modern point of view. The Greek philosophers were content to leave their views in the form of assertions, however,

An iron–nickel meteorite, showing the Widmanstätten structure of nickel (lighter color) in iron (darker color). (Leon Lewandowski)

**Figure 1-7**
An image of the surface of silicon constructed with a scanning tunneling microscope. The silicon atoms (shown in yellow) are mostly arranged in a regular pattern. (John Ozcomert, Michael Trenary)

and lacked the essentially modern scientific view that theories must be tested and refined by experiment.

In modern times, we are so used to speaking of atoms that we rarely stop to consider the experimental evidence for their existence. Twentieth-century science has developed a number of sophisticated techniques to measure the properties of single atoms, and powerful microscopes even allow us to observe them (Fig. 1-7). Well before single atoms were detected, however, chemists could speak with confidence about their existence and the ways in which they combine to form molecules. Moreover, although the absolute masses of single atoms of oxygen and hydrogen were not measured until the early 20th century, some 50 years earlier chemists could assert (correctly) that the *ratio* of the two masses was close to 16:1. The chemical evidence for the existence of atoms and for the scale of relative atomic masses provides a fascinating and important story, both in its own right and as an illustration of the way in which science progresses. This section presents that story.

### The Law of Conservation of Mass

The first key steps toward understanding the microscopic nature of matter were taken during the 18th century in the course of studies of heat and combustion. It had been observed that an organic material (such as wood) left a solid residue of ash when burned, and a metal heated in air was transformed into a "calx," which we now call an oxide. The explanation for this phenomenon that was popular in the first part of the 18th century was that a property called "phlogiston" was driven out of wood or metal by the heat of a fire. From the modern perspective this seems quite absurd, because the ash weighs less than the original wood, whereas the calx weighs more than the metal. At the time, however, the principle of conservation of mass was not yet established, and people saw no reason why the mass of a material should not change upon heating.

Further progress could be made only through the careful use of the chemical balance to determine the changes in mass[1] that occur in chemical reactions. The balance had been known since antiquity, but had been used principally as an assayer's tool and for verifying the masses of coins or commodities in commerce. The great French chemist Antoine Lavoisier used the balance to demonstrate that the sum of the masses of the products of a chemical reaction equals the sum of the masses of the reactants. He heated mercury in a sealed flask containing air. After several days a red substance, mercury(II) oxide, was produced. The gas remaining in the flask was reduced in mass and could no longer support life or combustion; a candle was extinguished by it, and animals suffocated when forced to breathe it. We now know that this residual gas was nitrogen, and that the oxygen in the air had reacted with the mercury. Lavoisier then took a carefully weighed amount of the red oxide of mercury and heated it strongly (Fig. 1-8). He weighed both the mercury and the gas that were produced, and showed that their combined mass was the same as

---

[1] Chemists sometimes use the term "weight" in place of "mass." Strictly speaking, weight and mass are not the same. The mass of a body is an invariant quantity, but its weight is the force exerted upon it by gravitational attraction (usually by the earth). Newton's second law relates the two ($w = m \times g$, where $g$ is the acceleration due to gravity). As $g$ varies from place to place on the earth's surface, so does the weight of a body. In chemistry we deal mostly with ratios, which are the same for masses and weights. We shall use the term "mass" exclusively, but "weight" is still in colloquial chemical use.

that of the mercury(II) oxide with which he had started. After further experiments Lavoisier was able to state the **law of conservation of mass:**

> In every chemical operation an equal quantity of matter exists before and after the operation.

Lavoisier was the first to observe that a chemical reaction is analogous to an algebraic equation. We would write his second reaction as

$$2\,HgO \longrightarrow 2\,Hg + O_2$$

although at that time the identity of the gas (oxygen) was not known.

### The Law of Definite Proportions

Rapid progress ensued as chemists began to make accurate determinations of the masses of reactants and products. A controversy arose between two schools of thought, led by the French chemists Claude Berthollet and Joseph Proust. Berthollet believed that the proportions (by mass) of the elements in a particular compound were not fixed but could vary over a certain range. Water, for example, rather than containing 11.1% by mass of hydrogen, might have somewhat less or more than this mass percentage. Proust disagreed, arguing that any apparent variation was due to impurities and experimental errors. He also stressed the difference between homogeneous mixtures and chemical compounds. Through his careful work, Proust demonstrated the fundamental **law of definite proportions:**

> In a given chemical compound, the proportions by mass of the elements that compose it are fixed, independent of the origin of the compound or its mode of preparation.

Pure sodium chloride contains 60.66% chlorine by mass, whether we obtain it from salt mines, or by crystallizing it from waters of the oceans or inland salt seas, or by synthesizing it from its elements, sodium and chlorine.[2]

The law of definite proportions was a crucial step in the development of modern chemistry, and by 1808 Proust's conclusions had become widely accepted. We now recognize that this law is not strictly true in all cases. Certain solids exist over a small range of compositions and are called **nonstoichiometric compounds.** An example is wüstite, which has the nominal chemical formula FeO (with 77.73% iron by mass) but whose composition in fact ranges continuously from $Fe_{0.95}O$ (with 76.8% iron) down to $Fe_{0.85}O$ (74.8% iron), depending on the method of preparation. Such compounds, called **berthollides** in honor of Berthollet, are discussed further in Section 17-5. All gaseous compounds, however, obey the law of definite proportions. This illustrates a common pattern of scientific progress: experimental observation of parallel behavior leads to the establishment of a law, or principle. More

**Figure 1-8**

When the red solid mercury(II) oxide is heated, it decomposes to mercury and oxygen. Note the drops of liquid mercury condensing on the side of the test tube.

---

[2] This statement needs some qualification. As we shall see in the next section, many elements have several *isotopes,* which are species whose atoms have almost identical chemical properties but different masses. Natural variation in isotope abundances leads to small variations in the mass proportions of elements in a compound, and larger variations can be induced by artificial isotopic enrichment.

accurate studies may then reveal exceptions to the general principle. Explanation of the exceptions leads to deeper understanding.

### The Atomic Theory of Dalton

The English scientist John Dalton was by no means the first person to propose the existence of atoms; as we have seen, such ideas date back to Greek times (the word "atom" comes from Greek *a-* [not] plus *tomos* [cut]: "not divisible"). Dalton's major contribution was to marshal the evidence for their existence. He showed that the mass relationships found by Lavoisier and Proust could be interpreted most simply by postulating the existence of atoms of the various elements.

In 1808 Dalton published *A New System of Chemical Philosophy,* in which the following five statements comprise the **atomic theory of matter:**

1. Matter consists of indivisible atoms.
2. All of the atoms of a given chemical element are identical in mass and in all other properties.
3. Different chemical elements have different kinds of atoms; in particular, their atoms have different masses.
4. Atoms are indestructible and retain their identities in chemical reactions.
5. A compound forms from its elements through the combination of atoms of unlike elements in small whole-number ratios.

Dalton's fourth postulate is clearly related to the law of conservation of mass, and the fifth is an attempt to explain the law of definite proportions. Suppose that one rejects the atomic theory and adopts a view in which compounds are subdivisible without limit. What, then, ensures the constancy of composition of a substance such as sodium chloride? On the other hand, if each sodium atom in sodium chloride is matched by one chlorine atom, then the constancy of composition can be understood. For this argument about why there should be a law of definite proportions, it does not matter how small the atoms of sodium and chlorine are. It is important merely that there be some lower bound to the subdivisibility of matter, because the moment we put in such a lower bound, arithmetic steps in. Matter becomes countable, and the units of counting are simply atoms. Believing in the law of definite proportions as an established experimental fact, Dalton *postulated* the atom.

### The Law of Multiple Proportions

The composition of a compound is shown by its **chemical formula.** The symbol "$H_2O$" for water indicates that the substance water contains two atoms of hydrogen for each atom of oxygen. It is now known that in water the atoms in each group of three (two H and one O) are linked together by attractive forces strong enough to keep the group together for a reasonable period of time. Such a group is called a **molecule.** In the absence of knowledge about a compound's molecules, the numerical subscripts in the chemical formula simply give the relative proportions of the elements in the compound. How do we know that these are the true proportions? The determination of chemical formulas (and the accompanying determination of relative atomic masses) was one of the major accomplishments of 19th-century chemistry, building on the atomic hypothesis of Dalton.

The simplest kind of compound we can apprehend is one in which two elements combine, contributing equal numbers of atoms to the union to form **diatomic mole-**

**cules,** which are made up of two atoms apiece. It was well known to 18th- and 19th-century chemists, however, that two elements often combine in different proportions, forming more than one compound.

For example, the elements carbon (C) and oxygen (O) combine under different conditions to form two different compounds. Let us call them A and B. Analysis shows that A contains 1.333 grams (g) of oxygen per 1.000 g of carbon, and that B contains 2.667 g of oxygen per 1.000 g of carbon. Although at this point we know nothing about the chemical formulas of the two oxides of carbon, we can immediately say that compound A contains half as many oxygen atoms per carbon atom in its molecules as does compound B. The evidence for this is that the ratio of the masses of oxygen in A and B, for a fixed mass of carbon in each, is 1.333 : 2.667, or 1 : 2. If the formula of compound A were CO, then the formula of compound B would have to be $CO_2$, $C_2O_4$, $C_3O_6$, or some other multiple of $CO_2$. If compound A were $CO_2$, then compound B would be $CO_4$ or $C_2O_8$, and so on. At this point, we cannot say which of these (or an infinite number of other possibilities) are true formulas of the molecules of compounds A and B, but we do know that the number of oxygen atoms per carbon atom in the two compounds is the *quotient of integers*.

Consider another example. Arsenic (As) and sulfur (S) combine to form two sulfides, A and B, in which the masses of sulfur per 1.000 g of arsenic are respectively 0.428 g and 0.642 g. The ratio of these sulfur masses is 0.428 : 0.642 = 0.667 ≈ 2 : 3. We conclude that *if* the formula of compound A is a multiple of AsS, then the formula of compound B must be a multiple of $As_2S_3$.

These two examples illustrate the **law of multiple proportions:**

> When two elements form a series of compounds, the masses of one element that combine with a fixed mass of the other element are in the ratio of small integers to each other.

In our first example, the ratio of the masses of oxygen in the two compounds, for a given mass of carbon, was 1 : 2. In the second example, the ratio of the masses of sulfur in the two compounds, for a given mass of arsenic, was 2 : 3. Today we know that the carbon oxides are CO (carbon monoxide) and $CO_2$ (carbon dioxide), and the arsenic sulfides are $As_4S_4$ and $As_2S_3$ (Fig. 1-9). Dalton could not have known this, however, because he had no information from which to decide how many atoms of

**Figure 1-9**

The yellow mineral orpiment, $As_2S_3$, and the red mineral realgar, $As_4S_4$, are two relatively abundant sources of arsenic, often occurring in the same sample. (Julius Weber)

carbon and oxygen are in one molecule of the carbon–oxygen compounds, or how many atoms of arsenic and sulfur are in the arsenic–sulfur compounds.

### Example 1-1

Chlorine (Cl) and oxygen form four different binary compounds. Analysis gives these results:

| Compound | Mass of O Combined with 1.0000 g of Cl |
|:---:|:---:|
| A | 0.22564 g |
| B | 0.90255 g |
| C | 1.3539  g |
| D | 1.5795  g |

(a) Show that the law of multiple proportions holds for these compounds.
(b) If the formula of compound A is a multiple of $Cl_2O$, then determine the formulas of compounds B, C, and D.

### Solution

(a) Form ratios by dividing each mass of oxygen by the smallest, which is 0.22564 g:

$$0.22564 \text{ g} : 0.22564 \text{ g} = 1.0000 \text{ for compound A}$$

$$0.90255 \text{ g} : 0.22564 \text{ g} = 4.0000 \text{ for compound B}$$

$$1.3539 \text{ g} : 0.22564 \text{ g} = 6.0003 \text{ for compound C}$$

$$1.5795 \text{ g} : 0.22564 \text{ g} = 7.0001 \text{ for compound D}$$

The ratios are whole numbers to a high degree of precision, and the law of multiple proportions is satisfied. Ratios of whole numbers would also have satisfied that law.
(b) If compound A has a formula that is some multiple of $Cl_2O$, then compound B is $Cl_2O_4$ (or $ClO_2$, or $Cl_3O_6$, and so forth) because it is four times richer in oxygen than compound A. Similarly, compound C, which is six times richer in oxygen than compound A, is $Cl_2O_6$ (or $ClO_3$, or $Cl_3O_9$, and so forth), and compound D, which is seven times richer in oxygen than compound A, is $Cl_2O_7$ (or a multiple thereof).

*Related Problems:* 7, 8, 9, 10

Dalton made a sixth assumption to resolve the dilemma of the absolute number of atoms present in a molecule, and he called it the "rule of greatest simplicity." He maintained that if two elements form only a single compound, its molecules will have the simplest possible formula: AB (see Fig. 1-10). Thus, he assumed that when hydrogen and oxygen combine to form water, the reaction is

$$H + O \longrightarrow HO$$

# SAUNDERS INVITES YOU TO BE A PART OF OUR LATEST INNOVATION IN CHEMISTRY!

**SAUNDERS COLLEGE PUBLISHING** *a division of Harcourt Brace College Publishers* • The Public Ledger Building, Suite 1250 • 150 South Independence Mall West • Philadelphia, PA 19106

**Figure 1-10**
Dalton's chemical symbols and some chemical equations he would have written. Note that some of his molecular formulas were incorrect, such as HO instead of $H_2O$. Modern names for these substances are given in parentheses.

Dalton was wrong, as we now know, and the correct reaction is

$$2\,H_2 + O_2 \longrightarrow 2\,H_2O$$

## The Law of Combining Volumes

At this time a French chemist, Joseph Gay-Lussac, conducted some important experiments on the volumes of gases that react with one another to form new gases. He discovered the **law of combining volumes:**

> The volumes of two reacting gases (at the same temperature and pressure) are in the ratio of simple integers. Moreover, the ratio of the volume of each product gas to the volume of either reacting gas is the ratio of simple integers.

We consider three examples:

2 volumes of hydrogen + 1 volume of oxygen $\longrightarrow$ 2 volumes of water vapor
1 volume of nitrogen + 1 volume of oxygen $\longrightarrow$ 2 volumes of nitrogen oxide
3 volumes of hydrogen + 1 volume of nitrogen $\longrightarrow$ 2 volumes of ammonia

Gay-Lussac did not theorize on his experimental findings, but shortly after their publication an Italian chemist, Amedeo Avogadro, used them to formulate an important hypothesis.

### Avogadro's Hypothesis

In 1811 Avogadro stated a postulate that has become known as **Avogadro's hypothesis:**

> Equal volumes of different gases (at the same temperature and pressure) contain equal numbers of particles.

The question immediately arose: are "particles" of the elements the same as Dalton's atoms? Avogadro took the point of view that they were not; rather, elements could exist as diatomic molecules. With his hypothesis Avogadro could explain Gay-Lussac's law of combining volumes (Fig. 1-11). Thus, the reactions we wrote out in words become

$$2 H_2 + O_2 \longrightarrow 2 H_2O$$

$$N_2 + O_2 \longrightarrow 2 NO$$

$$3 H_2 + N_2 \longrightarrow 2 NH_3$$

all of whose coefficients are proportional to the volumes of the reactant and product gases in Gay-Lussac's experiments, and whose chemical formulas agree with modern results. Dalton, on the other hand, would have written

$$H + O \longrightarrow OH$$

$$N + O \longrightarrow NO$$

$$H + N \longrightarrow NH$$

all of whose coefficients disagree with Gay-Lussac's observations of their relative volumes.

In addition to predicting correct molecular formulas, Avogadro's hypothesis gives correct results for the relative atomic masses of the elements. Analysis by chemists during the 18th century had revealed that 1 g of hydrogen combines fully with 8 g of oxygen to make 9 g of water. If Dalton's formula for water, HO, were correct, then an atom of oxygen would have to weigh eight times as much as an atom of hydrogen; that is, Dalton's assumption requires the **relative atomic mass** of oxygen to be 8 on a scale where the relative atomic mass of hydrogen is set at 1. From Avogadro's hypothesis, however, each water molecule contains twice as many atoms of hydrogen as oxygen, so to achieve the experimental mass relationship, each oxygen atom must have twice as large a relative atomic mass. This gives a relative atomic mass of 16 for oxygen, a result consistent with modern measurements.

One would think that Dalton would have welcomed Avogadro's brilliant hypothesis, but he did not. Dalton and others remained firm in their conviction that elements could not exist as diatomic molecules. One reason for this was the belief that molecules were held together by a force called "affinity," which expressed the attraction of opposites, just as we think of the attraction between positive and nega-

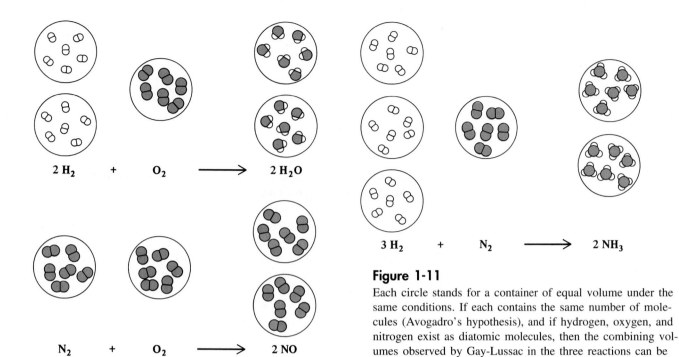

$$2 \, H_2 \quad + \quad O_2 \quad \longrightarrow \quad 2 \, H_2O$$

$$N_2 \quad + \quad O_2 \quad \longrightarrow \quad 2 \, NO$$

$$3 \, H_2 \quad + \quad N_2 \quad \longrightarrow \quad 2 \, NH_3$$

**Figure 1-11**

Each circle stands for a container of equal volume under the same conditions. If each contains the same number of molecules (Avogadro's hypothesis), and if hydrogen, oxygen, and nitrogen exist as diatomic molecules, then the combining volumes observed by Gay-Lussac in the three reactions can be understood.

tive electric charges. If this were true, why should two *like* atoms be held together in a molecule? Moreover, if like atoms somehow did hold together in pairs, why should they not aggregate further to form molecules with three, four, or six atoms and so forth? As a result of loyalty to the affinity theory, Avogadro's reasoning did not attract the attention it deserved. Confusion reigned, because different chemists adopted different chemical formulas for molecules. A textbook published by August Kekulé in 1861 gave 19 different chemical formulas for acetic acid!

It was 50 years before an Italian chemist, Stanislao Cannizzaro, presented a paper at the First International Chemical Congress in 1860 in Karlsruhe, Germany, that convinced others to accept Avogadro's approach. Cannizzaro analyzed many gaseous compounds and showed that their chemical formulas could be established with a consistent scheme that used Avogadro's hypothesis but avoided any extra assumptions about molecular formulas. Gaseous hydrogen, oxygen, and nitrogen (and fluorine, chlorine, bromine, and iodine as well) indeed turn out to consist of diatomic molecules under ordinary conditions.

## 1-4 BUILDING BLOCKS OF THE ATOM

In the original Greek conception, carried over into Dalton's time, the atom was considered the ultimate and indivisible building block of matter. At the end of the 19th century this notion began to yield to the view that atoms were themselves composed of smaller particles. Scientists carried the process of analysis (Sec. 1-2) down to the subatomic level.

## Electrons

One important piece of evidence for the existence of particles smaller than atoms came from studies of the effects of large electric fields on atoms and molecules. When a gas is placed in a glass tube between two conducting plates, and a large potential difference is imposed across the plates, current passes through the gas, suggesting that the electric field has broken down the atoms into new species that carry charge. (The same effect occurs when an electrical discharge from a lightning bolt passes through air.) As the amount of gas in the tube is lowered, current continues to flow; the mysterious invisible current carriers, called **cathode rays,** travel in straight lines and produce a luminous spot where they impinge upon the glass tube. Several decades of research revealed that cathode rays could be deflected by both magnetic and electric fields and could heat a piece of metal foil in the tube until it glowed. Influenced by Maxwell's recent electromagnetic theory of light, one school of physicists believed the cathode rays to be a strange form of invisible light, while another group considered them to be a stream of negatively charged particles.

In 1897 the British physicist J. J. Thomson carried out a series of experiments that resolved the controversy by proving that cathode rays are negatively charged particles, called **electrons.** His key experiment is depicted schematically in Figure 1-12. In a highly evacuated tube, a beam of cathode rays passed between a pair of plates that could be charged positively and negatively to establish an electric field perpendicular to the cathode-ray trajectory. For the arrangement shown in Figure 1-12, the cathode rays were deflected downward (indicating that they carried a negative charge), and the deviation could be measured accurately from the displacement of the luminous spot on a screen at the end of the tube. A magnetic field was established in the same region by passing an electric current through a pair of coils, its direction perpendicular to both that of the electric field and that of the cathode rays. The magnetic field also produced a deflection of the rays that could be made to oppose the deflection caused by the electric field. By varying the strengths of the two fields, Thomson could cause the cathode rays to pass undeviated through the tube. Under these conditions the beam experienced two equal but opposing forces. One, due to the electric field $E$, was

$$f_E = Ee$$

and the other, due to the magnetic field $H$, was

$$f_H = Hev$$

where $e$ is the electric charge on the supposed particles and $v$ is their velocity. The velocity of the cathode rays was therefore

$$v = \frac{E}{H}$$

When the magnetic field was turned off, only the electric field acted on the beam, giving it a displacement

$$s = \frac{1}{2}at^2$$

away from the central axis of the tube at the point where it emerged from the field between the plates (Fig. 1-12b). The particles acquired the acceleration $a$ toward the

(a)

(b)

**Figure 1-12**

Thomson's apparatus to measure the electron charge-to-mass ratio, $e/m_e$. Electrons (cathode rays) stream across the tube from left to right. The electric field alone deflects the beam down, and the magnetic field alone deflects it up. By adjusting the two field strengths, Thomson achieved a condition of zero net deflection.

positive plate in the time $t$ it took for them to traverse the length of the condenser $\ell$. Because $v = \ell/t$, $t$ and then $a$ can be calculated. From Newton's second law,

$$f_E = m_e a = Ee$$

Successively substituting the above expressions for $a$, $t$, and $v$ leaves only measurable quantities:

$$\frac{e}{m_e} = \frac{a}{E} = \frac{2s}{t^2 E} = \frac{2sv^2}{\ell^2 E} = \frac{2sE}{\ell^2 H^2}$$

From this, the charge-to-mass ratio for the electron can be determined. The currently accepted value is $e/m_e = 1.7588196 \times 10^{11}\ \text{C kg}^{-1}$, where charge is measured in coulombs and mass in kilograms. (See Appendix B for a full discussion of units.)

Thomson's experiment allowed determination of only the *ratio* of the charge to the mass of the electron. The actual value of the electric charge was measured in 1906 by the American physicist Robert Millikan with his student H. A. Fletcher. In their elegant experiment (Fig. 1-13), tiny drops of oil became charged by collisions with electrons or ionized gas molecules in the surrounding air. A charged oil drop

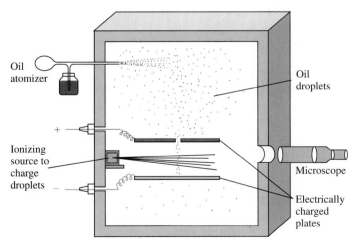

**Figure 1-13**

Millikan's apparatus to measure the charge on an electron, $e$. By adjusting the electric field strength between the charged plates, Millikan could halt the fall of negatively charged oil drops and determine their net charge.

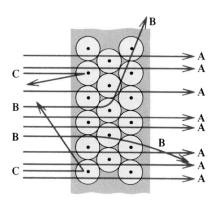

(b)

## Figure 1-14

(a) Flashes of light mark the arrival of alpha particles at the detector screen. In the Rutherford experiment, the rate of hits on the screen varied from around 20 per minute at high angles to nearly 132,000 per minute at low angles. (b) The interpretation of Rutherford's experiment. Most of the alpha particles pass through the space between nuclei and undergo only small deflections (A). A few pass close to a nucleus and are more strongly deflected (B). Some are even scattered backward (C). The nucleus is far smaller proportionately than the dots shown.

(with charge $Q$ and mass $M$) situated in an electric field between two plates was subject to two forces: the force of gravity $-Mg$, causing it to fall, and a force $QE$ from the electric field, causing it to rise. By adjusting the electric field to balance the two forces, and independently determining the drops' masses, $M$, from their falling speeds in the absence of an electric field, Millikan showed that the charge $Q$ was always an integral multiple of the same basic charge, $1.59 \times 10^{-19}$ C. He suggested that the different oil drops carried integral numbers of a fundamental charge, which he took to be the charge of a single electron. More accurate modern measurements led to the value $e = 1.6021773 \times 10^{-19}$ C. Together with the $e/m_e$ ratio found by Thomson, this gives $m_e = 9.109390 \times 10^{-31}$ kg for the electron mass.

### The Nucleus

Along with the electrons pulled off atoms by an electric field, positively charged particles also form. These **canal rays** were observed in cathode ray tubes but traveled in the opposite direction and were thousands of times more massive than the electron. This led Thomson to propose an atomic model in which the atom was regarded as consisting of electrons imbedded in a jelly-like fluid of positive charge. In this ''plum pudding'' model, Thomson supposed that each atom had an integral number of electrons $Z$ whose charge was exactly balanced by the jelly-like positive charge, making the atom electrically neutral.

In 1911 Ernest Rutherford made a startling discovery. He and his students had been studying the properties of alpha particles emitted from radium, an element recently isolated by Marie and Pierre Curie. Rutherford's experiments consisted in bombarding thin gold foil ($6 \times 10^{-5}$ cm thick) with these particles and observing their deflections via the scintillations they produced on a ZnS fluorescent screen (Fig. 1-14). Almost all of the alpha particles passed straight through the foil, but occasionally one was found to have been deflected through a large angle. Even less frequently, a particle was observed to be scattered backwards! Rutherford was astounded, because the alpha particles were relatively massive and fast-moving. In his words, ''It was almost as incredible as if you fired a 15-inch shell at a piece of tissue paper and it came back and hit you.'' He and his students studied the frequency with which such large deflections occurred. They concluded that most of the mass in the gold foil was concentrated in very dense, extremely small, positively charged particles that they called **nuclei.** By analyzing the trajectories of the scattered particles, they estimated the radius of the gold nucleus to be less than $10^{-12}$ cm and the positive charge on each nucleus to be approximately $+100e$ (the actual value is $+79e$).

Rutherford proposed a model of the atom in which the nucleus possesses a net charge of $+Ze$, with $Z$ electrons surrounding the nucleus out to a distance of about $10^{-8}$ cm. In the Rutherford picture, a gold atom has 79 electrons (each with a charge of $-1e$) arranged about a nucleus of charge $+79e$. Nearly all of the volume of an atom is occupied by its electrons; nearly all of its mass is concentrated in the nucleus. The Rutherford theory continues to be the accepted model of atomic structure. The properties of a given chemical element arise from the charge $+Ze$ on its nucleus and the presence of $Z$ electrons around the nucleus. This integer $Z$ is called the **atomic number** of the element. Atomic numbers are given inside the back cover of this book.

### Protons, Neutrons, and Isotopes

The smallest and simplest nucleus is that of the hydrogen atom—the **proton.** It has a positive charge of exactly the same magnitude as the charge on the electron, but its mass is $1.67262 \times 10^{-27}$ kg, which is 1836 times greater than the electron mass. Nuclei of other elements contain $Z$ times the charge on the proton, but their masses are greater than $Z$ times the proton mass. The helium nucleus, for example, has atomic number $Z = 2$, but its mass is approximately four times the proton mass. In 1920 Rutherford suggested the existence in nuclei of an uncharged particle, the **neutron,** with a mass close to that of the proton. In this case, the nucleus consists of $Z$ protons and $N$ neutrons. The **mass number** $A$ is defined as $Z + N$.

A nuclear species (**nuclide**) is characterized by its atomic number $Z$ (that is, the nuclear charge in units of $e$, or the number of protons in the nucleus) and its mass number $A$ (the sum of protons plus neutrons in the nucleus). We denote an atom that contains such a nuclide with the symbol ${}_{Z}^{A}X$, where X is the chemical symbol for the element. The atomic number $Z$ is sometimes omitted because it is implied by the chemical symbol for the element. Thus, ${}_{1}^{1}H$ (or ${}^{1}H$) is a hydrogen atom, and ${}_{6}^{12}C$ (or ${}^{12}C$) is a carbon atom whose nucleus contains six protons and six neutrons. **Isotopes** are nuclides of the same chemical species (that is, having the same $Z$) with different mass numbers $A$ and therefore different numbers of neutrons in the nucleus. The nuclides of hydrogen, deuterium, and tritium, represented by ${}_{1}^{1}H$, ${}_{1}^{2}H$, and ${}_{1}^{3}H$, respectively, are all members of the family of isotopes that belong to the element hydrogen.

### Example 1-2

Radon-222 (${}^{222}Rn$) has recently received publicity because its presence in basements may increase the number of cancer cases in the general population, especially among smokers. State the number of electrons, protons, and neutrons that make up an atom of ${}^{222}Rn$.

### Solution

From the table at the back of the book, the atomic number of radon is 86, so the nucleus contains 86 protons and $222 - 86 = 136$ neutrons. The atom has 86 electrons to balance the positive charge on the nucleus.

*Related Problems:* 15, 16, 17, 18

## 1-5 WEIGHING AND COUNTING MOLECULES

As pointed out in Section 1-3, elemental analysis allowed 19th-century chemists to conclude that oxygen atoms weigh 16 times more than hydrogen atoms. In the early 20th century, the accurate determination of atomic and molecular masses was further advanced by the work of J. J. Thomson, F. W. Aston, and others who developed the technique of **mass spectrometry.** It is the chemist's most accurate method for determining relative atomic and molecular masses. In a mass spectrometer (Fig. 1-15), one or more electrons are removed from each such atom or molecule. The

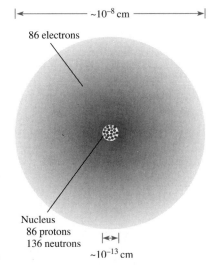

A schematic picture of a ${}^{222}Rn$ atom, with a dense, positively charged nucleus surrounded by a cloud of electrons. In a real atom, the nucleus would be far smaller proportionately than that shown.

**Figure 1-15**

A simplified representation of a modern mass spectrometer. Gas molecules are ionized and accelerated by the electric field between the accelerating plates. The ion beam passes into a magnetic field, where it is separated into components, each containing particles having a characteristic charge-to-mass ratio. Here the spectrometer is adjusted to detect the less strongly deflected $^{22}Ne^+$ ions. By changing the magnitude of the electric or magnetic field, one can move the beam of $^{20}Ne^+$ ions from B to A so that it can be detected.

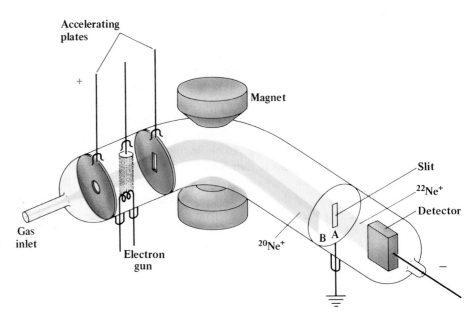

resulting positively charged species, called ions, are accelerated by an electric field and then passed through a magnetic field. The extent of curvature of the particle trajectories depends on the ratio of their charges to their masses, just as in Thomson's experiments on cathode rays (electrons), described in Section 1-4. This allows species of different mass to be separated and detected. Early experiments in mass spectrometry demonstrated, for example, a mass ratio of 16:1 for oxygen relative to hydrogen, confirming by *physical* techniques a relationship deduced originally on *chemical* grounds.

Because many elements have more than one naturally occurring isotope, however, the relationship between the chemist's and the physicist's relative atomic masses was not always so simple. The atomic mass of chlorine determined by chemical means was 35.45 (relative to an oxygen atomic mass of 16), but instead of showing a single peak in the mass spectrum corresponding to a relative atomic mass of 35.45, chlorine showed two peaks, with relative masses near 35 and 37. Approximately three quarters of all chlorine atoms appear to have relative atomic mass 35 (the $^{35}Cl$ atoms), and one quarter have relative atomic mass 37 (the $^{37}Cl$ atoms). Naturally occurring chlorine is thus a mixture of two isotopes with different masses but very nearly identical chemical properties. Elements in nature usually are mixtures, after all. Dalton's second assumption (that all atoms of a given element are identical in mass) is thus shown to be wrong in most cases. Because the *chemical* properties of different isotopes are so similar, their existence was not discovered until the development of the mass spectrometer.

Until 1900 chemists worked with a scale of relative atomic masses in which the average relative atomic mass of hydrogen was set at 1. At about that time, they changed to a scale in which the average relative atomic mass of naturally occurring oxygen (a mixture of $^{16}O$, $^{17}O$, and $^{18}O$) was set at 16. In 1961, by international agreement, the atomic mass scale was further revised, with the adoption of exactly

12 as the relative atomic mass of $^{12}C$. There are two stable isotopes of carbon: $^{12}C$ and $^{13}C$ ($^{14}C$ and other isotopes of carbon are unstable, and of very low terrestrial abundance). Natural carbon contains 98.892% $^{12}C$ and 1.108% $^{13}C$ by mass. The relative atomic masses of the elements as found in nature can be obtained as averages over the masses of the isotopes of each element, weighted by their observed fractional abundances. If an element consists of $n$ isotopes, of which the $i$th isotope has a mass $A_i$ and a fractional abundance $p_i$, then the average relative atomic mass of the element in nature (its chemical relative atomic mass) will be

$$A = A_1 p_1 + A_2 p_2 + \cdots + A_n p_n \equiv \sum_{i=1}^{n} A_i p_i \qquad (1\text{-}1)$$

The relative atomic mass of a nuclide is close to but (except for $^{12}C$) not exactly equal to its mass number.

## Example 1-3

Calculate the chemical relative atomic mass of carbon, taking the relative atomic mass of $^{13}C$ to be 13.003354 on the $^{12}C$ scale.

### Solution

Set up the following table:

| Isotope | Isotopic Mass $\times$ Abundance |
|---------|----------------------------------|
| $^{12}C$ | $12.000000 \times 0.98892 = 11.867$ |
| $^{13}C$ | $13.003354 \times 0.01108 = \underline{0.144}$ |
| | Chemical relative atomic mass = 12.011 |

*Related Problems:* 19, 20, 21, 22

The number of significant figures in a table of chemical or natural relative atomic masses (see inside the back cover of this book) is limited not only by the accuracy of the mass spectrometric data but also by any variability in the natural abundances of the isotopes. If lead from one mine has a relative atomic mass of 207.18 and lead from another has 207.23, there is no way a result more precise than 207.2 can be obtained. In fact, geochemists now are able to use small variations in the $^{16}O:^{18}O$ isotopic abundance ratio as a "thermometer" to deduce the temperatures at which different oxygen-containing rocks were formed in the earth's crust over geologic time scales. They also find anomalies in the oxygen isotopic compositions of certain meteorites, implying that their origins may lie outside our solar system.

*Relative atomic masses have no units* because they are ratios of two masses measured in whatever units we choose (grams, kilograms, pounds, and so forth). The **relative molecular mass** of a compound is the sum of the relative atomic masses of the elements that constitute it, each one multiplied by the number of atoms of that element in a molecule. The formula of water is $H_2O$, for example, so its relative molecular mass is relative molecular mass of water

$$= 2 \text{ (relative atomic mass of H)} + 1 \text{ (relative atomic mass of O)}$$
$$= 2(1.0079) + 1(15.9994) = 18.0152$$

## Avogadro's Number

What are the actual masses (in grams) of individual atoms and molecules? To answer this, it is necessary to establish a connection between the macroscopic scale of masses used in the laboratory and the microscopic scale of the masses of individual atoms and molecules. The link between the two is provided by **Avogadro's number** ($N_0$), defined as the number of atoms in exactly 12 g of $^{12}C$. We shall later consider some of the many experimental methods that have been developed to determine the numerical value of $N_0$. Its currently accepted value is

$$N_0 = 6.022137 \times 10^{23}$$

The mass of a single $^{12}C$ atom is then found by dividing exactly 12 g by $N_0$:

$$\text{mass of a } ^{12}C \text{ atom} = \frac{12.0000 \text{ g}}{6.022137 \times 10^{23}} = 1.992648 \times 10^{-23} \text{ g}$$

This is truly a very small mass, reflecting the very large number of atoms in a 12-g sample of carbon.

Avogadro's number is defined relative to the $^{12}C$ atom because that isotope is the basis for the modern scale of relative atomic masses. We can apply it to other substances as well, in a particularly simple fashion. Consider sodium, which has a relative atomic mass of 22.98977. A sodium atom is 22.98977/12 times as heavy as a $^{12}C$ atom. If $N_0$ atoms of $^{12}C$ have a mass of 12 g, then the mass of $N_0$ atoms of sodium must be

$$\frac{22.98977}{12} (12 \text{ g}) = 22.98977 \text{ g}$$

The mass, in grams, of $N_0$ atoms of *any* element is numerically equal to the relative atomic mass of that element. The same conclusion applies to molecules. From the relative molecular mass of water calculated earlier, the mass of $N_0$ molecules of water is 18.0152 g.

## Example 1-4

One of the heaviest atoms found in nature is $^{238}U$. It has a relative atomic mass of 238.0508 on a scale in which 12 is the atomic mass of $^{12}C$. Calculate the mass (in grams) of one atom of $^{238}U$.

### Solution

Because $N_0$ atoms of $^{238}U$ have a mass of 238.0508 g and $N_0$ is $6.022137 \times 10^{23}$, one atom must have a mass equal to

$$\frac{238.0508 \text{ g}}{6.022137 \times 10^{23}} = 3.952929 \times 10^{-22} \text{ g}$$

*Related Problems:* 23, 24

## The Mole

Atoms and molecules have very small masses, so chemical experiments with measurable amounts of substances involve many of these atoms and molecules. It is convenient to group atoms or molecules in counting units of $N_0 = 6.022137 \times 10^{23}$ to measure the **chemical amount** of a substance. One of these counting units is called a **mole** (abbreviated ''mol''; from Latin *moles,* meaning ''heap'' or ''pile''). One mole of a substance is the amount that contains Avogadro's number of atoms, molecules, or other entities. In other words, 1 mol of $^{12}C$ contains $N_0$ $^{12}C$ atoms, 1 mol of water contains $N_0$ water molecules, and so forth. We must be careful in some cases, because a phrase such as ''1 mol of oxygen'' is ambiguous. We should refer instead to ''1 mol of $O_2$'' if there are $N_0$ oxygen *molecules,* and ''1 mol of O'' if there are $N_0$ oxygen *atoms.*

The mass of one mole of atoms of an element—the **molar mass,** with units of grams per mole—is numerically equal to the dimensionless relative atomic mass of that element, and the same relationship holds between the molar mass of a compound and its relative molecular mass. Thus, the relative molecular mass of water is 18.0152, and its molar mass is $18.0152 \text{ g mol}^{-1}$.

To determine the chemical amount, in moles, of a given substance, we use the chemist's most powerful tool, the laboratory balance. If a sample of iron weighs 8.232 g, then

$$\text{chemical amount of iron} = \frac{\text{number of grams of iron}}{\text{molar mass of iron}}$$

$$= \frac{8.232 \text{ g Fe}}{55.847 \text{ g mol}^{-1}}$$

$$= 0.1474 \text{ mol Fe}$$

where the molar mass of iron was obtained from a table of relative atomic masses. The calculation can be turned around, as well. Suppose a certain chemical amount, 0.2000 mol, of water is needed in a chemical reaction. We have

$$\text{(chemical amount of water)} \times \text{(molar mass of water)} = \text{mass of water}$$

$$(0.2000 \text{ mol H}_2\text{O}) \times (18.015 \text{ g mol}^{-1}) = 3.603 \text{ g H}_2\text{O}$$

Thus, we simply measure out 3.603 g of water. In both cases, the molar mass is the conversion factor between the substance's mass and its chemical amount in moles.

Although numbers of moles are frequently determined by weighing, it is still preferable to think of a mole as a fixed number of particles (Avogadro's number) rather than as a fixed mass. The term ''mole'' is thus analogous to a term such as ''dozen'': one dozen pennies weigh 26 g, substantially less than the mass of one dozen nickels, 60 g; but each group contains 12 coins. Figure 1-16 shows mole quantities of several substances.

.......................................................................................................

## Example 1-5

Nitrogen dioxide ($NO_2$) is a major component of urban air pollution. For a sample containing 4.000 g of $NO_2$, calculate (a) the chemical amount (the number of moles) of $NO_2$ and (b) the number of molecules of $NO_2$.

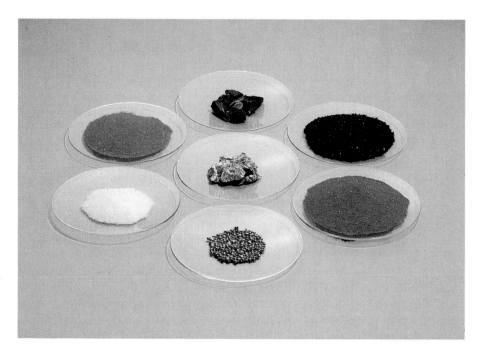

**Figure 1-16**
One-mole quantities of several substances. Clockwise from top: graphite (C), potassium permanganate ($KMnO_4$), copper sulfate pentahydrate ($CuSO_4 \cdot 5H_2O$), copper (Cu), sodium chloride (NaCl), and potassium dichromate ($K_2Cr_2O_7$). Antimony (Sb) is at the center. (Leon Lewandowski)

### Solution

(a) From the tabulated molar masses of nitrogen (14.007 g mol$^{-1}$) and oxygen (15.999 g mol$^{-1}$), the molar mass of $NO_2$ is

$$14.007 \text{ g mol}^{-1} + (2 \times 15.999 \text{ g mol}^{-1}) = 46.005 \text{ g mol}^{-1}$$

The chemical amount of $NO_2$ is then

$$\text{moles } NO_2 = \frac{4.000 \text{ g } NO_2}{46.005 \text{ g mol}^{-1}} = 0.08695 \text{ mol } NO_2$$

(b) To convert from moles to number of molecules, multiply by Avogadro's number:

$$\text{molecules } NO_2 = (0.08695 \text{ mol } NO_2) \times 6.0221 \times 10^{23} \text{ mol}^{-1}$$
$$= 5.236 \times 10^{22} \text{ molecules } NO_2$$

*Related Problems:* 29, 30

### Density and Molecular Size

The **density** of a sample is the ratio of its mass to its volume:

$$\text{density} = \frac{\text{mass}}{\text{volume}} \tag{1-2}$$

The base unit of mass in the International System of Units (discussed in Appendix B) is the kilogram (kg), but this is often inconveniently large for practical purposes in chemistry. Frequently the gram is used instead; moreover, the gram is the standard unit for expressing molar masses. Several units for volume are in frequent use. The base SI unit of the cubic meter ($m^3$) is also quite unwieldy for laboratory purposes (1 $m^3$ of water weighs 1000 kg, or 1 metric ton). For volume we shall therefore use the liter (1 L = $10^{-3}$ $m^3$) and the cubic centimeter, which is identical to the milliliter (1 $cm^3$ = 1 mL = $10^{-3}$ L = $10^{-6}$ $m^3$). Figure 1-17 shows some laboratory equipment used to measure volumes of liquids, and Table 1-1 lists the densities of some substances in units of grams per cubic centimeter.

The density of a substance is not a fixed, invariant quantity but depends on the conditions of pressure and temperature that exist at the time of measurement. For some substances (especially gases and liquids), the volume may be more convenient to measure than the mass, and when the density is known, it provides the conversion factor between volume and mass. For example, near room temperature liquid benzene ($C_6H_6$) has a density of 0.8765 g $cm^{-3}$. Suppose that 0.2124 L of benzene is measured into a container. The mass of benzene is then the product of volume and density:

$$0.2124 \text{ L} \times (1 \times 10^3 \text{ cm}^3 \text{ L}^{-1}) \times (0.8765 \text{ g cm}^{-3}) = 186.2 \text{ g}$$

Dividing this by the molar mass of benzene (78.114 g $mol^{-1}$) gives the corresponding chemical amount, 2.384 mol.

Knowing the density and molar mass of a substance, we readily compute its **molar volume,** the volume occupied by one mole of substance:

$$V_m = \frac{\text{molar mass (g mol}^{-1})}{\text{density (g cm}^{-3})} = \text{molar volume (cm}^3 \text{ mol}^{-1})$$

Near 0°C, for example, ice has a density of 0.92 g $cm^{-3}$, so the molar volume of solid water under these conditions is

$$V_m = \frac{18.0 \text{ g mol}^{-1}}{0.92 \text{ g cm}^{-3}} = 20 \text{ cm}^3 \text{ mol}^{-1}$$

The molar volume of a gas is much larger than this. For $O_2$ at room conditions, the data in Table 1-1 give a molar volume of 24,600 $cm^3$ $mol^{-1}$ = 24.6 L $mol^{-1}$, which is more than 1000 times larger than the molar volume just computed for ice under the same conditions of temperature and pressure. How can we interpret this fact on a microscopic level? We note also that the volumes of liquids and solids do not shift very much with changes in temperature or pressure, but gas volumes are quite sensitive to these changes. A hypothesis to explain such observations is that molecules in liquids and solids are close enough to touch one another, but in a gas they are separated by large distances. If this hypothesis is correct (and it has been borne out by further study), then the sizes of the molecules themselves can be estimated from the volume per molecule in the liquid or solid state. The volume per molecule is the molar volume divided by Avogadro's number; for ice this gives

$$\text{volume per H}_2\text{O molecule} = \frac{20 \text{ cm}^3 \text{ mol}^{-1}}{6.02 \times 10^{23} \text{ mol}^{-1}} = 3.3 \times 10^{-23} \text{ cm}^3$$

This corresponds to a cube with edges about $3.2 \times 10^{-8}$ cm long. We conclude from this and other density measurements that the characteristic size of atoms and small molecules is on the order of $10^{-8}$ cm. Avogadro's number provides the link between laboratory-scale measurements and the masses and volumes of single atoms and molecules.

**Figure 1-17**

Some laboratory equipment, used to contain and dispense measured volumes of liquids. From left, a buret, a volumetric flask, a graduated cylinder, and a pipet.

| Table 1-1 | Densities of Some Substances[†] |
|---|---|
| **Substance** | **Density (g cm$^{-3}$)** |
| Hydrogen | 0.000082 |
| Oxygen | 0.00130 |
| Water | 1.00 |
| Magnesium | 1.74 |
| Sodium chloride | 2.16 |
| Quartz | 2.65 |
| Aluminum | 2.70 |
| Iron | 7.86 |
| Copper | 8.96 |
| Silver | 10.5 |
| Lead | 11.4 |
| Mercury | 13.5 |
| Gold | 19.3 |
| Platinum | 21.4 |

[†]These densities are measured at room temperature and at average atmospheric pressure near sea level.

## 1-6 CHEMICAL FORMULAS AND CHEMICAL EQUATIONS

Chemists use balanced **chemical equations** to describe the reactants and products taking part in a chemical reaction. The chemical formulas that appear in such equations indicate the relative numbers of atoms of the elements in a given substance. Now we distinguish between two types of formulas: the molecular formula and the empirical formula. The **molecular formula** of a substance specifies the number of atoms of each element in one molecule of that substance. Thus, the molecular formula of carbon dioxide is $CO_2$; each molecule of carbon dioxide contains one atom of carbon and two atoms of oxygen. The molecular formula of glucose is $C_6H_{12}O_6$; each glucose molecule contains 6 atoms of carbon, 6 of oxygen, and 12 of hydrogen. Molecular formulas can be defined for all gaseous substances and for those liquids or solids that, like glucose, possess well-defined molecular subunits.

By contrast, the **empirical formula** of a compound is the simplest formula that gives the correct relative numbers of atoms of each kind in a compound. The empirical formula of glucose, for example, is $CH_2O$, indicating that the numbers of atoms of carbon, hydrogen, and oxygen are in a ratio of $1:2:1$. When a molecular formula is known, it is clearly preferable because it conveys more information. In some solids and liquids, however, distinct small molecules do not exist, and the only meaningful chemical formula is an empirical one. Solid cobalt(II) chloride, which has the empirical formula $CoCl_2$, is an example. There are strong attractive forces between a cobalt atom and two adjoining chlorine atoms in solid cobalt(II) chloride, but it is not possible to distinguish the forces *within* such a "molecule" of $CoCl_2$ from those operating *between* it and a neighbor; the latter are equally strong. The solid is, in effect, a single giant molecule. Consequently, cobalt(II) chloride is represented with an empirical formula and referred to by a **formula unit** of $CoCl_2$, rather than by "a molecule of $CoCl_2$."

In some cases, small molecules are incorporated into a solid structure, and the chemical formula is written to show this fact explicitly. Thus, cobalt and chlorine form not only the anhydrous salt $CoCl_2$ just mentioned, but also the hexahydrate $CoCl_2 \cdot 6H_2O$, in which six water molecules are incorporated per $CoCl_2$ formula unit (Fig. 1-18). The dot in this formula is used to set off a well-defined molecular component of the solid, such as water.

**Figure 1-18**
When cobalt(II) chloride crystallizes from solution, it brings with it six water molecules per formula unit, giving a red solid with the empirical formula $CoCl_2 \cdot 6H_2O$. This solid melts at 86°C; above approximately 110°C it loses some water and forms a lavender solid. (Leon Lewandowski)

## Empirical Formula and Percentage Composition

The empirical formula of a compound and the percentage composition by mass of the elements that constitute it are related simply through the mole concept. For example, the empirical formula of ethylene (molecular formula $C_2H_4$) is $CH_2$. Its composition by mass is calculated from the masses of carbon and hydrogen in 1 mol of $CH_2$ formula units:

$$\text{mass of C} = 1 \text{ mol C} \times (12.011 \text{ g mol}^{-1}) = 12.011 \text{ g}$$

$$\text{mass of H} = 2 \text{ mol H} \times (1.00794 \text{ g mol}^{-1}) = 2.0159 \text{ g}$$

Adding these gives a total mass of 14.027 g. The mass percentages of carbon and hydrogen in the compound are then found by dividing each of their masses by this total mass and multiplying by 100%, giving 85.628% C and 14.372% H.

We can reverse this procedure and determine the empirical formula from the elemental analysis of a compound, as illustrated by the following example.

## Example 1-6

A 60.00-g sample of a dry-cleaning fluid was analyzed and found to contain 10.80 g C, 1.36 g H, and 47.84 g Cl. Determine the empirical formula of the compound, using a table of atomic masses.

## Solution

The chemical amounts of the elements in the sample are

For carbon: $\dfrac{10.80 \text{ g C}}{12.011 \text{ g mol}^{-1}} = 0.8992 \text{ mol C}$

For hydrogen: $\dfrac{1.36 \text{ g H}}{1.008 \text{ g mol}^{-1}} = 1.35 \text{ mol H}$

For chlorine: $\dfrac{47.84 \text{ g Cl}}{35.453 \text{ g mol}^{-1}} = 1.349 \text{ mol Cl}$

The ratio of the chemical amount of carbon to that of chlorine (or hydrogen) is $0.8992 : 1.349 = 0.6666$, which is very close to $2:3$. The numbers of moles are then in a ratio of $2:3:3$, so the empirical formula is $C_2H_3Cl_3$. Further measurements would be necessary to find the actual molecular mass and the correct *molecular* formula from among $C_2H_3Cl_3$, $C_4H_6Cl_6$, or any higher multiples, $(C_2H_3Cl_3)_n$.

*Related Problems:* 41, 42, 43, 44, 45, 46

A **hydrocarbon** is a compound containing only carbon and hydrogen. Its empirical formula can be determined by using the combustion train shown in Figure 1-19. In this device, a weighed amount of the hydrocarbon is burned completely in oxygen, yielding carbon dioxide and water. The masses of water and carbon dioxide can then be determined, and from these data the empirical formula is found, as in the following example.

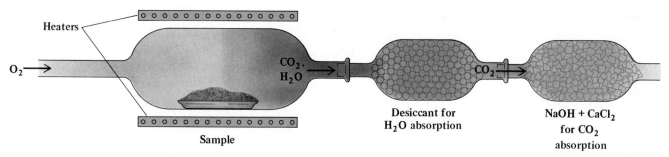

Heaters

$O_2 \rightarrow$

$CO_2$, $H_2O$ $\rightarrow$

$CO_2 \rightarrow$

Sample

Desiccant for
$H_2O$ absorption

NaOH + $CaCl_2$
for $CO_2$
absorption

## Figure 1-19

A combustion train to determine amounts of carbon and hydrogen in hydrocarbons. A weighed sample is burned in a flow of oxygen to form water and carbon dioxide. These products of combustion pass over a desiccant such as magnesium perchlorate, $Mg(ClO_4)_2$, which absorbs the water. In a second stage, the carbon dioxide passes on and is absorbed on finely divided particles of sodium hydroxide, NaOH, mixed with calcium chloride, $CaCl_2$. The changes in mass of the two absorbers give the amounts of water and carbon dioxide produced.

## Example 1-7

A certain compound, used as a welding fuel, contains carbon and hydrogen only. Burning a small sample of it completely in oxygen gives 3.38 g of $CO_2$, 0.692 g of water, and no other products. What is the empirical formula of the compound?

### Solution

We first compute the chemical amounts of the $CO_2$ and $H_2O$. Because all of the carbon has been converted to $CO_2$, and all of the hydrogen to water, the chemical amounts of C and H in the unburned gas can be determined:

$$\text{moles of C} = \text{moles of } CO_2 = \frac{3.38 \text{ g}}{44.01 \text{ g mol}^{-1}} = 0.0768 \text{ mol}$$

$$\text{moles of H} = 2(\text{moles of } H_2O) = 2\left(\frac{0.692 \text{ g}}{18.02 \text{ g mol}^{-1}}\right) = 0.0768 \text{ mol}$$

Note that the number of moles of water is multiplied by 2 to find the number of moles of hydrogen atoms, because each water molecule contains two hydrogen atoms. Because the compound contains equal chemical amounts (numbers of moles) of carbon and hydrogen, its empirical formula is CH. Its molecular formula may be CH, or $C_2H_2$, or $C_3H_3$, and so on.

*Related Problems:* 47, 48

The molecular formula is some whole-number multiple of the empirical formula. To determine the molecular formula, it is necessary to know the approximate molar mass of the compound under study. From Avogadro's hypothesis, the ratio of molar masses of two gaseous compounds is the same as the ratio of their densities, provided that those densities are measured at the same temperature and pressure. (This is true because a given volume contains the same number of molecules of the two gases.) The welding gas from Example 1-7 has density 1.06 g $L^{-1}$ at 25°C and atmospheric pressure. Under the same conditions, gaseous oxygen (which exists as diatomic $O_2$ molecules with molar mass 32.0 g mol$^{-1}$) has a density of 1.31 g $L^{-1}$. The approximate molar mass of the welding gas is then

$$\text{molar mass of welding gas} = \frac{1.06 \text{ g L}^{-1}}{1.31 \text{ g L}^{-1}}(32.0 \text{ g mol}^{-1}) = 25.9 \text{ g mol}^{-1}$$

The molar mass corresponding to the *empirical* formula CH is 13.0 g mol$^{-1}$. Because 25.9 g mol$^{-1}$ is approximately twice this, there must be two CH units per molecule, so the molecular formula is $C_2H_2$. The gas is acetylene (Fig. 1-20).

## Writing Balanced Chemical Equations

Chemical reactions combine elements into compounds, decompose compounds back into elements, and transform existing compounds into new compounds. Because atoms are indestructible in chemical reactions, the same number of atoms (or moles of atoms) of each element must be present before and after any reaction. The conservation of matter in a chemical change is represented in a balanced chemical equation for that process.

An equation can be balanced using stepwise reasoning. Consider the decomposition of ammonium nitrate ($NH_4NO_3$) upon gentle heating to dinitrogen oxide ($N_2O$) and water. The *unbalanced* equation for this process is

$$NH_4NO_3 \longrightarrow N_2O + H_2O$$

The formulas on the left side of the arrow denote **reactants,** and those on the right side denote **products.** This equation is unbalanced because there are 3 mol of O atoms on the left (and 4 of H) but only 2 mol of O atoms and 2 of H atoms on the right. To balance the equation, first assign 1 as the coefficient of one species, usually the one containing the most elements—in this case, $NH_4NO_3$. Next, seek out the elements that appear in only one other place in the equation and assign coefficients to balance the numbers of their atoms. Here nitrogen appears in only one other place ($N_2O$), and a coefficient of 1 for the $N_2O$ ensures that there are 2 mol of N atoms on each side of the equation. Hydrogen appears in $H_2O$, so its coefficient is 2 to balance the 4 mol of H atoms on the left side. This gives

$$NH_4NO_3 \longrightarrow N_2O + 2\,H_2O$$

Finally, verify that the last element, oxygen, is also balanced because there are 3 mol of O atoms on each side. The coefficients of 1 in front of the $NH_4NO_3$ and $N_2O$ are omitted by convention.

As a second example, consider the reaction in which butane ($C_4H_{10}$) is burned in oxygen to form carbon dioxide and water:

$$\_\,C_4H_{10} + \_\,O_2 \longrightarrow \_\,CO_2 + \_\,H_2O$$

Spaces have been left for the coefficients that tell the number of moles of each reactant and product. Start with 1 mol of butane, $C_4H_{10}$. It contains 4 mol of carbon atoms and thus must produce 4 mol of carbon dioxide molecules, because there is nowhere else the carbon can go. The coefficient of $CO_2$ is therefore 4. In the same way, the 10 mol of hydrogen *atoms* must form 5 mol of water *molecules,* because each water molecule contains two hydrogen atoms; thus, the coefficient of the $H_2O$ is 5:

$$C_4H_{10} + \_\,O_2 \longrightarrow 4\,CO_2 + 5\,H_2O$$

Four moles of $CO_2$ contains 8 mol of oxygen atoms, and 5 mol of $H_2O$ contains 5 mol of oxygen atoms, giving a total of 13 mol of oxygen atoms. Thirteen moles of oxygen atoms are equivalent to $\frac{13}{2}$ mol of oxygen molecules, so the coefficient of $O_2$ is $\frac{13}{2}$. The balanced equation is

**Figure 1-20**

An oxyacetylene torch is used to weld metals in the high-temperature flame produced by burning acetylene in oxygen. (Bethlehem Steel/Phototake)

$$C_4H_{10} + \tfrac{13}{2} O_2 \longrightarrow 4\,CO_2 + 5\,H_2O$$

There is nothing wrong with fractions such as $\tfrac{13}{2}$ in a balanced equation, because fractions of moles are perfectly meaningful. It is often the custom to eliminate any such fractions, however. Here, multiplying all coefficients in the equation by 2 gives

$$2\,C_4H_{10} + 13\,O_2 \longrightarrow 8\,CO_2 + 10\,H_2O$$

Let us summarize the steps in balancing a chemical equation:

1. Assign 1 as the coefficient of one species. The best choice is the most complicated species, with the largest number of elements.
2. Identify, in sequence, elements that appear in only one chemical species whose coefficient is not yet determined. Choose that coefficient to balance the number of moles of atoms of that element. Continue until all coefficients are identified.
3. If desired, multiply the whole equation by the smallest integer that will eliminate any fractions.

This method of balancing equations "by inspection" works in many, but not all, cases. In Section 10-1, we present techniques for balancing certain more complex chemical equations.

## Example 1-8

The Hargreaves process is an industrial procedure for making sodium sulfate ($Na_2SO_4$) for use in papermaking. The starting materials are sodium chloride (NaCl), sulfur dioxide ($SO_2$), water, and oxygen; hydrogen chloride (HCl) is generated as a byproduct. Write a balanced chemical equation for this process.

### Solution

The unbalanced equation is

$$\underline{\quad}\,NaCl + \underline{\quad}\,SO_2 + \underline{\quad}\,H_2O + \underline{\quad}\,O_2 \longrightarrow \underline{\quad}\,Na_2SO_4 + \underline{\quad}\,HCl$$

Begin by assigning a coefficient of 1 to the $Na_2SO_4$ because it is the most complex species, composed of three different elements. There are 2 mol of Na atoms on the right, giving 2 as the coefficient of the NaCl. By the same argument, the coefficient of the $SO_2$ must be 1 to balance the 1 mol of sulfur on the right. This gives

$$2\,NaCl + SO_2 + \underline{\quad}\,H_2O + \underline{\quad}\,O_2 \longrightarrow Na_2SO_4 + \underline{\quad}\,HCl$$

Next we note that there are 2 mol of Cl atoms on the left (reactant) side, so the coefficient of the HCl must be 2. Hydrogen is the next element to balance, with 2 mol on the right side and therefore a coefficient of 1 for the $H_2O$:

$$2\,NaCl + SO_2 + H_2O + \underline{\quad}\,O_2 \longrightarrow Na_2SO_4 + 2\,HCl$$

Finally, the oxygen atoms must be balanced. There are 4 mol of oxygen atoms on the right, but the left contains 2 mol from the $SO_2$ and 1 from the $H_2O$; therefore, 1 mol of oxygen *atoms* must come from the $O_2$. The coefficient of the $O_2$ is therefore $\tfrac{1}{2}$:

$$2\,NaCl + SO_2 + H_2O + \tfrac{1}{2} O_2 \longrightarrow Na_2SO_4 + 2\,HCl$$

Multiply all coefficients in the equation by 2, giving

$$4\,NaCl + 2\,SO_2 + 2\,H_2O + O_2 \longrightarrow 2\,Na_2SO_4 + 4\,HCl$$

In balancing the equation, oxygen was considered last because it appears in several places on the left side of the equation.

*Related Problems:* 53, 54

## 1-7 MASS RELATIONSHIPS IN CHEMICAL REACTIONS

A balanced chemical equation makes quantitative statements about the relative masses of the substances reacting. The chemical equation for the combustion of butane

$$2\,C_4H_{10} + 13\,O_2 \longrightarrow 8\,CO_2 + 10\,H_2O$$

can be interpreted as saying either that

2 molecules of $C_4H_{10}$ + 13 molecules of $O_2 \longrightarrow$
$\qquad\qquad\qquad$ 8 molecules of $CO_2$ + 10 molecules of $H_2O$

or that

2 mol of $C_4H_{10}$ + 13 mol of $O_2 \longrightarrow$ 8 mol of $CO_2$ + 10 mol of $H_2O$

Multiplying the molar mass of each substance in the reaction by the number of moles represented in the balanced equation gives

116.3 g of $C_4H_{10}$ + 416.0 g of $O_2 \longrightarrow$ 352.1 g of $CO_2$ + 180.2 g of $H_2O$

The study of the relationships between masses of reactants and products is called **stoichiometry** (from the Greek *stoicheion,* "element," + *metron,* "measure"). Stoichiometry is fundamental to all of chemistry.

The coefficients in a balanced chemical equation give "chemical conversion factors" between the chemical amounts of substances consumed in or produced by a reaction. If 6.16 mol of butane reacts according to the preceding equation, the amounts of $O_2$ consumed and $CO_2$ generated are

$$\text{moles } O_2 = 6.16 \text{ mol } C_4H_{10} \times \left(\frac{13 \text{ mol } O_2}{2 \text{ mol } C_4H_{10}}\right) = 40.0 \text{ mol } O_2$$

$$\text{moles } CO_2 = 6.16 \text{ mol } C_4H_{10} \times \left(\frac{8 \text{ mol } CO_2}{2 \text{ mol } C_4H_{10}}\right) = 24.6 \text{ mol } CO_2$$

For most practical purposes we shall be interested in the *masses* of reactants and products, because those are the quantities that are directly measured. In this case, the molar masses (calculated from a table of atomic masses) are used to convert the chemical amount of a substance (in moles) to its mass (in grams), as illustrated by the following example.

## Example 1-9

Calcium hypochlorite, $Ca(OCl)_2$, is used as a bleaching agent (Fig. 1-21). It is produced from sodium hydroxide, calcium hydroxide, and chlorine according to the overall equation

**Figure 1-21**
A solution of calcium hypochlorite, $Ca(OCl)_2$, is a powerful bleaching agent. A few drops of a solution quickly destroy the color of the blue dye on the cloth (*right*). (Julius Weber)

$$2\,NaOH + Ca(OH)_2 + 2\,Cl_2 \longrightarrow Ca(OCl)_2 + 2\,NaCl + 2\,H_2O$$

How many grams of chlorine and sodium hydroxide react with 1067 g of $Ca(OH)_2$, and how many grams of calcium hypochlorite are produced?

### Solution

The chemical amount of $Ca(OH)_2$ consumed is

$$\frac{1067\text{ g }Ca(OH)_2}{74.09\text{ g mol}^{-1}} = 14.40\text{ mol }Ca(OH)_2$$

where the molar mass of $Ca(OH)_2$ has been obtained from the molar masses of calcium, oxygen, and hydrogen as

$$40.08 + 2(15.999) + 2(1.0079) = 74.09\text{ g mol}^{-1}$$

According to the balanced equation, 1 mol of $Ca(OH)_2$ reacts with 2 mol of NaOH and 2 mol of $Cl_2$ to produce 1 mol of $Ca(OCl)_2$. If 14.40 mol of $Ca(OH)_2$ reacts, then

$$\text{moles NaOH} = 14.40\text{ mol }Ca(OH)_2 \left(\frac{2\text{ mol NaOH}}{1\text{ mol }Ca(OH)_2}\right)$$
$$= 28.80\text{ mol NaOH}$$

$$\text{moles }Cl_2 = 14.40\text{ mol }Ca(OH)_2 \left(\frac{2\text{ mol }Cl_2}{1\text{ mol }Ca(OH)_2}\right)$$
$$= 28.80\text{ mol }Cl_2$$

$$\text{moles }Ca(OCl)_2 = 14.40\text{ mol }Ca(OH)_2 \left(\frac{1\text{ mol }Ca(OCl)_2}{1\text{ mol }Ca(OH)_2}\right)$$
$$= 14.40\text{ mol }Ca(OCl)_2$$

From the chemical amounts and molar masses of reactants and products, the desired masses are found:

$$\text{mass NaOH reacting} = (28.80\text{ mol})(40.00\text{ g mol}^{-1}) = 1152\text{ g}$$

$$\text{mass }Cl_2\text{ reacting} = (28.80\text{ mol})(70.91\text{ g mol}^{-1}) = 2042\text{ g}$$

$$\text{mass }Ca(OCl)_2\text{ produced} = (14.40\text{ mol})(142.98\text{ g mol}^{-1}) = 2059\text{ g}$$

*Related Problems:* 55, 56, 57, 58

In calculations such as that in Example 1-9, a known mass of one substance takes part in a reaction, and we need to calculate the masses of one or more other reactants or products. Figure 1-22 summarizes the three-step process used. With experience, it is possible to write down the answers in a shorthand form so that all

### Figure 1-22

The steps in a stoichiometric calculation. In a typical calculation, the mass of one reactant or product is known, and the masses of one or more other reactants or products are to be calculated using the balanced chemical equation and a table of relative atomic masses.

three conversions are carried out at the same time. The chemical amount of NaOH reacting in the preceding example can be written as

$$\left(\frac{1067 \text{ g Ca(OH)}_2}{74.10 \text{ g mol}^{-1}}\right) \times \left(\frac{2 \text{ mol NaOH}}{1 \text{ mol Ca(OH)}_2}\right) \times 40.00 \text{ g mol}^{-1} = 1152 \text{ g NaOH}$$

It is better at first, however, to follow a stepwise procedure for such calculations.

## Limiting Reactant

In the cases considered so far, the reactants were present in the exact ratios necessary for them to be completely consumed in forming products. This is not the usual case, however. Suppose arbitrary amounts of reactants are mixed and allowed to react. The one that is used up first is the **limiting reactant,** and a portion of the other reactants remains after completion of the reaction; they are present **in excess.** An increase in the amount of the limiting reactant will lead to an increase in the amount of product formed. This is not true of the other reactants. In an industrial process, the limiting reactant is frequently the most expensive one, to ensure that none of it is wasted. For instance, the silver nitrate used in preparing silver chloride for photographic film by the reaction

$$AgNO_3 + NaCl \longrightarrow AgCl + NaNO_3$$

is far more expensive than the sodium chloride (ordinary salt). Thus, it makes sense to carry out the reaction with an excess of sodium chloride, to ensure that as much of the silver nitrate as possible reacts to form products.

There is a systematic method to find the limiting reactant and determine the maximum possible amounts of products. Take each reactant in turn, assume that it is used up completely in the reaction, and calculate the mass of one of the products that will be formed. Whichever reactant gives the *smallest* mass of this product is the limiting reactant. Once it has reacted fully, no further product can be formed.

## Example 1-10

Sulfuric acid ($H_2SO_4$) forms in the chemical reaction

$$2 SO_2 + O_2 + 2 H_2O \longrightarrow 2 H_2SO_4$$

Suppose 400 g $SO_2$, 175 g $O_2$, and 125 g $H_2O$ are mixed and the reaction proceeds until one of the reactants is used up. Which is the limiting reactant? What mass of $H_2SO_4$ is produced, and what masses of the other reactants remain?

## Solution

The chemical amount of each reactant originally present is calculated by dividing each mass by the corresponding molar mass:

$$\frac{400 \text{ g SO}_2}{64.06 \text{ g mol}^{-1}} = 6.24 \text{ mol SO}_2$$

$$\frac{175 \text{ g O}_2}{32.00 \text{ g mol}^{-1}} = 5.47 \text{ mol O}_2$$

$$\frac{125 \text{ g H}_2\text{O}}{18.02 \text{ g mol}^{-1}} = 6.94 \text{ mol H}_2\text{O}$$

If all the $SO_2$ reacted, it would give

$$6.24 \text{ mol } SO_2 \times \left( \frac{2 \text{ mol } H_2SO_4}{2 \text{ mol } SO_2} \right) = 6.24 \text{ mol } H_2SO_4$$

If all the $O_2$ reacted, it would give

$$5.47 \text{ mol } O_2 \times \left( \frac{2 \text{ mol } H_2SO_4}{1 \text{ mol } O_2} \right) = 10.94 \text{ mol } H_2SO_4$$

Finally, if all the water reacted, it would give

$$6.94 \text{ mol } H_2O \times \left( \frac{2 \text{ mol } H_2SO_4}{2 \text{ mol } H_2O} \right) = 6.94 \text{ mol } H_2SO_4$$

In this case $SO_2$ is the limiting reactant, because the computation based on its amount gives the smallest amount of product (6.24 mol $H_2SO_4$). Oxygen and water are present in excess. After the reaction the amount of each that remains is the original amount minus the amount reacting:

$$\text{moles } O_2 = 5.47 \text{ mol } O_2 - \left( 6.24 \text{ mol } SO_2 \times \frac{1 \text{ mol } O_2}{2 \text{ mol } SO_2} \right)$$

$$= 5.47 - 3.12 \text{ mol } O_2 = 2.35 \text{ mol } O_2$$

$$\text{moles } H_2O = 6.94 \text{ mol } H_2O - \left( 6.24 \text{ mol } SO_2 \times \frac{2 \text{ mol } H_2O}{2 \text{ mol } SO_2} \right)$$

$$= 6.94 - 6.24 \text{ mol } H_2O = 0.70 \text{ mol } H_2O$$

The masses of reactants and products after the reaction are

$$\text{mass } H_2SO_4 \text{ produced} = (6.24 \text{ mol})(98.07 \text{ g mol}^{-1}) = 612 \text{ g}$$

$$\text{mass } O_2 \text{ remaining} = (2.35 \text{ mol})(32.00 \text{ g mol}^{-1}) = 75 \text{ g}$$

$$\text{mass } H_2O \text{ remaining} = (0.70 \text{ mol})(18.02 \text{ g mol}^{-1}) = 13 \text{ g}$$

The total mass at the end is $612 + 13 + 75 = 700$ g, which is, of course, equal to the total mass originally present, $400 + 175 + 125 = 700$ g, as required by the law of conservation of mass.

*Related Problems:* 69, 70

## Percentage Yield

The amounts of products calculated so far have been **theoretical yields,** determined by assuming that the reaction goes cleanly and completely. The **actual yield** of a product (the amount present after separating it from other products and reactants and purifying it) is usually less than the theoretical yield. There are several possible reasons for this. The reaction may stop short of completion so that reactants remain unreacted. There may be competing reactions that give other products and so reduce the yield of the desired one. Finally, in the process of separation and purification some of the product is invariably lost, although that amount can be reduced by careful experimental techniques. The ratio of the actual yield to the theoretical yield (multiplied by 100%) gives the **percentage yield** for that product in the reaction.

## Example 1-11

The sulfide ore of zinc (ZnS) is reduced to elemental zinc by ''roasting'' it (heating it in air) to give ZnO and then heating the ZnO with carbon monoxide. The two reactions can be written as

$$ZnS + \tfrac{3}{2} O_2 \longrightarrow ZnO + SO_2$$

$$ZnO + CO \longrightarrow Zn + CO_2$$

Suppose 5.32 kg of ZnS is treated in this way and 3.30 kg of pure Zn is obtained. Calculate the theoretical yield of Zn and its actual percentage yield.

## Solution

From the molar mass of ZnS (97.46 g mol$^{-1}$), the chemical amount of ZnS initially present is

$$\frac{5320 \text{ g ZnS}}{97.46 \text{ g mol}^{-1}} = 54.6 \text{ mol ZnS}$$

Because each mole of ZnS gives 1 mol of ZnO in the first chemical equation, and each mole of ZnO then gives 1 mol of Zn, the theoretical yield of Zn is 54.6 mol. In grams, this is

$$54.6 \text{ mol Zn} \times 65.39 \text{ g mol}^{-1} = 3570 \text{ g Zn}$$

The ratio of actual yield to theoretical yield, multiplied by 100%, gives the percentage yield of Zn:

$$\% \text{ yield} = \left( \frac{3.30 \text{ kg}}{3.57 \text{ kg}} \right) \times 100\% = 92.4\%$$

*Related Problems:* 71, 72

It is clearly desirable to achieve as high a percentage yield of product as possible, in order to reduce consumption of raw materials. In some synthetic reactions (especially in organic chemistry), the final product is the result of many successive reactions. In such processes the yields in the individual steps must be quite high if the synthetic method is to be a practical success. Suppose, for example, that ten consecutive reactions must be carried out to reach the product, and each has a percentage yield of only 50% (a fractional yield of 0.5). The overall yield is the product of the fractional yields of the steps:

$$\underbrace{(0.5) \times (0.5) \times \cdots \times (0.5)}_{10 \text{ terms}} = (0.5)^{10} = 0.001$$

This overall percentage yield of 0.1% makes the process useless for synthetic purposes. If all the individual percentage yields could be increased to 90%, however, the overall yield would be $(0.9)^{10} = 0.35$, or a 35% overall yield. This is much more reasonable and might make the process a feasible one.

## Titanium

Metallic titanium and its alloys (especially those with aluminum and vanadium) combine the advantages of high strength and light weight, and so are widely used in the aerospace industry for the bodies and engines of airplanes. The major natural source for titanium is the ore rutile, containing titanium dioxide ($TiO_2$).

(a) An intermediate in the preparation of elemental titanium from $TiO_2$ is a volatile chloride of titanium (boiling point 136°C) that contains 25.24% titanium by mass. Determine the empirical formula of this compound.

(b) At 136°C and atmospheric pressure, the density of this gaseous chloride is 5.6 g $L^{-1}$. Under the same conditions, the density of gaseous nitrogen ($N_2$, molar mass 28.0 g $mol^{-1}$) is 0.83 g $L^{-1}$. Determine the molecular formula of this compound.

(c) The titanium chloride dealt with in parts (a) and (b) is produced by the reaction of chlorine with a hot mixture of titanium dioxide and coke (carbon), with carbon dioxide generated as a byproduct. Write a balanced chemical equation for this reaction.

(d) What mass of chlorine is needed to produce 79.2 g of the titanium chloride?

(e) The titanium chloride then reacts with liquid magnesium at 900°C to give titanium and magnesium chloride ($MgCl_2$). Write a balanced chemical equation for this step in the refining of titanium.

Titanium sponge, ingot, and mill products after removal from a furnace. (TIMET)

(f) Suppose the reaction chamber for part (e) contains 351 g of the titanium chloride and 63.2 g of liquid magnesium. Which is the limiting reactant? What maximum mass of titanium could result?

(g) Isotopic analysis of the titanium from a particular ore gave these results:

| Isotope | Relative Mass | Abundance (%) |
|---------|---------------|---------------|
| $^{46}Ti$ | 45.952633 | 7.93 |
| $^{47}Ti$ | 46.95176 | 7.28 |
| $^{48}Ti$ | 47.947948 | 73.94 |
| $^{49}Ti$ | 48.947867 | 5.51 |
| $^{50}Ti$ | 49.944789 | 5.34 |

Calculate the mass of a single $^{48}Ti$ atom and the *average* mass of the titanium atoms in this ore sample.

### Answers

(a) $TiCl_4$

(b) $TiCl_4$

(c) $TiO_2 + C + 2 Cl_2 \rightarrow TiCl_4 + CO_2$

(d) 59.2 g

(e) $TiCl_4 + 2 Mg \rightarrow Ti + 2 MgCl_2$

(f) Mg; 62.3 g

(g) $7.961949 \times 10^{-23}$ g, $7.950 \times 10^{-23}$ g

After studying this chapter and working the problems that follow, you should be able to

1. Describe in operational terms how to distinguish among mixtures, compounds, and elements (Sec. 1-2).
2. Outline Dalton's atomic theory of matter and describe its experimental basis.
3. Describe the reasoning that permits chemical formulas to be determined by purely chemical means (Sec. 1-3, problems 7–10).
4. Describe the experiments that led to the discovery of the electron and the measurement of its mass and charge, and those that revealed the nature of the nucleus (Sec. 1-4).
5. State the numbers of protons, neutrons, and electrons in particular atoms (Sec. 1-4, problems 15–18).
6. Given the atomic masses and natural abundances of the isotopes of an element, calculate its chemical atomic mass (Sec. 1-5, problems 19–20).
7. Interconvert the mass, chemical amount, number of molecules, and (using the density) volume of a substance (Sec. 1-5, problems 25–34).
8. Given the percentages by mass of the elements in a compound, determine its empirical formula (Sec. 1-6, problems 41–46).
9. Use ratios of gas densities to estimate molecular mass and determine molecular formulas (Sec. 1-6, problems 49–52).
10. Balance simple chemical equations (Sec. 1-6, problems 53–54).
11. Given the mass of a reactant or product in a chemical reaction, use a balanced chemical equation to calculate the masses of other reactants consumed and other products formed (Sec. 1-7, problems 55–64).
12. Given a set of initial masses of reactants and a balanced chemical equation, determine the limiting reactant and calculate the masses of reactants and products after the reaction has gone to completion (Sec. 1-7, problems 69–70).
13. Determine the percentage yield of a reaction from its calculated theoretical yield and its measured actual yield (Sec. 1-7, problems 71–72).

## PROBLEMS

### The Composition of Matter

1. Classify the following materials as substances or mixtures: table salt, wood, mercury, air, water, seawater, sodium chloride, mayonnaise. If they are mixtures, subclassify them as homogeneous or heterogeneous; if they are substances, subclassify them as compounds or elements.
2. Classify the following materials as substances or mixtures: absolute (pure) alcohol, milk (as purchased in a store), copper wire, rust, barium bromide, concrete, baking soda, baking powder. If they are mixtures, subclassify them as homogeneous or heterogeneous; if they are substances, subclassify them as compounds or elements.
3. A 17th-century chemist wrote of the "simple bodies which enter originally into the composition of mixtures and into which these mixtures resolve themselves or may be finally resolved." What is being discussed?
4. Since 1800, nearly 200 sincere but erroneous reports of the discovery of new chemical elements have been made. Why

have mistaken reports of new elements been so numerous? Why is it relatively easy to prove that a material is *not* a chemical element but difficult to prove absolutely that a material *is* an element?

### The Atomic Theory of Matter

5. A sample of ascorbic acid (vitamin C) is synthesized in the laboratory. It contains 30.0 g of carbon and 40.0 g of oxygen. Another sample of ascorbic acid, isolated from lemons (an excellent source of the vitamin), contains 12.7 g of carbon. Compute the mass of oxygen (in grams) in the second sample.
6. A sample of a compound synthesized and purified in the laboratory contains 25.0 g of hafnium and 31.5 g of tellurium. The identical compound is discovered in a rock formation. A sample from the rock formation contains 0.125 g of hafnium. Determine how much tellurium is in the sample from the rock formation.

7. Nitrogen (N) and silicon (Si) form two binary compounds with the following compositions:

| Compound | Mass % N | Mass % Si |
|----------|----------|-----------|
| 1 | 33.28 | 66.72 |
| 2 | 39.94 | 60.06 |

(a) Compute the mass of silicon that combines with 1.0000 g of nitrogen in each case.

(b) Show that these compounds satisfy the law of multiple proportions. If the second compound has the formula $Si_3N_4$, what is the formula of the first compound?

8. Iodine (I) and fluorine (F) form a series of binary compounds with the following compositions:

| Compound | Mass % I | Mass % F |
|----------|----------|----------|
| 1 | 86.979 | 13.021 |
| 2 | 69.007 | 30.993 |
| 3 | 57.191 | 42.809 |
| 4 | 48.829 | 51.171 |

(a) Compute in each case the mass of fluorine that combines with 1.0000 g of iodine.

(b) By figuring out small whole-number ratios among the four answers in part (a), show that these compounds satisfy the law of multiple proportions.

9. Vanadium and oxygen form a series of compounds with the following compositions:

| Mass % V | Mass % O |
|----------|----------|
| 76.10 | 23.90 |
| 67.98 | 32.02 |
| 61.42 | 38.58 |
| 56.02 | 43.98 |

What are the relative numbers of atoms of oxygen in the compounds for a given mass of vanadium?

10. Tungsten (W) and chlorine (Cl) form a series of compounds with the following compositions:

| Mass % W | Mass % Cl |
|----------|-----------|
| 72.17 | 27.83 |
| 56.45 | 43.55 |
| 50.91 | 49.09 |
| 46.36 | 53.64 |

If a molecule of each compound contains only one tungsten atom, what are the formulas for the four compounds?

11. A liquid compound containing only hydrogen and oxygen is placed in a flask. Two electrodes are dipped into the liquid, and an electric current is passed between them. Gaseous hydrogen forms at one electrode and gaseous oxygen at the other. After a time, 14.4 mL of hydrogen has evolved at the negative terminal and 14.4 mL of oxygen has evolved at the positive terminal.

(a) Assign a chemical formula to the compound in the cell.

(b) Explain why more than one formula is possible as the answer to part (a).

12. A sample of liquid $N_2H_4$ is decomposed to give gaseous $N_2$ and gaseous $H_2$. The two gases are separated, and the nitrogen occupies 13.7 mL at room conditions of pressure and temperature. Determine the volume of the hydrogen under the same conditions.

13. Pure nitrogen dioxide ($NO_2$) forms when dinitrogen oxide ($N_2O$) and oxygen ($O_2$) are mixed in the presence of a certain catalyst. What volumes of $N_2O$ and oxygen are needed to produce 4.0 L of $NO_2$ if all gases are held at the same conditions of temperature and pressure?

14. Gaseous methanol ($CH_3OH$) reacts with oxygen ($O_2$) to produce water vapor and carbon dioxide. What volumes of water vapor and carbon dioxide will be produced from 2.0 L of methanol if all gases are held at the same temperature and pressure conditions?

## Building Blocks of the Atom

15. The isotope of plutonium used for nuclear fission is $^{239}Pu$. Determine (a) the ratio of the number of neutrons in a $^{239}Pu$ nucleus to the number of protons and (b) the number of electrons in a single Pu atom.

16. The last "missing" element from the first six periods was promethium, which was finally discovered in 1947 among the fission products of uranium. Determine (a) the ratio of the number of neutrons in a $^{145}Pm$ nucleus to the number of protons and (b) the number of electrons in a single Pm atom.

17. The americium isotope $^{241}Am$ is used in smoke detectors. Describe the composition of a neutral atom of this isotope in terms of protons, neutrons, and electrons.

18. In 1982, the production of a single atom of $^{266}_{109}Une$ (unnilennium-266) was reported. Describe the composition of a neutral atom of this isotope in terms of protons, neutrons, and electrons.

## Weighing and Counting Molecules

19. The natural abundances and isotopic masses of the element silicon (Si) relative to $^{12}C = 12.00000$ are

| Isotope | % Abundance | Isotopic Mass |
|---------|-------------|---------------|
| $^{28}Si$ | 92.21 | 27.97693 |
| $^{29}Si$ | 4.70 | 28.97649 |
| $^{30}Si$ | 3.09 | 29.97376 |

Calculate the atomic mass of naturally occurring silicon.

20. The natural abundances and isotopic masses of the element neon (Ne) are

| Isotope | % Abundance | Isotopic Mass |
|---------|-------------|---------------|
| $^{20}Ne$ | 90.00 | 19.99212 |
| $^{21}Ne$ | 0.27 | 20.99316 |
| $^{22}Ne$ | 9.73 | 21.99132 |

Calculate the atomic mass of naturally occurring neon.

21. Only two isotopes of boron (B) occur in nature; their atomic masses and abundances are given in the following table. Complete the table by computing the relative atomic mass of $^{11}B$ to four significant figures, taking the tabulated relative atomic mass of natural boron as 10.811.

| Isotope | % Abundance | Atomic Mass |
|---------|-------------|-------------|
| $^{10}B$ | 19.61 | 10.013 |
| $^{11}B$ | 80.39 | ? |

22. Over half of all the atoms in naturally occurring zirconium are $^{90}Zr$. The other four stable isotopes of zirconium have the following relative atomic masses and abundances:

| Isotope | % Abundance | Atomic Mass |
|---------|-------------|-------------|
| $^{91}Zr$ | 11.27 | 90.9056 |
| $^{92}Zr$ | 17.17 | 91.9050 |
| $^{94}Zr$ | 17.33 | 93.9063 |
| $^{96}Zr$ | 2.78 | 95.9083 |

Compute the relative atomic mass of $^{90}Zr$ to four significant figures, using the tabulated relative atomic mass 91.224 for natural zirconium.

23. Compute the mass, in grams, of a single iodine atom if the relative atomic mass of iodine is 126.90447 on the accepted scale of atomic masses (based on 12 as the relative atomic mass of $^{12}C$).

24. Determine the mass, in grams, of exactly 100 million atoms of fluorine if the relative atomic mass of fluorine is 18.998403 on a scale on which exactly 12 is the relative atomic mass of $^{12}C$.

25. Compute the relative molecular masses of the following compounds on the $^{12}C$ scale.
    (a) $P_4O_{10}$
    (b) $BrCl$
    (c) $Ca(NO_3)_2$
    (d) $KMnO_4$
    (e) $(NH_4)_2SO_4$

26. Compute the relative molecular masses of the following compounds on the $^{12}C$ scale.
    (a) $[Ag(NH_3)_2]Cl$
    (b) $Ca_3[Co(CO_3)_3]_2$
    (c) $OsO_4$
    (d) $H_2SO_4$
    (e) $Ca_3Al_2(SiO_4)_3$

27. Suppose that a person counts out gold atoms at the rate of one each second for the entire span of an 80-year life. Has the person counted enough atoms to be detected with an ordinary balance? Explain.

28. A gold atom has a diameter of $2.88 \times 10^{-10}$ m. Suppose the atoms in 1.00 mol of gold atoms are arranged just touching their neighbors in a single straight line. Determine the length of the line.

29. The vitamin A molecule has the formula $C_{20}H_{30}O$, and a molecule of vitamin $A_2$ has the formula $C_{20}H_{28}O$. Determine how many moles of vitamin $A_2$ contain the same number of atoms as 1.000 mol of vitamin A.

30. Arrange the following in order of increasing mass: 1.06 mol $SF_4$; 117 g $CH_4$; $8.7 \times 10^{23}$ molecules of $Cl_2O_7$; $417 \times 10^{23}$ atoms of argon (Ar).

31. Mercury is traded by the "flask," a unit that has a mass of 34.5 kg. Determine the volume of a flask of mercury if the density of mercury is 13.6 g cm$^{-3}$.

32. Gold costs $400 per troy ounce, and one troy ounce equals 31.1035 g. Determine the cost of 10.0 cm$^3$ of gold if the density of gold is 19.32 g cm$^{-3}$ at room conditions.

33. Aluminum oxide ($Al_2O_3$) occurs in nature as a mineral called corundum, which is noted for its hardness and its resistance to attack by acids. Its density is 3.97 g cm$^{-3}$. Calculate the number of atoms of aluminum in 15.0 cm$^3$ of corundum.

34. Calculate the number of atoms of silicon (Si) in 415 cm$^3$ of the colorless gas disilane at 0°C and atmospheric pressure, where its density is 0.00278 g cm$^{-3}$. The molecular formula of disilane is $Si_2H_6$.

## Chemical Formulas and Chemical Equations

35. A newly synthesized compound has the molecular formula $ClF_2O_2PtF_6$. Compute, to four significant figures, the mass percentage of each of the four elements in this compound.

36. Acetaminophen is the generic name of the pain reliever in Tylenol and some other headache remedies. The compound has the molecular formula $C_8H_9NO_2$. Compute, to four significant figures, the mass percentage of each of the four elements in acetaminophen.

37. Arrange the following compounds from left to right in order of increasing percentage by mass of hydrogen: $H_2O$, $C_{12}H_{26}$, $N_4H_6$, LiH.

38. Arrange the following compounds from left to right in order of increasing percentage by mass of fluorine: HF, $C_6HF_5$, BrF, $UF_6$.

39. "Q-gas" is a mixture of 98.70% helium and 1.30% butane ($C_4H_{10}$) by mass. It is used as a filling for gas-flow Geiger counters. Compute the mass percentage of hydrogen in Q-gas.

40. A pharmacist prepares an antiulcer medicine by mixing 286 g of $Na_2CO_3$ with water, adding 150 g of glycine ($C_2H_5NO_2$), and stirring continuously at 40°C until a firm mass results. She heats the mass gently until all of the water has been driven away. No other chemical changes occur in this step.

Compute the mass percentage of carbon in the resulting white crystalline medicine.

41. Zinc phosphate is used as a dental cement. A 50.00-mg sample is broken down into its constituent elements and gives 16.58 mg of oxygen, 8.02 mg of phosphorus, and 25.40 mg of zinc. Determine the empirical formula of zinc phosphate.

42. Bromoform is 94.85% bromine, 0.40% hydrogen, and 4.75% carbon by mass. Determine its empirical formula.

43. Fulgurites are the products of the melting that occurs when lightning strikes the earth. Microscopic examination of a sand fulgurite reveals that it is a globule with variable composition, containing some grains of the definite chemical composition Fe 46.01%, Si 53.99%. Determine the empirical formula of these grains.

44. A sample of a ''suboxide'' of cesium gives up 1.6907% of its mass as gaseous oxygen when gently heated, leaving pure cesium behind. Determine the empirical formula of this binary compound.

45. Barium and nitrogen form two binary compounds containing 90.745% and 93.634% barium, respectively. Determine the empirical formulas of these two compounds.

46. Carbon and oxygen form no fewer than five different binary compounds. The mass percentages of carbon in the five compounds are as follows: A, 27.29; B, 42.88; C, 50.02; D, 52.97; and E, 65.24. Determine the empirical formulas of the five compounds.

47. A sample of 1.000 g of a compound containing carbon and hydrogen reacts with oxygen at elevated temperature to yield 0.692 g of $H_2O$ and 3.381 g of $CO_2$.
    (a) Calculate the masses of C and H in the sample.
    (b) Does the compound contain any other elements?
    (c) What are the mass percentages of C and H in the compound?
    (d) What is the empirical formula of the compound?

48. Burning a compound of calcium, carbon, and nitrogen in oxygen in a combustion train generates calcium oxide (CaO), carbon dioxide ($CO_2$), nitrogen dioxide ($NO_2$), and no other substances. A small sample gives 2.389 g of CaO, 1.876 g of $CO_2$, and 3.921 g of $NO_2$. Determine the empirical formula of the compound.

49. The empirical formula of a gaseous fluorocarbon is $CF_2$. At a certain temperature and pressure, a 1-liter volume holds 8.93 g of this fluorocarbon, while under the same conditions the 1-liter volume holds only 1.70 g of gaseous fluorine ($F_2$). Determine the molecular formula of this compound.

50. At its boiling point (280°C) and at atmospheric pressure, phosphorus has a gas density of 2.7 g $L^{-1}$. Under the same conditions nitrogen has a gas density of 0.62 g $L^{-1}$. How many atoms of phosphorus are there in one phosphorus molecule in the vapor under these conditions?

51. A gaseous binary compound has a vapor density that is 1.94 times that of oxygen at the same temperature and pressure. When 1.39 g of the gas is burned in an excess of oxygen, 1.21 g of water is formed, taking off all the hydrogen originally present.
    (a) Estimate the molecular mass of the gaseous compound.
    (b) How many H atoms are there in a molecule of the compound?
    (c) What is the maximum possible value of the atomic mass of the second element in the compound?
    (d) Are other values possible for the atomic mass of the second element? Use a table of atomic masses to identify the element that best fits the data.
    (e) What is the molecular formula of the compound?

52. A gaseous binary compound has a vapor density that is 2.53 times that of nitrogen at 100°C and atmospheric pressure. When 8.21 g of the gas reacts with $AlCl_3$ at 100°C, 1.62 g of gaseous nitrogen is produced, taking off all the nitrogen originally present.
    (a) Estimate the molecular mass of the gaseous compound.
    (b) How many N atoms are there in a molecule of the compound?
    (c) What is the maximum possible value of the atomic mass of the second element?
    (d) Are other values possible for the atomic mass of the second element? Use a table of atomic masses to identify the element that best fits the data.
    (e) What is the molecular formula of the compound?

53. Balance the following chemical equations.
    (a) $H_2 + N_2 \rightarrow NH_3$
    (b) $K + O_2 \rightarrow K_2O_2$
    (c) $PbO_2 + Pb + H_2SO_4 \rightarrow PbSO_4 + H_2O$
    (d) $BF_3 + H_2O \rightarrow B_2O_3 + HF$
    (e) $KClO_3 \rightarrow KCl + O_2$
    (f) $CH_3COOH + O_2 \rightarrow CO_2 + H_2O$
    (g) $K_2O_2 + H_2O \rightarrow KOH + O_2$
    (h) $PCl_5 + AsF_3 \rightarrow PF_5 + AsCl_3$

54. Balance the following chemical equations.
    (a) $Al + HCl \rightarrow AlCl_3 + H_2$
    (b) $NH_3 + O_2 \rightarrow NO + H_2O$
    (c) $Fe + O_2 + H_2O \rightarrow Fe(OH)_2$
    (d) $HSbCl_4 + H_2S \rightarrow Sb_2S_3 + HCl$
    (e) $Al + Cr_2O_3 \rightarrow Al_2O_3 + Cr$
    (f) $XeF_4 + H_2O \rightarrow Xe + O_2 + HF$
    (g) $(NH_4)_2Cr_2O_7 \rightarrow N_2 + Cr_2O_3 + H_2O$
    (h) $NaBH_4 + H_2O \rightarrow NaBO_2 + H_2$

## Mass Relationships in Chemical Reactions

55. For each of the following chemical reactions, calculate the mass of the underlined reactant that is required to produce 1.000 g of the underlined product.
    (a) $\underline{Mg} + 2 HCl \rightarrow \underline{H_2} + MgCl_2$
    (b) $\underline{2\,CuSO_4} + 4\,KI \rightarrow 2\,CuI + \underline{I_2} + 2\,K_2SO_4$
    (c) $\underline{NaBH_4} + 2\,H_2O \rightarrow NaBO_2 + 4\,\underline{H_2}$

56. For each of the following chemical reactions, calculate the

mass of the underlined product that is produced from 1.000 g of the underlined reactant.

(a) $\underline{CaCO_3} + H_2O \rightarrow \underline{Ca(OH)_2} + CO_2$

(b) $\underline{C_3H_8} + 5\,O_2 \rightarrow 3\,\underline{CO_2} + 4\,H_2O$

(c) $2\,\underline{MgNH_4PO_4} \rightarrow \underline{Mg_2P_2O_7} + 2\,NH_3 + H_2O$

57. An 18.6-g sample of $\underline{K_2CO_3}$ was treated in such a way that all of its carbon was captured in the compound $K_2Zn_3[Fe(CN)_6]_2$. Compute the mass (in grams) of this product.

58. A chemist dissolves 1.406 g of pure platinum (Pt) in an excess of a mixture of hydrochloric and nitric acids and then, after a series of subsequent steps involving several other chemicals, isolates a compound of molecular formula $Pt_2C_{10}H_{18}N_2S_2O_6$. Determine the maximum possible yield of this compound.

59. Disilane ($Si_2H_6$) is a gas that reacts with oxygen to give silica ($SiO_2$) and water. Calculate the mass of silica that would form if 25.0 cm$^3$ of disilane (with a density of $2.78 \times 10^{-3}$ g cm$^{-3}$) reacted with excess oxygen.

60. Tetrasilane ($Si_4H_{10}$) is a liquid with a density of 0.825 g cm$^{-3}$. It reacts with oxygen to give silica ($SiO_2$) and water. Calculate the mass of silica that would form if 25.0 cm$^3$ of tetrasilane reacted completely with excess oxygen.

61. Cryolite ($Na_3AlF_6$) is used in the production of aluminum from its ores. It is made by the reaction

$$6\,NaOH + Al_2O_3 + 12\,HF \longrightarrow 2\,Na_3AlF_6 + 9\,H_2O$$

Calculate the mass of cryolite that can be prepared by the complete reaction of 287 g of $Al_2O_3$.

62. Carbon disulfide ($CS_2$) is a liquid that is used in the production of rayon and cellophane. It is manufactured from methane and elemental sulfur via the reaction

$$CH_4 + 4\,S \longrightarrow CS_2 + 2\,H_2S$$

Calculate the mass of $CS_2$ that can be prepared by the complete reaction of 67.2 g of sulfur.

63. Potassium nitrate ($KNO_3$) is used as a fertilizer for certain crops. It is produced through the reaction

$$4\,KCl + 4\,HNO_3 + O_2 \longrightarrow 4\,KNO_3 + 2\,Cl_2 + 2\,H_2O$$

Calculate the minimum mass of KCl required to produce 567 g of $KNO_3$. What mass of $Cl_2$ will be generated as well?

64. Elemental phosphorus can be prepared from calcium phosphate via the overall reaction

$$2\,Ca_3(PO_4)_2 + 6\,SiO_2 + 10\,C \longrightarrow$$
$$6\,CaSiO_3 + P_4 + 10\,CO$$

Calculate the minimum mass of $Ca_3(PO_4)_2$ required to produce 69.8 g of $P_4$. What mass of $CaSiO_3$ is generated as a byproduct?

*65. An element X has a dibromide with the empirical formula $XBr_2$ and a dichloride with the empirical formula $XCl_2$. The dibromide is completely converted to the dichloride when it is heated in a stream of chlorine according to the equation

$$XBr_2 + Cl_2 \longrightarrow XCl_2 + Br_2$$

When 1.500 g of $XBr_2$ is treated, 0.890 g of $XCl_2$ results.

(a) Calculate the atomic mass of the element X.

(b) By reference to a list of the atomic masses of the elements, identify the element X.

*66. An element A has a triiodide with the formula $AI_3$ and a trichloride with the formula $ACl_3$. The triiodide is quantitatively converted to the trichloride when it is heated in a stream of chlorine, according to the equation

$$AI_3 + \tfrac{3}{2}Cl_2 \longrightarrow ACl_3 + \tfrac{3}{2}I_2$$

If 0.8000 g of $AI_3$ is treated, 0.3776 g of $ACl_3$ is obtained.

(a) Calculate the atomic mass of the element A.

(b) Identify the element A.

*67. A mixture consisting of only sodium chloride (NaCl) and potassium chloride (KCl) weighs 1.0000 g. When the mixture is dissolved in water and an excess of silver nitrate is added, all of the chloride ions associated with the original mixture are precipitated as insoluble silver chloride (AgCl). The mass of the silver chloride is found to be 2.1476 g. Calculate the mass percentages of sodium chloride and potassium chloride in the original mixture.

*68. A mixture of aluminum and iron weighing 9.62 g reacts with hydrogen chloride in aqueous solution according to the parallel reactions

$$2\,Al + 6\,HCl \longrightarrow 2\,AlCl_3 + 3\,H_2$$
$$Fe + 2\,HCl \longrightarrow FeCl_2 + H_2$$

A 0.738-g quantity of hydrogen is evolved when the metals react completely. Calculate the mass of iron in the original mixture.

69. When ammonia is mixed with hydrogen chloride (HCl), the white solid ammonium chloride ($NH_4Cl$) is produced. Suppose 10.0 g of ammonia is mixed with the same mass of hydrogen chloride. What substances will be present after the reaction has gone to completion, and what will their masses be?

70. The poisonous gas hydrogen cyanide (HCN) is produced by the high-temperature reaction of ammonia with methane ($CH_4$). Hydrogen is also produced in this reaction.

(a) Write a balanced chemical equation for the reaction that occurs.

(b) Suppose 500.0 g of methane is mixed with 200.0 g of ammonia. Calculate the masses of the substances present after the reaction is allowed to proceed to completion.

71. The iron oxide $Fe_2O_3$ reacts with carbon monoxide, CO, to give iron and carbon dioxide:

$$Fe_2O_3 + 3\,CO \longrightarrow 2\,Fe + 3\,CO_2$$

The reaction of 433.2 g of $Fe_2O_3$ with excess CO yields

254.3 g of iron. Calculate the theoretical yield of iron (assuming complete reaction) and its percentage yield.

72. Titanium dioxide, $TiO_2$, reacts with carbon and chlorine to give gaseous $TiCl_4$:

$$TiO_2 + 2 C + 2 Cl_2 \longrightarrow TiCl_4 + 2 CO$$

The reaction of 7.39 kg of titanium dioxide with excess C and $Cl_2$ gives 14.24 kg of titanium tetrachloride. Calculate the theoretical yield of $TiCl_4$ (assuming complete reaction) and its percentage yield.

## Additional Problems

73. Soft-wood chips weighing 17.2 kg are placed in an iron vessel and mixed with 150.1 kg of water and 22.43 kg of sodium hydroxide. A steel lid seals the vessel, which is then placed in an oven at 250°C for 6 hours. Much of the wood fiber decomposes under these conditions; the vessel and lid do not react. (a) Classify each of the materials mentioned as a substance or mixture. Subclassify the substances as elements or compounds.
(b) Determine the mass of the contents of the iron vessel after the reaction.

*74. In a reproduction of the Millikan oil-drop experiment, a student obtains the following values for the charges on nine different oil droplets.

| | | |
|---|---|---|
| $6.563 \times 10^{-19}$ C | $13.13 \times 10^{-19}$ C | $19.71 \times 10^{-19}$ C |
| $8.204 \times 10^{-19}$ C | $16.48 \times 10^{-19}$ C | $22.89 \times 10^{-19}$ C |
| $11.50 \times 10^{-19}$ C | $18.08 \times 10^{-19}$ C | $26.18 \times 10^{-19}$ C |

(a) Based on these data alone, what is your best estimate of the number of electrons on each of the above droplets? (*Hint:* Begin by considering differences in charges between adjacent data points, and see what groups these fall into.)
(b) Based on these data alone, what is your best estimate of the charge on the electron?
(c) Is it conceivable that the actual charge is half the charge you calculated in (b)? What evidence would help you decide one way or the other?

75. A rough estimate of the radius of a nucleus is provided by the formula $r = kA^{1/3}$, where $k$ is approximately $1.3 \times 10^{-13}$ cm and $A$ is the mass number of the nucleus. Estimate the density of the nucleus of $^{127}I$ (which has a nuclear mass of $2.1 \times 10^{-22}$ g), in grams per cubic centimeter. Compare with the density of solid iodine, 4.93 g cm$^{-3}$.

76. In a neutron star, gravity causes the electrons to combine with protons to form neutrons. A typical neutron star has a mass half that of the sun, compressed into a sphere of radius 20 km. If such a neutron star contains $6.0 \times 10^{56}$ neutrons, calculate its density in grams per cubic centimeter. Compare this with the density inside a $^{232}Th$ nucleus, in which 142 neutrons and 90 protons occupy a sphere of radius $9.1 \times 10^{-13}$ cm. Take

the mass of a neutron to be $1.675 \times 10^{-24}$ g and that of a proton to be $1.673 \times 10^{-24}$ g.

77. Dalton's 1808 version of the atomic theory of matter included five general statements (Sec. 1-3). According to modern understanding, four of those statements require amendment or extension. List the modifications that have been made to four of the five original postulates.

78. Naturally occurring rubidium (Rb) consists of two isotopes: $^{85}Rb$ (atomic mass 84.9117) and $^{87}Rb$ (atomic mass 86.9092). The atomic mass of the isotope mixture found in nature is 85.4678. Calculate the percent abundances of the two isotopes in rubidium.

*79. The inhabitants of a planet in a distant galaxy measure mass in units of "margs," where 1 marg = 4.8648 g. Their scale of atomic masses is based on the isotope $^{32}S$ (atomic mass on earth = 31.972), so they define one "elom" of $^{32}S$ as the amount of sulfur in exactly 32 margs of $^{32}S$. Furthermore, they define $N_{or}$, or "Ordagova's number" (after the well-known scientist Oedema Ordagova), as the number of atoms of sulfur in exactly 32 margs of $^{32}S$.
(a) Calculate the numerical value of $N_{or}$ to four significant figures.
(b) Calculate the atomic mass of phosphorus in margs per elom to five significant figures, assuming that the abundances of phosphorus isotopes are the same as those found on earth.

80. At one time, it was thought that indium formed a chloride of formula $InCl_2$. More recent work shows that the compound in question is actually $In_3[In_2Cl_9]$. Determine the percentage by mass of indium according to each formula, and explain how the error could be made.

81. Human parathormone has the impressive molecular formula $C_{691}H_{898}N_{125}O_{164}S_{11}$. Compute the mass percentages of all of the elements in this compound.

82. A white oxide of tungsten is 79.2976% tungsten by mass. A blue tungsten oxide also contains exclusively tungsten and oxygen, but it is 80.8473% tungsten by mass. Determine the empirical formulas of white tungsten oxide and blue tungsten oxide.

83. A dark brown binary compound contains oxygen and a metal. It is 13.38% oxygen by mass. Heating it moderately drives off some of the oxygen and gives a red binary compound that is 9.334% oxygen by mass. Strong heating drives off more oxygen and gives still another binary compound that is only 7.168% oxygen by mass.
(a) Compute the mass of oxygen that is combined with 1.000 g of the metal in each of these three oxides.
(b) Assume that the empirical formula of the first compound is $MO_2$ (where M stands for the metal). Give the empirical formulas of the second and third compounds.
(c) Name the metal.

84. A binary compound of nickel and oxygen contains 78.06% nickel by mass. Is this a stoichiometric or a nonstoichiometric compound? Explain.

**85.** Two binary oxides of the element manganese contain, respectively, 30.40% and 36.81% oxygen by mass. Calculate the empirical formulas of the two oxides.

**\*86.** A sample of a gaseous binary compound of boron and chlorine weighing 2.842 g occupies 0.153 L. This sample is decomposed to give 0.664 g of solid boron and enough gaseous chlorine ($Cl_2$) to occupy 0.688 L at the same temperature and pressure. Determine the molecular formula of the compound.

**87.** A possible practical way to eliminate oxides of nitrogen (such as $NO_2$) from automobile exhaust gases employs cyanuric acid, $C_3N_3(OH)_3$. When heated to the relatively low temperature of 625°F, cyanuric acid converts to gaseous isocyanic acid (HNCO). Isocyanic acid reacts with $NO_2$ in the exhaust to form nitrogen, carbon dioxide, and water, all of which are normal constituents of the air.
(a) Write balanced equations for these two reactions.
(b) If the process just described became practical, how much cyanuric acid (in kilograms) would be required to absorb the $1.7 \times 10^{10}$ kg of $NO_2$ generated annually in auto exhaust in the United States?

**88.** Aspartame (molecular formula $C_{14}H_{18}N_2O_5$) is a sugar substitute in soft drinks. Under certain conditions, one mole of aspartame reacts with two moles of water to give one mole of aspartic acid (molecular formula $C_4H_7NO_4$), one mole of methanol (molecular formula $CH_3OH$), and one mole of phenylalanine. Determine the molecular formula of phenylalanine.

**89.** 3′-Methylphthalanilic acid is used commercially as a "fruit set" to prevent premature drop of apples, pears, cherries, and peaches from the tree. It is 70.58% carbon, 5.13% hydrogen, 5.49% nitrogen, and 18.80% oxygen. If eaten, the fruit set reacts with water in the body to split into an innocuous product that contains carbon, hydrogen, and oxygen only and m-toluidine ($NH_2C_6H_4CH_3$), which causes anemia and kidney damage. Compute the mass of the fruit set that would produce 5.23 g of m-toluidine.

**90.** Aluminum carbide ($Al_4C_3$) reacts with water to produce gaseous methane ($CH_4$). Calculate the mass of methane formed from 63.2 g of $Al_4C_3$.

**91.** Citric acid ($C_6H_8O_7$) is made by fermentation of sugars such as sucrose ($C_{12}H_{22}O_{11}$) in air. Oxygen is consumed and water generated as a byproduct.
(a) Write a balanced equation for the overall reaction that occurs in the manufacture of citric acid from sucrose.
(b) What mass of citric acid is made from 15.0 kg of sucrose?

**92.** A sample that contains only $SrCO_3$ and $BaCO_3$ weighs 0.800 g. When it is dissolved in excess acid, 0.211 g of carbon dioxide is liberated. What percentage of $SrCO_3$ did the sample contain? Assume all the carbon originally present is converted to carbon dioxide.

**93.** A sample of a substance with the empirical formula $XBr_2$ weighs 0.5000 g. When it is dissolved in water and all of its

bromide is converted to insoluble AgBr by addition of an excess of silver nitrate, the mass of the resulting AgBr is found to be 1.0198 g. The chemical reaction is

$$XBr_2 + 2\,AgNO_3 \longrightarrow 2\,AgBr + X(NO_3)_2$$

(a) Calculate the molecular mass (i.e., formula mass) of $XBr_2$.
(b) Calculate the atomic mass of X and give its name and symbol.

**94.** A newspaper article about the danger of global warming from the accumulation of greenhouse gases such as carbon dioxide states that "reducing driving your car by 20 miles a week would prevent release of over 1000 pounds of $CO_2$ per year into the atmosphere." Is this a reasonable statement? Assume that gasoline is octane (molecular formula $C_8H_{18}$) and that it is burned completely to $CO_2$ and $H_2O$ in the engine of your car. Facts (or reasonable guesses) about your car's gas mileage, the density of octane, and other factors will also be needed.

**95.** In the Solvay process for producing sodium carbonate ($Na_2CO_3$), the following reactions occur in sequence:

$$NH_3 + CO_2 + H_2O \longrightarrow NH_4HCO_3$$
$$NH_4HCO_3 + NaCl \longrightarrow NaHCO_3 + NH_4Cl$$
$$2\,NaHCO_3 \xrightarrow{\text{heat}} Na_2CO_3 + H_2O + CO_2$$

How many metric tons of sodium carbonate would be produced per metric ton of $NH_3$ if the process were 100% efficient? 1 metric ton = 1000 kg.

**96.** A yield of 3.00 g of $KClO_4$ is obtained from the (unbalanced) reaction

$$KClO_3 \longrightarrow KClO_4 + KCl$$

when 4.00 g of the reactant is used. What is the percentage yield of the reaction?

**97.** An industrial-scale process for making acetylene consists of the following sequence of operations:

$$\underset{\text{limestone}}{CaCO_3} \longrightarrow \underset{\text{lime}}{CaO} + \underset{\text{carbon dioxide}}{CO_2}$$

$$CaO + 3\,C \longrightarrow \underset{\text{calcium carbide}}{CaC_2} + \underset{\text{carbon monoxide}}{CO}$$

$$CaC_2 + 2\,H_2O \longrightarrow \underset{\substack{\text{calcium} \\ \text{hydroxide}}}{Ca(OH)_2} + \underset{\text{acetylene}}{C_2H_2}$$

What is the percentage yield of the overall process if 2.32 metric tons of $C_2H_2$ is produced from 10.0 metric tons of limestone? 1 metric ton = 1000 kg.

**98.** Silicon nitride ($Si_3N_4$), a valuable ceramic, is made by the direct combination of silicon and nitrogen at high temperature. How much silicon must react with excess nitrogen to prepare 125 g of silicon nitride if the yield of the reaction is 95.0%?

# CHAPTER

2

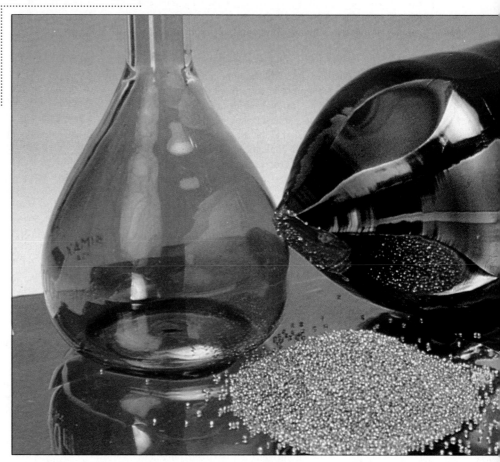

*Samples of bromine and silicon (top) and of copper (below) on a reflecting sheet of aluminum.*

# Chemical Periodicity and Inorganic Reactions

hapter 1 revealed how the law of conservation of mass, through the mole concept, determines quantitative mass relationships in chemical reactions. That discussion assumed prior knowledge of the chemical formulas of the reactants and products in each equation. The far more open-ended question of what compounds are found in nature (or can be made in the laboratory) and what types of reactions they undergo now arises. Why are some elements and compounds fiercely reactive and others inert? Why are there compounds with chemical formulas $H_2O$ and $NaCl$, but never $H_3O$ or $NaCl_2$? Why are helium and the other noble gases monatomic, when molecules of hydrogen and chlorine are diatomic? These are questions that can be answered on various levels, and they will be addressed throughout this book.

As discussed in Section 1-4, every atom contains electrons. The details of the behavior of electrons in atoms are complex, and that discussion is deferred to Chapter 13. The first concern in this chapter is to note certain recurring similarities in the physical and chemical properties of the elements and to explore the origin of these similarities in the numbers and arrangements of electrons in the elements' atoms. Much chemical reactivity involves the transfer or sharing of electrons among atoms, processes that can be described with the Lewis electron-dot model. The second concern in this chapter is to show how to set down and interpret these useful diagrams.

In the course of the chapter a system of inorganic nomenclature is developed and several important classes of chemical reactions are identified and described. The goal is to lay a foundation for organizing the vast body of chemical information in a rational way.

## 2-1 THE PERIODIC TABLE

By the late 1860s more than 60 chemical elements had been identified. Much was known about the descriptive chemistry of these elements, and certain groupings had been suggested based on similarities in physical or chemical properties. Elements were classified as **metals** or **nonmetals,** for example, depending on the presence (or absence) of a characteristic metallic luster, their good (or poor) ability to conduct electricity and heat, and their malleability (or brittleness). The classification of an element as a metal or nonmetal was generally straightforward, but certain elements (antimony, arsenic, boron, silicon, and tellurium) resembled metals in some respects and nonmetals in others and were therefore called **semimetals** or **metalloids.** Close scrutiny of other properties of the elements—in particular the empirical formulas of their binary compounds with chlorine (their *chlorides*), with oxygen (their *oxides*), and with hydrogen (their *hydrides*), allowed the establishment of smaller groups of three to six chemically similar elements. With some changes, this classification of elements into groups has had significance to this day.

For example, the **alkali metals** (lithium, sodium, potassium, rubidium, and cesium) are all relatively soft metals (Fig. 2-1) with low melting points that form

**Figure 2-1**

Elemental sodium is a very soft metal that can be cut easily with a knife. It melts at a temperature of only 97.8°C (just below the normal boiling point of water). Sodium is stored under oil because it reacts rapidly with water vapor in the air. (Leon Lewandowski)

(a)                                                                (b)

**Figure 2-2**

(*a*) When a tiny piece of sodium is added to water, it reacts rapidly to give hydrogen and sodium hydroxide. The hydrogen appears as small gas bubbles that cause the piece of sodium to skitter over the surface of the water. An indicator that turns red as the sodium hydroxide forms has been added to the water. (*b*) The parallel reaction of potassium with water is more vigorous. Its heat ignites both the hydrogen that forms and unreacted potassium, causing a burst of flames as they burn in air. (*a*, Leon Lewandowski)

1:1 compounds with chlorine, having chemical formulas such as NaCl and RbCl. The alkali metals react with water to liberate hydrogen; potassium, rubidium, and cesium do this so vigorously that they must be handled with particular caution, as the heat of their reaction with water causes the hydrogen generated to burst into flame (Fig. 2-2). A second group of elements, the **alkaline-earth metals** (beryllium, magnesium, calcium, strontium, barium, and radium), react in a 1:2 atom ratio with chlorine, giving compounds such as $MgCl_2$ and $CaCl_2$.

Calcium carbonate, $CaCO_3$, occurs in two crystalline forms: calcite (*left*) and aragonite (*right*). Limestone is an aggregate of small crystals of the two, together with a minor percentage of impurities such as $MgCO_3$ and $SiO_2$. (Leon Lewandowki)

Brass-colored crystals of iron pyrites ($FeS_2$), or fool's gold, surrounded by a white feldspar matrix. Pyrite minerals are an abundant and important source of elemental sulfur. Its luster makes iron pyrites appear metallic, but although malleable gold—a true metal—can be cut with a knife, the brittle fool's gold shatters. (Julius Weber)

Of the nonmetallic elements, the **chalcogens** (oxygen, sulfur, selenium, and tellurium) form 1:1 compounds with alkaline-earth metals but 2:1 compounds with the alkali metals; that is, the alkaline-earth chalcogenides have formulas such as CaO and BaS, whereas the alkali-metal chalcogenides have formulas such as $Li_2O$ and $Na_2S$. Chlorine is a member of a family of elements called **halogens** (fluorine, chlorine, bromine, and iodine). These differ significantly in physical properties (the first two are gases at room temperature, bromine is a liquid, and iodine a solid; Fig. 2-3), but their chemical behaviors are similar. Any alkali-metal element will combine with any halogen in 1:1 proportion to form a compound such as LiF or RbI, called an alkali halide. Alkaline-earth metals react with halogens in 1:2 proportion, giving alkaline-earth halides such as $CaF_2$ and $MgBr_2$.

Other elements fall into three additional groups that are somewhat less clearly defined in their chemical and physical properties than those already mentioned. The first of these contains a semimetal (boron) and four metals (aluminum, gallium, indium, and thallium). All form 1:3 chlorides (such as $GaCl_3$) and 2:3 oxides (such as $Al_2O_3$). Another group includes a nonmetal as well as semimetals and metals. It consists of the elements carbon, silicon, germanium, tin, and lead. All of these

**Figure 2-3**
Chlorine, bromine, and iodine differ in color and physical state, being respectively gas, liquid, and solid at room temperature. As members of the halogen family, these elements have closely related chemical properties. (Leon Lewandowski)

Borax, $Na_2B_4O_5(OH)_4 \cdot 8H_2O$, is mined on a scale of millions of tons per year and is the major source of boron. Although boron is not an abundant element in the earth's crust, it occurs in a small number of extensive deposits (associated with past volcanic activity) and can be mined economically. This photo shows an open-pit borax mine and processing plant at Boron, California. (Phototake)

**Figure 2-4**
Graphite (*a*), diamond (*b*), and buckminsterfullerene (*c*) are three different forms of carbon. Graphite is a soft, grayish-black, flaky solid that conducts electricity; diamond consists of very hard, transparent crystals that are nonconductors of electricity; buckminsterfullerene is a newly discovered molecular form of carbon with the formula $C_{60}$. All three forms arise in nature and are now extensively synthesized. (*a*, *b*, Julius Weber; *c*, Donald Huffman, photo by W. Krätschmer)

elements form 1:4 chlorides (such as $SiCl_4$), 1:4 hydrides (such as $GeH_4$), and 1:2 oxides (such as $SnO_2$). In this group, tin and lead are metals with low melting points, and silicon and germanium are semimetals. Elemental carbon, the nonmetal, exists in several forms: graphite, diamond, and the recently discovered fullerenes (Fig. 2-4). A seventh and final group includes nitrogen, phosphorus, arsenic, antimony, and bismuth. With hydrogen and oxygen these elements form binary compounds with empirical formulas such as $PH_3$ and $N_2O_5$. The hydrides become increasingly unstable as their molar masses increase, and $BiH_3$ can only be kept below $-45°C$. A similar trend exists for the oxides, and $Bi_2O_5$ has never been obtained in pure form. The lighter members of this group are clearly nonmetals (nitrogen and

(*a*)    (*b*)    (*c*)

phosphorus); bismuth is clearly a metal; and arsenic and antimony are classed as semimetals.

The seven groups just outlined reflect modern understanding of the relationships among the elements. To a 19th-century chemist some of the associations would have been evident (such as those in the alkali-metal group), but others would have been much less obvious (such as placing nitrogen and bismuth in the same group). Moreover, there are other elements that do not fall into these seven groups, and several were well known in the 19th century. Silver, for example, resembles the alkali metals in forming a $1:1$ chloride (AgCl) and a $2:1$ oxide ($Ag_2O$), but its melting point is much higher (962°C, compared with 39°C for rubidium, for example) and it does not react with water. It therefore required considerable chemical judgment to determine which elements really belonged together in groups.

The next step started with the observation that there was a connection between group properties and atomic masses. When the elements known at the time were written in order of increasing atomic mass, repetitive sequences of the groups appeared. An element of the halogen group, for example, was always followed by one of the alkali-metal group and preceded by one of the chalcogens. When the series of elements was written so as to begin a new row with each alkali metal, elements of the same groups were automatically assembled in columns in a **periodic table of the elements.**

When the periodic table was first introduced in 1869–70 by the German chemist Lothar Meyer and (independently) the Russian Dmitri Mendeleyev (Fig. 2-5), one third of the naturally occurring chemical elements had not yet been discovered,

## TABELLE II

| REIHEN | GRUPPE I. — $R^2O$ | GRUPPE II. — RO | GRUPPE III. — $R^2O^3$ | GRUPPE IV. $RH^4$ $RO^2$ | GRUPPE V. $RH^3$ $R^2O^5$ | GRUPPE VI. $RH^2$ $RO^3$ | GRUPPE VII. RH $R^2O^7$ | GRUPPE VIII. — $RO^4$ |
|---|---|---|---|---|---|---|---|---|
| I | H = I | | | | | | | |
| 2 | Li = 7 | Be = 9,4 | B = II | C = I2 | N = I4 | O = I6 | F = I9 | |
| 3 | Na = 23 | Mg = 24 | Al = 27,3 | Si = 28 | P = 31 | S = 32 | Cl = 35,5 | |
| 4 | K = 39 | Ca = 40 | — = 44 | Ti = 48 | V = 51 | Cr = 52 | Mn = 55 | Fe = 56, Co = 59, Ni = 59, Cu = 63. |
| 5 | (Cu = 63) | Zn = 65 | — = 68 | — = 72 | As = 75 | Se = 78 | Br = 80 | |
| 6 | Rb = 85 | Sr = 87 | ?Yt = 88 | Zr = 90 | Nb = 94 | Mo = 96 | — = I00 | Ru = I04, Rh = I04, Pd = I06, Ag = I08. |
| 7 | (Ag = I08) | Cd = II2 | In = II3 | Sn = II8 | Sb = I22 | Te = I25 | J = I27 | |
| 8 | Cs = I33 | Ba = I37 | ?Di = I38 | ?Ce = I40 | — | — | — | — — — — |
| 9 | (—) | — | — | — | — | — | — | |
| I0 | — | — | ?Er = I78 | ?La = I80 | Ta = I82 | W = I84 | — | Os = I95, Ir = I97, Pt = I98, Au = I99. |
| II | (Au = I99) | Hg = 200 | Tl = 204 | Pb = 207 | Bi = 208 | — | — | |
| I2 | — | — | — | Th = 23I | — | U = 240 | — | — — — — |

## Figure 2-5

Dmitri Mendeleyev published this early version of the periodic table in 1872. The letter R is used as a generic symbol for the elements of a group (German *Gruppe*). Note how Mendeleyev coped with the problem of elements such as manganese (Mn) that do not fit well in the pattern of the seven main groups, displaying them together with the main-group elements. The gap below boron was later filled by scandium, that below aluminum by gallium, and that below silicon by germanium.

Rayleigh, discovered a gaseous element in 1894 that did not fit into the previous classification scheme. A century earlier, Henry Cavendish had noted the existence of a residual gas when oxygen and nitrogen are removed from air, but its importance had not been realized. Together with William Ramsay, Rayleigh isolated the gas and named it argon. Ramsay then studied a gas that was present in natural gas deposits and discovered that it was helium, an element whose presence in the sun had been noted earlier in the spectrum of sunlight but that had not previously been known on earth (Fig. 2-6). Rayleigh and Ramsay postulated the existence of a new group of elements, and in 1898 other members of the series (neon, krypton, and xenon) were isolated. These elements are referred to as **noble gases** or sometimes as inert gases because of their relative inertness toward chemical combination.

Elements in the periodic table are ordered by atomic number, the key property that determines the chemical behavior of the element. As discussed in Section 1-4, the atomic number is equal to the positive charge of the nucleus and to the number of electrons in a neutral atom. Hydrogen, the lightest element, has one electron in each atom and atomic number 1. Helium, with atomic number 2, has two electrons, and the neutral atoms of each successive element in the periodic table have one more electron than those of the preceding element (balanced by a correspondingly higher nuclear charge).

The modern periodic table, shown in Figure 2-7 and inside the front cover of this book, places elements in **groups** (arranged vertically) and **periods** (arranged horizontally). The properties of the eight groups of **representative elements** (''main-group'' elements) have been sketched already. In addition, there are ten groups (and three periods) of **transition-metal elements;** a period of elements from atomic number 57 through 71 called the rare-earth or **lanthanide elements;** and a period of elements from atomic number 89 through 103 called the **actinides,** all of which are unstable and most of which must be produced artificially. The lanthanide and actinide elements are usually placed below the rest of the table to conserve space. The groups of representative elements are numbered (with Roman numerals) from I to VIII, with the letter A sometimes added to differentiate them from the transition-metal groups, which are labeled from IB to VIIIB. In this book we shall use group numbers exclusively for the representative elements (dropping the ''A''), and we shall refer to transition-metal elements by the first element in the corresponding group. For example, the elements in the carbon group (C, Si, Ge, Sn, Pb) are designated as Group IV, and the elements chromium (Cr), molybdenum (Mo), and tungsten (W) as the chromium group.[1]

Physical as well as chemical properties of the elements vary systematically across the periodic table. Important physical properties include melting points and boiling points, thermal and electrical conductivity, density, atomic size, and the energy change involved in adding an electron to and removing one from a neutral atom. Appendix F presents numerical values for most of these properties. In general, the elements on the left side of the table (especially in the later periods) are metallic solids and good conductors of electricity. On the right side (especially in earlier periods) they are poor conductors of electricity and are generally found in the gaseous state at room temperature. In between, the semimetals form a zigzag line of

**Figure 2-6**

When an electrical discharge is passed through a tube filled with helium gas at low pressure, the helium atoms emit a characteristic violet light. The same characteristic emitted light proves the presence of helium in the sun. (Leon Lewandowski)

---

[1]Recently, several international organizations recommended a new system of group designation in which the main-group elements make up Groups 1, 2, and 13 through 18, and the transition-metal elements fill Groups 3 through 12.

PERIOD

**Figure 2-7**

The modern periodic table of the elements. Below each symbol is the element's year of discovery; elements with no dates have been known since ancient times. Above each symbol is the atomic number. The color coding indicates the relative abundance by mass of the elements in the world (the atmosphere, oceans and freshwater bodies, and the earth's crust to a depth of 40 km). Oxygen alone comprises almost 50% of this mass, and silicon comprises more than 25%.

division between metallic and nonmetallic elements (see inside the front cover of this book). One of the major accomplishments of modern chemistry is its ability to account for these systematic variations (and many individual exceptions as well) both qualitatively and quantitatively.

## 2-2 IONS AND IONIC COMPOUNDS

The periodic table is a useful way to describe systematically the properties of the elements, and we shall return to it frequently. Chapter 13 will show how the structure of the periodic table arises from the application of quantum mechanics to the electronic structures of atoms. Here we confine our discussion to a simple and helpful model called the **Lewis electron-dot model.** It was proposed by the American chemist G. N. Lewis in 1916, well before modern quantum mechanics was fully developed. It works equally well for ions and ionic compounds and for the compounds we shall shortly identify as covalent.

The Lewis model begins by recognizing that not all the electrons in an atom participate in chemical bonding. Electrons appear to occupy a set of **shells** surrounding the nucleus, and those in inner shells (called **core electrons**) are not significantly involved in the formation of bonds between atoms. The outermost, partially filled shell (called the **valence shell**) contains the electrons that need to be included in most descriptions of chemical bonding, the **valence electrons.** A filled shell possesses great chemical stability. Progress through the elements in order of increasing atomic number reveals a filled shell whenever a noble-gas element such as helium, neon, or argon is reached. The additional electron in atoms of the first element past a noble-gas element (that is, the outermost electron in an alkali-metal atom) is the first occupant of a new shell, so an alkali-metal atom has one valence electron. With the exception of helium, the number of valence electrons in a neutral atom of a main-group element (those in Groups I through VIII) is equal to the element's group number in the periodic table.

The main-group elements that follow a series of transition-metal elements require some special attention. Atoms of bromine, for example, have 17 more electrons than atoms of argon, the preceding noble gas. We still say that bromine has 7 valence electrons (like chlorine), not 17. The reason is that in the fourth, fifth, and sixth periods the ten electrons added in the course of the transition-metal series (although they are very important for the bonding of those elements) have become *core* electrons by the time the end of the transition-metal series is reached. The bonding properties of an element such as bromine resemble those of the lighter elements in its group.

The Lewis model represents valence electrons with dots; core electrons are not shown. The first four dots are displayed singly around the four sides of the symbol of the element. If there are more than four valence electrons, their dots are paired with those already present. The result is a **Lewis dot symbol** for that atom. The Lewis notation for the elements of the first two periods is

$$\cdot \text{H} \qquad\qquad\qquad\qquad\qquad\qquad\qquad\qquad\qquad :\text{He}$$

$$\cdot \text{Li} \quad \cdot \text{Be} \cdot \quad \cdot \dot{\text{B}} \cdot \quad \cdot \dot{\underset{\cdot}{\text{C}}} \cdot \quad :\dot{\underset{\cdot}{\text{N}}} \cdot \quad \cdot \ddot{\underset{\cdot\cdot}{\text{O}}} \cdot \quad :\ddot{\underset{\cdot\cdot}{\text{F}}} \cdot \quad :\ddot{\underset{\cdot\cdot}{\text{Ne}}} :$$

## The Formation of Ionic Compounds

A positively charged **ion** (called a **cation**) forms when an atom loses one or more electrons, and a negatively charged ion (called an **anion**) forms when an atom adds electrons. The creation of ions is indicated by removing dots from or adding them to the Lewis dot symbol and also by writing the net electric charge of the ion as a right superscript. For example:

| Na $\cdot$ | Na$^+$ | $\cdot$ Ca $\cdot$ | Ca$^{2+}$ |
|---|---|---|---|
| Sodium atom | Sodium ion | Calcium atom | Calcium ion |
| $:\ddot{\text{F}} \cdot$ | $:\ddot{\text{F}}:^-$ | $\cdot \dot{\text{S}} \cdot$ | $:\ddot{\text{S}}:^{2-}$ |
| Fluorine atom | Fluoride ion | Sulfur atom | Sulfide ion |

Special stability results when an atom, by either losing or gaining electrons, forms an ion whose outermost shell has the same number of electrons as the outermost shell of a noble-gas atom. Except for hydrogen and helium, whose valence shells are completed with two electrons, atoms of the first few periods of the periodic table have a maximum of eight electrons in their valence shells. We say that a chlorine ion ($: \ddot{C}l : ^-$) or an argon atom ($: \ddot{A}r :$) has a completed **octet** in its valence shell.

The tendency of atoms to achieve valence octets describes much chemical reactivity. Atoms of elements in Groups I and II achieve an octet by losing electrons to form cations; atoms of elements in Groups VI and VII do so by gaining electrons to form anions. Reactions of the metallic elements on the left side of the periodic table with the nonmetallic elements on the right side always transfer just enough electrons to form ions with completed octets. The following equations, in which $e^-$ stands for an electron, use Lewis symbols to show the formation first of a cation and an anion and then of an **ionic compound.**

$$\text{Na} \cdot \longrightarrow \text{Na}^+ + e^- \qquad \text{Loss of a valence electron}$$

$$e^- + : \ddot{C}l \cdot \longrightarrow : \ddot{C}l : ^- \qquad \text{Gain of a valence electron}$$

$$\text{Na}^+ + : \ddot{C}l : ^- \longrightarrow \text{Na}^+ : \ddot{C}l : ^- \qquad \text{Combination to form the compound NaCl}$$

Another example is the formation of $CaBr_2$:

$$\cdot \text{Ca} \cdot + 2 : \ddot{B}r \cdot \longrightarrow \text{Ca}^{2+} + 2 : \ddot{B}r : ^- \longrightarrow \text{Ca}^{2+}(\text{Br}^-)_2$$

| Neutral atoms not having octets | Positive ion with octet | Negative ions with octets | Ionic compound |

The model predicts a 1:1 compound between Na and Cl and a 1:2 compound between Ca and Br, in agreement with experiment.

Ionic compounds, except those with $OH^-$ as the anion, are often called **salts** by analogy with NaCl, common table salt. They are solids at room conditions and generally have high melting and boiling points (for example, NaCl melts at 801°C and boils at 1413°C). Solid ionic compounds usually conduct electricity poorly, but their melts (the molten liquids) conduct well.

## Names and Formulas of Ionic Compounds

The combination of cations with anions results in ionic compounds. Each one's name consists of the name of the cation followed by that of the anion. Ions can be either monatomic or polyatomic; the latter are also referred to as molecular ions.

A monatomic cation bears the name of the parent element. We have already encountered such examples as the sodium ion ($Na^+$) and the calcium ion ($Ca^{2+}$); ions of the other elements in Groups I and II are named in the same way. The transition metals and the metallic elements of Groups III, IV, and V differ from the Group I and II metals in that they often form several stable ions in compounds and in solution. Although calcium compounds never contain $Ca^{3+}$ ions (always $Ca^{2+}$), the element iron forms both the $Fe^{2+}$ and $Fe^{3+}$ ions, and thallium forms both the $Tl^+$ and $Tl^{3+}$ ions. When a metal forms ions of more than one charge, we distinguish them by placing a Roman numeral in parentheses after the name of the metal:

| | | | | | |
|---|---|---|---|---|---|
| $Cu^+$ | copper(I) ion | $Fe^{2+}$ | iron(II) ion | $Sn^{2+}$ | tin(II) ion |
| $Cu^{2+}$ | copper(II) ion | $Fe^{3+}$ | iron(III) ion | $Sn^{4+}$ | tin(IV) ion |

Three compounds of lead and oxygen. Lead dioxide ($PbO_2$, *left*) contains $Pb^{4+}$ ions; litharge (PbO, *right*) contains $Pb^{2+}$ ions; minium ($Pb_3O_4$, *top*) contains both $Pb^{2+}$ and $Pb^{4+}$ ions. (Leon Lewandowski)

An earlier method for distinguishing between such pairs of ions used the suffixes *-ous* and *-ic* added to the root of the (usually Latin) name of the metal to indicate the ions of lower and higher charge, respectively. Thus, $Fe^{2+}$ was called the ferrous ion and $Fe^{3+}$ the ferric ion. This method, although still sometimes used, is not recommended for systematic nomenclature and will not appear again in this book.

A few polyatomic cations are of importance in inorganic chemistry. These include the ammonium ion, $NH_4^+$ (obtained by adding $H^+$ to ammonia); the hydronium ion, $H_3O^+$ (obtained by adding $H^+$ to water); and the particularly interesting molecular ion formed by mercury: $Hg_2^{2+}$, the mercury(I) ion. This species must be carefully distinguished from $Hg^{2+}$, the mercury(II) ion. The Roman numeral I in parentheses means in this case that the average charge on each of the two mercury atoms is $+1$. Compounds with the empirical formulas HgCl and HgBr correspond to $Hg_2Cl_2$ and $Hg_2Br_2$.

A monatomic anion is named by adding the suffix *-ide* to the first portion of the name of the element. Thus, chlor*ine* becomes the chlor*ide* ion, and ox*ygen* becomes the ox*ide* ion. The other monatomic anions of Groups V, VI, and VII are named similarly. Many polyatomic anions exist, and the naming of these species is more complex. The names of the oxoanions (each contains oxygen in combination with a second element) are derived by adding the ending *-ate* to the stem of the name of that second element. Some elements form two oxoanions. The *-ate* ending is then used for the oxoanion with the larger number of oxygen atoms (e.g., $NO_3^-$, nit*rate*), and the ending *-ite* is added for the name of the anion with the smaller number (e.g., $NO_2^-$, nit*rite*). For elements such as chlorine, which form more than two oxoanions, we use the additional prefixes *per-* (largest number of oxygen atoms) and *hypo-* (smallest number of oxygen atoms). An oxoanion containing hydrogen as a third element includes that word in its name. The $HCO_3^-$ oxoanion, for example, is called the hydrogen carbonate ion in preference to its common (nonsystematic) name, "bicarbonate ion," and $HSO_4^-$, often called "bisulfate ion," is better designated as the hydrogen sulfate ion. Table 2-2 lists some of the most important anions. It is

## Table 2-2   Formulas and Names of Some Common Anions

| | | | |
|---|---|---|---|
| $F^-$ | fluoride | $CO_3^{2-}$ | carbonate |
| $Cl^-$ | chloride | $HCO_3^-$ | hydrogen carbonate |
| $Br^-$ | bromide | $NO_2^-$ | nitrite |
| $I^-$ | iodide | $NO_3^-$ | nitrate |
| $H^-$ | hydride | $SiO_4^{4-}$ | silicate |
| $O^{2-}$ | oxide | $PO_4^{3-}$ | phosphate |
| $S^{2-}$ | sulfide | $HPO_4^{2-}$ | hydrogen phosphate |
| $O_2^{2-}$ | peroxide | $H_2PO_4^-$ | dihydrogen phosphate |
| $O_2^-$ | superoxide | $SO_3^{2-}$ | sulfite |
| $OH^-$ | hydroxide | $SO_4^{2-}$ | sulfate |
| $CN^-$ | cyanide | $HSO_4^-$ | hydrogen sulfate |
| $CNO^-$ | cyanate | $ClO^-$ | hypochlorite |
| $SCN^-$ | thiocyanate | $ClO_2^-$ | chlorite |
| $MnO_4^-$ | permanganate | $ClO_3^-$ | chlorate |
| $CrO_4^{2-}$ | chromate | $ClO_4^-$ | perchlorate |
| $Cr_2O_7^{2-}$ | dichromate | | |

important to be able to recognize and name the ions from that table, bearing in mind that the electric charge is an essential part of the formula.

The composition of an ionic compound is determined by overall charge neutrality: the total positive charge on the cations must exactly balance the total negative charge on the anions. The following names and formulas of ionic compounds illustrate this point.

| | | |
|---|---|---|
| Tin(II) bromide | One 2+ cation, two 1− anions | $SnBr_2$ |
| Potassium permanganate | One 1+ cation, one 1− anion | $KMnO_4$ |
| Ammonium sulfate | Two 1+ cations, one 2− anion | $(NH_4)_2SO_4$ |
| Iron(II) dihydrogen phosphate | One 2+ cation, two 1− anions | $Fe(H_2PO_4)_2$ |

## Example 2-1

Give the chemical formulas of (a) calcium cyanide and (b) copper(II) phosphate.

### Solution

(a) Calcium cyanide is composed of $Ca^{2+}$ and $CN^-$ ions. For the overall charge to be 0, there must be two $CN^-$ ions for each $Ca^{2+}$ ion. Thus, the chemical formula of calcium cyanide is $Ca(CN)_2$.

(b) The ions present in this compound are $Cu^{2+}$ and $PO_4^{3-}$. To ensure charge neutrality, there must be three $Cu^{2+}$ ions (total charge +6) and two $PO_4^{3-}$ ions (total charge −6) per formula unit. Thus, the chemical formula of copper(II) phosphate is $Cu_3(PO_4)_2$.

*Related Problems:* 11, 12

## 2-3 COVALENT COMPOUNDS AND THEIR LEWIS STRUCTURES

Elements in Groups III through V of the periodic table (especially in the first two periods) have a lesser tendency to form ions than those at the left and right sides of the table. Consider the simplest stable compound of carbon and hydrogen: methane ($CH_4$). Unlike ionic compounds, this substance is a gas at room temperature, not a solid. Cooling methane to low temperatures condenses it to a solid in which the $CH_4$ molecules retain their identities. Methane dissolves in water to a slight extent, but it does not ionize. Thus, it is not useful to think of methane as an ionic substance made up of $C^{4-}$ and $H^+$ ions (or $C^{4+}$ and $H^-$ ions). It is a nonionic compound.

The Lewis electron-dot model can describe the bonding in molecules of nonionic substances as well as in ionic compounds. Electrons are not transferred from one atom to another in a nonionic compound, but are *shared* between atoms to form **covalent bonds.** Hydrogen and chlorine combine, for example, to form the **covalent compound** hydrogen chloride. This can be indicated with a **Lewis structure** for the molecule of the product, in which the valence electrons from each atom are redistributed so that one electron from the hydrogen atom and one from the chlorine atom are now shared by the two atoms. The two dots representing this electron pair are placed between the symbols for the two elements:

$$H \cdot + \cdot \overset{\cdot\cdot}{\underset{\cdot\cdot}{Cl}} : \longrightarrow H : \overset{\cdot\cdot}{\underset{\cdot\cdot}{Cl}} :$$

The basic rule that governs the writing of Lewis structures is the **octet rule:** whenever possible, the electrons in a covalent compound are distributed in such a way that each main-group element (except hydrogen) is surrounded by eight electrons (an *octet* of electrons). Hydrogen has two electrons in such a structure. When the octet rule is satisfied, the atom attains the special stability of a noble-gas shell. In the structure for HCl just shown, the H nucleus, through sharing, is close to two valence electrons (like the noble gas, He), and the Cl has eight valence electrons near it (like Ar). Electrons that are shared between two atoms are counted as contributing to the filling of the valence shell of each atom.

A shared pair of electrons can also be represented by a short line (—):

$$H\!-\!\ddot{\underset{..}{Cl}}:$$

The unshared electron pairs around the chlorine atom in the Lewis structure are called **lone pairs,** and they make no contribution to the bond between the atoms. Lewis structures of some simple covalent compounds are

$$NH_3 \qquad H_2O \qquad CH_4$$

Lewis structures indicate how bonds connect the atoms in a molecule, but they do not show the molecule's spatial geometry. The ammonia molecule is not planar but pyramidal, for example, with the nitrogen atom at the apex. The water molecule is bent rather than straight. Three-dimensional geometry can be indicated by ball-and-stick models such as those in Figure 2-8.

In some molecules we must suppose that more than one pair of electrons are shared by two atoms. In the oxygen molecule, each atom has six valence electrons, so for each to achieve an octet configuration, *two* pairs of electrons must be shared, making a **double bond** between the atoms:

$$\ddot{O}::\ddot{O} \qquad or \qquad O\!=\!O$$

Similarly, the $N_2$ molecule has a **triple bond,** involving three shared electron pairs:

$$:N:::N: \qquad or \qquad :N\!\equiv\!N:$$

A carbon–carbon bond can involve the sharing of one, two, or three electron pairs. A progression from single to triple bonding is found in the hydrocarbons ethane ($C_2H_6$), ethylene ($C_2H_4$), and acetylene ($C_2H_2$):

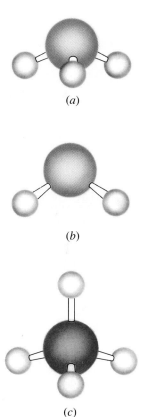

**Figure 2-8**
Molecules of three familiar substances, drawn in ball-and-stick fashion. The sizes of the balls have been reduced somewhat to show the bonds more clearly, but the balls' sizes relative to one another are correct. (*a*) Ammonia, $NH_3$. (*b*) Water, $H_2O$. (*c*) Methane, $CH_4$.

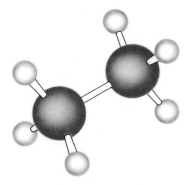

Ethane, $C_2H_6$, can be burned in oxygen as a fuel, and if strongly heated it reacts to form hydrogen and ethylene.

Ethylene, $C_2H_4$, is the largest-volume organic (carbon-containing) chemical produced.

Acetylene, $C_2H_2$, has a triple bond that makes it highly reactive.

Carbon monoxide, CO, is a colorless, odorless, and toxic gas produced by the incomplete burning of hydrocarbons in air. It is used in the production of elemental metals from their oxide ores.

| Table 2-3 | Average Bond Lengths | | | | |
|-----------|------|-----|------|-----|------|
| C—C | 1.54 | N—N | 1.45 | C—H | 1.10 |
| C=C | 1.34 | N=N | 1.25 | N—H | 1.01 |
| C≡C | 1.20 | N≡N | 1.10 | O—H | 0.96 |
| C—O | 1.43 | N—O | 1.43 | C—N | 1.47 |
| C=O | 1.20 | N=O | 1.18 | C≡N | 1.16 |

All values are in units of $10^{-10}$ meters.

Multiple bonding to attain an octet most often involves the elements carbon, nitrogen, oxygen, and, to a lesser degree, sulfur. Double and triple bonds are shorter than a single bond between the same pair of atoms (see Table 2-3).

## Formal Charges

In **homonuclear** diatomic molecules (in which both atoms are the same, as in $H_2$ and $Cl_2$) the electrons are equally shared between the two atoms, and for such molecules the covalency is nearly ideal.

Consider, however, a molecule of carbon monoxide (CO). Its Lewis structure has a triple bond:

$$: C : : : O :$$

This uses the ten valence electrons (four from the C and six from the O) and gives each atom an octet. If the six bonding electrons were shared equally, the carbon atom would own five valence electrons (one *more* than its group number) and the oxygen atom would own five electrons (one *less* than its group number). Equal sharing implies that formally the carbon atom must gain an electron, attaining a **formal charge** of $-1$, and the oxygen atom must lose an electron, attaining a formal charge of $+1$:

$$\overset{\ominus}{} : C : : : O : \overset{\oplus}{}$$

Carbon monoxide is a covalent compound, and the assignment of formal charges does not make it ionic. We emphasize that the equal sharing of electrons in the bonds of such **heteronuclear** diatomic molecules is without experimental basis and is simply a postulate. It has a useful purpose in cases in which two or more Lewis structures are possible for the same molecule. When this happens, the structure with the smallest formal charges is the preferred one, and it generally gives the best description of bonding in the molecule.

The formal charge on an atom in a Lewis structure is simple to calculate. If the valence electrons were removed from an atom, it would have a positive charge equal to its group number in the periodic table (elements in Group VI, the chalcogens, have six valence electrons and therefore a charge of $+6$ when those electrons are removed). From this positive charge, subtract the number of lone-pair valence electrons possessed by the atom in the Lewis structure, and then subtract one half of the number of bonding electrons shared by it:

formal charge = group number − number of electrons in lone pairs
$- \frac{1}{2}$(number of electrons in bonding pairs)

## Example 2-2

Compute the formal charges on the atoms in the following Lewis structure, which represents the azide ion ($N_3^-$):

$$[\ddot{N}{=}N{=}\ddot{N}]^-$$

### Solution

Nitrogen is in Group V. Hence, each N atom contributes five valence electrons to the bonding, and the negative charge on the ion contributes one more electron. The Lewis structure correctly represents 16 electrons.

Each of the terminal nitrogen atoms has four electrons in lone pairs and four bonding electrons (which comprise a double bond) associated with it. Therefore,

$$\text{formal charge}_{(\text{terminal N})} = 5 - 4 - \tfrac{1}{2}(4) = -1$$

The nitrogen atom in the center of the structure has no electrons in lone pairs. Its entire octet lies in the eight bonding electrons:

$$\text{formal charge}_{(\text{central N})} = 5 - 0 - \tfrac{1}{2}(8) = +1$$

The sum of the three formal charges is $-1$, which is the true overall charge on this polyatomic ion. Failure of this check indicates an error in either the Lewis structure or the arithmetic.

*Related Problems:* 15, 16

## Drawing Lewis Structures

In drawing Lewis structures, we shall assume that the molecular "skeleton" (that is, a plan of the bonding of specific atoms to other atoms) is known. In this respect, it helps to know that hydrogen and fluorine are always terminal atoms in Lewis structures, bonded to only one other atom. A systematic procedure for drawing Lewis structures can then be used, as expressed by the following rules:

1. Count up the total number of valence electrons available by first using the group numbers to add up the valence electrons from all the atoms present. If the species is a negative ion, *add* the absolute value of the total charge; if it is a positive ion, *subtract* the value of the charge.
2. Calculate the total number of electrons that would be needed if each atom had its *own* noble-gas shell of electrons around it (two for hydrogen, eight for carbon and heavier elements).
3. Subtract the number in step 1 from the number in step 2. This is the number of shared (or bonding) electrons present.
4. Assign two bonding electrons (one pair) to each bond in the molecule or ion.
5. If bonding electrons remain, assign them in pairs by making some of the bonds double or triple bonds. In some cases there may be more than one way to do this. In general, double bonds form only between atoms of the elements C, N, O, and S. Triple bonds are usually restricted to either C or N.
6. Assign the remaining electrons as lone pairs to the atoms, giving octets to all atoms except hydrogen.

**7.** Determine the formal charge on each atom and write it next to that atom. Check that the formal charges add to give the correct total charge on the molecule or polyatomic ion.

The last rule identifies structures that are undesirable because they imply large separations of negative from positive charge, and catches inadvertent errors (e.g., the wrong number of dots).

The use of these rules is illustrated by the following example.

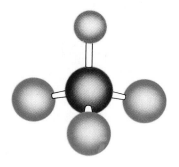

**Figure 2-9**
Phosphoryl chloride, $POCl_3$, is a reactive compound used synthetically to introduce phosphorus into organic molecules.

## Example 2-3

Write a Lewis electron-dot structure for phosphoryl chloride, $POCl_3$ (Fig. 2-9). Assign formal charges to all of the atoms.

### Solution

The first step is to calculate the total number of valence electrons available in the molecule. For $POCl_3$ it is

$$5 \text{ (from P)} + 6 \text{ (from O)} + [3 \times 7 \text{ (from Cl)}] = 32$$

Next, calculate how many electrons would be necessary if each atom were to have its own noble-gas shell of electrons around it. Because there are five atoms in the present case (none of them hydrogen), 40 electrons would be required. From the difference of these numbers ($40 - 32 = 8$), each atom can achieve an octet only if eight electrons are shared between pairs of atoms. Eight electrons correspond to four electron pairs, so each of the four linkages in $POCl_3$ must be a single bond. (If the number of shared electron pairs were *larger* than the number of bonds, double or triple bonds would be present.)

The other 24 valence electrons are assigned as lone pairs to the atoms in such a way that each achieves an octet configuration. The resulting Lewis structure is

Formal charges are already indicated in this diagram. Phosphorus has the group number 5, and it shares eight electrons with no lone-pair electrons, so

$$\text{formal charge on P} = 5 - 4 = +1$$

Oxygen has the group number 6 with six lone-pair electrons and two shared electrons, so

$$\text{formal charge on O} = 6 - 6 - \tfrac{1}{2}(2) = -1$$

All three chlorine atoms have zero formal charge, computed by

$$\text{formal charge on Cl} = 7 - 6 - \tfrac{1}{2}(2) = 0$$

*Related Problems:* 21, 22, 23, 24

## Resonance Structures

For certain molecules or molecular ions, two or more equivalent Lewis structures can be written. An example is ozone ($O_3$), for which there are two possible Lewis structures:

Ozone, $O_3$, is a pale-blue gas with a pungent odor that condenses to a deep-blue liquid below $-112°C$.

These structures suggest that one O—O bond is a single bond and the other is double, so the molecule would be asymmetric. In fact, the two O—O bond lengths are found experimentally to be identical, with a length of $1.28 \times 10^{-10}$ m, intermediate between the O—O single bond length in $H_2O_2$ ($1.49 \times 10^{-10}$ m) and the O=O double bond length in $O_2$ ($1.21 \times 10^{-10}$ m). The Lewis structure picture fails, but it can be patched up by saying that the true structure is a **resonance hybrid** of the two Lewis structures, in which each of the bonds is intermediate between a single and a double bond. This is diagrammed by connecting the structures with a double-headed arrow:

The term "resonance" does not mean that the molecule oscillates back and forth from one structure to the other. Rather, the true structure is a hybrid that *simultaneously* includes features of each of the contributing Lewis structures. This awkwardness can be avoided through a treatment of bonding using molecular orbitals, as shown in Chapter 14.

## Example 2-4

Draw three resonance forms for the nitrate ion, $NO_3^-$ (Fig. 2-10), and estimate the bond lengths.

### Solution

In $NO_3^-$ there are 24 valence electrons. For each atom to have its own octet, $4 \times 8 = 32$ electrons would be required. Therefore, $32 - 24 = 8$ electrons must be shared between atoms, implying a total of four bonding pairs. These can be distributed in one double and two single bonds, leading to the equivalent resonance structures

The two singly bonded oxygen atoms carry formal charges of $-1$, and that of the

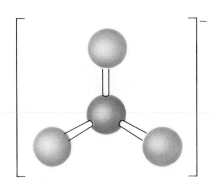

**Figure 2-10**
The nitrate ion, $NO_3^-$, has a symmetric planar structure.

nitrogen atom is +1. The bond lengths should all be equal and should lie between the values given in Table 2-3 for N—O ($1.43 \times 10^{-10}$ m) and N=O ($1.18 \times 10^{-10}$ m).

*Related Problems:* 27, 28, 29, 30, 31, 32

## Breakdown of the Octet Rule

Lewis structures with the octet rule are useful in predicting the types of molecules that will be stable under ordinary conditions of temperature and pressure. For example, we can write a simple Lewis structure for water ($H_2O$),

$$H : \overset{..}{\underset{..}{O}} : H$$

in which each atom has a noble-gas configuration. This is impossible to do for OH or for $H_3O$, which suggests that these species are either unstable or highly reactive.

There are several situations in which the octet rule is *not* satisfied.

*Case 1: Odd-Electron Molecules*  The electrons in a Lewis structure that satisfies the octet rule must occur in pairs—bonding pairs or lone pairs. Any molecule that has an odd number of electrons cannot satisfy the octet rule. Most stable molecules have even numbers of electrons, but a few have odd numbers. An example is nitrogen oxide (NO), a stable (though reactive) molecule that is an important factor in air pollution. It has 11 electrons, and the best electron-dot structure for it is

$$: \overset{.}{N} : : \overset{..}{O}$$

in which only the oxygen atom has a noble-gas configuration. The stability of NO contradicts the octet rule.

*Case 2: Octet-Deficient Molecules*  Some molecules are stable even though they have too few electrons to achieve an octet. For example, the standard rules for $BF_3$ would lead to the Lewis structure

$$: \overset{..}{\underset{..}{F}} : \quad \textcircled{\tiny +1}$$
$$\overset{..}{:}F — B = F$$
$$\textcircled{\tiny -1}$$

but experimental evidence strongly indicates that there are no double bonds in $BF_3$ (fluorine never forms double bonds). Moreover, the placement of a positive formal charge on fluorine is never correct. A more nearly correct structure is

$$: \overset{..}{\underset{..}{F}} :$$
$$: \overset{..}{\underset{..}{F}} — B — \overset{..}{\underset{..}{F}} :$$

Although this Lewis structure denies an octet to the boron atom, it does at least assign zero formal charges to all atoms.

*Case 3: Valence Shell Expansion*  Lewis structures become more complex in the compounds of elements from the third and subsequent periods. Sulfur, for example,

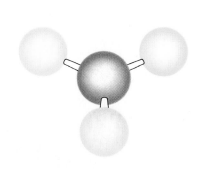

Boron trifluoride, $BF_3$, is a highly reactive gas that condenses to a liquid at $-100°C$. Its major use is in speeding up a large class of reactions involving carbon compounds.

forms some compounds that are readily described by Lewis structures that give closed shells to all atoms. An example is hydrogen sulfide ($H_2S$), which is analogous to water in its Lewis structure. In sulfur hexafluoride ($SF_6$), however, the central sulfur is bonded to six fluorine atoms. This cannot be described by a Lewis diagram unless more than eight electrons are allowed around the sulfur atom, a process called **valence shell expansion.** The resulting Lewis structure is

$$\begin{array}{c} \ddot{\phantom{.}} \\ :\!\overset{..}{F}\!: \\ :\!F\!\diagdown \;\mid\; \diagup\!F\!: \\ S \\ :\!F\!\diagup \;\mid\; \diagdown\!F\!: \\ :\!\overset{..}{F}\!: \end{array}$$

The fluorine atoms have octets, but the central sulfur atom shares a total of 12 electrons.

In the standard procedure for writing Lewis structures, the need for valence shell expansion is signaled when the number of shared electrons is not sufficient to place a bonding pair between each pair of atoms that are supposed to be bonded. In $SF_6$, for example, 48 electrons are available but 56 are needed to form separate octets on seven atoms. This means that $56 - 48 = 8$ electrons would be shared. Four electron pairs are not sufficient to make even single bonds between the central S atom and the six terminal F atoms. In this case we still follow rule 4 (assign one bonding pair to each bond in the molecule or ion) even though this will use more than eight electrons. Rule 5 becomes irrelevant, because there are no extra shared electrons. Rule 6 is now replaced with a new rule:

**6′.** Assign lone pairs to the *terminal* atoms to give them octets. If any electrons still remain, assign them to the central atoms as lone pairs.

The effect of rule 6′ is to abandon the octet rule for the central atom but preserve it for the terminal atoms.

Sulfur hexafluoride, $SF_6$, is an extremely stable, dense, and unreactive gas. It is used as an insulator in high-voltage generators and switches.

## Example 2-5

Write a Lewis structure for the linear $I_3^-$ (triiodide) ion.

### Solution

There are 7 valence electrons from each iodine atom plus 1 from the overall ion charge, giving a total of 22. Because $3 \times 8 = 24$ electrons would be needed in separate octets, $24 - 22 = 2$ are shared according to the original rules. Two electrons are not sufficient to make two different bonds, however, so valence expansion is necessary.

A pair of electrons is placed in each of the two bonds, and rule 6′ is used to complete the octets of the two terminal I atoms. This leaves

$$:\!\overset{..}{\underset{..}{I}}\!-\!I\!-\!\overset{..}{\underset{..}{I}}\!:$$

At this stage, 16 valence electrons have been used. The remaining 6 are placed as lone pairs on the central I atom. A formal charge of $-1$ then resides on this atom:

$$:\!\overset{..}{\underset{..}{I}}\!-\!\overset{\ominus}{\overset{..}{\underset{..}{I}}}\!-\!\overset{..}{\underset{..}{I}}\!:$$

Note the valence expansion on the central atom: it shares or owns a total of ten electrons rather than the eight required by adherence to the octet rule.

*Related Problems:* 33, 34

## 2-4 OXIDATION NUMBERS AND INORGANIC NOMENCLATURE

In Sections 2-1, 2-2, and 2-3 we used the periodic table and the Lewis model to describe the transfer and sharing of electrons in chemical compounds. Before we move on (in Section 2-5) to consider the types of reactions these compounds undergo, we pause to discuss the systematic nomenclature of inorganic compounds. In Section 2-2, ionic compounds were named by following the name of the positive ion with the name of the negative ion. If a metal atom can form more than one stable ion, the ionic charge is given by a Roman numeral enclosed in parentheses; for example, iron(III) represents the $Fe^{3+}$ ion. This Roman numeral notation can be generalized to apply to compounds with covalent character as well, giving rise to a quantity called *oxidation number*. Oxidation numbers are used in naming compounds, in classifying types of reactions (Section 2-5), and in exploring the systematic chemistry of the elements.

### Oxidation Number

The terms "ionic" and "covalent" have been used up to now as though they represent the complete transfer of one or more electrons from one atom to another or the equal sharing of pairs of electrons between atoms in a molecule. In actuality, both types of bonding are idealizations that rarely apply exactly. It is possible to make isolated positive ions by giving gaseous atoms sufficient energy to ionize them, of course, and to make isolated negative ions of many elements by allowing gaseous atoms to capture slow-moving electrons. In each case the ions carry an electric charge that is numerically equal to the charge on the electron. But if these positive and negative ions come together, the gaseous diatomic molecules they form will not be 100% ionic. Even an NaF(*g*) molecule is only about 88% ionic, because the electron distribution that surrounds the two nuclei is distorted. This is referred to as polarization and may be thought of as a small degree of covalent interaction between the two ions.

At the opposite limit, some ionic character is inevitably present even in molecules that appear to be ideally covalent. Formal charge is an artificial way to make this evident: if electrons are assumed to be shared equally between bonded atoms, then molecules such as CO must have formal charges placed on atoms that would suggest ionic interactions between them. As we have stressed, formal charges are not real, but their appearance is evidence that the simple picture of perfect covalent behavior is inadequate, and if the separation of formal charge is too large, the Lewis structure is inadequate to describe the molecule. Even when no formal charges are present, there is still an ionic contribution to bonding in "covalent" compounds. With sophisticated theoretical calculations it is possible to obtain an electron density profile for any diatomic molecule. For a heteronuclear diatomic molecule such as LiH (Fig. 2-11*a*), an examination of the density contours shows that there is a small

**(a) LiH**

**(b) F₂**

## Figure 2-11

Electron contour diagrams for a heteronuclear diatomic molecule (LiH) and a homonuclear diatomic molecule (F₂). The red dots indicate the positions of the nuclei, and the blue lines represent lines over which the electron density is constant. Thus, density falls from the nuclei to the outer parts of the molecules. The skewing of the electron distribution to the right in LiH is an indication of the net negative charge on the H atom in this molecule. The calculation of such contour diagrams will be considered more fully in Chapter 14.

net positive charge on the lithium atom that is balanced by an equal and opposite negative charge on the hydrogen atom, resulting in partial ionic character in the molecule. In a homonuclear diatomic molecule such as $F_2$ (Fig. 2-11$b$) there is, *on average,* no net charge on either atom; even here, however, a contribution to the bonding comes from "resonant structures" (in the Lewis sense) in which the two atoms bear opposite charge. There is an "ionic" contribution to the bonding of every covalent molecule.

Using actual electron density contours would seem only to complicate the problem of accounting for how electrons hold atoms together. What is needed, however, is not the actual electron distribution in a molecule but rather enough information about that distribution to account for the transfer of electrons from one molecule or ion to another in oxidation–reduction reactions. We also need a way to name different binary compounds of the same elements. For these purposes it is not necessary to have detailed knowledge of the exact electron distributions in molecules, whether they are chiefly ionic or covalent. We can assign convenient fictitious charges to the atoms in a molecule and call them **oxidation numbers,** making certain the law of charge conservation is strictly observed. Oxidation numbers are chosen so that in ionic compounds the oxidation number coincides with the charge on the ion. The following simple conventions are useful:

1. The oxidation numbers of the atoms in a neutral molecule must add up to zero, and those in an ion must add up to the charge on the ion.
2. Alkali metal atoms have oxidation number +1, alkaline earth atoms +2, in their compounds.
3. Fluorine always has oxidation number −1 in its compounds. The other halogens have oxidation number −1 in their compounds except those with oxygen and with other halogens, where they can have positive oxidation numbers.
4. Hydrogen is assigned oxidation number +1 in its compounds except in metal hydrides such as LiH, where convention 2 takes precedence and hydrogen has oxidation number −1.
5. Oxygen is assigned oxidation number −2 in compounds. There are two exceptions: in compounds with fluorine, convention 3 takes precedence, and in compounds that contain O—O bonds, conventions 2 and 4 take precedence. Thus, the oxidation number of oxygen in $OF_2$ is +2; in peroxides (such as $H_2O_2$ and $Na_2O_2$) it is −1. In superoxides (such as $KO_2$) oxygen's oxidation number is $-\frac{1}{2}$.

Convention 1 is fundamental because it guarantees charge conservation: the total number of electrons must remain constant in chemical reactions. This rule also makes the oxidation numbers of the neutral atoms of elements all zero. Conventions 2 to 5 are based on the principle that in ionic compounds the oxidation number should equal the charge on the ion. Note that fractional oxidation numbers, though not common, are allowed and in fact necessary to be consistent with this set of conventions.

With the preceding conventions in hand, chemists can assign oxidation numbers to the atoms in the vast majority of compounds. Apply conventions 2 through 5 as listed above, noting the exceptions given; then assign oxidation numbers to the other elements in such a way that convention 1 is always obeyed. Note that convention 1 applies not only to free ions but to the components that make up ionic solids.

Chlorine has oxidation number $-1$ not only as a free $Cl^-$ ion but in the ionic solid AgCl and in covalent $CH_3Cl$. It is important to recognize common ionic species (especially molecular ions) and to know the total charges they carry. Table 2-2 listed the names and formulas of many common anions. Inspection of the table reveals that several elements exhibit different oxidation numbers in different compounds. In $Ag_2S$ sulfur appears as the sulfide ion and has oxidation number $-2$, but in $Ag_2SO_4$ it appears as part of a sulfate ($SO_4^{2-}$) ion. In this case,

$$\text{(oxidation number of S)} + [4 \times \text{(oxidation number of O)}] = \text{total charge on ion}$$

$$x + [4(-2)] = -2$$

$$\text{oxidation number of S} = x = +6$$

A convenient way to indicate the oxidation number of an atom is to write it directly above the corresponding symbol in the formula of the compound:

$$\overset{+1\,-2}{N_2O} \qquad \overset{+1\,-1}{LiH} \qquad \overset{0}{O_2} \qquad \overset{+6\,-2}{SO_4^{2-}}$$

## Example 2-6

Assign oxidation numbers to the atoms in the following chemical compounds and ions: $NaCl$, $ClO^-$, $Fe_2(SO_4)_3$, $SO_2$, $I_2$, $KMnO_4$, $CaH_2$.

### Solution

$\overset{+1\ -1}{NaCl}$ — From conventions 2 and 3.

$\overset{+1\ -2}{ClO^-}$ — From conventions 1 and 5.

$\overset{+3\ +6\ -2}{Fe_2(SO_4)_3}$ — From conventions 1 and 5. This is solved by recognizing the presence of sulfate ($SO_4^{2-}$) groups.

$\overset{+4\ -2}{SO_2}$ — From conventions 1 and 5.

$\overset{0}{I_2}$ — From convention 1. $I_2$ is an element.

$\overset{+1\ +7\ -2}{KMnO_4}$ — From conventions 1, 2, and 5.

$\overset{+2\ -1}{CaH_2}$ — From conventions 1 and 2 (metal hydride case).

Related Problems: 35, 36

**Figure 2-12**
Several oxides of manganese. They are arranged in order of increasing oxidation number of the Mn, counterclockwise from bottom left: MnO, $Mn_3O_4$, $Mn_2O_3$, and $MnO_2$. A compound of still higher oxidation state, $Mn_2O_7$, is a dark red liquid that explodes easily. (Leon Lewandowski)

Oxidation numbers must not be confused with the formal charges described in Section 2-3. They resemble formal charges to the extent that both are assigned, by arbitrary conventions, to symbols in formulas for specific purposes. The purposes differ, however. Formal charges are used solely to identify preferred Lewis structures. Oxidation numbers are used in nomenclature, in identifying oxidation–reduction reactions, and in exploring trends in chemical reactivity across the periodic table. Oxidation numbers are often not the same as the true charges on atoms.

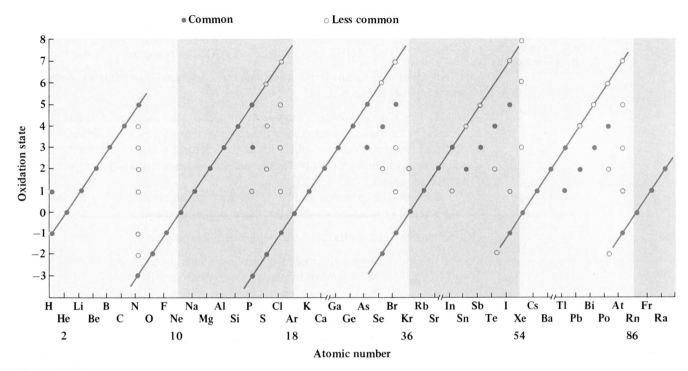

**Figure 2-13**

Important oxidation states of the main-group elements.

In $KMnO_4$ or $Mn_2O_7$, for example, Mn has the oxidation number +7, but the actual net charge on the atom is much less than this. A high oxidation state usually indicates significant covalent character in the bonding of that compound. The oxides of manganese with lower oxidation states (Fig. 2-12) have more ionic character.

Figure 2-13 shows the most common oxidation states of the elements of the main groups. The strong diagonal lines for these elements reflect the stability of closed electron octets. For example, the oxidation states +6 and −2 for sulfur would correspond to losing or gaining enough electrons to attain a noble-gas octet. The octet-based Lewis model is not useful for compounds of the transition elements, which are discussed further in Chapter 16.

## Naming Binary Covalent Compounds

How do we name nonionic (covalent) compounds? If a pair of elements forms only one compound, begin with the name of the element appearing first in the chemical formula, followed by the second element with the suffix -*ide* added to its root. This is quite analogous to the naming of ionic compounds. Just as NaBr is sodium bromide, so the following names designate typical covalent compounds:

| | | | |
|---|---|---|---|
| HBr | hydrogen bromide | $H_2S$ | hydrogen sulfide |
| $BeCl_2$ | beryllium chloride | BN | boron nitride |

A number of well-established nonsystematic names continue to be used (even the strictest chemist does not call water ''hydrogen oxide''!). They include

| Table 2-4 | Prefixes Used for Naming Binary Covalent Compounds | |
| --- | --- |
| **Number** | **Prefix** |
| 1 | *mono-* |
| 2 | *di-* |
| 3 | *tri-* |
| 4 | *tetra-* |
| 5 | *penta-* |
| 6 | *hexa-* |
| 7 | *hepta-* |
| 8 | *octa-* |
| 9 | *nona-* |
| 10 | *deca-* |

| $H_2O$ | water | $PH_3$ | phosphine |
| --- | --- | --- | --- |
| $NH_3$ | ammonia | $AsH_3$ | arsine |
| $N_2H_4$ | hydrazine | $COCl_2$ | phosgene |

If a pair of elements forms more than one compound, two methods can be used to distinguish between them.

**1.** Use Greek prefixes (Table 2-4) to specify the number of atoms of each element in the molecular formula of the compound (*di-* for two, *tri-* for three, and so forth). If the compound is a solid without well-defined molecules, name the empirical formula in this way. The prefix for one (*mono-*) is omitted except in the case of carbon monoxide.

**2.** Write the oxidation number of the first-named element in Roman numerals and place it in parentheses after the name of that element.

Applying the two methods to the oxides of nitrogen gives

| $N_2O$ | dinitrogen oxide | nitrogen(I) oxide |
| --- | --- | --- |
| $NO$ | nitrogen oxide | nitrogen(II) oxide |
| $N_2O_3$ | dinitrogen trioxide | nitrogen(III) oxide |
| $NO_2$ | nitrogen dioxide | nitrogen(IV) oxide |
| $N_2O_4$ | dinitrogen tetraoxide | nitrogen(IV) oxide |
| $N_2O_5$ | dinitrogen pentaoxide | nitrogen(V) oxide |

The first method has some advantages over the second. It distinguishes between $NO_2$ (nitrogen dioxide) and $N_2O_4$ (dinitrogen tetraoxide), two distinct compounds that would both be called nitrogen(IV) oxide under the second system of nomenclature. Two of these oxides have common (nonsystematic) names that may be encountered elsewhere: $N_2O$ is often called nitrous oxide, and $NO$ is called nitric oxide.

## 2-5 INORGANIC REACTION CHEMISTRY

At the heart of chemistry lie chemical reactions, processes that transform combinations of substances into other substances with widely differing properties. It is chemistry that allows the remarkable transformation of sodium, a soft silvery metal, and chlorine, a poisonous yellow-green gas, into sodium chloride, a high-melting white crystalline solid that is used to preserve and season food (Fig. 2-14). In this section, several important types of chemical reactions are discussed and classified. The concepts of electron transfer and electron sharing (developed earlier in the chapter in connection with Lewis structures and oxidation states) will prove useful in describing the types of reactions that can occur.

### Dissolution and Precipitation Reactions

Potassium sulfate is an ionic solid whose solubility in water at 25°C is 120 g L$^{-1}$. Because the resulting solution is a good conductor of electricity, $K_2SO_4$ is called a **strong electrolyte.** Its aqueous solution contains ions ($K^+$ and $SO_4^{2-}$ in this case) that can move when an electric field is applied (Fig. 2-15). The chemical equation for this **dissolution reaction** is written as

$$K_2SO_4(s) \longrightarrow 2\,K^+(aq) + SO_4^{2-}(aq)$$

**Figure 2-14**

The reaction between sodium and chlorine proceeds very slowly until the addition of a drop of water sets it off. Sparks then fly, the reaction vessel becomes hot, and a white powder of sodium chloride forms on the walls of the flask.

where $(s)$ stands for solid and $(aq)$ for aqueous, indicating that the ions are surrounded by water molecules in solution. (Other solvents can be used as well, but water is the exclusive concern here.) Different compounds dissolve to different extents in water. Only a small amount (0.0025 g) of solid barium sulfate dissolves per liter of water at 25°C according to the reaction

$$BaSO_4(s) \longrightarrow Ba^{2+}(aq) + SO_4^{2-}(aq)$$

Solubility differences among ionic compounds are explored more extensively in Chapter 7.

The near-total insolubility of barium sulfate suggests that if a large enough amount of aqueous barium ion were mixed with aqueous sulfate ion, the reverse reaction would occur and solid barium sulfate would appear. Of course, it is not possible to prepare a beaker containing ions of one charge only. Ions of opposite charge must be present to maintain overall charge neutrality. It is possible, however, to prepare one beaker that contains a soluble barium compound in water (such as barium chloride) and a second beaker that contains a soluble sulfate compound in water (such as potassium sulfate). Mixing the two solutions (Fig. 2-16) then produces solid barium sulfate through the reaction

$$Ba^{2+}(aq) + SO_4^{2-}(aq) \longrightarrow BaSO_4(s)$$

which is called a **precipitation reaction.**

Such precipitation reactions are sometimes written as

$$BaCl_2(aq) + K_2SO_4(aq) \longrightarrow BaSO_4(s) + 2\,KCl(aq)$$

which suggests ionic exchange, in which the two anions exchange places. This is somewhat misleading, however, because $BaCl_2$, $K_2SO_4$, and $KCl$ are all dissociated into ions in aqueous solution. It is more nearly correct to write

$$Ba^{2+}(aq) + 2\,Cl^-(aq) + 2\,K^+(aq) + SO_4^{2-}(aq) \longrightarrow$$
$$BaSO_4(s) + 2\,K^+(aq) + 2\,Cl^-(aq)$$

The potassium and chloride ions appear on both sides of the equation. They are **spectator ions,** which assure charge neutrality but do not take part directly in the chemical reaction. Omitting such spectator ions from the balanced chemical equation leads to the **net ionic equation**

$$Ba^{2+}(aq) + SO_4^{2-}(aq) \longrightarrow BaSO_4(s)$$

A net ionic equation includes only the ions (and molecules) that actually take part in the reaction.

In dissolution and precipitation reactions, ions retain their identities and, in particular, oxidation states do not change. The ions simply exchange the positions they had in a solid (surrounded by other ions) for new positions in solution (surrounded by solvent molecules), or they undergo the reverse process.

## Example 2-7

An aqueous solution of sodium carbonate is mixed with an aqueous solution of calcium chloride, and a white precipitate immediately forms. Write a net ionic equation to account for this precipitate.

**Figure 2-15**

An aqueous solution of potassium sulfate conducts electricity. When metallic plates (electrodes) charged by a battery are put in the solution, positive ions $(K^+)$ migrate toward the negative plate, and negative ions $(SO_4^{2-})$ toward the positive plate.

**Figure 2-16**

A solution of potassium sulfate is being added to one of barium chloride. A cloud of white solid barium sulfate is formed; the potassium chloride remains in solution. (Leon Lewandowski)

## Solution

Aqueous sodium carbonate contains $Na^+(aq)$ and $CO_3^{2-}(aq)$ ions, and aqueous calcium chloride contains $Ca^{2+}(aq)$ and $Cl^-(aq)$ ions. Mixing the two solutions places $Na^+(aq)$ and $Cl^-(aq)$ ions and also $Ca^{2+}(aq)$ and $CO_3^{2-}(aq)$ ions in contact for the first time. The precipitate forms by the reaction

$$Ca^{2+}(aq) + CO_3^{2-}(aq) \longrightarrow CaCO_3(s)$$

because the other combination of ions leads to sodium chloride, a compound that is known to be soluble in water.

*Related Problems:* 41, 42

Acetic acid, $CH_3COOH$, is a common weak organic acid that is found in vinegar as a 3 to 5% solution by mass.

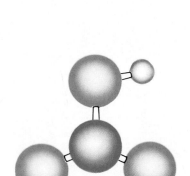

Nitric acid, $HNO_3$, is manufactured from ammonia.

## Acid–Base Reactions

Acids and bases have been known and characterized since early times. Acids (from Latin *acidus,* ''sharp'') are recognized by their sharp, sour taste: acetic acid, $CH_3COOH$ (the main constituent of vinegar), is one example. The ashes of plants yield substances called bases that taste bitter and neutralize the effects of acids. Many acids contain oxygen; examples are nitric acid ($HNO_3$), sulfuric acid ($H_2SO_4$), and acetic acid. In fact, in Lavoisier's time it was thought that oxygen was an essential constituent of any acid. It was not until hydrochloric acid (HCl) was analyzed by the British chemist Sir Humphry Davy in 1810 that it was recognized that oxygen was not an essential element in an acid. From then on, the role of *hydrogen* in the properties of acids was increasingly stressed. Table 2-5 lists the names and formulas of a number of important acids.

Acids and bases can be dissolved in solvents other than water, but we confine our attention to aqueous solution. In pure water, small but equal amounts of hydrogen ions ($H^+$) and hydroxide ions ($OH^-$) are present.[2] These arise from the partial ionization of water:

$$H_2O(\ell) \longrightarrow H^+(aq) + OH^-(aq)$$

where the symbol ($\ell$) is used to designate the liquid state, just as we have used ($s$) for the solid and shall use ($g$) for the gas. Based on the work of the Swedish chemist Svante Arrhenius, we define an **acid** as a substance that, when dissolved in water, increases the amount of hydrogen ion over that present in pure water. Gaseous hydrogen chloride reacts with water to give hydrochloric acid:

$$HCl(g) \longrightarrow H^+(aq) + Cl^-(aq)$$

A **base** is defined as a substance that, when dissolved, leads to an increase in the amount of hydroxide ion over that present in pure water. Sodium hydroxide dissolves extensively in water according to the equation

$$NaOH(s) \longrightarrow Na^+(aq) + OH^-(aq)$$

and is a strong base. Ammonia is another base, as shown by the products of its reaction with water:

[2]As we shall see further in Chapter 6, a hydrogen ion in aqueous solution is closely held by a water molecule and is better represented as a hydronium ion, $H_3O^+(aq)$. For the time being, we employ the simpler notation $H^+(aq)$.

$$NH_3(aq) + H_2O(\ell) \longrightarrow NH_4^+(aq) + OH^-(aq)$$

When an acidic solution is mixed with a basic solution, a **neutralization reaction** occurs:

$$H^+(aq) + OH^-(aq) \longrightarrow H_2O(\ell)$$

This is the reverse of the water ionization reaction shown earlier. If the spectator ions are put back into the equation it reads, for example,

$$\underset{\text{Acid}}{HCl} + \underset{\text{Base}}{NaOH} \longrightarrow \underset{\text{Water}}{H_2O} + \underset{\text{Salt}}{NaCl}$$

showing that a salt can be defined as the product (other than water) of the reaction of an acid with a base. It is usually preferable, however, to omit the spectator ions and to indicate explicitly only the reacting ions.

The Lewis structure model can be used to describe a more general kind of acid–base reaction. A **Lewis base** is any species that donates lone-pair electrons, and a **Lewis acid** is any species that accepts such electron pairs. The Arrhenius acids and bases considered so far fit this description (with the Lewis acid, $H^+$, acting as an acceptor toward various Lewis bases such as $NH_3$ and $OH^-$, the electron pair donors). Other reactions that do *not* involve hydrogen ions can still be considered Lewis acid–base reactions. An example is the reaction between electron-deficient $BF_3$ and electron-rich $NH_3$:

Here ammonia, the Lewis base, donates lone-pair electrons to $BF_3$, the Lewis acid or electron acceptor. The bond that forms is a **coordinate covalent bond,** in which both electrons are supplied by a lone pair on the Lewis base.

Octet-deficient compounds involving elements of Group III such as boron and aluminum are often strong Lewis acids, because Group III atoms can achieve octet configurations by forming coordinate covalent bonds. Atoms and ions from Groups V through VII have the necessary lone pairs to act as Lewis bases. Compounds of

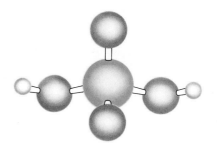

Sulfuric acid, $H_2SO_4$, is the industrial chemical produced on the largest scale in the world, in amounts exceeding 100 million tons per year.

Hydrochloric acid, HCl, is used in the pickling of steel and other metals to remove oxide layers on the surface.

| Table 2-5 Names of Common Acids | | |
|---|---|---|
| **Binary Acids** | **Oxoacids** | **Organic Acids** |
| HF, hydrofluoric acid | $H_2CO_3$, carbonic acid | HCOOH, formic acid |
| HCl, hydrochloric acid | $H_3PO_4$, phosphoric acid | $CH_3COOH$, acetic acid |
| HCN, hydrocyanic acid[†] | $HNO_2$, nitrous acid | $C_6H_5COOH$, benzoic acid |
| $H_2S$, hydrosulfuric acid | $HNO_3$, nitric acid | HOOC—COOH, oxalic acid |
| | $H_2SO_3$, sulfurous acid | |
| | $H_2SO_4$, sulfuric acid | |
| | HClO, hypochlorous acid | |
| | $HClO_2$, chlorous acid | |
| | $HClO_3$, chloric acid | |
| | $HClO_4$, perchloric acid | |

[†]Contains three elements but is named as a binary acid.

main-group elements from the later periods can also act as Lewis acids through valence expansion. In such reactions, the central atom accepts a share in additional lone pairs beyond the eight electrons needed to satisfy the octet rule. For example, $SnCl_4$ is a Lewis acid that accepts electrons from chloride ion lone pairs:

$$SnCl_4(\ell) + 2\, Cl^-(aq) \longrightarrow [SnCl_6]^{2-}(aq)$$

After the reaction, each tin atom is surrounded by 12 rather than 8 valence electrons.

Like dissolution and precipitation reactions, acid–base reactions involve no change in oxidation number. They differ, however, in that they involve changes in the sharing of electrons, with the Lewis acid attracting electrons from the lone pairs of the Lewis base to form a covalent bond.

### Oxidation–Reduction Reactions

We turn now to a third class of reactions, those in which the oxidation states of some of the participants in the reaction undergo changes. These are called **oxidation–reduction** (or **redox**) **reactions.** A prototype redox reaction is that of magnesium with oxygen (Fig. 2-17). When this reaction is carried to completion, the product is magnesium oxide:

$$2\, Mg(s) + O_2(g) \longrightarrow 2\, MgO(s)$$

Magnesium is **oxidized** in this process; it *gives up* electrons as its oxidation number *increases* from 0 (in elemental Mg) to +2 (in MgO). A second species must always be present to *accept* these electrons. This second species is said to be **reduced.** In this case it is oxygen, as the oxidation number *decreases* from 0 to −2. The transfer of electrons ($e^-$) can be indicated with arrows:

$$2\, \overset{0}{Mg} + \overset{0}{O}_2 \longrightarrow 2\, \overset{+2\ -2}{MgO}$$
$$2 \times 2e^- = 2 \times 2e^-$$

The arrows point *away from* the species being oxidized (giving up electrons) and *toward* the species being reduced (accepting electrons). The electron "bookkeeping" beneath the equation ensures that the same number of electrons are taken up by oxygen as are given up by magnesium:

2 Mg atoms × 2 electrons per Mg atom =
        2 oxygen atoms per formula unit × 2 electrons per oxygen atom

Originally, the term "oxidation" referred only to reactions with oxygen. It now is used to describe any process in which the oxidation number of a species increases, even if oxygen is not involved in the reaction. For example, when calcium combines with chlorine to form calcium chloride,

$$\overset{0}{Ca}(s) + \overset{0}{Cl}_2(g) \longrightarrow \overset{+2\ -1}{CaCl_2}$$
$$2e^- = 2 \times 1e^-$$

the calcium has been oxidized and the chlorine has been reduced.

**Figure 2-17**

Magnesium burning in air gives off an extremely bright light. This characteristic led to the incorporation of magnesium into the flash powder used in early photography. Magnesium powder is still used in fireworks for the same reason. (Leon Lewandowski)

Oxidation–reduction reactions are among the most important in chemistry and in modern life. The burning of coal, natural gas, and gasoline (for heat and power) are examples of redox reactions, as are the recovery of metals such as iron and aluminum from their oxide ores and the production of chemicals such as sulfuric acid from sulfur, air, and water. Many of these reactions will be discussed throughout this book. Chapter 10, for example, describes redox reactions in electrochemical cells.

## Example 2-8

Determine whether the following equations represent oxidation–reduction reactions.
(a) $SnCl_2(s) + Cl_2(g) \rightarrow SnCl_4(\ell)$
(b) $CaCO_3(s) \rightarrow CaO(s) + CO_2(g)$
(c) $2\,H_2O_2(\ell) \rightarrow 2\,H_2O(\ell) + O_2(g)$

## Solution

(a) Tin increases its oxidation number from $+2$ to $+4$, and the oxidation number of chlorine in $Cl_2$ is reduced from 0 to $-1$. Some other chlorine atoms are unchanged in oxidation number, but this is still a redox reaction.
(b) The oxidation state of Ca remains at $+2$, that of O at $-2$, and that of C at $+4$. Thus, this is not a redox reaction.
(c) The oxidation number of oxygen in $H_2O_2$ is $-1$. This changes to $-2$ in the product $H_2O$ and 0 in $O_2$. This is a redox reaction in which the same element is both oxidized and reduced.

*Related Problems:* 51, 52, 53, 54

## 2-6 CHEMISTRY OF THE HYDRIDES AND OXIDES

The hydrides and oxides of the elements vary systematically in chemical and physical properties through the periodic table. The reactions of these compounds illustrate the classes of chemical reaction described in Section 2-5.

### Chemistry of the Hydrides

We have mentioned the role played by hydrogen in the properties of acids. Can it be concluded that all binary compounds containing hydrogen are acids? Not at all! In fact, the hydrides show quite varied behavior across the periodic table.

Hydrogen forms compounds with the alkali metals and the heavier alkaline-earth metals that are called *saline* (salt-like) hydrides. For example, at high pressures and high temperatures hydrogen reacts with liquid sodium to form sodium hydride:

$$2\,Na(\ell) + H_2(g) \longrightarrow 2\,NaH(s)$$

The saline hydrides are ionic solids that consist of metal cations and *negatively* charged hydride ions ($H^-$). Sodium hydride is a white crystalline solid that resembles sodium chloride in appearance but has quite different chemical properties. Sodium hydride reacts vigorously with water, for example, to yield hydrogen and a

**Figure 2-18**
Calcium hydride reacts vigorously with water to generate hydrogen.

basic solution. The reaction of calcium hydride with water is a convenient laboratory source for hydrogen (Fig. 2-18):

$$CaH_2(s) + 2\,H_2O(\ell) \longrightarrow 2\,H_2(g) + Ca^{2+}(aq) + 2\,OH^-(aq)$$

The saline hydrides are strong bases because they react completely with water, as shown, to give hydroxide ions.

At elevated temperatures hydrogen reacts with most of the nonmetallic elements from Groups III through VII to form binary covalent compounds. A gaseous mixture of hydrogen and nitrogen, for example, reacts at elevated temperatures to form ammonia:

$$N_2(g) + 3\,H_2(g) \longrightarrow 2\,NH_3(g)$$

Most covalent hydrides can be described by Lewis structures in which each hydrogen atom shares two electrons to attain a configuration resembling that of the noble gas helium. A fascinating exception, which the Lewis model is unable to describe, is provided by the boron hydrides, the simplest of which has the formula $B_2H_6$ (Fig. 2-19). Two of the hydrogen atoms in this molecule form bridge bonds in which two electrons are shared by three atoms—the two boron atoms and the bridging hydrogen atom. Many higher boron hydrides (*boranes*) are known as well, some of which have very complex basket-like or cage-like structures.

Covalent compounds of hydrogen can be acidic, basic, or neither. Hydrides of elements in Groups V through VII all have lone pairs as well as hydrogen ions and so, in principle, can act as bases or as acids. The hydrogen halides are acids, as we have seen, and ammonia is a base. Water itself acts in some reactions as a base and in others as an acid and is therefore referred to as **amphoteric.** Phosphine ($PH_3$), hydrides of carbon such as methane ($CH_4$), and heavier Group IV and Group V hydrides dissolve to only small extents in water and exhibit little acid *or* base character. Boron hydrides react with water

$$B_2H_6(g) + 6\,H_2O(\ell) \longrightarrow 2\,B(OH)_3(aq) + 6\,H_2(g)$$

to give the weak acid $B(OH)_3$, boric acid (Fig. 2-20). Boric acid functions not by ionizing directly, as $H_2SO_4$ and most other oxoacids do,

$$H_2SO_4(aq) \longrightarrow H^+(aq) + HSO_4^-(aq)$$

but rather by taking $OH^-$ from water:

$$B(OH)_3(aq) + H_2O(\ell) \longrightarrow B(OH)_4^-(aq) + H^+(aq)$$

Although the net effect is to increase the amount of $H^+$ in solution, the mechanisms differ. The reaction of boric acid with water is a good example of a Lewis acid–base reaction: the Lewis acid, $B(OH)_3$, forms a coordinate covalent bond with the Lewis base, $OH^-$.

**Figure 2-19**
Diborane, $B_2H_6$, is a much-studied compound, both because it has an unusual structure (in which hydrogen atoms are simultaneously bonded to two different boron atoms) and because it is useful as an intermediate in chemical synthesis.

### Example 2-9

Complete and balance the following chemical equations, and state whether each is an example of a redox, acid–base, dissolution, or precipitation reaction.
(a) $H_2(g) + Cl_2(g) \rightarrow$
(b) $H_2S(aq) \rightarrow$
(c) $RbH(s) + H_2O(\ell) \rightarrow$

## Solution

(a) $H_2(g) + Cl_2(g) \rightarrow 2\,HCl(g)$
This is a redox reaction, because the oxidation states of H and Cl change from 0 to $+1$ and $-1$, respectively.
(b) $H_2S(aq) \rightarrow H^+(aq) + HS^-(aq)$
This is an acid–base reaction.
(c) $RbH(s) + H_2O(\ell) \rightarrow Rb^+(aq) + OH^-(aq) + H_2(g)$
This is a redox reaction, because the hydride $H^-$ ion changes oxidation state from $-1$ to 0 while one of the water hydrogen atoms changes from $+1$ to 0. It is also an acid–base reaction, with the hydride ion acting as a Lewis base and water as an acid that furnishes a hydrogen ion to form $H_2$.

*Related Problems:* 55, 56

**Figure 2-20**
Boric acid, $B(OH)_3$, is a planar, octet-deficient molecule that is a good Lewis acid. It forms a coordinate co-valent bond with the Lewis base $OH^-$ to give the ion $B(OH)_4^-$.

## Chemistry of the Oxides

The binary compounds of oxygen are just as varied as the hydrides in their chemical properties. Alkali-metal oxides (such as $Na_2O$) and alkaline-earth oxides (such as CaO) are ionic compounds composed of metal ions and oxide ions ($O^{2-}$). However, only lithium and the alkaline-earth elements from beryllium through strontium form primarily the "expected" oxides through reactions such as

$$4\,Li(s) + O_2(g) \longrightarrow 2\,Li_2O(s)$$

Sodium and barium take on more oxygen to form the peroxides $Na_2O_2$ and $BaO_2$ through reactions such as

$$Ba(s) + O_2(g) \longrightarrow BaO_2(s)$$

Potassium, rubidium, and cesium are even more reactive than sodium and form the superoxides $KO_2$, $RbO_2$, and $CsO_2$ through reactions such as

$$K(s) + O_2(g) \longrightarrow KO_2(s)$$

Oxides of the metals that form peroxides or superoxides are readily prepared by decomposition of their carbonates on heating, as with

$$Na_2CO_3(s) \longrightarrow Na_2O(s) + CO_2(g)$$

Oxides, peroxides, and superoxides react in different ways with water. A soluble oxide reacts with water to give a strongly basic solution:

$$Na_2O(s) + H_2O(\ell) \longrightarrow 2\,Na^+(aq) + 2\,OH^-(aq)$$

A less soluble oxide such as MgO dissolves to only a limited extent in water. It is still a base, however, and dissolves readily in acidic solution according to

$$MgO(s) + 2\,H^+(aq) \longrightarrow Mg^{2+}(aq) + H_2O(\ell)$$

A peroxide also gives a basic solution when it reacts with water, but hydrogen peroxide is generated, as well:

$$Na_2O_2(s) + 2\,H_2O(\ell) \longrightarrow 2\,Na^+(aq) + 2\,OH^-(aq) + H_2O_2(aq)$$

Hydrogen peroxide, $H_2O_2$, is a pale-blue liquid that freezes at $-0.43°C$ and boils at $150°C$. It can act as either an oxidizing agent or a reducing agent in acidic or basic solutions.

Superoxides react with water to liberate oxygen as well as hydroxide ion and hydrogen peroxide:

$$2\,KO_2(s) + 2\,H_2O(\ell) \longrightarrow 2\,K^+(aq) + 2\,OH^-(aq) + H_2O_2(aq) + O_2(g)$$

The reactions of these compounds with carbon dioxide are also of interest. Group I and II oxides give carbonates through reactions such as

$$CaO(s) + CO_2(g) \longrightarrow CaCO_3(s)$$

Peroxides and superoxides also give carbonates but liberate additional oxygen:

$$2\,Na_2O_2(s) + 2\,CO_2(g) \longrightarrow 2\,Na_2CO_3(s) + O_2(g)$$

$$4\,KO_2(s) + 2\,CO_2(g) \longrightarrow 2\,K_2CO_3(s) + 3\,O_2(g)$$

These two reactions are used in respiratory support systems for firefighters and the crews of submarines and spacecraft. For every two volumes of $CO_2$ that are breathed out, one volume of $O_2$ is regenerated by the first reaction. The superoxides are even better; they regenerate three volumes of $O_2$ for every two volumes of $CO_2$ exhaled.

Nonmetals burn in oxygen (Fig. 2-21) to form oxides that are intermediate in character between covalent and ionic. In most cases several oxidation states occur. Carbon monoxide forms when carbon is burned in insufficient oxygen, whereas carbon dioxide results when oxygen is in excess. There are at least six oxides of chlorine and six of nitrogen.

Many binary oxides can be considered as anhydrides of acids and bases, obtained by dehydration. An **acid anhydride** is obtained by removing water from an oxoacid until only the oxide remains; thus, $CO_2$ is the anhydride of carbonic acid ($H_2CO_3$).

## Example 2-10

What is the acid anhydride of phosphoric acid ($H_3PO_4$)?

### Solution

If the formula unit $H_3PO_4$ (which contains an odd number of hydrogen atoms) is doubled, $H_6P_2O_8$ is obtained. Subtraction of $3\,H_2O$ from this gives $P_2O_5$, which is the empirical formula of tetraphosphorus decaoxide ($P_4O_{10}$). This compound is the acid anhydride of phosphoric acid.

*Related Problems:* 57, 58

The oxides of most of the nonmetals are acid anhydrides, which react with an excess of water to form acidic solutions. Examples are

$$N_2O_5(s) + H_2O(\ell) \longrightarrow 2\,HNO_3(aq) \longrightarrow 2\,H^+(aq) + 2\,NO_3^-(aq)$$

$$SO_3(g) + H_2O(\ell) \longrightarrow H_2SO_4(aq) \longrightarrow H^+(aq) + HSO_4^-(aq)$$

Silica ($SiO_2$) is the acid anhydride of the very weak silicic acid $H_2SiO_3$, a gelatinous material that is insoluble in water but readily dissolves in strongly basic aqueous solutions according to

$$H_2SiO_3(s) + 2\,OH^-(aq) \longrightarrow SiO_3^{2-}(aq) + 2\,H_2O(\ell)$$

**Figure 2-21**

Molten sulfur burns in oxygen to give sulfur dioxide. The higher oxide, sulfur trioxide, requires the presence of platinum for its formation.

| I | II | III | IV | V | VI | VII |
|---|---|---|---|---|---|---|
| $Li_2O$ | BeO | $B_2O_3$ | $CO_2$ | $N_2O_5$ | $(O_2)$ | $OF_2$ |
| $Na_2O$ | MgO | $Al_2O_3$ | $SiO_2$ | $P_4O_{10}$ | $SO_3$ | $Cl_2O_7$ |
| $K_2O$ | CaO | $Ga_2O_3$ | $GeO_2$ | $As_2O_5$ | $SeO_3$ | $Br_2O_7$ |
| $Rb_2O$ | SrO | $In_2O_3$ | $SnO_2$ | $Sb_2O_5$ | $TeO_3$ | $I_2O_7$ |
| $Cs_2O$ | BaO | $Tl_2O_3$ | $PbO_2$ | $Bi_2O_5$ | $PoO_3$ | $At_2O_7$ |

Increasing basicity (vertical, left side)

Increasing acidity (vertical, right side)

⟵ Increasing basicity

**Figure 2-22**

Among the oxides of the main-group elements, acidity tends to increase from left to right and from bottom to top in the periodic table. Oxygen difluoride, however, has only weakly acidic properties.

Oxides of Group I and II metals are **base anhydrides,** obtained by removing water from the corresponding hydroxides. Calcium oxide, CaO, is the base anhydride of calcium hydroxide, $Ca(OH)_2$. The removal of water from $Ca(OH)_2$ is the reverse of the addition of water to the oxide:

$$CaO(s) + H_2O(\ell) \longrightarrow Ca(OH)_2(s)$$

$$Ca(OH)_2(s) \longrightarrow CaO(s) + H_2O(\ell)$$

The base anhydride of NaOH is $Na_2O$. Oxides of metals in the middle groups of the periodic table (III through V) lie on the border between ionic and covalent behavior and are frequently amphoteric (they can act as acid or base). An example is aluminum oxide ($Al_2O_3$), which dissolves to only a limited extent in water but much more readily in either acids or bases:

Acting as a base:   $Al_2O_3(s) + 6\,H^+(aq) \longrightarrow 2\,Al^{3+}(aq) + 3\,H_2O(\ell)$

Acting as an acid:   $Al_2O_3(s) + 2\,OH^-(aq) + 3\,H_2O(\ell) \longrightarrow 2\,Al(OH)_4^-(aq)$

Figure 2-22 summarizes the acid–base character of the main-group oxides in the first periods.

Although oxoacids and hydroxides are Arrhenius acids and bases [releasing $H^+(aq)$ or $OH^-(aq)$ into aqueous solution], acid and base anhydrides do not fall into this classification; they contain neither $H^+$ nor $OH^-$. Acid anhydrides are still acids in the Lewis sense, however (they accept electron pairs), and base anhydrides are bases in the Lewis sense (their $O^{2-}$ ions donate electron pairs). The reaction between an acid anhydride and a base anhydride is then a Lewis acid–base reaction. An example of such a reaction is

$$CaO(s) + CO_2(g) \longrightarrow CaCO_3(s)$$

Here, the Lewis base CaO donates an electron pair (one of the lone pairs of the oxygen atom) to the Lewis acid ($CO_2$) to form a coordinate covalent bond in the $CO_3^{2-}$ ion. Similar Lewis acid–base reactions can be written for other acid–base anhydride pairs. Sulfur trioxide, for example, reacts with metal oxides to form sulfates:

$$MgO(s) + SO_3(g) \longrightarrow MgSO_4(s)$$

Note that these reactions are not redox reactions (oxidation numbers do not change). They are clearly not dissolution or precipitation reactions, nor are they acid–base reactions in the Arrhenius sense. They can be usefully classified as acid–base reactions in the Lewis sense, however.

---

### Example 2-11

Complete and balance the following chemical equations, and state whether each is an example of redox, acid–base (including Lewis acid–base), dissolution, or precipitation.
(a) $Rb(s) + O_2(g) \rightarrow$
(b) $Na_2O(s) + HCl(aq) \rightarrow$
(c) $K_2O(s) + CO_2(g) \rightarrow$

### Solution

(a) Rubidium forms superoxides upon reaction with oxygen in the redox reaction

$$Rb(s) + O_2(g) \longrightarrow RbO_2(s)$$

(b) This is an acid–base reaction of a strong acid with a base anhydride. The product is the highly soluble NaCl:

$$Na_2O(s) + 2\,HCl(aq) \longrightarrow 2\,Na^+(aq) + 2\,Cl^-(aq) + H_2O(\ell)$$

(c) This is a Lewis acid–base reaction:

$$K_2O(s) + CO_2(g) \longrightarrow K_2CO_3(s)$$

*Related Problems:* 61, 62

---

**PRACTICE EXERCISE**

### Oxides and Peroxides

Consider the three compounds $KO_2$, $BaO_2$, and $TiO_2$. Each contains two oxygen atoms per metal atom, but the oxygen occurs in different forms in the three compounds.
(a) The oxygen in $TiO_2$ occurs as $O^{2-}$ ions. Give the Lewis dot symbol for this ion.

On the right is a sample of rutile, $TiO_2$, which is the primary ore of titanium. Behind it is a large quartz crystal through which slender rutile hairs penetrate. Crystals of this type are called Venus hairstone. (Michael Dalton/Fundamental Photographs)

How many valence electrons does this ion have? What is the chemical name for $TiO_2$?

(b) Recall that Group II elements form stable $+2$ ions. By referring to Table 2-2, identify the oxygen-containing ion in $BaO_2$ and give the name of the compound. Draw a Lewis structure for the oxygen-containing ion, showing formal charge. Is the bond in this ion a single or a double bond?

(c) Recall that Group I elements form stable $+1$ ions. By referring to Table 2-2, identify the oxygen-containing ion in $KO_2$ and give the name of the compound. Show that the oxygen-containing ion is an odd-electron species. Draw the best Lewis structure you can for it.

(d) Write balanced chemical equations for the reactions of $BaO_2$ with strong acid and with carbon dioxide, and identify the species that are oxidized and reduced.

**Answers**

(a) The ion $:\ddot{O}:^{2-}$ has eight valence electrons (an octet). $TiO_2$ is titanium(IV) oxide.

(b) The ion in $BaO_2$ must be the peroxide ion ($O_2^{2-}$), and the compound is barium peroxide. The Lewis structure of the peroxide ion is

$$\overset{\ominus}{:}\ddot{O}:\ddot{O}:\overset{\ominus}{}$$

and the O—O bond is a single bond.

(c) The ion in $KO_2$ must be the superoxide ion ($O_2^-$), and the compound is potassium superoxide. The superoxide ion has 13 valence electrons. The best Lewis structures one can draw are the resonance structures

$$\cdot\ddot{O}:\ddot{O}:^{\ominus} \quad \text{and} \quad ^{\ominus}:\ddot{O}:\ddot{O}\cdot$$

in which only one of the oxygen atoms attains an octet electron configuration.

(d) $$2\,BaO_2(s) + 4\,H^+(aq) \rightarrow 2\,Ba^{2+}(aq) + 2\,H_2O(\ell) + O_2(g)$$

The oxygen atoms in the peroxide ion (oxidation state $-1$) are either oxidized to $O_2$ (oxidation state $0$) or reduced to $H_2O$ (oxidation state $-2$).

$$2\,BaO_2(s) + 2\,CO_2(g) \longrightarrow 2\,BaCO_3(s) + O_2(g)$$

Again, oxygen is reduced to the $-2$ oxidation state (in $BaCO_3$) and oxidized to $0$ (in $O_2$).

After studying this chapter and working the problems that follow, you should be able to

**CONCEPTS & SKILLS**

1. Describe the structure of the periodic table, and predict chemical and physical properties of an element based on those of others in its group or period (Section 2-1, problems 1–2).

2. Use the principle of charge balance to write chemical formulas for ionic compounds (Section 2-2, problems 7–14).

3. Given a molecular formula, draw a Lewis structure for the molecule (Section 2-3, problems 21–24).

4. Assign formal charges and identify resonance structures for a given Lewis structure (Section 2-3, problems 27–32).

5. Assign oxidation numbers to atoms in compounds (Section 2-4, problems 35–36).

6. Name inorganic compounds, given their chemical formulas, and write chemical formulas for named inorganic compounds (Section 2-4, problems 37–40).

7. Classify reactions as dissolution, precipitation, acid–base, and oxidation–reduction (Section 2-5).

8. Describe the chemical behavior of the hydrides and oxides (Section 2-6, problems 55–62).

## PROBLEMS

### The Periodic Table

1. Before the element scandium was discovered in 1879, it was known as "eka-boron." Predict the properties of scandium from averages of the corresponding properties of its neighboring elements in the periodic table (which follow). Compare your predictions with the observed values in Appendix F.

| Element | Symbol | Melting Point (°C) | Boiling Point (°C) | Density (g cm$^{-3}$) |
|---------|--------|--------------------|--------------------|-----------------------|
| Calcium | Ca | 839 | 1484 | 1.55 |
| Titanium | Ti | 1660 | 3287 | 4.50 |
| Scandium | Sc | ? | ? | ? |

2. The element technetium (Tc) is not found in nature but has been produced artificially through nuclear reactions. Use the data in the following table for several neighboring elements to estimate the melting point, boiling point, and density of technetium. Compare your predictions with the observed values in Appendix F.

| Element | Symbol | Melting Point (°C) | Boiling Point (°C) | Density (g cm$^{-3}$) |
|---------|--------|--------------------|--------------------|-----------------------|
| Manganese | Mn | 1244 | 1962 | 7.2 |
| Molybdenum | Mo | 2610 | 5560 | 10.2 |
| Rhenium | Re | 3180 | 5627 | 20.5 |
| Ruthenium | Ru | 2310 | 3900 | 12.3 |

3. Use the group structure of the periodic table to predict the empirical formulas for the binary compounds that hydrogen forms with the elements antimony, bromine, tin, and selenium.

4. Use the group structure of the periodic table to predict the empirical formulas for the binary compounds that hydrogen forms with the elements germanium, fluorine, tellurium, and bismuth.

### Ions and Ionic Compounds

5. For each of the following atoms or ions, give the Lewis dot symbol and state the total number of electrons, the number of valence electrons, and the number of core electrons.
   (a) Rn          (c) $Se^{2-}$
   (b) $Sr^+$       (d) $Sb^-$

6. For each of the following atoms or ions, give the Lewis dot symbol and state the total number of electrons, the number of valence electrons, and the number of core electrons.
   (a) $Ra^{2+}$     (c) $Bi^{2-}$
   (b) Br           (d) $Ga^+$

7. Give the name and formula of an ionic compound involving only the elements in each pair that follows. Write Lewis symbols for the elements both before and after chemical combination.
   (a) Chlorine and cesium
   (b) Calcium and astatine
   (c) Aluminum and sulfur
   (d) Potassium and tellurium

8. Give the name and formula of an ionic compound involving only the elements in each pair that follows. Write Lewis symbols for the elements both before and after chemical combination.
   (a) Gallium and bromine
   (b) Strontium and polonium
   (c) Magnesium and iodine
   (d) Lithium and selenium

9. Give systematic names to the following compounds.

(a) $Al_2O_3$

(b) $Rb_2Se$

(c) $(NH_4)_2S$

(d) $Ca(NO_3)_2$

(e) $Cs_2SO_4$

(f) $KHCO_3$

10. Give systematic names to the following compounds.

(a) $KNO_2$

(b) $Sr(MnO_4)_2$

(c) $MgCr_2O_7$

(d) $NaH_2PO_4$

(e) $BaCl_2$

(f) $NaClO_3$

11. Write the chemical formulas for the following compounds.

(a) Silver cyanide

(b) Calcium hypochlorite

(c) Potassium chromate

(d) Gallium oxide

(e) Potassium superoxide

(f) Barium hydrogen carbonate

12. Write the chemical formulas for the following compounds.

(a) Cesium sulfite

(b) Strontium thiocyanate

(c) Lithium hydride

(d) Sodium peroxide

(e) Ammonium dichromate

(f) Rubidium hydrogen sulfate

13. Trisodium phosphate (TSP) is a heavy-duty cleaning agent. Write its chemical formula. What would be the systematic name for this ionic compound?

14. Monoammonium phosphate is a compound made up of $NH_4^+$ and $H_2PO_4^-$ ions that is employed as a flame retardant (its use for this purpose was first suggested by Gay-Lussac in 1821). Write its chemical formula. What is the systematic chemical name of this compound?

## Covalent Compounds and Their Lewis Structures

15. Assign formal charges to all atoms in the following Lewis structures.

(a) $SO_4^{2-}$

(b) $S_2O_3^{2-}$

(c) $SbF_3$

(d) $SCN^-$

16. Assign formal charges to all atoms in the following Lewis structures.

(a) $ClO_4^-$

(b) $SO_2$

(c) $BrO_2^-$

(d) $NO_3^-$

17. Determine the formal charges on all the atoms in the following Lewis structures:

$$H-N=O \qquad \text{and} \qquad H-O=N$$

Which one would probably be favored as the structure of the molecule HNO?

18. Determine the formal charges on all the atoms in the following Lewis structures:

$$:Cl-Cl-O: \qquad \text{and} \qquad :Cl-O-Cl:$$

Which one would probably be favored as the structure of the molecule $Cl_2O$?

19. In each of the following Lewis structures, Z represents a main-group element. Name the group to which Z belongs in each case, and give an example of such a compound or ion that actually exists.

(a)

(b)

(c)

(d)

20. In each of the following Lewis structures, Z represents a main-group element. Name the group to which Z belongs in each case, and give an example of such a compound or ion that actually exists.

(a)

(b)

(c)

(d)

21. Draw Lewis electron-dot structures for the following species.
    (a) $AsH_3$
    (b) HOCl
    (c) $KrF_2$
    (d) $PO_2Cl_2^-$ (central P atom)
22. Draw Lewis electron-dot structures for the following species.
    (a) Methane
    (b) Carbon dioxide
    (c) Phosphorus trichloride
    (d) Chlorate ion
23. Urea is an important chemical fertilizer with the chemical formula $(H_2N)CO(NH_2)$. The carbon atom is bonded to both nitrogen atoms and to the oxygen atom. Draw urea's Lewis structure and use Table 2-3 to estimate its bond lengths.
24. Acetic acid is the active ingredient of vinegar. Its chemical formula is $CH_3COOH$, and the second carbon atom is bonded to the first carbon atom and to both oxygen atoms. Draw acetic acid's Lewis structure and use Table 2-3 to estimate its bond lengths.
25. Under certain conditions the stable form of sulfur consists of rings of eight sulfur atoms. Draw the Lewis structure for such a ring.
26. White phosphorus $(P_4)$ consists of four phosphorus atoms arranged at the corners of a tetrahedron. Draw the valence electrons on this structure to give a Lewis structure that satisfies the octet rule.

27. Draw Lewis electron-dot structures for the following species, indicating formal charges and resonance structures where applicable.
    (a) $H_3NBF_3$
    (b) $CH_3COO^-$ (acetate ion)
    (c) $HCO_3^-$ (hydrogen carbonate ion)
28. Draw Lewis electron-dot structures for the following species, indicating formal charges and resonance structures where applicable.
    (a) HNC (central N atom)
    (b) $SCN^-$ (thiocyanate ion)
    (c) $H_2CNN$ (the first N atom is bonded to the carbon and the second N)
29. Draw Lewis structures for the two resonance forms of the nitrite ion, $NO_2^-$. In what range do you expect the nitrogen–oxygen bond length to fall? (*Hint:* Use Table 2-3.)
30. Draw Lewis structures for the three resonance forms of the carbonate ion, $CO_3^{2-}$. In what range do you expect the carbon–oxygen bond length to fall? (*Hint:* Use Table 2-3.)
31. Methyl isocyanate, which was involved in the disaster in Bhopal, India, in 1984, has the chemical formula $CH_3NCO$.

Draw its Lewis structure, including resonance forms. (*Note:* The N atom is bonded to the two C atoms.)
32. Peroxyacetyl nitrate (PAN) is one of the prime irritants in photochemical smog. It has the formula $CH_3COOONO_2$, with the structure

$$CH_3-C{\overset{\textstyle O}{\underset{\textstyle O-O-NO_2}{\big\|}}}$$

Draw its Lewis structure, including resonance forms.
33. Draw Lewis structures for the following compounds. In the formula, the symbol of the central atom is given first. (*Hint:* The valence octet may be expanded for the central atom.)
    (a) $PF_5$
    (b) $SF_4$
    (c) $XeO_2F_2$
34. Draw Lewis structures for the following ions. In the formula, the symbol of the central atom is given first. (*Hint:* The valence octet may be expanded for the central atom.)
    (a) $BrO_4^-$
    (b) $PCl_6^-$
    (c) $XeF_3^+$

## Oxidation Numbers and Inorganic Nomenclature

35. Assign oxidation numbers to the atoms in each of the following species: $SrBr_2$, $Zn(OH)_4^{2-}$, $SiH_4$, $CaSiO_3$, $Cr_2O_7^{2-}$, $Ca_5(PO_4)_3F$, $KO_2$, CsH.
36. Assign oxidation numbers to the atoms in each of the following species: $NH_4NO_3$, $CaMgSiO_4$, $Fe(CN)_6^{4-}$, $B_2H_6$, $BaH_2$, $PbCl_2$, $Cu_2O(SO_4)$, $S_4O_6^{2-}$.
37. Write the chemical formula for each of the following compounds.
    (a) Silicon dioxide
    (b) Ammonium carbonate
    (c) Lead(IV) oxide
    (d) Diphosphorus pentaoxide
    (e) Calcium iodide
    (f) Iron(III) nitrate
38. Write the chemical formula for each of the following compounds.
    (a) Lanthanum(III) sulfide
    (b) Cesium sulfate
    (c) Dinitrogen trioxide
    (d) Iodine pentafluoride
    (e) Chromium(III) sulfate
    (f) Potassium permanganate
39. Give the systematic name for each of the following compounds.
    (a) $Cu_2S$ and CuS        (d) $ZrCl_4$
    (b) $Na_2SO_4$              (e) $Cl_2O_7$
    (c) $As_4O_6$               (f) $Ga_2O$
40. Give the systematic name for each of the following compounds.

(a) $Mg_2SiO_4$
(b) $Fe(OH)_2$ and $Fe(OH)_3$
(c) $As_2O_5$
(d) $(NH_4)_2HPO_4$
(e) $SeF_6$
(f) $Hg_2SO_4$

## Inorganic Reaction Chemistry

**41.** Rewrite the following balanced equations as net ionic equations.
(a) $NaCl(aq) + AgNO_3(aq) \rightarrow AgCl(s) + NaNO_3(aq)$
(b) $K_2CO_3(s) + 2\,HCl(aq) \rightarrow$
$$2\,KCl(aq) + CO_2(g) + H_2O(\ell)$$
(c) $2\,Cs(s) + 2\,H_2O(\ell) \rightarrow 2\,CsOH(aq) + H_2(g)$
(d) $2\,KMnO_4(aq) + 16\,HCl(aq) \rightarrow$
$$5\,Cl_2(g) + 2\,MnCl_2(aq) + 2\,KCl(aq) + 8\,H_2O(\ell)$$
**42.** Rewrite the following balanced equations as net ionic equations.
(a) $Na_2SO_4(aq) + BaCl_2(aq) \rightarrow BaSO_4(s) + 2\,NaCl(aq)$
(b) $6\,NaOH(aq) + 3\,Cl_2(g) \rightarrow$
$$NaClO_3(aq) + 5\,NaCl(aq) + 3\,H_2O(\ell)$$
(c) $Hg_2(NO_3)_2(aq) + 2\,KI(aq) \rightarrow Hg_2I_2(s) + 2\,KNO_3(aq)$
(d) $3\,NaOCl(aq) + KI(aq) \rightarrow$
$$NaIO_3(aq) + 2\,NaCl(aq) + KCl(aq)$$
**43.** Write a balanced equation for the acid–base reaction that leads to the production of each of the following salts. Name the acid, base, and salt.
(a) $CaF_2$
(b) $Rb_2SO_4$
(c) $Zn(NO_3)_2$
(d) $KCH_3COO$
**44.** Write a balanced equation for the acid–base reaction that leads to the production of each of the following salts. Name the acid, base, and salt.
(a) $Na_2SO_3$
(b) $Ca(C_6H_5COO)_2$
(c) $PbSO_4$
(d) $CuCl_2$
**45.** Hydrogen sulfide can be removed from natural gas by reaction with excess sodium hydroxide. Name the salt that is produced in this reaction. (*Note:* Hydrogen sulfide loses both of its hydrogen atoms in the course of this reaction.)
**46.** During the preparation of viscose rayon, cellulose is dissolved in a bath containing sodium hydroxide and later reprecipitated as rayon using a solution of sulfuric acid. Name the salt that is a byproduct of this process. Rayon production is in fact a significant commercial source for this salt.
**47.** An important step in many industrial processes is the slaking of lime, in which water is added to calcium oxide to make calcium hydroxide.
(a) Write the balanced equation for this process.
(b) Can this be considered a Lewis acid–base reaction? If so, what is the Lewis acid and what is the Lewis base?
**48.** Silica ($SiO_2$) is an impurity that must be removed from a metal oxide or sulfide ore when the ore is being reduced to elemental metal. To do this, lime ($CaO$) is added. It reacts with the silica to form a slag of calcium silicate ($CaSiO_3$), which can be separated and removed from the ore.
(a) Write the balanced equation for this process.
(b) Can this be considered a Lewis acid–base reaction? If so, what is the Lewis acid and what is the Lewis base?
**49.** Chemists working with fluorine and its compounds sometimes find it helpful to think in terms of acid–base reactions in which the fluoride ion ($F^-$) is donated and accepted.
(a) Would the acid in this system be the fluoride donor or fluoride acceptor?
(b) Identify the acid and base in each of these reactions:
$$ClF_3O_2 + BF_3 \longrightarrow ClF_2O_2 \cdot BF_4$$
$$TiF_4 + 2\,KF \longrightarrow K_2[TiF_6]$$
**50.** Researchers working with glasses often think of acid–base reactions in terms of oxide donors and oxide acceptors. The oxide ion is $O^{2-}$.
(a) In this system, is the base the oxide donor or the oxide acceptor?
(b) Identify the acid and base in each of these reactions:
$$2\,CaO + SiO_2 \longrightarrow Ca_2SiO_4$$
$$Ca_2SiO_4 + SiO_2 \longrightarrow 2\,CaSiO_3$$
$$Ca_2SiO_4 + CaO \longrightarrow Ca_3SiO_5$$
**51.** For each of the following balanced equations, write the oxidation number above the symbol of each atom that changes oxidation state in the course of the reactions.
(a) $2\,PF_2I(\ell) + 2\,Hg(\ell) \rightarrow P_2F_4(g) + Hg_2I_2(s)$
(b) $2\,KClO_3(s) \rightarrow 2\,KCl(s) + 3\,O_2(g)$
(c) $4\,NH_3(g) + 5\,O_2(g) \rightarrow 4\,NO(g) + 6\,H_2O(g)$
(d) $2\,As(s) + 6\,NaOH(\ell) \rightarrow 2\,Na_3AsO_3(s) + 3\,H_2(g)$
**52.** For each of the following balanced equations, write the oxidation number above the symbol of each atom that changes oxidation state in the course of the reaction.
(a) $N_2O_4(g) + KCl(s) \rightarrow NOCl(g) + KNO_3(s)$
(b) $H_2S(g) + 4\,O_2F_2(s) \rightarrow SF_6(g) + 2\,HF(g) + 4\,O_2(g)$
(c) $2\,POBr_3(s) + 3\,Mg(s) \rightarrow 2\,PO(s) + 3\,MgBr_2(s)$
(d) $4\,BCl_3(g) + 3\,SF_4(g) \rightarrow$
$$4\,BF_3(g) + 3\,SCl_2(\ell) + 3\,Cl_2(g)$$
**53.** Selenic acid ($H_2SeO_4$) is a powerful oxidizing acid that dissolves not only silver (as does the related acid $H_2SO_4$) but gold, through the reaction
$$2\,Au(s) + 6\,H_2SeO_4(aq) \longrightarrow$$
$$Au_2(SeO_4)_3(aq) + 3\,H_2SeO_3(aq) + 3\,H_2O(\ell)$$
Determine the oxidation numbers of the atoms in this equation. Which species is oxidized and which is reduced?
**54.** Diiodine pentaoxide oxidizes carbon monoxide to carbon dioxide under room conditions, yielding iodine as the second product:

$$I_2O_5(s) + 5\ CO(g) \longrightarrow I_2(s) + 5\ CO_2(g)$$

This can be used in an analytical method to measure the amount of carbon monoxide in a sample of air. Determine the oxidation numbers of the atoms in this equation. Which species is oxidized and which is reduced?

## Chemistry of the Hydrides and Oxides

55. Write balanced equations for the formation of barium hydride from the elements and the reactions of barium hydride with water, with hydrochloric acid, and with aqueous zinc sulfate.

56. Write balanced equations for the formation of hydrogen bromide from the elements and the reactions of hydrogen bromide with water, with potassium carbonate, and with magnesium.

57. Identify each of the following oxides as an acid or base anhydride. Write the chemical formula and give the name of the acid or base formed upon reaction with water.
    (a) $MgO$
    (b) $Cl_2O$
    (c) $SO_3$
    (d) $Cs_2O$

58. Write the chemical formula and give the name of the anhydride corresponding to each of the following acids or bases, and identify it as an acid or base anhydride.
    (a) $H_3AsO_4$
    (b) $H_2MoO_4$
    (c) $RbOH$
    (d) $H_2SO_3$

59. Tin(II) oxide is amphoteric. Write balanced chemical equations for its reactions with an aqueous solution of hydrochloric acid and with an aqueous solution of sodium hydroxide. (*Note:* The hydroxide complex ion of tin(II) is $[Sn(OH)_3]^-$.)

60. Zinc oxide is amphoteric. Write balanced chemical equations for its reactions with an aqueous solution of hydrochloric acid and with an aqueous solution of sodium hydroxide. (*Note:* The hydroxide complex ion of zinc is $[Zn(OH)_4]^{2-}$.)

61. Complete and balance each of the following chemical equations, and state whether each represents a redox, acid–base (including Lewis acid–base), dissolution, or precipitation reaction.
    (a) $H_2Te(aq) + H_2O(\ell) \rightarrow$
    (b) $SrO(s) + CO_2(g) \rightarrow$
    (c) $HI(aq) + CaCO_3(s) \rightarrow$
    (d) $Na_2O(s) + NH_4Br(s) \rightarrow$

62. Complete and balance each of the following chemical equations, and state whether each represents a redox, acid–base (including Lewis acid–base), dissolution, or precipitation reaction.
    (a) $HBr(aq) + ZnO(s) \rightarrow$
    (b) $Cs(s) + O_2(g) \rightarrow$
    (c) $MgO(s) + SO_3(g) \rightarrow$
    (d) $NaH(s) + H_2O(\ell) \rightarrow$

## Additional Problems

63. Trends in physical and chemical properties are not always smooth across the periodic table.
    (a) Estimate the boiling point of zinc, based on the boiling points of its neighbors copper (2567°C) and gallium (2403°C). Compare your estimate with the measured value of 907°C.
    (b) Gold and thallium are solids at room temperature, whereas mercury is a liquid. Can you draw a general conclusion about the boiling temperatures of elements in the zinc group?
    (c) Where would you predict the boiling point of cadmium to lie relative to those of silver and indium?

64. Indium forms a pale yellow oxide that is 82.712% indium and 17.288% oxygen by mass.
    (a) At the time of Mendeleyev, it was believed that the relative atomic mass of oxygen was 16 and the relative atomic mass of indium was 76. At that time, what formula was given to the pale yellow indium oxide?
    (b) Mendeleyev suggested that the correct atomic mass of indium was 114 ($\frac{3}{2}$ times 76). What formula was he simultaneously suggesting for the pale yellow oxide?

65. Only some of the elements in the seventh period have been discovered.
    (a) Recently scientists were able to prepare a few atoms of element 111. What elements should it resemble in chemical properties? Do you expect it to be a metal?
    (b) What other elements should resemble element 117? Predict the chemical formula of the compound of element 117 with potassium.

66. Which of the following do you expect to exist as stable ionic compounds? For each unstable compound, write the empirical formula of a stable compound containing the same ions. Name the five compounds.
    (a) $Na_3CO_3$          (d) $BaCN$
    (b) $MgCl$              (e) $Sr(OH)_2$
    (c) $(NH_4)_2SO_4$

67. Many important fertilizers are ionic compounds containing the elements nitrogen, phosphorus, and potassium, because these are frequently the limiting nutrients in soil for plant growth.
    (a) Write the chemical formulas for the following chemical fertilizers: ammonium phosphate, potassium nitrate, ammonium sulfate.
    (b) Calculate the mass percentage of nitrogen, phosphorus, and potassium for each of the compounds in part (a).

68. Two possible Lewis structures for sulfine ($H_2CSO$) are

    (a) Compute the formal charges on all atoms.

(b) Draw a Lewis structure for which all the atoms in sulfine have formal charges of zero.

**69.** There is persuasive evidence for the brief existence of the unstable molecule OPCl.

(a) Draw a Lewis structure for this molecule in which the octet rule is satisfied on all atoms and the formal charges on all atoms are zero.

(b) The compound OPCl reacts with oxygen to give $O_2PCl$. Draw a Lewis structure of $O_2PCl$ for which all formal charges are equal to zero. Draw a Lewis structure in which the octet rule is satisfied on all atoms.

**70.** The compound $SF_3N$ has been synthesized.

(a) Draw the Lewis structure of this molecule, supposing that the three F's and the N surround the S. Indicate the formal charges. Repeat, but assume that the three F's and the S surround the N.

(b) On the basis of the results in part (a), speculate about which arrangement is more likely to correspond to the actual molecular structure.

**71.** In nitryl chloride ($NO_2Cl$), the chlorine atom and the two oxygen atoms are bonded to a central nitrogen atom, and all the atoms lie in a plane. Draw the two electron-dot resonance structures that satisfy the octet rule and that together are consistent with the fact that the two nitrogen-oxygen bonds are equivalent.

**72.** The molecular ion $S_3N_3^-$ has the cyclic structure

All S—N bonds are equivalent.

(a) Give six equivalent resonance hybrid Lewis structures for this molecular ion.

(b) Compute the formal charges on all atoms in the molecular ion in each of the six Lewis structures.

(c) Determine the charge on each atom in the polyatomic ion, assuming that the true distribution of electrons is the *average* of the six Lewis structures arrived at in parts (a) and (b).

(d) An advanced calculation suggests that the actual charge resident on each N atom is $-0.375$, and that on each S atom is $+0.041$. Show that this result is consistent with the overall $-1$ charge on the molecular ion.

**\*73.** The two compounds nitrogen dioxide and dinitrogen tetraoxide are mentioned in Section 2-4.

(a) $NO_2$ is an odd-electron compound. Draw the best Lewis structures you can for it, recognizing that one atom cannot achieve an octet configuration. Use formal charges to decide whether that should be the (central) nitrogen atom or one of the oxygen atoms.

(b) Draw resonance structures for $N_2O_4$ that obey the octet rule. The two N atoms are bonded in this molecule.

**74.** Although magnesium and the alkaline-earth metals situated below it in the periodic table form ionic chlorides, beryllium chloride ($BeCl_2$) is a covalent compound.

(a) Follow the usual rules to write a Lewis structure for $BeCl_2$, in which each atom attains an octet configuration. Indicate formal charges.

(b) The Lewis structure that results from part (a) is an extremely unlikely one because of the double bonds and formal charges it shows. By relaxing the requirement of placing an octet on the beryllium atom, show how a Lewis structure without formal charges can be written.

**75.** (a) The first noble-gas compound, prepared by Neil Bartlett in 1962, was an orange-yellow ionic solid consisting of $XeF^+$ and $Pt_2F_{11}^-$ ions. Draw a Lewis structure for the $XeF^+$ ion.

(b) Shortly after the preparation of the ionic compound discussed in part (a), it was found that the irradiation of mixtures of xenon and fluorine with sunlight produced white crystalline $XeF_2$. Draw a Lewis structure for this molecule, allowing valence expansion on the central xenon atom.

**\*76.** Represent the bonding in $SF_2$ (F—S—F) with Lewis structures. Include the formal charges on all atoms. The dimer of this compound has the formula $S_2F_4$. It was isolated in 1980 and shown to have the structure $F_3S$—SF. Draw a possible Lewis structure to represent the bonding in the dimer, indicating the formal charges on all atoms. Is it possible to draw a Lewis structure for $S_2F_4$ in which all atoms have valence octets? Explain why or why not.

**\*77.** One method for making acetylene ($C_2H_2$) is the addition of calcium carbide ($CaC_2$) to water. Calcium hydroxide is generated as a byproduct.

(a) Write a balanced equation for the reaction that takes place, and show that no changes in oxidation state occur.

(b) Can this be described as a Lewis acid–base reaction? If so, which species is the Lewis acid and which is the Lewis base?

**\*78.** Sodium borohydride ($NaBH_4$) is a versatile and selective reducing agent that is made by the reaction of sodium hydride with diborane ($B_2H_6$).

(a) Write a balanced chemical equation for this synthetic reaction.

(b) Assuming that sodium remains as an $Na^+$ ion throughout the reaction, can this be described as a Lewis acid–base reaction? Name the Lewis acid and the Lewis base.

**79.** Bismuth forms an ion with the formula $Bi_5^{3+}$. Arsenic and fluorine form a complex ion $[AsF_6]^-$, with fluorine atoms arranged around a central arsenic atom. Assign oxidation numbers to each of the atoms in the bright yellow crystalline solid with the formula $Bi_5(AsF_6)_3 \cdot 2SO_2$.

**80.** A good method of preparing very pure oxygen on a small scale is the decomposition of $KMnO_4$ in a vacuum above $215°C$:

$$2\ KMnO_4(s) \longrightarrow K_2MnO_4(s) + MnO_2(s) + O_2(g)$$

Assign an oxidation number to each atom and verify that the

total number of electrons lost is equal to the total number gained.

**81.** One step in the Bessemer steelmaking process is the reaction of elemental silicon as an impurity with iron(II) oxide and calcium oxide to give metallic iron and calcium metasilicate ($CaSiO_3$). Write a balanced chemical equation for this reaction and indicate which species is oxidized and which is reduced.

**\*82.** (a) Determine the oxidation number of lead in each of the following oxides: $PbO$, $PbO_2$, $Pb_2O_3$, $Pb_3O_4$.
(b) The only known lead ions are $Pb^{2+}$ and $Pb^{4+}$. How can you reconcile this statement with your answer to part (a)?

**83.** The hydrogen xenate ion reacts with base according to the equation

$$2\ HXeO_4^-(aq) + 2\ OH^-(aq) \longrightarrow$$
$$XeO_6^{4-}(aq) + Xe(g) + O_2(g) + 2\ H_2O(\ell)$$

Give the oxidation numbers for all the atoms in this equation. Which species are being reduced? Which oxidized?

**84.** In some forms of the periodic table, hydrogen is placed in Group I; in others, it is placed in Group VII. Give arguments in favor of each location.

**\*85.** The structure of $B_4H_{10}$ can be indicated as

$$
\begin{array}{c}
H \\
| \\
H-B-H \\
H_2B\diagup \quad | \quad \diagdown BH_2 \\
\diagdown H-B-H \diagup \\
| \\
H
\end{array}
$$

with four bridging hydrogen atoms and six ordinary B—H single bonds. Show the distribution of electrons in this structure.

**86.** Can the reaction of $Na_2O$ with $SiO_2$ to give $Na_2SiO_3$ be considered a Lewis acid–base reaction? If so, which species is the Lewis acid and which the Lewis base? Use Lewis electron-dot structures to help you answer this question. The product, sodium metasilicate, is used as a substitute for phosphates in detergents.

**87.** A compound is said to be *hygroscopic* if it has a strong affinity for water. A very hygroscopic compound such as $P_4O_{10}$ will dehydrate many oxoacids to give the corresponding acid anhydrides (while, in turn, it forms a phosphorus oxoacid). Write a balanced chemical equation for the reaction between $P_4O_{10}$ and nitric acid. In this case, the oxoacid of phosphorus produced is $HPO_3$ rather than the fully hydrated $H_3PO_4$.

**88.** Oxygen difluoride does not react with water to give an oxoacid. Rather, its reaction is described by the equation

$$OF_2(g) + H_2O(\ell) \longrightarrow O_2(g) + 2\ HF(aq)$$

(a) Is this a redox reaction? If so, name the species being oxidized and the species being reduced.
(b) The oxoacid that might be predicted to form from the addition of water to $OF_2$ was first prepared in 1968 through the reaction of $F_2$ with water at very low temperatures and under special conditions. It is a white solid that melts to a pale yellow liquid at $-117°C$. Give its chemical formula and its name, based on the formula and name of the parallel oxoacid of chlorine.

**89.** Barium perxenate ($Ba_2XeO_6$) is a colorless salt of the perxenate ion.
(a) Americium perxenate has also been prepared. Write its chemical formula, given that the stable ion formed by americium is $Am^{3+}$.
(b) The corresponding oxoacid, perxenic acid, has not been prepared because it is spontaneously reduced in aqueous solution. Give its chemical formula.
(c) Give the chemical formula and name for the anhydride of perxenic acid.

**\*90.** Arsenic forms two important oxoacids with chemical formulas $H_3AsO_3$ and $H_3AsO_4$.
(a) Give the chemical formulas and names of the acid anhydrides of these two acids.
(b) Follow the usual naming procedure for an element that forms two oxoacids to determine the names of the oxoacids of arsenic.
(c) One of the two acid anhydrides has amphoteric properties and dissolves in strong acids as well as in strong bases. Which of the acid anhydrides do you expect to be amphoteric?
(d) Arsenic is in the same group as nitrogen. How do the chemical formulas of the oxoacids of arsenic and nitrogen differ?
(e) Can you explain the result in part (d) based on Lewis structures of the arsenic oxoacids? (*Hint:* Recall that, although second-period elements readily form multiple bonds, this is unusual in the fourth period.)

**91.** When coal that contains sulfur is burned to generate electricity, sulfur dioxide and sulfur trioxide are produced. Because these are air pollutants that contribute significantly to acid precipitation, their removal in the stacks of power plants would be very desirable. One possible method is to line the stacks with lime ($CaO$). What would be the products of lime's reaction with the sulfur oxides in the stacks?

**92.** Balance the following equations and classify them as representing dissolution, precipitation, acid–base (including Lewis acid–base), or redox reactions.
(a) $H_3PO_4(aq) + NaOH(aq) \rightarrow NaH_2PO_4(aq) + H_2O(\ell)$
(b) $Fe_3O_4(s) + C(s) \rightarrow Fe(s) + CO(g)$
(c) $Ca(NO_3)_2(aq) + K_3PO_4(aq) \rightarrow$
$$Ca_3(PO_4)_2(s) + KNO_3(aq)$$
(d) $K_2Cr_2O_7(aq) + HI(aq) \rightarrow$
$$CrI_3(s) + I_2(s) + KI(aq) + H_2O(\ell)$$

**93.** A certain element, M, is a main-group metal that reacts with chlorine to give a compound with the chemical formula $MCl_2$, and with oxygen to give the compound MO.
(a) To which group in the periodic table does element M belong?
(b) The chloride contains 44.7% chlorine by mass. Name the element M.

**94.** The last step in the industrial preparation of phosphoryl chloride (Example 2-3) is the reaction of tetraphosphorus decaoxide with phosphorus pentachloride to give $POCl_3$ as the sole product. Write a balanced chemical equation for this reaction.

**95.** If 5.00 g of sodium reacts with oxygen at elevated pressures to give a mixed peroxide–superoxide solid that weighs 9.25 g, calculate the fraction of total oxygen present as peroxide and the fraction present as superoxide in the solid.

**\*96.** An ionic compound used as a chemical fertilizer has the composition (by mass) 48.46% O, 23.45% P, 21.21% N, 6.87% H. Give the name and chemical formula of the compound, and draw Lewis structures for the two types of ions that make it up.

# CHAPTER

3

*Gaseous nitrogen dioxide is generated by the reaction of copper with nitric acid.*

# The Gaseous State

There are two major ways of classifying matter. The first emphasizes the *chemical* constitution of a material, identifying the substances present in a mixture and the elements present in a compound. Chapters 1 and 2 used this approach. The second scheme of classification considers the *physical state* of a material, labeling it a gas, a liquid, or a solid. We recognize gases, liquids, and solids quite easily because we are surrounded by important examples of all of them: air, water, and earth. Gases and liquids are fluid, but solids are rigid. A gas expands to fill any container it occupies; a liquid has a fixed volume but flows to conform to the shape of its container; a solid has a fixed volume *and* a fixed shape that resists deformation. These simple properties enable us to classify nearly all substances, at least under ordinary terrestrial conditions. There are also situations in which two or more states of matter coexist—for example, an ice cube with liquid water and water vapor in a closed container at the freezing temperature, or a solid precipitate at the bottom of an aqueous solution.

In this chapter and the next, our goal is to develop a deeper understanding of the states of matter and the transitions between them. This will allow us to later explore quantitatively the extents and rates of chemical reactions. We begin with gases, both because they have played such a central role throughout the history of chemistry and

88

because their physical properties are simpler than those of liquids and solids. In the next chapter we will turn to condensed states of matter.

## 3-1 THE CHEMISTRY OF GASES

The Greeks considered air to be one of the four fundamental elements in nature, and as early as the 17th century its physical properties (such as resistance to compression) were being studied. The chemical composition of air (Table 3-1) was not appreciated until late in the 18th century, when Lavoisier, Priestley, and others showed that air consists primarily of two different substances: oxygen and nitrogen. Oxygen was characterized by its ability to support life. It was recognized that, once a volume of oxygen was used up (by burning a candle in a closed container, for example), the nitrogen that remained could no longer keep animals alive. More than 100 years elapsed before a careful reanalysis of air showed that oxygen and nitrogen account for only about 99% of the total volume, with most of the remaining 1% taken up by a new gas, given the name "argon." The other noble gases (helium, neon, krypton, and xenon) are present in air to lesser extents.

Additional gases are found on the earth's surface. Methane ($CH_4$) is produced by bacterial processes, especially in swampy areas (hence its older name, "marsh gas"). It is a major constituent of natural-gas deposits formed over many millennia by decaying plant matter under the earth's surface. The recovery of methane from municipal landfills for use as a fuel gas is now a commercially feasible process. Gases also form when liquids evaporate. Water vapor in the air from the evaporation of liquid water is the best known example; it causes the humidity of the air.

Gases are also formed by chemical reactions. When heated, some solids decompose to give gaseous products. The decomposition of mercury(II) oxide to mercury and oxygen (Fig. 1-8) has already been mentioned:

$$2 \, HgO(s) \xrightarrow{\text{Heat}} 2 \, Hg(\ell) + O_2(g)$$

## Table 3-1  Composition of Air[†]

| Constituent | Formula | Fraction by Volume |
|---|---|---|
| Nitrogen | $N_2$ | 0.78110 |
| Oxygen | $O_2$ | 0.20953 |
| Argon | Ar | 0.00934 |
| Carbon dioxide | $CO_2$ | 0.00034 |
| Neon | Ne | $1.82 \times 10^{-5}$ |
| Helium | He | $5.2 \times 10^{-6}$ |
| Methane | $CH_4$ | $1.5 \times 10^{-6}$ |
| Krypton | Kr | $1.1 \times 10^{-6}$ |
| Hydrogen | $H_2$ | $5 \times 10^{-7}$ |
| Dinitrogen oxide | $N_2O$ | $3 \times 10^{-7}$ |
| Xenon | Xe | $8.7 \times 10^{-8}$ |

[†] Air also contains other constituents, the abundances of which are quite variable in the atmosphere. Examples are water ($H_2O$), 0–0.07; ozone ($O_3$), $0–7 \times 10^{-8}$; carbon monoxide (CO), $0–2 \times 10^{-8}$; nitrogen dioxide ($NO_2$), $0–2 \times 10^{-8}$; sulfur dioxide ($SO_2$), $0–1 \times 10^{-6}$.

Carbon dioxide, $CO_2$, is dissolved in aqueous solution to carbonate beverages. It also serves as an inexpensive acid through its reaction with water to form $H_2CO_3(aq)$. Solid carbon dioxide (Dry Ice) is used for refrigeration.

Sulfur dioxide, $SO_2$, is produced by burning sulfur in air.

This reaction was used by Joseph Priestley in his discovery of oxygen. Even earlier (1756) Joseph Black had shown that marble, which consists primarily of calcium carbonate ($CaCO_3$), dissociates on heating to give quicklime (CaO) and carbon dioxide:

$$CaCO_3(s) \xrightarrow{\text{Heat}} CaO(s) + CO_2(g)$$

Ammonium chloride ($NH_4Cl$) can be decomposed by heat to produce two gases, ammonia and hydrogen chloride:

$$NH_4Cl(s) \xrightarrow{\text{Heat}} NH_3(g) + HCl(g)$$

Some gas-forming reactions proceed explosively. The decomposition of nitroglycerine is a detonation in which all the products are gaseous:

$$4\,C_3H_5(NO_3)_3(\ell) \longrightarrow 6\,N_2(g) + 12\,CO_2(g) + O_2(g) + 10\,H_2O(g)$$

Several elements react with oxygen to form gaseous oxides (Section 2-6). Carbon dioxide, for example, forms during animal respiration but is also produced by burning coal, oil, and other materials containing compounds of carbon. Oxides of sulfur are produced by burning elemental sulfur, and those of nitrogen from elemental nitrogen:

$$S(s) + O_2(g) \longrightarrow SO_2(g)$$
$$2\,SO_2(g) + O_2(g) \longrightarrow 2\,SO_3(g)$$
$$N_2(g) + O_2(g) \longrightarrow 2\,NO(g)$$
$$2\,NO(g) + O_2(g) \longrightarrow 2\,NO_2(g)$$

The role of these compounds in air pollution will be considered further in Section 15-4.

The actions of acids on certain ionic solids are another important class of reactions that form gases. Carbon dioxide is produced from the reaction of acids with carbonates (Fig. 3-1):

$$CaCO_3(s) + 2\,HCl(g) \longrightarrow CaCl_2(s) + CO_2(g) + H_2O(\ell)$$

**Figure 3-1**

The calcium carbonate in a piece of chalk reacts with an aqueous solution of hydrochloric acid, producing bubbles of carbon dioxide.

Other examples of this type of reaction are

$$Na_2S(s) + 2\,HCl(g) \longrightarrow 2\,NaCl(s) + H_2S(g)$$

$$K_2SO_3(s) + 2\,HCl(g) \longrightarrow 2\,KCl(s) + SO_2(g) + H_2O(\ell)$$

$$NaCl(s) + H_2SO_4(\ell) \longrightarrow NaHSO_4(s) + HCl(g)$$

In these reactions, the metal ions (sodium, calcium, and potassium) play no direct role, and parallel reactions can be written for other metal ions.

The gases just mentioned show extremely varied chemical behavior. Some, such as HCl and $SO_3$, are very reactive, acidic, and corrosive, whereas others, such as $N_2O$ and nitrogen, are much less reactive. Although the chemical properties of gases vary significantly, their physical properties are much simpler to understand. At sufficiently low densities the physical properties of all gases behave in the same way and approach those of an "ideal" gas, the subject of the following sections.

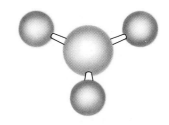

Sulfur trioxide, $SO_3$, is a corrosive gas produced by the further oxidation of sulfur dioxide.

## 3-2 PRESSURE AND TEMPERATURE OF GASES

Three useful properties for describing gases are volume ($V$), pressure ($P$), and temperature ($T$). Volume is self-evident and needs no comment; as stated in Section 1-5, we shall use the liter as a unit of volume. Pressure and temperature require a little more care to define.

### Pressure

The force exerted by a gas on a unit area of the walls of its container is called the **pressure** of that gas. We do not often stop to think that the air around us exerts a pressure on us and on everything else at the earth's surface. This fact was demonstrated in an ingenious experiment by Evangelista Torricelli (1608–1647), an Italian scientist who had been an assistant to Galileo. He sealed a long glass tube at one end and filled it with mercury. He then covered the open end with his thumb, turned the tube upside down, and immersed it in a dish of liquid mercury (Fig. 3-2a), taking care that no air leaked in. The mercury in the tube fell, leaving a nearly perfect vacuum at the closed end, but it did not all run out of the tube. It stopped when its top was about 76 cm above the level of the mercury in the dish. Torricelli showed that the exact height varied somewhat from day to day and from place to place.

This simple device, called a **barometer,** works like a balance, one arm of which is loaded with the mass of mercury in the tube and the other with a column of air, of the same cross-sectional area $A$, that extends to the top of the earth's atmosphere, approximately 150 km up (Fig. 3-2b). The height of the mercury column adjusts itself so that its mass and that of the air column (and thus the two forces on the surface of the mercury in the dish) become equal. Day-to-day changes in the height of the column occur as the force exerted by the atmosphere varies with the weather. A barometer at a high elevation has a portion of the atmosphere below it and gives a lower reading than one at sea level, if the effect of the weather is excluded.

The relationship between Torricelli's invention and the pressure of the atmosphere can be understood from Newton's laws of motion. His second law stated that

$$\text{force} = \text{mass} \times \text{acceleration}$$

$$F = ma$$

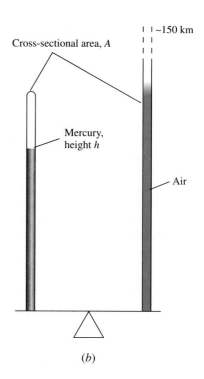

**Figure 3-2**

(*a*) In Torricelli's barometer, the top of the mercury in the tube is approximately 76 cm higher than that in the open beaker. (*b*) The mass of mercury in height *h* exactly balances that of a column of air of the same diameter extending to the top of the atmosphere.

where the acceleration of a body (*a*) is the rate at which its velocity changes. All bodies attract one another because of gravity, and the force of attraction imparts an acceleration to each body. The standard acceleration due to the earth's gravitational field (usually denoted by *g* instead of *a*) is $g = 9.80665$ m s$^{-2}$. Pressure is the force per unit area, or the total force *F* divided by the area *A*:

$$P = \frac{F}{A} = \frac{mg}{A}$$

Because the volume of mercury in the barometer is $V = Ah$,

$$P = \frac{mg}{A} = \frac{mg}{V/h} = \frac{mgh}{V}$$

If the density is stated as $\rho = m/V$, this equation can be rewritten

$$P = \rho g h \tag{3-1}$$

With this equation we can calculate the pressure exerted by the atmosphere. The density of mercury at 0°C, in SI units (see Appendix B), is

$$\rho = 13.5951 \text{ g cm}^{-3} = 1.35951 \times 10^4 \text{ kg m}^{-3}$$

and the height of the mercury column under ordinary atmospheric conditions near sea level is close to 0.76 m (760 mm). Let us use exactly this value for the height in our computation:

$$P = \rho g h = (1.35951 \times 10^4 \text{ kg m}^{-3})(9.80665 \text{ m s}^{-2})(0.760000 \text{ m})$$
$$= 1.01325 \times 10^5 \text{ kg m}^{-1} \text{ s}^{-2}$$

**Table 3-2   Units for Pressure**

| Unit | Definition or Relationship |
|------|---------------------------|
| pascal (Pa) | $1 \text{ kg m}^{-1}\text{s}^{-2}$ |
| bar | $1 \times 10^5 \text{ Pa}$ |
| atmosphere (atm) | 101,325 Pa |
| torr | 1/760 atm |
| 760 mm Hg (at 0°C) | 1 atm |
| 14.6960 pounds per square inch (psi, lb in$^{-2}$) | 1 atm |

Pressure can be expressed in various units. The natural SI unit is the $\text{kg m}^{-1}\text{s}^{-2}$, which arose in the preceding calculation and is called the *pascal* (Pa). One **standard atmosphere** (1 atm) is defined as exactly $1.01325 \times 10^5$ Pa. The standard atmosphere is a useful unit because the pascal is inconveniently small and because ''atmospheric pressure'' is important as a standard of reference. Pressures must be converted to pascals if calculations are to be carried out entirely in SI units.

Unfortunately, pressure units in common use have proliferated. Although we shall work primarily with the standard atmosphere, it is important to be able to recognize other units and convert among them. Weather reports express atmospheric pressure in terms of the height (in millimeters or inches) of the column of mercury it can support. A pressure of 1 atm supports a 760-mm column of mercury at 0°C, so we often speak of 1 atm pressure as 760 mm (of Hg). Because the density of mercury depends slightly on temperature, for accurate work it is necessary to specify the temperature and make the proper corrections to the density. A more precise term is the *torr*, defined as 1 torr = 1/760 atm (or 760 torr = 1 atm) at *any* temperature. Only at 0°C do the torr and the millimeters of Hg coincide. These and other units of pressure are summarized in Table 3-2.

## Boyle's Law

In the 17th century, Robert Boyle, an English natural philosopher and theologian, studied the properties of confined gases. He noted that a gas tends to spring back to its original volume after being compressed or expanded. These properties are much like those of metal springs, which were being investigated by his collaborator Robert Hooke. Boyle's experiments on the compression and expansion of air were published in 1662 in a monograph, ''The Spring of the Air and Its Effects.''

Boyle worked with a simple piece of apparatus: a J-tube in which air was trapped at the closed end by a column of mercury (Fig. 3-3). If the difference in height $h$ between the two mercury levels in such a tube is 0, then the pressure of the air in the closed part exactly balances that of the atmosphere, so the pressure $P$ is 1 atm. Adding mercury to the open end of the tube increases the pressure of the confined air by 1 mm Hg for every millimeter by which the level of mercury on the open side of the tube exceeds the level on the closed side. Expressed in atmospheres, the pressure is

$$P = 1 \text{ atm} + \frac{h \text{ (in mm)}}{760 \text{ mm atm}^{-1}}$$

**Figure 3-3**

(*a*) Boyle's J-tube. When the heights of mercury on the two sides of the tube are the same (*left*), the pressure of the confined gas must equal that of the atmosphere—1 atm, or 760 mm Hg. After mercury has been added (*right*), the pressure of the gas is increased by the number of millimeters of mercury in the height difference *h*. The compression of the gas causes it to occupy a smaller volume. (*b*) The same principle applies in a manometer, used to measure the pressure of a gas. The glass tube at the bottom is connected to the gas sample, and the difference in height between the levels of the mercury in the two arms is read off on a calibrated scale. (*b*, Leon Lewandowski)

The volume of the confined air can be read off from the previously calibrated tube. Boyle discovered that the product of pressure and volume, *PV*, is a constant *C* at a given temperature for a sample of gas:

$$PV = C \qquad \text{(fixed temperature and fixed amount of gas)} \qquad \textbf{(3-2)}$$

This result is known as **Boyle's law.**

The equation $P = C/V$ is the equation of a hyperbola, shown in Figure 3-4*a*. Many curves that look like hyperbolas in fact are not. How are we to know whether Boyle's experiments are best described by $PV = C$ and not some more complicated law? There are two ways of plotting the data to answer this question. The first is to rewrite Boyle's law in the form

$$P = \frac{C}{V} = C\left(\frac{1}{V}\right)$$

and to note that $P$ is proportional to $1/V$. A graph of $P$ against $1/V$ (rather than $V$) should be a straight line passing through the origin with slope $C$ (see Fig. 3-4$b$). Alternatively, because $PV = C$, a plot of $PV$ against $P$ should be independent of $P$, as shown in Figure 3-4$c$.

It is important to remember that the constant $C$ depends on temperature and on the amount (number of moles) of gas used. At 0°C and for 1 mol of gas (e.g., 31.999 g $O_2$, 28.013 g $N_2$, or 2.0159 g $H_2$), an extrapolation of $PV$ to zero pressure (Fig. 3-4$c$) gives a limiting value of 22.414 L atm for $C$, so that

$$PV = 22.414 \text{ L atm} \qquad \text{(for one mole of gas at 0°C)}$$

If the pressure is 1.00 atm, the volume should be 22.4 L; if $P$ is 4.00 atm, $V$ should be 22.414/4.00 = 5.60 L. Boyle's law is an idealization that is satisfied exactly by all gases at *very* low pressures; for real gases near 1 atm pressure, small corrections may be necessary for highly accurate work. For high-pressure gases (beyond 50 to 100 atm), substantial corrections are needed.

## Example 3-1

The long cylinder of a bicycle pump has a volume of 1131 cm$^3$ and is filled with air at a pressure of 1.02 atm. The outlet valve is sealed shut and the pump handle is pushed down until the volume of the air is 517 cm$^3$. Compute the pressure inside the pump, and express it in atmospheres and pounds per square inch.

## Solution

Note that the temperature and amount of gas are not stated in this problem, so the value of 22.414 L atm cannot be used for the constant $C$. All that is necessary, however, is to assume that the temperature does not change as the pump handle is pushed down. If $P_1$ and $P_2$ are the initial and final pressures, and $V_1$ and $V_2$ the initial and final volumes, then

$$P_1 V_1 = P_2 V_2$$

because the temperature and amount of air in the pump do not change. Substitution gives

$$(1.02 \text{ atm})(1131 \text{ cm}^3) = P_2(517 \text{ cm}^3)$$

which can be solved for $P_2$:

$$P_2 = 2.23 \text{ atm}$$

In pounds per square inch (Table 3-2), this pressure is

$$P_2 = 2.23 \text{ atm} \times 14.696 \text{ psi atm}^{-1} = 32.8 \text{ psi}$$

*Related Problems:* 11, 12

(a)

(b)

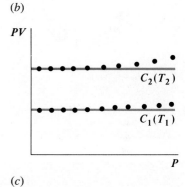

(c)

**Figure 3-4**

Three ways of depicting Boyle's law. Small deviations from the law (especially apparent in parts $b$ and $c$) arise for real gases at higher pressures or smaller volumes.

## Temperature and Charles's Law

**Temperature** is one of those elusive properties that all of us think we understand but that are very difficult to pin down in a quantitative fashion. We have an instinctive feeling (through the sense of touch) of *hot* and *cold;* for example, water at its freezing point is colder than at its boiling point and thus should have a lower temperature. On the Celsius scale of temperature, the freezing point is assigned a temperature of 0°C and the boiling point (at 1 atm pressure) is 100°C. On the Fahrenheit scale the same two temperatures are 32°F and 212°F.

Assigning two fixed points in this way does not resolve the problem of defining a temperature scale. Ether, for example, boils at atmospheric pressure somewhere between 0 and 100°C, but what temperature should be assigned to its boiling point? Further arbitrary choices are certainly not the answer. The problem is that temperature is not a mechanical quantity like pressure, and thus is more difficult to define.

There are mechanical properties that *depend* on temperature, however. For example, as liquid mercury is heated from 0°C to 100°C its volume increases by 1.82%. This change in volume could be used as the basis for the definition of a temperature scale. Temperature could be *assumed* to be linearly related to the volume of mercury so that temperatures could be read off from the volume of mercury in a tube (a mercury thermometer). The problem with this definition, of course, is that it is tied to the properties of a single substance, mercury. If the volume of some other substance is measured on the mercury scale, it will *not* be a linear function of temperature. Why should mercury be chosen rather than another substance? How can temperature be defined below the freezing point or above the boiling point of mercury? To answer this question, let us examine the behavior of gases upon heating or cooling.

Boyle observed that the product of the pressure and volume of a confined gas changes upon heating and thus depends on temperature (Fig. 3-5), but the first quantitative experiments were carried out by the French scientist Jacques Charles more than a century later. The crucial physical content of Charles's observations is that, at sufficiently low pressures, *all* gases expand by the same relative amount if they have equal initial and final temperatures. For example, heating a sample of $N_2$ from the freezing point of water to the boiling point causes the gas to expand to 1.366 times its original volume; the same 36.6% increase in volume is found for $O_2$, $CO_2$, and other gases. (In contrast, liquids and solids vary widely in thermal expansion.) This universal behavior suggests that temperature can be expressed as a linear function of gas volume. We therefore write

$$t = c\left(\frac{V}{V_0} - 1\right)$$

where $V$ is the volume of the gas at temperature $t$, $V_0$ is its volume at the freezing point of water, and $c$ is a constant, the same for all gases. We have written the linear equation in this form to ensure that the freezing point of water (when $V = V_0$) will be at $t = 0$, corresponding to the zero of the Celsius scale of temperature. From the measured fractional increase in $V$ to the boiling point (taken to be 100°C), the value of $c$ can be determined. In 1802, Gay-Lussac reported a value for $c$ of 267°C. Subsequent experiments have refined this result to give $c = 273.15$°C. The definition of temperature (in degrees Celsius) is then

$$t = 273.15°C\left(\frac{V}{V_0} - 1\right)$$

**Figure 3-5**

When a balloon is cooled with liquid nitrogen, its volume shrinks drastically as the gas inside cools. (Leon Lewandowski)

The temperature of a gas can be measured by taking a gas sample at low pressure and comparing its volume with that at the freezing point of water. For many gases atmospheric pressure is sufficiently low, but for highly accurate temperature determinations it is necessary to use lower pressures or to apply small corrections.

Once temperature has been defined in this way, we can return to mercury and measure its *actual* changes in volume with temperature. The result found is almost, but not quite, linear. For example, if a mercury thermometer is calibrated to match the gas thermometer at 0°C and 100°C and if the scale in between is divided evenly into 100 parts to mark off degrees, a small error will result from the use of this thermometer. A temperature of 40.00°C on the gas thermometer will be read as 40.11°C on the mercury thermometer, which indicates that the volume of liquid mercury does not vary exactly linearly with temperature.

The preceding equation can be rewritten to express the gas volume in terms of the temperature. The result is

$$V = V_0\left(1 + \frac{t}{273.15°C}\right) \qquad (3\text{-}3)$$

In words, the volume of a gas varies linearly with its temperature. This is the most common statement of **Charles's law,** but it is misleading in a way because the linearity is built in through the definition of temperature. The key observation is the universal nature of the constant 273.15°C, which is the same for all gases at low pressures. Writing Charles's law in this form suggests a very interesting lower limit to the temperatures that can be obtained. Negative temperatures on the Celsius scale correspond to temperatures below the freezing point of water and of course are meaningful. But what would happen as $t$ approached −273.15°C? The volume would then approach 0 (Fig. 3-6), and if $t$ could go below this value the volume would become negative, clearly an impossible result. We therefore surmise that $t = -273.15°C$ is a fundamental limit below which the temperature cannot be low-

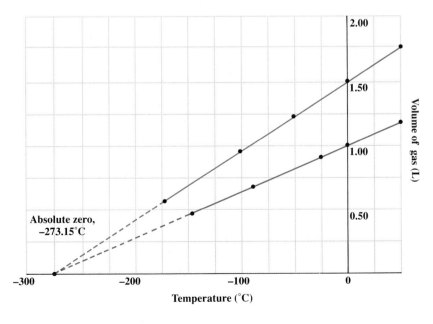

**Figure 3-6**

The volume of a sample of a gas is a function of its temperature under conditions of constant pressure. The observed straight-line response of volume to temperature illustrates Charles's law. A particular sample of a gas (red line) has a volume of 1.0 L at a temperature of 0°C. Another sample of a gas (blue line) held at the same pressure takes up more volume at 0°C but shrinks faster as cooled. Its *percentage* change in volume is the same as that of the first sample for every degree of temperature change. Extrapolation of the trends (dashed lines) predicts that the volumes of the samples go to zero at a temperature of −273.15°C. Similar observations are made irrespective of the chemical identity of the gas.

ered. All real gases condense to a liquid or solid before they reach this **absolute zero** of temperature, but more rigorous arguments show that *no* substance (gas, liquid, or solid) can be cooled below $-273.15°C$. In fact, it becomes increasingly difficult to cool a substance as absolute zero is approached. The coldest temperatures reached to date are on the order of several microdegrees (1 microdegree $= 10^{-6}\,°C$) above absolute zero.

The absolute zero of temperature is a compellingly logical choice as the zero point of a temperature scale. The easiest way to create such a new scale is to add 273.15 to the Celsius temperature. This gives the **Kelvin temperature scale:**

$$T\ (\text{Kelvin}) = 273.15 + t\ (\text{Celsius}) \qquad \textbf{(3-4)}$$

The capital $T$ signifies that this is an absolute scale, whose unit is the *kelvin* (K). Thus, a temperature of 25.00°C corresponds to $273.15 + 25.00 = 298.15$ K. On this scale, Charles's law takes the form

$$V \propto T \qquad \text{(fixed pressure and fixed amount of gas)}$$

where the proportionality constant is determined by the pressure and the amount of gas present. The ratios of volumes occupied at two different temperatures by a fixed amount of gas at fixed pressure follow:

$$\frac{V_1}{V_2} = \frac{T_1}{T_2}$$

## Example 3-2

A scientist studying the properties of hydrogen at low temperature takes a volume of 2.50 L of hydrogen at atmospheric pressure and a temperature of 25.00°C and cools the gas at constant pressure to $-200.00°C$. Predict the volume of the hydrogen.

### Solution

The first step is always to convert temperatures to kelvins:

$$t_1 = \quad 25.00°C \implies T_1 = 273.15 + 25.00 = 298.15\ \text{K}$$
$$t_2 = -200.00°C \implies T_2 = 273.15 - 200.00 = 73.15\ \text{K}$$

The ratio in Charles's law is:

$$\frac{V_1}{T_1} = \frac{2.50\,\text{L}}{298.15\,\text{K}} = \frac{V_2}{T_2} = \frac{V_2}{73.15\,\text{K}}$$

$$V_2 = \frac{(73.15\,\text{K})(2.50\,\text{L})}{298.15\,\text{K}} = 0.613\ \text{L}$$

*Related Problems:* 13, 14, 15, 16, 17, 18

## 3-3 THE IDEAL GAS LAW

Several properties of gases at low pressures have been deduced to this point. From Boyle's law,

$$V \propto \frac{1}{P} \qquad \text{(at constant temperature, fixed amount of gas)}$$

and from Charles's law,

$$V \propto T \qquad \text{(at constant pressure, fixed amount of gas)}$$

where $T$ is the absolute temperature in kelvins. In addition, it is clear that

$$V \propto n \qquad \text{(at constant temperature and pressure)}$$

where $n$ is the chemical amount of substance (in moles), because at a given temperature and pressure, doubling the amount of *any* substance (gas, liquid, or solid) must double the volume.

These three statements may be combined in the form

$$V \propto \frac{nT}{P}$$

A proportionality constant called $R$ converts this proportionality to an equation:

$$V = R\frac{nT}{P} \qquad \text{or} \qquad PV = nRT \tag{3-5}$$

This is the **ideal gas law,** an empirical law that holds approximately for all gases near atmospheric pressure and that becomes increasingly accurate as the pressure is decreased.

Situations frequently arise in which a gas undergoes a change that takes it from some initial condition (described by $P_1$, $V_1$, $T_1$, and $n_1$) to a final condition (described by $P_2$, $V_2$, $T_2$, and $n_2$). Because $R$ is a constant,

$$\frac{P_1 V_1}{n_1 T_1} = \frac{P_2 V_2}{n_2 T_2} \tag{3-6}$$

This is a useful alternative form of the ideal gas law.

### Example 3-3

A weather balloon filled with helium has a volume of $1.0 \times 10^4$ L at 1.00 atm and 30°C. It rises to an altitude at which the pressure has dropped to 0.60 atm and the temperature is −20°C. What is the volume of the balloon then? Assume that the balloon stretches in such a way that the pressure inside stays close to the pressure outside.

### Solution

Because the amount of helium does not change, we can set $n_1$ equal to $n_2$ and cancel it out of Equation 3-6, giving

**Figure 3-7**

In a hot-air balloon, the volume remains nearly constant (the balloon is rigid) and the pressure is nearly constant as well (unless the balloon rises very high). Thus, $n$ is inversely proportional to $T$; as the air inside the balloon is heated, its amount decreases and its density falls. This reduced density gives the balloon its lift. (Jim Marshall/Picture Perfect USA™)

$$\frac{P_1 V_1}{T_1} = \frac{P_2 V_2}{T_2}$$

Solving this for the only unknown quantity, $V_2$, gives

$$V_2 = V_1 \left(\frac{P_1}{P_2}\right)\left(\frac{T_2}{T_1}\right)$$

$$= 1.0 \times 10^4\,\text{L}\left(\frac{1.00\,\text{atm}}{0.60\,\text{atm}}\right)\left(\frac{253\,\text{K}}{303\,\text{K}}\right)$$

$$= 1.4 \times 10^4\,\text{L} = 14{,}000\,\text{L}$$

Note that temperatures must be converted to kelvins before the ideal gas law is used.

*Related Problems:* 19, 20, 21, 22

The approach outlined in Example 3-3 can be applied when other combinations of the four variables ($P$, $V$, $T$, and $n$) remain constant (Fig. 3-7). For example, when $n$ and $T$ remain constant, they can be cancelled from the equation and Boyle's law, which was used in solving Example 3-1, results. When $n$ and $P$ remain constant, this relationship reduces to Charles's law, used in Example 3-2.

We are not quite finished. For a given gas, $PV/nT$ is a constant, called $R$, but it has not yet been established whether $R$ is a *universal* constant or whether it changes from one gas to another (e.g., from oxygen to nitrogen to hydrogen). To deal with this issue, we invoke Avogadro's hypothesis, whose validity is established by the success of the laws of chemical combination considered in Chapter 1. According to this hypothesis, fixed volumes of different gases at the same temperature and pressure contain the same number of molecules (and therefore of moles). In other words, for fixed $V$, $P$, and $T$, the value of $n$ is fixed as well and is independent of the particular gas studied. Therefore, $R$ is a **universal gas constant** that is the same for all gaseous substances.

The numerical value of $R$ depends on the units chosen for $P$ and $V$. At the freezing point of water ($T = 273.15$ K) the product $PV$ for 1 mol of any gas approaches the value 22.414 L atm at low pressures; hence, $R$ has the value

$$R = \frac{PV}{nT} = \frac{22.414\,\text{L atm}}{(1.0000\,\text{mol})(273.15\,\text{K})} = 0.082058\,\text{L atm mol}^{-1}\,\text{K}^{-1}$$

If $P$ is measured in SI units of pascals (kg m$^{-1}$ s$^{-2}$) and $V$ in cubic meters, then $R$ has the value

$$R = \frac{(1.01325 \times 10^5\,\text{kg m}^{-1}\,\text{s}^{-2})(22.414 \times 10^{-3}\,\text{m}^3)}{(1.0000\,\text{mol})(273.15\,\text{K})}$$

$$= 8.3145\,\text{kg m}^2\,\text{s}^{-2}\,\text{mol}^{-1}\,\text{K}^{-1}$$

Because 1 kg m$^2$ s$^{-2}$ is defined to be one *joule* (the SI unit of energy, abbreviated J), the gas constant is

$$R = 8.3145\,\text{J mol}^{-1}\,\text{K}^{-1}$$

## Example 3-4

What mass of helium is needed to fill the weather balloon from Example 3-3?

### Solution

First, solve the ideal gas law for $n$:

$$n = \frac{PV}{RT}$$

If $P$ is expressed in atmospheres and $V$ in liters, the value $R = 0.08206 \text{ L atm mol}^{-1}\text{ K}^{-1}$ must be used.

$$n = \frac{(1.00 \text{ atm})(1.00 \times 10^4 \text{ L})}{(0.08206 \text{ L atm mol}^{-1}\text{ K}^{-1})(303.15 \text{ K})} = 402 \text{ mol}$$

Because the molar mass of helium is $4.00 \text{ g mol}^{-1}$, the mass of helium required is

$$(402 \text{ mol})(4.00 \text{ g mol}^{-1}) = 1610 \text{ g} = 1.61 \text{ kg}$$

*Related Problems:* 23, 24

## Chemical Calculations for Gases

One of the most important applications of the gas laws in chemistry is in the calculation of volumes of gases consumed or produced in chemical reactions. If the conditions of pressure and temperature are known, then the ideal gas law can be used to convert between chemical amount and gas volume. Instead of working with the mass of the gas taking part, we can then use its volume, which is easier to measure. This is illustrated by the following example.

## Example 3-5

Concentrated nitric acid acts on copper to give nitrogen dioxide and dissolved copper ions (Fig. 3-8) according to the balanced chemical equation

$$Cu(s) + 4 \text{ H}^+(aq) + 2 \text{ NO}_3^-(aq) \longrightarrow 2 \text{ NO}_2(g) + Cu^{2+}(aq) + 2 \text{ H}_2\text{O}(\ell)$$

Suppose that 6.80 g of copper is consumed in this reaction, and that the $NO_2$ is collected at a pressure of 0.970 atm and a temperature of 45°C. Calculate the volume of $NO_2$ produced.

### Solution

The first step (as in Fig. 1-22) is to convert from the mass of the known reactant or product (in this case, 6.80 g of Cu) to its chemical amount by using the molar mass of copper, $63.55 \text{ g mol}^{-1}$:

$$\frac{6.80 \text{ g Cu}}{63.55 \text{ g mol}^{-1}} = 0.107 \text{ mol Cu}$$

**Figure 3-8**

When concentrated nitric acid is added to a copper penny, the copper is oxidized and an aqueous solution of blue copper(II) nitrate forms. In addition, some of the nitrate ion is reduced to brown gaseous nitrogen dioxide, which bubbles off. (Leon Lewandowski)

Next, the chemical amount of $NO_2$ generated in the reaction is found by forming the ''chemical conversion factor'' 2 mol of $NO_2$ per 1 mol of Cu from the balanced equation and multiplying the chemical amount of Cu by it:

$$0.107 \text{ mol Cu} \times \left(\frac{2 \text{ mol } NO_2}{1 \text{ mol Cu}}\right) = 0.214 \text{ mol } NO_2$$

Finally, the ideal gas law is used to find the volume from the chemical amount (recall that the temperature must first be expressed in kelvins by adding 273.15):

$$V = \frac{nRT}{P} = \frac{(0.214 \text{ mol})(0.08206 \text{ L atm mol}^{-1} \text{ K}^{-1})(273 + 45 \text{ K})}{0.970 \text{ atm}}$$

$$= 5.76 \text{ L}$$

Therefore, 5.76 L of $NO_2$ is produced under these conditions.

*Related Problems:* 25, 26, 27, 28, 29, 30, 31, 32

## 3-4 MIXTURES OF GASES

Suppose a mixture of gases occupies a container at a certain temperature. We define the **partial pressure** of each gas as the pressure that gas would exert if it alone were present in the container. **Dalton's law** (named after its discoverer, John Dalton) then states that the total pressure measured, $P_{tot}$, is the sum of the partial pressures of the individual gases (Fig. 3-9). This law holds under the same conditions as the ideal gas law itself: it is approximate at moderate pressures, but becomes increasingly more accurate as the pressure is lowered.

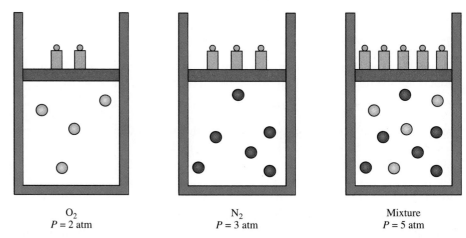

|  $O_2$  |  $N_2$  |  Mixture  |
| :---: | :---: | :---: |
| $P = 2$ atm | $P = 3$ atm | $P = 5$ atm |

**Figure 3-9**

According to Dalton's law, the total pressure of a gas mixture (indicated here by the number of weights holding the piston in place) is the sum of the pressures exerted by the individual gases.

For a gas mixture at low pressure, the partial pressure of one component, A, is

$$P_A = n_A \frac{RT}{V}$$

The total pressure is the sum of the partial pressures:

$$P_{tot} = P_A + P_B + P_C + \cdots = (n_A + n_B + n_C + \cdots) \frac{RT}{V} = n_{tot} \frac{RT}{V} \quad (3\text{-}7)$$

where $n_{tot}$ is the total number of moles in the gas mixture. Dividing the first equation by the second gives

$$\frac{P_A}{P_{tot}} = \frac{n_A}{n_{tot}} \quad \text{or} \quad P_A = \frac{n_A}{n_{tot}} P_{tot}$$

We define

$$X_A = \frac{n_A}{n_{tot}}$$

as the **mole fraction** of A in the mixture—the number of moles of A divided by the total number of moles present. Then

$$P_A = X_A P_{tot} \quad (3\text{-}8)$$

The partial pressure of any component in a mixture of ideal gases is the total pressure multiplied by the mole fraction of that component. Note that the volume fractions in Table 3-1 are the same as mole fractions.

## Example 3-6

When $NO_2$ is cooled to room temperature, some of it reacts to form a *dimer,* $N_2O_4$, through the reaction

$$2\,NO_2 \longrightarrow N_2O_4$$

Suppose 15.2 g of $NO_2$ is placed in a 10.0-L flask at high temperature and the flask is cooled to 25°C. The total pressure is measured to be 0.500 atm. What partial pressures and mole fractions of $NO_2$ and $N_2O_4$ are present?

### Solution

Initially there is 15.2 g/46.01 g $mol^{-1}$ = 0.330 mol $NO_2$, and therefore the same number of moles of nitrogen atoms. If at 25°C there are $n_{NO_2}$ moles of $NO_2$ and $n_{N_2O_4}$ moles of $N_2O_4$, then, because the total number of moles of nitrogen atoms is unchanged,

$$n_{NO_2} + 2\,n_{N_2O_4} = 0.330 \text{ mol} \quad \textbf{(a)}$$

To find a second relation between $n_{NO_2}$ and $n_{N_2O_4}$, use Dalton's law:

$$P_{NO_2} + P_{N_2O_4} = 0.500 \text{ atm}$$

$$\frac{RT}{V}n_{NO_2} + \frac{RT}{V}n_{N_2O_4} = 0.500 \text{ atm}$$

$$n_{NO_2} + n_{N_2O_4} = 0.500\frac{V}{RT}$$

$$= \frac{(0.500 \text{ atm})(10.0 \text{ L})}{(0.08206 \text{ L atm mol}^{-1} \text{ K}^{-1})(298 \text{ K})}$$

$$n_{NO_2} + n_{N_2O_4} = 0.204 \text{ mol} \qquad \textbf{(b)}$$

Subtracting Equation b from Equation a gives

$$n_{N_2O_4} = 0.126 \text{ mol}$$

$$n_{NO_2} = 0.078 \text{ mol}$$

From these results, $NO_2$ has a mole fraction of 0.38 and a partial pressure of (0.38)(0.500 atm) = 0.19 atm, and $N_2O_4$ has a mole fraction of 0.62 and a partial pressure of (0.62)(0.500 atm) = 0.31 atm.

*Related Problems:* 33, 34, 35, 36, 37, 38

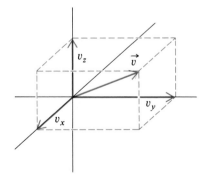

**Figure 3-10**

The path of a molecule in a box.

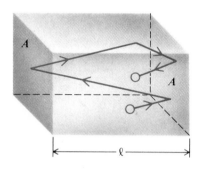

**Figure 3-11**

The velocity is shown by an arrow of length $v$. It can be separated into three components, $v_x$, $v_y$, and $v_z$, along the three Cartesian coordinate axes.

## 3-5 THE KINETIC THEORY OF GASES

The ideal gas law summarizes certain physical properties of gases at low pressures. It is an empirical law, the consequence of experimental observations, but its simplicity and generality prompt us to ask whether it has some underlying microscopic explanation that involves the properties of atoms and molecules in a gas. Such an explanation must allow other properties of gases at low pressures to be predicted, and must explain why real gases deviate from the ideal gas law to small but measurable extents. Such a theory was developed in the 19th century, notably by the physicists Rudolf Clausius, James Clerk Maxwell, and Ludwig Boltzmann. The **kinetic theory of gases** is one of the great milestones of science, and its success provides strong evidence for the atomic theory of matter discussed in Chapter 1.

This section involves a type of reasoning that differs from the one we have used so far. Instead of proceeding from experimental observations to empirical laws, we shall begin with a model and, using basic laws of physics with mathematical reasoning, show how this model helps to explain the measured properties of gases. Boyle's law is identified with the result of the model, and a connection is established between temperature and the distribution of speeds in a group of gas molecules.

The underlying assumptions of the kinetic theory of gases are very simple:

1. A pure gas consists of a large number of identical molecules separated by distances that are great compared with their size.
2. The gas molecules are constantly moving in random directions with a distribution of speeds.
3. The molecules exert no forces on one another between collisions, so between collisions they move in straight lines with constant velocities.

Vessel wall

$m\mathrm{v}_y$

$\overrightarrow{m\mathrm{v}}$

$-m\mathrm{v}_x$

$m\mathrm{v}_x$

$m\mathrm{v}_y$

$\overrightarrow{m\mathrm{v}}$

Point of impact

Before

After

## Figure 3-12

An elastic collision of a molecule with a wall. The component of the molecule's momentum perpendicular to the wall reverses sign, from $m v_x$ to $-m v_x$. The component parallel to the wall, $m v_y$, is unchanged. The total momentum is shown by the red arrow. Although its direction is changed by the collision, its length is not, reflecting the fact that molecular speeds are unaffected by elastic wall collisions.

**4.** The collisions of molecules with the walls of the container are elastic; no energy is lost during a collision.

## The Meaning of Temperature

We first use the kinetic theory of gases to find a relationship among pressure, volume, and the motions of molecules in an ideal gas. Comparison with the ideal gas law ($PV = nRT$) gives a deeper understanding of the meaning of temperature.

Suppose a container has the shape of a rectangular box of length $\ell$, with end faces each of which has area $A$ (see Fig. 3-10). A single molecule that is moving with speed $u$ in some direction is placed in the box. It is important to distinguish between **speed** and **velocity.** The velocity of a molecule specifies both the rate at which it is moving (its speed, in meters per second) and the direction of motion. As shown in Figure 3-11, the velocity can be indicated by an arrow (vector) $\vec{v}$, which has a length equal to the speed $u$ and which points in the direction of motion of the molecule. The velocity can also be represented by its components along three coordinate axes: $v_x$, $v_y$, and $v_z$. These are related to the speed $u$ by the Pythagorean theorem:

$$u^2 = v_x^2 + v_y^2 + v_z^2$$

The **momentum** of a molecule, $\vec{p}$, is its velocity multiplied by its mass. When the molecule collides elastically with a wall of the box, such as one of the end faces of area $A$, the $y$ and $z$ components of the velocity, $v_y$ and $v_z$, are unchanged, but the $x$ component (perpendicular to $A$) reverses sign (Fig. 3-12). The change in the $x$ component of the momentum of the molecule, $\Delta p_{x,\mathrm{mol}}$, is[1]

$$\Delta p_{x,\mathrm{mol}} = \text{final momentum} - \text{initial momentum}$$
$$= m(-v_x) - m v_x = -2\,m v_x$$

However, the total momentum of the system (molecule plus box) must be conserved, so this momentum change is balanced by an equal and opposite momentum change given to the wall:

$$\Delta p_{x,\mathrm{wall}} = 2\,m v_x$$

After this collision with the wall, the molecule reverses direction, strikes the opposite face of the box, and then again approaches the original face. In between, it may strike the top, bottom, and sides. These collisions do not change $v_x$, however, so they

[1]The capital Greek delta ($\Delta$) is used to indicate the change in a quantity—the final value minus the initial value.

do not affect the time between collisions with the original end face (see Fig. 3-10). The distance traveled in the $x$ direction is $2\ell$ and the magnitude of the velocity component in this direction is $v_x$, so the time elapsed between collisions with this end face is

$$\Delta t = \frac{2\ell}{v_x}$$

The momentum transferred to the wall per second is the momentum change per collision divided by $\Delta t$:

$$\frac{\Delta p_{x,\text{wall}}}{\Delta t} = \frac{2mv_x}{2\ell/v_x} = \frac{mv_x^2}{\ell}$$

From Newton's second law, this is the force exerted on the original face by the repeatedly colliding molecule:

$$f = ma = m\frac{\Delta v}{\Delta t} = \frac{\Delta p}{\Delta t} = \frac{mv_x^2}{\ell}$$

Suppose now that a large number, $N$, of molecules of mass $m$ are moving independently in the box with $x$ components of velocity $v_{x1}, v_{x2}, v_{x3}$, and so forth. Then the total force exerted on the face by the $N$ molecules is the sum of the forces exerted by the individual molecules:

$$F = \frac{mv_{x1}^2}{\ell} + \frac{mv_{x2}^2}{\ell} + \cdots + \frac{mv_{xN}^2}{\ell} = \frac{Nm}{\ell}\overline{v_x^2}$$

where

$$\overline{v_x^2} = \frac{1}{N}(v_{x1}^2 + v_{x2}^2 + \cdots + v_{xN}^2)$$

Here $\overline{v_x^2}$ is the average of the square of the $x$ component of the velocity of the $N$ molecules, obtained by summing $v_x^2$ for the $N$ molecules and dividing by $N$. The pressure is the total force on the wall divided by the area $A$, so

$$P = \frac{F}{A} = \frac{Nm}{A\ell}\overline{v_x^2}$$

Because $A\ell$ is the volume $V$ of the box, we conclude that

$$PV = Nm\,\overline{v_x^2}$$

The gas molecules have no preferred direction of motion, so $\overline{v_x^2}$, $\overline{v_y^2}$, and $\overline{v_z^2}$ should be equal to each other. We conclude that

$$\overline{u^2} = \overline{v_x^2} + \overline{v_y^2} + \overline{v_z^2} = 3\overline{v_x^2}$$

so

$$PV = \tfrac{1}{3}Nm\overline{u^2} \qquad\qquad (3\text{-}9)$$

where $\overline{u^2}$ is the **mean-square speed** of the gas molecules. From the ideal gas law,

$$PV = nRT$$

so we conclude that

$$\tfrac{1}{3}Nm\overline{u^2} = nRT$$

With this equation our major goal has been achieved: a relation between speeds of molecules and temperature. It can be simplified a little, however. The equation has the number of molecules, $N$, on the left and the number of moles, $n$, on the right. Because $N$ is just $n$ multiplied by Avogadro's number $N_0$, we can divide both sides by $n$ to find

$$\tfrac{1}{3}N_0 m\overline{u^2} = RT \tag{3-10}$$

Let us examine this equation in two ways. First, we note that the **kinetic energy** of a molecule of mass $m$ moving at speed $u$ is equal to $\tfrac{1}{2}mu^2$, so the *average* kinetic energy of $N_0$ molecules (1 mol) is $\tfrac{1}{2}N_0 m\overline{u^2}$. This is exactly the left side of the equation, with the factor $\tfrac{1}{2}$ instead of $\tfrac{1}{3}$:

$$\text{kinetic energy per mole} = \tfrac{1}{2}N_0 m\overline{u^2} = \tfrac{3}{2} \times (\tfrac{1}{3}N_0 m\overline{u^2}) = \tfrac{3}{2}RT$$

The average kinetic energy of the molecules of a gas depends only on the temperature and is independent of the mass of the molecules or their density.

A second way to look at the equation is to recall that if $m$ is the mass of a single molecule, then $N_0 m$ is the mass of 1 mol of molecules—the molar mass, abbreviated $\mathcal{M}$. Solve the equation for the mean-square speed:

$$\overline{u^2} = \frac{3RT}{\mathcal{M}} \tag{3-11}$$

The higher the temperature and the lighter the molecules, the greater the mean-square speed.

## Distribution of Molecular Speeds

We define the **root-mean-square speed** $u_{\text{rms}}$ as the square root of the mean-square speed $3RT/\mathcal{M}$:

$$u_{\text{rms}} = \sqrt{\overline{u^2}} = \sqrt{\frac{3RT}{\mathcal{M}}}$$

This equation should be used only when all the terms in it are expressed in self-consistent units, such as SI units. The value used for $R$ will then be

$$R = 8.3145 \text{ J mol}^{-1}\text{ K}^{-1} = 8.3145 \text{ kg m}^2\text{ s}^{-2}\text{ mol}^{-1}\text{ K}^{-1}$$

Molar masses $\mathcal{M}$ must be converted to *kilograms* per mole for use in the equation. The final result will then be expressed in the expected SI unit of speed, meters per second.

## Example 3-7

Calculate $u_{\text{rms}}$ for (a) a helium atom, (b) an oxygen molecule, and (c) a xenon atom at 298 K.

## Solution

Because the factor $3RT$ will appear in all three expressions for $u_{rms}$, let us calculate it first:

$$3RT = (3)(8.3145 \text{ kg m}^2 \text{ s}^{-2} \text{ mol}^{-1} \text{ K}^{-1})(298 \text{ K})$$

$$= 7.43 \times 10^3 \text{ kg m}^2 \text{ s}^{-2} \text{ mol}^{-1}$$

The molar masses of He, $O_2$, and Xe are $4.00 \text{ g mol}^{-1}$, $32.00 \text{ g mol}^{-1}$, and $131.3 \text{ g mol}^{-1}$, respectively. Convert them to $4.00 \times 10^{-3} \text{ kg mol}^{-1}$, $32.00 \times 10^{-3} \text{ kg mol}^{-1}$, and $131.3 \times 10^{-3} \text{ kg mol}^{-1}$, and insert them together with the value for $3RT$ into the formula for $u_{rms}$:

$$u_{rms}(\text{He}) = \sqrt{\frac{7.43 \times 10^3 \text{ kg m}^2 \text{ s}^{-2} \text{ mol}^{-1}}{4.00 \times 10^{-3} \text{ kg mol}^{-1}}} = 1360 \text{ m s}^{-1}$$

$$u_{rms}(O_2) = \sqrt{\frac{7.43 \times 10^3 \text{ kg m}^2 \text{ s}^{-2} \text{ mol}^{-1}}{32.00 \times 10^{-3} \text{ kg mol}^{-1}}} = 482 \text{ m s}^{-1}$$

$$u_{rms}(\text{Xe}) = \sqrt{\frac{7.43 \times 10^3 \text{ kg m}^2 \text{ s}^{-2} \text{ mol}^{-1}}{131.3 \times 10^{-3} \text{ kg mol}^{-1}}} = 238 \text{ m s}^{-1}$$

At the same temperature the He, $O_2$, and Xe molecules all have the same average kinetic energy; lighter molecules move faster to compensate for their lesser masses. The average molecule moves along quite rapidly at room temperature. These rms speeds convert to 3050, 1080, and 532 mph, respectively.

*Related Problems:* 41, 42, 43, 44

The root-mean-square speed gives some idea of the typical speed of molecules in a gas, but it would be useful to have a more complete picture of the entire distribution of molecular speeds. In particular, we should like to know the fraction of molecules, $\Delta N/N$, that have speeds between $u$ and $u + \Delta u$. This fraction is given by the speed distribution function $f(u)$:

$$\frac{\Delta N}{N} = f(u) \, \Delta u$$

The function $f(u)$ was determined independently by Maxwell and Boltzmann for a gas of molecules of mass $m$ at temperature $T$. It is called the **Maxwell–Boltzmann speed distribution** and has the form

$$f(u) = 4\pi \left( \frac{m}{2\pi k_B T} \right)^{3/2} u^2 \exp \left( -mu^2/2k_B T \right) \qquad \text{(3-12)}$$

where **Boltzmann's constant** $k_B$ is equal to $R/N_0$. This distribution is plotted in Figure 3-13 for several temperatures. As the temperature is raised, the entire distribution of molecular speeds shifts toward higher values. Very few molecules have either very low or very high speeds, so $f(u)$ is small in these limits and it has a maximum at some intermediate speed. The speed distribution of the molecules in a gas has been measured experimentally by an apparatus such as the one in Figure 3-14, giving results consistent with the predictions of Maxwell and Boltzmann.

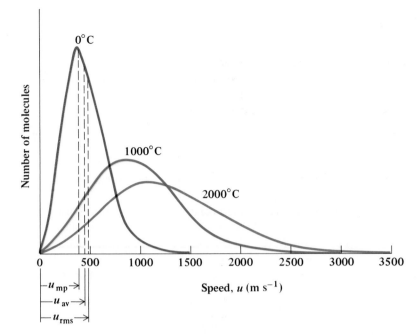

**Figure 3-13**
The Maxwell–Boltzmann distribution of molecular speeds in nitrogen at three temperatures. The peak in each curve gives the most probable speed $u_{mp}$, which is slightly smaller than the root-mean-square speed $u_{rms}$. The average speed $u_{av}$ (obtained simply by adding the speeds and dividing by the number of molecules in the sample) lies in between. All three measures give comparable estimates of typical molecular speeds and show how these speeds increase with temperature.

Another way to think of $f(u)$ is as a *probability distribution:* $f(u)\,\Delta u$ is the probability that any given molecule will have a speed between $u$ and $u + \Delta u$. The **most probable speed** $u_{mp}$ is the speed at which $f(u)$ has its maximum. For the Maxwell–Boltzmann distribution function this is

$$u_{mp} = \sqrt{\frac{2k_BT}{m}} = \sqrt{\frac{2RT}{\mathcal{M}}}$$

Another important quantity is the **average speed** $\bar{u}$, which is

$$\bar{u} = \sqrt{\frac{8k_BT}{\pi m}} = \sqrt{\frac{8RT}{\pi\mathcal{M}}}$$

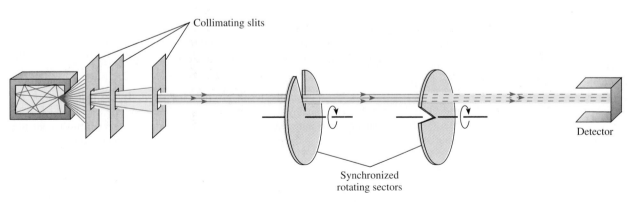

**Figure 3-14**

This device can be used to measure the distribution of molecular speeds experimentally. Only those molecules with the correct speed to pass through *both* rotating sectors will reach the detector, where they can be counted. By changing the rate of rotation of the sectors, the speed distribution can be determined.

We have already seen that

$$u_{rms} = \sqrt{\frac{3k_B T}{m}} = \sqrt{\frac{3RT}{\mathcal{M}}}$$

These three speeds are close to each other, but not equal. They stand in a ratio:

$$u_{mp} : \bar{u} : u_{rms} = 1.000 : 1.128 : 1.225$$

For each temperature there is a unique distribution curve that defines temperature in the kinetic theory of gases. Unless the molecules in a gas have a speed distribution that conforms to such a curve, temperature has no meaning for the gas. Temperature is a parameter of a system of gaseous molecules only when their speed distribution is represented by the Maxwell–Boltzmann function. Consider a closed container filled with molecules whose speed distribution is not "Maxwellian." Such a situation is possible (for example, just after an explosion) but it cannot persist for very long. Any distribution of molecular speeds other than a Maxwell–Boltzmann distribution quickly becomes such a distribution through collisions of molecules, which exchange energy. Once attained, the Maxwell–Boltzmann distribution persists indefinitely. The gas molecules have come to **thermal equilibrium** with one another, and we can speak of a system as having a temperature only if the condition of thermal equilibrium exists.

## 3-6 A Deeper Look . . . Applications of the Kinetic Theory

The kinetic theory we have developed can be applied to several important properties of gases. A study of the rates at which atoms and molecules collide with a wall and with one another helps to explain phenomena ranging from isotope separation based on gaseous diffusion to gas-phase chemical kinetics.

### Wall Collisions, Effusion, and Diffusion

Let us call $Z_w$ the rate of collisions of gas molecules with a section of wall of area $A$. A full mathematical calculation of $Z_w$ requires the use of integral calculus and solid geometry. We present instead some simple physical arguments to demonstrate the form of the dependence of this rate on the properties of the gas.

First of all, $Z_w$ should be proportional to the area $A$, because doubling the area will double the number of collisions with the wall. Second, $Z_w$ should be proportional to the average molecular speed $\bar{u}$, because if molecules move twice as fast they will collide twice as often with a given wall area. Finally, the wall collision rate should be proportional to the number density $N/V$, because if there are twice as many molecules in a given volume they will collide twice as often with the wall. All of these arguments are consistent with the kinetic theory of gases and are confirmed by the full mathematical analysis. We conclude that

$$Z_w \propto \frac{N}{V} \bar{u} A$$

Note that the units of both sides are inverse seconds, as they should be for a rate. The proportionality constant, calculated from a complete analysis of the directions from which molecules impinge on the wall, turns out to have the value $\frac{1}{4}$, so the wall collision rate is

$$Z_w = \frac{1}{4} \frac{N}{V} \bar{u} A = \frac{1}{4} \frac{N}{V} \sqrt{\frac{8RT}{\pi \mathcal{M}}} A \qquad \text{(3-13)}$$

using the result given earlier for $\bar{u}$. This simple equation has many uses. It sets an upper limit upon the rate at which a gas may react with a solid, for example. It also permits calculation of the rate at which molecules effuse through a small hole in the wall of a vessel.

### Example 3-8

Calculate the number of collisions that oxygen molecules make per second on $1.00$ cm$^2$ of the surface of the vessel containing them, if the pressure is $1.00 \times 10^{-6}$ atm and the temperature is 25°C (298 K).

## Solution

First compute the quantities that appear in the equation for $Z_w$:

$$\frac{N}{V} = \frac{N_0 n}{V} = \frac{N_0 P}{RT}$$

$$= \frac{(6.022 \times 10^{23}\ \text{mol}^{-1})(1.00 \times 10^{-6}\ \text{atm})}{(0.08206\ \text{L atm mol}^{-1}\ \text{K}^{-1})(298\ \text{K})}$$

$$= 2.46 \times 10^{16}\ \text{L}^{-1} = 2.46 \times 10^{19}\ \text{m}^{-3}$$

$$A = 1.00\ \text{cm}^2 = 1.00 \times 10^{-4}\ \text{m}^2$$

$$\bar{u} = \sqrt{\frac{8RT}{\pi \mathcal{M}}}$$

$$= \sqrt{\frac{8(8.3145\ \text{J mol}^{-1}\ \text{K}^{-1})(298\ \text{K})}{\pi(32.00 \times 10^{-3}\ \text{kg mol}^{-1})}}$$

$$= 444\ \text{m s}^{-1}$$

Then the collision rate is

$$Z_w = \frac{1}{4}\frac{N}{V}\bar{u}A$$

$$= \tfrac{1}{4}(2.46 \times 10^{19}\ \text{m}^{-3})(444\ \text{m s}^{-1})(1.00 \times 10^{-4}\ \text{m}^2)$$

$$= 2.73 \times 10^{17}\ \text{s}^{-1}$$

*Related Problems: 47, 48*

The formula just given for the rate of collisions with a wall can be used to explain **Graham's law of effusion.** In 1846 Thomas Graham showed experimentally that the rate of effusion of a gas through a small hole into a vacuum (Fig. 3-15) is inversely proportional to the square root of its molar mass, assuming the different gases are studied at the same temperature and pressure. If $T$ and $P$ are held constant, the number density $N/V$ does not change, and the only dependence on the gas studied is the $1/\sqrt{\mathcal{M}}$ factor in Equation 3-13, exactly as observed by Graham. This agreement provides additional support for the kinetic theory of gases.

Graham's law can be applied to the effusion of a mixture of two gases through a small hole. The ratio of the rates of effusion of the two species, A and B, is

$$\frac{\text{rate of effusion of A}}{\text{rate of effusion of B}} = \frac{\dfrac{1}{4}\dfrac{N_A}{V}\sqrt{\dfrac{8RT}{\pi \mathcal{M}_A}}A}{\dfrac{1}{4}\dfrac{N_B}{V}\sqrt{\dfrac{8RT}{\pi \mathcal{M}_B}}A}$$

$$= \frac{N_A}{N_B}\sqrt{\frac{\mathcal{M}_B}{\mathcal{M}_A}} \qquad \textbf{(3-14)}$$

This ratio is equal to the ratio of the numbers of molecules of

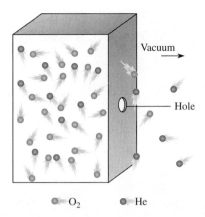

## Figure 3-15

A small hole in the box permits molecules to effuse out into a vacuum. The lighter particles (here, helium atoms) effuse at greater rates than the heavier oxygen molecules because their speeds are greater on the average.

the two species coming out through the hole in a given short time interval. The gas that emerges is enriched in the lighter component because lighter molecules effuse more rapidly than heavier ones. If B is heavier than A, the **enrichment factor** is $\sqrt{\mathcal{M}_B/\mathcal{M}_A}$ for the lighter species, A.

## Example 3-9

Calculate the enrichment factors from effusion for a mixture of $^{235}\text{UF}_6$ and $^{238}\text{UF}_6$, uranium hexafluoride with two different uranium isotopes (the atomic mass of $^{235}\text{U}$ is 235.04, and that of $^{238}\text{U}$ is 238.05). The atomic mass of fluorine is 19.00.

## Solution

The two molar masses are

$$\mathcal{M}(^{238}\text{UF}_6) = 238.05 + 6(19.00) = 352.05\ \text{g mol}^{-1}$$

$$\mathcal{M}(^{235}\text{UF}_6) = 235.04 + 6(19.00) = 349.04\ \text{g mol}^{-1}$$

$$\text{enrichment factor} = \sqrt{\frac{\mathcal{M}(^{238}\text{UF}_6)}{\mathcal{M}(^{235}\text{UF}_6)}} = \sqrt{\frac{352.05\ \text{g mol}^{-1}}{349.04\ \text{g mol}^{-1}}}$$

$$= 1.0043$$

*Related Problems: 49, 50*

Graham's law of effusion holds true only if the opening in the vessel is small enough, and the pressure low enough, so that most molecules follow straight-line trajectories through the opening without undergoing collisions with one another. A related, but more complex, phenomenon is **gaseous diffusion**

**Figure 3-16**
The gaseous diffusion plant for separation of uranium isotopes in uranium hexafluoride at Oak Ridge National Laboratories. If the uranium is to be used as fuel in nuclear power plants, a total enrichment is necessary over many stages, to take the gas from 0.7% to 3% $^{235}U$. (Oak Ridge National Laboratories)

through a porous barrier. This differs from effusion in that molecules undergo many collisions with one another and with the barrier during their passage through it. Interestingly, the diffusion rate is also inversely proportional to the square root of the molar mass of the gas, although the reasons for this dependence are not those outlined for the effusion process. If a mixture of gases is placed in contact with a porous barrier, the gas passing through is enriched in the lighter component, A, by a factor of $\sqrt{\mathcal{M}_B/\mathcal{M}_A}$, and the gas remaining behind is enriched in the heavier component.

During World War II, it became necessary to separate the more easily fissionable isotope $^{235}U$ from $^{238}U$ in order to develop the atomic bomb. Because $^{235}U$ has a natural abundance of only 0.7%, its isolation in nearly pure form appeared a daunting task. The procedure that was adopted was designed to allow uranium to react with fluorine to form the relatively volatile compound $UF_6$ (boiling point 56°C), which can be enriched in $^{235}U$ by passing it through a porous barrier. As shown in Example 3-9, however, each passage gives an enrichment factor of only 1.0043, so many such barriers in succession were necessary to provide sufficient enrichment of the $^{235}U$ component. Such a gaseous diffusion chamber was constructed in a very short time at the Oak Ridge National Laboratories, and the enriched $^{235}U$ was used in the Hiroshima atomic bomb. Similar methods are still used today to separate uranium isotopes (Fig. 3-16).

**Example 3-10**

How many diffusion stages are required if $^{235}U$ is to be enriched from 0.70% to 7.0% by means of the gaseous $UF_6$ diffusion process?

**Solution**

From Example 3-9, the enrichment factor per stage is 1.0043, so that for the first stage there is the relation

$$\left(\frac{N\,^{235}UF_6}{N\,^{238}UF_6}\right)_1 = \left(\frac{N\,^{235}UF_6}{N\,^{238}UF_6}\right)_0 (1.0043)$$

where the subscripts 0 and 1 denote the concentrations initially and after the first stage. For the ratio of the numbers after $n$ stages, therefore,

$$\left(\frac{N\,^{235}UF_6}{N\,^{238}UF_6}\right)_n = \left(\frac{N\,^{235}UF_6}{N\,^{238}UF_6}\right)_0 (1.0043)^n$$

$$\frac{7.0}{93.0} = \frac{0.70}{99.30}(1.0043)^n$$

$$10.677 = (1.0043)^n$$

This equation can be solved by taking logarithms of both sides (see Appendix C):

$$\log_{10}(10.677) = n \log_{10}(1.0043)$$

$$1.02845 = 0.0018635\,n$$

$$n = 552 \text{ stages}$$

*Related Problems:* 51, 52

## Frequency of Molecular Collisions

Collisions lie at the very heart of chemistry, because chemical reactions can occur only when molecules collide with one another. The kinetic theory of gases lets us estimate the frequency

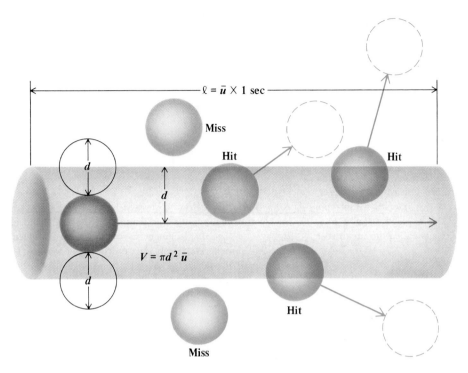

**Figure 3-17**

An average molecule (red) sweeps out a cylinder of volume $\pi d^2 \bar{u}$ in one second. It will collide with any molecules whose centers lie within the cylinder. From this, the rate of its collisions with other molecules can be calculated.

of molecular collisions and the average distance traveled by a molecule between collisions.

Molecules are assumed to be approximately spherical, with diameters, $d$, on the order of $10^{-10}$ m (Section 1-5). We initially suppose that a given molecule moves through a gas of stationary molecules of the same diameter. Such a molecule "sweeps out" a cylinder with cross-sectional area $A = \pi d^2$ (see Fig. 3-17). It collides with any molecule whose center lies inside the cylinder, within a distance $d$ of the center of the moving molecule. In one second, the length of such a cylinder is $\bar{u}\,\Delta t = \bar{u} \times 1$ s, where $\bar{u}$ is the average molecular speed. It does not matter that the moving molecule is scattered in another direction when it collides with other molecules; these little cylinders are merely joined together, end to end, to effect a cylinder whose length is still $\bar{u} \times 1$ s.

In one second, the moving molecule therefore sweeps out a volume

$$V = \pi d^2 \bar{u}$$

If $N/V$ is the number of molecules per unit volume in the gas (the number density of the gas), then the number of collisions per second is

$$\text{collision rate} = Z_1 = \frac{N}{V} V = \frac{N}{V} \pi d^2 \bar{u} \qquad \text{(approximately)}$$

Actually, this equation is only approximate, because the other molecules are not standing still but are moving. A full calculation gives an extra factor of $\sqrt{2}$:

$$Z_1 = \sqrt{2}\,\frac{N}{V}\,\pi d^2 \bar{u}$$

Inserting the result for $\bar{u}$ from the kinetic theory gives

$$Z_1 = 4\frac{N}{V}d^2 \sqrt{\frac{\pi RT}{\mathcal{M}}} \qquad \textbf{(3-15)}$$

**Example 3-11**

Calculate the collision frequency for (a) a molecule in a sample of oxygen at 1.00 atm pressure and 25°C and (b) a molecule of hydrogen in a region of interstellar space where the number density is $1.0 \times 10^{10}$ molecules per cubic meter and the temperature is 30 K. Take the diameter of $O_2$ to be $2.92 \times 10^{-10}$ m and that of $H_2$ to be $2.34 \times 10^{-10}$ m.

**Solution**

(a) $\dfrac{N}{V} = \dfrac{N_0 P}{RT} = \dfrac{(6.022 \times 10^{23}\ \text{mol}^{-1})(1.00\ \text{atm})}{(0.08206\ \text{L atm mol}^{-1}\ \text{K}^{-1})(298\ \text{K})}$

$\qquad = 2.46 \times 10^{22}\ \text{L}^{-1} = 2.46 \times 10^{25}\ \text{m}^{-3}$

In Example 3-8 we found $\bar{u}$ to be 444 m s$^{-1}$. Thus, the collision frequency is

$Z_1 = \sqrt{2}\ \pi(2.46 \times 10^{25}\ \text{m}^{-3})$
$\qquad\qquad \times (2.92 \times 10^{-10}\ \text{m})^2(444\ \text{m s}^{-1})$

$\qquad = 4.14 \times 10^9\ \text{s}^{-1}$

This is the average number of collisions undergone by each molecule per second.

(b) An analogous calculation gives

$$Z_1 = 1.4 \times 10^{-6} \text{ s}^{-1}$$

In other words, the average molecule under these conditions waits $7.3 \times 10^5$ s, or 8.5 days, between collisions.

## Mean Free Path and Diffusion

$Z_1$ is the rate at which a given molecule collides with other molecules. Its inverse, $Z_1^{-1}$, is therefore a measure of the average time between collisions. During this interval a molecule travels an average distance $\bar{u}Z_1^{-1}$. This average distance traveled by a molecule between collisions is called the **mean free path** $\lambda$.

$$\lambda = \bar{u}Z_1^{-1} = \frac{\bar{u}}{\sqrt{2}\dfrac{N}{V}\pi d^2 \bar{u}} = \frac{1}{\sqrt{2}\,\pi d^2 N/V} \qquad \text{(3-16)}$$

The mean free path, unlike the collision frequency, is independent of the molar mass. It must be large compared to the molecular diameter to result in ideal gas behavior. For a molecule of diameter $3 \times 10^{-10}$ m, the mean free path at 25°C and atmospheric pressure is $1 \times 10^{-7}$ m, or 300 times larger.

Molecules in a gas move in straight lines only over rather short distances before being deflected by collisions and changing direction (Fig. 3-18). Because each one follows a zigzag course, molecules take a longer time to achieve a given displacement from their starting points than if there were no collisions. This helps to explain the slowness of diffusion in gases. Recall that the speed of a molecule is on the order of $1 \times 10^3$ m s$^{-1}$. If an odorous gas were released in one part of a room and the molecules followed straight-line trajectories, the odor would reach other points in the room almost instantaneously. Instead, there is a time lapse due to the irregular paths (''random walks'') followed by the molecules.

We can describe diffusion in a gas by means of averaged quantities such as the mean-square displacement $\overline{\Delta r^2} = \overline{\Delta x^2} + \overline{\Delta y^2} + \overline{\Delta z^2}$ (analogous to the mean-square speed considered earlier). If there are no gas currents to perturb the motion of the gas molecules (a rather strong condition, requiring strict isolation from the surroundings), then $\overline{\Delta r^2}$ is found to be proportional to the time elapsed, $t$:

$$\overline{\Delta r^2} = 6Dt \qquad \text{(3-17)}$$

The proportionality constant is $6D$, where $D$ is the **diffusion constant** of the molecules. The root-mean-square displacement $\sqrt{\overline{\Delta r^2}}$ is therefore equal to $\sqrt{6Dt}$.

The diffusion constant has units of m$^2$ s$^{-1}$. It is proportional to the mean free path $\lambda$ and to the mean molecular speed $\bar{u}$, but the proportionality constant is difficult to work out.

## Figure 3-18

A gas molecule follows a straight-line path for only a short time before undergoing a collision, so its overall path is a zigzag one. The displacement $\Delta r$ of a particular molecule in time $t$ is shown. The mean-square displacement, $\overline{\Delta r^2}$, is equal to $6Dt$, where $D$ is the diffusion constant of the molecules in the gas.

For the simplest case, a single-component gas, it has the value $3\pi/16$, so

$$D = \frac{3\pi}{16}\lambda\bar{u} = \frac{3\pi}{16}\sqrt{\frac{8RT}{\pi\mathcal{M}}}\frac{1}{\sqrt{2}\,\pi d^2\,N/V}$$

$$= \frac{3}{8}\sqrt{\frac{RT}{\pi\mathcal{M}}}\frac{1}{d^2\,N/V} \qquad \text{(3-18)}$$

## Example 3-12

Calculate the mean free path and the diffusion constant for the molecules in Example 3-11.

### Solution

(a) The mean free path is

$$\lambda = \frac{1}{\sqrt{2}\,\pi(2.92 \times 10^{-10}\text{ m})^2(2.46 \times 10^{25}\text{ m}^{-3})}$$

$$= 1.07 \times 10^{-7}\text{ m}$$

The diffusion constant is

$$D = \frac{3\pi}{16}\lambda\bar{u} = \frac{3\pi}{16}(1.07 \times 10^{-7}\text{ m})(444\text{ m s}^{-1})$$

$$= 2.80 \times 10^{-5}\text{ m}^2\text{ s}^{-1}$$

(b) An analogous calculation for an average molecule of $H_2$ in interstellar space gives

$$\lambda = 4.1 \times 10^8\text{ m}$$

For comparison, the distance from earth to moon is $3.8 \times 10^8$ m. The diffusion constant is

$$D = 1.4 \times 10^{11}\text{ m}^2\text{ s}^{-1}$$

*Related Problems:* 53, 54

The ideal gas law, $PV = nRT$, is a particularly simple example of an **equation of state**—an equation relating the pressure, temperature, chemical amount, and volume to one another. Such equations of state can be obtained from either theory or experiment. They are useful not only for ideal gases but for real gases, liquids, and solids.

Real gases follow the ideal gas equation of state only at sufficiently low densities. Deviations appear in a variety of forms. Boyle's law, $PV = C$, is no longer satisfied at high pressures, and Charles's law, $V \propto T$, begins to break down when the temperature becomes low. Avogadro's hypothesis also does not hold strictly for real gases at moderate pressures. At atmospheric pressure it is quite well satisfied for most gases, but for some (e.g., water vapor and ammonia) there are deviations of 1 to 2%. These deviations are revealed in the **compressibility factor** $z$:

$$z = \frac{PV}{nRT} \tag{3-19}$$

When $z$ differs from 1 (Fig. 3-19), the ideal gas law is inadequate. A more accurate equation of state is needed.

### The van der Waals Equation of State

One of the earliest and most important improvements on the ideal gas equation of state was proposed in 1873 by the Dutch physicist Johannes van der Waals. The **van der Waals equation of state** has the form

$$\left( P + a\frac{n^2}{V^2} \right)(V - nb) = nRT \tag{3-20}$$

$$P = \frac{nRT}{V - nb} - a\frac{n^2}{V^2}$$

Comparison with the ideal gas law shows two modifications arising from the forces between molecules, which are repulsive at short distances and attractive at large distances. Because of repulsive forces, molecules cannot overlap. They exclude other molecules from the volumes they occupy; in this way the effective volume available to a given molecule is not $V$ but $V - nb$, where $b$ is a constant describing the volume excluded per mole of molecules. Attractive forces hold pairs or groups of molecules together. Any tendency to cluster together reduces the effective number of independent molecules in the gas and therefore reduces the rate of collisions made with the walls of the container. Fewer wall collisions means a lower pressure. Van der Waals argued that this effect, because it depends on attractions between *pairs* of molecules, should be proportional to the *square* of the number of molecules per unit volume ($N^2/V^2$) or, equivalently, proportional to $n^2/V^2$. This effect reduces the pressure by an amount $a(n/V)^2$, where $a$ is a positive constant that depends on the strength of the attractive forces. Putting these two corrections into the ideal gas law produces the van der Waals equation just presented.

The constants $a$ and $b$ are obtained by fitting experimental $PVT$ data for real gases (see Table 3-3). They have the units

**Figure 3-19**

A plot of $z = PV/nRT$ against pressure shows deviations from the ideal gas law quite clearly. For an ideal gas, the plot is the straight horizontal line. The behavior of several real gases at 25°C (*a*) and of nitrogen at several temperatures (*b*) deviate as shown.

(*a*)

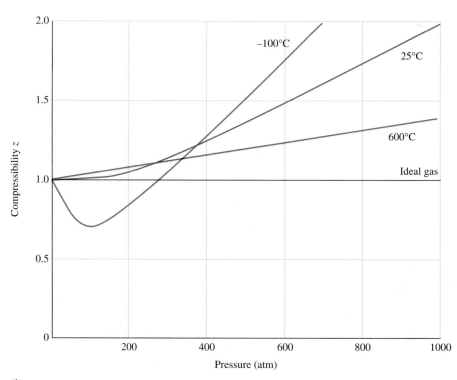

(*b*)

$$a: \quad \text{atm L}^2 \text{ mol}^{-2}$$

$$b: \quad \text{L mol}^{-1}$$

when $R$ has the units L atm mol$^{-1}$ K$^{-1}$.

## Example 3-13

A sample consists of 8.00 kg of gaseous nitrogen and fills a 100-L flask at 300°C. What is the pressure of the gas, using the van der Waals equation of state? What pressure would be predicted by the ideal gas equation?

### Solution

Because the molar mass of $N_2$ is 28 g mol$^{-1}$,

$$n = \frac{8.00 \times 10^3 \text{ g}}{28.0 \text{ g mol}^{-1}} = 286 \text{ mol}$$

The temperature (in kelvins) is $T = 300 + 273 = 573$ K, and the volume $V$ is 100 L. Using $R = 0.08206$ L atm mol$^{-1}$ K$^{-1}$ and the van der Waals constants for nitrogen given in Table 3-3, we calculate $P = 151 - 11 = 140$ atm. If the ideal gas law is used instead, a pressure of 134 atm is calculated. This illustrates the magnitude of deviations from the ideal gas law at higher pressures.

*Related Problems:* 55, 56, 57, 58

The effects of the two van der Waals parameters can be seen clearly by writing the compressibility factor for this equation of state:

$$z = \frac{PV}{nRT} = \frac{V}{V - nb} - \frac{a}{RT}\frac{n}{V} = \frac{1}{1 - bn/V} - \frac{a}{RT}\frac{n}{V}$$

Respulsive forces (through $b$) increase $z$ above 1, while attractive forces (through $a$) reduce $z$. At low particle densities, $n/V$ is small. After the approximation

**Table 3-3  Van der Waals Constants of Several Gases**

| Name | Formula | $a$ <br> (atm L$^2$ mol$^{-2}$) | $b$ <br> (L mol$^{-1}$) |
|---|---|---|---|
| Ammonia | $NH_3$ | 4.170 | 0.03707 |
| Argon | Ar | 1.345 | 0.03219 |
| Carbon dioxide | $CO_2$ | 3.592 | 0.04267 |
| Hydrogen | $H_2$ | 0.2444 | 0.02661 |
| Hydrogen chloride | HCl | 3.667 | 0.04081 |
| Methane | $CH_4$ | 2.253 | 0.04278 |
| Nitrogen | $N_2$ | 1.390 | 0.03913 |
| Nitrogen dioxide | $NO_2$ | 5.284 | 0.04424 |
| Oxygen | $O_2$ | 1.360 | 0.03183 |
| Sulfur dioxide | $SO_2$ | 6.714 | 0.05636 |
| Water | $H_2O$ | 5.464 | 0.03049 |

$$\frac{1}{1 - bn/V} \cong 1 + b\frac{n}{V} + \cdots$$

is made, the compressibility factor becomes

$$z \cong 1 + \left(b - \frac{a}{RT}\right)\frac{n}{V} + \cdots = 1 + B(T)\frac{n}{V} + \cdots$$

The temperature at which the **second virial coefficient** $B(T)$ multiplying the density $n/V$ vanishes is called the Boyle temperature, $T_B$. In the van der Waals model it has the value

$$T_B = \frac{a}{Rb}$$

At temperatures above $T_B$, repulsive forces dominate and $z > 1$; at temperatures below $T_B$, attractive forces dominate and $z < 1$ (at low enough density). A gas at the Boyle temperature is still not ideal, but deviations from $z = 1$ do not appear until higher densities.

The constant $b$ is the volume excluded by 1 mol of molecules and should therefore be close to the volume per mole in the liquid state, where molecules are essentially in contact with each other. For example, the density of liquid nitrogen is $0.808$ g cm$^{-3}$. One mole of $N_2$ weighs 28.0 g, so

$$\text{molar volume of } N_2(\ell) = \frac{28.0 \text{ g mol}^{-1}}{0.808 \text{ g cm}^{-3}} = 34.7 \text{ cm}^3 \text{ mol}^{-1}$$

$$= 0.0347 \text{ L mol}^{-1}$$

This is reasonably close to the van der Waals $b$ parameter of $0.03913$ L mol$^{-1}$, obtained by fitting the equation of state to $PVT$ data for nitrogen.

### Intermolecular Forces

The attractive force parameter $a$ and the excluded volume per mole, $b$, characterize the forces between the atoms or molecules in a gas. Real atoms are not impenetrable, hard spheres, however. As two atoms are brought together, they attract each other until the distance between their centers, $R$, becomes short enough, at which point they repel each other with increasing force as they are pushed together. This can be described by a **potential energy curve** $V(R)$ such as that shown for argon in Figure 3-20. If two molecules can lower their energy by moving closer together, then a net attractive force exists between them; but if they can lower their energy only by moving apart, there is a net repulsive force. The force changes from attractive to repulsive at the minimum of $V(R)$. Accurate potential energy curves for atoms are obtained by fitting $V(R)$ to measured properties of real gases such as the second virial coefficient $B(T)$. Additional information is provided by experiments that scatter beams of atoms from one another (Fig. 1-1).

For many purposes the detailed shape of the potential is less important than aspects such as the depth and location of the potential minimum. A simple expression that is frequently used to model these interactions is the **Lennard–Jones potential:**

$$V_{\text{LJ}}(R) = 4\epsilon\left[\left(\frac{\sigma}{R}\right)^{12} - \left(\frac{\sigma}{R}\right)^6\right] \tag{3-21}$$

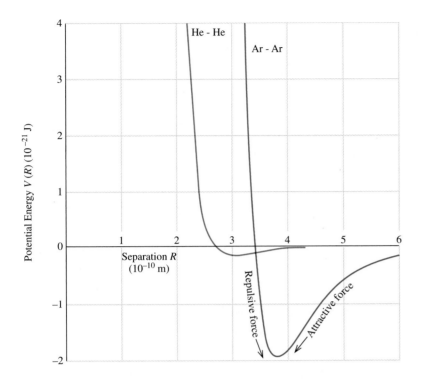

**Figure 3-20**
Potential energy curves $V(R)$ for pairs of helium atoms (red) and pairs of argon atoms (blue). Where $V$ decreases with increasing $R$, the atoms repel each other; where $V$ increases with $R$, the atoms attract.

where $\epsilon$ is the depth and $\sigma$ is the distance at which $V(R)$ passes through 0. This potential has an attractive part, proportional to $R^{-6}$, and a repulsive part, proportional to $R^{-12}$. The minimum in this potential is located at $2^{1/6}\sigma$, or $1.22\sigma$. Table 3-4 lists Lennard–Jones parameters for a number of atoms. Note that the depth and range of the potential increase for the heavier noble-gas atoms. Molecules such as $N_2$ that are nearly spherical can also be treated approximately with Lennard–Jones potentials. The two parameters $\epsilon$ and $\sigma$, like the van der Waals parameters $a$ and $b$, are simple ways of characterizing the interactions between molecules in real gases.

**Table 3-4  Lennard–Jones Parameters for Atoms and Molecules**

| Substance | $\sigma$ (m) | $\epsilon$ (J) |
|---|---|---|
| He | $2.56 \times 10^{-10}$ | $1.41 \times 10^{-22}$ |
| Ne | $2.75 \times 10^{-10}$ | $4.92 \times 10^{-22}$ |
| Ar | $3.40 \times 10^{-10}$ | $1.654 \times 10^{-21}$ |
| Kr | $3.60 \times 10^{-10}$ | $2.36 \times 10^{-21}$ |
| Xe | $4.10 \times 10^{-10}$ | $3.06 \times 10^{-21}$ |
| $H_2$ | $2.93 \times 10^{-10}$ | $5.11 \times 10^{-22}$ |
| $O_2$ | $3.58 \times 10^{-10}$ | $1.622 \times 10^{-21}$ |
| CO | $3.76 \times 10^{-10}$ | $1.383 \times 10^{-21}$ |
| $N_2$ | $3.70 \times 10^{-10}$ | $1.312 \times 10^{-21}$ |
| $CH_4$ | $3.82 \times 10^{-10}$ | $2.045 \times 10^{-21}$ |

## PRACTICE EXERCISE

A space shuttle taking off. (NASA/ Phototake)

### Ammonium Perchlorate

Ammonium perchlorate ($NH_4ClO_4$) is a solid rocket fuel used in space shuttles. When heated above 200°C, it decomposes to a variety of gaseous products, of which the most important are nitrogen, chlorine, oxygen, and water vapor.

(a) Write a balanced chemical equation for the decomposition of $NH_4ClO_4$, assuming the products just listed are the only ones generated.

(b) The sudden appearance of hot gaseous products in a small initial volume leads to rapid increases in pressure and temperature, which give the rocket its thrust. What total pressure of gas would be produced at 800°C by igniting $7.00 \times 10^5$ kg of $NH_4ClO_4$ (a typical charge of the booster rockets in the space shuttle) and allowing it to expand to fill a volume of 6400 m³ ($6.40 \times 10^6$ L)? Use the ideal gas law.

(c) Calculate the mole fraction of chlorine and its partial pressure in the mixture of gases produced.

(d) The van der Waals equation applies strictly to pure real gases, not to mixtures. For a mixture like the one resulting from the reaction of part (a), it may still be possible to define effective $a$ and $b$ parameters to relate total pressure, volume, temperature, and total chemical amount. Suppose the gas mixture has $a = 4.00$ atm $L^2$ mol$^{-2}$ and $b = 0.0330$ L mol$^{-1}$. Recalculate the pressure of the gas mixture in part (b), using the van der Waals equation.

(e) Calculate and compare the root-mean-square speeds of water and chlorine molecules under the conditions of part (b).

(f) The gas mixture from part (b) cools and expands until it reaches a temperature of 200°C and a pressure of 3.20 atm. Calculate the volume occupied by the gas mixture after this expansion has occurred. Assume ideal gas behavior.

**Answers**

(a) $2\,NH_4ClO_4(s) \rightarrow N_2(g) + Cl_2(g) + 2\,O_2(g) + 4\,H_2O(g)$

(b) 328 atm

(c) $X_{Cl_2} = \frac{1}{8} = 0.125$ (exactly); $P_{Cl_2} = 41.0$ atm

(d) 318 atm

(e) $u_{rms}(H_2O) = 1220$ m s$^{-1}$; $u_{rms}(Cl_2) = 614$ m s$^{-1}$

(f) $2.89 \times 10^5$ m³

## CONCEPTS & SKILLS

After studying this chapter and working the problems that follow, you should be able to

1. Write chemical equations for several reactions that lead to gas formation (Section 3-1, problems 1–4).

2. Describe how pressure and temperature are defined and measured (Section 3-2, problems 5–10).

3. Use the ideal gas law to relate pressure, volume, temperature, and chemical amount of an ideal gas, and to do stoichiometric calculations involving gases (Section 3-3, problems 19–32).

4. Use Dalton's law to calculate partial pressures in gas mixtures (Section 3-4, problems 33–38).

5. Use the Maxwell–Boltzmann distribution of molecular speeds to calculate root-mean-square, most probable, and average speeds of molecules in a gas (Section 3-5, problems 41–44).

6. Describe the connection between temperature and the speeds or kinetic energies of the molecules in a gas (Section 3-5, problems 45–46).

7. Calculate the rate of collisions of molecules with a wall, and from that determine the effusion rate of a gas through a small hole of known area (Section 3-6, problems 47–48).

8. Calculate the enrichment factor for lighter molecules when a gas consisting of a mixture of light and heavy molecules effuses through a small aperture in a vessel wall (Section 3-6, problems 49–52).

9. Calculate the collision frequency, the mean free path, and the diffusion constant for gases (Section 3-6, problems 53–54).

10. Use the van der Waals equation to relate the pressure, volume, temperature, and chemical amount of a nonideal gas (Section 3-7, problems 55–58).

11. Discuss how forces between atoms and molecules vary with distance (Section 3-7).

## PROBLEMS

### The Chemistry of Gases

1. Solid ammonium hydrosulfide ($NH_4HS$) decomposes entirely to gases when it is heated. Write a chemical equation representing this change.

2. Solid ammonium carbamate ($NH_4CO_2NH_2$) decomposes entirely to gases when it is heated. Write a chemical equation representing this change.

3. Ammonia ($NH_3$) is an important and useful gas. Suggest a way to generate it from ammonium bromide ($NH_4Br$). Include a balanced chemical equation.

4. Hydrogen cyanide (HCN) is a poisonous gas. Explain why solutions of potassium cyanide (KCN) should never be acidified. Include a balanced chemical equation.

### Pressure and Temperature of Gases

5. Suppose a barometer were designed using water (with a density of $1.00 \ \mathrm{g \ cm^{-3}}$) rather than mercury as its fluid. What would be the height of the column of water balancing 1.00 atm pressure?

6. A vessel that contains a gas has two pressure gauges attached to it. One contains liquid mercury, the other an oil such as dibutylphthalate. The difference in levels of mercury in the two arms of the mercury gauge is observed to be 9.50 cm. Given:

$$\text{density of mercury} = 13.60 \ \mathrm{g \ cm^{-3}}$$

$$\text{density of oil} = 1.045 \ \mathrm{g \ cm^{-3}}$$

$$\text{acceleration due to gravity} = 9.806 \ \mathrm{m \ s^{-2}}$$

(a) What is the pressure of the gas?

(b) What is the difference in heights of the oil in the two arms of the oil pressure gauge?

7. Calcium dissolved in the ocean is used by marine organisms to form $CaCO_3(s)$ in skeletons and shells. When the organisms die, their remains fall to the bottom. The amount of calcium carbonate that can be dissolved in seawater depends on the pressure. At great depths, where the pressure exceeds about 414 atm, the shells slowly redissolve. This prevents all the world's calcium from being tied up as insoluble $CaCO_3(s)$ at the bottom of the sea. Estimate the depth (in feet) of water that exerts a pressure great enough to dissolve seashells.

8. Suppose that the atmosphere were perfectly uniform, with a density throughout equal to that of air at 0°C, $1.3 \ \mathrm{g \ L^{-1}}$. Calculate the thickness of such an atmosphere that would cause a pressure of exactly one standard atmosphere at the earth's surface.

9. The "critical pressure" of mercury is 172.00 MPa. Above this pressure mercury cannot be liquefied, no matter what the temperature. Express this pressure in atmospheres and in bars (1 bar = $10^5$ Pa).

10. Experimental studies of solid surfaces and the chemical reactions that occur on them require very low gas pressures to avoid surface contamination. High-vacuum apparatus for such experiments can routinely reach pressures of $5 \times 10^{-10}$ torr. Express this pressure in atmospheres and in pascals.

11. Some nitrogen is held in a 2.00-L tank at a pressure of 3.00 atm. The tank is connected to a 5.00-L tank that is completely empty (evacuated), and a valve is opened to connect

the two. No temperature change occurs in the process. Determine the total pressure in this two-tank system after the nitrogen stops flowing.

12. The Stirling engine, a heat engine invented by a Scottish minister, has been considered for use in automobile engines because of its efficiency. In such an engine, a gas goes through a four-step cycle of (1) expansion at constant $T$, (2) cooling at constant $V$, (3) compression at constant $T$ to its original volume, and (4) heating at constant $V$ to its original temperature. Suppose the gas starts at a pressure of 1.23 atm, and the volume of the gas changes from 0.350 L to 1.31 L during its expansion at constant $T$. Calculate the pressure of the gas at the end of this step in the cycle.

13. The absolute temperature of a 4.00-L sample of gas is doubled at constant pressure. Determine the volume of the gas after this change.

14. The Celsius temperature of a 4.00-L sample of gas is doubled from 20.0°C to 40.0°C at constant pressure. Determine the volume of the gas after this change.

15. The gill is an obscure unit of volume. If some $H_2(g)$ has a volume of 17.4 gills at 100°F, what volume would it have if the temperature were reduced to 0°F, assuming that its pressure stayed constant?

16. A gas originally at a temperature of 26.5°C is cooled at constant pressure. Its volume decreases from 5.40 L to 5.26 L. Determine its new temperature in degrees Celsius.

17. Calcium carbide ($CaC_2$) reacts with water to produce acetylene ($C_2H_2$) according to the equation

$$CaC_2(s) + 2\ H_2O(\ell) \longrightarrow Ca(OH)_2(s) + C_2H_2(g)$$

A certain mass of $CaC_2$ reacts completely with water to give 64.5 L of $C_2H_2$ at 50°C and $P = 1.00$ atm. If the same mass of $CaC_2$ reacts completely at 400°C and $P = 1.00$ atm, what volume of $C_2H_2$ will be collected at the higher temperature?

18. A convenient laboratory source for oxygen of very high purity is the decomposition of potassium permanganate at 230°C:

$$2\ KMnO_4(s) \longrightarrow K_2MnO_4(s) + MnO_2(s) + O_2(g)$$

Suppose 3.41 L of oxygen is needed at atmospheric pressure and a temperature of 20°C. What volume of oxygen should be collected at 230°C and the same pressure in order to give this volume when cooled?

## The Ideal Gas Law

19. A bicycle tire is inflated to a gauge pressure of 30.0 psi at a temperature of $t = 0$°C. What will its gauge pressure be at 32°C if the tire is considered nonexpandable? (*Note:* the gauge pressure is the *difference* between the tire pressure and atmospheric pressure, 14.7 psi.)

20. The pressure of a poisonous gas inside a sealed container is 1.47 atm at 20°C. If the barometric pressure is 0.96 atm, to what temperature (in degrees Celsius) must the container and its contents be cooled so that the container can be opened with no risk of gas spurting out?

21. A 20.6-L sample of "pure" air is collected in Greenland at a temperature of −20.0°C and a pressure of 1.01 atm and is forced into a 1.05-L bottle for shipment to Europe for analysis.
(a) Compute the pressure inside the bottle just after it is filled.
(b) Compute the pressure inside the bottle as it is opened in the 21.0°C comfort of the European laboratory.

22. Iodine heptafluoride ($IF_7$) can be made at elevated temperatures by the reaction

$$I_2(g) + 7\ F_2(g) \longrightarrow 2\ IF_7(g)$$

Suppose 63.6 L of gaseous $IF_7$ is made by this reaction at 300°C and a pressure of 0.459 atm. Calculate the volume this gas will occupy if heated to 400°C at a pressure of 0.980 atm.

23. According to a reference handbook, "The weight of one liter [of $H_2Te(g)$ is] 6.234 g." Why is this information nearly valueless? Assume that $H_2Te(g)$ is an ideal gas, and calculate the temperature (in degrees Celsius) at which this statement is true if the pressure is 1.00 atm.

24. A scuba diver's tank contains 0.30 kg of oxygen ($O_2$) compressed into a volume of 2.32 L.
(a) Use the ideal gas law to estimate the gas pressure inside the tank at 5°C, and express it in atmospheres and in pounds per square inch.
(b) What volume would this oxygen occupy at 30°C and a pressure of 0.98 atm?

25. Hydrogen is produced by the complete reaction of 6.24 g of sodium with an excess of gaseous hydrogen chloride.
(a) Write a balanced chemical equation for the reaction that occurs.
(b) How many liters of hydrogen will be produced at a temperature of 50.0°C and a pressure of 0.850 atm?

26. Aluminum reacts with excess aqueous hydrochloric acid to produce hydrogen.
(a) Write a balanced chemical equation for the reaction. (*Hint:* Water-soluble $AlCl_3$ is the stable chloride of aluminum.)
(b) Calculate the mass of pure aluminum that will furnish 10.0 L of hydrogen at a pressure of 0.750 atm and a temperature of 30.0°C.

27. The classic method for manufacturing hydrogen chloride—which is still in use today to a small extent—involves the reaction of sodium chloride with excess sulfuric acid at elevated temperatures. The overall equation for this process is

$$NaCl(s) + H_2SO_4(\ell) \longrightarrow NaHSO_4(s) + HCl(g)$$

What volume of hydrogen chloride is produced at 550°C and a pressure of 0.97 atm from the reaction of 2500 kg of sodium chloride?

28. In 1783 the French physicist Jacques Charles supervised and took part in the first human flight in a hydrogen balloon. Such

balloons rely for their buoyancy on the low density of hydrogen relative to air. In Charles's balloon ascent, the hydrogen was produced (together with iron(II) sulfate) from the action of aqueous sulfuric acid on iron filings.
(a) Write a balanced chemical equation for this reaction.
(b) What volume of hydrogen is produced at 300 K and a pressure of 1.0 atmosphere when 300 kg of sulfuric acid is consumed in this reaction?
(c) What would be the radius of a spherical balloon filled by the gas in part (b)?

**29.** Potassium chlorate decomposes when heated, giving oxygen and potassium chloride:

$$2\,KClO_3(s) \longrightarrow 2\,KCl(s) + 3\,O_2(g)$$

A test tube holding 87.6 g of $KClO_3$ is heated and the reaction goes to completion. What volume of $O_2$ will be evolved if it is collected at a pressure of 1.04 atm and a temperature of 13.2°C?

**30.** Elemental chlorine was first produced by Carl Wilhelm Scheele in 1774 using the reaction of pyrolusite ($MnO_2$) with sulfuric acid and sodium chloride:

$$4\,NaCl(s) + 2\,H_2SO_4(\ell) + MnO_2(s) \longrightarrow$$
$$2\,Na_2SO_4(s) + MnCl_2(s) + 2\,H_2O(\ell) + Cl_2(g)$$

Calculate the minimum mass of $MnO_2$ required to generate 5.32 L of gaseous chlorine, measured at a pressure of 0.953 atm and a temperature of 33°C.

**31.** Elemental sulfur can be recovered from gaseous hydrogen sulfide ($H_2S$) through the reaction

$$2\,H_2S(g) + SO_2(g) \longrightarrow 3\,S(s) + 2\,H_2O(\ell)$$

(a) What volume of $H_2S$ (in liters at 0°C and 1.00 atm) is required to produce 2.00 kg (2000 g) of sulfur by this process?
(b) What minimum mass and volume (at 0°C and 1.00 atm) of $SO_2$ are required to produce 2.00 kg of sulfur by this reaction?

**32.** When ozone ($O_3$) is placed in contact with dry, powdered KOH at −15°C, the red-brown solid potassium ozonide ($KO_3$) forms, according to the balanced equation

$$5\,O_3(g) + 2\,KOH(s) \longrightarrow 2\,KO_3(s) + 5\,O_2(g) + H_2O(s)$$

Calculate the volume of ozone needed (at a pressure of 0.134 atm and −15°C) to produce 4.69 g of $KO_3$.

## Mixtures of Gases

**33.** Sulfur dioxide reacts with oxygen in the presence of platinum to give sulfur trioxide:

$$2\,SO_2(g) + O_2(g) \longrightarrow 2\,SO_3(g)$$

Suppose that at one stage in the reaction 26.0 mol $SO_2$, 83.0 mol $O_2$, and 17.0 mol $SO_3$ are present in the reaction vessel at a total pressure of 0.950 atm. Calculate the mole fraction of $SO_3$ and its partial pressure.

**34.** The synthesis of ammonia from the elements is carried out at high pressures and temperatures.

$$N_2(g) + 3\,H_2(g) \longrightarrow 2\,NH_3(g)$$

Suppose that at one stage in the reaction 13 mol $NH_3$, 31 mol $N_2$, and 93 mol $H_2$ are present in the reaction vessel at a total pressure of 210 atm. Calculate the mole fraction of $NH_3$ and its partial pressure.

**35.** The atmospheric pressure at the surface of Mars is $5.92 \times 10^{-3}$ atm. The Martian atmosphere is 95.3% $CO_2$ and 2.7% $N_2$ by volume, with small amounts of other gases also present. Compute the mole fraction and partial pressure of $N_2$ in the atmosphere of Mars.

**36.** The atmospheric pressure at the surface of Venus is 90.8 atm. The Venusian atmosphere is 96.5% $CO_2$ and 3.5% $N_2$ by volume, with small amounts of other gases also present. Compute the mole fraction and partial pressure of $N_2$ in the atmosphere of Venus.

**37.** A gas mixture at room temperature contains 10.0 mol of CO and 12.5 mol of $O_2$.
(a) Compute the mole fraction of CO in the mixture.
(b) The mixture is then heated, and the CO starts to react with the $O_2$ to give $CO_2$:

$$CO(g) + \tfrac{1}{2}\,O_2(g) \longrightarrow CO_2(g)$$

At a certain point in the heating, 3.0 mol of $CO_2$ is present. Determine the mole fraction of CO in the new mixture.

**38.** A gas mixture contains 4.5 mol $Br_2$ and 33.1 mol $F_2$.
(a) Compute the mole fraction of $Br_2$ in the mixture.
(b) The mixture is heated above 150°C and starts to react to give $BrF_5$:

$$Br_2(g) + 5\,F_2(g) \longrightarrow 2\,BrF_5(g)$$

At a certain point in the reaction, 2.2 mol of $BrF_5$ is present. Determine the mole fraction of $Br_2$ in the mixture at that point.

**39.** The partial pressure of water vapor in saturated air at 20°C is 0.0230 atm.
(a) How many molecules of water are in 1.00 cm$^3$ of saturated air at 20°?
(b) What volume of saturated air at 20°C contains 0.500 mol of water?

**40.** The partial pressure of oxygen in a mixture of oxygen and hydrogen is 0.200 atm, and that of hydrogen is 0.800 atm.
(a) How many molecules of oxygen are in a 1.500-L container of this mixture at 40°C?
(b) If a spark is introduced into the container, how many grams of water will be produced?

## The Kinetic Theory of Gases

**41.** (a) Compute the root-mean-square speed of $H_2$ molecules in hydrogen at a temperature of 300 K.

(b) Repeat for $SF_6$ molecules in gaseous sulfur hexafluoride at 300 K.

**42.** Researchers recently reported the first optical atomic trap. In this device, beams of laser light replace the physical walls of conventional containers. The laser beams are tightly focused. They briefly (for 0.5 s) exert enough pressure to confine 500 sodium atoms in a volume of $1.0 \times 10^{-15}$ m$^3$. The temperature of this gas is 0.00024 K, the lowest temperature ever reached for a gas. Compute the root-mean-square speed of the atoms in this confinement.

**43.** Compare the root-mean-square speed of helium atoms near the surface of the sun, where the temperature is approximately 6000 K, with that of helium atoms in an interstellar cloud, where the temperature is 100 K.

**44.** The "escape velocity" necessary for objects to leave the gravitational field of the earth is 11.2 km s$^{-1}$. Calculate the ratio of the escape velocity to the root-mean-square speed of helium, argon, and xenon atoms at 2000 K. Does your result help to explain the low abundance of the light gas helium in the atmosphere? Explain.

**45.** Chlorine dioxide ($ClO_2$) is used for bleaching wood pulp. In a gaseous sample held at thermal equilibrium at a particular temperature, 35.0% of the molecules have speeds exceeding 400 m s$^{-1}$. If the sample is heated slightly, will the percentage of molecules with speeds in excess of 400 m s$^{-1}$ then be greater than or less than 35%? Explain.

**46.** The $ClO_2$ from problem 45 is heated further until it explodes, yielding $Cl_2$, $O_2$, and other gaseous products. The mixture is then cooled until the original temperature is reached. Is the percentage of *chlorine* molecules with speeds in excess of 400 m s$^{-1}$ greater than or less than 35%? Explain.

## Applications of the Kinetic Theory

**47.** A spherical bulb with a volume of 500 cm$^3$ is evacuated to a negligibly small residual gas pressure and then closed off. One hour later the pressure in the vessel is found to be $1.00 \times 10^{-7}$ atm due to the fact that the bulb has a tiny hole in it. Assume that the surroundings are at atmospheric pressure, $T = 300$ K, and the average molar mass of molecules in the atmosphere is 28.8 g mol$^{-1}$. Calculate the radius of the hole in the vessel wall, assuming it to be circular.

**48.** A 200-cm$^3$ vessel contains hydrogen gas at a temperature of 25°C and a pressure of 0.990 atm. Unfortunately, the vessel has a tiny hole in its wall, and over a period of 1.0 hour the pressure drops to 0.989 atm. What is the radius of the hole (assumed to be circular)?

**49.** Methane ($CH_4$) effuses through a small opening in the side of a container at the rate of $1.30 \times 10^{-8}$ mol s$^{-1}$. An unknown gas effuses through the same opening at the rate of $5.42 \times 10^{-9}$ mol s$^{-1}$ when maintained at the same temperature and pressure as the methane. Determine the molar mass of the unknown gas.

**50.** Equal chemical amounts of two gases, fluorine and bromine

pentafluoride, are mixed. Determine the ratio of the rates of effusion of the two gases through a very small opening in their container.

**51.** Calculate the theoretical number of stages that would be needed to enrich $^{235}U$ to 95% purity by means of the barrier diffusion process, using $^{235}UF_6$ and $^{238}UF_6$ as the gaseous compounds. The natural abundance of $^{238}U$ is 99.27%, and that of $^{235}U$ is 0.72%. Take the relative atomic masses of $^{235}U$ and $^{238}U$ to be 235.04 and 238.05, respectively.

**52.** A mixture of $H_2$ and He at 300 K effuses from a very tiny hole in the vessel that contains it. What is the mole fraction of $H_2$ in the original gas mixture if 3.00 times as many He atoms as $H_2$ molecules escape from the orifice in unit time? If the same mixture is to be separated by a barrier-diffusion process, how many stages are necessary to achieve $H_2$ of 99.9% purity?

**53.** At what pressure does the mean free path of Kr atoms ($d = 3.16 \times 10^{-10}$ m) become comparable with the diameter of the 1-L spherical vessel that contains them at 300 K? Calculate the diffusion constant at this pressure.

**54.** At what pressure does the mean free path of Kr atoms ($d = 3.16 \times 10^{-10}$ m) become comparable with the diameter of a Kr atom if $T = 300$ K? Calculate the diffusion constant at this pressure. Assume that krypton obeys the ideal gas law even at these high pressures.

## Real Gases

**55.** Oxygen is supplied to hospitals and chemical laboratories under pressure in large steel cylinders. Typically, such a cylinder has an internal volume of 28.0 L and contains 6.80 kg of oxygen. Use the van der Waals equation to estimate the pressure inside such a cylinder at 20°C. Express it in atmospheres and in pounds per square inch.

**56.** Steam at high pressures and temperatures is used to generate electrical power in utility plants. A large utility boiler has a volume of 2500 m$^3$ and contains 140 metric tons (1 metric ton = $10^3$ kg) of steam at a temperature of 540°C. Use the van der Waals equation to estimate the pressure of the steam under these conditions, in atmospheres and in pounds per square inch.

**57.** Using (a) the ideal gas law and (b) the van der Waals equation, calculate the pressure exerted by 50.0 g of carbon dioxide in a 1.00-L vessel at 25°C. Do attractive or repulsive forces dominate?

**58.** When 60.0 g of methane ($CH_4$) is placed in a 1.00-L vessel, the pressure is measured to be 130 atm. Calculate the temperature of the gas using (a) the ideal gas law and (b) the van der Waals equation. Do attractive or repulsive forces dominate?

## Additional Problems

**59.** The earth is approximately a sphere of radius 6370 km. Taking the average barometric pressure on the earth's surface to be 730 mm (Hg), estimate the total mass of the earth's atmosphere.

*60. After a flood fills a basement to a depth of 9.0 ft and completely saturates the surrounding earth, the owner buys an electric pump and quickly pumps the water out of the basement. Suddenly a basement wall collapses, the structure is severely damaged, and mud oozes in. Explain this event by estimating the difference between the outside pressure at the base of the basement walls and the pressure inside the drained basement. Assume that the mud has a density of 4.9 g cm$^{-3}$. Report the answer both in atmospheres and in pounds per square inch.

61. The density of mercury is 13.5955 g cm$^{-3}$ at 0.0°C, but only 13.5094 g cm$^{-3}$ at 35°C. Suppose that a mercury barometer is read on a hot summer day when the temperature is 35°C. The column of mercury is 760.0 mm long. Correct for the expansion of the mercury and compute the true pressure in atmospheres.

62. When a gas is cooled at constant pressure, it is found that the volume decreases linearly:

$$V = 209.4 \text{ L} + \left(0.456\frac{\text{L}}{°\text{F}}\right) \times t_F$$

where $t_F$ is the temperature in degrees Fahrenheit. From this relationship, estimate the absolute zero of temperature in degrees Fahrenheit.

63. Amontons's law relates pressure to absolute temperature. Consider the ideal gas law and then write a statement of Amontons's law in a form that is analogous to the statements of Charles's law and Boyle's law in the text.

64. A gas has a density of 2.94 g L$^{-1}$ at 50°C and $P = 1.00$ atm. What will be its density at 150°C? Calculate the molar mass of the gas, assuming it obeys the ideal gas law.

65. A lighter-than-air balloon contains 1005 mol of helium at 1.00 atm and 25.0°C.
(a) Compute the difference between the mass of the helium it contains and the mass of the air it displaces, assuming the molar mass of air to be 29.0 g mol$^{-1}$.
(b) The balloon now ascends to an altitude of 10 miles, where the temperature is −80.0°C. The walls of the balloon are elastic enough that the pressure inside it equals the pressure outside. Repeat the calculation of part (a).

66. Baseball reporters say that long fly balls that would have carried for home runs in July "die" in the cool air of October and are caught. The idea behind this observation is that a baseball carries better when the air is less dense. Dry air is a mixture of gases with an effective molar mass of 29.0 g mol$^{-1}$.
(a) Compute the density of dry air on a July day when the temperature is 95.0°F and the pressure is 1.00 atm.
(b) Compute the density of dry air on an October evening when the temperature is 50.0°F and the pressure is 1.00 atm.
(c) Suppose that the humidity on the July day is 100%, so that the air is saturated with water vapor. Is the density of this

hot, moist air less than, equal to, or greater than the density of the hot, dry air computed in part (a)? In other terms, does high humidity favor the home run?

67. Sulfuric acid reacts with sodium chloride to produce gaseous hydrogen chloride according to the equation

$$NaCl(s) + H_2SO_4(\ell) \longrightarrow NaHSO_4(s) + HCl(g)$$

A 10.0-kg mass of NaCl reacts completely with sulfuric acid to give a certain volume of HCl($g$) at 50°C and $P = 1.00$ atm. If the same volume of hydrogen chloride is collected at 500°C and $P = 1.00$ atm, what mass of NaCl has reacted?

68. Exactly 1.0 lb of Hydrone, an alloy of sodium with lead, yields (at 0.0°C and 1.00 atm) 2.6 ft$^3$ of hydrogen when it is treated with water. All of the sodium reacts according to the equation

$$2 \text{ Na}_{\text{in alloy}} + 2 \text{ H}_2O(\ell) \longrightarrow 2 \text{ NaOH}(aq) + H_2(g)$$

and the lead does not react with water. Compute the percentage by mass of sodium in the alloy.

69. A sample of limestone (calcium carbonate, CaCO$_3$) is heated at 950 K until it is completely converted to calcium oxide (CaO) and CO$_2$. The CaO is then all converted to calcium hydroxide by addition of water, yielding 8.47 kg of solid Ca(OH)$_2$. Calculate the volume of CO$_2$ produced in the first step, assuming it to be an ideal gas at 950 K and a pressure of 0.976 atm.

70. A gas exerts a pressure of 0.740 atm in a certain container. Suddenly a chemical change occurs that consumes half of the molecules originally present and forms two new molecules for every three consumed. Determine the new pressure in the container if the volume of the container and the temperature are unchanged.

71. The following arrangement of flasks is set up. Assuming no temperature change, determine the final pressure inside the

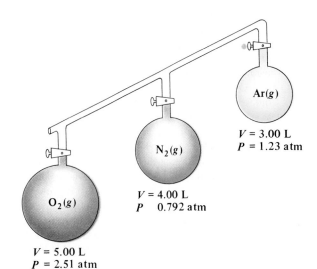

Ar($g$)
V = 3.00 L
P = 1.23 atm

N$_2$($g$)
V = 4.00 L
P  0.792 atm

O$_2$($g$)
V = 5.00 L
P = 2.51 atm

system after all stopcocks are opened. The connecting tube has zero volume.

**\*72.** A mixture of $CS_2(g)$ and excess $O_2(g)$ in a 10.0-L reaction vessel at 100.0°C is under a pressure of 3.00 atm. When the mixture is ignited by a spark, it explodes. The vessel successfully contains the explosion, in which all of the $CS_2(g)$ reacts to give $CO_2(g)$ and $SO_2(g)$. The vessel is cooled back to its original temperature of 100.0°C, and the pressure of the mixture of the two product gases and the unreacted $O_2(g)$ is found to be 2.40 atm. Calculate the mass (in grams) of $CS_2(g)$ originally present.

**73.** Acetylene reacts with hydrogen in the presence of a catalyst to form ethane according to the equation

$$C_2H_2(g) + 2\ H_2(g) \longrightarrow C_2H_6(g)$$

The pressure of a mixture of acetylene and an excess of hydrogen decreases from 0.100 atm to 0.042 atm in a vessel of a given volume after the catalyst is introduced, and the temperature is restored to its initial value after the reaction reaches completion. What was the mole fraction of acetylene in the original mixture?

**74.** Refer to the atomic trap described in problem 42.
(a) Assume ideal gas behavior to compute the pressure exerted on the "walls" of the optical bottle in this experiment.
(b) In this gas, the mean free path (the average distance traveled by the sodium atoms between collisions) is 3.9 m. Compare this to the mean free path of the atoms in gaseous sodium at room conditions.

**75.** Deuterium ($^2H$), when heated to sufficiently high temperature, undergoes a nuclear fusion reaction that results in the production of helium. The reaction proceeds rapidly at a temperature $T$, at which the average kinetic energy of the deuterium atoms is $8 \times 10^{-16}$ J. (At this temperature, deuterium molecules dissociate completely into deuterium atoms.)
(a) Calculate $T$ in kelvins. (Atomic mass of $^2H$ = 2.015.)
(b) For the fusion reaction to occur with ordinary hydrogen atoms, the average energy of the atoms must be about $32 \times 10^{-16}$ J. By what factor does the average speed of the $^1H$ atoms differ from that of the $^2H$ atoms of part (a)?

**76.** Molecules of oxygen of the following isotopic composition are separated in an oxygen enrichment plant: $^{16}O^{16}O$, $^{16}O^{17}O$, $^{16}O^{18}O$, $^{17}O^{17}O$, $^{17}O^{18}O$, $^{18}O^{18}O$.
(a) Compare the average translational kinetic energy of the lightest and heaviest molecular oxygen species at 200°C and at 400°C.
(b) Compare the average speeds of the lightest and heaviest molecular oxygen species at the same two temperatures.

**\*77.** What is the probability that an $O_2$ molecule in a sample of oxygen at 300 K has a speed between $5.00 \times 10^2$ m s$^{-1}$ and $5.10 \times 10^2$ m s$^{-1}$? (*Hint:* Try approximating the area under the Maxwell–Boltzmann distribution by small rectangles.)

**\*78.** Molecules in a spherical container make wall collisions in a great circle plane of the sphere. All paths traveled between collisions are equal in length.

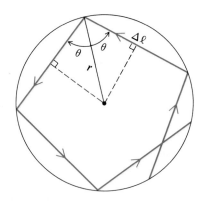

(a) Relate the path length $\Delta\ell$ to the angle $\theta$ and the sphere radius $r$.
(b) Express the component of the molecular momentum transferred to the wall in a collision in terms of the molecular mass $m$, the speed $u$, and the angle $\theta$.
(c) Using parts (a) and (b), calculate the average force exerted on the wall by one molecule.
(d) Derive the relation $PV = \frac{1}{3} Nm\overline{u^2}$ for a spherical container. This is the same relation found for a rectangular box in Section 3-5.

**79.** A vessel with a small hole in its wall is filled with oxygen to a pressure of 1.00 atm at 25°C. In a one-minute period, 3.25 g of oxygen effuses out through the hole into a vacuum. The vessel is evacuated and filled with an unknown gas at the same pressure and volume. In this case, 5.39 g of the unknown gas effuses out in one minute. Calculate the molar mass of the unknown gas.

**80.** A cylindrical storage tank for natural gas (mostly methane, $CH_4$) with a 20-ft radius and a 50-ft height is filled to a pressure of 2000 psi at 20°C. A small leak of 1.0-mm$^2$ area develops at one of the welds. Calculate the mass of $CH_4$, in grams, that leaks out of the tank in one day. What fraction of the total gas escapes per day?

**81.** A Thermos bottle (Dewar vessel) has an evacuated space between its inner and outer walls to diminish the rate of transfer of thermal energy to or from the bottle's contents. For good insulation the mean free path of the residual gas (air; average molecular mass = 29) should be at least ten times the distance between the inner and outer walls, which is about 1.0 cm. What should be the maximum residual gas pressure in the evacuated space if $T$ = 300 K? Take an average diameter of $d = 3.1 \times 10^{-10}$ m for the molecules in the air.

**82.** A tanker truck carrying liquid ammonia overturns, releasing ammonia vapor into the air.
(a) Approximating ammonia, oxygen, and nitrogen as spheres of equal diameter ($3 \times 10^{-10}$ m), estimate the diffusion constant of ammonia in air at atmospheric pressure and 20°C.
(b) Calculate the time required for a 100-m root-mean-square displacement of ammonia from the truck, and express

this time in everyday units (seconds, minutes, hours, days, or years). The actual time for the ammonia to travel this distance is far shorter because of the existence of air currents (even when there is no wind).

83. Molecules of $UF_6$ are approximately 175 times more massive than $H_2$ molecules; however, Avogadro's number of $H_2$ molecules confined at a set temperature exert the same pressure on the walls of the container as the same number of $UF_6$ molecules. Explain how this is possible.

84. The number density of atoms (chiefly hydrogen) in interstellar space is about 10 per cubic centimeter, and the temperature is about 100 K.
    (a) Calculate the pressure of the gas in interstellar space, and express it in atmospheres.
    (b) Under these conditions, an atom of hydrogen collides with another atom once every $1 \times 10^9$ seconds (that is, once every 30 years). By using the root-mean-square speed, estimate the distance traveled by a hydrogen atom between collisions. Compare this distance with the distance from the earth to the sun (150 million km).

85. A sample of 2.00 mol of argon is confined at low pressure in a volume at a temperature of 50°C. Describe quantitatively the effects of each of the following changes on the pressure, the average energy per atom in the gas, the root-mean-square speed, the rate of collisions with a given area of wall, the frequency of Ar–Ar collisions, and the mean free path:

(a) The temperature is decreased to $-50°C$.
(b) The volume is doubled.
(c) The amount of argon is increased to 3.00 mol.

86. By assuming that the collision diameter of a $CH_4$ molecule is given by its Lennard–Jones $\sigma$ parameter (Table 3-4), estimate the rate at which methane molecules collide with one another at 25°C and a pressure of (a) 1.00 atm; (b) $1.0 \times 10^{-7}$ atm.

*87. The van der Waals constant $b$ is related to the volume excluded per mole of molecules, so it should be proportional to $N_0\sigma^3$, where $\sigma$ is the distance parameter in the Lennard-Jones potential.
    (a) Make a plot of $b$ against $N_0\sigma^3$ for Ar, $H_2$, $CH_4$, $N_2$, and $O_2$, using data from Tables 3–3 and 3–4. Do you see an overall correlation between the two?
    (b) The van der Waals constant $a$ has dimensions of pressure times the square of the molar volume. Rewrite the units of $a$ in terms of energy, length, and number of moles, and suggest a relationship between $a$ and some combination of the constants $\epsilon$, $\sigma$, and $N_0$. Make a plot of $a$ against this combination of constants for the gases of part (a). Do you see an overall correlation in this case?

*88. Take the derivative of the Lennard-Jones potential to express the force exerted on one atom by another in terms of the distance $R$ between them. Calculate the forces (in joules per meter) on a pair of interacting argon atoms at distances of 3.0, 3.4, 3.8, and $4.2 \times 10^{-10}$ m. Is the force attractive or repulsive at each of these distances?

## CUMULATIVE PROBLEMS

89. A gaseous hydrocarbon, in a volume of 25.4 L at 400 K and a pressure of 3.40 atm, reacts in an excess of oxygen to give 47.4 g of $H_2O$ and 231.6 g of $CO_2$. Determine the molecular formula of the hydrocarbon.

90. A sample of a gaseous binary compound of boron and chlorine weighing 2.842 g occupies 0.153 L at 0°C and 1.00 atm pressure. This sample is decomposed to give solid boron and gaseous chlorine ($Cl_2$). The chlorine occupies 0.688 L at the same temperature and pressure. Determine the molecular formula of the compound.

91. A mixture of calcium carbonate, $CaCO_3$, and barium carbonate, $BaCO_3$, weighing 5.40 g reacts fully with hydrochloric

acid, $HCl(aq)$, to generate 1.39 L of $CO_2(g)$, measured at 50°C and 0.904 atm pressure. Calculate the percentages by mass of $CaCO_3$ and $BaCO_3$ in the original mixture.

92. A solid sample of $Rb_2SO_3$ weighing 6.24 g reacts with 1.38 L of gaseous HBr, measured at 75°C and 0.953 atm pressure. The solid RbBr, extracted from the reaction mixture and purified, has a mass of 7.32 g.
    (a) What is the limiting reactant?
    (b) What is the theoretical yield of RbBr, assuming complete reaction?
    (c) What is the actual percentage yield of product?

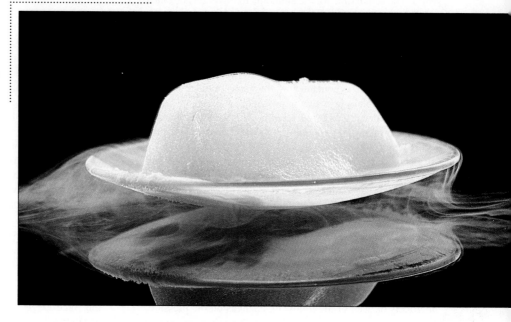

# CHAPTER

# 4

*Sublimation of Dry Ice ($CO_2$). Solid $CO_2$ changes directly from solid to gas. (Matt Meadows/Peter Arnold, Inc.)*

# Condensed Phases and Solutions

I n a gas the distances between molecules are so large that either their interactions can be ignored or collisions between isolated pairs of molecules can be considered. In a condensed phase (liquid or solid), by contrast, a molecule interacts simultaneously with many neighboring molecules and no such simplifications apply. In this chapter we consider the phase transitions undergone by a gas as it liquefies and freezes, and investigate the dynamic processes that prevail when two or more phases coexist. We then turn to the properties of solutions and mixtures, examining the preparation of solutions and their reaction stoichiometry. We consider the effects of small concentrations of dissolved species on phase transitions and explore some of the types of phase coexistence that arise in multicomponent systems. Applications are made to such varied processes as the determination of molar mass, the desalinization of water, and the separation of substances via distillation.

## 4-1 PHASE EQUILIBRIUM AND PHASE DIAGRAMS

Suppose 1 mol of gaseous sulfur dioxide is compressed at a temperature fixed at 30.0°C. The volume is measured at each pressure, and a graph of volume against pressure is constructed (Fig. 4-1). At low pressures the graph shows the inverse

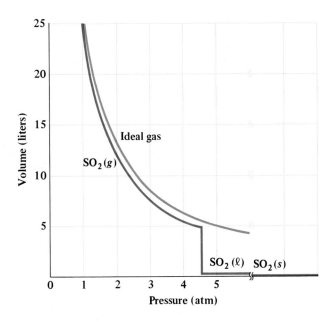

**Figure 4-1**

As one mole of $SO_2$ is compressed at a constant temperature of 30°C, the volume at first falls somewhat below its ideal gas value. Then, at 4.52 atm, the volume falls abruptly as the gas condenses to a liquid. At a much higher pressure, a further transition to the solid takes place.

dependence ($V \propto 1/P$) predicted by the ideal gas law. As the pressure increases, deviations appear because the gas is not ideal. At this temperature, attractive forces dominate; therefore, the volume falls below its ideal gas value and approaches 4.74 L (rather than 5.50 L) as the pressure approaches 4.52 atm.

For pressures up to 4.52 atm this behavior is quite regular and can be described by the van der Waals equation. At 4.52 atm, something surprising takes place: the volume drops abruptly by a factor of 100 and remains low as the pressure is increased further. What has happened? The gas has been liquefied solely by the application of pressure. If the compression of $SO_2$ is continued, another abrupt (but small) change in volume will occur as the liquid freezes to a solid. Condensed phases also arise when the temperature of a gas is lowered at constant pressure. If steam (water vapor) is cooled at 1 atm pressure, it condenses to liquid water at 100°C and freezes to solid ice at 0°C. Liquids and solids form at low temperatures once the attractive forces between molecules become strong enough to dominate the random effects of kinetic energy.

## Phase Equilibrium

Liquids and solids, like gases, are **phases**—samples of matter that are uniform throughout in both chemical constitution and physical state. Two or more phases can coexist. Suppose a small quantity of liquid water is put in an evacuated flask, with the temperature held at 25°C by placing the system in a constant-temperature bath. A pressure gauge is used to monitor changes in the pressure of water vapor inside the flask. Immediately after the water enters the flask, the pressure of water vapor begins to rise from zero. It increases with time and gradually levels off at a value of 0.03126 atm, which is the **vapor pressure** of water at 25°C. The contents of the flask have reached **equilibrium**, a condition in which no further changes in macroscopic properties occur as long as the system remains isolated. This passage toward

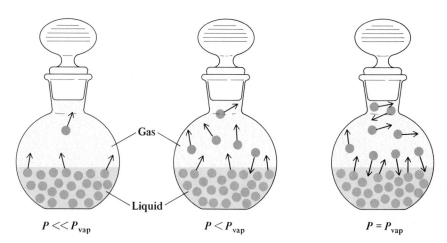

**Figure 4-2**

The approach to equilibrium in evaporation and condensation. Initially the pressure above the liquid is very low, and many more molecules leave the liquid surface than return to it. As time passes, more molecules fill the gas phase until the equilibrium vapor pressure $P_{vap}$ is approached; the rates of evaporation and condensation then become equal.

equilibrium is a spontaneous process, occurring in a closed system without any external influence. If some of the water vapor that has formed is removed, additional water evaporates from the liquid to reestablish the same vapor pressure, $P_{vap}(H_2O) = 0.03126$ atm.

What is happening on a microscopic scale to cause this spontaneous movement of the system toward equilibrium? According to the kinetic theory, the molecules of water in the liquid are in a constant state of thermal motion. Some of those near the surface are moving fast enough to escape the attractive forces holding them in the liquid; this process of **evaporation** causes the pressure of the water vapor to increase. As the number of molecules in the vapor phase increases, the reverse process begins to occur: molecules in the vapor strike the surface of the liquid and some are captured, leading to **condensation.** As the pressure of the gas increases, the rate of condensation increases until it balances the rate of evaporation from the surface (Fig. 4-2). Once this occurs, there is no further net flow of matter from one phase to the other; the system has reached **phase equilibrium,** characterized by a particular value of the water vapor pressure. Water molecules continue to evaporate from the surface of the liquid, but other water molecules return to the liquid from the vapor at an equal rate. A similar phase equilibrium is established between an ice cube and liquid water at the freezing point.

The vapor pressure of the water is independent of the size and shape of the container. If the experiment is duplicated in a larger flask, then a greater *amount* of water evaporates on the way to equilibrium, but the final pressure in the flask at 25°C is still 0.03126 atm as long as some liquid water is present. If the experiment is repeated at a temperature of 30.0°C, everything happens as just described except that the pressure in the space above the water reaches 0.04187 atm. A higher temperature corresponds to a larger average kinetic energy for the water molecules. A new balance between the rates of evaporation and condensation is struck, but at a higher

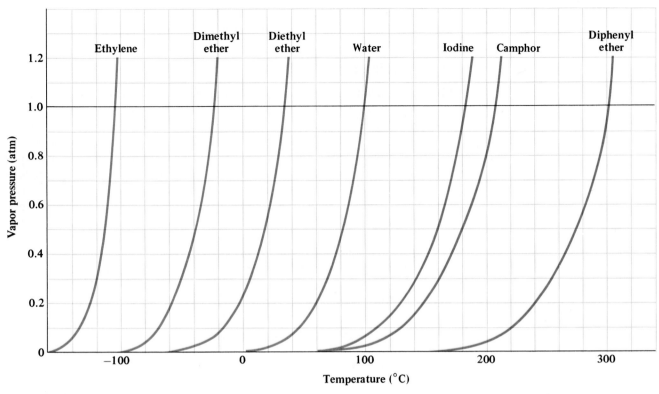

**Figure 4-3**
The vapor pressure of a solid or liquid depends strongly on temperature. The temperature at which the vapor pressure becomes 1 atm defines the normal boiling point of a liquid and the normal sublimation point of a solid.

vapor pressure. The vapor pressure of water, and of all other substances, increases with increasing temperature (Fig. 4-3 and Table 4-1).

Phase equilibrium is a *dynamic* process that is quite different from the static equilibrium achieved as a marble rolls to a stop after being spun into a bowl. In the equilibrium between liquid water and water vapor, the partial pressure levels off, not because evaporation and condensation stop, but because at equilibrium their rates become equal. One further feature of equilibrium that can be seen from the liquid–vapor system is that its properties are independent of the direction from which it is approached. If we inject enough water vapor into the empty flask so that initially the pressure of the vapor is *above* the vapor pressure of liquid water, $P_{vap}(H_2O)$, then liquid water will condense until the same equilibrium vapor pressure is achieved (0.03126 atm at 25°C). Of course, if we do not use enough water vapor to exceed a pressure of 0.03126 atm, all the water will remain in the vapor phase and two-phase equilibrium will not occur.

The presence of water vapor above an aqueous solution has an important practical consequence. If a reaction in aqueous solution generates gases, these gases are "wet," containing water vapor at a partial pressure given by the equilibrium vapor pressure of water at the temperature of the experiment. The amount of gas generated is determined not by the total pressure but by the partial pressure of the gas. Dalton's

| Table 4-1 | Vapor Pressure of Water at Various Temperatures |
|---|---|
| **Temperature (°C)** | **Vapor Pressure (atm)** |
| 15.0 | 0.01683 |
| 17.0 | 0.01912 |
| 19.0 | 0.02168 |
| 21.0 | 0.02454 |
| 23.0 | 0.02772 |
| 25.0 | 0.03126 |
| 30.0 | 0.04187 |
| 50.0 | 0.1217 |

law (Section 3-4) must be used to subtract the partial pressure of water as listed in Table 4-1. This correction is significant in quantitative work (see problems 1–4).

## Phase Transitions

Six **phase transitions** occur among the three states of matter (Fig. 4-4). Typically, solids melt to give liquids when they are heated, and liquids boil to give gases. **Boiling** is an extension of evaporation (in which the vapor escapes from the surface only). In boiling, bubbles of gaseous substance form actively throughout the body of a liquid and rise up to escape at the surface. Only when the vapor pressure of a liquid exceeds the external pressure under which it is being held can the liquid start to boil. The **boiling point** is the temperature at which the vapor pressure of a liquid equals the external pressure. The external pressure influences boiling points quite strongly; for example, water boils at 25°C if the external pressure is reduced below 0.03126 atm (recall that the vapor pressure of water at 25°C is just this number) but requires a temperature of 121°C to boil under an external pressure of 2.0 atm. At high elevations the pressure of the atmosphere is lower than 1 atm, so water boils at a temperature less than 100°C and food cooks more slowly in boiling water than it would at a lower elevation. By contrast, the use of a pressure cooker increases the boiling temperature and speeds the cooking of food. The **normal boiling point** is defined as the temperature at which the vapor pressure of the liquid equals 1 atm. Figure 4-3 shows that in general a lower normal boiling point implies a higher vapor pressure at any fixed temperature (and therefore a more volatile liquid).

**Melting** is the conversion of a solid to the liquid state. The **normal melting point** of a solid is the temperature at which solid and liquid are in equilibrium under a pressure of 1 atm. The normal melting point of ice is 0.00°C, so liquid water and ice coexist indefinitely (are in equilibrium) at this temperature at a pressure of 1 atm. If the temperature is lowered by even a small amount, then all of the water eventually freezes; if the temperature is infinitesimally raised, all of the ice eventually melts. The qualifying term "normal" is often omitted in talking about melting points because they depend only weakly on pressure.

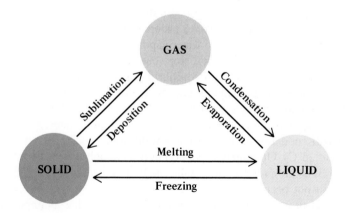

### Figure 4-4

Direct transitions among all three states of matter not only are possible but are observed in everyday life.

**Figure 4-5**

When sugar is heated, it melts and simultaneously decomposes to a dark-colored caramelized mixture. (Leon Lewandowski)

It is sometimes possible to overshoot a phase boundary, with the new phase appearing only after some delay. An example is the superheating of a liquid. Liquid water, for example, can reach a temperature somewhat above 100°C if heated rapidly. When vaporization of a superheated liquid does occur, it can be quite violent, with liquid thrown out of the container. To avoid this superheating in the laboratory, boiling chips (pieces of porous fired clay) may be added to the liquid. They help to initiate boiling as soon as the normal boiling point is reached, by providing sites where gas bubbles can form. Undercooling of liquids below their freezing points is also possible. In careful experiments, undercooled liquid water has been studied at temperatures below −30°C (at atmospheric pressure).

Many materials react chemically when heated, before they have a chance to melt or boil. Substances that change their chemical identities before changing state do not have normal melting or boiling points. For example, sucrose (table sugar) melts, but quickly begins to darken and eventually chars (Fig. 4-5). A temperature hot enough to overcome the intermolecular attractions in sugar is also sufficient to break apart the sugar molecules themselves.

## Phase Diagrams

If the temperature of a substance is held constant and its pressure is changed, a phase transition will be observed at some particular pressure between two of its three phases: solid, liquid, and gas. Repeating the process at many different temperatures gives the data necessary to draw the **phase diagram** of that substance—a plot of pressure against temperature that shows the stable state for every pressure–temperature combination. Figure 4-6 shows the phase diagrams for carbon dioxide and water. A great deal of information can be read from such diagrams. For each substance there is a unique combination of pressure and temperature, called the **triple point** (marked ''T''), at which the gas, liquid, and solid phases coexist in equilibrium. Extending from the triple point are three lines, each denoting the conditions for the equilibrium coexistence of two phases. Along line TA, solid and gas are in equilibrium; along TB, solid and liquid; and along TC, liquid and gas. The areas between lines encompass combinations of pressure and temperature where only one phase exists.

The gas–liquid coexistence curve extends upward in temperature and pressure from the triple point. This line, stretching from T to C in the phase diagrams, is the

**Figure 4-6**

Phase diagrams for (a) carbon dioxide and (b) water. The pressures and temperatures are not drawn to scale.

(a) Carbon dioxide

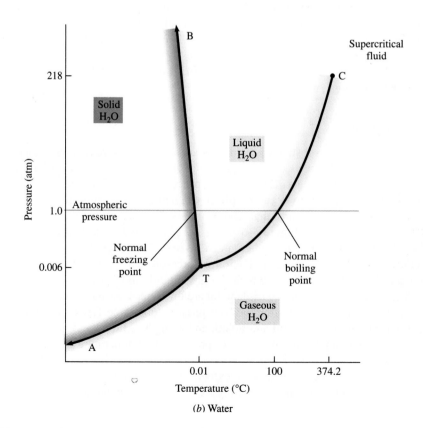

(b) Water

vapor pressure curve of the liquid substance, portions of which were shown in Figure 4-3. The gas–liquid coexistence curve does not continue indefinitely, but instead terminates at the **critical point** (points C in Fig. 4-6). Along this coexistence curve there is an abrupt, discontinuous change in the density and other properties from one side to the other. The differences between the properties of the liquid and the gas become smaller as the critical point is approached, and disappear altogether at that point. If the substance is placed in a closed container and is gradually heated, a **meniscus** is observed at the boundary between liquid and gas (Fig. 4-7); at the critical point this meniscus disappears. For pressures above the critical pressure (218 atm for water) it is no longer possible to identify a particular state as gas or liquid. A substance beyond its critical point is called a **supercritical fluid** because the term "fluid" includes both gases and liquids.

The liquid–solid coexistence curve does not terminate, as the gas–liquid curve does at the critical point, but continues to indefinitely high pressures. In practice, such a curve is almost vertical because very large changes in pressure are necessary to change the freezing temperature of a liquid. For most substances, this curve inclines slightly to the right (see Fig. 4-6a): an increase in pressure increases the freezing point of the liquid. Another way of saying this is that at constant temperature an increase in pressure leads to the formation of a phase with higher density (smaller volume), and for most substances the solid is denser than the liquid. Water and a few other substances are anomalous (see Fig. 4-6b); for them, the liquid–solid coexistence curve slopes up initially to the *left,* showing that an increase in pressure causes the solid to melt. This anomaly is related to the densities of the liquid and solid phases: ice is less dense than water (that is why ice cubes float on water), so when ice is compressed at 0°C, it melts.

For most substances, including water (Fig. 4-6b), atmospheric pressure occurs somewhere between the triple-point pressure and the critical pressure, so in our ordinary experience all three phases—gas, liquid, and solid—are observed. For a few substances the triple-point pressure lies *above P* = 1 atm, however, and under atmospheric conditions there is a direct transition called **sublimation** from solid to gas, without an intermediate liquid state. Carbon dioxide is such a substance (see Fig. 4-6a); its triple-point pressure is 5.117 atm (the triple-point temperature is −56.57°C). Solid $CO_2$ (Dry Ice) sublimes directly to gaseous $CO_2$ at atmospheric

**Figure 4-7**
When a gas and liquid coexist, the interface between them is clearly visible as a meniscus. The meniscus is useful for reading the volume of a liquid in a buret. It disappears as the critical point is reached. (Leon Lewandowski)

Sublimation of solid carbon dioxide (Dry Ice). The white clouds are drops of water vapor (moisture in the air) that condense at the low temperatures near the solid surface. Gaseous carbon dioxide itself is transparent.

pressure. In this respect it differs from ordinary ice, which melts before it evaporates and sublimes only at pressures below its triple-point pressure, 0.0060 atm. This fact is used in freeze-drying, a process in which foods are frozen and then put in a vacuum chamber at a pressure of less than 0.0060 atm. The ice crystals that formed on freezing then sublime, leaving a dried food that can be reconstituted by adding water.

Many substances exhibit more than one solid phase as the temperature and pressure are varied. At ordinary pressures the stable state of carbon is graphite, a rather soft, black solid, but at high enough pressures the hard, transparent diamond form is stable. Below 13.2°C (and at atmospheric pressure) elemental tin undergoes a slow transformation from the metallic white form to a powdery gray form, a process referred to as "tin disease." This has caused the destruction of tin organ pipes in unheated churches. No fewer than nine solid forms of ice are known, some of which exist only over a very limited range of temperatures and pressures.

### Example 4-1

Consider a sample of $CO_2$ held at point B in Figure 4-6a.
(a) What phase(s) is (are) present at equilibrium?
(b) Suppose the $CO_2$ originally held at point B is heated at constant pressure. What happens?
(c) Describe a procedure to convert all the $CO_2$ to gas without changing the temperature.

### Solution

(a) Because point B lies on the liquid–solid coexistence curve, both liquid and solid phases are present.
(b) An increase in temperature at constant pressure corresponds to a movement to the right in the figure, and the solid $CO_2$ melts to a pure liquid.
(c) If the pressure is lowered sufficiently at constant temperature (below the triple-point pressure of 5.1 atm, for example) the $CO_2$ will be converted completely to gas.

*Related Problems:* 9, 10

## 4-2 INTERMOLECULAR FORCES IN LIQUIDS

The same forces that make gases nonideal (Section 3-7) are responsible for the condensation of gases to form liquids. The stronger the attractive forces are, the greater the stability of the liquid and the higher the normal boiling point $T_b$. (Melting points depend more sensitively than do boiling points on molecular shapes and on details of the molecular interactions, and so vary less systematically with the strength of the attractive forces.) Ionic liquids generally have the strongest attractions, due to the Coulomb interaction of charged ions, and have very high boiling points. Molten NaCl, for example, boils at 1686 K under atmospheric pressure. At the opposite extreme, the normal boiling point of helium is only 4.2 K. Within a series of related compounds, those of higher molar mass tend to have higher normal boiling points. Thus, from helium to xenon, normal boiling points increase (see Fig.

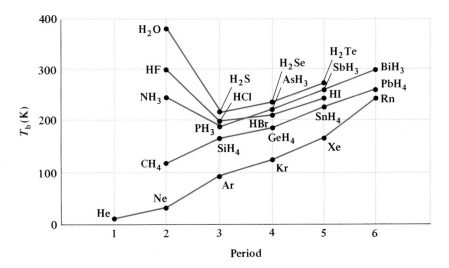

**Figure 4-8**

Trends in the boiling points of hydrides of some main-group elements and of the noble gases.

4-8) along with the strengths of the attractive forces among the noble gases (represented by the well depth $\epsilon$ in Table 3-4).

Between the noble-gas and ionic liquids falls a class of liquids called **polar liquids.** Recall from Section 2-4 that the electrons in a molecule such as HCl are distributed around the two nuclei to give a net small positive charge on one end (the H atom, in this case) and an equal and opposite small negative charge on the other end (the Cl atom). In liquid HCl, the molecules arrange themselves so that, as much as possible, the H atom of one molecule is close to the Cl atoms of neighboring molecules. Intermolecular forces in such dipolar liquids are weaker than those in ionic liquids (where the charge is transferred completely from one atom to another to make two ions) but stronger than those in **nonpolar liquids** such as $N_2$, where the two ends of the molecule are identical and the bonding is fully covalent. As Figure 4-8 shows, HCl has a higher boiling point than argon (a nonpolar fluid of atoms with nearly the same molar mass) because of its polar nature.

### Hydrogen Bonds

Figure 4-8 shows the normal boiling points of several series of hydrides. As just stated, the boiling point usually increases with increasing molar mass in a series of related compounds. The dramatic deviations from these systematic trends shown by HF, $NH_3$, and especially $H_2O$ indicate the strength and importance of the special type of bond that is common to these cases, a **hydrogen bond.** Such a bond forms when a hydrogen atom bonded to an O, N, or F atom (atoms that strongly attract an additional electron) also interacts with the lone electron pair of another such atom nearby. Figure 4-9 shows the interaction of a pair of water molecules to form a dimer in the gas phase. The hydrogen bond that forms is weaker than an ordinary O—H covalent bond but significantly stronger than most intermolecular interactions. Like most hydrogen bonds, that in water is linear but asymmetric, with the

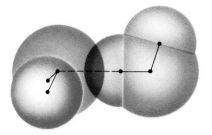

**Figure 4-9**

A single hydrogen bond between water molecules forms a dimer. This bond is far weaker than a covalent bond but still strong enough to resist dissociation at room temperature. The shared hydrogen atom at the center approaches the neighboring oxygen atom quite closely.

hydrogen atom closer to and more strongly bound to one of the oxygen atoms. It is indicated as O—H $\cdots$ O.

Like HCl or $H_2S$ molecules, water molecules are polar, with a small positive charge on each hydrogen atom and a small negative charge on each oxygen. In the liquid, these molecules orient themselves so that the hydrogen atoms are close to oxygen atoms on neighboring molecules. The hydrogen atom in a bond such as O—H is only slightly shielded by electrons because, unlike all other elements, it has no electrons other than valence electrons. As a result, it can approach very close to the lone-pair electrons on a neighboring oxygen atom, causing a strong electrostatic (Coulomb) interaction between the two. In addition, a small amount of covalent bonding arises from the sharing of electrons between the two oxygen atoms and the intervening hydrogen atom. These effects combine to make the interaction unusually strong.

## Water

Water makes up about 0.023% of the total mass of the earth. About $1.4 \times 10^{21}$ kg of it is distributed above, on, and below the earth's surface. The volume of this vast amount of water is about 1.4 billion $km^3$. Most of the earth's water (97.7%) is contained in the oceans, with about 1.9% in the form of ice or snow and most of the remainder (a very small fraction of the total) available as fresh water in lakes, rivers, underground sources, and atmospheric water vapor. A small but important fraction is bound to cations in certain minerals, such as clays and hydrated crystalline salts. More than 80% of the surface of the earth is covered with water—as ice and snow near the poles, as relatively pure water in lakes and rivers, and as a salt solution in the world's oceans. The unusual properties of water, which come from its network of hydrogen bonds, have profound effects on life on earth. Figure 4-8 compared the boiling points of water and hydrides that lack hydrogen bonds. An extrapolation of the trends from the latter compounds would give a boiling point for "water without hydrogen bonds" near 150 K (−123°C). Life as we know it would not be possible under these circumstances.

If all possible hydrogen bonds form in a mole ($N_0$ molecules) of pure water, then every oxygen atom is surrounded tetrahedrally by four hydrogen atoms: its own two and two from neighboring molecules. The hydrogen bonds number $2N_0$ and form a three-dimensional network. In ice, this full complement of hydrogen bonds does form. The result is an array of interlocking six-membered rings of water molecules (Fig. 4-10), echoed macroscopically in the characteristic six-fold symmetry of snowflakes.

Water has its maximum density at 4°C (Fig. 4-11), and it expands upon freezing (recall also Fig. 4-6b). This unusual behavior, which is seen in very few other liquids, also is due to hydrogen bonds. When ice melts, some of the hydrogen bonds that maintain the open structure of Figure 4-10 break and the structure partially collapses, producing a liquid with a smaller volume (higher density). The reverse process, a sudden expansion of water on freezing, can cause bursting of water pipes and freeze–thaw cracking of rocks and concrete. Such expansion has many beneficial effects as well, however. If ice were denser than water, the winter ice that forms at the surface of a lake would sink to the bottom and the lake would freeze from the bottom up. Instead, the ice remains at the surface and the water near the bottom achieves a stable wintertime temperature near 4°C, which allows the lake's fish to survive.

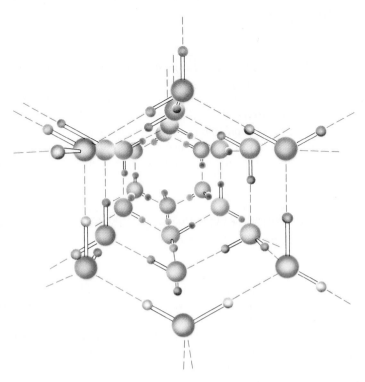

**Figure 4-10**

The structure of ice is quite open. Each water molecule has only four nearest neighbors, with which it interacts by means of hydrogen bonds (blue dashed lines).

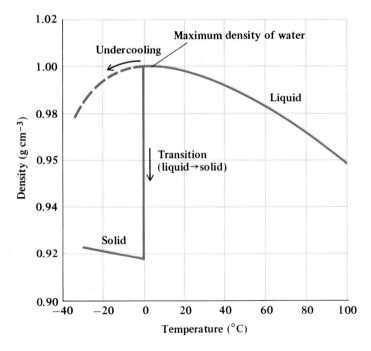

**Figure 4-11**

The density of water rises to a maximum as it is cooled to 3.98°C, then starts to decrease slowly. Undercooled water (water chilled below its freezing point but not yet converted to ice) continues the smooth decrease in density. When liquid water freezes, the density drops abruptly.

Water easily dissolves many ionic compounds and is the medium of choice for much solution-phase inorganic chemistry. However, large classes of compounds, such as hydrocarbons, dissolve to only very small extents in water. In general, water follows the rule "Like dissolves like." Substances such as ethyl alcohol, which can form hydrogen bonds, have appreciable solubilities in water, whereas polar molecules without hydrogen bonds dissolve to lesser extents. Chemical reactions involving nonpolar molecules such as hydrocarbons require a medium other than water.

## Example 4-2

Predict the order of increase in the normal boiling points of the following substances: $F_2$, HBr, $NH_4Cl$, and HF.

### Solution

As an ionic substance, $NH_4Cl$ should have the highest boiling point of the three (measured value: 520°C). HF should have a higher boiling point than HBr, because its molecules form hydrogen bonds (Fig. 4-8) that are stronger than the dipolar interactions in HBr (measured: 20°C for HF, −67°C for HBr). Fluorine, $F_2$, is nonpolar and contains light atoms, and so should have the lowest boiling point of the four substances (measured: −188°C).

*Related Problems:* 13, 14, 15, 16

## 4-3  COMPOSITION OF SOLUTIONS

Homogeneous systems that contain two or more substances are called **solutions.** Usually a solution is thought of as a liquid that contains some dissolved substance, such as a solid or gas. This is the sense in which the term is used throughout most of this chapter, but solutions of one solid in another are also common (an example is an alloy of gold and silver). In fact, any homogeneous system of two or more substances (liquid, solid, or gas) is a solution. The major component is usually called the **solvent,** and minor components are the **solutes.** Several measures can be used to specify the composition of a solution. **Mass percentage** (colloquially called weight percentage) is frequently employed in everyday life and is defined as the percentage by mass of a given substance in the solution. In chemistry, the most useful measures of composition are *mole fraction, molarity,* and *molality.*

The mole fraction of a substance in a mixture is the number of moles of that substance divided by the total number of moles present. This term was introduced in the discussion of gas mixtures and Dalton's law (Section 3-4). In a binary mixture containing $n_1$ mol of species 1 and $n_2$ mol of species 2, the mole fractions $X_1$ and $X_2$ are

$$X_1 = \frac{n_1}{n_1 + n_2}$$

$$X_2 = \frac{n_2}{n_1 + n_2} = 1 - X_1$$

The mole fractions of all the species present must add up to 1. When it is possible to make a clear distinction between solvent and solutes, the label 1 denotes the solvent, and higher numbers are given to the solutes. If comparable amounts of two liquids such as water and alcohol are mixed, however, the assignment of the labels 1 and 2 is arbitrary.

The **concentration** of a substance is the number of moles per unit volume. The SI units moles per cubic meter are of inconvenient size for chemical work, so the **molarity,** defined as the number of moles of solute per liter of solution, is used instead:

$$\text{molarity} = \frac{\text{moles solute}}{\text{liters solution}} = \text{mol L}^{-1}$$

"M" is the abbreviation for "moles per liter." A 0.1 M (read "0.1 molar") solution of HCl has 0.1 mol of HCl (dissociated into ions, of course) per liter of solution. Molarity is the most common way of specifying the compositions of dilute solutions. For accurate measurements it has the disadvantage that it depends slightly on temperature. If a solution is heated or cooled, its volume changes, so the number of moles of solute per liter of solution also changes.

The **molality,** in contrast, involves a ratio of masses and is independent of temperature. It is defined as the number of moles of solute per kilogram of *solvent*:

$$\text{molality} = \frac{\text{moles solute}}{\text{kilograms solvent}} = \text{mol kg}^{-1}$$

Because water has a density of 1.00 g cm$^{-3}$ at 20°C, 1.00 liter of water weighs $1.00 \times 10^3$ g, or 1.00 kg. It follows that in a dilute aqueous solution the number of moles of solute per liter is approximately the same as the number of moles per kilogram of water, so that molarity and molality have nearly equal values. For nonaqueous solutions, and concentrated aqueous solutions, this approximate equality fails.

## Example 4-3

A solution is prepared by dissolving 22.4 g of $MgCl_2$ in 0.200 L of water. Taking the density of pure water to be 1.00 g cm$^{-3}$ and the density of the resulting solution to be 1.089 g cm$^{-3}$, calculate the mole fraction, molarity, and molality of $MgCl_2$ in this solution.

### Solution

We are given the *mass* of the $MgCl_2$ and the *volume* of the water. The chemical amounts are

$$\text{moles } MgCl_2 = \frac{22.4 \text{ g}}{95.22 \text{ g mol}^{-1}} = 0.235 \text{ mol}$$

$$\text{moles water} = \frac{(0.200 \text{ L})(1000 \text{ cm}^3 \text{ L}^{-1})(1.00 \text{ g cm}^{-3})}{18.02 \text{ g mol}^{-1}}$$

$$= 11.1 \text{ mol}$$

$$\text{mole fraction } MgCl_2 = \frac{0.235 \text{ mol}}{(11.1 + 0.235) \text{ mol}} = 0.0207$$

To calculate the molarity, we must first determine the volume of solution. Its mass is 200 g water + 22.4 g MgCl = 222.4 g, and its density is 1.089 g cm$^{-3}$, so the volume is

$$\text{volume solution} = \frac{222.4 \text{ g}}{1.089 \text{ g cm}^{-3}} = 204 \text{ cm}^3 = 0.204 \text{ L}$$

$$\text{molarity MgCl}_2 = \frac{0.235 \text{ mol MgCl}_2}{0.204 \text{ L}} = 1.15 \text{ M}$$

$$\text{molality MgCl}_2 = \frac{0.235 \text{ mol MgCl}_2}{0.200 \text{ kg H}_2\text{O}} = 1.18 \text{ mol kg}^{-1}$$

*Related Problems:* 23, 24

In Example 4-3, it is important to note that, although the molarity of MgCl$_2$ is 1.15 M, the molarity of Cl$^-$ ions in the solution is twice as large, or 2.30 M, because each formula unit of MgCl$_2$ dissociates to give two Cl$^-$ ions.

## Example 4-4

A 9.386 M aqueous solution of sulfuric acid has a density of 1.5091 g cm$^{-3}$. Calculate the molality, the percentage by mass, and the mole fraction of sulfuric acid in this solution.

### Solution

It is convenient to choose one liter of the solution, whose mass is

$$(1000 \text{ cm}^3)(1.5091 \text{ g cm}^{-3}) = 1509.1 \text{ g} = 1.5091 \text{ kg}$$

One liter contains 9.386 mol H$_2$SO$_4$, or

$$9.386 \text{ mol H}_2\text{SO}_4 \times 98.08 \text{ g mol}^{-1} = 920.6 \text{ g H}_2\text{SO}_4$$

The mass of water in this liter of solution is then obtained by subtraction:

$$\text{mass of water in 1 L of solution} = 1.5091 - 0.9206 \text{ kg} = 0.5885 \text{ kg}$$

The molality is now directly obtained as

$$\text{molality H}_2\text{SO}_4 = \frac{9.386 \text{ mol H}_2\text{SO}_4}{0.5885 \text{ kg H}_2\text{O}} = 15.95 \text{ mol kg}^{-1}$$

and the mass percentage is

$$\text{mass percentage H}_2\text{SO}_4 = \frac{0.9206 \text{ kg}}{1.5091 \text{ kg}} \times 100\% = 61.00\%$$

The chemical amount of water is

$$\text{moles H}_2\text{O} = \frac{588.5 \text{ g}}{18.02 \text{ g mol}^{-1}} = 32.66 \text{ mol}$$

so that the mole fraction of H$_2$SO$_4$ is

$$\text{mole fraction } H_2SO_4 = X_2 = \frac{9.386 \text{ mol}}{9.386 + 32.66 \text{ mol}} = 0.2232$$

*Related Problems:* 25, 26

## Preparation of Solutions

Examples 4-3 and 4-4 show that if a known mass of solute is added to a known volume of solvent, the molarity can be calculated only if the density of the resulting solution is known. If one liter of solvent is used, for example, the volume of the resulting solution is less than one liter in some cases and greater in others. If a solution is to have a given molarity, it is clearly inconvenient to need to know the solution density. This is avoided in practice by dissolving the measured amount of solute in a smaller amount of solvent, then adding solvent continuously until the desired total volume is reached. For accurate work, a carefully calibrated **volumetric flask** is used (Fig. 4-12).

On occasion it may be necessary to prepare a dilute solution of specified concentration from a more concentrated solution of known concentration by adding pure solvent to it. Suppose that the initial concentration (molarity) is $c_i$ and the initial solution volume is $V_i$. The chemical amount of solute is $(c_i \text{ mol L}^{-1})(V_i \text{ L}) = c_i V_i$ mol. This does not change upon dilution to a final solution volume $V_f$ because only solvent, and not solute, is being added. Thus, $c_i V_i = c_f V_f$, and the final molarity is

$$c_f = \frac{\text{moles solute}}{\text{final solution volume}} = \frac{c_i V_i}{V_f} \qquad \textbf{(4-1)}$$

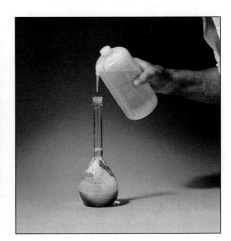

## Figure 4-12
To prepare a solution of ammonium chromate, $(NH_4)_2CrO_4$, with accurately known concentration, weigh out an amount of the solid, transfer it to a volumetric flask, dissolve it in somewhat less than the required amount of water, and dilute to the total volume marked on the neck of the flask. (Leon Lewandowski)

This equation can be used either to calculate the final concentration after dilution to a given final volume, or to determine what final volume should be used to obtain a given concentration.

## Example 4-5

(a) Describe how to prepare 0.500 L of a 0.100 M aqueous solution of potassium hydrogen carbonate ($KHCO_3$).
(b) Describe how to dilute this solution to a final concentration of 0.0400 M $KHCO_3$.

### Solution

(a)    moles solute = $(0.500 \text{ L})(0.100 \text{ mol L}^{-1}) = 0.0500 \text{ mol}$

grams solute = $(0.0500 \text{ mol})(100.12 \text{ g mol}^{-1}) = 5.01 \text{ g}$

because 100.12 is the molar mass of $KHCO_3$. We would therefore dissolve 5.01 g of $KHCO_3$ in a small amount of water and dilute the solution to 0.500 L.
(b) Rearranging Equation 4-1 gives

$$V_f = \frac{c_i}{c_f} V_i$$

$$= \left( \frac{0.100 \text{ mol L}^{-1}}{0.0400 \text{ mol L}^{-1}} \right)(0.500 \text{ L}) = 1.25 \text{ L}$$

To achieve this, the solution from part (a) is diluted to a total volume of 1.25 L by adding water.

*Related Problems:* 29, 30

## 4-4 THE STOICHIOMETRY OF REACTIONS IN SOLUTION

The great majority of all chemical reactions that occur on the earth's surface, whether in living organisms or among inorganic substances, take place in aqueous solution. When a chemical reaction is carried out between substances in solution, the same requirements of stoichiometry apply that were discussed in Section 1-7, in the sense that the conservation laws (embodied in balanced chemical equations) are always in force. It is now necessary to modify our use of them, however. Instead of a conversion between masses and chemical amounts (using the molar mass as a conversion factor), the conversion is now between *solution volumes* and chemical amounts, with the concentration as the conversion factor.

For instance, consider the reaction that is used commercially to prepare elemental bromine from its salts in solution:

$$2 \text{ Br}^-(aq) + \text{Cl}_2(aq) \longrightarrow 2 \text{ Cl}^-(aq) + \text{Br}_2(aq)$$

Suppose there is 50.0 mL of a 0.0600 M solution of NaBr. What volume of a 0.0500 M solution of $Cl_2$ is needed to react completely with the $Br^-$? To answer this, find the chemical amount of bromide ion present:

$$0.0500 \text{ L} \times (0.0600 \text{ mol L}^{-1}) = 3.00 \times 10^{-3} \text{ mol Br}^-$$

Next, use the chemical conversion factor 1 mol of $Cl_2$ per 2 mol of $Br^-$ to find

$$\text{moles } Cl_2 \text{ reacting} = 3.00 \times 10^{-3} \text{ mol } Br^- \left( \frac{1 \text{ mol } Cl_2}{2 \text{ mol } Br^-} \right)$$

$$= 1.50 \times 10^{-3} \text{ mol } Cl_2$$

Finally, find the necessary volume of aqueous chlorine:

$$\frac{1.50 \times 10^{-3} \text{ mol}}{0.0500 \text{ mol L}^{-1}} = 3.00 \times 10^{-2} \text{ L solution}$$

The reaction requires $3.00 \times 10^{-2}$ L, or 30.0 mL, of the $Cl_2$ solution. In practice, an excess of $Cl_2$ solution would be used to ensure more nearly complete conversion of the bromide ion to bromine.

The chloride ion concentration after completion of the reaction might also be of interest. Because each mole of bromide ion that reacts gives one mole of chloride ion in the products, the chemical amount of $Cl^-$ produced is $3.00 \times 10^{-3}$ mol. The final volume of the solution is 0.0800 L, so the final concentration of $Cl^-$ is

$$[Cl^-] = \frac{3.00 \times 10^{-3} \text{ mol}}{0.0800 \text{ L}} = 0.0375 \text{ M}$$

where square brackets around a chemical symbol are introduced to refer to molarity.

## Example 4-6

When potassium dichromate is added to concentrated hydrochloric acid, it reacts according to the chemical equation

$$K_2Cr_2O_7(s) + 14 \text{ HCl}(aq) \longrightarrow$$

$$2 \text{ K}^+(aq) + 2 \text{ Cr}^{3+}(aq) + 8 \text{ Cl}^-(aq) + 7 \text{ H}_2O(\ell) + 3 \text{ Cl}_2(g)$$

producing a mixed solution of chromium(III) chloride and potassium chloride and evolving gaseous chlorine. Suppose that 6.20 g of $K_2Cr_2O_7$ reacts with concentrated HCl and that the final volume of the solution is 100.0 mL. Calculate the final concentration of $Cr^{3+}(aq)$ and the chemical amount of chlorine produced.

## Solution

The first step is to convert the mass of $K_2Cr_2O_7$ to its chemical amount:

$$\frac{6.20 \text{ g } K_2Cr_2O_7}{294.19 \text{ g mol}^{-1}} = 0.0211 \text{ mol } K_2Cr_2O_7$$

The balanced chemical equation states that 1 mol of $K_2Cr_2O_7$ reacts to give 2 mol of $Cr^{3+}$ and 3 mol of $Cl_2$. Use of these two chemical conversion factors gives

$$\text{moles } Cr^{3+} = 0.0211 \text{ mol } K_2Cr_2O_7 \left( \frac{2 \text{ mol } Cr^{3+}}{1 \text{ mol } K_2Cr_2O_7} \right)$$

$$= 0.0422 \text{ mol } Cr^{3+}$$

$$\text{moles } Cl_2 = 0.0211 \text{ mol } K_2Cr_2O_7 \left( \frac{3 \text{ mol } Cl_2}{1 \text{ mol } K_2Cr_2O_7} \right)$$

$$= 0.0633 \text{ mol } Cl_2$$

(a)

(b)

**Figure 4-13**

(a) Addition of a small amount of potassium permanganate from a buret gives a dash of purple color to the $Fe^{2+}$ solution, which disappears quickly as the permanganate ion reacts to give $Fe^{3+}$ and $Mn^{2+}$.

(b) When all the $Fe^{2+}$ ions have been consumed, additional drops of permanganate solution give a pale purple color to the solution.

Because the final volume of the solution is 0.100 L, the concentration of $Cr^{3+}(aq)$ is

$$[Cr^{3+}] = \frac{0.0422 \text{ mol}}{0.100 \text{ L}} = 0.422 \text{ M}$$

*Related Problems:* 33, 34

### Titration

One of the most important techniques in analytical chemistry is **titration**—the addition of a carefully measured volume of one solution, of known concentration, to a second solution containing a different substance in unknown concentration, with which it will react quantitatively. Completion of the reaction is signaled by a change in some physical property, such as the color of the reacting mixture or the color of an indicator that has been added to it. Titration enables chemists to determine the amount of a substance present in a sample. A titration can be used, for example, to determine the amount (or concentration) of $Fe^{2+}$ ions in a solution. The reaction of potassium permanganate in acidic solution with $Fe^{2+}$ is quantitative:

$$MnO_4^-(aq) + 5\,Fe^{2+}(aq) + 8\,H^+(aq) \longrightarrow Mn^{2+}(aq) + 5\,Fe^{3+}(aq) + 4\,H_2O(\ell)$$

A solution of potassium permanganate of accurately known concentration is prepared and placed in a buret—a glass tube with a delivery valve (a stopcock) and markings for measurement of solution volumes. The volumes of solution dispensed from it can be read very accurately.

The titration is begun by opening the stopcock of the buret and letting a small volume of permanganate solution run into the flask that contains the $Fe^{2+}$ solution. A dash of purple colors the solution (Fig. 4-13a) but rapidly disappears as the permanganate ion reacts with $Fe^{2+}$ ions to give the nearly colorless $Mn^{2+}$ and $Fe^{3+}$ products. The addition of incremental volumes of permanganate solution is continued until the $Fe^{2+}$ is almost completely converted to $Fe^{3+}$. At this stage, the addition of just one drop of $KMnO_4$ imparts a pale purple color to the reaction mixture (Fig. 4-13b) and signals the completion of the reaction. The point at which the first permanent color change occurs is called the **endpoint** of the titration, and the volume of the titrant $KMnO_4$ solution is calculated by subtracting the initial reading of the solution meniscus in the buret from the final volume reading.

Suppose that the permanganate solution has a concentration of 0.09625 M, and 26.34 mL (0.02634 L) of it is added to reach the endpoint. The chemical amount of $Fe^{2+}$ in the original solution is calculated in two steps:

$$\text{amount of } MnO_4^- \text{ reacting} = 0.02634 \text{ L} \times 0.09625 \text{ mol L}^{-1}$$

$$= 2.535 \times 10^{-3} \text{ mol } MnO_4^-$$

From the balanced chemical equation, each mole of $MnO_4^-$ used causes the oxidation of 5 mol of $Fe^{2+}$, so

$$\text{amount of } Fe^{2+} \text{ reacting} = 2.535 \times 10^{-3} \text{ mol } MnO_4^- \times \left(\frac{5 \text{ mol } Fe^{2+}}{1 \text{ mol } MnO_4^-}\right)$$

$$= 1.268 \times 10^{-2} \text{ mol } Fe^{2+}$$

Many analytical procedures are indirect and involve additional preliminary reactions of the sample before the titration can be carried out. For example, a soluble calcium salt will not take part in a redox reaction with potassium permanganate. However, adding ammonium oxalate to the solution containing $Ca^{2+}$ causes the quantitative precipitation of calcium oxalate:

$$Ca^{2+}(aq) + C_2O_4^{2-}(aq) \longrightarrow CaC_2O_4(s)$$

After the precipitate is filtered and washed, it is dissolved in sulfuric acid to form oxalic acid:

$$CaC_2O_4(s) + 2\,H^+(aq) \longrightarrow Ca^{2+}(aq) + H_2C_2O_4(aq)$$

Finally, the oxalic acid is titrated with permanganate solution of accurately known concentration, taking advantage of the redox reaction

$$2\,MnO_4^-(aq) + 5\,H_2C_2O_4(aq) + 6\,H^+(aq) \longrightarrow$$

$$2\,Mn^{2+}(aq) + 10\,CO_2(g) + 8\,H_2O(\ell)$$

In this way, the quantity of calcium can be determined indirectly by reactions that involve precipitation, acid–base, and redox steps.

Oxalic acid, $H_2C_2O_4$, is an acid that can give up two hydrogen ions in aqueous solution to form the oxalate ion, $C_2O_4^{2-}$.

## Acid–Base Titration

The permanganate titration has the special advantage that a striking color change takes place at the endpoint, when $MnO_4^-$ has been added in very slight excess. In most reactions, however, no such sharp distinction in color is observed at the endpoint, and in such cases it is necessary to add a small amount of an **indicator,** a dye that changes color when the reaction is complete (see Section 6-2). This is the case with most acid–base titrations. Phenolphthalein is such an indicator, undergoing a change from colorless to pink when a solution changes from acidic to basic. The concentration of acetic acid in an aqueous solution can be determined by adding a few drops of a phenolphthalein solution and then titrating it with a solution of sodium hydroxide of accurately known concentration. At the first permanent appearance of a pink color, the stopcock of the buret is closed. At this point the reaction

$$CH_3COOH(aq) + OH^-(aq) \longrightarrow CH_3COO^-(aq) + H_2O(\ell)$$

has gone stoichiometrically to completion.

## Example 4-7

A sample of vinegar is to be analyzed for its acetic acid content. A volume of 50.0 mL is measured out and titrated with a solution of 1.306 M NaOH; 31.66 mL of that titrant is required to reach the phenolphthalein endpoint. Calculate the concentration of acetic acid in the vinegar (in moles per liter).

## Solution

The chemical amount of NaOH reacting is found by multiplying the volume of NaOH solution (31.66 mL = 0.03166 L) by its concentration (1.306 M):

$$0.03166\,L \times 1.306\,mol\,L^{-1} = 4.135 \times 10^{-2}\,mol\,NaOH$$

Because 1 mol of acetic acid reacts with 1 mol of $OH^-(aq)$, the chemical amount of acetic acid originally present must also have been $4.135 \times 10^{-2}$ mol. Its concentration was then

$$[CH_3COOH] = \frac{4.135 \times 10^{-2} \text{ mol}}{0.0500 \text{ L}} = 0.827 \text{ M}$$

*Related Problems:* 41, 42

## 4-5  COLLIGATIVE PROPERTIES OF SOLUTIONS

We return now from the chemical reactions of solutions to such physical properties as their vapor pressures and phase diagrams. Consider first a solution made by dissolving a nonvolatile solute in a solvent. By ''nonvolatile'' we mean that the vapor pressure of the solute above the solution is negligible. An example is a solution of sucrose (cane sugar) in water, in which the vapor pressure of sucrose above the solution is zero. The case of a volatile solute is taken up in Section 4-6.

The *solvent* vapor pressure is not zero and changes with the composition of the solution at a fixed temperature. If the mole fraction of solvent ($X_1$) is 1, then the vapor pressure is $P_1^\circ$, the vapor pressure of pure solvent at the temperature of the experiment. When $X_1$ approaches zero (giving pure solute), the vapor pressure $P_1$ of the solvent must go to zero also, because solvent is no longer present. As the mole fraction $X_1$ changes from 1 to 0, $P_1$ drops from $P_1^\circ$ to 0. What is the shape of the curve?

The French chemist François-Marie Raoult found that for some solutions a plot of solvent vapor pressure against solvent mole fraction can be fitted closely by a straight line (see Fig. 4-14). Solutions that conform to this straight-line relationship obey the simple equation

$$P_1 = X_1 P_1^\circ \tag{4-2}$$

which is known as **Raoult's law.** Such solutions are called **ideal solutions.** Other solutions deviate from straight-line behavior and are called **nonideal solutions.**

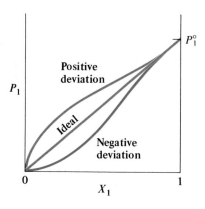

**Figure 4-14**

In an ideal solution, a graph of solvent vapor pressure $P_1$ versus mole fraction of solvent, $X_1$, is a straight line. Nonideal solutions behave differently; examples of positive and negative deviations from the ideal solution are shown. The vapor pressure of pure solvent is $P_1^\circ$.

They may show positive deviations (with vapor pressures higher than those predicted by Raoult's law) or negative deviations (with lower vapor pressures). On a molecular level, negative deviations arise when the solute attracts solvent molecules especially strongly, reducing their tendency to escape into the vapor phase. Positive deviations arise in the opposite case, when solvent and solute molecules are not strongly attracted to each other. Even nonideal solutions with nondissociating solutes approach Raoult's law as $X_1$ approaches 1, however, just as all real gases obey the ideal gas law at sufficiently low densities.

Raoult's law forms the basis for four properties of dilute solutions that are called **colligative properties** (from Latin *colligare*, "to collect together") because they depend on the collective effect of the *number* of dissolved particles rather than on the *nature* of the particular particles involved. These four properties are

1. The lowering of the vapor pressure of a solution relative to pure solvent
2. The elevation of the boiling point
3. The depression of the freezing point
4. The phenomenon of osmotic pressure

## Vapor-Pressure Lowering

Because $X_1 = 1 - X_2$ for a two-component solution, Raoult's law can be rewritten as

$$\Delta P_1 = P_1 - P_1^\circ = X_1 P_1^\circ - P_1^\circ = -X_2 P_1^\circ \qquad \text{(4-3)}$$

so that the *change* in vapor pressure of the solvent (indicated by the capital Greek delta, $\Delta$) is proportional to the mole fraction of solute. The negative sign implies **vapor-pressure lowering;** the vapor pressure is always lower above a dilute solution than it is above the pure solvent.

## Example 4-8

At 25°C the vapor pressure of pure benzene is $P_1^\circ = 0.1252$ atm. Suppose 6.40 g of naphthalene, $C_{10}H_8$ (molar mass 128.17 g mol$^{-1}$), is dissolved in 78.0 g of benzene (molar mass 78.0 g mol$^{-1}$). Calculate the vapor pressure of benzene over the solution, assuming ideal behavior.

### Solution

The chemical amount of solvent, $n_1$, is 1.00 mol in this case (because 78.0 g = 1.00 mol benzene has been used). The chemical amount of solute is $n_2 = 6.40$ g/128.17 g mol$^{-1}$ = 0.0499 mol $C_{10}H_8$, so the mole fraction $X_1$ is

$$X_1 = \frac{n_1}{n_1 + n_2} = \frac{1.00 \text{ mol}}{1.00 + 0.0499 \text{ mol}} = 0.9525$$

From Raoult's law, the vapor pressure of benzene above the solution is

$$P_1 = X_1 P_1^\circ = 0.9525 \times 0.1252 \text{ atm} = 0.1193 \text{ atm}$$

*Related Problems:* 43, 44

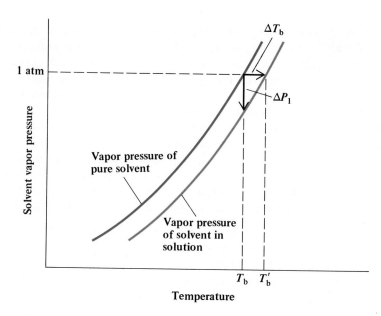

**Figure 4-15**

The vapor pressure of the solvent above a dilute solution is lower than that of the pure solvent at all temperatures. As a result, in order for the solution to boil (i.e., for the vapor pressure to reach 1 atm), a higher temperature is required for the solution than for the pure solvent. This amounts to an elevation of the boiling point.

### Boiling-Point Elevation

The normal boiling point of a pure liquid or a solution is the temperature at which the vapor pressure reaches 1 atm. Because a dissolved solute reduces the vapor pressure, the temperature of the solution must be increased to make it boil. That is, the boiling point of a solution is higher than that of the pure solvent. This phenomenon, referred to as **boiling-point elevation,** provides an alternative method for determining molar masses.

The vapor-pressure curve of a dilute solution lies slightly below that for the pure solvent. In Figure 4-15, $\Delta P_1$ is the decrease of vapor pressure at $T_b$, and $\Delta T_b$ is the change in temperature necessary to hold the vapor pressure at 1 atm (in other words, $\Delta T_b = T_b' - T_b$ is the increase in boiling point caused by addition of solute to the pure solvent). For small concentrations of nondissociating solutes, the two curves are parallel, so

$$-\frac{\Delta P_1}{\Delta T_b} = \text{slope of curve} = S$$

$$\Delta T_b = -\frac{\Delta P_1}{S} = \frac{X_2 P_1^\circ}{S}$$

$$= \frac{1}{S}\left(\frac{n_2}{n_1 + n_2}\right) \quad \text{(from Raoult's law, with } P_1^\circ = 1 \text{ atm)}$$

The constant $S$ is a property of the pure solvent only, because it is the slope of the

**Table 4-2  Boiling-Point Elevation and Freezing-Point Depression Constants**

| Solvent | Formula | $T_b$ (°C) | $K_b$ (K kg mol$^{-1}$) | $T_f$ (°C) | $K_f$ (K kg mol$^{-1}$) |
|---|---|---|---|---|---|
| Acetic acid | $CH_3COOH$ | 118.1 | 3.07 | 17 | 3.9 |
| Benzene | $C_6H_6$ | 80.1 | 2.53 | 5.5 | 4.9 |
| Carbon tetrachloride | $CCl_4$ | 76.7 | 5.03 | −22.9 | 32 |
| Diethyl ether | $C_4H_{10}O$ | 34.7 | 2.02 | −116.2 | 1.8 |
| Ethanol | $C_2H_5OH$ | 78.4 | 1.22 | −114.7 | 1.9 |
| Naphthalene | $C_{10}H_8$ | — | — | 80.5 | 6.8 |
| Water | $H_2O$ | 100.0 | 0.512 | 0.0 | 1.86 |

vapor pressure curve $-\Delta P_1/\Delta T_b$ near 1 atm pressure. In other words, $S$ is independent of the solute species involved.

For very dilute solutions, $n_1 \gg n_2$, and this may be simplified to

$$\Delta T_b = \frac{1}{S}\frac{n_2}{n_1} = \frac{1}{S}\left(\frac{m_2/\mathcal{M}_2}{m_1/\mathcal{M}_1}\right)$$

where $m_1$ and $m_2$ are the masses of solvent and solute (in grams) and $\mathcal{M}_1$ and $\mathcal{M}_2$ are their molar masses in grams per mole. Because $\mathcal{M}_1$, like $S$, is a property of the solvent only, it is convenient to combine the two and define a new constant $K_b$ through

$$K_b = \frac{\mathcal{M}_1}{(1000 \text{ g kg}^{-1})S}$$

Then

$$\Delta T_b = K_b\left(\frac{m_2/\mathcal{M}_2}{m_1/(1000 \text{ g kg}^{-1})}\right)$$

Because $m_1$ is measured in grams, $m_1/(1000 \text{ g kg}^{-1})$ is the number of kilograms of solvent; also, $m_2/\mathcal{M}_2$ is the number of moles of solute. The expression in parentheses is therefore the molality ($m$) of the solution.

$$\Delta T_b = K_b m \qquad\qquad \textbf{(4-4)}$$

For a given solvent, $K_b$ is obtained by measuring the boiling-point elevations for dilute solutions of known molality (i.e., containing a known amount of solute with known molar mass). Table 4-2 gives values of $K_b$ for a number of solvents. Once $K_b$ has been found, it can be used either to predict boiling-point elevations for solutes of known molar mass or to determine molar masses from measured boiling-point elevations, as illustrated in part (b) of the following example.

## Example 4-9

(a) When 5.50 g of biphenyl ($C_{12}H_{10}$) is dissolved in 100.0 g of benzene, the boiling point increases by 0.903°C. Calculate $K_b$ for benzene.
(b) When 6.30 g of an unknown hydrocarbon is dissolved in 150.0 g of benzene,

the boiling point of the solution increases by 0.597°C. What is the molar mass of the unknown substance?

## Solution

(a) Because the molar mass of biphenyl is $154.2 \text{ g mol}^{-1}$, 5.50 g contains $5.50 \text{ g}/154.2 \text{ g mol}^{-1} = 0.0357 \text{ mol}$. The molality $m$ is

$$m = \frac{\text{mol solvent}}{\text{kg solvent}} = \frac{0.0357 \text{ mol}}{0.1000 \text{ kg}} = 0.357 \text{ mol kg}^{-1}$$

$$K_b = \frac{\Delta T_b}{m} = \frac{0.903 \text{ K}}{0.357 \text{ mol kg}^{-1}} = 2.53 \text{ K kg mol}^{-1} \qquad \text{for benzene}$$

(b) Solving $\Delta T_b = K_b m$ for $m$ gives

$$m = \frac{\Delta T_b}{K_b} = \frac{0.597 \text{ K}}{2.53 \text{ K kg mol}^{-1}} = 0.236 \text{ mol kg}^{-1}$$

The chemical amount of solute is the product of the molality of the solution and the mass of the solvent, $m_1$:

$$n_2 = (0.236 \text{ mol kg}^{-1}) \times (0.1500 \text{ kg}) = 0.0354 \text{ mol}$$

Finally, the molar mass of the solute is its mass divided by its chemical amount:

$$\text{molar mass of solute} = \mathcal{M}_2 = \frac{m_2}{n_2} = \frac{6.30 \text{ g}}{0.0354 \text{ mol}} = 178 \text{ g mol}^{-1}$$

The unknown hydrocarbon might be anthracene ($C_{14}H_{10}$), which has a molar mass of $178.24 \text{ g mol}^{-1}$.

*Related Problems:* 45, 46, 47, 48

So far, only nondissociating solutes have been considered. Colligative properties depend on the total number of moles per liter of dissolved species present. If a solute dissociates (as sodium chloride dissolves to furnish $Na^+$ and $Cl^-$ ions in aqueous solution), then the molality $m$ to be used is the *total* molality. One mole of NaCl dissolves to furnish 2 mol of ions, so the total molality and the boiling-point elevation are twice as large as they would be if NaCl molecules were present in solution. One mole of $Ca(NO_3)_2$ dissolves to furnish 3 mol of ions (1 mol of $Ca^{2+}$ ions and 2 mol of $NO_3^-$ ions), giving three times the boiling-point elevation. The corresponding vapor-pressure lowering is greater, as well. Ions are not exactly like neutral molecules in solution, however, and nonideal behavior appears at lower concentrations in solutions that contain ions.

## Example 4-10

Lanthanum (III) chloride ($LaCl_3$) is a salt that is completely dissociated into ions in dilute aqueous solution,

$$LaCl_3(s) \longrightarrow La^{3+}(aq) + 3 \, Cl^-(aq)$$

yielding 4 mol of ions per mole of $LaCl_3$. Suppose 0.2453 g of $LaCl_3$ is dissolved in 10.00 g of $H_2O$. What is the boiling point of the solution at atmospheric pressure, assuming ideal solution behavior?

### Solution

The molar mass of $LaCl_3$ is 245.3 g mol$^{-1}$.

$$\text{moles of } LaCl_3 = \frac{0.2453 \text{ g}}{245.3 \text{ g mol}^{-1}}$$

$$= 1.000 \times 10^{-3} \text{ mol}$$

$$\text{total molality} = m = \frac{(4)(1.000 \times 10^{-3}) \text{ mol of ions}}{0.0100 \text{ kg solvent}}$$

$$= 0.400 \text{ mol kg}^{-1}$$

This is inserted into the formula for the boiling-point elevation:

$$\Delta T_b = K_b m = (0.512 \text{ K kg mol}^{-1})(0.400 \text{ mol kg}^{-1}) = 0.205 \text{ K}$$

$$T_b = 100.205°C$$

The actual boiling point is slightly lower than this because the solution is non-ideal.

### Freezing-Point Depression

The phenomenon of **freezing-point depression** is analogous to that of boiling-point elevation. Here we consider only cases in which the first solid that crystallizes from solution is the pure solvent. If solute crystallizes out with solvent, the situation is more complicated.

Pure solid solvent coexists at equilibrium with a certain vapor pressure of solvent, determined by the temperature. Solvent in solution likewise coexists with a certain vapor pressure of solvent. If solid solvent and the solvent in solution are to coexist, they must have the *same* vapor pressure. This means that the freezing temperature of a solution can be identified as the temperature at which the vapor-pressure curve of the pure solid solvent intersects that of the solution (see Fig. 4-16). As solute is added to the solution, the vapor pressure of the solvent falls and the freezing point, the temperature at which the first crystals of pure solvent begin to appear, drops. The difference $\Delta T_f = T_f' - T_f$ is therefore negative, and a freezing-point depression is observed.

The change in temperature $\Delta T_f$ is once again proportional to the change in vapor pressure $\Delta P_1$. For small enough concentrations of solute, the freezing-point depression is related to the total molality $m$ (by analogy with the case of boiling-point elevation) through

$$\Delta T_f = T_f' - T_f = -K_f m \qquad \textbf{(4-5)}$$

where $K_f$ is a positive constant that depends only on the properties of the solvent (Table 4-2). The phenomenon of freezing-point depression is responsible for the fact

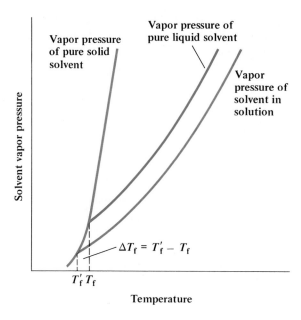

**Figure 4-16**

The vapor pressure of solvent above a dilute solution, compared with that above pure liquid and solid solvent. The depression of the freezing point from $T_f$ to $T_f'$ is shown.

that seawater, containing dissolved salts, has a slightly lower freezing point than fresh water. Concentrated salt solutions have still lower freezing points. Salt spread on an icy road reduces the freezing point of the ice, and the ice melts.

Measurements of the drop in the freezing point, like those of elevation of the boiling point, can be used to determine molar masses of unknown substances. If a substance dissociates in solution, the *total* molality of all species present (ionic or neutral) must be used in the calculation.

### Example 4-11

The chemical amounts of the major dissolved species in a 1.000-L sample of seawater follow. Estimate the freezing point of the seawater, assuming $K_f = 1.86$ K kg mol$^{-1}$ for water.

| | | | |
|---|---|---|---|
| $Na^+$ | 0.458 mol | $Cl^-$ | 0.533 mol |
| $Mg^{2+}$ | 0.052 mol | $SO_4^{2-}$ | 0.028 mol |
| $Ca^{2+}$ | 0.010 mol | $HCO_3^-$ | 0.002 mol |
| $K^+$ | 0.010 mol | $Br^-$ | 0.001 mol |
| Neutral species | 0.001 mol | | |

#### Solution

Because water has a density of 1.00 g cm$^{-3}$, 1.00 L of water weighs 1.00 kg. For dilute *aqueous* solutions, the number of moles per kilogram of solvent (the molality $m$) is therefore approximately equal to the number of moles per liter. The total molality, obtained by adding the individual species molalities just given, is $m = 1.095$ mol kg$^{-1}$. Then

$$\Delta T = -K_f m = -(1.86 \text{ K kg mol}^{-1})(1.095 \text{ mol kg}^{-1}) = -2.04 \text{ K}$$

The seawater should freeze at approximately $-2°C$. Nonideal solution effects make the actual freezing point slightly higher than this.

*Related Problems:* 49, 50

Both freezing-point depression and boiling-point elevation can be used to determine whether a species of known molar mass dissociates in solution (see Fig. 4-17), as the following example shows.

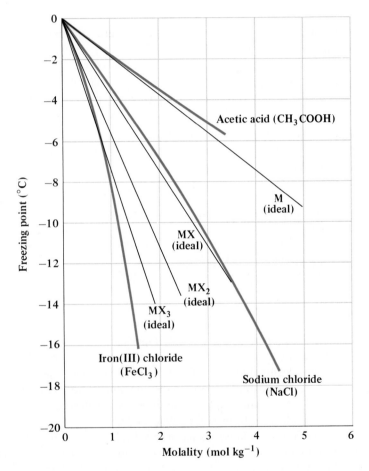

**Figure 4-17**

The heavy colored lines give the observed depression of the freezing point of water by acetic acid, NaCl, and FeCl$_3$ as the molality of the solutions increases. Straight black lines sketch the predicted ideal behavior for one through four moles of particles per mole in solution. The experimental curve for NaCl (which gives *two* moles of dissolved particles) stays close to the ideal straight line for MX; the experimental curve for FeCl$_3$ (which gives *four* moles of dissolved particles) stays fairly close to the ideal straight line for MX$_3$. The pattern suggests that acetic acid dissolves to give one mole of particles per mole of solute. As the molalities of the solutions increase, the observed freezing-point depressions deviate in varying ways from the straight lines.

## Example 4-12

When 0.494 g of $K_3Fe(CN)_6$ is dissolved in 100.0 g of water, the freezing point is found to be $-0.093°C$. How many ions are present for each formula unit of $K_3Fe(CN)_6$ dissolved?

### Solution

The total molality of all species in solution is

$$m = \frac{-\Delta T_f}{K_f} = \frac{0.093 \text{ K}}{1.86 \text{ K kg mol}^{-1}} = 0.050 \text{ mol kg}^{-1}$$

Because the molar mass of $K_3Fe(CN)_6$ is 329.25 g mol$^{-1}$, the total molality if *no* dissociation had taken place would be

$$\frac{\left(\dfrac{0.494 \text{ g}}{329.25 \text{ g mol}^{-1}}\right)}{0.100 \text{ kg}} = 0.0150 \text{ mol kg}^{-1}$$

This is between one fourth and one third of the measured total molality in solution, so each $K_3Fe(CN)_6$ must dissociate into three to four ions. In fact, the dissociation that takes place is

$$K_3Fe(CN)_6(s) \longrightarrow 3 \text{ K}^+(aq) + [Fe(CN)_6]^{3-}(aq)$$

Deviations from ideal solution behavior have reduced the effective total molality from 0.060 to 0.050 mol kg$^{-1}$.

*Related Problems:* 53, 54

## Osmotic Pressure

The fourth colligative property is particularly important in cellular biology, because it plays a vital role in the transport of molecules across cell membranes. Such membranes are **semipermeable,** allowing small molecules (such as water) to pass through but blocking the passage of large molecules such as proteins and carbohydrates. A semipermeable membrane (e.g., common cellophane) can be used to separate small solvent molecules from large solute molecules.

Suppose a solution is contained in an inverted tube, the lower end of which is covered by a semipermeable membrane. This solution has a solute concentration of $c$ moles per liter. When the end of the tube is inserted in a beaker of pure solvent (Fig. 4-18), solvent flows from the beaker into the tube. The volume of the solution increases, and the solvent rises in the tube until, at equilibrium, it reaches a height $h$ above the solvent in the beaker. The pressure on the solution side of the membrane is greater than the atmospheric pressure on the surface of the pure solvent by an amount given by the **osmotic pressure, $\pi$:**

$$\pi = \rho g h \tag{4-6}$$

where $\rho$ is the density of the solution (1.00 g cm$^{-3}$ for a dilute aqueous solution) and $g$ the acceleration due to gravity (9.807 m s$^{-2}$).

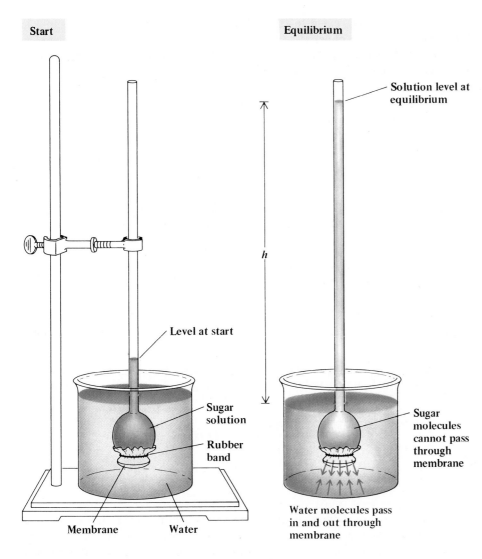

**Start**

Level at start

Sugar
solution

Rubber
band

Membrane          Water

**Equilibrium**

Solution level at
equilibrium

$h$

Sugar
molecules
cannot pass
through
membrane

Water molecules pass
in and out through
membrane

**Figure 4-18**

In this device to measure osmotic
pressure, the semipermeable membrane allows solvent, but not solute,
molecules to pass through. This results in a net flow of solvent into the
tube until equilibrium is achieved,
with the level of solution at a height
$h$ above the solvent in the beaker.
Once this happens, the solvent molecules pass through the membrane at
the same rate in both directions.

For example, a height $h$ of 0.17 m corresponds to an osmotic pressure for a
dilute aqueous solution of

$$\pi = [(1.00 \text{ g cm}^{-3})(10^{-3} \text{ kg g}^{-1})(10^{6} \text{ cm}^3 \text{ m}^{-3})](9.807 \text{ m s}^{-2})(0.17 \text{ m})$$

$$= 1.7 \times 10^3 \text{ kg m}^{-1} \text{ s}^{-2} = 1.7 \times 10^3 \text{ Pa}$$

$$\pi \text{ (atm)} = \frac{1.7 \times 10^3 \text{ Pa}}{1.013 \times 10^5 \text{ Pa atm}^{-1}} = 0.016 \text{ atm}$$

This illustrates how accurately very small osmotic pressures can be measured.

In 1887 Jacobus van't Hoff discovered an important relationship among osmotic pressure $\pi$, concentration $c$, and absolute temperature $T$:

$$\pi = cRT \qquad \qquad \textbf{(4-7)}$$

$R$ is the gas constant, equal to 0.08206 L atm mol$^{-1}$ K$^{-1}$ if $\pi$ is expressed in atmospheres and $c$ in moles per liter. Because $c = n/V$, where $n$ is the chemical amount of solute and $V$ is the volume of the solution, van't Hoff's equation can be rewritten as

$$\pi V = nRT$$

which bears a striking similarity to the ideal gas law. With this relationship the molar mass of a dissolved substance can be determined from the osmotic pressure of its solution.

### Example 4-13

A chemist dissolves 2.00 g of a protein in 0.100 L water. The osmotic pressure is 0.021 atm at 25°C. What is the approximate molar mass of the protein?

### Solution

The concentration in moles per liter is

$$c = \frac{\pi}{RT} = \frac{0.021 \text{ atm}}{(0.08206 \text{ L atm mol}^{-1} \text{ K}^{-1})(298 \text{ K})}$$

$$= 8.6 \times 10^{-4} \text{ mol L}^{-1}$$

Now 2.00 g dissolved in 0.100 L gives the same concentration as 20.0 g in 1.00 L. Therefore, $8.6 \times 10^{-4}$ mol of protein must weigh 20.0 g, and the molar mass is

$$\mathcal{M} = \frac{20.0 \text{ g}}{8.6 \times 10^{-4} \text{ mol}} = 23,000 \text{ g mol}^{-1}$$

*Related Problems:* 55, 56

When a carrot is immersed in salt water (*left*), the osmotic pressure outside the cells of the vegetable is higher than that inside. Water flows out into the solution, causing the carrot to shrink. A carrot left in pure water (*right*) does not shrivel.

Osmotic pressure is particularly useful for measuring molar masses of large molecules such as proteins, whose solubilities may be low. In the case given in Example 4-13, for instance, the height difference $h$ is 22 cm, an easily measured quantity. By contrast, the other three colligative properties in this example would show very small effects:

vapor-pressure lowering $= 4.8 \times 10^{-7}$ atm

boiling-point elevation $= 0.00044$ K

freezing-point depression $= 0.0016$ K

All of these changes are too small for accurate measurement. As with the other techniques, the *total* number of moles of solute species determines the osmotic pressure if dissociation occurs.

Osmosis has other important uses. In some parts of the world potable water is a precious commodity. It can be obtained much more economically by desalinizing brackish waters, through a process called **reverse osmosis,** than by distillation. When an ionic solution in contact with a semipermeable membrane has a pressure applied to it that exceeds its osmotic pressure, water of quite high purity passes through. Reverse osmosis is also used to control water pollution.

## 4-6 MIXTURES AND DISTILLATION

The preceding section described the properties of solutions of nonvolatile solutes in liquid solvents. The concept of an ideal solution can be extended to mixtures of two or more components, each of which is volatile. An ideal solution is one in which the vapor pressure of *each* species present is proportional to its mole fraction in solution over the whole range of mole fraction:

$$P_i = X_i P_i^\circ$$

where $P_i^\circ$ is the vapor pressure (at a given temperature) of pure substance $i$, $X_i$ is its mole fraction in solution, and $P_i$ is its partial vapor pressure over the solution. This is a generalization of Raoult's law to each component of a solution.

For an ideal mixture of two volatile substances, the vapor pressure of component 1 is

$$P_1 = X_1 P_1^\circ$$

and that of component 2 is

$$P_2 = X_2 P_2^\circ = (1 - X_1)P_2^\circ$$

The vapor pressures for such an ideal solution are shown in Figure 4-19, together with typical vapor pressures for a nonideal solution.

### Henry's Law

At sufficiently low mole fraction $X_2$, the vapor pressure of component 2 (even in a nonideal solution) is proportional to $X_2$:

$$P_2 = k_2 X_2 \qquad\qquad \textbf{(4-8)}$$

where $k_2$ is a constant. For $X_1$ small ($X_2$ near 1),

$$P_1 = k_1 X_1 = k_1(1 - X_2)$$

This linear vapor pressure of what is called the solute (because it is present at small mole fraction) is known as **Henry's law:** the vapor pressure of a volatile dissolved substance is proportional to the mole fraction of that substance in solution. When-

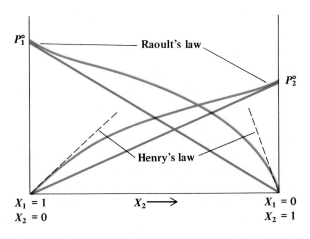

$X_1 = 1$      $X_2 \longrightarrow$      $X_1 = 0$
$X_2 = 0$                              $X_2 = 1$

### Figure 4-19

Vapor pressures above a mixture of two volatile liquids. Both ideal behavior (blue lines) and nonideal behavior (red curves) are shown. Positive deviations from ideal solution behavior are illustrated here, although negative deviations are observed for other nonideal solutions. Raoult's and Henry's laws are shown as dilute solution limits for the nonideal mixture.

ever Raoult's law is valid for a solvent, Henry's law is valid for the solute (Fig. 4-19).

One familiar application of Henry's law is in the carbonation of beverages. If the partial pressure of $CO_2$ above a solution is increased, the amount dissolved in the solution increases proportionately. When the beverage can is opened, dissolved gas bubbles out of solution in response to the lower $CO_2$ pressure outside. Henry's law is important in biology (gases such as oxygen dissolve in blood and other bodily fluids) and in environmental chemistry (volatile pollutants can move between bodies of water and the atmosphere).

### Example 4-14

The Henry's law constant for oxygen dissolved in water is $4.34 \times 10^4$ atm at 25°C. If the partial pressure of oxygen in air is 0.20 atm under ordinary atmospheric conditions, calculate the concentration (in moles per liter) of dissolved oxygen in water that is in equilibrium with air at 25°C.

### Solution

Henry's law is used to calculate the mole fraction of oxygen in water:

$$X_{O_2} = \frac{P_{O_2}}{k_{O_2}} = \frac{0.20 \text{ atm}}{4.34 \times 10^4 \text{ atm}} = 4.6 \times 10^{-6}$$

Next, the mole fraction is converted to molarity. One liter of water weighs 1000 g, so it contains

$$\frac{1000 \text{ g H}_2\text{O}}{18.02 \text{ g mol}^{-1}} = 55.5 \text{ mol water}$$

Because $X_{O_2}$ is so small, $n_{H_2O} + n_{O_2}$ is very close to $n_{H_2O}$, and it can be written

$$X_{O_2} = \frac{n_{O_2}}{n_{H_2O} + n_{O_2}} \approx \frac{n_{O_2}}{n_{H_2O}}$$

$$4.6 \times 10^{-6} = \frac{n_{O_2}}{55.5 \text{ mol}}$$

Thus, the chemical amount of oxygen in one liter of water is

$$n_{O_2} = (4.6 \times 10^{-6})(55.5 \text{ mol}) = 2.6 \times 10^{-4} \text{ mol}$$

and the concentration of dissolved $O_2$ is $2.6 \times 10^{-4}$ M.

*Related Problems:* 59, 60

### Distillation

The vapor pressures of the pure components of an ideal solution usually differ, and for this reason such a solution has a composition different from that of the vapor phase with which it is in equilibrium. This can best be seen in an example.

Hexane ($C_6H_{14}$) and heptane ($C_7H_{16}$) form a nearly ideal solution over the whole range of mole fractions. At 25°C, the vapor pressure of pure hexane is $P_1^\circ =$

0.198 atm, and that of pure heptane is $P_2^\circ = 0.0600$ atm. Suppose a solution contains 4.00 mol of hexane and 6.00 mol of heptane, so that its mole fractions are $X_1 = 0.400$ and $X_2 = 0.600$. The vapor in equilibrium with this ideal solution has partial pressures

$$P_{\text{hexane}} = P_1 = X_1 P_1^\circ = (0.400)(0.198 \text{ atm}) = 0.0792 \text{ atm}$$

$$P_{\text{heptane}} = P_2 = X_2 P_2^\circ = (0.600)(0.0600 \text{ atm}) = 0.0360 \text{ atm}$$

From Dalton's law, the total pressure is the sum of these partial pressures:

$$P_{\text{total}} = P_1 + P_2 = 0.1152 \text{ atm}$$

If $X_1'$ and $X_2'$ are the mole fractions in the vapor, then

$$X_1' = \frac{0.0792 \text{ atm}}{0.1152 \text{ atm}} = 0.688$$

$$X_2' = 1 - X_1' = \frac{0.0360 \text{ atm}}{0.1152 \text{ atm}} = 0.312$$

The liquid and the vapor with which it is in equilibrium have different compositions (Fig. 4-20), and the vapor is enriched in the more volatile component.

Suppose some of this vapor were removed and condensed to a liquid. The vapor in equilibrium with this new solution would be still richer in the more volatile component, and the process could be continued further (Fig. 4-20). This progression underlies the technique of separating a mixture into its pure components by **fractional distillation,** a process in which the components are successively evaporated and recondensed. What has been described so far corresponds to a constant-temperature process, but actual distillation is carried out at constant total pressure. The vapor pressure–mole fraction plot is transformed into a boiling temperature–mole fraction plot, shown in Figure 4-21. Note that the component with the lower vapor pressure (2) has the higher boiling point, $T_b^2$. If the temperature of a solution of a certain composition is raised until it touches the liquid line in the plot, the vapor in equilibrium with the solution is richer in the more volatile component 1. Its composition lies at the intersection of the horizontal constant-temperature line and the equilibrium vapor curve.

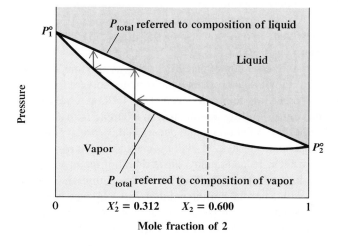

**Figure 4-20**

The composition of the vapor above a solution differs from the composition of the liquid with which it is in equilibrium. Here the upper (straight) line is the total pressure of the vapor in equilibrium with an ideal solution having mole fraction $X_2$ of component 2. By moving horizontally from that line to a point of intersection with the lower curve, we can locate the mole fraction $X_2'$ of component 2 in the vapor (red arrow). Subsequent condensations and vaporizations are shown by blue arrows.

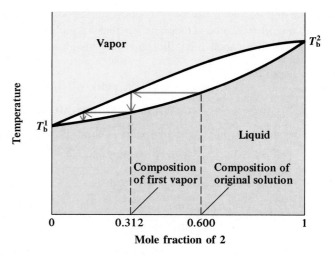

**Figure 4-21**

The boiling point of an ideal solution varies with the composition of the solution. The upper curve is the boiling temperature referred to the vapor composition, and the lower curve is the boiling temperature referred to the liquid composition. The vapors boiling off of a solution that has a 0.600 mole fraction of component 2 are enriched in the more volatile component 1 to the extent that their mole fraction of component 2 is only 0.312 (red arrow). The subsequent blue arrows show the further steps used in obtaining nearly pure component 1 by fractional distillation.

A liquid can be vaporized in different ways. It can simply be boiled until it is entirely vaporized and the final composition of the vapor is the same as that of the original liquid. It is clear that such a mixture boils over a range of temperatures, rather than at a single $T_b$ like a pure liquid. Alternatively, if the boiling is stopped midway, the vapor fraction that has boiled off can be collected and recondensed. The resulting liquid (i.e., the condensate) will be richer in component 1 than the original solution was. By repeating the process again and again, mixtures successively richer in 1 will be obtained. This is the principle behind the **distillation column** (Fig. 4-22). Throughout the length of the tube, such evaporations and recondensations take place, and this allows mixtures to be separated into their constituent substances. Such a process is used to separate nitrogen and oxygen in air (Fig. 4-23); the air is liquefied and then distilled, with the lower-boiling nitrogen ($T_b = -196°C$) vaporizing before the oxygen ($T_b = -183°C$).

Nonideal solutions may have more complicated behavior. A mixture showing large *negative* deviations from Raoult's law (one in which solute–solvent forces are strongly attractive) will show a boiling-point *maximum* (Fig. 4-24a). A solution at the maximum is called a **maximum-boiling azeotrope;** an example is that formed by the $H_2O$/HCl system. The boiling-point maximum occurs in this case at 108.58°C and 1 atm pressure for a composition of 20.22% HCl by mass. A mixture showing large *positive* deviations from ideal behavior may show a boiling-point *minimum* (Fig. 4-24b) and a corresponding **minimum-boiling azeotrope.** Ethyl alcohol and water form such an azeotrope with a normal boiling point of 78.17°C and a composition of 4% water by mass. In this case, attractive forces between ethyl alcohol molecules and between water molecules are stronger than those between ethyl alco-

**Figure 4-22**

In a distillation column, temperature decreases with height in the column. The less volatile components condense and fall back to the flask, but the more volatile ones continue up the column into the water-cooled condenser, where they condense and are recovered in the receiver.

**Figure 4-23**

This Pennsylvania plant produces more than 1000 tons per day of gases from air (nitrogen, oxygen, and argon). Atmospheric air is liquefied and separated by distillation in these towers. (Praxair, Inc.)

(a) Constant boiling maximum

(b) Constant boiling minimum

**Figure 4-24**

Dependence of boiling temperature on mole fraction for maximum- and minimum-boiling azeotropes. The coordinates are not to scale.

**Figure 4-25**
A natural opal. (Julius Weber)

hol and water, so the solution boils at a lower temperature than either pure component. An azeotrope behaves like a single-component fluid in that it boils at a well-defined temperature and the solution and vapor have the same composition. A mixture of two substances that has an azeotrope cannot be separated by fractional distillation into two pure substances, but only into one pure substance and a mixture with the azeotropic composition. A mixture of 50% ethanol and water, for example, can be distilled to obtain pure water and an azeotropic mixture containing 4% water and 96% ethanol. The last 4% of water cannot be removed by distillation to obtain pure ethanol.

## 4-7 COLLOIDAL SUSPENSIONS

A **colloid** is a mixture of two or more substances in which one phase is suspended as a large number of very small particles in a second phase. The dispersed substance and the background medium may be any combination of gas, liquid, or solid. Examples of colloids include aerosol sprays (liquid suspended in gas), smoke (solid particles in air), milk (fat droplets and solids in water), mayonnaise (water droplets in oil), paint (solid pigment particles in oil for oil-based paints, or pigment and oil dispersed in water for latex paints). Colloidal particles are larger than single molecules but are too small to be seen by the eye; their dimensions typically range from $10^{-9}$ to $10^{-6}$ m. Their presence can be seen most dramatically in the way in which they scatter light; a familiar example is the passage of light from a movie projector through a suspension of small dust particles in air. The gemstone opal has remarkable optical properties that arise from colloidal water suspended in solid silicon dioxide (Fig. 4-25).

Although some colloids settle out into two separate phases if left standing long enough, others persist indefinitely; a suspension of gold particles prepared by Michael Faraday in 1857 shows no apparent settling to date. In many colloids, the particles have net positive or negative charges on their surfaces, balanced by an opposite charge of ions in solution. The settling out of such colloids is speeded by dissolving salts in the solution, a process called **flocculation.** The salts reduce the repulsive electrostatic forces between the suspended particles, causing aggregation and sedimentation (Fig. 4-26). Flocculation occurs in river deltas; when river water

**Figure 4-26**
When a salt is added to a colloidal dispersion (*a*), the repulsive forces between the colloidal particles are reduced and aggregation occurs (*b*). Eventually the aggregated particles fall to the bottom of the container as a low-density sediment (*c*).

(*a*) Dispersion

(*b*) Aggregation

(*c*) Sedimentation

(a)

(b)

**Figure 4-27**
(a) This colloidal suspension of $PbCrO_4$ appears cloudy. (b) After flocculation, the precipitate settles to the bottom. (Leon Lewandowski)

containing suspended clay particles meets the salt water of the ocean, the clay settles out as open, low-density sediments. Flocculating agents are deliberately added to paints so that the pigment will settle in a loosely packed sediment. When the paint is stirred, the pigment is redispersed through the medium. In the absence of such agents, the suspended particles tend to settle in compact sediments that are difficult to resuspend.

In some cases the formation of a colloid is not desirable, as in the precipitation of a solid from solution (Chapter 2). Especially with metal sulfides, the solid precipitate may appear as a colloidal suspension with particles small enough to pass through ordinary filter paper (Fig. 4-27). If this happens, precipitated solid can be separated out only by flocculation or centrifugation or by forcing the suspension through a membrane such as cellophane that permits passage of only the small solvent molecules.

Suspended particles are in a constant state of motion called **Brownian motion** after Robert Brown, a Scottish botanist who used a microscope to observe the motion of pollen particles in water. Brownian motion results from the constant random buffeting of the particles by solvent molecules. In 1905 Albert Einstein showed how the motion of Brownian particles could be described on a microscopic level; his work provided one of the most striking and convincing verifications of the molecular hypothesis and of the kinetic theory of matter, and led to a fairly accurate determination of Avogadro's number.

## Maple Syrup

**PRACTICE EXERCISE**

The sap in a maple tree can be described approximately as a 3.0% (by mass) solution of sucrose ($C_{12}H_{22}O_{11}$) in water. Sucrose does not dissociate to any significant extent in aqueous solution.
(a) At 20°C, the density of sap is 1.010 g cm$^{-3}$. Calculate the molarity of sucrose in sap.
(b) A typical maple tree yields about 12 gal of sap per year. Calculate how many grams of sucrose are contained in this volume of sap (1 gal = 3.785 L).
(c) The rising of sap in trees is caused largely by osmosis; the concentration of dissolved sucrose in sap is higher than that of the groundwater outside the tree.

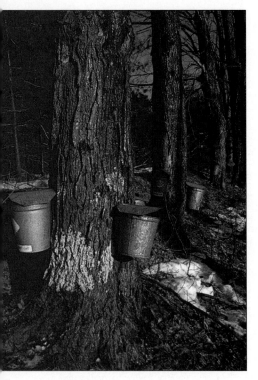

Tapping syrup from sugar maple trees. (Ted Levin/Phototake)

Calculate the osmotic pressure of a sap solution and the height to which the sap should rise above the ground on a day when the temperature is 20°C. Approximate the groundwater as pure (although in fact it typically contains 0.01 to 0.03 M dissolved species). Express the answer in meters and in feet (1 m = 3.28 ft).

(d) To produce maple syrup from sap, one boils the sap to reduce its water content. Calculate the normal boiling point of a sap solution.

(e) Maple syrup is the concentrated sap solution that results when most of the water is boiled off. The syrup has a composition of approximately 64% (by mass) sucrose and 36% water, with flavoring components present in small concentrations. If the density of the maple syrup is 1.31 g cm$^{-3}$, calculate the mole fraction, molarity, and molality of sucrose in maple syrup.

(f) What volume (in gallons) of maple syrup can be obtained from the sap in one typical tree?

(g) In the presence of vanadium(V) oxide, dinitrogen tetraoxide oxidizes sucrose to oxalic acid ($H_2C_2O_4$) according to the equation

$$C_{12}H_{22}O_{11}(aq) + 9\,N_2O_4(s) \longrightarrow 6\,H_2C_2O_4(aq) + 18\,NO(g) + 5\,H_2O(\ell)$$

Calculate the mass of $N_2O_4$ that will react completely with 7.00 L of the sap solution from part (a), and give the concentration of oxalic acid that results.

**Answers**
(a) 0.089 M
(b) $1.4 \times 10^3$ g sucrose
(c) $\pi = 2.1$ atm; height = 22 m = 72 ft
(d) 100.046°C
(e) Mole fraction = 0.086; molarity = 2.4 M; molality = 5.2 mol kg$^{-1}$
(f) 0.56 gal
(g) $5.2 \times 10^2$ g $N_2O_4$;   0.53 M

## CONCEPTS & SKILLS

After studying this chapter and working the problems that follow, you should be able to

1. Discuss the evidence that phase equilibrium is a dynamic process at the molecular level (Section 4-1).

2. Sketch the pressure–temperature phase diagram for a typical substance, and identify the lines, areas, and singular points (Section 4-1).

3. Describe the effects of different kinds of intermolecular forces on phase transitions (Section 4-2, problems 13–20).

4. Express the concentration of a solute in solution in units of mass percentage, molarity, molality, and mole fraction (Section 4-3, problems 23–28).

5. Describe how a solution of a given molarity is prepared, and the effect on molarity of dilution (Section 4-3, problems 29–32).

6. Calculate the chemical amounts of substances reacting during a solution-phase reaction such as titration (Section 4-4, problems 33–42).

7. Calculate the molar mass of a nonvolatile solute from the changes it causes in the colligative properties (vapor-pressure lowering, boiling-point elevation, freezing-point lowering, or osmotic pressure) of its dilute solution (Section 4-5, problems 43–58).

**8.** Discuss the meaning of Henry's law and use it to calculate the solubilities of gases in liquids (Section 4-6, problems 59–62).

**9.** Relate the total pressure and composition of the vapor in equilibrium with an ideal two-component solution to the composition of the solution and the vapor pressures of its pure components (Section 4-6, problems 63–66).

**10.** Explain how distillation is used to separate the volatile components of a binary liquid solution (Section 4-6).

**11.** Describe the physical properties of a colloidal suspension (Section 4-7).

**PROBLEMS**

## Phase Equilibrium and Phase Diagrams

**1.** Calcium carbide reacts with water to produce acetylene ($C_2H_2$) and calcium hydroxide. The acetylene is collected over water at 40.0°C under a total pressure of 0.9950 atm. The vapor pressure of water at this temperature is 0.0728 atm. Calculate the mass of acetylene per liter of "wet" acetylene collected in this way, assuming ideal gas behavior.

**2.** A metal reacts with aqueous hydrochloric acid to produce hydrogen. The hydrogen is collected over water at 25°C under a total pressure of 0.9900 atm. The vapor pressure of water at this temperature is 0.0313 atm. Calculate the mass of hydrogen per liter of "wet" hydrogen above the water, assuming ideal gas behavior.

**3.** Carbon dioxide is liberated by the reaction of aqueous hydrochloric acid with calcium carbonate:

$$CaCO_3(s) + 2\,H^+(aq) \longrightarrow$$
$$Ca^{2+}(aq) + CO_2(g) + H_2O(\ell)$$

A volume of 722 mL of $CO_2(g)$ is collected over water at 20°C and a total pressure of 0.9963 atm. At this temperature, water has a vapor pressure of 0.0231 atm. Calculate the mass of calcium carbonate that has reacted, assuming no losses of carbon dioxide.

**4.** When an excess of sodium hydroxide is added to an aqueous solution of ammonium chloride, gaseous ammonia is produced:

$$NaOH(aq) + NH_4Cl(aq) \longrightarrow$$
$$NaCl(aq) + NH_3(g) + H_2O(\ell)$$

Suppose 3.68 g of ammonium chloride reacts in this way at 30°C and a total pressure of 0.9884 atm. At this temperature, the vapor pressure of water is 0.0419 atm. Calculate the volume of ammonia saturated with water vapor that will be produced under these conditions, assuming no leaks or other losses of gas.

**5.** High in the Andes, an explorer notes that the water being heated for tea boils vigorously at a temperature of 194°F. Use Figure 4-3 to estimate the atmospheric pressure at the altitude of the camp. What fraction of the earth's atmosphere lies below the level of the explorer's camp?

**6.** The total pressure in a pressure cooker filled with water rises to 4.0 atm when it is heated, and this pressure is maintained by the periodic operation of a relief valve. Explain how the use of a pressure cooker speeds the cooking of food.

**7.** At its melting point (624°C), the density of solid plutonium is 16.24 g cm$^{-3}$. The density of liquid plutonium is 16.66 g cm$^{-3}$. A small sample of liquid plutonium at 625°C is strongly compressed. Predict what phase changes, if any, will occur.

**8.** Phase changes occur between different solid forms as well as from solid to liquid, liquid to gas, and solid to gas. When white tin at 1.00 atm is cooled below 13.2°C, it spontaneously changes (over a period of weeks) to gray tin. The density of gray tin is less than the density of white tin (5.75 g cm$^{-3}$ versus 7.31 g cm$^{-3}$). Some white tin is compressed to a pressure of 2.00 atm. At this pressure, should the temperature be higher or lower than 13.2°C for the conversion to gray tin to take place? Explain your reasoning. (*Hint:* See the discussion of the effect of density on the slope of the liquid–solid coexistence curve.)

**9.** The following table lists several important points on the pressure–temperature diagram of nitrogen:

|  | **P (atm)** | **T (K)** |
|---|---|---|
| Triple point | 0.123 | 63.15 |
| Critical point | 33.40 | 126.2 |
| Normal boiling point | 1.0 | 77.35 |
| Normal melting point | 1.0 | 63.29 |

The density of $N_2(s)$ is 1.03 g cm$^{-3}$, and that of $N_2(\ell)$ is 0.808 g cm$^{-3}$. Use this information to sketch the phase diagram of nitrogen.

**10.** Some water starts out at a temperature of 298 K and a pressure of 1 atm. It is compressed to 512 atm at constant temperature and then heated to 750 K at constant pressure. Next it is decompressed back to 1 atm at 750 K and finally cooled to 400 K at constant pressure.

(a) What was the state (solid, liquid, or gas) of the water at the start of the experiment?

(b) What is the state (solid, liquid, or gas) of the water at the end of the experiment?

(c) Did any phase transitions occur during the four steps described? If so, at what temperature and pressure did they occur? (*Hint:* Trace out the changes on the phase diagram of water [see Fig. 4-6*b*].)

11. The vapor pressure of solid acetylene at −84.0°C is 760 torr.
    (a) Does the triple-point temperature lie above or below −84.0°C? Explain.
    (b) Suppose a sample of solid acetylene is held under an external pressure of 0.80 atm and heated from 10 K to 300 K. What phase change(s), if any, occur?

12. The triple point of hydrogen occurs at a temperature of 13.8 K and a pressure of 0.069 atm.
    (a) What is the vapor pressure of solid hydrogen at 13.8 K?
    (b) Suppose a sample of solid hydrogen is held under an external pressure of 0.030 atm and heated from 5 K to 300 K. What phase change(s), if any, occur?

## Intermolecular Forces in Liquids

13. Under room conditions, fluorine and chlorine are gases, bromine is a liquid, and iodine is a solid. Explain the origin of this trend in the physical state of the halogens.

14. The later halogens form pentafluorides: $ClF_5$, $BrF_5$, and $IF_5$. At 0°C, one of these is a solid, one a liquid, and one a gas. Specify which is which, and explain your reasoning.

15. List the following substances in order of increasing normal boiling points $T_b$, and explain your reasoning: NO, $NH_3$, Ne, RbCl.

16. List the following substances in order of increasing normal boiling points $T_b$, and explain your reasoning: $SO_2$, He, HF, $CaF_2$, Ar.

17. As a vapor, methanol exists to an extent as a tetramer, $(CH_3OH)_4$, in which four $CH_3OH$ molecules are held together by hydrogen bonds. Propose a reasonable structure for this tetramer.

18. Hypofluorous acid (HOF) is the simplest possible compound that allows comparison between fluorine and oxygen in their abilities to form hydrogen bonds. Although F attracts electrons more strongly than O, solid HOF unexpectedly contains no H···F hydrogen bonds! Draw a proposed structure for chains of HOF molecules in the crystalline state. The bond angle in HOF is 101°.

19. Hydrazine ($N_2H_4$) is used as a reducing agent and in the manufacture of rocket fuels. How do you expect its boiling point to compare with that of ethylene ($C_2H_4$)?

20. Hydrogen peroxide ($H_2O_2$) is a major industrial chemical that is produced on a scale approaching $10^9$ kg per year. It is widely used as a bleach and in chemical manufacturing processes. How do you expect its boiling point to compare with those of fluorine ($F_2$) and hydrogen sulfide ($H_2S$), two substances with molar masses comparable to that of hydrogen peroxide?

## Composition of Solutions

21. A patient has a "cholesterol count" of 214. Like many blood-chemistry measurements, this result has the units mg $dL^{-1}$.
    (a) Determine the molar concentration of cholesterol in this patient's blood, taking the molar mass of cholesterol to be 386.64 g $mol^{-1}$.
    (b) Estimate the molality of cholesterol in the patient's blood.
    (c) If 214 is a typical cholesterol reading among men in the United States, determine the volume of such blood required to furnish 8.10 g of cholesterol.

22. In many states, a person is legally intoxicated if his or her blood has a concentration of 0.1 g (or more) of ethyl alcohol ($C_2H_5OH$) per deciliter. Express this "threshold concentration" in mol $L^{-1}$.

23. A solution of hydrochloric acid in water is 38.00% hydrochloric acid by mass. Its density is 1.1886 g $cm^{-3}$ at 20°C. Compute its molarity, mole fraction, and molality at this temperature.

24. A solution of acetic acid and water contains 205.0 g $L^{-1}$ of acetic acid and 820.0 g $L^{-1}$ of water.
    (a) Compute the density of the solution.
    (b) Compute the molarity, molality, mole fraction, and mass percentage of acetic acid in this solution.
    (c) Take the acetic acid as the solvent, and do the same for water as the solute.

25. A 6.0835 M aqueous solution of acetic acid ($C_2H_4O_2$) has a density of 1.0438 g $cm^{-3}$. Compute its molality.

26. A 1.241 M aqueous solution of $AgNO_3$ (used to prepare silver chloride photographic emulsions) has a density of 1.171 g $cm^{-3}$. Compute its molality.

27. Water is slightly soluble in liquid nitrogen. At −196°C (the boiling point of liquid nitrogen), the mole fraction of water in a saturated solution is $1.00 \times 10^{-5}$. Compute the mass of water that can dissolve in 1.00 kg of boiling liquid nitrogen.

28. Some water dissolves in liquid methane at −161°C to give a solution in which the mole fraction of water is $6.0 \times 10^{-5}$. Determine the mass of water dissolved in 1.00 L of this solution if the density of the solution is 0.78 g $cm^{-3}$.

29. Concentrated phosphoric acid as sold for use in the laboratory is usually 90% $H_3PO_4$ by mass (the rest is water). Such a solution contains 12.2 mol of $H_3PO_4$ per liter of solution at 25°C.
    (a) Compute the density of this solution.
    (b) What volume of this solution should be used in mixing 2.00 L of a 1.00 M phosphoric acid solution?

30. A perchloric acid solution is 60.0% $HClO_4$ by mass. It is simultaneously 9.20 M at 25°C.
    (a) Compute the density of this solution.
    (b) What volume of this solution should be used in mixing 1.00 L of a 1.00 M perchloric acid solution?

31. Suppose 25.0 g of solid NaOH is added to 1.50 L of an aqueous solution that is already 2.40 M in NaOH. Then water is

added until the final volume is 4.00 L. Determine the concentration of the NaOH in the resulting solution.

32. Suppose 0.400 L of a solution of 0.0700 M nitric acid is added to 0.800 L of a solution of 0.0300 M nitric acid, giving a total volume of 1.200 L. Calculate the concentration (molarity) of nitric acid in the resulting solution.

## Stoichiometry of Reactions in Solution

33. When treated with acid, lead(IV) oxide is reduced to a lead(II) salt, with the liberation of oxygen:

$$2\,PbO_2(s) + 4\,HNO_3(aq) \longrightarrow$$
$$2\,Pb(NO_3)_2(aq) + 2\,H_2O(\ell) + O_2(g)$$

What volume of a 7.91 M solution of nitric acid is just sufficient to react with 15.9 g of lead(IV) oxide according to this equation?

34. Phosphoric acid is made industrially by the reaction of fluorapatite, $Ca_5(PO_4)_3F$, in phosphate rock with sulfuric acid:

$$Ca_5(PO_4)_3F(s) + 5\,H_2SO_4(aq) + 10\,H_2O(\ell) \longrightarrow$$
$$3\,H_3PO_4(aq) + 5(CaSO_4\cdot2H_2O) + HF(aq)$$

What volume of 6.3 M phosphoric acid is generated by the reaction of 2.2 metric tons (2200 kg) of fluorapatite?

35. The carbon dioxide produced (together with hydrogen) from the industrial-scale oxidation of methane in the presence of nickel is removed from the gas mixture in a scrubber containing an aqueous solution of potassium carbonate:

$$CO_2(g) + H_2O(\ell) + K_2CO_3(aq) \longrightarrow 2\,KHCO_3(aq)$$

Calculate the volume of carbon dioxide (at 50°C and 1.00 atm pressure) that will react with 187 L of a 1.36 M potassium carbonate solution.

36. Nitrogen oxide can be generated on a laboratory scale by the reaction of dilute sulfuric acid with aqueous sodium nitrite:

$$6\,NaNO_2(aq) + 3\,H_2SO_4(aq) \longrightarrow$$
$$4\,NO(g) + 2\,HNO_3(aq) + 2\,H_2O(\ell) + 3\,Na_2SO_4(aq)$$

What volume of 0.646 M aqueous $NaNO_2$ should be used in this reaction to generate 5.00 L of nitrogen oxide at a temperature of 20°C and a pressure of 0.970 atm?

37. Phosphorus trifluoride is a highly toxic gas that reacts slowly with water to give a mixture of phosphorous acid and hydrofluoric acid.
    (a) Write a balanced chemical equation for this reaction.
    (b) Determine the concentration (in moles per liter) of each of the acids that result from the reaction of 1.94 L of phosphorus trifluoride (measured at 25°C and 0.970 atm pressure) with water to give a solution volume of 872 mL.

38. Phosphorus pentachloride reacts violently with water to give a mixture of phosphoric acid and hydrochloric acid.
    (a) Write a balanced chemical equation for this reaction.
    (b) Determine the concentration (in moles per liter) of each of the acids that result from the complete reaction of 1.22 L

of phosphorus pentachloride (measured at 215°C and 0.962 atm pressure) with enough water to give a solution volume of 697 mL.

39. Potassium dichromate in acidic solution is used to titrate a solution of iron(II) ions, with which it reacts according to

$$Cr_2O_7^{2-}(aq) + 6\,Fe^{2+}(aq) + 14\,H^+(aq) \longrightarrow$$
$$2\,Cr^3(aq) + 6\,Fe^{3+}(aq) + 7\,H_2O(\ell)$$

A potassium dichromate solution is prepared by dissolving 5.134 g of $K_2Cr_2O_7$ in water and diluting to a total volume of 1.000 L. A total of 34.26 mL of this solution is required to reach the endpoint in a titration of a 500.0-mL sample containing $Fe^{2+}(aq)$. Determine the concentration of $Fe^{2+}$ in the original solution.

40. Cerium(IV) ions are strong oxidizing agents in acidic solution, oxidizing arsenious acid to arsenic acid according to the equation

$$2\,Ce^{4+}(aq) + H_3AsO_3(aq) + H_2O(\ell) \longrightarrow$$
$$2\,Ce^{3+}(aq) + H_3AsO_4(aq) + 2\,H^+(aq)$$

A sample of $As_2O_3$ weighing 0.217 g is dissolved in basic solution and then acidified to make $H_3AsO_3$. Its titration with a solution of acidic cerium(IV) sulfate requires 21.47 mL. Determine the original concentration of $Ce^{4+}(aq)$ in the titrating solution.

41. To determine the concentration of a solution of nitric acid, a 100.0-mL sample is placed in a flask and titrated with a 0.1279 M solution of potassium hydroxide. A volume of 37.85 mL is required to reach the phenolphthalein endpoint. Calculate the concentration of nitric acid in the original sample.

42. The concentration of aqueous ammonia in a cleaning solution is determined by titration with hydrochloric acid. A volume of 23.18 mL of 0.8381 M HCl is needed to titrate a 50.0-mL sample of the ammonia solution to a methyl red endpoint. Calculate the concentration of ammonia in the cleaning solution.

## Colligative Properties of Solutions

43. The vapor pressure of pure acetone ($CH_3COCH_3$) at 30°C is 0.3270 atm. Suppose 15.0 g of benzophenone, $C_{13}H_{10}O$, is dissolved in 50.0 g of acetone. Calculate the vapor pressure of acetone above the resulting solution.

44. The vapor pressure of diethyl ether (molar mass, 74.12 g $mol^{-1}$) at 30°C is 0.8517 atm. Suppose 1.800 g of maleic acid, $C_4H_4O_4$, is dissolved in 100.0 g of diethyl ether at 30°C. Calculate the vapor pressure of diethyl ether above the resulting solution.

45. Pure toluene ($C_7H_8$) has a normal boiling point of 110.60°C. A solution of 7.80 g of anthracene ($C_{14}H_{10}$) in 100.0 g of toluene has a boiling point of 112.06°C. Calculate $K_b$ for toluene.

46. When 2.62 g of the nonvolatile solid anthracene, $C_{14}H_{10}$, is dissolved in 100.0 g of cyclohexane, $C_6H_{12}$, the boiling point

of the cyclohexane is raised by 0.41°C. Calculate $K_b$ for cyclohexane.

47. When 39.8 g of a nondissociating, nonvolatile sugar is dissolved in 200.0 g of water, the boiling point of the water is raised by 0.30°C. Estimate the molar mass of the sugar.

48. When 2.60 g of a substance that contains only indium and chlorine is dissolved in 50.0 g of tin(IV) chloride, the normal boiling point of the tin(IV) chloride is raised from 114.1°C to 116.3°C. If $K_b = 9.43$ K kg mol$^{-1}$ for $SnCl_4$, what are the approximate molar mass and the probable molecular formula of the solute?

49. The Rast method for determining molar masses employs camphor as the solvent. Camphor melts at 178.4°C, and its large $K_f$ (37.7 K kg mol$^{-1}$) makes it especially useful for accurate work. A sample of an unknown substance that weighs 0.840 g lowers the freezing point of 25.0 g of camphor to 170.8°C. What is its molar mass?

50. Barium chloride has a freezing point of 962°C and a freezing-point depression constant of 108 K kg mol$^{-1}$. If 12 g of an unknown substance dissolved in 562 g of barium chloride gives a solution with a freezing point of 937°C, compute the molar mass of the unknown, assuming no dissociation takes place.

51. Ice cream is made by freezing a liquid mixture that, as a first approximation, can be considered a solution of sucrose ($C_{12}H_{22}O_{11}$) in water. Estimate the temperature at which the first ice crystals begin to appear in a mix that consists of 34% (by mass) sucrose in water. As ice crystallizes out, the remaining solution becomes more concentrated. What happens to its freezing point?

52. The solution to problem 51 shows that to make homemade ice cream, temperatures ranging downward from −3°C are needed. Ice cubes from a freezer have a temperature of about −12°C (+10°F), which is cold enough, but contact with the warmer ice cream mixture causes them to melt to liquid at 0°C, which is too warm. To obtain a liquid that is cold enough, salt (NaCl) is dissolved in water, and ice is added to the salt water. The salt lowers the freezing point of the water enough so that it can freeze the liquid inside the ice cream maker. The instructions for an ice cream maker say to add one part salt to eight parts water (by mass). What is the freezing point of this solution (in degrees Celsius and degrees Fahrenheit)? Assume that the NaCl dissociates fully into ions and that the solution is ideal.

53. An aqueous solution is 0.8402 molal in $Na_2SO_4$. It has a freezing point of −4.218°C. Determine the effective number of particles arising from each $Na_2SO_4$ formula unit in this solution.

54. The freezing-point depression constant of pure $H_2SO_4$ is 6.12 K kg mol$^{-1}$. When 2.3 g of ethanol ($C_2H_5OH$) is dissolved in 1.00 kg of pure sulfuric acid, the freezing point of the solution is 0.92 K lower than the freezing point of pure sulfuric acid. Determine how many particles are formed as one molecule of ethanol goes into solution in sulfuric acid.

55. A 200-mg sample of a purified compound of unknown molar mass is dissolved in benzene and diluted with that solvent to a volume of 25.0 cm$^3$. The resulting solution is found to have an osmotic pressure of 0.0105 atm at 300 K. What is the molar mass of the unknown compound?

56. Suppose 2.37 g of a protein is dissolved in water and diluted to a total volume of 100.0 mL. The osmotic pressure of the resulting solution is 0.0319 atm at 20°C. What is the molar mass of the protein?

57. A polymer of large molar mass is dissolved in water at 15°C, and the resulting solution rises to a final height of 15.2 cm above the level of the pure water, as water molecules pass through a semipermeable membrane into the solution. If the solution contains 4.64 g polymer per liter, calculate the molar mass of the polymer.

58. Suppose 0.125 g of a protein is dissolved in 10.0 cm$^3$ of ethyl alcohol ($C_2H_5OH$), whose density at 20°C is 0.789 g cm$^{-3}$. The solution rises to a height of 26.3 cm in an osmometer (an apparatus for measuring osmotic pressure). What is the approximate molar mass of the protein?

## Mixtures and Distillation

59. The Henry's law constant at 25°C for carbon dioxide dissolved in water is $1.65 \times 10^3$ atm. If a carbonated beverage is bottled under a $CO_2$ pressure of 5.0 atm:
(a) Calculate the chemical amount of carbon dioxide dissolved per liter of water under these conditions, using 1.00 g cm$^{-3}$ as the density of water.
(b) Explain what happens on a microscopic level after the bottle cap is removed.

60. The Henry's law constant at 25°C for nitrogen dissolved in water is $8.57 \times 10^4$ atm, that for oxygen is $4.34 \times 10^4$ atm, and that for helium is $1.7 \times 10^5$ atm.
(a) Calculate the number of moles of nitrogen and oxygen dissolved per liter of water in equilibrium with air at 25°C. Use Table 3-1.
(b) Air is dissolved in blood and other bodily fluids. As a deep-sea diver descends, the pressure increases and the concentration of dissolved air in the blood increases. If the diver returns to the surface too quickly, gas bubbles out of solution within the body so rapidly that it can cause a dangerous condition called the bends. Use Henry's law to show why divers sometimes use a combination of helium and oxygen in their breathing tanks in place of compressed air.

61. At 25°C, some water is added to a sample of gaseous methane ($CH_4$) at 1.00 atm pressure in a closed vessel, and shaken until as much methane as possible dissolves. Then 1.00 kg of the solution is removed and boiled to expel the methane, yielding a volume of 3.01 L of $CH_4(g)$ at 0°C and 1.00 atm. Determine the Henry's law constant for methane in water.

62. When exactly the procedure of problem 61 is carried out using benzene ($C_6H_6$) in place of water, the volume of meth-

ane that results is 0.510 L at 0°C and 1.00 atm. Determine the Henry's law constant for methane in benzene.

63. At 20°C, the vapor pressure of toluene is 0.0289 atm and the vapor pressure of benzene is 0.0987 atm. Equal chemical amounts (equal numbers of moles) of toluene and benzene are mixed and form an ideal solution. Compute the mole fraction of benzene in the vapor in equilibrium with this solution.

64. At 90°C, the vapor pressure of toluene is 0.534 atm and the vapor pressure of benzene is 1.34 atm. Benzene (0.400 mol) is mixed with toluene (0.900 mol) to form an ideal solution. Compute the mole fraction of benzene in the vapor in equilibrium with this solution.

65. At 40°C, the vapor pressure of pure carbon tetrachloride ($CCl_4$) is 0.293 atm and the vapor pressure of pure dichloroethane ($C_2H_4Cl_2$) is 0.209 atm. A nearly ideal solution is prepared by mixing 30.0 g of carbon tetrachloride with 20.0 g of dichloroethane.
    (a) Calculate the mole fraction of $CCl_4$ in the solution.
    (b) Calculate the total vapor pressure of the solution at 40°C.
    (c) Calculate the mole fraction of $CCl_4$ in the vapor in equilibrium with the solution.

66. At 300 K, the vapor pressure of pure benzene ($C_6H_6$) is 0.1355 atm and the vapor pressure of pure n-hexane ($C_6H_{14}$) is 0.2128 atm. Mixing 50.0 g of benzene with 50.0 g of n-hexane gives a solution that is nearly ideal.
    (a) Calculate the mole fraction of benzene in the solution.
    (b) Calculate the total vapor pressure of the solution at 300 K.
    (c) Calculate the mole fraction of benzene in the vapor in equilibrium with the solution.

## Additional Problems

67. You boil a small quantity of water inside a 5.0-L metal can. When the can is filled with water vapor (and all air has been expelled), you quickly seal the can with a screw-on cap and remove it from the source of heat. It cools to 60°C, and most of the steam condenses to liquid water. Determine the pressure inside the can. (*Hint:* Refer to Figure 4-3.)

*68. Butane-fueled cigarette lighters, which give hundreds of lights each, typically contain 4 to 5 g of butane ($C_4H_{10}$) that is confined in a 10-mL plastic container and exerts a pressure of 3.0 atm at room temperature (25°C). The butane boils at −0.5°C under normal pressure. Butane lighters have been known to explode in use, inflicting serious injury. A person hoping to end such accidents suggests that less butane be placed in the lighters so that the pressure inside each does not exceed 1.0 atm. Estimate how many grams of butane would be contained in such a lighter.

69. A cooling bath is prepared in a laboratory by mixing chunks of solid $CO_2$ with ethanol. $CO_2(s)$ sublimes at −78.5°C to $CO_2(g)$. Ethanol freezes at −114.5°C and boils at +78.4°C. State the temperature of this cooling bath, and describe what

will be seen when it is prepared under ordinary laboratory conditions.

70. Oxygen melts at 54.8 K and boils at 90.2 K at atmospheric pressure. At the normal boiling point, the density of the liquid is 1.14 g $cm^{-3}$ and the vapor can be approximated as an ideal gas. The critical point is defined by $T_c = 154.6$ K, $P_c = 49.8$ atm, and (density) $\rho_c = 0.436$ g $cm^{-3}$. The triple point is defined by $T_t = 54.4$ K, $P_t = 0.0015$ atm, a liquid density of 1.31 g $cm^{-3}$, and a solid density of 1.36 g $cm^{-3}$. At 130 K, the vapor pressure of the liquid is 17.25 atm. Use this information to construct a phase diagram showing $P$ versus $T$ for oxygen. You need not draw the diagram to scale, but you should give numerical labels to as many points as possible on both axes.

71. It can be shown that if a gas obeys the van der Waals equation, its critical temperature, pressure, and molar volume are given by

$$T_c = \frac{8a}{27Rb} \qquad P_c = \frac{a}{27b^2} \qquad \left(\frac{V}{n}\right)_c = 3b$$

where $a$ and $b$ are the van der Waals constants of the gas. Use the van der Waals constants for oxygen, carbon dioxide, and water (from Table 3-3) to estimate the critical-point properties of these substances. Compare with the observed values given in Figure 4-6 and in problem 70.

72. Each increase in pressure of 100 atm decreases the melting point of water by about 1.0°C.
    (a) Estimate the temperature at which liquid water freezes under a pressure of 400 atm.
    (b) One possible explanation of why a skate moves smoothly over ice is that the pressure exerted by the skater on the ice lowers its freezing point and causes it to melt. The pressure exerted by an object is the force (its mass times the acceleration of gravity, 9.8 m $s^{-2}$) divided by the area of contact. Calculate the change in freezing point of ice when a skater with a mass of 75 kg stands on a blade of area $8.0 \times 10^{-5}$ $m^2$ in contact with the ice. Is this sufficient to explain the ease of skating at a temperature of, for example, −5°C (23°F)?

*73. (a) Sketch the phase diagram of *temperature* versus *molar volume* for carbon dioxide, indicating the region of each of the phases (gas, liquid, solid) and the coexistence regions for two phases.
    (b) Liquid water has a maximum at 4°C in its curve of density against temperature at $P = 1$ atm, and the solid is less dense than the liquid. What happens if you try to draw a phase diagram of $T$ versus molar volume for water?

74. The critical temperature of HCl is 51°C, lower than those of both HF, 188°C, and HBr, 90°C. Explain this by analyzing the nature of the intermolecular forces in each case.

75. The normal boiling points of the fluorides of the second-period elements are as follows: LiF, 1676°C; $BeF_2$, 1175°C; $BF_3$, −100°C; $CF_4$, −128°C; $NF_3$, −129°C; $OF_2$, −145°C; $F_2$, −188°C. Describe the nature of the intermolecular forces in this series of liquids, and account for the trends in boiling point.

**76.** Potassium iodide (KI) is used to make photographic emulsions. Suppose 0.157 g of KI(s) is added to 57.6 g of a solution that contains 4.40% KI by mass. Determine the molality of KI in the resulting solution.

**77.** Veterinarians use Donovan's solution to treat skin diseases in animals. The solution is prepared by mixing 1.00 g of $AsI_3(s)$, 1.00 g of $HgI_2(s)$, and 0.900 g of $NaHCO_3(s)$ in enough water to make a total volume of 100.0 mL.
(a) Compute the total mass of iodine per liter of Donovan's solution, in grams per liter.
(b) You need a lot of Donovan's solution to treat an outbreak of rash in an elephant herd. You have plenty of mercury(II) iodide and sodium hydrogen carbonate, but the only arsenic(III) iodide you can find is 1.50 L of a 0.100 M aqueous solution. Explain how to prepare 3.50 L of Donovan's solution starting with these materials.

**78.** Relative solubilities of salts in liquid ammonia can differ significantly from those in water. Thus, silver bromide is soluble in ammonia, but barium bromide is not (the reverse of the situation in water).
(a) Write a balanced equation for the reaction of an ammonia solution of barium nitrate with an ammonia solution of silver bromide. Silver nitrate is soluble in liquid ammonia.
(b) What volume of a 0.50 M solution of silver bromide will react completely with 0.215 L of a 0.076 M solution of barium nitrate in ammonia?
(c) What mass of barium bromide will precipitate from the reaction in part (b)?

**\*79.** Suppose 150 mL of a 10.00% by mass solution of sodium chloride (density = 1.0726 g cm$^{-3}$) is acidified with sulfuric acid and then treated with an excess of $MnO_2(s)$. Under these conditions, all of the chlorine is liberated as $Cl_2(g)$. The chlorine is collected without loss and reacts with excess $H_2(g)$ to form HCl(g). The HCl(g) is dissolved in enough water to make 250 mL of solution. Compute the molarity of this solution.

**\*80.** The amount of ozone in a mixture of gases can be determined by passing the mixture through an acidic aqueous solution of potassium iodide, where the ozone reacts according to

$$O_3(g) + 3\,I^-(aq) + H_2O(\ell) \longrightarrow$$
$$O_2(g) + I_3^-(aq) + 2\,OH^-(aq)$$

to form the triiodide ion $I_3^-$. The amount of triiodide produced is then determined by titrating with thiosulfate solution:

$$I_3^-(aq) + 2\,S_2O_3^{2-}(aq) \longrightarrow 3\,I^-(aq) + S_4O_6^{2-}(aq)$$

A small amount of starch solution is added as an indicator because it forms a deep-blue complex with the triiodide solution. Disappearance of the blue color thus signals the completion of the titration. Suppose 53.2 L of a gas mixture at a temperature of 18°C and a total pressure of 0.993 atm is passed through a solution of potassium iodide until the ozone in the mixture has reacted completely. The solution requires 26.2 mL of a 0.1359 M solution of thiosulfate ion to titrate to

the endpoint. Calculate the mole fraction of ozone in the original gas sample.

**81.** The vapor pressure of pure liquid $CS_2$ is 0.3914 atm at 20°C. When 40.0 g of rhombic sulfur is dissolved in 1.00 kg of $CS_2$, the vapor pressure of $CS_2$ falls to 0.3868 atm. Determine the molecular formula for the sulfur molecules dissolved in $CS_2$.

**82.** The expressions for boiling-point elevation and freezing-point depression apply accurately to *dilute* solutions only. A saturated aqueous solution of NaI (sodium iodide) in water has a boiling point of 144°C. The mole fraction of NaI in the solution is 0.390. Compute the molality of this solution. Compare the boiling-point elevation predicted by the expression in this chapter to the elevation actually observed.

**83.** You take a bottle of soft drink out of your refrigerator. The contents are liquid and stay liquid, even when you shake them. Presently you remove the cap, and the liquid freezes solid. Offer a possible explanation for this observation.

**84.** Mercury(II) chloride ($HgCl_2$) freezes at 276.1°C and has a freezing-point depression constant $K_f$ of 34.3 K kg mol$^{-1}$. When 1.36 g of solid mercury(I) chloride (empirical formula HgCl) is dissolved in 100 g of $HgCl_2$, the freezing point is lowered by 0.99°C. Calculate the molar mass of the dissolved solute species, and give its molecular formula.

**\*85.** The vapor pressure of an aqueous solution of $CaCl_2$ at 25°C is 0.02970 atm. The vapor pressure of pure water at the same temperature is 0.03126 atm. Estimate the freezing point of the solution.

**86.** Ethylene glycol ($CH_2OHCH_2OH$) is used in antifreeze because, when mixed with water, it lowers the freezing point below 0°C. What mass percentage of ethylene glycol in water must be used to reduce the freezing point of the mixture to −5.0°C, assuming ideal solution behavior?

**87.** A new compound has the empirical formula $GaCl_2$. This surprises some chemists who, based on the position of gallium in the periodic table, expect a chloride of gallium to have the formula $GaCl_3$ or possibly GaCl. They suggest that the "$GaCl_2$" is really $Ga[GaCl_4]$, in which the bracketed group behaves as a unit with a −1 charge. Suggest experiments to test this hypothesis.

**\*88.** Suppose two beakers are placed in a small closed container at 25°C. One contains 400 mL of a 0.100 M aqueous solution of NaCl; the second contains 200 mL of a 0.250 M aqueous solution of NaCl. Small amounts of water evaporate from both solutions. As time passes, the volume of solution in the second beaker gradually increases, and that in the first gradually decreases. Why? If we wait long enough, what will the final volumes and concentrations be?

**\*89.** The walls of erythrocytes (red blood cells) are permeable to water. In a salt solution, they shrivel (lose water) when the outside salt concentration is high, and swell (take up water) when the outside salt concentration is low. In an experiment at 25°C, an aqueous solution of NaCl that has a freezing point of −0.406°C causes erythrocytes neither to swell nor to shrink, indicating that the osmotic pressure of their contents is equal to that of the NaCl solution. Calculate the osmotic pressure of the solution inside the erythrocytes under these conditions, assuming that its molarity and molality are equal.

**90.** Silver dissolves in molten lead. Compute the osmotic pressure of a 0.010 M solution of silver in lead at 423°C. Compute the height of a column of molten lead ($\rho = 11.4$ g cm$^{-3}$) to which this pressure corresponds.

**91.** Henry's law is important in environmental chemistry, where it predicts the distribution of pollutants between water and the atmosphere. Benzene ($C_6H_6$) emitted in wastewater streams, for example, can pass into the air, where it is degraded by processes induced by light from the sun. The Henry's law constant for benzene in water at 25°C is 301 atm. Calculate the partial pressure of benzene vapor in equilibrium with a solution of 2.0 g of benzene per 1000 L of water. How many benzene molecules are present in each cubic centimeter?

**\*92.** Refer to the data of problem 64. Calculate the mole fraction of toluene in a mixture of benzene and toluene that boils at 90°C under atmospheric pressure.

**93.** What is the difference between a solution and a colloidal suspension? Give examples of each, and show how, in some cases, it may be difficult to classify a mixture as one or the other.

**CUMULATIVE PROBLEMS**

**94.** The tungsten filament in an incandescent light bulb ordinarily operates at a temperature of about 2500°C. At this temperature, the vapor pressure of solid tungsten is $7.0 \times 10^{-9}$ atm. Estimate the number of gaseous tungsten atoms per cubic centimeter under these conditions.

**95.** At 25°C, the equilibrium vapor pressure of water is 0.03126 atm. A humidifier is placed in a 100-m$^3$ room and is operated until the air becomes saturated with water vapor. Assuming that no water vapor leaks out of the room and that initially there was no water vapor in the air, calculate the number of grams of water that have passed into the air.

**96.** The air over an unknown liquid is saturated with the vapor of that liquid at 25°C and a total pressure of 0.980 atm. Suppose a 6.00-L sample of the saturated air is collected, and the vapor of the unknown liquid is removed from the sample by cooling and condensation. The pure air remaining occupies a volume of 3.75 L at $-50$°C and 1.000 atm. Calculate the vapor pressure of the unknown liquid at 25°C.

**97.** Other things being equal, ionic character in compounds of metals decreases with increasing oxidation number. Rank the following compounds from lowest to highest normal boiling points: $AsF_5$, $SbF_3$, $SbF_5$, $F_2$.

**98.** A student prepares a solution by dissolving 1.000 mol of $Na_2SO_4$ in water. She accidentally leaves the container uncovered, and comes back the next week to find only a white, solid residue. The mass of the residue is 322.2 g. Determine the chemical formula of this residue.

**99.** Complete combustion of 2.40 g of a compound of carbon, hydrogen, and oxygen yielded 5.46 g $CO_2$ and 2.23 g $H_2O$. When 8.69 g of the compound was dissolved in 281 g of water, the freezing point of the solution was found to be $-0.97$°C. What is the molecular formula of the compound?

# CHAPTER

5

*As solid sodium hydrogen carbonate (NaHCO₃) dissolves in water, a heterogeneous equilibrium is established with bubbles of carbon dioxide. (Tom Carroll)*

# Chemical Equilibrium

In Chapter 1 we discussed the stoichiometry of chemical reactions and showed how to calculate the amounts of products formed when the amounts of reactants are known, assuming the reaction goes to completion so that one of the reactants (the limiting reactant) is used up. In fact, however, many reactions do not proceed to completion, but rather approach an equilibrium state in which unconsumed reactants are mixed with the products but no further change takes place. The extent to which a given reaction occurs depends on the nature of the dynamic equilibrium between those reactants and products. In this chapter we examine the way in which this equilibrium is established, and show how to use equilibrium constants to calculate amounts of substances present at equilibrium, especially for gas-phase reactions. Then, in Chapters 6 and 7, we shall focus on two particularly important types of chemical equilibria in aqueous solution—those involving acids and bases, and those between ionic solids and their dissolved ions.

## 5-1 CHEMICAL REACTIONS AND EQUILIBRIUM

Section 4-1 described the phase equilibrium between liquid water and water vapor. It can be written as a chemical equation:

$$H_2O(\ell) \rightleftharpoons H_2O(g)$$

## Figure 5-1

The equilibrium between $N_2O_4$ and $NO_2$ depends on temperature. The tube on the right, held in an ice bath at 0°C, contains mostly $N_2O_4$. Its color is pale because only $NO_2$ is colored. The deeper color in the tube on the left, which is held at 50°C, reflects the increased $NO_2$ present in equilibrium at the higher temperature. The tubes contain the same masses of substance, distributed in different ways between $NO_2$ and $N_2O_4$.

The double arrows ($\rightleftarrows$) emphasize the dynamic nature of phase equilibrium: liquid water evaporates to form water vapor, and at the same time vapor condenses to give liquid. An analogous dynamic description applies to a chemical equilibrium, in which bonds are broken or formed. An example is the equilibrium between nitrogen dioxide ($NO_2$) and its dimer, dinitrogen tetraoxide ($N_2O_4$) (briefly considered in Example 3-6). This equilibrium is expressed by the chemical equation

$$2\,NO_2(g) \rightleftarrows N_2O_4(g)$$

Here the double arrows emphasize that molecules of $NO_2$ can react to form $N_2O_4$, and $N_2O_4$ molecules can break apart into pairs of $NO_2$ molecules. At equilibrium, both reactions occur at the same rate. Because $NO_2$ is a brown gas but $N_2O_4$ is colorless, the equilibrium between them can be studied by observing the color of a tube containing the two gases. At high temperatures, $NO_2$ predominates and a brown color results; as the temperature is lowered, the partial pressure of $N_2O_4$ increases and the color fades (Fig. 5-1).

If the tube is filled with a known amount of $NO_2$ at high temperature and then cooled, the total pressure drops below that given by the ideal gas law for $NO_2$ alone. The reason is that, for each molecule of $N_2O_4$ formed, *two* molecules of $NO_2$ are consumed, decreasing the total number of molecules and therefore the total pressure. In Example 3-6, the partial pressures of $NO_2$ and $N_2O_4$ present *at equilibrium* were calculated from the amount of $NO_2$ originally present and the total pressure. After a series of such experiments, in which the initial and final temperatures are the same but the amount of $NO_2$ placed into the flask is changed, an interesting relationship emerges between the partial pressures of $NO_2$ and $N_2O_4$ present at equilibrium. A plot of $P_{N_2O_4}$ against $P_{NO_2}$ has a simple parabolic form (Fig. 5-2), so

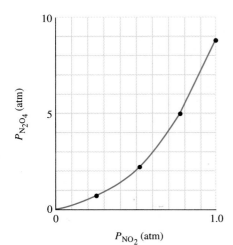

## Figure 5-2

At a constant temperature of 25°C, a graph of $N_2O_4$ partial pressure at equilibrium versus $NO_2$ partial pressure has the form of a parabola.

High temperature

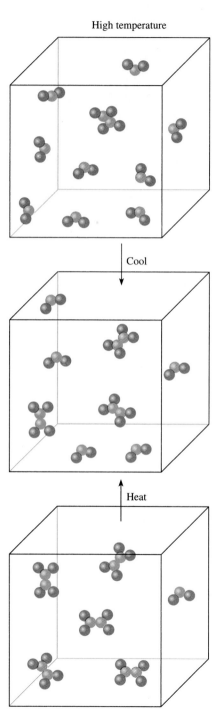

Cool

Heat

Low temperature

$$P_{N_2O_4} = K(P_{NO_2})^2$$

where $K$ is an **equilibrium constant** whose value depends only on the final temperature at which equilibrium is established. For the $N_2O_4$–$NO_2$ equilibrium, $K$ decreases as the temperature $T$ increases, so the amount of $N_2O_4$ present for a given partial pressure of $NO_2$ falls with increasing temperature as the dimer dissociates at elevated temperatures. This parabolic relationship can be written in the form

$$\frac{P_{N_2O_4}}{(P_{NO_2})^2} = K$$

$K$ has the numerical value 8.8 at $t = 25°C$, provided the partial pressures of $N_2O_4$ and $NO_2$ are expressed in atmospheres.

Chemical equilibrium, like the equilibrium between phases, is a dynamic process. When the initial partial pressures of the reactants are high, collisions between their molecules cause product molecules to form. Once the partial pressures of the products have increased sufficiently, the reverse reaction (forming ''reactants'' from ''products'') begins to occur. As the equilibrium state is approached, the forward and backward rates of reaction become equal and there is no further net change in reactant or product partial pressures. Just as the equilibrium between liquid water and water vapor is a dynamic process on the molecular scale, with evaporation and condensation taking place simultaneously, so the chemical equilibrium between $NO_2$ and $N_2O_4$ occurs through the continuous formation of molecules of $N_2O_4$ from $NO_2$ molecules and their dissociation back into $NO_2$ molecules with equal rates. Chemical equilibrium is not a static condition, although macroscopic properties such as partial pressures do stop changing when equilibrium is attained. Rather, it is the consequence of a dynamic balance between forward and backward reactions.

The same equilibrium state is reached whether one starts with the reactants or with the products. Suppose the system is prepared not with pure $NO_2$ at high temperature, but with an equal mass of pure $N_2O_4$ at low temperature (Fig. 5-3). Upon heating the $N_2O_4$ to 25°C, dissociation to $NO_2$ begins, and it continues until the same equilibrium state is reached as would be reached starting from $NO_2$. This can be used to test whether a system is truly in equilibrium or whether the reaction is just so slow that changes in concentration are unmeasurably small, even though the system is far from equilibrium. If the same state is reached from either reactants or products, it may be concluded that it is a true equilibrium state.

There are four fundamental aspects of equilibrium states:

1. They display no macroscopic evidence of change.
2. They are reached through spontaneous processes.
3. They show a dynamic balance of forward and reverse processes.
4. They are the same regardless of direction of approach.

### Figure 5-3

A particular equilibrium state is reached at a given temperature, regardless of whether the initial state consists of some mass of $NO_2$ molecules that dimerize as the gas is cooled from a high temperature, or of the same mass of $N_2O_4$ molecules that dissociate as the gas is heated from a low temperature.

## The Law of Mass Action

The relationship between the concentrations of $NO_2$ and $N_2O_4$ at equilibrium

$$\frac{P_{N_2O_4}}{(P_{NO_2})^2} = K$$

is a particular example of a general law that was first stated in approximate form in 1864 by two Norwegians, C. M. Guldberg (a mathematician) and his brother-in-law P. Waage (a chemist). This law states that the partial pressures of reacting gases present at equilibrium are not independent of each other but are related through a temperature-dependent equilibrium constant. Such relationships can be determined experimentally for other gas-phase chemical reactions. The reaction between hydrogen and nitrogen to produce ammonia

$$N_2(g) + 3\,H_2(g) \rightleftharpoons 2\,NH_3(g)$$

is characterized by the equilibrium law

$$\frac{(P_{NH_3})^2}{(P_{N_2})(P_{H_2})^3} = K$$

where $K$ is equal to $6.8 \times 10^5$ at 25°C and $1.9 \times 10^{-4}$ at 400°C, with pressures expressed in atmospheres.

The results of these (and other) measurements of the partial pressures of reactants and products present at equilibrium can be summarized by the following law:

> In a gas-phase chemical reaction in which $a$ moles of species A and $b$ moles of species B react to form $c$ moles of species C and $d$ moles of species D,
>
> $$aA + bB \rightleftharpoons cC + dD$$
>
> the partial pressures of gases at equilibrium are related through
>
> $$\frac{P_C^c P_D^d}{P_A^a P_B^b} = K(T)$$
>
> provided all species are present as low-pressure gases.

It is straightforward to generalize this law to the case in which different numbers of reactant or product species take part. The partial pressures of *products* appear in the numerator of the expression for the equilibrium constant, and those of *reactants* appear in the denominator. Each partial pressure is raised to a power equal to its coefficient in the balanced chemical equation. In Section 5-5 the law will be extended to include solutions and heterogeneous equilibria involving liquids and solids; this more general form is called the **law of mass action.**

The equilibrium expressions derived from this law hold for low-density gas mixtures, no matter how complex. For example, the quadratic relationship between the partial pressures of $NO_2$ and $N_2O_4$ at equilibrium holds for a mixture containing many other species besides. $N_2$, $O_2$, and $NO$ may all be present, and additional equilibrium laws will relate their partial pressures to those of $NO_2$ and $N_2O_4$; however, the relationship between $P_{N_2O_4}$ and $P_{NO_2}$ and the numerical value of $K$ will be unchanged. Equilibrium laws are not affected by the presence of other species or by simultaneous equilibria, provided the temperature is unchanged. At high enough densities, when the gas mixture is not ideal and Dalton's law begins to fail, deviations from the equilibrium law appear.

The magnitude of the equilibrium constant gives direct information about the nature of the equilibrium. If $K$ is very large, then at equilibrium the partial pressures of the products will be large compared with those of the reactants. In this case only, stoichiometry can be used to calculate the chemical amounts or the masses of product formed. If $K$ is very small, the reactant pressures will be large compared to product pressures, and the extent of reaction will be very small. For intermediate values of $K$, both reactants and products will be present in significant proportions at equilibrium.

## Example 5-1

Write equilibrium expressions for the following gas-phase chemical equilibria.
(a) $2\,NOCl(g) \rightleftarrows 2\,NO(g) + Cl_2(g)$
(b) $CO(g) + \frac{1}{2}O_2(g) \rightleftarrows CO_2(g)$

### Solution

(a)
$$\frac{(P_{NO})^2(P_{Cl_2})}{(P_{NOCl})^2} = K$$

The powers of 2 come from the factors of 2 in the balanced equation.

(b)
$$\frac{P_{CO_2}}{(P_{CO})(P_{O_2})^{1/2}} = K$$

Fractional powers appear in the equilibrium expression whenever they are present in the balanced equation.

*Related Problems:* 1, 2, 3, 4

## The Concept of Activity

The alert reader may have noticed that units have not been assigned to equilibrium constants, even though the expressions to which they have been set equal are the ratios of product and reactant pressures that *do* have units. The concept of **activity** resolves this apparent inconsistency.

A full treatment of activity requires thermodynamic results that will be derived in Chapter 9. At this point let us merely state the two major results. First, the equilibrium expressions written hold strictly only if the partial pressures $P_i$ that appear in them are replaced by *activities, $a_i$.* Second, the activity of a dilute gas is given by

$$a_i = \frac{P_i}{P_{ref}}$$

where $P_i$ is the partial pressure of gas $i$ and $P_{ref}$ is a reference pressure, chosen to equal one atmosphere (or 760 torr, or 101,325 Pa, depending upon which pressure unit is used). The equilibrium expression for the reaction

$$aA + bB \rightleftharpoons cC + dD$$

then has the form

$$\frac{a_C^c a_D^d}{a_A^a a_B^b} = \frac{(P_C/P_{ref})^c (P_D/P_{ref})^d}{(P_A/P_{ref})^a (P_B/P_{ref})^b} = K$$

The activities, being ratios of pressures, are dimensionless, so this expression is consistent with a dimensionless equilibrium constant $K$. Collecting the terms involving $P_{ref}$ on the right side of the equation gives

$$\frac{P_C^c P_D^d}{P_A^a P_B^b} = K(P_{ref})^{-a-b+c+d}$$

If pressures are expressed in atmospheres, then $P_{ref} = 1$ atm and the right-hand side has the same numerical value as $K$, with $P_{ref}$ factors serving only to make the equation dimensionally correct. If some other unit is chosen for pressures, the $P_{ref}$ factors no longer have a numerical value of unity. The convention followed in this book is to express all pressures in atmospheres. The $P_{ref}$ factors will not be explicitly included because their value is unity with these choices of pressure unit and reference pressure.

## 5-2 CALCULATING EQUILIBRIUM CONSTANTS

If the equilibrium partial pressures of all the reactants and products can be measured, the equilibrium constant can be calculated by writing the equilibrium expression and substituting the experimental values (in atmospheres) into it. In many cases it is not practical to directly measure the equilibrium partial pressure of each separate reactant and product. Despite this, the equilibrium constant can usually be derived from other available data, although the determination is less direct, as in the following example.

### Example 5-2

Phosgene, $COCl_2$, forms from CO and $Cl_2$ according to the equilibrium

$$CO(g) + Cl_2(g) \rightleftharpoons COCl_2(g)$$

At 600°C, a gas mixture of CO and $Cl_2$ is prepared that has initial partial pressures (before reaction) of 0.60 atm for CO and 1.10 atm for $Cl_2$. After the reaction mixture has reached equilibrium, the partial pressure of $COCl_2(g)$ at this temperature is measured to be 0.10 atm. Calculate the equilibrium constant for this reaction.

### Solution

Only the equilibrium partial pressure of $COCl_2(g)$ and the *initial* partial pressures of the other two gases are given. To find the equilibrium constant, it is necessary to determine the equilibrium partial pressures of CO and $Cl_2$. To do this, we set up a simple table:

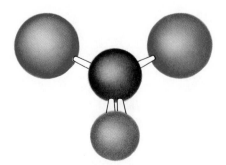

Phosgene, $COCl_2$, is a chemical intermediate used in making polyurethanes for foams and surface coatings.

|  | CO(g) | + | Cl₂(g) | ⇌ | COCl₂(g) |
|---|---|---|---|---|---|
| Initial partial pressure (atm) | 0.60 | | 1.10 | | 0 |
| Change in partial pressure (atm) | ? | | ? | | +0.10 |
| Equilibrium partial pressure (atm) | ? | | ? | | 0.10 |

Note that the first two lines must add to give the third. We use the relationships built into the balanced chemical equation to fill in the blanks in the table. Because this is the only reaction taking place, every mole of $COCl_2$ produced consumes exactly 1 mol of CO and exactly 1 mol of $Cl_2$. According to the ideal gas equation, the partial pressures of gases are proportional to their chemical amounts (in moles) as long as the volume and temperature are held fixed. Therefore, the *change* in partial pressure of each gas must be proportional to the change in its chemical amount as the mixture goes toward equilibrium. If the partial pressure of $COCl_2$ *increases* by 0.10 atm through reaction, the partial pressures of CO and $Cl_2$ must each *decrease* by 0.10 atm. Inserting these values into the table gives

|  | CO(g) | + | Cl₂(g) | ⇌ | COCl₂(g) |
|---|---|---|---|---|---|
| Initial partial pressure (atm) | 0.60 | | 1.10 | | 0 |
| Change in partial pressure (atm) | −0.10 | | −0.10 | | +0.10 |
| Equilibrium partial pressure (atm) | 0.50 | | 1.00 | | 0.10 |

where the last line was obtained by adding the first two.

We now insert the equilibrium partial pressures into the equilibrium expression to calculate the equilibrium constant:

$$K = \frac{P_{COCl_2}}{(P_{CO})(P_{Cl_2})} = \frac{(0.10)}{(0.50)(1.00)} = 0.20$$

*Related Problems:* 9, 10, 11, 12

For a reaction in which some of the coefficients in the balanced equation are not equal to 1, deriving expressions for the changes in the partial pressures of the products and reactants requires care. Consider the combustion of ethane at constant volume:

$$2\,C_2H_6(g) + 7\,O_2(g) \rightleftharpoons 4\,CO_2(g) + 6\,H_2O(g)$$

If the initial partial pressures are all 1.00 atm and there is a net reaction from left to right in this equation, the table is

|  | 2 C₂H₆(g) | + | 7 O₂(g) | ⇌ | 4 CO₂(g) | + | 6 H₂O(g) |
|---|---|---|---|---|---|---|---|
| Initial partial pressure | 1.00 | | 1.00 | | 1.00 | | 1.00 |
| Change in partial pressure | −2y | | −7y | | +4y | | +6y |
| Equilibrium partial pressure | 1.00 − 2y | | 1.00 − 7y | | 1.00 + 4y | | 1.00 + 6y |

If the $C_2H_6$ partial pressure decreases by $2y$ atm, that of $O_2$ must decrease by $7y$, because the coefficients in the balanced equation are 2 and 7. The changes in partial pressures of products will have the opposite sign (positive instead of negative), because as reactants disappear, products appear. Their magnitudes are determined according to the coefficients in the balanced equation.

## Relationships Among Equilibrium Expressions

Suppose a reaction is written in two opposing directions:

$$2 H_2(g) + O_2(g) \rightleftharpoons 2 H_2O(g) \qquad \frac{P_{H_2O}^2}{P_{H_2}^2 P_{O_2}} = K_1$$

$$2 H_2O(g) \rightleftharpoons 2 H_2(g) + O_2(g) \qquad \frac{P_{H_2}^2 P_{O_2}}{P_{H_2O}^2} = K_2$$

In the first reaction, hydrogen and oxygen combine to form water vapor, whereas in the second, water vapor dissociates into hydrogen and oxygen. The equilibrium constants $K_1$ and $K_2$ are clearly the inverse of each other, so their product $K_1 K_2$ equals 1. This is true quite generally: the equilibrium constant for a reverse reaction is the reciprocal of the equilibrium constant for the corresponding forward reaction.

What happens if a balanced chemical equation is multiplied by a constant? In the preceding example, multiplying the first equation by $\frac{1}{2}$ gives

$$H_2(g) + \tfrac{1}{2} O_2(g) \rightleftharpoons H_2O(g)$$

This is a perfectly satisfactory way to write the equation for the chemical reaction; it says that 1 mol of hydrogen reacts with $\frac{1}{2}$ mol of oxygen to yield 1 mol of water vapor. The corresponding equilibrium expression is

$$\frac{P_{H_2O}}{P_{H_2} P_{O_2}^{1/2}} = K_3$$

Comparison with the expression for $K_1$ shows that

$$K_3 = K_1^{1/2}$$

When a balanced chemical equation is multiplied by a constant factor, the corresponding equilibrium constant is raised to a power equal to that factor.

A further operation that can be carried out on chemical equations is to add two equations to give a third. In this case, the equilibrium constant for the third equation is the product of the equilibrium constants for the first two. For example, at 25°C,

$$2 BrCl(g) \rightleftharpoons Cl_2(g) + Br_2(g) \qquad \frac{P_{Cl_2} P_{Br_2}}{P_{BrCl}^2} = K_1 = 0.45$$

$$Br_2(g) + I_2(g) \rightleftharpoons 2 IBr(g) \qquad \frac{P_{IBr}^2}{P_{Br_2} P_{I_2}} = K_2 = 0.051$$

Adding the two chemical equations gives

$$2 BrCl(g) + Br_2(g) + I_2(g) \rightleftharpoons 2 IBr(g) + Cl_2(g) + Br_2(g)$$

Removing $Br_2(g)$ from both sides leaves

$$2 BrCl(g) + I_2(g) \rightleftharpoons 2 IBr(g) + Cl_2(g) \qquad \frac{P_{IBr}^2 P_{Cl_2}}{P_{BrCl}^2 P_{I_2}} = K_3$$

and, by inspection, $K_3 = K_1 K_2 = (0.45)(0.051) = 0.023$.

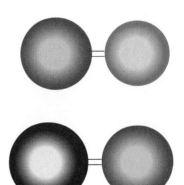

Bromine chloride, BrCl, and iodine bromide, IBr, are two examples of *interhalogens,* combinations of two or more halogen elements.

If a second equation is *subtracted* from the first, the resulting equilibrium constant is that of the first *divided* by that of the second (subtracting a reaction is the same as adding the reverse reaction). Applied to chemical equations, the operations of addition and subtraction transform into multiplication and division of the equilibrium expressions and equilibrium constants.

## Example 5-3

The concentrations of the oxides of nitrogen are monitored in air-pollution reports. At 25°C, the equilibrium constant for the reaction

$$NO(g) + \tfrac{1}{2} O_2(g) \rightleftharpoons NO_2(g)$$

is

$$\frac{P_{NO_2}}{P_{NO} P_{O_2}^{1/2}} = K_1 = 1.3 \times 10^6$$

and that for

$$\tfrac{1}{2} N_2(g) + \tfrac{1}{2} O_2(g) \rightleftharpoons NO(g)$$

is

$$\frac{P_{NO}}{P_{N_2}^{1/2} P_{O_2}^{1/2}} = K_2 = 6.5 \times 10^{-16}$$

Find the equilibrium constant $K_3$ for the reaction

$$N_2(g) + 2 O_2(g) \rightleftharpoons 2 NO_2(g)$$

### Solution

Adding the chemical equations for the first two reactions gives

$$\tfrac{1}{2} N_2(g) + O_2(g) \rightleftharpoons NO_2(g)$$

The equilibrium constant for this reaction, $K_3'$, is just the product of $K_1$ and $K_2$, or $K_1 K_2$. The constant sought, $K_3$, is defined by a chemical equation that is twice this, so $K_3'$ must be raised to the power 2 (i.e., squared) to give $K_3$:

$$K_3 = (K_3')^2 = (K_1 K_2)^2 = 7.1 \times 10^{-19}$$

*Related Problems:* 13, 14, 15, 16

## 5-3 CALCULATING GAS-PHASE EQUILIBRIA

The law of mass action gives the *form* of the expression for the equilibrium constant for an arbitrary gas-phase reaction. Such expressions can then be used to predict the outcomes of chemical reactions. How to approach the problem depends on the type of experimental data available. One frequently encountered type of calculation involves determining the partial pressures of reactants and products at equilibrium, given initial partial pressures of the reactants. This is illustrated by the following example.

## Example 5-4

Suppose $H_2(g)$ and $I_2(g)$ are sealed in a flask at $T = 400$ K with partial pressures $P_{H_2} = 1.320$ atm and $P_{I_2} = 1.140$ atm. At this temperature $H_2$ and $I_2$ do not react rapidly to form $HI(g)$, although after a long enough time they would produce $HI(g)$ at its equilibrium partial pressure. Suppose, instead, that the gases are heated in the sealed flask to 600 K, a temperature at which they speedily reach equilibrium:

$$H_2(g) + I_2(g) \rightleftharpoons 2 HI(g)$$

The equilibrium constant for the reaction is 92.6 at 600 K:

$$\frac{P_{HI}^2}{P_{H_2} P_{I_2}} = 92.6$$

(a) What are the equilibrium values of $P_{H_2}$, $P_{I_2}$, and $P_{HI}$ at 600 K?
(b) What percentage of the $I_2$ originally present has reacted when equilibrium is reached?

## Solution

(a) Suppose $H_2$ and $I_2$ did *not* react at 600 K. From the ideal gas law at constant volume, their partial pressures would be

$$P_{H_2}^{\circ} = 1.320 \text{ atm} \times \left(\frac{600 \text{ K}}{400 \text{ K}}\right) = 1.980 \text{ atm}$$

$$P_{I_2}^{\circ} = 1.140 \text{ atm} \times \left(\frac{600 \text{ K}}{400 \text{ K}}\right) = 1.710 \text{ atm}$$

Of course, these gases *do* react, and the extent of the reaction can be calculated. To do this, set up a table as in Example 5-2:

|  | $H_2(g)$ | + | $I_2(g)$ | $\rightleftharpoons$ | $2 HI(g)$ |
|---|---|---|---|---|---|
| Initial partial pressure (atm) | 1.980 | | 1.710 | | 0 |
| Change in partial pressure (atm) | $-x$ | | $-x$ | | $+2x$ |
| Equilibrium partial pressure (atm) | $1.980 - x$ | | $1.710 - x$ | | $2x$ |

If the partial pressure of $H_2$ decreases by $x$ atm as the reaction proceeds, then the partial pressure of $I_2$ must also decrease by $x$ atm because each mole of $H_2$ reacts with 1 mol of $I_2$. By similar reasoning, the partial pressure of HI increases by $2x$ atm: 2 mol of HI forms from each mole of $H_2$. Inserting the equilibrium partial pressures into the equilibrium expression results in the equation

$$\frac{(2x)^2}{(1.980 - x)(1.710 - x)} = 92.6$$

Multiplying and collecting terms gives

$$88.6x^2 - 341.694x + 313.525 = 0$$

Solving for $x$ using the quadratic formula (Appendix C) gives

$$x = \frac{-(-341.694) \pm \sqrt{(341.694)^2 - 4(88.6)(313.525)}}{2(88.6)}$$

$$= 1.5044 \text{ atm or } 2.3522 \text{ atm}$$

The second root is unphysical because it leads to negative answers for the equilibrium partial pressures of the $H_2(g)$ and $I_2(g)$. Discarding it leaves

$$P_{HI} = 2 \times 1.5044 \text{ atm} = 3.0088 \text{ atm}$$

$$P_{H_2} = 1.980 \text{ atm} - 1.5044 \text{ atm} = 0.4756 \text{ atm}$$

$$P_{I_2} = 1.710 \text{ atm} - 1.5044 \text{ atm} = 0.2056 \text{ atm}$$

It is a good idea to check such results by inserting the calculated equilibrium partial pressures back into the equilibrium expression to make sure that the known value of $K$ comes out.

$$\text{Check:} \quad \frac{(3.0088)^2}{(0.4756)(0.2056)} = 92.6$$

As the final step, round off each answer to the correct number of significant digits: $P_{HI} = 3.01$ atm, $P_{H_2} = 0.48$ atm, $P_{I_2} = 0.21$ atm. Rounding off sooner makes the $K$ calculated in the check differ somewhat from 92.6.

(b) The fraction of $I_2$ that has *not* reacted is the ratio of the chemical amount of $I_2$ present at the end to that present at the beginning. Because neither volume nor temperature changes during the reaction, this equals the ratio of the final partial pressure of $I_2$ (0.2056 atm) to its initial partial pressure (1.710 atm):

$$\text{percentage unreacted} = \frac{0.2056 \text{ atm}}{1.710 \text{ atm}} \times 100\% = 12\%$$

The percentage that *has* reacted is then 88%.

*Related Problems:* 17, 18, 19, 20, 21, 22

If one or more of the products of a reaction are present initially, together with reactants, they must be incorporated into the equilibrium calculation, as illustrated by the following example.

## Example 5-5

Hydrogen is made from natural gas (methane) for immediate consumption in industrial processes, such as ammonia production. The first step is the "steam reforming of methane":

$$CH_4(g) + H_2O(g) \rightleftharpoons CO(g) + 3 H_2(g)$$

The equilibrium constant for this reaction is $1.8 \times 10^{-7}$ at 600 K. Gaseous $CH_4$, $H_2O$, and CO are introduced into an evacuated container at 600 K, and their initial partial pressures (before reaction) are 1.40 atm, 2.30 atm, and 1.60 atm, respectively. Determine the partial pressure of $H_2(g)$ that will result at equilibrium.

## Solution

The equilibrium expression is

$$\frac{(P_{CO})(P_{H_2})^3}{(P_{CH_4})(P_{H_2O})} = K$$

Set up the table of partial pressures:

|  | CH$_4$(g) | + | H$_2$O(g) | $\rightleftarrows$ | CO(g) | + | 3 H$_2$(g) |
|---|---|---|---|---|---|---|---|
| Initial partial pressure | 1.40 |  | 2.30 |  | 1.60 |  | 0 |
| Change in partial pressure | $-y$ |  | $-y$ |  | $+y$ |  | $+3y$ |
| Equilibrium partial pressure | $1.40 - y$ |  | $2.30 - y$ |  | $1.60 + y$ |  | $3y$ |

Insert the equilibrium partial pressures into the equilibrium expression:

$$\frac{(1.60 + y)(3y)^3}{(1.40 - y)(2.30 - y)} = 1.8 \times 10^{-7}$$

If this equation were expanded by multiplying through by the denominator, a polynomial equation of fourth order in $y$ would result, for which the quadratic formula from the preceding problem would be useless. How can the equation be solved?

The equilibrium constant in this case is quite small, so the extent of reaction will also be small. This suggests that $y$ will be a small number relative to the partial pressures of the gases present initially. Try the approximation that $y$ can be ignored where it is added to a number of order one; that is, replace $1.60 + y$ with 1.60 in the preceding equation, and make the same approximation for the two terms in the denominator. When $y$ *multiplies* something, as in the $(3y)^3$ term, of course it cannot be set equal to zero. The result of this is the approximate equation

$$\frac{(1.60)(3y)^3}{(1.40)(2.30)} = 1.8 \times 10^{-7}$$

$$y^3 = 1.34 \times 10^{-8}$$

The cube roots of both sides give

$$y = 2.38 \times 10^{-3}$$

This value is indeed small compared with 1.60, 1.40, and 2.30, and so its neglect is justified. At equilibrium, then,

$$P_{H_2} = 3y = 7.1 \times 10^{-3} \text{ atm}$$

*Related Problems:* 23, 24, 25, 26

Approximation methods are of great importance in solving equilibrium problems and will arise frequently in Chapters 6 and 7. Appendix C discusses strategies for attacking such problems.

## Concentrations of Gases in Equilibrium Calculations

Sometimes the data for a calculation of gas-phase equilibrium are given in terms of concentrations rather than partial pressures. In such a case all concentrations can be converted to partial pressures before the calculations are carried out, or the equilibrium expression can be rewritten in terms of concentration variables. The concentration [A] of an ideal gas is related to its partial pressure $P_A$ through

$$[A] = \frac{n_A}{V} = \frac{P_A}{RT}$$

which can be written

$$P_A = RT[A]$$

Such a relation can be substituted for each species appearing in the equilibrium expression. It is best to put the factors of $P_{ref} = 1$ atm back into the equilibrium expression in order to examine the units of the resulting equations. For the general gas-phase reaction

$$aA(g) + bB(g) \rightleftharpoons cC(g) + dD(g)$$

the equilibrium expression is

$$\frac{(RT[C]/P_{ref})^c(RT[D]/P_{ref})^d}{(RT[A]/P_{ref})^a(RT[B]/P_{ref})^b} = K$$

Rearranging gives

$$\frac{[C]^c[D]^d}{[A]^a[B]^b} = K(RT/P_{ref})^{+a+b-c-d}$$

This expression relates concentrations of gas-phase species at equilibrium. If $a + b - c - d = 0$ (i.e., if there is no change in the total number of moles of gases in the reaction mixture), the right-hand side of the equilibrium expression reduces to $K$.

## Example 5-6

At elevated temperatures, $PCl_5$ dissociates extensively according to

$$PCl_5(g) \rightleftharpoons PCl_3(g) + Cl_2(g)$$

At 300°C, the equilibrium constant for this reaction is $K = 11.5$. The concentrations of $PCl_3$ and $Cl_2$ at equilibrium in a container at 300°C are both 0.0100 mol L$^{-1}$. Calculate [PCl$_5$].

### Solution

Two moles of gases are produced for each mole of gas consumed, so $RT/P_{ref}$ must be raised to the power $1 - 2 = -1$. Hence,

$$\frac{[PCl_3][Cl_2]}{[PCl_5]} = K\left(\frac{RT}{P_{ref}}\right)^{-1} = K\left(\frac{P_{ref}}{RT}\right)$$

$$= 11.5 \times \frac{1 \text{ atm}}{(0.08206 \text{ L atm mol}^{-1}\text{ K}^{-1})(573 \text{ K})} = 0.245\frac{\text{mol}}{\text{L}}$$

Solving this equation for $[PCl_5]$ gives

$$[PCl_5] = \frac{[PCl_3][Cl_2]}{0.245 \text{ mol L}^{-1}} = \frac{(0.0100 \text{ mol L}^{-1})(0.0100 \text{ mol L}^{-1})}{0.245 \text{ mol L}^{-1}}$$

$$= 4.08 \times 10^{-4} \text{ mol L}^{-1}$$

*Related Problems:* 27, 28

## 5-4 THE DIRECTION OF CHANGE

The law of mass action not only gives information about the nature of the equilibrium state; it also tells the direction in which a reaction will proceed spontaneously when reactants and products are mixed together with arbitrary partial pressures, and it describes the response of an equilibrium system to an external perturbation.

### The Reaction Quotient

The **reaction quotient** $Q$ for the general gas-phase reaction is defined as

$$Q = \frac{P_C^c P_D^d}{P_A^a P_B^b}$$

in which the partial pressures do *not* necessarily have their equilibrium values. The distinction between $Q$ and $K$ is crucial. The equilibrium constant $K$ is determined by the partial pressures of reactants and products *at equilibrium* and is a constant, dependent only on the temperature. The reaction quotient $Q$ depends on the actual instantaneous partial pressures, whatever they may be; thus, $Q$ changes with time. As the reaction approaches equilibrium, $Q$ approaches $K$. The initial partial pressures $P_A^\circ$, $P_B^\circ$, $P_C^\circ$, and $P_D^\circ$ give an initial reaction quotient $Q_0$, whose magnitude relative to $K$ determines the direction in which the reaction will proceed spontaneously toward equilibrium. If $Q_0$ is *less* than $K$, then $Q$ must *increase* as time goes on. This requires an increase in the product partial pressures and a decrease in reactant partial pressures; in other words, the reaction proceeds from left to right. If $Q_0$ is *greater* than $K$, similar reasoning shows that the reaction will proceed from right to left, with $Q$ *decreasing* with time until it becomes equal to $K$ (Fig. 5-4).

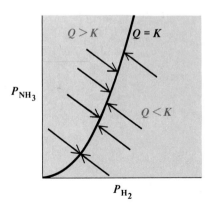

### Figure 5-4

If nitrogen and hydrogen are mixed in 1:3 proportions together with some ammonia, they react according to the chemical equation

$$N_2(g) + 3 H_2(g) \rightleftharpoons 2 NH_3(g)$$

(a) If the initial reaction quotient $Q_0$ is less than $K$, it increases with time; if it is greater than $K$, it decreases. (b) The pathways from part a are shown here in a graph of the partial pressure of ammonia against that of hydrogen (under conditions where there are three moles of $H_2$ for each mole of $N_2$). As the reaction goes toward equilibrium, the partial pressures approach a parabolic curve representing partial pressures that can coexist at chemical equilibrium.

## Example 5-7

The reaction between nitrogen and hydrogen to produce ammonia

$$N_2(g) + 3\,H_2(g) \rightleftharpoons 2\,NH_3(g)$$

is essential in making nitrogen-containing fertilizers. This reaction has an equilibrium constant equal to $1.9 \times 10^{-4}$ at 400°C. Suppose that 0.10 mol of $N_2$, 0.040 mol of $H_2$, and 0.020 mol of $NH_3$ are sealed in a 1.00-L vessel at 400°C. In which direction will the reaction proceed?

### Solution

The initial pressures $P_i = n_i\,RT/V$ are readily calculated to be

$$P_{N_2} = 5.5 \text{ atm} \qquad P_{H_2} = 2.2 \text{ atm} \qquad P_{NH_3} = 1.1 \text{ atm}$$

The initial numerical value of $Q$ is therefore

$$Q_0 = \frac{(P^{\circ}_{NH_3})^2}{P^{\circ}_{N_2}\,(P^{\circ}_{H_2})^3} = \frac{(1.1)^2}{(5.5)(2.2)^3} = 2.1 \times 10^{-2}$$

Because $Q_0 > K$, the reaction will proceed from right to left and ammonia will dissociate until equilibrium is reached.

*Related Problems:* 29, 30, 31, 32

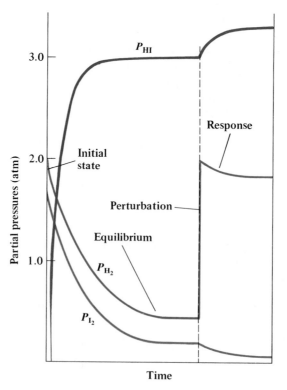

**Figure 5-5**

Partial pressures versus time for the equilibrium

$$H_2(g) + I_2(g) \rightleftharpoons 2\,HI(g)$$

The left part of the figure shows the attainment of equilibrium starting from the initial conditions of Example 5-4. Then the equilibrium state is abruptly perturbed by an increase in the partial pressure of $H_2$ to 2.000 atm. In accordance with Le Châtelier's principle, the system responds (Example 5-8) in such a way as to decrease the partial pressure of $H_2$—that is, to counteract the perturbation that moved it away from equilibrium in the first place.

### External Effects and Le Châtelier's Principle

Suppose a system at equilibrium is perturbed by some external stress such as a change in volume or temperature or a change in the partial pressure of one of the reactants or products. How will the system respond? The qualitative answer is embodied in a principle stated by Henri Le Châtelier in 1884:

> A system in equilibrium that is subjected to a stress will react in a way that tends to counteract the stress.

**Le Châtelier's principle** provides a way to predict the direction of change of a system under an external perturbation.

As a simple example, consider what happens when a small quantity of a reactant is added to an equilibrium mixture. The addition of reactant lowers the reaction quotient $Q$ below $K$ and a net reaction takes place in the forward direction, partially converting reactants to products, until $Q$ again equals $K$. The system partially counteracts the stress (the increase in the quantity of one of the reactants) and attains a new equilibrium state. If one of the *products* is added to an equilibrium mixture, $Q$ temporarily becomes *greater* than $K$ and a net *back* reaction occurs, partially counteracting the imposed stress by reducing the concentration of products (Fig. 5-5).

### Example 5-8

An equilibrium gas mixture of $H_2(g)$, $I_2(g)$, and $HI(g)$ at 600 K has

$$P_{H_2} = 0.4756 \text{ atm} \qquad P_{I_2} = 0.2056 \text{ atm} \qquad P_{HI} = 3.009 \text{ atm}$$

This is essentially the final equilibrium state of Example 5-4. Enough $H_2$ is added to increase its partial pressure to 2.000 atm at 600 K before any reaction takes place. The mixture then once again reaches equilibrium at 600 K. What are the final partial pressures of the three gases?

### Solution

Set up the usual table, in which "initial" now means the moment after the addition of the new $H_2$ but before it reacts further.

| | $H_2(g)$ + | $I_2(g)$ ⇌ | $2\,HI(g)$ |
|---|---|---|---|
| Initial partial pressure (atm) | 2.000 | 0.2056 | 3.009 |
| Change in partial pressure (atm) | $-x$ | $-x$ | $+2x$ |
| Equilibrium partial pressure (atm) | $2.000 - x$ | $0.2056 - x$ | $3.009 + 2x$ |

From Le Châtelier's principle it follows that net reaction to consume $H_2$ will follow addition of $H_2$, and this fact has been used in assigning a negative sign to the change in the partial pressure of $H_2$ in the table.

Substitution of the equilibrium partial pressures into the equilibrium law gives

$$\frac{(3.009 + 2x)^2}{(2.000 - x)(0.2056 - x)} = 92.6$$

Expansion of this expression results in the quadratic equation

$$88.60x^2 - 216.275x + 29.023 = 0$$

which can be solved to give

$$x = 0.1425 \quad \text{or} \quad 2.299$$

The second root would lead to negative partial pressures of $H_2$ and $I_2$ and is therefore physically impossible. Substitution of the first root into the expressions from the table gives

$$P_{H_2} = 2.000 - 0.1425 = 1.86 \text{ atm}$$

$$P_{I_2} = 0.2056 - 0.1425 = 0.063 \text{ atm}$$

$$P_{HI} = 3.009 + 2(0.1425) = 3.29 \text{ atm}$$

$$\text{Check:} \quad \frac{(3.29)^2}{(1.86)(0.063)} = 92.4$$

When HI is made from the elements, iodine is a much more expensive reactant than hydrogen. It therefore makes sense to add hydrogen to the reaction mixture (as in Example 5-8) to ensure more complete reaction of the iodine. If one of the products is removed from an equilibrium mixture, the reaction will also occur in the forward direction to compensate partially by increasing the partial pressures of products. Most industrial operations are designed in such a way that products can be removed continuously to achieve high overall yields, even for reactions with small equilibrium constants.

Le Châtelier's principle also predicts the effect of a change in volume on gas-phase equilibrium. A decrease in the volume of a gaseous system causes an increase in the total pressure, and the system responds, if possible, in such a way as to reduce the total pressure. In the equilibrium

$$2 \, P_2(g) \rightleftharpoons P_4(g)$$

for example, the reaction shifts in the forward direction when the volume is decreased. This occurs because every two molecules of $P_2$ consumed produce only one molecule of $P_4$, thus reducing the total pressure and partially compensating for the external stress arising from the change in volume. An *increase* in volume, on the other hand, favors reactants over products in this system, and some $P_4$ dissociates to form $P_2$ (Fig. 5-6). If there is no difference in the total numbers of gas-phase molecules on the two sides of the equation, then a change in volume has no effect on the equilibrium.

This effect of changing the volume of an equilibrium reacting mixture can also be understood by using the reaction quotient. For the phosphorus equilibrium just described, the reaction quotient is

$$Q = \frac{P_{P_4}}{(P_{P_2})^2}$$

Initially, $Q$ equals $K$. Suppose the volume is then decreased by a factor of 2; because the temperature is unchanged, this initially increases each partial pressure by a factor of 2. Because there are two powers of the pressure in the denominator and only one in the numerator, this decreases $Q$ by a factor of 2, making it lower than $K$. Reaction must then occur in the forward direction until $Q$ again equals $K$.

**Figure 5-6**

An equilibrium mixture of $P_2$ and $P_4$ (*center*) is compressed (*left*). Some $P_2$ molecules combine to give $P_4$ molecules, to reduce the total number of molecules and thus the total pressure. If the volume is increased (*right*), some $P_4$ molecules dissociate to pairs of $P_2$ molecules to increase the total number of molecules and the pressure exerted by those molecules.

When the volume of a system is decreased, its total pressure increases. Another way to increase the total pressure is to add an inert gas such as argon to the reaction mixture without changing the total volume. In this case, the effect on the equilibrium is entirely different. Because the partial pressures of the reactant and product gases are unchanged by an inert gas, adding argon at constant volume has no effect on the position of the equilibrium.

A final way to perturb an equilibrium system is to change its temperature. Chemical reactions are either **endothermic** (taking up heat from the surroundings) or **exothermic** (giving off heat). Raising the temperature of an equilibrium mixture by adding heat causes reaction to occur in such a way as to absorb some of the added heat. The equilibrium in an endothermic reaction shifts from left to right, while that in an exothermic reaction shifts from right to left, with "products" reacting to give "reactants." Equivalently, we can describe the shifts in terms of the effect of temperature on equilibrium constants. The equilibrium constant for an endothermic reaction increases with increasing temperature, while that for an exothermic reaction decreases with increasing temperature. The reaction of $NO_2$ to give $N_2O_4$ (discussed in Section 5-1) is an exothermic process, with a smaller equilibrium constant at higher temperatures.

As an application of Le Châtelier's principle, consider the reaction

$$N_2(g) + 3 H_2(g) \rightleftharpoons 2 NH_3(g)$$

which is the basis of the industrial synthesis of ammonia. Because this reaction is exothermic, the yield of ammonia is increased by working at as low a temperature as possible (Fig. 5-7a). At too low a temperature, however, the reaction is very slow, so a compromise temperature near 500°C is usually employed. Because the number of moles of gas decreases as the reaction occurs, the yield of product is enhanced by decreasing the volume of the reaction vessel. Typically, total pressures of 150 to 300 atm are used (Fig. 5-7b), although some plants work at up to 900 atm of pressure. Even at high pressures, however, the yield of ammonia is usually only 15% to 20% because the equilibrium constant is so small. To overcome this, ammonia plants employ a cyclic process in which the gas mixture is cooled so that the ammonia liquefies (its boiling point is much higher than those of nitrogen and hydrogen) and is removed. Continuous removal of products helps to drive the reaction to completion. We will discuss the industrial production of ammonia in Chapter 21.

Le Châtelier's principle is a qualitative way of summarizing certain properties of equilibrium states, and in this regard it is a precursor of the more complete and quantitative thermodynamic analysis presented in Chapters 8 and 9. At that point,

(a)

(b)

**Figure 5-7**

(a) The equilibrium mole percentage of ammonia in a 1:3 mixture of $N_2$ and $H_2$ varies with temperature; low temperatures favor high yields of $NH_3$. The data shown correspond to a fixed total pressure of 300 atm. (b) At a fixed temperature (here, 500°C), the yield of $NH_3$ increases with increasing total pressure.

heat effects in chemical reactions and the effect of temperature on equilibrium will be discussed in greater depth.

## 5-5 HETEROGENEOUS EQUILIBRIA

We turn now to equilibria that involve solids, liquids, and dissolved species as well as gases. Recall the phase equilibrium between liquid water and water vapor from Section 4-1:

$$H_2O(\ell) \rightleftharpoons H_2O(g) \qquad\qquad P_{H_2O} = K$$

As long as *some* liquid water is in the container, the pressure of water vapor at 25°C will be 0.03126 atm. The position of this equilibrium is not affected by the amount of liquid water present, and the liquid does not appear in the equilibrium expression. The reason is that the concentration of liquid water is a constant at a given temperature and pressure. An analogous situation occurs in the equilibrium between solid iodine and iodine dissolved in aqueous solution:

$$I_2(s) \rightleftharpoons I_2(aq) \qquad\qquad [I_2] = K$$

The position of the equilibrium (given by the concentration of $I_2$ dissolved at a given temperature) is independent of the amount of solid present, as long as there is some; the pure solid does not appear in the equilibrium law. A dissolved species—here, $I_2(aq)$—enters through its *concentration,* expressed in moles per liter.

Another example involves a chemical reaction, the decomposition of calcium carbonate:

$$CaCO_3(s) \rightleftharpoons CaO(s) + CO_2(g) \qquad\qquad P_{CO_2} = K$$

If calcium carbonate is heated, it decomposes into calcium oxide (lime) and carbon dioxide (Fig. 5-8); the reverse reaction is favored at sufficiently high pressures of carbon dioxide. The equilibrium can be studied experimentally, and it is found that at any given temperature the pressure of $CO_2(g)$ is constant, independent of the amounts of $CaCO_3(s)$ and $CaO(s)$, as long as some of each is present (Fig. 5-9). The two pure solids do not enter the equilibrium expression, and the equilibrium constant reduces simply to the partial pressure of carbon dioxide.

The more general form of the **law of mass action** can now be stated:

1. Gases enter equilibrium expressions as partial pressures, in atmospheres.
2. Dissolved species enter as concentrations, in moles per liter.
3. Pure solids and pure liquids do not appear in equilibrium expressions; neither does a solvent taking part in a chemical reaction, provided the solution is dilute.
4. Partial pressures and concentrations of products appear in the numerator, and those of reactants in the denominator, each raised to a power equal to its coefficient in the balanced chemical equation.

Equilibrium expressions written in this way are valid to reasonable accuracy when the pressures of gases do not exceed several atmospheres and the concentrations of solutes do not exceed 0.1 M. At higher pressures or concentrations, where gases and solutions depart seriously from ideality, more advanced methods of treatment are required.

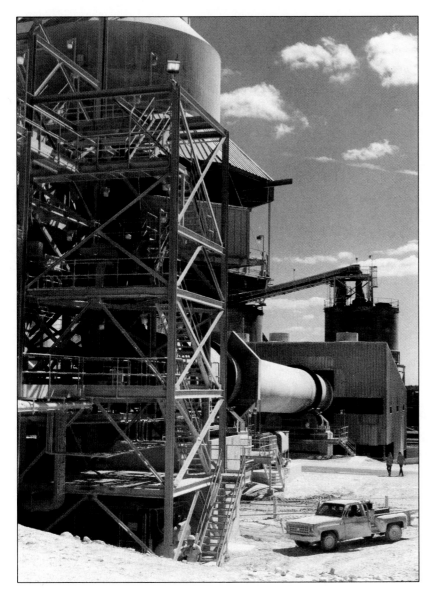

**Figure 5-8**
The conversion of limestone to lime through the reaction

$$CaCO_3(s) \longrightarrow CaO(s) + CO_2(g)$$

is carried out at high temperatures in giant rotary lime kilns. Lime is used to make chemicals as well as in building materials such as glass and cement. (National Lime Association)

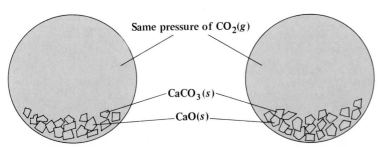

Much $CaCO_3(s)$, little $CaO(s)$          Little $CaCO_3(s)$, much $CaO(s)$

**Figure 5-9**
As long as both $CaO(s)$ and $CaCO_3(s)$ are present at equilibrium in a closed container, the partial pressure of $CO_2(g)$ at a fixed temperature does not depend on the amounts of the two solids present.

## Example 5-9

Hypochlorous acid (HOCl) is produced by bubbling chlorine through an agitated suspension of mercury(II) oxide in water. The chemical equation for this process is

$$2\,Cl_2(g) + 2\,HgO(s) + H_2O(\ell) \rightleftharpoons HgO\cdot HgCl_2(s) + 2\,HOCl(aq)$$

Write the equilibrium expression for this reaction.

### Solution

$$\frac{[HOCl]^2}{P_{Cl_2}^2} = K$$

The HgO and HgO·HgCl$_2$ do not appear because they are solids, and water does not appear because it is the solvent. Chlorine, as a gas, enters in the form of its activity, numerically equal to its partial pressure in atmospheres. The activity of HOCl, a dissolved species, is numerically equal to its concentration in moles per liter. Both the concentration of HOCl and the partial pressure of Cl$_2$ are raised to the second power because their coefficients are 2 in the balanced chemical equation.

*Related Problems:* 41, 42, 43, 44

## Example 5-10

Graphite is added to a vessel that contains only CO$_2$(g) at a certain high temperature. The pressure rises as a reaction occurs that produces CO(g):

$$C(s) + CO_2(g) \rightleftharpoons 2\,CO(g) \qquad \frac{(P_{CO})^2}{P_{CO_2}} = K = 4.2$$

and the total pressure reaches an equilibrium value of 1.40 atm. Calculate the equilibrium partial pressure of the CO.

### Solution

There is no purpose in setting up a table of partial pressures, because there is no information on the initial partial pressure of CO$_2$. Instead, call the equilibrium partial pressure of CO $y$, so $1.40 - y$ is the equilibrium partial pressure of CO$_2$. The equilibrium expression then becomes

$$\frac{y^2}{1.40 - y} = 4.2$$

which is solved to give $y = 1.1$. The partial pressure of CO is 1.1 atm at equilibrium.

*Related Problems:* 49, 50

## 5-6 EXTRACTION AND SEPARATION PROCESSES

An important type of equilibrium involves partitioning a solute species between two immiscible solvent phases. Such equilibria are used in many separation processes in chemical research and in industry.

Suppose two immiscible liquids, such as water and carbon tetrachloride, are put in a container. "Immiscible" means mutually insoluble; these liquids separate into two phases with the less dense liquid, in this case water, lying on top of the other liquid. A visible boundary, the meniscus, separates the two phases. If a solute such as iodine is added to the mixture and the vessel is shaken to distribute the iodine through the container (Fig. 5-10), the iodine is partitioned between the two phases at equilibrium with a characteristic concentration ratio, the **partition coefficient** $K$. This is the equilibrium constant for the process

$$I_2(aq) \rightleftharpoons I_2(CCl_4)$$

and can be written as

$$\frac{[I_2]_{CCl_4}}{[I_2]_{aq}} = K$$

in which $[I_2]_{CCl_4}$ and $[I_2]_{aq}$ are the concentrations (in moles per liter) of $I_2$ in the $CCl_4$ and aqueous phases, respectively. At 25°C, $K$ has the value 85 for this equilibrium. The fact that $K$ is greater than 1 shows that iodine is more soluble in $CCl_4$ than in water. If iodide ion ($I^-$) is dissolved in the water, then it can react with iodine to form the triiodide ion $I_3^-$:

$$I_2(aq) + I^-(aq) \longrightarrow I_3^-(aq)$$

This consumes $I_2(aq)$ and, by Le Châtelier's principle, causes more $I_2$ in the first equilibrium to move from the $CCl_4$ phase to the aqueous phase.

**Extraction** takes advantage of the partitioning of a solute between two immiscible solvents to remove that solute from one solvent into another. Suppose iodine is present as a contaminant in water that also contains other solutes which are insoluble in carbon tetrachloride. In such a case, most of the iodine could be removed by shaking the aqueous solution with $CCl_4$, allowing the two phases to separate, and then pouring off the water layer from the heavier layer of carbon tetrachloride. The greater the equilibrium constant for the partition of a solute from the original solvent into the extracting solvent, the more complete such a separation will be.

**Figure 5-10**

Iodine is dissolved in water and poured on top of carbon tetrachloride in a separatory funnel (*left*). After the funnel is shaken (*right*), the iodine reaches a partition equilibrium between the upper (aqueous) phase and the lower ($CCl_4$) phase. The deeper color in the lower phase indicates that iodine dissolves preferentially in the denser $CCl_4$ phase. (Leon Lewandowski)

### Example 5-11

An aqueous solution has an iodine concentration of $2.00 \times 10^{-3}$ M. Calculate the percentage of iodine remaining in the aqueous phase after extraction of 0.100 L of this aqueous solution with 0.050 L of $CCl_4$ at 25°C.

### Solution

The chemical amount of $I_2$ present is

$$(2.00 \times 10^{-3} \text{ mol L}^{-1})(0.100 \text{ L}) = 2.00 \times 10^{-4} \text{ mol}$$

Suppose that $y$ mol remains in the aqueous phase and $2.00 \times 10^{-4} - y$ mol passes into the $CCl_4$ phase. Then

$$\frac{[I_2]_{CCl_4}}{[I_2]_{aq}} = K = 85$$

$$= \frac{(2.00 \times 10^{-4} - y)/0.050}{y/0.100}$$

$$= \frac{2(2.00 \times 10^{-4} - y)}{y}$$

Note that the volumes of the two solvents used are unchanged because they are immiscible. Solving for $y$ gives

$$y = 4.6 \times 10^{-6} \, \text{mol}$$

The fraction remaining in the aqueous phase is $4.6 \times 10^{-6}/2.0 \times 10^{-4} = 0.023$, or 2.3%. Additional extractions could be carried out to remove more of the $I_2$ from the aqueous phase.

One extraction process used industrially on a large scale is the purification of sodium hydroxide for use in the manufacture of rayon. The sodium hydroxide produced by electrolysis (see Section 22-1) typically contains 1% sodium chloride and 0.1% sodium chlorate as impurities. If a concentrated aqueous solution of sodium hydroxide is extracted with liquid ammonia, the NaCl and $NaClO_3$ are partitioned into the ammonia phase in preference over the aqueous phase. The heavier aqueous phase is added to the top of an extraction vessel filled with ammonia, and equilibrium is reached as droplets of it settle through the ammonia phase to the bottom. This procedure reduces impurity concentrations in the sodium hydroxide solution to about 0.08% NaCl and 0.0002% $NaClO_3$.

## Chromatographic Separations

Partition equilibria are the basis of an important class of separation techniques called **chromatography.** This word comes from the Greek root *chroma,* meaning "color," and was chosen because the original chromatographic separations involved colored substances. The technique can be applied to a wide variety of mixtures of substances.

Chromatography is a continuous extraction process in which solute species are exchanged between two phases. One, the mobile phase, moves with respect to the other, stationary phase. The partition ratio $K$ of a solute A between the stationary and mobile phases is

$$\frac{[A]_{stationary}}{[A]_{mobile}} = K$$

As the mobile phase containing solute passes over the stationary phase, the solute molecules move between the two phases. True equilibrium is never fully established because the motion of the fluid phase continually brings fresh solvent into contact with the stationary phase. Nevertheless, the partition coefficient $K$ provides a guide to the behavior of a particular solute. The greater $K$ is, the more time the solute

Paper chromatography separates a mixture of watercolor paints into their component colors. The solvent rises through the paper, drawing different colors along at different rates.

spends in the stationary phase and therefore the slower its progress through the separation system. Solutes with different values of $K$ are separated by their different rates of travel.

Different types of chromatography use different mobile and stationary phases; Table 5-1 (page 198) lists some of the most important. **Column chromatography** (Fig. 5-11) uses a tube packed with a porous material, frequently a silica gel on which water has been adsorbed. Water is therefore the stationary liquid phase in this case. Other solvents such as pyridine or benzene are used in the mobile phase; in some cases it is most efficient to use different solvents in succession to separate the

(a)                                   (b)

## Figure 5-11

(a) In a column chromatograph, the top of the column is loaded with a mixture of solutes to be separated (green). (b) With addition of solvent, the different solutes travel at different rates, giving rise to bands. The separate fractions can be collected in different flasks for use or analysis.

**Table 5-1   Chromatographic Separation Techniques[†]**

| Name | Mobile Phase | Stationary Phase |
|------|--------------|------------------|
| Gas–liquid | Gas | Liquid adsorbed on a porous solid in a tube |
| Gas–solid | Gas | Porous solid in a tube |
| Column | Liquid | Liquid adsorbed on a porous solid in a tubular column |
| Paper | Liquid | Liquid held in the pores of a thick paper |
| Thin layer | Liquid | Liquid or solid; solid is held on a glass plate and liquid may be adsorbed on it |
| Ion exchange | Liquid | Solid (finely divided ion-exchange resin) in a tubular column |

[†]Adapted from D. A. Skoog and D. M. West, *Analytical Chemistry* (Saunders College Publishing, Philadelphia, 1980), Table 18-1.

components of a solute mixture. As the solute fractions reach the bottom of the column, they are separated for analysis or use. Column chromatography is important in industry because it is easily increased from laboratory to production scale.

**Gas–liquid chromatography** (Fig. 5-12) is one of the most important separation techniques for modern chemical research. The stationary phase is again a liquid adsorbed on a porous solid, but the mobile phase is now gaseous. The sample is vaporized and passes through the column, carried in a stream of an inert gas such as helium or nitrogen. The residence time in the column depends on the partition coefficient of the solute species, allowing an efficient separation of mixtures. The

## Figure 5-12

In a gas–liquid chromatograph, the sample is vaporized and passes through a column, carried in a stream of an inert gas such as helium or nitrogen. The residence time of any substance on the column depends on its partition coefficient from the vapor to the liquid in the column. A species leaving the column at a given time can be detected by a variety of techniques. The result is a gas chromatogram, with a peak corresponding to each substance in the mixture.

solute leaving the column at a given time can be detected by a variety of techniques that produce a **gas chromatogram** with a peak corresponding to each solute species in the mixture. Gas–liquid chromatography is widely employed for separating the products of organic reactions. It can also be used to determine the purity of substances, because even very small amounts of impurities appear clearly as separate peaks in the chromatogram. The technique is important in the separation and identification of trace amounts of possibly toxic substances in environmental and biological samples. Amounts on the order of parts per trillion ($10^{-12}$ g in a 1-g sample) can be detected and identified.

## Sulfuric Acid

Sulfuric acid is produced in larger volume than any other chemical and has a tremendous number of applications ranging from fertilizer manufacture to metal treatment and chemical synthesis. Its production, properties, and uses are discussed in detail in Chapter 21.

The modern industrial production of sulfuric acid involves three steps, for which the balanced chemical equations are

**1.** $S(s) + O_2(g) \rightleftharpoons SO_2(g)$
**2.** $SO_2(g) + \frac{1}{2} O_2(g) \rightleftharpoons SO_3(g)$
**3.** $SO_3(g) + H_2O(\ell) \rightleftharpoons H_2SO_4(\ell)$

(a) Write an equilibrium expression for each of these steps, with equilibrium constants $K_1$, $K_2$, and $K_3$.
(b) If these reactions could be carried out at 25°C, the equilibrium constants would be $3.9 \times 10^{52}$, $2.6 \times 10^{12}$, and $2.2 \times 10^{14}$. Write a balanced equation for the overall reaction and calculate its equilibrium constant at 25°C.
(c) Although the products of all three equilibria are strongly favored at 25°C [part (b)], reactions 1 and 2 occur too slowly to be practical; they must be carried out at elevated temperatures. At 700°C, the partial pressures of $SO_2$, $O_2$, and $SO_3$ in an equilibrium mixture are measured to be 2.23 atm, 1.14 atm, and 6.26 atm, respectively. Calculate $K_2$ at 700°C.
(d) At 300°C, $K_2$ has the value $1.3 \times 10^4$. Suppose some $SO_3$ is introduced into an evacuated vessel at an initial partial pressure of 0.89 atm. Calculate the partial pressure of $SO_2$ that will be reached at equilibrium, assuming that only reaction 2 takes place under these conditions. (*Hint:* $K_2$ is large enough that you can try the approximation that the fraction of $SO_3$ dissociated is very small.)
(e) Some $SO_2$ is placed in a flask and heated with oxygen to 600°C, at which point $K_2 = 9.5$. At equilibrium, 62% of it has reacted to give $SO_3$. Calculate the partial pressure of oxygen at equilibrium in this reaction mixture.
(f) Equal chemical amounts of $SO_2$, $O_2$, and $SO_3$ are mixed and heated to 600°C, where their total pressure before reaction is 0.090 atm. Will reaction 2 occur from right to left or from left to right? Will the total pressure increase or decrease during the course of the reaction?
(g) Reactions 1 and 3 are both exothermic. State the effects on equilibria 1 and 3 of increasing the temperature and of decreasing the volume. (*Note:* A change in volume has little effect on liquids and solids taking part in a reaction.)

A pile of sulfur awaiting conversion into sulfuric acid. (© 1991 Stefan Schulhof)

**Answers**

(a)
$$\frac{P_{SO_2}}{P_{O_2}} = K_1 \qquad \frac{P_{SO_3}}{P_{SO_2}P_{O_2}^{1/2}} = K_2 \qquad \frac{1}{P_{SO_3}} = K_3$$

(b) $S(s) + \frac{3}{2}O_2(g) + H_2O(\ell) \rightleftharpoons H_2SO_4(\ell); \quad K = K_1K_2K_3 = 2.2 \times 10^{79}$

(c) $K_2$ is 2.63.

(d) The $SO_2$ partial pressure is $2.1 \times 10^{-3}$ atm.

(e) The $O_2$ partial pressure is 0.029 atm.

(f) Left to right; pressure will decrease.

(g) Increasing the temperature will shift both equilibria to the left. Decreasing the volume will not affect equilibrium 1 and will shift equilibrium 3 to the right.

## CONCEPTS & SKILLS

After studying this chapter and working the problems that follow, you should be able to

1. Set up an expression relating the partial pressures of the reactants and products that are in equilibrium in a chemical reaction (Section 5-1, problems 1–4).
2. Calculate equilibrium constants from experimental measurements of partial and total pressures (Section 5-2, problems 7–12).
3. Calculate the equilibrium partial pressures of all species involved in a gas-phase chemical reaction from the initial pressure(s) of the reactants (Section 5-3, problems 17–26).
4. Relate concentrations to partial pressures in equilibrium calculations (Section 5-3, problems 27–28).
5. Determine the direction in which a chemical reaction will proceed spontaneously by calculating its reaction quotient (Section 5-4, problems 29–30).
6. State Le Châtelier's principle and give several applications (Section 5-4, problems 35–40).
7. Formulate and use the equilibrium expression for heterogeneous reactions (Section 5-5, problems 41–54).
8. Use the law of mass action to explain the distribution of a solute between two immiscible solvents (Section 5-6, problems 55–58).
9. Outline the basis for the separation of compounds by partition chromatography (Section 5-6).

## PROBLEMS

### Chemical Reactions and Equilibrium

1. Write equilibrium expressions for the following gas-phase reactions.
   (a) $2H_2(g) + O_2(g) \rightleftharpoons 2H_2O(g)$
   (b) $Xe(g) + 3F_2(g) \rightleftharpoons XeF_6(g)$
   (c) $2C_6H_6(g) + 15O_2(g) \rightleftharpoons 12CO_2(g) + 6H_2O(g)$
2. Write equilibrium expressions for the following gas-phase reactions.

   (a) $2Cl_2(g) + O_2(g) \rightleftharpoons 2Cl_2O(g)$
   (b) $N_2(g) + O_2(g) + Br_2(g) \rightleftharpoons 2NOBr(g)$
   (c) $C_3H_8(g) + 5O_2(g) \rightleftharpoons 3CO_2(g) + 4H_2O(g)$
3. At a moderately elevated temperature, phosphoryl chloride ($POCl_3$) can be produced in the vapor phase from the gaseous elements. Write a balanced chemical equation and an equilibrium expression for this system. Note that gaseous phosphorus consists of $P_4$ molecules at moderate temperatures.

**4.** If confined at high temperature, ammonia and oxygen quickly react and come to equilibrium with their products, water vapor and nitrogen oxide. Write a balanced chemical equation and an equilibrium expression for this system.

**5.** An important step in the industrial production of hydrogen is the reaction of carbon monoxide with water:

$$CO(g) + H_2O(g) \rightleftharpoons CO_2(g) + H_2(g)$$

(a) Use the law of mass action to write the equilibrium expression for this reaction.

(b) At 500°C, the equilibrium constant for this reaction is 3.9. Suppose that the equilibrium partial pressures of CO and $H_2O$ are both 0.10 atm and that of $CO_2$ is 0.70 atm. Calculate the equilibrium partial pressure of $H_2(g)$.

**6.** Phosgene ($COCl_2$) is an important intermediate in the manufacture of certain plastics. It is produced by the reaction

$$CO(g) + Cl_2(g) \rightleftharpoons COCl_2(g)$$

(a) Use the law of mass action to write the equilibrium expression for this reaction.

(b) At 600°C, the equilibrium constant for this reaction is 0.20. Calculate the partial pressure of phosgene in equilibrium with a mixture of CO (at 0.0020 atm) and $Cl_2$ (at 0.00030 atm).

## Calculating Equilibrium Constants

**7.** At 454 K, $Al_2Cl_6(g)$ reacts to form $Al_3Cl_9(g)$ according to the equation

$$3\ Al_2Cl_6(g) \rightleftharpoons 2\ Al_3Cl_9(g)$$

In an experiment at this temperature, the equilibrium partial pressure of $Al_2Cl_6(g)$ is 1.00 atm and the equilibrium partial pressure of $Al_3Cl_9(g)$ is $1.02 \times 10^{-2}$ atm. Compute the equilibrium constant of the above reaction at 454 K.

**8.** At 298 K, $F_3SSF(g)$ decomposes partially to $SF_2(g)$. At equilibrium, the partial pressure of $SF_2(g)$ is $1.1 \times 10^{-4}$ atm and the partial pressure of $F_3SSF$ is 0.0484 atm.

(a) Write a balanced equilibrium equation to represent this reaction.

(b) Compute the equilibrium constant corresponding to the equation you wrote.

**9.** The compound 1,3-di-*t*-butylcyclohexane exists in two forms that are known as the chair and boat conformations because their molecular structures resemble those objects. Equilibrium exists between the two forms, represented by the equation

$$chair \rightleftharpoons boat$$

At 580 K, 6.42% of the molecules are in the chair form. Calculate the equilibrium constant for the reaction as written above.

**10.** At 248°C and a total pressure of 1.000 atm, the fractional dissociation of $SbCl_5$ is 0.718 for the reaction

$$SbCl_5(g) \rightleftharpoons SbCl_3(g) + Cl_2(g)$$

This means that 718 out of every 1000 molecules of $SbCl_5$ originally present have dissociated. Calculate the equilibrium constant.

**11.** Sulfuryl chloride ($SO_2Cl_2$) is a colorless liquid that boils at 69°C. Above this temperature, the vapors dissociate into sulfur dioxide and chlorine:

$$SO_2Cl_2(g) \rightleftharpoons SO_2(g) + Cl_2(g)$$

This reaction is slow at 100°C, but it is accelerated by the presence of some $FeCl_3$ (which does not affect the final position of the equilibrium). In an experiment, 3.174 g of $SO_2Cl_2(\ell)$ and a small amount of solid $FeCl_3$ are put into an evacuated 1.000-L flask, which is then sealed and heated to 100°C. The total pressure in the flask at that temperature is found to be 1.30 atm.

(a) Calculate the partial pressure of each of the three gases present.

(b) Calculate the equilibrium constant at this temperature.

**12.** A certain amount of $NOBr(g)$ is sealed in a flask, and the temperature is raised to 350 K. The following equilibrium is established:

$$NOBr(g) \rightleftharpoons NO(g) + \tfrac{1}{2}Br_2(g)$$

The total pressure in the flask when equilibrium is reached at this temperature is 0.675 atm, and the vapor density is $2.219\ \text{g L}^{-1}$.

(a) Calculate the partial pressure of each species.

(b) Calculate the equilibrium constant at this temperature.

**13.** At a certain temperature, the value of the equilibrium constant for the reaction

$$CS_2(g) + 3\ O_2(g) \rightleftharpoons CO_2(g) + 2\ SO_2(g)$$

is $K_1$. How is $K_1$ related to the equilibrium constant $K_2$ for the related equilibrium

$$\tfrac{1}{3}CS_2(g) + O_2(g) \rightleftharpoons \tfrac{1}{3}CO_2(g) + \tfrac{2}{3}SO_2(g)$$

at the same temperature?

**14.** At 25°C, the equilibrium constant for the reaction

$$6\ ClO_3F(g) \rightleftharpoons 2\ ClF(g) + 4\ ClO(g) + 7\ O_2(g) + 2\ F_2(g)$$

is 32.6. Calculate the equilibrium constant at 25°C for the reaction

$$\tfrac{1}{3}ClF(g) + \tfrac{2}{3}ClO(g) + \tfrac{7}{6}O_2(g) + \tfrac{1}{3}F_2(g) \rightleftharpoons ClO_3F(g)$$

**15.** Suppose that $K_1$ and $K_2$ are the respective equilibrium constants for the two reactions

$$XeF_6(g) + H_2O(g) \rightleftharpoons XeOF_4(g) + 2\ HF(g)$$

$$XeO_4(g) + XeF_6(g) \rightleftharpoons XeOF_4(g) + XeO_3F_2(g)$$

Give the equilibrium constant for the reaction

$$XeO_4(g) + 2\ HF(g) \rightleftharpoons XeO_3F_2(g) + H_2O(g)$$

in terms of $K_1$ and $K_2$.

16. At 1330 K, germanium(II) oxide (GeO) and tungsten(VI) oxide ($W_2O_6$) are both gases. The following two equilibria are established simultaneously:

$$2\ GeO(g) + W_2O_6(g) \rightleftharpoons 2\ GeWO_4(g)$$

$$GeO(g) + W_2O_6(g) \rightleftharpoons GeW_2O_7(g)$$

The equilibrium constants for the two are respectively $7.0 \times 10^3$ and $38 \times 10^3$. Compute $K$ for the reaction

$$GeO(g) + GeW_2O_7(g) \rightleftharpoons 2\ GeWO_4(g)$$

## Calculating Gas-Phase Equilibria

17. The dehydrogenation of benzyl alcohol to make the flavoring agent benzaldehyde is an equilibrium process described by the equation

$$C_6H_5CH_2OH(g) \rightleftharpoons C_6H_5CHO(g) + H_2(g)$$

At 523 K, the value of its equilibrium constant is $K = 0.558$.
(a) Suppose 1.20 g of benzyl alcohol is placed in a 2.00-L vessel and heated to 523 K. What is the partial pressure of benzaldehyde when equilibrium is attained?
(b) What fraction of benzyl alcohol is dissociated into products at equilibrium?

18. Isopropyl alcohol can dissociate into acetone and hydrogen:

$$(CH_3)_2CHOH(g) \rightleftharpoons (CH_3)_2CO(g) + H_2(g)$$

At 179°C, the equilibrium constant for this dehydrogenation reaction is 0.444.
(a) If 10.00 g of isopropyl alcohol is placed in a 10.00-L vessel and heated to 179°C, what is the partial pressure of acetone when equilibrium is attained?
(b) What fraction of isopropyl alcohol is dissociated at equilibrium?

19. A weighed quantity of $PCl_5(s)$ is sealed in a 100.0-cm$^3$ glass bulb to which a pressure gauge is attached. The bulb is heated to 250°C, and the gauge shows that the pressure in the bulb rises to 0.895 atm. At this temperature, the solid $PCl_5$ is all vaporized and also partially dissociated into $Cl_2(g)$ and $PCl_3(g)$ according to the equation

$$PCl_5(g) \rightleftharpoons PCl_3(g) + Cl_2(g)$$

At 250°C, $K = 2.15$ for this reaction. Assume that the contents of the bulb are at equilibrium and calculate the partial pressure of the three different chemical species in the vessel.

20. Suppose 93.0 g of HI(g) is placed in a glass vessel and heated to 1107 K. At this temperature, equilibrium is quickly established between HI(g) and its decomposition products, $H_2(g)$ and $I_2(g)$:

$$2\ HI(g) \rightleftharpoons H_2(g) + I_2(g)$$

The equilibrium constant at 1107 K is 0.0259, and the total pressure at equilibrium is observed to equal 6.45 atm. Calculate the equilibrium partial pressures of HI(g), $H_2(g)$, and $I_2(g)$.

21. The equilibrium constant at 350 K for the reaction

$$Br_2(g) + I_2(g) \rightleftharpoons 2\ IBr(g)$$

has a value of 322. Bromine at an initial partial pressure of 0.0500 atm is mixed with iodine at an initial partial pressure of 0.0400 atm and held at 350 K until equilibrium is reached. Calculate the equilibrium partial pressure of each of the gases.

22. The equilibrium constant for the reaction of fluorine and oxygen to form oxygen difluoride ($OF_2$) is 40.1 at 298 K:

$$F_2(g) + \tfrac{1}{2}O_2(g) \rightleftharpoons OF_2(g)$$

Suppose some $OF_2$ is introduced into an evacuated container at 298 K and allowed to dissociate until its partial pressure reaches an equilibrium value of 1.00 atm. Calculate the equilibrium partial pressures of $F_2$ and $O_2$ in the container.

23. At 25°C, the equilibrium constant for the reaction

$$N_2(g) + O_2(g) \rightleftharpoons 2\ NO(g)$$

is $4.2 \times 10^{-31}$. Suppose a container is filled with nitrogen (at an initial partial pressure of 0.41 atm), oxygen (at an initial partial pressure of 0.59 atm), and nitrogen oxide (at an initial partial pressure of 0.22 atm). Calculate the partial pressures of all three gases after equilibrium is reached at this temperature.

24. At 25°C, the equilibrium constant for the reaction

$$2\ NO_2(g) \rightleftharpoons 2\ NO(g) + O_2(g)$$

is $5.9 \times 10^{-13}$. Suppose a container is filled with nitrogen dioxide at an initial partial pressure of 0.89 atm. Calculate the partial pressures of all three gases after equilibrium is reached at this temperature.

*25. The equilibrium constant for the synthesis of ammonia

$$N_2(g) + 3\ H_2(g) \rightleftharpoons 2\ NH_3(g)$$

has the value $K = 6.78 \times 10^5$ at 25°C. Calculate the equilibrium partial pressures of $N_2(g)$, $H_2(g)$, and $NH_3(g)$ at 25°C if the total pressure is 1.00 atm and the H:N atom ratio in the system is 3:1. (*Hint:* Try the approximation that $P_{N_2}$ and $P_{H_2} \ll P_{NH_3}$ and see if the resulting equations are simplified.)

*26. At 400°C, $K = 3.19 \times 10^{-4}$ for the reaction in problem 25. Repeat the calculation for $P_{N_2}$, $P_{H_2}$, and $P_{NH_3}$, assuming the same total pressure and composition. (*Hint:* Try the approximation that $P_{NH_3} \ll P_{N_2}$ and $P_{H_2}$ and see if the resulting equations are simplified.)

27. Calculate the concentration of phosgene ($COCl_2$) that will be present at 600°C in equilibrium with carbon monoxide (at a concentration of $2.3 \times 10^{-4}$ mol L$^{-1}$) and chlorine (at a con-

centration of $1.7 \times 10^{-2}$ mol L$^{-1}$). (Use the data of problem 6.)

**28.** The reaction

$$SO_2Cl_2(g) \rightleftharpoons SO_2(g) + Cl_2(g)$$

has an equilibrium constant at 100°C of 2.40. Calculate the concentration of $SO_2$ that will be present at 100°C in equilibrium with $SO_2Cl_2$ (at a concentration of $3.6 \times 10^{-4}$ mol L$^{-1}$) and chlorine (at a concentration of $6.9 \times 10^{-3}$ mol L$^{-1}$).

## The Direction of Change

**29.** Some $Al_2Cl_6$ (at a partial pressure of 0.473 atm) is placed in a closed container at 454 K with some $Al_3Cl_9$ (at a partial pressure of $1.02 \times 10^{-2}$ atm). Enough argon is added to raise the total pressure to 1.00 atm.
(a) Calculate the initial reaction quotient for the reaction

$$3\,Al_2Cl_6(g) \rightleftharpoons 2\,Al_3Cl_9(g)$$

(b) As the gas mixture reaches equilibrium, will there be net production or consumption of $Al_3Cl_9$? (Use the data given in problem 7.)

**30.** Some $SF_2$ (at a partial pressure of $2.3 \times 10^{-4}$ atm) is placed in a closed container at 298 K with some $F_3SSF$ (at a partial pressure of 0.0484 atm). Enough argon is added to raise the total pressure to 1.000 atm.
(a) Calculate the initial reaction quotient for the decomposition of $F_3SSF$ to $SF_2$.
(b) As the gas mixture reaches equilibrium, will there be net formation or dissociation of $F_3SSF$? (Use the data given in problem 8.)

**31.** The progress of the reaction

$$H_2(g) + Br_2(g) \rightleftharpoons 2\,HBr(g)$$

can be monitored visually by following changes in the color of the reaction mixture ($Br_2$ is reddish brown, and $H_2$ and HBr are colorless). A gas mixture is prepared at 700 K, in which 0.40 atm is the initial partial pressure of both $H_2$ and $Br_2$ and 0.90 atm is the initial partial pressure of HBr. The color of this mixture then fades as the reaction progresses toward equilibrium. Give a condition that must be satisfied by the equilibrium constant $K$ (for example, it must be greater than or smaller than a given number).

**32.** Recall from our discussion of the $NO_2$–$N_2O_4$ equilibrium that $NO_2$ has a brownish color. At elevated temperatures, $NO_2$ reacts with CO according to

$$NO_2(g) + CO(g) \rightleftharpoons NO(g) + CO_2(g)$$

The other three gases taking part in this reaction are colorless. When a gas mixture is prepared at 500 K, in which 3.4 atm is the initial partial pressure of both $NO_2$ and CO, and 1.4 atm is the partial pressure of both NO and $CO_2$, the brown color of the mixture is observed to fade as the reaction progresses

toward equilibrium. Give a condition that must be satisfied by the equilibrium constant $K$ (for example, it must be greater than or smaller than a given number).

**33.** At $T = 1200$°C, the reaction

$$P_4(g) \rightleftharpoons 2\,P_2(g)$$

has an equilibrium constant $K = 0.612$.
(a) Suppose the initial partial pressure of $P_4$ is 5.00 atm and that of $P_2$ is 2.00 atm. Calculate the reaction quotient $Q$ and state whether the reaction proceeds to the right or to the left as equilibrium is approached.
(b) Calculate the partial pressures at equilibrium.
(c) If the volume of the system is then increased, will there be net formation or net dissociation of $P_4$?

**34.** At $T = 100$°C, the reaction

$$SO_2Cl_2(g) \rightleftharpoons SO_2(g) + Cl_2(g)$$

has an equilibrium constant $K = 2.4$.
(a) Suppose the initial partial pressure of $SO_2Cl_2$ is 1.20 atm, and $P_{SO_2} = P_{Cl_2} = 0$. Calculate the reaction quotient $Q$ and state whether the reaction proceeds to the right or to the left as equilibrium is approached.
(b) Calculate the partial pressures at equilibrium.
(c) If the volume of the system is then decreased, will there be net formation or net dissociation of $SO_2Cl_2$?

**35.** Explain the effect of each of the following stresses on the position of the following equilibrium:

$$3\,NO(g) \rightleftharpoons N_2O(g) + NO_2(g)$$

The reaction as written is exothermic.
(a) $N_2O(g)$ is added to the equilibrium mixture without change of volume or temperature.
(b) The volume of the equilibrium mixture is reduced at constant temperature.
(c) The equilibrium mixture is cooled.
(d) Gaseous argon (which does not react) is added to the equilibrium mixture while both the total gas pressure and the temperature are kept constant.
(e) Gaseous argon is added to the equilibrium mixture without changing the volume.

**36.** Explain the effect of each of the following stresses on the position of the equilibrium

$$SO_3(g) \rightleftharpoons SO_2(g) + \tfrac{1}{2}O_2(g)$$

The reaction as written is endothermic.
(a) $O_2(g)$ is added to the equilibrium mixture without changing volume or temperature.
(b) The mixture is compressed at constant temperature.
(c) The equilibrium mixture is cooled.
(d) An inert gas is pumped into the equilibrium mixture while the total gas pressure and the temperature are kept constant.
(e) An inert gas is added to the equilibrium mixture without changing the volume.

37. In a gas-phase reaction, it is observed that the equilibrium yield of products is increased by lowering the temperature and by reducing the volume.
    (a) Is the reaction exothermic or endothermic?
    (b) Is there a net increase or a net decrease in the number of gas molecules in the reaction?
38. The equilibrium constant of a gas-phase reaction increases as temperature is increased. When the nonreacting gas neon is admitted to a mixture of reacting gases (holding the temperature and the total pressure fixed and increasing the volume of the reaction vessel), the product yield is observed to decrease.
    (a) Is the reaction exothermic or endothermic?
    (b) Is there a net increase or a net decrease in the number of gas molecules in the reaction?
39. The most extensively used organic compound in the chemical industry is ethylene ($C_2H_4$). The two equations

$$C_2H_4(g) + Cl_2(g) \rightleftharpoons C_2H_4Cl_2(g)$$

$$C_2H_4Cl_2(g) \rightleftharpoons C_2H_3Cl(g) + HCl(g)$$

represent the way in which vinyl chloride ($C_2H_3Cl$) is synthesized for eventual use in polymeric plastics (polyvinyl chloride, PVC). The byproduct of the reaction, HCl, is now most cheaply made by this and similar reactions, rather than by direct combination of $H_2$ and $Cl_2$. Heat is given off in the first reaction and taken up in the second. Describe how you would design an industrial process to maximize the yield of vinyl chloride.
40. Methanol is made via the exothermic reaction

$$CO(g) + 2 H_2(g) \longrightarrow CH_3OH(g)$$

Describe how you would control the temperature and pressure to maximize the yield of methanol.

## Heterogeneous Equilibria

41. Using the law of mass action, write the equilibrium expression for each of the following reactions.
    (a) $8 H_2(g) + S_8(s) \rightleftharpoons 8 H_2S(g)$
    (b) $C(s) + H_2O(\ell) + Cl_2(g) \rightleftharpoons COCl_2(g) + H_2(g)$
    (c) $CaCO_3(s) \rightleftharpoons CaO(s) + CO_2(g)$
    (d) $3 C_2H_2(g) \rightleftharpoons C_6H_6(\ell)$
42. Using the law of mass action, write the equilibrium expression for each of the following reactions.
    (a) $3 C_2H_2(g) + 3 H_2(g) \rightleftharpoons C_6H_{12}(\ell)$
    (b) $CO_2(g) + C(s) \rightleftharpoons 2 CO(g)$
    (c) $CF_4(g) + 2 H_2O(\ell) \rightleftharpoons CO_2(g) + 4 HF(g)$
    (d) $K_2NiF_6(s) + TiF_4(s) \rightleftharpoons K_2TiF_6(s) + NiF_2(s) + F_2(g)$
43. Using the law of mass action, write the equilibrium expression for each of the following reactions.
    (a) $Zn(s) + 2 Ag^+(aq) \rightleftharpoons Zn^{2+}(aq) + 2 Ag(s)$
    (b) $VO_4^{3-}(aq) + H_2O(\ell) \rightleftharpoons VO_3(OH)^{2-}(aq) + OH^-(aq)$
    (c) $2 As(OH)_6^{3-}(aq) + 6 CO_2(g) \rightleftharpoons$
    $$As_2O_3(s) + 6 HCO_3^-(aq) + 3 H_2O(\ell)$$

44. Using the law of mass action, write the equilibrium expression for each of the following reactions.
    (a) $6 I^-(aq) + 2 MnO_4^-(aq) + 4 H_2O(\ell) \rightleftharpoons$
    $$3 I_2(aq) + 2 MnO_2(s) + 8 OH^-(aq)$$
    (b) $2 Cu^{2+}(aq) + 4 I^-(aq) \rightleftharpoons 2 CuI(s) + I_2(aq)$
    (c) $\frac{1}{2} O_2(g) + Sn^{2+}(aq) + 3 H_2O(\ell) \rightleftharpoons$
    $$SnO_2(s) + 2 H_3O^+(aq)$$
45. $N_2O_4$ is soluble in the solvent cyclohexane; however, dissolution does not prevent $N_2O_4$ from breaking down to give $NO_2$ according to the equation

$$N_2O_{4(cyclohexane)} \rightleftharpoons 2 NO_{2(cyclohexane)}$$

An effort to compare this solution equilibrium to the similar equilibrium in the gas gave the following actual experimental data at 20°C:

| $[N_2O_4]$ (mol L$^{-1}$) | $[NO_2]$ (mol L$^{-1}$) |
|---|---|
| $0.190 \times 10^{-3}$ | $2.80 \times 10^{-3}$ |
| $0.686 \times 10^{-3}$ | $5.20 \times 10^{-3}$ |
| $1.54 \times 10^{-3}$ | $7.26 \times 10^{-3}$ |
| $2.55 \times 10^{-3}$ | $10.4 \times 10^{-3}$ |
| $3.75 \times 10^{-3}$ | $11.7 \times 10^{-3}$ |
| $7.86 \times 10^{-3}$ | $17.3 \times 10^{-3}$ |
| $11.9 \times 10^{-3}$ | $21.0 \times 10^{-3}$ |

(a) Graph the *square* of the concentration of $NO_2$ versus the concentration of $N_2O_4$.
(b) Compute the average equilibrium constant of this reaction.
46. $NO_2$ is soluble in carbon tetrachloride ($CCl_4$). As it dissolves, it dimerizes to give $N_2O_4$ according to the equation

$$2 NO_{2(CCl_4)} \rightleftharpoons N_2O_{4(CCl_4)}$$

A study of this equilibrium gave the following experimental data at 20°C:

| $[N_2O_4]$ (mol L$^{-1}$) | $[NO_2]$ (mol L$^{-1}$) |
|---|---|
| $0.192 \times 10^{-3}$ | $2.68 \times 10^{-3}$ |
| $0.721 \times 10^{-3}$ | $4.96 \times 10^{-3}$ |
| $1.61 \times 10^{-3}$ | $7.39 \times 10^{-3}$ |
| $2.67 \times 10^{-3}$ | $10.2 \times 10^{-3}$ |
| $3.95 \times 10^{-3}$ | $11.0 \times 10^{-3}$ |
| $7.90 \times 10^{-3}$ | $16.6 \times 10^{-3}$ |
| $11.9 \times 10^{-3}$ | $21.4 \times 10^{-3}$ |

(a) Graph the concentration of $N_2O_4$ versus the *square* of the concentration of $NO_2$.
(b) Compute the average equilibrium constant of this reaction.
47. At 298 K, the equilibrium constant for the reaction

$$Fe_2O_3(s) + 3 H_2(g) \rightleftharpoons 2 Fe(s) + 3 H_2O(\ell)$$

is $4.0 \times 10^{-6}$, and that for

$$CO_2(g) + H_2(g) \rightleftharpoons CO(g) + H_2O(\ell)$$

is $3.2 \times 10^{-4}$. Suppose some solid $Fe_2O_3$, solid Fe, and liquid $H_2O$ are brought into equilibrium with $CO(g)$ and $CO_2(g)$ in a closed container at 298 K. Calculate the ratio of the partial pressure of $CO(g)$ to that of $CO_2(g)$ at equilibrium.

**48.** A sample of ammonium carbamate placed in a glass vessel at 25°C undergoes the reaction

$$NH_4OCONH_2(s) \rightleftharpoons 2\,NH_3(g) + CO_2(g)$$

The total pressure of gases in equilibrium with the solid is found to be 0.115 atm.
(a) Calculate the partial pressures of $NH_3$ and $CO_2$.
(b) Calculate the equilibrium constant at 25°C.

**49.** The equilibrium constant for the reaction

$$NH_3(g) + HCl(g) \rightleftharpoons NH_4Cl(s)$$

at 340°C is $K = 4.0$.
(a) If the partial pressure of ammonia is $P_{NH_3} = 0.80$ atm and solid ammonium chloride is present, what is the equilibrium partial pressure of hydrogen chloride at 340°C?
(b) An excess of solid $NH_4Cl$ is added to a container filled with ammonia at 340°C and a pressure of 1.50 atm. Calculate the pressures of $NH_3(g)$ and $HCl(g)$ reached at equilibrium.

**50.** The equilibrium constant for the reaction

$$H_2(g) + I_2(s) \rightleftharpoons 2\,HI(g)$$

at 25°C is $K = 0.345$.
(a) If the partial pressure of hydrogen is $P_{H_2} = 1.00$ atm and solid iodine is present, what is the equilibrium partial pressure of hydrogen iodide, $P_{HI}$, at 25°C?
(b) An excess of solid $I_2$ is added to a container filled with hydrogen at 25°C and a pressure of 4.00 atm. Calculate the pressures of $H_2(g)$ and $HI(g)$ reached at equilibrium.

**51.** The equilibrium constant for the "water gas" reaction

$$C(s) + H_2O(g) \rightleftharpoons CO(g) + H_2(g)$$

is $K = 2.6$ at a temperature of 1000 K. Calculate the reaction quotient $Q$ for each of the following conditions, and state which direction the reaction shifts in coming to equilibrium.
(a) $P_{H_2O} = 0.600$ atm; $P_{CO} = 1.525$ atm; $P_{H_2} = 0.805$ atm
(b) $P_{H_2O} = 0.724$ atm; $P_{CO} = 1.714$ atm; $P_{H_2} = 1.383$ atm

**52.** The equilibrium constant for the reaction

$$H_2S(g) + I_2(g) \rightleftharpoons 2\,HI(g) + S(s)$$

at 110°C is equal to 0.0023. Calculate the reaction quotient $Q$ for each of the following conditions and determine whether solid sulfur is consumed or produced as the reaction comes to equilibrium.
(a) $P_{I_2} = 0.461$ atm; $P_{H_2S} = 0.05$ atm; $P_{HI} = 0.0$ atm
(b) $P_{I_2} = 0.461$ atm; $P_{H_2S} = 0.05$ atm; $P_{HI} = 9.0$ atm

**53.** Pure solid $NH_4HSe$ is placed in an evacuated container at 24.8°C. Eventually, the pressure above the solid reaches the equilibrium pressure 0.0184 atm due to the reaction

$$NH_4HSe(s) \rightleftharpoons NH_3(g) + H_2Se(g)$$

(a) Calculate the equilibrium constant of this reaction at 24.8°C.
(b) In a different container, the partial pressure of $NH_3(g)$ in equilibrium with $NH_4HSe(s)$ at 24.8°C is 0.0252 atm. What is the partial pressure of $H_2Se(g)$?

**54.** The total pressure of the gases in equilibrium with solid sodium hydrogen carbonate at 110°C is 1.648 atm, corresponding to the reaction

$$2\,NaHCO_3(s) \rightleftharpoons Na_2CO_3(s) + H_2O(g) + CO_2(g)$$

($NaHCO_3$ is used in dry chemical fire extinguishers because the products of this decomposition reaction smother the fire.)
(a) Calculate the equilibrium constant at 110°C.
(b) What is the partial pressure of water vapor in equilibrium with $NaHCO_3(s)$ at 110°C if the partial pressure of $CO_2(g)$ is 0.800 atm?

## Extraction and Separation Processes

**55.** An aqueous solution, initially $1.00 \times 10^{-2}$ M in iodine ($I_2$), is shaken with an equal volume of an immiscible organic solvent, $CCl_4$. The iodine distributes itself between the aqueous and $CCl_4$ layers, and when equilibrium is reached at 27°C, the concentration of $I_2$ in the aqueous layer is $1.30 \times 10^{-4}$ M. Calculate the partition coefficient $K$ at 27°C for the reaction

$$I_2(aq) \rightleftharpoons I_2(CCl_4)$$

**56.** An aqueous solution, initially $2.50 \times 10^{-2}$ M in iodine ($I_2$), is shaken with an equal volume of an immiscible organic solvent, $CS_2$. The iodine distributes itself between the aqueous and $CS_2$ layers, and when equilibrium is reached at 25°C, the concentration of $I_2$ in the aqueous layer is $4.16 \times 10^{-5}$ M. Calculate the partition coefficient $K$ at 25°C for the reaction

$$I_2(aq) \rightleftharpoons I_2(CS_2)$$

**57.** Benzoic acid ($C_6H_5COOH$) dissolves in water to the extent of 2.00 g $L^{-1}$ at 15°C and in diethyl ether to the extent of $6.6 \times 10^2$ g $L^{-1}$ at the same temperature.
(a) Calculate the equilibrium constants at 15°C for the two reactions

$$C_6H_5COOH(s) \rightleftharpoons C_6H_5COOH(aq)$$

and

$$C_6H_5COOH(s) \rightleftharpoons C_6H_5COOH(ether)$$

(b) From your answers to part (a), calculate the partition coefficient $K$ for the reaction

$$C_6H_5COOH(aq) \rightleftharpoons C_6H_5COOH(ether)$$

**58.** Citric acid ($C_6H_8O_7$) dissolves in water to the extent of 1300 g $L^{-1}$ at 15°C and in diethyl ether to the extent of 22 g $L^{-1}$ at the same temperature.
(a) Calculate the equilibrium constants at 15°C for the two reactions

$$C_6H_8O_7(s) \rightleftharpoons C_6H_8O_7(aq)$$

and

$$C_6H_8O_7(s) \rightleftharpoons C_6H_8O_7(ether)$$

(b) From your answers to part (a), calculate the partition coefficient $K$ for the reaction

$$C_6H_8O_7(aq) \rightleftharpoons C_6H_8O_7(ether)$$

## Additional Problems

**59.** At 298 K, unequal amounts of $BCl_3(g)$ and $BF_3(g)$ were mixed in a container. The gases reacted to form $BFCl_2(g)$ and $BClF_2(g)$. When equilibrium was finally reached, the four gases were present in these relative chemical amounts: $BCl_3(90)$, $BF_3(470)$, $BClF_2(200)$, $BFCl_2(45)$.
(a) Determine the equilibrium constants at 298 K of the two reactions

$$2\, BCl_3(g) + BF_3(g) \rightleftharpoons 3\, BFCl_2(g)$$

$$BCl_3(g) + 2\, BF_3(g) \rightleftharpoons 3\, BClF_2(g)$$

(b) Determine the equilibrium constant of the reaction

$$BCl_3(g) + BF_3(g) \rightleftharpoons BFCl_2 + BClF_2(g)$$

and explain why knowing this equilibrium constant really adds nothing to what you knew in part (a).
**60.** Methanol can be synthesized by means of the equilibrium reaction

$$CO(g) + 2\, H_2(g) \rightleftharpoons CH_3OH(g)$$

for which the equilibrium constant at 225°C is $6.08 \times 10^{-3}$. Assume that the ratio of the pressures of $CO(g)$ and $H_2(g)$ is 1:2. What values should they have if the partial pressure of methanol is to be 0.500 atm?
**61.** At equilibrium at 425.6°C, a sample of *cis*-1-methyl-2-ethylcyclopropane is 73.6% converted into the *trans* form:

$$cis \rightleftharpoons trans$$

(a) Compute the equilibrium constant $K$ for this reaction.
(b) Suppose that 0.525 mol of the *cis* compound is placed in a 15.00-L vessel and heated to 425.6°C. Compute the equilibrium partial pressure of the *trans* compound.
**62.** The equilibrium constant for the reaction

$$(CH_3)_3COH(g) \rightleftharpoons (CH_3)_2CCH_2(g) + H_2O(g)$$

is 2.42 at 450 K.
(a) A pure sample of the reactant, which is named "*t*-butanol," is confined in a container of fixed volume at a temperature of 450 K and at an original pressure of 0.100 atm. Calculate the fraction of this starting material that is converted to products at equilibrium.

(b) A second sample of the reactant is confined, this time at an original pressure of 5.00 atm. Again, calculate the fraction of the starting material that is converted to products at equilibrium.
**63.** The hydrogenation of pyridine to piperidine

$$C_5H_5N(g) + 3\, H_2(g) \rightleftharpoons C_5H_{11}N(g)$$

is an equilibrium process whose equilibrium constant is given by the equation

$$\log_{10} K = -20.281 + \frac{10,560\ K}{T}$$

(a) Calculate the value of $K$ at $T = 500$ K.
(b) If the partial pressure of hydrogen is 1.00 atm, what fraction of the nitrogen is in the form of pyridine molecules at $T = 500$ K?
**64.** Acetic acid in the vapor phase consists of both monomeric and dimeric forms in equilibrium:

$$2\, CH_3COOH(g) \rightleftharpoons (CH_3COOH)_2(g)$$

At 110°C, the equilibrium constant for this reaction is 3.72.
(a) Calculate the partial pressure of the dimer when the total pressure is 0.725 atm at equilibrium.
(b) What percentage of the acetic acid is dimerized under these conditions?
**\*65.** Repeat the calculation of problem 26 with a total pressure of 100 atm. (*Note:* Here it is necessary to solve the equation by successive approximations. See Appendix C.)
**\*66.** At 900 K, the equilibrium constant for the reaction

$$\tfrac{1}{2} O_2(g) + SO_2(g) \rightleftharpoons SO_3(g)$$

has the value $K = 0.587$. What will be the equilibrium partial pressure of $O_2(g)$ if a sample of $SO_3$ weighing 0.800 g is heated to 900 K in a quartz-glass vessel whose volume is 100.0 cm³? (*Note:* Here it is necessary to solve the equation by successive approximations. See Appendix C.)
**67.** At 300°C, the equilibrium constant for the reaction

$$PCl_5(g) \rightleftharpoons PCl_3(g) + Cl_2(g)$$

is $K = 11.5$.
(a) Calculate the reaction quotient $Q$ if initially $P_{PCl_3} = 2.0$ atm, $P_{Cl_2} = 6.0$ atm, and $P_{PCl_5} = 0.10$ atm. State whether the reaction proceeds to the right or to the left as equilibrium is approached.
(b) Calculate $P_{PCl_3}$, $P_{Cl_2}$, and $P_{PCl_5}$ at equilibrium.
(c) If the volume of the system is then increased, will the amount of $PCl_5$ present increase or decrease?
**68.** Write the equilibrium expression for the oxidation of gold in a basic solution containing cyanide ions, a reaction used in the recovery of gold from mine tailings:

$$4\, Au(s) + 8\, CN^-(aq) + O_2(g) + 2\, H_2O(\ell) \rightleftharpoons$$
$$4\, Au(CN)_2^-(aq) + 4\, OH^-(aq)$$

*69. At 298 K, chlorine is only slightly soluble in water. Thus, under a pressure of 1.00 atm of $Cl_2(g)$, 1.00 L of water at equilibrium dissolves just 0.091 mol of $Cl_2$.

$$Cl_2(g) \rightleftharpoons Cl_2(aq)$$

In such solutions, the $Cl_2(aq)$ concentration is 0.061 M and the concentrations of $Cl^-(aq)$ and $HOCl(aq)$ are both 0.030 M. These two additional species are formed by the equilibrium

$$Cl_2(aq) + H_2O(\ell) \rightleftharpoons H^+(aq) + Cl^-(aq) + HOCl(aq)$$

There are no other Cl-containing species. Compute the equilibrium constants $K_1$ and $K_2$ for the two reactions.

70. At 400°C, the reaction

$$BaO_2(s) + 4\,HCl(g) \rightleftharpoons BaCl_2(s) + 2\,H_2O(g) + Cl_2(g)$$

has an equilibrium constant equal to $K_1$. How is $K_1$ related to the equilibrium constant $K_2$ of the reaction

$$2\,Cl_2(g) + 4\,H_2O(g) + 2\,BaCl_2(s) \rightleftharpoons \\ 8\,HCl(g) + 2\,BaO_2(s)$$

at 400°C?

71. Ammonium hydrogen sulfide, a solid, decomposes to give $NH_3(g)$ and $H_2S(g)$. At 25°C, some $NH_4HS(s)$ is placed in an evacuated container. A portion of it decomposes, and the total pressure at equilibrium is 0.659 atm. Extra $NH_3(g)$ is then injected into the container and, when equilibrium is reestablished, the partial pressure of $NH_3(g)$ is 0.750 atm.
(a) Compute the equilibrium constant for the decomposition of ammonium hydrogen sulfide.
(b) Determine the final partial pressure of $H_2S(g)$ in the container.

72. The equilibrium constant for the reaction

$$KOH(s) + CO_2(g) \rightleftharpoons KHCO_3(s)$$

is $6 \times 10^{15}$ at 25°C. Suppose 7.32 g of KOH and 9.41 g of $KHCO_3$ are placed in a closed evacuated container and allowed to reach equilibrium. Calculate the pressure of $CO_2(g)$ at equilibrium.

73. The equilibrium constant for the reduction of nickel(II) oxide to nickel at 754°C is 255.4, corresponding to the reaction

$$NiO(s) + CO(g) \rightleftharpoons Ni(s) + CO_2(g)$$

If the total pressure of the system at 754°C is 2.50 atm, calculate the partial pressures of $CO(g)$ and $CO_2(g)$.

74. Both glucose (corn sugar) and fructose (fruit sugar) taste sweet, but fructose tastes sweeter. Each year in the United States, tons of corn syrup destined to sweeten food are treated to convert glucose as fully as possible to the sweeter fructose. The reaction is an equilibrium:

$$glucose \rightleftharpoons fructose$$

(a) A 0.2564 M solution of pure glucose is treated at 25°C with an enzyme (catalyst) that causes the preceding equilibrium to be reached quickly. The final concentration of fructose is 0.1175 M. In another experiment at the same temperature, a 0.2666 M solution of pure fructose is treated with the same enzyme, and the final concentration of glucose is 0.1415 M. Compute an average equilibrium constant for the preceding reaction.
(b) At equilibrium under these conditions, what percentage of glucose is converted to fructose?

75. Polychlorinated biphenyls (PCBs) are a major environmental problem. These oily substances have many uses, but they resist breakdown by bacterial action when spilled in the environment and, being fat-soluble, can accumulate to dangerous concentrations in the fatty tissues of fish and animals. One little-appreciated complication in controlling the problem is that there are *209* different PCBs, all now in the environment. They are generally similar, but their solubilities in fats differ considerably. The best measure of this is $K_{ow}$, the equilibrium constant for the partition of a PCB between the fat-like solvent octanol and water.

$$PCB(aq) \rightleftharpoons PCB(octanol)$$

An equimolar mixture of PCB-2 and PCB-11 in water is treated with an equal volume of octanol. Determine the ratio between the amounts of PCB-2 and PCB-11 in the water at equilibrium. At room temperature, $K_{ow}$ is $3.98 \times 10^4$ for PCB-2 and $1.26 \times 10^5$ for PCB-11.

76. Refer to the data in problems 57 and 58. Suppose 2.00 g of a solid consisting of 50.0% benzoic acid and 50.0% citric acid by mass is added to 100.0 mL of water and 100.0 mL of diethyl ether, and the whole assemblage is shaken. When the immiscible layers are separated and the solvents are removed by evaporation, two solids result. Calculate the percentage (by mass) of the major component in each solid.

77. At 25°C, the partition coefficient for the equilibrium

$$I_2(aq) \rightleftharpoons I_2(CCl_4)$$

has the value $K = 85$. To 0.100 L of an aqueous solution, which is initially $2 \times 10^{-3}$ M in $I_2$, we add 0.025 L of $CCl_4$. The mixture is shaken in a separatory funnel and allowed to separate into two phases, and the $CCl_4$ phase is withdrawn.
(a) Calculate the fraction of the $I_2$ remaining in the aqueous phase.
(b) Suppose the remaining aqueous phase is shaken with another 0.025 L of $CCl_4$ and again separated. What fraction of the $I_2$ from the original aqueous solution is *now* in the aqueous phase?
(c) Compare your answer with that of Example 5-11, in which the same total amount of $CCl_4$ (0.050 L) was used in a *single* extraction. For a given total amount of extracting solvent, which is the more efficient way to remove iodine from water?

## CUMULATIVE PROBLEMS

**78.** A 4.72-g mass of methanol, $CH_3OH$, is placed in an evacuated 1.00-L flask and heated to 250°C. It vaporizes and then reaches the following equilibrium:

$$CH_3OH(g) \rightleftharpoons CO(g) + 2\,H_2(g)$$

A tiny hole forms in the side of the container, allowing a small amount of gas to effuse out. Analysis of the escaping gases shows that the rate of escape of the hydrogen is 33 times the rate of escape of the methanol. Calculate the equilibrium constant for the preceding reaction at 250°C.

**79.** The reaction

$$P_4(g) \rightleftharpoons 2\,P_2(g)$$

is endothermic and begins to occur at moderate temperatures. (a) In which direction do you expect deviations to occur from Boyle's law, $P \propto 1/V$ (constant $T$), for gaseous $P_4$?

(b) In which direction do you expect deviations to occur from Charles's law, $V \propto T$ (constant $P$), for gaseous $P_4$?

**80.** At 25°C the equilibrium constant for the reaction

$$CaSO_4(s) + 2\,H_2O(g) \rightleftharpoons CaSO_4 \cdot 2H_2O(s)$$

is $1.6 \times 10^3$. Over what range of relative humidities do you expect $CaSO_4(s)$ to be converted to $CaSO_4 \cdot 2H_2O$? (*Note:* The relative humidity is the partial pressure of water vapor divided by its equilibrium vapor pressure and multiplied by 100%. Use Table 4-1.)

**81.** In an extraction process, a solute species is partitioned between two immiscible solvents. Suppose the two solvents are water and carbon tetrachloride. State which phase will have the higher concentration of each of the following solutes: (a) $CH_3OH$, (b) $C_2Cl_6$, (c) $Br_2$, (d) NaCl. Explain your reasoning.

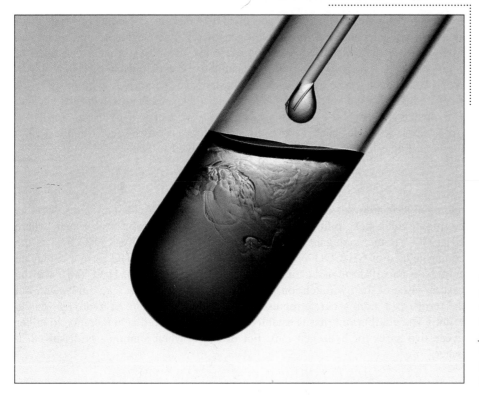

*Cobalt chloride: chemical equilibrium in HCl. A solution of $Co(H_2O)_6^{2+}$ + $4\,Cl^- \rightleftharpoons CoCl_4^{2-} + 6\,H_2O$ shows a concentration of pink $Co(H_2O)_6^{2+}$ ions formed by the addition of water. The $CoCl_4^{2-}$ solution has greater density. (Richard Megna/Fundamental Photographs, New York)*

# Acid–Base Equilibria

**W**e have established acid–base reactions as one of the major categories of chemical action (Chapter 2) and discussed the use of acid–base titrations (Chapter 4). Now the ideas about chemical equilibrium developed in Chapter 5 are applied to an especially important aspect of acid–base chemistry: calculating how far such reactions proceed before reaching equilibrium.

Although many acid–base reactions take place in the gaseous and solid states and in nonaqueous solutions, in this chapter the focus is on acid–base reactions in aqueous solutions. Such solutions play roles in everyday life. Vinegar, orange juice, and battery fluid are familiar acidic aqueous solutions; basic solutions are produced when such common products as borax, baking soda, and antacids are dissolved in water. Acid–base chemistry will appear many times throughout the rest of this book. We will consider the effects of acidity on the dissolution of solids in Chapter 7, on redox reactions in electrochemical cells in Chapter 10, and on the rates of reaction in Chapter 11. Chapter 21 discusses the manufacture and uses of the workhorse industrial acids (nitric, phosphoric, and sulfuric). Chapter 23 points out the central biochemical roles of amino acids and nucleic acids. The reactions of acids and bases lie at the heart of all chemistry.

## 6-1 PROPERTIES OF ACIDS AND BASES

In Section 2-5, where the nature of acids and bases was discussed, it was stated that acids increase the concentration of hydrogen ion ($H^+$) above that in pure water, and bases increase the concentration of hydroxide ion ($OH^-$). A broader definition of acids and bases, which will be useful in quantitative calculations in this chapter, was proposed independently by Johannes Brønsted and Thomas Lowry in 1923. A **Brønsted–Lowry acid** is defined as a substance that can donate a hydrogen ion, and a **Brønsted–Lowry base** is a substance that can accept a hydrogen ion. In a Brønsted–Lowry acid–base reaction, hydrogen ions are transferred from the acid to the base. For example, when acetic acid is dissolved in water,

$$\underset{\text{Acid}_1}{CH_3COOH(aq)} + \underset{\text{Base}_2}{H_2O(\ell)} \rightleftharpoons \underset{\text{Acid}_2}{H_3O^+(aq)} + \underset{\text{Base}_1}{CH_3COO^-(aq)}$$

hydrogen ions are transferred from acetic acid to water. Acids and bases occur as **conjugate acid–base pairs.** $CH_3COOH$ and $CH_3COO^-$ form such a pair, where $CH_3COO^-$ is the conjugate base of $CH_3COOH$ (equivalently, $CH_3COOH$ is the conjugate acid of $CH_3COO^-$). In the same way, $H_3O^+$ and $H_2O$ form a conjugate acid–base pair. [Throughout this chapter, the hydronium ion, $H_3O^+(aq)$, will be used in place of the hydrogen ion, $H^+(aq)$. Chemically the two are equivalent, but the former is a more accurate representation of the true nature of hydrogen ions in water.] The equilibrium that is established may be pictured as the competition between two bases for hydrogen ions. For example, when ammonia is dissolved in water,

$$\underset{\text{Acid}_1}{H_2O(\ell)} + \underset{\text{Base}_2}{NH_3(aq)} \rightleftharpoons \underset{\text{Acid}_2}{NH_4^+(aq)} + \underset{\text{Base}_1}{OH^-(aq)}$$

the two bases $NH_3$ and $OH^-$ compete for hydrogen ions.

One advantage of the Brønsted–Lowry approach is that it is not limited to aqueous solutions. An example with liquid ammonia as the solvent is

$$\underset{\text{Acid}_1}{HCl(\text{in } NH_3)} + \underset{\text{Base}_2}{NH_3(\ell)} \rightleftharpoons \underset{\text{Acid}_2}{NH_4^+(\text{in } NH_3)} + \underset{\text{Base}_1}{Cl^-(\text{in } NH_3)}$$

The $NH_3$ acts as a base even though hydroxide ion, $OH^-$, is not present. The Arrhenius scheme that was introduced in Chapter 2 has been extended and freed from its limitation to aqueous solution.

Some molecules and ions function either as acids or bases depending on reaction conditions, and are called **amphoteric.** The most common example is water itself. Water acts as an acid in donating a hydrogen ion to $NH_3$ (its conjugate base here is $OH^-$) and as a base in accepting a hydrogen ion from $CH_3COOH$ (its conjugate acid here is $H_3O^+$). In the same way, the hydrogen carbonate ion can act as an acid

$$HCO_3^-(aq) + H_2O(\ell) \rightleftharpoons H_3O^+(aq) + CO_3^{2-}(aq)$$

or as a base:

$$HCO_3^-(aq) + H_2O(\ell) \rightleftharpoons H_2CO_3(aq) + OH^-(aq)$$

### Autoionization of Water

What happens when water acts as both acid and base in the same reaction? The resulting equilibrium is

$$H_2O(\ell) + H_2O(\ell) \rightleftharpoons H_3O^+(aq) + OH^-(aq)$$
$$\text{Acid}_1 \qquad \text{Base}_2 \qquad\quad \text{Acid}_2 \qquad \text{Base}_1$$

or

$$2\,H_2O(\ell) \rightleftharpoons H_3O^+(aq) + OH^-(aq)$$

This reaction is responsible for the small but measurable **autoionization** of water. The equilibrium expression that results is

$$[H_3O^+][OH^-] = K_w$$

where $K_w$, the ion product constant for water, has the numerical value $1.0 \times 10^{-14}$ at 25°C. Note that $H_2O(\ell)$, as the solvent, does not enter the equilibrium expression. The temperature dependence of $K_w$ is given in Table 6-1; all problems in this chapter are assumed to refer to 25°C unless otherwise stated.

Pure water contains no ions other than $H_3O^+$ and $OH^-$, and because there is overall electrical neutrality, there must be an equal number of ions of each type. Using the equilibrium expression, this gives

$$[H_3O^+] = [OH^-] = y$$
$$y^2 = 1.0 \times 10^{-14}$$
$$y = 1.0 \times 10^{-7}$$

so that in pure water at 25°C the concentrations of both $H_3O^+$ and $OH^-$ are $1.0 \times 10^{-7}$ M.

**Table 6-1 Temperature Dependence of $K_w$**

| $t$ (°C) | $K_w$ | pH of Water |
|---|---|---|
| 0 | $0.114 \times 10^{-14}$ | 7.47 |
| 10 | $0.292 \times 10^{-14}$ | 7.27 |
| 20 | $0.681 \times 10^{-14}$ | 7.08 |
| 25 | $1.01 \times 10^{-14}$ | 7.00 |
| 30 | $1.47 \times 10^{-14}$ | 6.92 |
| 40 | $2.92 \times 10^{-14}$ | 6.77 |
| 50 | $5.47 \times 10^{-14}$ | 6.63 |
| 60 | $9.61 \times 10^{-14}$ | 6.51 |

## Strong Acids and Bases

An aqueous acidic solution contains an excess of $H_3O^+$ over $OH^-$ ions. A **strong acid** is one that ionizes essentially completely in aqueous solution. When the strong acid HCl (hydrochloric acid) is put in water, the reaction that occurs is

$$HCl(aq) + H_2O(\ell) \longrightarrow H_3O^+(aq) + Cl^-(aq)$$

A single rather than a double arrow is used here to indicate that the reaction is essentially complete. Another strong acid is perchloric acid ($HClO_4$). If 0.10 mol of either of these acids is dissolved in enough water to make 1.0 L of solution, a 0.10 M concentration of $H_3O^+(aq)$ results. The acid–base properties of solutions are determined by their concentrations of $H_3O^+(aq)$, so these two strong acids have the same apparent strength in water despite differences in their intrinsic abilities to donate hydrogen ions. Water is said to have a **leveling effect** on a certain group of acids ($HCl$, $HBr$, $HI$, $H_2SO_4$, $HNO_3$, and $HClO_4$) in that they all behave as strong acids when water is the solvent; the reactions of these acids with water all lie so far to the right at equilibrium that the differences between the acids are undetectably small. The concentration of $H_3O^+$ in a 0.10 M solution of *any* strong acid that donates one hydrogen ion per molecule is simply 0.10 M; the $OH^-$ concentration is

$$[OH^-] = \frac{K_w}{[H_3O^+]} = \frac{1.0 \times 10^{-14}}{0.10} = 1.0 \times 10^{-13}\ \text{M}$$

A **strong base** is defined in an analogous fashion, as a base that reacts essentially completely to give $OH^-(aq)$ ion when put in water. The amide ion ($NH_2^-$) and the hydride ion ($H^-$) are both strong bases. For every mole per liter of either of these ions that is added to water, one mole per liter of $OH^-(aq)$ forms. The other products are $NH_3$ and $H_2$, respectively (see Fig. 2-18):

$$H_2O(\ell) + NH_2^-(aq) \longrightarrow NH_3(aq) + OH^-(aq)$$

$$H_2O(\ell) + H^-(aq) \longrightarrow H_2(aq) + OH^-(aq)$$

The important base sodium hydroxide, a solid, increases the $OH^-$ concentration in water when it dissolves:

$$NaOH(s) \longrightarrow Na^+(aq) + OH^-(aq)$$

For every mole of NaOH that dissolves in water, one mole of $OH^-(aq)$ forms, making NaOH a strong base. Strong bases are leveled in aqueous solution in the same way that strong acids are leveled. If 0.10 mol of NaOH or $NH_2^-$ or $H^-$ is put into enough water to make 1.0 L of solution, then in every case

$$[OH^-] = 0.10 \text{ M}$$

$$[H_3O^+] = \frac{1.0 \times 10^{-14}}{0.10} = 1.0 \times 10^{-13} \text{ M}$$

The $OH^-$ contribution from the autoionization of water is negligible in these cases, as was the contribution of $H_3O^+$ from that source in the case of the 0.10 M HCl or $HClO_4$ solution. Of course, if only a very small amount of strong acid or base were added to pure water (for example, $10^{-7}$ mol $L^{-1}$), the autoionization of water would have to be taken into account.

## The pH Function

In aqueous solution, the concentration of hydronium ion can range from 10 M to $10^{-15}$ M. It is convenient to compress this enormous range by introducing a logarithmic acidity scale, called **pH** and defined by

$$pH = -\log_{10} [H_3O^+]$$

Pure water at 25°C has $[H_3O^+] = 1.0 \times 10^{-7}$ M, so

$$pH = -\log_{10}(1.0 \times 10^{-7}) = -(-7.00) = 7.00$$

A 0.10 M solution of HCl has $[H_3O^+] = 0.10$ M, so

$$pH = -\log_{10}(0.10) = -\log_{10}(1.0 \times 10^{-1}) = -(-1.00) = 1.00$$

and at 25°C a 0.10 M solution of NaOH has

$$pH = -\log_{10}\left(\frac{1.0 \times 10^{-14}}{0.10}\right) = -\log_{10}(1.0 \times 10^{-13}) = -(-13.00) = 13.00$$

As these examples show, calculating the pH is especially easy when the concentration of $H_3O^+$ is exactly a power of 10, because the logarithm is then just the power to which 10 is raised. In other cases, a calculator is needed. If the pH is given, the concentration of $H_3O^+$ can be calculated by raising 10 to the power $(-pH)$.

The most commonly encountered $H_3O^+$ concentrations are less than 1 M, so the pH function is defined with a negative sign in order to produce a number that is

**Figure 6-1**

A simple pH meter with a digital read-out.

positive in most cases. A *high* pH means a *low* concentration of $H_3O^+$, and vice versa. At 25°C,

| | |
|---|---|
| pH < 7 | Acidic solution |
| pH = 7 | Neutral solution |
| pH > 7 | Basic solution |

At other temperatures the pH of water differs from 7.00 (see Table 6-1). A change of one pH unit implies a change of a factor of 10 (that is, one order of magnitude) in the concentrations of $H_3O^+$ and $OH^-$. The pH is most directly measured with a **pH meter** (Fig. 6-1). The mechanism by which pH meters operate will be considered in Chapter 10. Figure 6-2 shows the pH values for several common fluids.

## Example 6-1

(a) A solution is prepared by dissolving 0.23 mol of NaH($s$) in 2.8 L of water. Calculate its pH.
(b) The pH of some orange juice at 25°C is 2.85. Calculate $[H_3O^+]$ and $[OH^-]$.

### Solution

(a) Because $H^-$ is a strong base, it reacts to give 0.23 mol of $OH^-$:

$$NaH(s) + H_2O(\ell) \longrightarrow Na^+(aq) + OH^-(aq) + H_2(g)$$

The concentration is

$$[OH^-] = \frac{0.23 \text{ mol}}{2.8 \text{ L}} = 8.2 \times 10^{-2} \text{ M}$$

$$[H_3O^+] = \frac{1.0 \times 10^{-14}}{8.2 \times 10^{-2}} = 1.22 \times 10^{-13} \text{ M}$$

The pH is then

$$pH = -\log_{10}(1.22 \times 10^{-13}) = 12.91$$

(b)
$$pH = 2.85 = -\log_{10}[H_3O^+]$$

$$[H_3O^+] = 10^{-2.85}$$

**Figure 6-2**

Many everyday materials are acidic or basic aqueous solutions with a wide range of pH values.

This can be evaluated by using a calculator to give

$$[H_3O^+] = 1.4 \times 10^{-3} \text{ M}$$

$$[OH^-] = \frac{1.0 \times 10^{-14}}{1.4 \times 10^{-3}} = 7.1 \times 10^{-12}$$

*Related Problems:* 5, 6, 7, 8

## 6-2 ACID AND BASE STRENGTH

Acids are classified as strong or weak, depending on whether they are fully or partially ionized in solution. In a **weak acid,** the transfer of hydrogen ions to water does not proceed to completion. A weak acid such as acetic acid is thus also a **weak electrolyte;** its aqueous solutions conduct electricity less well than a strong acid of the same concentration because fewer ions are present. A weak acid produces smaller colligative properties than a strong acid (recall the effect of dissolved acetic acid on the freezing point of water in Fig. 4-17).

The Brønsted–Lowry theory helps to establish a quantitative scale for acid strength. The ionization of an acid (symbolized by ''HA'') in aqueous solution can be written as

$$HA(aq) + H_2O(\ell) \rightleftharpoons H_3O^+(aq) + A^-(aq)$$

where $A^-$ is the conjugate base of HA. The equilibrium expression for this chemical reaction is

$$\frac{[H_3O^+][A^-]}{[HA]} = K_a$$

where the subscript ''a'' stands for ''acid.'' For example, if the symbol $A^-$ refers to the cyanide ion ($CN^-$), we write

$$\frac{[H_3O^+][CN^-]}{[HCN]} = K_a$$

Hydrogen cyanide, HCN, is a highly toxic gas that dissolves in water to form equally toxic solutions of hydrocyanic acid.

where $K_a$ is the **acid ionization constant** for hydrogen cyanide in water and has a numerical value of $6.17 \times 10^{-10}$ at 25°C. Table 6-2 gives values of $K_a$ for a number of important acids. The useful quantity $pK_a = -\log_{10} K_a$ is also given. The acid ionization constant is a quantitative measure of the strength of the acid in a given solvent (here, water). A strong acid has $K_a$ greater than 1, so [HA] in the denominator is small and the acid ionizes almost completely. In a weak acid, $K_a$ is smaller than 1 and the ionized species have low concentrations; reaction with water proceeds to a limited extent before equilibrium is reached.

The strength of a base is inversely related to the strength of its conjugate acid; the weaker the acid, the stronger its conjugate base, and vice versa. To see this, note that the equation representing the ionization of a base such as ammonia in water can be written as

$$\underset{\text{Acid}_1}{H_2O(\ell)} + \underset{\text{Base}_2}{NH_3(aq)} \rightleftharpoons \underset{\text{Acid}_2}{NH_4^+(aq)} + \underset{\text{Base}_1}{OH^-(aq)}$$

which gives an equilibrium expression of the form

$$\frac{[\text{NH}_4^+][\text{OH}^-]}{[\text{NH}_3]} = K_b$$

where the subscript "b" on $K_b$ stands for "base." Because $[\text{OH}^-]$ and $[\text{H}_3\text{O}^+]$ are related through the water autoionization equilibrium expression

$$[\text{OH}^-][\text{H}_3\text{O}^+] = K_w$$

| Table 6-2   Ionization Constants of Acids at 25°C | | | | |
|---|---|---|---|---|
| **Acid** | **HA** | **A$^-$** | **$K_a$** | **p$K_a$** |
| Hydriodic | HI | I$^-$ | $\sim 10^{11}$ | $\sim -11$ |
| Hydrobromic | HBr | Br$^-$ | $\sim 10^{9}$ | $\sim -9$ |
| Perchloric | HClO$_4$ | ClO$_4^-$ | $\sim 10^{7}$ | $\sim -7$ |
| Hydrochloric | HCl | Cl$^-$ | $\sim 10^{7}$ | $\sim -7$ |
| Chloric | HClO$_3$ | ClO$_3^-$ | $\sim 10^{3}$ | $\sim -3$ |
| Sulfuric (1) | H$_2$SO$_4$ | HSO$_4^-$ | $\sim 10^{2}$ | $\sim -2$ |
| Nitric | HNO$_3$ | NO$_3^-$ | $\sim 20$ | $\sim -1.3$ |
| Hydronium ion | H$_3$O$^+$ | H$_2$O | 1 | 0.0 |
| Iodic | HIO$_3$ | IO$_3^-$ | $1.6 \times 10^{-1}$ | 0.80 |
| Oxalic (1) | H$_2$C$_2$O$_4$ | HC$_2$O$_4^-$ | $5.9 \times 10^{-2}$ | 1.23 |
| Sulfurous (1) | H$_2$SO$_3$ | HSO$_3^-$ | $1.54 \times 10^{-2}$ | 1.81 |
| Sulfuric (2) | HSO$_4^-$ | SO$_4^{2-}$ | $1.2 \times 10^{-2}$ | 1.92 |
| Chlorous | HClO$_2$ | ClO$_2^-$ | $1.1 \times 10^{-2}$ | 1.96 |
| Phosphoric (1) | H$_3$PO$_4$ | H$_2$PO$_4^-$ | $7.52 \times 10^{-3}$ | 2.12 |
| Arsenic (1) | H$_3$AsO$_4$ | H$_2$AsO$_4^-$ | $5.0 \times 10^{-3}$ | 2.30 |
| Chloroacetic | CH$_2$ClCOOH | CH$_2$ClCOO$^-$ | $1.4 \times 10^{-3}$ | 2.85 |
| Hydrofluoric | HF | F$^-$ | $6.6 \times 10^{-4}$ | 3.18 |
| Nitrous | HNO$_2$ | NO$_2^-$ | $4.6 \times 10^{-4}$ | 3.34 |
| Formic | HCOOH | HCOO$^-$ | $1.77 \times 10^{-4}$ | 3.75 |
| Benzoic | C$_6$H$_5$COOH | C$_6$H$_5$COO$^-$ | $6.46 \times 10^{-5}$ | 4.19 |
| Oxalic (2) | HC$_2$O$_4^-$ | C$_2$O$_4^{2-}$ | $6.4 \times 10^{-5}$ | 4.19 |
| Hydrazoic | HN$_3$ | N$_3^-$ | $1.9 \times 10^{-5}$ | 4.72 |
| Acetic | CH$_3$COOH | CH$_3$COO$^-$ | $1.76 \times 10^{-5}$ | 4.75 |
| Propionic | CH$_3$CH$_2$COOH | CH$_3$CH$_2$COO$^-$ | $1.34 \times 10^{-5}$ | 4.87 |
| Pyridinium ion | HC$_5$H$_5$N$^+$ | C$_5$H$_5$N | $5.6 \times 10^{-6}$ | 5.25 |
| Carbonic (1) | H$_2$CO$_3$ | HCO$_3^-$ | $4.3 \times 10^{-7}$ | 6.37 |
| Sulfurous (2) | HSO$_3^-$ | SO$_3^{2-}$ | $1.02 \times 10^{-7}$ | 6.91 |
| Arsenic (2) | H$_2$AsO$_4^-$ | HAsO$_4^{2-}$ | $9.3 \times 10^{-8}$ | 7.03 |
| Hydrosulfuric | H$_2$S | HS$^-$ | $9.1 \times 10^{-8}$ | 7.04 |
| Phosphoric (2) | H$_2$PO$_4^-$ | HPO$_4^{2-}$ | $6.23 \times 10^{-8}$ | 7.21 |
| Hypochlorous | HClO | ClO$^-$ | $3.0 \times 10^{-8}$ | 7.53 |
| Hydrocyanic | HCN | CN$^-$ | $6.17 \times 10^{-10}$ | 9.21 |
| Ammonium ion | NH$_4^+$ | NH$_3$ | $5.6 \times 10^{-10}$ | 9.25 |
| Carbonic (2) | HCO$_3^-$ | CO$_3^{2-}$ | $4.8 \times 10^{-11}$ | 10.32 |
| Arsenic (3) | HAsO$_4^{2-}$ | AsO$_4^{3-}$ | $3.0 \times 10^{-12}$ | 11.53 |
| Hydrogen peroxide | H$_2$O$_2$ | HO$_2^-$ | $2.4 \times 10^{-12}$ | 11.62 |
| Phosphoric (3) | HPO$_4^{2-}$ | PO$_4^{3-}$ | $2.2 \times 10^{-13}$ | 12.67 |
| Water | H$_2$O | OH$^-$ | $1.0 \times 10^{-14}$ | 14.00 |

the $K_b$ expression can be written as

$$K_b = \frac{[NH_4^+]K_w}{[NH_3][H_3O^+]} = \frac{K_w}{K_a}$$

where $K_a$ is the acid ionization constant for $NH_4^+$, the conjugate acid of the base $NH_3$. This general relationship between the $K_b$ of a base and the $K_a$ of its conjugate acid shows that $K_b$ need not be tabulated separately from $K_a$, because the two are related through

$$K_w = K_a K_b$$

It is also clear that if $K_a$ is large (so that the acid is strong), then $K_b$ is small (the conjugate base is weak). Figure 6-3 summarizes the strengths of conjugate acid–base pairs.

If two bases compete for hydrogen ions, the stronger base is favored in the equilibrium that is reached. The stronger acid donates hydrogen ions to the stronger

**Figure 6–3**

The relative strengths of some acids and their conjugate bases.

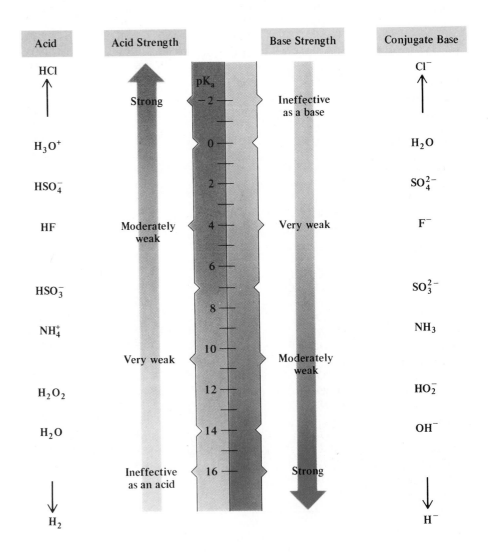

base, producing a weaker acid and a weaker base. To see this, consider the equilibrium

$$\text{HF}(aq) + \text{CN}^-(aq) \rightleftharpoons \text{HCN}(aq) + \text{F}^-(aq)$$
$$\text{Acid}_1 \qquad \text{Base}_2 \qquad\qquad \text{Acid}_2 \qquad \text{Base}_1$$

with equilibrium constant

$$\frac{[\text{HCN}][\text{F}^-]}{[\text{HF}][\text{CN}^-]} = K$$

The two bases $\text{F}^-$ and $\text{CN}^-$ compete for hydrogen ions. The overall reaction is obtained by starting with one acid ionization reaction

$$\text{HF}(aq) + \text{H}_2\text{O}(\ell) \rightleftharpoons \text{H}_3\text{O}^+(aq) + \text{F}^-(aq)$$

$$\frac{[\text{H}_3\text{O}^+][\text{F}^-]}{[\text{HF}]} = K_\text{a} = 6.6 \times 10^{-4}$$

and *subtracting* from it a second acid ionization reaction:

$$\text{HCN}(aq) + \text{H}_2\text{O}(\ell) \rightleftharpoons \text{H}_3\text{O}^+(aq) + \text{CN}^-(aq)$$

$$\frac{[\text{H}_3\text{O}^+][\text{CN}^-]}{[\text{HCN}]} = K_\text{a}' = 6.17 \times 10^{-10}$$

When the second equation is subtracted from the first, the corresponding equilibrium constants must be divided. The numerical value of $K$ is

$$K = \frac{K_\text{a}}{K_\text{a}'} = \frac{6.6 \times 10^{-4}}{6.17 \times 10^{-10}} = 1.1 \times 10^6$$

Because HCN is a weaker acid than HF, $K_\text{a}'$ is smaller than $K_\text{a}$, and $K$ is larger than 1. The equilibrium described by $K$ lies strongly to the right, and there is a net donation of $\text{H}^+$ from the stronger acid (HF) to the stronger base ($\text{CN}^-$), producing the weaker acid (HCN) and the weaker base ($\text{F}^-$). The magnitudes of acid ionization constants can be used to predict the direction of net hydrogen ion transfer in reactions between acids and bases in aqueous solution.

## Indicators

An **indicator** is a soluble dye that changes color noticeably over a fairly narrow range of pH. The typical indicator is a weak organic acid that has a different color from its conjugate base (Fig. 6-4). Litmus changes from red to blue as its acid form

Hydrogen fluoride, HF, is a colorless liquid that boils at 19.5°C. It dissolves in water to give solutions of hydrofluoric acid, a weak acid.

**Figure 6–4**

Color differences in four indicators: bromophenol red, thymolphthalein, phenolphthalein, and bromocresol green. In each case, the acidic form is on the left and the basic form is on the right. (Leon Lewandowski)

is converted to base. Good indicators have such intense colors that only a few drops of a dilute indicator solution must be added to the solution being studied. The very low concentration of indicator molecules has almost no effect on the pH of the solution. The color changes of the indicator reflect the effects of the *other* acids and bases present in the solution.

If the acid form of a given indicator is represented as HIn and the conjugate base form as In$^-$, their acid–base equilibrium is

$$HIn(aq) + H_2O(\ell) \rightleftharpoons H_3O^+(aq) + In^-(aq) \qquad \frac{[H_3O^+][In^-]}{[HIn]} = K_a$$

where $K_a$ is the acid ionization constant for the indicator. This expression can be rearranged to give

$$\frac{[H_3O^+]}{K_a} = \frac{[HIn]}{[In^-]}$$

**Figure 6–5**

Indicators change their colors at very different pH values, so the best choice of indicator depends on the particular experimental conditions.

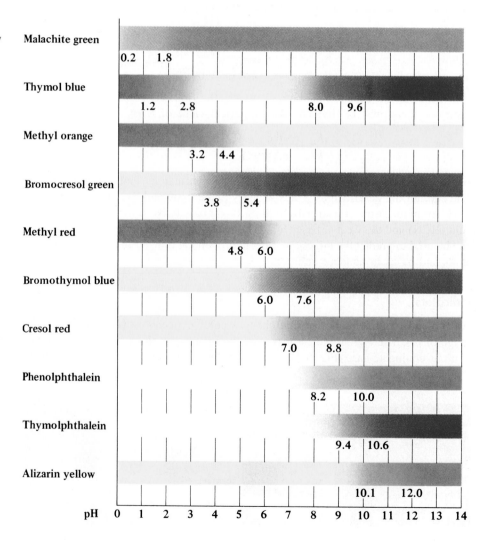

| | pH | | | | | | | | | | | | | | |
|---|---|---|---|---|---|---|---|---|---|---|---|---|---|---|---|
| Malachite green | | 0.2 | 1.8 | | | | | | | | | | | | |
| Thymol blue | | | 1.2 | 2.8 | | | | 8.0 | 9.6 | | | | | | |
| Methyl orange | | | | 3.2 | 4.4 | | | | | | | | | | |
| Bromocresol green | | | | 3.8 | 5.4 | | | | | | | | | | |
| Methyl red | | | | | 4.8 | 6.0 | | | | | | | | | |
| Bromothymol blue | | | | | | 6.0 | 7.6 | | | | | | | | |
| Cresol red | | | | | | | 7.0 | 8.8 | | | | | | | |
| Phenolphthalein | | | | | | | | 8.2 | 10.0 | | | | | | |
| Thymolphthalein | | | | | | | | | 9.4 | 10.6 | | | | | |
| Alizarin yellow | | | | | | | | | | 10.1 | 12.0 | | | | |

pH  0  1  2  3  4  5  6  7  8  9  10  11  12  13  14

If the concentration of hydronium ion $[H_3O^+]$ is large relative to $K_a$, this ratio is large, and $[HIn]$ is large compared with $[In^-]$. The solution has the color of the acid form of the indicator because most of the indicator molecules are in the acid form. Litmus, for example, has a $K_a$ near $10^{-7}$. If the pH is 5, then

$$\frac{[H_3O^+]}{K_a} = \frac{10^{-5}}{10^{-7}} = 100$$

Thus, approximately 100 times as many indicator molecules are in the acid form as in the base form, and the solution is red.

As the concentration of hydronium ion is reduced, more molecules of acid indicator ionize to give the base form. When $[H_3O^+]$ is near $K_a$, almost equal amounts of the two forms are present, and the color is a mixture of the colors of the two indicator states (violet for litmus). A further decrease in $[H_3O^+]$ to a value much smaller than $K_a$ then leads to a predominance of the base form, with the corresponding color being observed.

Different indicators have different values for $K_a$ and thus show color changes at different pH values (Fig. 6-5). The weaker an indicator is as an acid, the higher is the pH at which the color change takes place. Such color changes occur over a range of 1 to 2 pH units. Methyl red, for example, is red when the pH is below 4.8 and yellow above 6.0; shades of orange are seen for intermediate pH values. This limits the accuracy to which the pH can be determined through the use of indicators. However, in Section 6-5 it will be seen that this does not affect the analytical determination of acid or base concentrations through titration, provided that an appropriate indicator is employed.

Many natural dyes found in fruits, vegetables, and flowers act as pH indicators by changing color with changes in acidity (Fig. 6-6). A particularly striking example is cyanidin, which is responsible both for the red color of poppies and the blue color

(*a*)                          b)                          (*c*)

**Figure 6–6**
Red cabbage extract is a natural pH indicator. (*a*) In acidic solution it is red. (*b*) Upon addition of base, several colors (including green and blue) appear as different regions of the mixture have different pH. (*c*) Once a uniformly basic pH is established, the color of the cabbage extract is blue. (Leon Lewandowski)

of cornflowers. The sap of the poppy is sufficiently acidic to turn cyanidin red, but the sap of the cornflower is basic and makes the dye blue. Related natural dyes called anthocyanins contribute to the colors of raspberries, strawberries, and blackberries.

## 6-3 EQUILIBRIA INVOLVING WEAK ACIDS AND BASES

Weak acids and bases react only partially with water, so to calculate the pH of their solutions we use $K_a$ or $K_b$ and the laws of chemical equilibrium. The calculations that result follow the pattern of Example 5-4 for gas equilibria. In that case, the initial gas-phase pressures $P°$ were known, and the pressures of products resulting from partial reaction were calculated. Now the initial concentration of acid or base is known, and the concentrations of products resulting from its partial reaction with water are calculated.

### Weak Acids

A weak acid has a $K_a$ smaller than 1. Values of the $pK_a$ start at zero for the strongest weak acid and range upward. (If the $pK_a$ is greater than 14, the compound is ineffective as an acid in aqueous solution.) When a weak acid is dissolved in water, the original concentration is almost always known, but partial reaction with water consumes some HA and generates $A^-$ and $H_3O^+$:

$$HA(aq) + H_2O(\ell) \rightleftharpoons H_3O^+(aq) + A^-(aq)$$

To calculate the amounts of $H_3O^+$, $A^-$, and HA at equilibrium, methods analogous to those of Chapter 5 are used, with partial pressures replaced by concentrations. A new feature here is that one of the products ($H_3O^+$) has a second source, the auto-ionization of the solvent, water. In the cases of usual interest this second effect is small and can be neglected; however, it is a good idea to verify at the end of each calculation that the $[H_3O^+]$ from the acid ionization alone exceeds $10^{-7}$ M by at least one order of magnitude. If this is not true, the more complete method of analysis given in Section 6-7 must be used.

### Example 6-2

Acetic acid ($CH_3COOH$) has a $K_a$ of $1.76 \times 10^{-5}$ at 25°C. Suppose 1.000 mol is dissolved in enough water to give 1.000 L of solution. Calculate the pH and the fraction of acetic acid ionized at equilibrium.

### Solution

The initial concentration of acetic acid is 1.000 M. If $y$ mol $L^{-1}$ ionizes, then

$$CH_3COOH(aq) + H_2O(\ell) \rightleftharpoons H_3O^+(aq) + CH_3COO^-(aq)$$

| | $CH_3COOH$ | $H_3O^+$ | $CH_3COO^-$ |
|---|---|---|---|
| Initial concentration (M) | 1.000 | ≈0 | 0 |
| Change in concentration (M) | $-y$ | $+y$ | $+y$ |
| Equilibrium concentration (M) | $1.000 - y$ | $y$ | $y$ |

Note that the $H_3O^+$ initially present from the ionization of water has been ignored, because it is small compared to $y$ for all but the weakest acids or the most dilute solutions. The equilibrium expression states that

$$\frac{[H_3O^+][CH_3COO^-]}{[CH_3COOH]} = \frac{y^2}{1.000 - y} = K_a = 1.76 \times 10^{-5}$$

This equation could be solved using the quadratic formula, as was done in Chapter 5. A quicker way is based on the fact that acetic acid is a weak acid, so that only a small fraction is ionized at equilibrium. The approximation that $y$ is small relative to 1.000 (that is, the final equilibrium concentration of $CH_3COOH$ is close to its initial concentration) is reasonable. This gives

$$\frac{y^2}{1.000 - y} \approx \frac{y^2}{1.000} = 1.76 \times 10^{-5}$$

$$y = 4.20 \times 10^{-3}$$

so that

$$[H_3O^+] = y = 4.20 \times 10^{-3} \text{ M}$$

The approximations should now be checked. First, $y$ is indeed much smaller than the original concentration of 1.000 M (by a factor of 200), so the neglect of it in the denominator is justified. Second, the concentration of $H_3O^+$ from acetic acid ($4.2 \times 10^{-3}$ M) is large compared with $10^{-7}$ M, so the neglect of water ionization is also justified. Therefore,

$$pH = -\log_{10}(4.2 \times 10^{-3}) = 2.38$$

The fraction ionized is the ratio of the concentration of $CH_3COO^-$ present at equilibrium to the concentration of $CH_3COOH$ present in the first place:

$$\frac{y}{1.000} = y = 4.2 \times 10^{-3}$$

The percentage of the acetic acid that is ionized is 0.42%. Fewer than one in a hundred of the molecules of this typical weak acid dissociate in this solution.

*Related Problems:* 19, 20, 21, 22

Approximation methods of the type used in this example were also employed in Example 5-5 and are discussed more extensively in Appendix C. As the solution of a weak acid becomes more dilute, an increasing fraction of it ionizes, as shown by the following example.

## Example 6-3

Suppose that 0.00100 mol of acetic acid is used instead of the 1.000 mol in the preceding example. Calculate the pH and the percentage of acetic acid ionized.

### Solution

Once again the assumption to be made (and checked later) is that the contribution of the ionization of water to $[H_3O^+]$ is negligible. Hence,

$$[CH_3COOH] = 0.00100 - y$$

$$[CH_3COO^-] = [H_3O^+] = y$$

$$\frac{y^2}{0.00100 - y} = 1.76 \times 10^{-5}$$

In this case, $y$ is *not* small relative to 0.00100, because a substantial fraction of the acetic acid molecules ionizes. There are two alternatives for solving the problem in this case.

The first is to solve the quadratic equation that results from multiplying out the equilibrium expression:

$$y^2 + 1.76 \times 10^{-5}y - 1.76 \times 10^{-8} = 0$$

Use of the quadratic formula gives

$$y = 1.24 \times 10^{-4} \text{ M} = [H_3O^+] = [CH_3COO^-]$$

Because $y \gg 10^{-7}$ M, the neglect of water ionization is justified.

$$\text{pH} = -\log_{10}(1.24 \times 10^{-4}) = 3.91$$

$$\text{percentage ionized} = \frac{1.24 \times 10^{-4} \text{ M}}{0.00100 \text{ M}} \times 100\% = 12.4\%$$

The other alternative is to use successive approximations (Appendix C). Begin by neglecting $y$ in the denominator (relative to 0.00100), which gives

$$\frac{y^2}{0.00100} \approx 1.76 \times 10^{-5}$$

$$y \approx 1.33 \times 10^{-4} \text{ M}$$

This is just what was done in Example 6-2. Now, however, continue by reinserting this *approximate* value of $y$ into the *denominator* and recalculating $y$:

$$\frac{y^2}{0.00100 - 0.000133} = 1.76 \times 10^{-5}$$

$$y^2 = 1.53 \times 10^{-8}$$

$$y = 1.23 \times 10^{-4} \text{ M}$$

A final iteration (inserting this new value of $y$ into the denominator) gives $1.24 \times 10^{-4}$ M, the same value produced by the quadratic formula. The iteration process is terminated when two successive results for $y$ agree to the desired number of significant figures.

*Related Problems:* 23, 24

The method of successive approximations is often faster to apply than the quadratic formula. It is important to remember that the accuracy of a result is limited both by the accuracy of the input data (values of $K_a$ and initial concentrations) and by the fact that solutions are not ideal. It is therefore pointless to calculate equilibrium concentrations to any higher degree of accuracy than 1% to 3%.

## Weak Bases

The definition and description of weak acids, their $K_a$'s, and the production of $H_3O^+(aq)$ ion when they are dissolved in water can be translated to apply to weak bases, their $K_b$'s, and the production of $OH^-(aq)$ ion. A **weak base** such as ammonia reacts only partially with water to produce $OH^-(aq)$:

$$H_2O(\ell) + NH_3(aq) \rightleftharpoons NH_4^+(aq) + OH^-(aq)$$

$$\frac{[NH_4^+][OH^-]}{[NH_3]} = K_b = 1.8 \times 10^{-5}$$

The $K_b$ of a weak base is smaller than 1, and the weaker the base, the smaller the $K_b$. If the $K_b$ of a compound is smaller than $1 \times 10^{-14}$, that compound is ineffective as a base in aqueous solution.

The analogy to weak acids continues in the calculation of the aqueous equilibria of weak bases, as the following example shows.

## Example 6-4

Calculate the pH of a solution made by dissolving 0.0100 mol of $NH_3$ in enough water to give 1.000 L of solution at 25°C. The $K_b$ for ammonia is $1.8 \times 10^{-5}$.

### Solution

Set up the table of the changes in concentrations that occur as the reaction goes to equilibrium, neglecting the small contribution to $[OH^-]$ from the autoionization of water:

|  | $H_2O(\ell) + NH_3(aq) \rightleftharpoons NH_4^+(aq) + OH^-(aq)$ | | |
| --- | --- | --- | --- |
| Initial concentration (M) | 0.0100 | 0 | $\approx 0$ |
| Change in concentration (M) | $-y$ | $+y$ | $+y$ |
| Equilibrium concentration (M) | $0.100 - y$ | $y$ | $y$ |

Substitution into the equilibrium expression gives

$$\frac{y^2}{0.0100 - y} = K_b = 1.8 \times 10^{-5}$$

which can be solved for $y$ by either the quadratic formula or the method of successive approximations:

$$y = 4.15 \times 10^{-4} \text{ M} = [OH^-]$$

The product of $[H_3O^+]$ and $[OH^-]$ is always $K_w$. Hence,

$$[H_3O^+] = \frac{K_w}{[OH^-]} = \frac{1.0 \times 10^{-14}}{4.15 \times 10^{-4}} = 2.4 \times 10^{-11}$$

$$pH = -\log_{10}(2.4 \times 10^{-11}) = 10.62$$

The pH is greater than 7, as expected for a solution of a base.

*Related Problems:* 27, 28

## Hydrolysis

Most of the acids considered up to now have been uncharged species with the general formula HA. In the Brønsted–Lowry picture, however, there is no reason why the acid should be an electrically neutral molecule. When $NH_4Cl$, a salt, dissolves in water, $NH_4^+$ ions are present. These ionize partially by transferring hydrogen ions to water, a straightforward Brønsted–Lowry acid–base reaction:

$$NH_4^+(aq) + H_2O(\ell) \rightleftharpoons H_3O^+(aq) + NH_3(aq)$$
$$\text{Acid}_1 \qquad \text{Base}_2 \qquad\qquad \text{Acid}_2 \qquad \text{Base}_1$$

$$\frac{[H_3O^+][NH_3]}{[NH_4^+]} = K_a = 5.6 \times 10^{-10}$$

$NH_4^+$ acts as an acid here, just as acetic acid did in Examples 6-2 and 6-3. Because $K_a$ is much smaller than 1, only a small amount of $H_3O^+$ is generated. $NH_4^+$ is a weak acid, but it is nonetheless an acid, and a solution of ammonium chloride has a pH below 7 (Fig. 6-7).

   **Hydrolysis** is the general term given to the reaction of a substance with water, and it is applied in particular to a reaction in which the pH changes from 7 upon dissolving a salt (in this case, $NH_4Cl$) in water. There is no reason to treat hydrolysis in any special way. The hydrolysis that takes place when $NH_4Cl$ dissolves in water can be completely described as a Brønsted–Lowry reaction in which water acts as a base and $NH_4^+$ acts as an acid to give a pH below 7. In parallel fashion, a solution becomes basic when a salt whose anion is a weak base is dissolved. This, too, is a case of hydrolysis, and once again it is simply another Brønsted–Lowry acid–base reaction, with water acting now as an acid (a hydrogen ion donor).

   Two salts that give basic solutions are sodium acetate and sodium fluoride. When these salts dissolve in water, they furnish acetate ($CH_3COO^-$) and fluoride ($F^-$) ions, respectively, both of which act as Brønsted–Lowry bases,

$$H_2O(\ell) + CH_3COO^-(aq) \rightleftharpoons CH_3COOH(aq) + OH^-(aq)$$

$$H_2O(\ell) + F^-(aq) \rightleftharpoons HF(aq) + OH^-(aq)$$

causing the $OH^-$ concentration to increase and giving a pH above 7.

**Figure 6–7**

Proof that ammonium chloride is an acid. (*a*) A solution of sodium hydroxide, with the pink color of phenolphthalein indicating its basic character. (*b, c*) As aqueous ammonium chloride is added, a neutralization reaction takes place and the solution turns colorless from the bottom up. (Leon Lewandowski)

(*a*)

(*b*)

(*c*)

## Example 6-5

Suppose 0.100 mol of $NaCH_3COO$ is dissolved in enough water to make 1.00 L of solution. What is the pH of the solution?

### Solution

The $K_a$ for $CH_3COOH$, the conjugate acid of $CH_3COO^-$, is $1.76 \times 10^{-5}$. Therefore, $K_b$ for the acetate ion is

$$\frac{[CH_3COOH][OH^-]}{[CH_3COO^-]} = K_b = \frac{K_w}{K_a} = 5.7 \times 10^{-10}$$

The equilibrium here is

$$H_2O(\ell) + CH_3COO^-(aq) \rightleftharpoons CH_3COOH(aq) + OH^-(aq)$$

|  |  |  |  |
|---|---|---|---|
| Initial concentration (M) | 0.100 | 0 | $\approx 0$ |
| Change in concentration (M) | $-y$ | $+y$ | $+y$ |
| Equilibrium concentration (M) | $0.100 - y$ | $y$ | $y$ |

Substitution into the equilibrium expression gives

$$\frac{y^2}{0.100 - y} = K_b = 5.7 \times 10^{-10}$$

$$y = 7.5 \times 10^{-6} \,\text{M} = [OH^-]$$

Note that $y$ is small relative to 0.100 and fairly large relative to $10^{-7}$.

$$[H_3O^+] = \frac{K_w}{[OH^-]} = \frac{1.0 \times 10^{-14}}{7.5 \times 10^{-6}} = 1.3 \times 10^{-9} \,\text{M}$$

$$\text{pH} = 8.89$$

*Related Problems:* 33, 34

Hydrolysis does not occur with all ions, only with those that are conjugate acids of weak bases or conjugate bases of weak acids. Chloride ion is the conjugate base of the strong acid HCl and consequently is ineffective as a base, unlike $F^-(aq)$ and $CH_3COO^-(aq)$. Its interaction with water would therefore scarcely change the $OH^-(aq)$ concentration. For this reason, a solution of NaCl is neutral, while one of NaF is slightly basic.

## 6-4 BUFFER SOLUTIONS

A **buffer solution** is any solution that maintains an approximately constant pH despite small additions of acid or base. Typically, a buffer solution contains a weak acid and its conjugate weak base in approximately equal concentrations. Buffer solutions play important roles in controlling the solubility of ions in solution and in maintaining the pH in biochemical and physiological processes. Many life processes are sensitive to pH and require regulation within a small range of $H_3O^+$ and $OH^-$

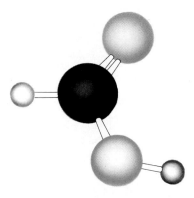

Formic acid, HCOOH, is the simplest carboxylic acid, with only a hydrogen atom attached to the —COOH group.

concentrations. Organisms have built-in buffers to protect them against large changes in pH. Human blood, for example, has a pH near 7.4 that is maintained by a combination of carbonate, phosphate, and protein buffer systems. A blood pH below 7.0 or above 7.8 leads quickly to death.

## Calculations of Buffering Action

Consider a typical weak acid, formic acid (HCOOH), and its conjugate base, formate ion (HCOO$^-$). The latter can be obtained by dissolving a salt such as sodium formate (NaHCOO) in water. The acid–base equilibrium established between these species is given by

$$HCOOH(aq) + H_2O(\ell) \rightleftharpoons H_3O^+(aq) + HCOO^-(aq)$$

with an acid ionization constant

$$\frac{[H_3O^+][HCOO^-]}{[HCOOH]} = K_a = 1.77 \times 10^{-4}$$

In Section 6-2 the pH of a solution containing only a weak acid (such as HCOOH) or only a weak base (such as HCOO$^-$) was considered. Suppose now that the weak acid and its conjugate base are *both* present initially. The resulting calculations resemble closely the calculation of Example 5-7, in which equilibrium was established starting from a mixture in which both reactants and products were present.

---

### Example 6-6

Suppose 1.00 mol of HCOOH and 0.500 mol of NaHCOO are added to water and diluted to 1.00 L. Calculate the pH of the solution.

#### Solution

| | $\mathbf{HCOOH}(aq) + \mathbf{H_2O}(\ell) \rightleftharpoons \mathbf{H_3O^+}(aq) +$ | | $\mathbf{HCOO^-}(aq)$ |
|---|---|---|---|
| Initial concentration (M) | 1.00 | $\approx 0$ | 0.500 |
| Change in concentration (M) | $-y$ | $+y$ | $+y$ |
| Equilibrium concentration (M) | $1.00 - y$ | $y$ | $0.500 + y$ |

The equilibrium expression is

$$\frac{y(0.500 + y)}{1.00 - y} = K_a = 1.77 \times 10^{-4}$$

Because $y$ is likely to be small relative to 1.00 and to 0.500, the expression is approximated as

$$\frac{y(0.500)}{1.00} \approx 1.77 \times 10^{-4}$$

$$y = 3.54 \times 10^{-4} \text{ M} = [H_3O^+]$$

A glance verifies that $y$ is indeed small relative to 1.00 and 0.500. Then

$$pH = -\log_{10}(3.54 \times 10^{-4}) = 3.45$$

*Related Problems:* 35, 36

To see how buffer solutions work, let us write the equilibrium expression for the ionization of a weak acid HA in the form

$$[H_3O^+] = K_a \frac{[HA]}{[A^-]}$$

The concentration of hydronium ion depends on the ratio of the concentration of the weak acid to the concentration of its conjugate base. The key to effective buffer action is to keep both of these concentrations nearly equal and fairly large. Adding a small amount of base to an effective buffer takes away only a few percent of the HA molecules by converting them into $A^-$ ions, and adds only a few percent to the amount of $A^-$ that was originally present. The ratio $[HA]/[A^-]$ decreases, but only very slightly. Added acid consumes a small fraction of the base $A^-$ to generate a bit more HA. The ratio $[HA]/[A^-]$ now increases, but again the change is only slight. Because the concentration of $H_3O^+$ is tied directly to this ratio, it also changes only slightly. The following example illustrates buffer action quantitatively.

## Example 6-7

Suppose 0.10 mol of a strong acid such as HCl is added to the solution in Example 6-6. Calculate the pH of the resulting solution.

### Solution

The strong acid HCl ionizes essentially completely in dilute aqueous solution. Initially, assume that *all* of the hydrogen ions from HCl are taken up by formate ions, giving formic acid, after which some formic acid ionizes back to formate ion and $H_3O^+$. This is simply a way of looking at the route by which equilibrium is approached, and not a statement of the sequence of reactions that actually occurs. The position of the final equilibrium does not depend on the route by which it is attained.

Because 0.10 mol of HCl reacts with an equal number of moles of $HCOO^-$, the concentrations of $HCOO^-$ and HCOOH *before* ionization are

$$[HCOO^-]_0 = 0.50 - 0.10 = 0.40 \text{ M}$$

$$[HCOOH]_0 = 1.00 + 0.10 = 1.10 \text{ M}$$

The chart to calculate the concentrations at equilibrium is

| $HCOOH(aq) + H_2O(\ell) \rightleftharpoons H_3O^+(aq) + HCOO^-(aq)$ | | |
|---|---|---|
| Initial concentration (M)          | 1.10       | $\approx 0$ | 0.40       |
| Change in concentration (M)        | $-y$       | $+y$        | $+y$       |
| Equilibrium concentration (M)      | $1.10 - y$ | $y$         | $0.40 + y$ |

The equilibrium expression then becomes

$$\frac{y(0.40 + y)}{1.10 - y} = 1.77 \times 10^{-4}$$

Because $y$ is again likely to be small relative to both 0.40 and 1.10,

$$y \approx \left(\frac{1.10}{0.40}\right)(1.77 \times 10^{-4}) = 4.9 \times 10^{-4}$$

$$pH = 3.31$$

Even though 0.10 mol of a strong acid was added, the pH changed only slightly, from 3.45 to 3.31. By contrast, the same amount of acid added to a liter of pure water would change the pH from 7 to 1.

*Related Problems:* 37, 38

Note that this problem was solved by performing first a stoichiometric (limiting reactant) calculation and then an equilibrium calculation. A similar calculation can be performed if a strong base such as $OH^-$ is added instead of a strong acid. The base reacts with formic acid to produce formate ions. Adding 0.10 mol of $OH^-$ to the $HCOOH/HCOO^-$ buffer of Example 6-6 increases the pH only to 3.58. In the absence of the buffer system, the same base would raise the pH to 13.00.

In any buffer there is competition between the tendency of the acid to donate hydrogen ions to water (increasing the acidity) and the tendency of the base to accept hydrogen ions from water (increasing the basicity). The resulting pH depends on the magnitude of $K_a$. If $K_a$ is large relative to $10^{-7}$, the acid ionization will win out and acidity will increase, as in the $HCOOH/HCOO^-$ buffer. A basic buffer (with pH > 7) can be made by working with an acid–base pair with $K_a$ smaller than $10^{-7}$. In this case, the net reaction leads to production of $OH^-$, and $K_b$ must be used to determine the equilibrium state. A typical example is the $NH_4^+/NH_3$ buffer made by mixing ammonium chloride with ammonia.

## Example 6-8

Calculate the pH of a solution made by adding 0.100 mol of $NH_4Cl$ and 0.200 mol of $NH_3$ to water and diluting to 1.000 L. $K_a$ for $NH_4^+$ is $5.6 \times 10^{-10}$.

### Solution

Because $K_a \ll 10^{-7}$ for $NH_4^+$ (equivalently, $K_b \gg 10^{-7}$ for $NH_3$), the net reaction is production of $OH^-$ ions. Therefore, the equilibrium is written to show the net transfer of hydrogen ions from water to $NH_3$:

| | $H_2O(\ell) +$ | $NH_3(aq)$ | $\rightleftarrows$ $NH_4^+(aq)$ | $+$ $OH^-(aq)$ |
|---|---|---|---|---|
| Initial concentration (M) | | 0.200 | 0.100 | $\approx 0$ |
| Change in concentration (M) | | $-y$ | $+y$ | $+y$ |
| Equilibrium concentration (M) | | $0.200 - y$ | $0.100 + y$ | $y$ |

The equilibrium expression is

$$\frac{y(0.100 + y)}{0.200 - y} = K_b = \frac{K_w}{K_a} = 1.8 \times 10^{-5}$$

If $y$ is small relative to the original concentrations of both $NH_3$ and $NH_4^+$, then

$$\frac{y(0.100)}{0.200} \approx 1.8 \times 10^{-5}$$

$$y \approx \left(\frac{0.200}{0.100}\right)(1.8 \times 10^{-5}) = 3.6 \times 10^{-5} \ll 0.100, \, 0.200$$

$$[\text{H}_3\text{O}^+] = \frac{1.0 \times 10^{-14}}{[\text{OH}^-]} = \frac{1.0 \times 10^{-14}}{3.6 \times 10^{-5}} = 2.8 \times 10^{-10} \,\text{M}$$

$$\text{pH} = -\log_{10}(2.8 \times 10^{-10}) = 9.55$$

## Designing Buffers

Control of pH is vital in synthetic and analytical chemistry, just as it is in living organisms. Procedures that work well at a pH of 5 may fail when the concentration of hydronium ion in the solution is raised tenfold to make the pH 4. Fortunately, by the proper choice of a weak acid and the ratio in which it is mixed with its conjugate base, it is possible to prepare buffer solutions that maintain the pH close to any desired value. Consider now how to choose the best conjugate acid–base system and how to calculate the required acid–base ratio.

In Examples 6-6, 6-7, and 6-8 the equilibrium concentrations of acid and base in the buffer systems were close to the initial concentrations. When this is the case, the calculation of pH is simplified greatly, because for either an acidic *or* a basic buffer,

$$K_a = \frac{[\text{H}_3\text{O}^+][\text{A}^-]}{[\text{HA}]} \approx \frac{[\text{H}_3\text{O}^+][\text{A}^-]_0}{[\text{HA}]_0}$$

$$[\text{H}_3\text{O}^+] = \frac{[\text{HA}]_0}{[\text{A}^-]_0} K_a$$

$$\text{pH} \approx \text{p}K_a - \log_{10}\frac{[\text{HA}]_0}{[\text{A}^-]_0}$$

where the last equation was obtained by taking the logarithm and changing sign. It can be verified that this simple equation gives the correct result for the three preceding examples. However, some care must be used in employing it because it is only approximate. It will be valid only if *both* $[\text{H}_3\text{O}^+]$ and $[\text{OH}^-]$ (and therefore the extent of ionization) are small relative to $[\text{HA}]_0$ and $[\text{A}^-]_0$.

This expression relating pH to $\text{p}K_a$ can be used to design buffers with a given pH. An optimal buffer is one in which the acid and its conjugate base are as nearly equal in concentration as possible; if the differences are too great, the buffer becomes less resistant to the effects of additional acid or base. To select a buffer system, an acid with a $\text{p}K_a$ as close as possible to the desired pH should be chosen. The concentrations of acid and conjugate base can then be adjusted to give exactly the desired pH.

## Example 6-9

Design a buffer system with pH 4.60.

### Solution

From Table 6-2, the $pK_a$ for acetic acid is 4.75, so the $CH_3COOH/CH_3COO^-$ buffer is a suitable one. The concentrations required to give the desired pH are related by

$$pH = 4.60 = pK_a - \log_{10} \frac{[CH_3COOH]_0}{[CH_3COO^-]_0}$$

$$\log_{10} \frac{[CH_3COOH]_0}{[CH_3COO^-]_0} = pK_a - pH = 4.75 - 4.60 = 0.15$$

$$\frac{[CH_3COOH]_0}{[CH_3COO^-]_0} = 10^{0.15} = 1.4$$

Such a ratio could be established by dissolving 0.100 mol of sodium acetate and 0.140 mol of acetic acid in water and diluting to 1.00 L, or 0.200 mol of $NaCH_3COO$ and 0.280 mol of $CH_3COOH$ in the same volume, and so on. As long as the ratio of the concentrations is 1.4, the solution will be buffered at approximately pH 4.60.

*Related Problems:* 39, 40, 41, 42

As the preceding example shows, in the determination of the pH the *absolute* concentrations of acid and conjugate base in a buffer are much less important than is their ratio. However, the absolute concentrations do affect the capacity of the solution to resist changes in pH induced by added acid or base. The higher the concentrations of buffering species, the smaller the change in pH when a fixed amount of a strong acid or base is added. In Example 6-6, for instance, a change in buffer concentrations from 1.00 M and 0.500 M to 0.500 M and 0.250 M does not alter the original pH of 3.45, because the ratio of acid to base concentrations is unchanged. The pH after 0.100 mol of HCl is added does change, to the value 3.15 rather than

**Figure 6–8**

Addition of a given volume of base to buffered and unbuffered solutions causes a much greater change in the pH of the unbuffered solution. Of the two buffered solutions, the one with higher buffer concentration resists pH changes more effectively.

3.31. The buffer at lower concentration is less resistant to pH change (Fig. 6-8). The buffering capacity of *any* buffer is exhausted if enough strong acid (or strong base) is added to use up, through chemical reaction, the original amount of weak base (or weak acid).

## 6-5 ACID–BASE TITRATION CURVES

Section 4-3 described an acid–base titration as the addition of carefully metered volumes of a basic solution of known concentration to an acidic solution of un-known concentration (or the addition of acid to base) to reach an endpoint. The endpoint is signaled by the color change of an indicator or by a sudden rise or fall in pH, although the pH of the reaction mixture changes continuously over the course of an acid–base titration. A graph of the pH versus the volume $V$ of titrating solution is a **titration curve.** Its shape depends on the value of $K_a$ and on the concentrations of the acid and base reacting. The concepts of acid–base equilibria provide tools to predict the exact shapes of titration curves when these quantities are all known. The same concepts allow $K_a$ and the unknown concentration to be calculated from an experimental titration curve. Here we examine three categories of titration: strong acid reacting with strong base, weak acid reacting with strong base, and strong acid reacting with weak base. Titrations of a weak acid with a weak base (and the reverse) are not useful for analytical purposes.

### Titrations Involving a Strong Acid and a Strong Base

The addition of a strong base to a strong acid (or the reverse) is the simplest type of titration. The chemical reaction is the neutralization:

$$H_3O^+(aq) + OH^-(aq) \longrightarrow 2 H_2O(\ell)$$

Suppose a solution of 100.0 mL (0.1000 L) of 0.1000 M HCl is titrated with 0.1000 M NaOH. What does the titration curve look like? Of course, the curve can be measured experimentally, giving the result in Figure 6-9. A theoretical curve can also be constructed by figuring out the pH of the reaction mixture at many different points during the addition of the NaOH solution, and plotting the results. The following example describes the latter procedure.

**1.  $V = 0$ mL NaOH added**
Initially, $[H_3O^+] = 0.1000$ M, so the pH is 1.000. The chemical amount of $H_3O^+$ present initially is

$$n_{H_3O^+} = [H_3O^+](\text{volume}) = (0.1000 \text{ mol L}^{-1})(0.1000 \text{ L})$$
$$= 1.000 \times 10^{-2} \text{ mol}$$

**2.  $V = 30.00$ mL NaOH added**
30.00 mL of 0.1000 M NaOH solution contains

$$(0.1000 \text{ mol L}^-)(0.03000 \text{ L}) = 3.000 \times 10^{-3} \text{ mol OH}^-$$

This reacts with (and neutralizes) an equal chemical amount of the $H_3O^+$ ion present initially, and reduces $n_{H_3O^+}$ to

$$n_{H_3O^+} = (1.000 \times 10^{-2} - 3.000 \times 10^{-3})\text{mol} = 7.00 \times 10^{-3} \text{ mol}$$

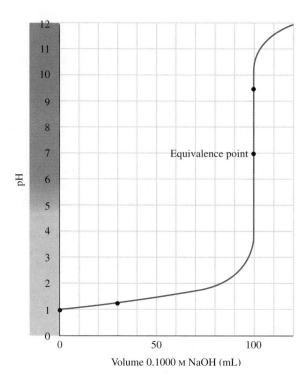

**Figure 6–9**

A titration curve for the titration of a strong acid by a strong base. The curve shown is for 100.0 mL of 0.1000 M HCl titrated with NaOH.

In addition (and this is very important to remember), the volume of the titration mixture has increased from 100.0 to 130.0 mL (that is, from 0.1000 to 0.1300 L). The *concentration* of $H_3O^+$ at this point in the titration is

$$[H_3O^+] = \frac{n_{H_3O^+}}{V_{tot}} = \frac{7.00 \times 10^{-3}\ \text{mol}}{0.1300\ \text{L}} = 0.0538\ \text{M}$$

$$pH = 1.27$$

### 3.  $V = 100.00$ mL NaOH added

This is called the **equivalence point,** that point in the titration at which the chemical amount of base added equals the chemical amount of acid originally present. The volume of base added up to the equivalence point is the **equivalent volume,** $V_e$. At the equivalence point in the titration of a strong acid with a strong base, the concentrations of $OH^-$ and $H_3O^+$ must be equal and the pH 7.0, due to the autoionization of water. At this point the solution is simply a nonhydrolyzing salt (in this case, NaCl) in water. The pH is 7 at the equivalence point *only* in the titration of a strong acid with a strong base (or vice versa). The pH at the equivalence point differs from 7 if a weak acid or weak base takes part in the titration.

### 4.  $V = 100.05$ mL NaOH added

Beyond the equivalence point, $OH^-$ is added to a neutral unbuffered solution. The $OH^-$ concentration can be found from the chemical amount of $OH^-$ added after the equivalence point has been reached. The volume beyond the equivalence point at this stage is 0.05 mL, or $5 \times 10^{-5}$ L. (This is the volume of approximately one drop of solution let in from the buret.) The chemical amount of $OH^-$ in this volume is

$$(0.1000 \text{ mol L}^{-1})(5 \times 10^{-5} \text{ L}) = 5 \times 10^{-6} \text{ mol}$$

Meanwhile, the total volume of the titration mixture is

0.1000 L HCl solution + 0.10005 L NaOH solution = 0.20005 L solution

so that the concentration of $OH^-$ is

$$[OH^-] = \frac{\text{moles } OH^-}{\text{total volume}} = \frac{5 \times 10^{-5} \text{ L}}{0.20005 \text{ L}} = 2.5 \times 10^{-5} \text{ M}$$

$$[H_3O^+] = 4 \times 10^{-10} \text{ M}; \quad pH = 9.4$$

As the titration curve shows (Fig. 6-9), the pH increases dramatically in the immediate vicinity of the equivalence point: $[H_3O^+]$ changes by four orders of magnitude between 99.98 mL and 100.02 mL NaOH! Any indicator whose color changes between pH = 5.0 and pH = 9.0 therefore signals the endpoint of the titration to an accuracy of ±0.02 mL in 100.0 mL, or ±0.02%. The titration endpoint (the experimentally measured volume at which the indicator changes color) is then almost identical to the equivalence point (the theoretical volume at which the chemical amount of added base equals that of acid present originally).

The titration of a strong base by a strong acid is entirely parallel. In this case the pH starts at a higher value and *drops* through a pH of 7 at the equivalence point. The roles of acid and base, and of $H_3O^+$ and $OH^-$, are reversed in the equations already given.

## Titration of Weak Acids and Bases

We turn now to the titration of a weak acid with a strong base (the titration of a weak base with a strong acid is analogous). The equivalence point has the same meaning as for a strong acid titration. At the equivalence point, the chemical amount of base added (in volume $V_e$) is equal to the chemical amount of acid originally present (in volume $V_0$), so that once again

$$c_0 V_0 = c_t V_e$$

where $c_0$ is the original weak acid concentration and $c_t$ is the $OH^-$ concentration in the titrating solution.

The calculation of the titration curve differs from the strong acid–strong base case in that now equilibrium (reflected in the $K_a$ of the weak acid) enters the picture. As an example, consider the titration of 100.0 mL of a 0.1000 M solution of acetic acid ($CH_3COOH$) with 0.1000 M NaOH. For this titration,

$$V_e = \frac{c_0}{c_t} V_0 = \left(\frac{0.1000 \text{ M}}{0.1000 \text{ M}}\right)(100.0 \text{ mL})$$

$$= 100.0 \text{ mL}$$

There are four distinct ranges in the titration; each corresponds to a type of calculation that has been considered already. Figure 6-10 shows pH versus volume for this titration, and the following paragraphs outline the calculation of four typical points on the curve.

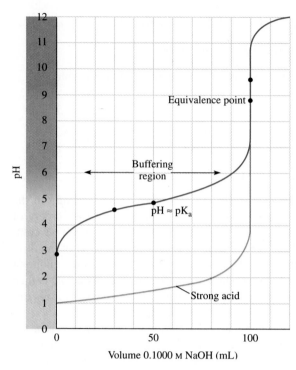

**Figure 6-10**

A titration curve for the titration of a weak acid by a strong base. The curve shown is for 100.0 mL of 0.1000 M $CH_3COOH$ titrated with NaOH. For comparison, the thin red line shows the titration curve for a strong acid of the same amount and concentration.

## 1. $V = 0$ mL NaOH added

Before any NaOH is added (that is, at the beginning of the titration), the problem is simply the ionization of a weak acid, which was considered in Section 6-3. A calculation analogous to those of Examples 6-2 and 6-3 gives a pH of 2.88.

## 2. $0 < V < V_e$

In this range the acid has been partially neutralized by added NaOH solution. Because $OH^-$ is a stronger base than acetate ion, it reacts almost completely with the acid originally present:

$$CH_3COOH(aq) + OH^-(aq) \rightleftharpoons H_2O(\ell) + CH_3COO^-(aq)$$

$$K = \frac{1}{K_b} = \frac{K_a}{K_w} = 2 \times 10^9 \gg 1$$

As a specific example, suppose that 30.00 mL of 0.1000 M NaOH has been added. The 30.00 mL NaOH contains

$$(0.1000 \text{ mol L}^-)(0.03000 \text{ L}) = 3.000 \times 10^{-3} \text{ mol OH}^-$$

and the original solution contained

$$(0.1000 \text{ mol L}^{-1})(0.1000 \text{ L}) = 1.000 \times 10^{-2} \text{ mol CH}_3\text{COOH}$$

The neutralization reaction generates one $CH_3COO^-$ ion for every $OH^-$ ion added. Hence, $3.000 \times 10^{-3}$ mol of $CH_3COO^-$ ion is generated. The amount of acetic acid that remains unreacted is

$$1.000 \times 10^{-2} - 3.000 \times 10^{-3} = 7.00 \times 10^{-3} \text{ mol}$$

Because the total volume is now 130.0 mL (or 0.1300 L), the nominal concentrations after reaction are

$$[CH_3COOH] \approx \frac{7.00 \times 10^{-3} \text{ mol}}{0.1300 \text{ L}} = 5.38 \times 10^{-2} \text{ M}$$

$$[CH_3COO^-] \approx \frac{3.00 \times 10^{-3} \text{ mol}}{0.1300 \text{ L}} = 2.31 \times 10^{-2} \text{ M}$$

This is nothing other than a buffer solution containing acetic acid at a concentration of $5.38 \times 10^{-2}$ M and sodium acetate at a concentration of $2.31 \times 10^{-2}$ M. Because the $K_a$ for acetic acid is larger than $10^{-7}$, hydronium ion (not hydroxide ion) predominates, and this is an acidic buffer. The pH can be found from the procedure used in Example 6-6 or, more approximately,

$$pH \approx pK_a - \log_{10} \frac{[CH_3COOH]_0}{[CH_3COO^-]_0} = 4.75 - \log_{10} \frac{5.38 \times 10^{-2}}{2.31 \times 10^{-2}} = 4.38$$

This region of the titration shows clearly the buffering action of a mixture of a weak acid with its conjugate base. At the half-equivalence point $V = V_e/2$, $[CH_3COOH]_0 = [CH_3COO^-]_0$, which corresponds to an equimolar buffer; at this point pH $\approx pK_a$. On either side of this point the pH rises relatively slowly as the NaOH solution is added.

### 3. $V = V_e$

At the equivalence point, $c_t V_e$ mol of $OH^-$ has been added to an equal chemical amount of $CH_3COOH$. An identical solution could have been prepared simply by adding $c_t V_e = 1.000 \times 10^{-2}$ mol of the base $CH_3COO^-$ (in the form of $NaCH_3COO$) to 0.2000 L of water. The pH at the equivalence point corresponds to the hydrolysis of $CH_3COO^-$:

$$H_2O(\ell) + CH_3COO^-(aq) \rightleftharpoons CH_3COOH(aq) + OH^-(aq)$$

as considered in Example 6-5. At the equivalence point the pH is 8.73.

Note that in the titration of a weak acid by a strong base the equivalence point comes not at pH 7 but at a higher (more basic) value. By the same token, the equivalence point in the titration of a weak base by a strong acid occurs at a pH lower than 7.

### 4. $V > V_e$

Beyond the equivalence point, $OH^-(aq)$ is added to a solution of the base $CH_3COO^-$. The $[OH^-]$ comes almost entirely from the hydroxide ion added beyond the equivalence point; very little comes from the reaction of the $CH_3COO^-$ with water. Beyond $V_e$, the pH for the titration of a weak acid by a strong base is very close to that for a strong acid by a strong base.

The equivalent volume $V_e$ is readily determined in the laboratory by using an indicator that changes color near pH 8.7, the pH at the equivalence point of the acetic acid titration. A suitable choice would be phenolphthalein, which changes from colorless to red over a pH range from 8.2 to 10.0. The slope of pH versus volume of strong base is less steep near the equivalence point for a weak acid than it is for a strong acid, making determinations of the equivalent volume (and of the original weak acid concentration) somewhat less accurate.

The use of titrations to determine concentrations and ionization constants for unknown acids and bases is illustrated by the following example.

## Example 6-10

A volume of 50.00 mL of a weak acid of unknown concentration is titrated with a 0.1000 M solution of NaOH. The equivalence point is reached after 39.30 mL of NaOH solution has been added. At the half-equivalence point (19.65 mL) the pH is 4.85. Calculate the original concentration of the acid and its ionization constant $K_a$.

### Solution

The chemical amount of acid originally present, $c_0 V_0$, is equal to the chemical amount of base added at the equivalence point, $c_t V_e$, so that

$$c_0 = \frac{V_e}{V_0} c_t = \frac{39.50 \text{ mL}}{50.00 \text{ mL}} \times 0.100 \text{ M} = 0.0786 \text{ M}$$

This is the original concentration of acid.

At the half-equivalence point,

$$pH = 4.85 \approx pK_a$$

$$K_a = 10^{-4.85} = 1.4 \times 10^{-5}$$

The unknown acid could be propionic acid, $CH_3CH_2COOH$ (see Table 6-2).

*Related Problems:* 53, 54

## 6-6 POLYPROTIC ACIDS

So far, only **monoprotic acids** have been considered. They are acids whose molecules are capable of donating only a single hydrogen ion to acceptor molecules. **Polyprotic acids** can donate two or more hydrogen ions to acceptors. Sulfuric acid is a familiar and important example. It reacts in two stages—first

$$H_2SO_4(aq) + H_2O(\ell) \longrightarrow H_3O^+(aq) + HSO_4^-(aq)$$

to give the hydrogen sulfate ion, and then

$$HSO_4^-(aq) + H_2O(\ell) \longrightarrow H_3O^+(aq) + SO_4^{2-}(aq)$$

The hydrogen sulfate ion is amphoteric, meaning that it is a base in the first reaction (with conjugate acid $H_2SO_4$) and an acid in the second (with conjugate base $SO_4^{2-}$). In its first ionization $H_2SO_4$ is a strong acid, but the product of that ionization ($HSO_4^-$) is itself only a weak acid. The $H_3O^+$ produced in a solution of $H_2SO_4$ comes primarily from the first ionization, and the solution has a pH close to that of a monoprotic strong acid of the same concentration. When this solution reacts with a strong base, however, its neutralizing power is twice that of a monoprotic acid of the same concentration, because each mole of sulfuric acid can react with and neutralize two moles of hydroxide ion.

## Weak Polyprotic Acids

Weak polyprotic acids ionize in two or more stages. Examples are carbonic acid ($H_2CO_3$), formed from solvated $CO_2$ (carbonated water; Fig. 6-11), and phosphoric acid ($H_3PO_4$). Carbonic acid can give up one hydrogen ion to form $HCO_3^-$ (hydrogen carbonate ion) or two hydrogen ions to form $CO_3^{2-}$ (carbonate ion). Phosphoric acid ionizes in three stages, giving successively $H_2PO_4^-$, $HPO_4^{2-}$, and $PO_4^{3-}$.

Two simultaneous equilibria are involved in the ionization of a diprotic acid such as $H_2CO_3$:[1]

$$H_2CO_3(aq) + H_2O(\ell) \rightleftharpoons H_3O^+(aq) + HCO_3^-(aq)$$

$$\frac{[H_3O^+][HCO_3^-]}{[H_2CO_3]} = K_{a1} = 4.3 \times 10^{-7}$$

and

$$HCO_3^-(aq) + H_2O(\ell) \rightleftharpoons H_3O^+(aq) + CO_3^{2-}(aq)$$

$$\frac{[H_3O^+][CO_3^{2-}]}{[HCO_3^-]} = K_{a2} = 4.8 \times 10^{-11}$$

**Figure 6–11**

The indicator bromothymol blue is blue in basic solution *(left)*. When Dry Ice (solid carbon dioxide) is placed in the bottom of the cylinder *(right)*, it dissolves to form carbonic acid. This causes the indicator to change to its colorless form. (Leon Lewandowski)

Two important observations can be made at the outset:

1. The $[H_3O^+]$ in the two ionization equilibria are one and the same.
2. $K_{a2}$ is invariably smaller than $K_{a1}$ because the negative charge left behind by the loss of a hydrogen ion in the first ionization causes the second hydrogen ion to be more tightly bound.

Exact calculations of simultaneous equilibria can be complex. They simplify considerably when the original acid concentration is not too small and the ionization constants $K_{a1}$ and $K_{a2}$ differ substantially in magnitude (by a factor of 100 or more). The latter condition is almost always satisfied. Under such conditions, the two equilibria can be treated sequentially, as in the following example.

## Example 6-11

Calculate the concentrations at equilibrium of $H_2CO_3$, $HCO_3^-$, $CO_3^{2-}$, and $H_3O^+$ in a saturated aqueous solution of $CO_2$, in which the original concentration of $H_2CO_3$ is 0.034 M.

### Solution

The $H_3O^+$ arises both from the ionization of $H_2CO_3$ and from the subsequent ionization of $HCO_3^-$, but because $K_{a2} \ll K_{a1}$ it is reasonable to ignore the contribution of $[H_3O^+]$ from the second ionization (as well as from the autoionization of water). These approximations will be checked later in the calculation.

If $y$ mol $L^{-1}$ of $H_2CO_3$ ionizes, the following approximations apply:

[1] An accurate description of carbonic acid and solvated $CO_2$ is somewhat more complicated than indicated here. In fact, most of the dissolved $CO_2$ remains as $CO_2(aq)$, and only a small fraction actually reacts with water to give $H_2CO_3(aq)$. However, we shall indicate by $[H_2CO_3]$ the total concentration of both of these species. Approximately 0.034 mol of $CO_2$ dissolves per liter of water at 25°C and atmospheric pressure.

| $H_2CO_3(aq) + H_2O(\ell) \rightleftarrows H_3O^+(aq) + HCO_3^-(aq)$ | | |
| --- | --- | --- |
| Initial concentration (M)       0.034 | $\approx 0$ | 0 |
| Change in concentration (M)       $-y$ | $+y$ | $+y$ |
| Equilibrium concentration (M)    $0.034 - y$ | $y$ | $y$ |

where equating both $[HCO_3^-]$ and $[H_3O^+]$ to $y$ involves the assumption that the subsequent ionization of $HCO_3^-$ has only a small effect on its concentration. Then the first ionization equilibrium can be written as

$$\frac{[H_3O^+][HCO_3^-]}{[H_2CO_3]} = K_{a1}$$

$$\frac{y^2}{0.034 - y} = 4.3 \times 10^{-7}$$

Solving this equation for $y$ gives

$$y = 1.2 \times 10^{-4}\ \text{M} = [H_3O^+] = [HCO_3^-]$$

$$[H_2CO_3] = 0.034 - y = 0.034\ \text{M}$$

The second equilibrium can be written as

$$\frac{[H_3O^+][CO_3^{2-}]}{[HCO_3^-]} = K_{a2} = \frac{(1.2 \times 10^{-4})[CO_3^{2-}]}{1.2 \times 10^{-4}} = 4.8 \times 10^{-11}$$

$$[CO_3^{2-}] = 4.8 \times 10^{-11}\ \text{M}$$

The concentration of the base produced in the second ionization, $[CO_3^{2-}]$, is numerically equal to $K_{a2}$.

The assumptions must now be checked. Because

$$[CO_3^{2-}] = 4.8 \times 10^{-11}\ \text{M} \ll 1.2 \times 10^{-4}\ \text{M} = [HCO_3^-]$$

we were justified in ignoring the effect of the second ionization on the concentrations of $HCO_3^-$ and $H_3O^+$. The additional concentration of $H_3O^+$ furnished by $HCO_3^-$ is only $4.8 \times 10^{-11}$ M. Finally, $[H_3O^+]$ is much larger than $10^{-7}$ M, so the neglect of water autoionization was also justified.

*Related Problems:* 55, 56

For a triprotic acid such as $H_3PO_4$, the concentration of the base ($PO_4^{3-}$) resulting from the third ionization could have been calculated.

An analogous procedure applies to the reactions of a base that can accept two or more hydrogen ions. In a solution of sodium carbonate ($Na_2CO_3$), for example, the carbonate ion reacts with water to form first $HCO_3^-$ and then $H_2CO_3$:

$$H_2O(\ell) + CO_3^{2-}(aq) \rightleftharpoons HCO_3^-(aq) + OH^-(aq)$$

$$\frac{[OH^-][HCO_3^-]}{[CO_3^{2-}]} = K_{b1} = \frac{K_w}{K_{a2}} = 2.1 \times 10^{-4}$$

$$H_2O(\ell) + HCO_3^-(aq) \rightleftharpoons H_2CO_3(aq) + OH^-(aq)$$

$$\frac{[OH^-][H_2CO_3]}{[HCO_3^-]} = K_{b2} = \frac{K_w}{K_{a1}} = 2.3 \times 10^{-8}$$

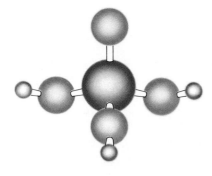

Phosphoric acid, $H_3PO_4$, is a triprotic acid that is used in the manufacture of phosphate fertilizers and in the food industry.

In this case $K_{b1} \gg K_{b2}$, so essentially all the $OH^-$ arises from the first reaction. The ensuing calculation of the concentrations of species present is just like that in Example 6-11.

## Effect of pH on Solution Composition

Changing the pH of a solution shifts the positions of all acid–base equilibria, including those involving polyprotic acids. Acid–base equilibrium expressions and equilibrium constants are used to calculate the amount of the change. For example, the two equilibria that apply for $H_2CO_3$–$HCO_3^-$–$CO_3^{2-}$ solutions can be written as

$$\frac{[HCO_3^-]}{[H_2CO_3]} = \frac{K_{a1}}{[H_3O^+]} \qquad \frac{[CO_3^{2-}]}{[HCO_3^-]} = \frac{K_{a2}}{[H_3O^+]}$$

At a given pH, the right-hand sides are known and the relative amounts of the three carbonate species can be calculated. This is illustrated by the following example.

........................................................................................................

### Example 6-12

Calculate the fractions of carbonate present as $H_2CO_3$, $HCO_3^-$, and $CO_3^{2-}$ at pH 10.00.

### Solution

At this pH, $[H_3O^+] = 1.0 \times 10^{-10}$ M, and the values of $K_{a1}$ and $K_{a2}$ for the preceding equations are used to find

$$\frac{[HCO_3^-]}{[H_2CO_3]} = \frac{4.3 \times 10^{-7}}{1.0 \times 10^{-10}} = 4.3 \times 10^3$$

$$\frac{[CO_3^{2-}]}{[HCO_3^-]} = \frac{4.8 \times 10^{-11}}{1.0 \times 10^{-10}} = 0.48$$

It is most convenient to rewrite these ratios with the same species (say, $HCO_3^-$) in the denominator. The first equation becomes

$$\frac{[H_2CO_3]}{[HCO_3^-]} = \frac{1}{4.3 \times 10^3} = 2.3 \times 10^{-4}$$

The fraction of each species present is obtained by dividing the species' concentration by the sum of the three concentrations. For $H_2CO_3$ this gives

$$\text{fraction } H_2CO_3 = \frac{[H_2CO_3]}{[H_2CO_3] + [HCO_3^-] + [CO_3^{2-}]}$$

This can be simplified by dividing numerator and denominator by $[HCO_3^-]$ and substituting in the ratios that have already been calculated:

$$\text{fraction } H_2CO_3 = \frac{[H_2CO_3]/[HCO_3^-]}{([H_2CO_3]/[HCO_3^-]) + 1 + ([CO_3^{2-}]/[HCO_3^-])}$$

$$= \frac{2.3 \times 10^{-4}}{2.3 \times 10^{-4} + 1 + 0.48} = 1.6 \times 10^{-4}$$

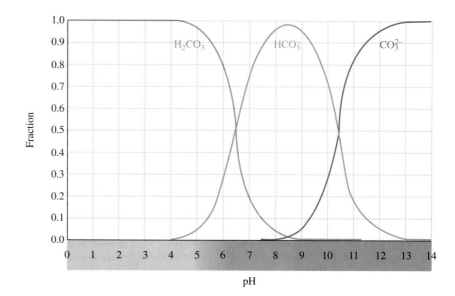

**Figure 6–12**

The equilibrium fractions of $H_2CO_3$, $HCO_3^-$, and $CO_3^{2-}$ that are present in aqueous solution at different values of the pH.

In the same way the fractions of $HCO_3^-$ and $CO_3^{2-}$ present are calculated to be 0.68 and 0.32, respectively.

*Related Problems:* 59, 60

If the calculation of Example 6-12 is repeated at a series of different pH values, the graph shown in Figure 6-12 is obtained. At high pH, $CO_3^{2-}$ predominates, and at low pH, $H_2CO_3$ is the major species. At intermediate pH (near pH 8, the approximate pH of seawater), the hydrogen carbonate ion $HCO_3^-$ is most prevalent. The variation in composition of sedimentary rocks from different locations can be traced back to the effect of pH on solution composition. Sediments containing carbonates were formed from alkaline (high-pH) lakes and oceans in which $CO_2$ was present mainly as $CO_3^{2-}$ ion. Sediments deposited from waters with intermediate pH are hydrogen carbonates or mixtures between carbonates and hydrogen carbonates. An example of the latter is trona ($2Na_2CO_3 \cdot NaHCO_3 \cdot 2H_2O$), an ore from the western United States that is an important source of both carbonates of sodium. Acidic waters did not deposit carbonates, but instead released $CO_2(g)$ to the atmosphere.

## 6-7 A Deeper Look . . . Exact Treatment of Acid–Base Equilibria

Sections 6-3 and 6-4 outlined methods for calculating equilibria involving weak acids, bases, and buffer solutions. There it was assumed that the amount of hydronium ion (or hydroxide ion) resulting from the ionization of water could be neglected in comparison with that produced by the ionization of dissolved acids or bases. In this section, that approximation is replaced by an exact treatment of acid–base equilibria (exact, that is, in the context of the mass-action law). This approach leads to somewhat more complicated equations, but it serves several purposes. It is of practical importance in cases (such as very weak acids or bases, or very dilute solutions) in which the previous approximations no longer hold. It includes as special cases the various aspects of acid–base equilibrium considered earlier. And finally, it provides a foundation for treating amphoteric equilibrium later in this section.

Consider a general case in which the initial concentration of a weak acid HA is called $c_a$, and the initial concentration of its conjugate base (which for simplicity is assumed to come from the salt NaA) is $c_b$. In solution there will be five dissolved species: HA, $A^-$, $Na^+$, $H_3O^+$, and $OH^-$. It is necessary to write down and solve five independent equations that relate the equilibrium concentrations of these species to the initial concentrations $c_a$ and $c_b$, and to $K_a$, the acid ionization constant of HA. The first equation is simply

$$[Na^+] = c_b \qquad \text{(a)}$$

reflecting the fact that the $Na^+(aq)$ from the dissolved salt is a spectator ion that does not take part in the acid–base equilibrium. Next there are two equilibrium relations:

$$[H_3O^+][OH^-] = K_w \qquad \text{(b)}$$

$$\frac{[H_3O^+][A^-]}{[HA]} = K_a \qquad \text{(c)}$$

The fourth relation is one of stoichiometry, or conservation of ''A-material'':

$$c_a + c_b = [HA] + [A^-] \qquad \text{(d)}$$

The original A-material was introduced either as acid or as base, with a total concentration of $c_a + c_b$. When equilibrium is reached, some redistribution has doubtless occurred, but the *total* concentration, $[HA] + [A^-]$, must be the same. The fifth and final relation results from charge balance; the solution must be electrically neutral, so the total amount of positive charge must be equal to the total amount of negative charge:

$$[Na^+] + [H_3O^+] = [A^-] + [OH^-] \qquad \text{(e)}$$

These five independent equations completely determine the five unknown concentrations. To solve them, begin by substituting (a) into (e) and solving for $[A^-]$:

$$[A^-] = c_b + [H_3O^+] - [OH^-] \qquad \text{(e}'\text{)}$$

Next, insert (e′) into (d) and solve for [HA]:

$$[HA] = c_a + c_b - [A^-] = c_a - [H_3O^+] + [OH^-] \quad \text{(d}'\text{)}$$

Next, substitute both (d′) and (e′) into (c) to find

$$\frac{[H_3O^+](c_b + [H_3O^+] - [OH^-])}{(c_a - [H_3O^+] + [OH^-])} = K_a \qquad \text{(c}'\text{)}$$

There are now two things that can be done with the general equation (c′). First, the exact solution for $[H_3O^+]$ can be set up, a procedure that is useful for very weak acids or bases or for very dilute solutions. Alternatively, (c′) can be reduced in various limits to cases already considered.

For the exact solution, eliminate $[OH^-]$ in (c′) by using (b). This gives

$$\frac{[H_3O^+]\left(c_b + [H_3O^+] - \dfrac{K_w}{[H_3O^+]}\right)}{\left(c_a - [H_3O^+] + \dfrac{K_w}{[H_3O^+]}\right)} = K_a$$

The numerator and denominator are multiplied by $[H_3O^+]$ and

the fraction is cleared by moving the denominator to the right side:

$$[H_3O^+](c_b[H_3O^+] + [H_3O^+]^2 - K_w)$$
$$= K_a(c_a[H_3O^+] - [H_3O^+]^2 + K_w)$$

This can be rewritten as

$$[H_3O^+]^3 + (c_b + K_a)[H_3O^+]^2 - (K_w + c_a K_a)[H_3O^+] - K_a K_w = 0$$

This is a cubic equation for $[H_3O^+]$, which, although more complicated than the quadratic equations encountered so far, can nonetheless be solved with a calculator. The concentrations of $OH^-$, $A^-$, and HA can then be found by successive substitutions into (b), (e′), and (d′).

Alternatively, the general equation (c′) can be examined in various limits. In an acidic buffer (as in Examples 6-6 and 6-7), if it can be assumed that $[H_3O^+] \gg [OH^-]$, then (c′) simplifies to

$$\frac{[H_3O^+](c_b + [H_3O^+])}{(c_a - [H_3O^+])} = K_a$$

which is exactly the equation used in those examples. If, additionally, $c_b = 0$, the weak-acid ionization limit of Examples 6-2 and 6-3 is reached. In a basic buffer, on the other hand, if it can be assumed that $[OH^-] \gg [H_3O^+]$, then (c′) simplifies to

$$\frac{(K_w/[OH^-])(c_b - [OH^-])}{(c_a + [OH^-])} = K_a$$

Here (b) was used to substitute for $[H_3O^+]$ where it multiplies the whole expression. This can be rewritten as

$$\frac{[OH^-](c_a + [OH^-])}{(c_b - [OH^-])} = \frac{K_w}{K_a} = K_b$$

which is exactly the equation used in Example 6-8. If no acid is present initially ($c_a = 0$), the weak-base ionization limits of Examples 6-4 and 6-5 are reached. The general approach therefore includes all of the previous calculations as special cases.

Unless conditions require the use of the exact solution, approximate equations are preferable because they are easier to apply and provide greater physical insight. If a calculation (ignoring water autoionization) of the ionization of a weak acid gives a concentration of $H_3O^+$ smaller than $10^{-6}$ M, or if a calculation of base ionization gives a concentration of $OH^-$ smaller than $10^{-6}$ M, the more exact treatment is needed. For buffer solutions, a pH near 7 does not necessarily mean that water ionization is important, unless the acid or base concentration becomes very small.

## Example 6-13

Calculate the pH of a $1.00 \times 10^{-5}$ M solution of HCN($aq$). The $K_a$ of HCN($aq$) is $6.17 \times 10^{-10}$.

## Solution

Suppose the autoionization of water is ignored and the method of Examples 6-2 and 6-3 is used. This gives $[H_3O^+] = 7.9 \times 10^{-8}$ M, which of course makes no sense, because it is *lower* than the concentration of hydronium ion in pure water. HCN is a very weak acid, but it is nonetheless an acid, not a base.

It is necessary, therefore, to use the exact cubic equation for $[H_3O^+]$, inserting into it the proper coefficients and taking $c_a = 1.00 \times 10^{-5}$ and $c_b = 0$. This gives

$$[H_3O^+]^3 + 6.17 \times 10^{-10}[H_3O^+]^2 - 1.617 \times 10^{-14}[H_3O^+] - 6.17 \times 10^{-24} = 0$$

Unfortunately, there is no method as simple as the quadratic formula to solve a cubic equation. The easiest way to solve this equation is to try a series of values for $[H_3O^+]$ on the left side, varying them to obtain a result as close as possible to 0 (see Appendix C). It is safe to assume that the final answer will be slightly larger than $1 \times 10^{-7}$, so the initial guesses should be of that magnitude. Carrying out the procedure gives

$$[H_3O^+] = 1.27 \times 10^{-7} \text{ M} \qquad pH = 6.90$$

*Related Problems:* 61, 62

## Amphoteric Equilibria

A second situation in which an exact analysis of acid–base equilibrium is useful occurs when an amphoteric species is dissolved in water. The hydrogen carbonate ion ($HCO_3^-$) is amphoteric because it can act as an acid in the equilibrium

$$HCO_3^-(aq) + H_2O(\ell) \rightleftharpoons H_3O^+(aq) + CO_3^{2-}(aq)$$

$$\frac{[H_3O^+][CO_3^{2-}]}{[HCO_3^-]} = K_{a2} = 4.8 \times 10^{-11}$$

or as a base in the equilibrium

$$H_2O(\ell) + HCO_3^-(aq) \rightleftharpoons H_2CO_3(aq) + OH^-(aq)$$

$$\frac{[OH^-][H_2CO_3]}{[HCO_3^-]} = K_{b2} = 2.3 \times 10^{-8}$$

If sodium hydrogen carbonate ($NaHCO_3$) is dissolved in water, a competition occurs between the tendencies of $HCO_3^-$ to accept hydrogen ions and to donate them. Because $K_{b2} > K_{a2}$, there should be more production of $OH^-$ than of $H_3O^+$, so the solution should be basic.

In an exact treatment of this equilibrium, there are six unknown concentrations—those of $Na^+$, $H_2CO_3$, $HCO_3^-$, $CO_3^{2-}$, $OH^-$, and $H_3O^+$. Two equilibrium equations were already presented, and a third relates $[OH^-]$ and $[H_3O^+]$ to $K_w$. If $[HCO_3^-]_0$ is the original concentration of $NaHCO_3$, then from stoichiometry

$$[HCO_3^-]_0 = [HCO_3^-] + [H_2CO_3] + [CO_3^{2-}]$$

because the total amount of carbonate material is conserved; any reduction in $[HCO_3^-]$ must be compensated by a corresponding increase in either $[H_2CO_3]$ or $[CO_3^{2-}]$. Next the principle of conservation of charge is used. The positively charged species present are $Na^+$ and $H_3O^+$, and the negatively charged species are $HCO_3^-$, $CO_3^{2-}$, and $OH^-$. Because there is overall charge neutrality,

$$[Na^+] + [H_3O^+] = [HCO_3^-] + 2[CO_3^{2-}] + [OH^-]$$

where the coefficient 2 for $[CO_3^{2-}]$ arises because each carbonate ion is doubly charged. In addition, the $Na^+$ concentration is unchanged, so

$$[Na^+] = [HCO_3^-]_0$$

In principle, these six equations could be solved simultaneously to calculate the exact $[H_3O^+]$ for an arbitrary initial concentration of $HCO_3^-$. The result is complex and gives little physical insight. Instead, we give only a simpler, approximate solution, which is sufficient in the cases considered here. Subtracting the carbonate balance equation from the charge balance equation gives

$$[H_3O^+] = [CO_3^{2-}] - [H_2CO_3] + [OH^-]$$

The three equilibrium expressions are used to rewrite this as

$$[H_3O^+] = K_{a2}\frac{[HCO_3^-]}{[H_3O^+]} - \frac{[H_3O^+][HCO_3^-]}{K_{a1}} + \frac{K_w}{[H_3O^+]}$$

$[CO_3^{2-}]$ and $[H_2CO_3]$ have been eliminated in favor of $[HCO_3^-]$.

Multiplying by $K_{a1}[H_3O^+]$ gives

$$K_{a1}[H_3O^+]^2 + [HCO_3^-][H_3O^+]^2 = K_{a1}K_{a2}[HCO_3^-] + K_{a1}K_w$$

$$[H_3O^+]^2 = \frac{K_{a1}K_{a2}[HCO_3^-] + K_{a1}K_w}{K_{a1} + [HCO_3^-]}$$

This equation still contains two unknown quantities, $[H_3O^+]$ and $[HCO_3^-]$. Because both $K_{a2}$ and $K_{b2}$ are small, $[HCO_3^-]$ should be close to its original value, $[HCO_3^-]_0$. If $[HCO_3^-]$ is set equal to $[HCO_3^-]_0$, this becomes

$$[H_3O^+]^2 \approx \frac{K_{a1}K_{a2}[HCO_3^-]_0 + K_{a1}K_w}{K_{a1} + [HCO_3^-]_0}$$

which can be solved for $[H_3O^+]$. In many cases of interest, $[HCO_3^-]_0 \gg K_{a1}$, and $K_{a2}[HCO_3^-]_0 \gg K_w$. When this is so, the expression simplifies to

$$[H_3O^+]^2 \approx K_{a1}K_{a2}$$

$$[H_3O^+] \approx \sqrt{K_{a1}K_{a2}}$$

$$pH \approx \frac{1}{2}(pK_{a1} + pK_{a2})$$

and so the pH of such a solution is the average of the $pK_a$ values for the two ionizations.

## Example 6-14

What is the pH of a solution that is 0.100 M in $NaHCO_3$?

### Solution

First, the two assumptions are checked:

$$[HCO_3^-]_0 = 0.100 \gg 4.3 \times 10^{-7} = K_{a1}$$

$$[HCO_3^-]_0 K_{a2} = 4.8 \times 10^{-12} \gg 1.0 \times 10^{-14} = K_w$$

so both are satisfied. Therefore,

$$[H_3O^+] = \sqrt{K_{a1}K_{a2}} = 4.5 \times 10^{-9} \text{ M}$$

$$pH = 8.34$$

and the solution is basic, as expected.

## Titration of a Polyprotic Acid

A polyprotic acid shows more than one equivalence point. The first equivalence point occurs when the volume $V_{e1}$ of base added is sufficient to remove one hydrogen ion from each acid molecule; $V_{e2}$ is the volume sufficient to remove two hydrogen ions from each, and so forth. A diprotic acid shows two equivalence points, and a triprotic acid, three. The equivalent volumes are related to each other by

$$V_{e1} = \frac{1}{2} V_{e2} = \frac{1}{3} V_3$$

Figure 6-13 shows a titration curve for triprotic phosphoric acid. The three equivalence points are at 100.0, 200.0, and 300.0 mL. The calculation of pH as a function of the volume of added base presents no new complications beyond those already considered. The initial pH is given by a calculation analogous to that of Example 6-11, and the pH in the flat regions between equivalence points is obtained by a buffer calculation like that for a diprotic acid. For example, the pH after addition of 50.0 mL of base is that of an equimolar $H_3PO_4$/$H_2PO_4^-$ buffer (subsequent ionization of $H_2PO_4^-$ can be ig-

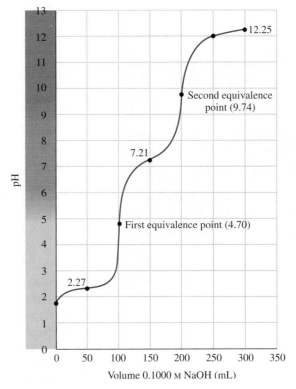

### Figure 6–13

A titration curve for the titration of a polyprotic acid (phosphoric acid) by a strong base. The curve shown is for 100.0 mL of 0.1000 M $H_3PO_4$ titrated with NaOH. No clear third equivalence point is seen at 300 mL because $K_a$ for $HPO_4^{2-}$ is not much greater than $K_a$ for $H_2O$ in aqueous solution.

nored). Finally, the pH at the first equivalence point is that for a solution of $NaH_2PO_4$ and uses the amphoteric equilibria equations presented earlier in this section ($PO_4^{3-}$ can be ignored in this case). The pH at the second equivalence point is an amphoteric equilibrium in which $HPO_4^{2-}$ is in equilibrium with $H_2PO_4^-$ and with $PO_4^{3-}$.

## Acid Precipitation

Acid precipitation is a major environmental problem throughout the industrialized world. One major source is the burning of fossil fuels containing sulfur (coal, oil, and natural gas). The sulfur dioxide released into the air dissolves in water or, more seriously, may be oxidized further to sulfur trioxide. The $SO_3$ dissolves in water to form sulfuric acid:

$$SO_3(g) + H_2O(\ell) \longrightarrow H_2SO_4(aq)$$

**PRACTICE EXERCISE**

The effect of acid rain on a stand of fir trees on Mount Mitchell, North Carolina. (John Shaw/Bruce Coleman, Inc.)

The net effect is to increase the acidity of the rain, which damages trees, kills fish in lakes, dissolves stone, and corrodes metal.

(a) A sample of rainwater is tested for acidity by using two indicators. Addition of methyl orange to half of the sample gives a yellow color, and addition of methyl red to the other half gives a red color. Estimate the pH of the sample.

(b) The pH in acid rain can range down to 3 or even lower in heavily polluted areas. Calculate the concentrations of $H_3O^+$ and $OH^-$ in a raindrop at pH 3.30 at 25°C.

(c) When $SO_2$ dissolves in water to form sulfurous acid, $H_2SO_3(aq)$, that acid can donate a hydrogen ion to water. Write a balanced chemical equation for this reaction, and identify the stronger Brønsted–Lowry acid and base in the equation.

(d) Ignore the further ionization of $HSO_3^-$, and calculate the pH of a solution whose initial concentration of $H_2SO_3$ is $4.0 \times 10^{-4}$ M. (*Hint:* Use the quadratic equation in this case.)

(e) Now suppose that all the dissolved $SO_2$ from part (d) has been oxidized further to $SO_3$, so that $4.0 \times 10^{-4}$ mol of $H_2SO_4$ is dissolved per liter. Calculate the pH in this case. [*Hint:* Because the first ionization of $H_2SO_4$ is that of a strong acid, the concentration of $H_3O^+$ can be written as $4.0 \times 10^{-4}$ plus the unknown amount of dissociation from $HSO_4^-(aq)$.]

(f) Lakes have a natural buffering capacity, especially in regions where there is limestone that gives rise to dissolved calcium carbonate. Write an equation for the effect of a small amount of acid rain containing sulfuric acid if it falls into a lake containing carbonate ($CO_3^{2-}$) ions. Discuss how the lake will resist further pH changes. What happens if a large excess of acid rain is deposited?

(g) A sample of 1.00 L of rainwater known to contain only sulfurous (and not sulfuric) acid is titrated with 0.0100 M NaOH. The equivalence point of the $H_2SO_3$/$HSO_3^-$ titration is reached after 31.6 mL has been added. Calculate the original concentration of sulfurous acid in the sample, again ignoring any effect of $SO_3^{2-}$ on the equilibria.

(h) Calculate the pH at the half-equivalence point, after 15.8 mL has been added. (*Hint:* Use the quadratic equation.)

**Answers**

(a) 4.4 to 4.8

(b) $[H_3O^+] = 5.0 \times 10^{-4}$ M;   $[OH^-] = 2.0 \times 10^{-11}$ M

(c) $H_2SO_3(aq) + H_2O(\ell) \rightleftharpoons HSO_3^-(aq) + H_3O^+(aq)$. The stronger acid is $H_3O^+$, and the stronger base is $HSO_3^-$.

(d) The pH is 3.41.

(e) The pH is 3.11.

(f) The $H_3O^+$ in the sulfuric acid solution reacts according to $H_3O^+(aq) + CO_3^{2-}(aq) \rightleftharpoons HCO_3^-(aq) + H_2O(\ell)$. The $HSO_4^-$ in the sulfuric acid reacts according to $HSO_4^-(aq) + CO_3^{2-}(aq) \rightleftharpoons SO_4^{2-}(aq) + HCO_3^-(aq)$. This gives rise to a $HCO_3^-$/$CO_3^{2-}$ buffer that can resist further changes in pH. An excess of acid rain overwhelms the buffer and leads to the formation of $H_2CO_3$.

(g) $3.16 \times 10^{-4}$ M

(h) The pH is 3.81.

**CONCEPTS & SKILLS**

After studying this chapter and working the problems that follow, you should be able to

1. Define an acid and a base in the Brønsted–Lowry system and give several examples of their reaction with a solvent (Section 6-1, problems 1–4).
2. Define the pH function and convert between pH and $[H_3O^+]$ (Section 6-1, problems 5–8).
3. State the relationship between the ionization constant for an acid and that for its conjugate base (Section 6-2, problems 13–14).
4. Explain how indicators allow the pH of a solution to be estimated (Section 6-2, problems 17–18).
5. Formulate the equilibrium expression for the ionization of a weak acid or base, and use it to determine the pH and fraction ionized (Section 6-3, problems 19–28).
6. Explain the behavior of a buffer solution. Calculate its pH from the concentrations of its conjugate acid–base pair (Section 6-4, problems 35–38).
7. Design a buffer system that will regulate the pH to a particular value (Section 6-4, problems 39–42).
8. Calculate the pH at any stage in the titration of a strong acid or base by a strong base or acid (Section 6-5, problems 43–44).
9. Calculate the pH at any stage in the titration of a weak acid or base by a strong base or acid (Section 6-5, problems 45–48).
10. Calculate the concentrations of all the species present in a solution of a weak polyprotic acid (Section 6-6, problems 55–58).
11. Outline the procedure for the exact treatment of acid–base equilibrium and use it to find the pH of a very dilute solution of a weak acid or base (Section 6-7, problems 61–62).
12. Calculate the pH at selected points in the titration of a polyprotic acid (Section 6-7, problems 63–64).

### Properties of Acids and Bases

1. Which of the following can act as Brønsted–Lowry acids? Give the formula of the conjugate Brønsted–Lowry base for each of them.
   (a) $Cl^-$
   (b) $HSO_4^-$
   (c) $NH_4^+$
   (d) $NH_3$
   (e) $H_2O$
2. Which of the following can act as Brønsted–Lowry bases? Give the formula of the conjugate Brønsted–Lowry acid for each of them.
   (a) $F^-$
   (b) $SO_4^{2-}$
   (c) $O^{2-}$
   (d) $OH^-$
   (e) $H_2O$
3. Lemon juice contains citric acid ($C_6H_8O_7$). What species serves as a base when lemon juice is mixed with baking soda (sodium hydrogen carbonate) during the preparation of some lemon cookies?
4. A treatment recommended in case of accidental swallowing of ammonia-containing cleanser is to drink large amounts of diluted vinegar. Write an equation for the chemical reaction on which this procedure depends.
5. The concentration of $H_3O^+$ in a sample of wine is $2.0 \times 10^{-4}$ M. Calculate the pH of the wine.
6. The concentration of $OH^-$ in a solution of household bleach is $3.6 \times 10^{-2}$ M. Calculate the pH of the bleach.
7. The pH of normal human urine is in the range of 5.5 to 6.5. Compute the range of the $H_3O^+$ concentration and the range of the $OH^-$ concentration in normal urine.
8. The pH of normal human blood is in the range of 7.35 to 7.45. Compute the range of the concentration of $H_3O^+$ and the range of the $OH^-$ concentration in normal blood.
9. The $pK_w$ of seawater at 25°C is 13.776. This differs from the usual $pK_w$ of 14.00 at this temperature because dissolved salts make seawater a nonideal solution. If the pH in seawater is

8.00, what are the concentrations of $H_3O^+$ and $OH^-$ in seawater at 25°C?

10. At body temperature (98.6°F = 37.0°C), $K_w$ has the value $2.4 \times 10^{-14}$. If the pH of blood is 7.4 under these conditions, what are the concentrations of $H_3O^+$ and $OH^-$?

11. When placed in water, potassium starts to react instantly and continues to react with great vigor. On the basis of this information, select the better of the following two equations to represent the reaction.

$$2 \text{ K}(s) + 2 \text{ H}_2\text{O}(\ell) \longrightarrow 2 \text{ KOH}(aq) + \text{H}_2(g)$$

$$2 \text{ K}(s) + 2 \text{ H}_3\text{O}^+(aq) \longrightarrow 2 \text{ K}^+(aq) + \text{H}_2(g) + 2 \text{ H}_2\text{O}(\ell)$$

State the reason for your choice.

12. Molecules of $t$-butyl chloride, $(CH_3)_3CCl$, react very slowly when mixed with water at low pH to give $t$-butyl alcohol, $(CH_3)_3COH$. When the pH is raised, the reaction takes place rapidly. Write an equation or equations to explain these facts.

## Acid and Base Strength

13. Ephedrine ($C_{10}H_{15}ON$) is a base that is used in nasal sprays as a decongestant.
    (a) Write an equation for its equilibrium reaction with water.
    (b) The $K_b$ for ephedrine is $1.4 \times 10^{-4}$. Calculate the $K_a$ for its conjugate acid.
    (c) Is ephedrine a weaker or a stronger base than ammonia?

14. Niacin ($C_5H_4NCOOH$), one of the B vitamins, is an acid.
    (a) Write an equation for its equilibrium reaction with water.
    (b) The $K_a$ for niacin is $1.5 \times 10^{-5}$. Calculate the $K_b$ for its conjugate base.
    (c) Is the conjugate base of niacin a stronger or a weaker base than pyridine, $C_5H_5N$?

15. Use the data in Table 6-2 to determine the equilibrium constant for the reaction

$$\text{HClO}_2(aq) + \text{NO}_2^-(aq) \rightleftharpoons \text{HNO}_2(aq) + \text{ClO}_2^-(aq)$$

Identify the stronger Brønsted–Lowry acid and the stronger Brønsted–Lowry base.

16. Use the data in Table 6-2 to determine the equilibrium constant for the reaction

$$\text{HPO}_4^{2-} + \text{HCO}_3^- \rightleftharpoons \text{PO}_4^{3-} + \text{H}_2\text{CO}_3$$

Identify the stronger Brønsted–Lowry acid and the stronger Brønsted–Lowry base.

17. (a) Which is the stronger acid—the acidic form of the indicator bromocresol green or the acidic form of methyl orange?
    (b) A solution is prepared in which bromocresol green is green and methyl orange is orange. Estimate the pH of this solution.

18. (a) Which is the stronger base—the basic form of the indicator cresol red or the basic form of thymolphthalein?
    (b) A solution is prepared in which cresol red is red and thymolphthalein is colorless. Estimate the pH of this solution.

## Equilibria Involving Weak Acids and Bases

19. Aspirin is acetylsalicylic acid, $HC_9H_7O_4$, which has a $K_a$ of $3.0 \times 10^{-4}$. Calculate the pH of a solution made by dissolving 0.65 g of acetylsalicylic acid in water and diluting to 50.0 mL.

20. Vitamin C is ascorbic acid ($HC_6H_7O_6$), for which $K_a$ is $8.0 \times 10^{-5}$. Calculate the pH of a solution made by dissolving a 500-mg tablet of pure vitamin C in water and diluting to 100 mL.

21. (a) Calculate the pH of a 0.20 M solution of benzoic acid at 25°C.
    (b) How many moles of acetic acid must be dissolved per liter of water to obtain the same pH as that from part (a)?

22. (a) Calculate the pH of a 0.35 M solution of propionic acid at 25°C.
    (b) How many moles of formic acid must be dissolved per liter of water to obtain the same pH as that from part (a)?

23. Iodic acid ($HIO_3$) is fairly strong for a weak acid, having a $K_a$ equal to 0.16 at 25°C. Compute the pH of a 0.100 M solution of $HIO_3$.

24. At 25°C, the $K_a$ of pentafluorobenzoic acid ($C_6F_5COOH$) is 0.033. Suppose 0.100 mol of pentafluorobenzoic acid is dissolved in 1.00 L of water. What is the pH of this solution?

25. Papaverine hydrochloride ($papH^+Cl^-$) is a drug used as a muscle relaxant. It is a weak acid. At 25°C, a 0.205 M solution of $papH^+Cl^-$ has a pH of 3.31. Compute the $K_a$ of the $papH^+$ ion.

26. The unstable weak acid 2-germaacetic acid ($GeH_3COOH$) is derived structurally from acetic acid ($CH_3COOH$) by having a germanium atom replace one of the carbon atoms. At 25°C, a 0.050 M solution of 2-germaacetic acid has a pH of 2.42. Compute the $K_a$ of 2-germaacetic acid and compare it with that of acetic acid.

27. Morphine is a weak base for which $K_b$ is $8 \times 10^{-7}$. Calculate the pH of a solution made by dissolving 0.0400 mol of morphine in water and diluting to 600 mL.

28. Methylamine is a weak base for which $K_b$ is $4.4 \times 10^{-4}$. Calculate the pH of a solution made by dissolving 0.070 mol of methylamine in water and diluting to 800.0 mL.

29. The pH at 25°C of an aqueous solution of hydrofluoric acid, HF ($K_a = 6.6 \times 10^{-4}$), is 2.13. Calculate the concentration of HF in this solution, in moles per liter.

30. The pH at 25°C of an aqueous solution of sodium cyanide (NaCN) is 11.50. ($K_a$ of HCN is $4.93 \times 10^{-10}$.) Calculate the concentration of $CN^-$ in this solution, in moles per liter.

31. You have 50.00 mL of a solution that is 0.100 M in acetic acid, and you neutralize it by adding 50.00 mL of a solution that is 0.100 M in sodium hydroxide. The pH of the resulting solution is not 7.00. Explain why. Is the pH of the solution greater than or less than 7?

32. A 75.00-mL portion of a solution that is 0.0460 M in $HClO_4$ is treated with 150.00 mL of 0.0230 M KOH($aq$). Is the pH of

the resulting mixture greater than, less than, or equal to 7.0? Explain.

33. Suppose a 0.100 M solution of each of the following substances is prepared. Rank the pH of the resulting solutions from lowest to highest: $NH_4Br$, NaOH, KI, $NaCH_3COO$, HCl.

34. Suppose a 0.100 M solution of each of the following substances is prepared. Rank the pH of the resulting solutions from lowest to highest: KF, $NH_4I$, HBr, NaCl, LiOH.

## Buffer Solutions

35. "Tris" is short for tris(hydroxymethyl)aminomethane. This weak base is widely used in biochemical research for the preparation of buffers. It offers low toxicity and a $pK_b$ (5.92 at 25°C) that is convenient for the control of pH in clinical applications. A buffer is prepared by mixing 0.050 mol of tris with 0.025 mol of HCl in a volume of 2.00 L. Compute the pH of the solution.

36. "Bis" is short for bis(hydroxymethyl)aminomethane. It is a weak base that is closely related to tris (see problem 35) and has similar properties and uses. Its $pK_b$ is 8.8 at 25°C. A buffer is prepared by mixing 0.050 mol of bis with 0.025 mol of HCl in a volume of 2.00 L (the same proportions as in the preceding problem). Compute the pH of the solution.

37. (a) Calculate the pH in a solution prepared by dissolving 0.050 mol of acetic acid and 0.020 mol of sodium acetate in water and adjusting the volume to 500 mL.
(b) Suppose 0.010 mol of NaOH is added to the buffer from part (a). Calculate the pH of the solution that results.

38. Sulfanilic acid ($NH_2C_6H_4SO_3H$) is used in manufacturing dyes. It ionizes in water according to the equilibrium equation

$$NH_2C_6H_4SO_3H(aq) + H_2O(\ell) \rightleftharpoons$$
$$NH_2C_6H_4SO_3^-(aq) + H_3O^+(aq)$$
$$K_a = 5.9 \times 10^{-4}$$

A buffer is prepared by dissolving 0.20 mol of sulfanilic acid and 0.13 mol of sodium sulfanilate ($NaNH_2C_6H_4SO_3$) in water and diluting to 1.00 L.
(a) Compute the pH of the solution.
(b) Suppose 0.040 mol of HCl is added to the buffer. Calculate the pH of the solution that results.

39. A physician wishes to prepare a buffer solution at pH = 3.82 that efficiently resists changes in pH yet contains only small concentrations of the buffering agents. Determine which one of the following weak acids, together with its sodium salt, would probably be best to use: *m*-chlorobenzoic acid, $K_a = 1.04 \times 10^{-4}$; *p*-chlorocinnamic acid, $K_a = 3.89 \times 10^{-5}$; 2,5-dihydroxybenzoic acid, $K_a = 1.08 \times 10^{-3}$; or acetoacetic acid, $K_a = 2.62 \times 10^{-4}$. Explain.

40. Suppose you were designing a buffer system for imitation blood, and wanted the buffer to maintain the blood at the realistic pH of 7.40. All other things being equal, which

buffer system would be preferable—$H_2CO_3/HCO_3^-$ or $H_2PO_4^-/HPO_4^{2-}$? Explain.

41. You have at your disposal an ample quantity of a solution of 0.0500 M NaOH and 500 mL of a solution of 0.100 M formic acid (HCOOH). How much of the NaOH solution should be added to the acid solution to produce a buffer of pH 4.00?

42. You have at your disposal an ample quantity of a solution of 0.100 M HCl and 400 mL of a solution of 0.0800 M NaCN. How much of the HCl solution should be added to the NaCN solution to produce a buffer of pH 9.60?

## Acid–Base Titration Curves

43. Suppose 100.0 mL of a 0.3750 M solution of the strong base $Ba(OH)_2$ is titrated with a 0.4540 M solution of the strong acid $HClO_4$. The neutralization reaction is

$$Ba(OH)_2(aq) + 2\,HClO_4(aq) \longrightarrow$$
$$Ba(ClO_4)_2(aq) + 2\,H_2O(\ell)$$

Compute the pH of the titration solution before any acid is added, when the titration is 1.00 mL short of the equivalence point, when the titration is at the equivalence point, and when the titration is 1.00 mL past the equivalence point. (*Note:* Each mole of $Ba(OH)_2$ gives *two* moles of $OH^-$ in solution.)

44. A sample containing 26.38 mL of 0.1439 M HBr is titrated with a solution of NaOH having a molarity of 0.1219 M. Compute the pH of the titration solution before any base is added, when the titration is 1.00 mL short of the equivalence point, when the titration is at the equivalence point, and when the titration is 1.00 mL past the equivalence point.

45. A sample containing 50.00 mL of 0.1000 M hydrazoic acid ($HN_3$) is being titrated with 0.1000 M sodium hydroxide. Compute the pH before any base is added, after the addition of 25.00 mL of the base, after the addition of 50.00 mL of the base, and after the addition of 51.00 mL of the base.

46. A sample of 50.00 mL of 0.1000 M aqueous solution of chloroacetic acid, $CH_2ClCOOH$ ($K_a = 1.4 \times 10^{-3}$), is titrated with a 0.1000 M NaOH solution. Calculate the pH at the following stages in the titration, and plot the titration curve: 0, 5.00, 25.00, 49.00, 49.90, 50.00, 50.10, and 55.00 mL NaOH.

47. The base ionization constant of ethylamine ($C_2H_5NH_2$) in aqueous solution is $K_b = 6.41 \times 10^{-4}$ at 25°C. Calculate the pH for the titration of 40.00 mL of a 0.1000 M solution of ethylamine with 0.1000 M HCl at the following volumes of added HCl: 0, 5.00, 20.00, 39.90, 40.00, 40.10, and 50.00 mL.

48. Ammonia is a weak base with a $K_b$ of $1.8 \times 10^{-5}$. A 140.0-mL sample of a 0.175 M solution of aqueous ammonia is titrated with a 0.106 M solution of the strong acid HCl. The reaction is

$$NH_3(aq) + HCl(aq) \longrightarrow NH_4^+(aq) + Cl^-(aq)$$

Compute the pH of the titration solution before any acid is added, when the titration is at the half-equivalence point, when the titration is at the equivalence point, and when the titration is 1.00 mL past the equivalence point.

49. Sodium benzoate, the sodium salt of benzoic acid, is used as a food preservative. A sample containing solid sodium benzoate mixed with sodium chloride is dissolved in 50.0 mL of 0.500 M HCl, giving an acidic solution (benzoic acid mixed with HCl). This mixture is then titrated with 0.393 M NaOH. After the addition of 46.50 mL of the NaOH solution, the pH is found to be 8.2. At this point, the addition of one more drop (0.02 mL) of NaOH raises the pH to 9.3. Calculate the mass of sodium benzoate ($NaC_6H_5COO$) in the original sample. (*Hint:* At the equivalence point, the *total* number of moles of acid [here HCl] equals the *total* number of moles of base [here, both NaOH and $NaC_6H_5COO$]).

50. An antacid tablet (such as Tums or Rolaids) weighs 1.3259 g. The only acid-neutralizing ingredient in this brand of antacid is $CaCO_3$. When placed in 12.07 mL of 1.070 M HCl, the tablet fizzes merrily as $CO_2(g)$ is given off. After all of the $CO_2$ has left the solution, an indicator is added, followed by 11.74 mL of 0.5310 M NaOH. The indicator shows that, at this point, the solution is definitely basic. Addition of 5.12 mL of 1.070 M HCl makes the solution acidic again. Then, 3.17 mL of the 0.5310 M NaOH brings the titration exactly to an endpoint, as signaled by the indicator. Compute the percentage by mass of $CaCO_3$ in the tablet.

51. What is the mass of diethylamine, $(C_2H_5)_2NH$, in 100.0 mL of an aqueous solution if it requires 15.90 mL of 0.0750 M HCl to titrate it to the equivalence point? What is the pH at the equivalence point if $K_b = 3.09 \times 10^{-4}$? What would be a suitable indicator for the titration?

52. A chemist who works in the process laboratory of the Athabasca Alkali Company makes frequent analyses for ammonia recovered from the Solvay process for making sodium carbonate. What is the pH at the equivalence point if she titrates the aqueous ammonia solution (approximately 0.10 M) with a strong acid of comparable concentration? Select an indicator that would be suitable for the titration.

53. If 50.00 mL of a 0.200 M solution of the weak base *N*-ethylmorpholine ($C_6H_{13}NO$) is mixed with 8.00 mL of 1.00 M HCl and then diluted to a final volume of 100.0 mL with water, the result is a buffer with a pH of 7.0. Compute the $K_b$ of *N*-ethylmorpholine.

54. The sodium salt of cacodylic acid, a weak acid, has the formula $NaO_2As(CH_3)_2 \cdot 3H_2O$. Its molar mass is 214.02 g $mol^{-1}$. A solution is prepared by mixing 21.40 g of this substance with enough water to make 1.000 L of solution. Then 50.00 mL of the sodium cacodylate solution is mixed with 29.55 mL of 0.100 M HCl and enough water to bring the volume to a total of 100.00 mL. The pH of the solution is 6.00. Determine the $K_a$ of cacodylic acid.

## Polyprotic Acids

55. Arsenic acid ($H_3AsO_4$) is a weak triprotic acid. Given the three acid ionization constants from Table 6-2 and an initial concentration of arsenic acid (before ionization) of 0.1000 M, calculate the equilibrium concentrations of $H_3AsO_4$, $H_2AsO_4^-$, $HAsO_4^{2-}$, $AsO_4^{3-}$, and $H_3O^+$.

56. Phthalic acid ($H_2C_8H_4O_4$, abbreviated $H_2Ph$) is a diprotic acid. Its ionization in water at 25°C takes place in two steps:

$$H_2Ph(aq) + H_2O(\ell) \rightleftharpoons H_3O^+(aq) + HPh^-(aq)$$
$$K_{a1} = 1.26 \times 10^{-3}$$

$$HPh^-(aq) + H_2O(\ell) \rightleftharpoons H_3O^+(aq) + Ph^{2-}(aq)$$
$$K_{a2} = 3.10 \times 10^{-6}$$

If 0.0100 mol of phthalic acid is dissolved per liter of water, calculate the equilibrium concentrations of $H_2Ph$, $HPh^-$, $Ph^{2-}$, and $H_3O^+$.

57. A solution as initially prepared contains 0.050 mol $L^{-1}$ of phosphate ion ($PO_4^{3-}$) at 25°C. Given the three acid ionization constants from Table 6-2, calculate the equilibrium concentrations of $PO_4^{3-}$, $HPO_4^{2-}$, $H_2PO_4^-$, $H_3PO_4$, and $OH^-$.

58. Oxalic acid ionizes in two stages in aqueous solution:

$$H_2C_2O_4(aq) + H_2O(\ell) \rightleftharpoons H_3O^+(aq) + HC_2O_4^-(aq)$$
$$K_{a1} = 5.9 \times 10^{-2}$$

$$HC_2O_4^-(aq) + H_2O(\ell) \rightleftharpoons H_3O^+(aq) + C_2O_4^{2-}(aq)$$
$$K_{a2} = 6.4 \times 10^{-5}$$

Calculate the equilibrium concentrations of $C_2O_4^{2-}$, $HC_2O_4^-$, $H_2C_2O_4$, and $OH^-$ in a 0.10 M solution of sodium oxalate ($Na_2C_2O_4$).

59. The pH of a normal raindrop is 5.60. Compute the concentrations of $H_2CO_3(aq)$, $HCO_3^-(aq)$, and $CO_3^{2-}(aq)$ in this raindrop, if the total concentration of dissolved carbonates is $1.0 \times 10^{-5}$ mol $L^{-1}$.

60. The pH of a drop of acid rain is 4.00. Compute the concentrations of $H_2CO_3(aq)$, $HCO_3^-(aq)$, and $CO_3^{2-}(aq)$ in the acid raindrop, if the total concentration of dissolved carbonates is $3.6 \times 10^{-5}$ mol $L^{-1}$.

## Exact Treatment of Acid–Base Equilibria

61. Thiamine hydrochloride (vitamin $B_1$ hydrochloride, $HC_{12}H_{17}ON_4SCl_2$) is a weak acid with $K_a = 3.4 \times 10^{-7}$. Suppose $3.0 \times 10^{-5}$ g of thiamine hydrochloride is dissolved in 1.00 L of water. Calculate the pH of the resulting solution. (*Hint:* This is a sufficiently dilute solution that the autoionization of water cannot be neglected.)

62. A sample of vinegar contains 40.0 g of acetic acid ($CH_3COOH$) per liter of solution. Suppose 1.00 mL is removed and diluted to 1.00 L, and 1.00 mL of *that* solution is removed and diluted to 1.00 L. Calculate the pH of the resulting solution. (*Hint:* This is a sufficiently dilute solution that the autoionization of water cannot be neglected.)

63. At 25°C, 50.00 mL of a 0.1000 M solution of maleic acid, a diprotic acid whose ionization constants are $K_{a1} = 1.42 \times 10^{-2}$ and $K_{a2} = 8.57 \times 10^{-7}$, is titrated with a 0.1000 M NaOH solution. Calculate the pH at the following volumes of added base: 0, 5.00, 25.00, 50.00, 75.00, 99.90, 100.00, and 105.00 mL.

**64.** Quinine ($C_{20}H_{24}O_2N_2$) is a water-soluble base that ionizes in two stages, with $K_{b1} = 3.31 \times 10^{-6}$ and $K_{b2} = 1.35 \times 10^{-10}$, at 25°C. Calculate the pH during the titration of an aqueous solution of 1.622 g of quinine in 100.00 mL of water as a function of the volume of added 0.1000 M HCl solution at the following volumes: 0, 25.00, 50.00, 75.00, 99.90, 100.00, and 105.00 mL.

## Additional Problems

**65.** Calculate the concentrations of $H_3O^+$ and $OH^-$ at 25°C in the following:
(a) Orange juice (pH 2.8)
(b) Tomato juice (pH 3.9)
(c) Milk (pH 4.1)
(d) Borax solution (pH 8.5)
(e) Household ammonia (pH 11.9)

**66.** Try to choose which of the following is the pH of a $6.44 \times 10^{-10}$ M $Ca(OH)_2(aq)$ solution, without doing any written calculations.
(a) 4.81
(b) 5.11
(c) 7.00
(d) 8.89
(e) 9.19

**67.** $Cl_2(aq)$ reacts with $H_2O(\ell)$ as follows:

$$Cl_2(aq) + 2 H_2O(\ell) \Longrightarrow$$
$$H_3O^+(aq) + Cl^-(aq) + HOCl(aq)$$

For an experiment to succeed, $Cl_2(aq)$ must be present, but the amount of $Cl^-(aq)$ in the solution must be minimized. For this purpose, should the pH of the solution be high, low, or neutral? Explain.

**68.** Use the data in Table 6-2 to determine the equilibrium constant for the reaction

$$H_2PO_4^-(aq) + 2 CO_3^{2-}(aq) \Longrightarrow PO_4^{3-}(aq) + 2 HCO_3^-(aq)$$

**69.** Urea ($NH_2CONH_2$) is a component of urine. It is a very weak base, having an estimated $pK_b$ of 13.8 at room temperature.
(a) Write the formula of the conjugate acid of urea.
(b) Compute the equilibrium concentration of urea in a solution that contains no urea but starts out containing 0.15 mol $L^{-1}$ of the conjugate acid of urea.

**70.** Exactly 1.0 L of solution of acetic acid gives the same color with methyl red as 1.0 L of a solution of hydrochloric acid. Which solution will neutralize the greater amount of 0.10 M $NaOH(aq)$? Explain.

**\*71.** The $K_a$ for acetic acid drops from $1.76 \times 10^{-5}$ at 25°C to $1.63 \times 10^{-5}$ at 50°C. Between the same two temperatures, $K_w$ increases from $1.00 \times 10^{-14}$ to $5.47 \times 10^{-14}$. At 50°C, the density of a 0.10 M solution of acetic acid is 98.81% of its density at 25°C. Will the pH of a 0.10 M solution of acetic acid in water increase, decrease, or remain the same when it is heated from 25°C to 50°C? Explain.

**72.** Calculate the pH of a solution that is prepared by dissolving 0.23 mol of hydrofluoric acid (HF) and 0.57 mol of hypochlorous acid (HClO) in water and diluting to 3.60 L. Also, calculate the equilibrium concentrations of HF, $F^-$, HClO, and $ClO^-$. (*Hint:* The pH will be determined by the stronger acid of this pair.)

**73.** For each of the following compounds, indicate whether a 0.100 M aqueous solution is acidic (pH < 7), basic (pH > 7), or neutral (pH = 7): HCl, $NH_4Cl$, $KNO_3$, $Na_3PO_4$, $NaCH_3COO$.

**\*74.** Calculate $[H_3O^+]$ in a solution that contains 0.100 mol of $NH_4CN$ per liter.

$$NH_4^+(aq) + H_2O(\ell) \Longrightarrow H_3O^+(aq) + NH_3(aq)$$
$$K_a = 5.6 \times 10^{-10}$$

$$HCN(aq) + H_2O(\ell) \Longrightarrow H_3O^+(aq) + CN^-(aq)$$
$$K_a = 6.17 \times 10^{-10}$$

**75.** A buffer solution is prepared by mixing 1.00 L of 0.050 M pentafluorobenzoic acid ($C_6F_5COOH$) and 1.00 L of 0.060 M sodium pentafluorobenzoate ($NaC_6F_5COO$). The $K_a$ of this weak acid is 0.033. Determine the pH of the buffer solution.

**76.** A chemist needs to prepare a buffer solution with pH = 10.00, and has both $Na_2CO_3$ and $NaHCO_3$ in pure crystalline form. What mass of each should be dissolved in 1.00 L of solution if the combined mass of the two salts is to be 10.0 g?

**77.** Which of these procedures would *not* make a pH = 4.75 buffer?
(a) Mix 50.0 mL of 0.10 M acetic acid and 50.0 mL of 0.10 M sodium acetate.
(b) Mix 50.0 mL of 0.20 M acetic acid and 50.0 mL of 0.10 M NaOH.
(c) Start with 50.0 mL of 0.20 M acetic acid and add a solution of strong base until the pH equals 4.75.
(d) Start with 50.0 mL of 0.20 M HCl and add a solution of strong base until the pH equals 4.75.
(e) Start with 100.0 mL of 0.20 M sodium acetate and add 50.0 mL of 0.20 M HCl.

**78.** It takes 4.71 mL of 0.0410 M NaOH to titrate a 50.00-mL sample of flat (no $CO_2$) GG's Cola to a pH of 4.9. At this point, the addition of one more drop (0.02 mL) of NaOH raises the pH to 6.1. The only acid in GG's Cola is phosphoric acid. Compute the concentration of phosphoric acid in this cola. Assume that the 4.71 mL of base removes only the first hydrogen from the $H_3PO_4$; that is, assume that the reaction is

$$H_3PO_4(aq) + OH^-(aq) \longrightarrow H_2O(\ell) + H_2PO_4^-(aq)$$

**79.** A 0.1000 M solution of a weak acid, HA, requires 50.00 mL of 0.1000 M NaOH to titrate it to its equivalence point. The pH of the solution is 4.50 when only 40.00 mL of the base has been added.
(a) Calculate the ionization constant $K_a$ of the acid.
(b) Calculate the pH of the solution at the equivalence point.

**80.** The chief chemist of Victory Vinegar Works, Ltd., interviews two chemists for employment. He states, "Quality control requires that our high-grade vinegar contain 5.00 ± 0.01% acetic acid by mass. How would you analyze our product to ensure that it meets this specification?"

Anne Dalton says, "I would titrate a 50.00-mL sample of the vinegar with 1.000 M NaOH, using phenolphthalein to detect the equivalence point to within ±0.02 mL of base."

Charlie Cannizzarro says, "I would use a pH meter to determine the pH to ±0.01 pH units and interface it with a computer to print out the mass percentage of acetic acid."

Which candidate did the chief chemist hire? Why?

81. Phosphonocarboxylic acid

$$H-O-\overset{\overset{\displaystyle :O:}{\|}}{\underset{\underset{\displaystyle H-O:}{|}}{P}}-\overset{\overset{\displaystyle H}{|}}{\underset{\underset{\displaystyle H}{|}}{C}}-\overset{\overset{\displaystyle :O:}{\|}}{C}-O-H$$

effectively inhibits the replication of the herpes virus. Structurally, it is a combination of phosphoric acid and acetic acid. It can donate three protons. The equilibrium constant values are $K_{a1} = 1.0 \times 10^{-2}$, $K_{a2} = 7.8 \times 10^{-6}$, and $K_{a3} = 2.0 \times 10^{-9}$. Enough phosphonocarboxylic acid is added to blood (pH 7.40) to make its total concentration $1.0 \times 10^{-5}$ M. The pH of the blood does not change. Determine the concentrations of all four forms of the acid in this mixture.

82. Egg whites contain dissolved carbon dioxide and water, which react together to give carbonic acid ($H_2CO_3$). In the few days after an egg is laid, it loses carbon dioxide through its shell. Does the pH of the egg white increase or decrease during this period?

83. If you breathe too rapidly (hyperventilate), the concentration of dissolved $CO_2$ in your blood drops. What effect does this have on the pH of the blood?

84. A reference book states that a saturated aqueous solution of potassium hydrogen tartrate is a buffer with a pH of 3.56. Write two chemical equations that show the buffer action of this solution. (Tartaric acid is a diprotic acid with the formula $H_2C_4H_4O_6$. Potassium hydrogen tartrate is $KHC_4H_4O_6$.)

*85. Glycine, the simplest amino acid, has both an acid group and a basic group in its structure ($H_2N-CH_2-COOH$). In aqueous solution it exists predominantly as a self-neutralized species called a zwitterion ($H_3\overset{+}{N}-CH_2-COO^-$). The zwitterion therefore behaves both as an acid and as a base, according to the equilibria at 25°C:

$$H_3\overset{+}{N}-CH_2-COO^-(aq) + H_2O(\ell) \rightleftharpoons$$
$$H_2N-CH_2-COO^-(aq) + H_3O^+(aq)$$
$$K_a = 1.7 \times 10^{-10}$$

$$H_3\overset{+}{N}-CH_2-COO^-(aq) + H_2O(\ell) \rightleftharpoons$$
$$H_3\overset{+}{N}-CH_2-COOH(aq) + OH^-(aq)$$
$$K_b = 2.2 \times 10^{-12}$$

Calculate the pH of a 0.10 M aqueous solution of glycine at 25°C. (*Hint:* You may need to take account of the autoionization of water.)

## CUMULATIVE PROBLEMS

86. Use the data from Table 6-1, together with Le Châtelier's principle, to decide whether the autoionization of water is exothermic or endothermic.

87. Baking soda (sodium hydrogen carbonate, $NaHCO_3$) is used in baking because it reacts with acids in foods to form carbonic acid ($H_2CO_3$), which in turn decomposes to water and carbon dioxide. In a batter, the carbon dioxide appears as gas bubbles that cause the bread or cake to rise.
(a) A rule of thumb in cooking is that $\frac{1}{2}$ teaspoon baking soda is neutralized by 1 cup of sour milk. The acid component of sour milk is lactic acid ($HC_3H_5O_3$). Write an equation for the neutralization reaction.
(b) If the density of baking soda is 2.16 g cm$^{-3}$, calculate the concentration of lactic acid in the sour milk, in moles per liter. Take 1 cup = 236.6 mL = 48 teaspoons.
(c) Calculate the volume of carbon dioxide that is produced at 1 atm pressure and 350°F (177°C) from the reaction of $\frac{1}{2}$ teaspoon of baking soda.

88. Boric acid, $B(OH)_3$, is an acid that acts differently from the usual Brønsted–Lowry acids. It reacts with water according to

$$B(OH)_3(aq) + 2 H_2O(\ell) \rightleftharpoons B(OH)_4^-(aq) + H_3O^+(aq)$$
$$K_a = 5.8 \times 10^{-10}$$

(a) Draw Lewis structures for $B(OH)_3$ and $B(OH)_4^-$. Can these be described as Lewis acids or Lewis bases?
(b) Calculate the pH of a 0.20 M solution of $B(OH)_3(aq)$.

89. At 40°C and 1.00 atm pressure, a gaseous monoprotic acid has a density of 1.05 g L$^{-1}$. After 1.85 g of this gas is dissolved in water and diluted to 450 mL, the pH is measured to be 5.01. Determine the $K_a$ of this acid and use Table 6-2 to identify it.

90. At 25°C, the Henry's law constant for carbon dioxide dissolved in water is $1.8 \times 10^3$ atm. Calculate the pH of water saturated with $CO_2(g)$ at 25°C in Denver, where the barometric pressure is 0.833 atm.

*A precipitate of lead iodine.*

# Dissolution and Precipitation Equilibria

D issolution and precipitation reactions, by which solids pass into and out of solution, constituted one of the major types of chemical reactions discussed in Chapter 2. They rank alongside acid–base reactions in practical importance. The dissolution and reprecipitation of solids permit chemists to isolate single products from reaction mixtures or to purify impure solid samples. Understanding the mechanisms of these reactions helps engineers to prevent scale from forming in boilers and doctors to reduce the incidence of painful kidney stones. Dissolution and precipitation control the formation of mineral deposits and profoundly affect the ecologies of rivers, lakes, and oceans.

In this chapter we explore some quantitative aspects of the equilibria that govern dissolution and precipitation. We begin with the case of a single solid salt partially dissolved in water and then move on to the effects of other salts, of acids and bases, and of ligands that can bind to metal ions to form complex ions. Our theme throughout this chapter is the manipulation of equilibria to control the solubilities of ionic solids.

## 7-1 THE NATURE OF SOLUBILITY EQUILIBRIA

Often the major task in synthetic chemistry is not making a compound, but purifying it. Side reactions can generate significant amounts of impurities; other impurities enter with the starting materials or are put in deliberately to make the reaction go faster. Running a reaction may take only hours, but the *work-up* (separation of crude product) and subsequent purification may require weeks. **Recrystallization,** one of the most powerful methods for purifying solids, relies on differences between the solubilities of the desired substance and its contaminants. An impure product is dissolved and reprecipitated, repeatedly if necessary, with careful control of the factors that influence solubility. Manipulating solubility requires an understanding of the equilibria that exist between an undissolved substance and its solution.

In recrystallization, a solution begins to deposit a compound when it is brought to the point of saturation with respect to that compound. A **saturated solution** is one in which a dissolution–precipitation equilibrium exists between the solid substance and its dissolved form. For example, the equilibrium between solid iodine and iodine dissolved in $CCl_4$ can be written as

$$I_2(s) \rightleftharpoons I_2(\text{in } CCl_4)$$

In dissolution, the solvent attacks the solid and **solvates** it at the level of individual particles. In this example, $I_2$–$CCl_4$ attractions replace the $I_2$–$I_2$ interactions formerly present in the solid, as the iodine molecules go off into the bulk of the $CCl_4$. In precipitation, the reverse occurs: solute-to-solute attractions are reestablished as the solute leaves the solution. Often, solute-to-solvent attractions persist right through the process of precipitation, and solvent incorporates itself into the solid. When lithium sulfate ($Li_2SO_4$) precipitates from water, it brings with it into the solid one molecule of water per formula unit:

$$2\,Li^+(aq) + SO_4^{2-}(aq) + H_2O(\ell) \longrightarrow Li_2SO_4 \cdot H_2O(s)$$

Such loosely bound solvent is known as solvent of crystallization (Fig. 1-18). Dissolving and then reprecipitating a compound may thus furnish material that has a different chemical formula and a different mass.

Solubility equilibria resemble the equilibria that volatile liquids (or solids) establish with their vapors in a closed container. In both cases, particles from a condensed phase tend to escape and spread through a larger, but limited, volume. In both cases, equilibrium is a dynamic compromise in which the rate of escape of particles from the condensed phase is equal to their rate of return. In a vaporization–condensation equilibrium, we assumed that the vapor above the condensed phase was an ideal gas. The analogous starting assumption for a dissolution–precipitation reaction is that the solution above the undissolved solid is an ideal solution.

Le Châtelier's principle applies to these equilibria, as it does to all equilibria. One way to exert a stress on a solubility equilibrium is to change the amount of solvent. Adding solvent reduces the concentration of dissolved substance; more solid then tends to dissolve to restore the concentration of the dissolved substance to its equilibrium value. If an excess of solvent is added so that all of the solid dissolves, then obviously the solubility equilibrium ceases to exist, and the solution is **unsaturated.** In a vaporization–condensation equilibrium this corresponds to the complete evaporation of the condensed phase. Removing solvent from an already saturated solution forces additional solid to precipitate in order to maintain a constant concentration. A volatile solvent is often removed by simply letting a solution stand uncovered until the solvent evaporates. When conditions are right, the solid forms as crystals on the bottom and sides of the container (Fig. 7-1).

Dissolution–precipitation reactions frequently come to equilibrium slowly. Days or even weeks of shaking a solid in contact with a solvent may be required before the solution becomes saturated. Moreover, solutions sometimes become **su-**

**Figure 7-1**
The beaker contained an aqueous solution of $K_2PtCl_4$, a substance used as a starting material in the synthesis of anticancer drugs. It was loosely covered to keep out dust, and allowed to stand. As the water evaporated, the solution became saturated and deposited solid $K_2PtCl_4$ in the form of long, needle-like red crystals. (Leon Lewandowski)

persaturated, a condition in which the concentration of dissolved solid *exceeds* its equilibrium value. The delay, then, is in forming, not dissolving, the solid. A supersaturated solution may persist for months or years and require extraordinary measures to be brought to equilibrium, although the tendency toward equilibrium is always there. The generally sluggish approach to equilibrium in dissolution–precipitation reactions is quite the opposite of the rapid rates at which acid–base reactions reach equilibrium.

## The Solubility of Salts

The **solubility** of a substance in a solvent is defined as the greatest amount (expressed either in grams or in moles) that will dissolve in equilibrium in a specified volume of solvent at a particular temperature. Although solvents other than water are used in many applications, aqueous solutions are the most important and are the exclusive concern here. Salts show a wide range of solubilities in water (Fig. 7-2). Silver perchlorate ($AgClO_4$) dissolves to the remarkable extent of about 5570 g (or almost 27 mol) per liter of water at 25°C, but at the same temperature only about 0.0018 g (or $1.3 \times 10^{-5}$ mol) of silver chloride (AgCl) dissolves per liter of water. Many salts with even lower solubilities are known. Solubilities often depend strongly on temperature (Fig. 7-3). Most dissolution reactions for ionic solids are endothermic (heat is absorbed), so by Le Châtelier's principle the solubility increases with increasing temperature. Those dissolution reactions that are exothermic (such as for $CaSO_4$) show the opposite behavior.

Although all ionic compounds dissolve to some extent in water, those having solubilities (at 25°C) of less than 0.1 g $L^{-1}$ are called *insoluble*. Those having solubilities of more than 10 g $L^{-1}$ are *soluble,* and the intermediate cases (0.1 to 10 g $L^{-1}$) are said to be *slightly soluble.* Fortunately, it is not necessary to memorize long lists of solubility data. Table 7-1 lists some generalizations concerning groups of salts and gives enough factual data to support good predictions about precipita-

**Figure 7-2**
Vastly different quantities of different compounds will dissolve in 1 L of water at 20°C. Clockwise from the front are borax, potassium permanganate, lead(II) chloride, sodium phosphate decahydrate, calcium oxide, and potassium dichromate. (Leon Lewandowski)

### Table 7-1  Solubilities of Ionic Compounds in Water

| Anion | Soluble[†] | Slightly Soluble | Insoluble |
|---|---|---|---|
| $NO_3^-$ (nitrate) | All | — | — |
| $CH_3COO^-$ (acetate) | Most | — | $Be(CH_3COO)_2$ |
| $ClO_3^-$ (chlorate) | All | — | — |
| $ClO_4^-$ (perchlorate) | Most | $KClO_4$ | — |
| $F^-$ (fluoride) | Group I, $AgF$, $BeF_2$ | $SrF_2$, $BaF_2$, $PbF_2$ | $MgF_2$, $CaF_2$ |
| $Cl^-$ (chloride) | Most | $PbCl_2$ | $AgCl$, $Hg_2Cl_2$ |
| $Br^-$ (bromide) | Most | $PbBr_2$, $HgBr_2$ | $AgBr$, $Hg_2Br_2$ |
| $I^-$ (iodide) | Most | — | $AgI$, $Hg_2I_2$, $PbI_2$, $HgI_2$ |
| $SO_4^{2-}$ (sulfate) | Most | $CaSO_4$, $Ag_2SO_4$, $Hg_2SO_4$ | $SrSO_4$, $BaSO_4$, $PbSO_4$ |
| $S^{2-}$ (sulfide) | Groups I and II, $(NH_4)_2S$ | — | Most |
| $CO_3^{2-}$ (carbonate) | Group I, $(NH_4)_2CO_3$ | — | Most |
| $SO_3^{2-}$ (sulfite) | Group I, $(NH_4)_2SO_3$ | — | Most |
| $PO_4^{3-}$ (phosphate) | Group I, $(NH_4)_3PO_4$ | — | Most |
| $OH^-$ (hydroxide) | Group I, $Ba(OH)_2$ | $Sr(OH)_2$, $Ca(OH)_2$ | Most |

[†] Soluble compounds are defined as those that dissolve to the extent of 10 or more grams per liter; slightly soluble compounds, 0.1 to 10 grams per liter; and insoluble compounds, less than 0.1 gram per liter at room temperature.

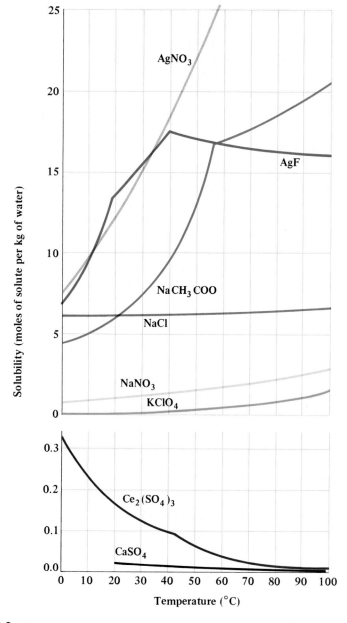

**Figure 7-3**
Most solubilities increase with increasing temperature, but some decrease. Note that the changes are not always smooth because different solid hydrates form over different temperature ranges.

**Figure 7-4**
A yellow precipitate of lead iodide forms when a drop of KI solution is added to a solution of $Pb(NO_3)_2$.

tion or dissolution in thousands of situations of practical importance. Knowing the solubilities of ionic substances, even in these qualitative terms, provides a way to predict the courses of numerous reactions. For example, when a solution of KI is added to one of $Pb(NO_3)_2$, $K^+$ and $NO_3^-$ ions are brought into contact, as are $Pb^{2+}$ and $I^-$ ions. From the table, $KNO_3$ is a soluble salt but $PbI_2$ is insoluble, and so a precipitate of $PbI_2$ will appear (Fig. 7-4).

## 7-2 IONIC EQUILIBRIA

When an ionic solid such as CsCl dissolves in water, it breaks up into ions that move apart from each other and become solvated by water molecules (*aquated,* or *hydrated;* Fig. 7-5). The aquated ions are shown in the chemical equation for the solubility equilibrium. Thus,

$$CsCl(s) \rightleftharpoons Cs^+(aq) + Cl^-(aq)$$

shows that the dissolved particles are ions. For a highly soluble salt (such as CsCl), the concentrations of the ions in a saturated aqueous solution are so large that the solution is very nonideal. There is much association among the ions in solution, resulting in temporary pairs of oppositely charged ions and in larger clusters, as well. In such cases, the simple type of equilibrium expression that was developed in Section 5-5 does not apply; therefore, we restrict our attention to sparingly soluble and ''insoluble'' salts, for which the concentrations in a saturated solution are 0.1 mol $L^{-1}$ or less—low enough that the interactions among the solvated ions are relatively small.

The sparingly soluble salt silver chloride, for example, establishes the following equilibrium when placed in water:

$$AgCl(s) \rightleftharpoons Ag^+(aq) + Cl^-(aq)$$

The equilibrium law for this reaction (written by following the rules for heterogeneous equilibria from Section 5-5) is

$$[Ag^+][Cl^-] = K_{sp}$$

where the subscript ''sp,'' standing for **solubility product,** distinguishes the $K$ as referring to the dissolution of a slightly soluble ionic solid in water. At 25°C, $K_{sp}$ has the numerical value $1.6 \times 10^{-10}$ for silver chloride. In the AgCl solubility product expression, the concentrations of the two ions produced are raised to the first power because their coefficients are 1 in the chemical equation. Solid AgCl does not appear in the equilibrium expression; the amount of pure solid AgCl does not affect the equilibrium as long as *some* is present. If *no* solid is present, then the product of the two ion concentrations is no longer constrained by the solubility product expression.

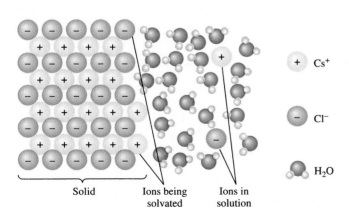

**Figure 7-5**

The dissolution of the ionic solid CsCl in water. Note the different solvations of positive and negative ions. Water molecules arrange themselves to point the positively charged hydrogen atoms toward the negatively charged $Cl^-$ ions, and the negatively charged oxygen atoms point toward the positively charged $Cs^+$ ions.

+  $Cs^+$

−  $Cl^-$

$H_2O$

Solid      Ions being      Ions in
           solvated       solution

## Solubility and $K_{sp}$

The molar solubility of a salt in water is not the same as its solubility product constant, but a simple relation often exists between them. For example, let us define $S$ as the molar solubility of AgCl($s$) in water at 25°C. Then, from stoichiometry, $[Ag^+] = [Cl^-] = S$ is the molarity of either ion at equilibrium. Hence,

$$[Ag^+][Cl^-] = S^2 = K_{sp} = 1.6 \times 10^{-10}$$

Taking the square roots of both sides of the equation gives

$$S = 1.26 \times 10^{-5}\,\text{M}$$

which rounds off to $1.3 \times 10^{-5}$ M. This is the molar solubility of AgCl in water. It is converted to a gram solubility by multiplying by the molar mass of AgCl:

$$(1.26 \times 10^{-5}\,\text{mol L}^{-1})(143.3\,\text{g mol}^{-1}) = 1.8 \times 10^{-3}\,\text{g L}^{-1}$$

Therefore, $1.8 \times 10^{-3}$ g of AgCl dissolves per liter of water at 25°C.

Solubility product constants (like solubilities) can be sensitive to temperature. At 100°C, the $K_{sp}$ for silver chloride is $2.2 \times 10^{-8}$; hot water dissolves about 12 times as much silver chloride as does water at 25°C. Refer to Table 7-2 for the solubility product constants at 25°C of a number of important sparingly soluble salts.

### Example 7-1

The $K_{sp}$ of calcium fluoride is $3.9 \times 10^{-11}$. Calculate the concentrations of calcium and fluoride ions in a saturated solution of CaF$_2$ at 25°C, and determine the solubility of CaF$_2$ in grams per liter.

### Solution

The solubility equilibrium is

$$CaF_2(s) \rightleftharpoons Ca^{2+}(aq) + 2\,F^-(aq)$$

and the expression for the solubility product is

$$[Ca^{2+}][F^-]^2 = K_{sp}$$

The concentration of fluoride ion is squared because it has a coefficient of 2 in the chemical equation.

If $S$ mol of CaF$_2$ dissolves in 1 L, the equilibrium concentration of Ca$^{2+}$ will be $[Ca^{2+}] = S$. The concentration of F$^-$ will be $[F^-] = 2S$, because each mole of CaF$_2$ produces *two* moles of fluoride ions. Therefore,

$$[Ca^{2+}][F^-]^2 = S \times (2S)^2 = 4S^3 = K_{sp} = 3.9 \times 10^{-11}$$

Solving for $S^3$ gives

$$S^3 = \frac{1}{4}(3.9 \times 10^{-11})$$

Taking the cube roots of both sides of this equation gives

$$S = 2.1 \times 10^{-4}$$

The equilibrium concentrations are therefore

$$[Ca^{2+}] = S = 2.1 \times 10^{-4} \text{ M}$$

$$[F^-] = 2S = 4.3 \times 10^{-4} \text{ M}$$

Because the molar mass of $CaF_2$ is 78.1 g mol$^{-1}$, the gram solubility is

$$\text{gram solubility} = (2.1 \times 10^{-4} \text{ mol L}^{-1})(78.1 \text{ g mol}^{-1}) = 0.017 \text{ g L}^{-1}$$

*Related Problems:* 7, 8, 9, 10, 11, 12

It is possible to reverse the procedure just outlined, of course, and determine the value of $K_{sp}$ from measured solubilities, as the following example illustrates.

## Example 7-2

Silver chromate ($Ag_2CrO_4$) is a red solid that dissolves in water to the extent of 0.029 g L$^{-1}$ at 25°C. Estimate its $K_{sp}$ and compare your estimate with the value in Table 7-2.

### Table 7-2 Solubility Product Constants $K_{sp}$ at 25°C

**Fluorides**

| | |
|---|---|
| $BaF_2$ | $[Ba^{2+}][F^-]^2 = 1.7 \times 10^{-6}$ |
| $CaF_2$ | $[Ca^{2+}][F^-]^2 = 3.9 \times 10^{-11}$ |
| $MgF_2$ | $[Mg^{2+}][F^-]^2 = 6.6 \times 10^{-9}$ |
| $PbF_2$ | $[Pb^{2+}][F^-]^2 = 3.6 \times 10^{-8}$ |
| $SrF_2$ | $[Sr^{2+}][F^-]^2 = 2.8 \times 10^{-9}$ |

**Chlorides**

| | |
|---|---|
| $AgCl$ | $[Ag^+][Cl^-] = 1.6 \times 10^{-10}$ |
| $CuCl$ | $[Cu^+][Cl^-] = 1.0 \times 10^{-6}$ |
| $Hg_2Cl_2$ | $[Hg_2^{2+}][Cl^-]^2 = 2 \times 10^{-18}$ |

**Bromides**

| | |
|---|---|
| $AgBr$ | $[Ag^+][Br^-] = 7.7 \times 10^{-13}$ |
| $CuBr$ | $[Cu^+][Br^-] = 4.2 \times 10^{-8}$ |
| $Hg_2Br_2$ | $[Hg_2^{2+}][Br^-]^2 = 1.3 \times 10^{-21}$ |

**Iodides**

| | |
|---|---|
| $AgI$ | $[Ag^+][I^-] = 1.5 \times 10^{-16}$ |
| $CuI$ | $[Cu^+][I^-] = 5.1 \times 10^{-12}$ |
| $PbI_2$ | $[Pb^{2+}][I^-]^2 = 1.4 \times 10^{-8}$ |
| $Hg_2I_2$ | $[Hg_2^{2+}][I^-]^2 = 1.2 \times 10^{-28}$ |

**Hydroxides**

| | |
|---|---|
| $AgOH$ | $[Ag^+][OH^-] = 1.5 \times 10^{-8}$ |
| $Al(OH)_3$ | $[Al^{3+}][OH^-]^3 = 3.7 \times 10^{-15}$ |
| $Fe(OH)_3$ | $[Fe^{3+}][OH^-]^3 = 1.1 \times 10^{-36}$ |
| $Fe(OH)_2$ | $[Fe^{2+}][OH^-]^2 = 1.6 \times 10^{-14}$ |
| $Mg(OH)_2$ | $[Mg^{2+}][OH^-]^2 = 1.2 \times 10^{-11}$ |
| $Mn(OH)_2$ | $[Mn^{2+}][OH^-]^2 = 2 \times 10^{-13}$ |
| $Zn(OH)_2$ | $[Zn^{2+}][OH^-]^2 = 4.5 \times 10^{-17}$ |

**Iodates**

| | |
|---|---|
| $AgIO_3$ | $[Ag^+][IO_3^-] = 3.1 \times 10^{-8}$ |
| $CuIO_3$ | $[Cu^+][IO_3^-] = 1.4 \times 10^{-7}$ |
| $Pb(IO_3)_2$ | $[Pb^{2+}][IO_3^-]^2 = 2.6 \times 10^{-13}$ |

**Carbonates**

| | |
|---|---|
| $Ag_2CO_3$ | $[Ag^+]^2[CO_3^{2-}] = 6.2 \times 10^{-12}$ |
| $BaCO_3$ | $[Ba^{2+}][CO_3^{2-}] = 8.1 \times 10^{-9}$ |
| $CaCO_3$ | $[Ca^{2+}][CO_3^{2-}] = 8.7 \times 10^{-9}$ |
| $PbCO_3$ | $[Pb^{2+}][CO_3^{2-}] = 3.3 \times 10^{-14}$ |
| $MgCO_3$ | $[Mg^{2+}][CO_3^{2-}] = 4.0 \times 10^{-5}$ |
| $SrCO_3$ | $[Sr^{2+}][CO_3^{2-}] = 1.6 \times 10^{-9}$ |

**Chromates**

| | |
|---|---|
| $Ag_2CrO_4$ | $[Ag^+][CrO_4^{2-}] = 1.9 \times 10^{-12}$ |
| $BaCrO_4$ | $[Ba^{2+}][CrO_4^{2-}] = 2.1 \times 10^{-10}$ |
| $PbCrO_4$ | $[Pb^{2+}][CrO_4^{2-}] = 1.8 \times 10^{-14}$ |

**Oxalates**

| | |
|---|---|
| $CuC_2O_4$ | $[Cu^{2+}][C_2O_4^{2-}] = 2.9 \times 10^{-8}$ |
| $FeC_2O_4$ | $[Fe^{2+}][C_2O_4^{2-}] = 2.1 \times 10^{-7}$ |
| $MgC_2O_4$ | $[Mg^{2+}][C_2O_4^{2-}] = 8.6 \times 10^{-5}$ |
| $PbC_2O_4$ | $[Pb^{2+}][C_2O_4^{2-}] = 2.7 \times 10^{-11}$ |
| $SrC_2O_4$ | $[Sr^{2+}][C_2O_4^{2-}] = 5.6 \times 10^{-8}$ |

**Sulfates**

| | |
|---|---|
| $BaSO_4$ | $[Ba^{2+}][SO_4^{2-}] = 1.1 \times 10^{-10}$ |
| $CaSO_4$ | $[Ca^{2+}][SO_4^{2-}] = 2.4 \times 10^{-5}$ |
| $PbSO_4$ | $[Pb^{2+}][SO_4^{2-}] = 1.1 \times 10^{-8}$ |

## Solution

The molar solubility is calculated from the gram solubility and the molar mass of silver chromate ($331.73$ g mol$^{-1}$):

$$\text{molar solubility} = \frac{0.029 \text{ g L}^{-1}}{331.73 \text{ g mol}^{-1}} = 8.74 \times 10^{-5} \text{ mol L}^{-1}$$

An extra significant digit is carried in this intermediate result to avoid round-off errors. Because each mole that dissolves gives *two* moles of silver ion, the concentration of $Ag^+(aq)$ is

$$[Ag^+] = 2 \times 8.74 \times 10^{-5} \text{ M} = 1.75 \times 10^{-4} \text{ M}$$

and that of $CrO_4^{2-}$ is simply $8.74 \times 10^{-5}$ M. The solubility product constant is then

$$K_{sp} = [Ag^+]^2[CrO_4^{2-}] = (1.75 \times 10^{-4})^2 \times (8.74 \times 10^{-5})$$
$$= 2.7 \times 10^{-12}$$

This is fairly close to the tabulated value, $1.9 \times 10^{-12}$.

*Related Problems:* 13, 14, 15, 16

Computing $K_{sp}$ from a solubility, and solubility from $K_{sp}$, is valid if the solution is ideal and if there are no side reactions that reduce the concentrations of the ions after they enter solution. If such reactions are present, they cause higher solubilities than are predicted from the $K_{sp}$ expression. For example, the solubility computed for $PbSO_4(s)$ at 25°C from its $K_{sp}$ is 0.032 g L$^{-1}$, whereas that measured experimentally is 0.0425 g L$^{-1}$. The difference is caused by the presence of species other than $Pb^{2+}(aq)$ in solution, such as $PbOH^+(aq)$. Further discussion of these side reactions is deferred to Section 7-6.

## 7-3 PRECIPITATION AND THE SOLUBILITY PRODUCT

So far, we have considered only cases in which a single slightly soluble salt attains equilibrium with its component ions in water. The relative concentrations of the cations and anions in such solutions echo their relative numbers of moles in the original salt. Thus, when AgCl is dissolved, equal numbers of moles of $Ag^+(aq)$ and $Cl^-(aq)$ ions result, and when $Ag_2SO_4$ is dissolved, twice as many moles of $Ag^+(aq)$ ions as $SO_4^{2-}(aq)$ are produced. A solubility product relationship such as

$$[Ag^+][Cl^-] = K_{sp}$$

is more general than this, however, and continues in force even if the relative chemical amounts of the two ions in solution differ from those in the pure solid compound. Such a situation often results when two solutions are mixed to give a precipitate or when another salt is present that contains an ion common to the salt under consideration.

### Precipitation from Solution

Suppose a solution is prepared by mixing one soluble salt, such as $AgNO_3$, with a solution of a second, such as NaCl. Will a precipitate of very slightly soluble silver chloride form? To reach an answer, the reaction quotient $Q$ that was defined in

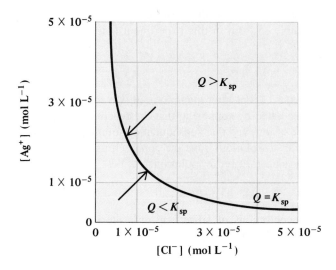

**Figure 7-6**

Some solid silver chloride is in contact with a solution containing $Ag^+(aq)$ and $Cl^-(aq)$ ions. If a solubility equilibrium exists, then the product $Q$ of the concentrations of the ions $[Ag^+] \times [Cl^-]$ is a constant, $K_{sp}$ (curved line). When $Q$ exceeds $K_{sp}$, solid silver chloride tends to precipitate until equilibrium is attained. When $Q$ is less than $K_{sp}$, additional solid tends to dissolve. If no solid is present, $Q$ remains less than $K_{sp}$.

connection with gaseous equilibria (Section 5-4) is used. The initial reaction quotient $Q_0$, when the mixing of the solutions is complete but before any reaction occurs, is

$$Q_0 = [Ag^+]_0[Cl^-]_0$$

If $Q_0 < K_{sp}$, no solid silver chloride can appear. On the other hand, if $Q_0 > K_{sp}$, solid silver chloride precipitates until the reaction quotient $Q$ reaches $K_{sp}$ (Fig. 7-6).

## Example 7-3

An emulsion of silver chloride for photographic film is prepared by adding a soluble chloride salt to a solution of silver nitrate. Suppose 500 mL of a solution of $CaCl_2$ with a chloride ion concentration of $8.0 \times 10^{-6}$ M is added to 300 mL of a 0.0040 M solution of $AgNO_3$. Will a precipitate of $AgCl(s)$ have appeared when equilibrium is reached?

### Solution

The "initial concentrations" to be used in calculating $Q_0$ are those *before* reaction but *after* dilution through mixing the two solutions. The initial concentration of $Ag^+(aq)$ after dilution from 300 mL to 800 mL of solution is

$$[Ag^+]_0 = 0.0040 \text{ M} \times \left(\frac{300 \text{ mL}}{800 \text{ mL}}\right) = 0.0015 \text{ M}$$

and that of $Cl^-(aq)$ is

$$[Cl^-]_0 = 8.0 \times 10^{-6} \text{ M} \times \left(\frac{500 \text{ mL}}{800 \text{ mL}}\right) = 5.0 \times 10^{-6} \text{ M}$$

The initial reaction quotient is

$$Q_0 = [Ag^+]_0 [Cl^-]_0 = (0.0015)(5.0 \times 10^{-6}) = 7.5 \times 10^{-9}$$

Because $Q_0 > K_{sp}$, a precipitate of silver chloride appears at equilibrium, although there may be too little to detect visually. Another possible precipitate, calcium nitrate, is far too soluble to form in this experiment (see Table 7-1).

*Related Problems:* 17, 18, 19, 20

The equilibrium concentrations of ions after the mixing of two solutions to give a precipitate are most easily calculated by supposing that the reaction first goes to completion (consuming one type of ion) and that subsequent dissolution of the solid restores some of that ionic species to solution—just the approach used in Example 6-7 for the addition of a strong acid to a buffer solution.

## Example 7-4

Calculate the equilibrium concentrations of silver and chloride ions resulting from the precipitation reaction of Example 7-3.

### Solution

In this case, the silver ion is clearly in excess, and so the chloride ion is the limiting reactant. If all of it were used up to make solid AgCl, the concentration of the remaining silver ion would be

$$[Ag^+] = 0.0015 - 5.0 \times 10^{-6} = 0.0015 \text{ M}$$

Set up the equilibrium as

|  | $AgCl(s) \rightleftharpoons Ag^+(aq) + Cl^-(aq)$ | |
| --- | --- | --- |
| Initial concentration (M) | 0.0015 | 0 |
| Change in concentration (M) | +y | +y |
| Equilibrium concentration (M) | 0.0015 + y | y |

so that the equilibrium expression is

$$(0.0015 + y)y = K_{sp} = 1.6 \times 10^{-10}$$

This quadratic equation can be solved by use of the quadratic formula, provided the calculator carries ten significant figures. It can be solved more easily by making the approximation that $y$ is much smaller than 0.0015, so that the equation simplifies to

$$0.0015y \approx 1.6 \times 10^{-10}$$
$$y \approx 1.1 \times 10^{-7} \text{ M} = [Cl^-]$$

The assumption about the size of $y$ was justified. The concentration of silver ion is

$$[Ag^+] = 0.0015 \text{ M}$$

*Related Problems:* 21, 22, 23, 24

## The Common-Ion Effect

Suppose a small amount of NaCl(s) is added to a saturated solution of AgCl. What happens? Sodium chloride is quite soluble in water and dissolves to form $Na^+(aq)$ and $Cl^-(aq)$ ions, raising the concentration of chloride ion. The quantity $Q_0 =$

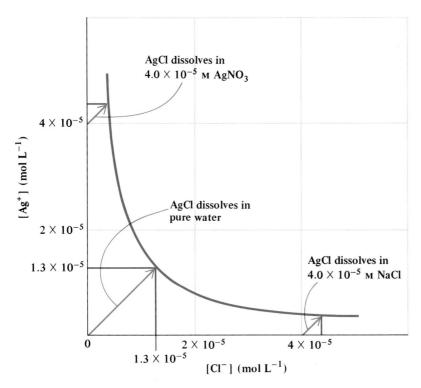

**Figure 7-7**

The presence of a dissolved common ion reduces the solubility of a salt in solution. As the AgCl dissolves, the concentrations of the ions follow the paths shown by the red arrows until they reach the blue equilibrium line. The molar solubilities are proportional to the lengths of the green lines: $1.3 \times 10^{-5}$ mol L$^{-1}$ for AgCl in pure water, but only $0.37 \times 10^{-5}$ mol L$^{-1}$ in either $4.0 \times 10^{-5}$ M AgNO$_3$ or $4.0 \times 10^{-5}$ M NaCl.

$[Ag^+][Cl^-]$ then exceeds the $K_{sp}$ of silver chloride, and silver chloride precipitates until the concentrations of Ag$^+$(aq) and Cl$^-$(aq) are sufficiently reduced that the solubility product expression once again is satisfied.

The same equilibrium may be approached from the other direction. The amount of AgCl(s) that can dissolve in a solution of sodium chloride is less than the amount that could dissolve in the same volume of pure water. Because $[Ag^+][Cl^-] = K_{sp}$, a graph of the equilibrium concentration of silver ion against chloride concentration has the form of a hyperbola (Fig. 7-7). The presence of excess Cl$^-$(aq) reduces the concentration of Ag$^+$(aq) permitted, and the solubility of AgCl(s) is reduced. In the same way, the prior presence of Ag$^+$(aq) in the solvent (for example, when an attempt is made to dissolve AgCl in water that already contains AgNO$_3$) reduces the amount of Cl$^-$(aq) permitted at equilibrium and also reduces the solubility of AgCl. This is referred to as the **common-ion effect:** if the solution and the solid salt to be dissolved in it have an ion in common, then the solubility of the salt is depressed.

Let us examine the quantitative consequences of the common-ion effect. Suppose an excess of AgCl(s) is added to 1.00 L of a 0.100 M NaCl solution and the solubility is again determined. If $S$ mol of AgCl dissolves per liter, the concentration of Ag$^+$(aq) will be $S$ mol L$^{-1}$, and that of Cl$^-$(aq) will be

$$[Cl^-] = 0.100 + S$$

because the chloride ion has two sources: the 0.100 M NaCl and the dissolution of AgCl. The expression for the solubility product is written as

$$[Ag^+][Cl^-] = S(0.100 + S) = K_{sp} = 1.6 \times 10^{-10}$$

The solubility of AgCl(s) in this solution must be smaller than it is in pure water, which is much smaller than 0.100. That is,

$$S < 1.3 \times 10^{-5} \ll 0.100$$

Thus, $(0.100 + S)$ can be approximated by 0.100 (as in Example 7-4), giving

$$(0.100)S \approx 1.6 \times 10^{-10}$$

$$S \approx 1.6 \times 10^{-9}$$

This is indeed much smaller than 0.100, so the approximation was a very good one. Therefore, at equilibrium,

$$[Ag^+] = S = 1.6 \times 10^{-9} \text{ M}$$

$$[Cl^-] = 0.100 \text{ M}$$

The gram solubility of AgCl here is

$$(1.6 \times 10^{-9} \text{ mol L}^{-1})(143.3 \text{ g mol}^{-1}) = 2.3 \times 10^{-7} \text{ g L}^{-1}$$

The solubility of AgCl in 0.100 M NaCl is lower than that in pure water by a factor of about 8000.

## Example 7-5

What is the gram solubility of $CaF_2(s)$ in a 0.100 M solution of NaF?

### Solution

Again, the molar solubility is denoted by $S$. The only source of the $Ca^{2+}(aq)$ in the solution at equilibrium is the dissolution of $CaF_2$, whereas the $F^-(aq)$ has two sources, the $CaF_2$ and the NaF. Hence,

| | $CaF_2(s) \rightleftharpoons$ | $Ca^{2+}(aq)$ + | $2 F^-(aq)$ |
|---|---|---|---|
| Initial concentration (M) | | 0 | 0.100 |
| Change in concentration (M) | | $+S$ | $+\ 2S$ |
| Equilibrium concentration (M) | | $S$ | $0.100 + 2S$ |

If $0.100 + 2S$ is approximated as 0.100, then

$$[Ca^{2+}][F^-]^2 = K_{sp}$$

$$S(0.100)^2 = 3.9 \times 10^{-11}$$

$$S = 3.9 \times 10^{-9}$$

Clearly,

$$2S = 7.8 \times 10^{-9} \ll 0.100$$

so the assumption was justified. The gram solubility of $CaF_2$ is

$$(3.9 \times 10^{-9} \text{ mol L}^{-1})(78.1 \text{ g mol}^{-1}) = 3.0 \times 10^{-7} \text{ g L}^{-1}$$

and the solubility in this case is reduced by a factor of 50,000.

*Related Problems:* 25, 26, 27, 28

## 7-4 THE EFFECTS OF pH ON SOLUBILITY

Some solids are only weakly soluble in water but dissolve readily in acidic solutions. Copper and nickel sulfides from ores, for example, can be brought into solution with strong acids, a fact that aids greatly in the separation and recovery of these valuable metals in their elemental forms. The effect of pH on solubility is shown dramatically in the damage done to buildings and monuments by acid precipitation (Fig. 7-8). Both marble and limestone are made up of small crystals of calcite ($CaCO_3$), which dissolves to only a limited extent in "natural" rain (with a pH of about 5.6) but dissolves much more extensively as the rainwater becomes more acidic. The reaction

$$CaCO_3(s) + H_3O^+(aq) \longrightarrow Ca^{2+}(aq) + HCO_3^-(aq) + H_2O(\ell)$$

causes this increase. This section examines the role of pH in solubility.

### Solubility of Hydroxides

One direct effect of pH on solubility occurs with the metal hydroxides. The concentration of $OH^-$ appears explicitly in the expression for the solubility product of such compounds. Thus, for the dissolution of $Zn(OH)_2(s)$,

$$Zn(OH)_2(s) \rightleftharpoons Zn^{2+}(aq) + 2\,OH^-(aq)$$

the solubility product expression is

$$[Zn^{2+}][OH^-]^2 = K_{sp} = 4.5 \times 10^{-17}$$

As the solution is made more acidic, the concentration of hydroxide ion decreases, causing an increase in the concentration of $Zn^{2+}(aq)$ ion. Zinc hydroxide is thus more soluble in acidic solution than in pure water.

### Example 7-6

Compare the solubility of $Zn(OH)_2$ in pure water with that in a solution buffered at pH 6.00.

### Solution

In pure water the usual solubility product calculation applies:

$$[Zn^{2+}] = S; \quad [OH^-] = 2S$$

$$S(2S)^2 = 4S^3 = K_{sp} = 4.5 \times 10^{-17}$$

$$S = 2.2 \times 10^{-6}\ \text{M} = [Zn^{2+}]$$

and so the solubility is $2.2 \times 10^{-6}$ mol $L^{-1}$, or $2.2 \times 10^{-4}$ g $L^{-1}$. Using

$$[OH^-] = 2S = 4.5 \times 10^{-6}\ \text{M}$$

the resulting solution is found to have pH 8.65.

In the second case, it is assumed that the solution is buffered sufficiently that the pH remains 6.00 after dissolution of the zinc hydroxide. Then

$$[OH^-] = 1.0 \times 10^{-8}\ \text{M}$$

**Figure 7-8**

The calcium carbonate in marble and limestone is very slightly soluble in neutral water. Its solubility is much greater in acidic water. Objects carved of these materials dissolve relatively rapidly in areas where rain, snow, or fog is acidified from air pollution. Shown here is the damage to one of the caryatids from the Erechtheum in the Acropolis in Athens. (Bruce Coleman, Inc.)

$$[Zn^{2+}] = \frac{K_{sp}}{[OH^-]^2} = \frac{4.5 \times 10^{-17}}{(1.0 \times 10^{-8})^2} = 0.45 \text{ M}$$

so that $0.45 \text{ mol L}^{-1}$, or $45 \text{ g L}^{-1}$, should dissolve in this case. When ionic concentrations are this high, the simple form of the solubility expression will likely break down, but the qualitative conclusion is still valid: $Zn(OH)_2$ is far more soluble at pH 6.00 than in pure water.

*Related Problems:* 31, 32

## Solubility of Salts of Bases

Metal hydroxides can be described as salts of a strong base, the hydroxide ion. The solubility of salts in which the anion is a different weak or strong base are also affected by pH. For example, consider a solution of a slightly soluble fluoride, such as calcium fluoride. The solubility equilibrium is

$$CaF_2(s) \rightleftharpoons Ca^{2+}(aq) + 2 F^-(aq) \qquad K_{sp} = 3.9 \times 10^{-11}$$

As the solution is made more acidic, some of the fluoride ion reacts with hydronium ion through

$$H_3O^+(aq) + F^-(aq) \rightleftharpoons HF(aq) + H_2O(\ell)$$

Because this is just the reverse of the acid ionization of HF, its equilibrium constant is the reciprocal of $K_a$ for HF, or $1/(3.5 \times 10^{-4}) = 2.9 \times 10^3$. As acid is added, the concentration of fluoride ion is reduced, so the calcium ion concentration must increase to maintain the solubility product equilibrium for $CaF_2$. As a result, the solubility of fluoride salts increases in acidic solution. The same applies to other ionic substances in which the anion is a weak or a strong base. By contrast, the solubility of a salt such as AgCl is only very slightly affected by a decrease in pH. The reason is that HCl is a strong acid, and so $Cl^-$ is ineffective as a base. The reaction

$$Cl^-(aq) + H_3O^+(aq) \longrightarrow HCl(aq) + H_2O(\ell)$$

occurs to a negligible extent in acidic solution.

## 7-5 A Deeper Look . . . Selective Precipitation of Ions

One way to analyze a mixture of ions in solution is to separate the mixture into its components by exploiting the differences in the solubilities of compounds containing the ions. To separate silver ion from lead ion, for example, a search is made for compounds of these elements that (1) have a common anion and (2) have widely different solubilities. The chlorides AgCl and $PbCl_2$ are two such compounds for which the solubility equilibria are

$$AgCl(s) \rightleftharpoons Ag^+(aq) + Cl^-(aq) \qquad K_{sp} = 1.6 \times 10^{-10}$$
$$PbCl_2(s) \rightleftharpoons Pb^{2+}(aq) + 2 Cl^-(aq) \qquad K_{sp} = 2.4 \times 10^{-4}$$

Lead chloride is far more soluble in water than is silver chloride. Consider a solution that is 0.10 M in both $Ag^+$ and $Pb^{2+}$. Is it possible to add enough $Cl^-$ to precipitate almost all the $Ag^+$ ions but leave all the $Pb^{2+}$ ions in solution? If so, a quantitative separation of the two species can be achieved.

In order for $Pb^{2+}$ to remain in solution, its reaction quotient must remain smaller than $K_{sp}$: $Q = [Pb^{2+}][Cl^-]^2 < K_{sp}$. Inserting $K_{sp}$ and the concentration of $Pb^{2+}$ gives

$$[Cl^-]^2 < \frac{K_{sp}}{[Pb^{2+}]} = \frac{2.4 \times 10^{-4}}{0.10} = 2.4 \times 10^{-3}$$

The square root of this is

$$[Cl^-] < 4.9 \times 10^{-2} \text{ M}$$

Thus, as long as the chloride ion concentration remains smaller than 0.049 M, no $PbCl_2$ should precipitate. To reduce the *silver* ion concentration in solution as far as possible (that is, to precipitate out as much silver chloride as possible), the chloride ion concentration should be kept as high as possible without exceeding 0.049 M. If exactly this concentration of $Cl^-(aq)$ is chosen, then at equilibrium,

$$[Ag^+] = \frac{K_{sp}}{[Cl^-]} = \frac{1.6 \times 10^{-10}}{0.049} = 3.3 \times 10^{-9}$$

At that concentration of $Cl^-$, the concentration of $Ag^+$ has been reduced to $3.3 \times 10^{-9}$ from the original concentration of 0.10 M. In other words, only about three $Ag^+$ ions in $10^8$ remain in solution, but all the $Pb^{2+}$ ions are left in solution. A nearly perfect separation of the two ionic species has been achieved.

This calculation gave the optimal theoretical separation factor for the two ions. In practice, it is necessary to keep the chloride concentration lower. If $[Cl^-]$ is ten times smaller, or 0.0049 M, about three $Ag^+$ ions in $10^7$ remain in solution with the $Pb^{2+}$. This is ten times more $Ag^+$ than if $[Cl^-] = 0.049$ M, but the separation of $Ag^+$ from $Pb^{2+}$ is still very good.

Figure 7-9 shows graphically how ions can be separated based on solubility. The relation between the concentration of $Pb^{2+}$ and $Cl^-$ ions in contact with solid $PbCl_2$

$$[Pb^{2+}] = \frac{K_{sp}}{[Cl^-]^2}$$

can be rewritten in a useful form by taking the base-10 logarithms of both sides:

$$\log_{10}[Pb^{2+}] = -2\log_{10}[Cl^-] + \log_{10}K_{sp}$$

That is, a graph of $\log_{10}[Pb^{2+}]$ against $\log_{10}[Cl^-]$ is a straight line with slope $-2$. The corresponding graph for the AgCl solubility equilibrium has slope $-1$.

$$\log_{10}[Ag^+] = -\log_{10}[Cl^-] + \log_{10}K_{sp}$$

If the concentration of $Cl^-$ and the initial concentrations of the metal ions correspond to a point that lies between the two lines in Figure 7-9, then AgCl precipitates but $PbCl_2$ does not.

## Metal Sulfides

Controlling the solubility of metal sulfides has important applications. According to Table 7-1, most metal sulfides are very slightly soluble in water; only a very small amount of a com-

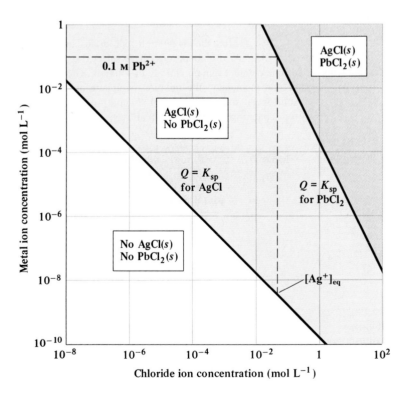

**Figure 7-9**

To separate a mixture of $Ag^+$ and $Pb^{2+}$ ions, a chloride ion concentration is selected that gives a $Pb^{2+}$ concentration below the equilibrium curve for $PbCl_2$, so all $Pb^{2+}$ remains in solution, but well above the equilibrium curve for $Ag^+$. As a result, nearly all the $Ag^+$ precipitates as AgCl. As the dashed lines show, $[Cl^-] = 0.049$ M is optimal if $[Pb^{2+}] = [Ag^+] = 0.1$ M.

pound such as $ZnS(s)$ will dissolve in water. Although it is tempting to write the resulting equilibrium as

$$ZnS(s) \rightleftharpoons Zn^{2+}(aq) + S^{2-}(aq) \qquad K_{sp} = ?$$

in analogy with the equilibria for other weakly soluble salts, this is misleading, because $S^{2-}$, like $O^{2-}$, is a very strong base (stronger than $OH^-$) and reacts almost quantitatively with water:

$$S^{2-}(aq) + H_2O(\ell) \longrightarrow HS^-(aq) + OH^-(aq)$$

Recent research has shown that $K_b$ for this reaction is on the order of $10^5$, and so essentially no $S^{2-}$ is present in aqueous solution. The net dissolution reaction is found by adding the two preceding equations,

$$ZnS(s) + H_2O(\ell) \rightleftharpoons Zn^{2+}(aq) + OH^-(aq) + HS^-(aq)$$

for which the equilibrium constant is

$$[Zn^{2+}][OH^-][HS^-] = K \approx 2 \times 10^{-25}$$

Table 7-3 gives values of the equilibrium constants for comparable reactions of other metal sulfides. As the pH decreases, the concentration of $OH^-$ decreases. At the same time, the concentration of $HS^-$ also decreases as the equilibrium

$$HS^-(aq) + H_3O^+(aq) \rightleftharpoons H_2S(aq) + H_2O(\ell)$$

shifts to the right upon addition of $H_3O^+$. If both $[OH^-]$ and $[HS^-]$ decrease, then $[Zn^{2+}]$ must increase in order to maintain a constant value for the product of the three concentrations. As a result, the solubility of $ZnS(s)$ increases as the pH of the solution decreases. Other metal sulfides behave the same way, becoming more soluble in acidic solution.

A quantitative calculation of metal sulfide solubility requires treating several simultaneous equilibria, as the following example illustrates.

### Example 7-7

In a solution that is saturated with $H_2S$, $[H_2S]$ is fixed at 0.1 M. Calculate the molar solubility of $FeS(s)$ in such a solution, if it is buffered at pH 3.0.

Hydrogen sulfide, $H_2S$, is a poisonous, foul-smelling gas. When dissolved in water, it gives a weak acid, hydrosulfuric acid.

**Solution**

If the pH is 3.0, then

$$[OH^-] = 1 \times 10^{-11} \text{ M}$$

In addition, the acid ionization of $H_2S$ must be considered:

$$H_2S(aq) + H_2O(\ell) \rightleftharpoons H_3O^+(aq) + HS^-(aq)$$

Substitution of $[H_2S] = 0.1$ M for a saturated solution and $[H_3O^+] = 1 \times 10^{-3}$ M (at pH 3.0) into the equilibrium expression for this reaction gives

$$\frac{[H_3O^+][HS^-]}{[H_2S]} = \frac{(1 \times 10^{-3})[HS^-]}{0.1} = K_a = 9.1 \times 10^{-8}$$

$$[HS^-] = 9 \times 10^{-6} \text{ M}$$

where $K_a$ came from Table 6-2. For the reaction

$$FeS(s) + H_2O(\ell) \rightleftharpoons Fe^{2+}(aq) + HS^-(aq) + OH^-(aq)$$

the equilibrium constant from Table 7-3 is

$$[Fe^{2+}][HS^-][OH^-] = 5 \times 10^{-19}$$

Substituting the values of $[HS^-]$ and $[OH^-]$ and solving for $[Fe^{2+}]$ give

$$[Fe^{2+}](9 \times 10^{-6})(1 \times 10^{-11}) = 5 \times 10^{-19}$$

$$[Fe^{2+}] = 6 \times 10^{-3} \text{ M}$$

Hence, $6 \times 10^{-3}$ mol of FeS dissolves per liter under these conditions.

*Related Problems:* 41, 42

By adjusting the pH through appropriate choice of buffers, as in Example 7-7, conditions can be selected so that metal ions of one element remain entirely in solution, whereas those of a second element in the mixture precipitate almost entirely as solid metal sulfide (Fig. 7-10). Such a procedure is important for separating metal ions in qualitative analysis.

**Table 7-3  Equilibrium Constants for Metal Sulfide Dissolution at 25°C**

| Metal Sulfide | $K^†$ |
|---|---|
| CuS | $5 \times 10^{-37}$ |
| PbS | $3 \times 10^{-28}$ |
| CdS | $7 \times 10^{-28}$ |
| SnS | $9 \times 10^{-27}$ |
| ZnS | $2 \times 10^{-25}$ |
| FeS | $5 \times 10^{-19}$ |
| MnS | $3 \times 10^{-14}$ |

$^†K$ is the equilibrium constant for the reaction $MS(s) + H_2O(\ell) \rightleftharpoons M^{2+}(aq) + OH^-(aq) + HS^-(aq)$.

**Figure 7-10**
These sulfides are insoluble at pH 1 and so can be separated out of a mixture containing other, more soluble sulfides. From left to right, they are PbS, $Bi_2S_3$, CuS, CdS, $Sb_2S_3$, $SnS_2$, $As_2S_3$, and HgS. (Leon Lewandowski)

........
## 7-6 COMPLEX IONS AND SOLUBILITY

Many transition-metal ions form **coordination complexes** in solution or in the solid state; these consist of a metal ion surrounded by a group of anions or neutral molecules called **ligands.** The interaction involves the sharing by the metal ion of a lone pair on each ligand molecule, giving a partially covalent bond with that ligand. Such complexes frequently have strikingly deep colors. When exposed to gaseous ammonia, greenish white crystals of copper sulfate ($CuSO_4$) give a deep-blue crystalline solid with the chemical formula $Cu(NH_3)_4SO_4$ (Fig. 7-11a). The anions in the solid are still sulfate ions ($SO_4^{2-}$), but the cations are now **complex ions,** or coordination complexes of the central $Cu^{2+}$ ion with four ammonia molecules, $Cu(NH_3)_4^{2+}$. The ammonia molecules coordinate to the copper ion through their lone-pair electrons (Fig. 7-12), acting as Lewis bases toward the metal ion, the Lewis acid. When the solid is dissolved in water, the deep blue color remains. This is evidence that the complex persists in water, because when ordinary $CuSO_4$ (without ammonia ligands) is dissolved in water, a much paler blue color results (Fig. 7-11b).

(a)                              (b)

**Figure 7-11**
(a) From left to right, crystals of $Cu(NH_3)_4SO_4$, $CuSO_4 \cdot 5H_2O$, and $CuSO_4$. (b) Aqueous solutions of copper sulfate containing (left) and not containing (right) ammonia. (Leon Lewandowski)

**Figure 7-12**
The structure of $Cu(NH_3)_4^{2+}$.

**Figure 7-13**
The structure of $Ag(NH_3)^+$.

Let us explore the effects of the formation of complex ions on equilibria in aqueous solutions, deferring to Chapter 16 a discussion of the microscopic structure and bonding in these complexes.

## Complex Ion Equilibria

When silver ions are dissolved in an aqueous ammonia solution, doubly coordinated silver–ammonia complexes, shown in Figure 7-13, form in two stepwise reactions:

$$Ag^+(aq) + NH_3(aq) \rightleftharpoons Ag(NH_3)^+(aq)$$

$$\frac{[Ag(NH_3)^+]}{[Ag^+][NH_3]} = K_1 = 2.1 \times 10^3$$

$$Ag(NH_3)^+(aq) + NH_3(aq) \rightleftharpoons Ag(NH_3)_2^+(aq)$$

$$\frac{[Ag(NH_3)_2^+]}{[Ag(NH_3)^+][NH_3]} = K_2 = 8 \times 10^3$$

If these two chemical equations are added (and their corresponding equilibrium laws are multiplied), the result is

$$Ag^+(aq) + 2\,NH_3(aq) \rightleftharpoons Ag(NH_3)_2^+(aq)$$

$$\frac{[Ag(NH_3)_2^+]}{[Ag^+][NH_3]^2} = K_f = K_1 K_2 = 1.7 \times 10^7$$

where $K_f$ is the **formation constant** of the full complex ion $Ag(NH_3)_2^+$. Table 7-4 gives formation constants for a representative selection of complex ions. The larger the formation constant $K_f$, the more stable the corresponding complex ion, for ions with the same number of ligands.

Because $K_1$ and $K_2$ of the $Ag(NH_3)_2^+$ complex ion are both large, a silver salt dissolved in water that contains a high concentration of ammonia will be primarily in the form of the complex ion $[Ag(NH_3)_2]^+$ at equilibrium.

## Example 7-8

Suppose 0.100 mol of $AgNO_3$ is dissolved in 1.00 L of a 1.00 M solution of $NH_3$. Calculate the concentrations of the $Ag^+$ and $Ag(NH_3)^+$ ions present at equilibrium.

### Solution

Suppose that most of the $Ag^+$ is present as $Ag(NH_3)_2^+$ (this will be checked later). Then

$$[Ag(NH_3)_2^+]_0 = 0.100 \text{ M}$$

$$[NH_3]_0 = 1.00 \text{ M} - (2 \times 0.100) \text{ M} = 0.80 \text{ M}$$

after each silver ion has become complexed with two ammonia molecules. The two stages of the dissociation of the $Ag(NH_3)_2^+$ ion are the reverses of the

## Table 7-4   Formation Constants of Coordination Complexes in Aqueous Solution

| | $K_f$ | $K_1$ | $K_2$ | $K_3$ | $K_4$ | $K_5$ | $K_6$ |
|---|---|---|---|---|---|---|---|
| **Ammines** | | | | | | | |
| $Ag(NH_3)_2^+$ | $1.7 \times 10^7$ | $2.1 \times 10^3$ | $8.2 \times 10^3$ | | | | |
| $Co(NH_3)_6^{2+}$ | $2.5 \times 10^4$ | $1.0 \times 10^2$ | 32 | 8.5 | 4.4 | 1.1 | 0.18 |
| $Cu(NH_3)_4^{2+}$ | $1.1 \times 10^{12}$ | $1.0 \times 10^4$ | $2 \times 10^3$ | $5 \times 10^2$ | 90 | | |
| $Ni(NH_3)_6^{2+}$ | $8 \times 10^{-7}$ | $5 \times 10^2$ | $1.3 \times 10^2$ | 40 | 12 | 3.3 | 0.8 |
| $Zn(NH_3)_4^{2+}$ | $5 \times 10^8$ | $1.5 \times 10^2$ | $1.8 \times 10^2$ | $2 \times 10^2$ | 90 | | |
| **Chlorides** | | | | | | | |
| $AgCl_2^-$ | $1.8 \times 10^5$ | $1.7 \times 10^3$ | $1.0 \times 10^2$ | | | | |
| $FeCl_4^-$ | 0.14 | 28 | 4.5 | 0.1 | $1.1 \times 10^{-2}$ | | |
| $HgCl_4^{2-}$ | $1.2 \times 10^{15}$ | $5.5 \times 10^6$ | $3 \times 10^6$ | 7 | 10 | | |
| $PbCl_4^{2-}$ | 25 | 40 | 1.5 | 0.8 | 0.5 | | |
| $SnCl_4^{2-}$ | 30 | 32 | 5.4 | 0.6 | 0.3 | | |
| **Hydroxides** | | | | | | | |
| $Co(OH)_3^-$ | $3 \times 10^{10}$ | $4 \times 10^4$ | 1 | $8 \times 10^5$ | | | |
| $Cu(OH)_4^{2-}$ | $3 \times 10^{18}$ | $1 \times 10^7$ | $5 \times 10^6$ | $2 \times 10^3$ | 30 | | |
| $Ni(OH)_3^-$ | $2 \times 10^{11}$ | $9 \times 10^4$ | $4 \times 10^3$ | $6 \times 10^2$ | | | |
| $Pb(OH)_3^-$ | $4 \times 10^{14}$ | $7 \times 10^7$ | $1.1 \times 10^3$ | $5 \times 10^3$ | | | |
| $Zn(OH)_4^{2-}$ | $5 \times 10^{14}$ | $2.5 \times 10^4$ | $8 \times 10^6$ | 70 | 33 | | |

complexation reactions, and so their equilibrium constants are the reciprocals of $K_2$ and $K_1$, respectively:

$$Ag(NH_3)_2^+(aq) \rightleftharpoons Ag(NH_3)^+(aq) + NH_3(aq)$$

$$\frac{[Ag(NH_3)^+][NH_3]}{[Ag(NH_3)_2^+]} = \frac{1}{K_2} = \frac{1}{8.2 \times 10^3}$$

$$Ag(NH_3)^+(aq) \rightleftharpoons Ag^+(aq) + NH_3(aq)$$

$$\frac{[Ag^+][NH_3]}{[Ag(NH_3)^+]} = \frac{1}{K_1} = \frac{1}{2.1 \times 10^3}$$

If $y$ mol $L^{-1}$ of $Ag(NH_3)_2^+$ dissociates at equilibrium according to the first equation,

| | $Ag(NH_3)_2^+(aq)$ | $\rightleftharpoons$ | $Ag(NH_3)^+(aq)$ | + | $NH_3(aq)$ |
|---|---|---|---|---|---|
| Initial concentration (M) | 0.100 | | 0 | | 0.80 |
| Change in concentration (M) | $-y$ | | $+y$ | | $+y$ |
| Equilibrium concentration (M) | $0.100 - y$ | | $y$ | | $0.80 + y$ |

then the first equilibrium expression becomes

$$\frac{y(0.80 + y)}{0.10 - y} = \frac{1}{K_2} = \frac{1}{8.2 \times 10^3}$$

$$y = 1.5 \times 10^{-5} \text{ M} = [Ag(NH_3)^+]$$

We can then calculate the concentration of free $Ag^+$ ions from the equilibrium law for the second step of the dissociation of the complex ion:

$$\frac{[Ag^+][NH_3]}{[Ag(NH_3)^+]} = \frac{1}{K_1}$$

$$\frac{[Ag^+](0.80)}{1.5 \times 10^{-5}} = \frac{1}{2.1 \times 10^3}$$

$$[Ag^+] = 9 \times 10^{-9} \text{ M}$$

It is clear that the original assumption was correct, and most of the silver present is tied up in the $Ag(NH_3)_2^+$ complex.

*Related Problems: 43, 44*

There is a close similarity between the working of Example 7-8 and an acid–base calculation. The first step (the assumption that the reaction goes to completion and is followed by a small amount of back dissociation) is analogous to the procedure for dealing with the addition of a small amount of a strong acid to a solution of weak base. The subsequent calculation of the successive dissociation steps resembles the calculation of polyprotic acid equilibria in Example 6-11. The only difference is that in complex ion equilibria it is conventional to work with formation constants, which are the inverses of the dissociation constants used in acid–base equilibria.

The formation of coordination complexes can have a large effect on the solubility of a compound in water. Silver bromide is only very weakly soluble in water,

$$AgBr(s) \rightleftharpoons Ag^+(aq) + Br^-(aq) \qquad K_{sp} = 7.7 \times 10^{-13}$$

but addition of thiosulfate ion ($S_2O_3^{2-}$) to the solution allows the complex ion $Ag(S_2O_3)_2^{3-}$ to form:

$$AgBr(s) + 2\,S_2O_3^{2-}(aq) \rightleftharpoons Ag(S_2O_3)_2^{3-}(aq) + Br^-(aq)$$

This greatly increases the solubility of the silver bromide (Fig. 7-14). The formation of this complex ion is an important step in the development of photographic images; thiosulfate ion is a component of the fixer that brings silver bromide into solution from the unexposed portion of the film.

## Example 7-9

Calculate the solubility of AgBr in a 1.00 M aqueous solution of ammonia.

### Solution

We tentatively assume that almost all the silver that dissolves is complexed as $Ag(NH_3)_2^+$ (this will be checked later). The overall reaction is then

$$AgBr(s) + 2\,NH_3(aq) \rightleftharpoons Ag(NH_3)_2^+(aq) + Br^-(aq)$$

Note that this is the sum of the two reactions

$$AgBr(s) \rightleftharpoons Ag^+(aq) + Br^-(aq) \qquad K_{sp} = 7.7 \times 10^{-13}$$
$$Ag^+(aq) + 2\,NH_3(aq) \rightleftharpoons Ag(NH_3)_2^+(aq) \qquad K_f = 1.7 \times 10^7$$

and so its equilibrium constant is the product $K_{sp}K_f = 1.3 \times 10^{-5}$.

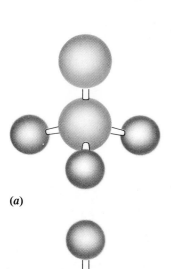

*(a)*

*(b)*

The thiosulfate ion *(a)* is related to the sulfate ion, $SO_4^{2-}$ *(b)*, by the replacement of one oxygen atom with one sulfur atom. It is prepared, however, by the reaction of elemental sulfur with the sulfite ion ($SO_3^{2-}$).

## Figure 7-14

An illustration of the effect of complex ion formation on solubility. Each test tube contains 2.0 g AgBr, but the one on the left also contains dissolved thiosulfate ion, which forms a complex ion with $Ag^+$. Almost none of the white solid AgBr has dissolved in pure water, but all of it has dissolved in the solution containing thiosulfate. (Leon Lewandowski)

If $S$ mol $L^{-1}$ of AgBr dissolves, then

$$S = [Br^-] \approx [Ag(NH_3)_2^+]$$

$$[NH_3] = 1.00 - 2S$$

because 2 mol of $NH_3$ is used up for each mole of complex formed. The equilibrium expression is

$$\frac{S^2}{(1.00 - 2S)^2} = K_{sp}K_f = 1.3 \times 10^{-5}$$

$$S = 3.6 \times 10^{-3} \text{ M} = [Ag(NH_3)_2^+] = [Br^-]$$

To check the original assumption, calculate the concentration of free silver ion:

$$[Ag^+] = \frac{K_{sp}}{[Br^-]} = \frac{K_{sp}}{S} = 2.1 \times 10^{-10} \ll [Ag(NH_3)_2^+]$$

verifying that almost all of the silver is complexed. The solubility is therefore $3.6 \times 10^{-3}$ mol $L^{-1}$, significantly greater than the solubility in pure water:

$$\text{solubility in pure water} = \sqrt{K_{sp}} = 8.8 \times 10^{-7} \text{ mol } L^{-1}$$

*Related Problems:* 47, 48

Another interesting effect of complex ions on solubilities is illustrated by the addition of iodide ion to a solution containing mercury(II) ion. After a moderate amount of iodide ion has been added, an orange precipitate forms (Fig. 7-15) through the reaction

$$Hg^{2+}(aq) + 2\,I^-(aq) \rightleftharpoons HgI_2(s)$$

With further addition of iodide ion, however, the orange solid redissolves because complex ions form:

$$HgI_2(s) + I^-(aq) \rightleftharpoons HgI_3^-(aq)$$

$$HgI_3^-(aq) + I^-(aq) \rightleftharpoons HgI_4^{2-}(aq)$$

## Figure 7-15

The "orange tornado" is a striking demonstration of the effects of complex ions on solubility. A solution is prepared with an excess of $I^-(aq)$ over $Hg^{2+}(aq)$ so that the $Hg^{2+}$ is complexed as $HgI_3^-$ and $HgI_4^{2-}$. A magnetic stirrer is used to create a vortex in the solution. Addition of a solution containing $Hg^{2+}$ down the center of the vortex then causes the orange solid $HgI_2$ to form in a layer at the edges of the vortex, giving the tornado effect. (Leon Lewandowski)

In the same way, silver chloride will dissolve in a concentrated solution of sodium chloride by forming soluble $AgCl_2^-$ complex ions. Complex ion formation affects solubility in the opposite direction from the common-ion effect of Section 7-3.

## Acidity and Amphoterism of Complex Ions

When dissolved in water, many metal ions increase the acidity of the solution. The iron(III) ion is an example: each dissolved $Fe^{3+}$ ion is strongly solvated by six water molecules, leading to a complex ion $Fe(H_2O)_6^{3+}$. This complex ion can act as a Brønsted–Lowry acid, donating hydrogen ions to the solvent, water:

$$Fe(H_2O)_6^{3+}(aq) + H_2O(\ell) \rightleftharpoons H_3O^+(aq) + Fe(H_2O)_5OH^{2+}(aq)$$
$$\text{Acid}_1 \qquad \text{Base}_2 \qquad \text{Acid}_2 \qquad \text{Base}_1$$

$$\frac{[H_3O^+][Fe(H_2O)_5OH^{2+}]}{[Fe(H_2O)_6^{3+}]} = K_a = 7.7 \times 10^{-3}$$

Metal ion hydrolysis fits into the general scheme of the Brønsted–Lowry acid–base reaction.

### Example 7-10

Calculate the pH of a solution that is 0.100 M in $Fe(NO_3)_3$.

### Solution

The iron(III) is present as $Fe(H_2O)_6^{3+}$, which reacts as a weak acid:

$$Fe(H_2O)_6^{3+} + H_2O(\ell) \rightleftharpoons Fe(H_2O)_5OH^{2+} + H_3O^+(aq)$$

with $K_a$ equal to $7.7 \times 10^{-3}$. If $y$ mol L$^{-1}$ of $[Fe(H_2O)_6]^{3+}$ reacts, then (neglecting the ionization of water itself)

$$[H_3O^+] = [Fe(H_2O)_5OH^{2+}] = y$$
$$[Fe(H_2O)_6^{3+}] = 0.100 - y$$

The equilibrium expression has the form

$$\frac{y^2}{0.100 - y} = 7.7 \times 10^{-3}$$

$$y = 2.4 \times 10^{-2} \text{ M} = [H_3O^+]$$

and so the pH is 1.62. Solutions of iron(III) salts are strongly acidic.

*Related Problems:* 51, 52

Another acceptable way to write the reaction that makes iron(III) solutions acidic is

$$Fe^{3+}(aq) + 2 H_2O(\ell) \rightleftharpoons H_3O^+(aq) + FeOH^{2+}(aq)$$

in which the specific mention of the six waters of hydration is now omitted. The $FeOH^{2+}$ complex ions are brown, but $Fe^{3+}$ ions are almost colorless. This reaction occurs to a sufficient extent to make a solution of $Fe(NO_3)_3$ in water pale brown.

| Table 7-5 | pH of 0.1 M Aqueous Metal Nitrate Solutions at 25°C | |
| --- | --- | --- |
| **Metal Nitrate** | | **pH** |
| $Fe(NO_3)_3$ | | 1.6 |
| $Pb(NO_3)_2$ | | 3.6 |
| $Cu(NO_3)_2$ | | 4.0 |
| $Zn(NO_3)_2$ | | 5.3 |
| $Ca(NO_3)_2$ | | 6.7 |
| $NaNO_3$ | | 7.0 |

**Figure 7-16**
The acidity of metal ions is illustrated by the vigorous reaction of anhydrous $AlCl_3$ with water to generate hydrated aluminum oxides, HCl(aq), and heat. The HCl turns the indicator to its red acid form. (Leon Lewandowski)

When strong acid is added, the equilibrium is driven back to the left, and the color fades. Table 7-5 gives values of the pH for 0.1 M solutions of several metal ions. Those that form strong complexes with hydroxide ion have low pH (Fig. 7-16), whereas those that do not form such complexes give neutral solutions (pH 7).

Different cations behave differently as water ligands are replaced by hydroxide ions in an increasingly basic solution. A particularly interesting example is $Zn^{2+}$. It forms a series of hydroxo complex ions:

$$Zn^{2+}(aq) + OH^-(aq) \rightleftharpoons ZnOH^+(aq)$$

$$ZnOH^+(aq) + OH^-(aq) \rightleftharpoons Zn(OH)_2(s)$$

$$Zn(OH)_2(s) + OH^-(aq) \rightleftharpoons Zn(OH)_3^-(aq)$$

$$Zn(OH)_3^-(aq) + OH^-(aq) \rightleftharpoons Zn(OH)_4^{2-}(aq)$$

In the Brønsted–Lowry theory, a polyprotic acid, $Zn(H_2O)_4^{2+}(aq)$, donates hydrogen ions in succession to make all the product ions. The second product, $Zn(OH)_2$, is amphoteric; it can react as either acid or base. It is only slightly soluble in pure water (its $K_{sp}$ is only $1.9 \times 10^{-17}$). If enough acid is added to solid $Zn(OH)_2$, the $OH^-$ ligands are removed, forming the soluble $Zn^{2+}$ ion; if enough base is added, $OH^-$ ligands attach to form the soluble $Zn(OH)_4^{2-}$ (zincate) ion. Thus, $Zn(OH)_2$ is soluble in strongly acidic *or* strongly basic solutions, but is only slightly soluble at intermediate pH values (Fig. 7-17). This amphoterism can be used to separate $Zn^{2+}$ from other cations that do not share the property. For example, $Mg^{2+}$ adds a maximum of two $OH^-$ ions to form $Mg(OH)_2$, a sparingly soluble hydroxide. Further addition of $OH^-$ does not lead to the formation of new complex ions. If a mixture of $Mg^{2+}$ and $Zn^{2+}$ ions is made sufficiently basic, the $Mg^{2+}$ precipitates as $Mg(OH)_2$, but the zinc remains in solution as $Zn(OH)_4^{2-}$, allowing the two to be separated. In the same way, aluminum is separated from iron industrially by dissolving solid $Al(OH)_3$ in strong base as $Al(OH)_4^-(aq)$ while $Fe(OH)_3$ remains as a precipitate.

**Figure 7-17**

Zinc hydroxide is insoluble in water (*center*) but dissolves readily in acid (*left*) and base (*right*). The indicator used is bromocresol red, which turns from red to yellow in acidic solution. (Leon Lewandowski)

## Carbonate Minerals

The carbonates are among the most abundant and important minerals in the earth's crust. When these minerals come into contact with fresh water or seawater, solubility equilibria are established that greatly affect the chemistry of the natural waters. Calcium carbonate ($CaCO_3$), the most important natural carbonate, makes up limestone and other forms of rock such as marble. Other carbonate minerals include dolomite, $CaMg(CO_3)_2$, and magnesite, $MgCO_3$. These compounds are sufficiently soluble that their solutions are nonideal, and so calculations based on solubility product expressions are only approximate.

**PRACTICE EXERCISE**

(a) The rare mineral nesquehonite contains $MgCO_3$ together with water of hydration. A sample containing 21.7 g of nesquehonite is acidified and heated, and the volume of $CO_2(g)$ produced is measured to be 3.51 L at 0°C and P = 1 atm. Assuming all the carbonate has reacted to form $CO_2$, give the chemical formula for nesquehonite.

(b) Write a chemical equation and a solubility product expression for the dissolution of dolomite in water.

Carbonate minerals. Calcite (*left*) and aragonite (*middle*) are both $CaCO_3$, and smithsonite (*right*) is $ZnCO_3$. (Julius Weber)

(c) In a sufficiently basic solution, the carbonate ion does not react significantly with water to form hydrogen carbonate ion. Calculate the solubility (in grams per liter) of limestone (calcium carbonate) in a 0.10 M solution of sodium hydroxide. Use the $K_{sp}$ from Table 7-2.

(d) In a strongly basic 0.10 M solution of $Na_2CO_3$, the concentration of $CO_3^{2-}$ is 0.10 M. What is the gram solubility of limestone in this solution? Compare your answer with that for part (c).

(e) In a mountain lake having a pH of 8.1, the total concentration of carbonate species, $[CO_3^{2-}] + [HCO_3^-]$, is measured to be $9.6 \times 10^{-4}$ M, whereas the concentration of $Ca^{2+}$ is $3.8 \times 10^{-4}$ M. Calculate the concentration of $CO_3^{2-}$ in this lake, using $K_a = 4.8 \times 10^{-11}$ for the acid ionization of $HCO_3^-$ to $CO_3^{2-}$. Is the lake unsaturated, saturated, or supersaturated with respect to $CaCO_3$?

(f) Will acid rainfall into the lake increase or decrease the solubility of limestone rocks in the lake's bed?

(g) Seawater contains a high concentration of $Cl^-$ ions, which form weak complexes such as $CaCl^+$ with calcium ions. Does the presence of such complexes increase or decrease the equilibrium solubility of $CaCO_3$ in seawater?

**Answers**

(a) $MgCO_3 \cdot 3H_2O$

(b) $CaMg(CO_3)_2(s) \rightleftarrows Ca^{2+}(aq) + Mg^{2+}(aq) + 2\,CO_3^{2-}(aq)$
$[Ca^{2+}][Mg^{2+}][CO_3^{2-}]^2 = K_{sp}$

(c) $9.3 \times 10^{-3}$ g $L^{-1}$

(d) $8.7 \times 10^{-6}$ g $L^{-1}$, smaller than in part (c) because of the common-ion effect

(e) $5.8 \times 10^{-6}$ M. $Q = 2.2 \times 10^{-9} < K_{sp} = 8.7 \times 10^{-9}$, and so the lake is slightly less than saturated.

(f) Increase

(g) Increase

## CONCEPTS & SKILLS

After studying this chapter and working the problems that follow, you should be able to

1. Discuss the dynamical processes that lead to solubility equilibria (Section 7-1).

2. Relate the solubilities of sparingly soluble salts in water to their solubility product constants (Section 7-2, problems 7–16).

3. Use the reaction quotient to predict whether a precipitate will form when two solutions are mixed, and then calculate the equilibrium concentrations that will result (Section 7-3, problems 17–24).
4. Calculate the solubility of a sparingly soluble salt in a solution that contains a given concentration of a common ion (Section 7-3, problems 25–30).
5. Determine the dependence on pH of the solubility of a salt of a weak base (Section 7-4, problems 33–34).
6. Specify the optimum conditions for the separation of two elements on the basis of the differing solubilities of their ionic compounds (Section 7-5, problems 35–42).
7. Calculate the concentrations of molecular and ionic species in equilibrium with complex ions (Section 7-6, problems 43–46).
8. Determine the effect of complex ion formation on the solubility of sparingly soluble salts involving a common cation (Section 7-6, problems 47–48).
9. Calculate the pH of aqueous solutions containing metal cations (Section 7-6, problems 49–54).

## PROBLEMS

### The Nature of Solubility Equilibria

1. Gypsum has the formula $CaSO_4 \cdot 2H_2O$. Plaster of paris has the chemical formula $CaSO_4 \cdot \frac{1}{2}H_2O$. In making wall plaster, water is added to plaster of paris, and the mixture then hardens into solid gypsum. How much water (in liters, at a density of $1.00 \text{ kg L}^{-1}$) should be added to 25.0 kg of plaster of paris to turn it into gypsum, assuming no loss from evaporation?
2. A 1.00-g sample of magnesium sulfate is dissolved in water, and the water is then evaporated away until the residue is bone dry. If the temperature of the water is kept between 48°C and 69°C, the solid that remains weighs 1.898 g. If the experiment is repeated with the temperature held between 69°C and 100°C, however, the solid has a mass of 1.150 g. Determine how many waters of crystallization per $MgSO_4$ there are in each of these two solids.
3. The following graph shows the solubility of KBr in water in

units of grams of KBr per 100 g of $H_2O$. If 80 g of KBr is added to 100 g of water at 10°C and the mixture is heated slowly, at what temperature will the last KBr dissolve?
4. Figure 7-3 shows the solubility of $AgNO_3$ in water in units of moles of $AgNO_3$ per kilogram of $H_2O$. If 255 g of $AgNO_3$ is added to 100 g of water at 95°C and cooled slowly, at what temperature will the solution become saturated?

### Ionic Equilibria

5. Iron(III) sulfate, $Fe_2(SO_4)_3$, is a yellow compound that is used as a coagulant in water treatment. Write a balanced chemical equation and a solubility product expression for its dissolution in water.
6. Lead antimonate, $Pb_3(SbO_4)_2$, is used as an orange pigment in oil-base paints and in glazes. Write a balanced chemical equation and a solubility product expression for its dissolution in water.
7. Thallium(I) iodate ($TlIO_3$) is only slightly soluble in water. Its $K_{sp}$ at 25°C is $3.07 \times 10^{-6}$. Estimate the solubility of thallium(I) iodate in water in units of grams per 100.0 mL of water.
8. Thallium thiocyanate (TlSCN) is only slightly soluble in water. Its $K_{sp}$ at 25°C is $1.82 \times 10^{-4}$. Estimate the solubility of thallium thiocyanate in units of grams per 100.0 mL of water.
9. Potassium perchlorate, $KClO_4$, has a $K_{sp}$ at 25°C of $1.07 \times 10^{-2}$. Compute its solubility in grams per liter of solution.
10. Ammonium hexachloroplatinate(IV), $(NH_4)_2(PtCl_6)$, is one of the few sparingly soluble ammonium salts. Its $K_{sp}$ at 20°C is $5.6 \times 10^{-6}$. Compute its solubility in grams per liter of solution.

11. The solubility product constant of mercury(I) iodide is $1.2 \times 10^{-28}$ at 25°C. Estimate the concentration of $Hg_2^{2+}$ and $I^-$ in equilibrium with solid $Hg_2I_2$.

12. The solubility product constant of $Hg_2Cl_2$ is $2 \times 10^{-18}$ at 25°C. Estimate the concentration of $Hg_2^{2+}$ and $Cl^-$ in equilibrium with solid $Hg_2Cl_2$ at 25°C.

13. The solubility of silver chromate ($Ag_2CrO_4$) in 500 mL of water at 25°C is 0.0129 g. Calculate its solubility product constant.

14. At 25°C, 400 mL of water can dissolve 0.00896 g of lead iodate, $Pb(IO_3)_2$. Calculate $K_{sp}$ for lead iodate.

15. At 100°C, water dissolves $1.8 \times 10^{-2}$ g of AgCl per liter. Compute the $K_{sp}$ of AgCl at this temperature.

16. A mass of 0.017 g of silver dichromate ($Ag_2Cr_2O_7$) will dissolve in 300 mL of water at 25°C. Calculate the solubility product constant $K_{sp}$ of silver dichromate.

## Precipitation and the Solubility Product

17. A solution of barium chromate ($BaCrO_4$) is prepared by dissolving $6.3 \times 10^{-3}$ g of this yellow solid in 1.00 L of hot water. Will solid barium chromate precipitate upon cooling to 25°C, according to the solubility product expression? Explain.

18. A solution is prepared by dissolving 0.090 g of $PbI_2$ in 1.00 L of hot water and cooling the solution to 25°C. Will solid precipitate result from this process, according to the solubility product expression? Explain.

19. A solution is prepared by mixing 250.0 mL of $2.0 \times 10^{-3}$ M $Ce(NO_3)_3$ and 150.0 mL of $1.0 \times 10^{-2}$ M $KIO_3$ at 25°C. Determine whether $Ce(IO_3)_3(s)$ ($K_{sp} = 1.9 \times 10^{-10}$) tends to precipitate from this mixture.

20. Suppose 100.0 mL of a 0.0010 M $CaCl_2$ solution is added to 50.0 mL of a $6.0 \times 10^{-5}$ M NaF solution at 25°C. Determine whether $CaF_2(s)$ ($K_{sp} = 3.9 \times 10^{-11}$) tends to precipitate from this mixture.

21. Suppose 50.0 mL of a 0.0500 M solution of $Pb(NO_3)_2$ is mixed with 40.0 mL of a 0.200 M solution of $NaIO_3$ at 25°C. Calculate the $[Pb^{2+}]$ and $[IO_3^-]$ when the mixture comes to equilibrium. At this temperature, $K_{sp}$ for $Pb(IO_3)_2$ is $2.6 \times 10^{-13}$.

22. Silver iodide (AgI) is used in place of silver chloride for the fastest photographic film because it is more sensitive to light, and so can form an image in a very short exposure time. A silver iodide emulsion is prepared by adding 6.60 L of 0.10 M NaI solution to 1.50 L of 0.080 M $AgNO_3$ solution at 25°C. Calculate the concentration of silver ion remaining in solution when the mixture comes to equilibrium, and its chemical amount relative to the amount present initially.

23. When 50.0 mL of 0.100 M $AgNO_3$ and 30.0 mL of 0.0600 M $Na_2CrO_4$ are mixed, a precipitate of silver chromate ($Ag_2CrO_4$) is formed. The solubility product $K_{sp}$ of silver chromate in water at 25°C is $1.9 \times 10^{-12}$. Calculate the $[Ag^+]$ and $[CrO_4^{2-}]$ remaining in solution at equilibrium.

24. When 40.0 mL of 0.0800 M $Sr(NO_3)_2$ and 80.0 mL of 0.0500 M KF are mixed, a precipitate of strontium fluoride ($SrF_2$) is formed. The solubility product $K_{sp}$ of strontium fluoride in water at 25°C is $2.8 \times 10^{-9}$. Calculate the $[Sr^{2+}]$ and $[F^-]$ remaining in solution at equilibrium.

25. Calculate the solubility (in mol $L^{-1}$) of $CaF_2(s)$ at 25°C in a 0.040 M aqueous solution of NaF.

26. Calculate the mass of AgCl that can dissolve in 100 mL of 0.150 M NaCl solution.

27. The solubility product of nickel(II) hydroxide, $Ni(OH)_2$, at 25°C is $K_{sp} = 1.6 \times 10^{-16}$.
(a) Calculate the molar solubility of $Ni(OH)_2$ in pure water at 25°C.
(b) Calculate the molar solubility of $Ni(OH)_2$ in 0.100 M NaOH.

28. Silver arsenate ($Ag_3AsO_4$) is a slightly soluble salt having a solubility product of $K_{sp} = 1.0 \times 10^{-22}$ at 25°C for the equilibrium

$$Ag_3AsO_4(s) \rightleftharpoons 3\,Ag^+(aq) + AsO_4^{3-}(aq)$$

(a) Calculate the molar solubility of silver arsenate in pure water at 25°C.
(b) Calculate the molar solubility of silver arsenate in 0.10 M $AgNO_3$.

29. A saturated solution of $Mg(OH)_2$ at 25°C is prepared by equilibrating solid $Mg(OH)_2$ with water. Concentrated NaOH is then added until the solubility of $Mg(OH)_2$ is 0.0010 times that in $H_2O$ alone. (Ignore the change in volume resulting from the addition of NaOH.) The solubility product $K_{sp}$ of $Mg(OH)_2$ is $1.2 \times 10^{-11}$ at 25°C. Calculate the concentration of hydroxide ion in the solution after the addition of the NaOH.

30. A saturated solution of $BaF_2$ at 25°C is prepared by equilibrating solid $BaF_2$ with water. Powdered NaF is then dissolved in the solution until the solubility of $BaF_2$ is 1.0% of that in $H_2O$ alone. The solubility product $K_{sp}$ of $BaF_2$ is $1.7 \times 10^{-6}$ at 25°C. Calculate the concentration of fluoride ion in the solution after addition of the powdered NaF.

## The Effects of pH on Solubility

31. Compare the molar solubility of AgOH in pure water with that in a solution buffered at pH 7.00. Note the difference between the two: when AgOH is dissolved in pure water, the pH does not remain at 7.

32. Compare the molar solubility of $Mg(OH)_2$ in pure water with that in a solution buffered at pH 9.00.

33. For each of the following ionic compounds, state whether the solubility will increase, decrease, or remain unchanged as a solution at pH 7 is made acidic.
(a) $PbI_2$
(b) AgOH
(c) $Ca_3(PO_4)_2$

34. For each of the following ionic compounds, state whether the

solubility will increase, decrease, or remain unchanged as a solution at pH 7 is made acidic.

(a) $SrCO_3$    (b) $Hg_2Br_2$    (c) $MnS$

## Selective Precipitation of Ions

**35.** An aqueous solution at 25°C is 0.10 M in both $Mg^{2+}$ and $Pb^{2+}$ ions. We wish to separate the two kinds of metal ions by taking advantage of the different solubilities of their oxalates, $MgC_2O_4$ and $PbC_2O_4$.

(a) What is the highest possible oxalate ion concentration that allows only one solid oxalate salt to be present at equilibrium? Which ion is present in the solid—$Mg^{2+}$ or $Pb^{2+}$?

(b) What fraction of the less soluble ion still remains in solution under the conditions of part (a)?

**36.** An aqueous solution at 25°C is 0.10 M in $Ba^{2+}$ and 0.50 M in $Ca^{2+}$ ions. We wish to separate the two by taking advantage of the different solubilities of their fluorides, $BaF_2$ and $CaF_2$.

(a) What is the highest possible fluoride ion concentration that allows only one solid fluoride salt to be present at equilibrium? Which ion is present in the solid—$Ba^{2+}$ or $Ca^{2+}$?

(b) What fraction of the less soluble ion still remains in solution under the conditions of part (a)?

**37.** The cations in an aqueous solution that contains 0.100 M $Hg_2(NO_3)_2$ and 0.0500 M $Pb(NO_3)_2$ are to be separated by taking advantage of the difference in the solubilities of their iodides. $K_{sp}(PbI_2) = 1.4 \times 10^{-8}$ and $K_{sp}(Hg_2I_2) = 1.2 \times 10^{-28}$. What should be the concentration of iodide ion for the best separation? In the "best" separation, one of the cations should remain entirely in solution and the other should precipitate as fully as possible.

**38.** The cations in an aqueous solution that contains 0.150 M $Ba(NO_3)_2$ and 0.0800 M $Ca(NO_3)_2$ are to be separated by taking advantage of the difference in the solubilities of their sulfates. $K_{sp}(BaSO_4) = 1.1 \times 10^{-10}$ and $K_{sp}(CaSO_4) = 2.4 \times 10^{-5}$. What should be the concentration of sulfate ion for the best separation?

**39.** Calculate the $[Zn^{2+}]$ in a solution that is in equilibrium with $ZnS(s)$ and in which $[H_3O^+] = 1.0 \times 10^{-5}$ M and $[H_2S] = 0.10$ M.

**40.** Calculate the $[Cd^{2+}]$ in a solution that is in equilibrium with $CdS(s)$ and in which $[H_3O^+] = 1.0 \times 10^{-3}$ M and $[H_2S] = 0.10$ M.

**41.** What is the highest pH at which 0.10 M $Fe^{2+}$ will remain entirely in a solution that is saturated with $H_2S$ at a concentration of $[H_2S] = 0.10$ M? At this pH, what would be the concentration of $Pb^{2+}$ in equilibrium with solid PbS in this solution?

**42.** What is the highest pH at which 0.050 M $Mn^{2+}$ will remain entirely in a solution that is saturated with $H_2S$ at a concentration of $[H_2S] = 0.10$ M? At this pH, what would be the concentration of $Cd^{2+}$ in equilibrium with solid CdS in this solution?

## Complex Ions and Solubility

**43.** Suppose 0.10 mol of $Cu(NO_3)_2$ and 1.50 mol of $NH_3$ are dissolved in water and diluted to a total volume of 1.00 L. Calculate the concentrations of $Cu(NH_3)_4^{2+}$ and of $Cu^{2+}$ at equilibrium.

**44.** The formation constant of the $TlCl_4^-$ complex ion is $1 \times 10^{18}$. Suppose 0.15 mol of $Tl(NO_3)_3$ is dissolved in 1.00 L of a 0.50 M solution of NaCl. Calculate the concentration at equilibrium of $TlCl_4^-$ and of $Tl^{3+}$.

**45.** The organic compound "18-crown-6" binds alkali metals in aqueous solution by wrapping around and enfolding the ion. It presents a niche that nicely accommodates the $K^+$ ion but is too small for $Rb^+$ ion and too large for the $Na^+$ ion. The values of the equilibrium constants show this:

$$Na^+(aq) + \text{18-crown-6}(aq) \rightleftharpoons Na\text{-crown}^+(aq)$$
$$K = 6.6$$

$$K^+(aq) + \text{18-crown-6}(aq) \rightleftharpoons K\text{-crown}^+(aq)$$
$$K = 111.6$$

$$Rb^+(aq) + \text{18-crown-6}(aq) \rightleftharpoons Rb\text{-crown}^+(aq)$$
$$K = 36$$

An aqueous solution is initially 0.0080 M in 18-crown-6(aq) and also 0.0080 M in $K^+(aq)$. Compute the equilibrium concentration of free $K^+$. ("Free" means not tied up with the 18-crown-6.) Compute the concentration of free $Na^+$ if the solution contains 0.0080 M $Na^+(aq)$ instead of $K^+(aq)$.

**46.** The organic compound 18-crown-6 (see preceding problem) also binds strongly with the alkali metal ions in methanol.

$$K^+ + \text{18-crown-6} \rightleftharpoons [\text{complex}]^+$$

In methanol solution, the equilibrium constant is $1.41 \times 10^6$. A similar reaction with $Cs^+$ has an equilibrium constant of only $2.75 \times 10^4$. A solution is made (in methanol) containing 0.020 mol $L^{-1}$ each of $K^+$ and $Cs^+$. It also contains 0.30 mol $L^{-1}$ of 18-crown-6. Compute the equilibrium concentrations of both the uncomplexed $K^+$ and the uncomplexed $Cs^+$.

**47.** Will silver chloride dissolve to a significantly greater extent in a 1.00 M NaCl solution than in pure water, due to the possible formation of $AgCl_2^-$ ions? Use data from Tables 7-2 and 7-4 to provide a quantitative answer to this question. What will happen in a 0.100 M NaCl solution?

**48.** Calculate how many grams of silver chloride will dissolve in 1.0 L of a 1.0 M $NH_3$ solution, through formation of the complex ion $Ag(NH_3)_2^+$.

**49.** The pH of a 0.2 M solution of $CuSO_4$ is 4.0. Write chemical equations to explain why a solution of this salt is neither basic [from the reaction of $SO_4^{2-}(aq)$ with water] nor neutral, but acidic.

**50.** Will a 0.05 M solution of $FeCl_3$ be acidic, basic, or neutral? Explain your answer by writing chemical equations to describe any reactions taking place.

**51.** The acid ionization constant for $Co(H_2O)_6^{2+}(aq)$ is $3 \times 10^{-10}$. Calculate the pH of a 0.10 M solution of $Co(NO_3)_2$.

**52.** The acid ionization constant for $Fe(H_2O)_6^{2+}(aq)$ is $3 \times 10^{-6}$. Calculate the pH of a 0.10 M solution of $Fe(NO_3)_2$, and compare it with the pH of the corresponding iron(III) nitrate solution from Example 7-10.

**53.** A 0.15 M aqueous solution of the chloride salt of the complex ion $Pt(NH_3)_4^{2+}$ is found to be weakly acidic with a pH of 4.92. This is initially puzzling because the $Cl^-$ ion in water is not acidic and $NH_3$ in water is *basic,* not acidic. Finally, it is suggested that the $Pt(NH_3)_4^{2+}$ ion as a group donates hydrogen ions. Compute the $K_a$ of this acid, assuming that just one hydrogen ion is donated.

**54.** The pH of a 0.10 M solution of $Ni(NO_3)_2$ is 5.0. Calculate the acid ionization constant of $Ni(H_2O)_6^{2+}(aq)$.

**\*55.** $K_{sp}$ for $Pb(OH)_2$ is $4.2 \times 10^{-15}$, and $K_f$ for $Pb(OH)_3^-$ is $4 \times 10^{14}$. Suppose a solution whose initial concentration of $Pb^{2+}(aq)$ is 1.00 M is brought to pH 13.0 by addition of solid NaOH. Will solid $Pb(OH)_2$ precipitate, or will the lead be dissolved as $Pb(OH)_3^-(aq)$? What will be $[Pb^{2+}]$ and $[Pb(OH)_3^-]$ at equilibrium? Repeat the calculation for an initial $Pb^{2+}$ concentration of 0.050 M. (*Hint:* One way to solve this problem is to *assume* that $Pb(OH)_2(s)$ is present and calculate $[Pb^{2+}]$ and $[Pb(OH)_3^-]$ that would be in equilibrium with the solid. If the sum of these is less than the original $[Pb^{2+}]$, the remainder can be assumed to have precipitated. If not, there is a contradiction and we must assume that *no* $Pb(OH)_2(s)$ is present. In this case we can calculate $[Pb^{2+}]$ and $[Pb(OH)_3^-]$ directly from $K_f$.)

**\*56.** $K_{sp}$ for $Zn(OH)_2$ is $4.5 \times 10^{-17}$, and $K_f$ for $Zn(OH)_4^{2-}$ is $5 \times 10^{14}$. Suppose a solution whose initial concentration of $Zn^{2+}(aq)$ is 0.010 M is brought to pH 14.0 by addition of solid NaOH. Will solid $Zn(OH)_2$ precipitate, or will the zinc be dissolved as $Zn(OH)_4^{2-}(aq)$? What will be $[Zn^{2+}]$ and $[Zn(OH)_4^{2-}]$ at equilibrium? Repeat the calculation at pH 13 for an initial $Zn^{2+}$ concentration of 0.10 M. See the hint in problem 55.

## Additional Problems

**57.** Write a chemical equation for the dissolution of mercury(I) chloride in water, and give its solubility product expression.

**\*58.** Magnesium ammonium phosphate has the formula $MgNH_4PO_4 \cdot 6H_2O$. It is only slightly soluble in water (its $K_{sp}$ is $2.3 \times 10^{-13}$). Write a chemical equation and the corresponding equilibrium law for the dissolution of this compound in water.

**59.** Soluble barium compounds are poisonous, but barium sulfate is routinely ingested as a suspended solid in a "barium cocktail" to improve the contrast in x-ray images. Calculate the concentration of dissolved barium per liter of water in equilibrium with solid barium sulfate.

**60.** A saturated aqueous solution of silver perchlorate ($AgClO_4$) contains 84.8% by mass $AgClO_4$, but a saturated solution of $AgClO_4$ in 60% aqueous perchloric acid contains only 5.63%

by mass $AgClO_4$. Explain this large difference, using chemical equations.

**61.** Suppose 140 mL of 0.0010 M $Sr(NO_3)_2$ is mixed with enough 0.0050 M NaF to make 1.00 L of solution. Will $SrF_2(s)$ ($K_{sp} = 2.8 \times 10^{-9}$) precipitate at equilibrium? Explain.

**62.** The concentration of calcium ion in a town's supply of drinking water is 0.0020 M. (This water is referred to as hard water because it contains such a large concentration of $Ca^{2+}$.) Suppose the water is to be fluoridated by the addition of NaF for the purpose of reducing tooth decay. What is the maximum concentration of fluoride ion that can be achieved in the water before precipitation of $CaF_2$ begins? Will the water supply attain the level of fluoride ion recommended by the U.S. Public Health Service, about $5 \times 10^{-5}$ M (1 mg fluorine per liter)?

**63.** The two solids $CuBr(s)$ and $AgBr(s)$ are only very slightly soluble in water. $K_{sp}(CuBr) = 4.2 \times 10^{-8}$ and $K_{sp}(AgBr) = 7.7 \times 10^{-13}$. Some $CuBr(s)$ and $AgBr(s)$ are both mixed into a quantity of water that is then stirred until it is saturated with respect to both solutes. Next, a small amount of KBr is added and dissolves completely. Compute the ratio of $[Cu^+]$ to $[Ag^+]$ after the system reestablishes equilibrium.

**\*64.** The two salts $BaCl_2$ and $Ag_2SO_4$ are both far more soluble in water than either $BaSO_4$ ($K_{sp} = 1.1 \times 10^{-10}$) or $AgCl$ ($K_{sp} = 1.6 \times 10^{-10}$) at 25°C. Suppose 50.0 mL of 0.040 M $BaCl_2(aq)$ is added to 50.0 mL of 0.020 M $Ag_2SO_4(aq)$. Calculate the concentrations of $SO_4^{2-}(aq)$, $Cl^-(aq)$, $Ba^{2+}(aq)$, and $Ag^+(aq)$ that remain in solution at equilibrium.

**65.** The Mohr method is a technique for determining the amount of chloride ion in an unknown sample. It is based on the difference in solubility between silver chloride ($AgCl$; $K_{sp} = 1.6 \times 10^{-10}$) and silver chromate ($Ag_2CrO_4$; $K_{sp} = 1.9 \times 10^{-12}$). In using this method, one adds a small amount of chromate ion to a solution with unknown chloride concentration. By measuring the volume of $AgNO_3$ added before the appearance of the red silver chromate, one can determine the amount of $Cl^-$ originally present. Suppose we have a solution that is 0.100 M in $Cl^-$ and 0.00250 M in $CrO_4^{2-}$. If we add 0.100 M $AgNO_3$ solution drop by drop, will $AgCl$ or $Ag_2CrO_4$ precipitate first? When $Ag_2CrO_4(s)$ first appears, what fraction of the $Cl^-$ that was originally present remains in solution?

**66.** Oxide ion, like sulfide ion, is a strong base. Write an equation for the dissolution of CaO in water and give its equilibrium constant expression. Write the corresponding equation for the dissolution of CaO in an aqueous solution of a strong acid, and relate its equilibrium constant to the previous one.

**67.** Water that has been saturated with magnesia (MgO) at 25°C has a pH of 10.16. Write a balanced chemical equation for the equilibrium between $MgO(s)$ and the ions it furnishes in aqueous solution, and calculate the equilibrium constant at 25°C. What is the solubility, in moles per liter, of MgO in water?

**68.** To 1.00 L of a 0.100 M $AgNO_3$ solution is added an excess of sodium chloride. Then 1.00 L of 0.500 M $NH_3(aq)$ is added. Finally, sufficient nitric acid is added until the pH of the resulting solution is 1.0. Write balanced equations for the reactions that take place (if any) at each of the three steps in this process.

**69.** Only about 0.16 mg of AgBr(s) will dissolve in 1.0 L of water (this volume of solid is smaller than the head of a pin). In a solution of ammonia that contains 0.10 mol ammonia per liter of water, there are about 555 water molecules for every molecule of ammonia. However, more than 400 times as much AgBr (68 mg) will dissolve in this solution as in plain water. Explain how such a tiny change in the composition of the solution can have such a large effect on the solubility of AgBr.

**\*70.** (a) Calculate the solubility of calcium oxalate ($CaC_2O_4$) in 1.0 M oxalic acid ($H_2C_2O_4$) at 25°C, using the two acid ionization constants for oxalic acid from Table 6-2 and the solubility product $K_{sp} = 2.6 \times 10^{-9}$ for $CaC_2O_4$.
(b) Calculate the solubility of calcium oxalate in pure water at 25°C.
(c) Account for the difference between the results of (a) and (b).

**\*71.** When 6 M HCl is added to solid CdS, some of the solid dissolves to give the complex ion $CdCl_4^{2-}(aq)$.
(a) Write a balanced equation for the reaction that occurs.
(b) Use data from Tables 6-2 and 7-3 and the formation constant of $CdCl_4^{2-}$ ($K_f = 8 \times 10^2$) to calculate the equilibrium constant for the reaction of part (a).
(c) What is the molar solubility of CdS per liter of 6 M HCl?

**\*72.** Using data from Table 7-4, calculate the concentrations of $Hg^{2+}(aq)$, $HgCl^+(aq)$, and $HgCl_2(aq)$ that result when 1.00 L of a 0.100 M $Hg(NO_3)_2$ solution is mixed with an equal volume of a 0.100 M $HgCl_2$ solution. (*Hint:* Use the analogy with amphoteric equilibria discussed in Section 6-7.)

**\*73.** Calculate the concentration of $Cu^{2+}(aq)$ in a solution that contains 0.020 mol of $CuCl_2$ and 0.100 mol of NaCN in 1.0 L.

$$Cu^{2+}(aq) + 4\,CN^-(aq) \rightleftharpoons Cu(CN)_4^{2-}(aq)$$
$$K = 2.0 \times 10^{30}$$

(*Hint:* Do not overlook the reaction of $CN^-$ with water to give HCN.)

**74.** An aqueous solution of $K_2[Pt(OH)_6]$ has a pH greater than 7. Explain this fact by writing an equation showing the $Pt(OH)_6^{2-}$ ion acting as a Brønsted–Lowry base and accepting a hydrogen ion from water.

**75.** In Example 7-10 we included only the first acid dissociation $K_{a1}$ of a 0.100 M aqueous solution of $Fe(H_2O)_6^{3+}$. Subsequent dissociation can also occur, with $K_{a2} = 2.0 \times 10^{-5}$, to give $Fe(H_2O)_4(OH)_2^+$.
(a) Calculate the concentration of $Fe(H_2O)_4(OH)_2^+$ at equilibrium. Does the pH change significantly when this second dissociation is taken into account?
(b) We can describe the same reaction as the dissociation of a complex ion $Fe(OH)_2^+$ to $Fe^{3+}$ and two $OH^-$ ions. Calculate $K_f$, the formation constant for $Fe(OH)_2^+$.

**76.** The volume of a certain saturated solution is greater than the sum of the volumes of the water and salt from which it is made. Predict the effect of increased pressure on the solubility of this salt.

**77.** Codeine has the molecular formula $C_{18}H_{21}NO_3$. It is soluble in water to the extent of 1.00 g per 120 mL of water at room temperature and 1.00 g per 60 mL of water at 80°C. Compute the molal solubility (in mol $kg^{-1}$) at both temperatures, taking the density of water to be fixed at 1.00 g $cm^{-3}$. Is the dissolution of codeine in water endothermic or exothermic?

**78.** Suppose 1.44 L of a saturated solution of strontium carbonate ($SrCO_3$) in boiling water at 100°C is prepared. The solution is then strongly acidified and shaken to drive off all the gaseous

$CO_2$ that forms. The volume of this gas (at a temperature of 100°C and a partial pressure of 0.972 atm) is measured to be 0.20 L (200 mL).
(a) Calculate the molar solubility of $SrCO_3$ in water at 100°C.
(b) Estimate the solubility product constant $K_{sp}$ of $SrCO_3$ at this temperature.
(c) Explain why the actual $K_{sp}$ at this temperature may be lower than you predicted in part (b).

**79.** A buffer is prepared by adding 50.0 mL of 0.15 M $HNO_3(aq)$ to 100.0 mL of 0.12 M NaHCOO(aq) (sodium formate). Calculate the solubility of $CaF_2(s)$ in this solution.

8

*The reaction of sodium with water is highly exothermic.*

# Thermodynamic Processes and Thermochemistry

T hermodynamics, in which a few apparently simple laws summarize a rich variety of observed behavior, is one of the surest and most powerful conceptual tools we possess in science. The central feature of thermodynamics is the universality of its basic laws, and the beauty of the subject is that so many physical conclusions can be deduced from those few laws. The laws of thermodynamics cannot themselves be derived or proved. Instead, they are generalizations that have been drawn from a vast number of observations of the behavior of matter. The history of thermodynamics, like those of many fields of science, has been fraught with misconceptions. As we look back upon the beginnings of the discipline in the 19th century, it appears to have been agonizingly slow to develop. But it *has* developed, and its laws are the pillars upon which much of modern science rests. Most of thermodynamics is in finished form today, but it is still being applied to research in the forefront of science, on systems ranging from black holes in distant parts of the universe to the growth and development of the biological cell.

Thermodynamics is an *operational* science, concerned with macroscopic properties that, in principle, can be measured. Its goal is to predict what types of chemical and physical processes are possible and under what conditions, and to calculate quantitatively the properties of the equilibrium state that ensues when a process is carried out. For example, with thermodynamics the following kinds of questions can be raised.

1. If hydrogen and nitrogen are mixed, will they tend to react? If so, in what percentage yield will they produce ammonia?
2. What effect will a particular change in temperature or pressure have on the extent of their reaction?
3. How can the conditions for a process be optimized so as to maximize its efficiency?

Thermodynamics is an immensely practical subject. The knowledge that a particular chemical process is impossible under certain conditions can save a great deal of time spent vainly trying to make it occur. Thermodynamics can also suggest ways in which conditions can be changed to make a process possible.

The power of thermodynamics lies in its generality: it rests upon no particular model of the structure of matter. In fact, if the entire atomic theory of matter were to be overthrown (a *very* unlikely event!), the foundations of thermodynamics would be unshaken. At the same time, thermodynamics has some important limitations. It is a tool that enables the prediction of certain macroscopic properties, but it cannot explain why a substance has a particular property. It can determine whether a process can occur, but it cannot say how rapidly it will occur. Thermodynamic arguments show that diamond is an unstable substance at atmospheric pressure and will eventually revert to graphite, but they cannot state how long the process will take.

## 8-1 SYSTEMS, STATES, AND PROCESSES

Terms that arise in thermodynamics must be defined precisely. Many apparent contradictions can be traced to loose or inaccurate use of language. Some terms have everyday meanings that are different from their thermodynamic usage. This section contains a brief introduction to the language of thermodynamics.

A **system** is a real or imaginary portion of the universe that is confined by certain boundaries or mathematical constraints. For example, the gas confined in a closed box clearly can be considered a system, but the gas in a particular cubic centimeter of space in the middle of a room may also be considered a system. In the first case the boundaries are physical ones; in the second they are conceptual or mathematical. The rest of the universe that lies outside of a system is called the **surroundings.** A **closed system** is one whose boundaries prevent the flow of matter

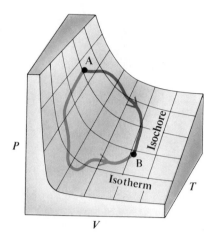

**Figure 8-1**

The *PVT* surface of 1 mol of ideal gas. Each point on the surface represents a combination of pressure, volume, and temperature allowed by the equation of state of the gas. Along an isotherm (*T* constant), the pressure varies inversely with volume; along an isochore (*V* constant), it varies linearly with temperature. Two processes are shown connecting states A and B along paths that satisfy the equation of state at every point.

**Figure 8-2**

Stages in an irreversible expansion of a gas from an initial state (*a*) of volume $V_1$ to a final state (*c*) of volume $V_2$. In the intermediate stage shown (*b*), the gas is not in equilibrium; because of turbulence, pressure and temperature cannot be defined.

into and out of it (they are impermeable), whereas in an **open system** the boundaries permit such flow. The amount of matter in an open system can change with time.

Thermodynamics is concerned with macroscopic properties of systems and the ways in which they change. Such properties are of two kinds, *extensive* and *intensive*. To distinguish between them, consider the following "thought experiment." Place a thin wall through the middle of a system, dividing it into two subsystems each of which, like the system itself, is characterized by certain properties. An **extensive property** of the system is one that can be written as the *sum* of the corresponding properties of the subsystems. Volume, mass, and energy are typical extensive properties; the volume of a system, for example, is the sum of the volumes of its subsystems. An **intensive property** of the system is one that is the *same* as the corresponding properties of each of the subsystems. Temperature and pressure are typical intensive properties; if a system at 298 K is divided in half, the temperature of each half will still be 298 K.

A **thermodynamic state** is a macroscopic condition of a system whose properties are *uniquely determined* and *independent of time.* A system characterized by a particular thermodynamic state is said to have reached **equilibrium.** As was stated in Chapters 4 and 5, the same equilibrium state can be reached from different directions. The thermodynamic state of a system composed of a given quantity of substance (pure by definition) is fixed when any two of its independent properties are given. Thus, the specification of *P* and *T* for 1 mol of a gas fixes not merely the volume *V* but all other properties of the material, such as the energy *E*. These relationships can be indicated by three-dimensional plots of any property as a function of two independent properties. For 1 mol of an ideal gas, a plot can be made of pressure against volume and temperature, for example (Fig. 8-1). The resulting surface is the equation of state of the substance (see Section 3-7). The points on such a surface (A, B, . . .) represent thermodynamic states of the system characterized by particular values of *P*, *V*, and *T* and therefore of all other macroscopic properties.

A **thermodynamic process** leads to a change in the thermodynamic state of a system. Such a process may be a *physical* process (e.g., a change in the pressure of a gaseous system or the boiling of a liquid) or a *chemical* process, in which there is a change in the distribution of matter among different chemical species (e.g., the reaction of solid $CaCO_3$, at 900 K and 1 atm pressure, to give solid $CaO$ and gaseous $CO_2$ at the same temperature and pressure).

Many conditions of a system do not correspond to any thermodynamic state. For example, suppose a gas is confined by a piston in a cylinder with volume $V_1$ (thermodynamic state A). If the piston is abruptly pulled out to increase the volume to $V_2$ (Fig. 8-2), chaotic gas currents arise as the molecules begin to move into the larger volume. These intermediate stages are not thermodynamic states, because such properties as density and temperature are changing rapidly both through space and in time. Eventually the currents die out and the system approaches a new

(*a*)

(*b*)

(*c*)

— Piston

thermodynamic state, B. States A and B are thermodynamic states, but the conditions in between cannot be described by only a few macroscopic variables and so are not thermodynamic states. Such a process is called **irreversible;** it cannot be represented as a path on a thermodynamic surface (as in Fig. 8-1), because the intermediate stages do not correspond to points on the equation-of-state surface.

By contrast, a **reversible process** is one that proceeds through a continuous series of thermodynamic states and thus can be indicated as a path on the equation-of-state surface. Such a process is an idealization, because true equilibrium is reached only after an infinite length of time and therefore such a process could never occur in a finite time. However, if the process is carried out slowly enough and in small enough steps, the real (irreversible) process can be considered an approximation to the limiting reversible process. There is an infinite number of reversible paths between any two thermodynamic states A and B; two of them are shown in Figure 8-1. The term "reversible" is used because an infinitesimal change in external conditions suffices to reverse the direction of motion of the system. For example, if a gas is expanded by pulling out a piston very slowly, only a tiny change in the forces exerted from the outside is required to change the direction of motion of the piston and begin to compress the gas.

Certain properties of a system, called **state functions,** are uniquely determined by the thermodynamic state of the system. Volume, temperature, and pressure are examples of state functions, and it will shortly be seen that the energy $E$ is also a state function. The Greek letter $\Delta$ (delta) is used to indicate *changes* in state functions in a thermodynamic process. Thus, $\Delta V = V_{\text{final}} - V_{\text{initial}}$ (or $V_f - V_i$) is the change in volume between initial and final states, and $\Delta E = E_{\text{final}} - E_{\text{initial}}$ is the corresponding change in energy (note that these changes are taken to be the final-state property minus the initial-state property). Because $\Delta E$ (or $\Delta V$ or $\Delta T$) depends only on the initial and final states, the same change, $\Delta E$, will be measured no matter which path (reversible or irreversible) is followed between any given pair of thermodynamic states; *the change in any state function is independent of path.* The converse statement is also true: if the change in a property of a system is independent of path, the property is a state function. Figure 8-3 illustrates two different paths that connect a given initial state and a common final state.

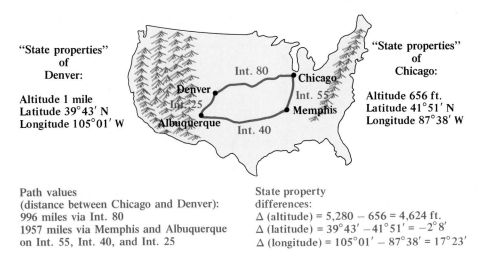

"State properties"
of
Denver:

Altitude 1 mile
Latitude 39°43′ N
Longitude 105°01′ W

"State properties"
of
Chicago:

Altitude 656 ft.
Latitude 41°51′ N
Longitude 87°38′ W

**Figure 8-3**

Differences in state properties (such as the difference in altitude between two points) are independent of the path followed. Other properties (such as the total distance traveled) depend on the particular path.

Path values
(distance between Chicago and Denver):
996 miles via Int. 80
1957 miles via Memphis and Albuquerque
on Int. 55, Int. 40, and Int. 25

State property
differences:
$\Delta$ (altitude) $= 5{,}280 - 656 = 4{,}624$ ft.
$\Delta$ (latitude) $= 39°43′ - 41°51′ = -2°8′$
$\Delta$ (longitude) $= 105°01′ - 87°38′ = 17°23′$

## 8-2 ENERGY, WORK, AND HEAT

The first law of thermodynamics, which is stated in the next section, relates the energy change in a thermodynamic process to the amount of work done on the system and the amount of heat transferred to the system. To understand the significance of this law, it is first necessary to examine the ways in which amounts of heat and work are measured. You will see that heat and work are simply different means by which energy is transferred into or out of a system.

### Work

The mechanical definition of **work** is the product of the external force on a body times the distance through which the force acts. If a body moves in a straight line from point $r_i$ to $r_f$ with a constant force $F$ applied along the direction of the path, the work done on the body is

$$w = F(r_f - r_i) \qquad \text{(force along direction of path)}$$

What is the relationship between work and energy in mechanical systems? Suppose a block of mass $M$ is moving with initial velocity $v_i$ along a frictionless surface. If a constant force $F$ is exerted on it in the direction of its motion, it will experience a constant acceleration $a = F/M$. After a time $t$, the block's velocity will have increased from $v_i$ to $v_f$ and its position will have changed from $r_i$ to $r_f$. The work done on the block is

$$w = F(r_f - r_i) = Ma(r_f - r_i)$$

The distance traveled, $r_f - r_i$, is given by the average velocity, in this case $(v_i + v_f)/2$, multiplied by the elapsed time $t$:

$$r_f - r_i = \left(\frac{v_i + v_f}{2}\right)t$$

When the acceleration is constant, it is equal to the change in velocity, $v_f - v_i$, divided by the elapsed time:

$$a = (v_f - v_i)/t$$

Substituting both of these results into the expression for the work done,

$$w = M\left(\frac{v_f - v_i}{t}\right)\left(\frac{v_i + v_f}{2}\right)t$$

$$= \frac{M}{2}(v_f - v_i)(v_f + v_i)$$

$$= \frac{M}{2}v_f^2 - \frac{M}{2}v_i^2$$

The expression on the right side is the change in kinetic energy, $\frac{1}{2}Mv^2$, of the block. For this idealized example with a frictionless surface, the work done is equal to the change in energy (in this case, kinetic) of the block.

As a second example, consider the work done in lifting an object in a gravitational field. To raise a mass $M$ from an initial height $h_i$ to a final height $h_f$, an upward

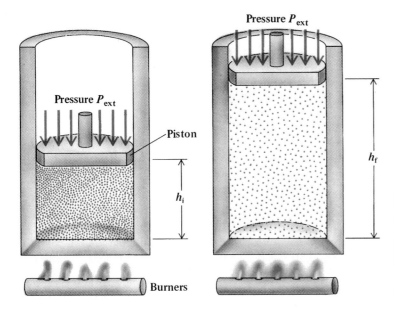

**Figure 8-4**

As the gas inside this cylinder is heated, it expands, pushing the piston against the pressure $P_{ext}$ exerted by the gas outside. As the piston is displaced over a distance $h_f - h_i = \Delta h$, the volume of the cylinder increases by an amount $A\,\Delta h$, where $A$ is the surface area of the piston.

force sufficient to counteract the downward force of gravity $Mg$ must be exerted. The work done on the object in this case is

$$w = Mg(h_f - h_i) = Mg\,\Delta h$$

This is the change in *potential* energy $Mgh$ of the object, showing once again a connection between mechanical work done and a change in energy.

One important kind of mechanical work in chemistry is **pressure–volume work,** which results when a gas is compressed or expanded under the influence of an outside pressure, such as occurs in the internal combustion engine of an automobile. Imagine that a gas has pressure $P_i$ and is confined in a cylinder by a frictionless piston of cross-sectional area $A$ and negligible mass (Fig. 8-4). The force exerted on the piston face by the gas is $F_i = P_i A$, because pressure is force divided by area. If there is a gas on the other side of the piston with pressure $P_{ext}$ ("ext" for "external"), then if $P_{ext} = P_i$ the piston will experience no net force; if $P_{ext}$ is increased, the gas will be compressed, and if it is decreased, the gas will expand. Consider first the case in which the external force is less than the initial force exerted by the gas, $P_i A$. Then the gas will expand and lift the piston from $h_i$ to $h_f$. The work in this case is

$$w = -F_{ext}(h_f - h_i)$$

The negative sign is inserted because the force from the gas outside opposes the expansion of the gas inside the cylinder. This is rewritten as

$$w = -P_{ext}A\,\Delta h$$

The product $A\,\Delta h$ is the volume change of the system, $\Delta V$, so the work is

$$w = -P_{ext}\,\Delta V \qquad \textbf{(8-1)}$$

For an expansion, $\Delta V > 0$, so $w < 0$ and the system does work; it pushes back the surroundings. If the gas had been compressed (by making $P_{ext}$ larger than $P_i$), work would have been done *on* the system. As before, $w = -P_{ext} \Delta V$, but now $\Delta V < 0$, so $w > 0$. If there is no volume change, $\Delta V = 0$ and no pressure–volume work is done. Finally, if there is no mechanical link to the surroundings (that is, if $P_{ext} = 0$), then once again no pressure–volume work can be performed.

If the pressure $P_{ext}$ is expressed in pascals and the volume in cubic meters, their product is in joules (J). These are the SI units. For many purposes, however, it is more convenient to express pressures in atmospheres and volumes in liters, resulting in liter-atmospheres (L atm) as a unit of work. The two work units are related by

$$1 \text{ L atm} = (10^{-3} \text{ m}^3)(1.01325 \times 10^5 \text{ kg m}^{-1} \text{ s}^{-2}) = 101.325 \text{ J}$$

## Example 8-1

A cylinder confines 2.00 L of gas under a pressure of 1.00 atm. The external pressure is also 1.00 atm. The gas is heated slowly, with the piston sliding freely to maintain the pressure of the gas close to 1.00 atm. Suppose the heating continues until a final volume of 3.50 L is reached. Calculate the work done on the gas, and express it in joules.

### Solution

This is an expansion of a system from 2.00 L to 3.50 L against a constant external pressure of 1.00 atm. The work done on the system is then

$$w = -P_{ext} \Delta V = -(1.00 \text{ atm})(3.50 \text{ L} - 2.00 \text{ L}) = -1.50 \text{ L atm}$$

Conversion to joules gives

$$w = (-1.50 \text{ L atm})(101.325 \text{ J L}^{-1} \text{ atm}^{-1}) = -152 \text{ J}$$

Because $w$ is negative, we can say either that $-152$ J of work is done *on* the gas or that $+152$ J of work is done *by* the gas.

*Related Problems:* 1, 2

## Heat

So far, two types of energy have been considered: the kinetic energy of a moving object and the potential energy of an object in a gravitational field. A third type of energy, less visible but equally important, is the *internal* energy of an object due to the relative motions and positions of the atoms that compose it (Fig. 8-5). Gas molecules are in a constant state of motion (Section 3-5) even when no overall gas flow is taking place in the container; the same is true of molecules in liquids and solids. If a piece of hot metal is plunged into a container of water, the temperature of the water rises as its molecules begin to move faster, corresponding to an increase in the internal energy of the water. The act of transferring this type of energy between objects is called **heat.** When a hot body is brought into contact with a colder body, the two temperatures change until they become equal. This process is sometimes described as the ''flow'' of heat from the hotter to the colder body. Although this picture is useful, it is somewhat misleading because it implies that heat is a substance that is contained in matter. Instead, heat (like work) is a way in which energy is exchanged between a system and its surroundings.

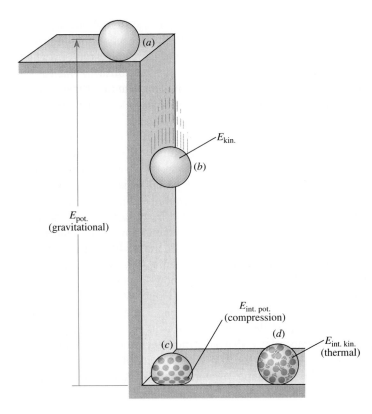

**Figure 8-5**

A ball being dropped from a height exhibits four kinds of energy. (*a*) It begins with a high *potential energy* associated with its attraction to the earth's surface through the force of gravity. (*b*) As it falls, this potential energy is converted to *kinetic energy* of motion of the ball. (*c*) After impact with the ground, the molecules near the surface of the ball are pushed against one another, increasing the *internal potential energy* of the ball. (*d*) As the ball stops bouncing, the molecules readjust their positions, but they move a little faster. The ball has a higher *internal kinetic energy* and a higher temperature now than just before impact with the ground.

How can amounts of heat be measured? One simple way is to use an **ice calorimeter** (Fig. 8-6), which consists of a bath containing ice and water, well insulated to prevent heat transfer to the surroundings. If heat is transferred to the bath from the system, some of the ice melts. Because a given mass of water has a smaller volume than the same mass of ice, the total volume of the ice–water mixture decreases as heat enters. If twice as much heat is transferred, twice as much ice will melt and the volume change will be twice as great. The amount of heat transferred is determined from the change in volume of the contents of the calorimeter. Heat transferred from the ice bath *into* the system causes water to freeze and *increases* the total volume of the ice bath.

An alternative way to measure heat is to take advantage of the fact that if heat is transferred to or removed from a substance in a single phase at constant pressure, the temperature changes in a reproducible way. The **specific heat capacity** of a material is the amount of heat required to raise the temperature of a one-gram mass by 1°C. If twice as much heat is transferred, twice as great a temperature change will be seen (provided the specific heat capacity itself does not change appreciably with temperature). The temperature change of a fixed amount of a given substance can thus be used as a measure of the quantity of heat transferred to or from it. This is described by

$$q = Mc_s \, \Delta T \qquad (8\text{-}2)$$

where $q$ is the heat transferred to a body of mass $M$ with specific heat capacity $c_s$ in order to cause a temperature change of $\Delta T$.

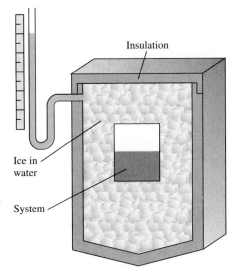

**Figure 8-6**

An ice calorimeter. As the ice melts, the volume of the ice–water mixture decreases, an effect that can be read off the scale on the left.

**Figure 8-7**

The falling weight turns a paddle that does work on the system (the water), causing an increase in its temperature.

Because heat, like work, is a form in which energy is transferred, the appropriate unit for it is also the joule. Historically, however, the connections among work, heat, and energy were not appreciated until the middle of the 19th century, by which time a separate unit for heat, the calorie, was already well established. One calorie was defined as the amount of heat required to raise the temperature of one gram of water from 14.5°C to 15.5°C (or, in other words, the specific heat capacity of water, $c_s$, at 15°C was *defined* as 1.00 cal $K^{-1}$ $g^{-1}$).

The qualitative equivalence of heat and work as means of energy transfer was suggested in 1798 by Benjamin Thompson, Count Rumford. In the course of his work as military advisor to the King of Bavaria, Thompson observed that the quantity of heat produced in the boring of cannons was proportional to the amount of work done in the process. Moreover, the operation could be continued indefinitely, demonstrating that heat was not a substance contained in the metal of the cannon. More quantitative measurements were carried out by the German physician Julius Mayer and by the English physicist James Joule. In the 1840s these scientists showed that the temperature of a substance could be raised by doing work on the substance as well as by adding heat to it. Figure 8-7 shows an apparatus in which a paddle, driven by a falling weight, churns the water in a tank. Work is performed on the water, and the temperature increases. The work done is $-Mg\,\Delta h$, where $\Delta h$ is the (negative) change in the height of the weight and $M$ is its mass. If all of this work goes to increase the water temperature (that is, if no heat enters the container or leaks out to the surroundings), then the specific heat capacity of the water in *joules* per gram per degree is equal to the quantity of work done divided by the product of the mass of water and its temperature rise.

### Example 8-2

Suppose a 10.00-kg mass drops through a height difference of 3.00 m, and the resulting work is used to turn a paddle in 200.0 g of water, initially at 15.00°C. The final water temperature is found to be 15.35°C. Assuming that the work done is used entirely to raise the water temperature, calculate the conversion factor between joules and calories.

### Solution

The total work done is

$$w = -Mg\,\Delta h = -(10.00\text{ kg})(9.806\text{ m s}^{-2})(-3.00\text{ m}) = 294\text{ J}$$

The heat (in calories) required to raise the water temperature by the same amount is

$$q = Mc_s\,\Delta T = (200.0\text{ g})(1.000\text{ cal K}^{-1}\text{ g}^{-1})(0.35\text{ K}) = 70\text{ cal}$$

Because the work done has the same effect on the water as direct transfer of heat, these two expressions can be set equal to each other, giving

$$70\text{ calories} = 294\text{ joules}$$

$$1\text{ calorie} \approx 4.2\text{ joules}$$

*Related Problems:* 3, 4

These and other experiments eliminate the need for the calorie as an independent unit, and the calorie is now *defined* as

$$1 \text{ cal} = 4.184 \text{ J (exactly)}$$

In this book the joule is used as the primary unit for heat and energy. However, much of the chemical literature continues to use the calorie as the unit of heat, so it is important to be familiar with both units.

The definition of the specific heat capacity as the amount of heat necessary to raise the temperature of 1 g of material by 1 K is a little imprecise, because in fact the amount of heat required depends on whether the process is carried out at constant volume or at constant pressure. This difference is especially important for gases. Two specific heat capacities must be defined: the specific heat capacity at constant volume, and that at constant pressure. Table 8-1 lists specific heat capacities at constant pressure for a number of common substances.

The specific heat capacity refers to 1 g of material. More generally, the **heat capacity** of any object is defined as the amount of heat necessary to raise the temperature of the object by 1 K. In thermodynamics the *molar* heat capacities are particularly useful: $c_V$ is the amount of heat required to raise the temperature of 1 mol of substance by 1 K at constant volume, and $c_P$ is the corresponding amount required at constant pressure. If the total heat transferred to $n$ mol at constant volume is $q_V$, then

$$q_V = nc_V(T_2 - T_1) = nc_V \,\Delta T \qquad \text{(8-3)}$$

If an amount $q_P$ is transferred at constant pressure, then

$$q_P = nc_P \,\Delta T \qquad \text{(8-4)}$$

provided that $c_V$ and $c_P$ do not change significantly between the initial and final temperatures.

When two objects at different temperatures are brought into contact with each other, energy (in the form of heat) is exchanged between them until they reach a common temperature. If they are insulated from their surroundings, the amount of heat $q_2$ taken up by the cooler object is equal to $-q_1$, the amount of heat given up by the hotter object. As always, the convention followed is that energy transferred to an object has a positive sign, so $q_2$ is positive when $q_1$ is negative.

## Example 8-3

A piece of iron weighing 72.4 g is heated to 100.0°C and plunged into 100.0 g of water that is initially at 10.0°C in a Styrofoam cup calorimeter (Fig. 8-8). Calculate the final temperature that is reached, assuming no heat loss to the surroundings.

### Solution

Because the data involve masses, it is easier to work with specific heat capacities (Table 8-1) than with molar heat capacities. If $t_f$ is the final temperature (in degrees Celsius), then the equation for heat balance gives

| Table 8-1 | Specific Heat Capacities at Constant Pressure (at 25°C) | |
| --- | --- |
| **Substance** | **Specific Heat ($J \ K^{-1} \ g^{-1}$)** |
| $Hg(\ell)$ | 0.140 |
| $Cu(s)$ | 0.385 |
| $Fe(s)$ | 0.449 |
| $SiO_2(s)$ | 0.739 |
| $CaCO_3(s)$ | 0.818 |
| $O_2(g)$ | 0.917 |
| $H_2O(\ell)$ | 4.18 |

**Figure 8-8**
A Styrofoam cup calorimeter. As the piece of metal cools, it releases heat to the water. The amount of heat released can be determined from the temperature change of the water.

$$M_1(c_{s1})\,\Delta T_1 = -M_2(c_{s2})\,\Delta T_2$$

$$(100.0\text{ g }H_2O)(4.18\text{ J }°C^{-1}\,g^{-1})(t_f - 10.0°C) =$$
$$(72.4\text{ g Fe})(0.449\text{ J }°C^{-1}\,g^{-1})(t_f - 100.0°C)$$

This is a linear equation for the unknown temperature $t_f$, and its solution is

$$418t_f - 4180 = -32.51t_f + 3251$$

$$t_f = 16.5°C$$

Note that specific heat capacities are numerically the same whether expressed in $J\,K^{-1}\,g^{-1}$ or $J\,(°C)^{-1}\,g^{-1}$ (because the degree Celsius and the kelvin have the same size). Converting 10.0°C and 100.0°C to kelvins and using specific heat capacities in units of $J\,K^{-1}\,g^{-1}$ gives $t_f = 289.7$ K, an equivalent answer.

*Related Problems:* 9, 10

## 8-3 THE FIRST LAW OF THERMODYNAMICS

Both heat and work are forms in which energy is transferred into and out of a system; they can be thought of as energy in transit. If the energy change is caused by *mechanical* contact of the system with its surroundings, work is done; if it is caused by *thermal* contact (leading to equalization of temperatures), heat is transferred. In many processes both heat and work cross the boundary of a system, and that system's internal energy change is the sum of the two contributions. This statement, called the **first law of thermodynamics,** takes the mathematical form[1]

$$\Delta E = q + w \tag{8-5}$$

A system cannot be thought to "contain" work or heat, because both work and heat refer not to states of the system but to *processes* that transform one state into another. In the Joule experiment of Figure 8-7, the work done on the water (the system) by the falling weight increased the water's temperature. Work was performed on the system without heat being transferred, and from the second law, $\Delta E = w$. The same change in state of the system can be brought about by transferring heat to the system without work being done, so that $\Delta E = q$. Because $q$ and $w$ depend on the particular process (or path) connecting the states, they are not state functions. However, their sum, $\Delta E = q + w$, is independent of path, and so internal energy is a function of state. The fundamental physical content of the first law of thermodynamics is the following observation:

---

[1] Some books, especially older ones, define work as positive when it is done *by* the system. The reason is that many engineering applications focus on the work done by a particular heat engine, and so it is helpful to define that quantity as positive. Thus, the work is given by

$$w = P_{ext}\,\Delta V \quad \text{(older convention)}$$

and the first law reads

$$\Delta E = q - w \quad \text{(older convention)}$$

Although this book does not use the older convention, you should check which convention is employed in other books.

Although $q$ and $w$ depend individually on the path followed between a given pair of states, their sum does not.

As was already asserted, the laws of thermodynamics cannot be derived or proved; they are the results of countless experiments on a tremendous variety of substances. It is not possible even to "check" the first law by independently measuring $\Delta E$, $w$, and $q$, because no "energy gauges" exist to determine energy changes $\Delta E$. But what *can* be done is to measure $w$ and $q$ for a series of different processes connecting the same initial and final states, and to find that their sum $q + w$ is always the same. This reliance upon the first law is what defines the energy change $\Delta E$.

In any process, the heat *added to* the system is *removed from* the surroundings, and so

$$q_{sys} = -q_{surr}$$

In the same way, the work done *on* the system is done *by* the surroundings, and so

$$w_{sys} = -w_{surr}$$

Adding these two and invoking the first law give

$$\Delta E_{sys} = -\Delta E_{surr}$$

Thus, the energy changes of system and surroundings have the same magnitude but opposite signs. The total energy of the universe (system plus surroundings) is then

$$\Delta E_{univ} = \Delta E_{sys} + \Delta E_{surr} = 0 \qquad \textbf{(8-6)}$$

Our conclusion is that in any process, the total energy of the universe remains unchanged; energy is conserved. This is another statement of the first law of thermodynamics.

## Heat Capacities of Ideal Gases

The first law of thermodynamics has been stated in very general terms. Before it can be used to make specific predictions about a substance, certain properties of that substance, such as its equation of state and its heat capacities, must be known. Thermodynamic arguments alone cannot predict the actual values of such properties. Experimental data or theoretical calculations can, however, be used together with thermodynamics to determine those properties.

An example is the pair of molar heat capacities $c_V$ and $c_P$ for an ideal monatomic gas, which can be calculated from the results of the kinetic theory of gases and the ideal gas equation of state. From Section 3-5, the average translational kinetic energy of $n$ mol of an ideal gas is

$$E_k = \tfrac{3}{2}nRT$$

As will be seen in Chapter 15, the rotations and vibrations of diatomic or polyatomic molecules make additional contributions to the energy. In a *monatomic* gas, however, these other contributions are not present, and changes in the total energy $\Delta E$ measured in thermodynamics can be equated to changes in the translational kinetic energy. If $n$ mol of a monatomic gas is taken from a temperature $T_1$ to a temperature $T_2$, the energy change is

$$\Delta E = \tfrac{3}{2}nR(T_2 - T_1) = \tfrac{3}{2}nR\,\Delta T$$

Note that the pressure and volume do not affect $E$ explicitly (except through temperature changes), and so this result is independent of the change in pressure or volume of the gas.

Now consider an ideal-gas temperature change at constant volume from the point of view of thermodynamics. If the volume does not change (the gas is heated in a container with rigid walls), the pressure–volume work $w$ must be zero, and therefore

$$\Delta E = q_V = nc_V\,\Delta T \qquad \text{(ideal gas)}$$

where $q_V$ is the heat transferred at constant volume. Equating this thermodynamic relation with the previous expression for $\Delta E$ (from kinetic theory) shows that

$$c_V = \tfrac{3}{2}R \qquad \text{(monatomic ideal gas)} \tag{8-7}$$

Similarly, the molar heat capacity at constant pressure, $c_P$, is calculated by examining the heating of a monatomic ideal gas at constant pressure from temperature $T_1$ to $T_2$. Experimentally, such a process can be carried out by placing the gas in a cylinder with a piston that moves out as the gas is heated, keeping the gas pressure equal to the outside pressure. In this case,

$$\Delta E = \tfrac{3}{2}nR\,\Delta T = nc_V\,\Delta T$$

still holds (because the energy change depends only on the temperatures for an ideal gas), but we now have

$$\Delta E = q_P + w$$

because the work is no longer zero. The work for a constant-pressure process is easily calculated from

$$w = -P\,\Delta V = -P(V_2 - V_1)$$

and the heat transferred is

$$q_P = nc_P\,\Delta T$$

Because $w$ is negative, $q_P$ is larger than $q_V$ by the amount of work done by the gas as it expands. This gives

$$\Delta E = q + w$$

$$nc_V\,\Delta T = nc_P\,\Delta T - P(V_2 - V_1)$$

From the ideal gas law, $PV_1 = nRT_1$ and $PV_2 = nRT_2$, so that

$$nc_V\,\Delta T = nc_P\,\Delta T - nR\,\Delta T$$

$$c_V = c_P - R$$

$$c_P = c_V + R$$

For a monatomic ideal gas, this shows that $c_P = \tfrac{5}{2}R$. It is important to use the proper units for $R$ in these expressions for $c_V$ and $c_P$. If heat is to be measured in joules, $R$ must be expressed as

$$R = 8.315 \text{ J K}^{-1}\text{ mol}^{-1}$$

For a diatomic or polyatomic ideal gas, $c_V$ is greater than $\frac{3}{2}R$, because energy can be stored in rotational and vibrational motions of the molecules; a greater amount of heat must be transferred to achieve a given temperature change. However, it is still true that

$$c_P = c_V + R \qquad \text{(any ideal gas)} \qquad \textbf{(8-8)}$$

and it is also still true that energy changes depend only on the temperature change; so, for a small temperature change $\Delta T$,

$$\Delta E = nc_V \Delta T \qquad \text{(any ideal gas)} \qquad \textbf{(8-9)}$$

## Example 8-4

Suppose that 1.00 kJ of heat is transferred to 2.00 mol of argon (at 298 K, 1 atm). What will the final temperature $T_f$ be if the heat is transferred (a) at constant volume, or (b) at constant pressure? Calculate the energy change $\Delta E$ in each case.

### Solution

Because argon is a monatomic, approximately ideal gas,

$$c_V = \tfrac{3}{2}R = 12.47 \text{ J K}^{-1} \text{ mol}^{-1}$$
$$c_P = \tfrac{5}{2}R = 20.79 \text{ J K}^{-1} \text{ mol}^{-1}$$

At constant volume,

$$q_V = nc_V \Delta T$$
$$1000 \text{ J} = (2.00 \text{ mol})(12.47 \text{ J K}^{-1} \text{ mol}^{-1})\, \Delta T$$
$$\Delta T = 40.1 \text{ K}; \quad T_f = 298 + 40.1 = 338 \text{ K}$$
$$\Delta E = nc_V \Delta T = q_V = 1000 \text{ J}$$

At constant pressure,

$$q_P = nc_P \Delta T$$
$$1000 \text{ J} = (2.00 \text{ mol})(20.79 \text{ J K}^{-1} \text{ mol}^{-1})\, \Delta T$$
$$\Delta T = 24.0 \text{ K}; \quad T_f = 298 + 24.0 = 322 \text{ K}$$
$$\Delta E = nc_V \Delta T = (2.00 \text{ mol})(12.47 \text{ J K}^{-1} \text{ mol}^{-1})(24 \text{ K}) = 600 \text{ J}$$

Note that the expression for $\Delta E$ involves $c_V$ even though the process is carried out at constant pressure. The difference of 400 J between the input $q_P$ and $\Delta E$ is the work done by the gas as it expands.

*Related Problems:* 21, 22

## Heat and Work for Ideal Gases

Calculations of heat and work can now be carried out for a variety of processes involving an ideal gas. To illustrate the fact that $q$ and $w$ depend individually on the path followed but their sum does not, consider the expansion of 1.00 mol of an ideal

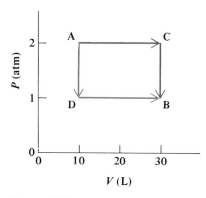

**Figure 8-9**

States A and B of a system are connected by two different ideal-gas processes, one passing through state C and the other through state D.

monatomic gas following two different paths. The system begins at state A ($P_A = 2.00$ atm, $V_A = 10.0$ L) and reaches a final state, B ($P_B = 1.00$ atm, $V_B = 30.0$ L), via either of two paths shown in Figure 8-9. Along path ACB (red arrows) the system is first heated at constant pressure ($P_{ext} = P_A = 2$ atm) until the volume has tripled; then it is cooled at constant volume until the pressure is halved. Along path ADB (blue arrows) the system is cooled at constant volume until the pressure is halved and then heated at constant pressure ($P_{ext} = P_B = 1$ atm) until the volume has tripled.

The calculations of heat and work for each step are of the type already performed, and are straightforward. Thus,

$$w_{AC} = -P_{ext} \Delta V = -P_A(V_B - V_A) \qquad w_{CB} = 0 \text{ because } V_C = V_B$$

$$q_{AC} = q_P = nc_P \Delta T = \tfrac{5}{2}nR(T_C - T_A)$$

$$q_{CB} = q_V = nc_V \Delta T = \tfrac{3}{2}nR(T_B - T_C)$$

From the ideal gas law, $nRT_A = P_AV_A$, $nRT_B = P_BV_B$, and $nRT_C = P_CV_C = P_AV_B$ (because $P_A = P_C$ and $V_B = V_C$). Using these relations and summing over the two steps give

$$w_{ACB} = w_{AC} + w_{CB} = -P_A(V_B - V_A) = -40.0 \text{ L atm} = -4050 \text{ J}$$

$$q_{ACB} = q_{AC} + q_{CB} = \tfrac{5}{2}nR(T_C - T_A) + \tfrac{3}{2}nR(T_B - T_C)$$

$$= \tfrac{5}{2}P_A(V_B - V_A) + \tfrac{3}{2}V_B(P_B - P_A)$$

$$= (100.0 - 45.0) \text{ L atm} = 5570 \text{ J}$$

The sum of these is $\Delta E = w_{ACB} + q_{ACB} = 1520$ J. (This could also have been obtained by using the ideal gas law to calculate the initial and final temperatures $T_A$ and $T_B$.)

The corresponding calculation for path ADB gives

$$w_{ADB} = -2030 \text{ J} \qquad \text{and} \qquad q_{ADB} = 3550 \text{ J}$$

Even though both the work and the heat have changed, their sum is still 1520 J, an illustration of the first law of thermodynamics.

## 8-4 THERMOCHEMISTRY

The energy changes considered so far have arisen from direct mechanical work on a system (as in the paddle wheel driven by a falling weight) or from the establishment of thermal contact between two systems at different temperatures. In chemistry, an important additional source of energy change in a system comes from the heat given off or taken up in the course of a chemical reaction. The study of these heat effects is referred to as **thermochemistry.**

### Enthalpy

Suppose some reacting species are sealed in a small closed container (called a bomb) and the container is placed in a calorimeter like the one in Figure 8-10. As the molecules react chemically, heat is given off or taken up and the change in temperature of the calorimetric fluid is measured. Because the container is sealed tightly, its volume is constant and no pressure–volume is work done. Therefore, the change in

**Figure 8-10**

The combustion calorimeter is also called a "bomb calorimeter"; the combustion reaction in it is carried out at a fixed volume. (Leon Lewandowski)

internal energy is equal to the measured heat absorbed in the chemical reaction at constant volume:

$$\Delta E = q_V$$

Such experiments at constant volume are often inconvenient or difficult to carry out. They require the use of a well-constructed reaction vessel that can resist the large pressure changes that occur in many chemical reactions. Most experiments in chemistry are done under constant (atmospheric) *pressure* conditions rather than at constant volume, however, and so it would be desirable to relate the heat transferred at constant pressure, $q_P$, to some state property analogous to the internal energy $E$.

If the work done is entirely pressure–volume work, and if the external pressure is held constant, then

$$\Delta E = q_P + w = q_P - P_{ext}\,\Delta V$$

If the external pressure is now assumed to be equal to the internal pressure of the system $P$, then

$$\Delta E = q_P - P\,\Delta V$$

$$q_P = \Delta E + P\,\Delta V$$

Because $P$ is constant, $P\,\Delta V = \Delta(PV)$, and this equation becomes

$$q_P = \Delta(E + PV)$$

The combination $E + PV$ appearing on the right side is now defined as the **enthalpy** $H$:

$$H = E + PV \qquad \text{(8-10)}$$

and so

$$q_P = \Delta(E + PV) = \Delta H$$

Because $E$, $P$, and $V$ are state functions, $H$ must be one, as well. Heat transfer at constant pressure has therefore been related to a state function.

It is important to remember that

$$\Delta H = q_P = \Delta E + P\,\Delta V \quad \text{(constant pressure)}$$

holds only at constant pressure. If the pressure changes, the more general relationship

$$\Delta H = \Delta E + \Delta(PV)$$

must be used. Like the energy, the enthalpy change is determined by the initial and final states and is independent of the particular path along which the process is carried out. This is always true for a state function.

For an ideal gas process,

$$\Delta H = \Delta E + \Delta(PV) = nc_V\,\Delta T + nR\,\Delta T$$

$$\Delta H = nc_P\,\Delta T \quad \text{(ideal gas)} \tag{8-11}$$

because $c_P = c_V + R$. This result holds for any ideal gas process and shows that enthalpy changes, like energy changes, depend only on the temperature difference between initial and final states.

**Enthalpies of Reaction**

When carbon monoxide is burned in oxygen to carbon dioxide,

$$CO(g) + \tfrac{1}{2}O_2(g) \longrightarrow CO_2(g)$$

heat is given off. Because this heat is transferred out of the reaction vessel (the system) and into the calorimeter, it has a *negative* sign. Careful calorimetric measurements show that 1.000 mol of CO reacted completely with 0.500 mol of $O_2$, at 25°C and a constant pressure of 1 atm, leads to an enthalpy change of

$$\Delta H = q_P = -2.830 \times 10^5 \text{ J} = -283.0 \text{ kJ}$$

The kilojoule (kJ), equal to $10^3$ J, is used because most enthalpy changes for chemical reactions lie in the range of thousands of joules per mole.

When heat is given off by a reaction ($\Delta H$ is negative), the reaction is said to be **exothermic** (Fig. 8-11). Reactions in which heat is taken up ($\Delta H$ positive) are called **endothermic** (Fig. 8-12). An obvious example of an endothermic reaction is the preceding reaction written in the opposite direction:

$$CO_2(g) \longrightarrow CO(g) + \tfrac{1}{2}O_2(g) \qquad \Delta H = +283.0 \text{ kJ}$$

If the direction of a chemical reaction is reversed, the enthalpy change reverses sign. Heat is required to convert $CO_2$ to CO and $O_2$ at constant pressure. The decomposition of $CO_2$ into CO and $O_2$ is very difficult to carry out in the laboratory, but the reverse reaction presents no practical obstacle. Thermodynamics allows prediction of the $\Delta H$ of the decomposition reaction with complete confidence, even if a calorimetric experiment is never actually carried out.

## Figure 8-11

The thermite reaction

$$2\,Al(s) + Fe_2O_3(s) \longrightarrow 2\,Fe(s) + Al_2O_3(s)$$

is among the most exothermic of all reactions, liberating 16 kJ of heat for every gram of Al that reacts. The iron(III) oxide and aluminum are mixed in finely divided form. Like the combustion of paper, the reaction requires a source of ignition, but then continues on its own.

(a)

(b)

## Figure 8-12

(a) When mixed in a flask, the two solids $Ba(OH)_2 \cdot 8H_2O(s)$ and $NH_4NO_3(s)$ undergo an acid–base reaction:

$$Ba(OH)_2 \cdot 8H_2O(s) + 2\,NH_4NO_3(s) \longrightarrow Ba(NO_3)_2(aq) + 2\,NH_3(aq) + 10H_2O(\ell)$$

(b) The water produced dissolves excess ammonium nitrate in an endothermic reaction. The dissolution absorbs so much heat that the water on the surface of the wet wooden block freezes to the bottom of the flask, and the block can be lifted up with the flask.

297

**Figure 8-13**

Red phosphorus reacts exothermically in liquid bromine. The rising gases are a mixture of the product $PBr_3$ and unreacted bromine that has boiled off.

The convention is used that a **reaction enthalpy** written after a balanced chemical equation refers to the enthalpy change that accompanies the complete conversion of stoichiometric amounts of reactants to products; the numbers of moles of reactants and products are given by the coefficients in the equation. The preceding equation shows the enthalpy change when 1 mol of $CO_2$ is converted to 1 mol of CO and $\frac{1}{2}$ mol of $O_2$. If this equation is multiplied by a factor of 2, the enthalpy change must also be doubled because twice as many moles are then involved (enthalpy, like energy, is an *extensive* property).

$$2\, CO_2(g) \longrightarrow 2\, CO(g) + O_2(g) \qquad \Delta H = +566.0 \text{ kJ}$$

The molar amounts need not be integers, as the following example illustrates.

## Example 8-5

Red phosphorus reacts with liquid bromine in an exothermic reaction (Fig. 8-13):

$$2\, P(s) + 3\, Br_2(\ell) \longrightarrow 2\, PBr_3(g) \qquad \Delta H = -243 \text{ kJ}$$

Calculate the enthalpy change when 2.63 g of phosphorus reacts with an excess of bromine in this way.

### Solution

First, convert from grams of phosphorus to moles, using the molar mass of phosphorus, 30.97 g $mol^{-1}$:

$$\text{moles P} = \frac{2.63 \text{ g P}}{30.97 \text{ g mol}^{-1}} = 0.0849 \text{ mol}$$

Given that an enthalpy change of $-243$ kJ is associated with 2 mol of P, it is readily seen that the enthalpy change associated with 0.0849 mol is

$$\Delta H = 0.0849 \text{ mol P} \times \left(\frac{-243 \text{ kJ}}{2 \text{ mol P}}\right) = -10.3 \text{ kJ}$$

*Related Problems:* 23, 24, 25, 26

The enthalpy change for the reaction of 1 mol of carbon monoxide with oxygen was stated to be $-283.0$ kJ. In a second experiment, the heat evolved when 1 mol of carbon (graphite) is burned in oxygen to carbon dioxide at 25°C is readily measured to be

$$C(s,\text{gr}) + O_2(g) \longrightarrow CO_2(g) \qquad \Delta H = -393.5 \text{ kJ}$$

Suppose the enthalpy change for the reaction

$$C(s,\text{gr}) + \tfrac{1}{2} O_2(g) \longrightarrow CO(g) \qquad \Delta H = ?$$

is wanted. This reaction cannot be carried out simply in the laboratory. If 1 mol of graphite is heated with $\frac{1}{2}$ mol of oxygen, almost half of the carbon burns to $CO_2(g)$ and the remainder is left as carbon at equilibrium. But the heat that would be evolved if it were possible to perform the experiment can be calculated from the fact that $\Delta H$ is independent of the path followed from reactants to products (Fig. 8-14). A path is chosen in which 1 mol of C is burned with $O_2$ to $CO_2$ (with $\Delta H = -393.5$ kJ), and to this is added the calculated enthalpy change for the hypothetical process in which $CO_2$ is converted to CO and $O_2$ ($\Delta H = +283.0$ kJ). The total $\Delta H$

$C(s) + O_2(g)$     $C(s) + \frac{1}{2}O_2(g) + \frac{1}{2}O_2(g)$

Reactants

$\Delta H = -110.5$ kJ

Products    $CO(g) + \frac{1}{2}O_2(g)$

$\Delta H = -393.5$ kJ

$\Delta H = +283.0$ kJ

$CO_2(g)$

**Figure 8-14**

Because enthalpy is a state property, the enthalpy change for the reaction of carbon with oxygen to give carbon monoxide (red arrow) can be determined through measurements along a path that is less direct but easier to study (blue arrows). The enthalpy change sought is the sum of the enthalpy change to burn carbon to carbon dioxide and that to convert carbon dioxide to carbon monoxide and oxygen.

is the algebraic sum of the two known enthalpy changes, $-393.5$ kJ $+ 283.0$ kJ $= -110.5$ kJ. To see this more clearly, we write out the reactions:

$$C(s,\text{gr}) + O_2(g) \longrightarrow CO_2(g) \qquad\qquad \Delta H_1 = -393.5 \text{ kJ}$$

$$\underline{CO_2(g) \longrightarrow CO(g) + \tfrac{1}{2}O_2(g) \qquad\qquad \Delta H_2 = +283.0 \text{ kJ}}$$

$$C(s,\text{gr}) + \tfrac{1}{2}O_2(g) \longrightarrow CO(g) \qquad\qquad \Delta H = \Delta H_1 + \Delta H_2 = -110.5 \text{ kJ}$$

> If two or more chemical equations are added to give another chemical equation, the corresponding enthalpies of reaction must be added.

This is known as **Hess's law** and derives from the fact that enthalpy is a state function.

The corresponding energy change $\Delta E$ might be desired for this reaction. The quantity $\Delta E$ is simple to calculate because of the relation

$$\Delta H = \Delta E + \Delta(PV)$$

or

$$\Delta E = \Delta H - \Delta(PV)$$

The gases can be assumed to obey the ideal gas law, so that

$$\Delta(PV) = \Delta(nRT) = RT\,\Delta n_g$$

because the temperature is constant at 25°C. Here $\Delta n_g$ is the change in the number of moles of *gas* in the reaction as written:

$$\Delta n_g = \text{total moles of product gases} - \text{total moles of reactant gases}$$

$$= 1 \text{ mol} - \tfrac{1}{2} \text{ mol} = \tfrac{1}{2} \text{ mol}$$

(Graphite is a solid, and its volume is negligible compared with the volumes of the gases.) Hence,

$$\Delta(PV) = RT\,\Delta n_g = (8.315 \text{ J K}^{-1}\text{ mol}^{-1})(298 \text{ K})(\tfrac{1}{2}\text{ mol})$$

$$= 1.24 \times 10^3 \text{ J} = 1.24 \text{ kJ}$$

**Table 8-2    Enthalpy Changes of Fusion and Vaporization$^\dagger$**

| Substance | $\Delta H_{fus}$ (kJ mol$^{-1}$) | $\Delta H_{vap}$ (kJ mol$^{-1}$) |
|---|---|---|
| $NH_3$ | 5.65 | 23.35 |
| HCl | 1.992 | 16.15 |
| CO | 0.836 | 6.04 |
| $CCl_4$ | 2.5 | 30.0 |
| $H_2O$ | 6.007 | 40.66 |
| NaCl | 28.8 | 170 |

$^\dagger$The enthalpy changes are measured at the normal melting point and the normal boiling point, respectively.

Note that $R$ must be expressed in J K$^{-1}$ mol$^{-1}$ if a result in joules is desired. Therefore,

$$\Delta E = -110.5 \text{ kJ} - 1.24 \text{ kJ} = -111.7 \text{ kJ}$$

For reactions in which only liquids and solids are involved, or those in which the number of moles of gas does not change, the enthalpy and energy changes are almost equal and their difference can be neglected.

Phase changes are not chemical reactions, but they can be considered in the same context. Heat must be transferred to ice in order to transform it to water, and so the phase change is endothermic, with $\Delta H$ positive:

$$H_2O(s) \longrightarrow H_2O(\ell) \qquad \Delta H_{fus} = +6.007 \text{ kJ}$$

Here $\Delta H_{fus}$ is the **molar enthalpy of fusion,** the heat that must be transferred at constant pressure to melt one mole of substance. When a liquid freezes, the reaction is reversed, and an equal amount of heat is given off to the surroundings; that is, $\Delta H_{freez} = -\Delta H_{fus}$. The vaporization of one mole of liquid at constant pressure and temperature requires an amount of heat called the **molar enthalpy of vaporization** $\Delta H_{vap}$,

$$H_2O(\ell) \longrightarrow H_2O(g) \qquad \Delta H_{vap} = +40.7 \text{ kJ}$$

whereas the condensation of a liquid from a vapor is an exothermic process, with $\Delta H_{cond} = -\Delta H_{vap}$. Table 8-2 lists enthalpies of fusion and vaporization.

### Example 8-6

To vaporize 100.0 g of carbon tetrachloride at its normal boiling point, 349.9 K, and $P = 1$ atm, 19.5 K of heat is required. Calculate $\Delta H_{vap}$ for $CCl_4$, and compare it with $\Delta E$ for the same process.

### Solution

The molar mass of $CCl_4$ is 153.8 g mol$^{-1}$, and so the chemical amount in 100.0 g is

$$\frac{100.0 \text{ g } CCl_4}{153.8 \text{ g mol}^{-1}} = 0.6502 \text{ mol } CCl_4$$

The enthalpy change for one mole of $CCl_4$ is then

$$\left(\frac{19.5 \text{ kJ}}{0.6502 \text{ mol } CCl_4}\right) \times 1.00 \text{ mol } CCl_4 = +30.0 \text{ kJ} = \Delta H_{vap}$$

The energy change is then

$$\Delta E = \Delta H_{vap} - \Delta(PV) = \Delta H_{vap} - RT \Delta n_g$$

Inserting $T = 349.9$ K and $\Delta n_g = 1$ (because there is an increase of one mole of gaseous products for each mole of liquid that is vaporized) gives

$$\Delta E = 30.0 \text{ kJ} - (8.315 \text{ J K}^{-1} \text{ mol}^{-1})(349.9 \text{ K})(1.00 \text{ mol})(10^{-3} \text{ kJ J}^{-1})$$

$$= (30.0 - 2.9) \text{ kJ} = +27.1 \text{ kJ mol}^{-1}$$

Thus, of the 30.0 kJ of energy transferred from the surroundings in the form of heat, 27.1 kJ is used to increase the internal energy of the molecules ($\Delta E$), and 2.9 kJ is used to expand the resulting vapor, $\Delta(PV)$.

## Standard-State Enthalpies

The **standard enthalpy** of a reaction (written $\Delta H°$) is the enthalpy change when all reactants and products are at 25°C and 1 atm pressure. Absolute enthalpies of substances, like absolute energies, cannot be measured or calculated. Only *changes* in enthalpy are subject to experimental determination. Just as altitudes are measured relative to a standard altitude (sea level), it is necessary to adopt a reference state for the enthalpies of substances. By definition, the **standard enthalpy of formation** $\Delta H_f°$ of a compound is the enthalpy change for the reaction that produces one mole of the compound from its elements in their stable states, all at 25°C and 1 atm pressure. For example, the standard enthalpy of formation of liquid water is the enthalpy change for the reaction

$$H_2(g) + \tfrac{1}{2} O_2(g) \longrightarrow H_2O(\ell) \qquad \Delta H° = -285.83 \text{ kJ}$$

$$\Delta H_f°(H_2O(\ell)) = -285.83 \text{ kJ mol}^{-1}$$

Here the superscript ° indicates standard-state conditions, and the subscript "f" stands for *formation*. In choosing a reference state, it is necessary to specify the stable state of the element, because some elements occur in more than one form at 25°C and 1 atm pressure. For example, carbon exists as graphite and diamond. Graphite is the thermodynamically stable form of carbon at 25°C and 1 atm, and diamond is unstable with respect to it. The reaction

$$C(dia) \longrightarrow C(gr)$$

should occur as written, given sufficient (very, very long!) time. (At extremely high pressures, however, diamond is the stable form of carbon.) Therefore, graphite is the reference state of carbon that is used in determining standard enthalpies of formation of compounds of carbon.[2]

---

[2] There is one exception to this choice of reference state. The reference state of phosphorus is taken to be white phosphorus, rather than the more stable red or black form, because the latter are less well characterized and less reproducible in their properties.

The $\Delta H_f^\circ$ for an *element* that is already in its standard state is clearly zero, because no further change is needed to bring it to standard-state conditions. The standard enthalpy of formation of a mole of *atoms* of an element is frequently a large positive quantity, however (the reaction is endothermic):

$$\tfrac{1}{2}\,H_2(g) \longrightarrow H(g) \qquad\qquad \Delta H^\circ = +217.96 \text{ kJ}$$

$$\Delta H_f^\circ(H(g)) = 217.96 \text{ kJ mol}^{-1}$$

For dissolved species, the reference state is defined as an ideal solution with a concentration of 1 M (this is obtained in practice by extrapolating the dilute solution behavior up to this concentration). A special comment is in order on the standard enthalpies of formation of ions. When a strong electrolyte dissolves in water, both positive and negative ions form; it is impossible to produce one without the other. It is therefore also impossible to measure the enthalpy change of formation of ions of only one charge. Only the sum of the enthalpies of formation of the positive and negative ions is accessible to calorimetric experiments. A further arbitrary choice is therefore made, namely that $\Delta H_f^\circ$ of $H^+(aq)$ is set to zero.

The reason for introducing standard enthalpies of formation is that it is then necessary only to tabulate $\Delta H_f^\circ$ for compounds (a long list at that), rather than $\Delta H^\circ$ for all the conceivable reactions those compounds may undergo with one another (a vastly longer list!). Appendix D gives a short table of standard enthalpies of formation at 25°C. The following example shows how they can be used to determine enthalpy changes for reactions carried out at 25°C and 1 atm pressure.

## Example 8-7

Using Appendix D, calculate $\Delta H^\circ$ for the following reaction at 25°C and 1 atm pressure:

$$2\,NO(g) + O_2(g) \longrightarrow 2\,NO_2(g)$$

### Solution

Because enthalpy is a function of state, $\Delta H$ can be calculated along any convenient path. In particular, two steps can be chosen for which $\Delta H$ is found easily. In step 1, the reactants are decomposed into the elements in their standard states:

$$2\,NO(g) \longrightarrow N_2(g) + O_2(g) \quad \Delta H_1 = -2\,\Delta H_f^\circ(NO)$$

The minus sign appears because the process chosen is the reverse of the formation of NO; the factor of 2 is present because 2 mol of NO is involved. Because oxygen is already an element in its standard state, it does not need to be changed (equivalently, $\Delta H_f^\circ(O_2(g))$ is 0).

In step 2, the elements are combined to form products:

$$N_2(g) + 2\,O_2(g) \longrightarrow 2\,NO_2(g) \quad \Delta H_2 = 2\,\Delta H_f^\circ(NO_2)$$

The enthalpy change of the overall reaction is then the sum of these two enthalpies:

$$\Delta H^\circ = \Delta H_1 + \Delta H_2 = -2\,\Delta H_f^\circ(NO) + 2\,\Delta H_f^\circ(NO_2)$$

$$= -(2 \text{ mol})(90.25 \text{ kJ mol}^{-1}) + (2 \text{ mol})(33.18 \text{ kJ mol}^{-1}) = -114.14 \text{ kJ}$$

*Related Problems:* 35, 36, 37, 38

The general pattern should be clear from this example. The change $\Delta H°$ for a reaction at atmospheric pressure and 25°C is the sum of the $\Delta H_f°$ for the *products* (multiplied by their coefficients in the balanced chemical equation) minus the sum of the $\Delta H_f°$ for the *reactants* (also multiplied by their coefficients). For a general reaction of the form

$$aA + bB \longrightarrow cC + dD$$

the standard enthalpy change is

$$\Delta H° = c \, \Delta H_f°(C) + d \, \Delta H_f°(D) - a \, \Delta H_f°(A) - b \, \Delta H_f°(B)$$

## Bond Enthalpies

Chemical reactions between molecules require existing bonds to break and new ones to form with the atoms arranged differently. Chemists have developed methods to study highly reactive intermediates, species in which bonds have been broken and not yet re-formed. For example, a hydrogen atom can be removed from a methane molecule,

$$CH_4(g) \longrightarrow CH_3(g) + H(g)$$

leaving two fragments, neither of which has a stable valence electron structure in the Lewis electron-dot picture. Both will go on to react rapidly with other molecules or fragments and eventually form the stable products of the reaction. During the brief presence of these reactive species, however, many of their properties can be measured.

An important quantity that has been measured is the enthalpy change when a bond is broken in the gas phase, called the **bond enthalpy.** This is invariably positive, because heat must be added to a collection of stable molecules to break their bonds. For example, the bond enthalpy of a C—H bond in methane is 438 kJ $mol^{-1}$, the measured standard enthalpy change for the reaction

$$CH_4(g) \longrightarrow CH_3(g) + H(g) \qquad \Delta H° = +438 \text{ kJ}$$

in which 1 mol of C—H bonds is broken, one for each molecule of methane. Bond enthalpies are fairly reproducible from one compound to another. Each of the following gas-phase reactions involves the breaking of a C—H bond:

$$C_2H_6(g) \longrightarrow C_2H_5(g) + H(g) \qquad \Delta H° = +410 \text{ kJ}$$

$$CHF_3(g) \longrightarrow CF_3(g) + H(g) \qquad \Delta H° = +429 \text{ kJ}$$

$$CHCl_3(g) \longrightarrow CCl_3(g) + H(g) \qquad \Delta H° = +380 \text{ kJ}$$

$$CHBr_3(g) \longrightarrow CBr_3(g) + H(g) \qquad \Delta H° = +377 \text{ kJ}$$

The approximate constancy of the measured enthalpy changes (all lie within 8% of their average value) suggests that the C—H bonds in all five molecules are rather similar. Because such bond enthalpies are reproducible from one molecule to another, it is useful to tabulate *average* bond enthalpies from measurements on a series of compounds (Table 8-3, p. 304). Any given bond enthalpy will differ somewhat from those shown, but in most cases the deviations are small.

The bond enthalpies of Table 8-3 can be used, together with enthalpies of atomization of the elements from the same table, to estimate standard enthalpies of formation $\Delta H_f°$ for molecules in the gas phase, and enthalpy changes $\Delta H°$ for gas-phase reactions. This is illustrated by the following example.

| Table 8-3 Average Bond Enthalpies | | | | | | | | | | |
|---|---|---|---|---|---|---|---|---|---|---|
| | **Molar Enthalpy of Atomization (kJ mol$^{-1}$)†** | **Bond Enthalpy (kJ mol$^{-1}$)‡** | | | | | | | | |
| | | H— | C— | C= | C≡ | N— | N= | N≡ | O— | O= |
| H | 218.0 | 436 | 413 | | | 391 | | | 463 | |
| C | 716.7 | 413 | 348 | 615 | 812 | 292 | 615 | 891 | 351 | 728 |
| N | 472.7 | 391 | 292 | 615 | 891 | 161 | 418 | 945 | | |
| O | 249.2 | 463 | 351 | 728 | | | | | 139 | 498 |
| S | 278.8 | 339 | 259 | 477 | | | | | | |
| F | 79.0 | 563 | 441 | | | 270 | | | 185 | |
| Cl | 121.7 | 432 | 328 | | | 200 | | | 203 | |
| Br | 111.9 | 366 | 276 | | | | | | | |
| I | 106.8 | 299 | 240 | | | | | | | |

† From Appendix D.
‡ From L. Pauling, *The Nature of the Chemical Bond,* 3rd ed. (Cornell University Press, Ithaca, New York, 1960).

## Example 8-8

Estimate the standard enthalpy of formation of dichlorodifluoromethane, $CCl_2F_2(g)$ (Fig. 8-15). This compound is also known as Freon-12 and has been used as a refrigerant because of its low reactivity and high volatility. It and other, related chlorofluorocarbons (CFCs) are being phased out because of their role in depleting the ozone layer in the outer atmosphere, as discussed in Section 15-4.

### Solution

The standard enthalpy of formation of $CCl_2F_2(g)$ is the enthalpy change for the process in which it is formed from the elements in their standard states at 25°C:

$$C(s,gr) + Cl_2(g) + F_2(g) \longrightarrow CCl_2F_2(g)$$

This reaction can be replaced by a hypothetical two-step process: all the species appearing on the left are atomized and then the atoms are combined to make $CCl_2F_2$.

$$C(s,gr) + Cl_2(g) + F_2(g) \longrightarrow C(g) + 2\,Cl(g) + 2\,F(g)$$
$$C(g) + 2\,Cl(g) + 2\,F(g) \longrightarrow CCl_2F_2(g)$$

The enthalpy change $\Delta H_1$ for the first step is the sum of the atomization energies:

$$\Delta H_1 = \Delta H_f^\circ(C(g)) + 2\,\Delta H_f^\circ(Cl(g)) + 2\,\Delta H_f^\circ(F(g))$$
$$= 716.7 + 2(121.7) + 2(79.0) = 1118 \text{ kJ}$$

The enthalpy change $\Delta H_2$ for the second step can be estimated from the bond enthalpies of Table 8-3. This step involves the formation (with release of heat and therefore negative enthalpy change) of two C—Cl and two C—F bonds per molecule. The net $\Delta H$ for this step is then

$$\Delta H_2 \approx -[2(328) + 2(441)] = -1538 \text{ kJ}$$

$$\Delta H_1 + \Delta H_2 = -1538 + 1118 = -420 \text{ kJ}$$

**Figure 8-15**
Dichlorodifluoromethane, $CCl_2F_2$, also known as Freon-12.

This $\Delta H_f^{\circ}$, $-420\,\text{kJ}\,\text{mol}^{-1}$, compares fairly well with the experimental value, $-477\,\text{kJ}\,\text{mol}^{-1}$. In general, much better agreement than this is not to be expected, because tabulated bond enthalpies are only average values.

*Related Problems:* 45, 46, 47, 48

## 8-5 REVERSIBLE GAS PROCESSES

In this section we consider several types of reversible processes that can be carried out using ideal gases. Recall from Section 8-1 that a true reversible process is an idealization; it is a process in which the system proceeds with infinitesimal speed through a series of equilibrium states. The external pressure $P_{\text{ext}}$ therefore can never differ by more than an infinitesimal amount from the pressure $P$ of the gas itself. The heat, work, and energy and enthalpy changes for ideal gases at constant volume (called **isochoric processes**) and at constant pressure **(isobaric processes)** have already been considered. In this section, *isothermal* (constant-temperature) and *adiabatic* ($q = 0$) processes are examined. The ultimate goal (to be realized in Chapter 9) is to understand the form of the law of mass action for gas-phase equilibrium.

### Isothermal Processes

An **isothermal process** is one carried out at constant temperature. This is accomplished by placing the system in a large reservoir (bath) at fixed temperature and allowing heat to be transferred as required between system and reservoir. The reservoir is large enough that its temperature is almost unchanged by this heat transfer. In Section 8-3, $E$ for an ideal gas was shown to depend only on temperature, and therefore $\Delta E = 0$ for any isothermal gas process. From the first law it follows that

$$w = -q \qquad \text{(isothermal process, ideal gas)}$$

In a *reversible* process, $P_{\text{ext}} = P_{\text{gas}} \equiv P$. However, the relation $w = -P_{\text{ext}}\,\Delta V$ from Section 8-2 cannot be used to calculate the work, because that expression applies only if the external pressure remains constant as the volume changes. In the reversible isothermal expansion of an ideal gas,

$$P_{\text{ext}} = P = \frac{nRT}{V}$$

by Boyle's law, and the pressure falls as the volume is increased from $V_1$ to $V_2$, as shown by the solid line in Figure 8-16 (p. 306). In this case, the work is calculated by approximating the process as a *series* of expansions by small amounts $\delta V$, during each of which $P_{\text{ext}}$ is held constant at $P_i$, with $i = 1, 2, 3, \ldots$ labeling the step. The work done in step $i$ is $-P_i\,\delta V$, and so the total work done is the sum of the work in all steps:

$$w = -P_1\,\delta V - P_2\,\delta V - \cdots$$

The summation is the sum of the areas of the rectangles in the figure. As the step size $\delta V$ is made smaller, it approaches the infinitesimal volume change that is written $dV$, and the corresponding work becomes

$$dw = -P_{\text{ext}}\,dV$$

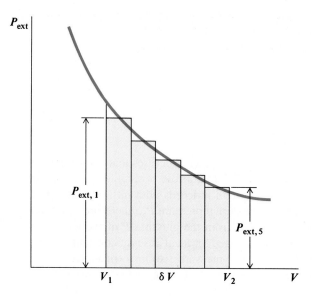

**Figure 8-16**

The area under the graph of external pressure against volume can be approximated as the sum of the areas of the rectangles shown.

In this limit the sum of the areas of the rectangles approaches the area under the graph of $P$ versus $V$. This limiting sum is the *integral* of $P_{ext}$ from $V_1$ to $V_2$ and is written symbolically using an integral sign:

$$w = -\int_{V_1}^{V_2} P\, dV \qquad \text{(reversible process)}$$

This is a compact way of saying that the work $w$ for a reversible expansion process is (with a minus sign) the area under the graph of $P$ plotted against $V$ from $V_1$ to $V_2$.

Mathematical methods of defining and calculating integrals are beyond the scope of this book, but Appendix C gives some supplementary information about several integrals that are important in chemistry, and it should be read by students unfamiliar with calculus. The main idea is the relationship between a thermodynamic quantity (in this case, the work) and the area under a curve.

In the present case the ideal gas law is used to write this area as

$$w = -nRT \int_{V_1}^{V_2} \frac{1}{V}\, dV$$

This integral is the area under a graph of $1/V$ against $V$ (a hyperbola) from $V_1$ to $V_2$. It defines the *natural logarithm* function, symbolized "ln" (see Appendix C). In particular,

$$w = -nRT \ln \frac{V_2}{V_1}$$

$$q = -w = nRT \ln \frac{V_2}{V_1}$$

$$\Delta E = 0 \quad \text{because} \quad \Delta T = 0$$

$$\Delta H = \Delta E + \Delta(PV) = \Delta E + \Delta(nRT) = 0 \qquad \textbf{(8-12)}$$

The enthalpy change, like the energy change, depends only on temperature for an ideal gas.

The fact that $V_2$ is greater than $V_1$ implies that $w < 0$ and $q > 0$; in an isothermal expansion, the system does work and heat must be transferred to it. In an isothermal compression, the reverse is true: the heat withdrawn is completely converted to work done on the system.

## Example 8-9

Calculate the heat and the work associated with a process in which 5.00 mol of gas expands reversibly at constant temperature $T = 298$ K from a pressure of 10.0 atm to 1.00 atm.

### Solution

At constant $T$ and $n$,

$$\frac{V_2}{V_1} = \frac{P_1}{P_2} = \frac{10.0 \text{ atm}}{1.00 \text{ atm}} = 10.0$$

Thus,

$$w = -(5.00 \text{ mol})(8.315 \text{ J K}^{-1} \text{ mol}^{-1})(298 \text{ K}) \ln 10.0$$
$$= -2.85 \times 10^4 \text{ J} = -28.5 \text{ kJ}$$

$$q = -w = 28.5 \text{ kJ}$$

*Related Problems:* 51, 52

## Adiabatic Processes

An **adiabatic process** is one in which there is no transfer of heat into or out of the system. This is accomplished by placing insulation around the system to prevent heat flow.

$$q = 0$$

$$\Delta E = w$$

Consider a very small adiabatic change. The volume changes by an amount $dV$ and the temperature by an amount $dT$. Now $E$ depends only on temperature for an ideal gas, and so

$$dE = nc_V \, dT$$

As always, the work is given by $-P_{ext} \, dV$. Setting these equal gives

$$nc_V \, dT = -P_{ext} \, dV$$

In other words, the temperature change $dT$ is related to the volume change $dV$ in such a process.

If the process is *reversible* as well as adiabatic, so that $P_{ext} \approx P$, the ideal gas law can be used to write

$$nc_V \, dT = -P \, dV = -\frac{nRT}{V} \, dV$$

The equation is simplified by dividing both sides through by $nT$, making the left side depend only on $T$ and the right side only on $V$:

$$\frac{c_V}{T}\,dT = -\frac{R}{V}\,dV$$

Suppose now that the change is not infinitesimal but large. How are temperature and volume related in this case? If a series of such infinitesimal changes is added together, the result is an integral of both sides of the equation from the initial state (specified by $T_1$ and $V_1$) to the final state (specified by $T_2$ and $V_2$):

$$c_V \int_{T_1}^{T_2} \frac{1}{T}\,dT = -R \int_{V_1}^{V_2} \frac{1}{V}\,dV$$

Here $c_V$ has been assumed to be approximately independent of $T$ over the range from $T_1$ to $T_2$. Evaluating the integrals gives

$$c_V \ln \frac{T_2}{T_1} = -R \ln \frac{V_2}{V_1} = R \ln \frac{V_1}{V_2}$$

A more useful form results from taking antilogarithms of both sides:

$$\left(\frac{T_2}{T_1}\right)^{c_V} = \left(\frac{V_1}{V_2}\right)^{R} = \left(\frac{V_1}{V_2}\right)^{c_P - c_V}$$

The last step used the fact that $R = c_P - c_V$. Thus,

$$\left(\frac{T_2}{T_1}\right) = \left(\frac{V_1}{V_2}\right)^{c_P/c_V - 1} = \left(\frac{V_1}{V_2}\right)^{\gamma - 1}$$

where $\gamma = c_P/c_V$ is the ratio of specific heats. This can be rearranged to give

$$T_1 V_1^{\gamma - 1} = T_2 V_2^{\gamma - 1} \tag{8-13}$$

In many situations, the initial thermodynamic state and therefore $T_1$ and $V_1$ are known. If the final volume $V_2$ is known, $T_2$ can be calculated; if $T_2$ is known, $V_2$ can be calculated.

In some cases, only the final *pressure*, $P_2$, of an adiabatic process is known. In this case the ideal gas law gives

$$\frac{P_1 V_1}{T_1} = \frac{P_2 V_2}{T_2}$$

Multiplying this by Equation 8-13 gives

$$P_1 V_1^{\gamma} = P_2 V_2^{\gamma} \tag{8-14}$$

This can be used to calculate $V_2$ from a known $P_2$.

Once the pressure, temperature, and volume of the final state are known, the energy and enthalpy changes and the work done are straightforward to calculate:

$$\Delta E = nc_V(T_2 - T_1) = w \qquad \text{(reversible adiabatic process for ideal gas)}$$

$$\Delta H = \Delta E + \Delta(PV) = \Delta E + (P_2 V_2 - P_1 V_1)$$

or, more simply,

$$\Delta H = nc_P \, \Delta T$$

## Example 8-10

Suppose 5.00 mol of an ideal monatomic gas at an initial temperature of 298 K and pressure of 10.0 atm is expanded adiabatically and reversibly until the pressure has dropped to 1.00 atm. Calculate the final volume and temperature, the energy and enthalpy changes, and the work done.

### Solution

The initial volume is

$$V_1 = \frac{nRT_1}{P_1} = \frac{(5.00\text{ mol})(0.08206\text{ L atm K}^{-1}\text{ mol}^{-1})(300\text{ K})}{(10.0\text{ atm})} = 12.3\text{ L}$$

and the heat capacity ratio for a monatomic gas is

$$\gamma = \frac{c_P}{c_V} = \frac{\frac{5}{2}R}{\frac{3}{2}R} = \frac{5}{3}$$

For a reversible adiabatic process,

$$\frac{P_1}{P_2}V_1^\gamma = V_2^\gamma$$

$$(10.0)(12.3)^{5/3} = V_2^{5/3}$$

$$V_2 = (12.3)(10.0)^{3/5} = 49.0\text{ L}$$

The final temperature can now be calculated from the ideal gas law:

$$T_2 = \frac{P_2 V_2}{nR} = 119\text{ K}$$

From this, the work done and the energy change can be found,

$$w = \Delta E = nc_V\,\Delta T$$

$$= (5.00\text{ mol})(\tfrac{3}{2} \times 8.315\text{ J K}^{-1}\text{ mol}^{-1})(119\text{ K} - 300\text{ K})$$

$$= -11{,}300\text{ J}$$

as well as the enthalpy change:

$$\Delta H = nc_P\,\Delta T = (5.00\text{ mol})(\tfrac{5}{2} \times 8.315\text{ J K}^{-1}\text{ mol}^{-1})(119\text{ K} - 300\text{ K})$$

$$= -18{,}800\text{ J}$$

*Related Problems:* 53, 54

Figure 8-17 (p. 310) compares the adiabatic expansion of this example to the isothermal expansion of Example 8-9. Note that the initial states were the same in the two cases, as were the final pressures. However, the final volume is larger by more than a factor of 2 for the isothermal expansion, and the work output for the adiabatic case is only 40% of the output from the isothermal expansion. Because $\gamma > 1$, the adiabatic line falls off more rapidly with increasing volume than the isothermal line. Because no heat is transferred in an adiabatic expansion, the work comes from the internal energy of the gas, and thus the temperature drops.

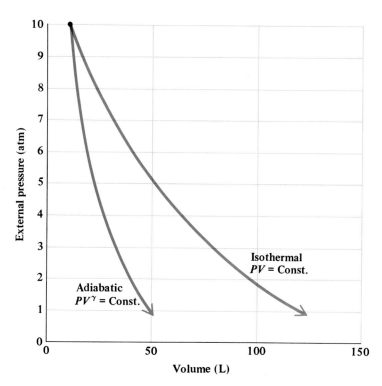

**Figure 8-17**
A comparison of reversible isothermal and adiabatic expansions.

## 8-6  A Deeper Look . . . Heat Engines and Refrigerators

A **cyclic process** is one in which a system is taken from some initial state through a series of other states and finally back to the original state. The thermodynamics of such processes has many practical applications, ranging from the operation of internal combustion engines in automobiles to the principles that underlie cooling and heating systems. Beneath these applications is a fundamental goal: the demonstration of a law of nature that provides an unfailing criterion for the direction of spontaneous change.

### The Carnot Cycle

One particularly important cyclic process is called the **Carnot cycle** after Sadi Carnot, an officer in Napoleon's French Army Corps of Engineers. Carnot devised this cycle during his theoretical investigations of the efficiency of conversion of heat to work in engines. The modern steam engine developed by James Watt had become a promising source of power, and armies needed reliable and efficient ways to transport cannons, construct bridges, and haul materiel. In the course of his study of this immensely practical problem, Carnot discovered some of

the fundamental principles underlying thermodynamics, although his classic paper "On the Motive Power of Heat" was not appreciated until a quarter century later, when it was cited by William Thompson (later Lord Kelvin).

In a Carnot cycle (Fig. 8-18), a system traverses two isothermal and two adiabatic paths to return to its original state. Each path is carried out reversibly (that is, in thermal equilibrium with internal and external forces nearly balanced at every step). As the system proceeds from state A to state C through state B, the system performs work (Fig. 8-18a). Of course, work performed *on* a system in a reversible process was defined to be

$$w_{ABC} = -\int_{V_A}^{V_C} P\,dV$$
(Path ABC)

and it is clearly negative for this (expansion) process. If the system then returns from state C to A through state D,

$$w_{CDA} = -\int_{V_C}^{V_A} P\,dV$$
(Path CDA)

*(a)* Expansion

*(b)* Compression

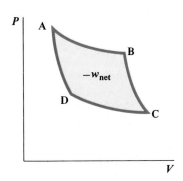

*(c)* Complete cycle

## Figure 8-18

Stages of the Carnot cycle. The net work done per cycle is the area enclosed by the curve ABCDA.

which is now a positive quantity for this (compression) process; work is performed upon the system (Fig. 8-18*b*). In the course of the cycle, the work performed by the system is the area under curve ABC, whereas that performed on the system is the (smaller) area under curve ADC. The net result of the whole cycle is that work is performed by the system, and the amount of this work is the difference between the two areas, which is the area enclosed by the cycle (Fig. 8-18*c*). As in any cyclic

process, the overall energy change $\Delta E$ is zero, and so the net work equals the negative of the total heat added to the system.

We have not yet specified the material contained in the system. Assume initially that it is an ideal gas, for which the results from Section 8-5 apply directly, with $T_h$ defined to be the higher temperature in the cycle and $T_l$ the lower.

**Path AB: Isothermal Expansion (temperature $T_h$)**

$$w_{AB} = -q_{AB} = -nRT_h \ln\left(\frac{V_B}{V_A}\right)$$

**Path BC: Adiabatic Expansion**

$$q_{BC} = 0$$
$$w_{BC} = nc_V(T_l - T_h) = -nc_V(T_h - T_l)$$

**Path CD: Isothermal Compression (temperature $T_l$)**

$$w_{CD} = -q_{CD} = -nRT_l \ln\left(\frac{V_D}{V_C}\right)$$
$$= nRT_l \ln\left(\frac{V_C}{V_D}\right)$$

**Path DA: Adiabatic Compression**

$$q_{DA} = 0$$
$$w_{DA} = nc_V(T_h - T_l)$$

The net work done on the system is

$$w_{net} = w_{AB} + w_{BC} + w_{CD} + w_{DA}$$
$$= -nRT_h \ln\left(\frac{V_B}{V_A}\right) - nc_V(T_h - T_l)$$
$$+ nRT_l \ln\left(\frac{V_C}{V_D}\right) + nc_V(T_h - T_l)$$
$$= -nRT_h \ln\left(\frac{V_B}{V_A}\right) + nRT_l \ln\left(\frac{V_C}{V_D}\right)$$

This can be simplified by noting that $V_B$ and $V_C$ lie on one adiabatic path, and $V_A$ and $V_D$ lie on another. In Section 8-5, the relation for a reversible adiabatic process was found:

$$\frac{T_2}{T_1} = \left(\frac{V_1}{V_2}\right)^{\gamma - 1}$$

Hence,

$$\frac{T_h}{T_l} = \left(\frac{V_C}{V_B}\right)^{\gamma - 1} \qquad \text{for path BC}$$

and

$$\frac{T_h}{T_l} = \left(\frac{V_D}{V_A}\right)^{\gamma - 1} \qquad \text{for path DA}$$

Equating these expressions gives

$$\left(\frac{V_C}{V_B}\right)^{\gamma-1} = \left(\frac{V_D}{V_A}\right)^{\gamma-1}$$

or

$$\frac{V_C}{V_B} = \frac{V_D}{V_A} \quad \text{and} \quad \frac{V_B}{V_A} = \frac{V_C}{V_D}$$

Hence, the net work done in one passage around the Carnot cycle is

$$w_{net} = -nR(T_h - T_l)\ln\frac{V_B}{V_A} \tag{8-15}$$

## Heat Engines

The Carnot cycle is an idealized model for a heat engine. When a certain amount of heat, $q_{AB}$, is added to the system at the higher temperature $T_h$, a net amount of work $-w_{net}$ is obtained from the system. In addition, some heat $q_{CD}$ is discharged at the lower temperature, but this energy is "degraded" and is no longer available for use in the engine. The **efficiency** $\epsilon$ of such an engine is the ratio of the negative of the net work done on the system, $-w_{net}$, to the heat added along the high-temperature isothermal path:

$$\epsilon = \frac{-w_{net}}{q_{AB}}$$

It is the net work that is available for the performance of useful mechanical tasks such as turning electrical generators or dynamos, but it is the heat $q_{AB}$ absorbed at the higher temperature $T_h$ that must be "paid for" in terms of coal or oil consumed to supply it. The efficiency $\epsilon$ must be maximized to get out the most work possible for the lowest cost.

For the ideal gas Carnot cycle, the efficiency is easily calculated:

$$\epsilon = \frac{-w_{net}}{q_{AB}} = \frac{nR(T_h-T_l)\ln(V_B/V_A)}{nRT_h\ln(V_B/V_A)}$$

$$= \frac{T_h-T_l}{T_h} = 1 - \frac{T_l}{T_h} \tag{8-16}$$

This result places a fundamental limit on the efficiency with which heat can be converted to mechanical work. Only if the high temperature $T_h$ were infinite or the low temperature $T_l$ were 0 would it be possible to have a heat engine that operates with 100% efficiency. To maximize efficiency, the greatest possible temperature difference should be used. Although this result has been derived only for the ideal gas, it will be shown later in this section that it applies to *any* reversible engine operating between two temperatures. For a *real* engine (which must operate irreversibly) the actual efficiency must be lower than this ideal Carnot efficiency, or **thermodynamic efficiency.**

## Example 8-11

Suppose a heat engine absorbs 10.0 kJ of heat from a high-temperature source at $T_h = 450$ K and discards heat to a low-temperature reservoir at $T_l = 350$ K. Calculate the thermodynamic efficiency $\epsilon$ of conversion of heat to work; the amount of work performed, $-w_{net}$; and the amount of heat discharged at $T_l$, $q_{CD}$.

### Solution

$$\epsilon = \frac{T_h - T_l}{T_h} = \frac{450\text{ K} - 350\text{ K}}{450\text{ K}} = 0.222$$

so that the engine can be, at most, 22.2% efficient. Because $\epsilon = -w_{net}/q_{AB}$,

$$w_{net} = -\epsilon q_{AB} = -(0.222)(10.0\text{ kJ}) = -2.22\text{ kJ}$$

Because $\Delta E$ for the whole cycle is 0,

$$\Delta E = 0 = q_{AB} + q_{CD} + w_{net}$$

$$q_{CD} = -q_{AB} - w_{net} = -10.0 + 2.22 = -7.8\text{ kJ}$$

Therefore, 7.8 kJ is discharged at 350 K. This heat must be removed from the vicinity of the engine by a cooling system, as otherwise it will cause $T_l$ to increase and reduce the efficiency of the engine.

*Related Problems:* 55, 56

## Refrigerators and Heat Pumps

If the direction in which the Carnot cycle is traversed is reversed and the system is taken along path ADCBA, then net work is done on the system, heat is absorbed from the low-temperature reservoir and discharged to the high-temperature reservoir, and a **refrigeration machine** results. The same Carnot equation applies in this case:

$$\frac{T_h - T_l}{T_h} = -\frac{w_{net}}{q_{AB}}$$

In the case of a refrigerator, there is no direct interest in the heat $q_{AB}$ that is discharged at the higher temperature. It must, of course, be removed to prevent $T_h$ from increasing and reducing the operating efficiency (for this reason, adequate air ventilation at the back of a refrigerator is necessary). What is of interest is how much heat ($q_{CD}$) will be absorbed from the interior of the refrigerator to keep its contents cold for a given amount of net work done on the system during a refrigeration cycle. It is necessary merely to solve the Carnot equation for $q_{AB}$ and substitute it into

$$q_{AB} + q_{CD} + w_{net} = \Delta E = 0$$

to obtain

$$q_{CD} = w_{net}\left(\frac{T_1}{T_h - T_1}\right)$$

### Example 8-12

Suppose the interior of a refrigerator is to be maintained at 275 K while the temperature of the room is 303 K. Suppose $w_{net} = 1.00$ kJ of work is done in each cycle. How much heat is withdrawn from the refrigerator per cycle, and how much is discharged into the room at $T_h$?

### Solution

The heat withdrawn from the refrigerator per cycle is

$$q_{CD} = w_{net}\left(\frac{T_1}{T_h - T_1}\right)$$

$$= 1.00 \text{ kJ}\left(\frac{275 \text{ K}}{303 \text{ K} - 275 \text{ K}}\right) = 9.8 \text{ kJ}$$

An interesting fact emerges. The Carnot cycle, operating as a refrigerator, is very efficient: the expenditure of 1.00 kJ of work permits the removal of 9.8 kJ of heat from the lower-temperature reservoir. For a refrigerator, the *smaller* the temperature difference, the more efficient the operation. The heat discharged at $T_h = 303$ K is

$$q_{AB} = -q_{CD} - w_{net} = -9.8 \text{ kJ} - 1.00 \text{ kJ} = -10.8 \text{ kJ}$$

*Related Problem: 57*

A **heat pump** is a thermodynamic engine operated ''in reverse'' like a refrigerator, in that it absorbs heat from a low-temperature thermal reservoir and discharges it to a thermal reservoir at high temperature. Its principal use is to heat buildings, and so now the focus is on the heat delivered at the higher temperature. A heat pump removes heat from the earth or the surrounding atmosphere and delivers it to the inside of the building in order to keep the building warm. In principle, a heat pump can be quite efficient because the amount of work done (and paid for in the form of electricity) can be small relative to the amount of heat delivered into the house; each kilojoule of work causes the transfer of several kilojoules of heat. By contrast, in an ordinary electrical heating system the electrical work is converted directly to heat, and the amount of heat produced is the same as the amount of work done. Heat pumps are fairly widely used, but their higher initial cost partially offsets their greater economy of operation.

Real refrigerators and heat pumps do not use ideal gases as their working fluids, nor do they follow a strict Carnot cycle. Instead, they make use of the heat absorbed when a volatile liquid is vaporized and the heat liberated when the vapor condenses in a cyclic operation. A nontoxic fluid such as $CCl_2F_2$, known as Freon-12 (b.p. $-29.8°C$), has commonly been used in refrigerators, although other working fluids are now under investigation (because of the damage chlorofluorocarbons do to the ozone layer). Inside the refrigerator the liquid absorbs heat and is vaporized. It is then compressed by a compressor and cooled by the air outside the refrigerator until it has recondensed to a liquid. When the thermostat shows that additional cooling is needed, the liquid Freon is again allowed to enter the tubes inside the refrigerator to absorb heat and be vaporized.

### Efficiency of General Carnot Engines

In this section, all Carnot engines operating reversibly between two temperatures $T_h$ and $T_1$ are shown to have the same efficiency:

$$\epsilon = \frac{-w_{net}}{q_h} = \frac{T_h - T_1}{T_h} \qquad (8\text{-}17)$$

In other words, the efficiency that was calculated for an ideal gas applies equally to any other working fluid. To demonstrate this, the *contrary* is assumed, and that assumption is shown to lead to a contradiction with experience; the assumption is therefore deemed false.

Assume that there *are* two reversible machines operating between the same two temperatures $T_h$ and $T_1$, one of which has an efficiency $\epsilon_1$ that is greater than the efficiency $\epsilon_2$ of the other. The two machines are adjusted so that the total work output is the same for both. The more efficient machine is run as a heat engine so that it produces mechanical work $-w_1$. This work is used to operate the other machine in the opposite sense (as a heat pump), so that $w_2 = -w_1$. The net work input to the *combined* machines is then 0, because the work produced by the first machine is used to run the second.

Now examine what happens to heat in this situation. Because engine 1 is more efficient than engine 2, and because the work is the same for both, engine 1 must withdraw less heat from the hot reservoir at $T_h$ than is discharged into the same reservoir by engine 2; there is a net transfer of heat *into* the high-temperature reservoir. For the combined engines, $\Delta E_{tot} = w_{tot} = 0$; so $q_{tot}$ must also be 0 and a net transfer of heat *out of* the low-temperature reservoir must therefore occur. What has been devised is an apparatus that can transfer heat from a low-temperature reservoir to a high-temperature reservoir with no net expenditure of work.

*But this is impossible.* All of our experience shows that one cannot make a device that permits transfer of heat from a cold body to a hot body without doing work. In fact, exactly the opposite is seen: heat flows spontaneously from hotter to colder bodies. Our experience can be summarized in the statement

It is impossible to construct a device that will transfer heat from a cold reservoir to a hot reservoir in a continuous cycle with no net expenditure of work.

This is one form of the **second law of thermodynamics,** as stated by Rudolf Clausius.

Let us return to our argument. An assumption was made that led to a conclusion that contradicted experience. Therefore, the original assumption must have been wrong, and there *cannot* be two reversible engines operating between the same two temperatures with different efficiencies. In other words, all Carnot engines must have the same efficiency, which has already been calculated to be

$$\epsilon = \frac{T_{\mathrm{h}} - T_{\mathrm{l}}}{T_{\mathrm{h}}}$$

for the ideal gas. For any substance, whether ideal or not, undergoing a Carnot cycle,

$$\epsilon = \frac{-w_{\mathrm{net}}}{q_{\mathrm{h}}} = \frac{q_{\mathrm{h}} + q_{\mathrm{l}}}{q_{\mathrm{h}}} = \frac{T_{\mathrm{h}} - T_{\mathrm{l}}}{T_{\mathrm{h}}}$$

The Carnot cycle forms the basis for a *thermodynamic* scale of temperature. Because $\epsilon = 1 - (T_{\mathrm{l}}/T_{\mathrm{h}})$, the Carnot efficiencies determine temperature ratios and thereby establish a temperature scale. In practice, the difficulty of operating real engines close to the reversible limit makes this procedure rather impractical; instead, real gases at low pressures are used to define and determine temperatures.

The last two terms of the preceding equation can be rewritten as

$$1 + \frac{q_{\mathrm{l}}}{q_{\mathrm{h}}} = 1 - \frac{T_{\mathrm{l}}}{T_{\mathrm{h}}}$$

and simplified to

$$\frac{q_{\mathrm{h}}}{T_{\mathrm{h}}} + \frac{q_{\mathrm{l}}}{T_{\mathrm{l}}} = 0 \qquad \textbf{(8-18)}$$

This simple equation has profound importance, because it contains the essence of the second law of thermodynamics. To see this, consider a *general* reversible cyclic process, and draw a series of closely spaced adiabatics, as shown in Figure 8-19. (An adiabatic is a representation on the phase diagram of thermodynamic states connected by a reversible adiabatic process.) Now replace each segment along the given cycle (ABDCA) with a series of alternating isothermal and adiabatic segments. It is clear that by taking more and more closely spaced adiabatics, a path can be obtained that is arbitrarily close to the desired curve.

Now follow the evolution of $\Sigma_i(q_i/T_i)$ along this curve. The key observation is that the contributions $q_i/T_i$ appear in pairs: for each $q_{\mathrm{h},i}/T_{\mathrm{h},i}$ from an isothermal segment along the ABC path, there is a $q_{\mathrm{l},i}/T_{\mathrm{l},i}$ along the CDA path. Any given pair forms two sides of a Carnot cycle, with the other two sides determined by adiabatics, and so

$$\frac{q_{\mathrm{h},i}}{T_{\mathrm{h},i}} + \frac{q_{\mathrm{l},i}}{T_{\mathrm{l},i}} = 0$$

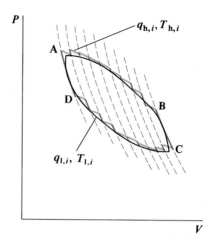

**Figure 8-19**
A general cyclic process (ABCDA) can be approximated to arbitrary accuracy by the sum of a series of Carnot cycles.

Summing over $i$ shows that

$$\sum_i \left( \frac{q_{\mathrm{h},i}}{T_{\mathrm{h},i}} + \frac{q_{\mathrm{l},i}}{T_{\mathrm{l},i}} \right) = 0$$

Because this summation follows the original reversible path to an arbitrary accuracy (if the lengths of the segments are made short enough), it follows that

$$\int \frac{1}{T} \, dq_{\mathrm{rev}} = 0$$

for *any* closed, reversible path. Although $q$ is not a state function, $q_{\mathrm{rev}}/T$ *is* a state function because, like energy and enthalpy, its total change is 0 for any process that begins and ends in the same state. In Chapter 9 the function $q_{\mathrm{rev}}/T$ is identified with the change in *entropy* of a system when it undergoes a process reversibly at a temperature $T$.

As we have stated, the principles of thermodynamics are based on observations of nature and are not subject to rigorous proof. We have accomplished something quite significant, however. From the premise (based on physical observation) that heat cannot be transferred from a low-temperature body to a high-temperature body without expenditure of work, we have derived the result that $\int (1/T) \, dq_{\mathrm{rev}}$ is independent of path and is therefore a state function. Some might say, "If we have to make some assumption anyway, why don't we just assume that $\int (1/T) \, dq_{\mathrm{rev}}$ is path-independent to start with?" This is certainly possible, and is practiced in many presentations of thermodynamics. The approach followed here has been different; we have preferred to base our assumptions directly on physical observation rather than on abstract mathematical axioms.

## Methanol as a Gasoline Substitute

Methanol ($CH_3OH$) is employed as a substitute for gasoline in certain high-performance vehicles. To design engines that will run on it, its thermochemistry must be understood.

(a) The methanol in an automobile engine must be in the gas phase before it can react. Calculate the heat (in kilojoules) that must be added to 1.00 kg of liquid methanol to raise its temperature from 25.0°C to its normal boiling point, 65.0°C. The molar heat capacity of liquid methanol is 81.6 J $K^{-1}$ $mol^{-1}$.

(b) Once the methanol has reached its boiling point, it must be vaporized. The molar enthalpy of vaporization for methanol is 38 kJ $mol^{-1}$. How much heat must be added to vaporize 1.00 kg of methanol?

(c) Once it is in the vapor phase, the methanol can react with oxygen in the air according to

$$CH_3OH(g) + \tfrac{3}{2}O_2(g) \longrightarrow CO_2(g) + 2\,H_2O(g)$$

A methanol-powered bus. (Vanessa Vick/Photo Researchers)

Use average bond enthalpies to estimate the enthalpy change in this reaction, for one mole of methanol reacting.

(d) Use data from Appendix D to calculate the actual enthalpy change in this reaction, assuming it to be the same at 65°C as at 25°C.

(e) Calculate the heat released when 1.00 kg of gaseous methanol is burned in air at constant pressure. Use the more accurate result of part (d) rather than that of part (c).

(f) Calculate the *difference* between the change in enthalpy and the change in internal energy when 1.00 kg of gaseous methanol is oxidized to gaseous $CO_2$ and $H_2O$ at 65°C.

(g) Suppose now that the methanol is burned inside the cylinder of an automobile. Taking the radius of the cylinder to be 4.0 cm and the distance moved by the piston during one stroke to be 12 cm, calculate the work done on the gas per stroke as it expands against an external pressure of 1.00 atm. Express your answer in liter-atmospheres and in joules.

### Answers

(a) 102 kJ

(b) $1.2 \times 10^3$ kJ

(c) −508 kJ

(d) −676.49 kJ

(e) $2.11 \times 10^4$ kJ

(f) $\Delta H - \Delta E = 43.9$ kJ

(g) −0.60 L atm = −61 J

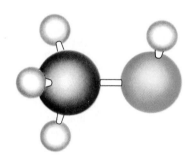

Methanol, $CH_3OH$.

---

After studying this chapter and working the problems that follow, you should be able to

1. Give precise definitions for the terms thermodynamic system, open system, closed system, thermodynamic state, reversible and irreversible process (Section 8-1).

2. Define and give examples of properties that are state functions of a system (Section 8-1).

3. Calculate the work done on an ideal gas when it is compressed reversibly (Section 8-2, problems 1–2).

4. Give a physical interpretation to the concept of heat, and calculate the change in temperature of a given quantity of a substance from its heat capacity and the amount of heat transferred to it (Section 8-2, problems 5–8).

5. Calculate the final temperature reached when two substances of different mass, heat capacity, and temperature are placed in thermal contact (Section 8-2, problems 9–12).

6. Calculate the amounts of heat and work and the change in energy of an ideal gas during expansions and compressions (Section 8-3, problems 15–22).

7. Calculate the energy and enthalpy changes for chemical reactions from the standard molar enthalpies of formation of reactants and products (Section 8-4, problems 35–40).

8. Use bond enthalpies to estimate enthalpies of formation of gaseous compounds (Section 8-4, problems 45–48).

9. Calculate the heat absorbed and work done by an ideal gas when it expands reversibly and either isothermally or adiabatically (Section 8-5, problems 51–54).

10. Calculate thermodynamic efficiencies of heat engines, heat pumps, and refrigerators. In each case, relate the work done per cycle to the heat absorbed and to that given off (Section 8-6, problems 55–58).

## PROBLEMS

### Energy, Work, and Heat

1. Some nitrogen for use in synthesizing ammonia is heated slowly, maintaining the external pressure close to the internal pressure of 50.0 atm, until its volume has increased from 542 L to 974 L. Calculate the work done on the nitrogen as it is heated, and express it in joules.

2. The gas mixture inside one of the cylinders of an automobile engine expands against a constant external pressure of 0.98 atm, from an initial volume of 150 mL (at the end of the compression stroke) to a final volume of 800 mL. Calculate the work done on the gas mixture during this process, and express it in joules.

3. When a ball of mass $m$ is dropped through a height difference $\Delta h$, its potential energy changes by the amount $mg\,\Delta h$ where $g$ is the acceleration of gravity, equal to $9.81 \text{ m s}^{-2}$. Suppose that when the ball hits the ground, all that energy is converted to heat, increasing the temperature of the ball. If the specific heat capacity of the material in the ball is $0.850 \text{ J K}^{-1} \text{ g}^{-1}$, calculate the height from which the ball must be dropped to increase the temperature of the ball by 1.00°C.

4. During his honeymoon in Switzerland, James Joule is said to have used a thermometer to measure the temperature difference between the water at the top and at the bottom of a waterfall. Take the height of the waterfall to be $\Delta h$ and the acceleration of gravity $g$ to be $9.81 \text{ m s}^{-2}$. Assuming that all

the potential energy change $mg\,\Delta h$ of a mass $m$ of water is used to heat that water by the time it reaches the bottom, calculate the temperature difference between the top and the bottom of a waterfall 100 m high. Take the specific heat capacity of water to be $4.18 \text{ J K}^{-1} \text{ g}^{-1}$.

5. The specific heat capacities of $Li(s)$, $Na(s)$, $K(s)$, $Rb(s)$, and $Cs(s)$ at 25°C are 3.57, 1.23, 0.756, 0.363, and 0.242 $\text{J K}^{-1} \text{ g}^{-1}$, respectively. Compute the molar heat capacities of these elements and identify any periodic trend. If there is a trend, use it to predict the molar heat capacity of francium, $Fr(s)$.

6. The specific heat capacities of $F_2(g)$, $Cl_2(g)$, $Br_2(g)$, and $I_2(g)$ are 0.824, 0.478, 0.225, and 0.145 $\text{J K}^{-1} \text{ g}^{-1}$, respectively. Compute the molar heat capacities of these elements and identify any periodic trend. If there is a trend, use it to predict the molar heat capacity of astatine, $At_2(g)$.

7. The specific heat capacities of the metals nickel, zinc, rhodium, tungsten, gold, and uranium at 25°C are 0.444, 0.388, 0.243, 0.132, 0.129, and 0.116 $\text{J K}^{-1} \text{ g}^{-1}$, respectively. Calculate the molar heat capacities of these six metals. Note how closely the molar heat capacities for these metals, which were selected at random, cluster about a value of $25 \text{ J K}^{-1} \text{ mol}^{-1}$. The rule of Dulong and Petit states that the molar heat capacities of the metallic elements are approximately $25 \text{ J K}^{-1} \text{ mol}^{-1}$.

8. Use the empirical rule of Dulong and Petit stated in problem

7 to estimate the specific heat capacities of vanadium, gallium, and silver.

9. Suppose 61.0 g of hot metal, which is initially at 120.0°C, is plunged into 100.0 g of water that is initially at 20.00°C. The metal cools down and the water heats up until they reach a common temperature of 26.39°C. Calculate the specific heat capacity of the metal, using $4.18 \text{ J K}^{-1} \text{ g}^{-1}$ as the specific heat capacity of the water.

10. A piece of zinc at 20.0°C that weighs 60.0 g is dropped into 200.0 g of water at 100.0°C. The specific heat capacity of zinc is $0.389 \text{ J K}^{-1} \text{ g}^{-1}$, and that of water near 100°C is $4.22 \text{ J K}^{-1} \text{ g}^{-1}$. Calculate the final temperature reached by the zinc and the water.

11. Very early in the study of the nature of heat it was observed that if two bodies of equal mass but different temperatures are placed in thermal contact, their specific heat capacities depend inversely on the change in temperature each undergoes on reaching its final temperature. Write a mathematical equation in modern notation to express this fact.

12. Iron pellets weighing 17.0 g at a temperature of 92.0°C are mixed in an insulated container with 17.0 g of water at a temperature of 20.0°C. The specific heat capacity of water is ten times greater than that of iron. What is the final temperature inside the container?

13. In their *Memoir on Heat,* published in 1783, Lavoisier and Laplace reported, ''The heat necessary to melt ice is equal to three quarters of the heat that can raise the same mass of water from the temperature of the melting ice to that of boiling water'' (English translation). Use this 18th-century observation to compute the amount of heat (in joules) needed to melt 1.00 g of ice. Assume that heating 1.00 g of water requires 4.18 J of heat for each 1.00°C throughout the range from 0°C to 100°C.

14. Galen, the great physician of antiquity, suggested scaling temperature from a reference point defined by mixing equal masses of ice and boiling water in an insulated container. Imagine that this is done with the ice at 0.00°C and the water at 100.0°C. Assume that the heat capacity of the container is negligible and that it takes 333.4 J of heat to melt 1.000 g of ice at 0.00°C to water at 0.00°C. Compute Galen's reference temperature in degrees Celsius.

## The First Law of Thermodynamics

15. If 0.500 mol of neon at 1.00 atm and 273 K expands against a constant external pressure of 0.100 atm until the gas pressure reaches 0.200 atm and the temperature reaches 210 K, calculate the work done on the gas, the internal energy change, and the heat absorbed by the gas.

16. Hydrogen behaves as an ideal gas at temperatures above 200 K and at pressures below 50 atm. Suppose 6.00 mol of hydrogen is initially contained in a 100-L vessel at a pressure of 2.00 atm. The average molar heat capacity of hydrogen at constant pressure, $c_P$, is $29.3 \text{ J K}^{-1} \text{ mol}^{-1}$ in the temperature range of this problem. The gas is cooled reversibly at constant pressure from its initial state to a volume of 50 L. Calculate the following quantities for this process.
(a) Temperature of the gas in the final state, $T_2$
(b) Work done on the gas, $w$, in joules
(c) Internal energy change of the gas, $\Delta E$, in joules
(d) Heat absorbed by the gas, $q$, in joules

17. Suppose 2.00 mol of an ideal, monatomic gas is initially at a pressure of 3.00 atm and a temperature $T = 350$ K. It is expanded *irreversibly* and *adiabatically* $(q = 0)$ against a constant external pressure of 1.00 atm until the volume has doubled.
(a) Calculate the final volume.
(b) Calculate $w$, $q$, and $\Delta E$ for this process, in joules.
(c) Calculate the final temperature of the gas.

18. Consider the free, isothermal (constant $T$) expansion of an ideal gas. ''Free'' means that the external force is zero, perhaps because a stopcock has been opened and the gas is allowed to expand into a vacuum. Calculate $\Delta E$ for this irreversible process. Show that $q = 0$ so that the expansion is also adiabatic $(q = 0)$ for an ideal gas. This is analogous to a classic experiment first carried out by Joule.

19. If 6.00 mol of argon in a 100-L vessel initially at 300 K is compressed adiabatically $(q = 0)$ and irreversibly until a final temperature of 450 K is reached, calculate the energy change of the gas, the heat added to the gas, and the work done on the gas.

20. A gas expands against a constant external pressure of 2.00 atm until its volume has increased from 6.00 L to 10.00 L. During this process it absorbs 500 J of heat from the surroundings.
(a) Calculate the energy change of the gas $\Delta E$.
(b) Calculate the work $w$ done on the gas in an irreversible adiabatic $(q = 0)$ process connecting the same initial and final states.

21. A chemical system is sealed in a strong, rigid container at room temperature and then heated vigorously.
(a) State whether $\Delta E$, $q$, and $w$ of the system are positive, negative, or zero during the heating process.
(b) Next, the container is cooled to its original temperature. Determine the signs of $\Delta E$, $q$, and $w$ for the cooling process.
(c) Designate heating as step 1 and cooling as step 2. Determine the signs of $(\Delta E_1 + \Delta E_2)$, $(q_1 + q_2)$, and $(w_1 + w_2)$ if possible.

22. A battery harnesses a chemical reaction to extract energy in the form of useful electrical work.
(a) A certain battery runs a toy truck and becomes partially discharged. In the process, it performs a total of 117.0 J of work on its immediate surroundings. It also gives off 3.0 J of heat, which the surroundings absorb. No other work or heat is exchanged with the surroundings. Compute $q$, $w$, and $\Delta E$ of the battery, making sure each quantity has the proper sign.
(b) The same battery is now recharged exactly to its original condition. This requires 210.0 J of electrical work from an

outside generator. Determine $q$ for the battery in this process. Explain why $q$ has the sign that it does.

## Thermochemistry

23. For each of the following reactions, the enthalpy change written is that measured when the numbers of moles of reactants and products taking part in the reaction are as given by their coefficients in the equation. Calculate the enthalpy change when 1.00 *gram* of the underlined substance is consumed or produced.

    (a) $4\,Na(s) + O_2(g) \rightarrow 2\,\underline{Na_2O}(s)$      $\Delta H = -828\,kJ$

    (b) $CaMg(CO_3)_2(s) \rightarrow CaO(s) + \underline{MgO}(s) + 2\,CO_2(g)$
             $\Delta H = +302\,kJ$

    (c) $H_2(g) + 2\,\underline{CO}(g) \rightarrow H_2O_2(\ell) + 2\,C(s)$
             $\Delta H = +33.3\,kJ$

24. For each of the following reactions, the enthalpy change given is that measured when the numbers of moles of reactants and products taking part in the reaction are as given by their coefficients in the equation. Calculate the enthalpy change when 1.00 *gram* of the underlined substance is consumed or produced.

    (a) $Ca(s) + \underline{Br_2}(\ell) \rightarrow CaBr_2(s)$     $\Delta H = -683\,kJ$

    (b) $6\,Fe_2O_3(s) \rightarrow 4\,\underline{Fe_3O_4}(s) + O_2(g)$    $\Delta H = +472\,kJ$

    (c) $2\,\underline{NaHSO_4}(s) \rightarrow 2\,NaOH(s) + 2\,SO_2(g) + O_2(g)$
             $\Delta H = +806\,kJ$

25. Liquid bromine dissolves readily in aqueous NaOH:

    $$Br_2(\ell) + 2\,NaOH(aq) \longrightarrow$$
    $$NaBr(aq) + NaOBr(aq) + H_2O(\ell)$$

    Suppose $2.88 \times 10^{-3}$ mol of $Br_2(\ell)$ is sealed in a glass capsule that is then immersed in a solution containing excess $NaOH(aq)$. The capsule is broken, the mixture is stirred, and a measured 121.3 J of heat evolves. In a separate experiment, simply breaking an empty capsule and stirring the solution in the same way evolves 2.34 J of heat. Compute the heat evolved as 1.00 mol of $Br_2(\ell)$ dissolves in excess $NaOH(aq)$.

26. A chemist mixes 1.00 g of $CuCl_2$ with an excess of $(NH_4)_2HPO_4$ in dilute aqueous solution. He measures the evolution of 670 J of heat as the two substances react to give $Cu_3(PO_4)_2(s)$. Compute the $\Delta H$ that would result from the reaction of 1.00 mol of $CuCl_2$ with an excess of $(NH_4)_2HPO_4$.

27. Calculate the enthalpy change when 2.38 g of carbon monoxide (CO) vaporizes at its normal boiling point. Use data from Table 8-2.

28. Molten sodium chloride is used for making elemental sodium and chlorine. Suppose the electrical power to a vat containing 56.2 kg of molten sodium chloride is cut off, and the salt crystallizes (without changing its temperature). Calculate the enthalpy change, using data from Table 8-2.

29. Suppose an ice cube weighing 36.0 g at a temperature of $-10°C$ is placed in 360 g of water at a temperature of $20°C$. Calculate the temperature after thermal equilibrium is

reached, assuming no heat loss to the surroundings. The enthalpy of fusion of ice is $\Delta H_{fus} = 6.007\,kJ\,mol^{-1}$, and the molar heat capacities $c_P$ of ice and water are 38 and 75 $J\,K^{-1}\,mol^{-1}$, respectively.

30. You have a supply of ice at $0.0°C$ and a glass containing 150 g of water at $25°C$. The enthalpy of fusion for ice is $\Delta H_{fus} = 333\,J\,g^{-1}$, and the specific heat capacity of water is 4.18 $J\,K^{-1}\,g^{-1}$. How many grams of ice must be added to the glass (and melted) to reduce the temperature of the water to $0°C$?

31. The measured enthalpy change for burning ketene ($CH_2CO$)

    $$CH_2CO(g) + 2\,O_2(g) \longrightarrow 2\,CO_2(g) + H_2O(g)$$

    is $\Delta H_1 = -981.1\,kJ$ at $25°C$. The enthalpy change for burning methane

    $$CH_4(g) + 2\,O_2(g) \longrightarrow CO_2(g) + 2\,H_2O(g)$$

    is $\Delta H_2 = -802.3\,kJ$ at $25°C$. Calculate the enthalpy change at $25°C$ for the reaction

    $$2\,CH_4(g) + 2\,O_2(g) \longrightarrow CH_2CO(g) + 3\,H_2O(g)$$

32. Given the following two reactions and corresponding enthalpy changes,

    $$CO(g) + SiO_2(s) \longrightarrow SiO(g) + CO_2(g)$$
    $$\Delta H = +520.9\,kJ$$

    $$8\,CO_2(g) + Si_3N_4(s) \longrightarrow 3\,SiO_2(s) + 2\,N_2O(g) + 8\,CO(g)$$
    $$\Delta H = +461.05\,kJ$$

    compute the $\Delta H$ of the reaction

    $$5\,CO_2(g) + Si_3N_4(s) \longrightarrow$$
    $$3\,SiO(g) + 2\,N_2O(g) + 5\,CO(g)$$

33. The enthalpy change to make diamond from graphite is $1.88\,kJ\,mol^{-1}$. Which gives off more heat when burned—a pound of diamonds or a pound of graphite? Explain.

34. The enthalpy change of combustion of monoclinic sulfur to $SO_2(g)$ is $-9.376\,kJ\,g^{-1}$. Under the same conditions, the rhombic form of sulfur has an enthalpy change of combustion to $SO_2(g)$ of $-9.293\,kJ\,g^{-1}$. Compute the $\Delta H$ of the reaction

    $$S(monoclinic) \longrightarrow S(rhombic)$$

    per gram of sulfur reacting.

35. Calculate the standard enthalpy change $\Delta H°$ at $25°C$ for the reaction

    $$N_2H_4(\ell) + 3\,O_2(g) \longrightarrow 2\,NO_2(g) + 2\,H_2O(\ell)$$

    using the standard enthalpies of formation ($\Delta H_f°$) of reactants and products at $25°C$ from Appendix D.

36. Using the data in Appendix D, calculate $\Delta H°$ for each of the following processes.

    (a) $2\,NO(g) + O_2(g) \rightarrow 2\,NO_2(g)$

    (b) $C(s) + CO_2(g) \rightarrow 2\,CO(g)$

    (c) $2\,NH_3(g) + \frac{7}{2}\,O_2(g) \rightarrow 2\,NO_2(g) + 3\,H_2O(g)$

    (d) $C(s) + H_2O(g) \rightarrow CO(g) + H_2(g)$

37. Zinc is commonly found in nature in the form of the mineral sphalerite (ZnS). A step in the smelting of zinc is the roasting of sphalerite with oxygen to produce zinc oxide:

$$2\,ZnS(s) + 3\,O_2(g) \longrightarrow 2\,ZnO(s) + 2\,SO_2(g)$$

(a) Calculate the standard enthalpy change $\Delta H°$ for this reaction, using data from Appendix D.
(b) Calculate the heat absorbed when 3.00 metric tons (1 metric ton = $10^3$ kg) of sphalerite is roasted under constant-pressure conditions.

38. The thermite process (see Fig. 8-11) is used for welding railway track together. In this reaction, aluminum reduces iron(III) oxide to metallic iron:

$$2\,Al(s) + Fe_2O_3(s) \longrightarrow 2\,Fe(s) + Al_2O_3(s)$$

Igniting a small charge of barium peroxide mixed with aluminum triggers the reaction of a mixture of aluminum powder and iron(III) oxide; the molten iron produced flows into the space between the steel rails that are to be joined.
(a) Calculate the standard enthalpy change $\Delta H°$ for this reaction, using data from Appendix D.
(b) Calculate the heat given off when 3.21 g of iron(III) oxide is reduced by aluminum at constant pressure.

39. The dissolution of calcium chloride in water

$$CaCl_2(s) \longrightarrow Ca^{2+}(aq) + 2\,Cl^-(aq)$$

is used in first-aid hot packs. In these packs, an inner pouch containing the salt is broken, allowing the salt to dissolve in the surrounding water.
(a) Calculate the standard enthalpy change $\Delta H°$ for this reaction, using data from Appendix D.
(b) Suppose 20.0 g of $CaCl_2$ is dissolved in 0.100 L of water at 20.0°C. Calculate the temperature reached by the solution, assuming it to be an ideal solution with a heat capacity close to that of 100 g of pure water (418 J K$^{-1}$).

40. Ammonium nitrate dissolves in water according to the reaction

$$NH_4NO_3(s) \longrightarrow NH_4^+(aq) + NO_3^-(aq)$$

(a) Calculate the standard enthalpy change $\Delta H°$ for this reaction, using data from Appendix D.
(b) Suppose 15.0 g of $NH_4NO_3$ is dissolved in 0.100 L of water at 20.0°C. Calculate the temperature reached by the solution, assuming it to be an ideal solution with a heat capacity close to that of 100 g of pure water (418 J K$^{-1}$).
(c) From a comparison with the results of problem 39, can you suggest a practical application of this dissolution reaction?

41. The standard enthalpy change of combustion (to $CO_2(g)$ and $H_2O(\ell)$) at 25°C of the organic liquid cyclohexane ($C_6H_{12}(\ell)$) is $-3923.7$ kJ mol$^{-1}$. Determine the $\Delta H_f°$ of $C_6H_{12}(\ell)$. Use data from Appendix D.

42. The standard enthalpy change of combustion (to $CO_2(g)$ and $H_2O(\ell)$) at 25°C of the organic liquid cyclohexene ($C_6H_{10}(\ell)$) is $-3731.7$ kJ mol$^{-1}$. Determine the $\Delta H_f°$ of $C_6H_{10}(\ell)$.

43. A sample of pure solid naphthalene ($C_{10}H_8$) weighing 0.6410 g is burned completely with oxygen to $CO_2(g)$ and $H_2O(\ell)$ in a constant-volume calorimeter at 25°C. The amount of heat evolved is observed to be 25.79 kJ.
(a) Write and balance the chemical equation for the combustion reaction.
(b) Calculate the standard change in internal energy ($\Delta E°$) for the combustion of 1.000 mol of naphthalene to $CO_2(g)$ and $H_2O(\ell)$.
(c) Calculate the standard enthalpy change ($\Delta H°$) for the same reaction as in part (b).
(d) Calculate the standard enthalpy of formation per mole of naphthalene, using data for the standard enthalpies of formation of $CO_2(g)$ and $H_2O(\ell)$ from Appendix D.

44. A sample of solid benzoic acid ($C_6H_5COOH$) that weighs 0.800 g is burned in an excess of oxygen to $CO_2(g)$ and $H_2O(\ell)$ in a constant-volume calorimeter at 25°C. The temperature rise is observed to be 2.15°C. The heat capacity of the calorimeter and its contents is known to be 9382 J K$^{-1}$.
(a) Write and balance the equation for the combustion of benzoic acid.
(b) Calculate the standard change in internal energy ($\Delta E°$) for the combustion of 1.000 mol of benzoic acid to $CO_2(g)$ and $H_2O(\ell)$ at 25°C.
(c) Calculate the standard enthalpy change ($\Delta H°$) for the same reaction as in part (b).
(d) Calculate the standard enthalpy of formation per mole of benzoic acid, using data for the standard enthalpies of formation of $CO_2(g)$ and $H_2O(\ell)$ from Appendix D.

45. A second chlorofluorocarbon used as a refrigerant and in aerosols (besides that discussed in Example 8-8) is $CCl_3F$. Use the atomization enthalpies and average bond enthalpies from Table 8-3 to estimate the standard enthalpy of formation ($\Delta H_f°$) of this compound in the gas phase.

46. The compound $CF_3CHCl_2$ (with a C—C bond) has been proposed as a substitute for $CCl_3F$ and $CCl_2F_2$ because it decomposes more quickly in the atmosphere and is much less liable to reduce the concentration of ozone in the stratosphere. Use the atomization enthalpies and average bond enthalpies from Table 8-3 to estimate the standard enthalpy of formation ($\Delta H_f°$) of $CF_3CHCl_2$ in the gas phase.

47. Propane has the structure $H_3C$—$CH_2$—$CH_3$. Use average bond enthalpies from Table 8-3 to estimate the change in enthalpy $\Delta H°$ for the reaction

$$C_3H_8(g) + 5\,O_2(g) \longrightarrow 3\,CO_2(g) + 4\,H_2O(g)$$

48. Use average bond enthalpies from Table 8-3 to estimate the change in enthalpy $\Delta H°$ for the reaction

$$C_2H_4(g) + H_2(g) \longrightarrow C_2H_6(g)$$

Refer to the molecular structures on page 58.

49. The reaction

$$BBr_3(g) + BCl_3(g) \longrightarrow BBr_2Cl(g) + BCl_2Br(g)$$

has a $\Delta H$ very close to 0. Sketch the Lewis structures of the four compounds, and explain why $\Delta H$ is so small.

**50.** At 381 K, the following reaction takes place:

$$Hg_2Cl_4(g) + Al_2Cl_6(g) \longrightarrow 2\,HgAlCl_5(g)$$

$$\Delta H = +10\ kJ$$

(a) Offer an explanation for the very small $\Delta H$ for this reaction in terms of the known structures of the compounds

(b) Explain why the small $\Delta H$ in this reaction is evidence against

as the structure of $Hg_2Cl_4(g)$.

## Reversible Gas Processes

**51.** If 2.00 mol of an ideal gas at 25°C expands isothermally and reversibly from 9.00 to 36.00 L, calculate the work done on the gas and the heat absorbed by the gas in the process. What are the changes in energy ($\Delta E$) and in enthalpy ($\Delta H$) of the gas in the process?

**52.** If 54.0 g of argon at 400 K is compressed isothermally and reversibly from a pressure of 1.50 atm to 4.00 atm, calculate the work done on the gas and the heat absorbed by the gas in the process. What are the changes in energy ($\Delta E$) and in enthalpy ($\Delta H$) of the gas?

**53.** Suppose 2.00 mol of a monatomic ideal gas ($c_V = \frac{3}{2}R$) is expanded adiabatically and reversibly from a temperature $T = 300$ K, where the volume of the system is 20.0 L, to a volume of 60.0 L. Calculate the final temperature of the gas, the work done on the gas, and the energy and enthalpy changes.

**54.** Suppose 2.00 mol of an ideal gas is contained in a heat-insulated cylinder with a moveable frictionless piston. Initially the gas is at 1.00 atm and 0°C. The gas is compressed reversibly to 2.00 atm. The molar heat capacity at constant pressure, $c_P$, equals 29.3 J K$^{-1}$ mol$^{-1}$. Calculate the final temperature of the gas, the change in its internal energy $\Delta E$, and the work done on the gas.

## Heat Engines and Refrigerators

**55.** A thermodynamic engine operates cyclically and reversibly between two temperature reservoirs, absorbing heat from the high-temperature bath at 450 K and discharging heat to the low-temperature bath at 300 K.

(a) What is the thermodynamic efficiency of the engine?

(b) How much heat is discarded to the low-temperature bath if 1500 J of heat is absorbed from the high-temperature bath during each cycle?

(c) How much work does the engine perform in one cycle of operation?

**56.** In each cycle of its operation, a thermal engine absorbs 1000 J of heat from a large heat reservoir at 400 K and discharges heat to another large heat sink at 300 K. Calculate:

(a) The thermodynamic efficiency of the heat engine, operated reversibly

(b) The quantity of heat discharged to the low-temperature sink each cycle

(c) The maximum amount of work the engine can perform each cycle

**57.** A scientist needs a refrigeration machine to maintain a chemical reaction at a temperature of −10°C. How much work must be performed on the system during each cycle of its operation if 800 J of heat is to be withdrawn from the −10°C reservoir each cycle and discharged to the room at +20°C? How much heat is discharged to the room at +20°C each cycle? Assume that the machine operates at 100% of its theoretical efficiency.

**58.** A cyclic thermodynamic process is invented for the purpose of converting heat to mechanical work. Its inventor claims that his machine has an efficiency of 45% and operates from a source of heat at 600°C. Under what circumstances might his claim be true?

## Additional Problems

**59.** At one time, it was thought that the molar mass of indium was near 76 g mol$^{-1}$. By referring to the law of Dulong and Petit (problem 7), show how the measured specific heat of metallic indium, 0.233 J K$^{-1}$ g$^{-1}$, makes this value unlikely.

**60.** The following table shows how the specific heat at constant pressure of liquid helium changes with temperature. Note the sharp increase over this temperature range:

Temperature (K):

| 1.80 | 1.85 | 1.90 | 1.95 | 2.00 | 2.05 | 2.10 | 2.15 |

$c_s$ (J K$^{-1}$ g$^{-1}$):

| 2.81 | 3.26 | 3.79 | 4.42 | 5.18 | 6.16 | 7.51 | 9.35 |

Estimate how much heat it takes at constant pressure to raise the temperature of 1.00 g of He($\ell$) from 1.8 K to 2.15 K. (*Hint:* For each temperature interval of 0.05 K, take the average, $c_s$, as the sum of the values at the ends of the interval divided by 2.)

**61.** Imagine that 2.00 mol of argon, confined by a moveable, frictionless piston in a cylinder at a pressure of 1.00 atm and a temperature of 398 K, is cooled to 298 K. Argon gas may be considered ideal, and its molar heat capacity at constant pressure is $c_P = (5/2)R$ where $R = 8.315 \text{ J K}^{-1} \text{ mol}^{-1}$. Calculate:

(a) The work done on the system, $w$
(b) The heat absorbed by the system, $q$
(c) The energy change of the system, $\Delta E$
(d) The enthalpy change of the system, $\Delta H$

62. Suppose 1.00 mol of ice at $-30°C$ is heated at atmospheric pressure until it is converted to steam at 140°C. Calculate $q$, $w$, $\Delta H$, and $\Delta E$ for this process. For ice, water, and steam, $c_P$ is 38, 75, and 36 J K$^{-1}$ mol$^{-1}$, respectively, and can be taken to be approximately independent of temperature. $\Delta H_{fus}$ for ice is 6.007 kJ mol$^{-1}$, and $\Delta H_{vap}$ for water is 40.66 kJ mol$^{-1}$. Use the ideal gas law for steam, and assume that the volume of 1 mol of ice or water is negligible relative to that of 1 mol of steam.

63. The gas inside a cylinder expands against a constant external pressure of 1.00 atm from a volume of 5.00 L to a volume of 13.00 L. In doing so, it turns a paddle immersed in 1.00 L of water. Calculate the temperature rise of the water, assuming no loss of heat to the surroundings or frictional losses in the mechanism. Take the density of water to be 1.00 g cm$^{-3}$ and its specific heat to be 4.18 J K$^{-1}$ g$^{-1}$.

64. Suppose 1.000 mol of argon (assumed to be an ideal gas) is confined in a strong, rigid container of volume 22.41 L at 273.15 K. The system is heated until 3.000 kJ (3000 J) of heat has been added. The molar heat capacity of the gas does not change during the heating and equals 12.47 J K$^{-1}$ mol$^{-1}$.

(a) Calculate the original pressure inside the vessel (in atmospheres).
(b) Determine $q$ for the system during the heating process.
(c) Determine $w$ for the system during the heating process.
(d) Compute the temperature of the gas after the heating, in degrees Celsius. Assume the container has zero heat capacity.
(e) Compute the pressure (in atmospheres) inside the vessel after the heating.
(f) Compute $\Delta E$ of the gas during the heating process.
(g) Compute $\Delta H$ of the gas during the heating process.
(h) The correct answer to part (g) exceeds 3.000 kJ. The increase in enthalpy (which at one time was misleadingly called the "heat content") in this system exceeds the amount of heat actually added. Why is this not a violation of the law of conservation of energy?

65. When glucose, a sugar, reacts fully with oxygen, carbon dioxide and water are produced:

$$C_6H_{12}O_6(s) + 6\,O_2(g) \longrightarrow 6\,CO_2(g) + 6\,H_2O(\ell)$$
$$\Delta H° = -2820 \text{ kJ}$$

Suppose a person weighing 50 kg (mostly water, with specific heat capacity 4.18 J K$^{-1}$ g$^{-1}$) eats a candy bar containing 14.3 g of glucose. If all the glucose reacted with oxygen and the heat produced were used entirely to increase the person's body temperature, what temperature increase would result? (In fact, most of the heat produced is lost to the surroundings before such a temperature increase occurs.)

66. In walking a kilometer, you use about 100 kJ of energy. This energy comes from the oxidation of foods, which is about 30% efficient. How much energy do you save by walking a kilometer instead of driving a car that gets 8.0 km L$^{-1}$ of gasoline (19 miles per gallon)? The density of gasoline is 0.68 g cm$^{-3}$ and its enthalpy of combustion is $-48$ kJ g$^{-1}$.

67. Liquid helium and liquid nitrogen are both used as coolants; He($\ell$) boils at 4.21 K, and N$_2$($\ell$) boils at 77.35 K. The specific heat of liquid helium near its boiling point is 4.25 J K$^{-1}$ g$^{-1}$, and the specific heat of liquid nitrogen near its boiling point is 1.95 J K$^{-1}$ g$^{-1}$. The enthalpy of vaporization of He($\ell$) is 25.1 J g$^{-1}$, and the enthalpy of vaporization of N$_2$($\ell$) is 200.3 J g$^{-1}$ (these data are calculated from the values in Appendix F). Discuss which liquid is the better coolant (on a per-gram basis) *near* its boiling point, and which is better *at* its boiling point.

68. When 1.00 g of potassium chlorate (KClO$_3$) is dissolved in 50.0 g of water in a Styrofoam calorimeter of negligible heat capacity, the temperature drops from 25.00 to 23.36°C. Calculate $q$ for the water and $\Delta H°$ for the process.

$$KClO_3(s) \longrightarrow K^+(aq) + ClO_3^-(aq)$$

The specific heat of water is 4.184 J K$^{-1}$ g$^{-1}$.

69. The enthalpy of combustion and the standard enthalpy of formation of a fuel can be determined by measuring the temperature change in a calorimeter when a weighed amount of the fuel is burned in oxygen.

(a) Write a balanced chemical equation for the combustion of isooctane, C$_8$H$_{18}$($\ell$), to CO$_2$($g$) and H$_2$O($\ell$). Isooctane is a component of gasoline and is used as a reference standard in determining the "octane rating" of a fuel mixture.
(b) Suppose 0.542 g of isooctane is placed in a fixed-volume (bomb) calorimeter, which contains 750 g of water, initially at 20.450°C, surrounding the reaction compartment. The heat capacity of the calorimeter itself (excluding the water) has been measured to be 48 J K$^{-1}$ in a separate calibration. After the combustion of the isooctane is complete, the water temperature is measured to be 28.670°C. Taking the specific heat of water to be 4.184 J K$^{-1}$ g$^{-1}$, calculate $\Delta E$ for the combustion of 0.542 g of isooctane.
(c) Calculate $\Delta E$ for the combustion of 1 mol of isooctane.
(d) Calculate $\Delta H$ for the combustion of 1 mol of isooctane.
(e) Calculate $\Delta H_f°$ for the isooctane.

70. The enthalpy change to form 1 mol of Hg$_2$Br$_2$(s) from the elements at 25°C is $-206.77$ kJ mol$^{-1}$, and that of HgBr($g$) is 96.23 kJ mol$^{-1}$. Compute the enthalpy change for the decomposition of 1 mol of Hg$_2$Br$_2$(s) to 2 mol of HgBr($g$):

$$Hg_2Br_2(s) \longrightarrow 2\,HgBr(g)$$

*71. The gas most commonly used in welding is acetylene (C$_2$H$_2$($g$)). When acetylene is burned in oxygen, the reaction is

$$C_2H_2(g) + \tfrac{5}{2}O_2(g) \longrightarrow 2\,CO_2(g) + H_2O(g)$$

(a) Using data from Appendix D, calculate $\Delta H°$ for this reaction.

(b) Calculate the total heat capacity of 2.00 mol of $CO_2(g)$ and 1.00 mol of $H_2O(g)$, using $c_P(CO_2(g)) = 37$ J K$^{-1}$ mol$^{-1}$ and $c_P(H_2O(g)) = 36$ J K$^{-1}$ mol$^{-1}$.

(c) When this reaction is carried out in an open flame, almost all the heat produced in part (a) goes to raise the temperature of the products. Calculate the maximum flame temperature that is attainable in an open flame burning acetylene in oxygen. The actual flame temperature would be lower than this because heat is lost to the surroundings.

*72. The enthalpy of reaction changes somewhat with temperature. Suppose we wish to calculate $\Delta H$ for a reaction at a temperature $T$ that is different from 298 K. To do this, we can replace the direct reaction at $T$ with a three-step process. In the first step, the temperature of the reactants is changed from $T$ to 298 K. $\Delta H$ for this step can be calculated from the molar heat capacities of the reactants, which are assumed to be independent of temperature. In the second step, the reaction is carried out at 298 K with an enthalpy change $\Delta H°$. In the third step, the temperature of the products is changed from 298 K to $T$. The sum of these three enthalpy changes is $\Delta H$ for the reaction at temperature $T$.

An important process contributing to air pollution is the chemical reaction

$$SO_2(g) + \tfrac{1}{2}O_2(g) \longrightarrow SO_3(g)$$

For $SO_2(g)$ the heat capacity $c_P$ is 39.9, for $O_2(g)$ it is 29.4, and for $SO_3(g)$ it is 50.7 J K$^{-1}$ mol$^{-1}$. Calculate $\Delta H$ for the preceding reaction at 500 K, using the enthalpies of formation at 298.15 K from Appendix D.

*73. At the top of the compression stroke in one of the cylinders of an automobile engine (that is, at the minimum gas volume), the volume of the gas–air mixture is 150 mL, the temperature is 600 K, and the pressure is 12.0 atm. The ratio of the number of moles of octane vapor to the number of moles of air in the combustion mixture is 1.00:80.0. What is the maximum temperature attained in the gas if octane burns explosively before the power stroke of the piston (gas expansion) begins? The gases may be considered to be ideal, and their heat capacities at constant pressure (assumed to be temperature-independent) are

$$c_P(C_8H_{18}(g)) = 327 \text{ J K}^{-1}\text{ mol}^{-1}$$

$$c_P(O_2(g)) = 35.2 \text{ J K}^{-1}\text{ mol}^{-1}$$

$$c_P(N_2(g)) = 29.8 \text{ J K}^{-1}\text{ mol}^{-1}$$

$$c_P(CO_2(g)) = 45.5 \text{ J K}^{-1}\text{ mol}^{-1}$$

$$c_P(H_2O(g)) = 38.9 \text{ J K}^{-1}\text{ mol}^{-1}$$

The enthalpy of formation of $C_8H_{18}(g)$ at 600 K is $-57.4$ kJ mol$^{-1}$.

74. Initially, 46.0 g of oxygen is at a pressure of 1.00 atm and a temperature of 400 K. It expands adiabatically and reversibly until the pressure is reduced to 0.60 atm, and then it is compressed isothermally and reversibly until the volume returns to its original value. Calculate the final pressure and temperature of the oxygen, the work done and heat added to the oxygen in this process, and the energy change $\Delta E$. Take $c_P(O_2) = 29.4$ J K$^{-1}$ mol$^{-1}$.

75. A young chemist buys a ''one-lung'' motorcycle but, before learning how to drive it, wants to understand the processes that occur in its engine. The manual says the cylinder has a radius of 5.00 cm, a piston stroke of 12.00 cm, and a (volume) compression ratio of 8:1. If a mixture of gasoline vapor (taken to be $C_8H_{18}$) and air in mole ratio 1:62.5 is drawn into the cylinder at 80°C and 1.00 atm, calculate:

(a) The temperature of the compressed gases just before the spark plug ignites them. (Assume the gases are ideal, the compression is adiabatic, and the average heat capacity of the mixture of gasoline vapor and air is $c_P = 35$ J K$^{-1}$ mol$^{-1}$.)

(b) The volume of the compressed gases just before ignition.

(c) The pressure of the compressed gases just before ignition.

(d) The maximum temperature of the combustion products, assuming combustion is completed before the piston begins its downstroke. Take $\Delta H_f°(C_8H_{18}) = -57.4$ kJ mol$^{-1}$.

(e) The temperature of the exhaust gases, assuming the expansion stroke to be adiabatic.

76. In Section 8-6, a heat engine based on the Carnot cycle was considered in which isothermal and adiabatic steps alternate. Other types of cycle are possible, such as one in which constant pressure and adiabatic steps alternate. Suppose 1.000 mol of an ideal monatomic gas is taken reversibly through the following cycle:

state A ($P_A = 2.000$ atm, $V_A = 10.00$ L)

$\xrightarrow[\text{Const }P]{}$ state B ($P_B = 2.000$ atm, $V_B = 20.00$ L)

$\xrightarrow[\text{Adiabatic}]{}$ state C ($P_C = 1.000$ atm, $V_C = 30.31$ L)

$\xrightarrow[\text{Const }P]{}$ state D ($P_D = 1.000$ atm, $V_D = 15.16$ L)

$\xrightarrow[\text{Adiabatic}]{}$ state A

Calculate $w$ and $q$ for each step and the efficiency $\epsilon$ of the engine.

77. Suppose a refrigerator is placed in the middle of a closed room, its door is opened, and it is plugged in and allowed to run. What will happen to the temperature of the room? Why?

**78.** Suppose 32.1 g of $ClF_3(g)$ and 17.3 g of $Li(s)$ are mixed and allowed to react at atmospheric pressure and 25°C until one of the reactants is used up, producing $LiCl(s)$ and $LiF(s)$. Calculate the amount of heat evolved.

**79.** Calculate $\Delta H_f^\circ$ and $\Delta E_f^\circ$ for the formation of silane, $SiH_4(g)$, from its elements at 298 K, if 250 cm$^3$ of the gaseous compound at $T = 298$ K and $P = 0.658$ atm is burned in a constant-volume gas calorimeter in an excess of oxygen and causes the evolution of 9.757 kJ of heat. The combustion reaction is

$$SiH_4(g) + 2\,O_2(g) \longrightarrow SiO_2(s, \text{quartz}) + 2\,H_2O(\ell)$$

and the formation of silane from its elements is

$$Si(s) + 2\,H_2(g) \longrightarrow SiH_4(g)$$

**80.** Use Le Châtelier's principle and data from Appendix D to predict the direction in which each of the following equilibria will shift if the temperature is increased:

$$ZnCl_2(s) \rightleftharpoons Zn^{2+}(aq) + 2\,Cl^-(aq)$$

$$2\,N_2O(g) + N_2O_4(g) \rightleftharpoons 6\,NO(g)$$

$$HCOOH(aq) + H_2O(\ell) \rightleftharpoons H_3O^+(aq) + COOH^-(aq)$$

**81.** By considering the nature of the intermolecular forces in each case, rank the following substances from smallest to largest enthalpy of vaporization: $KBr$, $Ar$, $NH_3$, and $He$. Explain your reasoning.

**82.** (a) Draw Lewis structures for $O_2$, $CO_2$, $H_2O$, $CH_4$, $C_8H_{18}$, and $C_2H_5OH$. In $C_8H_{18}$, the eight carbon atoms form a chain with single bonds; in $C_2H_5OH$ the two carbon atoms are bonded to one another.

Using average bond enthalpies from Table 8-3, compute the enthalpy change in each of the following reactions, if 1 mol of each carbon compound is burned and all reactants and products are in the gas phase.

(b) $CH_4 + 2\,O_2 \rightarrow CO_2 + 2\,H_2O$
(burning methane, or natural gas)
(c) $C_8H_{18} + \frac{25}{2}\,O_2 \rightarrow 8\,CO_2 + 9\,H_2O$
(burning octane, in gasoline)
(d) $C_2H_5OH + 3\,O_2 \rightarrow 2\,CO_2 + 3\,H_2O$
(burning ethanol, in gasohol)

**83.** (a) Draw a Lewis structure for carbonic acid, $H_2CO_3$, with a central carbon atom bonded to the three oxygen atoms.
(b) Carbonic acid is unstable in aqueous solution and converts to dissolved carbon dioxide. Use bond enthalpies to estimate the enthalpy change for the reaction

$$H_2CO_3 \longrightarrow H_2O + CO_2$$

**\*84.** A supersonic nozzle is a cone-shaped object with a small hole in the end through which a gas is forced. As it moves through the nozzle opening, the gas expands in a manner that can be approximated as reversible and adiabatic. Such nozzles are used in molecular beams (Section 11-6) and in supersonic aircraft engines to provide thrust, because as the gas cools, its random thermal energy is converted into directed motion of the molecules with average velocity $\bar{v}$. Very little thermodynamic work is done because the external pressure is low, so the net effect is to convert thermal energy to net translational motion of the gas molecules. Suppose the gas in the nozzle is helium; its pressure is 50 atm and its temperature is 400 K before it begins its expansion.
(a) What are the average speed and the average velocity of the molecules *before* the expansion?
(b) What will be the temperature of the gas after its pressure has dropped to 1.0 atm in the expansion?
(c) What is the average velocity of the molecules at this point in the expansion?

*The reaction between sodium and chlorine proceeds very slowly until the addition of a drop of water sets it off.*

# Spontaneous Change and Equilibrium

A spontaneous change is one that, given enough time, occurs by itself without outside intervention. One of the most striking features of spontaneous change in nature is that it has a direction (Fig. 9-1). For instance, we observe a gas expanding into a vacuum but we never see the reverse process, a gas compressing itself spontaneously into a small part of its container. We measure heat flowing from

**Figure 9-1**
A bullet hitting a steel plate at a speed of 1600 feet per second melts as its kinetic energy is converted to heat, and sprays in all directions. These three photographs make sense only in the order shown; the reverse process is unmistakably unlikely. (Palm Press)

a hot body to a cold one when they are brought into thermal contact, but we never detect heat flowing spontaneously in the opposite direction. One of the goals of thermodynamics is to account for this *directionality* of spontaneous change.

It is clear that the first law of thermodynamics does not help in this regard. Energy is conserved both in a forward process and in its reverse; nothing in the first law indicates a preference for one direction of change over the other. In this chapter we define a new state function, entropy, and show how the examination of changes in entropy can lead to general predictions about the direction of spontaneous change and the nature of the equilibrium state that is ultimately achieved. We explore the connection between entropy, a macroscopic property, and the extent of disorder present on atomic scales. We then consider another state function, the Gibbs free energy, and demonstrate its connection to entropy and enthalpy in chemical reactions and phase equilibria.

## 9-1 ENTROPY, SPONTANEOUS CHANGE, AND THE SECOND LAW

It has already been shown that the amount of heat, $q$, transferred to a system is not a state function, because different paths connecting the same two states involve different total amounts of heat. If the heat added is divided by the temperature at each stage along a reversible path, however, the sum depends only on the initial and final states. In this section, this remarkable fact (which first emerged in Section 8-6) is illustrated and used to derive a law for the direction of spontaneous change in nature.

### Entropy and Reversible Processes

For a reversible process occurring at constant temperature, the **entropy change** $\Delta S$ of the system is defined as the amount of heat transferred to the system divided by the temperature:

$$\Delta S = \frac{q_{rev}}{T} \qquad \textbf{(9-1)}$$

The units of entropy are joules per kelvin. The reason for this definition, as well as its extension to irreversible processes or those in which the temperature changes, will be discussed further in this section. In Section 8-5, the heat transferred when a gas is expanded (or compressed) isothermally and reversibly from volume $V_1$ to $V_2$ was shown to be

$$q_{rev} = nRT \ln \left( \frac{V_2}{V_1} \right)$$

and therefore

$$\Delta S = nR \ln \left( \frac{V_2}{V_1} \right) \qquad \text{(constant } T\text{)} \qquad \textbf{(9-2)}$$

The entropy of a gas increases during an isothermal expansion and decreases during a compression.

Another type of constant-temperature process is a phase transition such as the melting of a solid at constant pressure. This occurs reversibly at the normal freezing temperature $T_f$ because an infinitesimal change in external conditions (e.g., a lowering of the temperature) serves to reverse the process. The reversible heat when 1 mol of substance melts is $q_{rev} = \Delta H_{fus}$, and

$$\Delta S_{fus} = \frac{q_{rev}}{T_f} = \frac{\Delta H_{fus}}{T_f} \qquad \textbf{(9-3)}$$

The entropy increases when a solid melts or a liquid vaporizes, and it decreases when the phase transition occurs in the opposite direction.

### Example 9-1

Calculate the entropy change when 3.00 mol of benzene vaporizes reversibly at its normal boiling point of 80.1°C. The molar enthalpy of vaporization of benzene at this temperature is 30.8 kJ mol$^{-1}$.

### Solution

The entropy change when 1 mol of benzene is vaporized at 80.1°C ( = 353.25 K) is

$$\Delta S_{vap} = \frac{\Delta H_{vap}}{T_b} = \frac{30,800 \text{ J mol}^{-1}}{353.25 \text{ K}} = +87.2 \text{ J K}^{-1} \text{ mol}^{-1}$$

When 3.00 mol is vaporized, the entropy change is three times as great:

$$\Delta S = (3.00 \text{ mol})(+87.2 \text{ J K}^{-1} \text{ mol}^{-1}) = +262 \text{ J K}^{-1}$$

*Related Problems:* 1, 2

Remarkably, most liquids have about the same molar entropy of vaporization. **Trouton's rule** states the magnitude of this entropy change:

$$\Delta S_{vap} = 88 \pm 5 \text{ J K}^{-1} \text{ mol}^{-1} \qquad \text{(9-4)}$$

Note that the $\Delta S_{vap}$ of benzene, calculated in Example 9-1, is within this range. The constancy of $\Delta S_{vap}$ means that $\Delta H_{vap}$ and $T_b$, which vary widely from substance to substance, must do so in the same proportion. The rule allows enthalpies of vaporization to be estimated from boiling temperatures. Exceptions exist, however; water has a molar entropy of vaporization of $109 \text{ J K}^{-1} \text{ mol}^{-1}$.

Now consider a reversible process in which the temperature changes. In this case, each small increment of heat transferred is divided by the temperature at which the transfer takes place; these are added to obtain the result. In the limit of a continuous process, the entropy change is then defined by an integral:

$$\Delta S = \sum_i \frac{q_i^{rev}}{T_i} = \int_A^B \frac{1}{T} \, dq_{rev}$$

For a reversible *adiabatic* process, $q = 0$ and therefore $\Delta S = 0$. Such a process is also called **isentropic** (that is, the entropy is constant).

In a reversible *isochoric* process, the volume is held constant and the system is heated or cooled by contact with a reservoir that differs from it in temperature by an infinitesimal amount $dT$. The heat transferred in this case is

$$dq_{rev} = nc_V \, dT$$

and the entropy change of the system as it is heated from $T_1$ to $T_2$ is

$$\Delta S = \int_{T_1}^{T_2} \frac{1}{T} \, dq_{rev} = \int_{T_1}^{T_2} \frac{nc_V}{T} \, dT$$

If $c_V$ is independent of $T$ over the temperature range of interest, it can be removed from the integral, giving the result

$$\Delta S = nc_V \int_{T_1}^{T_2} \frac{1}{T} \, dT = nc_V \ln\left(\frac{T_2}{T_1}\right) \qquad \text{(constant } V) \qquad \text{(9-5)}$$

The analogous result for the entropy change of the system in a reversible *isobaric* process (constant *pressure*) is

$$\Delta S = \int_{T_1}^{T_2} \frac{nc_P}{T} \, dT = nc_P \ln\left(\frac{T_2}{T_1}\right) \qquad \text{(constant } P) \qquad \text{(9-6)}$$

Entropy always increases with increasing temperature.

The following example illustrates the fact that the entropy is a state function, for which changes are independent of the path followed.

## Example 9-2

(a) Calculate the entropy change for the process described in Example 8-9: 5.00 mol of argon expands reversibly at a constant temperature of 298 K from a pressure of 10.0 atm to 1.00 atm.

(b) Calculate the entropy change for the same initial and final states as in part (a) but along a different path. First the 5.00 mol of argon expands reversibly and *adiabatically* between the same two pressures. This is the path followed in Example 8-10; it causes the temperature to fall to 118.6 K. Then the gas is heated at constant pressure back to 298 K.

## Solution

(a) At constant temperature, the entropy change is

$$\Delta S = nR \ln \left(\frac{V_2}{V_1}\right) = nR \ln \left(\frac{P_1}{P_2}\right)$$

$$= (5.00 \text{ mol})(8.315 \text{ J K}^{-1} \text{ mol}^{-1}) \ln 10.0$$

$$= +95.7 \text{ J K}^{-1}$$

(b) For the adiabatic part of this path, the entropy change is 0. When the gas is then heated reversibly at constant pressure from 118.6 K to 298 K, the entropy change is

$$\Delta S = nc_P \ln \left(\frac{T_2}{T_1}\right)$$

$$= (5.00 \text{ mol})(\tfrac{5}{2} \times 8.315 \text{ J K}^{-1} \text{ mol}^{-1}) \ln \frac{298 \text{ K}}{118.6 \text{ K}}$$

$$= +95.7 \text{ J K}^{-1}$$

This is the same as the result from part (a), an illustration of the fact that the entropy is a state function. By contrast, the amounts of heat for the two paths are different: 28.5 kJ and 18.6 kJ.

*Related Problems:* 5, 6

## Irreversible Processes

In an irreversible process, the temperature is undefined except at the beginning and end, and the previous expression does not apply. To calculate the entropy change associated with such a process, a way must be devised to connect the initial and final states by a *reversible* path. With a little ingenuity it is always possible to do this; in fact, there are infinitely many reversible paths from which to choose. Because entropy is a state function, the resulting entropy change will apply to the irreversible path as well.

The work performed *by* a system ($-w$) as it undergoes an irreversible expansion is always less than when the expansion is carried out reversibly. To see this, return to the definition of work done *on* a system:

$$w = -\int P_{ext} \, dV$$

During an expansion, $P_{ext}$ must be less than $P$, the pressure of the gas. For a reversible expansion $P_{ext}$ is only infinitesimally smaller (so the system is always very close to equilibrium), but for an irreversible expansion $P_{ext}$ is measurably smaller. Therefore, the area under a graph of $P_{ext}$ plotted against $V$ is less than that of a graph of $P$ against $V$ (Fig. 9-2), so that

$$-w_{irrev} = \int P_{ext}\, dV < \int P\, dV = -w_{rev}$$

and the work performed by the system, $-w_{irrev}$, is algebraically less than $-w_{rev}$. A reversible process is always more efficient than an irreversible one.

If the reversible and irreversible processes have the same initial and final states, then $\Delta E$ is the same for both.

$$\Delta E = w_{irrev} + q_{irrev} = w_{rev} + q_{rev}$$

But because

$$-w_{irrev} < -w_{rev}$$

we must have

$$w_{irrev} > w_{rev}$$

and

$$q_{irrev} < q_{rev}$$

The heat absorbed is a *maximum* when the process is carried out reversibly.
An example illustrates these inequalities.

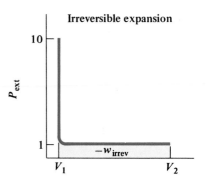

**Figure 9-2**

Work done by a system in reversible and irreversible expansions between the same initial and final states. The work performed is greater for the reversible process.

### Example 9-3

Calculate the heat absorbed and the work done by a system of 5.00 mol of an ideal gas as it expands irreversibly at constant temperature $T = 298$ K from a pressure of 10.0 atm to 1.00 atm. The external pressure is held constant at 1.00 atm.

#### Solution

The initial volume $V_1$ is

$$V_1 = \frac{nRT}{P_1} = \frac{(5.00 \text{ mol})(0.08206 \text{ L atm K}^{-1} \text{ mol}^{-1})(298 \text{ K})}{10.0 \text{ atm}} = 12.2 \text{ L}$$

The final volume is 10 times this, or

$$V_2 = 122 \text{ L}$$

For a constant external pressure,

$$w_{irrev} = -P_{ext}\, \Delta V = -(1.00 \text{ atm})(122 \text{ L} - 12.2 \text{ L}) = -110 \text{ L atm}$$

$$= -11.1 \text{ kJ}$$

At constant $T$, $\Delta E = 0$, however, so that

$$q_{irrev} = -w_{irrev} = 11.1 \text{ kJ}$$

In Example 8-9, a reversible expansion between the same two states was carried out, with the result that

$$w_{\mathrm{rev}} = -28.5 \text{ kJ}$$

This demonstrates that

$$-w_{\mathrm{irrev}} < -w_{\mathrm{rev}}$$

and

$$q_{\mathrm{irrev}} < q_{\mathrm{rev}}$$

## The Second Law

It is always true that

$$q_{\mathrm{rev}} > q_{\mathrm{irrev}}$$

Dividing this expression by $T$, the temperature at which the heat is transferred, gives

$$\frac{q_{\mathrm{rev}}}{T} > \frac{q_{\mathrm{irrev}}}{T}$$

The left-hand side is the entropy change $\Delta S$,

$$\Delta S = \frac{q_{\mathrm{rev}}}{T}$$

so

$$\Delta S > \frac{q_{\mathrm{irrev}}}{T}$$

The last two equations can be combined as

$$\Delta S \geq \frac{q}{T}$$

where the equality applies only to a reversible process.

Now consider an *isolated* system. In this case there is no transfer of heat into or out of the system, and $q = 0$. For an isolated system, then, $\Delta S \geq 0$. The universe itself (i.e., a system plus its surroundings) is clearly an isolated system. It follows, therefore, that

1. In a *reversible* process, the total entropy of a system plus its surroundings is unchanged.
2. In an *irreversible* process, the total entropy of a system plus its surroundings must increase.
3. A process for which $\Delta S_{\mathrm{total}} < 0$ is impossible.

These statements are one formulation of the **second law of thermodynamics.** As we have seen, a reversible process is an idealized limit of a series of irreversible steps. Any real process (one occurring in finite time) must have some degree of irreversibility, and therefore in any real process the entropy of the universe increases. This provides a very powerful criterion with which to describe the direction of a spontaneous process. If the total entropy change ($\Delta S_{\mathrm{sys}} + \Delta S_{\mathrm{surr}}$) is positive, the process can occur spontaneously. If it is negative (as it is for the reverse process), then according to the second law, the process is impossible.

## Example 9-4

Calculate $\Delta S_{tot} = \Delta S_{sys} + \Delta S_{surr}$ for the reversible and irreversible isothermal expansions of Examples 8-9 and 9-3.

### Solution

For the reversible expansion,

$$q_{rev} = 28.5 \text{ kJ}$$

$$\Delta S_{sys} = \frac{q_{rev}}{T} = \frac{28,500 \text{ J}}{298 \text{ K}} = +95.7 \text{ J K}^{-1}$$

The surroundings give up the same amount of heat at the same temperature. Hence,

$$\Delta S_{surr} = \frac{-28,500 \text{ J}}{298 \text{ K}} = -95.7 \text{ J K}^{-1}$$

$\Delta S_{tot} = 95.7 - 95.7 = 0$ for the reversible process.

For the irreversible expansion, it is still true that

$$\Delta S_{sys} = +95.7 \text{ J K}^{-1}$$

because $S$ is a function of state and the initial and final states are the same as for the reversible expansion. From Example 9-3, however, only 11.1 kJ of heat is given up by the surroundings in this case.[1] Hence,

$$\Delta S_{surr} = \frac{-11,100 \text{ J}}{298 \text{ K}} = -37.2 \text{ J K}^{-1}$$

and $\Delta S_{tot} = 95.7 - 37.2 = 58.5 \text{ J K}^{-1} > 0$ for the irreversible process.

*Related Problems:* 9, 10

## 9-2 ENTROPY AND IRREVERSIBILITY: A STATISTICAL INTERPRETATION

The concept of entropy may appear rather mysterious at this point. What is entropy *physically,* and why should the reversible heat divided by the temperature be related to the spontaneity of processes in nature? These are deep questions that we can only begin to answer here. To do so, we must step outside the confines of classical thermodynamics, which is concerned only with macroscopic properties, and examine the microscopic molecular basis for the second law. Such an approach, called **statistical thermodynamics,** shows that spontaneous change in nature can be understood by using probability theory to predict and explain the behavior of the many atoms and molecules that make up a macroscopic sample of matter.

---

[1] How can the heat from Example 9-3, which is irreversible from the perspective of the system, be reversible from the perspective of the surroundings? This can be accomplished by enclosing the gas in a material (such as a metal) that can efficiently transfer heat to and from the surroundings and thus remain close to equilibrium, at the same time that the gas itself is far from equilibrium due to the gas currents that occur during the irreversible expansion.

(a)

(b)

**Figure 9-3**

The free expansion of a gas into a vacuum. (a) The stopcock is closed with all gas in the left bulb. (b) With the stopcock open, half the gas is in each bulb.

## Irreversibility

Consider a particularly simple spontaneous process: the free adiabatic expansion of 1 mol of an ideal gas into a vacuum (Fig. 9-3). The gas is initially held in the left bulb in volume $V/2$, while the right bulb is evacuated. After the stopcock is opened, the gas expands to fill the entire volume $V$. As shown in problem 18 in Chapter 8, such a free adiabatic expansion of an ideal gas is also isothermal because no work is done, and $\Delta E = 0$. The entropy change can be evaluated from the heat transferred along a *reversible* path connecting the same two states. The entropy change for such a reversible isothermal expansion with $n = 1$ mol is

$$\Delta S_{sys} = nR \ln \frac{V_f}{V_i} = R \ln \frac{V}{V/2} = R \ln 2$$

for the present example. Because the state of the surroundings is unchanged, $\Delta S_{tot} = R \ln 2$, which is positive, as it must be for a spontaneous process. Note that it would have been incorrect to attempt to calculate the entropy change of the system from $q_{irrev}/T$.

Now examine the same free expansion from a microscopic point of view. If the path of some particular molecule is followed, it can be seen that after it spends some time on the left side it will cross to the right, then back to the left, and so forth. If enough time elapses, the molecule will eventually spend equal amounts of time on the two sides, because there is no reason for it to prefer one side to the other. Put in other terms, the *probability* that the molecule is on, say, the left side at any given instant will be 1/2, the same as the probability that it is on the right side. This concept of probability provides the key to understanding spontaneous change. The gas expands to fill the whole volume because it is overwhelmingly probable that it will do so. In the same way, a gas is never observed to compress spontaneously into a smaller volume, because such a process is overwhelmingly improbable.

Exactly how improbable is such a spontaneous compression of 1 mol of gas? The probability that one particular molecule is on the left side at a given time is 1/2. A second specific molecule may be on either the left or the right, and so the probability that both are on the left is $1/2 \times 1/2 = 1/4$. On average, four given molecules will be on the left $1/2 \times 1/2 \times 1/2 \times 1/2 = 1/16$ of the time (Fig. 9-4). Continuing this argument for all $N_0 = 6.0 \times 10^{23}$ molecules leads to the probability that all $N_0$ molecules are on the left:

$$\frac{1}{2} \times \frac{1}{2} \times \cdots \times \frac{1}{2} = \left(\frac{1}{2}\right)^{6.0 \times 10^{23}}$$

This can be rewritten in scientific notation as 1 divided by 10 raised to the power $a$ (or, equivalently, as $10^{-a}$):

$$\frac{1}{10^a} = \left(\frac{1}{2}\right)^{6.0 \times 10^{23}} = \frac{1}{2^{6.0 \times 10^{23}}}$$

A calculator can handle numbers this large (or this small) only if logarithms are used. Taking the base-10 logarithms of both denominators gives

$$a = \log 2^{(6.0 \times 10^{23})} = 6.0 \times 10^{23} \log 2$$

$$= (6.0 \times 10^{23})(0.30)$$

$$= 1.8 \times 10^{23}$$

The probability that all the molecules will be on the left is 1 in $10^{1.8 \times 10^{23}}$.

This is a vanishingly small probability, because $10^{1.8 \times 10^{23}}$ is an unimaginably large number. It is vastly larger than the number $1.8 \times 10^{23}$ (which is a large number already). To realize this, consider how such numbers would be written. The number $1.8 \times 10^{23}$ is fairly straightforward to write out. It contains 22 zeros:

$$180,000,000,000,000,000,000,000$$

The number $10^{1.8 \times 10^{23}}$ is 1 followed by $1.8 \times 10^{23}$ zeros. Written out, such a number would more than fill all the books in the world. Put in other terms, $1.8 \times 10^{23}$ corresponds to the number of molecules in about 5 cm$^3$ of water, but $10^{1.8 \times 10^{23}}$ is far larger than the number of molecules in the entire universe!

The reason that a violation of the second law of thermodynamics is never seen is that such a process is overwhelmingly improbable. Nothing in the laws of mechanics prevents a gas from compressing spontaneously, but the event is so improbable, it is not seen. The directionality of spontaneous change is a consequence of the large numbers of molecules in the macroscopic systems treated by thermodynamics. Clearly, if there were only 6 molecules, instead of $6 \times 10^{23}$, it would not be surprising to find them all on the left side at a given time. In fact, the probability that 6 molecules are on the left side is $(1/2)^6 = 1/64$, so there is 1 chance in 64 that this will occur. Irreversibility in nature results from the random, statistical behavior of large numbers of molecules.

## Example 9-5

Calculate the probability of a spontaneous compression of 1.00 mol of gas by 0.01%—that is, the probability that all the molecules will be found in a volume $V' = 0.9999V$ at a certain time.

### Solution

In this case, the probability that a given molecule is in $V'$ is not 1/2 but 0.9999 (the probability that it is in the remainder of $V$ is 0.0001). The probability that all $N_0$ molecules are in $V'$ is

$$(0.9999)^{N_0} = 1/10^a$$

$$a = -N_0 \log (0.9999) = (6.0 \times 10^{23})(4.3 \times 10^{-5})$$

$$= 2.6 \times 10^{19}$$

The chance of such a compression occurring spontaneously is 1 in $10^{2.6 \times 10^{19}}$, which is still vanishingly small. Thus, a spontaneous compression of even a fraction of a percent will not be seen.

*Related Problems:* 11, 12

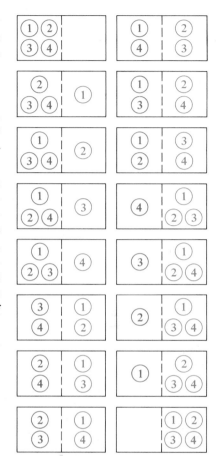

**Figure 9-4**

The 16 possible states of a system of four molecules that may occupy either side of a container. In only one of these are all four molecules on the left side.

## Entropy and the Number of Microstates

Spontaneous processes occur when constraints are removed from a system. In the free expansion of a gas, the molecules are initially confined in one part of the container. After the constraint is removed (the stopcock is opened), they are free to stay where they were, but they also have other possibilities open to them—the number of states available to the system increases.

If the volume is doubled, the number of positions available to a given molecule is doubled also. Therefore, the number of states available to the molecule should be proportional to the volume $V$:

$$\text{number of states available per molecule} = cV$$

where $c$ is a proportionality constant. The state of a *two*-molecule system is given jointly by the states of the molecules in it, so the number of microscopic states available is just the product of the number of states for each molecule, $(cV) \times (cV) = (cV)^2$. For an $N$-molecule system,

$$\text{microscopic states available} = \Omega = (cV)^N$$

The term "microscopic state," or **microstate,** is used to specify a particular distribution of molecules among the positions and momenta accessible to them, and the basic assumption is made that each microstate is equally likely to be occupied by the system. The symbol $\Omega$ denotes the number of microstates of a system.

In an irreversible process, the number of microstates available to the system increases. It is reasonable to expect some connection between the number of microstates $\Omega$ and the entropy $S$. When one increases, the other should as well. An important difference is that the entropy, because it is an extensive variable, is proportional to $N$, but $\Omega$ depends on $N$ in the *power* to which $cV$ is raised. Doubling $N$ doubles $S$, but leads to $\Omega$ being squared. Therefore, $S$ is related directly to the *natural logarithm* of $\Omega$:

$$S = k \ln \Omega = k \ln (cV)^N = Nk \ln (cV)$$

where $k$ is some proportionality constant. When $N$ doubles, the logarithm of $\Omega$ doubles also.

Given this relation, it is possible to calculate the entropy change for the free expansion of 1 mol ($N_0$ atoms) of gas from a volume $V/2$ to $V$:

$$\Delta S \text{ (microscopic)} = N_0 k \ln (cV) - N_0 k \ln (cV/2)$$

$$= N_0 k \ln \left( \frac{cV}{cV/2} \right) = N_0 k \ln 2$$

Note that the constant $c$ has dropped out. The entropy change for this process from the macroscopic view was found to be

$$\Delta S \text{ (thermodynamic)} = R \ln 2$$

If $k$ is identified as $R/N_0$, which is **Boltzmann's constant** $k_B$, then the two expressions are identical. The relation that results is

$$S = k_B \ln \Omega \qquad (9\text{-}7)$$

which was originally discovered by the Austrian physicist Ludwig Boltzmann in the late 19th century (Fig. 9-5). It is impossible to overstate its importance, for it provides the link between the microscopic world of atoms and molecules and the macroscopic world of bulk matter and thermodynamics. Although this equation was obtained here for one specific example, in fact it holds quite generally.

**Figure 9-5**

The fundamental relationship between entropy ($S$) and the number of microstates ($W$) was derived by Ludwig Boltzmann in 1868. On his tombstone in Vienna is carved the equation he obtained, $S = k \log W$. We would write "ln" instead of "log" for the natural logarithm. (D. Oxtoby)

## Entropy and Disorder

The entropy of an isolated system increases with the number of microscopic states available to the molecules. Another common way to describe this phenomenon is to relate the entropy to disorder: an ordered system has low entropy because its molecules are constrained to occupy only certain positions in space. As constraints are removed and the molecules are free to occupy more locations, the disorder increases and the entropy rises. An isolated system is not observed to become ordered spontaneously because the number of possible states, each equally probable, is overwhelmingly greater than the number of ordered ones.

The melting of a solid is a clear example of this. In a solid, the atoms or molecules are constrained to stay near their equilibrium positions in the crystal lattice, whereas in the liquid they can move far away from these fixed (ordered) positions. More states are available to the molecules of the liquid, corresponding to an increase in entropy: upon melting of the substance, $\Delta S_{sys}$ is positive. In the same way, when a liquid evaporates, the disorder and the entropy increase because the number of microstates available to the molecules increases enormously. Although the molecules in a liquid remain at the bottom of their container, those in a gas move throughout the container. They are in a greater state of disorder, so the entropy change $\Delta S_{sys}$ on vaporization is also positive. This change, $\Delta S_{vap}$, is particularly great for water ($109 \ J \ K^{-1}$) compared to the Trouton's rule value ($88 \ J \ K^{-1}$) because hydrogen bonding in liquid water orders that state and gives it a relatively low entropy.

Although the concepts of order and disorder are useful and are related qualitatively to entropy, the more fundamental connection is to the number of microstates available. In some cases (such as in certain phase transformations between solid phases), it is hard to say which phase is more ordered. A calculation of the relative number of microstates available, however, correctly predicts the sign of the entropy change in a transformation from one such phase to the other.

## 9-3 ABSOLUTE ENTROPIES AND THE THIRD LAW

In thermodynamic processes, only *changes* in entropy, $\Delta S$, are measured, just as only changes in energy, $\Delta E$, or enthalpy, $\Delta H$, arise. It is nevertheless useful to define *absolute* entropies relative to some reference state. An important experimental observation that simplifies the choice of reference state is the following:

> In any thermodynamic process involving only pure phases in their equilibrium states, the entropy change $\Delta S$ approaches zero as $T$ approaches 0 K.

This observation is the **Nernst heat theorem,** named after its discoverer, Walther Nernst. It immediately suggests a choice of reference state: the entropy of any pure element in its equilibrium state is defined to approach zero as $T$ approaches 0 K. From the Nernst result, the entropy change for any chemical reaction, including one in which elements react to give a compound, approaches zero at 0 K. This has as a consequence the **third law of thermodynamics:**

> The entropy of any substance (element or compound) in its equilibrium state approaches zero at the absolute zero of temperature.

Absolute zero can never actually be reached, so a small extrapolation is needed to make use of this result.

The third law, like the two laws that precede it, is a macroscopic law based on experimental measurements. It is, however, consistent with the microscopic interpretation of the entropy presented in Section 9-2. In particular, from quantum mechanics and statistical thermodynamics, as the temperature approaches absolute zero the number of microstates available to a substance at equilibrium falls rapidly toward 1, and so the absolute entropy defined as $k_B \ln \Omega$ should approach 0. We have stated that the entropy of a substance *in its equilibrium state* approaches zero at 0 K. In practice, equilibrium may be very difficult to achieve at low temperatures as particle motion slows. In solid CO, for example, molecules remain randomly oriented (CO or OC) as the crystal is cooled, even though the equilibrium state at low temperatures would correspond to a definite orientation of each molecule. Because a molecule is so slow to reorient at low temperatures, such a crystal may not reach its equilibrium state in a measurable period of time. A nonzero entropy measured at low temperatures indicates that the system is not in equilibrium.

## Standard-State Entropies

Because the entropy of any substance in its equilibrium state is zero at absolute zero, its entropy at any other temperature $T$ is given by the entropy change as it is heated from 0 K to $T$. If heat is added at constant pressure,

$$\Delta S = n \int_{T_1}^{T_2} \frac{c_P}{T} \, dT$$

and so $S_T$, the absolute entropy of 1 mol of substance at temperature $T$, is

$$S_T = \int_0^T \frac{c_P}{T} \, dT$$

It is necessary merely to measure $c_P$ as a function of temperature and determine the area under a plot of $c_P/T$ versus $T$ from 0 K to any desired temperature. If a substance melts, boils, or undergoes some other phase change before reaching the temperature $T$, the entropy change for that process must be added to $\int (c_P/T) \, dT$.

In particular, the absolute molar entropy $S°$ at 298.15 K and 1 atm pressure (Fig. 9-6) is

$$S° = \int_0^{298.15} \frac{c_P}{T} \, dT + \Delta S \quad \text{(phase changes between 0 and 298.15 K)} \quad \textbf{(9-8)}$$

Standard molar entropies $S°$ are tabulated for a number of elements and compounds in Appendix D. If $c_P$ is measured in J K$^{-1}$ mol$^{-1}$, then the entropy $S°$ will have the same units. For dissolved ions, the arbitrary convention $S°(H^+(aq)) = 0$ is applied (just as for the standard enthalpy of formation of $H^+$ discussed in Section 8-4). For this reason, some $S°$ values are negative for aqueous ions—an impossibility for substances.

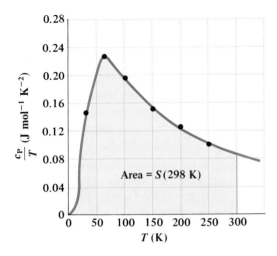

**Figure 9-6**

A graph of $c_P/T$ versus $T$ for platinum. The black dots represent experimental measurements. The area up to any temperature (here, 298 K) is the molar entropy at that temperature.

Tabulated standard molar entropies are used to calculate entropy changes in chemical reactions at 25°C and 1 atm, just as standard enthalpies of formation are combined to obtain enthalpies of reaction according to Hess's law (Section 8-4).

### Example 9-6

Using the table of standard molar entropies in Appendix D, calculate $\Delta S°$ for the chemical reaction

$$N_2(g) + 2\,O_2(g) \longrightarrow 2\,NO_2(g)$$

with reactants and products at a temperature of 25°C and a pressure of 1 atm.

### Solution

From the table,

$$S°(N_2(g)) = 191.50\ \mathrm{J\ K^{-1}\ mol^{-1}}$$

$$S°(O_2(g)) = 205.03\ \mathrm{J\ K^{-1}\ mol^{-1}}$$

$$S°(NO_2(g)) = 239.95\ \mathrm{J\ K^{-1}\ mol^{-1}}$$

The entropy change for the reaction is the sum of the entropies of the products, minus the sum of entropies of the reactants, each multiplied by its coefficient in the balanced chemical equation:

$$\Delta S° = 2S°(NO_2(g)) - S°(N_2(g)) - 2S°(O_2(g))$$

$$= (2\ \mathrm{mol})(239.95\ \mathrm{J\ K^{-1}\ mol^{-1}}) - (1\ \mathrm{mol})(191.50\ \mathrm{J\ K^{-1}\ mol^{-1}})$$
$$- (2\ \mathrm{mol})(205.03\ \mathrm{J\ K^{-1}\ mol^{-1}})$$

$$= -121.66\ \mathrm{J\ K^{-1}}$$

The factors of 2 multiply $S°$ for $NO_2$ and $O_2$ because 2 mol of each appears in the chemical equation. Note that standard molar entropies, unlike standard molar enthalpies of formation $\Delta H_f°$, are not 0 for elements at 25°C. The negative $\Delta S°$ results because this is the entropy change of the system only. The surroundings must undergo a positive entropy change in such a way that $\Delta S_{\mathrm{tot}} \geq 0$.

*Related Problems:* 19, 20, 21, 22

## 9-4 THE GIBBS FREE ENERGY

In Section 9-1, the change in entropy of a system plus its surroundings (that is, the total change of entropy $\Delta S_{tot}$) was shown to provide a criterion for whether a process is spontaneous, reversible, or impossible:

$$\Delta S_{tot} > 0 \quad \text{spontaneous}$$
$$\Delta S_{tot} = 0 \quad \text{reversible}$$
$$\Delta S_{tot} < 0 \quad \text{impossible}$$

Although $\Delta S_{tot}$ is a completely general criterion for assessing the spontaneity or impossibility of a process, it requires calculation of the entropy change for the surroundings as well as for the system. It is desirable to have a state function that gives the feasibility of a process without reference to the surroundings.

For the special case of processes at constant temperature and pressure (the most important in chemistry), such a state function exists. To identify it, note first that at constant pressure, $\Delta H_{sys} = q_P$, and so the heat transferred to the surroundings is $-q_P = -\Delta H_{sys}$. If the surroundings remain at constant temperature during the process, the transfer of heat must have the same effect on them as would a reversible transfer of the same amount of heat. Their entropy change is then

$$\Delta S_{surr} = \frac{-\Delta H_{sys}}{T}$$

The total entropy change is

$$\Delta S_{tot} = \Delta S_{sys} + \Delta S_{surr} = \Delta S_{sys} - \frac{\Delta H_{sys}}{T}$$
$$= \frac{-(\Delta H_{sys} - T\,\Delta S_{sys})}{T}$$

Because the temperature $T$ is constant, we can rewrite this as

$$\Delta S_{tot} = \frac{-\Delta(H_{sys} - TS_{sys})}{T}$$

We define the **Gibbs free energy** $G$ as

$$G = H - TS \tag{9-9}$$

so that

$$\Delta S_{tot} = \frac{-\Delta G_{sys}}{T}$$

Because the absolute temperature $T$ is always positive, $\Delta S_{tot}$ and $\Delta G_{sys}$ must have the opposite sign for processes occurring at constant $T$ and $P$. It follows that

$$\Delta G_{sys} < 0 \quad \text{spontaneous processes}$$
$$\Delta G_{sys} = 0 \quad \text{reversible processes}$$
$$\Delta G_{sys} > 0 \quad \text{nonspontaneous processes}$$

for processes carried out at constant temperature and pressure. If $\Delta G_{sys} > 0$ for a process, then $\Delta G_{sys} < 0$ for the *reverse* process, and that reverse process will occur spontaneously.

## The Gibbs Free Energy and Phase Transitions

As a simple example of the usefulness of the Gibbs free energy, consider the freezing of 1 mol of a liquid such as water to form ice:

$$H_2O(\ell) \longrightarrow H_2O(s)$$

Examine first what happens when this process is carried out at the ordinary freezing point of water under atmospheric pressure, 273.15 K. The measured enthalpy change (the heat absorbed at constant pressure) is

$$\Delta H_{273} = q_P = -6007 \text{ J}$$

At $T_f = 273.15$ K, the freezing of water takes place reversibly (i.e., the system remains close to equilibrium as it freezes). Therefore, the entropy change is

$$\Delta S_{273} = \frac{q_{rev}}{T_f} = \frac{-6007 \text{ J}}{273.15 \text{ K}} = -21.99 \text{ J K}^{-1}$$

The Gibbs free energy change is

$$\Delta G_{273} = \Delta H_{273} - T\Delta S_{273} = -6007 \text{ J} - (273.15 \text{ K})(-21.99 \text{ J K}^{-1}) = 0$$

At the normal freezing point the Gibbs free energy change is 0, because the freezing of water under these conditions is an equilibrium process.

Now consider what happens as the water is cooled below 273.15 K to 263.15 K (−10.00°C). Water under these conditions is *undercooled*; it can exist as a liquid only for a limited period before freezing. The Gibbs free energy change can now be calculated as water freezes at this lower temperature.

Assume that $\Delta H$ and $\Delta S$ for the freezing process do not depend on temperature. If that is the case,

$$\Delta G_{263} = -6007 \text{ J} - (263.15 \text{ K})(-21.99 \text{ J K}^{-1}) = -220 \text{ J}$$

An exact calculation that takes into account the fact that $\Delta H$ and $\Delta S$ *do* depend slightly on temperature leads to

$$\Delta G_{263} = -213 \text{ J}$$

for the process. Because $\Delta G < 0$, the undercooled water will freeze spontaneously at 263.15 K. At a temperature *greater* than $T_f$, $\Delta G$ is greater than 0, showing that freezing of the liquid would be impossible. This is in accord with our experience of nature: water does not freeze at atmospheric pressure if the temperature is held above 273.15 K; rather, the reverse process occurs, and ice melts.

Writing the Gibbs free energy change as $\Delta G = \Delta H - T\Delta S$ shows that a negative value of $\Delta G$ (and therefore a spontaneous process) is favored by a negative value of $\Delta H$ and a positive value of $\Delta S$. In the freezing of a liquid, $\Delta H$ is negative. However, $\Delta S$ for freezing is also negative, rather than positive (in the language of Section 9-2, the solid is more ordered than the liquid and so has a lower entropy). Whether or not a liquid freezes depends on the competition between two factors: an

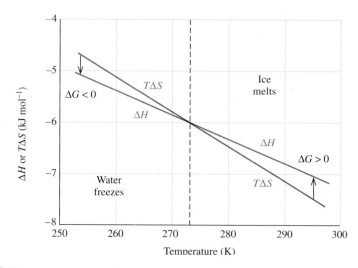

**Figure 9-7**

Plots of $\Delta H$ and $T\Delta S$ versus temperature for the freezing of water. At 273.15 K the two curves cross, so that at this temperature $\Delta G = 0$ and ice and water coexist. Below this temperature, the freezing of water to ice is spontaneous; above it, the reverse process, the melting of ice to water, is spontaneous.

enthalpy effect that favors freezing and an entropy effect that disfavors it (Fig. 9-7). At low temperatures (below $T_f$) the former dominates and the liquid freezes spontaneously, but at higher temperatures (above $T_f$) the latter dominates and freezing does not occur. At $T_f$ the Gibbs free energies of the two phases are equal ($\Delta G = 0$) and the phases coexist at equilibrium. Similar types of analysis can be carried out for other phase transitions, such as the condensation of a gas to a liquid.

## 9-5 THE GIBBS FREE ENERGY AND CHEMICAL EQUILIBRIUM

The change in the Gibbs free energy provides a criterion for the spontaneity of any process occurring at constant temperature and pressure (Fig. 9-8). In this section, free energy changes in chemical reactions are examined and the relationship between free energy and chemical equilibrium is considered.

### Standard-State Free Energies

The change in the Gibbs free energy for a chemical reaction carried out at constant temperature is

$$\Delta G = \Delta H - T\Delta S \tag{9-10}$$

where $\Delta H$ is the enthalpy change in the reaction (considered in Section 8-4) and $\Delta S$ is the entropy change of the system (Section 9-3). Although we cannot know the *absolute* value of the Gibbs free energy of a substance (just as we cannot know the absolute value of its energy $E$) it is convenient to define a **standard molar Gibbs**

**free energy of formation** $\Delta G_f^\circ$ that is analogous to the standard molar enthalpy of formation $\Delta H_f^\circ$ introduced in Section 8-4. The $\Delta G_f^\circ$ of a compound is the change in Gibbs free energy for the reaction in which 1 mol of the compound in its standard state is formed from its elements in their standard states. For example, $\Delta G_f^\circ$ for $CO_2(g)$ is given by the Gibbs free energy change for the reaction

$$C(s,gr) + O_2(g) \longrightarrow CO_2(g) \qquad \Delta G_f^\circ = ?$$

The $\Delta G_f^\circ$ can be calculated from $\Delta H$ and $\Delta S$ for this reaction. Here $\Delta H$ is simply $\Delta H_f^\circ$ for $CO_2(g)$ because graphite and oxygen are elements in their standard states,

$$\Delta H = \Delta H_f^\circ(CO_2) = -393.51 \text{ kJ}$$

and $\Delta S$ can be obtained from the absolute entropies of the substances involved at 25°C and 1 atm pressure (both elements and compounds, because the absolute entropy $S^\circ$ of an element is not zero in its standard state).

$$\Delta S^\circ = S^\circ(CO_2) - S^\circ(C) - S^\circ(O_2) = 213.63 - 5.74 - 205.03 \text{ J K}^{-1}$$

$$= +2.86 \text{ J K}^{-1}$$

The $\Delta G_f^\circ$ for $CO_2$ is then

$$\Delta G_f^\circ = \Delta H_f^\circ - T\,\Delta S^\circ = -393.51 \text{ kJ} - (298.15 \text{ K})(2.86 \text{ J K}^{-1})(10^{-3} \text{ kJ J}^{-1})$$

$$= -394.36 \text{ kJ}$$

From our definition, $\Delta G_f^\circ = 0$ for an *element* that is already in its standard state.

Because $G$ is a state function, chemical equations can be added together, with their $\Delta G_f^\circ$'s combined the way they were for changes in enthalpy and entropy. A table of values of $\Delta G_f^\circ$ (such as that in Appendix D) can be used to calculate Gibbs free energy changes for a wide variety of chemical reactions under standard-state conditions.

## Example 9-7

Calculate $\Delta G^\circ$ for the following reaction, using tabulated values for $\Delta G_f^\circ$ from Appendix D.

$$3 \text{ NO}(g) \longrightarrow \text{N}_2\text{O}(g) + \text{NO}_2(g)$$

### Solution

$$\Delta G^\circ = \Delta G_f^\circ(\text{N}_2\text{O}) + \Delta G_f^\circ(\text{NO}_2) - 3\,\Delta G_f^\circ(\text{NO})$$

$$= (1 \text{ mol})(104.18 \text{ kJ mol}^{-1}) + (1 \text{ mol})(51.29 \text{ kJ mol}^{-1})$$
$$- (3 \text{ mol})(86.55 \text{ kJ mol}^{-1})$$

$$= -104.18 \text{ kJ}$$

## The Equilibrium Constant

In chemical equilibria, the partial pressures of gases can differ from 1 atm, so it is important to investigate how the Gibbs free energy changes with pressure at constant temperature. If the pressure of an ideal gas is changed from $P_1$ to $P_2$ with the temperature held constant, the free energy change is

**Figure 9-8**

The dissolution of hydrogen chloride in water

$$\text{HCl}(g) \longrightarrow \text{HCl}(aq)$$

is a spontaneous process, with $\Delta G^\circ = -35.9$ kJ. In this demonstration, the upper flask is filled with gaseous hydrogen chloride, and a small amount of water is injected into it. As the hydrogen chloride dissolves spontaneously, its pressure drops. The resulting partial vacuum in the upper flask draws water up the tube from the lower flask, allowing more hydrogen chloride to dissolve. The change is so fast that a vigorous fountain of water plays into the upper flask. The free energy change of the process appears as work, raising the water. (Leon Lewandowski)

$$\Delta G = \Delta(H - TS) = \Delta H - T\Delta S = -T\Delta S$$

because $\Delta H = 0$ when the pressure of an ideal gas is changed at constant temperature. The entropy change for an ideal gas in an isothermal process was calculated in Section 9-1:

$$\Delta S = nR \ln\left(\frac{V_2}{V_1}\right) = nR \ln\left(\frac{P_1}{P_2}\right) = -nR \ln\left(\frac{P_2}{P_1}\right)$$

so

$$\Delta G = nRT \ln\left(\frac{P_2}{P_1}\right) \qquad\qquad \textbf{(9-11)}$$

Consider now a mixture of gases that react chemically, such as the NO, $N_2O$, and $NO_2$ given in Example 9-7:

$$3\,NO(g) \rightleftharpoons N_2O(g) + NO_2(g)$$

If all of the partial pressures are 1 atm, then $\Delta G$ for this reaction is just $\Delta G°$ at 25°C, but if the pressures differ from 1 atm, $\Delta G$ must be calculated from a three-step process (Fig. 9-9). In step 1, the partial pressure of the reactant (in this case, 3 mol of NO) is changed from its initial value, $P_{NO}$, to the reference pressure $P_{ref} = 1$ atm:

$$\Delta G_1 = 3RT \ln\left(\frac{P_{ref}}{P_{NO}}\right) = RT \ln\left(\frac{P_{ref}}{P_{NO}}\right)^3$$

In step 2, the reaction is carried out with all reactants and products at partial pressures of $P_{ref} = 1$ atm:

$$\Delta G_2 = \Delta G°$$

In step 3, the partial pressures of the products (in this case, 1 mol of $N_2O$ and 1 mol of $NO_2$) are changed from $P_{ref} = 1$ atm to $P_{N_2O}$ and $P_{NO_2}$:

$$\Delta G_3 = RT \ln\left(\frac{P_{N_2O}}{P_{ref}}\right) + RT \ln\left(\frac{P_{NO_2}}{P_{ref}}\right) = RT \ln\left[\left(\frac{P_{N_2O}}{P_{ref}}\right)\left(\frac{P_{NO_2}}{P_{ref}}\right)\right]$$

**Figure 9-9**

A three-step process (red arrows) to calculate $\Delta G$ of a reaction for which reactants and products are not in their standard states.

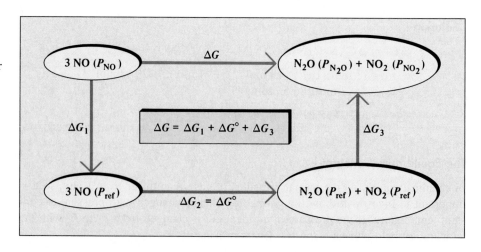

The overall Gibbs free energy change $\Delta G$ is the sum of the free energy changes for the three steps in the path:

$$\Delta G = \Delta G_1 + \Delta G_2 + \Delta G_3$$

$$= \Delta G° + RT \ln \left[ \frac{(P_{N_2O}/P_{ref})(P_{NO_2}/P_{ref})}{(P_{NO}/P_{ref})^3} \right]$$

The combination of partial pressures appearing in brackets in this expression is just the reaction quotient $Q$ introduced in Chapter 5, so that

$$\Delta G = \Delta G° + RT \ln Q$$

When a chemical reaction has reached equilibrium, $\Delta G = 0$; that is, there is no further spontaneous tendency for the reaction to occur in either the forward or reverse direction. In addition, $Q$ becomes equal to $K$. Under these conditions the preceding equation becomes

$$-\Delta G° = RT \ln K \qquad\qquad \text{(9-12)}$$

For the general reaction

$$a\text{A} + b\text{B} \longrightarrow c\text{C} + d\text{D}$$

the result, obtained in the same way, is

$$-\Delta G° = RT \ln \left[ \frac{(P_C/P_{ref})^c(P_D/P_{ref})^d}{(P_A/P_{ref})^a(P_B/P_{ref})^b} \right] = RT \ln K \qquad \text{(at equilibrium)}$$

We have accomplished two very important things. First, we have used thermodynamics to derive the mass action law (that is, the *form* of the equilibrium expression), a law that was introduced previously on purely empirical grounds. Note how the *activity* of a gas (the ratio of its pressure to the reference pressure $P_{ref}$) enters naturally in the thermodynamic approach, and how the equilibrium constant is a dimensionless quantity. Second, we have shown how that equilibrium constant can be calculated from $\Delta G°$ (that is, from $\Delta H°$ and $\Delta S°$) so that the *extent* of any equilibrium chemical reaction can be deduced from calorimetric data alone.

Away from equilibrium,

$$\Delta G = \Delta G° + RT \ln Q = -RT \ln K + RT \ln Q$$

$$\Delta G = RT \ln (Q/K) \qquad\qquad \text{(9-13)}$$

If the reaction quotient $Q$ is *less* than $K$, $\Delta G < 0$ and the reaction will proceed spontaneously as written, from left to right. If $Q > K$, then $\Delta G > 0$ and the *reverse* reaction (right to left) will occur spontaneously until equilibrium is reached. The reaction quotient criterion for the direction of reaction is a consequence of the second law of thermodynamics.

The expression connecting the standard Gibbs free energy change and the equilibrium constant can be rewritten as

$$\ln K = \frac{-\Delta G°}{RT} = \frac{\Delta S°}{R} - \frac{\Delta H°}{RT}$$

so that

$$K = \exp\left[\frac{-\Delta G^\circ}{RT}\right] = \exp\left[\frac{\Delta S^\circ}{R}\right]\exp\left[\frac{-\Delta H^\circ}{RT}\right]$$

Here $K$ is large (favoring the products) if $\Delta S^\circ$ is positive and large and if $\Delta H^\circ$ is negative and large. In other words, an increase in randomness or disorder ($\Delta S^\circ > 0$) and a decrease in enthalpy ($\Delta H^\circ < 0$) favor spontaneity. If $\Delta H^\circ$ and $\Delta S^\circ$ have the *same* sign, the value of $K$ will be a compromise between one effect that raises $K$ and the other that lowers it.

## Example 9-8

The $\Delta G^\circ$ of the chemical reaction

$$3\,NO(g) \longrightarrow N_2O(g) + NO_2(g)$$

was calculated in Example 9-7. Now calculate the equilibrium constant of this reaction at 25°C.

### Solution

The standard free energy change for the conversion of 3 mol of NO to 1 mol of $N_2O$ and 1 mol of $NO_2$ was found to be $-104.18$ kJ. To keep the units of the calculation correct, the $\Delta G^\circ$ is rewritten as $-104.18$ kJ mol$^{-1}$, where "per mole" signifies "per mole of the reaction as it is written," that is, per 3 mol of NO, 1 mol of $N_2O$, and 1 mol of $NO_2$, the chemical amounts in the balanced equation. Substitution gives

$$\ln K = \frac{-\Delta G^\circ}{RT} = \frac{-(-104{,}180\,\text{J mol}^{-1})}{(8.315\,\text{J K}^{-1}\,\text{mol}^{-1})(298.15\,\text{K})} = 42.03$$

$$K = e^{42.03} = 1.8 \times 10^{18}$$

*Related Problems:* 39, 40, 41, 42

The conversion of NO($g$) to $N_2O(g)$ plus $NO_2(g)$ is spontaneous under standard conditions. The forward reaction under these conditions is scarcely observed because it is so slow, but its equilibrium constant can nonetheless be calculated! Such calculations often have enormous practical impact. For example, this calculation shows that a plan to exploit the reaction to cut down the amount of NO in cooled automobile exhaust would not be thermodynamically doomed at the outset. The fundamental reaction tendency is there; success of the plan would hinge on making the reaction go faster.

Thermodynamics also applies to processes occurring in solution. Although the result is not derived here, the Gibbs free energy change for $n$ mol of a solute, as an ideal (dilute) solution changes in concentration from $c_1$ to $c_2$ mol L$^{-1}$, is

$$\Delta G = nRT \ln\left(\frac{c_2}{c_1}\right)$$

If the reference state for the solute is defined to be an ideal solution with a concentration $c_{\text{ref}} = 1$ M, then the same arguments used before for gas-phase reactions lead to expressions of the forms

$$\Delta G = \Delta G^\circ + RT \ln \left[ \frac{([C]/c_{ref})^c ([D]/c_{ref})^d}{([A]/c_{ref})^a ([B]/c_{ref})^b} \right]$$

and

$$\Delta G^\circ = -RT \ln K$$

for reactions involving dissolved species. Tables of standard free energies for solutes in aqueous solution at 25°C allow solute-phase equilibrium constants to be calculated.

If a chemical reaction involves both gases and dissolved species, the tabulated free energies should be used only if all gas concentrations are given as partial pressures expressed in atmospheres and all dissolved species concentrations are expressed in moles per liter. A pure solid or liquid is already in its standard state, and so its activity is unity and does not appear in the equilibrium expression.

## Temperature Dependence of Equilibrium Constants

Often the value of an equilibrium constant is needed for a reaction at a temperature other than 25°C. At least over a limited temperature range, $\Delta H$ and $\Delta S$ do not vary much with temperature (Fig. 9-10). To the extent that their temperature dependence may be neglected, it is evident that $\ln K$ is a linear function of $1/T$:

$$\ln K = -\frac{\Delta G}{RT} = -\frac{\Delta H^\circ}{RT} + \frac{\Delta S^\circ}{R} \tag{9-14}$$

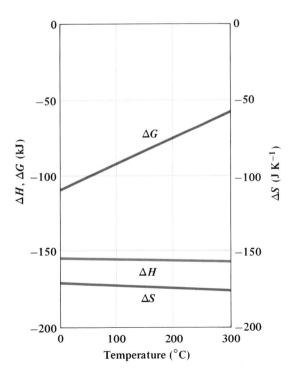

### Figure 9-10

The entropy change of the reaction

$$3 \, NO(g) \longrightarrow N_2O(g) + NO_2(g)$$

varies less than 5% between 0°C and 300°C; the enthalpy of reaction is even closer to constancy. The free energy change in the reaction shifts greatly over the temperature range, however.

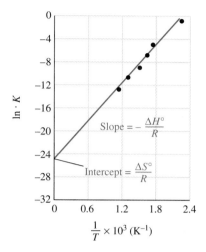

**Figure 9-11**

The temperature dependence of the equilibrium constant for the reaction

$$N_2(g) + 3\,H_2(g) \rightleftharpoons 2\,NH_3(g)$$

Experimental data are shown by points.

A graph of $\ln K$ against $1/T$ is approximately a straight line with slope $-\Delta H°/R$ and intercept $\Delta S°/R$ (Fig. 9-11). If the value of $K$ is known for one temperature and $\Delta H°$ is also known, $K$ can be calculated for other temperatures.

This can be seen in another way. Let $K_1$ and $K_2$ be the equilibrium constants for a reaction at temperatures $T_1$ and $T_2$, respectively. Then

$$\ln K_2 = -\frac{\Delta H°}{RT_2} + \frac{\Delta S°}{R}$$

$$\ln K_1 = -\frac{\Delta H°}{RT_1} + \frac{\Delta S°}{R}$$

Subtracting the second equation from the first gives

$$\ln\left(\frac{K_2}{K_1}\right) = -\frac{\Delta H°}{R}\left[\frac{1}{T_2} - \frac{1}{T_1}\right] \tag{9-15}$$

This is known as the **van't Hoff equation** after the Dutch chemist Jacobus van't Hoff. Given $\Delta H°$ and $K$ at one temperature, the equation can be used to calculate $K$ at another temperature, to the approximation that $\Delta H°$ and $\Delta S°$ are constant. Alternatively, it can be used to determine $\Delta H°$ without the use of a calorimeter if the values of $K$ are known at several temperatures.

The effect of a temperature change on the equilibrium constant depends on the sign of $\Delta H°$. If $\Delta H°$ is negative (the reaction is exothermic, giving off heat), then increasing the temperature *reduces K*. If $\Delta H°$ is positive (the reaction is endothermic, taking up heat), then increasing the temperature *increases K*. These observations can be related to Le Châtelier's principle. An increase in temperature is caused by transfer of heat to a system. A system at equilibrium responds in such a way as to take up some of that heat through chemical reaction—in the backward direction for an exothermic reaction (decreasing $K$) or in the forward direction for an endothermic reaction (increasing $K$).

## Example 9-9

Calculate $K$ for the equilibrium of Example 9-8 at $T = 400$ K, assuming $\Delta H°$ to be approximately independent of temperature over the range from 298 to 400 K.

### Solution

The first step is to calculate $\Delta H°$ for the reaction. Appendix D provides data to calculate

$$\Delta H° = -155.52 \text{ kJ}$$

From the van't Hoff equation,

$$\ln\left(\frac{K_{400}}{K_{298}}\right) = -\frac{\Delta H°}{R}\left[\frac{1}{400\text{ K}} - \frac{1}{298\text{ K}}\right]$$

$$= \frac{155{,}520 \text{ J mol}^{-1}}{8.315 \text{ J K}^{-1}\text{ mol}^{-1}}\left[\frac{1}{400\text{ K}} - \frac{1}{298\text{ K}}\right] = -16.01$$

$$\frac{K_{400}}{K_{298}} = e^{-16.01} = 1.1 \times 10^{-7}$$

Taking $K_{298}$ to be $1.8 \times 10^{18}$ (from Example 9-8) gives

$$K_{400} = (1.8 \times 10^{18})(1.1 \times 10^{-7}) = 2.0 \times 10^{11}$$

Because the reaction is exothermic, an increase in temperature reduces the equilibrium constant.

An alternative way to do this calculation would be to determine both $\Delta H^\circ$ and $\Delta S^\circ$ and from them to calculate $\Delta G$ at 400 K.

*Related Problems: 49, 50*

Dissolution reactions can be either endothermic or exothermic; this is reflected in the positive and negative slopes of the solubilities in Figure 7-3. At first glance, one might think that the entropy would increase whenever a solid dissolved in water, but this is not necessarily true. Although solute molecules become more disordered as they move from the crystal into the water, they also change the hydrogen-bonding network in the liquid. In some cases (such as with small, multiply charged positive ions) the water becomes more ordered in the solution and its entropy decreases. This can make the overall entropy change of dissolution negative, disfavoring the process.

## Vapor Pressure

Suppose pure liquid water is in equilibrium with its vapor at temperature $T$:

$$H_2O(\ell) \rightleftharpoons H_2O(g) \qquad P_{H_2O(g)} = K$$

The temperature dependence of $K$ (and therefore of the vapor pressure $P_{H_2O(g)}$) is a special case of the van't Hoff equation. If $\Delta H_{vap}$ and $\Delta S_{vap}$ are approximately independent of temperature, then from the van't Hoff equation,

$$\ln\left(\frac{K_2}{K_1}\right) = \ln\left(\frac{P_2}{P_1}\right) = -\frac{\Delta H_{vap}}{R}\left[\frac{1}{T_2} - \frac{1}{T_1}\right]$$

where $P_2$ and $P_1$ are the vapor pressures at temperatures $T_2$ and $T_1$. From this equation, the vapor pressure at any given temperature can be estimated if its value at some other temperature is known and if the enthalpy of vaporization is also known.

At the normal boiling point of a substance, $T_b$, the vapor pressure is 1 atm. If $T_1$ is taken to correspond to $T_b$ and $T_2$ to some other temperature $T$, the van't Hoff equation is

$$\ln P = -\frac{\Delta H_{vap}}{R}\left[\frac{1}{T} - \frac{1}{T_b}\right] \qquad (9\text{-}16)$$

where $P$ is the vapor pressure at temperature $T$, expressed in atmospheres.

## Example 9-10

The $\Delta H_{vap}$ for water is 40.66 kJ mol$^{-1}$ at the normal boiling point, $T_b = 373$ K. Assuming $\Delta H_{vap}$ and $\Delta S_{vap}$ are approximately independent of temperature from 50 to 100°C, estimate the vapor pressure of water at 50°C (323 K).

**Solution**

$$\ln P_{323} = \frac{-40660 \text{ J mol}^{-1}}{8.314 \text{ J K}^{-1} \text{ mol}^{-1}} \left[ \frac{1}{323 \text{ K}} - \frac{1}{373 \text{ K}} \right] = -2.03$$

$$P_{323} = 0.13 \text{ atm}$$

This differs slightly from the experimental value, 0.1217 atm, because $\Delta H_{vap}$ *does* change with temperature, an effect that was neglected in the approximate calculation.

*Related Problems:* 51, 52

## 9-6 A Deeper Look . . . Thermodynamics and Colligative Properties

In Chapter 4 we considered several colligative properties of dilute solutions, including the phenomena of boiling-point elevation, freezing-point depression, and osmotic pressure. A deeper understanding of all of them can be achieved through thermodynamics.

The colligative properties involve the free energy of the *solvent* in a dilute solution. Suppose $n_1$ mol of solvent is added to an ideal solution with solvent mole fraction $X_1$. Either $n_1$ is small enough or the solution volume is large enough that addition of this solvent does not appreciably change $X_1$. The free energy change $\Delta G$ is determined by using the three-step process shown in Figure 9-12. In step 1, $n_1$ mol of solvent is brought into the vapor phase with pressure $P_1^\circ$, the equilibrium

vapor pressure of pure solvent. Because solvent and vapor are at equilibrium at this pressure, $\Delta G_1 = 0$ for this step. In step 2, the pressure is changed from $P_1^\circ$ to $P_1$, the solvent vapor pressure above the solution. The change in free energy with pressure for gases, calculated in Section 9-5, is

$$\Delta G_2 = n_1 RT \ln \left( \frac{P_1}{P_1^\circ} \right)$$

In step 3, the gaseous solvent at pressure $P_1$ is brought into solution. $\Delta G_3$ is also 0 because the solvent in the solution is in equilibrium with the vapor at pressure $P_1$. We conclude that

$$\Delta G = \Delta G_2 = n_1 RT \ln \left( \frac{P_1}{P_1^\circ} \right)$$

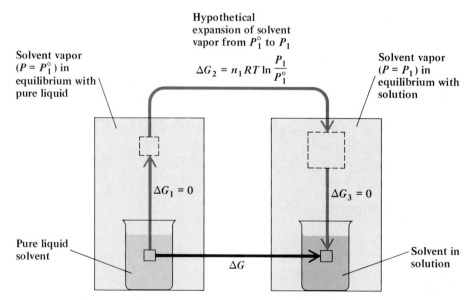

**Figure 9-12**

The three-step process to bring $n_1$ mol of solvent into a large volume of solution.

Hypothetical expansion of solvent vapor from $P_1^\circ$ to $P_1$

$$\Delta G_2 = n_1 RT \ln \frac{P_1}{P_1^\circ}$$

Solvent vapor $(P = P_1^\circ)$ in equilibrium with pure liquid

Solvent vapor $(P = P_1)$ in equilibrium with solution

$\Delta G_1 = 0$

$\Delta G_3 = 0$

Pure liquid solvent

Solvent in solution

$\Delta G$

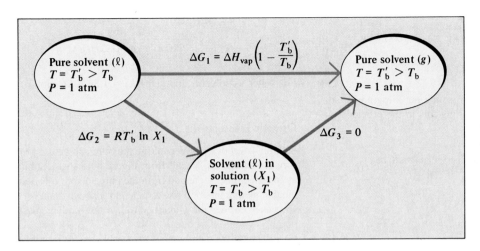

**Figure 9-13**

Two ways to vaporize 1 mol of solvent to 1 atm pressure. Path 1 (blue arrow) is irreversible, whereas path 2 + 3 (red arrows) is reversible. The overall change in Gibbs free energy is the same for the two paths.

From Raoult's law (which provides the *definition* of an ideal solution),

$$P_1 = X_1 P_1^\circ$$

so that

$$\Delta G = n_1 RT \ln X_1 \qquad \textbf{(9-17)}$$

Because $X_1 < 1$, the free energy decreases when the solvent is transferred to the solution.

### Boiling-Point Elevation and Freezing-Point Depression

Consider a solution of a nonvolatile solute with mole fraction $X_2$ in a solvent with mole fraction $X_1 = 1 - X_2$. At the normal boiling point $T_b'$ of the *solution,* the vapor pressure of the solvent above the solution is 1 atm. Then $T_b'$ differs from $T_b$, the boiling point of the pure solvent, by an amount that we now calculate using thermodynamics.

Consider the free energy change when 1 mol of pure solvent (at temperature $T_b'$) is vaporized to a gas at 1 atm pressure. This can be carried out in two different ways (Fig. 9-13). On one hand, the solvent can be vaporized directly (step 1). Because $T_b' \neq T_b$ (the system is above its equilibrium boiling point), $\Delta G_1$ differs from 0:

$$\Delta G_1 = \Delta H_{vap} - T_b' \Delta S_{vap}$$

$$= \Delta H_{vap}\left(1 - \frac{T_b'}{T_b}\right)$$

because $\Delta S_{vap} = \Delta H_{vap}/T_b$. The second path involves two steps. In step 2, 1 mol of solvent is transferred to the solution at mole fraction $X_1$:

$$\Delta G_2 = RT_b' \ln X_1 = RT_b' \ln (1 - X_2)$$

In step 3, 1 mol of solvent is then vaporized from the solution to the vapor at 1 atm pressure. For this step $\Delta G_3$ is 0, because this is the normal boiling point of the solution.

The results of the two paths must be equal, and so

$$\Delta G_1 = \Delta G_2 + \Delta G_3$$

$$\Delta H_{vap}\left(1 - \frac{T_b'}{T_b}\right) = RT_b' \ln (1 - X_2)$$

This expression relates the boiling point of the solution, $T_b'$, to the mole fraction of solute, $X_2$, and the enthalpy of vaporization of the pure solvent. If the solution is dilute, $X_2$ is small and the logarithm can be approximated as[2]

$$\ln (1 - X_2) \approx -X_2$$

In this case $T_b'$ will be close to $T_b$, so that it can be replaced by $T_b$ on the right side. This leaves

$$\frac{\Delta H_{vap}}{T_b}(T_b - T_b') = RT_b(-X_2)$$

$$\Delta T_b = T_b' - T_b = \frac{RT_b^2}{\Delta H_{vap}}X_2$$

The solute mole fraction is now converted to molality. First,

$$X_2 = \frac{n_2}{n_1 + n_2} \approx \frac{n_2}{n_1}$$

where $n_1$ and $n_2$ are the numbers of moles of solvent and solute, respectively. Because very dilute solutions are considered, $n_1 \gg n_2$. If $m_1$ and $m_2$ are the masses (in grams) of solvent and solute, and if $\mathcal{M}_1$ and $\mathcal{M}_2$ are their *molar* masses, then

---

[2]This well-known approximation results from an algebraic series. It can be checked readily with a calculator for small values of $X_2$.

$$X_2 = \frac{m_2/\mathcal{M}_2}{m_1/\mathcal{M}_1} = \frac{m_2/\mathcal{M}_2}{(m_1/1000 \text{ g kg}^{-1})} \times \frac{\mathcal{M}_1}{1000 \text{ g kg}^{-1}}$$

$$= m \times \frac{\mathcal{M}_1}{1000 \text{ g kg}^{-1}}$$

where $m$ is the *molality* (moles of solute per kilogram of solvent), defined in Chapter 4. Inserting this in the previous expression for $\Delta T_b$ gives

$$\Delta T_b = \frac{RT_b^2 \mathcal{M}_1}{(1000 \text{ g kg}^{-1}) \Delta H_{vap}} m = K_b m$$

The constant

$$K_b = \frac{RT_b^2 \mathcal{M}_1}{(1000 \text{ g kg}^{-1}) \Delta H_{vap}} \qquad \text{(9-18)}$$

depends only on properties of the pure solvent (normal boiling point, molar mass, and enthalpy of vaporization) and is independent of the nature of the solute.

This result has the same form as the expression for boiling-point elevation given in Chapter 4 from an empirical point of view. Thermodynamics shows why the boiling-point elevation has the form that it does, and relates the constant $K_b$ to independently measurable experimental properties of the solvent.

....................................................................

**Example 9-11**

Water's $\Delta H_{vap}$ is 40.66 kJ mol$^{-1}$, and its normal boiling point is 373.15 K. Calculate $K_b$ for water.

**Solution**

$$K_b = \frac{(8.315 \text{ J K}^{-1} \text{ mol}^{-1})(373.15 \text{ K})^2(18.02 \text{ g mol}^{-1})}{(1000 \text{ g kg}^{-1})(40660 \text{ J mol}^{-1})}$$

$$= 0.5130 \text{ K kg mol}^{-1}$$

This is close to the experimentally determined value of $K_b$ from Chapter 4, 0.512 K kg mol$^{-1}$.

*Related Problem:* 53

....................................................................

The freezing point depression of a solution is analogous. Here the phase change is the freezing of solvent to form its pure solid, and so the enthalpy change involved is $-\Delta H_{fus}$ rather than $\Delta H_{vap}$. Because of the negative sign, $\Delta T_f$ is negative and the freezing point is lowered. Reasoning parallel to that just given for boiling-point elevation leads to

$$\Delta T_f = T_f' - T_f = -K_f m$$

where

$$K_f = \frac{RT_f^2 \mathcal{M}_1}{(1000 \text{ g kg}^{-1}) \Delta H_{fus}} \qquad \text{(9-19)}$$

For water as solvent, this gives $K_f = 1.86$ K kg mol$^{-1}$.

## Osmotic Pressure

Section 4-5 showed how osmotic pressure can be used to measure molar masses. Recall from Figure 4-18 that a solution in an inverted tube, separated from the pure solvent by a semipermeable membrane, will rise to a height $h$ above the level of the solvent. Consider the situation from a thermodynamic standpoint.

Because the membrane is permeable to solvent molecules, the free energy per mole of solvent must be the same on both sides of the membrane at equilibrium, as otherwise there would be a net flow of solvent molecules through the membrane. The free energy of the solvent on the solution side is changed by two effects: the dilution to a mole fraction $X_1 < 1$ by the presence of solute, and the increase in external hydrostatic pressure by an amount $\Pi$ due to the rise of the level of solution in the tube. These two effects must cancel each other and give $\Delta G = 0$ at equilibrium.

The first effect has already been discussed; it changes the free energy for $n_1$ mol of solvent in the solution by $n_1 RT \ln X_1$. Because $X_1 < 1$, this term is negative; it must be balanced by a second term, which results from an increase in the free energy as the pressure on the system is increased. To calculate this second term, examine the effect of pressure on the free energy of pure solvent occupying a volume $V$. The general relationship for $G$ is

$$G = H - TS = E + PV - TS$$

If the pressure is changed at constant temperature,

$$\Delta G = \Delta E + \Delta(PV) - T\Delta S$$

$$= q + w + \Delta(PV) - T\Delta S$$

If this pressure change is carried out reversibly, $q = T\Delta S$ and

$$\Delta G = w + \Delta(PV)$$

A liquid is rather incompressible; even a large increase in external pressure leads to only a small change in its volume. If the volume is taken to be a constant, $\Delta(PV)$ becomes $V\Delta P$, and $w \approx 0$ because a volume change is necessary for work to be done. The free energy change of a liquid under pressure is thus approximately

$$\Delta G = V\Delta P$$

The pressure change $\Delta P$ is the osmotic pressure $\Pi$, the additional hydrostatic pressure on the solution side of the membrane. Adding this to the dilution contribution to the free energy per mole of solvent in solution gives

$$\Delta G = 0 = n_1 RT \ln X_1 + \Pi V$$

Therefore,

$$\Pi V = -n_1 RT \ln (1 - X_2)$$

For a dilute solution, where $\ln (1 - X_2) \approx -X_2$,

$$\Pi V \approx -n_1 RT(-X_2) \approx n_1 RT \frac{n_2}{n_1} = n_2 RT$$

and

$$\Pi = (n_2/V)RT = cRT \qquad \text{(9-20)}$$

where $c$ is the solute concentration. This is the osmotic pressure law of Section 4-5, derived here for ideal solutions by means of thermodynamics.

## Purifying Nickel

Impure nickel, obtained from the smelting of its sulfide ores in a blast furnace, can be converted to metal of 99.9% to 99.99% purity by the Mond process, which relies on the equilibrium

$$\text{Ni}(s) + 4\,\text{CO}(g) \rightleftharpoons \text{Ni(CO)}_4(g)$$

The standard enthalpy of formation of nickel tetracarbonyl, $\text{Ni(CO)}_4(g)$, is $-602.9 \text{ kJ mol}^{-1}$, and its absolute entropy $S°$ is $410.6 \text{ J K}^{-1} \text{ mol}^{-1}$.

(a) Predict (without referring to a table) whether the entropy change of the system (the reacting atoms and molecules) is positive or negative in this process.

(b) At a temperature where this reaction is spontaneous, predict whether the entropy change of the surroundings is positive or negative.

(c) Use the data in Appendix D to calculate $\Delta H°$ and $\Delta S°$ for this reaction.

(d) At what temperature is $\Delta G = 0$ (and $K = 1$) for this reaction?

(e) The first step in the Mond process is the equilibration of impure nickel with CO and $\text{Ni(CO)}_4$ at about 50°C. In this step, the equilibrium constant should be as large as possible to draw as much nickel as possible into the vapor-phase complex. Calculate the equilibrium constant for the preceding reaction at 50°C.

(f) In the second step of the Mond process, the gases are removed from the reaction chamber and heated to about 230°C. At high enough temperatures, the sign of $\Delta G$ is reversed and the reaction occurs in the opposite direction, depositing pure nickel. In this step, the equilibrium constant should be as small as possible. Calculate the equilibrium constant for the preceding reaction at 230°C.

(g) The Mond process relies on the volatility of $\text{Ni(CO)}_4$ for its success. Under room conditions this compound is a liquid, but it boils at 42.2°C with an enthalpy of vaporization of $29.0 \text{ kJ mol}^{-1}$. Calculate the entropy of vaporization of $\text{Ni(CO)}_4$ and compare it with that predicted by Trouton's rule.

(h) A recently developed variation of the Mond process carries out the first step at higher pressures and at a temperature of 150°C. Estimate the maximum pressure of $\text{Ni(CO)}_4(g)$ that can be attained before the gas will liquefy at this temperature [that is, calculate the vapor pressure of $\text{Ni(CO)}_4(\ell)$ at 150°C].

A huge reducer kiln at the heart of a nickel carbonyl refinery in Wales. (INCO)

**Answers**
(a) Negative
(b) Positive
(c) $\Delta H° = -160.8 \text{ kJ}$;   $\Delta S° = -409.5 \text{ J K}^{-1}$
(d) $392.7 \text{ K} = 119.5°C$

(e) $4.0 \times 10^4$

(f) $2.0 \times 10^{-5}$

(g) $92.0 \ J \ K^{-1} \ mol^{-1}$, close to the Trouton's rule value of $88 \ J \ K^{-1} \ mol^{-1}$

(h) 16.7 atm

## CONCEPTS & SKILLS

After studying this chapter and working the problems that follow, you should be able to

1. Calculate the entropy change for gaseous systems undergoing reversible isothermal, isochoric, and isobaric processes (Section 9-1, problems 1–8).
2. Calculate the entropy change for the system and the surroundings when irreversible processes are carried out (Section 9-1, problems 9–10).
3. Provide a statistical interpretation of the change in entropy that occurs when a gas undergoes a volume change (Section 9-2, problems 11–16).
4. Describe how the absolute entropy of a substance can be obtained (Section 9-3).
5. Calculate standard-state entropy changes for chemical reactions (Section 9-3, problems 19–22).
6. Define the Gibbs free energy function, and state the criterion it provides for the spontaneity of a process (Section 9-4).
7. Calculate the change in Gibbs free energy for reversible and spontaneous phase transformations (Section 9-4, problems 27–30).
8. Relate the equilibrium constant of a reaction to its standard Gibbs free energy change (Section 9-5, problems 39–42).
9. Relate the change in the equilibrium constant of a reaction with temperature to its standard enthalpy change (Section 9-5, problems 45–52).
10. Calculate the change in the Gibbs free energy of a solution when it is diluted and relate this to the colligative properties of dilute solutions (Section 9-6, problems 53–54).

## PROBLEMS

### Entropy, Spontaneous Change, and the Second Law

1. Tungsten melts at 3410°C and has an enthalpy change of fusion of $35.4 \ kJ \ mol^{-1}$. Calculate the entropy of fusion of tungsten.

2. Tetraphenylgermane, $(C_6H_5)_4Ge$, has a melting point of 232.5°C, and its enthalpy increases by $106.7 \ J \ g^{-1}$ during fusion. Calculate the molar enthalpy of fusion and molar entropy of fusion of tetraphenylgermane.

3. The normal boiling point of acetone is 56.2°C. Use Trouton's rule to estimate its molar enthalpy of vaporization.

4. The molar enthalpy of vaporization of liquid hydrogen chloride is $16.15 \ kJ \ mol^{-1}$. Use Trouton's rule to estimate its normal boiling point.

5. If 4.00 mol of hydrogen $(c_P = 28.8 \ J \ K^{-1} \ mol^{-1})$ is expanded reversibly and isothermally at 400 K from an initial volume of 12.0 L to a final volume of 30.0 L, calculate $\Delta E$, $q$, $w$, $\Delta H$, and $\Delta S$ for the gas.

6. Suppose 60.0 g of hydrogen bromide, $HBr(g)$, is heated reversibly from 300 K to 500 K at a constant volume of 50.0 L and then allowed to expand isothermally and reversibly until the original pressure is reached. Using $c_P(HBr(g)) = 29.1 \ J \ K^{-1} \ mol^{-1}$, calculate $\Delta E$, $q$, $w$, $\Delta H$, and $\Delta S$ for this process. Assume that HBr is an ideal gas under these conditions.

7. Exactly 1 mol of ice is heated reversibly at atmospheric pressure from −20 to 0°C, melted reversibly at 0°C, and then heated reversibly at atmospheric pressure to 20°C. $\Delta H_{fus} = 6007 \ J \ mol^{-1}$, $c_P(ice) = 38 \ J \ K^{-1} \ mol^{-1}$, and $c_P(water) = 75 \ J \ K^{-1} \ mol^{-1}$. Calculate $\Delta S$ for the system, the surroundings, and the universe for this process.

**8.** Suppose 1.00 mol of water at 25°C is flash-evaporated by allowing it to fall into an iron crucible maintained at 150°C. Calculate $\Delta S$ for the water, $\Delta S$ for the iron crucible, and $\Delta S_{tot}$, if $c_P(H_2O(\ell)) = 75.4$ J K$^{-1}$ mol$^{-1}$ and $c_P(H_2O(g)) = 36.0$ J K$^{-1}$ mol$^{-1}$. Take $\Delta H_{vap} = 40.68$ kJ mol$^{-1}$ for water at its boiling point of 100°C.

**9.** In Example 8-3, a process was considered in which 72.4 g of iron initially at 100.0°C was added to 100.0 g of water initially at 10.0°C, and an equilibrium temperature of 16.5°C was reached. Take $c_P(Fe)$ to be 25.1 J K$^{-1}$ mol$^{-1}$ and $c_P(H_2O)$ to be 75.3 J K$^{-1}$mol$^{-1}$, independent of temperature. Calculate $\Delta S$ for the iron, $\Delta S$ for the water, and $\Delta S_{tot}$ in this process.

**10.** Iron has a heat capacity of 25.1 J K$^{-1}$ mol$^{-1}$, approximately independent of temperature between 0 and 100°C.
(a) Calculate the enthalpy and entropy change of 1.00 mol of iron as it is cooled at atmospheric pressure from 100°C to 0°C.
(b) A piece of iron weighing 55.85 g and at 100°C is placed in a large reservoir of water held at 0°C. It cools irreversibly until its temperature equals that of the water. Assuming the water reservoir is large enough that its temperature remains very close to 0°C, calculate the entropy changes for the iron and the water, and the total entropy change in this process.

## Entropy and Irreversibility: A Statistical Interpretation

**11.** (a) How many "microstates" are there for the numbers that come up on a pair of dice?
(b) What is the probability that a role of a pair of dice will show two sixes?

**12.** (a) Suppose a volume is divided into three equal parts. How many microstates can be written for all possible ways of distributing four molecules among the three parts?
(b) What is the probability that all four molecules are in the leftmost third of the volume at the same time?

**13.** When $H_2O(\ell)$ and $D_2O(\ell)$ are mixed, the following reaction occurs spontaneously:

$$H_2O(\ell) + D_2O(\ell) \longrightarrow 2\ HOD(\ell)$$

There is little difference between the enthalpy of an O—H bond and that of an O—D bond. What is the main driving force for this reaction?

**14.** The two gases $BF_3(g)$ and $BCl_3(g)$ are mixed in equal molar amounts. All B—F bonds have about the same bond enthalpy, as do all B—Cl bonds. Explain why the mixture tends to react to form $BF_2Cl(g)$ and $BCl_2F(g)$.

**15.** Two large glass bulbs of identical volume are connected by means of a stopcock. One bulb initially contains 1.00 mol of $H_2$; the other, 1.00 mol of He. The stopcock is opened and the gases are allowed to mix and reach equilibrium. What is the probability that all of the $H_2$ in the first bulb will diffuse into the second bulb and all of the He gas in the second bulb will diffuse into the first bulb?

**16.** A mixture of 2.00 mol of nitrogen and 1.00 mol of oxygen is in thermal equilibrium in a 100-L container at 25°C. Calculate the probability that at a given time all the nitrogen will be found in the left half of the container and all the oxygen in the right half.

**17.** Predict the sign of the system's entropy change in each of the following processes.
(a) Sodium chloride melts.
(b) A building is demolished.
(c) A volume of air is divided into three separate volumes of nitrogen, oxygen, and argon, each at the same pressure and temperature as the original air.

**18.** Predict the sign of the system's entropy change in each of the following processes.
(a) A computer is constructed from iron, copper, carbon, silicon, gallium, and arsenic.
(b) A container holding a compressed gas develops a leak and the gas enters the atmosphere.
(c) Solid carbon dioxide (Dry Ice) sublimes to gaseous carbon dioxide.

## Absolute Entropies and the Third Law

**19.** (a) Use data from Appendix D to calculate the standard entropy change at 25°C for the reaction

$$N_2H_4(\ell) + 3\ O_2(g) \longrightarrow 2\ NO_2(g) + 2\ H_2O(\ell)$$

(b) Suppose the hydrazine ($N_2H_4$) is in the gaseous, rather than liquid, state. Will the entropy change for its reaction with oxygen be higher or lower than that calculated in part (a)? (*Hint:* Entropies of reaction can be added when chemical equations are added, in the same way that Hess's law allows enthalpies to be added.)

**20.** (a) Use data from Appendix D to calculate the standard entropy change at 25°C for the reaction

$$CH_3COOH(g) + NH_3(g) \longrightarrow$$
$$CH_3NH_2(g) + CO_2(g) + H_2(g)$$

(b) Suppose that 1.00 mol each of solid acetamide, $CH_3CONH_2(s)$, and of water, $H_2O(\ell)$, react to give the same products. Will the standard entropy change be larger or smaller than that calculated for the reaction in part (a)?

**21.** The alkali metals react with chlorine to give salts:

$$2\ Li(s) + Cl_2(g) \longrightarrow 2\ LiCl(s)$$
$$2\ Na(s) + Cl_2(g) \longrightarrow 2\ NaCl(s)$$
$$2\ K(s) + Cl_2(g) \longrightarrow 2\ KCl(s)$$
$$2\ Rb(s) + Cl_2(s) \longrightarrow 2\ RbCl(s)$$
$$2\ Cs(s) + Cl_2(g) \longrightarrow 2\ CsCl(s)$$

Using the data in Appendix D, compute $\Delta S°$ of each reaction and identify a periodic trend, if any.

**22.** All of the halogens react directly with $H_2(g)$ to give binary compounds. The reactions are

$$F_2(g) + H_2(g) \longrightarrow 2\,HF(g)$$

$$Cl_2(g) + H_2(g) \longrightarrow 2\,HCl(g)$$

$$Br_2(g) + H_2(g) \longrightarrow 2\,HBr(g)$$

$$I_2(g) + H_2(g) \longrightarrow 2\,HI(g)$$

Using the data in Appendix D, compute $\Delta S°$ of each reaction and identify a periodic trend, if any.

**23.** The dissolution of calcium chloride

$$CaCl_2(s) \longrightarrow Ca^{2+}(aq) + 2\,Cl^-(aq)$$

is a spontaneous process at 25°C, even though the standard entropy change of the preceding reaction is negative ($\Delta S° = -44.7\ J\ K^{-1}$). What conclusion can you draw about the change in entropy of the surroundings in this process?

**24.** Quartz, $SiO_2(s)$, does not spontaneously decompose to silicon and oxygen at 25°C in the reaction

$$SiO_2(s) \longrightarrow Si(s) + O_2(g)$$

even though the standard entropy change of the reaction is large and positive ($\Delta S° = +182.02\ J\ K^{-1}$). Explain.

**25.** Use the microscopic interpretation of entropy from Section 9-2 to explain why the entropy change of the system in problem 24 is positive.

**26.** (a) Why is the entropy change of the system negative for the reaction in problem 23, when the ions become dispersed through a large volume of solution? (*Hint:* Think about the role of the solvent, water.)
(b) Use Appendix D to calculate $\Delta S°$ for the corresponding dissolution of $CaF_2(s)$. Explain why this value is even more negative than that given in problem 23.

## The Gibbs Free Energy

**27.** The molar enthalpy of fusion of solid ammonia is $5.65\ kJ\ mol^{-1}$, and the molar entropy of fusion is $28.9\ J\ K^{-1}\ mol^{-1}$.
(a) Calculate the Gibbs free energy change for the melting of 1.00 mol of ammonia at 170 K.
(b) Calculate the Gibbs free energy change for the conversion of 3.60 mol of solid ammonia to liquid ammonia at 170 K.
(c) Will ammonia melt spontaneously at 170 K?
(d) At what temperature are solid and liquid ammonia in equilibrium at a pressure of 1 atm?

**28.** Solid tin exists in two forms: white and gray. For the transformation

$$Sn(s,\ white) \longrightarrow Sn(s,\ gray)$$

the enthalpy change is $-2.1\ kJ$ and the entropy change is $-7.4\ J\ K^{-1}$.

(a) Calculate the Gibbs free energy change for the conversion of 1.00 mol of white tin to gray tin at $-30°C$.
(b) Calculate the Gibbs free energy change for the conversion of 2.50 mol of white tin to gray tin at $-30°C$.
(c) Will white tin convert spontaneously to gray tin at $-30°C$?
(d) At what temperature are white and gray tin in equilibrium at a pressure of 1 atm?

**29.** Ethanol's enthalpy of vaporization is $38.7\ kJ\ mol^{-1}$ at its normal boiling point, 78°C. Calculate $q$, $w$, $\Delta E$, $\Delta S_{sys}$, and $\Delta G$ when 1.00 mol of ethanol is vaporized reversibly at 78°C and 1 atm. Assume that the vapor is an ideal gas, and neglect the volume of liquid ethanol relative to that of its vapor.

**30.** Suppose 1.00 mol of superheated ice melts to liquid water at 25°C. Assume the specific heats of ice and liquid water have the same value and are independent of temperature. The enthalpy change for the melting of ice at 0°C is $6007\ J\ mol^{-1}$. Calculate $\Delta H$, $\Delta S_{sys}$, and $\Delta G$ for this process.

## The Gibbs Free Energy and Chemical Equilibrium

**31.** At 1200°C, the reduction of iron oxide to elemental iron and oxygen is not spontaneous:

$$2\,Fe_2O_3(s) \longrightarrow 4\,Fe(s) + 3\,O_2(g) \quad \Delta G = +840\ kJ$$

Show how this process can be made to proceed if all the oxygen generated reacts with carbon:

$$C(s) + O_2(g) \longrightarrow CO_2(g) \qquad \Delta G = -400\ kJ$$

This observation is the basis for the smelting of iron ore with coke to extract metallic iron.

**32.** The primary medium for free-energy storage in living cells is adenosine triphosphate (ATP). Its formation from adenosine diphosphate (ADP) is not spontaneous:

$$ADP^{3-}(aq) + HPO_4^{2-}(aq) + H^+(aq) \longrightarrow$$
$$ATP^{4-}(aq) + H_2O(\ell) \qquad \Delta G = +34.5\ kJ$$

Cells couple ATP production with the metabolism of glucose (a sugar):

$$C_6H_{12}O_6(aq) + 6\,O_2(g) \longrightarrow$$
$$6\,CO_2(g) + 6\,H_2O(\ell) \qquad \Delta G = -2872\ kJ$$

The reaction of 1 molecule of glucose leads to the formation of 38 molecules of ATP from ADP. Show how the coupling makes this reaction spontaneous. What fraction of the free energy released in the oxidation of glucose is stored in the ATP?

**33.** A process at constant $T$ and $P$ can be described as spontaneous if $\Delta G < 0$ and nonspontaneous if $\Delta G > 0$. Over what range of temperatures is each of the following processes spontaneous? Assume that all gases are at a pressure of 1 atm. (*Hint:* Use Appendix D to calculate $\Delta H$ and $\Delta S$ [assumed independent of temperature and equal to $\Delta H°$ and $\Delta S°$, respectively] and then use the definition of $\Delta G$.)

(a) The rusting of iron, a complex reaction that can be approximated as

$$4\,Fe(s) + 3\,O_2(g) \longrightarrow 2\,Fe_2O_3(s)$$

(b) The preparation of $SO_3(g)$ from $SO_2(g)$, a step in the manufacture of sulfuric acid:

$$SO_2(g) + \tfrac{1}{2}\,O_2(g) \longrightarrow SO_3(g)$$

(c) The production of the anesthetic dinitrogen oxide through the decomposition of ammonium nitrate:

$$NH_4NO_3(s) \longrightarrow N_2O(g) + 2\,H_2O(g)$$

**34.** Follow the same procedure used in problem 33 to determine the range of temperatures over which each of the following processes is spontaneous.
(a) The preparation of the poisonous gas phosgene:

$$CO(g) + Cl_2(g) \longrightarrow COCl_2(g)$$

(b) The laboratory-scale production of oxygen from the decomposition of potassium chlorate:

$$2\,KClO_3(s) \longrightarrow 2\,KCl(s) + 3\,O_2(g)$$

(c) The reduction of iron(II) oxide (wüstite) by coke (carbon), a step in the production of iron in a blast furnace:

$$FeO(s) + C(s,\,gr) \longrightarrow Fe(s) + CO(g)$$

**35.** Explain why it is possible to reduce tungsten(VI) oxide ($WO_3$) to metal with hydrogen at an elevated temperature. At what temperature does the equilibrium constant of the reaction taking place in this conversion become unity? Use the data of Appendix D.

**36.** Tungsten(VI) oxide can also be reduced to tungsten by heating it with carbon in an electric furnace:

$$2\,WO_3(s) + 3\,C(s) \longrightarrow 2\,W(s) + 3\,CO_2(g)$$

(a) Calculate the standard free energy change ($\Delta G°$) for this reaction, and comment on the feasibility of the process at room conditions.
(b) What must be done to make the process thermodynamically feasible, assuming $\Delta H$ and $\Delta S$ do not vary much with temperature?

**37.** Although the process of dissolution of helium gas in water is favored in terms of energy, helium is only very slightly soluble in water. What keeps this gas from dissolving in great quantities in water?

**38.** At 298.15 K, the solubility of rhombic sulfur in water is slight, only $1.9 \times 10^{-8}$ mol sulfur per liter of water.
(a) Compute $\Delta G°$ for the dissolution of sulfur in water.
(b) Despite the fact that the $\Delta G°$ computed in part (a) is positive, sulfur placed in water at 298.15 K still spontaneously dissolves to the extent just given. Explain.

**39.** Calculate $\Delta G°$ and the equilibrium constant $K$ at 25°C for the reaction

$$2\,NH_3(g) + \tfrac{7}{2}\,O_2(g) \Longrightarrow 2\,NO_2(g) + 3\,H_2O(g)$$

using data in Appendix D.

**40.** Write a reaction for the dehydrogenation of gaseous ethane ($C_2H_6$) to acetylene ($C_2H_2$). Calculate $\Delta G°$ and the equilibrium constant for this reaction at 25°C, using data from Appendix D.

**41.** Use the thermodynamic data from Appendix D to calculate equilibrium constants at 25°C for the following reactions.
(a) $SO_2(g) + \tfrac{1}{2}\,O_2(g) \rightleftarrows SO_3(g)$
(b) $3\,Fe_2O_3(s) \rightleftarrows 2\,Fe_3O_4(s) + \tfrac{1}{2}\,O_2(g)$
(c) $CuCl_2(s) \rightleftarrows Cu^{2+}(aq) + 2\,Cl^-(aq)$
Write an equilibrium expression in each case.

**42.** Use the thermodynamic data from Appendix D to calculate equilibrium constants at 25°C for the following reactions.
(a) $H_2(g) + N_2(g) + 2\,O_2(g) \rightleftarrows 2\,HNO_2(g)$
(b) $Ca(OH)_2(s) \rightleftarrows CaO(s) + H_2O(g)$
(c) $Zn^{2+}(aq) + 4\,NH_3(aq) \rightleftarrows Zn(NH_3)_4^{2+}(aq)$
Write an equilibrium expression in each case.

**43.** One way to manufacture ethanol is by the reaction

$$C_2H_4(g) + H_2O(g) \Longrightarrow C_2H_5OH(g)$$

The $\Delta H_f°$ of $C_2H_4(g)$ is 52.3 kJ mol$^{-1}$; of $H_2O(g)$, $-241.8$ kJ mol$^{-1}$; and of $C_2H_5OH(g)$, $-235.3$ kJ mol$^{-1}$. Without doing detailed calculations, suggest the conditions of pressure and temperature that will maximize the yield of ethanol at equilibrium.

**44.** Dimethyl ether ($CH_3OCH_3$) is a good substitute for environmentally harmful propellants in aerosol spray cans. It is produced by the dehydration of methanol:

$$2\,CH_3OH(g) \Longrightarrow CH_3OCH_3(g) + H_2O(g)$$

Describe reaction conditions that favor the equilibrium production of this valuable chemical. As a basis for your answer, compute $\Delta H°$ and $\Delta S°$ of the reaction from the data in Appendix D.

**45.** The equilibrium constant at 25°C for the reaction

$$2\,NO_2(g) \Longrightarrow N_2O_4(g)$$

is 6.8. At 200°C, the equilibrium constant is $1.21 \times 10^{-3}$. Calculate the enthalpy change ($\Delta H$) for this reaction, assuming that $\Delta H$ and $\Delta S$ of the reaction are constant over the temperature range from 25 to 200°C.

**46.** The ionization constant of chloroacetic acid ($ClCH_2COOH$) in water is $1.528 \times 10^{-3}$ at 0°C and $1.230 \times 10^{-3}$ at 40°C. Calculate the enthalpy of ionization of the acid in water, assuming that $\Delta H$ and $\Delta S$ are constant over this range of temperature.

**47.** The equilibrium constant for the reaction

$$\tfrac{1}{2}\,Cl_2(g) + \tfrac{1}{2}\,F_2(g) \Longrightarrow ClF(g)$$

is measured to be $9.3 \times 10^9$ at 298 K and $3.3 \times 10^7$ at 398 K.
(a) Calculate $\Delta G°$ at 298 K for the reaction.
(b) Calculate $\Delta H°$ and $\Delta S°$, assuming the enthalpy and en-

tropy changes to be independent of temperature between 298 and 398 K.

**48.** The autoionization constant of water ($K_w$) is $1.139 \times 10^{-15}$ at 0°C and $9.614 \times 10^{-14}$ at 60°C.

(a) Calculate the enthalpy of autoionization of water.

$$2\,H_2O(\ell) \rightleftharpoons H_3O^+(aq) + OH^-(aq)$$

(b) Calculate the entropy of autoionization of water.

(c) At what temperature will the pH of pure water be 7.00, from these data?

**49.** For the synthesis of ammonia from its elements,

$$3\,H_2(g) + N_2(g) \rightleftharpoons 2\,NH_3(g)$$

the equilibrium constant $K = 5.9 \times 10^5$ at 298 K, and $\Delta H° = -92.2$ kJ. Calculate the equilibrium constant for the reaction at 600 K, assuming no change in $\Delta H°$ and $\Delta S°$ between 298 K and 600 K.

**50.** The reaction of $Cu_2O(s)$ with $H_2(g)$ to yield $Cu(s)$ is represented by the chemical equation

$$Cu_2O(s) + H_2(g) \rightleftharpoons 2\,Cu(s) + H_2O(\ell)$$

Calculate the equilibrium constant for this reaction at 373 K, assuming $\Delta H$ and $\Delta S$ are approximately independent of temperature. Use data from Appendix D.

**51.** The vapor pressure of ammonia at −50°C is 0.4034 atm; at 0°C, it is 4.2380 atm.

(a) Calculate the molar enthalpy of vaporization ($\Delta H_{vap}$) of ammonia.

(b) Calculate the normal boiling temperature of $NH_3(\ell)$.

**52.** The vapor pressure of butyl alcohol ($C_4H_9OH$) at 70.1°C is 0.1316 atm; at 100.8°C, it is 0.5263 atm.

(a) Calculate the molar enthalpy of vaporization ($\Delta H_{vap}$) of butyl alcohol.

(b) Calculate the normal boiling point of butyl alcohol.

## Thermodynamics and Colligative Properties

**53.** The enthalpy of vaporization $\Delta H_{vap}$ of benzene is 30.82 kJ mol$^{-1}$ at its normal boiling temperature, 80.2°C. Calculate the boiling point of a solution of 2.00 g of anthracene ($C_{14}H_{10}$) in 100.0 g of benzene.

**54.** The enthalpy of fusion of camphor ($C_{10}H_{16}O$) is 6.815 kJ mol$^{-1}$ at its melting point, 178.4°C. Calculate the freezing point of a solution consisting of 0.250 g of tetraphenyl methane, ($C_6H_5)_4C$, in 10.00 g of camphor.

## Additional Problems

**55.** Ethanol ($CH_3CH_2OH$) has a normal boiling point of 78.4°C and a molar enthalpy of vaporization of 38.74 kJ mol$^{-1}$. Calculate the molar entropy of vaporization of ethanol and compare it with the prediction of Trouton's rule.

**56.** A quantity of ice is mixed with a quantity of hot water in a sealed, rigid, insulated container. The insulation prevents heat exchange between the ice–water mixture and the surroundings. The contents of the container soon reach equilibrium. Tell whether the total *energy* of the contents decreases, remains the same, or increases in this process. Make a similar statement about the total *entropy* of the contents. Explain your answers.

**57.** (a) If 2.60 mol of $O_2(g)$ ($c_P = 29.4$ J K$^{-1}$ mol$^{-1}$) is compressed reversibly and adiabatically from an initial pressure of 1.00 atm and 300 K to a final pressure of 8.00 atm, calculate $\Delta S$ for the gas.

(b) Suppose a different path from that in part (a) is used. The gas is first heated at constant pressure to the same final temperature and then compressed reversibly and isothermally to the same final pressure. Calculate $\Delta S$ for this path and show that it is equal to that found in part (a).

**58.** One mole of a monatomic ideal gas begins in a state with $P = 1.00$ atm and $T = 300$ K. It is expanded reversibly and adiabatically until the volume has doubled; then it is expanded irreversibly and isothermally into a vacuum until the volume has doubled again; then it is heated reversibly at constant volume to 400 K. Finally it is compressed reversibly and isothermally until a final state with $P = 1.00$ atm and $T = 400$ K is reached. Calculate $\Delta S_{sys}$ for this process. (*Hint:* There are two ways to solve this problem—an easy way and a hard way.)

***59.** The motion of air masses through the atmosphere can be approximated as adiabatic (because air is a poor conductor of heat) and reversible (because pressure differences in the atmosphere are small). To a good approximation, air can be treated as an ideal gas with average molar mass 29 g mol$^{-1}$ and average heat capacity 29 J K$^{-1}$ mol$^{-1}$.

(a) Show that the displacement of the air masses occurs at constant entropy ($\Delta S = 0$).

(b) Suppose the average atmospheric pressure near the earth's surface is $P_0$ and the temperature is $T_0$. The air is displaced upward until its temperature is $T$ and its pressure is $P$. Determine the relationship between $P$ and $T$. (*Hint:* Consider the process as occurring in two steps: first a cooling from $T_0$ to $T$ at constant pressure, and then an expansion from $P_0$ to $P$ at constant temperature. Equate the sum of the two entropy changes to $\Delta S_{tot} = 0$.)

(c) In the lower atmosphere, the dependence of pressure on height $h$ above the earth's surface can be approximated as

$$\ln(P/P_0) = -\mathcal{M}gh/RT$$

where $\mathcal{M}$ is the molar mass (kg mol$^{-1}$), $g$ the acceleration due to gravity (9.8 m s$^{-2}$), and $R$ the gas constant. If the air temperature at sea level near the equator is 38°C ($\sim$100°F), calculate the air temperature at the summit of Mount Kilimanjaro, 5.9 km above sea level. For further discussion of this problem, see L. K. Nash, *J. Chem. Educ.* 61:23, 1984.

**60.** Calculate the entropy change that results from mixing 54.0 g of water at 273 K with 27.0 g of water at 373 K in a vessel whose walls are perfectly insulated from the surroundings. Consider the specific heat of water to be constant over the temperature range from 273 K to 373 K and to have the value $4.18 \, \text{J K}^{-1} \text{g}^{-1}$.

**61.** Problem 10 asked for the entropy change when a piece of iron is cooled by immersion in a reservoir of water at 0°C.
(a) Repeat problem 10(b), supposing that the iron is cooled to 50°C in a large water reservoir held at that temperature before being placed in the 0°C reservoir.
(b) Repeat the calculation supposing that four water reservoirs at 75°C, 50°C, 25°C, and 0°C are used.
(c) As more reservoirs are used, what happens to $\Delta S$ for the iron, for the water, and for the universe? How would you attempt to carry out a reversible cooling of the iron?

**62.** Problem 42 in Chapter 3 described an optical atomic trap. In one experiment, a gas of 500 sodium atoms is confined in a volume of $1000 \, \mu m^3$. The temperature of the system is 0.00024 K. Compute the probability that, by chance, these 500 slowly moving sodium atoms will all congregate in the left half of the available volume. Express your answer in scientific notation.

**\*63.** Suppose we have several different ideal gases, $i = 1, 2, 3, \ldots, N$, each occupying its own volume $V_i$, all at the same pressure and temperature. The boundaries between the volumes are removed so that the gases mix at constant temperature in the total volume $V = \Sigma_i V_i$.
(a) Using the microscopic interpretation of entropy, show that

$$\Delta S = -nR \sum_i X_i \ln X_i$$

for this process, where $n$ is the total number of moles of gas and $X_i$ is the mole fraction of gas $i$.
(b) Calculate the entropy change when 50 g each of $O_2(g)$, $N_2(g)$, and $Ar(g)$ are mixed at 1 atm and 0°C.
(c) Using Table 3-1, calculate the entropy change when 100 L of air (assumed to be a mixture of ideal gases) at 1 atm and 25°C is separated into its component gases at the same pressure and temperature.

**\*64.** The $N_2O$ molecule has the structure N—N—O. In an ordered crystal of $N_2O$, the molecules are lined up in a regular fashion, with the orientation of each determined by its position in the crystal. In a random crystal (formed on rapid freezing), each molecule has two equally likely orientations.
(a) Calculate the number of microstates available to a random crystal of $N_0$ (Avogadro's number) of molecules.
(b) Calculate the entropy change when 1.00 mol of a random crystal is converted to an ordered crystal.

**65.** By examining the following graphs, predict which element—copper or gold—has the higher absolute entropy at a temperature of 200 K.

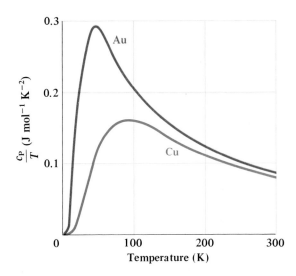

**\*66.** Consider the process described in problem 17 in Chapter 8. Use the results from that problem to do the following.
(a) Calculate $\Delta S$ for the system, the surroundings, and the universe.
(b) If the absolute entropy per mole of the gas *before* the expansion is $158.2 \, \text{J K}^{-1} \text{mol}^{-1}$, calculate $\Delta G_{\text{sys}}$ for the process.

**67.** Two different crystalline forms of sulfur are the rhombic form and the monoclinic form. At atmospheric pressure, rhombic sulfur undergoes a transition to monoclinic when it is heated above 368.5 K:

$$S(s, \text{rhombic}) \longrightarrow S(s, \text{monoclinic})$$

(a) What is the sign of the entropy change ($\Delta S$) for this transition?
(b) $|\Delta H|$ for this transition is $400 \, \text{J mol}^{-1}$. Calculate $\Delta S$ for the transition.

**\*68.** Use data from Appendix D to estimate the temperature at which $I_2(g)$ and $I_2(s)$ are in equilibrium at a pressure of 1 atm. Can this equilibrium actually be achieved? Refer to Appendix F for data on iodine.

**69.** The $\Delta G_f^\circ$ of $Si_3N_4(s)$ is $-642.6 \, \text{kJ mol}^{-1}$. Use this fact and the data in Appendix D to compute $\Delta G^\circ$ of the reaction

$$3 \, CO_2(g) + Si_3N_4(s) \longrightarrow$$
$$3 \, SiO_2(\text{quartz}) + 2 \, N_2(g) + 3 \, C(s, \text{gr})$$

**70.** The compound $Pt(NH_3)_2I_2$ comes in two forms, the *cis* and the *trans*, which differ in their molecular structure. The following data are available:

|       | $\Delta H_f^\circ$ (kJ mol$^{-1}$) | $\Delta G_f^\circ$ (kJ mol$^{-1}$) |
|-------|------------------------------------|------------------------------------|
| *cis*   | $-286.56$ | $-130.25$ |
| *trans* | $-316.94$ | $-161.50$ |

Combine these data with data from Appendix D to compute the standard entropies ($S°$) of both of these compounds at 25°C.

**71.** (a) Use data from Appendix D to calculate $\Delta H°$ and $\Delta S°$ at 25°C for the reaction

$$2\,CuCl_2(s) \rightleftharpoons 2\,CuCl(s) + Cl_2(g)$$

(b) Calculate $\Delta G$ at 590 K, assuming $\Delta H°$ and $\Delta S°$ are independent of temperature.

(c) Careful high-temperature measurements show that when this reaction is carried out at 590 K, $\Delta H_{590}$ is 158.36 kJ and $\Delta S_{590}$ is 177.74 J K$^{-1}$. Use these facts to compute an improved value of $\Delta G_{590}$ for this reaction. Determine the percentage error in $\Delta G_{590}$ that comes from using the 298-K values in place of 590-K values in this case.

**72.** (a) The normal boiling point of carbon tetrachloride ($CCl_4$) is 76.5°C. A student looks up the standard enthalpies of formation of $CCl_4(\ell)$ and of $CCl_4(g)$ in Appendix D. They are listed as $-135.44$ and $-102.9$ kJ mol$^{-1}$, respectively. By subtracting the first from the second, she computes $\Delta H°$ for the vaporization of $CCl_4$ to be 32.5 kJ mol$^{-1}$. But Table 8-2 states that the $\Delta H_{vap}$ of $CCl_4$ is 30.0 kJ mol$^{-1}$. Explain the discrepancy.

(b) Calculate the molar entropy change of vaporization ($\Delta S_{vap}$) of $CCl_4$ at 76.5°C.

**73.** The typical potassium ion concentration in the fluid outside a cell is 0.0050 M, whereas that inside a muscle cell is 0.15 M.

(a) What is the spontaneous direction of motion of ions through the cell wall?

(b) In *active transport*, cells use free energy stored in ATP (see problem 32) to move ions in the direction opposite their spontaneous direction of flow. Calculate the cost in free energy to move 1.00 mol of K$^+$ ions through the cell wall by active transport. Assume no change in K$^+$ concentrations during this process.

**74.** Barium nitride vaporizes slightly at high temperature as it undergoes the dissociation

$$Ba_3N_2(s) \rightleftharpoons 3\,Ba(g) + N_2(g)$$

At 1000 K, the equilibrium constant is $4.5 \times 10^{-19}$. At 1200 K, the equilibrium constant is $6.2 \times 10^{-12}$.

(a) Estimate $\Delta H°$ for this reaction.

(b) The equation is rewritten as

$$2\,Ba_3N_2(s) \rightleftharpoons 6\,Ba(g) + 2\,N_2(g)$$

Now the equilibrium constant is $2.0 \times 10^{-37}$ at 1000 K and $3.8 \times 10^{-23}$ at 1200 K. Estimate $\Delta H°$ of *this* reaction.

**75.** Although iodine is not very soluble in pure water, it dissolves readily in water that contains I$^-$($aq$) ion, thanks to the reaction

$$I_2(aq) + I^-(aq) \rightleftharpoons I_3^-(aq)$$

The equilibrium constant of this reaction was measured as a function of temperature with these results:

| $t$: | 3.8°C | 15.3°C | 25.0°C | 35.0°C | 50.2°C |
|------|-------|--------|--------|--------|--------|
| $K$: | 1160  | 841    | 689    | 533    | 409    |

(a) Plot ln $K$ on the $y$ axis as a function of $1/T$, the reciprocal of the absolute temperature.

(b) Estimate the $\Delta H°$ of this reaction.

**76.** The breaking of the O—O bond in peroxydisulfuryl difluoride ($FO_2SOOSO_2F$) gives $FO_2SO$:

$$(FO_2SO)_2 \rightleftharpoons 2\,FO_2SO$$

The compound on the left of this equation is a colorless liquid that boils at 67.1°C. Its vapor, when heated to about 100°C, turns brown as the product of the reaction forms. Suppose that, in a sample of the vapor, the intensity of the brown color doubles between 100°C and 110°C and that the total pressure increases only by the 2.7% predicted for an ideal gas. Estimate $\Delta H°$ for the preceding reaction.

**\*77.** Trouton's rule states that $\Delta S_{vap}$ is close to 88 J K$^{-1}$ mol$^{-1}$ for a wide variety of substances. Imagine that this rule holds exactly. Show that then

$$\frac{\Delta T_b}{T_b} = C X_2$$

where $\Delta T_b$ is the boiling point elevation, $T_b$ the boiling point of the pure solvent, and $X_2$ the mole fraction of solute. Show that the constant $C$ is independent of the solvent *and* solute chosen, and calculate its numerical value. Assume that the solution is dilute and the solute is nonvolatile.

## CUMULATIVE PROBLEMS

**78.** When a gas undergoes a reversible adiabatic expansion, its entropy remains constant even though the volume increases. Explain how this can be consistent with the microscopic interpretation of entropy developed in Section 9-2. (*Hint:* Consider what happens to the distribution of velocities in the gas.)

**79.** Calculate the equilibrium pressure (in atmospheres) of $O_2(g)$

over a sample of pure NiO($s$) in contact with pure Ni($s$) at 25°C. The NiO($s$) decomposes according to the equation

$$NiO(s) \longrightarrow Ni(s) + \tfrac{1}{2}\,O_2(g)$$

Use data from Appendix D.

**80.** $\Delta G_f°$ of SrSO$_4$($s$) is $-1340.9$ kJ mol$^{-1}$. Use this, together with information from Appendix D, to estimate the molar solubility of $SrSO_4$ in water at 25°C.

81. The triple bond in the $N_2$ molecule is very strong, but at high enough temperatures even it breaks down. At 5000 K, when the total pressure exerted by a sample of nitrogen is 1.00 atm, $N_2(g)$ is 0.65% dissociated at equilibrium:

$$N_2(g) \rightleftharpoons 2 N(g)$$

At 6000 K with the same total pressure, the proportion of $N_2(g)$ dissociated at equilibrium rises to 11.6%. Use the van't Hoff equation to estimate the $\Delta H$ of this reaction.

82. The sublimation pressure of solid $NbI_5$ is the pressure of gaseous $NbI_5$ present in equilibrium with the solid. It is given by the empirical equation

$$\log P = -6762/T + 8.566$$

The vapor pressure of liquid $NbI_5$, on the other hand, is given by

$$\log P = -4653/T + 5.43$$

In these two equations, $T$ is the absolute temperature in kelvins and $P$ is the pressure in atmospheres.

(a) Determine the enthalpy and entropy of sublimation of $NbI_5(s)$.

(b) Determine the enthalpy and entropy of vaporization of $NbI_5(\ell)$.

(c) Calculate the normal boiling point of $NbI_5(\ell)$.

(d) Calculate the triple point temperature and pressure of $NbI_5$. (*Hint:* At the triple point of a substance, the liquid and solid are in equilibrium and must have the same vapor pressure. If they did not, vapor would continually escape from the phase with the higher vapor pressure and collect in the phase with the lower vapor pressure.)

83. Snow and ice sublime spontaneously when the partial pressure of water vapor is below the equilibrium vapor pressure of ice. At 0°C, the vapor pressure of ice is 0.0060 atm (the triple-point pressure of water). Taking the enthalpy of sublimation of ice to be 50.0 kJ mol$^{-1}$, calculate the partial pressure of water vapor below which ice will sublime spontaneously at −15°C.

*84. Calculate the minimum work necessary to obtain 1.0 L of pure water by applying reverse osmosis to a large volume of seawater. Use the molality of seawater given in Example 4-11, and treat seawater as an ideal solution. (*Hint:* For processes at constant pressure and temperature, the reversible work is equal to $\Delta G$.)

# 10

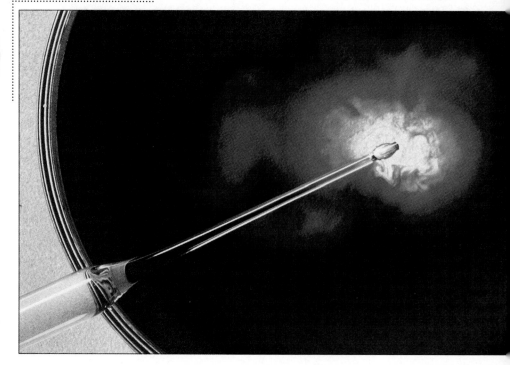

*When aqueous sodium chloride is added to a deep purple solution of potassium permanganate, an oxidation-reduction reaction occurs, generating a light pink solution of MnCl₂ as well as gaseous chlorine.*

# Electrochemistry

E lectrochemistry links the chemistry of oxidation–reduction reactions to the physics of charge flow. It provides an important way of utilizing the free energy available in spontaneous chemical reactions to perform useful work, and of using energy to carry out reactions that would otherwise not be possible. Electrochemistry is concerned with such practical problems as the storage of energy in batteries and the efficient conversion of energy from readily available sources (such as solar or chemical energy) to forms that are useful for technological applications. In this chapter we discuss redox reactions in aqueous solution and the way in which electrochemical cells operate, and examine the relationship between the current produced and reaction stoichiometry. We then apply thermodynamic principles to the operation of such cells, and show how these principles can help to guide the search for new electrochemical technologies.

## 10-1 BALANCING OXIDATION–REDUCTION EQUATIONS

Recall from Section 2-5 that oxidation–reduction reactions are those in which the oxidation states of some of the participants in a reaction undergo changes. Consider the reaction between copper and an aqueous solution of silver nitrate:

$$Cu(s) + 2\,Ag^+(aq) \longrightarrow Cu^{2+}(aq) + 2\,Ag(s)$$

The nitrate ions are spectators that do not take part in the reaction, and so they are omitted in the net equation. Two electrons are transferred from each reacting copper atom to a pair of silver ions. Copper (the electron donor) is oxidized, and silver ion (the electron acceptor) is reduced. This oxidation–reduction reaction occurs when a piece of copper is placed in an aqueous solution of silver nitrate or any other soluble silver salt (Fig. 10-1). Metallic silver immediately begins to plate out on the copper, the concentration of silver ion decreases, and blue $Cu^{2+}(aq)$ appears in solution and increases in concentration as time passes.

It is useful to consider this chemical equation as representing the sum of oxidation and reduction **half-reactions,** in which electrons ($e^-$) appear explicitly. The oxidation of copper is written as

$$Cu(s) \longrightarrow Cu^{2+}(aq) + 2\,e^-$$

and the reduction of silver ion as

$$Ag^+(aq) + e^- \longrightarrow Ag(s)$$

In the overall equation, electrons must not appear explicitly. Thus, the second equation must be multiplied by 2 before being added to the first, so that the electrons cancel out on both sides. As before, this gives

$$Cu(s) + 2\,Ag^+(aq) \longrightarrow Cu^{2+}(aq) + 2\,Ag(s)$$

In Section 1-6, the chemical equation for the burning of butane in oxygen (a redox reaction) was balanced through logical reasoning, taking into account the fact that the number of moles of atoms of each element is the same before and after the reaction. Many redox reactions are more difficult to balance than this, so it is useful to have a systematic procedure. In this section we outline a procedure based on half-reactions, applying it to reactions that occur in acidic or basic aqueous solution. In these reactions, water and $H_3O^+$ (acidic solution) or $OH^-$ (basic solution) may take part either as reactants or as products, so it is necessary to *complete* the corresponding equations as well as to balance them.

As an example, let us complete and balance the chemical equation for the dissolution of copper(II) sulfide in aqueous nitric acid (Fig. 10-2):

$$CuS(s) + NO_3^-(aq) \longrightarrow Cu^{2+}(aq) + SO_4^{2-}(aq) + NO(g)$$

**Figure 10-1**

When a piece of copper screen is inserted into a solution of silver nitrate, silver forms in a tree-like structure, and the solution turns blue as $Cu^{2+}$ ions form.

**Figure 10-2**

Copper(II) sulfide reacts with concentrated nitric acid to liberate nitrogen oxide and produces a solution of copper(II) sulfate, which displays the characteristic blue color of copper(II) ions in water. (Leon Lewandowski)

**Step 1** *Write two unbalanced half-equations, one for the species that is oxidized and its product and one for the species that is reduced and its product.*

Here the unbalanced half-reaction involving CuS is

$$CuS \longrightarrow Cu^{2+} + SO_4^{2-}$$

The unbalanced half-reaction involving $NO_3^-$ is

$$NO_3^- \longrightarrow NO$$

**Step 2** *Insert coefficients to make the numbers of atoms of all elements except oxygen and hydrogen equal on the two sides of each equation.*

In the present case copper, sulfur, and nitrogen are already balanced in the two half-equations, so this step is already completed.

**Step 3** *Balance oxygen by adding $H_2O$ to one side of each half-equation.*

$$CuS + 4\,H_2O \longrightarrow Cu^{2+} + SO_4^{2-}$$

$$NO_3^- \longrightarrow NO + 2\,H_2O$$

**Step 4** *Balance hydrogen. For an acidic solution, add $H_3O^+$ to the side of each half-equation that is "deficient" in hydrogen, and an equal amount of $H_2O$ to the other side. For a basic solution, add $H_2O$ to the side of each half-equation that is "deficient" in hydrogen, and an equal amount of $OH^-$ to the other side.*

Note that this step does not disrupt the oxygen balance achieved in Step 3. In the present case (acidic solution) the result is

$$CuS + 12\,H_2O \longrightarrow Cu^{2+} + SO_4^{2-} + 8\,H_3O^+$$

$$NO_3^- + 4\,H_3O^+ \longrightarrow NO + 6\,H_2O$$

**Step 5** *Balance charge by inserting $e^-$ (electrons) as a reactant or product in each half-equation.*

$$CuS + 12\,H_2O \longrightarrow Cu^{2+} + SO_4^{2-} + 8\,H_3O^+ + 8\,e^- \qquad \text{(oxidation)}$$

$$NO_3^- + 4\,H_3O^+ + 3\,e^- \longrightarrow NO + 6\,H_2O \qquad \text{(reduction)}$$

**Step 6** *Multiply the two half-equations by numbers chosen to make the number of electrons given off by the oxidation equal the number taken up by the reduction. Then add the two half-equations, canceling electrons. If $H_3O^+$, $OH^-$, or $H_2O$ appears on both sides of the final equation, cancel out the duplications.*

Here the oxidation half-equation must be multiplied by 3 (so that 24 electrons are produced) and the reduction half-equation by 8 (so that the same 24 electrons are consumed):

$$3\,CuS + 36\,H_2O \longrightarrow 3\,Cu^{2+} + 3\,SO_4^{2-} + 24\,H_3O^+ + 24\,e^-$$

$$\underline{8\,NO_3^- + 32\,H_3O^+ + 24\,e^- \longrightarrow 8\,NO + 48\,H_2O \qquad\qquad\qquad}$$

$$3\,CuS + 8\,NO_3^- + 8\,H_3O^+ \longrightarrow 3\,Cu^{2+} + 3\,SO_4^{2-} + 8\,NO + 12\,H_2O$$

This procedure allows the balancing of equations that are difficult to balance just by inspection. For basic solutions, remember to add $H_2O$ and $OH^-$, rather than $H_3O^+$ and $H_2O$, at step 4.

## Example 10-1

Balance the following equation, which represents a reaction that takes place in basic aqueous solution:

$$Ag(s) + HS^-(aq) + CrO_4^{2-}(aq) \longrightarrow Ag_2S(s) + Cr(OH)_3(s)$$

### Solution

**Step 1**
$$Ag + HS^- \longrightarrow Ag_2S$$
$$CrO_4^{2-} \longrightarrow Cr(OH)_3$$

**Step 2**
$$2\,Ag + HS^- \longrightarrow Ag_2S$$

The other half-reaction is unchanged.

**Step 3** $H_2O$ is now added to the second half-reaction to balance oxygen:

$$CrO_4^{2-} \longrightarrow Cr(OH)_3 + H_2O$$

**Step 4** The right side of the silver half-reaction is deficient by one hydrogen. Add one $H_2O$ to the right and one $OH^-$ to the left:

$$2\,Ag + HS^- + OH^- \longrightarrow Ag_2S + H_2O$$

In the chromium half-reaction, the left side is deficient by five hydrogens, and so $5\,H_2O$ is added to that side and $5\,OH^-$ to the right side:

$$CrO_4^{2-} + 4\,H_2O \longrightarrow Cr(OH)_3 + 5\,OH^-$$

(Notice that the $H_2O$ on the right cancelled out one of the five on the left, leaving four.)

**Step 5** Electrons are added to the right side of the silver half-reaction and to the left side of the chromium half-reaction to balance charge:

$$2\,Ag + HS^- + OH^- \longrightarrow Ag_2S + H_2O + 2\,e^- \qquad \text{(oxidation)}$$
$$CrO_4^{2-} + 4\,H_2O + 3\,e^- \longrightarrow Cr(OH)_3 + 5\,OH^- \qquad \text{(reduction)}$$

**Step 6** The gain and loss of electrons are equalized. The first equation is multiplied by 3 so that it absorbs six electrons, and the second is multiplied by 2 so that it liberates six electrons.

$$6\,Ag + 3\,HS^- + 3\,OH^- \longrightarrow 3\,Ag_2S + 3\,H_2O + 6\,e^- \qquad \text{(oxidation)}$$
$$2\,CrO_4^{2-} + 8\,H_2O + 6\,e^- \longrightarrow 2\,Cr(OH)_3 + 10\,OH^- \qquad \text{(reduction)}$$

$$\overline{6\,Ag + 3\,HS^- + 2\,CrO_4^{2-} + 5\,H_2O \longrightarrow 3\,Ag_2S + 2\,Cr(OH)_3 + 7\,OH^-}$$

*Related Problems:* 1, 2, 3, 4

## Example 10-2

Balance the following equation for the reaction of arsenic(III) sulfide with aqueous chloric acid:

$$As_2S_3(s) + ClO_3^-(aq) \longrightarrow H_3AsO_4(aq) + SO_4^{2-}(aq) + Cl^-(aq)$$

### Solution

Step 1
$$As_2S_3 \longrightarrow H_3AsO_4 + SO_4^{2-}$$
$$ClO_3^- \longrightarrow Cl^-$$

Step 2
$$As_2S_3 \longrightarrow 2\,H_3AsO_4 + 3\,SO_4^{2-}$$

Step 3
$$As_2S_3 + 20\,H_2O \longrightarrow 2\,H_3AsO_4 + 3\,SO_4^{2-}$$
$$ClO_3^- \longrightarrow Cl^- + 3\,H_2O$$

Step 4   The chloric acid makes this an acidic solution, and so $H_3O^+$ and $H_2O$ are used:

$$As_2S_3 + 54\,H_2O \longrightarrow 2\,H_3AsO_4 + 3\,SO_4^{2-} + 34\,H_3O^+$$
$$ClO_3^- + 6\,H_3O^+ \longrightarrow Cl^- + 9\,H_2O$$

Step 5
$$As_2S_3 + 54\,H_2O \longrightarrow 2\,H_3AsO_4 + 3\,SO_4^{2-} + 34\,H_3O^+ + 28\,e^-$$
$$ClO_3^- + 6\,H_3O^+ + 6\,e^- \longrightarrow Cl^- + 9\,H_2O$$

Step 6   We must find the least common multiple of 28 and 6. This is 84, and we therefore need to multiply the first equation by 84/28 = 3 and the second equation by 84/6 = 14:

$$3\,As_2S_3 + 162\,H_2O \longrightarrow 6\,H_3AsO_4 + 9\,SO_4^{2-} + 102\,H_3O^+ + 84\,e^-$$
$$\underline{14\,ClO_3^- + 84\,H_3O^+ + 84\,e^- \longrightarrow 14\,Cl^- + 126\,H_2O}$$
$$3\,As_2S_3 + 14\,ClO_3^- + 36\,H_2O \longrightarrow 6\,H_3AsO_4 + 9\,SO_4^{2-} + 14\,Cl^- + 18\,H_3O^+$$

## Disproportionation

An important type of redox reaction, **disproportionation,** occurs when a single substance is both oxidized and reduced. Example 2-8(c) provided such a reaction:

$$2\,\overset{-1}{H_2O_2}(\ell) \longrightarrow 2\,\overset{-2}{H_2}O(\ell) + \overset{0}{O_2}(g)$$

The oxygen in hydrogen peroxide ($H_2O_2$) is in an intermediate oxidation state of $-1$; some of it is oxidized to $O_2$ and some is reduced to $H_2O$. The balancing of equations for disproportionation reactions is described in the following example.

## Example 10-3

Balance the following equation for the reaction that occurs when chlorine is dissolved in basic solution:

$$Cl_2(g) \longrightarrow ClO_3^-(aq) + Cl^-(aq)$$

## Solution

**Step 1** We solve this by writing the $Cl_2$ on the left sides of *two* half-equations:

$$Cl_2 \longrightarrow ClO_3^-$$

$$Cl_2 \longrightarrow Cl^-$$

**Step 2**

$$Cl_2 \longrightarrow 2\,ClO_3^-$$

$$Cl_2 \longrightarrow 2\,Cl^-$$

**Step 3** The first half-equation becomes

$$Cl_2 + 6\,H_2O \longrightarrow 2\,ClO_3^-$$

**Step 4** Now the first half-equation becomes

$$Cl_2 + 12\,OH^- \longrightarrow 2\,ClO_3^- + 6\,H_2O$$

**Step 5**    $Cl_2 + 12\,OH^- \longrightarrow 2\,ClO_3^- + 6\,H_2O + 10\,e^-$    (oxidation)

$Cl_2 + 2\,e^- \longrightarrow 2\,Cl^-$                        (reduction)

**Step 6** Multiply the second equation by 5 and add:

$$Cl_2 + 12\,OH^- \longrightarrow 2\,ClO_3^- + 6\,H_2O + 10\,e^-$$

$$\underline{5\,Cl_2 + 10\,e^- \longrightarrow 10\,Cl^-}$$

$$6\,Cl_2 + 12\,OH^- \longrightarrow 2\,ClO_3^- + 10\,Cl^- + 6\,H_2O$$

Dividing this equation by 2 gives

$$3\,Cl_2 + 6\,OH^- \longrightarrow ClO_3^- + 5\,Cl^- + 3\,H_2O$$

*Related Problems:* 7, 8

# 10-2 ELECTROCHEMICAL CELLS

The reaction of copper with aqueous silver ions shown in Figure 10-1 is clearly spontaneous and therefore irreversible, and hence $\Delta G < 0$, although at this point its magnitude is unknown. Since no work is performed, the first law of thermodynamics requires that the entire energy change appear as the evolution of heat. This same reaction can be carried out quite differently without ever bringing the two reactants into *direct* contact with each other if a galvanic cell (a battery) is constructed from them. A copper strip is partially immersed in a solution of $Cu(NO_3)_2$ and a silver strip in a solution of $AgNO_3$, as Figures 10-3 and 10-4 illustrate. The two solutions are connected by a **salt bridge,** which is an inverted U-shaped tube containing a solution of a salt such as $NaNO_3$. The ends of the bridge are stuffed with porous plugs that prevent the two solutions from mixing but allow ions to pass through them. The two metal strips are connected by wires that lead to an **ammeter,** an instrument that measures the direction and magnitude of electric current through it.

As copper is oxidized on the left side, $Cu^{2+}$ ions enter the solution, and the electrons that are released pass through the external circuit from left to right, as shown by the deflection of the ammeter needle. The electrons enter the silver strip,

**Figure 10-3**

A metallic copper anode reacts to give a blue solution containing copper(II) ions as silver ions plate out on a silver cathode in a galvanic cell. (Leon Lewandowski)

and at the metal–solution interface they are picked up by $Ag^+$ ions, which plate out as atoms on the surface of the silver. This process would lead to an increase of positive charge in the left-hand beaker and to a decrease in the right-hand one were it not for the salt bridge; the bridge permits a net flow of positive ions through it into the right-hand beaker and of negative ions into the left-hand beaker, preserving charge neutrality in each.

As discussed in Section 10-1, this oxidation–reduction reaction is composed of two separate half-reactions. The oxidation half-reaction in the left-hand beaker is

$$Cu(s) \longrightarrow Cu^{2+}(aq) + 2\ e^-$$

and the reduction half-reaction in the right-hand beaker is

$$Ag^+(aq) + e^- \longrightarrow Ag(s)$$

Following Michael Faraday, chemists call the site at which oxidation occurs in an electrochemical cell the **anode** and the site at which reduction occurs the **cathode.** In the galvanic cell just discussed, copper is the anode (because it is oxidized) and silver is the cathode (because $Ag^+$ is reduced). Electrons flow in the external circuit from anode to cathode. In solution both positive and negative ions are free to move. In an electrochemical cell, negative ions (anions) move toward the anode, and positive ions (cations) move toward the cathode. We shall adopt a convention for galvanic cells in which the anode is shown on the left and the cathode on the right, so that electrons flow through the external circuit from left to right. Schematically the copper–silver galvanic cell is represented as

$$Cu|Cu^{2+} \parallel Ag^+|Ag$$

with the anode on the left and the cathode on the right, and with the metal–solution interface represented by | and the salt bridge by ‖.

**Figure 10-4**

In the galvanic cell of Figure 10-3, charged particles move when the circuit is completed. Electrons flow from the copper to the silver electrode through the wire. In solution, anions migrate toward the copper electrode and cations move toward the silver electrode.

The net chemical reaction that occurs in this simple galvanic cell is the same one that takes place when a copper strip is placed in an aqueous solution of silver nitrate, but there is an essential difference in the process. Because the reaction components are separated into two compartments while electrical continuity is preserved, the direct transfer of electrons from copper atoms to silver ions is avoided, and they are forced to travel through an external circuit (wire) before they ultimately accomplish the same net effect. The current of electrons through the wire can be used for a variety of purposes. For example, if a light bulb were placed in the electric circuit, the current passing through it would cause the bulb to light. The electrochemical cell would have allowed chemical energy to be converted to heat and radiant energy. Alternatively, the light bulb could be replaced by a small electric motor and the energy change of the chemical reaction used to perform mechanical work.

## Galvanic and Electrolytic Cells

What causes an electric current to flow in a galvanic cell? There must be an electrical **potential difference,** $\Delta\mathscr{E}$, between two points in the circuit to cause electrons to flow, just as a difference in gravitational potential between two points on the earth's surface causes water to flow downhill. This electrical potential difference, or **cell voltage,** can be measured with an instrument called a **voltmeter** inserted in the external circuit. The voltage measured in a galvanic cell depends on the magnitude of the current passing through the cell, and the voltage falls if the current becomes too large. The intrinsic cell voltage (that at zero current) can be measured by placing a variable voltage source in the external circuit in such a way that its potential difference $\Delta\mathscr{E}_{ext}$ *opposes* the intrinsic potential difference $\Delta\mathscr{E}$ of the electrochemical cell. The net potential difference is then

$$\Delta\mathscr{E}_{net} = \Delta\mathscr{E} - \Delta\mathscr{E}_{ext}$$

$\Delta\mathscr{E}$ can be measured by adjusting $\Delta\mathscr{E}_{ext}$ until $\Delta\mathscr{E}_{net}$ becomes 0, at which point the current through the circuit falls to 0 as well. If $\Delta\mathscr{E}_{ext}$ is held just below $\Delta\mathscr{E}$, the net potential difference becomes small and the cell operation is close to reversible, with only a small current and a slow rate of reaction at the electrodes.

If the opposing external voltage is increased *above* the natural potential difference of the cell, the electrons reverse direction and move toward the copper electrode. Copper ions in solution accept electrons and deposit as copper metal, and silver metal dissolves and furnishes additional $Ag^+$ ions. The net reaction occurring is then the reverse of the spontaneous reaction, namely,

$$2\,Ag(s) + Cu^{2+}(aq) \longrightarrow 2\,Ag^+ + Cu(s)$$

An electrochemical cell that operates spontaneously is called a **galvanic cell** (or voltaic cell). Such a cell converts chemical energy to electrical energy, which can be used to perform work. A cell in which an opposing external potential causes the reaction to occur in the direction opposite the spontaneous direction is called an **electrolytic cell;** such a cell uses electrical energy to carry out chemical reactions that would otherwise not occur. When a cell is changed into an electrolytic cell by the addition of an external potential source that reverses the direction of electron flow, there is also a reversal in the sites of the anode and the cathode. In the electrolytic cell, oxidation takes place at the silver electrode, which therefore becomes the anode, and the copper electrode becomes the cathode.

## Example 10-4

The final step in the production of magnesium from seawater is the electrolysis of molten magnesium chloride, in which the overall reaction is

$$Mg^{2+} + 2\,Cl^- \longrightarrow Mg(\ell) + Cl_2(g)$$

Write equations for the half-reactions occurring at the anode and at the cathode, and indicate the direction in which electrons flow through the external circuit.

### Solution

The anode is the site at which oxidation takes place—that is, where electrons are given up. The anode half-reaction must be

$$2\,Cl^- \longrightarrow Cl_2(g) + 2\,e^-$$

At the cathode, reduction takes place and the electrons are taken up. The cathode half-reaction is

$$Mg^{2+} + 2\,e^- \longrightarrow Mg(\ell)$$

Electrons move from the anode, where chlorine is liberated, through the external circuit to the cathode, where molten magnesium is produced.

*Related Problems:* 9, 10

## Faraday's Laws

The dual aspect of the electrochemical cell (galvanic and electrolytic) was realized shortly after the cell's discovery in 1800 by Alessandro Volta. Volta constructed a "battery" consisting of a number of platelets of silver and zinc that were separated from one another by porous strips of paper saturated with a salt solution. By 1807, Sir Humphry Davy had prepared elemental sodium and potassium by using a battery to electrolyze their respective hydroxides. The underlying scientific basis of the electrochemical cell was unclear, however. Michael Faraday's research showed a direct quantitative relationship between the amounts of substances that react at the cathode and the anode and the total electric charge that passes through the cell. This observation is the substance of **Faraday's laws,** which we state as

> **1.** The mass of a given substance that is produced or consumed at an electrode is proportional to the quantity of electric charge passed through the cell.
> **2.** Equivalent masses[1] of different substances are produced or consumed at an electrode by the passage of a given quantity of electric charge through the cell.

These laws, which summarize the stoichiometry of electrochemical processes, were discovered by Michael Faraday in 1833, more than half a century before the electron was discovered and the atomic basis of electricity was understood.

[1] The equivalent mass of an element in a redox reaction, or of a compound containing that element, is its molar mass divided by the number of moles of electrons transferred per mole of substance in the corresponding half-reaction.

The charge $e$ on a single electron (expressed in coulombs) has been very accurately determined to be

$$e = 1.6021773 \times 10^{-19}\,\text{C}$$

so that the quantity of charge represented by 1 mol of electrons is

$$Q = (6.022137 \times 10^{23}\,\text{mol}^{-1})(1.602177 \times 10^{-19}\,\text{C}) = 96{,}485.31\,\text{C mol}^{-1}$$

This quantity of charge is called the **Faraday constant** (symbol $\mathscr{F}$):

$$\mathscr{F} = 96{,}485.31\,\text{C mol}^{-1}$$

**Electric current** is the amount of charge flowing through a circuit per unit time. If $Q$ is the magnitude of the charge in coulombs and $t$ is the time in seconds that it takes to pass a point in the circuit, then the current $I$ is

$$I = \frac{Q}{t} \qquad\qquad \textbf{(10-1)}$$

where the units for $I$ are amperes (A) or coulombs per second. A current of $I$ amperes flowing for $t$ seconds causes $It$ coulombs of charge to pass through the circuit. The chemical amount, in moles, of electrons is

$$\frac{It}{96{,}485\,\text{C mol}^{-1}} = \text{moles of electrons}$$

From the number of moles of electrons that pass through a circuit, the number of moles (and therefore the number of grams) of substances reacting at the electrodes in the electrochemical cell can be calculated. Suppose a zinc–silver galvanic cell is constructed in which the anode half-reaction is

$$\text{Zn}(s) \longrightarrow \text{Zn}^{2+}(aq) + 2\,e^{-}$$

and the cathode half-reaction is

$$\text{Ag}^{+}(aq) + e^{-} \longrightarrow \text{Ag}(s)$$

Each mole of electrons that passes through the cell arises from the oxidation of $\frac{1}{2}$ mol of $\text{Zn}(s)$ (because each Zn atom gives up two electrons) and reduces 1 mol of silver ions. From the molar masses of silver and zinc, we calculate that $65.38/2 = 32.69$ g of zinc is dissolved at the anode and 107.87 g of silver is deposited at the cathode. The same relationships hold if the cell is operated as an electrolytic cell, but in that case silver is dissolved and zinc is deposited.

## Example 10-5

An electrolytic cell is constructed in which the silver ions in silver chloride are reduced to silver at the cathode and copper is oxidized to $\text{Cu}^{2+}(aq)$ at the anode. A current of 0.500 A is passed through the cell for 101 minutes. Calculate the mass of copper dissolved and the mass of silver deposited.

### Solution

$$t = (101\,\text{min})(60\,\text{s min}^{-1}) = 6.06 \times 10^{3}\,\text{s}$$

The chemical amount of electrons passed through the circuit during this time is

$$\frac{(0.500\,\text{C s}^{-1})(6.06 \times 10^{3}\,\text{s})}{96{,}485\,\text{C mol}^{-1}} = 3.14 \times 10^{-2}\,\text{mol}\,e^{-}$$

The half-cell reactions are

$$AgCl(s) + e^- \longrightarrow Ag(s) + Cl^-(aq) \qquad \text{(cathode)}$$

$$Cu(s) \longrightarrow Cu^{2+}(aq) + 2\,e^- \qquad \text{(anode)}$$

and so the masses of silver deposited and copper dissolved are

$$(3.14 \times 10^{-2} \text{ mol } e^-) \times \left(\frac{1 \text{ mol Ag}}{1 \text{ mol } e^-}\right) \times (107.87 \text{ g mol}^{-1})$$

$$= 3.39 \text{ g Ag deposited}$$

$$(3.14 \times 10^{-2} \text{ mol } e^-) \times \left(\frac{1 \text{ mol Cu}}{2 \text{ mol } e^-}\right) \times (63.55 \text{ g mol}^{-1})$$

$$= 0.998 \text{ g Cu dissolved}$$

*Related Problems:* 11, 12, 13, 14

## 10-3 THE GIBBS FREE ENERGY AND CELL VOLTAGE

Chapters 8 and 9 discussed the pressure–volume work associated with expanding and compressing gases. In electrochemistry a different kind of work, **electrical work,** is fundamental. If an amount of charge, $Q$, moves through a potential difference $\Delta\mathscr{E}$, the electrical work is

$$w_{\text{elec}} = -Q\,\Delta\mathscr{E} \tag{10-2}$$

The minus sign appears in this equation because the same thermodynamic convention is followed as in Chapters 8 and 9—namely, that work done by the system (here, an electrochemical cell) has a negative sign. Because work is measured in joules and charge in coulombs, $\Delta\mathscr{E}$ has units of joules per coulomb. One volt (abbreviated V) is 1 joule per coulomb. Because the total charge $Q$ is the current $I$ multiplied by the time $t$ in seconds during which the current is flowing, this equation for the electrical work can also be written

$$w_{\text{elec}} = -It\,\Delta\mathscr{E}$$

The potential difference $\Delta\mathscr{E}$ is positive for a galvanic cell, so $w_{\text{elec}}$ is negative in this case, and net electrical work is performed by the galvanic cell. In an electrolytic cell, in contrast, $\Delta\mathscr{E}$ is negative and $w_{\text{elec}}$ is positive, corresponding to net electrical work done on the system by an external source such as an electric generator.

### Example 10-6

A 6.00-V battery delivers a steady current of 1.25 A for a period of 1.50 hours. Calculate the total charge $Q$, in coulombs, that passes through the circuit and the electrical work done *by* the battery.

### Solution

The total charge is

$$Q = It = (1.25 \text{ C s}^{-1})(1.50 \text{ hr})(3600 \text{ s hr}^{-1}) = 6750 \text{ C}$$

The electrical work is

$$w_{elec} = -Q\,\Delta\mathscr{E} = -(6750\text{ C})(6.00\text{ J C}^{-1}) = -4.05 \times 10^4\text{ J}$$

This is the work done *on* the battery, and so the work done *by* the battery is the negative of this, or $+40.5$ kJ.

Thermodynamics demonstrates a fundamental relationship between the change in free energy, $\Delta G$, of a spontaneous chemical reaction at constant temperature and pressure, and the maximum electrical work that such a reaction is capable of producing:

$$-w_{elec,max} = |\Delta G| \qquad \text{(at constant } T \text{ and } P\text{)} \qquad \textbf{(10-3)}$$

To show this, recall the definition of the Gibbs free energy function $G$:

$$G = H - TS = E + PV - TS$$

For processes at constant pressure $P$ and constant temperature $T$ (the usual case in electrochemical cells),

$$\Delta G = \Delta E + P\,\Delta V - T\,\Delta S$$

From the first law of thermodynamics,

$$\Delta E = q + w$$

or

$$\Delta E = q + w_{elec} - P\,\Delta V$$

because there are now two kinds of work, electrical work $w_{elec}$ and pressure–volume work $-P_{ext}\,\Delta V = -P\,\Delta V$. Combining this with the equation for the change in free energy gives

$$\Delta G = q + w_{elec} - P\,\Delta V + P\,\Delta V - T\,\Delta S = q + w_{elec} - T\,\Delta S$$

If the condition of reversibility is imposed upon the galvanic cell, then

$$q = q_{rev} = T\,\Delta S$$

and the free energy change is

$$\Delta G = w_{elec,rev}$$

If the same cell were to be operated irreversibly (i.e., if a large current were permitted to flow), less electrical work would be produced. The maximum electrical work is produced by the galvanic cell when it is operated reversibly.

If $n$ mol of electrons (or $n\mathscr{F}$ coulombs of charge) passes through the external circuit of the galvanic cell when it is operated reversibly, and if $\Delta\mathscr{E}$ is the reversible cell voltage, then

$$\Delta G = w_{elec} = -Q\,\Delta\mathscr{E} = -n\mathscr{F}\,\Delta\mathscr{E} \qquad \text{(reversible)}$$

Electrical work is produced by an electrochemical cell only if $\Delta G < 0$ (or when $\Delta\mathscr{E} > 0$, which amounts to the same thing). This relationship provides a direct way to determine free energy changes from measurements of cell voltage.

## Standard States and Cell Voltages

Recall from Chapters 8 and 9 that the *standard state* of a substance means a temperature of 25°C and a pressure of 1 atm. In addition, the standard state of a solute is that for which its concentration in ideal solution is 1 M.[2] The standard free energy change $\Delta G°$ for a reaction in which all reactants and products are in their standard states can be calculated from a table of standard free energies of formation $\Delta G_f°$ of the substances taking part in the reaction (see Appendix D). For those reactions that can be carried out in electrochemical cells, the standard free energy change $\Delta G°$ is related to a **standard cell voltage** $\Delta \mathscr{E}°$ by

$$\Delta G° = -n\mathscr{F}\,\Delta \mathscr{E}° \qquad\qquad\text{(10-4)}$$

Here $\Delta \mathscr{E}°$ is the cell voltage (potential difference) of a galvanic cell in which reactants and products are in their standard states (gases at 1 atm pressure, solutes at 1 M concentration, metals in their pure stable states, and the temperature at 25°C). This standard cell voltage is an intrinsic electrical property of the cell, which can be calculated from the standard free energy change $\Delta G°$ by means of Equation 10-4. Conversely, $\Delta G°$ (and, hence, equilibrium constants for cell reactions) can be determined by measuring the standard cell voltage $\Delta \mathscr{E}°$ of a reaction in which $n$ mol of electrons passes through the external circuit.

### Example 10-7

A $Zn^{2+}|Zn$ half-cell is connected to a $Cu^{2+}|Cu$ half-cell to make a galvanic cell, in which $[Zn^{2+}] = [Cu^{2+}] = 1.00$ M. The cell voltage at 25°C is measured to be $\Delta \mathscr{E}° = 1.10$ V, and Cu is observed to plate out as the reaction proceeds. Calculate $\Delta G°$ for the chemical reaction that takes place in the cell, for 1.00 mol of zinc dissolved.

### Solution

The reaction is

$$Zn(s) + Cu^{2+}(aq) \longrightarrow Zn^{2+}(aq) + Cu(s)$$

because Cu is a product. For the reaction as written, in which 1 mol of $Zn(s)$ and 1 mol of $Cu^{2+}(aq)$ react, 2 mol of electrons passes through the external circuit, so $n = 2$. Therefore,

$$\Delta G° = -n\mathscr{F}\,\Delta \mathscr{E}° = -(2.00\text{ mol})(96{,}485\text{ C mol}^{-1})(1.10\text{ V})$$
$$= -2.12 \times 10^5 \text{ J} = -212 \text{ kJ}$$

*Related Problems:* 19, 20

## Half-Cell Voltages

We could tabulate all the conceivable galvanic cells and their standard voltages, but the list would be very long. To avoid this, the half-cell potentials $\mathscr{E}°$ are tabulated; they can be combined to obtain the standard cell voltage $\Delta \mathscr{E}°$ for any complete cell.

---

[2] Recall that the standard state of a solute has unit *activity,* which can differ significantly from 1 M concentration for a real electrolyte solution. In what follows, we shall nevertheless refer to 1 M concentration as the standard state.

To see this, return to the standard cell just considered, made up of $Zn^{2+}|Zn$ and $Cu^{2+}|Cu$ half-cells. Each half-cell reaction is written as a reduction:

$$Zn^{2+}(aq) + 2\,e^- \longrightarrow Zn(s)$$

$$Cu^{2+}(aq) + 2\,e^- \longrightarrow Cu(s)$$

Each half-reaction has a certain tendency to occur. If **reduction potentials** $\mathscr{E}°(Zn^{2+}|Zn)$ and $\mathscr{E}°(Cu^{2+}|Cu)$ are associated with the two half-cells, the magnitude and sign of each of these potentials is related to the tendency of the reaction to occur as written. The more positive the half-cell reaction's potential, the greater its tendency to take place as a reduction. Of course, a single half-cell reduction reaction cannot take place in isolation, because a source of electrons is required. This can be achieved only by combining two half-cells. The reaction with the more positive (or algebraically greater) reduction potential occurs as a reduction, and that with the less positive potential is forced to run in reverse (as an oxidation) to supply electrons. The net cell potential (voltage) is then the *difference* between the individual half-cell reduction potentials. In the preceding example, copper is observed to plate out on the cathode, and so its potential must be more positive than that of zinc:

$$\Delta\mathscr{E}° = \mathscr{E}°(Cu^{2+}|Cu) - \mathscr{E}°(Zn^{2+}|Zn) = 1.10\ V$$

When two half-cells are combined to make a galvanic cell, reduction occurs in the half-cell with the algebraically greater potential (making it the cathode), and oxidation occurs in the other half-cell (making it the anode). We therefore write

$$\Delta\mathscr{E}° = \mathscr{E}°(cathode) - \mathscr{E}°(anode) \qquad \textbf{(10-5)}$$

for a galvanic cell. Cell voltages and half-cell reduction potentials are *intensive* quantities, independent of the amount of matter in a cell or the amount reacting. Half-cell reduction potentials are therefore *not* multiplied by numbers of moles reacting (as Gibbs free energies would be) to obtain the overall cell potential.

In thermodynamics, only energy differences $\Delta E$ are measurable; absolute energies are not. Therefore, energies (or enthalpies or free energies) are defined relative to a reference state for which these quantities were arbitrarily set at 0. The same reasoning applies to half-cells: because only differences are measured, we are free to define a reference potential for a particular half-cell and measure other half-cell potentials relative to it. The convention used is to define $\mathscr{E}°$ for the half-cell reaction of $H_2(g)$ and $H_3O^+(aq)$ to be 0 at all temperatures, when the gas pressure at the electrode is 1 atm and the $H_3O^+(aq)$ concentration in solution is 1 M (Fig. 10-5).

$$2\,H_3O^+(aq) + 2\,e^- \longrightarrow H_2(g) + 2\,H_2O(\ell) \qquad \mathscr{E}° = 0\ V \text{ (by definition)}$$

All other half-cell potentials are then determined from this reference potential by combining their standard half-cells with the standard $H_3O^+|H_2$ half-cell in a galvanic cell and measuring the cell voltage. The magnitude and sign of that cell voltage give the standard potential of the half-cell in question, its reaction being written as an oxidation or a reduction according to whether the electrode behaves as the anode or the cathode of the galvanic cell.

For example, if a $Cu^{2+}(1\ M)|Cu$ half-cell is connected to the standard $H_3O^+(1\ M)|H_2$ half-cell, copper is observed to plate out, and so the copper half-cell is the cathode and the hydrogen half-cell is the anode. The observed cell voltage is 0.34 V; thus,

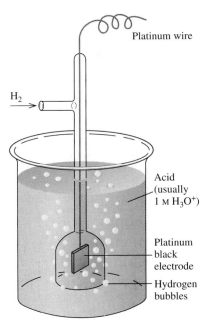

Platinum wire

$H_2$

Acid (usually 1 M $H_3O^+$)

Platinum black electrode

Hydrogen bubbles

**Figure 10-5**

Hydrogen is a gas at room conditions, and electrodes cannot be constructed from it directly. In the hydrogen half-cell shown here, a piece of platinum covered with a fine coating of platinum black is dipped into the solution, and a stream of hydrogen is passed over the surface. The platinum itself does not react but provides a support surface for the reaction of the $H_2$ and water to give $H_3O^+$, or the reverse. It also provides the necessary electrical connection to carry away or supply electrons.

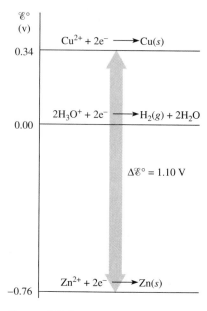

**Figure 10-6**

Sometimes standard reduction potentials are represented to scale. In this diagram, the distance between a pair of standard half-cell potentials is proportional to the $\Delta \mathscr{E}°$ generated by the electrochemical cell that combines the two. The half-reaction with the higher reduction potential forces the other half-reaction to occur in the reverse direction.

$$\Delta \mathscr{E}° = \mathscr{E}°(\text{cathode}) - \mathscr{E}°(\text{anode})$$
$$= \mathscr{E}°(Cu^{2+}|Cu) - \mathscr{E}°(H_3O^+|H_2) = \mathscr{E}°(Cu^{2+}|Cu) - 0$$
$$= 0.34 \text{ V}$$

The standard $Cu^{2+}|Cu$ half-cell potential is therefore 0.34 V on a scale in which the standard $H_3O^+|H_2$ half-cell potential is 0:

$$Cu^{2+}(1 \text{ M}) + 2\,e^- \longrightarrow Cu(s) \qquad \mathscr{E}° = 0.34 \text{ V}$$

When a $Zn^{2+}(1 \text{ M})|Zn$ half-cell is connected to the standard hydrogen half-cell, zinc dissolves; its half-cell is the anode because oxidation occurs in it. The measured cell voltage is 0.76 V, and so

$$\Delta \mathscr{E}° = 0.76 \text{ V} = \mathscr{E}°(\text{cathode}) - \mathscr{E}°(\text{anode}) = 0 - \mathscr{E}°(Zn^{2+}|Zn)$$

The $Zn^{2+}(1 \text{ M})|Zn$ half-cell reduction potential is therefore $-0.76$ V:

$$Zn^{2+}(1 \text{ M}) + 2\,e^- \longrightarrow Zn(s) \qquad \mathscr{E}° = -0.76 \text{ V}$$

If the zinc and copper half-cells are combined, the copper half-cell will be the cathode because it has the more positive half-cell reduction potential. The galvanic cell voltage under standard-state conditions (Fig. 10-6) will be

$$\Delta \mathscr{E}° = \mathscr{E}°(\text{cathode}) - \mathscr{E}°(\text{anode}) = 0.34 \text{ V} - (-0.76 \text{ V}) = 1.10 \text{ V}$$

in agreement with the measured value.

## Example 10-8

An aqueous solution of potassium permanganate ($KMnO_4$) has a deep purple color. In aqueous acidic solution, the permanganate ion can be reduced to the pale-pink manganese(II) ion ($Mn^{2+}$). Under standard conditions, the reduction potential of an $MnO_4^-|Mn^{2+}$ half-cell is $\mathscr{E}° = 1.49$ V. Suppose this half-cell is combined with a $Zn^{2+}|Zn$ half-cell in a galvanic cell, with $[Zn^{2+}] = [MnO_4^-] = [Mn^{2+}] = [H_3O^+] = 1$ M. (a) Write equations for the reactions at the anode and the cathode. (b) Write a balanced equation for the overall cell reaction. (c) Calculate the standard cell potential difference, $\Delta \mathscr{E}°$.

### Solution

(a) Because $\mathscr{E}°(MnO_4^-|Mn^{2+}) = 1.49$ V is algebraically greater than $\mathscr{E}°(Zn^{2+}|Zn) = -0.76$ V, permanganate ions will be reduced at the cathode. The balanced half-cell reaction requires the presence of $H_3O^+$ ions and water, giving

$$MnO_4^-(aq) + 8\,H_3O^+(aq) + 5\,e^- \longrightarrow Mn^{2+}(aq) + 12\,H_2O(\ell)$$

The half-equation for the oxidation of Zn at the anode is

$$Zn(s) \longrightarrow Zn^{2+}(aq) + 2\,e^-$$

(b) In the overall reaction, the number of electrons taken up at the cathode must equal the number released at the anode, and so the first equation must be multiplied by 2 and the second by 5. Adding the two gives the overall reaction:

$$2\,MnO_4^-(aq) + 16\,H_3O^+(aq) + 5\,Zn(s) \longrightarrow$$
$$2\,Mn^{2+}(aq) + 24\,H_2O(\ell) + 5\,Zn^{2+}(aq)$$

(c) The galvanic cell potential is the difference between the standard reduction potential for permanganate (at the cathode) and that for zinc (at the anode):

$$\Delta \mathscr{E}° = \mathscr{E}°(MnO_4^-|Mn^{2+}) - \mathscr{E}°(Zn^{2+}|Zn) = 1.49 - (-0.76) = 2.25 \text{ V}$$

Note that the half-cell potentials are not multiplied by their coefficients (2 and 5) before subtraction. Half-cell potentials are *intensive* properties of a galvanic cell, and so are independent of the amount of the reacting species.

*Related Problems:* 21, 22

Appendix E summarizes the standard reduction potentials for a large number of half-reactions. The reactions in that table are arranged in order of decreasing reduction potentials—that is, with the most positive at the top and the most negative at the bottom. In any galvanic cell, the half-cell that is listed higher in the table will act as the cathode (if both half-cells are in the standard state).

A strong **oxidizing agent** is a chemical species that is itself easily reduced. Such species are marked by large positive reduction potentials and appear at the top left in Appendix E. Fluorine has the largest reduction potential listed, and fluorine molecules are extremely eager to accept electrons to become fluoride ions. Other strong oxidizing agents include hydrogen peroxide ($H_2O_2$) and solutions of permanganate ion ($MnO_4^-$). A strong **reducing agent,** on the other hand, is very easily oxidized, so its corresponding reduction potential is large and negative. Such reducing agents appear at the lower right in Appendix E. The alkali and alkaline earth metals are especially good reducing agents.

Oxygen itself is a good oxidizing agent in acidic solution at pH 0 because it has a fairly high reduction potential:

$$O_2 + 4 H_3O^+ + 4 e^- \longrightarrow 6 H_2O \qquad \mathscr{E}° = 1.229 \text{ V}$$

Ozone ($O_3$) is still stronger, because its high free energy relative to oxygen provides an additional driving force for oxidation reactions, as shown by its high half-cell reduction potential in acidic aqueous solution:

$$O_3 + 2 H_3O^+ + 2 e^- \longrightarrow O_2 + 3 H_2O \qquad \mathscr{E}° = 2.07 \text{ V}$$

Because of its great oxidizing power, ozone is used commercially as a bleach for wood pulp and as a disinfectant and sterilizing agent for water, where it oxidizes algae and organic impurities but leaves no undesirable residue. In basic aqueous solution, the half-reactions of oxygen-containing species involve hydroxide ions instead of hydronium ions, and the reduction potentials are correspondingly changed. At pH 14 (standard basic conditions),

$$O_2 + 2 H_2O + 4 e^- \longrightarrow 4 OH^- \qquad \mathscr{E}° = 0.401 \text{ V}$$

$$O_3 + H_2O + 2 e^- \longrightarrow O_2 + 2 OH^- \qquad \mathscr{E}° = 1.24 \text{ V}$$

Both oxygen and ozone are less effective oxidizing agents in basic than in acidic solution.

## Adding and Subtracting Half-Cell Reactions

When two half-cells are combined to form a galvanic cell, their half-cell potentials are *not* multiplied by the coefficients that appear in the overall balanced chemical equation. It often happens, however, that two half-cells are to be combined to obtain *another half-cell.* For example, suppose we wish to know the standard half-cell potential corresponding to the reaction

$$Cu^{2+}(aq) + e^- \longrightarrow Cu^+(aq)$$

when the standard half-cell potentials are known for the reactions

$$Cu^{2+}(aq) + 2\,e^- \longrightarrow Cu(s) \qquad \mathscr{E}_1^\circ = \mathscr{E}^\circ(Cu^{2+}|Cu) = 0.340\ V$$

$$Cu^+(aq) + e^- \longrightarrow Cu(s) \qquad \mathscr{E}_2^\circ = \mathscr{E}^\circ(Cu^+|Cu) = 0.522\ V$$

Could we not simply subtract the second half-reaction from the first and, correspondingly, the second half-cell potential from the first half-cell potential, to obtain

$$Cu^{2+}(aq) + e^- \longrightarrow Cu^+(aq) \qquad \mathscr{E}_3^\circ = \mathscr{E}^\circ(Cu^{2+}|Cu^+) = -0.182\ V?$$

The answer is no. Instead, the *free energy change* for each half-reaction must be calculated from

$$\Delta G_{hc}^\circ = -n_{hc}\mathscr{F}\mathscr{E}^\circ$$

and combined before calculating $\mathscr{E}^\circ$ for their resultant. Thus, for the free energy change for the difference of the two half-reactions,

$$\Delta G_3^\circ = \Delta G_1^\circ - \Delta G_2^\circ = -n_1\mathscr{F}\mathscr{E}_1^\circ + n_2\mathscr{F}\mathscr{E}_2^\circ$$

Setting this equation equal to $-n_3\mathscr{F}\mathscr{E}_3^\circ$ gives

$$-n_3\mathscr{F}\mathscr{E}_3^\circ = -n_1\mathscr{F}\mathscr{E}_1^\circ + n_2\mathscr{F}\mathscr{E}_2^\circ$$

$$\mathscr{E}_3^\circ = \frac{n_1\mathscr{E}_1^\circ - n_2\mathscr{E}_2^\circ}{n_3}$$

The correct standard half-cell potential for the half-cell reaction

$$Cu^{2+}(aq) + e^- \longrightarrow Cu^+(aq)$$

is

$$\mathscr{E}^\circ(Cu^{2+}|Cu^+) = \frac{(2\ mol)(0.340\ V) - (1\ mol)(0.522\ V)}{1\ mol} = 0.158\ V$$

This same procedure using free energies can be used to calculate overall cell voltages from half-cell potentials. Here, however, the number of moles of electrons $n_1$ released at the anode is equal to the number $n_2$ taken up at the cathode, and is the number $n_3$ implied by the overall chemical equation. The electrons then cancel out and the result reduces to the simple form found before:

$$\Delta\mathscr{E}^\circ = \mathscr{E}^\circ(cathode) - \mathscr{E}^\circ(anode)$$

### Reduction Potential Diagrams and Disproportionation

We can summarize the half-reactions of copper in a **reduction potential diagram** of the form

$$Cu^{2+} \xrightarrow{\ 0.158\ V\ } Cu^+ \xrightarrow{\ 0.522\ V\ } Cu$$
$$\underset{0.340\ V}{\rule{4cm}{0.4pt}}$$

The line connecting each pair of species stands for the full half-reaction, written as a reduction and balanced by the addition of electrons (and, where necessary, water and $H_3O^+$ or $OH^-$). The number over each line is the corresponding reduction potential in volts.

Recall from Example 10-3 that *disproportionation* is the process in which a single substance is both oxidized and reduced. Reduction potential diagrams enable us to determine which species are stable with respect to disproportionation. A species can disproportionate if and only if a reduction potential that lies immediately to its right is larger than one that appears immediately to its left. To demonstrate this, note that a disproportionation reaction is the sum of a reduction half-reaction and an oxidation half-reaction. The driving force is the *difference* between the two reduction potentials, and if this difference is positive the disproportionation occurs spontaneously. In the case of $Cu^+$, the disproportionation reaction is

$$2\,Cu^+ \longrightarrow Cu^{2+} + Cu(s) \qquad \Delta\mathscr{E}° = 0.522 - 0.158 = 0.364\ \text{V}$$

Because $\Delta\mathscr{E}° > 0$, $\Delta G° < 0$ and the reaction occurs spontaneously, although the rate may be slow.

## Example 10-9

Hydrogen peroxide, $H_2O_2$, is a possible product of the reduction of oxygen in acidic solution:

$$O_2 + 2\,H_3O^+ + 2\,e^- \longrightarrow H_2O_2 + 2\,H_2O \qquad \mathscr{E}_3° = ?$$

It can then be further reduced to water:

$$H_2O_2 + 2\,H_3O^+ + 2\,e^- \longrightarrow 4\,H_2O \qquad \mathscr{E}_2° = 1.77\ \text{V}$$

(a) Use the half-cell potential just given for the reduction of $H_2O_2$, together with that given earlier,

$$O_2 + 4\,H_3O^+ + 4\,e^- \longrightarrow 6\,H_2O \qquad \mathscr{E}_1° = 1.229\ \text{V}$$

to calculate the standard half-cell potential for the reduction of $O_2$ to $H_2O_2$ in acidic solution.
(b) Write a reduction potential diagram for $O_2$, $H_2O_2$, and $H_2O$.
(c) Is $H_2O_2$ stable with respect to disproportionation in acidic solution?

## Solution

(a) The desired half-cell reaction is obtained by *subtracting* the reaction with potential $\mathscr{E}_2°$ from that with potential $\mathscr{E}_1°$. The half-cell potentials are not subtracted, however, but combined as described earlier in this section. Taking $n_1 = 4$, $n_2 = 2$, and $n_3 = 2$ gives

$$\mathscr{E}_3° = \frac{n_1\mathscr{E}_1° - n_2\mathscr{E}_2°}{n_3}$$

$$= \frac{(4\ \text{mol})(1.229\ \text{V}) - (2\ \text{mol})(1.77\ \text{V})}{2\ \text{mol}} = 0.69\ \text{V}$$

(b) The reduction potential diagram is obtained by omitting the electrons, water, and $H_3O^+$ from the corresponding half-equations:

$$O_2 \xrightarrow{\ 0.69\ \text{V}\ } H_2O_2 \xrightarrow{\ 1.77\ \text{V}\ } H_2O$$
$$\underset{1.229\ \text{V}}{\underline{\phantom{O_2 \xrightarrow{\ 0.69\ \text{V}\ } H_2O_2 \xrightarrow{}}}}$$

(c) $H_2O_2$ is thermodynamically unstable to disproportionation in acidic solution, because the half-cell potential to its right (1.77 V) is higher than that to its left (0.69 V). The disproportionation of $H_2O_2$ is also spontaneous in neutral solution, but is slow enough that aqueous solutions of hydrogen peroxide can be stored for a long time without deteriorating, as long as they are kept out of the light.

*Related Problems:* 31, 32, 33, 34

## 10-4 CONCENTRATION EFFECTS AND THE NERNST EQUATION

Cells in which concentrations or pressures differ from their standard-state values are often of interest. It is straightforward to use the thermodynamic principles of Chapter 9 to determine the effects of concentration and pressure on the cell voltage. There it was shown that the free energy change is related to the reaction quotient $Q$ through

$$\Delta G = \Delta G° + RT \ln Q$$

Combining this equation with

$$\Delta G = -n\mathscr{F}\,\Delta\mathscr{E}$$

and

$$\Delta G° = -n\mathscr{F}\,\Delta\mathscr{E}°$$

gives

$$-n\mathscr{F}\,\Delta\mathscr{E} = -n\mathscr{F}\,\Delta\mathscr{E}° + RT \ln Q$$

and

$$\Delta\mathscr{E} = \Delta\mathscr{E}° - \frac{RT}{n\mathscr{F}} \ln Q \tag{10-6}$$

a relationship known as the **Nernst equation.**

The Nernst equation can be rewritten in terms of common (base-10) logarithms by using the fact that

$$\ln Q \approx 2.303 \log_{10} Q$$

At 25°C (298.15 K), the combination of constants $2.303\,RT/\mathscr{F}$ becomes

$$2.303\frac{RT}{\mathscr{F}} = (2.303)\frac{(8.315\ \text{J K}^{-1}\ \text{mol}^{-1})(298.15\ \text{K})}{96485\ \text{C mol}^{-1}}$$

$$= 0.0592\ \text{J C}^{-1} = 0.0592\ \text{V}$$

because 1 joule per coulomb is 1 volt. The Nernst equation then becomes

$$\Delta\mathscr{E} = \Delta\mathscr{E}° - \frac{0.0592\ \text{V}}{n} \log_{10} Q \qquad \text{(at 25°C)} \tag{10-7}$$

which is its most familiar form. Here $n$ is the number of moles of electrons transferred in the overall chemical reaction as written. In a galvanic cell made from zinc, aluminum, and their ions, for example,

$$3\,Zn^{2+}(aq) + 6\,e^- \longrightarrow 3\,Zn(s)$$

$$2\,Al(s) \longrightarrow 2\,Al^{3+}(aq) + 6\,e^-$$

$$3\,Zn^{2+}(aq) + 2\,Al(s) \longrightarrow 2\,Al^{3+}(aq) + 3\,Zn(s)$$

there is a net transfer of $n = 6$ mol of electrons through the external circuit for the equation as written.

The Nernst equation applies to half-cells in exactly the same way as it does to complete electrochemical cells. For any half-cell potential at 25°C,

$$\mathscr{E} = \mathscr{E}° - \frac{0.0592\ V}{n_{hc}} \log_{10} Q_{hc}$$

where $n_{hc}$ is the number of electrons appearing in the half-reaction and $Q_{hc}$ is the reaction quotient for the half-cell reaction written as a reduction. The electrons themselves do not appear in $Q_{hc}$. In the $Zn^{2+}|Zn$ case, the half-reaction written as a reduction is

$$Zn^{2+}(aq) + 2\,e^- \longrightarrow Zn(s)$$

so that $n_{hc} = 2$ and $Q_{hc} = 1/[Zn^{2+}]$. If two half-cells in which reactants and products are not in their standard states are combined in a galvanic cell, the one with the more positive value of $\mathscr{E}$ will be the cathode, where reduction takes place, and the cell voltage will be

$$\Delta\mathscr{E} = \mathscr{E}(\text{cathode}) - \mathscr{E}(\text{anode})$$

## Example 10-10

Suppose the $Zn|Zn^{2+}\|MnO_4^-|Mn^{2+}$ cell from Example 10-8 is operated at pH 2.00 with $[MnO_4^-] = 0.12$ M, $[Mn^{2+}] = 0.0010$ M, and $[Zn^{2+}] = 0.015$ M. Calculate the cell voltage $\Delta\mathscr{E}$ at 25°C.

### Solution

Recall that the overall equation for this cell is

$$2\,MnO_4^-(aq) + 5\,Zn(s) + 16\,H_3O^+(aq) \longrightarrow$$
$$2\,Mn^{2+}(aq) + 5\,Zn^{2+}(aq) + 24\,H_2O(\ell)$$

From Example 10-8, for every 5 mol of Zn oxidized (or 2 mol of $MnO_4^-$ reduced), 10 mol of electrons passes through the external circuit, so $n = 10$. A pH of 2.00 corresponds to a hydronium ion concentration of 0.010 M. Putting this and the other concentrations into the expression for the reaction quotient and substituting both $Q$ and the value of $\Delta\mathscr{E}°$ (from Example 10-8) into the Nernst equation give

$$\Delta\mathscr{E} = 2.25\ V - \frac{0.0592\ V}{10} \log_{10} \frac{[Mn^{2+}]^2\,[Zn^{2+}]^5}{[MnO_4^-]^2\,[H_3O^+]^{16}}$$

$$= 2.25\ V - \frac{0.0592\ V}{10} \log_{10} \frac{(0.0010)^2(0.015)^5}{(0.12)^2(0.010)^{16}}$$

$$= 2.25\ V - \frac{0.0592\ V}{10} \log_{10}(5.3 \times 10^{18}) = 2.14\ V$$

*Related Problems:* 35, 36

## Measuring Equilibrium Constants

Electrochemistry provides a convenient and accurate way to measure equilibrium constants for many solution-phase reactions. For an overall cell reaction,

$$\Delta G^\circ = -n\mathscr{F}\,\Delta\mathscr{E}^\circ$$

In addition, $\Delta G^\circ$ is related to the equilibrium constant $K$ through

$$\Delta G^\circ = -RT \ln K$$

so

$$RT \ln K = n\mathscr{F}\,\Delta\mathscr{E}^\circ$$

$$\ln K = \frac{n\mathscr{F}}{RT}\,\Delta\mathscr{E}^\circ$$

$$\log_{10} K = \frac{n}{0.0592\text{ V}}\,\Delta\mathscr{E}^\circ \qquad \text{(at 25°C)} \tag{10-8}$$

The same result can be obtained in a slightly different way. Return to the Nernst equation, which reads (at 25°C)

$$\Delta\mathscr{E} = \Delta\mathscr{E}^\circ - \frac{0.0592\text{ V}}{n}\log_{10} Q$$

Suppose all species are present under standard-state conditions; then $Q = 1$ initially and the cell voltage $\Delta\mathscr{E}$ is the standard cell voltage $\Delta\mathscr{E}^\circ$. As the reaction takes place, $Q$ increases and $\Delta\mathscr{E}$ decreases. As reactants are used up and products are formed, the cell voltage approaches 0. Eventually, equilibrium is reached and $\Delta\mathscr{E}$ becomes 0. At that point,

$$\Delta\mathscr{E}^\circ = \frac{0.0592\text{ V}}{n}\log_{10} K$$

which is the same as the relationship just obtained. This relation lets us calculate equilibrium constants from standard cell voltages, as the following example illustrates.

### Example 10-11

Calculate the equilibrium constant for the redox reaction

$$2\,\text{MnO}_4^-(aq) + 5\,\text{Zn}(s) + 16\,\text{H}_3\text{O}^+(aq) \longrightarrow$$
$$2\,\text{Mn}^{2+}(aq) + 5\,\text{Zn}^{2+}(aq) + 24\,\text{H}_2\text{O}(\ell)$$

at 25°C using the cell voltage calculated in Example 10-8.

### Solution

From Example 10-8, $\Delta\mathscr{E}^\circ = 2.25$ V, and from the analysis there and in Example 10-10, $n = 10$ for the preceding reaction. Therefore,

$$\log_{10} K = \frac{n}{0.0592\text{ V}}\,\Delta\mathscr{E}^\circ = \frac{10}{0.0592\text{ V}}(2.25\text{ V}) = 380$$

$$K = 10^{380}$$

This overwhelmingly large equilibrium constant reflects the strength of permanganate ion as an oxidizing agent and of zinc as a reducing agent. It means that for all practical purposes no $MnO_4^-$ ions are present at equilibrium.

*Related Problems:* 43, 44

The foregoing example illustrates how equilibrium constants for overall cell reactions can be determined electrochemically. It was an instance of a redox equilibrium, but related procedures can be used to measure the solubility product constants of sparingly soluble ionic compounds or the ionization constants of weak acids and bases. Suppose, for example, that the solubility product constant of AgCl is to be determined by means of an electrochemical cell. One half-cell contains solid AgCl and Ag metal in equilibrium with a known concentration of $Cl^-(aq)$ (established with 0.00100 M NaCl, for example) so that an unknown but definite concentration of $Ag^+(aq)$ is present. A silver electrode is used so that the half-cell reaction involved is either the reduction of $Ag^+(aq)$ or the oxidation of Ag. This is, in effect, an $Ag^+|Ag$ half-cell whose potential is to be determined. The second half-cell can be any whose potential is accurately known, and its choice is a matter of convenience. In the following example, the second half-cell is a standard $H_3O^+|H_2$ half-cell.

### Example 10-12

A galvanic cell is constructed using a standard hydrogen half-cell (with platinum electrode) and a half-cell containing silver and silver chloride:

$$Pt|H_2(1\ atm)|H_3O^+(1\ M)\|Cl^-(1.00 \times 10^{-3}\ M) + Ag^+(?\ M)|AgCl|Ag$$

The $H_2|H_3O^+$ half-cell is observed to be the anode, and the measured cell voltage is $\Delta\mathscr{E} = 0.397$ V. Calculate the silver ion concentration in the cell and the $K_{sp}$ of AgCl at 25°C.

### Solution

The half-cell reactions are

$$H_2(g) + 2\ H_2O(\ell) \longrightarrow 2\ H_3O^+(aq) + 2\ e^- \quad \text{(anode)}$$

$$2\ Ag^+(aq) + 2\ e^- \longrightarrow 2\ Ag(s) \quad \text{(cathode)}$$

$$\Delta\mathscr{E}° = \mathscr{E}°(\text{cathode}) - \mathscr{E}°(\text{anode}) = 0.800 - 0.000\ V = 0.800\ V$$

Note that $n = 2$ for the overall cell reaction, and the reaction quotient simplifies to

$$Q = \frac{[H_3O^+]^2}{[Ag^+]^2 P_{H_2}} = \frac{1}{[Ag^+]^2}$$

because $[H_3O^+] = 1$ M and $P_{H_2} = 1$ atm. The Nernst equation is

$$\Delta\mathscr{E} = \Delta\mathscr{E}° - \frac{0.0592\ V}{n} \log_{10} Q$$

$$\log_{10} Q = \frac{n}{0.0592\ V}(\Delta\mathscr{E}° - \Delta\mathscr{E}) = \frac{2}{0.0592\ V}(0.800 - 0.397\ V) = 13.6$$

$$Q = 10^{13.6} = 4 \times 10^{13} = 1/[Ag^+]^2$$

This can be solved for the silver ion concentration $[Ag^+]$ to give

$$[Ag^+] = 1.6 \times 10^{-7} \text{ M}$$

so that

$$[Ag^+][Cl^-] = (1.6 \times 10^{-7})(1.00 \times 10^{-3})$$

$$K_{sp} = 1.6 \times 10^{-10}$$

*Related Problems:* 49, 50

## pH Meters

The voltage of a galvanic cell is sensitive to the pH if one of its electrodes is a $Pt|H_2$ electrode that dips into a solution of variable pH. A simple cell to measure pH is

$$Pt|H_2(1 \text{ atm})|H_3O^+(\text{variable}) \| H_3O^+(1 \text{ M})|H_2(1 \text{ atm})|Pt$$

If the half-cell reactions are written as

$$H_2(1 \text{ atm}) + 2 \text{ H}_2O(\ell) \longrightarrow 2 \text{ H}_3O^+(\text{var}) + 2\,e^- \qquad \text{(anode)}$$

$$2 \text{ H}_3O^+(1 \text{ M}) + 2\,e^- \longrightarrow H_2(1 \text{ atm}) + 2 \text{ H}_2O(\ell) \qquad \text{(cathode)}$$

then $n = 2$ and $Q = [H_3O^+(\text{variable})]^2$ because the other concentrations and gas pressures are unity. From the Nernst equation,

$$\Delta \mathscr{E} = \Delta \mathscr{E}^\circ - \frac{0.0592 \text{ V}}{n} \log_{10} Q$$

and because $\Delta \mathscr{E}^\circ = 0$,

$$\Delta \mathscr{E} = -\frac{0.0592 \text{ V}}{2} \log_{10} [H_3O^+]^2$$

$$= -0.0592 \text{ V} \log_{10} [H_3O^+] = (0.0592 \text{ V}) \text{ pH}$$

The measured cell voltage is proportional to the pH.

What we have described is a simple **pH meter.** It is inconvenient, though, to have to bubble hydrogen gas through the unknown and reference half-cells, and so a more portable and miniaturized pair of electrodes is employed to replace the hydrogen half-cells. In a typical commercial pH meter, two electrodes are dipped into the solution of unknown pH. One of these, the **glass electrode,** usually consists of an AgCl-coated silver electrode in contact with an HCl solution of known (e.g., 1.0 M) concentration in a thin-walled glass bulb. A pH-dependent potential develops across this thin glass membrane when the glass electrode is immersed in a solution of different, unknown $[H_3O^+]$. The second half-cell is frequently a **saturated calomel electrode,** consisting of a platinum wire in electrical contact with a paste of liquid mercury, calomel $(Hg_2Cl_2(s))$, and a saturated solution of KCl. The overall cell (Fig. 10-7) can be represented as

$$Ag|AgCl|Cl^- + H_3O^+(1.0 \text{ M})|glass|H_3O^+(\text{var}) \| Cl^-(\text{sat})|Hg_2Cl_2(s)|Hg|Pt$$

whose half-reactions are

$$2 \text{ Ag}(s) + 2 \text{ Cl}^-(1.0 \text{ M}) \longrightarrow 2 \text{ AgCl}(s) + 2\,e^- \qquad \text{(anode)}$$

$$H_3O^+(1.0 \text{ M}) \longrightarrow H_3O^+(\text{var})$$

$$Hg_2Cl_2(s) + 2\,e^- \longrightarrow 2\,Hg(\ell) + 2\,Cl^-(sat) \qquad \text{(cathode)}$$

The first and third half-reactions have half-cell potentials $\mathscr{E}(AgCl|Cl^-|Ag)$ and $\mathscr{E}(Hg_2Cl_2|Cl^-|Hg)$ that can be combined and called $\Delta\mathscr{E}(ref)$ because they make a constant contribution to the cell voltage. The second reaction is the source of a variable potential in the cell, corresponding to the free energy of dilution of $H_3O^+$ from a concentration of 1.0 M to an unknown and variable concentration, and its potential exists across the thin glass membrane of the glass electrode. The Nernst equation for the cell can therefore be written as

$$\Delta\mathscr{E} = \Delta\mathscr{E}(ref) - \frac{0.0592\text{ V}}{1} \log_{10} \frac{[H_3O^+(var)]}{1.00}$$

$$= \Delta\mathscr{E}(ref) + (0.0592\text{ V})\,pH$$

and the pH of the unknown solution is

$$pH = \frac{\Delta\mathscr{E} - \Delta\mathscr{E}(ref)}{0.0592\text{ V}}$$

The glass electrode has a number of advantages. It responds only to changes in $[H_3O^+]$ and does so over a wide range of pH. It is unaffected by strong oxidizing agents that would make a hydrogen electrode unreliable. Highly colored solutions that would render acid–base indicators useless do not interfere with the glass electrode. Finally, the glass electrode can be miniaturized to permit insertion into individual living cells, and therefore finds wide use in biology.

Other types of electrodes have been designed that are sensitive to the concentrations of ions other than $H_3O^+$. The simplest example of such an **ion-selective**

**Figure 10-7**

A pH meter consists of a glass electrode (*left*) and a calomel electrode (*right*), both of which dip into a solution of unknown hydronium ion concentration.

Voltmeter

Platinum wire

Silver wire coated with AgCl

Saturated solution of KCl and $Hg_2Cl_2$

Thin glass membrane

Solid calomel ($Hg_2Cl_2(s)$)

HCl (1.0 M)

Mercury

Solution of unknown pH

Porous plug

Batteries vary in size and chemistry. Shown here are an automobile lead-storage battery, rechargeable nickel–cadmium cells, alkaline cells, and zinc–carbon dry cells. (Richard Megna/Fundamental Photographs, New York)

Positive electrode

Graphite cathode

Zinc anode

Paper separator

Moist paste of $ZnCl_2$ plus $NH_4Cl$

Negative electrode

$MnO_2$ + graphite

## Figure 10-8

In a Leclanché dry cell, electrons are released to an external circuit at the anode and enter the cell again at the cathode, where the reduction of $MnO_2$ occurs.

**electrode** is a metal wire, which can be used to detect the concentration of the corresponding metal ion in solution. Silver and copper wires can be used reproducibly in this way to determine $Ag^+$ and $Cu^{2+}$ concentrations, respectively. Still other electrodes have been developed to detect specific ions. For example, glasses of chemically modified composition are used to construct membrane electrodes to determine potassium and sodium ions or halogen ions.

## 10-5 BATTERIES AND FUEL CELLS

The electrochemistry of galvanic cells leads to a wide variety of applications in industry and in everyday life. In this section we focus on two of the most important: batteries to store energy and fuel cells to convert chemical energy to electrical energy.

### Batteries

The origin of the **battery** is lost in history, but it is believed that Persian artisans must have used some form of battery to gold-plate jewelry as long ago as the second century B.C. The modern development of the battery began with Alessandro Volta in 1800. Volta constructed a stack of alternating zinc and silver disks, between pairs of which were inserted disks of paper saturated with a salt solution. This device generated a potential difference across its ends that could cause a shock if the stack contained enough disks. In Volta's invention, a large number of galvanic cells were arranged in series (cathode to anode) so that their individual voltages added together. Such an arrangement is, strictly speaking, a **battery of cells,** but modern usage makes no distinction between a single cell and a voltaic pile, and the word "battery" has come to mean the cell itself. Cells that are discarded when their electrical energy has been spent are called **primary** cells, and those that can be recharged are **secondary** cells.

The most familiar primary cell is the **Leclanché cell** (also called a *zinc–carbon dry cell*) used for flashlights, portable radios, and a host of other purposes. Each year more than 5 billion such dry cells are used worldwide, and estimates place the quantity of zinc consumed for this purpose at more than 30 metric tons per day. The "dry cell" is not really dry at all. Rather, its electrolyte is a moist powder containing ammonium chloride and zinc chloride. Figure 10-8 is a cutaway illustration of a dry cell, which consists of a zinc shell for an anode (negative pole) and an axial graphite rod for a cathode (positive pole), with the rod surrounded by a densely packed layer of graphite and manganese dioxide. Each of these components performs an interesting and essential function. At the zinc anode, oxidation takes place:

$$Zn(s) \longrightarrow Zn^{2+}(aq) + 2\,e^- \qquad \text{(anode)}$$

The moist salt mixture permits ions to transport through the cell a net charge equal to that carried by electrons in the external circuit, just as a salt bridge does in the cells we have considered up to now. Manganese dioxide is the ultimate electron acceptor and is reduced to $Mn_2O_3$ by electrons that pass to it from the graphite rod through the graphite particles:

$$2\,MnO_2(s) + 2\,NH_4^+(aq) + 2\,e^- \longrightarrow$$
$$Mn_2O_3(s) + 2\,NH_3(aq) + H_2O(\ell) \qquad \text{(cathode)}$$

The ingenious device of mixing powdered graphite with powdered $MnO_2$ greatly increases the effective surface area of the cathode, reduces the internal resistance of the cell, and enables currents of several amperes to flow. The overall cell reaction is

$$Zn(s) + 2\,MnO_2(s) + 2\,NH_4^+(aq) \longrightarrow$$
$$Zn^{2+}(aq) + Mn_2O_3(s) + 2\,NH_3(aq) + H_2O(\ell)$$

The cell components are hermetically sealed in a steel shell that is in contact with the zinc and acts as the negative terminal of the battery. A fresh zinc–carbon dry cell generates a potential difference of 1.5 V.

The Leclanché cell has the disadvantage that its concentrations change with time, and thus the voltage of the battery falls as it is used. In an **alkaline dry cell,** the ammonium chloride is replaced by potassium hydroxide, and the half-cell reactions become

$$Zn(s) + 2\,OH^-(aq) \longrightarrow Zn(OH)_2(s) + 2\,e^- \qquad \text{(anode)}$$

$$2\,MnO_2(s) + H_2O(\ell) + 2\,e^- \longrightarrow Mn_2O_3(s) + 2\,OH^-(aq) \qquad \text{(cathode)}$$

The overall reaction is then

$$Zn(s) + 2\,MnO_2(s) + H_2O(\ell) \longrightarrow Zn(OH)_2(s) + Mn_2O_3(s)$$

Because dissolved species do not appear in the overall reaction, concentrations do not change significantly and a steadier voltage results.

A third primary dry cell is the **zinc–mercuric oxide cell** depicted in Figure 10-9. It is frequently given the shape of a small button and is used in automatic cameras, hearing aids, digital calculators, and quartz–electric watches. This battery has an anode that is a mixture of mercury and zinc and a steel cathode in contact with solid mercury(II) oxide (HgO). The electrolyte is a 45% KOH solution that saturates an absorbent material. The anode half-reaction is the same as that in an alkaline dry cell,

$$Zn(s) + 2\,OH^-(aq) \longrightarrow Zn(OH)_2(s) + 2\,e^- \qquad \text{(anode)}$$

but the cathode half-reaction is now

$$HgO(s) + H_2O(\ell) + 2\,e^- \longrightarrow Hg(\ell) + 2\,OH^-(aq) \qquad \text{(cathode)}$$

The overall reaction is

$$Zn(s) + HgO(s) + H_2O(\ell) \longrightarrow Zn(OH)_2(s) + Hg(\ell)$$

This cell has a very stable output of 1.34 V, a fact that makes it especially valuable for use in communication equipment and scientific instruments.

## Rechargeable Batteries

In some batteries the electrodes can be regenerated after depletion by imposing an external potential across them that reverses the direction of current flow through the cell. These are called **secondary batteries,** and the process of reconstituting them to their original state is called "recharging." To recharge a run-down secondary battery, the voltage of the external source must be larger than that of the battery in its original state and, of course, opposite in polarity.

The **nickel–cadmium cell** (or *nicad battery;* Fig. 10-10) is used in hand-held electronic calculators and other cordless electric implements such as portable shavers. Its half-cell reactions during discharge are

**Figure 10-9**

A zinc–mercuric oxide dry cell, used in electric watches and cameras.

**Figure 10-10**

A rechargeable nickel–cadmium battery. (Eveready Battery Company)

**Figure 10-11**

In a lead–acid storage battery, anodes made of Pb alternate with cathodes of Pb coated with $PbO_2$. The electrolyte is sulfuric acid.

$$Cd(s) + 2\,OH^-(aq) \longrightarrow Cd(OH)_2(s) + 2\,e^- \qquad \text{(anode)}$$

$$2\,NiO(OH)(s) + 2\,H_2O(\ell) + 2\,e^- \longrightarrow 2\,Ni(OH)_2(s) + 2\,OH^-(aq) \quad \text{(cathode)}$$

$$Cd(s) + 2\,NiO(OH)(s) + 2\,H_2O(\ell) \longrightarrow Cd(OH)_2(s) + 2\,Ni(OH)_2(s)$$

This battery gives a fairly constant voltage of 1.4 V. When it is connected to an external voltage source, the preceding reactions are reversed as the battery is recharged.

One technically important secondary battery is the **lead–acid storage battery,** used in automobiles. A 12-V lead storage battery consists of six 2.0-V cells (Fig. 10-11) connected in series (cathode to anode) by an internal lead linkage and housed in a hard rubber or plastic case. In each cell the anode consists of metallic lead in porous form to maximize its contact area with the electrolyte. The cathode is of similar design, but its lead has been converted to lead dioxide. A sulfuric acid solution (37% by mass) serves as the electrolyte.

When the external circuit is completed, electrons are released from the anode to the external circuit and the resulting $Pb^{2+}$ ions precipitate on the electrode as insoluble lead sulfate. At the cathode, electrons from the external circuit reduce $PbO_2$ to water and $Pb^{2+}$ ions, which also precipitate as $PbSO_4$ on that electrode. The half-cell reactions are

$$Pb(s) + SO_4^{2-}(aq) \longrightarrow PbSO_4(s) + 2\,e^- \quad \text{(anode)}$$

$$PbO_2(s) + SO_4^{2-}(aq) + 4\,H_3O^+(aq) + 2\,e^- \longrightarrow PbSO_4(s) + 6\,H_2O(\ell)$$
$$\text{(cathode)}$$

$$Pb(s) + PbO_2(s) + 2\,SO_4^{2-}(aq) + 4\,H_3O^+(aq) \longrightarrow 2\,PbSO_4(s) + 6\,H_2O(\ell)$$

The anode and cathode are both largely converted to $PbSO_4(s)$ when the storage battery is fully discharged, and because sulfuric acid is a reactant, its concentration falls. Measuring the density of the electrolyte provides a quick way to estimate the state of charge of the battery.

When a voltage in excess of 12 V is applied across the terminals of the battery in the opposite direction, the half-cell reactions are reversed. This part of the cycle restores the battery to its initial state, ready to be used in another work-producing discharge half-cycle. Lead–acid storage batteries endure many thousands of cycles of discharge and charge before they ultimately fail because of the flaking off of $PbSO_4$ from the electrodes or the development of internal short circuits. In automobiles, they are usually not designed to undergo complete discharge–recharge cycles. Instead, a generator converts some of the kinetic energy of the vehicle to electrical energy for continuous or intermittent charging. About $1.8 \times 10^7$ J can be obtained in the discharge of an average automobile battery, and currents as large as 100 A are drawn for the short time needed to start the engine.

A drawback of the lead–acid storage battery is its low energy density, the amount of energy obtainable per kilogram of battery mass. This is not important when a battery is used to start a gasoline-powered automobile, but it precludes the battery's use in a vehicle driven by an electric motor. This difficulty of a low energy-to-mass ratio, which limits the range of vehicle operation before recharge is necessary, has spurred research by electrochemists to develop secondary batteries that have much higher energy densities.

One promising line of approach has been the development of rechargeable batteries that employ an alkali metal (lithium or sodium) as the anode and sulfur as the electron acceptor. Sulfur is a nonconductor of electricity, and so graphite is used as the cathode that conducts electrons to it. The elements must be in their liquid states, and so these batteries are high-temperature cells (sulfur melts at 112°C, lithium at 186°C, and sodium at 98°C). The sodium–sulfur cell (Fig. 10-12), for example, has an optimum operating temperature of 250°C. The half-cell reactions are

$$2\,Na \longrightarrow 2\,Na^+ + 2\,e^- \qquad (anode)$$

$$\underline{S + 2\,e^- \longrightarrow S^{2-} \qquad\qquad (cathode)}$$

$$2\,Na + S \longrightarrow 2\,Na^+ + S^{2-}$$

This is actually an oversimplification of the cathodic process, because sulfide ion forms polysulfides with sulfur:

$$S^{2-} + nS \longrightarrow S^{2-}_{n+1}$$

but the fundamental principles of the cell operation are the same.

What makes the sodium–sulfur cell possible is a remarkable property of a compound called beta-alumina, which has the composition $NaAl_{11}O_{17}$. It allows sodium ions to migrate through its structure very easily, but blocks the passage of polysulfide ions. Therefore, it can function as a semipermeable medium like the membranes used in osmosis (Section 4-4). Such an ion-conducting solid electrolyte is essential to prevent direct chemical reaction between sulfur and sodium. The lithium–sulfur battery operates on similar principles, and other solid electrolytes such as calcium fluoride, which permits ionic transport of fluoride ion, may find use in cells that use those elements.

The considerable promise of the alkali metal–sulfur cell lies in its high energy density, which makes it possible to construct lightweight batteries capable of generating large currents. Their use in electric cars is an attractive possibility. When the cost of petroleum-based fuels increases and their supply becomes less reliable, cells such as the alkali metal–sulfur battery may come into their own.

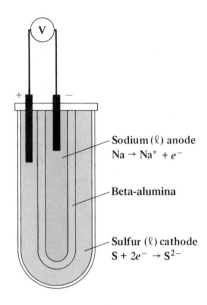

**Figure 10-12**

In a sodium–sulfur battery, $Na^+$ ions migrate through beta-alumina to the cathode to equalize the charge as electrons flow spontaneously from the anode to the cathode through the external circuit.

## Fuel Cells

A battery is a closed system that delivers electrical energy by electrochemical reactions. Once the chemicals originally present are consumed, the battery must be either recharged or discarded. In contrast, a **fuel cell** is designed for continuous operation, with reactants (fuel) being supplied and products removed continuously. It is an energy converter, transforming chemical energy into electrical energy. Fuel cells based on the reaction

$$2\,H_2(g) + O_2(g) \longrightarrow 2\,H_2O(\ell)$$

were used on the Gemini and Apollo space vehicles, for example, to help meet the electrical requirements of the missions (Fig. 10-13). After the product was purified by ion exchange, it was drunk by the crew.

Figure 10-14 represents the hydrogen–oxygen fuel cell schematically. The electrodes can be any nonreactive conductor (graphite, for example), whose function is to conduct electrons into and out of the cell and to facilitate electron exchange between the gases and the ions in solution. The electrolyte transports charge through the cell, and the ions dissolved in it participate in the half-reactions at each elec-

**Figure 10-13**
This hydrogen–oxygen fuel cell was used on U.S. space missions. (United Technologies)

trode. Acidic solutions present corrosion problems, and so an alkaline solution is preferable (1 M NaOH, for instance). The anode half-reaction is then

$$H_2(g) + 2\,OH^-(aq) \longrightarrow 2\,H_2O(\ell) + 2\,e^-$$

and the cathode reaction is

$$\tfrac{1}{2}\,O_2(g) + H_2O(\ell) + 2\,e^- \longrightarrow 2\,OH^-(aq)$$

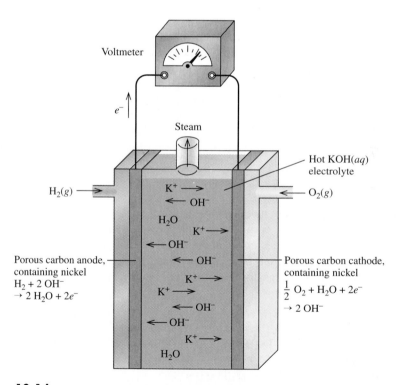

**Figure 10-14**
In a hydrogen–oxygen fuel cell, the two gases are fed in separately and are oxidized or reduced on the electrodes. A hot solution of potassium hydroxide between the electrodes completes the circuit, and the steam produced in the reaction evaporates from the cell continuously.

The standard reduction potentials at 25°C (1 M concentrations, 1 atm pressure) are

$$\mathscr{E}°(H_2|H_2O) = -0.828 \text{ V} \qquad \text{and} \qquad \mathscr{E}°(O_2|OH^-) = 0.401 \text{ V}$$

The overall cell reaction is the production of water,

$$H_2(g) + \tfrac{1}{2} O_2(g) \longrightarrow H_2O(\ell)$$

and the cell voltage is

$$\Delta\mathscr{E}° = \mathscr{E}°(\text{cathode}) - \mathscr{E}°(\text{anode}) = 0.401 - (-0.828) = 1.229 \text{ V}$$

The overall cell voltage does not depend on pH because the $OH^-$ has dropped out; the same voltage would be obtained in an acidic fuel cell.

Another practical fuel cell accomplishes the overall reaction

$$CO(g) + \tfrac{1}{2} O_2(g) \longrightarrow CO_2(g)$$

for which the half-reactions are

$$CO(g) + 3 H_2O(\ell) \longrightarrow CO_2(g) + 2 H_3O^+(aq) + 2\,e^- \qquad \text{(anode)}$$

$$\tfrac{1}{2} O_2(g) + 2 H_3O^+(aq) + 2\,e^- \longrightarrow 3 H_2O(\ell) \qquad\qquad \text{(cathode)}$$

The electrolyte is usually concentrated phosphoric acid, and the operating temperature is between 100° and 200°C. Platinum is the electrode material of choice because it facilitates the electron transfer reactions.

Natural gas (largely $CH_4$) and even fuel oil can be ''burned'' electrochemically to provide electrical energy by either of two approaches. They can be converted to CO and $H_2$ or to $CO_2$ and $H_2$ prior to use in a fuel cell by reaction with steam,

$$CH_4(g) + H_2O(g) \longrightarrow CO(g) + 3 H_2(g)$$

$$CO(g) + H_2O(g) \longrightarrow CO_2(g) + H_2(g)$$

at temperatures of about 500°C. Alternatively, they can be used directly in a fuel cell at higher temperatures (up to 750°C) with a molten alkali metal carbonate as the electrolyte. Either type of fuel cell is attractive as an electrochemical energy converter in regions where hydrocarbon fuels are readily available but large-scale power plants (fossil or nuclear) are remote.

The theoretical advantage of electrochemical fuel cells over more traditional fuel technology can be seen from a thermodynamic analysis. If a chemical reaction such as the oxidation of a fuel can be carried out electrochemically, the maximum (reversible) work obtainable is equal to the free energy change $\Delta G$ (Section 10-3):

$$-w_{\text{max}}(\text{fuel cell}) = |\Delta G|$$

In traditional fuel technology, the same fuel would be burned in air, producing an amount of heat $q_P = \Delta H$, the enthalpy of combustion. The heat would then be used to run a heat engine-generator system to produce electrical power. The efficiency of conversion of heat to work is limited by the laws of thermodynamics. If the heat is supplied at temperature $T_h$ and if the lower operating temperature is $T_l$, the maximum work obtainable (Section 8-6) is

$$-w_{\text{max}}(\text{heat engine}) = \epsilon|q_P| = \frac{T_h - T_l}{T_h}|\Delta H|$$

**Figure 10-15**

Rust and peeling paint on an iron hinge and handle. (Eric Kamp/ Phototake)

Because the magnitude of $\Delta H$ is generally comparable to that of $\Delta G$ for fuel oxidation reactions, the fuel cell will be more efficient because the factor $(T_h - T_l)/T_h$ for a thermal engine is much less than 1. In practice, fuel cells and heat engines must be operated irreversibly (to increase the rate of energy production), and the work obtained with both is less than $w_{max}$. The overall efficiency of practical heat engines rarely exceeds 30–35%, whereas that of fuel cells can be in the 60–70% range. This advantage of fuel cell technology is partially offset by the greater expense of constructing and maintaining fuel cells, however.

## 10-6 CORROSION AND ITS PREVENTION

The **corrosion** of metals is one of the most significant problems faced by advanced industrial societies (Fig. 10-15). It has been estimated that in the United States alone the annual cost of corrosion amounts to tens of *billions* of dollars. Effects of corrosion are both visible (the formation of rust on exposed iron surfaces) and invisible (the cracking and resulting loss of strength of metal beneath the surface). The mechanism of corrosion must be understood before processes can be developed for its prevention.

Although corrosion is a serious problem for many metals, we shall focus on the spontaneous electrochemical reactions of iron. Corrosion can be pictured as a "short-circuited" galvanic cell, in which some regions of the metal surface act as cathodes and others as anodes, and the electric "circuit" is completed by electron flow through the iron itself. These electrochemical cells form in parts of the metal where there are impurities or in regions that are subject to stress. The anode reaction is

$$Fe(s) \longrightarrow Fe^{2+}(aq) + 2\,e^-$$

Various cathode reactions are possible. In the absence of oxygen (for example, at the bottom of a lake), the corrosion reactions are

$$Fe(s) \longrightarrow Fe^{2+}(aq) + 2\,e^- \qquad \text{(anode)}$$
$$\underline{2\,H_2O(\ell) + 2\,e^- \longrightarrow 2\,OH^-(aq) + H_2(g) \qquad \text{(cathode)}}$$
$$Fe(s) + 2\,H_2O(\ell) \longrightarrow Fe^{2+}(aq) + 2\,OH^-(aq) + H_2(g)$$

These reactions are generally slow, however, and do not cause serious amounts of corrosion. Far more extensive corrosion takes place when the iron is in contact with both oxygen and water. In this case the cathode reaction is

$$\tfrac{1}{2}O_2(g) + 2\,H_3O^+(aq) + 2\,e^- \longrightarrow 3\,H_2O(\ell)$$

The $Fe^{2+}$ ions formed simultaneously at the anode migrate to the cathode, where they are further oxidized by $O_2$ to the $+3$ oxidation state to form rust ($Fe_2O_3 \cdot xH_2O$), a hydrated form of iron(III) oxide:

$$2\,Fe^{2+}(aq) + \tfrac{1}{2}O_2(g) + (6 + x)H_2O(\ell) \longrightarrow Fe_2O_3 \cdot xH_2O + 4\,H_3O^+(aq)$$

The hydronium ions produced in this reaction allow the corrosion cycle to continue.

When a portion of the paint that protects a piece of iron or steel is chipped off (Fig. 10-16), the exposed area acts as the cathode because it is open to the atmosphere (air and water) and is therefore rich in oxygen, whereas oxygen-poor areas *under* the paint act as anodes. Rust forms on the cathode (the visible, exposed region), and pitting (loss of metal through oxidation of iron and flow of metal ions

**Secondary reaction**

$$(6 + x)\,H_2O(\ell) + 2\,Fe^{2+}(aq) + \tfrac{1}{2}\,O_2(g) \rightarrow Fe_2O_3 \cdot xH_2O(s) + 4\,H_3O^+(aq)$$

Fe → Fe²⁺ + 2e⁻ ... Anode reaction ... ½ O₂ + 2 H₃O⁺ + 2e⁻ → 3 H₂O ... Cathode reaction

$$2\,Fe(s) + \tfrac{3}{2}\,O_2(g) + xH_2O(\ell) \rightarrow Fe_2O_3 \cdot xH_2O(s)$$
**Overall reaction**

**Figure 10-16**

The corrosion of iron. Note that pitting occurs in the anodic region, and rust appears in the cathodic region.

to the cathode) occurs at the anode. This pitting can lead to loss of structural strength in girders and other supports. The most serious harm done by corrosion is not the visible rusting but the damage done beneath the painted surface.

A number of factors speed corrosion. Dissolved salt provides an electrolyte that improves the flow of charge through solution; a well-known example is the rapid rusting of cars in areas where salt is spread on icy roads. Higher acidity also increases corrosion, as seen by the role of $H_3O^+$ as a reactant in the reduction process at the cathode. Acidity is enhanced by the presence of dissolved $CO_2$ (which produces $H_3O^+$ and $HCO_3^-$ ions) and by air pollution from oxides of sulfur, which leads to the formation of dissolved sulfuric acid in acid precipitation.

Corrosion of iron can be inhibited in a number of ways. Coatings of paint or plastics obviously protect the metal, but they can crack or suffer other damage, thereby localizing and accentuating the process. A very important method of protecting metals arises from the phenomenon of **passivation,** in which a thin metal oxide layer forms on the surface and prevents further electrochemical reactions. Some metals become passivated spontaneously upon exposure to air; aluminum, for example, reacts with oxygen to form a thin protective layer of $Al_2O_3$. Special paints designed for rust prevention contain potassium dichromate ($K_2Cr_2O_7$) and lead oxide ($Pb_3O_4$), which cause the superficial oxidation and passivation of iron. Stainless steel is an alloy of iron with chromium in which the chromium leads to passivation and prevents rusting.

A different way of preventing iron corrosion is to use a **sacrificial anode.** A comparison of the standard reduction potentials of iron and magnesium

$$Fe^{2+} + 2\,e^- \longrightarrow Fe(s) \qquad \mathscr{E}° = -0.41\ V$$
$$Mg^{2+} + 2\,e^- \longrightarrow Mg(s) \qquad \mathscr{E}° = -2.39\ V$$

shows that $Mg^{2+}$ is much harder to reduce than $Fe^{2+}$ or, conversely, that $Mg(s)$ is more easily oxidized than $Fe(s)$. A piece of magnesium in electrical contact with iron is oxidized in preference to the iron, and the iron is therefore protected. The magnesium is the sacrificial anode, and once it is consumed by oxidation it must be replaced. This method is used to protect ship hulls, bridges, and iron water pipes from corrosion. Magnesium plates are attached at regular intervals along a piece of buried pipe, and it is far easier to replace them periodically than to replace the entire pipe.

## 10-7 A Deeper Look . . . Electrolysis of Water and Aqueous Solutions

Until now we have considered primarily galvanic cells, in which chemical reactions are harnessed to produce electricity. In the reverse process, electrolysis, a source of electrical energy is used to drive a nonspontaneous reaction in the opposite direction to give products whose free energies are high enough that they may not ordinarily be present in nature. The recharging of a run-down secondary battery is an example. Electrolysis is used in the industrial production of metals, a subject discussed in Section 19-4. Here we examine only the electrolysis of water and aqueous solutions. Consider first the electrolysis of water between inert electrodes such as platinum, for which the half-cell reactions are

$$2\,H_3O^+(aq) + 2\,e^- \longrightarrow H_2(g) + 2\,H_2O(\ell) \qquad \text{(cathode)}$$
$$3\,H_2O(\ell) \longrightarrow \tfrac{1}{2}O_2(g) + 2\,H_3O^+(aq) + 2\,e^-$$
$$\text{(anode)}$$

---

$$H_2O(\ell) \longrightarrow H_2(g) + \tfrac{1}{2}O_2(g)$$

A practical problem immediately arises. The concentration of $H_3O^+(aq)$ and $OH^-(aq)$ ions in pure water at 25°C is only $1.0 \times 10^{-7}$ M, so the *rate* of electrolysis will be exceedingly small. This practical consideration is put aside for the moment, however, because it does not alter the *thermodynamic* analysis.

The potential $\mathscr{E}°$ for the cathode reaction is by definition 0 V, but because the $H_3O^+(aq)$ concentration in pure water is not 1 M but $1.0 \times 10^{-7}$ M, $\mathscr{E}$ differs from $\mathscr{E}°$ and equals

$$\mathscr{E}(\text{cathode}) = \mathscr{E}°(\text{cathode}) - \frac{0.0592\text{ V}}{n_{hc}} \log Q_{hc}$$

$$= 0.00 - \frac{0.0592\text{ V}}{2} \log \frac{P_{H_2}}{[H_3O^+]^2}$$

If $H_2(g)$ is produced at atmospheric pressure, this simplifies to

$$\mathscr{E}(\text{cathode}) = 0.00 - \frac{0.0592\text{ V}}{2} \log \frac{1}{(10^{-7})^2}$$

$$= -0.414\text{ V}$$

The anode half-reaction written as a reduction (i.e., in reverse) is

$$\tfrac{1}{2}O_2(g) + 2\,H_3O^+(aq) + 2\,e^- \longrightarrow 3\,H_2O(\ell)$$

A table of standard reduction potentials (Appendix E) gives $\mathscr{E}° = 1.229$ V. In the present case, the $H_3O^+(aq)$ concentration is $1.0 \times 10^{-7}$ M rather than 1 M, and so

$$\mathscr{E}(\text{anode}) = \mathscr{E}° - \frac{0.0592\text{ V}}{2} \log \frac{1}{(P_{O_2})^{1/2}[H_3O^+]^2}$$

$$= 1.229\text{ V} - \frac{0.0592\text{ V}}{2} \log \frac{1}{(10^{-7})^2}$$

$$= 0.815\text{ V}$$

if $P_{O_2} = 1$ atm. The overall cell voltage is

$$\Delta\mathscr{E} = \mathscr{E}(\text{cathode}) - \mathscr{E}(\text{anode}) = -0.414 - 0.815\text{ V}$$

$$= -1.229\text{ V}$$

A negative $\Delta\mathscr{E}$ means that the process does not occur spontaneously; it can be made to occur only by applying an external voltage sufficient to overcome the intrinsic negative voltage of the cell. In the electrolysis of water, this minimum external voltage, which is called the **decomposition potential** of water, is 1.229 V. When this potential difference has been imposed, an overall cell reaction will occur, given by the sum of the reduction and oxidation half-reactions:

$$H_2O(\ell) \longrightarrow \tfrac{1}{2}O_2(g) + H_2(g)$$

Hydrogen will bubble off at the cathode and oxygen at the anode.

Suppose an external voltage is now used to electrolyze an electrolyte solution instead of pure water. The resultant products depend upon the concentrations of the ions present and their half-cell potentials. In a 0.10 M NaCl solution, we could conceive of the following processes taking place:

Cathode:
$$Na^+(0.1\text{ M}) + e^- \longrightarrow Na(s)$$
$$2\,H_3O^+(10^{-7}\text{ M}) + 2\,e^- \longrightarrow H_2(g) + 2\,H_2O(\ell)$$

Anode:
$$Cl^-(0.1\text{ M}) \longrightarrow \tfrac{1}{2}Cl_2(g) + e^-$$
$$3\,H_2O(\ell) \longrightarrow \tfrac{1}{2}O_2(g) + 2\,H_3O^+(10^{-7}\text{ M}) + 2\,e^-$$

In each of these pairs of possible processes, which one will actually occur?

For the first half-reaction the reduction potential is

$$\mathscr{E}(Na^+|Na) = \mathscr{E}°(Na^+|Na) - \frac{0.0592\text{ V}}{1} \log \frac{1}{[Na^+]}$$

$$= -2.71 - 0.06 = -2.77\text{ V}$$

Because this result is algebraically less than the half-cell voltage $\mathscr{E}(H_3O^+(10^{-7}\text{ M})|H_2) = -0.414$ V for pure water, the reduction of $Na^+(aq)$ is impossible, and $H_2(g)$ is the cathode product.

For the third half-reaction, the reduction potential is

$$\mathscr{E}(Cl_2|Cl^-) = \mathscr{E}°(Cl_2|Cl^-) - \frac{0.0592\text{ V}}{1} \log \frac{[Cl^-]}{P_{Cl_2}^{1/2}}$$

$$= 1.36 + 0.06 = 1.42\text{ V}$$

Because this is more positive than the half-cell voltage

$\mathcal{E}(O_2,H_3O^+(10^{-7} \text{ M})|H_2O) = 0.815$ V for pure water, $Cl_2$ has a *greater* tendency to be reduced than $O_2$, and so $Cl^-$ has a *lesser* tendency to be oxidized than $H_2O$, and the anode product is $O_2(g)$. If we try to increase the external potential above 1.229 V, all that will happen is that water will be electrolyzed at a greater rate to produce hydrogen and oxygen. Sodium and chlorine will not appear as long as sufficient water is present.

Suppose now that 0.10 M NaI is substituted for the 0.10 M NaCl solution. Sodium ions still will not be reduced; however, $\mathcal{E}(I_2|I^-)$ is

$$\mathcal{E}(I_2|I^-) = \mathcal{E}°(I_2|I^-) - \frac{0.0592 \text{ V}}{1} \log [I^-]$$

$$= 0.535 + 0.059 = 0.594 \text{ V}$$

The half-cell potential for the reduction of iodine is *less* positive than the reduction potential of $O_2(g)$ in water at pH = 7 (0.815 V), so the oxidation of 0.10 M $I^-$ occurs in preference to the oxidation of water. The anode reaction is therefore

$$I^-(0.10 \text{ M}) \longrightarrow \tfrac{1}{2} I_2(s) + e^-$$

and the overall cell reaction is

$$H_3O^+(10^{-7} \text{ M}) + I^-(0.10 \text{ M}) \longrightarrow$$
$$\tfrac{1}{2} H_2(g) + \tfrac{1}{2} I_2(s) + H_2O(\ell)$$

The intrinsic cell voltage is

$$\Delta\mathcal{E} = \mathcal{E}(\text{cathode}) - \mathcal{E}(\text{anode})$$
$$= -0.414 - 0.594 = -1.008 \text{ V} < 0$$

When the decomposition potential of the solution (1.008 V) is exceeded, $H_2(g)$ and $I_2(s)$ begin to form. Of course, this causes $[I^-]$ to decrease and $\mathcal{E}(I_2|I^-)$ to increase. When the iodide ion concentration reaches about $2 \times 10^{-5}$ M, $\mathcal{E}(I_2|I^-)$ will have increased to 0.815 V and the external voltage required to maintain electrolysis will have increased to 1.229 V. At this point, water will start to be electrolyzed and oxygen will be produced at the anode.

Our results for the electrolysis of *neutral* aqueous solutions are summarized as follows:

1. A species can be reduced only if its reduction potential is algebraically greater than $-0.414$ V.
2. A species can be oxidized only if its reduction potential is algebraically smaller than 0.815 V.

In solutions with pH different from 7, these results must be modified, as shown by the following example.

.................................................................
**Example 10-13**

An aqueous 0.10 M solution of $NiCl_2$ is electrolyzed under 1 atm pressure. Determine the products formed at the anode and the cathode and the decomposition potential, if the pH is (a) 7.0; (b) 0.0.

**Solution**

The reduction of $Ni^{2+}$ at the cathode has the half-cell potential

$$Ni^{2+}(aq) + 2 e^- \longrightarrow Ni(s)$$

$$\mathcal{E}(Ni^{2+}|Ni) = \mathcal{E}°(Ni^{2+}|Ni) - \frac{0.0592 \text{ V}}{2} \log \frac{1}{[Ni^{2+}]}$$

$$= -0.23 - 0.03 = -0.26 \text{ V}$$

and the oxidation of $Cl^-$ at the anode has the *reduction* potential

$$\tfrac{1}{2} Cl_2(g) + e^- \longrightarrow Cl^-(aq)$$

$$\mathcal{E}(Cl_2|Cl^-) = \mathcal{E}°(Cl_2|Cl^-) - \frac{0.0592 \text{ V}}{1} \log \frac{[Cl^-]}{P_{Cl_2}^{1/2}}$$

$$= 1.36 \text{ V} - (0.0592 \text{ V}) \log 0.2 = 1.40 \text{ V}$$

(a) In neutral solution the cathode half-reaction is

$$Ni^{2+}(aq) + 2 e^- \longrightarrow Ni(s)$$

because $\mathcal{E}(Ni^{2+}|Ni) = -0.26 \text{ V} > -0.414$ V. The anode half-reaction is

$$3 H_2O(\ell) \longrightarrow 2 H_3O^+(10^{-7} \text{ M}) + \tfrac{1}{2} O_2(g) + 2 e^-$$

because $\mathcal{E}(Cl_2|Cl^-) = 1.40 \text{ V} > 0.815$ V (the reduction potential for this half-reaction). The cell voltage is

$$\Delta\mathcal{E} = \mathcal{E}(\text{cathode}) - \mathcal{E}(\text{anode}) = -0.26 - 0.815 \text{ V}$$
$$= -1.08 \text{ V}$$

and so the decomposition potential is 1.08 V.
(b) In 1.0 M acid solution (pH = 0.0), the anode half-reaction is still

$$3 H_2O(\ell) \longrightarrow 2 H_3O^+(1 \text{ M}) + \tfrac{1}{2} O_2(g) + 2 e^-$$

The reduction potential for this reaction is now $\mathcal{E}_{hc} = \mathcal{E}_{hc}° = 1.229$ V, which is still less than 1.40 V, the reduction potential for the competing reaction involving chlorine.

The cathode half-reaction now becomes

$$2 H_3O^+(1 \text{ M}) + 2 e^- \longrightarrow H_2(g) + 2 H_2O(\ell)$$

because $\mathcal{E}(H_3O^+|H_2) = 0.0 \text{ V} > -0.26$ V for $\mathcal{E}(Ni^{2+}|Ni)$. We now have

$$\Delta\mathcal{E} = \mathcal{E}(\text{cathode}) - \mathcal{E}(\text{anode}) = 0.00 - 1.229 = -1.229 \text{ V}$$

and so the decomposition potential of the solution is now 1.229 V, just as it is for pure water.

*Related Problems:* 63, 64

**PRACTICE EXERCISE**

Black crystals of manganite, MnOOH, an ore of manganese. (Julius Weber)

## Manganese

Manganese is the 12th most abundant element on the earth's surface. Its most important ore source is pyrolusite ($MnO_2$). The preparation and uses of manganese and its compounds (which range up to +7 in oxidation state) are intimately bound up with electrochemistry.

(a) Elemental manganese in a state of high purity can be prepared by electrolyzing aqueous solutions of $Mn^{2+}$. At which electrode (anode or cathode) does the Mn appear? Electrolysis is also used to make $MnO_2$ in high purity from $Mn^{2+}$ solutions. At which electrode does the $MnO_2$ appear?

(b) The Winkler method is an analytical procedure for determining the amount of oxygen dissolved in water. In the first step, $Mn(OH)_2(s)$ is oxidized by gaseous oxygen to $Mn(OH)_3(s)$ in basic aqueous solution. Write the oxidation and reduction half-equations for this step, and write the balanced overall equation. Then use Appendix E to calculate the standard voltage that would be measured if this reaction were carried out in an electrochemical cell.

(c) Calculate the equilibrium constant at 25°C for the reaction in part (b).

(d) In the second step of the Winkler method, the $Mn(OH)_3$ is acidified to give $Mn^{3+}$, and iodide ion is added. Will $Mn^{3+}$ spontaneously oxidize $I^-$? Write a balanced equation for its reaction with $I^-$, and use data from Appendix E to calculate its equilibrium constant. Titration of the $I_2$ produced completes the use of the Winkler method.

(e) Manganese(IV) is an even stronger oxidizing agent than manganese(III). It oxidizes zinc to $Zn^{2+}$ in the dry cell. Such a battery has a cell voltage of 1.5 V. Calculate the electrical work done by this battery in 1.00 hour if it produces a steady current of 0.70 A.

(f) Calculate the mass of zinc reacting in the process described in part (e).

(g) The reduction potential of permanganate ion (+7 oxidation state) in acidic aqueous solution is given by

$$MnO_4^-(aq) + 8\,H_3O^+(aq) + 5\,e^- \longrightarrow Mn^{2+}(aq) + 12\,H_2O(\ell)$$
$$\mathscr{E}° = 1.491\ \text{V}$$

whereas that of the analogous fifth-period species, pertechnetate ion, is

$$TcO_4^-(aq) + 8\,H_3O^+(aq) + 5\,e^- \longrightarrow Tc^{2+}(aq) + 12\,H_2O(\ell)$$
$$\mathscr{E}° = 0.500\ \text{V}$$

Which is the stronger oxidizing agent, permanganate ion or pertechnetate ion?

(h) A galvanic cell is made from two half-cells. In the first, a platinum electrode is immersed in a solution at pH 2.00 that is 0.100 M in both $MnO_4^-$ and $Mn^{2+}$. In the second, a zinc electrode is immersed in a 0.0100 M solution of $Zn(NO_3)_2$. Calculate the cell voltage that will be measured.

### Answers

(a) Mn appears at the cathode, and $MnO_2$ at the anode.

(b)
$$Mn(OH)_2(s) + OH^-(aq) \longrightarrow Mn(OH)_3(s) + e^- \qquad \text{(oxidation)}$$
$$O_2(g) + 2\,H_2O(\ell) + 4\,e^- \longrightarrow 4\,OH^-(aq) \qquad \text{(reduction)}$$
$$\overline{4\,Mn(OH)_2(s) + O_2(g) + 2\,H_2O(\ell) \longrightarrow 4\,Mn(OH)_3(s)}$$
$$\Delta\mathscr{E}° = 0.401 - (-0.40) = 0.80\ \text{V}$$

(c) $K = 1 \times 10^{54}$

(d) $Mn^{3+}$ will spontaneously oxidize $I^-$.

$$2\,Mn^{3+}(aq) + 2\,I^-(aq) \longrightarrow 2\,Mn^{2+}(aq) + I_2(s) \qquad K = 9 \times 10^{32}$$

(e) $3.8 \times 10^3\,J$

(f) 0.85 g Zn is oxidized.

(g) Permanganate ion

(h) 2.12 V

After studying this chapter and working the problems that follow, you should be able to

1. Balance equations for redox reactions in aqueous solution, using the half-reaction method (Section 10-1, problems 1–4).
2. Define the terms "anode" and "cathode" and give the convention that is used to represent a galvanic cell (Section 10-2, problems 9–10).
3. Use Faraday's laws to calculate the quantities of substances produced or consumed at the electrodes of electrochemical cells in relation to the total charge passing through the circuit (Section 10-2, problems 11–18).
4. Explain the relationship between the free energy change in a galvanic cell and the amount of electrical work (Section 10-3, problems 19–20).
5. Combine half-cell reactions and their standard reduction potentials to obtain the overall reactions and standard voltages of electrochemical cells (Section 10-3, problems 21–24).
6. Combine half-reactions and their standard reduction potentials to form other half-reactions and their standard reduction potentials (Section 10-3, problems 31–32).
7. Use reduction potential diagrams to determine strengths of oxidizing and reducing agents and stability toward disproportionation (Section 10-3, problems 33–34).
8. Apply the Nernst equation to calculate the voltage of a cell in which reactants and products are not in their standard states (Section 10-4, problems 35–42).
9. Calculate the value of an equilibrium constant from the voltage of an electrochemical cell (Section 10-4, problems 43–46).
10. Describe the principles that underlie the use of electrochemical cells as pH meters (Section 10-4, problems 47–48).
11. Discuss the electrochemistry of a primary battery and contrast it with that of a secondary battery and a fuel cell (Section 10-5, problems 51–58).
12. Discuss the electrochemical corrosion of metals and describe measures that may be used to minimize it (Section 10-6, problems 59–62).
13. Predict the products liberated at the anode and cathode of an electrolysis cell with a given aqueous electrolyte composition (Section 10-7, problems 63–64).

### Balancing Oxidation–Reduction Equations

1. Complete and balance the following equations for reactions taking place in acidic solution.

   (a) $VO_2^+(aq) + SO_2(g) \rightarrow VO^{2+}(aq) + SO_4^{2-}(aq)$

   (b) $Br_2(\ell) + SO_2(g) \rightarrow Br^-(aq) + SO_4^{2-}(aq)$

   (c) $Cr_2O_7^{2-}(aq) + Np^{4+}(aq) \rightarrow Cr^{3+}(aq) + NpO_2^{2+}(aq)$

   (d) $HCOOH(aq) + MnO_4^-(aq) \rightarrow CO_2(g) + Mn^{2+}(aq)$

   (e) $Hg_2HPO_4(s) + Au(s) + Cl^-(aq) \rightarrow$
   $$Hg(\ell) + H_2PO_4^-(aq) + AuCl_4^-(aq)$$

2. Complete and balance the following equations for reactions taking place in acidic solution.
   (a) $MnO_4^-(aq) + H_2S(aq) \rightarrow Mn^{2+}(aq) + SO_4^{2-}(aq)$
   (b) $Zn(s) + NO_3^-(aq) \rightarrow Zn^{2+}(aq) + NH_4^+(aq)$
   (c) $H_2O_2(aq) + MnO_4^-(aq) \rightarrow O_2(g) + Mn^{2+}(aq)$
   (d) $Sn(s) + NO_3^-(aq) \rightarrow Sn^{4+}(aq) + N_2O(g)$
   (e) $UO_2^{2+}(aq) + Te(s) \rightarrow U^{4+}(aq) + TeO_4^{2-}(aq)$

3. Complete and balance the following equations for reactions taking place in basic solution.
   (a) $Cr(OH)_3(s) + Br_2(aq) \rightarrow CrO_4^{2-}(aq) + Br^-(aq)$
   (b) $ZrO(OH)_2(s) + SO_3^{2-}(aq) \rightarrow Zr(s) + SO_4^{2-}(aq)$
   (c) $HPbO_2^-(aq) + Re(s) \rightarrow Pb(s) + ReO_4^-(aq)$
   (d) $HXeO_4^-(aq) \rightarrow XeO_6^{4-}(aq) + Xe(g)$
   (e) $N_2H_4(aq) + CO_3^{2-}(aq) \rightarrow N_2(g) + CO(g)$

4. Complete and balance the following equations for reactions taking place in basic solution.
   (a) $OCl^-(aq) + I^-(aq) \rightarrow IO_3^-(aq) + Cl^-(aq)$
   (b) $SO_3^{2-}(aq) + Be(s) \rightarrow S_2O_3^{2-}(aq) + Be_2O_3^{2-}(aq)$
   (c) $H_2BO_3^-(aq) + Al(s) \rightarrow BH_4^-(aq) + H_2AlO_3^-(aq)$
   (d) $O_2(g) + Sb(s) \rightarrow H_2O_2(aq) + SbO_2^-(aq)$
   (e) $Sn(OH)_6^{2-}(aq) + Si(s) \rightarrow HSnO_2^-(aq) + SiO_3^{2-}(aq)$

5. The following balanced equations represent reactions that occur in aqueous acid. Break them down into balanced oxidation and reduction half-equations.
   (a) $2\,H_3O^+(aq) + H_2O_2(aq) + 2\,Fe^{2+}(aq) \rightarrow$
   $$2\,Fe^{3+}(aq) + 4\,H_2O(\ell)$$
   (b) $H_3O^+(aq) + H_2O(\ell) + 2\,MnO_4^-(aq) + 5\,SO_2(aq) \rightarrow$
   $$2\,Mn^{2+}(aq) + 5\,HSO_4^-(aq)$$
   (c) $5\,ClO_2^-(aq) + 4\,H_3O^+(aq) \rightarrow$
   $$4\,ClO_2(g) + Cl^-(aq) + 6\,H_2O(\ell)$$

6. The following balanced equations represent reactions that occur in aqueous base. Break them down into balanced oxidation and reduction half-equations.
   (a) $4\,PH_3(g) + 4\,H_2O(\ell) + 4\,CrO_4^{2-}(aq) \rightarrow$
   $$P_4(s) + 4\,Cr(OH)_4^-(aq) + 4\,OH^-(aq)$$
   (b) $NiO_2(s) + 2\,H_2O(\ell) + Fe(s) \rightarrow$
   $$Ni(OH)_2(s) + Fe(OH)_2(s)$$
   (c) $CO_2(g) + 2\,NH_2OH(aq) \rightarrow$
   $$CO(g) + N_2(g) + 3\,H_2O(g)$$

7. Nitrous acid ($HNO_2$) disproportionates in acidic solution to nitrate ion ($NO_3^-$) and nitrogen oxide (NO). Write a balanced equation for this reaction.

8. Thiosulfate ion ($S_2O_3^{2-}$) disproportionates in acidic solution to give solid sulfur and aqueous hydrogen sulfite ion ($HSO_3^-$). Write a balanced equation for this reaction.

## Electrochemical Cells

9. Diagram the following galvanic cell, indicating the direction of flow of electrons in the external circuit and the motion of ions in the salt bridge.

   $$Pt(s)|Cr^{2+}(aq),\ Cr^{3+}(aq)\|Cu^{2+}(aq)|Cu(s)$$

   Write a balanced equation for the overall reaction in this cell.

10. Diagram the following galvanic cell, indicating the direction of flow of electrons in the external circuit and the motion of ions in the salt bridge.

    $$Ni(s)|Ni^{2+}(aq)\|HCl(aq)|H_2(g)|Pt(s)$$

    Write a balanced equation for the overall reaction in this cell.

11. A quantity of electricity equal to $6.95 \times 10^4$ C passes through an electrolytic cell that contains a solution of $Sn^{4+}(aq)$ ions. Compute the maximum chemical amount, in moles, of $Sn(s)$ that can be deposited at the cathode.

12. A quantity of electricity equal to $9.263 \times 10^4$ C passes through a galvanic cell that has an $Ni(s)$ anode. Compute the maximum chemical amount, in moles, of $Ni^{2+}(aq)$ that can be released into solution.

13. A galvanic cell is constructed that has a zinc anode immersed in a $Zn(NO_3)_2$ solution and a platinum cathode immersed in an NaCl solution equilibrated with $Cl_2(g)$ at 1 atm and 25°C. A salt bridge connects the two half-cells.
    (a) Write a balanced equation for the cell reaction.
    (b) A steady current of 0.800 A is observed to flow for a period of 25.0 min. How much charge passes through the circuit during this time? How many moles of electrons is this charge equivalent to?
    (c) Calculate the change in mass of the zinc electrode.
    (d) Calculate the volume of gaseous chlorine generated or consumed as a result of the reaction.

14. A galvanic cell consists of a cadmium cathode immersed in a $CdSO_4$ solution and a zinc anode immersed in a $ZnSO_4$ solution. A salt bridge connects the two half-cells.
    (a) Write a balanced equation for the cell reaction.
    (b) A current of 1.45 A is observed to flow for a period of 2.60 hours. How much charge passes through the circuit during this time? How many moles of electrons is this charge equivalent to?
    (c) Calculate the change in mass of the zinc electrode.
    (d) Calculate the change in mass of the cadmium electrode.

15. An acidic solution containing copper ions is electrolyzed, producing gaseous oxygen (from water) at the anode and copper at the cathode. For every 16.0 g of oxygen generated, 63.5 g of copper plates out. What is the oxidation state of the copper in the solution?

16. Michael Faraday reported that passing electricity through one solution liberated 1 mass of hydrogen at the cathode and 8 masses of oxygen at the anode. The same quantity of electricity liberated 36 masses of chlorine at the anode and 58 masses of tin at the cathode from a second solution. What were the oxidation states of hydrogen, oxygen, chlorine, and tin in these solutions?

17. Liquid potassium chloride, $KCl(\ell)$, is decomposed in an electrolytic cell to form potassium and chlorine. Liquid KCl consists of $K^+$ and $Cl^-$ ions.
    (a) Write balanced equations for the half-cell reactions at the anode and at the cathode and for the overall cell reaction.
    (b) If a current of 2.00 A is passed through the cell for a

period of 5.00 hours, calculate the mass of metal deposited and of gas liberated.

18. In the Hall–Héroult process for the electrolytic production of aluminum, $Al^{3+}$ ions from $Al_2O_3$ dissolved in molten cryolite ($Na_3AlF_6$) are reduced to $Al(\ell)$ while carbon (graphite) is oxidized to $CO_2$ by reaction with oxide ions.
   (a) Write balanced equations for the half-reactions at the anode and at the cathode and for the overall cell reaction.
   (b) If a current of 50,000 A is passed through the cell for a period of 24 hours, what mass of aluminum will be recovered?

## The Gibbs Free Energy and Cell Voltage

19. A $Ni|Ni^{2+}\|Ag^+|Ag$ galvanic cell is constructed in which the standard cell voltage is 1.03 V. Calculate the free energy change at 25°C when 1.00 g of silver plates out, if all concentrations remain at their standard value of 1 M throughout the process. What is the maximum electrical work done by the cell on its surroundings during this experiment?

20. A $Zn|Zn^{2+}\|Co^{2+}|Co$ galvanic cell is constructed in which the standard cell voltage is 0.48 V. Calculate the free energy change at 25°C per gram of zinc lost at the anode, if all concentrations remain at their standard value of 1 M throughout the process. What is the maximum electrical work done by the cell on its surroundings during this experiment?

21. A galvanic cell is constructed in which a $Br_2|Br^-$ half-cell is connected to a $Co^{2+}|Co$ half-cell.
   (a) By referring to Appendix E, write balanced chemical equations for the half-reactions at the anode and the cathode and for the overall cell reaction.
   (b) Calculate the cell voltage, assuming that all reactants and products are in their standard states.

22. A galvanic cell is constructed in which a $Pt|Fe^{2+}, Fe^{3+}$ half-cell is connected to a $Cd^{2+}|Cd$ half-cell.
   (a) By referring to Appendix E, write balanced chemical equations for the half-reactions at the anode and the cathode and for the overall cell reaction.
   (b) Calculate the cell voltage, assuming that all reactants and products are in their standard states.

23. In a galvanic cell, one half-cell consists of a zinc strip dipped into a 1.00 M solution of $Zn(NO_3)_2$. In the second half-cell, solid indium adsorbed on graphite is in contact with a 1.00 M solution of $In(NO_3)_3$. Indium is observed to plate out as the galvanic cell operates, and the initial cell voltage is measured to be 0.425 V at 25°C.
   (a) Write balanced equations for the half-reactions at the anode and the cathode.
   (b) Calculate the standard reduction potential of an $In^{3+}|In$ half-cell. Consult Appendix E for the reduction potential of the $Zn^{2+}|Zn$ electrode.

24. In a galvanic cell, one half-cell consists of gaseous chlorine bubbled over a platinum electrode at a pressure of 1.00 atm into a 1.00 M solution of NaCl. The second half-cell has a strip of solid gallium immersed in a 1.00 M $Ga(NO_3)_3$ solution. The initial cell voltage is measured to be 1.918 V at 25°C, and as the cell operates, the concentration of chloride ion is observed to increase.
   (a) Write balanced equations for the half-reactions at the anode and the cathode.
   (b) Calculate the standard reduction potential of a $Ga^{3+}|Ga$ half-cell. Consult Appendix E for the reduction potential of the $Cl_2|Cl^-$ electrode.

25. Would you expect powdered solid aluminum to act as an oxidizing agent or as a reducing agent?

26. Would you expect potassium perchlorate, $KClO_4(aq)$, in a concentrated acidic solution to act as an oxidizing agent or as a reducing agent?

27. Bromine is sometimes used in place of chlorine as a disinfectant in swimming pools. If the effectiveness of a chemical as a disinfectant depends solely on its strength as an oxidizing agent, do you expect bromine to be better or worse than chlorine as a disinfectant, at a given concentration?

28. Many bleaches, including chlorine and its oxides, oxidize dye compounds in cloth. Predict which of the following will be the strongest bleach at a given concentration and pH 0: $NaClO_3(aq)$, $NaClO(aq)$, $Cl_2(aq)$. How does the strongest chlorine-containing bleach compare in strength with ozone, $O_3(g)$?

29. Suppose you have the following reagents available at pH 0, atmospheric pressure, and 1 M concentration:

$$Co(s),\ Ag^+(aq),\ Cl^-(aq),\ Cr(s),\ BrO_3^-(aq),\ I_2(s)$$

   (a) Which is the strongest oxidizing agent?
   (b) Which is the strongest reducing agent?
   (c) Which reagent will reduce $Pb^{2+}(aq)$ while leaving $Cd^{2+}(aq)$ unreacted?

30. Suppose you have the following reagents available at pH 0, atmospheric pressure, and 1 M concentration:

$$Sc(s),\ Hg_2^{2+}(aq),\ Cr_2O_7^{2-}(aq),\ H_2O_2(aq),\ Sn^{2+}(aq),\ Ni(s)$$

   (a) Which is the strongest oxidizing agent?
   (b) Which is the strongest reducing agent?
   (c) Which reagent will oxidize $Fe(s)$ while leaving $Cu(s)$ unreacted?

31. (a) Use the data from Appendix E to calculate the half-cell potential $\mathscr{E}°$ for the half-reaction

$$Mn^{3+}(aq) + 3\,e^- \longrightarrow Mn(s)$$

   (b) Consider the disproportionation reaction

$$3\,Mn^{2+}(aq) \rightleftharpoons Mn(s) + 2\,Mn^{3+}(aq)$$

   Will $Mn^{2+}$ disproportionate in aqueous solution?

32. The following standard reduction potentials have been measured in aqueous solution at 25°C:

$$Tl^{3+} + e^- \longrightarrow Tl^{2+} \qquad \mathscr{E}° = -0.37\ V$$

$$Tl^{3+} + 2\,e^- \longrightarrow Tl^+ \qquad \mathscr{E}° = 1.25\ V$$

(a) Calculate the half-cell potential for the half-reaction

$$Tl^{2+} + e^- \longrightarrow Tl^+$$

(b) Consider the disproportionation reaction

$$2\,Tl^{2+}(aq) \rightleftharpoons Tl^{3+}(aq) + Tl^+(aq)$$

Will $Tl^{2+}$ disproportionate in aqueous solution?

33. The following reduction potentials are measured at pH 0:

$$BrO_3^- + 6\,H_3O^+ + 5\,e^- \longrightarrow \tfrac{1}{2}Br_2(\ell) + 9\,H_2O \quad \mathscr{E}° = 1.52\ V$$

$$Br_2(\ell) + 2\,e^- \longrightarrow 2\,Br^- \qquad\qquad\qquad \mathscr{E}° = 1.065\ V$$

(a) Will bromine disproportionate spontaneously in acidic solution?

(b) Which is the stronger reducing agent at pH 0: $Br_2(\ell)$ or $Br^-$?

34. The following reduction potentials are measured at pH 14:

$$ClO^- + H_2O + 2\,e^- \longrightarrow Cl^- + 2\,OH^- \quad \mathscr{E}° = 0.90\ V$$

$$ClO_2^- + H_2O + 2\,e^- \longrightarrow ClO^- + 2\,OH^- \quad \mathscr{E}° = 0.59\ V$$

(a) Will $ClO^-$ disproportionate spontaneously in basic solution?

(b) Which is the stronger reducing agent at pH 14: $ClO^-$ or $Cl^-$?

## Concentration Effects and the Nernst Equation

35. A galvanic cell is constructed that carries out the reaction

$$Pb^{2+}(aq) + 2\,Cr^{2+}(aq) \longrightarrow Pb(s) + 2\,Cr^{3+}(aq)$$

If the initial concentration of $Pb^{2+}(aq)$ is 0.15 M, that of $Cr^{2+}(aq)$ is 0.20 M, and that of $Cr^{3+}(aq)$ is 0.0030 M, calculate the initial voltage generated by the cell at 25°C.

36. A galvanic cell is constructed that carries out the reaction

$$2\,Ag(s) + Cl_2(g) \longrightarrow 2\,Ag^+(aq) + 2\,Cl^-(aq)$$

If the partial pressure of $Cl_2(g)$ is 1.00 atm, the initial concentration of $Ag^+(aq)$ is 0.25 M, and that of $Cl^-(aq)$ is 0.016 M, calculate the initial voltage generated by the cell at 25°C.

37. Calculate the reduction potential for a $Pt|Cr^{3+},Cr^{2+}$ half-cell in which $[Cr^{3+}]$ is 0.15 M and $[Cr^{2+}]$ is 0.0019 M.

38. Calculate the reduction potential for an $I_2(s)|I^-$ half-cell in which $[I^-]$ is $1.5 \times 10^{-6}$ M.

39. An $I_2(s)|I^-(1.00\ M)$ half-cell is connected to an $H_3O^+|H_2(1\ atm)$ half-cell in which the concentration of the hydronium ion is unknown. The measured cell voltage is 0.841 V and the $I_2|I^-$ half-cell is the cathode. What is the pH in the $H_3O^+|H_2$ half-cell?

40. A $Cu^{2+}(1.00\ M)|Cu$ half-cell is connected to a $Br_2(\ell)|Br^-$ half-cell in which the concentration of bromide ion is unknown. The measured cell voltage is 0.963 V, and the $Cu^{2+}|Cu$ half-cell is the anode. What is the bromide ion concentration in the $Br_2(\ell)|Br^-$ half-cell?

41. The following reaction occurs in an electrochemical cell:

$$3\,HClO_2(aq) + 2\,Cr^{3+}(aq) + 12\,H_2O(\ell) \longrightarrow$$
$$3\,HClO(aq) + Cr_2O_7^{2-}(aq) + 8\,H_3O^+(aq)$$

(a) Calculate $\Delta\mathscr{E}°$ for this cell.

(b) At pH 0, with $[Cr_2O_7^{2-}] = 0.80$ M, $[HClO_2] = 0.15$ M, and $[HClO] = 0.20$ M, the cell voltage is found to be 0.15 V. Calculate the concentration of $Cr^{3+}(aq)$ in the cell.

42. A galvanic cell is constructed in which the overall reaction is

$$Cr_2O_7^{2-}(aq) + 14\,H_3O^+(aq) + 6\,I^-(aq) \longrightarrow$$
$$2\,Cr^{3+}(aq) + 3\,I_2(s) + 21\,H_2O(\ell)$$

(a) Calculate $\Delta\mathscr{E}°$ for this cell.

(b) At pH 0, with $[Cr_2O_7^{2-}] = 1.5$ M and $[I^-] = 0.40$ M, the cell voltage is found to equal 0.87 V. Calculate the concentration of $Cr^{3+}(aq)$ in the cell.

43. By using the half-cell potentials in Appendix E, calculate the equilibrium constant at 25°C for the reaction in problem 41. Dichromate ion ($Cr_2O_7^{2-}$) is orange, and $Cr^{3+}$ is light green in aqueous solution. If 2.00 L of 1.00 M $HClO_2$ solution is added to 2.00 L of 0.50 M $Cr(NO_3)_3$ solution, what color will the resulting solution have?

44. By using the half-cell potentials in Appendix E, calculate the equilibrium constant at 25°C for the reaction

$$6\,Hg^{2+}(aq) + 2\,Au(s) \rightleftharpoons 3\,Hg_2^{2+}(aq) + 2\,Au^{3+}(aq)$$

If 1.00 L of a 1.00 M $Au(NO_3)_3$ solution is added to 1.00 L of a 1.00 M $Hg_2(NO_3)_2$ solution, calculate the concentrations of $Hg^{2+}$, $Hg_2^{2+}$, and $Au^{3+}$ at equilibrium.

45. The following standard reduction potentials have been determined for the aqueous chemistry of indium:

$$In^{3+}(aq) + 2\,e^- \longrightarrow In^+(aq) \qquad \mathscr{E}° = -0.40\ V$$

$$In^+(aq) + e^- \longrightarrow In(s) \qquad\qquad \mathscr{E}° = -0.21\ V$$

Calculate the equilibrium constant ($K$) for the disproportionation of $In^+(aq)$ at 25°C.

$$3\,In^+(aq) \rightleftharpoons 2\,In(s) + In^{3+}(aq)$$

46. Use data from Appendix E to compute the equilibrium constant for the reaction

$$Hg^{2+}(aq) + Hg(\ell) \rightleftharpoons Hg_2^{2+}(aq)$$

47. A galvanic cell consists of a $Pt|H_3O^+(1.00\ M)|H_2(g)$ cathode connected to a $Pt|H_3O^+(aq)|H_2(g)$ anode in which the concentration of $H_3O^+$ is unknown but is kept constant by the action of a buffer consisting of a weak acid, $HA(0.10\ M)$, mixed with its conjugate base, $A^-(0.10\ M)$. The measured cell voltage is $\Delta\mathscr{E} = 0.150$ V at 25°C, with a hydrogen pressure of 1.00 atm at both electrodes. Calculate the pH in the buffer solution, and from it determine the $K_a$ of the weak acid.

48. In a galvanic cell, the cathode consists of an $Ag^+(1.00\ M)|Ag$ half-cell. The anode is a platinum wire, with hydrogen bubbling over it at 1.00-atm pressure, that is immersed in a buffer

solution containing benzoic acid and sodium benzoate. The concentration of benzoic acid ($C_6H_5COOH$) is 0.10 M, and that of benzoate ion ($C_6H_5COO^-$) is 0.050 M. The overall cell reaction is then

$$Ag^+(aq) + \tfrac{1}{2}H_2(g) + H_2O(\ell) \longrightarrow Ag(s) + H_3O^+(aq)$$

and the measured cell voltage is 1.030 V. Calculate the pH in the buffer solution and determine the $K_a$ of benzoic acid.

49. A galvanic cell is constructed in which the overall reaction is

$$Br_2(\ell) + H_2(g) + 2\,H_2O(\ell) \longrightarrow 2\,Br^-(aq) + 2\,H_3O^+(aq)$$

(a) Calculate $\Delta\mathscr{E}°$ for this cell.
(b) Silver ions are added until AgBr precipitates at the cathode and $[Ag^+]$ reaches 0.060 M. The cell voltage is then measured to be 1.710 V at pH = 0 and $P_{H_2}$ = 1.0 atm. Calculate $[Br^-]$ under these conditions.
(c) Calculate the solubility product constant $K_{sp}$ for AgBr.

50. A galvanic cell is constructed in which the overall reaction is

$$Pb(s) + 2\,H_3O^+(aq) \longrightarrow Pb^{2+}(aq) + H_2(g) + 2\,H_2O(\ell)$$

(a) Calculate $\Delta\mathscr{E}°$ for this cell.
(b) Chloride ions are added until $PbCl_2$ precipitates at the anode and $[Cl^-]$ reaches 0.15 M. The cell voltage is then measured to be 0.22 V at pH = 0 and $P_{H_2}$ = 1.0 atm. Calculate $[Pb^{2+}]$ under these conditions.
(c) Calculate the solubility product constant $K_{sp}$ of $PbCl_2$.

## Batteries and Fuel Cells

51. Calculate the voltage $\Delta\mathscr{E}°$ of a lead–acid cell if all reactants and products are in their standard states. What will be the voltage if six such cells are connected in series?

52. Calculate the standard voltage of the zinc–mercuric oxide cell shown in Figure 10-9. (*Hint:* The easiest way to proceed is to calculate $\Delta G°$ for the corresponding overall reaction, and then find $\Delta\mathscr{E}°$ from it.) Take $\Delta G_f°(Zn(OH)_2(s)) = -553.5$ kJ $mol^{-1}$.

53. (a) What quantity of charge (in coulombs) is a fully charged 12-V lead–acid storage battery theoretically capable of furnishing if the spongy lead available for reaction at the anodes weighs 10 kg and there is excess $PbO_2$?
(b) What is the theoretical maximum amount of work (in joules) that can be obtained from this battery?

54. (a) What quantity of charge (in coulombs) is a fully charged 1.34-V zinc–mercuric oxide watch battery theoretically capable of furnishing if the mass of HgO in the battery is 0.50 g?
(b) What is the theoretical maximum amount of work (in joules) that can be obtained from this battery?

55. The concentration of the electrolyte, sulfuric acid, in a lead–acid storage battery diminishes as the battery is discharged. Is a discharged battery recharged by replacing the dilute $H_2SO_4$ with fresh, concentrated $H_2SO_4$? Explain.

56. One cold winter morning the temperature is well below 0°F. In trying to start your car, you run the battery down com-

pletely. Several hours later, you return to replace your fouled spark plugs and find that the liquid in the battery has now frozen even though the air temperature is actually a bit higher than it was in the morning. Explain how this can happen.

57. Consider the fuel cell that accomplishes the overall reaction

$$H_2(g) + \tfrac{1}{2}O_2(g) \longrightarrow H_2O(\ell)$$

If the fuel cell operates with 60% efficiency, calculate the amount of electrical work generated per gram of water produced. The gas pressures are constant at 1 atm, and the temperature is 25°C.

58. Consider the fuel cell that accomplishes the overall reaction

$$CO(g) + \tfrac{1}{2}O_2(g) \longrightarrow CO_2(g)$$

Calculate the maximum electrical work that could be obtained from the conversion of 1.00 mol of $CO(g)$ to $CO_2(g)$ in such a fuel cell operated with 100% efficiency at 25°C and with the pressure of each gas equal to 1 atm.

## Corrosion and Its Prevention

59. Two half-reactions proposed for the corrosion of iron in the absence of oxygen are

$$Fe(s) \longrightarrow Fe^{2+}(aq) + 2\,e^-$$
$$2\,H_2O(\ell) + 2\,e^- \longrightarrow 2\,OH^-(aq) + H_2(g)$$

Calculate the standard voltage generated by a galvanic cell running this pair of half-reactions. Is the overall reaction spontaneous under standard conditions? As the pH falls from 14, will the reaction become spontaneous?

60. In the presence of oxygen, the cathode half-reaction written in the preceding problem is replaced by

$$\tfrac{1}{2}O_2(g) + 2\,H_3O^+(aq) + 2\,e^- \longrightarrow 3\,H_2O(\ell)$$

but the anode half-reaction is unchanged. Calculate the standard cell voltage for *this* pair of reactions operating as a galvanic cell. Is the overall reaction spontaneous under standard conditions? As the water becomes more acidic, does the driving force for the rusting of iron increase or decrease?

61. Could sodium be used as a sacrificial anode to protect the iron hull of a ship?

62. If it is shown that titanium can be used as a sacrificial anode to protect iron, what conclusion can be drawn about the standard reduction potential of its half-reaction?

$$Ti^{3+}(aq) + 3\,e^- \longrightarrow Ti(s)$$

## Electrolysis of Water and Aqueous Solutions

63. An electrolytic cell consists of a pair of inert metallic electrodes in a solution buffered to pH = 5.0 and containing nickel sulfate ($NiSO_4$) at a concentration of 1.00 M. A current of 2.00 A is passed through the cell for 10.0 hours.
(a) What product is formed at the cathode?

(b) What is the mass of this product?

(c) If the pH is changed to pH = 1.0, what product will form at the cathode?

**64.** A 0.100 M neutral aqueous $CaCl_2$ solution is electrolyzed using platinum electrodes. A current of 1.50 A passes through the solution for 50.0 hours.

(a) Write the half-reactions occurring at the anode and at the cathode.

(b) What is the decomposition potential?

(c) Calculate the mass, in grams, of the product formed at the cathode.

## Additional Problems

**65.** The drain cleaner Drano consists of aluminum turnings mixed with sodium hydroxide. When it is added to water, the sodium hydroxide dissolves and releases heat. The aluminum reacts with water to generate bubbles of hydrogen and aqueous $Al(OH)_4^-$ ions. Write a balanced net ionic equation for this reaction.

**66.** Sulfur-containing compounds in the air tarnish silver, giving black $Ag_2S$. A practical method of cleaning tarnished silverware is to place the tarnished item in electrical contact with a piece of zinc and dip both into water containing a small amount of salt. Write balanced half-equations to represent what takes place.

**67.** A current passed through inert electrodes immersed in an aqueous solution of sodium chloride produces chlorate ion, $ClO_3^-(aq)$, at the anode and gaseous hydrogen at the cathode. Given this fact, write a balanced equation for the chemical reaction if gaseous hydrogen and aqueous sodium chlorate are mixed and allowed to react spontaneously until they reach equilibrium.

**68.** A galvanic cell is constructed by linking a $Co^{2+}|Co(s)$ half-cell to an $Ag^+|Ag(s)$ half-cell through a salt bridge and then connecting the cobalt and silver electrodes through an external circuit. When the circuit is closed, the cell voltage is measured to be 1.08 V and silver is seen to plate out while cobalt dissolves.

(a) Write the half-reactions that occur at the anode and at the cathode, and the balanced overall cell reaction.

(b) The cobalt electrode is weighed after 150 min of operation and is found to have decreased in mass by 0.36 g. By what amount has the silver electrode increased in mass?

(c) What is the average current drawn from the cell during this period?

**69.** The galvanic cell $Zn(s)|Zn^{2+}(aq)\|Ni^{2+}(aq)|Ni(s)$ is constructed using a completely immersed zinc electrode that weighs 32.68 g and a nickel electrode immersed in 575 mL of 1.00 M $Ni^{2+}(aq)$ solution. A steady current of 0.0715 A is drawn from the cell as the electrons move from the zinc electrode to the nickel electrode.

(a) Which reactant is the limiting reactant in this cell?

(b) How long does it take for the cell to be completely discharged?

(c) How much mass has the nickel electrode gained when the cell is completely discharged?

(d) What is the concentration of the $Ni^{2+}(aq)$ when the cell is completely discharged?

**70.** A newly discovered bacterium can reduce selenate ion, $SeO_4^{2-}(aq)$, to elemental selenium, $Se(s)$, in reservoirs. This is significant because the soluble selenate ion is potentially toxic, but elemental selenium is insoluble and harmless. Assume that water is oxidized to oxygen as the selenate ion is reduced. Compute the mass of oxygen produced if all the selenate in a $10^{12}$-L reservoir contaminated with 100 mg $L^{-1}$ of selenate ion is reduced to selenium.

**71.** Thomas Edison invented an electric meter that was nothing more than a simple coulometer, a device to measure the amount of electricity passing through a circuit. In this meter, a small, fixed fraction of the total current supplied to a household was passed through an electrolytic cell, plating out zinc at the cathode. Each month, the cathode could then be removed and weighed to determine the amount of electricity used. If 0.25% of a household's electricity passed through such a coulometer, and the cathode increased in mass by 1.83 g in a month, how many coulombs of electricity were used during that month?

**\*72.** The chief chemist of the Brite-Metal Electroplating Co. is required to certify that the rinse solutions that are discharged from the company's tin-plating process into the municipal sewer system contain no more than 10 ppm (parts per million) by mass of $Sn^{2+}$. The chemist devises the following analytical procedure to determine the concentration. At regular intervals, a 100-mL (100 g) sample is withdrawn from the waste stream and acidified to pH = 1.0. A starch solution and 10 mL of 0.10 M potassium iodide are added, and a 25.0-mA current is passed through the solution between platinum electrodes. Iodine appears as a product of electrolysis at the anode when the oxidation of $Sn^{2+}$ to $Sn^{4+}$ is practically complete, and signals its presence with the deep blue color of a complex formed with starch. What is the maximum duration of electrolysis to the appearance of the blue color that ensures that the concentration of $Sn^{2+}$ does not exceed 10 ppm?

**73.** Estimate the cost of the electrical energy needed to produce $1.5 \times 10^{10}$ kg (a year's supply for the world) of aluminum from $Al_2O_3(s)$ if electrical energy costs 10 cents per kilowatt-hour (1 kWh = 3.6 MJ = $3.6 \times 10^6$ J) and if the cell voltage is 5 V.

**74.** Titanium can be produced by electrolytic reduction from an anhydrous molten salt electrolyte that contains titanium(IV) chloride and a spectator salt that furnishes ions to make the electrolyte conduct electricity. The standard enthalpy of formation of $TiCl_4(\ell)$ is $-750$ kJ $mol^{-1}$ and the standard entropies of $TiCl_4(\ell)$, $Ti(s)$, and $Cl_2(g)$ are 253, 30, and 223 J $K^{-1}$ $mol^{-1}$. What minimum applied voltage will be necessary at 100°C?

**75.** A half-cell has a graphite electrode immersed in an acidic solution (pH 0) of $Mn^{2+}$ (concentration 1.00 M) in contact with solid $MnO_2$. A second half-cell has an acidic solution (pH 0) of $H_2O_2$ (concentration 1.00 M) in contact with a platinum electrode past which gaseous oxygen at a pressure of 1.00 atm is bubbled. The two half-cells are connected to form a galvanic cell.
(a) By referring to Appendix E, write balanced chemical equations for the half-reactions at the anode and the cathode and for the overall cell reaction.
(b) Calculate the cell voltage.

**76.** By considering these half-reactions and their standard reduction potentials,

$$Pt^{2+} + 2\,e^- \longrightarrow Pt \qquad \mathscr{E}° = 1.2 \text{ V}$$
$$NO_3^- + 4\,H_3O^+ + 3\,e^- \longrightarrow NO + 6\,H_2O \quad \mathscr{E}° = 0.96 \text{ V}$$
$$PtCl_4^{2-} + 2\,e^- \longrightarrow Pt + 4\,Cl^- \qquad \mathscr{E}° = 0.73 \text{ V}$$

account for the fact that platinum will dissolve in a mixture of hydrochloric acid and nitric acid (*aqua regia*) but will not dissolve in either acid alone.

**77.** (a) One method to reduce the concentration of unwanted $Fe^{3+}(aq)$ in a solution of $Fe^{2+}(aq)$ is to drop a piece of metallic iron into the storage container. Write the reaction that removes the $Fe^{3+}$, and compute its standard cell potential.
(b) By referring to problem 31, suggest a way to remove unwanted $Mn^{3+}(aq)$ from solutions of $Mn^{2+}(aq)$.

**78.** (a) Based only on the standard reduction potentials for the $Cu^{2+}|Cu^+$ and the $I_2(s)|I^-$ half-reactions, would you expect $Cu^{2+}(aq)$ to be reduced to $Cu^+(aq)$ by $I^-(aq)$?
(b) The formation of solid CuI plays a role in the interaction between $Cu^{2+}(aq)$ and $I^-(aq)$.

$$Cu^{2+}(aq) + I^-(aq) + e^- \rightleftharpoons CuI(s) \quad \mathscr{E}° = 0.86 \text{ V}$$

Taking into account this added information, do you expect $Cu^{2+}$ to be reduced by iodide ion?

**79.** In some old European churches, the stained-glass windows have so darkened from corrosion and age that hardly any light comes through. Microprobe analysis showed that tiny cracks and defects on the glass surface were enriched in insoluble Mn(III) and Mn(IV) compounds. From Appendix E, suggest a reducing agent and conditions that might successfully convert these compounds to soluble Mn(II) without simultaneously reducing Fe(III) (which gives the glass its colors) to Fe(II). Take $MnO_2$ as representative of the insoluble Mn(III) and Mn(IV) compounds.

**80.** (a) Calculate the half-cell potential for the reaction

$$O_2(g) + 4\,H_3O^+(aq) + 4\,e^- \longrightarrow 6\,H_2O(\ell)$$

at pH 7 with the oxygen pressure at 1 atm.
(b) Explain why aeration of solutions of $I^-$ leads to their decomposition. Write a balanced equation for the redox reaction that occurs.
(c) Will the same problem arise with solutions containing $Br^-$ or $Cl^-$? Explain.

(d) Will decomposition be favored or opposed by increasing acidity?

**81.** An engineer needs to prepare a galvanic cell that uses the reaction

$$2\,Ag^+(aq) + Zn(s) \longrightarrow Zn^{2+}(aq) + 2\,Ag(s)$$

and generates an initial voltage of 1.50 V. She has 0.010 M $AgNO_3(aq)$ and 0.100 M $Zn(NO_3)_2(aq)$ solutions as well as electrodes of metallic copper and silver, wires, containers, water, and a $KNO_3$ salt bridge. Sketch the cell. Clearly indicate the concentrations of all solutions.

**82.** Consider a galvanic cell for which the anode reaction is

$$Pb(s) \longrightarrow Pb^{2+}(1.0 \times 10^{-2} \text{ M}) + 2\,e^-$$

and the cathode reaction is

$$VO^{2+}(0.10 \text{ M}) + 2\,H_3O^+(0.10 \text{ M}) + e^- \longrightarrow$$
$$V^{3+}(1.0 \times 10^{-5} \text{ M}) + 3\,H_2O(\ell)$$

The measured cell voltage is 0.640 V.
(a) Calculate $\mathscr{E}°$ for the $VO^{2+}|V^{3+}$ half-reaction, using $\mathscr{E}°(Pb^{2+}|Pb)$ from Appendix E.
(b) Calculate the equilibrium constant ($K$) at 25°C for the reaction

$$Pb(s) + 2\,VO^{2+}(aq) + 4\,H_3O^+(aq) \rightleftharpoons$$
$$Pb^{2+}(aq) + 2\,V^{3+}(aq) + 6\,H_2O(\ell)$$

**83.** Suppose we construct a pressure cell in which the gas pressures differ in the two half-cells. Suppose such a cell consists of a $Cl_2(0.010 \text{ atm})|Cl^-(1 \text{ M})$ half-cell connected to a $Cl_2(0.50 \text{ atm})|Cl^-(1 \text{ M})$ half-cell. Determine which half-cell will be the anode, write the overall equation for the reaction, and calculate the cell voltage.

**84.** A student decides to measure the solubility of lead sulfate in water and sets up the electrochemical cell

$$Pb|PbSO_4|SO_4^{2-}(aq, 0.0500 \text{ M}) \| Cl^-(aq, 1.00 \text{ M})|AgCl|Ag$$

At 25°C he finds the cell voltage to be 0.546 V, and from Appendix E he finds

$$AgCl(s) + e^- \rightleftharpoons Ag(s) + Cl^-(aq) \qquad \mathscr{E}° = 0.222 \text{ V}$$

What does he find for the $K_{sp}$ of $PbSO_4$?

**85.** A wire is fastened across the terminals of the Leclanché cell in Figure 10-8. Indicate the direction of electron flow in the wire.

**86.** Overcharging a lead–acid storage battery can generate hydrogen. Write a balanced equation to represent the reaction taking place.

**87.** An ambitious chemist discovers an alloy electrode that is capable of catalytically converting ethanol reversibly to carbon dioxide at 25°C according to the half-reaction

$$C_2H_5OH(\ell) + 15\,H_2O(\ell) \longrightarrow$$
$$2\,CO_2(g) + 12\,H_3O^+(aq) + 12\,e^-$$

Believing that this discovery is financially important, the

chemist patents its composition and designs a fuel cell that may be represented as

$$Alloy|C_2H_5OH(\ell)|CO_2(g) + H_3O^+(1\ M)\|H_3O^+(1\ M)|O_2|Ni$$

(a) Write the half-reaction occurring at the cathode.
(b) Using data from Appendix D, calculate $\Delta\mathscr{E}°$ for the cell at 25°C.
(c) What is the $\mathscr{E}°$ value for the ethanol half-cell?

88. Iron or steel is often covered by a thin layer of a second metal to prevent rusting: tin cans consist of steel covered with tin, and galvanized iron is made by coating iron with a layer of zinc. If the protective layer is broken, however, iron will rust more readily in a tin can than in galvanized iron. Explain this observation by comparing the half-cell potentials of iron, tin, and zinc.

89. An electrolysis cell contains a solution of 0.10 M $NiSO_4$. The anode and cathode are both strips of Pt foil. Another electrolysis cell contains the same solution, but the electrodes are strips of Ni foil. In each case, a current of 0.10 A flows through the cell for 10 hours.
(a) Write a balanced equation for the chemical reaction that occurs at the anode in each cell.

(b) Calculate the mass, in grams, of the product formed at the anode in each cell. (The product may be a gas, a solid, or an ionic species in solution.)

90. A potential difference of 2.0 V is impressed across a pair of inert electrodes (e.g., platinum) that are immersed in a 0.050 M aqueous KBr solution. What are the products that form at the anode and the cathode?

*91. An aqueous solution is simultaneously 0.10 M in $SnCl_2$ and in $CoCl_2$.
(a) If the solution is electrolyzed, which metal will appear first?
(b) At what decomposition potential will that metal first appear?
(c) As the electrolysis proceeds, the concentration of the metal being reduced will drop and the voltage will change. How complete a separation of the metals using electrolysis is theoretically possible? In other words, at the point where the second metal begins to form, what fraction of the first metal is left in solution?

## CUMULATIVE PROBLEMS

92. A 1.0 M solution of NaOH is electrolyzed, generating $O_2(g)$ at the anode. A current of 0.15 A is passed through the cell for a period of 75 minutes. Calculate the volume of (wet) oxygen generated in this period if the temperature is held at 25°C and the total pressure is 0.985 atm. (*Hint:* Use the vapor pressure of water at this temperature from Table 4-1.)

93. Use standard entropies from Appendix D to predict whether the standard voltage of the $Cu|Cu^{2+}\|Ag^+|Ag$ cell (diagrammed in Figures 10-3 and 10-4) will increase or decrease if the temperature is raised above 25°C.

94. About 50,000 kJ of electrical energy is required to produce 1.0 kg of Al from its $Al(OH)_3$ ore. The major energy cost in recycling aluminum cans is the melting of the aluminum. The enthalpy of fusion of Al(s) is 10.7 kJ mol$^{-1}$. Compare the energy cost for making new aluminum with that for recycling.

95. Use the following half-reactions and their reduction potentials to calculate the $K_{sp}$ of AgBr:

$$Ag^+(aq) + e^- \longrightarrow Ag(s) \qquad \mathscr{E}° = 0.7996\ V$$

$$AgBr(s) + e^- \longrightarrow Ag(s) + Br^-(aq) \qquad \mathscr{E}° = 0.0713\ V$$

(b) Estimate the solubility of AgBr in 0.10 M NaBr(aq).

96. Amounts of iodine dissolved in aqueous solution, $I_2(aq)$, can be determined by titration with thiosulfate ion ($S_2O_3^{2-}$). The thiosulfate ion is oxidized to $S_4O_6^{2-}$ while the iodine is reduced to iodide ion. Starch is used as an indicator because it has a strong blue color in the presence of dissolved iodine.
(a) Write a balanced equation for this reaction.
(b) If 56.40 mL of 0.100 M $S_2O_3^{2-}$ solution is used to reach the endpoint of a titration of an unknown amount of iodine, calculate the number of moles of iodine originally present.
(c) Combine the appropriate half-cell potentials from Appendix E with thermodynamic data from Appendix D for the equilibrium

$$I_2(s) \rightleftharpoons I_2(aq)$$

to calculate the equilibrium constant at 25°C for the reaction in part (a).

*Rust and corrosion of iron on the seashore. (Diane Schiumo/Fundamental Photographs, New York)*

# Chemical Kinetics

Why do some chemical reactions proceed with lightning speed, when others require days, months, or even years to produce detectable amounts of products? How do catalysts increase the rates of chemical reactions? Why do small changes in temperature often have such large effects on the cooking rate of food? How does a study of the rate of a chemical reaction inform us about the way in

**403**

which molecules combine to form products? All of these questions involve chemical kinetics.

Chemical kinetics is the study of the rates and mechanisms of chemical reactions. If thermodynamics addresses *why* reactions take place (What is the driving force?), kinetics explains *how*. Thermodynamics is a very powerful tool for understanding the nature of chemical equilibrium, but it gives no information about the crucial question of how rapidly that equilibrium is achieved. The speeds with which different chemical reactions occur vary enormously, ranging from explosive rapidity to glacial sluggishness. A mixture of hydrogen and oxygen can exist almost indefinitely in a closed container at room temperature and pressure, despite the inherent tendency of the two gases to react spontaneously. The introduction of a spark or a tiny piece of platinum causes the elements to react explosively to form water. It is of enormous practical importance to understand how conditions can be modified to carry out chemical reactions with reasonable speed, under proper control.

Chemical kinetics is a complex subject, and our mastery of it is not nearly as complete as that of thermodynamics. There are many reactions whose equilibrium constants are known accurately, but whose rates and detailed reaction pathways remain poorly understood. This is particularly true of reactions in which many species participate in the overall process that leads from reactants to products. An example is the reaction

$$5\,Fe^{2+}(aq) + MnO_4^-(aq) + 8\,H_3O^+(aq) \longrightarrow$$
$$5\,Fe^{3+}(aq) + Mn^{2+}(aq) + 12\,H_2O(\ell)$$

The equilibrium constant for this reaction is straightforward to measure (from the voltage of a galvanic cell, for example), and from it the equilibrium concentrations can be calculated for arbitrary initial conditions. The exact way in which the reaction goes from reactants to products is considerably harder to determine. It certainly does *not* involve the simultaneous collision of five $Fe^{2+}$ ions and one $MnO_4^-$ ion with eight $H_3O^+$ ions, because such a collision would be exceedingly rare. Instead, it proceeds through a series of elementary steps involving two or at most three ions, such as

$$Fe^{2+} + MnO_4^- \longrightarrow Fe^{3+} + MnO_4^{2-}$$
$$MnO_4^{2-} + H_3O^+ \longrightarrow HMnO_4^- + H_2O$$
$$HMnO_4^- + Fe^{2+} \longrightarrow HMnO_4^{2-} + Fe^{3+}$$

Other postulated steps take the process to its final products. Some of these steps are slow, others fast; taken together they constitute the **reaction mechanism.** A major goal of chemical kinetics is to discern the way the rate depends on the concentrations of the reacting species, and to deduce the mechanism of a reaction from experimental knowledge of that rate.

## 11-1 RATES OF CHEMICAL REACTIONS

The speed of a reaction depends on many factors. Concentrations of reacting species certainly play a major role in speeding up or slowing down a particular reaction (Fig. 11-1). As we shall see in Section 11-4, many reactions are extremely sensitive to temperature, a fact that makes careful control of temperature critical for quantita-

**Figure 11-1**

The rate of reaction of zinc with aqueous sulfuric acid depends on the concentration of the acid. The dilute solution reacts slowly (*left*), and the more concentrated solution reacts rapidly (*right*). (Charles Steele)

tive measurements in chemical kinetics. Finally, the physical forms of the reactants are often crucial to the observed rate. An iron nail oxidizes only very slowly in dry air to iron oxide, but steel wool burns spectacularly in oxygen (Fig. 11-2). The quantitative study of heterogeneous reactions (those involving two or more phases, such as a solid and a gas) is difficult, so we begin with homogeneous reactions (those that take place entirely within the gas phase or solution). In Section 11-7 we return briefly to some important aspects of heterogeneous reactions.

## Measuring Reaction Rates

A kinetics experiment measures the rate at which the concentration of a substance taking part in a chemical reaction changes with time. How can a changing concentration be monitored experimentally? If the reaction is slow enough, we can allow it to proceed for a measured time and then abruptly "quench" (effectively stop) it by rapidly cooling the reaction mixture to a low enough temperature. At that low temperature there is time to carry out a chemical analysis for some particular reactant or product. This procedure is not useful for rapid reactions, especially those involving gas mixtures, which are difficult to cool quickly. An alternative is to use the absorption of light as a probe. As will be seen in Chapter 15, molecules differ in the wavelengths of light they absorb. If a wavelength is absorbed by only one particular reactant or product, then measuring the amount of light absorbed at that wavelength by the reaction mixture determines the concentration of the absorbing species at a series of times. Often a flash of light can also be used to initiate a very fast reaction.

The average rate of a reaction is analogous to the average speed of a car. If the average position of a car is noted at two different times, then

$$\text{average speed} = \frac{\text{distance traveled}}{\text{elapsed time}} = \frac{\text{change in location}}{\text{change in time}}$$

**Figure 11-2**

Steel wool burning in oxygen. (Leon Lewandowski)

In the same way, the **average reaction rate** is obtained by dividing the change in concentration of a reactant or product by the time interval over which that change occurs:

$$\text{average reaction rate} = \frac{\text{change in concentration}}{\text{change in time}}$$

If concentration is measured in mol L$^{-1}$ and time in seconds, then the rate of a reaction has units of mol L$^{-1}$ s$^{-1}$.

Consider a specific example. In the gas-phase reaction

$$NO_2(g) + CO(g) \longrightarrow NO(g) + CO_2(g)$$

$NO_2$ and CO are consumed as NO and $CO_2$ are produced. If a probe can measure the NO concentration, the average rate of reaction can be estimated from the ratio of the change in NO concentration $\Delta[NO]$ to the time interval $\Delta t$:

$$\text{average rate} = \frac{\Delta[NO]}{\Delta t} = \frac{[NO]_f - [NO]_i}{t_f - t_i}$$

This estimate depends on the time interval $\Delta t$ that is selected, because the rate at which NO is produced changes with time. From the data in Figure 11-3, the average rate of reaction during the first 50 s is

$$\text{average rate} = \frac{\Delta[NO]}{\Delta t} = \frac{(0.0160 - 0)\ \text{mol L}^{-1}}{(50 - 0)\ \text{s}} = 3.2 \times 10^{-4}\ \text{mol L}^{-1}\ \text{s}^{-1}$$

During the second 50 s, the average rate is $1.6 \times 10^{-4}$ mol L$^{-1}$ s$^{-1}$, and during the third 50 s it is $9.6 \times 10^{-5}$ mol L$^{-1}$ s$^{-1}$. Clearly, this reaction slows down as it progresses, and its average rate indeed depends on the time interval chosen. Figure 11-3 shows a graphical method for determining average rates. The average rate is the slope of the straight line connecting the concentrations at the initial and final points of a time interval.

The **instantaneous rate** of a reaction is obtained by considering smaller and smaller time increments $\Delta t$ (with correspondingly smaller values of $\Delta[NO]$). As $\Delta t$ approaches 0, the rate becomes the slope of the line tangent to the curve at time $t$ (see Fig. 11-3). This slope is written as the derivative of [NO] with respect to time:

$$\text{instantaneous rate} = \text{limit}_{\Delta t \to 0}\ \frac{[NO]_{t+\Delta t} - [NO]_t}{\Delta t} = \frac{d[NO]}{dt}$$

Throughout the rest of this book, the instantaneous rate will be referred to simply as the *rate*. The instantaneous rate of a reaction at the moment that it begins (at $t = 0$) is the **initial rate** of that reaction.

The rate of this sample reaction could just as well have been measured by monitoring changes in the concentration of $CO_2$, $NO_2$, or CO instead of NO. Because every molecule of NO produced is accompanied by one molecule of $CO_2$, the rate of increase of $CO_2$ concentration is the same as that of NO. The concentrations of the two reactants, $NO_2$ and CO, *decrease* at the same rate that the concentrations of the products increase, because the coefficients in the balanced equation are also both equal to 1. This is summarized as

$$\text{rate} = -\frac{d[NO_2]}{dt} = -\frac{d[CO]}{dt} = \frac{d[NO]}{dt} = \frac{d[CO_2]}{dt}$$

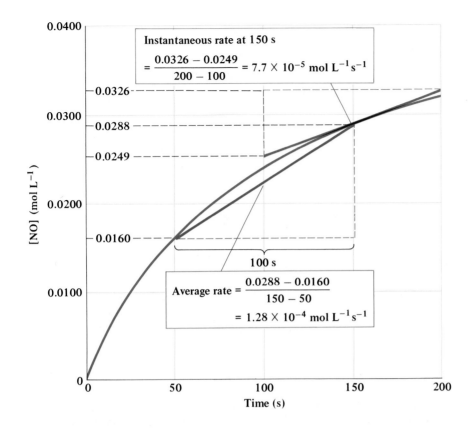

**Figure 11-3**

A graph of the concentration of NO against time in the reaction $NO_2 + CO \rightarrow NO + CO_2$. The average rate during the time interval from 50 to 150 seconds is obtained by dividing the change in NO concentration by the duration of the interval (green box). The instantaneous rate 150 seconds after the start of the reaction is found by calculating the slope of the line tangent to the curve at that point (red box).

Another gas-phase reaction is

$$2\,NO_2(g) + F_2(g) \longrightarrow 2\,NO_2F(g)$$

This equation states that two molecules of $NO_2$ disappear and two of $NO_2F$ appear for each molecule of $F_2$ that reacts. Thus, the $NO_2$ concentration changes twice as fast as the $F_2$ concentration; the $NO_2F$ concentration also changes twice as fast and has the opposite sign. We write the rate in this case as

$$\text{rate} = -\frac{1}{2}\frac{d[NO_2]}{dt} = -\frac{d[F_2]}{dt} = \frac{1}{2}\frac{d[NO_2F]}{dt}$$

The rate of change of concentration of each species is divided by its coefficient in the balanced chemical equation. Rates of change of reactants appear with negative signs, and those of products with positive signs. For the general reaction

$$a\mathrm{A} + b\mathrm{B} \longrightarrow c\mathrm{C} + d\mathrm{D}$$

the rate is

$$\text{rate} = -\frac{1}{a}\frac{d[\mathrm{A}]}{dt} = -\frac{1}{b}\frac{d[\mathrm{B}]}{dt} = \frac{1}{c}\frac{d[\mathrm{C}]}{dt} = \frac{1}{d}\frac{d[\mathrm{D}]}{dt} \qquad \textbf{(11-1)}$$

These relations hold true provided there are no transient intermediate species or, if there are intermediates, their concentrations are independent of time for most of the reaction period.

## 11-2 RATE LAWS

In discussing chemical equilibrium we stressed that both forward and reverse reactions can occur; once products are formed, they can react back to give the original reactants. The net rate is the difference:

$$\text{net rate} = \text{forward rate} - \text{reverse rate}$$

Strictly speaking, measurements of concentration give the net rate rather than simply the forward rate. Near the beginning of a reaction that starts from pure reactants, however, the concentrations of reactants are far higher than those of products, and the reverse rate can be neglected. In addition, many reactions go to "completion" ($K \gg 1$) and so have a measurable rate only in the forward direction, or else the experiment can be arranged in such a way that the products are removed as they are formed. In this section, the focus is on forward rates exclusively.

### Order of a Reaction

The forward rate of a chemical reaction depends on the concentrations of the reactants. As an example, consider the decomposition of dinitrogen pentaoxide ($N_2O_5$). This compound is a white solid that is stable below 0°C but decomposes when vaporized:

$$N_2O_5(g) \longrightarrow 2\,NO_2(g) + \tfrac{1}{2}\,O_2(g)$$

The rate of the reaction depends on the concentration of $N_2O_5(g)$. In fact, as Figure 11-4 shows, a graph of rate versus concentration is a straight line that can be extrapolated to pass through the origin, and so the rate can be written

$$\text{rate} = k[N_2O_5]$$

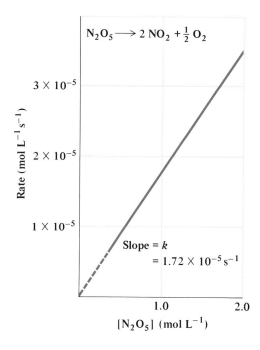

**Figure 11-4**

The rate of decomposition of $N_2O_5(g)$ at 25°C is proportional to its concentration. The slope of this line is equal to the rate constant $k$ for the reaction.

This relation between the rate of a reaction and concentration is called a **rate expression** or **rate law,** and the proportionality constant $k$ is called the **rate constant** for the reaction. Like an equilibrium constant, a rate constant is independent of concentration but depends on temperature, as we shall see with greater detail in Section 11-5.

For many (but not all) reactions with a single reactant, the rate is proportional to the concentration of that reactant raised to a power. That is, the rate expression for

$$a\text{A} \longrightarrow \text{products}$$

frequently has the form

$$\text{rate} = k[\text{A}]^n$$

It is important to note that the power $n$ in the rate expression has no direct relation to the coefficient $a$ in the balanced chemical equation. For the decomposition of ethane at high temperatures and low pressures,

$$\text{C}_2\text{H}_6(g) \longrightarrow 2\,\text{CH}_3(g)$$

the rate expression has the form

$$\text{rate} = k[\text{C}_2\text{H}_6]^2$$

so that $n = 2$ even though the coefficient in the chemical equation is 1.

The power to which the concentration is raised is called the **order** of the reaction with respect to that reactant. Thus, the decomposition of $\text{N}_2\text{O}_5$ is **first order,** whereas that of $\text{C}_2\text{H}_6$ is **second order.** Some processes are **zeroth order** over a range of concentrations. Because $[\text{A}]^0 = 1$, such reactions have rates that are independent of concentration:

$$\text{rate} = k \qquad \text{(zeroth-order kinetics)}$$

The order of a reaction does not have to be an integer; fractional powers are sometimes found. At 450 K, the decomposition of acetaldehyde ($\text{CH}_3\text{CHO}$) is described by the rate expression

$$\text{rate} = k[\text{CH}_3\text{CHO}]^{3/2}$$

The lesson from all this is that reaction order is an experimental property that cannot be predicted from the form of the chemical equation.

The following example illustrates how the order of a reaction can be deduced from experimental data.

## Example 11-1

At elevated temperatures, HI reacts according to the chemical equation

$$2\,\text{HI}(g) \longrightarrow \text{H}_2(g) + \text{I}_2(g)$$

At 443°C, the rate of the reaction increases with concentration of HI as follows:

| [HI] (mol $\text{L}^{-1}$) | 0.0050 | 0.010 | 0.020 |
|---|---|---|---|
| Rate (mol $\text{L}^{-1}\,\text{s}^{-1}$) | $7.5 \times 10^{-4}$ | $3.0 \times 10^{-3}$ | $1.2 \times 10^{-2}$ |

(a) Determine the order of the reaction and write the rate expression.
(b) Calculate the rate constant, and give its units.
(c) Calculate the reaction rate for a 0.0020 M concentration of HI.

### Solution

(a) The rate expressions at two different concentrations $[HI]_1$ and $[HI]_2$ are

$$rate_1 = k([HI]_1)^n$$

$$rate_2 = k([HI]_2)^n$$

After dividing the second equation by the first, the rate constant $k$ drops out, leaving the reaction order $n$ as the only unknown quantity.

$$\frac{rate_2}{rate_1} = \left(\frac{[HI]_2}{[HI]_1}\right)^n$$

We can now substitute any two sets of data into this equation and solve for $n$. Taking the first two sets with $[HI]_1 = 0.0050$ M and $[HI]_2 = 0.010$ M gives

$$\frac{3.0 \times 10^{-3}}{7.5 \times 10^{-4}} = \left(\frac{0.010}{0.0050}\right)^n$$

which simplifies to

$$4 = (2)^n$$

By inspection, $n = 2$, and so the reaction is second order in HI. In a case where the solution of the equation is less obvious, we can take the logarithms of both sides, giving (in this case)

$$\log_{10} 4 = n \log_{10} 2$$

$$n = \frac{\log_{10} 4}{\log_{10} 2} = \frac{0.602}{0.301} = 2$$

The rate expression has the form

$$rate = k[HI]^2$$

(b) The rate constant $k$ is calculated by inserting any of the sets of data into the rate expression. Taking the first set, for example, gives

$$7.5 \times 10^{-4} \text{ mol L}^{-1} \text{ s}^{-1} = k(0.0050 \text{ mol L}^{-1})^2$$

Solving for $k$ gives

$$k = 30 \text{ L mol}^{-1} \text{ s}^{-1}$$

(c) Finally, the rate is calculated for $[HI] = 0.0020$ M:

$$rate = k[HI]^2 = (30 \text{ L mol}^{-1} \text{ s}^{-1})(0.0020 \text{ mol L}^{-1})^2$$
$$= 1.2 \times 10^{-4} \text{ mol L}^{-1} \text{ s}^{-1}$$

So far, each reaction rate has depended only on a single concentration. In reality, many rates depend on the concentrations of two or more different chemical species, and the rate expression is written in a form such as

$$rate = -\frac{1}{a}\frac{d[A]}{dt} = k[A]^m[B]^n$$

Again the exponents $m$ and $n$ do not derive from the coefficients in the balanced equation for the reaction; they must be determined experimentally and are usually integers or half-integers.

The exponents $m, n, \ldots$ give the order of the reaction, just as in the simpler case where only one concentration appeared in the rate expression. For example, the preceding reaction is said to be $m$th order in A, meaning that a change in the concentration of A by a certain factor leads to a change in the rate by that factor raised to the $m$th power. The reaction is $n$th order in B, and the **overall reaction order** is $m + n$. For the reaction

$$H_2PO_2^-(aq) + OH^-(aq) \longrightarrow HPO_3^{2-}(aq) + H_2(g)$$

the experimentally determined rate expression is

$$\text{rate} = k[H_2PO_2^-][OH^-]^2$$

so that the reaction is said to be first order in $H_2PO_2^-(aq)$ and second order in $OH^-(aq)$, with an overall reaction order of 3. The units of $k$ depend on the reaction order. If all concentrations are expressed in mol $L^{-1}$ and if $p = m + n + \cdots$ is the overall reaction order, then $k$ has units of $mol^{-(p-1)} L^{p-1} s^{-1}$.

### Example 11-2

Use the preceding rate expression to determine the effect of the following changes on the rate of decomposition of $H_2PO_2^-(aq)$:
(a) Tripling the concentration of $H_2PO_2^-(aq)$ at constant pH
(b) Changing the pH from 13 to 14 at a constant concentration of $H_2PO_2^-(aq)$

### Solution

(a) Because the reaction is first order in $H_2PO_2^-(aq)$, tripling this concentration will triple the reaction rate.
(b) A change in pH from 13 to 14 corresponds to an increase in the $OH^-(aq)$ concentration by a factor of 10. Because the reaction is second order in $OH^-(aq)$ (i.e., this term is squared in the rate expression), this will increase the reaction rate by a factor of $10^2$, or 100.

*Related Problems:* 5, 6

Rate expressions that depend on more than one concentration are more difficult to obtain experimentally than those that depend on only one. One way to proceed is to find the instantaneous initial rates of reaction for several values of one of the concentrations, holding the other initial concentrations fixed from one run to the next. Then the experiment can be repeated, changing one of the other concentrations. The following example illustrates this procedure.

### Example 11-3

The reaction of $NO(g)$ with $O_2(g)$ gives $NO_2(g)$:

$$2\,NO(g) + O_2(g) \longrightarrow 2\,NO_2(g)$$

From the dependence of the initial rate ($-\frac{1}{2}\, d[NO]/dt$) on the initial concentrations of NO and $O_2$, determine the rate expression and the value of the rate constant.

| [NO] (mol L$^{-1}$) | [O$_2$] (mol L$^{-1}$) | Initial Rate (mol L$^{-1}$ s$^{-1}$) |
|---|---|---|
| $1.0 \times 10^{-4}$ | $1.0 \times 10^{-4}$ | $2.8 \times 10^{-6}$ |
| $1.0 \times 10^{-4}$ | $3.0 \times 10^{-4}$ | $8.4 \times 10^{-6}$ |
| $2.0 \times 10^{-4}$ | $3.0 \times 10^{-4}$ | $3.4 \times 10^{-5}$ |

### Solution

When [$O_2$] is multiplied by 3 (with [NO] constant), the rate is also multiplied by 3 (from $2.8 \times 10^{-6}$ to $8.4 \times 10^{-6}$), and so the reaction is first order in $O_2$. When [NO] is multiplied by 2 (with [$O_2$] constant), the rate is multiplied by

$$\frac{3.4 \times 10^{-5}}{8.4 \times 10^{-6}} \approx 4 = 2^2$$

so the reaction is second order in NO. Thus, the form of the rate expression is

$$\text{rate} = k[O_2][NO]^2$$

To evaluate $k$, any set of data can be inserted into the equation. From the first set,

$$2.8 \times 10^{-6} \text{ mol L}^{-1}\text{ s}^{-1} = k(1.0 \times 10^{-4} \text{ mol L}^{-1})(1.0 \times 10^{-4} \text{ mol L}^{-1})^2$$

$$k = 2.8 \times 10^{6} \text{ L}^2 \text{ mol}^{-2} \text{ s}^{-1}$$

*Related Problems:* 7, 8

### Integrated Rate Laws

Measuring an initial rate involves determining small changes in concentration $\Delta[A]$ that occur during a short time interval $\Delta t$. Sometimes it can be difficult to obtain sufficiently precise experimental data for these small changes. An alternative is to use an **integrated rate law,** which expresses the concentration of a species directly as a function of the time. For any given simple rate expression, a corresponding integrated rate law can be obtained.

As an example, consider again the reaction

$$N_2O_5(g) \longrightarrow 2\, NO_2(g) + \tfrac{1}{2}\, O_2(g)$$

whose rate law has been determined experimentally to be

$$\text{rate} = -\frac{d[N_2O_5]}{dt} = k[N_2O_5]$$

This is a first-order reaction. If we let [$N_2O_5$] = $c$, a function of time, we have

$$\frac{dc}{dt} = -kc$$

We seek a function whose slope at every time is proportional to the value of the function itself. This function can be found through the use of calculus. Separating the variables (with concentration $c$ on the left and time $t$ on the right) gives

$$\frac{1}{c}\,dc = -k\,dt$$

Integrating from an initial concentration $c_0$ at time $t = 0$ to a concentration $c$ at time $t$ (see Appendix C, Section C-5) gives

$$\int_{c_0}^{c}\frac{1}{c}\,dc = -k\int_{0}^{t}dt$$

$$\ln c - \ln c_0 = -kt$$

$$\ln(c/c_0) = -kt$$

$$c = c_0\,e^{-kt} \qquad (11\text{-}2)$$

The concentration will fall off exponentially with time. For a first-order reaction, a plot of $\ln c$ against $t$ should be a straight line with slope $-k$ (Fig. 11-5).

A useful concept in discussions of first-order reactions is the **half-life** $t_{1/2}$—the time it takes for the original concentration $c_0$ to be reduced to one-half its value, $c_0/2$. Setting $c = c_0/2$ gives

$$-kt_{1/2} = \ln\left(\frac{c}{c_0}\right) = \ln\left(\frac{c_0/2}{c_0}\right) = -\ln 2$$

$$t_{1/2} = \frac{\ln 2}{k} = \frac{0.6931}{k} \qquad (11\text{-}3)$$

If $k$ has units of $\text{s}^{-1}$, $(\ln 2)/k$ is the half-life in seconds. During each half-life, the concentration of A falls to half its value again (Fig. 11-6).

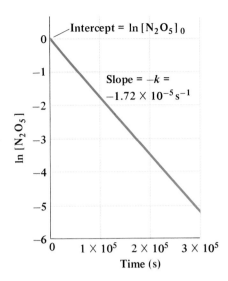

**Figure 11-5**

In a first-order reaction such as the decomposition of $N_2O_5$, a graph of the natural logarithm of the concentration against time is a straight line, the negative of whose slope gives the rate constant for the reaction.

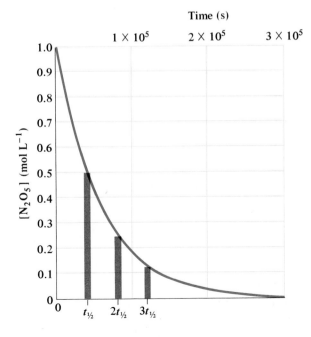

**Figure 11-6**

The same data as in Figure 11-5 are graphed in a concentration-versus-time picture. The half-life $t_{1/2}$ is the time it takes for the concentration to be reduced to one-half its initial value. In two half-lives, the concentration falls to one quarter of its initial value.

## Example 11-4

(a) What is the rate constant $k$ for the first-order decomposition of $N_2O_5(g)$ at 25°C if the half-life of $N_2O_5(g)$ at that temperature is $4.03 \times 10^4$ s?

(b) What percentage of the $N_2O_5$ molecules will *not* have reacted after one day?

### Solution

(a)
$$t_{1/2} = \frac{\ln 2}{k} = 4.03 \times 10^4 \text{ s}$$

Solving for the rate constant $k$ gives

$$k = \frac{\ln 2}{t_{1/2}} = \frac{0.6931}{4.03 \times 10^4 \text{ s}} = 1.72 \times 10^{-5} \text{ s}^{-1}$$

(b) From the integrated rate law for a first-order reaction,

$$\frac{c}{c_0} = e^{-kt}$$

Putting in the value for $k$ and setting $t$ to 1 day $= 8.64 \times 10^4$ s give

$$\frac{c}{c_0} = \exp\left[-(1.72 \times 10^{-5} \text{ s}^{-1})(8.64 \times 10^4 \text{ s})\right]$$
$$= e^{-1.49} = 0.226$$

Therefore, 22.6% of the molecules will not have reacted after one day.

*Related Problems:* 11, 12

Integrated rate laws can be obtained for reactions of other orders. The observed rate of the reaction

$$2\,NO_2 \longrightarrow 2\,NO + O_2$$

is second order in $[NO_2]$:

$$\text{rate} = -\frac{1}{2}\frac{d[NO_2]}{dt} = k[NO_2]^2$$

Writing $[NO_2] = c$ gives

$$\frac{dc}{dt} = -2kc^2$$

$$\frac{1}{c^2} dc = -2k\,dt$$

Integrating this from the initial concentration $c_0$ at time 0 to $c$ at time $t$ gives

$$\int_{c_0}^{c} \frac{1}{c^2} dc = -2k \int_0^t dt$$

$$-\frac{1}{c} + \frac{1}{c_0} = -2kt$$

$$\frac{1}{c} = \frac{1}{c_0} + 2kt \qquad\qquad \text{(11-4)}$$

For such a second-order reaction, a plot of $1/c$ against $t$ is linear (Fig. 11-7).

The concept of half-life has little use for second-order reactions. Setting $[NO_2]$ equal to $[NO_2]_0/2$ in the preceding equation and solving for $t$ give

$$\frac{2}{[NO_2]_0} = 2kt_{1/2} + \frac{1}{[NO_2]_0}$$

$$t_{1/2} = \frac{1}{2k[NO_2]_0}$$

Rather than being a constant, the half-life depends on the initial concentration.

## Example 11-5

The dimerization of tetrafluoroethylene ($C_2F_4$) to octafluorocyclobutane ($C_4F_8$) is second order in the reactant $C_2F_4$, and at 450 K its rate constant is $k = 0.0448$ L mol$^{-1}$ s$^{-1}$. If the initial concentration of $C_2F_4$ is 0.100 mol L$^{-1}$, what will its concentration be after 205 s?

## Solution

For a second-order reaction,

$$\frac{1}{c} - \frac{1}{c_0} = 2kt$$

Solving for the concentration $c$ after a time $t = 205$ s gives

$$\frac{1}{c} = (2)(0.0448 \text{ L mol}^{-1} \text{ s}^{-1})(205 \text{ s}) + \frac{1}{0.100 \text{ mol L}^{-1}} = 28.4 \text{ L mol}^{-1}$$

$$c = 3.53 \times 10^{-2} \text{ mol L}^{-1}$$

*Related Problems:* 15, 16

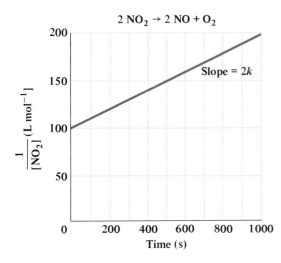

2 NO$_2$ → 2 NO + O$_2$

Slope = 2k

$\dfrac{1}{[NO_2]}$ (L mol$^{-1}$)

Time (s)

**Figure 11-7**

For a second-order reaction such as 2 NO$_2$ → 2 NO + O$_2$, a graph of the reciprocal of the concentration against time is a straight line with slope $2k$.

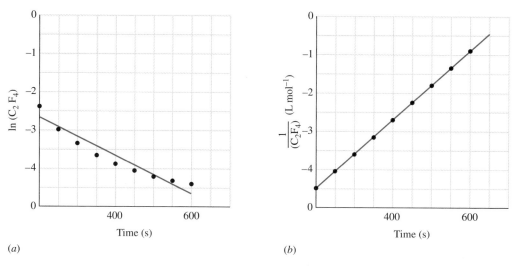

**Figure 11-8**

For the reaction in Example 11-5, (*a*) plotting the logarithm of the concentration of $C_2F_4$ against time tests for first-order kinetics, and (*b*) plotting the reciprocal of the concentration of $C_2F_4$ against time tests for second-order kinetics. It is clear that the assumption of first-order kinetics does not fit the data as well; no straight line will pass through the data.

In a real experimental study, of course, the reaction order may be unknown. In such a case, it is necessary to try several plots to find the one that best fits the data (Fig. 11-8).

## 11-3  REACTION MECHANISMS

Many reactions do not occur in a single step, but proceed through a series of steps. Each step is called an **elementary reaction** and is directly caused by the collisions of atoms, ions, or molecules. The rate expression for an overall reaction cannot be derived from the stoichiometry of the balanced equation, as we have emphasized, and must be determined experimentally. The rate of an elementary reaction, on the other hand, *is* directly proportional to the product of the concentrations of the reacting species, each raised to a power equal to its coefficient in the balanced elementary equation.

### Elementary Reactions

A **unimolecular** elementary reaction involves only a single reactant molecule. An example is the dissociation of energized $N_2O_5$ molecules in the gas phase:

$$N_2O_5^* \longrightarrow NO_2 + NO_3$$

The asterisk indicates that the $N_2O_5$ molecules have far more than thermal energy. This step is unimolecular and has the rate expression

$$\text{rate} = k[N_2O_5^*]$$

An important class of unimolecular reactions is the decay of radioactive nuclei, considered in Chapter 12.

The most common type of elementary reaction involves the collision of two atoms, ions, or molecules and is called **bimolecular.** An example is the reaction

$$NO(g) + O_3(g) \longrightarrow NO_2(g) + O_2(g)$$

The frequency at which a given NO molecule collides with ozone molecules is proportional to the concentration of ozone: if there are twice as many ozone molecules per unit volume, each NO molecule will undergo twice as many collisions as it moves through space, and will react twice as rapidly. The rate of collisions of *all* the NO molecules in the container is proportional to the concentration of NO as well, so the rate law of a bimolecular reaction like this one has the form

$$rate = k[NO][O_3]$$

A **termolecular** reaction step involves the simultaneous collision of three molecules, which is a much less likely event. An example is the recombination of iodine atoms in the gas phase to form iodine molecules. So much energy is released in forming the I—I bond that the molecule would simply fly apart as soon as it was formed if the event were a binary collision. A third atom or molecule is necessary to take away some of the excess energy. If iodine recombination takes place in the presence of a sufficiently high concentration of an inert gas such as argon, termolecular reactions

$$I + I + Ar \longrightarrow I_2 + Ar$$

occur in which the argon atom leaves with more kinetic energy than it had initially. The rate law for this termolecular reaction is

$$rate = k[I]^2[Ar]$$

Elementary reactions involving collisions of four or more molecules are not observed, and even termolecular collisions are rare if other pathways are possible.

Elementary reactions in liquid solvents involve encounters of solute species with one another. If the solution is ideal, the rates of these processes are proportional to the product of the concentrations of the solute species involved. Solvent molecules are always present and may affect the reaction, however, even though they do not appear in the rate expression because the solvent concentration cannot be varied appreciably. A reaction such as the recombination of iodine atoms occurs readily in a liquid. It appears to be second order with rate law

$$rate = k[I]^2$$

only because the third body involved is a solvent molecule. In the same way, a reaction between a solvent molecule and a solute molecule appears to be unimolecular, and only the concentration of solute molecules enters the rate expression for that step.

## Reaction Mechanisms

A **reaction mechanism** is a detailed series of elementary reactions, with their rates, that are combined to yield the overall reaction. It is often possible to write several reaction mechanisms, each of which is consistent with a given overall reaction. One of the goals of chemical kinetics is to use the observed rate of a reaction to choose among various conceivable reaction mechanisms.

The gas-phase reaction of nitrogen dioxide with carbon monoxide provides a good example of a reaction mechanism. The generally accepted mechanism at low temperatures has two steps, both bimolecular:

$$NO_2 + NO_2 \longrightarrow NO_3 + NO \qquad \text{(slow)}$$

$$NO_3 + CO \longrightarrow NO_2 + CO_2 \qquad \text{(fast)}$$

For any reaction mechanism, combining the steps must give the overall reaction. When each elementary step occurs the same number of times in the course of the reaction, the chemical equations can simply be added. (If one step occurs twice as often as the others, it must be multiplied by 2 before the elementary reactions are added.) In this case, we add the two chemical equations to give

$$2\,NO_2 + NO_3 + CO \longrightarrow NO_3 + NO + NO_2 + CO_2$$

Canceling out the $NO_3$ and one molecule of $NO_2$ from each side leads to

$$NO_2 + CO \longrightarrow NO + CO_2$$

A **reaction intermediate** (here, $NO_3$) is a chemical species that is formed and consumed in the reaction but does not appear in the overall balanced chemical equation. One of the major challenges in chemical kinetics is to identify intermediates, which are frequently so short-lived that they are difficult to detect directly.

## Example 11-6

Consider the following reaction mechanism:

$$Cl_2 \longrightarrow 2\,Cl$$

$$Cl + CHCl_3 \longrightarrow HCl + CCl_3$$

$$CCl_3 + Cl \longrightarrow CCl_4$$

(a) What is the molecularity of each elementary step?
(b) Write the overall equation for the reaction.
(c) Identify the reaction intermediate(s).

### Solution

(a) The first step is unimolecular, and the other two are bimolecular.
(b) Adding the three steps gives

$$Cl_2 + 2\,Cl + CHCl_3 + CCl_3 \longrightarrow 2\,Cl + HCl + CCl_3 + CCl_4$$

The two species that appear in equal amounts on both sides cancel out to leave

$$Cl_2 + CHCl_3 \longrightarrow HCl + CCl_4$$

(c) The two reaction intermediates are Cl and $CCl_3$.

*Related Problems:* 21, 22

## Kinetics and Chemical Equilibrium

There is a direct connection between the rates for the elementary steps in a chemical reaction mechanism and the overall equilibrium constant $K$. To explore it, consider the reaction

$$2\,NO(g) + 2\,H_2(g) \rightleftharpoons N_2(g) + 2\,H_2O(g)$$

This reaction is believed to occur in a three-step process involving $N_2O_2$ and $N_2O$ as intermediates:

$$NO + NO \underset{k_{-1}}{\overset{k_1}{\rightleftharpoons}} N_2O_2$$

$$N_2O_2 + H_2 \underset{k_{-2}}{\overset{k_2}{\rightleftharpoons}} N_2O + H_2O$$

$$N_2O + H_2 \underset{k_{-3}}{\overset{k_3}{\rightleftharpoons}} N_2 + H_2O$$

These elementary reactions are shown as equilibria, and so the reverse reactions (from products to reactants) are included here as well; $k_1$, $k_2$, and $k_3$ are the rate constants for the forward elementary steps, and $k_{-1}$, $k_{-2}$, and $k_{-3}$ are the rate constants for the corresponding reverse reactions.

We now invoke the principle of **detailed balance,** which states that at equilibrium the rate of *each* elementary process is balanced by (equal to) the rate of its reverse process. For the preceding mechanism we conclude that

$$k_1[NO]_{eq}^2 = k_{-1}[N_2O_2]_{eq}$$

$$k_2[N_2O_2]_{eq}[H_2]_{eq} = k_{-2}[N_2O]_{eq}[H_2O]_{eq}$$

$$k_3[N_2O]_{eq}[H_2]_{eq} = k_{-3}[N_2]_{eq}[H_2O]_{eq}$$

The equilibrium constants[1] $K_1$, $K_2$, and $K_3$ for the elementary reactions are equal to the ratio of the forward and reverse reaction rate constants:

$$K_1 = \frac{[N_2O_2]_{eq}}{[NO]_{eq}^2} = \frac{k_1}{k_{-1}}$$

$$K_2 = \frac{[N_2O]_{eq}[H_2O]_{eq}}{[N_2O_2]_{eq}[H_2]} = \frac{k_2}{k_{-2}}$$

$$K_3 = \frac{[N_2]_{eq}[H_2O]_{eq}}{[N_2O]_{eq}[H_2]_{eq}} = \frac{k_3}{k_{-3}}$$

The steps of the mechanism are now added together to obtain the overall reaction. Recall from Section 5-2 that when reactions are added, their equilibrium constants are multiplied. Therefore, the overall equilibrium constant $K$ is

---

[1] True equilibrium constants are dimensionless because they involve activities rather than partial pressures or concentrations. The convention in chemical kinetics is to use concentrations rather than activities, even for gaseous species. Therefore, the constants such as $K_1$, $K_2$, and $K_3$ that are introduced here (and that are *not* dimensionless) must be multiplied by the concentration of the reference state, $c_{ref} = RT/P_{ref}$, raised to the appropriate power in order to be made true, dimensionless equilibrium constants. Nevertheless, such constants as $K_1$, $K_2$, and $K_3$ are referred to as equilibrium constants in this section.

$$K = K_1 K_2 K_3 = \frac{k_1 k_2 k_3}{k_{-1} k_{-2} k_{-3}} = \frac{[N_2O_2]_{eq}[N_2O]_{eq}[H_2O]_{eq}[N_2]_{eq}[H_2O]_{eq}}{[NO]_{eq}^2[N_2O_2]_{eq}[H_2]_{eq}[N_2O]_{eq}[H_2]_{eq}}$$

$$= \frac{[H_2O]_{eq}^2[N_2]_{eq}}{[NO]_{eq}^2[H_2]_{eq}^2}$$

The concentrations of the intermediates $N_2O_2$ and $N_2O$ cancel out, giving the usual expression of the mass-action law.

This result can be generalized to any reaction mechanism. The product of the forward rate constants for the elementary reactions divided by the product of the reverse rate constants is always equal to the equilibrium constant of the overall reaction. If there are several possible mechanisms for a given reaction (which might involve intermediates other than $N_2O_2$ and $N_2O$), their forward and reverse rate constants will all be consistent in this way with the equilibrium constant of the overall reaction.

## 11-4 REACTION MECHANISMS AND RATE

In many reaction mechanisms, one step is significantly slower than all the others; it is called the **rate-determining step.** Because an overall reaction can occur only as fast as its slowest step, that step is crucial in determining the rate of the reaction. This is analogous to the flow of automobile traffic on a highway on which there is a slowdown at some point. The rate at which cars can complete a trip down the full length of the highway (in cars per minute) is approximately equal to the rate at which they pass through the bottleneck.

If the rate-determining step is the first one, the analysis is particularly simple. An example is the reaction

$$2\,NO_2 + F_2 \longrightarrow 2\,NO_2F$$

for which the experimental rate law is

$$\text{rate} = k_{obs}[NO_2][F_2]$$

A possible mechanism for the reaction is

$$NO_2 + F_2 \xrightarrow{k_1} NO_2F + F \qquad \text{(slow)}$$

$$NO_2 + F \xrightarrow{k_2} NO_2F \qquad \text{(fast)}$$

The first step is slow and determines the rate, $k_1[NO_2][F_2]$, in agreement with the observed rate expression. The subsequent fast step does not affect the reaction rate, because fluorine atoms react with $NO_2$ almost as soon as they are produced.

Mechanisms in which the rate-determining step occurs after one or more fast steps are often signaled by a reaction order greater than 2, by a nonintegral reaction order, or by an inverse concentration dependence on one of the species taking part in the reaction. An example is the reaction

$$2\,NO + O_2 \longrightarrow 2\,NO_2$$

for which the experimental rate law is

$$\text{rate} = k_{obs}[NO]^2[O_2]$$

One possible mechanism would be a single-step termolecular reaction of two NO molecules with one $O_2$ molecule. This would be consistent with the form of the rate expression, but termolecular collisions are quite rare, and if there is an alternative pathway it is usually followed.

One such alternative is the two-step mechanism

$$NO + NO \underset{k_{-1}}{\overset{k_1}{\rightleftharpoons}} N_2O_2 \qquad \text{(fast equilibrium)}$$

$$N_2O_2 + O_2 \xrightarrow{k_2} 2\,NO_2 \qquad \text{(slow)}$$

Because the slow step determines the overall rate, we can write

$$\text{rate} = k_2[N_2O_2][O_2]$$

The concentration of a reactive intermediate such as $N_2O_2$ cannot be varied at will, however. Because the $N_2O_2$ reacts only slowly with $O_2$, the reverse reaction (to 2 NO) is possible and must be taken into account. In fact, it is reasonable to assume that all of the elementary reactions that occur *before* the rate-determining step are in equilibrium, with the forward and reverse reactions occurring at the same rate. In this case, we have

$$\frac{[N_2O_2]}{[NO]^2} = \frac{k_1}{k_{-1}} = K_1$$

$$[N_2O_2] = K_1[NO]^2$$

$$\text{rate} = k_2 K_1 [NO]^2[O_2]$$

This result is consistent with the observed reaction order, with $k_2 K_1 = k_{obs}$.

........................................................................................................................

## Example 11-7

In basic aqueous solution the reaction

$$I^- + OCl^- \longrightarrow Cl^- + OI^-$$

follows a rate law that is consistent with the following mechanism:

$$OCl^-(aq) + H_2O(\ell) \underset{k_{-1}}{\overset{k_1}{\rightleftharpoons}} HOCl(aq) + OH^-(aq) \qquad \text{(fast equilibrium)}$$

$$I^-(aq) + HOCl(aq) \xrightarrow{k_2} HOI(aq) + Cl^-(aq) \qquad \text{(slow)}$$

$$OH^-(aq) + HOI(aq) \xrightarrow{k_3} H_2O(\ell) + OI^-(aq) \qquad \text{(fast)}$$

What rate law is predicted by this mechanism?

## Solution

The rate is determined by the slowest elementary step, the second one:

$$\text{rate} = k_2[I^-][HOCl]$$

However, the HOCl is in equilibrium with $OCl^-$ and $OH^-$ due to the first step:

$$\frac{[\text{HOCl}][\text{OH}^-]}{[\text{OCl}^-]} = K_1 = \frac{k_1}{k_{-1}}$$

Solving this for [HOCl] and inserting it into the previous expression gives the prediction

$$\text{rate} = k_2 K_1 \frac{[\text{I}^-][\text{OCl}^-]}{[\text{OH}^-]}$$

which is, in fact, the experimentally observed rate law.

*Related Problems:* 25, 26, 27, 28, 29, 30

The rate law of the foregoing example showed an inverse dependence on the concentration of $\text{OH}^-$ ion. Such a form is often a clue that a rapid equilibrium occurs in the first steps of a reaction, preceding the rate-determining step. Fractional orders of reaction provide a similar clue, as in the reaction of $H_2$ with $Br_2$ to form HBr,

$$H_2 + Br_2 \longrightarrow 2\,HBr$$

for which the initial reaction rate (before very much HBr builds up) is

$$\text{rate} = k_{\text{obs}}[H_2][Br_2]^{1/2}$$

How can such a fractional power appear? One reaction mechanism that predicts this rate law is

$$Br_2 + M \underset{k_{-1}}{\overset{k_1}{\rightleftharpoons}} Br + Br + M \qquad \text{(fast equilibrium)}$$

$$Br + H_2 \overset{k_2}{\longrightarrow} HBr + H \qquad \text{(slow)}$$

$$H + Br_2 \overset{k_3}{\longrightarrow} HBr + Br \qquad \text{(fast)}$$

Here M stands for a second molecule that does not react but that supplies the energy to break up the bromine molecules. For such a mechanism the reaction rate is determined by the slow step:

$$\text{rate} = k_2[Br][H_2]$$

However, [Br] is fixed by the establishment of equilibrium in the first reaction,

$$\frac{[Br]^2}{[Br_2]} = K_1 = \frac{k_1}{k_{-1}}$$

so that

$$[Br] = K_1^{1/2}[Br_2]^{1/2}$$

The rate expression predicted by this mechanism is thus

$$\text{rate} = k_2 K_1^{1/2}[H_2][Br_2]^{1/2}$$

This is in accord with the observed fractional power in the rate law. On the other hand, the simple bimolecular mechanism

$$H_2 + Br_2 \xrightarrow{k_1} 2\,HBr \qquad \text{(slow)}$$

predicts a rate law:

$$\text{rate} = k_1[H_2][Br_2]$$

This disagrees with the observed rate law, and so it can be ruled out as the major contributor to the measured rate.

It should be clear from our discussion that deducing a rate law from a proposed mechanism is relatively straightforward, but doing the reverse is much harder. In fact, several competing mechanisms often give rise to the same rate law, and only some independent type of measurement can help one to choose between them. A reaction mechanism can never be proved from an experimental rate law; it can only be disproved if it is inconsistent with the experimental behavior.

A classic example is the reaction

$$H_2 + I_2 \longrightarrow 2\,HI$$

for which the observed rate law is

$$\text{rate} = k_{obs}[H_2][I_2]$$

(Contrast this with the rate law already given for the analogous reaction of $H_2$ with $Br_2$.) This is one of the earliest and most extensively studied reactions in chemical kinetics, and until 1967 it was widely believed to occur as a one-step elementary reaction. At that time J. H. Sullivan investigated the effect of illuminating the reacting sample with light, which splits some of the $I_2$ molecules into iodine atoms. If the mechanism we proposed is correct, the effect of the light on the reaction should be small because it leads only to a small decrease in the $I_2$ concentration.

Instead, Sullivan observed a dramatic *increase* in the rate of reaction under illumination, which could be explained only by the participation of iodine *atoms* in the reaction mechanism. One such mechanism is

$$I_2 + M \underset{k_{-1}}{\overset{k_1}{\rightleftharpoons}} I + I + M \qquad \text{(fast equilibrium)}$$

$$H_2 + I + I \xrightarrow{k_2} 2\,HI \qquad \text{(slow)}$$

for which the rate law is

$$\text{rate} = k_2[H_2][I]^2 = k_2 K_1[H_2][I_2]$$

This mechanism gives the same rate law that is observed experimentally, but it is now consistent with the effect of light on the reaction. The other reaction mechanism also appears to contribute significantly to the overall rate.

This example illustrates the hazards of trying to determine reaction mechanisms from rate laws: several mechanisms can fit any given empirical rate law, and it is always possible that a new piece of information that suggests a different mechanism will be found. The problem is that, under ordinary conditions, reaction intermediates cannot be isolated and studied like the reactants and products. This state of affairs is changing, however, with the development of experimental techniques that allow the direct study of the transient intermediates that form in small concentration during the course of a chemical reaction.

## The Steady-State Approximation

In some reaction mechanisms there is no single step that is much slower than the others, and the methods discussed so far cannot be used to predict the rate law. In this case, the **steady-state approximation** is helpful, in which it is assumed that the concentrations of reactive intermediates remain nearly constant through most of the reaction.

To illustrate this approximation, examine the mechanism proposed by F. A. Lindemann for the dissociation of molecules in the gas phase. A molecule such as $N_2O_5$ undergoes collisions with neighboring molecules M, where M can stand for another $N_2O_5$ molecule or for an inert gas such as argon if the latter is present. Through such collisions the $N_2O_5$ molecule can become excited (or activated) to a state indicated by $N_2O_5^*$:

$$N_2O_5 + M \underset{k_{-1}}{\overset{k_1}{\rightleftharpoons}} N_2O_5^* + M$$

The reverse process, with rate constant $k_{-1}$, is also indicated because the activated molecule can be deactivated by collisions with other molecules. The second step is the unimolecular decomposition of $N_2O_5^*$:

$$N_2O_5^* \xrightarrow{k_2} NO_3 + NO_2$$

Subsequent reaction steps to form $O_2$ and $NO_2$ from $NO_3$ occur rapidly and do not affect the measured rate:

$$NO_3 + NO_2 \xrightarrow{k_3} NO + NO_2 + O_2 \qquad \text{(fast)}$$

$$NO_3 + NO \xrightarrow{k_4} 2 NO_2 \qquad \text{(fast)}$$

The $N_2O_5^*$ is a reactive intermediate; it is produced at a rate $k_1[N_2O_5][M]$ from collisions of $N_2O_5$ molecules with other molecules and is lost at a rate $k_{-1}[N_2O_5^*][M]$ due to deactivation and at a rate $k_2[N_2O_5^*]$ due to dissociation. The net rate of change of $[N_2O_5^*]$ is then

$$\frac{d[N_2O_5^*]}{dt} = k_1[N_2O_5][M] - k_{-1}[N_2O_5^*][M] - k_2[N_2O_5^*]$$

At the beginning of the reaction, $[N_2O_5^*]$ is 0, but this concentration builds up after a short time to a small value. The steady-state approximation consists of the assumption that after this short time the rates of production and loss of $N_2O_5^*$ become equal, and

$$\frac{d[N_2O_5^*]}{dt} = 0$$

The steady-state concentration of $[N_2O_5^*]$ persists practically unchanged throughout most of the course of the reaction.

Setting the net rate of change of the $N_2O_5^*$ concentration to 0 gives

$$\frac{d[N_2O_5^*]}{dt} = 0 = k_1[N_2O_5][M] - k_{-1}[N_2O_5^*][M] - k_2[N_2O_5^*]$$

Solving for $[N_2O_5^*]$ gives

$$[N_2O_5^*]\{k_2 + k_{-1}[M]\} = k_1[N_2O_5][M]$$

$$[N_2O_5^*] = \frac{k_1[N_2O_5][M]}{k_2 + k_{-1}[M]}$$

The rate of the overall reaction $N_2O_5 \rightarrow 2\,NO_2 + \frac{1}{2}\,O_2$ is

$$\text{rate} = \frac{1}{2}\frac{d[NO_2]}{dt} = k_2[N_2O_5^*] = \frac{k_1k_2[N_2O_5][M]}{k_2 + k_{-1}[M]}$$

This expression has two limiting cases:

1. *Low pressure*    When [M] is small enough, $k_2 \gg k_{-1}[M]$ and we can use the approximation

$$\text{rate} = k_1[N_2O_5][M] \qquad \text{(second order)}$$

This same result would be found by assuming the first step to be rate-determining.

2. *High pressure*    When [M] is large enough, $k_{-1}[M] \gg k_2$ and we can use the approximation

$$\text{rate} = \frac{k_1}{k_{-1}}k_2[N_2O_5] \qquad \text{(first order)}$$

This same result would be found by assuming the second step to be rate-determining.

The steady-state approximation is a more general approach than those considered earlier, and can be used when no single step is rate-determining.

## Chain Reactions

A **chain reaction** is one that proceeds through a series of elementary steps, some of which are repeated many times. Such chain reactions consist of three stages: (1) **initiation,** in which two or more reactive intermediates are generated; (2) **propagation,** in which products are formed but reactive intermediates are continuously regenerated; and (3) **termination,** in which two intermediates combine to give a stable product.

An example of a chain reaction is the reaction of methane with fluorine to give $CH_3F$ and $HF$:

$$CH_4(g) + F_2(g) \longrightarrow CH_3F(g) + HF(g)$$

Although in principle this reaction could occur through a one-step bimolecular process, that route turns out to be too slow to contribute significantly under normal reaction conditions. Instead, the mechanism involves a chain reaction of the following type:

$$
\begin{aligned}
CH_4 + F_2 &\longrightarrow CH_3 + HF + F && \text{(initiation)} \\
CH_3 + F_2 &\longrightarrow CH_3F + F && \text{(propagation)} \\
CH_4 + F &\longrightarrow CH_3 + HF && \text{(propagation)} \\
CH_3 + F + M &\longrightarrow CH_3F + M && \text{(termination)}
\end{aligned}
$$

In the initiation step, two reactive intermediates ($CH_3$ and F) are produced. During the propagation steps, these intermediates are not used up while reactants ($CH_4$ and $F_2$) are being converted to products ($CH_3F$ and HF). The propagation steps can be repeated again and again, until eventually two reactive intermediates come together in a termination step. As we shall see in Chapter 23, chain reactions are important in building up long-chain molecules called polymers.

The chain reaction just considered proceeds at a constant rate, because each propagation step both uses up and produces a reactive intermediate. The concentrations of the reactive intermediates remain approximately constant and are determined by the rates of chain initiation and termination. However, another type of chain reaction is possible in which the number of reactive intermediates increases during one or more propagation steps. This is called a **branching chain reaction.** An example is the reaction of oxygen with hydrogen. The mechanism is complex and can be initiated in various ways, leading to the formation of several reactive intermediates such as O, H, and OH. Some propagation steps are of the type already seen for $CH_4$ and $F_2$, such as

$$OH + H_2 \longrightarrow H_2O + H$$

in which one reactive intermediate (OH) is used up and one (H) is produced. Other propagation steps are branching, however:

$$H + O_2 \longrightarrow OH + O$$

$$O + H_2 \longrightarrow OH + H$$

In these steps, each reactive intermediate that is used up causes the generation of two others. This leads to a rapid growth in the number of reactive species, speeding the rate further and possibly causing an explosion. Branching chain reactions are critical in the fission of uranium (Chapter 12).

## 11-5 EFFECT OF TEMPERATURE ON REACTION RATES

The first four sections of this chapter discussed the experimental determination of rate laws and their relation to assumed mechanisms for chemical reactions. Little was said, however, about the actual magnitudes of rate constants (either for elementary reactions or for overall rates of multistep reactions), nor was the effect of temperature on reaction rates discussed. To consider these matters, it is necessary to establish the connection between collision rates and the rates of chemical reactions. Only gases, for which the kinetic theory of Chapter 3 is applicable, are considered here.

### Gas-Phase Reaction Rate Constants

In Section 3-6, the kinetic theory of gases was used to estimate the frequency of collisions in a gas between a given molecule and other molecules. In Example 3-11, this frequency was calculated to be $4.1 \times 10^9 \, s^{-1}$ under room conditions for a typical small molecule such as oxygen. If every collision led to reaction, the reaction would be practically complete in a period of about $10^{-9}$ s. Some reactions do proceed at almost this rate. An example is the bimolecular reaction between two $CH_3$ radicals to give ethane, $C_2H_6$,

$$2\,CH_3 \longrightarrow C_2H_6$$

for which the observed rate constant is $1 \times 10^{10}\,L\,mol^{-1}\,s^{-1}$. If the initial pressure of $CH_3$ were near 1 atm, then at 25°C the concentration initially would be about 0.04 M. According to the second-order integrated rate law from Section 11-2, after a period of $10^{-9}$ s the concentration would have dropped to 0.02 M. Much more commonly, however, reactions proceed at rates that are far lower—by factors of $10^{12}$ or more. The naive idea that "to collide is to react" clearly must be modified if such rates are to be understood.

A clue can be found in the observed temperature dependence of reaction rate constants. The rates of many reactions increase extremely rapidly with increases in temperature; typically, a 10°C rise in temperature may double the rate. In 1889, Svante Arrhenius suggested that rate constants vary exponentially with inverse temperature,

$$k = Ae^{-E_a/RT} \tag{11-5}$$

where $E_a$ is a constant with dimensions of energy and $A$ is a constant with the same dimensions as $k$. Taking the natural logarithm of this equation gives

$$\ln k = \ln A - \frac{E_a}{RT} \tag{11-6}$$

so that a plot of $\ln k$ against $1/T$ should be a straight line with slope $-E_a/R$ and intercept $\ln A$. Many rate constants do show just this kind of temperature dependence (Fig. 11-9).

.....................................................................................................................

## Example 11-8

The decomposition of hydroxylamine ($NH_2OH$) in the presence of oxygen follows the rate law

$$-\frac{d[NH_2OH]}{dt} = k_{obs}[NH_2OH][O_2]$$

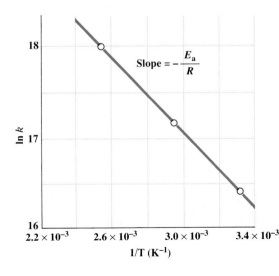

**Figure 11-9**

An Arrhenius plot of $\ln k$ against $1/T$ for the reaction of benzene vapor with oxygen atoms. An extrapolation to $1/T = 0$ gives the constant $\ln A$ from the intercept of this line.

where $k_{obs}$ is $0.237 \times 10^{-4}$ L mol$^{-1}$ s$^{-1}$ at 0°C and $2.64 \times 10^{-4}$ L mol$^{-1}$ s$^{-1}$ at 25°C. Calculate $E_a$ and the factor $A$ for this reaction.

### Solution

Let us write the Arrhenius equation at two different temperatures $T_1$ and $T_2$:

$$\ln k_1 = \ln A - \frac{E_a}{RT_1} \qquad \text{and} \qquad \ln k_2 = \ln A - \frac{E_a}{RT_2}$$

If the first equation is subtracted from the second, the term $\ln A$ cancels out, leaving

$$\ln k_2 - \ln k_1 = \ln \frac{k_2}{k_1} = -\frac{E_a}{R}\left(\frac{1}{T_2} - \frac{1}{T_1}\right)$$

which can be solved for $E_a$. In the present case, $T_1 = 273$ K and $T_2 = 298$ K, and so

$$\ln \frac{2.64 \times 10^{-4}}{0.237 \times 10^{-4}} = \frac{-E_a}{8.315 \text{ J K}^{-1} \text{ mol}^{-1}}\left(\frac{1}{298 \text{ K}} - \frac{1}{273 \text{ K}}\right)$$

$$2.410 = \frac{E_a}{8.315 \text{ J K}^{-1} \text{ mol}^{-1}}(3.07 \times 10^{-4} \text{ K}^{-1})$$

$$E_a = 6.52 \times 10^4 \text{ J mol}^{-1} = 65.2 \text{ kJ mol}^{-1}$$

Now that $E_a$ is known, the constant $A$ can be calculated by using data at either temperature. At 273 K,

$$\ln A = \ln k_1 + \frac{E_a}{RT}$$

$$= \ln(0.237 \times 10^{-4}) + \frac{6.52 \times 10^4 \text{ J mol}^{-1}}{8.315 \text{ J K}^{-1} \text{ mol}^{-1}(273 \text{ K})}$$

$$= -10.65 + 28.73 = 18.08$$

$$A = e^{18.08} = 7.1 \times 10^7 \text{ L mol}^{-1} \text{ s}^{-1}$$

A more accurate determination of $E_a$ and $A$ would use measurements at a series of temperatures rather than at only two, with a fit to a plot such as that in Figure 11-9.

*Related Problems:* 35, 36

Arrhenius believed that for molecules to react upon collision they must become "activated," and the parameter $E_a$ became known as the **activation energy.** His ideas were refined by scientists who followed, and in 1915 A. Marcelin pointed out that, although molecules make many collisions, not all collisions are reactive. Only those collisions for which the collision energy (i.e., the relative translational kinetic energy of the colliding molecules) exceeds some critical energy result in reaction. Thus, Marcelin gave a dynamic interpretation to the activation energy inferred from reaction rates.

The strong temperature dependence of rate constants, described by the Arrhenius law, can be related to the Maxwell-Boltzmann distribution of molecular energies (Fig. 11-10). If $E_a$ is the critical relative collision energy that a pair of molecules must have for a reaction to occur, only a small fraction of the molecules will possess this much energy (or more) when the temperature is sufficiently low. This fraction corresponds to the area under the Maxwell-Boltzmann distribution curve between $E_a$ and $\infty$. As the temperature is increased, the distribution function spreads out toward higher energies. The fraction of molecules that exceed the critical energy $E_a$ increases exponentially as $\exp(-E_a/RT)$, as Arrhenius's law and experiment require. The reaction rate is then proportional to $\exp(-E_a/RT)$, and both the strong temperature dependence and the order of magnitude of the experimental rate constants can be understood.

**Figure 11-10**

The Maxwell–Boltzmann distribution of molecular kinetic energies, showing the effect of temperature on the fraction of molecules having large enough kinetic energies to react. This figure shows only translational energy; internal vibrational and rotational energy also promote reactions.

## The Reaction Coordinate and the Activated Complex

Why should there be a critical collision energy $E_a$ for the reaction between two molecules? To understand this, let us consider the physical analogy of marbles rolling on a hilly surface. As a marble rolls up a hill, its potential energy increases and its kinetic energy decreases (it gets higher and slows down). If it can reach the top of the hill, it will fall down the other side, with its kinetic energy increasing and its potential energy decreasing. Not every marble will make it over the hill, however. If its initial speed (and therefore kinetic energy) is too small, a marble will roll part of the way up and then fall back. Only those marbles with initial kinetic energy higher than a critical threshold will pass over the hill.

We can translate this physical model into a description of molecular collisions and reactions. As two reactant molecules, atoms, or ions approach each other along a **reaction path,** their potential energy increases as the bonds within them distort. At some maximum potential energy they are associated in an unstable entity called an **activated complex** or **transition state.** The activated complex is the cross-over stage where the smooth ascent in potential energy as the reactants come together becomes a smooth descent as the product molecules separate. As in the case of the marbles, not all pairs of reacting bodies react. Only those pairs with sufficient kinetic energy can stretch bonds and rearrange atoms sufficiently to reach the transition state that separates reactants from products. If the barrier is too high, almost all colliding pairs of reactant molecules separate from each other without reacting. The height of the barrier is close to the measured activation energy for the reaction.

Figure 11-11 is a graph of potential energy versus position along the reaction path for the reaction

$$NO_2(g) + CO(g) \longrightarrow NO(g) + CO_2(g)$$

Two activation energies are shown in Figure 11-11: $E_{a,f}$ is the activation energy for the forward reaction, and $E_{a,r}$ is that for the reverse reaction, in which NO takes an oxygen atom from $CO_2$ to form $NO_2$ and CO. The difference between the two is $\Delta E$, the change in internal energy of the chemical reaction:

$$\Delta E = E_{a,f} - E_{a,r}$$

Although $\Delta E$ is a thermodynamic quantity, obtainable from calorimetric measurements, $E_{a,f}$ and $E_{a,r}$ must be found from the temperature dependence of the rate

**Figure 11-11**

The energy profile along the reaction coordinate for the reaction $NO_2 + CO \rightarrow NO + CO_2$. This direct reaction dominates the kinetics at high temperatures (above about 500 K).

constants for the forward and reverse reactions. In this reaction the forward and reverse activation energies are 132 and 358 kJ mol$^{-1}$, respectively, and $\Delta E$ from thermodynamics is $-226$ kJ mol$^{-1}$.

The activation energy for an elementary reaction is always positive (although in some cases it can be quite small) because there is always some energy barrier to surmount. Rates of elementary reactions therefore increase with increasing temperature. This is not necessarily true for rates of overall reactions consisting of more than one elementary reaction. These sometimes have "negative activation energies" so that the overall reaction rate slows at higher temperature. How can this be? Let us examine a specific example: the reaction of NO with oxygen,

$$2\,NO(g) + O_2(g) \longrightarrow 2\,NO_2(g)$$

for which the observed rate law is

$$\text{rate} = k_{obs}[NO]^2[O_2]$$

where $k_{obs}$ *decreases* with increasing temperature. In Section 11-4 we saw that this rate expression could be accounted for with a two-step mechanism. The first step is a rapid equilibrium (with equilibrium constant $K_1$) of two NO molecules with their dimer, $N_2O_2$. The second step is the slow reaction (rate constant $k_2$) of $N_2O_2$ with $O_2$ to form products. The overall rate constant is therefore the product of $k_2$ and $K_1$. Although $k_2$ is the rate constant for an elementary reaction (and thus increases with increasing temperature), $K_1$ is an *equilibrium* constant. Provided that the corresponding reaction is sufficiently exothermic (as it is in this case), $K_1$ will decrease so rapidly with increasing temperature that the product $k_2 K_1$ will decrease as well. This accounts for the observation of negative activation energies in some overall chemical reactions.

## 11-6  A Deeper Look . . . Reaction Dynamics

In this section, we take a closer look at the dynamics of gas-phase reactions, and then compare them with the corresponding dynamics in solution. The starting point is a result from Section 3-6, that the rate of collisions of a single molecule with other molecules of the same type is

$$Z_1 = \sqrt{2}\,\pi\,d^2\,\bar{u}\frac{N}{V} = 4\,d^2\sqrt{\frac{\pi RT}{\mathcal{M}}}\,\frac{N}{V}$$

Here $d$ is the molecular diameter, $\bar{u}$ the average speed, $\mathcal{M}$ the molar mass, and $N/V$ the number density of molecules in the gas. If there are $N$ molecules in the volume, the *total* number of collisions per unit time is $\frac{1}{2}N \times Z_1$. (The factor $\frac{1}{2}$ comes from the fact that the collision of two A molecules counts as only one collision—not as one collision for the first A molecule with the second, plus one for the second with the first.) The rate of collisions *per unit volume* is then this result divided by $V$, or

rate of collisions per unit volume = $Z_{AA}$

$$= 2d^2\sqrt{\frac{\pi RT}{\mathcal{M}}}\left(\frac{N}{V}\right)^2$$

The next step is to relate $Z_{AA}$ to the second-order rate constant for the reaction

$$A + A \longrightarrow \text{products}$$

If the activation energy for this reaction is $E_a$, then only a fraction $\exp(-E_a/RT)$ of these collisions will have sufficient energy to overcome the barrier leading to products. Each such *effective* collision of a pair of A molecules leads to a decrease in the number of A molecules in the reaction mixture by two, and so the rate of change of the number of A molecules per unit volume is

$$\frac{d(N/V)}{dt} = -2Z_{AA}e^{-E_a/RT}$$

$$= -2 \times 2\,d^2\sqrt{\frac{\pi RT}{\mathcal{M}}}e^{-E_a/RT}\left(\frac{N}{V}\right)^2$$

Rate constants involve the number of *moles* of A per unit volume, [A], not the number of molecules. The two are related by

$$N_0[A] = \frac{N}{V}$$

where $N_0$ is Avogadro's number. Substituting this equation and bringing a factor of $-\frac{1}{2}$ back to the left side (as in our usual definition of rate), we find

$$\text{rate} = -\frac{1}{2}\frac{d[A]}{dt} = 2\,d^2\,N_0\sqrt{\frac{\pi RT}{\mathcal{M}}}e^{-E_a/RT}[A]^2$$

The predicted rate constant can be identified as

$$k = 2\,d^2\,N_0\sqrt{\frac{\pi RT}{\mathcal{M}}}e^{-E_a/RT} \qquad (11\text{-}7)$$

How well does this simple theory agree with experiment? By fitting data on gas-phase elementary reaction rates to the Arrhenius form, the activation energy and the factor A can be obtained and the latter compared with the theory, once the molecular diameter is estimated. For the elementary reaction

$$2\,NOCl(g) \longrightarrow 2\,NO(g) + Cl_2(g)$$

the measured rate constant is 0.16 times the calculated rate constant. What this indicates is that not all collisions lead to reaction, even if the molecules have great enough relative kinetic energy. This result makes sense, because the relative orientations of the colliding molecules certainly should play a role in determining whether a particular collision results in a reaction. It seems obvious that, for a $Cl_2$ molecule to split off, the two NOCl molecules must approach each other in such a way that the chlorine atoms are close together (see Fig. 11-12). The calculated collision frequency must therefore be multiplied by a **steric factor** $P$ (0.16 in this case) to account for the fact that only a fraction of the collisions occur with the proper orientation to lead to reaction.

The collision-theory calculation of the rate constant can be extended to bimolecular reactions of two species, A and B. Comparison of observed and predicted rate constants leads to the values of $P$ shown in Table 11-1. As the colliding molecules become larger and more complex, $P$ becomes smaller, because a smaller fraction of collisions is effective in causing reaction. The steric factor is an empirical correction that can be predicted only in especially simple cases.

### Molecular Beams

The kinetics experiments considered so far have involved changing reactant concentrations and temperatures, which are macroscopic properties of a reacting mixture. An alternative

### Table 11-1  Steric Factors for Gas-Phase Reactions

| Reaction | Steric Factor $P$ |
|---|---|
| $2\,NOCl \rightarrow 2\,NO + Cl_2$ | 0.16 |
| $2\,NO_2 \rightarrow 2\,NO + O_2$ | $5.0 \times 10^{-2}$ |
| $2\,ClO \rightarrow Cl_2 + O_2$ | $2.5 \times 10^{-3}$ |
| $H_2 + C_2H_4 \rightarrow C_2H_6$ | $1.7 \times 10^{-6}$ |

Adapted from P. W. Atkins, *Physical Chemistry* (W. H. Freeman, New York, 1994), p. C30.

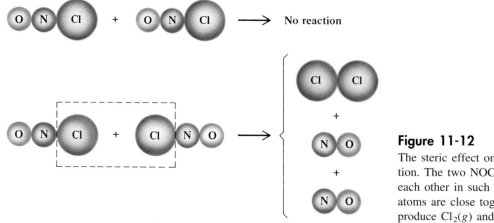

**Figure 11-12**

The steric effect on the probability of a reaction. The two NOCl molecules must approach each other in such a way that the two chlorine atoms are close together, if the encounter is to produce $Cl_2(g)$ and $NO(g)$.

way to study kinetics is to employ the crossed molecular beam. In this device, two beams of molecules are crossed in a chamber under high vacuum, and the reaction products are determined (Fig. 11-13).

Such molecular beams allow a high degree of selection of reactant molecules. For example, a velocity selector (Fig. 3-13) can be used to keep only those molecules that have velocities within a small range. This permits much finer tuning of reaction energies than can be achieved simply by changing the temperature. If electric and magnetic fields or laser light is used, molecules can be further selected according to how fast they rotate and how much they vibrate. The products can also be analyzed for velocity, and their angular distribution relative to the directions of the initial beams can be determined. This al-

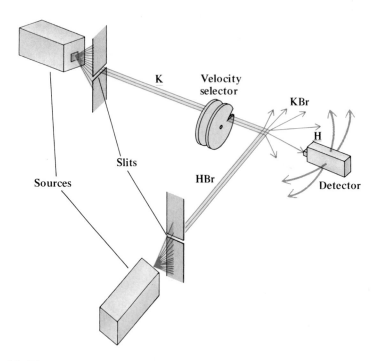

**Figure 11-13**

Diagram of a crossed molecular beam experiment. The reaction being studied is $K + HBr \rightarrow KBr + H$.

lows a much more detailed examination of the way in which molecules collide and react. For example, if the activated complex rotates many times before it finally breaks up, the angular distribution of products should be uniform, but if it breaks up before it has a chance to rotate, the angular distribution should be very nonuniform and should depend on the directions of the original beams. This type of measurement therefore gives information on the lifetime of an activated complex.

Molecular beams are limited to reactions that are carried out in the gas phase, for which well-defined beams of reactant molecules can be prepared. They cannot be used, for example, to study kinetics in liquid solvents. However, the detailed information that they give for certain reactions allows more precise testing of models and theories for the ways in which chemical reactions occur in the gas phase.

### Reaction Kinetics in Liquids

In a gas, atoms or molecules move in straight lines between occasional collisions (Fig. 3-18). In a liquid, the concept of collision has no meaning because molecules interact continuously with not one but many neighbors. Their trajectories can nonetheless be followed; these consist of rattling motions in a temporary "cage" formed by neighbors, superimposed upon a random and erratic diffusive displacement. In a time interval $t$, molecules undergo various net displacements $\Delta r$, but the average of the *square* of their displacements is found experimentally to be proportional to the time interval:

$$\overline{(\Delta r)^2} = 6Dt$$

This is the same law that was given for gases in Section 3-6, and it applies also to solids. The magnitude of the diffusion constant $D$ varies widely from one phase to another, however. Typical values of $D$ are on the order of $10^{-9}$ m$^2$ s$^{-1}$ for liquids (four orders of magnitude smaller than for a typical gas at atmospheric pressure), so that in a time interval of 1 s the mean-square displacement through diffusion is of order $10^{-8}$ m$^2$ and the root-mean-square displacement is of order $10^{-4}$ m (0.1 mm). For larger particles the diffusion coefficient can be much smaller and the resulting displacement smaller, as well.

The motion of molecules in a liquid has a significant effect on the kinetics of chemical reaction in solution. Molecules must diffuse together before they can react, and so their diffusion constants affect the rate of reaction. If the intrinsic reaction rate of two molecules that come into contact is fast enough (that is, if almost every encounter leads to reaction), then diffusion is the rate-limiting step. Such **diffusion-controlled reactions** have a maximum bimolecular rate constant on the order of $10^{10}$ L mol$^{-1}$ s$^{-1}$ in aqueous solution for the reaction of two neutral species. If the two species have opposite charges, the reaction rate can be even higher. One of the fastest known reactions in aqueous solution is the neutralization of hydronium ion ($H_3O^+$) by hydroxide ion ($OH^-$):

$$H_3O^+(aq) + OH^-(aq) \longrightarrow 2 H_2O(\ell)$$

for which the diffusion-controlled rate constant is greater than $10^{11}$ L mol$^{-1}$ s$^{-1}$.

### 11-7 KINETICS OF CATALYSIS

A **catalyst** is a substance that takes part in a chemical reaction and speeds it up, but itself undergoes no permanent chemical change. Catalysts therefore do not appear in the overall balanced chemical equation, but their presence very much affects the rate law, modifying and speeding existing pathways or, more commonly, providing completely new pathways by which a reaction can occur (Fig. 11-14). Catalysts exert significant effects on reaction rates even when they are present in very small amounts. In industrial chemistry, great effort is devoted to finding catalysts that will accelerate particular desired reactions without increasing the generation of undesired products.

Catalysis can be classified into two types: homogeneous and heterogeneous. In **homogeneous catalysis,** the catalyst is present in the same phase as the reactants, as when a gas-phase catalyst speeds up a gas-phase reaction, or a species dissolved in solution speeds up a reaction in solution. An example of homogeneous catalysis is the effect of chlorofluorocarbons and oxides of nitrogen on the depletion of ozone in the stratosphere, which will be examined in Section 15-4. A second example is the catalysis of the oxidation–reduction reaction

$$Tl^+(aq) + 2 Ce^{4+}(aq) \longrightarrow Tl^{3+}(aq) + 2 Ce^{3+}(aq)$$

**Figure 11-14**

The decomposition of hydrogen peroxide, $H_2O_2$, to water and oxygen is catalyzed by solid $MnO_2$. Here the water evolves as steam because of the heat given off in the reaction.

by silver ions in solution. The direct reaction of $Tl^+$ with a single $Ce^{4+}$ ion to give $Tl^{2+}$ as an intermediate is slow. The reaction can be speeded by adding $Ag^+$ ion, which takes part in a reaction mechanism of the form

$$Ag^+ + Ce^{4+} \underset{k_{-1}}{\overset{k_1}{\rightleftharpoons}} Ag^{2+} + Ce^{3+} \qquad \text{(fast)}$$

$$Tl^+ + Ag^{2+} \xrightarrow{k_2} Tl^{2+} + Ag^+ \qquad \text{(slow)}$$

$$Tl^{2+} + Ce^{4+} \xrightarrow{k_3} Tl^{3+} + Ce^{3+} \qquad \text{(fast)}$$

The $Ag^+$ ions are not permanently transformed by this reaction, because those used up in the first step are regenerated in the second; they play the role of catalyst in significantly speeding the rate of the overall reaction.

In **heterogeneous catalysis,** the catalyst is present as a distinct phase. The most important case is the catalytic action of certain solid surfaces on gas-phase and solution-phase reactions. A critical step in the production of sulfuric acid, for example, involves the use of a solid oxide of vanadium ($V_2O_5$) as catalyst. Many other

**Figure 11-15**

Platinum catalyzes the reaction $H_2$ + $C_2H_4$ by providing a surface that promotes the dissociation of $H_2$ to H atoms, which can then add to the $C_2H_4$ stepwise to give ethane, $C_2H_6$.

solid catalysts are used in industrial processes. One of the best studied of such reactions is the addition of hydrogen to ethylene to form ethane:

$$C_2H_4(g) + H_2(g) \longrightarrow C_2H_6(g)$$

The process occurs extremely slowly in the gas phase, but is catalyzed by a platinum surface (Fig. 11-15).

A familiar type of catalyst is used in the exhaust streams of automobile engines to reduce the emission of pollutants such as unburned hydrocarbons, carbon monoxide, and nitrogen oxides (Fig. 11-16). A **catalytic converter** is designed to simultaneously *oxidize* hydrocarbons and CO

$$CO, C_xH_y, O_2 \xrightarrow{\text{Catalyst}} CO_2, H_2O$$

and *reduce* nitrogen oxides:

$$NO, NO_2 \xrightarrow{\text{Catalyst}} N_2, O_2$$

Clearly, the best catalyst for one process may not be the best for another, so a combination of two catalysts is employed. The noble metals, although expensive, are particularly useful. Typically, platinum and rhodium are deposited on a fine honeycomb mesh of alumina ($Al_2O_3$), giving a large surface area that increases the contact time of the exhaust gas with the catalysts. The platinum serves primarily as an oxidation catalyst and the rhodium as a reduction catalyst. Catalytic converters can be poisoned with certain metals that block their active sites and reduce their effectiveness. Lead is one of the most serious such poisons; as a result, unleaded fuel must be employed in automobiles with catalytic converters.

A catalyst speeds up the rate of a reaction by increasing the Arrhenius factor $A$ or, more often, by lowering the activation energy $E_a$ by providing a new activated complex of lower potential energy (Fig. 11-17). A corollary of this fact is that the same catalyst will speed both the forward *and* reverse reactions, because it lowers both the forward and reverse activation energies equally. It is important to remember that a catalyst has no effect on the thermodynamics of the overall reaction. Because the free energy is a function of state, $\Delta G$ is independent of the path followed and therefore the equilibrium constant is not changed by the presence of a

**Figure 11-16**

A catalytic converter used to reduce automobile pollution, cut open to reveal the platinum, palladium, and rhodium catalysts. In this new design, a steel-alloy heating element raises the temperature to 400°C in seconds, activating the catalysts, and reducing the pollution emitted in the first minutes after the car is started. (Courtesy of Corning, Incorporated)

**Figure 11-17**

The most important way in which catalysts speed reactions is by reducing the activation energy.

catalyst. Reaction products that are not favored thermodynamically will still not form. The role of catalysts is to speed up the rate of production of products that *are* allowed by thermodynamics.

An **inhibitor** plays an opposite role to that of a catalyst. It slows the rate of a reaction, frequently by increasing the activation energy. Inhibitors are also important industrially because they can be used to reduce the rates of undesirable side reactions, allowing desired products to form in greater yield.

### Enzyme Catalysis

Many chemical reactions in living systems are carried out with enzymes acting as catalysts. An **enzyme** is a large protein molecule (typically of molar mass $20,000 \text{ g mol}^{-1}$ or more) which has a structure capable of carrying out a specific reaction or series of reactions. One or more reactant molecules (called **substrates**) bind to an enzyme at its **active sites.** These are regions on the surface of the enzyme whose structures and chemical properties are such that a given substrate will bind to them and its chemical transformation can be carried out (Fig. 11-18). Many enzymes are quite specific in their active sites. The enzyme urease, for example, catalyzes the hydrolysis of urea, $(NH_2)_2CO$,

$$H_3O^+(aq) + (NH_2)_2CO(aq) + H_2O(\ell) \xrightarrow{\text{Urease}} 2\,NH_4^+(aq) + HCO_3^-(aq)$$

but will not bind most other kinds of molecules, even those of similar structure. In some cases, however, a second species *does* bind to the enzyme and can act as an inhibitor, preventing the enzyme from carrying out its usual role as catalyst.

The kinetics of enzyme catalysis is indicated schematically by the reaction mechanism

$$E + S \underset{k_{-1}}{\overset{k_1}{\rightleftharpoons}} ES$$

$$ES \xrightarrow{k_2} E + P$$

Here E stands for the free enzyme, S for the substrate, ES for the complex they form when the substrate binds to the active site, and P for the product of the chemical transformation. The rate of formation of P can be derived for this mechanism by making a steady-state approximation for the concentration of the enzyme-substrate

**Figure 11-18**

The binding of a substrate to an enzyme, and the subsequent reaction of the substrate. As this schematic figure suggests, the size and shape of the active site play a role in determining which substrates bind. Equally important are the strengths of the intermolecular forces between nearby groups on the enzyme and substrate.

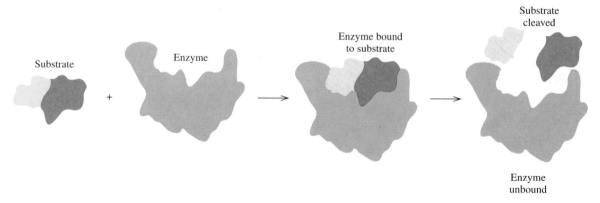

complex, [ES]. As shown in Section 11-4, this approximation involves setting the rate of change of [ES] to 0:

$$\frac{d[\text{ES}]}{dt} = 0 = k_1[\text{E}][\text{S}] - k_{-1}[\text{ES}] - k_2[\text{ES}]$$

This equation could be solved to find [ES] in terms of [E], but that is not quite what we want. Here [E] is the concentration of free enzyme (which does not have substrate bound to it), and [ES] is the concentration of bound enzyme. Their sum is $[\text{E}]_0$, the total amount of enzyme present:

$$[\text{E}]_0 = [\text{E}] + [\text{ES}]$$

It is $[\text{E}]_0$ that is accessible to experiment, so we replace [E] in the steady-state equation above with $[\text{E}]_0 - [\text{ES}]$. This gives

$$\frac{d[\text{ES}]}{dt} = 0 = k_1[\text{E}]_0[\text{S}] - k_1[\text{ES}][\text{S}] - k_{-1}[\text{ES}] - k_2[\text{ES}]$$

which can be solved for [ES] to give

$$[\text{ES}] = \frac{k_1[\text{E}]_0[\text{S}]}{k_1[\text{S}] + (k_{-1} + k_2)}$$

If $K_m$ is defined by

$$K_m = \frac{k_{-1} + k_2}{k_1}$$

then

$$[\text{ES}] = \frac{[\text{E}]_0[\text{S}]}{[\text{S}] + K_m}$$

The rate of formation of product is

$$\frac{d[\text{P}]}{dt} = k_2[\text{ES}] = \frac{k_2[\text{E}]_0[\text{S}]}{[\text{S}] + K_m} \qquad \textbf{(11-8)}$$

This is the **Michaelis-Menten equation,** which describes the kinetics of many enzyme-catalyzed reactions. Figure 11-19 shows a typical experimental curve for the

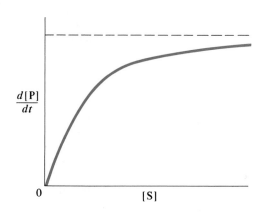

**Figure 11-19**

The dependence of the rate of an enzyme reaction on substrate concentration. The dashed line gives the rate attained for very high [S].

dependence of the rate of such a reaction on substrate concentration [S]. For low substrate concentrations the rate is linear in [S], but at high concentrations it levels off. This behavior is easily understood on physical grounds. When [S] is small, the more substrate is added, the greater is the rate of formation of product. As the amount of substrate becomes large, all the active sites of the enzyme molecules become bound to substrate, and further addition of S does not affect the rate of the reaction. A saturation effect is seen, and the rate levels off.

A revealing form is obtained by taking the inverses of both sides of the Michaelis–Menten equation, to give

$$\frac{1}{d[P]/dt} = \frac{1}{k_2[E]_0} + \frac{K_m}{k_2[E]_0[S]}$$

A plot of the inverse of the rate of reaction against the inverse of the substrate concentration should be a straight line whose slope and intercept allow $k_2[E]_0$ and $K_m$ to be determined.

Crystals of potassium hydrogen sulfite, $KHSO_3$, under polarized light. (Phillip A. Harrington/Fran Heyl Associates)

## PRACTICE EXERCISE

### Sulfite and Sulfate Kinetics

Sulfur dioxide dissolves in water droplets (fog, clouds, and rain) in the atmosphere and reacts according to the equation

$$SO_2(aq) + 2\,H_2O(\ell) \longrightarrow HSO_3^-(aq) + H_3O^+(aq)$$

The $HSO_3^-(aq)$ is then, much more slowly, oxidized by dissolved oxygen that is also in the droplets:

$$2\,HSO_3^-(aq) + O_2(aq) + 2\,H_2O(\ell) \longrightarrow 2\,SO_4^{2-}(aq) + 2\,H_3O^+(aq)$$

Although the second reaction has been studied for many years, only recently was the discovery made that it can proceed by the steps

$$2\,HSO_3^-(aq) + O_2(aq) \longrightarrow S_2O_7^{2-}(aq) + H_2O(\ell) \qquad \text{(fast)}$$

$$S_2O_7^{2-}(aq) + 3\,H_2O(\ell) \longrightarrow 2\,SO_4^{2-}(aq) + 2\,H_3O^+(aq) \qquad \text{(slow)}$$

The previously undetected intermediate, $S_2O_7^{2-}(aq)$, is well known in other reactions. It is the *disulfate* ion.

In an experiment at 25°C, a solution was mixed with the realistic initial concentrations of 0.270 M $HSO_3^-(aq)$ and 0.0135 M $O_2(aq)$. The initial pH was 3.90. The following table tells what happened in the solution, beginning at the moment of mixing.

| Time (s) | [HSO$_3^-$] (M) | [O$_2$] (M) | [S$_2$O$_7^{2-}$] (M) | [HSO$_4^-$] + [SO$_4^{2-}$] (M) |
|---|---|---|---|---|
| 0.000 | 0.270 | 0.0135 | 0.000 | 0.000 |
| 0.010 | 0.243 | 0.000 | $13.5 \times 10^{-3}$ | 0.000 |
| 10.0 | 0.243 | 0.000 | $11.8 \times 10^{-3}$ | $3.40 \times 10^{-3}$ |
| 45.0 | 0.243 | 0.000 | $7.42 \times 10^{-3}$ | $12.2 \times 10^{-3}$ |
| 90.0 | 0.243 | 0.000 | $4.08 \times 10^{-3}$ | $18.8 \times 10^{-3}$ |
| 150.0 | 0.243 | 0.000 | $1.84 \times 10^{-3}$ | $23.3 \times 10^{-3}$ |
| 450.0 | 0.243 | 0.000 | $0.034 \times 10^{-3}$ | $26.9 \times 10^{-3}$ |
| 600.0 | 0.243 | 0.000 | $0.005 \times 10^{-3}$ | $27.0 \times 10^{-3}$ |

(a) Determine the average rate of increase of the total of the concentrations of the sulfate plus hydrogen sulfate ions during the first 10 s of the experiment.

(b) Determine the average rate of disappearance of hydrogen sulfite ion during the first 0.010 s of the experiment.

(c) Explain why the hydrogen sulfite ion stops disappearing after 0.010 s.

(d) Plot the concentration of disulfate ion versus time on graph paper, and use the graph to estimate the instantaneous rate of disappearance of disulfate ion 90.0 s after the reaction starts.

(e) Determine the order with respect to the disulfate ion of the second step of the conversion, and the rate constant of that step.

(f) Determine the half-life of the second step of the conversion process.

(g) At 15°C, the rate constant of the second step of the conversion is only 62% of its value at 25°C. Compute the activation energy of the second step.

(h) The first step of the conversion occurs much faster when $1.0 \times 10^{-6}$ M $Fe^{2+}(aq)$ ion is added (but the rate of the second step is unaffected). What role does $Fe^{2+}$ play?

(i) Write a balanced equation for the overall reaction that gives sulfuric acid from $SO_2$ dissolved in water droplets in the air.

**Answers**

(a) $3.40 \times 10^{-4}$ mol $L^{-1}$ $s^{-1}$

(b) 2.7 mol $L^{-1}$ $s^{-1}$

(c) All of the oxygen is consumed, and so the first step of the process is over.

(d) $5.43 \times 10^{-5}$ mol $L^{-1}$ $s^{-1}$

(e) First order; $k = 0.0133$ $s^{-1}$

(f) 52 s

(g) 34 kJ $mol^{-1}$

(h) $Fe^{2+}$ acts as a catalyst.

(i) $2\,SO_2(aq) + O_2(aq) + 6\,H_2O(\ell) \rightarrow 2\,SO_4^{2-}(aq) + 4\,H_3O^+(aq)$

After studying this chapter and working the problems that follow, you should be able to

**CONCEPTS & SKILLS**

1. Describe experimental methods of measuring average and instantaneous rates (Section 11-1, problems 1–2).

2. Deduce rate laws and reaction orders from experimental measurements of the dependence of reaction rates on concentrations (Section 11-2, problems 5–8).

3. Use the integrated rate laws for first- and second-order reactions to calculate the concentrations remaining after a certain elapsed time (Section 11-2, problems 9–18).

4. Describe the relationship between the equilibrium constant for a reaction and the corresponding forward and reverse rate constants (Section 11-3, problems 23–24).

5. Deduce the rate law from a mechanism characterized by a single rate-determining step (Section 11-4, problems 25–30).

6. Use the steady-state approximation to deduce rate laws in cases where no single step is rate-determining (Section 11-4, problems 31–34).

7. Calculate Arrhenius factors and activation energies from measurements of the temperature dependence of rate constants (Section 11-5, problems 35–40).

8. Discuss the connection between activation energy and the energy distribution of molecules, and relate the forward and reverse activation energies to each other through thermodynamics (Section 11-5, problems 41–42).
9. Outline the quantitative calculation of rate constants, using collision theory of gases (Section 11-6, problems 43–44).
10. Describe several types of catalysts and their effects on chemical reactions (Section 11-7).
11. Relate the rate of an enzyme-catalyzed reaction to the concentrations of substrate and enzyme in the reaction mixture (Section 11-7, problems 45–46).

## PROBLEMS

### Rates of Chemical Reactions

1. Use Figure 11-3 to estimate graphically the instantaneous rate of production of NO at $t = 200$ s.
2. Use Figure 11-3 to estimate graphically the instantaneous rate of production of NO at $t = 100$ s.
3. Give three related expressions for the rate of the reaction

$$N_2(g) + 3\,H_2(g) \longrightarrow 2\,NH_3(g)$$

assuming that the concentrations of any intermediates are constant and that the volume of the reaction vessel does not change.

4. Give four related expressions for the rate of the reaction

$$2\,H_2CO(g) + O_2(g) \longrightarrow 2\,CO(g) + 2\,H_2O(g)$$

assuming that the concentrations of any intermediates are constant and that the volume of the reaction vessel does not change.

### Rate Laws

5. Nitrogen oxide reacts with hydrogen at elevated temperatures according to the following chemical equation:

$$2\,NO(g) + 2\,H_2(g) \longrightarrow N_2(g) + 2\,H_2O(g)$$

It is observed that, when the concentration of $H_2$ is cut in half, the rate of the reaction is also cut in half. When the concentration of NO is multiplied by 10, the rate of the reaction increases by a factor of 100.
(a) Write the rate expression for this reaction, and give the units of the rate constant $k$.
(b) If [NO] were multiplied by 3 and $[H_2]$ by 2, what change in the rate would be observed?

6. In the presence of vanadium oxide, $SO_2(g)$ reacts with an excess of oxygen to give $SO_3(g)$:

$$SO_2(g) + \tfrac{1}{2}O_2(g) \xrightarrow{\;V_2O_5\;} SO_3(g)$$

This reaction is an important step in the manufacture of sulfu-

ric acid. It is observed that tripling the $SO_2$ concentration increases the rate by a factor of 3, but tripling the $SO_3$ concentration *decreases* the rate by a factor of $1.7 \approx \sqrt{3}$. The rate is insensitive to the $O_2$ concentration as long as an excess of oxygen is present.
(a) Write the rate expression for this reaction, and give the units of the rate constant $k$.
(b) If $[SO_2]$ is multiplied by 2 and $[SO_3]$ by 4 but all other conditions are unchanged, what change in the rate will be observed?

7. In a study of the reaction of pyridine ($C_5H_5N$) with methyl iodide ($CH_3I$) in a benzene solution, the following set of initial reaction rates was measured at 25°C for different initial concentrations of the two reactants:

| $[C_5H_5N]$ (mol L$^{-1}$) | $[CH_3I]$ (mol L$^{-1}$) | Rate (mol L$^{-1}$ s$^{-1}$) |
|---|---|---|
| $1.00 \times 10^{-4}$ | $1.00 \times 10^{-4}$ | $7.5 \times 10^{-7}$ |
| $2.00 \times 10^{-4}$ | $2.00 \times 10^{-4}$ | $3.0 \times 10^{-6}$ |
| $2.00 \times 10^{-4}$ | $4.00 \times 10^{-4}$ | $6.0 \times 10^{-6}$ |

(a) Write the rate expression for this reaction.
(b) Calculate the rate constant $k$, and give its units.
(c) Predict the initial reaction rate for a solution in which $[C_5H_5N]$ is $5.0 \times 10^{-5}$ M and $[CH_3I]$ is $2.0 \times 10^{-5}$ M.

8. The rate for the oxidation of iron(II) by cerium(IV)

$$Ce^{4+}(aq) + Fe^{2+}(aq) \longrightarrow Ce^{3+}(aq) + Fe^{3+}(aq)$$

is measured at several different initial concentrations of the two reactants:

| $[Ce^{4+}]$ (mol L$^{-1}$) | $[Fe^{2+}]$ (mol L$^{-1}$) | Rate (mol L$^{-1}$ s$^{-1}$) |
|---|---|---|
| $1.1 \times 10^{-5}$ | $1.8 \times 10^{-5}$ | $2.0 \times 10^{-7}$ |
| $1.1 \times 10^{-5}$ | $2.8 \times 10^{-5}$ | $3.1 \times 10^{-7}$ |
| $3.4 \times 10^{-5}$ | $2.8 \times 10^{-5}$ | $9.5 \times 10^{-7}$ |

(a) Write the rate expression for this reaction.

(b) Calculate the rate constant $k$, and give its units.

(c) Predict the initial reaction rate for a solution in which $[Ce^{4+}]$ is $2.6 \times 10^{-5}$ M and $[Fe^{2+}]$ is $1.3 \times 10^{-5}$ M.

9. The reaction $SO_2Cl_2(g) \rightarrow SO_2(g) + Cl_2(g)$ is first order, with a rate constant of $2.2 \times 10^{-5}$ s$^{-1}$ at 320°C. The partial pressure of $SO_2Cl_2(g)$ in a sealed vessel at 320°C is 1.0 atm. How long will it take for the partial pressure of $SO_2Cl_2(g)$ to fall to 0.50 atm?

10. The reaction $FClO_2(g) \rightarrow FClO(g) + O(g)$ is first order, with a rate constant of $6.76 \times 10^{-4}$ s$^{-1}$ at 322°C.

(a) Calculate the half-life of the reaction at 322°C.

(b) If the initial partial pressure of $FClO_2$ in a container at 322°C is 0.040 atm, how long will it take to fall to 0.010 atm?

11. The decomposition of benzene diazonium chloride

$$C_6H_5N_2Cl \longrightarrow C_6H_5Cl + N_2$$

follows first-order kinetics with a rate constant of $4.3 \times 10^{-5}$ s$^{-1}$ at 20°C. If the initial partial pressure of $C_6H_5N_2Cl$ is 0.0088 atm, calculate its partial pressure after 10.0 hours.

12. At 600 K, the rate constant for the first-order decomposition of nitroethane

$$CH_3CH_2NO_2(g) \longrightarrow C_2H_4(g) + HNO_2(g)$$

is $1.9 \times 10^{-4}$ s$^{-1}$. A sample of $CH_3CH_2NO_2$ is heated to 600 K, at which point its initial partial pressure is measured to be 0.078 atm. Calculate its partial pressure after 3.0 hours.

13. Chloroethane decomposes at elevated temperatures according to the reaction

$$C_2H_5Cl(g) \longrightarrow C_2H_4(g) + HCl(g)$$

This reaction obeys first-order kinetics. After 340 s at 800 K, a measurement shows that the concentration of $C_2H_5Cl$ has decreased from 0.0098 mol L$^{-1}$ to 0.0016 mol L$^{-1}$. Calculate the rate constant $k$ at 800 K.

14. The isomerization reaction

$$CH_3NC \longrightarrow CH_3CN$$

obeys the first-order rate law

$$rate = -k[CH_3NC]$$

in the presence of an excess of argon. Measurements at 500 K reveal that in 520 s the concentration of $CH_3NC$ decreases to 71% of its original value. Calculate the rate constant $k$ of the reaction at 500 K.

15. At 25°C in $CCl_4$ solvent, the reaction

$$I + I \longrightarrow I_2$$

is second order in the concentration of the iodine atoms. The rate constant $k$ has been measured as $8.2 \times 10^9$ L mol$^{-1}$ s$^{-1}$. Suppose the initial concentration of I atoms is $1.00 \times 10^{-4}$ M. Calculate their concentration after $2.0 \times 10^{-6}$ s.

16. $HO_2$ is a highly reactive chemical species that plays a role in atmospheric chemistry. The rate of the gas-phase reaction

$$HO_2(g) + HO_2(g) \longrightarrow H_2O_2(g) + O_2(g)$$

is second order in $[HO_2]$, with a rate constant at 25°C of $1.4 \times 10^9$ L mol$^{-1}$ s$^{-1}$. Suppose some $HO_2$ with an initial concentration of $2.0 \times 10^{-8}$ M could be confined at 25°C. Calculate the concentration that would remain after 1.0 s, assuming no other reactions take place.

17. The rate for the reaction

$$OH^-(aq) + NH_4^+(aq) \longrightarrow H_2O(\ell) + NH_3(aq)$$

is first order in both $OH^-$ and $NH_4^+$ concentrations, and the rate constant $k$ at 20°C is $3.4 \times 10^{10}$ L mol$^{-1}$ s$^{-1}$. Suppose 1.00 L of a 0.0010 M NaOH solution is rapidly mixed with the same volume of 0.0010 M $NH_4Cl$ solution. Calculate the time (in seconds) required for the $OH^-$ concentration to decrease to a value of $1.0 \times 10^{-5}$ M.

18. The rate for the reaction

$$OH^-(aq) + HCN(aq) \longrightarrow H_2O(\ell) + CN^-(aq)$$

is first order in both $OH^-$ and HCN concentrations, and the rate constant $k$ at 25°C is $3.7 \times 10^9$ L mol$^{-1}$ s$^{-1}$. Suppose 0.500 L of a 0.0020 M NaOH solution is rapidly mixed with the same volume of a 0.0020 M HCN solution. Calculate the time (in seconds) required for the $OH^-$ concentration to decrease to a value of $1.0 \times 10^{-4}$ M.

## Reaction Mechanisms

19. Identify each of the following elementary reactions as unimolecular, bimolecular, or termolecular, and write the rate expression.

(a) $HCO + O_2 \rightarrow HO_2 + CO$

(b) $CH_3 + O_2 + N_2 \rightarrow CH_3O_2 + N_2$

(c) $HO_2NO_2 \rightarrow HO_2 + NO_2$

20. Identify each of the following elementary reactions as unimolecular, bimolecular, or termolecular, and write the rate expression.

(a) $BrONO_2 \rightarrow BrO + NO_2$

(b) $HO + NO_2 + Ar \rightarrow HNO_3 + Ar$

(c) $O + H_2S \rightarrow HO + HS$

21. Consider the following reaction mechanism:

$$H_2O_2 \longrightarrow H_2O + O$$

$$O + CF_2Cl_2 \longrightarrow ClO + CF_2Cl$$

$$ClO + O_3 \longrightarrow Cl + 2\,O_2$$

$$Cl + CF_2Cl \longrightarrow CF_2Cl_2$$

(a) What is the molecularity of each elementary step?

(b) Write the overall equation for the reaction.

(c) Identify the reaction intermediate(s).

**22.** Consider the following reaction mechanism:

$$NO_2Cl \longrightarrow NO_2 + Cl$$

$$Cl + H_2O \longrightarrow HCl + OH$$

$$OH + NO_2 + N_2 \longrightarrow HNO_3 + N_2$$

(a) What is the molecularity of each elementary step?
(b) Write the overall equation for the reaction.
(c) Identify the reaction intermediate(s).

**23.** The rate constant of the elementary reaction

$$BrO(g) + NO(g) \longrightarrow Br(g) + NO_2(g)$$

is $1.3 \times 10^{10}$ L mol$^{-1}$ s$^{-1}$ at 25°C, and its equilibrium constant is $5.0 \times 10^{10}$ at this temperature. Calculate the rate constant at 25°C of the elementary reaction

$$Br(g) + NO_2(g) \longrightarrow BrO(g) + NO(g)$$

**24.** The compound $IrH_3(CO)(P(C_6H_5)_3)_2$ exists in two forms, the meridional ("mer") and facial ("fac"). At 25° in a nonaqueous solvent, the reaction mer → fac has a rate constant of 2.33 s$^{-1}$, and the reaction fac → mer has a rate constant of 2.10 s$^{-1}$. What is the equilibrium constant of the mer-to-fac reaction at 25°C?

## Reaction Mechanism and Rate

**25.** Write the overall reaction and rate laws that correspond to the following reaction mechanisms. Be sure to eliminate intermediates from the answers.

(a) $A + B \underset{k_{-1}}{\overset{k_1}{\rightleftharpoons}} C + D$    (fast equilibrium)

   $C + E \overset{k_2}{\longrightarrow} F$    (slow)

(b)   $A \underset{k_{-1}}{\overset{k_1}{\rightleftharpoons}} B + C$    (fast equilibrium)

   $C + D \underset{k_{-2}}{\overset{k_2}{\rightleftharpoons}} E$    (fast equilibrium)

   $E \overset{k_3}{\longrightarrow} F$    (slow)

**26.** Write the overall reaction and the rate laws that correspond to the following reaction mechanisms. Be sure to eliminate intermediates from the answers.

(a) $2 A + B \underset{k_{-1}}{\overset{k_1}{\rightleftharpoons}} D$    (fast equilibrium)

   $D + B \overset{k_2}{\longrightarrow} E + F$    (slow)

   $F \overset{k_3}{\longrightarrow} G$    (fast)

(b) $A + B \underset{k_{-1}}{\overset{k_1}{\rightleftharpoons}} C$    (fast equilibrium)

   $C + D \underset{k_{-2}}{\overset{k_2}{\rightleftharpoons}} F$    (fast equilibrium)

   $F \overset{k_3}{\longrightarrow} G$    (slow)

**27.** HCl reacts with propene ($CH_3CHCH_2$) in the gas phase according to the overall reaction

$$HCl + CH_3CHCH_2 \longrightarrow CH_3CHClCH_3$$

The experimental rate expression is

$$rate = k[HCl]^3[CH_3CHCH_2]$$

Which, if any, of the following mechanisms are consistent with the observed rate expression?

(a) $HCl + HCl \rightleftharpoons H + HCl_2$    (fast equilibrium)
   $H + CH_3CHCH_2 \rightarrow CH_3CHCH_3$    (slow)
   $HCl_2 + CH_3CHCH_3 \rightarrow CH_3CHClCH_3 + HCl$    (fast)

(b) $HCl + HCl \rightleftharpoons H_2Cl_2$    (fast equilibrium)
   $HCl + CH_3CHCH_2 \rightleftharpoons CH_3CHClCH_3^*$

   (fast equilibrium)
   $CH_3CHClCH_3^* + H_2Cl_2 \rightarrow CH_3CHClCH_3 + 2 HCl$

   (slow)

(c) $HCl + CH_3CHCH_2 \rightleftharpoons H + CH_3CHClCH_2$

   (fast equilibrium)

   $H + HCl \rightleftharpoons H_2Cl$    (fast equilibrium)
   $H_2Cl + CH_3CHClCH_2 \rightarrow HCl + CH_3CHClCH_3$

   (slow)

**28.** Chlorine reacts with hydrogen sulfide in aqueous solution

$$Cl_2(aq) + H_2S(aq) \longrightarrow S(s) + 2 H^+(aq) + 2 Cl^-(aq)$$

in a second-order reaction that follows the rate expression

$$rate = k[Cl_2][H_2S]$$

Which, if any, of the following mechanisms are consistent with the observed rate expression?

(a) $Cl_2 + H_2S \rightarrow H^+ + Cl^- + Cl^+ + HS^-$    (slow)
   $Cl^+ + HS^- \rightarrow H^+ + Cl^- + S$    (fast)

(b) $H_2S \rightleftharpoons HS^- + H^+$    (fast equilibrium)
   $HS^- + Cl_2 \rightarrow 2 Cl^- + S + H^+$    (slow)

(c) $H_2S \rightleftharpoons HS^- + H^+$    (fast equilibrium)
   $H^+ + Cl_2 \rightleftharpoons H^+ + Cl^- + Cl^+$    (fast equilibrium)
   $Cl^+ + HS^- \rightarrow H^+ + Cl^- + S$    (slow)

**29.** Nitryl chloride is a reactive gas with a normal boiling point of $-16$°C. Its decomposition to nitrogen dioxide and chlorine is described by the equation

$$2 NO_2Cl \longrightarrow 2 NO_2 + Cl_2$$

The rate expression for this reaction has the form

$$rate = k[NO_2Cl]$$

Which, if any, of the following mechanisms are consistent with the observed rate expression?

(a) $NO_2Cl \rightarrow NO_2 + Cl$    (slow)
   $Cl + NO_2Cl \rightarrow NO_2 + Cl_2$    (fast)

(b) $2 NO_2Cl \rightleftharpoons N_2O_4 + Cl_2$    (fast equilibrium)
   $N_2O_4 \rightarrow 2 NO_2$    (slow)

(c) $2 NO_2Cl \rightleftharpoons ClO_2 + N_2O + ClO$    (fast equilibrium)
   $N_2O + ClO_2 \rightleftharpoons NO_2 + NOCl$    (fast equilibrium)
   $NOCl + ClO \rightarrow NO_2 + Cl_2$    (slow)

**30.** Ozone in the upper atmosphere is decomposed by nitrogen oxide through the reaction

$$O_3 + NO \longrightarrow O_2 + NO_2$$

The experimental rate expression for this reaction is

$$\text{rate} = k[O_3][NO]$$

Which, if any, of the following mechanisms are consistent with the observed rate expression?
(a) $O_3 + NO \rightarrow O + NO_3$               (slow)
    $O + O_3 \rightarrow 2\,O_2$                        (fast)
    $NO_3 + NO \rightarrow 2\,NO_2$                (fast)
(b) $O_3 + NO \rightarrow O_2 + NO_2$          (slow)
(c) $NO + NO \rightleftarrows N_2O_2$              (fast equilibrium)
    $N_2O_2 + O_3 \rightarrow NO_2 + 2\,O_2$   (slow)

**31.** Consider the mechanism of problem 25(a). Suppose *no* assumptions are made about the relative rates of the steps. By making a steady-state approximation for the concentration of the intermediate (C), express the rate of production of the product (F) in terms of the concentrations of A, B, D, and E. In what limit does this reduce to the result of problem 25(a)?

**32.** Consider the mechanism of problem 25(b). Suppose *no* assumptions are made about the relative rates of the steps. By making a steady-state approximation for the concentrations of the intermediates (C and E), express the rate of production of the product (F) in terms of the concentrations of A, B, and D. In what limit does this reduce to the result of problem 25(b)?

**33.** The mechanism for the decomposition of $NO_2Cl$ is

$$NO_2Cl \underset{k_{-1}}{\overset{k_1}{\rightleftarrows}} NO_2 + Cl$$

$$NO_2Cl + Cl \xrightarrow{k_2} NO_2 + Cl_2$$

By making a steady-state approximation for [Cl], express the rate of appearance of $Cl_2$ in terms of the concentrations of $NO_2Cl$ and $NO_2$.

**34.** A key step in the formation of sulfuric acid from dissolved $SO_2$ in acid precipitation is the oxidation of hydrogen sulfite ion by hydrogen peroxide:

$$HSO_3^-(aq) + H_2O_2(aq) \longrightarrow HSO_4^-(aq) + H_2O(\ell)$$

The mechanism involves peroxymonosulfurous acid, $SO_2OOH^-$:

$$HSO_3^-(aq) + H_2O_2(aq) \underset{k_{-1}}{\overset{k_1}{\rightleftarrows}} SO_2OOH^-(aq) + H_2O(\ell)$$

$$SO_2OOH^-(aq) + H_3O^+(aq) \xrightarrow{k_2} HSO_4^-(aq) + H_3O^+(aq)$$

By making a steady-state approximation for the reactive intermediate concentration, $[SO_2OOH^-(aq)]$, express the rate of formation of $HSO_4^-(aq)$ in terms of the concentrations of $HSO_3^-(aq)$, $H_2O_2(aq)$, and $H_3O^+(aq)$.

## Effect of Temperature on Reaction Rates

**35.** The rate of the elementary reaction

$$Ar + O_2 \longrightarrow Ar + O + O$$

has been studied as a function of temperature between 5000 and 18,000 K. The following data were obtained for the rate constant $k$:

| Temperature (K) | $k$ (L mol$^{-1}$ s$^{-1}$) |
|---|---|
| 5,000 | $5.49 \times 10^6$ |
| 10,000 | $9.86 \times 10^8$ |
| 15,000 | $5.09 \times 10^9$ |
| 18,000 | $8.60 \times 10^9$ |

(a) Calculate the activation energy of this reaction.
(b) Calculate the factor $A$ in the Arrhenius equation for the temperature dependence of the rate constant.

**36.** The gas-phase reaction

$$H + D_2 \longrightarrow HD + D$$

is the exchange of isotopes of hydrogen of atomic mass 1 (H) and 2 (D, deuterium). The following data were obtained for the rate constant $k$ of this reaction:

| Temperature (K) | $k$ (L mol$^{-1}$ s$^{-1}$) |
|---|---|
| 299 | $1.56 \times 10^4$ |
| 327 | $3.77 \times 10^4$ |
| 346 | $7.6 \times 10^4$ |
| 440 | $1.07 \times 10^6$ |
| 549 | $8.7 \times 10^6$ |
| 745 | $8.7 \times 10^7$ |

(a) Calculate the activation energy of this reaction.
(b) Calculate the factor $A$ in the Arrhenius equation for the temperature dependence of the rate constant.

**37.** The rate constant of the elementary reaction

$$BH_4^-(aq) + NH_4^+(aq) \longrightarrow BH_3NH_3(aq) + H_2(g)$$

is $k = 1.94 \times 10^{-4}$ L mol$^{-1}$ s$^{-1}$ at 30.0°C, and the reaction has an activation energy of 161 kJ mol$^{-1}$.
(a) Compute the rate constant of the reaction at a temperature of 40.0°C.
(b) After equal concentrations of $BH_4^-(aq)$ and $NH_4^+(aq)$ are mixed at 30.0°C, $1.00 \times 10^4$ s is required for half of them to be consumed. How long will it take to consume half of the reactants if an identical experiment is performed at 40.0°C?

**38.** Dinitrogen tetraoxide ($N_2O_4$) decomposes spontaneously at room temperature in the gas phase:

$$N_2O_4(g) \longrightarrow 2\,NO_2(g)$$

The rate law governing the disappearance of $N_2O_4$ with time is

$$-\frac{d[N_2O_4]}{dt} = k[N_2O_4]$$

At 30°C, $k = 5.1 \times 10^6 \text{ s}^{-1}$ and the activation energy for the reaction is 54.0 kJ mol$^{-1}$.

(a) Calculate the time (in seconds) required for the partial pressure of $N_2O_4(g)$ to decrease from 0.10 atm to 0.010 atm at 30°C.

(b) Repeat the calculation of part (a) at 300°C.

39. The activation energy for the isomerization reaction of $CH_3NC$ in problem 14 is 161 kJ mol$^{-1}$, and the reaction rate constant at 600 K is 0.41 s$^{-1}$.

(a) Calculate the Arrhenius factor $A$ for this reaction.

(b) Calculate the rate constant for this reaction at 1000 K.

40. Cyclopropane isomerizes to propylene according to a first-order reaction:

$$\text{cyclopropane} \longrightarrow \text{propylene}$$

The activation energy is $E_a = 272$ kJ mol$^{-1}$. At 500°C, the reaction rate constant is $6.1 \times 10^{-4} \text{ s}^{-1}$.

(a) Calculate the Arrhenius factor $A$ for this reaction.

(b) Calculate the rate constant for this reaction at 25°C.

41. The activation energy of the gas-phase reaction

$$OH(g) + HCl(g) \longrightarrow H_2O(g) + Cl(g)$$

is 3.5 kJ mol$^{-1}$, and the change in the internal energy in the reaction is $\Delta E = -66.8$ kJ mol$^{-1}$. Calculate the activation energy of the reaction

$$H_2O(g) + Cl(g) \longrightarrow OH(g) + HCl(g)$$

42. The compound HOCl is known, but the related compound HClO, with a different order for the atoms in the molecule, is not known. Calculations suggest that the activation energy for the conversion HOCl → HClO is 311 kJ mol$^{-1}$ and that for the conversion HClO → HOCl is 31 kJ mol$^{-1}$. Estimate $\Delta E$ for the reaction HOCl → HClO.

## Collision Theory of Gas-Phase Reactions

43. Use collision theory to estimate the preexponential factor in the rate constant for the elementary reaction

$$NOCl + NOCl \longrightarrow 2\,NO + Cl_2$$

at 25°C. Take the average diameter of an NOCl molecule to be $3.0 \times 10^{-10}$ m and use the steric factor $P$ from Table 11-1.

44. Use collision theory to estimate the preexponential factor in the rate constant for the elementary reaction

$$NO_2 + NO_2 \longrightarrow 2\,NO + O_2$$

at 500 K. Take the average diameter of an $NO_2$ molecule to be $2.6 \times 10^{-10}$ m and use the steric factor $P$ from Table 11-1.

## Kinetics of Catalysis

45. Certain bacteria use the enzyme penicillinase to decompose penicillin and render it inactive. The Michaelis–Menten con-

stants for this enzyme and substrate are $K_m = 5 \times 10^{-5}$ mol L$^{-1}$ and $k_2 = 2 \times 10^3 \text{ s}^{-1}$.

(a) What is the maximum rate of decomposition of penicillin if the enzyme concentration is $6 \times 10^{-7}$ M?

(b) At what substrate concentration will the rate of decomposition be half that calculated in part (a)?

46. The conversion of dissolved carbon dioxide in blood to $HCO_3^-$ and $H_3O^+$ is catalyzed by the enzyme carbonic anhydrase. The Michaelis–Menten constants for this enzyme and substrate are $K_m = 8 \times 10^{-5}$ mol L$^{-1}$ and $k_2 = 6 \times 10^5 \text{ s}^{-1}$.

(a) What is the maximum rate of reaction of carbon dioxide if the enzyme concentration is $5 \times 10^{-6}$ M?

(b) At what $CO_2$ concentration will the rate of decomposition be 30% of that calculated in part (a)?

## Additional Problems

47. Hemoglobin molecules in blood bind oxygen and carry it to cells, where it takes part in metabolism. The binding of oxygen

$$\text{hemoglobin}(aq) + O_2(aq) \longrightarrow \text{hemoglobin·}O_2(aq)$$

is first order in hemoglobin and first order in dissolved oxygen, with a rate constant of $4 \times 10^7$ L mol$^{-1}$ s$^{-1}$. Calculate the initial rate at which oxygen will be bound to hemoglobin if the concentration of hemoglobin is $2 \times 10^{-9}$ M and that of oxygen is $5 \times 10^{-5}$ M.

*48. Suppose 1.00 L of $9.95 \times 10^{-3}$ M $S_2O_3^{2-}$ is mixed with 1.00 L of $2.52 \times 10^{-3}$ M $H_2O_2$ at a pH of 7.0 and a temperature of 25°C. These species react by two competing pathways, represented by the balanced equations

$$S_2O_3^{2-} + 4\,H_2O_2 \longrightarrow 2\,SO_4^{2-} + H_2O + 2\,H_3O^+$$

$$2\,S_2O_3^{2-} + H_2O_2 + 2\,H_3O^+ \longrightarrow S_4O_6^{2-} + 4\,H_2O$$

At the instant of mixing, the thiosulfate ion ($S_2O_3^{2-}$) is observed to be disappearing at the rate of $7.9 \times 10^{-7}$ mol L$^{-1}$ s$^{-1}$. At the same moment, the $H_2O_2$ is disappearing at the rate of $8.8 \times 10^{-7}$ mol L$^{-1}$ s$^{-1}$.

(a) Compute the percentage of the $S_2O_3^{2-}$ that is, at that moment, reacting according to the first equation.

(b) It is observed that the hydronium ion concentration drops. Use the data and answer from part (a) to compute how many milliliters per minute of 0.100 M $H_3O^+$ must be added to keep the pH equal to 7.0.

49. $8.23 \times 10^{-3}$ mol of InCl(s) is placed in 1.00 L of 0.010 M HCl(aq) at 75°C. The InCl(s) dissolves quite quickly, and then the following reaction occurs:

$$3\,In^+(aq) \longrightarrow 2\,In(s) + In^{3+}(aq)$$

As this disproportionation proceeds, the solution is analyzed at intervals to determine the concentration of $In^+(aq)$ that remains.

| Time (s) | $[In^+]$ (mol $L^{-1}$) |
|----------|-------------------------|
| 0 | $8.23 \times 10^{-3}$ |
| 240 | $6.41 \times 10^{-3}$ |
| 480 | $5.00 \times 10^{-3}$ |
| 720 | $3.89 \times 10^{-3}$ |
| 1000 | $3.03 \times 10^{-3}$ |
| 1200 | $3.03 \times 10^{-3}$ |
| 10000 | $3.03 \times 10^{-3}$ |

(a) Plot ln $[In^+]$ versus time, and determine the apparent rate constant for this first-order reaction.
(b) Determine the half-life of this reaction.
(c) Determine the equilibrium constant $K$ for the reaction under the experimental conditions.

*50. A compound called di-$t$-butyl peroxide [abbreviation DTBP, formula $(CH_3)_3COOC(CH_3)_3$] decomposes to give acetone $[(CH_3)_2CO]$ and ethane $(C_2H_6)$:

$$(CH_3)_3COOC(CH_3)_3(g) \longrightarrow 2\ (CH_3)_2CO(g) + C_2H_6(g)$$

The *total* pressure of the reaction mixture changes with time, as shown by the following data at 147.2°C:

| Time (min) | $P_{tot}$ (atm) | Time (min) | $P_{tot}$ (atm) |
|------------|-----------------|------------|-----------------|
| 0 | 0.2362 | 26 | 0.3322 |
| 2 | 0.2466 | 30 | 0.3449 |
| 6 | 0.2613 | 34 | 0.3570 |
| 10 | 0.2770 | 38 | 0.3687 |
| 14 | 0.2911 | 40 | 0.3749 |
| 18 | 0.3051 | 42 | 0.3801 |
| 20 | 0.3122 | 46 | 0.3909 |
| 22 | 0.3188 | | |

(a) Calculate the partial pressure of DTBP at each time from these data. Assume that at time 0, DTBP is the only gas present.
(b) Are the data better described by a first-order or a second-order rate expression with respect to DTBP concentration?

51. The reaction of $OH^-$ with HCN in aqueous solution at 25°C has a forward rate constant $k_f$ of $3.7 \times 10^9$ L $mol^{-1}$ $s^{-1}$. Using this information and the measured acid ionization constant of HCN (see Table 6-2), calculate the rate constant $k_r$ in the first-order rate law

$$\text{rate} = k_r[CN^-]$$

for the transfer of hydrogen ions to $CN^-$ from surrounding water molecules:

$$H_2O(\ell) + CN^-(aq) \longrightarrow OH^-(aq) + HCN(aq)$$

52. Carbon dioxide reacts with ammonia to give ammonium carbamate, a solid. The reverse reaction also occurs:

$$CO_2(g) + 2\ NH_3(g) \rightleftharpoons NH_4OCONH_2(s)$$

The forward reaction is first order in $CO_2(g)$ and second order in $NH_3(g)$. Its rate constant is 0.238 $atm^{-2}$ $s^{-1}$ at 0.0°C (expressed in terms of partial pressures rather than concentrations). The reaction in the reverse direction is zero order, and its rate constant, at the same temperature, is $1.60 \times 10^{-7}$ atm $s^{-1}$. Experimental studies show that, at all stages in the progress of this reaction, the net rate is equal to the forward rate minus the reverse rate. Compute the equilibrium constant of this reaction at 0.0°C.

53. For the reactions

$$I + I + M \longrightarrow I_2 + M$$
$$Br + Br + M \longrightarrow Br_2 + M$$

the rate laws are

$$-\frac{d[I]}{dt} = k_1[I]^2[M]$$
$$-\frac{d[Br]}{dt} = k_{Br}[Br]^2[M]$$

The ratio $k_1/k_{Br}$ at 500°C is 3.0 when M is an Ar molecule. Initially, $[I]° = 2[Br]°$, while [M] is the same for both reactions and is much greater than $[I]°$. Calculate the ratio of the time required for [I] to decrease to half its initial value to the same time for [Br] at 500°C.

54. In some reactions there is a competition between kinetic control and thermodynamic control over product yields. Suppose compound A can undergo two elementary reactions to stable products:

$$A \underset{k_{-1}}{\overset{k_1}{\rightleftharpoons}} B \quad \text{or} \quad A \underset{k_{-2}}{\overset{k_2}{\rightleftharpoons}} C$$

For simplicity we assume first-order kinetics for both forward and reverse reactions. We take the numerical values $k_1 = 1 \times 10^8$ $s^{-1}$, $k_{-1} = 1 \times 10^2$ $s^{-1}$, $k_2 = 1 \times 10^9$ $s^{-1}$, and $k_{-2} = 1 \times 10^4$ $s^{-1}$.
(a) Calculate the equilibrium constant for the equilibrium

$$B \rightleftharpoons C$$

From this value, give the ratio of the concentration of B to that of C at equilibrium. This is an example of thermodynamic control.
(b) In the case of kinetic control, the products are isolated (or undergo additional reaction) before the back reactions can take place. Suppose the back reactions in the preceding example ($k_{-1}$ and $k_{-2}$) can be ignored. Calculate the concentration ratio of B to C reached in this case.

55. Compare and contrast the mechanisms for the two gas-phase reactions

$$H_2 + Br_2 \longrightarrow 2\ HBr$$
$$H_2 + I_2 \longrightarrow 2\ HI$$

56. In Section 11-4, the steady-state approximation was used to derive a rate expression for the decomposition of $N_2O_5(g)$:

$$\text{rate} = \frac{k_1 k_2 [M][N_2O_5]}{k_2 + k_{-1}[M]} = k_{\text{eff}}[N_2O_5]$$

At 300 K, with an excess of nitrogen present, the following values of $k_{\text{eff}}$ as a function of total pressure were found:

| P (atm) | $k_{\text{eff}}$ (s$^{-1}$) | P (atm) | $k_{\text{eff}}$ (s$^{-1}$) |
|---------|------------------|---------|------------------|
| 9.21    | 0.265            | 0.625   | 0.116            |
| 5.13    | 0.247            | 0.579   | 0.108            |
| 3.16    | 0.248            | 0.526   | 0.104            |
| 3.03    | 0.223            | 0.439   | 0.092            |
|         |                  | 0.395   | 0.086            |

Use the data to estimate the value of $k_{\text{eff}}$ at very high total pressure and the value of $k_1$ in L mol$^{-1}$ s$^{-1}$.

**57.** The decomposition of ozone by light can be described by the mechanism

$$O_3 + \text{light} \xrightarrow{k_1} O_2 + O$$

$$O + O_2 + M \xrightarrow{k_2} O_3 + M$$

$$O + O_3 \xrightarrow{k_3} 2\,O_2$$

with the overall reaction being

$$2\,O_3 + \text{light} \longrightarrow 3\,O_2$$

The rate constant $k_1$ depends on the light intensity and the type of light source used. By making a steady-state approximation for the concentration of oxygen atoms, express the rate of formation of $O_2$ in terms of the $O_2$, $O_3$, and M concentrations and the elementary rate constants. Show that only the ratio $k_3/k_2$, and not the individual values of $k_2$ and $k_3$, affects the rate.

**\*58.** In Section 11-4 we considered the following mechanism for the reaction of $Br_2$ with $H_2$:

$$Br_2 + M \underset{k_{-1}}{\overset{k_1}{\rightleftarrows}} Br + Br + M$$

$$Br + H_2 \xrightarrow{k_2} HBr + H$$

$$H + Br_2 \xrightarrow{k_3} HBr + Br$$

Although this is adequate for calculating the *initial* rate of reaction, before product HBr builds up, there is an additional process that can participate as the reaction continues:

$$HBr + H \xrightarrow{k_4} H_2 + Br$$

(a) Write an expression for the rate of change of [H].
(b) Write an expression for the rate of change of [Br].
(c) As hydrogen and bromine atoms are both short-lived species, we can make the steady-state approximation and set the rates from parts (a) and (b) to 0. Express the steady-state

concentrations [H] and [Br] in terms of concentrations of $H_2$, $Br_2$, HBr, and M. (*Hint:* Try adding the rate for part (a) to that for part (b).)
(d) Express the rate of production of HBr in terms of concentrations of $H_2$, $Br_2$, HBr, and M.

**59.** The gas-phase decomposition of acetaldehyde can be represented by the overall chemical equation

$$CH_3CHO \longrightarrow CH_4 + CO$$

It is thought to occur through the sequence of reactions

$$CH_3CHO \longrightarrow CH_3 + CHO$$

$$CH_3 + CH_3CHO \longrightarrow CH_4 + CH_2CHO$$

$$CH_2CHO \longrightarrow CO + CH_3$$

$$CH_3 + CH_3 \longrightarrow CH_3CH_3$$

Show that this reaction mechanism corresponds to a chain reaction, and identify the initiation, propagation, and termination steps.

**60.** Lanthanum(III) phosphate crystallizes as a hemihydrate, $LaPO_4 \cdot \frac{1}{2}H_2O$. When it is heated, it loses water to give anhydrous lanthanum(III) phosphate:

$$2(LaPO_4 \cdot \tfrac{1}{2}H_2O(s)) \longrightarrow 2\,LaPO_4(s) + H_2O(g)$$

This reaction is first order in the chemical amount of $LaPO_4 \cdot \frac{1}{2}H_2O$. The rate constant varies with temperature as follows:

| Temperature (°C) | k (s$^{-1}$) |
|------------------|--------------|
| 205              | $2.3 \times 10^{-4}$  |
| 219              | $3.69 \times 10^{-4}$ |
| 246              | $7.75 \times 10^{-4}$ |
| 260              | $12.3 \times 10^{-4}$ |

Compute the activation energy of this reaction.

**61.** The water in a pressure cooker boils at a temperature greater than 100°C because it is under pressure. At this higher temperature, the chemical reactions associated with the cooking of food take place at a greater rate.
(a) Some food cooks fully in 5 min in a pressure cooker at 112°C and in 10 min in an open pot at 100°C. Calculate the average activation energy for the reactions associated with the cooking of this food.
(b) How long will the same food take to cook in an open pot of boiling water in Denver, where the average atmospheric pressure is 0.818 atm and the boiling point of water is 94.4°C?

**62.** (a) A certain first-order reaction has an activation energy of 53 kJ mol$^{-1}$. It is run twice, first at 298 K and then at 308 K (10°C higher). All other conditions are identical. Show that, in the second run, the reaction occurs at double its rate in the first run.
(b) The same reaction is run twice more at 398 K and 408 K.

Show that the reaction goes 1.5 times as fast at 408 K as it does at 398 K.

*63. The gas-phase reaction between hydrogen and iodine

$$H_2(g) + I_2(g) \underset{k_r}{\overset{k_f}{\rightleftharpoons}} 2\,HI(g)$$

proceeds with a forward rate constant at 1000 K of $k_f = 240\ L\ mol^{-1}\ s^{-1}$ and an activation energy of 165 kJ $mol^{-1}$. By using this information and data from Appendix D, calculate the activation energy for the reverse reaction and the value of $k_r$ at 1000 K. Assume that $\Delta H$ and $\Delta S$ for the reaction are independent of temperature between 298 and 1000 K.

64. The following reaction mechanism has been proposed for a chemical reaction:

$$A_2 \underset{k_{-1}}{\overset{k_1}{\rightleftharpoons}} A + A \qquad \text{(fast equilibrium)}$$

$$A + B \underset{k_{-2}}{\overset{k_2}{\rightleftharpoons}} AB \qquad \text{(fast equilibrium)}$$

$$AB + CD \overset{k_3}{\longrightarrow} AC + BD \qquad \text{(slow)}$$

(a) Write a balanced equation for the overall reaction.
(b) Write the rate expression that corresponds to the preceding mechanism. Express the rate in terms of concentrations of reactants only ($A_2$, B, CD).
(c) Suppose that the first two steps in the preceding mecha-

nism are endothermic and the third one is exothermic. Will an increase in temperature increase the reaction rate constant, decrease it, or cause no change? Explain.

65. How would you describe the role of the $CF_2Cl_2$ in the reaction mechanism of problem 21?

66. In Section 11-7 we wrote a mechanism in which silver ions catalyze the reaction of $Tl^+$ with $Ce^{4+}$. Determine the rate law for this mechanism by making a steady-state approximation for the concentration of the reactive intermediate $Ag^{2+}$.

67. The rates of enzyme catalysis can be lowered by the presence of inhibitor molecules I, which bind to the active site of the enzyme. This adds the following additional step to the reaction mechanism considered in Section 11-7:

$$E + I \underset{k_{-3}}{\overset{k_3}{\rightleftharpoons}} EI \qquad \text{(fast equilibrium)}$$

Determine the effect of the presence of inhibitor at total concentration $[I]_0 = [I] + [EI]$ on the rate expression for formation of products derived at the end of this chapter.

68. The enzyme lysozyme kills certain bacteria by attacking a sugar called N-acetylglucosamine (NAG) in their cell walls. At an enzyme concentration of $2 \times 10^{-6}$ M, the maximum rate for substrate (NAG) reaction, found at high substrate concentration, is $1 \times 10^{-6}$ mol $L^{-1}\ s^{-1}$. The rate is reduced by a factor of 2 when the substrate concentration is reduced to $6 \times 10^{-6}$ M. Determine the Michaelis–Menten constants $K_m$ and $k_2$ for lysozyme.

## CUMULATIVE PROBLEMS

69. The rate of the gas-phase reaction

$$H_2 + I_2 \longrightarrow 2\,HI$$

is given by

$$\text{rate} = -\frac{d[I_2]}{dt} = k[H_2][I_2]$$

with $k = 0.0242$ L $mol^{-1}\ s^{-1}$ at 400°C. If the initial concentration of $H_2$ is 0.081 mol $L^{-1}$ and that of $I_2$ is 0.036 mol $L^{-1}$, calculate the initial rate at which heat is absorbed or emitted during the reaction. Assume that the enthalpy change at 400°C is the same as that at 25°C.

70. The rate of the reaction

$$2\,ClO_2(aq) + 2\,OH^-(aq) \longrightarrow$$
$$ClO_3^-(aq) + ClO_2^-(aq) + H_2O(\ell)$$

is given by

$$\text{rate} = k[ClO_2]^2[OH^-]$$

with $k = 230$ L$^2$ $mol^{-2}\ s^{-1}$ at 25°C. A solution is prepared that has initial concentrations $[ClO_2] = 0.020$ M, $[HCN] = 0.095$ M, and $[CN^-] = 0.17$ M. Calculate the initial rate of the reaction.

71. A gas mixture was prepared at 500 K with total pressure 3.26 atm and a mole fraction of 0.00057 of NO and 0.00026 of $O_3$. The elementary reaction

$$NO + O_3 \longrightarrow NO_2 + O_2$$

has a second-order rate constant of $7.6 \times 10^7$ L $mol^{-1}\ s^{-1}$ at this temperature. Calculate the initial rate of the reaction under these conditions.

72. The activation energy for the reaction

$$2\,NO_2 \longrightarrow 2\,NO + O_2$$

is $E_a = 111$ kJ $mol^{-1}$. Calculate the root-mean-square velocity of an $NO_2$ molecule at 400 K and compare it to the velocity of an $NO_2$ molecule with kinetic energy $E_a/N_0$.

*Atomic particle tracks in a bubble chamber. (Fermilab/Peter Arnold, Inc.)*

C H A P T E R

# 12

# Nuclear Chemistry

Our approach to chemistry so far has focused primarily on macroscopic or bulk properties of matter: masses of reactants and products in stoichiometry, pressures and concentrations in reacting mixtures at or away from equilibrium, and energy changes in physical and chemical processes. Until the end of the 19th century, chemistry was concerned almost exclusively with such macroscopic properties, and so it was appropriate for us to begin with such a viewpoint. Of course, some microscopic concepts have also appeared in our discussion. The laws of combining masses and volumes, macroscopic observations that they are, demanded interpretation in terms of atoms; the macroscopic relationship between energy and temperature for ideal gases was clarified through the kinetic theory of gases. Even so, our excursions into the microscopic world in the first 11 chapters were aimed largely at clarifying macroscopic experience, not at discovering their ultimate causes.

We now shift our point of view, and the next six chapters examine matter from a microscopic point of view, exploring the ways in which the building blocks introduced in Section 1-4 (proton, neutron, and electron) fit together to give matter its properties. In this chapter we consider the properties and reactions of the nucleus. In subsequent chapters we will move on to atoms, molecules, and the regular arrays of molecules that form solids.

## 12-1 MASS–ENERGY RELATIONSHIPS IN NUCLEI

Recall from Section 1-4 that almost all the mass of an atom is concentrated in a very small volume in the nucleus. The small size of the nucleus (which occupies less than one *trillionth* of the space in the atom) and the strong forces between the protons and neutrons that make it up largely isolate its behavior from the outside world of electrons and other nuclei. This greatly simplifies our analysis of nuclear chemistry, allowing us to examine single nuclei without being concerned about the atoms, ions, or molecules in which they may be found.

As defined in Section 1-4, a nuclide is characterized by the number of protons, $Z$, and the number of neutrons, $N$, it contains. The atomic number $Z$ determines the charge $+Ze$ on the nucleus and therefore decides the identity of the element; the sum $Z + N = A$ is the mass number of the nuclide and is the integer closest to the relative atomic mass of the nuclide. Nuclides are designated by the symbol $_Z^A X$, where X is the chemical symbol for the element.

The absolute masses of nuclides and of elementary particles such as the proton, neutron, and electron are far too small to express in kilograms or grams without the constant use of 10 raised to some large negative exponent. A unit of mass that is better sized for measuring on the submicroscopic scale is needed. An **atomic mass unit** (abbreviated u) is defined as exactly $\frac{1}{12}$ the mass of a single atom of $^{12}$C. Because 1 mol of $^{12}$C atoms weighs exactly 12 g, one atom weighs $12/N_0$ grams, and

$$1 \text{ u} = \frac{1}{12}\left(\frac{12 \text{ g mol}^{-1}}{6.022137 \times 10^{23} \text{ mol}^{-1}}\right)$$

$$= 1.660540 \times 10^{-24} \text{ g} = 1.660540 \times 10^{-27} \text{ kg}$$

The conversion factor between atomic mass units and grams is numerically equal to the inverse of Avogadro's number $N_0$, and the mass of a single atom in atomic mass units is numerically equal to the mass of 1 mol of atoms in grams. Thus, one atom of $^1$H has a mass of 1.007825 u because 1 mol of $^1$H has a mass of 1.007825 g. The *dalton* is a mass unit that is equivalent to the atomic mass unit and is used frequently in biochemistry.

Table 12-1 lists the masses of elementary particles and selected neutral atoms, obtained from mass spectrometry (Fig. 1-15). Note that each atomic mass includes the contribution from the surrounding $Z$ electrons as well as that from the nucleus.

### The Einstein Mass–Energy Relationship

Some nuclides are stable and can exist indefinitely, but others are unstable (radioactive) and decay spontaneously to form other nuclides in processes to be considered in Section 12-2. Figure 12-1 shows the distribution of the number of neutrons $N$ ($= A - Z$) against atomic number $Z$ (protons in the nucleus) for the known nuclides of the elements. For the light elements, $N/Z \approx 1$ for stable nuclides, and so nearly equal numbers of protons and neutrons are found in the nucleus. For the heavier elements, $N/Z > 1$ and there are progressively more neutrons than protons in the nucleus. One of the goals of nuclear chemistry is to understand the differing stabilities of the isotopes. Nuclear mass provides the key to this understanding.

## Table 12-1   Masses of Selected Elementary Particles and Atoms

| Elementary Particle | Symbol | Mass (u) | Mass (kg) |
|---|---|---|---|
| Electron, beta particle | $_{-1}^{0}e^{-}$ | 0.00054857990 | $9.109390 \times 10^{-31}$ |
| Positron | $_{1}^{0}e^{+}$ | 0.00054857990 | $9.109390 \times 10^{-31}$ |
| Proton | $_{1}^{1}p^{+}$ | 1.00727647 | $1.672623 \times 10^{-27}$ |
| Neutron | $_{0}^{1}n$ | 1.00866490 | $1.674929 \times 10^{-27}$ |

| Atom | Mass (u) | Atom | Mass (u) |
|---|---|---|---|
| $_{1}^{1}\text{H}$ | 1.00782505 | $_{11}^{23}\text{Na}$ | 22.989770 |
| $_{1}^{2}\text{H}$ | 2.0141079 | $_{12}^{24}\text{Mg}$ | 23.985045 |
| $_{1}^{3}\text{H}$ | 3.016049 | $_{14}^{30}\text{Si}$ | 29.973761 |
| $_{2}^{3}\text{He}$ | 3.01602930 | $_{15}^{30}\text{P}$ | 29.97832 |
| $_{2}^{4}\text{He}$ | 4.0026033 | $_{16}^{32}\text{S}$ | 31.972072 |
| $_{3}^{7}\text{Li}$ | 7.016005 | $_{17}^{35}\text{Cl}$ | 34.9688527 |
| $_{4}^{8}\text{Be}$ | 8.0053052 | $_{20}^{40}\text{Ca}$ | 39.962589 |
| $_{4}^{9}\text{Be}$ | 9.0121825 | $_{22}^{49}\text{Ti}$ | 48.947871 |
| $_{4}^{10}\text{Be}$ | 10.013535 | $_{35}^{81}\text{Br}$ | 80.91629 |
| $_{5}^{8}\text{B}$ | 8.024612 | $_{37}^{87}\text{Rb}$ | 86.909184 |
| $_{5}^{10}\text{B}$ | 10.012938 | $_{38}^{87}\text{Sr}$ | 86.908890 |
| $_{5}^{11}\text{B}$ | 11.009305 | $_{53}^{127}\text{I}$ | 126.904477 |
| $_{6}^{11}\text{C}$ | 11.011433 | $_{88}^{226}\text{Ra}$ | 226.025406 |
| $_{6}^{12}\text{C}$ | 12 exactly | $_{88}^{228}\text{Ra}$ | 228.03123 |
| $_{6}^{13}\text{C}$ | 13.003354 | $_{89}^{228}\text{Ac}$ | 228.03117 |
| $_{6}^{14}\text{C}$ | 14.003242 | $_{90}^{232}\text{Th}$ | 232.038054 |
| $_{7}^{14}\text{N}$ | 14.00307400 | $_{90}^{234}\text{Th}$ | 234.0436 |
| $_{8}^{16}\text{O}$ | 15.9949146 | $_{91}^{231}\text{Pa}$ | 231.035881 |
| $_{8}^{17}\text{O}$ | 16.999131 | $_{92}^{231}\text{U}$ | 231.0363 |
| $_{8}^{18}\text{O}$ | 17.9991594 | $_{92}^{234}\text{U}$ | 234.040947 |
| $_{9}^{19}\text{F}$ | 18.9984033 | $_{92}^{235}\text{U}$ | 235.043925 |
| $_{11}^{21}\text{Na}$ | 20.99764 | $_{92}^{238}\text{U}$ | 238.050786 |

Note from Table 12-1 that the neutron has a slightly greater mass than the proton. The neutron is stable inside a nucleus, but in free space it is unstable, decaying into a proton and an electron with a half-life of about 12 minutes:

$$_{0}^{1}n \longrightarrow {}_{1}^{1}p^{+} + {}_{-1}^{0}e^{-}$$

Here $_{0}^{1}n$ and $_{1}^{1}p^{+}$ represent the neutron and the proton, respectively. We have written this as a **balanced nuclear equation.** To check such an equation for balance, first verify that the total mass numbers (the superscripts) are equal on the two sides. Then verify that the nuclear charges are balanced by checking the sum of the subscripts.

Mass is not conserved in the reaction just written. By subtracting the mass of the reactant from the masses of the products (using Table 12-1), the change in mass is calculated to be

$$\Delta m = -1.395 \times 10^{-30}\,\text{kg} = -8.3985 \times 10^{-4}\,\text{u}$$

What has happened to the "lost" mass? It has been converted to energy—specifically, kinetic energy carried off by the electron and proton. According to

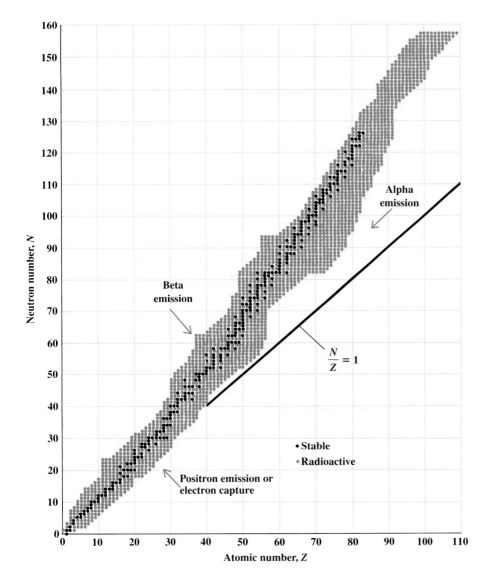

**Figure 12-1**

A plot of $N$ versus $Z$ for nuclides (represented by dots). Arrows show the directions of decay of unstable nuclides via alpha emission, beta emission, positron emission, or electron capture.

Einstein's special theory of relativity, a change in mass always accompanies a change in energy:

$$E = mc^2$$

or, better,

$$\Delta E = c^2 \, \Delta m \qquad \text{(12-1)}$$

where $c$ is the speed of light in a vacuum and $\Delta m$ is the change in mass (mass of products minus mass of reactants). To express $\Delta E$ in joules, $c$ must have units of meters per second, and $\Delta m$, units of kilograms. In the decay of one neutron, for example,

$$\Delta E = c^2 \, \Delta m = (2.9979 \times 10^8 \text{ m s}^{-1})^2 (-1.395 \times 10^{-30} \text{ kg}) = -1.254 \times 10^{-13} \text{ J}$$

For ordinary chemical reactions, the change in mass is negligibly small. In the combustion of 1 mol of carbon to $CO_2$, the energy change is about 400 kJ, corresponding to a change in mass of only $4 \times 10^{-9}$ g.

Changes in energy in nuclear reactions are nearly always expressed in more convenient energy units than the joule—namely, the **electron volt** (eV) and the **million electron volt** (MeV). The electron volt is defined as the energy given to an electron when it is accelerated through a potential difference of exactly 1 V:

$$\Delta E \text{ (joules)} = Q \text{ (coulombs)} \times \Delta \mathscr{E} \text{ (volts)}$$

$$1 \text{ eV} = 1.602177 \times 10^{-19} \text{ C} \times 1 \text{ V} = 1.602177 \times 10^{-19} \text{ J}$$

The million electron volt is 1 million times larger:

$$1 \text{ MeV} = 1.602177 \times 10^{-13} \text{ J}$$

A change in mass of 1 u = $1.660540 \times 10^{-27}$ kg corresponds to

$$\Delta E = c^2 \Delta m = (2.9979246 \times 10^8 \text{ m s}^{-1})^2 (1.660540 \times 10^{-27} \text{ kg})$$
$$= 1.492419 \times 10^{-10} \text{ J}$$

Converting to MeV,

$$\frac{1.492419 \times 10^{-10} \text{ J}}{1.602177 \times 10^{-13} \text{ J MeV}^{-1}} = 931.494 \text{ MeV}$$

We say that the **energy equivalent** of 1 u is 931.494 MeV. In the decay of a neutron,

$$\Delta E = (-8.3985 \times 10^{-4} \text{ u})(931.494 \text{ MeV u}^{-1}) = 0.782 \text{ MeV}$$

## Binding Energies of Nuclei

The **binding energy** $E_B$ of a nucleus is defined as the negative of the energy change $\Delta E$ that would occur if that nucleus were formed from its component protons and neutrons. For the $^4_2\text{He}$ nucleus, for example,

$$2\,^1_1p^+ + 2\,^1_0n \longrightarrow\; ^4_2\text{He}^{2+} \qquad\qquad \Delta E = ?$$

Binding energies of nuclei are calculated with Einstein's relationship and accurate data from a mass spectrometer. Most measurements, however, determine the mass of the atom with most or all of its electrons, not the mass of the bare nucleus (recall that it is the $^{12}$C *atom,* not the $^{12}$C nucleus, that is defined to have atomic mass 12). Therefore, we take the nuclear binding energy $E_B$ of the helium nucleus to be the negative of the energy change $\Delta E$ for the formation of an *atom* of helium from hydrogen *atoms* and neutrons.

$$2\,^1_1\text{H} + 2\,^1_0n \longrightarrow\; ^4_2\text{He}$$

The correction for the differences in binding energies of electrons (which of course are present both in the hydrogen atoms and in the atom being formed) is a very small one that need not concern us here.

### Example 12-1

Calculate the binding energy of $^4$He from the data in Table 12-1, and express it both in joules and in million electron volts (MeV).

**Solution**

The mass change is

$$\Delta m = m[{}^4_2\text{He}] - 2m[{}^1_1\text{H}] - 2m[{}^1_0 n]$$

$$= 4.0026033 - 2(1.00782505) - 2(1.0086649) = -0.0303766 \text{ u}$$

Einstein's relation then gives

$$\Delta E = (-0.0303766 \text{ u})(1.660540 \times 10^{-27} \text{ kg u}^{-1})(2.9979246 \times 10^8 \text{ m s}^{-1})^2$$

$$= -4.53346 \times 10^{-12} \text{ J}$$

$$E_B = 4.53346 \times 10^{-12} \text{ J}$$

If 1 *mole* of ${}^4_2\text{He}$ atoms were formed in this way, the energy change would be greater by a factor of Avogadro's number $N_0$, giving $\Delta E = -2.73 \times 10^{12} \text{ J mol}^{-1}$. This is an enormous quantity, seven orders of magnitude greater than typical energy changes in chemical reactions.

The energy change, in MeV, accompanying the formation of a ${}^4_2\text{He}$ atom is

$$\Delta E = (-0.0303766 \text{ u})(931.494 \text{ MeV u}^{-1}) = -28.2956 \text{ MeV}$$

*Related Problems:* 3, 4

**Figure 12-2**

The variation of the binding energy per nucleon with mass number. A total of 2212 nuclides are shown, of which 274 are stable. The greatest nuclear stability is seen in the vicinity of ${}^{56}\text{Fe}$. Note that ${}^4\text{He}$ (red circle) is unusually stable for its mass number.

There are four nucleons in the $_2^4$He nucleus, and so the binding energy *per nucleon* is

$$\frac{E_B}{4} = \frac{28.2956 \text{ MeV}}{4} = 7.07390 \text{ MeV}$$

The binding energy per nucleon is a direct measure of the stability of the nucleus. It varies with atomic number among the stable elements, increasing to a maximum of about 8.8 MeV for iron and nickel, as Figure 12-2 on page 453 illustrates.

## 12-2 NUCLEAR DECAY PROCESSES

Why are some nuclei **radioactive,** decaying spontaneously, when others are stable? Thermodynamics gives the criterion $\Delta G < 0$ for a process to be spontaneous at constant $T$ and $P$. The energy change $\Delta E$ in a nuclear decay process is so great that the free energy change is essentially equal to it, and the thermodynamic criterion for spontaneous nuclear reaction consequently simplifies to

$$\Delta E < 0 \qquad \text{or} \qquad \Delta m < 0$$

The equivalence of these criteria follows from Einstein's mass–energy relationship. Spontaneous transformations of one nucleus into others can occur only if the combined mass of products is less than the mass of the original nuclide.

Before we address the details of nuclear decay, it is necessary to mention **antiparticles.** Each subatomic particle is thought to have an antiparticle of the same mass but opposite charge. Thus, the antiparticle of an electron is a **positron** ($e^+$), and the antiparticle of a proton is a negatively charged particle called an **antiproton.** Antiparticles have only transient existences because matter and antimatter annihilate one another when they come together, emitting radiation carrying an equivalent amount of energy. In the case of an electron and a positron, two oppositely directed **gamma rays** ($\gamma$-rays; high-energy photons), called the **annihilation radiation,** are emitted. Another important particle–antiparticle pair is the neutrino (symbol $\nu$) and antineutrino ($\tilde{\nu}$), both of which appear to be massless, uncharged particles. (They differ only in parity, a quantum-mechanical symmetry upon reflection.) Neutrinos interact so weakly with matter that elaborate detectors are required to record the extremely rare events induced by them in selected nuclides.

### Beta Decay

If an unstable nuclide contains fewer protons than do stable isotopes of the same mass number, it is called "proton deficient." Such a nucleus can decay by transforming one of its neutrons into a proton and emitting a high-energy electron $_{-1}^0e^-$, also called a **beta particle,** and an antineutrino ($\tilde{\nu}$). The superscript 0 indicates a mass number of 0 because an electron contains neither protons nor neutrons and has a much smaller mass than a nucleus. The subscript $-1$ indicates the negative charge on the particle. It is not a true "atomic number," but writing the symbol in this way is helpful in balancing nuclear equations. The nuclide that results has the same mass number $A$, but its atomic number $Z$ is increased by 1 because a neutron has been transformed into a proton. Examples of **beta decay** of unstable nuclei are

$$_6^{14}\text{C} \longrightarrow {}_7^{14}\text{N} + {}_{-1}^0e^- + \tilde{\nu} \qquad\qquad {}_{11}^{24}\text{Na} \longrightarrow {}_{12}^{24}\text{Mg} + {}_{-1}^0e^- + \tilde{\nu}$$

The criterion $\Delta m < 0$ for spontaneous beta decay requires that

$$\underset{\text{(Parent nucleus)}}{m[^A Z]} > \underset{\text{(Daughter nucleus)}}{m[^A(Z + 1)]} + m[_{-1}^{\ 0}e^-]$$

because the antineutrino $\tilde{\nu}$ has zero mass. This inequality refers to the masses of the parent and daughter *nuclei* and the emitted electron. It must be reformulated in terms of neutral *atom* masses because the mass spectrometer yields masses of atoms rather than nuclei. To do this, add the mass of $Z$ electrons to both sides, giving a total of $Z + 1$ electrons on the right side (because the emitted $_{-1}^{\ 0}e^-$ is also counted). The inequality becomes

$$\underset{\text{(Parent atom)}}{m[^A Z]} > \underset{\text{(Daughter atom)}}{m[^A(Z + 1)]}$$

A direct comparison of atomic masses (Table 12-1) lets us decide whether beta emission can take place.

The condition that energy be released follows from the Einstein mass–energy relation:

$$\Delta E = c^2\{m[^A(Z + 1)] - m[^A Z]\} < 0$$

The liberated energy is carried off as kinetic energy by the beta particle (electron) and the antineutrino, because the nucleus produced is heavy enough that its recoil energy is small and can be neglected. The kinetic energy of the electron can fall anywhere in a continuous range from 0 up to $-\Delta E$ (Fig. 12-3), with the antineutrino carrying off the balance of the energy.

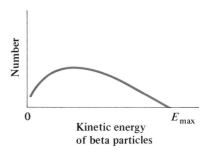

**Figure 12-3**

Emitted beta particles have a distribution of kinetic energies up to a cutoff value of $E_{max}$.

---

### Example 12-2

Calculate the maximum kinetic energy of the electron in the decay

$$_{6}^{14}\text{C} \longrightarrow {}_{7}^{14}\text{N} + {}_{-1}^{\ 0}e^- + \tilde{\nu}$$

### Solution

Table 12-1 gives the masses of the relevant atoms. From them we calculate

$$\begin{aligned}
\Delta m &= m[_{7}^{14}\text{N}] - m[_{6}^{14}\text{C}] \\
&= 14.003074 - 14.003242 \\
&= -0.000168 \text{ u}
\end{aligned}$$

$$\Delta E = (-1.68 \times 10^{-4}\,\text{u})(931.5\ \text{MeV u}^{-1}) = -0.156\ \text{MeV}$$

The maximum kinetic energy of the electron is 0.156 MeV.

*Related Problem:* 8

---

### Positron Emission

When a nucleus has too many protons for stability relative to the number of neutrons it contains, it may decay by emitting a positron. In this event, a proton is converted to a neutron, and a high-energy positron (symbolized $_{1}^{0}e^+$) and a neutrino ($\nu$) are emitted. As in beta decay, the mass number of the nuclide remains the same, but now the atomic number $Z$ *decreases* by 1. Examples of **positron emission** are

$$^{11}_{6}C \longrightarrow ^{11}_{5}B + ^{0}_{1}e^{+} + \nu \qquad\qquad ^{19}_{10}Ne \longrightarrow ^{19}_{9}F + ^{0}_{1}e^{+} + \nu$$

The positron has the same mass as an electron but has a positive charge. The mass criterion $\Delta m < 0$ is different from that for beta decay. For the bare nuclei the criterion is

$$\underset{\text{(Parent nucleus)}}{m[^{A}Z]} > \underset{\text{(Daughter nucleus)}}{m[^{A}(Z-1)]} + m[^{0}_{1}e^{+}]$$

Adding the mass of $Z$ electrons to both sides gives the mass of the neutral atom $^{A}Z$ on the left, but only $Z - 1$ electrons are needed on the right side to form a neutral atom $^{A}(Z-1)$. This leaves additional electron and positron masses on the right. These masses are the same, so the criterion for spontaneous positron emission is

$$\underset{\text{(Parent atom)}}{m[^{A}Z]} > \underset{\text{(Daughter atom)}}{m[^{A}(Z-1)]} + 2m[^{0}_{1}e^{+}]$$

The energy change in positron emission is

$$\Delta E = c^{2}\{m[^{A}(Z-1)] + 2m[^{0}_{1}e^{+}] - m[^{A}Z]\} < 0$$

The kinetic energy $(-\Delta E)$ is distributed between the positron and the neutrino.

## Example 12-3

Calculate the maximum kinetic energy of the positron emitted in the decay.

$$^{11}_{6}C \longrightarrow ^{11}_{5}B + ^{0}_{1}e^{+} + \nu$$

### Solution

The change in mass is

$$\begin{aligned}
\Delta m &= m[^{11}_{5}B] + 2m[^{0}_{1}e^{+}] - m[^{11}_{6}C] \\
&= 11.009305 + 2(0.0005486) - 11.011433 \\
&= -0.001031 \text{ u}
\end{aligned}$$

The energy change is then

$$\Delta E = (-0.001031 \text{ u})(931.5 \text{ MeV u}^{-1}) = -0.960 \text{ MeV}$$

The maximum kinetic energy of the positron is $+0.960$ MeV.

*Related Problem:* 7

## Electron Capture

If $m[^{A}Z]$ exceeds $m[^{A}(Z-1)]$ (both referring to neutral atoms) but by less than $1.0972 \times 10^{-3}$ u (equivalent to $2m[^{0}_{1}e^{+}]$), positron emission cannot occur. An alternative process that *is* allowed is **electron capture,** in which an electron outside the nucleus of the parent atom is captured by the nucleus and a proton is converted to a neutron. As in positron emission, the mass number is unchanged and the atomic number decreases by 1. In this case, however, the only particle emitted is a neutrino. An example is

$$\ce{^{231}_{92}U} + \ce{^{0}_{-1}e^-} \longrightarrow \ce{^{231}_{91}Pa} + \nu$$

The mass criterion for electron capture is simply that

$$\underset{\text{(Parent atom)}}{m[^AZ]} > \underset{\text{(Daughter atom)}}{m[^A(Z-1)]}$$

and so the energy change is

$$\Delta E = c^2\{m[^A(Z-1)] - m[^AZ]\} < 0$$

Electron capture is quite common for heavier neutron-deficient nuclei, for which the mass change with atomic number is too small for positron emission to be possible. In the preceding example,

$$\Delta m = m[\ce{^{231}_{91}Pa}] - m[\ce{^{231}_{92}U}] = 231.0359 - 231.0363 = -0.0004 \text{ u}$$

Because $\Delta m < 0$, this process can and does occur by electron capture, but because 0.0004 u is less than $2m[\ce{^0_1 e^+}]$, the analogous process could not occur through positron emission.

## Alpha Decay

The three forms of decay discussed so far all lead to changes in the atomic number $Z$ but not in the mass number $A$. Another process, **alpha decay,** involves the emission of an alpha particle ($\ce{^4_2He^{2+}}$ ion) and a decrease in mass number $A$ by 4 (the atomic number $Z$ and the neutron number $N$ each decrease by 2). An example is

$$\ce{^{238}_{92}U} \longrightarrow \ce{^{234}_{90}Th} + \ce{^4_2He}$$

Such alpha decay occurs chiefly for elements in the unstable region beyond bismuth ($Z = 83$) in the periodic table.

The criterion for alpha decay is once again $\Delta m < 0$. In the preceding example,

$$\Delta m = \underset{\text{(Atom)}}{m[\ce{^{234}_{90}Th}]} + \underset{\text{(Atom)}}{m[\ce{^4_2He}]} - \underset{\text{(Atom)}}{m[\ce{^{238}_{92}U}]}$$

$$= 234.0436 + 4.0026 - 238.0508 \text{ u} = -0.0046 \text{ u} < 0$$

$$\Delta E = (-0.0046 \text{ u})(931.5 \text{ MeV u}^{-1}) = -4.3 \text{ MeV}$$

Most of the energy is carried away by the lighter helium atom, with a small fraction appearing as recoil energy of the heavy thorium atom.

## Other Modes of Decay

The arrows in Figure 12-1 show the directions of the types of decay discussed so far. Nuclei whose values of $N/Z$ are too great move toward the line of stability by beta emission, while those whose values are too small undergo positron emission or electron capture. Nuclei that are simply too massive can move to lower values of both $N$ and $Z$ by alpha emission. The range of stable nuclides is limited by the interplay of attractive nuclear forces and repulsive Coulomb forces between protons in the nucleus. Scientists believe that it is unlikely that additional stable elements will be found, although some hope that long-lived radioactive nuclides exist at still higher Z.

When nuclei are very proton-deficient or very neutron-deficient, an excess particle may "boil off," that is, be ejected directly from the nucleus. These decay modes are called **neutron emission** and **proton emission,** respectively, and move nuclides down or to the left in Figure 12-1. Finally, certain unstable nuclei undergo spontaneous **fission,** in which they split into two nuclei of roughly equal size. Nuclear fission will be discussed in more detail in Section 12-5.

## Detecting and Measuring Radioactivity

Many methods have been developed to detect, identify, and quantitatively measure the products of nuclear reactions. Some are quite simple, but others require complex electronic instrumentation. Perhaps the simplest radiation detector is the **photographic emulsion,** first employed by Henri Becquerel, the discoverer of radioactivity. In 1896 Becquerel reported his observation that potassium uranyl sulfate ($K_2UO_2(SO_4)_2\cdot2H_2O$) could expose a photographic plate even in the dark. Such detectors are still used today in the film badges that are worn to monitor exposure to penetrating radiation. The degree of darkening of the film is proportional to the quantity of radiation received.

Rutherford and his students used a screen coated with zinc sulfide to detect the arrival of alpha particles by the pinpoint scintillations of light they produce. That simple device has been developed into the modern **scintillation counter.** Instead of a ZnS screen, the modern scintillation counter employs a crystal of sodium iodide, in which a small fraction of the $Na^+$ ions have been replaced by thallium ($Tl^+$) ions. The crystal emits a pulse of light when it absorbs a beta particle or a gamma ray, and a photomultiplier tube detects and counts the light pulses.

The **Geiger counter** (Figure 12-4) consists of a cylindrical tube, usually of glass, coated internally with metal to provide a negative electrode and with a wire

(a)

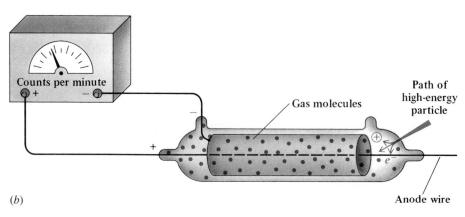

(b)

## Figure 12-4

(a) A Geiger counter measuring the emissions from a sample of carnotite, a uranium-containing mineral. (b) In a Geiger tube, radiation ionizes gas in the tube, freeing electrons that are accelerated to the anode wire in a cascade. Their arrival creates an electrical pulse. (a, David Davidson/Tom Stack and Associates) down the center for a positive electrode. The tube is filled to a total pressure of about 0.1 atm with a mixture of 90% argon and 10% ethyl alcohol vapor, and a potential difference of about 1000 V is applied across the electrodes. When a high-energy electron (beta particle) enters the tube, it produces positive ions and electrons. The light electrons are quickly accelerated toward the positively charged wire. As they advance, they encounter and ionize other neutral atoms. An avalanche of electrons builds up, and a large electron current flows into the central wire. This causes a drop in the potential difference, which is recorded, and the multiplicative electron discharge is quenched by the alcohol molecules. In this way single beta particles produce electrical pulses that can be amplified and counted. Portable Geiger counters are widely used in uranium prospecting and to measure radiation in workplaces.

## 12-3  KINETICS OF RADIOACTIVE DECAY

The decay of any given unstable nucleus is a random event and is independent of the number of surrounding nuclei that have decayed. When the number of nuclei is large, we can be confident that during any given period of time a definite fraction of the original number of nuclei will have undergone a transformation into another nuclear species. In other words, the rate of decay of a collection of nuclei is proportional to the number of nuclei present, showing that nuclear decay follows a first-order rate equation of the type discussed in Chapter 11. All the results developed there apply to the present situation; for example, the integrated rate law has the form

$$N = N_i \, e^{-kt}$$

where $N_i$ is the number of nuclei originally present at $t = 0$. The decay constant $k$ is related to a half-life $t_{1/2}$ through

$$t_{1/2} = \frac{\ln 2}{k} = \frac{0.6931}{k}$$

just as in the first-order gas-phase chemical kinetics of Section 11-2. The half-life is the time it takes for the nuclei in a sample to decay to one-half their initial number, and it can range from less than $10^{-21}$ s to more than $10^{24}$ years for unstable nuclei. Table 12-2 lists the half-lives and decay modes of some unstable nuclides.

There is one important practical difference between chemical kinetics and nuclear kinetics. In chemical kinetics, the concentration of a reactant or product is monitored over time, and the rate of a reaction is then found from the rate of change of that concentration. In nuclear kinetics, the rate of occurrence of decay events, $-dN/dt$, is measured directly with a Geiger counter or other radiation detector. This decay rate—the average disintegration rate in numbers of nuclei per unit time—is called the **activity.**

$$\text{activity} = A = -\frac{dN}{dt} = kN \qquad \qquad \textbf{(12-2)}$$

Because the activity is proportional to the number of nuclei $N$, it also decays exponentially with time:

$$A = A_i \, e^{-kt} \qquad \qquad \textbf{(12-3)}$$

**Table 12-2   Decay Characteristics of Some Radioactive Nuclei**

| Nuclide | $t_{1/2}$ | Decay Mode† | Daughter |
|---|---|---|---|
| $^3_1$H (tritium) | 12.26 years | $e^-$ | $^3_2$He |
| $^8_4$Be | $\sim 1 \times 10^{-16}$ s | $\alpha$ | $^4_2$He |
| $^{14}_6$C | 5730 years | $e^-$ | $^{14}_7$N |
| $^{22}_9$Na | 2.601 years | $e^+$ | $^{22}_{10}$Ne |
| $^{24}_9$Na | 15.02 hours | $e^-$ | $^{24}_{12}$Mg |
| $^{32}_{15}$P | 14.28 days | $e^-$ | $^{32}_{16}$S |
| $^{35}_{16}$S | 87.2 days | $e^-$ | $^{35}_{17}$Cl |
| $^{36}_{17}$Cl | $3.01 \times 10^5$ years | $e^-$ | $^{36}_{18}$Ar |
| $^{40}_{19}$K | $1.28 \times 10^9$ years | $\begin{cases} e^- (89.3\%) \\ \text{E.C. } (10.7\%) \end{cases}$ | $^{40}_{20}$Ca $^{40}_{18}$Ar |
| $^{59}_{26}$Fe | 44.6 days | $e^-$ | $^{59}_{27}$Co |
| $^{60}_{27}$Co | 5.27 years | $e^-$ | $^{60}_{28}$Ni |
| $^{90}_{38}$Sr | 29 years | $e^-$ | $^{90}_{39}$Y |
| $^{109}_{48}$Cd | 453 days | E.C. | $^{109}_{47}$Ag |
| $^{125}_{53}$I | 59.7 days | E.C. | $^{125}_{52}$Te |
| $^{131}_{53}$I | 8.041 days | $e^-$ | $^{131}_{54}$Xe |
| $^{127}_{54}$Xe | 36.41 days | E.C. | $^{127}_{53}$I |
| $^{137}_{57}$La | $\sim 6 \times 10^4$ years | E.C. | $^{137}_{56}$Ba |
| $^{222}_{86}$Rn | 3.824 days | $\alpha$ | $^{218}_{84}$Po |
| $^{226}_{88}$Ra | 1600 years | $\alpha$ | $^{222}_{86}$Rn |
| $^{232}_{90}$Th | $1.40 \times 10^{10}$ years | $\alpha$ | $^{228}_{88}$Ra |
| $^{235}_{92}$U | $7.04 \times 10^8$ years | $\alpha$ | $^{231}_{90}$Th |
| $^{238}_{92}$U | $4.468 \times 10^9$ years | $\alpha$ | $^{234}_{90}$Th |
| $^{239}_{93}$Np | 2.350 days | $e^-$ | $^{239}_{94}$Pu |
| $^{239}_{94}$Pu | $2.411 \times 10^4$ years | $\alpha$ | $^{235}_{92}$U |

†E.C. stands for electron capture; $e^+$, for positron emission; $e^-$, for beta emission; $\alpha$, for alpha emission.

A plot of $\ln A$ against time $t$ is linear with slope $-k = -(\ln 2)/t_{1/2}$, as Figure 12-5 shows. The activity $A$ is reduced to half its initial value in a time $t_{1/2}$. Once $A$ and $k$ are known, the number of nuclei $N$ at that time can be calculated from

$$N = \frac{A}{k} = \frac{A t_{1/2}}{\ln 2} = \frac{A t_{1/2}}{0.6931}$$

The S.I. unit of activity is the becquerel (Bq), defined as 1 radioactive disintegration per second. An older and much larger unit of activity is the curie (abbreviated Ci), which is defined as $3.7 \times 10^{10}$ disintegrations per second. The activity of 1 g of radium is 1 Ci.

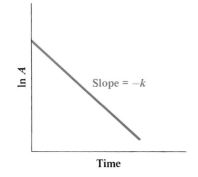

**Figure 12-5**

A graph of the logarithm of the activity of a radioactive nuclide against time is a straight line with slope $-k$. The decay rate can also be measured as the ratio of the activity to the number of atoms of the radioisotope, if the latter is known.

### Example 12-4

Tritium ($^3$H) decays by beta emission to $^3$He with a half-life of 12.26 years. A sample of a tritiated compound has an initial activity of 0.833 Bq. Calculate the number $N_i$ of tritium nuclei in the sample initially, the decay constant $k$, and the activity after 2.50 years.

### Solution

Convert the half-life to seconds:

$$t_{1/2} = (12.26 \text{ yr})(60 \times 60 \times 24 \times 365 \text{ s yr}^{-1}) = 3.866 \times 10^8 \text{ s}$$

The number of nuclei originally present was

$$N_i = \frac{A_i t_{1/2}}{\ln 2} = \frac{(0.833 \text{ s}^{-1})(3.866 \times 10^8 \text{ s})}{0.6931} = 4.65 \times 10^8 \text{ }^3\text{H nuclei}$$

The decay constant $k$ is calculated directly from the half-life:

$$k = \frac{\ln 2}{t_{1/2}} = \frac{0.6931}{3.866 \times 10^8 \text{ s}} = 1.793 \times 10^{-9} \text{ s}^{-1}$$

To find the activity after 2.50 years, convert this time to seconds ($7.884 \times 10^7$ s) and use

$$A = A_i e^{-kt}$$

$$= (0.833 \text{ Bq}) \exp\left[-(1.793 \times 10^{-9} \text{ s}^{-1})(7.884 \times 10^7 \text{ s})\right] = 0.723 \text{ Bq}$$

*Related Problems:* 21, 22

### Radioactive Dating

The decay of radioactive nuclides with known half-lives enables geochemists to measure the ages of rocks from their isotopic compositions. Suppose that a uranium-bearing mineral was deposited some 2 billion ($2.00 \times 10^9$) years ago and has remained geologically unaltered to the present time. The $^{238}\text{U}$ in the mineral decays with a half-life of $4.51 \times 10^9$ years to form a series of short-lived intermediates, ending in the stable lead isotope $^{206}\text{Pb}$ (Fig. 12-6). The fraction of uranium remaining after $2.00 \times 10^9$ years should be

$$\frac{N}{N_i} = e^{-kt} = e^{-0.6931 \, t/t_{1/2}} = \exp\left(\frac{-0.6931 \times 2.00 \times 10^9 \text{ yr}}{4.51 \times 10^9 \text{ yr}}\right) = 0.735$$

and the number of $^{206}\text{Pb}$ atoms should be approximately

$$(1 - 0.735)N_i = 0.265 \, N_i(^{238}\text{U})$$

The ratio of abundances

$$\frac{N(^{206}\text{Pb})}{N(^{238}\text{U})} = \frac{0.265}{0.735} = 0.361$$

therefore is determined by the time elapsed since the deposit was originally formed—in this case, $2.00 \times 10^9$ years. Of course, to calculate the age of the mineral, we work backward from the measured $N(^{206}\text{Pb})/N(^{238}\text{U})$ ratio.

To use the method, it is necessary to be certain that the stable nuclide generated ($^{206}\text{Pb}$ in this case) arises only from the parent species ($^{238}\text{U}$ here) and that neither lead nor uranium has left or entered the rock over the course of geologic time. If possible, it is desirable to measure the ratios of several different isotopically paired

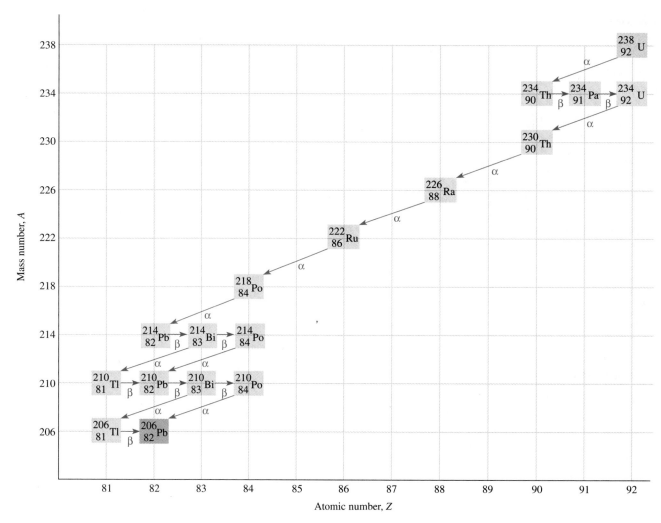

**Figure 12-6**

The radioactive nuclide $^{238}U$ decays via a series of alpha and beta emissions to the stable nuclide $^{206}Pb$.

species in the same rock sample. For example, $^{87}Rb$ decays to $^{87}Sr$ with a half-life of $4.9 \times 10^{10}$ years, and one mode of $^{40}K$ decay (with a half-life of $1.28 \times 10^{9}$ years) is to $^{40}Ar$. Each pair should ideally yield the same age. Detailed analysis of a large number of samples suggests that the oldest surface rocks on earth are about 3.8 billion years old. An estimate of 4.5 billion years for the age of the earth and solar system comes from indirect evidence involving an isotopic analysis of meteorites, which are believed to have formed at the same time.

A somewhat different type of dating uses measurements of $^{14}C$ decay, which covers the range of human history and prehistory back to about 30,000 years ago. This unstable species (with a half-life of 5730 years) is produced continuously in the atmosphere. Cosmic rays of very high energy cause nuclear reactions that produce neutrons. These can collide with $^{14}_{7}N$ nuclei to produce $^{14}_{6}C$ by the reaction

$$^{14}_{7}N + ^{1}_{0}n \longrightarrow ^{14}_{6}C + ^{1}_{1}H$$

The resulting $^{14}C$ enters the carbon reservoir on the earth's surface, mixing with stable $^{12}C$ as dissolved $H^{14}CO_3^-$ in the oceans, as $^{14}CO_2$ in the atmosphere, and in the tissues of plants and animals. This mixing, which is believed to have occurred at a fairly constant rate over the last 50,000 years, means that the $^{14}C$ in a living organism has a specific activity of close to 15.3 disintegrations per minute per gram of total carbon—that is, 0.255 Bq g$^{-1}$. When a plant or animal dies (for example, when a tree is cut down), the exchange of carbon with the surroundings stops, and the amount of $^{14}C$ in the sample falls exponentially with time. By measuring the $^{14}C$ activity remaining in an archaeological sample, we can estimate its age. This $^{14}C$ dating method, developed by W. F. Libby, has been calibrated against other dating techniques (such as counting the annual rings of bristlecone pines or examining the written records that may accompany a carbon-containing artifact) and has been found to be quite reliable over the time span for which it can be checked. This indicates an approximately constant rate of production of $^{14}C$ near the earth's surface over a period of thousands of years. The burning of fossil fuels in the last hundred years has increased the proportion of $^{12}C$ in living organisms and will cause difficulty in applying $^{14}C$ dating in the future.

## Example 12-5

A wooden implement has a specific activity of $^{14}C$ of 0.195 Bq g$^{-1}$. Estimate the age of the implement.

### Solution

The decay constant for $^{14}C$ is

$$k = \frac{0.6931}{5730 \text{ yr}} = 1.21 \times 10^{-4} \text{ yr}^{-1}$$

The initial specific activity was 0.255 Bq g$^{-1}$, and the measured activity now (after $t$ years) is 0.195 Bq g$^{-1}$, and so

$$A = A_i \, e^{-kt}$$

$$0.195 \text{ Bq g}^{-1} = 0.255 \text{ Bq g}^{-1} \, e^{-(1.21 \times 10^{-4})t}$$

$$\ln\left(\frac{0.195}{0.255}\right) = -(1.21 \times 10^{-4} \text{ yr}^{-1})t$$

$$t = 2200 \text{ yr}$$

The implement comes from a tree cut down approximately 2200 years ago.

*Related Problems:* 25, 26

## 12-4 RADIATION IN BIOLOGY AND MEDICINE

Radiation has both harmful and beneficial effects for living organisms. All forms of radiation cause damage in direct proportion to the amount of energy they deposit in cells and tissues. The damage takes the form of chemical changes in cellular molecules, which alter their functions and lead either to uncontrolled multiplication and growth of cells or to their death. Alpha particles lose their kinetic energy over very short distances in matter (typically 10 cm in air or 0.05 cm in water or tissues), producing intense ionization in their wakes until they accept electrons and are neutralized to harmless helium atoms. Radium, for example, is an alpha emitter that substitutes for calcium in bone tissue and destroys its capacity to produce both red and white blood cells. Beta particles, gamma rays, and x-rays have greater penetrating power than alpha particles and so present a radiation hazard even when their source is well outside an organism.

The amount of damage produced in tissue by any of these kinds of radiation is proportional to the number of particles or photons and to their energy. A given activity of tritium causes less damage than the same activity of $^{14}C$, because the beta particles from tritium have a maximum kinetic energy of 0.0179 MeV, whereas those from $^{14}C$ have an energy of 0.156 MeV. What is important is the amount of ionization produced or the quantity of energy deposited by radiation. For this purpose, several units are used. The *rad* (radiation *a*bsorbed *d*ose) is defined as the amount of radiation that deposits $10^{-2}$ J of energy per kilogram of tissue. The damage produced in human tissue depends upon still other factors, such as the nature of the tissue, the kind of radiation, the total radiation dose, and the dose rate. To take all these into account, the *rem* (*r*oentgen *e*quivalent in *m*an) has been defined to measure the effective dosages of radiation received by humans. A physical dose of 1 rad of beta or gamma radiation translates into a human dose of 1 rem. Alpha radiation is more toxic; a physical dose of 1 rad of alpha radiation equals about 10 rems.

Exposure to radiation is unavoidable. The average person in the United States receives about 100 millirems (mrem) annually from natural sources that include cosmic radiation and radioactive nuclides such as $^{40}K$ and $^{222}Rn$. Another 50 to 100 mrem (variable) comes from human activities (including dental and medical x-rays and airplane flights, which increase exposure to cosmic rays higher in the atmosphere). The safe level of exposure to radiation is a controversial issue among biologists; one group maintains that the effects of radiation are cumulative, another that a threshold dose is necessary for pathological change. The problem is made even more complex by the necessity to distinguish between tissue damage in an exposed individual and genetic damage, which may not become apparent for several generations. It is much easier to define the radiation level that, with high probability, will cause death in an exposed person. The $LD_{50}$ level in human beings (the level carrying a 50% probability that death will result within 30 days after a single exposure) is 500 rad.

## Example 12-6

The beta decay of $^{40}K$ that is a natural part of the body makes all human beings slightly radioactive. An adult weighing 70.0 kg contains about 170 g of potassium. The relative natural abundance of $^{40}K$ is 0.0118%, its half-life is $1.28 \times 10^9$ years, and its beta particles have an average kinetic energy of 0.55 MeV.
(a) Calculate the total activity of $^{40}K$ in this person.
(b) Determine (in rad per year) the annual radiation absorbed dose arising from this internal $^{40}K$.

### Solution

(a) First calculate the decay constant of $^{40}K$ in $s^{-1}$:

$$k = \frac{0.693}{t_{1/2}} = \frac{0.693}{(1.28 \times 10^9 \text{ yr})(365 \times 24 \times 60 \times 60 \text{ s yr}^{-1})}$$
$$= 1.72 \times 10^{-17} \text{ s}^{-1}$$

$$\text{number of } ^{40}\text{K atoms} = \frac{170 \text{ g}}{40.0 \text{ g mol}^{-1}}(1.18 \times 10^{-4})(6.02 \times 10^{23} \text{ mol}^{-1})$$
$$= 3.02 \times 10^{20}$$

$$A = -\frac{dN}{dt} = kN = (1.72 \times 10^{-17} \text{ s}^{-1})(3.02 \times 10^{20}) = 5.19 \times 10^3 \text{ s}^{-1}$$

(b) Each disintegration of $^{40}$K emits an average of 0.55 MeV of energy, and we assume that all of this energy is deposited within the body. From part (a), $5.19 \times 10^3$ disintegrations occur per second, and we know how many seconds are in a year. The total energy deposited per year is then

$$5.19 \times 10^3 \text{ s}^{-1} \times (60 \times 60 \times 24 \times 365 \text{ s yr}^{-1}) \times 0.55 \text{ MeV}$$
$$= 9.0 \times 10^{10} \text{ MeV yr}^{-1}$$

Next, because a rad is a centijoule (cJ) per kilogram of tissue, we express this answer in centijoules per year:

$$(9.0 \times 10^{10} \text{ MeV yr}^{-1})(1.602 \times 10^{-13} \text{ J MeV}^{-1})(10^2 \text{ cJ J}^{-1}) = 1.44 \text{ cJ yr}^{-1}$$

Each kilogram of body tissue receives 1/70.0 of this amount of energy per year, because the person weighs 70.0 kg. The dose is thus $21 \times 10^{-3}$ cJ kg$^{-1}$ yr$^{-1}$, which is equivalent to 21 mrad yr$^{-1}$. This is about a fifth of the annual background dosage received by a person.

*Related Problems:* 35, 36

Although radiation can do great harm, it confers great benefits in medical applications (Fig. 12-7). The diagnostic importance of x-ray imaging hardly needs mention. Both x-rays and gamma rays are used selectively in cancer therapy to destroy

**Figure 12-7**
An advanced generator for $^{99m}$Tc, an excited state of $^{99}$Tc that decays to the nuclear ground state by emission of gamma rays. This isotope is used in nuclear medicine to study the heart. The $^{99m}$Tc is taken up by heart tissue; a gamma-ray detector then provides an image of the heart. (Dupont Pharma)

**Figure 12-8**
A patient in a positron emission tomography (PET) scanner squeezes a handgrip to stimulate flow of blood to the heart. The electrodes on his chest record an electrocardiogram while an array of detectors around the scanner registers nuclear decays. (GTE)

malignant cells. The beta-emitting $^{131}$I nuclide finds use in the treatment of cancer of the thyroid because iodine is taken up preferentially by the thyroid gland. Heart pacemakers use the decay of tiny amounts of radioactive $^{238}$Pu, converted to electrical energy.

Positron emission tomography (PET) is an important diagnostic technique using radiation (Fig. 12-8). It employs radioisotopes such as $^{11}$C (half-life 20.3 min) or $^{15}$O (half-life 124 s) that emit positrons when they decay. These are incorporated (quickly, because of their short half-lives) into substances such as glucose, which are injected into the patient. By following the pattern of positron emission from the body, researchers can study blood flow and glucose metabolism in healthy and diseased individuals. Computer-reconstructed pictures of positron emissions from the brain are particularly useful, because the locations of glucose metabolism appear to differ between normal persons and patients suffering from ailments such as manic depression and schizophrenia.

Less direct benefits come from other applications. An example is the study of the mechanism of photosynthesis, in which carbon dioxide and water are combined to form glucose in the green leaves of plants.

$$6\,CO_2(g) + 6\,H_2O(\ell) \longrightarrow C_6H_{12}O_6(s) + 6\,O_2(g)$$

Exposure of plants to $CO_2$ containing a higher-than-normal proportion of $^{14}$C as a tracer enables scientists to follow the mechanism of the reaction. At intervals, the plants are analyzed to find what compounds contain $^{14}$C in their molecules and thereby to identify the intermediates in photosynthesis. Radioactive tracers are also widely used in medical diagnosis. The radioimmunoassay technique, invented by Nobel laureate Rosalind Yalow, determines the levels of drugs and hormones in body fluids.

## 12-5 NUCLEAR FISSION

The nuclear reactions considered so far have been spontaneous, first-order decays of unstable nuclides. By the 1920s, physicists and chemists were using particle accelerators to bombard samples with high-energy particles in order to *induce* nuclear reactions. One of the first results of this program was the identification in 1932 by James Chadwick (a student of Rutherford) of the neutron as a product of the reaction of alpha particles with light nuclides such as $^9$Be:

$$^4_2He + {}^9_4Be \longrightarrow {}^1_0n + {}^{12}_6C$$

Shortly after Chadwick's discovery, a group of physicists in Rome, led by Enrico Fermi, began to study the interaction of neutrons with the nuclei of various elements. The experiments produced a number of radioactive species, and it was evident that the absorption of a neutron increased the $N:Z$ ratio in target nuclei above the stability line (Fig. 12-1). One of the targets used was uranium, the heaviest naturally occurring element. Several radioactive products resulted, none of which had chemical properties characteristic of the elements between $Z = 86$ (radon) and $Z = 92$ (uranium). It appeared to the Italian scientists in 1934 that several new transuranic elements ($Z > 92$) had been synthesized, and an active period of investigation followed.

In 1938 in Berlin, Otto Hahn and Fritz Strassmann sought to characterize the supposed transuranic elements. To their bewilderment, they found instead that barium ($Z = 56$) appeared among the products of the bombardment of uranium by neutrons. Hahn informed his former colleague Lise Meitner, and she conjectured that the products of the bombardment of uranium by neutrons were not transuranic elements but fragments of uranium atoms resulting from a process she termed **fission.** The implication of this phenomenon—the possible release of enormous amounts of energy—was immediately evident. When the outbreak of World War II appeared imminent in the summer of 1939, Albert Einstein wrote to President Franklin Roosevelt to inform him of the possible military uses of fission and of his concern that Germany might develop a nuclear explosive. As a result, in 1942 President Roosevelt authorized the Manhattan District Project, an intense, coordinated effort by a large number of physicists, chemists, and engineers to make a fission bomb of unprecedented destructive power.

The operation of the first atomic bomb hinged on the fission of uranium in a chain reaction induced by absorption of neutrons. The two most abundant isotopes of uranium are $^{235}$U and $^{238}$U, whose natural relative abundances are 0.720% and 99.275%, respectively. Both undergo fission upon absorbing a neutron, the latter only with "fast" neutrons and the former with both "fast" and "slow" neutrons. In the early days of neutron research, it was not known that the absorption of neutrons by nuclei depends strongly upon neutron velocity. By accident, Fermi and his colleagues discovered that experiments conducted on a wooden table led to a much higher yield of radioactive products than those performed on a marble-topped table. Fermi then repeated the irradiation experiments with a block of paraffin wax interposed between the radium–beryllium neutron source and the target sample, with the startling result that the induced level of radioactivity was greatly enhanced. Within hours Fermi had found the explanation: the high-energy neutrons emitted from the radium–beryllium source were reduced to thermal energy by collision with the low-mass nuclei of the paraffin molecules. Because of their lower energies, their probability of reaction with $^{235}$U was greater and a higher yield was achieved. Hydrogen nuclei and the nuclei of other light elements such as $^{12}$C (in graphite) are very effective in reducing the energies of high-velocity neutrons and are called **moderators.**

The fission of $^{235}$U follows many different patterns, and some 34 elements have been identified among the fission products. In any single fission event two particular nuclides are produced together with two or three secondary neutrons; collectively, they carry away about 200 MeV of kinetic energy. Usually the daughter nuclei have different $Z$ and $A$ numbers, and so the fission process is asymmetric. Three of the many pathways are

$$_{0}^{1}n + _{92}^{235}U \longrightarrow \begin{cases} _{30}^{72}Zn + _{62}^{162}Sm + 2\,_{0}^{1}n \\ _{38}^{80}Sr + _{54}^{153}Xe + 3\,_{0}^{1}n \\ _{36}^{94}Kr + _{56}^{139}Ba + 3\,_{0}^{1}n \end{cases}$$

Figure 12-9 shows the distribution of the nuclides produced. The emission of more than one neutron per neutron absorbed in the fission process means that this is a branching chain reaction in which the number of neutrons grows exponentially with time (Fig. 12-10). Permitted to proceed unchecked, this reaction would quickly lead to the release of enormous quantities of energy. Neutrons can be lost by various processes, however, so it is not obvious that a self-propagating reaction will occur. On December 2, 1942, Fermi and his associates demonstrated that a self-sustaining neutron chain reaction occurred in a uranium "pile" with a graphite moderator. They limited its power output to $\frac{1}{2}$ J s$^{-1}$ by inserting cadmium control rods to absorb neutrons, thereby balancing the rates of neutron generation and loss.

Fermi's work made two developments possible: (1) the exploitation of nuclear fission for the controlled generation of energy in nuclear reactors and (2) the production of $^{239}$Pu, a slow- and fast-neutron fissionable isotope of plutonium, as an alternative to $^{235}$U for the construction of atomic bombs. For the sudden release of energy that is required in an explosive, it was necessary to obtain $^{235}$U or $^{239}$Pu in a state free of neutron-absorbing impurities. Both alternatives were pursued simulta-

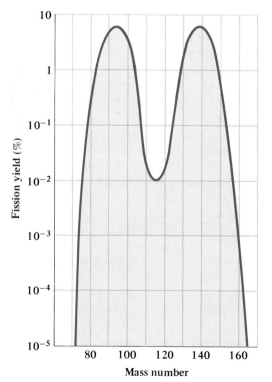

**Figure 12-9**

The distribution of nuclides produced in the fission of $^{235}$U has two peaks. Nuclei having mass numbers in the vicinity of $A = 95$ and $A = 139$ are formed with the highest yield; those with $A \approx 117$ are produced with lower probability.

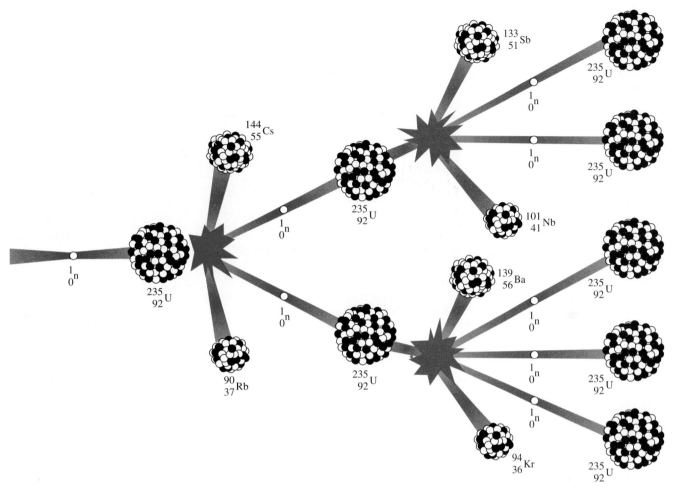

**Figure 12-10**

In a self-propagating nuclear chain reaction, the number of neutrons grows exponentially during fission.

neously. The first required enriching $^{235}U$ from its relative abundance of 0.72% in natural uranium. This was accomplished through gaseous diffusion (Section 3-6). For the second alternative, $^{239}Pu$ was recovered from the partially spent uranium fuel of large nuclear reactors by means of redox reactions, precipitation, and solvent extraction.

Although Enrico Fermi and his associates were the first scientists to demonstrate a self-sustaining nuclear chain reaction, a natural uranium fission reactor "went critical" about 1.8 billion years ago in a place now called Oklo, in the Gabon Republic of equatorial Africa. In 1972 French scientists discovered that the $^{235}U$ content of ore from a site in the open-pit mine at Oklo was only 0.7171%; the normal content of ore from other areas of the mine was 0.7207%. Although this deviation was not large, it was significant, and an investigation revealed that other elements were also present in the ore, in the exact proportions expected after nuclear fission. This discovery established that a self-sustaining nuclear reaction had occurred at Oklo. The geological age of the ore body was found to be about $1.8 \times 10^9$ years, and the original $^{235}U$ concentration is calculated to have been about 3%. From the size of the active ore mass and the depletion of $^{235}U$, it is estimated that the reactor generated about 15,000 megawatt-years ($5 \times 10^{17}$ J) of energy over a period of about 100,000 years.

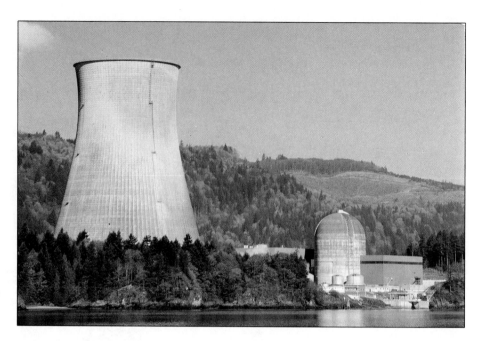

**Figure 12-11**

A nuclear power plant. The large structure on the left is a cooling tower; the containment building is the smaller building on the right with the domed top. (Jonathan Turk)

## Nuclear Power Reactors

Most **nuclear power reactors** in the United States (Fig. 12-11) use rods of $U_3O_8$ as fuel. The uranium is primarily $^{238}U$, but the amount of $^{235}U$ is enriched above natural abundance to a level of about 3%. The moderator used to slow the neutrons (to increase the efficiency of the fission) is ordinary water in most cases, so these reactors are called ''light-water'' reactors. The controlled release of energy by nuclear fission in power reactors demands a delicate balance between neutron generation and neutron loss. As mentioned earlier, this is accomplished by means of steel control rods containing $^{112}Cd$ or $^{10}B$, isotopes that have a very large neutron-capture probability. These rods are automatically inserted into or withdrawn from the fissioning system in response to a change in the neutron flux. As the nuclear reaction proceeds, the moderator (water) is heated and transfers its heat to a steam generator. The steam then goes to turbines that generate electricity (Fig. 12-12).

The power reactors discussed so far rely on the fission of $^{235}U$, an isotope in quite limited supply. An alternative is to convert the much more abundant $^{238}U$ to fissionable plutonium ($^{239}Pu$) by neutron bombardment:

$$^{238}_{92}U + ^{1}_{0}n \longrightarrow ^{239}_{93}Np + _{-1}^{0}e^- \longrightarrow ^{239}_{94}Pu + 2\,_{-1}^{0}e^-$$

Fissionable $^{233}U$ can also be made from thorium:

$$^{232}_{90}Th + ^{1}_{0}n \longrightarrow ^{233}_{91}Pa + _{-1}^{0}e^- \longrightarrow ^{233}_{92}U + 2\,_{-1}^{0}e^-$$

In a **breeder reactor,** in addition to heat being generated by fission, neutrons are absorbed in a blanket of uranium or thorium. This causes the preceding reactions to occur and generates additional fuel for the reactor to use. An advanced technology for breeder reactors uses liquid sodium instead of water as the coolant, allowing the use of faster neutrons than in water-cooled reactors. The faster neutrons cause more complete consumption of radioactive fuels, increasing efficiency and greatly reducing radioactive waste.

Liquid water under high pressure (carries heat to steam generator)

Steam

NUCLEAR REACTOR

Heat source

Steam generator

Steam turbine (generates electricity)

Condenser (steam from turbine is condensed by river water)

pump

pump

pump

80°F    RIVER    100°F

**Figure 12-12**

A schematic diagram of a pressurized-water nuclear power reactor.

The risks associated with the operation of nuclear reactors are small but not negligible, as the failure of the Three Mile Island reactor in 1979 and the disaster at Chernobyl in 1987 demonstrated. If a reactor has to be shut down quickly, there is danger of a meltdown, in which the heat from the continuing fission processes melts the uranium fuel. Coolant must be circulated until heat from the decay of short-lived isotopes has been dissipated. The Three Mile Island accident resulted in a partial meltdown because some water coolant pumps were inoperative and others were shut down too soon, causing damage to the core and a slight release of radioactivity into the environment. The Chernobyl disaster was caused by a failure of the water-cooling system and a meltdown. The rapid and uncontrolled nuclear reaction that took place set the graphite moderator on fire and caused the reactor building to rupture, spreading radioactive nuclides with an activity estimated at $2 \times 10^{20}$ Bq into the atmosphere. A major part of the problem was that the reactor at Chernobyl, unlike those in the United States, was not in a massive containment building.

The safe disposal of the radioactive wastes from nuclear reactors is an important and controversial matter. A variety of proposals have been made, including the burial of radioactive waste in deep mines on either a recoverable or a permanent basis, burial at sea, and launching the waste into outer space. The first alternative is the only one that appears credible. The essential requirement is that the disposal site(s) be stable with respect to possible earthquakes or invasion by underground water. Spent nuclear fuel can be encased in blocks of borosilicate glass, packed in metal containers, and buried in stable rock formations. For a nuclide such as $^{239}$Pu, whose half-life is 24,000 years, a storage site that is stable over 240,000 years is needed before the activity drops to 0.1% of its original value. Some shorter-lived isotopes are more hazardous over short periods of time, but their threat diminishes more quickly.

## 12-6 NUCLEAR FUSION AND NUCLEOSYNTHESIS

**Nuclear fusion** is the union of two light nuclides to form a heavier nuclide with the release of energy. Fusion processes are often called **thermonuclear reactions,** because they require that the colliding particles possess very high kinetic energies, corresponding to temperatures of millions of degrees, before they are initiated. They are the processes that occur in the sun and other stars. In 1939 Hans Bethe (and, independently, Carl von Weizsäcker) proposed that in normal stars (main sequence) the following reactions take place:

$$\, ^1_1\text{H} + \, ^1_1\text{H} \longrightarrow \, ^2_1\text{H} + \, ^0_1e^+ + \nu$$

$$\, ^2_1\text{H} + \, ^1_1\text{H} \longrightarrow \, ^3_2\text{He} + \gamma$$

$$\, ^3_2\text{He} + \, ^3_2\text{He} \longrightarrow \, ^4_2\text{He} + 2\, ^1_1\text{H}$$

In the first reaction, two high-velocity protons fuse to form a deuteron, with the emission of a positron and a neutrino that carry away (as kinetic energy) the additional 0.415 MeV of energy released. In the second reaction, a high-energy deuteron combines with a high-velocity proton to form a helium nucleus of mass 3 and a gamma ray. The third reaction completes the cycle with the formation of a normal helium nucleus ($^4_2$He) and the regeneration of two protons. Each of these reactions is exothermic, but up to 1.25 MeV is required to overcome the repulsive barrier between the positively charged nuclei. The overall result of the cycle is to convert hydrogen nuclei to helium nuclei, and the process is called **hydrogen burning.**

As such a star ages and accumulates helium, it begins to contract under the influence of its immense gravity. As it contracts, its helium core heats up, and when it reaches a temperature of about $10^8$ K, a stage of **helium burning** begins. The first reaction that occurs is

$$2\, ^4_2\text{He} \rightleftharpoons \, ^8_4\text{Be}$$

This reaction is written as an equilibrium because the $^8$Be quickly reverts to helium nuclei with a half-life of only $2 \times 10^{-16}$ s. Even with this short half-life, the $^8$Be nuclei are believed to occasionally react with alpha particles to form stable $^{12}$C:

$$\, ^8_4\text{Be} + \, ^4_2\text{He} \longrightarrow \, ^{12}_6\text{C}$$

The overall effect of the helium-burning phase of a star's life is to convert three helium nuclei to a carbon nucleus, just as helium was formed from four hydrogen nuclei in the hydrogen-burning phase. The density of the core of a star that is burning helium is on the order of $10^5$ g cm$^{-3}$.

This process of **nucleosynthesis** continues beyond the formation of $^{12}$C to produce $^{13}$N, $^{13}$C, $^{14}$N, $^{15}$O, $^{15}$N, and $^{16}$O. The stars in this stage are classified as red giants. Similar cycles occur until the temperature of a star core is about $4 \times 10^9$ K, its density is about $3 \times 10^6$ g cm$^{-3}$, and the nuclei are $^{56}$Fe, $^{59}$Co, and $^{60}$Ni. These are the nuclei that have the maximum binding energy per nucleon (see Fig. 12-2). It is thought that the synthesis of still heavier nuclei occurs in the immense explosions of supernovae.

Heavy elements can also be produced in particle accelerators, which accelerate ions to high speeds, causing collisions that generate the new elements. Technetium, for example, is not found in nature but was first produced in 1937 when high-energy deuterons were directed at a molybdenum source:

$$^{96}_{42}\text{Mo} + {}^{2}_{1}\text{H} \longrightarrow {}^{97}_{43}\text{Tc} + {}^{1}_{0}n$$

The first **transuranic element** was produced in 1940. Neptunium ($Z = 93$) results from the capture of a neutron by $^{238}\text{U}$, followed by beta decay. Subsequent work by Glenn Seaborg and others led to the production of plutonium ($Z = 94$) and heavier elements. In recent years, nuclides with $Z$ as high as 111 have been made, but in tiny quantities. These nuclides have very short half-lives.

Major efforts are now under way to achieve controlled nuclear fusion as a source of energy. One approach is based on the reaction between deuterium and tritium atoms,

$$^{2}_{1}\text{H} + {}^{3}_{1}\text{H} \longrightarrow {}^{4}_{2}\text{He} + {}^{1}_{0}n$$

with a predicted energy release of 17.6 MeV. The neutrons produced in this reaction can be used to make the needed tritium by means of the reaction

$$^{6}_{3}\text{Li} + {}^{1}_{0}n \longrightarrow {}^{3}_{1}\text{H} + {}^{4}_{2}\text{He}$$

in a lithium "blanket" that surrounds the region of the reaction. Temperatures on the order of $10^{8}$ K are required to initiate these reactions, by which point atoms are stripped of all their electrons and the system is a **plasma** of nuclei, electrons, and neutrons. There are two major problems: heating the plasma to this high temperature and confining it. In a laser fusion reactor, pellets of deuterium–tritium fuel are raised to the thermonuclear ignition temperature by bursts of light fired from lasers (Fig. 12-13). Contact between the plasma and the wall of the confining vessel would immediately result in prohibitive energy losses and extinguish the fusion reaction. A plasma can be contained in an intense toroidal (doughnut-shaped) magnetic field. Research reactors using this confinement method have progressed and are reaching the point where they produce more energy than they consume. Formidable problems still must be solved, including the development of materials for containment vessels that can withstand the corrosive effects of the intense x-ray and neutron radiation that is present.

The rewards of a workable nuclear fusion process would be great. Fusion produces neither the long-lived radioactive nuclides that accompany nuclear fission

*(a)*

*(b)*

**Figure 12-13**

*(a)* The target chamber of Nova, a 16-foot-diameter aluminum sphere inside of which ten powerful laser beams converge. *(b)* The target itself is a tiny capsule (1 mm in diameter) of deuterium mixed with tritium. A laser pulse one billionth of a second in duration heats the target hotter than the sun's core, raising its pressure to more than 100 million atm, to initiate nuclear fusion. (*a*, Phototake; *b*, Gary Stone/Lawrence Livermore National Laboratory)

(although tritium requires care in handling) nor the environmental pollutants released by the burning of fossil fuels. Although deuterium is present in only 1/6000 of the abundance of ordinary hydrogen, its separation from the latter by the electrolysis of water is readily accomplished, and the oceans contain a virtually unlimited quantity of deuterium.

## PRACTICE EXERCISE

Radon most commonly enters houses through the foundation or basement walls.

### Radon

Radioactive $^{222}$Rn and $^{220}$Rn form constantly from the decay of uranium and thorium in rocks and soil and, being gaseous, seep out of the ground. The radon isotopes decay fairly quickly, but their products, which are also radioactive, are then in the air and attach themselves to dust particles. Thus, airborne radioactivity can accumulate to worrisome levels in poorly ventilated basements in ground that is rich in uranium and thorium.

(a) Describe the composition of an atom of $^{222}$Rn and compare it with that of an atom of $^{220}$Rn.

(b) Although $^{222}$Rn is a decay product of $^{238}$U, $^{220}$Rn comes from $^{232}$Th. How many alpha particles are emitted in the formation of these radon isotopes from their uranium or thorium starting points? (*Hint:* Alpha decay changes the mass number $A$, but other decay processes do not.)

(c) Can alpha decay alone explain the formation of these radon isotopes from $^{238}$U and $^{232}$Th? If not, state what other types of decay must occur.

(d) Can $^{222}_{86}$Rn and $^{220}_{86}$Rn decay by alpha particle emission? Write balanced nuclear equations for these two decay processes, and calculate the changes in mass that would result. The masses of $^{222}$Rn and $^{220}$Rn atoms are 222.0175 and 220.01140 u, respectively; those of $^{218}$Po and $^{216}$Po are 218.0089 and 216.00192 u, respectively.

(e) Calculate the energy change in the alpha decay of one $^{220}$Rn nucleus, in million electron volts and in joules.

(f) The half-life of $^{222}$Rn is 3.82 days. Calculate the initial activity of $2.00 \times 10^{-8}$ g of $^{222}$Rn, in disintegrations per second.

(g) What will be the activity of the $^{222}$Rn from part (f) after 14 days?

(h) The half-life of $^{220}$Rn is 54 s. Are the health risks of exposure to a given amount of radon for a given short length of time greater or smaller for $^{220}$Rn than for $^{222}$Rn?

### Answers

(a) An atom of $^{222}$Rn has 86 electrons outside the nucleus. Inside the nucleus are 86 protons and $222 - 86 = 136$ neutrons. An atom of $^{220}$Rn has the same number of electrons and protons, but only 134 neutrons in its nucleus.

(b) Four alpha particles are produced to make $^{222}$Rn from $^{238}$U; three are produced to make $^{220}$Rn from $^{232}$Th.

(c) If $^{238}_{92}$U were to lose four alpha particles, $^{222}_{84}$Po would result instead of $^{222}_{84}$Rn. Two $_{-1}^{0}e^{-}$ beta particles must be ejected from the nucleus along the way to raise the atomic number to $Z = 86$. The same is true of the production of $^{220}$Rn from $^{232}$Th.

(d) $^{222}_{86}$Rn $\rightarrow$ $^{218}_{84}$Po $+$ $^{4}_{2}$He; $\Delta m = -0.0060$ u $< 0$; allowed

  $^{220}_{86}$Rn $\rightarrow$ $^{216}_{84}$Po $+$ $^{4}_{2}$He; $\Delta m = -0.0069$ u $< 0$; allowed

(e) $\Delta E = -6.4$ MeV $= -1.03 \times 10^{-12}$ J

(f) $A = 1.14 \times 10^{8}$ s$^{-1}$

(g) $A = 9.0 \times 10^{6}$ s$^{-1}$

(h) Greater

After you have studied this chapter and worked the problems that follow, you should be able to

1. Calculate the binding energy of a nucleus (Section 12-1, problems 3–6).
2. Write balanced nuclear equations for beta decay, positron emission, electron capture, and alpha decay processes and calculate the maximum kinetic energies of particles emitted (Section 12-2, problems 7–18).
3. Describe several methods that are used to detect the products of radioactive decay (Section 12-2).
4. Solve problems involving the half-life or decay constant of a radioactive sample and its activity (Section 12-3, problems 19–24).
5. Apply the kinetics of nuclear decay to the dating of rocks or artifacts (Section 12-3, problems 25–30).
6. Discuss the interactions of radiation with various kinds of matter, and the measurement of radiation dosage (Section 12-4, problems 33–36).
7. Describe the processes of nuclear fission and fusion, and calculate the amounts of energy released when they occur (Sections 12-5 and 12-6, problems 37–47).
8. Explain the benefits and risks associated with the use of nuclear reactions for power generation (Sections 12-5 and 12-6).

### Mass–Energy Relationships in Nuclei

1. Complete and balance the following equations for nuclear reactions that are thought to take place in stars:
   (a) $2\ {}^{12}_{6}\text{C} \rightarrow ? + {}^{1}_{0}n$
   (b) $? + {}^{1}_{1}\text{H} \rightarrow {}^{12}_{6}\text{C} + {}^{4}_{2}\text{He}$
   (c) $2\ {}^{3}_{2}\text{He} \rightarrow ? + 2\ {}^{1}_{1}\text{H}$
2. Complete and balance the following equations for nuclear reactions that are used in particle accelerators to make elements beyond uranium:
   (a) ${}^{4}_{2}\text{He} + {}^{253}_{99}\text{Es} \rightarrow ? + 2\ {}^{1}_{0}n$
   (b) ${}^{249}_{98}\text{Cf} + ? \rightarrow {}^{257}_{103}\text{Lr} + 2\ {}^{1}_{0}n$
   (c) ${}^{238}_{92}\text{U} + {}^{12}_{6}\text{C} \rightarrow {}^{244}_{98}\text{Cf} + ?$
3. Calculate the total binding energy, in both kJ per mole and MeV per atom, and the binding energy per nucleon of the following nuclides, using the data from Table 12-1.
   (a) ${}^{40}_{20}\text{Ca}$
   (b) ${}^{87}_{37}\text{Rb}$
   (c) ${}^{238}_{92}\text{U}$
4. Calculate the total binding energy, in both kilojoules per mole and MeV per atom, and the binding energy per nucleon of the following nuclides, using the data from Table 12-1.
   (a) ${}^{10}_{4}\text{Be}$
   (b) ${}^{35}_{17}\text{Cl}$
   (c) ${}^{49}_{22}\text{Ti}$
5. Use the data from Table 12-1 to predict which is more stable: four protons, four neutrons, and four electrons organized as two ${}^{4}\text{He}$ atoms or as one ${}^{8}\text{Be}$ atom. What is the mass difference?

6. Use the data from Table 12-1 to predict which is more stable: 16 protons, 16 neutrons, and 16 electrons organized as two ${}^{16}\text{O}$ atoms or as one ${}^{32}\text{S}$ atom. What is the mass difference?

### Nuclear Decay Processes

7. The nuclide ${}^{8}_{5}\text{B}$ decays by positron emission to ${}^{8}_{4}\text{Be}$. What is the energy released (in MeV)?
8. The nuclide ${}^{10}_{4}\text{Be}$ undergoes spontaneous radioactive decay to ${}^{10}_{5}\text{B}$ with emission of a beta particle. Calculate the maximum kinetic energy of the emitted beta particle.
9. Write balanced equations that represent the following nuclear reactions.
   (a) Beta emission by ${}^{39}_{17}\text{Cl}$
   (b) Positron emission by ${}^{22}_{11}\text{Na}$
   (c) Alpha emission by ${}^{224}_{88}\text{Ra}$
   (d) Electron capture by ${}^{82}_{38}\text{Sr}$
10. Write balanced equations that represent the following nuclear reactions.
    (a) Alpha emission by ${}^{155}_{70}\text{Yb}$
    (b) Positron emission by ${}^{26}_{14}\text{Si}$
    (c) Electron capture by ${}^{65}_{30}\text{Zn}$
    (d) Beta emission by ${}^{100}_{41}\text{Nb}$
11. The stable isotopes of neon are ${}^{20}\text{Ne}$, ${}^{21}\text{Ne}$, and ${}^{22}\text{Ne}$. Predict the nuclides formed when ${}^{19}\text{Ne}$ and ${}^{23}\text{Ne}$ decay.
12. The two stable isotopes of carbon are ${}^{12}\text{C}$ and ${}^{13}\text{C}$. Predict the nuclides formed when ${}^{11}\text{C}$ and ${}^{14}\text{C}$ decay. Is alpha emission by ${}^{14}\text{C}$ possible?
13. The free neutron is an unstable particle that decays into a

proton. What other particle is formed in neutron decay, and what is the maximum kinetic energy (in MeV) that it can possess?

**14.** The radionuclide $^{210}_{84}$Po decays by alpha emission to a daughter nuclide. The atomic mass of $^{210}_{84}$Po is 209.9829 u, and that of its daughter is 205.9745 u.

(a) Identify the daughter, and write the nuclear equation for the radioactive decay process.

(b) Calculate the total energy released per disintegration (in MeV).

(c) Calculate the kinetic energy of the emitted alpha particle.

**15.** The natural abundance of $^{30}$Si is 3.1%. Upon irradiation with neutrons, this isotope is converted to $^{31}$Si, which decays to the stable isotope $^{31}$P. This provides a way of introducing trace amounts of phosphorus into silicon in a much more uniform fashion than is possible by ordinary mixing of silicon and phosphorus and gives semiconductor devices the capability of handling much higher levels of power. Write balanced nuclear equations for the two steps in the preparation of $^{31}$P from $^{30}$Si.

**16.** The most convenient way to prepare the element polonium is to expose bismuth (which is 100% $^{209}$Bi) to neutrons. Write balanced nuclear equations for the two steps in the preparation of polonium.

**17.** One convenient source of neutrons is the reaction of an alpha particle from an emitter such as polonium ($^{210}$Po) with an atom of beryllium ($^{9}$Be). Write nuclear equations for the reactions that occur.

**18.** Three atoms of element 111 were produced in 1994 by bombarding $^{209}$Bi with $^{64}$Ni.

(a) Write a balanced equation for this nuclear reaction. What other species is produced?

(b) Write a balanced equation for the alpha decay process of this nuclide of element 111.

## Kinetics of Radioactive Decay

**19.** How many radioactive disintegrations occur per minute in a 0.0010-g sample of $^{209}$Po that has been freshly separated from its decay products? The half-life of $^{209}$Po is 103 years.

**20.** How many alpha particles are emitted per minute by a 0.0010-g sample of $^{238}$U that has been freshly separated from its decay products? Assume that each decay emits one alpha particle. The half-life of $^{238}$U is $4.47 \times 10^9$ years.

**21.** The nuclide $^{19}$O, prepared by neutron irradiation of $^{19}$F, has a half-life of 29 s.

(a) How many $^{19}$O atoms are in a freshly prepared sample if its decay rate is $2.5 \times 10^4$ s$^{-1}$?

(b) After 2.00 min, how many $^{19}$O atoms remain?

**22.** The nuclide $^{35}$S decays by beta emission with a half-life of 87.1 days.

(a) How many grams of $^{35}$S are in a sample that has a decay rate from that nuclide of $3.70 \times 10^2$ s$^{-1}$?

(b) After 365 days, how many grams of $^{35}$S remain?

**23.** Astatine is the rarest naturally occurring element, with $^{219}$At

appearing as the product of a very minor side branch in the decay of $^{235}$U (itself not a very abundant isotope). It is estimated that the mass of all the naturally occurring $^{219}$At in the upper kilometer of the earth's surface has a steady value of only 44 mg. Calculate the total activity (in disintegrations per second) caused by all the naturally occurring astatine in this part of the earth. The half-life of $^{219}$At is 54 s, and its atomic mass is 219.01 u.

**24.** Technetium has not been found in nature. It can be obtained readily as a product of uranium fission in nuclear power plants, however, and is now produced in quantities of many kilograms per year. One medical use relies on the tendency of $^{99m}$Tc (an excited nuclear state of $^{99}$Tc) to concentrate in abnormal heart tissue. Calculate the total activity (in disintegrations per second) caused by the decay of 1.0 $\mu$g of $^{99m}$Tc, which has a half-life of 6.0 hours.

**25.** The specific activity of $^{14}$C in the biosphere is 0.255 Bq g$^{-1}$. What is the age of a piece of papyrus from an Egyptian tomb if its beta counting rate is 0.153 Bq g$^{-1}$? The half-life of $^{14}$C is 5730 years.

**26.** The specific activity of an article found in the Lascaux Caves in France is 0.0375 Bq g$^{-1}$. Calculate the age of the article.

**27.** Over geological time, an atom of $^{238}$U decays to a stable $^{206}$Pb atom in a series of eight alpha emissions, each of which leads to the formation of one helium atom. A geochemist analyzes a rock and finds that it contains $9.0 \times 10^{-5}$ cm$^3$ of helium (at 0°C and atmospheric pressure) per gram and $2.0 \times 10^{-7}$ g of $^{238}$U per gram. Estimate the age of the mineral, given that $t_{1/2}$ of $^{238}$U is $4.47 \times 10^9$ years.

**28.** The isotope $^{232}$Th decays to $^{208}$Pb by the emission of six alpha particles, with a half-life of $1.39 \times 10^{10}$ years. Analysis of 1.00 kg of ocean sediment shows it to contain 7.4 mg of $^{232}$Th and $4.9 \times 10^{-3}$ cm$^3$ of gaseous helium at 0°C and atmospheric pressure. Estimate the age of the sediment, assuming no loss or gain of thorium or helium from the sediment since its formation and assuming that the helium arose entirely from the decay of thorium.

**29.** The half-lives of $^{235}$U and $^{238}$U are $7.04 \times 10^8$ years and $4.47 \times 10^9$ years, respectively, and the present abundance ratio is $^{238}$U/$^{235}$U = 137.7. It is thought that their abundance ratio was 1 at some time *before* our earth and solar system were formed about $4.5 \times 10^9$ years ago. Estimate how long ago the supernova occurred that supposedly produced all the uranium isotopes in equal abundance, including the two longest-lived isotopes, $^{238}$U and $^{235}$U.

**30.** Using the result of problem 29 and the accepted age of the earth, $4.5 \times 10^9$ yr, calculate the $^{238}$U/$^{235}$U ratio at the time the earth was formed.

## Radiation in Biology and Medicine

**31.** Write balanced equations for the decays of $^{11}$C and $^{15}$O, both of which are used in positron emission tomography to scan the uptake of glucose in the body.

**32.** Write balanced equations for the decays of $^{13}$N and $^{18}$F, two other radioisotopes that are used in positron emission tomography. What is the ultimate fate of the positrons?

33. The positrons emitted by $^{11}$C have a maximum kinetic energy of 0.99 MeV, and those emitted by $^{15}$O have a maximum kinetic energy of 1.72 MeV. Calculate the ratio of the number of millirems of radiation exposure caused by ingesting a given fixed chemical amount (equal numbers of atoms) of each of these radioisotopes.

34. Compare the relative health risks of contact with a given amount of $^{226}$Ra, which has a half-life of 1622 years and emits 4.78-MeV alpha particles, with contact with the same chemical amount of $^{14}$C, which has a half-life of 5730 years and emits beta particles with energies of up to 0.155 MeV.

35. The nuclide $^{131}$I undergoes beta decay with a half-life of 8.041 days. Large quantities of this nuclide were released into the environment in the Chernobyl accident. A victim of radiation poisoning has absorbed $5.0 \times 10^{-6}$ g (5.0 $\mu$g) of $^{131}$I.
    (a) Compute the activity, in becquerels, of the $^{131}$I in this person, taking the atomic mass of the nuclide to equal 131 g mol$^{-1}$.
    (b) Compute the radiation absorbed dose, in millirads, caused by this nuclide during the first *second* after its ingestion. Assume that beta particles emitted by $^{131}$I have an average kinetic energy of 0.40 MeV, that all of this energy is deposited within the victim's body, and that the victim weighs 60 kg.
    (c) Is this dose likely to be lethal? Remember that the $^{131}$I diminishes its activity as it decays.

36. The nuclide $^{239}$Pu undergoes alpha decay with a half-life of $2.411 \times 10^4$ years. An atomic energy worker breathes in $5.0 \times 10^{-6}$ g (5.0 $\mu$g) of $^{239}$Pu, which lodges permanently in a lung.
    (a) Compute the activity, in becquerels, of the $^{239}$Pu ingested, taking the atomic mass of the nuclide to be 239 g mol$^{-1}$.
    (b) Determine the radiation absorbed dose, in millirads, during the first *year* after its ingestion. Assume that alpha particles emitted by $^{239}$Pu have an average kinetic energy of 5.24 MeV, that all of this energy is deposited within the worker's body, and that the worker weighs 60 kg.
    (c) Is this dose likely to be lethal?

## Nuclear Fission and Fusion

37. Strontium-90 is one of the most hazardous products of atomic weapons testing because of its long half-life ($t_{1/2} = 28.1$ years) and its tendency to accumulate in the bones.
    (a) Write nuclear equations for the decay of $^{90}$Sr via the successive emission of two beta particles.
    (b) The atomic mass of $^{90}$Sr is 89.9073 u and that of $^{90}$Zr is 89.9043 u. Calculate the energy released per $^{90}$Sr atom, in MeV, in decaying to $^{90}$Zr.
    (c) What will be the initial activity of 1.00 g of $^{90}$Sr released into the environment, in disintegrations per second?

(d) What activity will the material from part (c) show after 100 years?

38. Plutonium-239 is the fissionable isotope produced in breeder reactors; it is also produced in ordinary nuclear plants and in weapons tests. It is an extremely poisonous substance with a half-life of 24,100 years.
    (a) Write an equation for the decay of $^{239}$Pu via alpha emission.
    (b) The atomic mass of $^{239}$Pu is 239.05216 u, and that of $^{235}$U is 235.04393 u. Calculate the energy released per $^{239}$Pu atom, in MeV, in decaying via alpha emission.
    (c) What will be the initial activity, in disintegrations per second, of 1.00 g of $^{239}$Pu buried in a disposal site for radioactive wastes?
    (d) What activity will the material from part (c) show after 100,000 years?

39. The three naturally occurring isotopes of uranium are $^{234}$U (half-life $2.5 \times 10^5$ years), $^{235}$U (half-life $7.0 \times 10^8$ years), and $^{238}$U (half-life $4.5 \times 10^9$ years). As time passes, will the average atomic mass of the uranium in a sample taken from nature increase, decrease, or remain constant?

40. Natural lithium consists of 7.42% $^6$Li and 92.58% $^7$Li. Much of the tritium ($^3_1$H) used in experiments with fusion reactions is made by the capture of neutrons by $^6$Li atoms.
    (a) Write a balanced nuclear equation for the process. What is the other particle produced?
    (b) After $^6$Li is removed from natural lithium, the remainder is sold for other uses. Is the molar mass of the leftover lithium greater or smaller than that of natural lithium?

41. Calculate the amount of energy released, in kilojoules per *gram* of uranium, in the fission reaction

$$^{235}_{92}\text{U} + ^{0}_{1}n \longrightarrow ^{94}_{36}\text{Kr} + ^{139}_{56}\text{Ba} + 3\,^{1}_{0}n$$

Use the atomic masses in Table 12-1. The atomic mass of $^{94}$Kr is 93.919 u, and that of $^{139}$Ba is 138.909 u.

42. Calculate the amount of energy released, in kilojoules per *gram* of deuterium ($^2$H), for the fusion reaction

$$^{2}_{1}\text{H} + ^{2}_{1}\text{H} \longrightarrow ^{4}_{2}\text{He}$$

Use the atomic masses in Table 12-1. Compare your answer with that from the preceding problem.

## Additional Problems

43. When an electron and a positron meet, they are replaced by two gamma rays, called the "annihilation radiation." Calculate the energies of these radiations, assuming that the kinetic energies of the particles are 0.

44. The nuclide $^{231}_{92}$U converts spontaneously to $^{231}_{91}$Pa.
    (a) Write two balanced nuclear equations for this conversion, one if it proceeds by electron capture and the other if it proceeds by positron emission.
    (b) Using the nuclidic masses in Table 16-1, calculate the change in mass for each process. Explain why electron cap-

ture can occur spontaneously in this case, but positron emission cannot.

**45.** The radioactive nuclide $^{64}_{29}Cu$ decays by beta emission to $^{64}_{30}Zn$ or by positron emission to $^{64}_{28}Ni$. The maximum kinetic energy of the beta particles is 0.58 MeV, and that of the positrons is 0.65 MeV. The mass of the neutral $^{64}_{29}Cu$ atom is 63.92976 u.
(a) Calculate the mass, in atomic mass units, of the neutral $^{64}_{30}Zn$ atom.
(b) Calculate the mass, in atomic mass units, of the neutral $^{64}_{28}Ni$ atom.

**46.** A puzzling observation that led to the discovery of isotopes was the fact that lead obtained from uranium-containing ores had an atomic mass lower by two full atomic mass units than lead obtained from thorium-containing ores. Explain this result, using the fact that decay of radioactive uranium and thorium to stable lead occurs via alpha and beta emission.

**47.** By 1913 the elements radium, actinium, thorium, and uranium had all been discovered, but element 91, between thorium and uranium in the periodic table, was not yet known. The approach used by Meitner and Hahn was to look for the parent that decays to form actinium. Alpha and beta emission are the most important decay pathways among the heavy radioactive elements. What elements would decay to actinium by each of these two pathways? If radium salts show no sign of actinium, what does this suggest about the parent of actinium? What is the origin of the name of element 91, discovered by Meitner and Hahn in 1918?

**48.** Working in Rutherford's laboratory in 1932, Cockcroft and Walton bombarded a lithium target with 700-keV protons and found that the following reaction occurred:

$$^{7}_{3}Li + ^{1}_{1}H \longrightarrow 2 \, ^{4}_{2}He$$

Each of the alpha particles was found to have a kinetic energy of 8.5 MeV. This research provided the first experimental test of Einstein's $\Delta E = c^2 \, \Delta m$ relationship. Discuss. Using the atomic masses from Table 12-1, calculate the value of $c$ needed to account for this result.

**49.** (a) Calculate the binding energy per nucleon in $^{30}_{15}P$.
(b) The radioactive decay of the $^{30}_{15}P$ occurs through positron emission. Calculate the maximum kinetic energy carried off by the positron.
(c) The half-life for this decay is 150 s. Calculate the rate constant $k$ and the fraction remaining after 450 s.

**50.** Selenium-82 undergoes *double* beta decay:

$$^{82}_{34}Se \longrightarrow ^{82}_{36}Kr + 2 \, ^{0}_{-1}e^- + 2 \, \tilde{\nu}$$

This low-probability process occurs with a half-life of $3.5 \times 10^{27}$ s, one of the longest half-lives ever measured. Estimate the activity in an 82.0-g (1.00 mol) sample of this isotope. How many $^{82}Se$ nuclei decay in a day?

**51.** Gallium citrate, which contains the radioactive nuclide $^{67}Ga$, is used in medicine as a tumor-seeking agent. Gallium-67 decays with a half-life of 77.9 hours. How much time is required for it to decay to 5.0% of its initial activity?

**52.** The nuclide $^{241}Am$ is used in smoke detectors. As it decays (with a half-life of 458 years), the emitted alpha particles ionize the air. When combustion products enter the detector, the number of ions changes and with it the conductivity of the air, setting off an alarm. If the activity of $^{241}Am$ in the detector is $3 \times 10^4$ Bq, calculate the mass of $^{241}Am$ present.

**53.** The half-life of $^{14}C$ is $t_{1/2} = 5730$ years, and 1.00 g of modern wood charcoal has an activity of 0.255 Bq.
(a) Calculate the number of $^{14}C$ atoms per gram of carbon in modern wood charcoal.
(b) Calculate the fraction of carbon atoms in the biosphere that are $^{14}C$.

**54.** Carbon-14 is produced in the upper atmosphere by the reaction

$$^{14}_{7}N + ^{1}_{0}n \longrightarrow ^{14}_{6}C + ^{1}_{1}H$$

where the neutrons come from nuclear processes induced by cosmic rays. It is estimated that the steady-state $^{14}C$ activity in the biosphere is $1.1 \times 10^{19}$ Bq.
(a) Estimate the total mass of carbon in the biosphere, using the data in problem 53.
(b) The earth's crust has an average carbon content of 250 parts per million by mass, and the total crustal mass is $2.9 \times 10^{25}$ g. Estimate the fraction of the carbon in the earth's crust that is part of the biosphere. Speculate on the whereabouts of the rest of the carbon in the earth's crust.

**55.** Analysis of a rock sample shows that it contains 0.42 mg of $^{40}Ar$ for every 1.00 mg of $^{40}K$. Assuming that all the argon resulted from decay of the potassium and that neither element has left or entered the rock since its formation, estimate the age of the rock. (*Hint:* Use data from Table 12-2.) Note that not all the $^{40}K$ decays to $^{40}Ar$.

**\*56.** Cobalt-60 and iodine-131 are used in treatments for some types of cancer. Cobalt-60 decays with a half-life of 5.27 years, emitting beta particles with a maximum energy of 0.32 MeV. Iodine-131 decays with a half-life of 8.04 days, emitting beta particles with a maximum energy of 0.60 MeV.
(a) Suppose a fixed small number of moles of each of these isotopes were to be ingested and remain in the body indefinitely. What is the *ratio* of the number of millirems of total lifetime radiation exposure that would be caused by the two radioisotopes?
(b) Now suppose that the contact with each of these isotopes is for a fixed short period, such as 1 hour. What is the ratio of millirems of radiation exposure for the two in this case?

**57.** Boron is used in control rods in nuclear power reactors because it is a good neutron absorber. When the isotope $^{10}B$ captures a neutron, an alpha particle (helium nucleus) is emitted. What other atom is formed? Write a balanced equation.

**\*58.** The average energy released in the fission of a $^{235}U$ nucleus is about 200 MeV. Suppose the conversion of this energy to electrical energy is 40% efficient. What mass of $^{235}U$ is converted to its fission products in a year's operation of a 1000-megawatt nuclear power station? Recall that 1 W is 1 J s$^{-1}$.

**59.** The energy released by a bomb is sometimes expressed in tons of TNT. When one ton of TNT (trinitrotoluene) explodes, $4 \times 10^9$ J of energy is released. The fission of one mole of uranium releases approximately $2 \times 10^{13}$ J of energy. Calculate the energy released by the fission of 1.2 kg of uranium in a small atomic bomb. Express your answer in tons of TNT.

**60.** The solar system abundances of the elements Li, Be, and B are four to seven orders of magnitude lower than those of the elements that immediately follow them: C, N, and O. Explain.

**\*61.** The sun's distance from earth is approximately $1.50 \times 10^8$ km and the earth's radius is 6371 km. The earth receives radiant energy from hydrogen burning in the sun at a rate of $0.135$ J s$^{-1}$ cm$^{-2}$. Using the data of Table 12-1, calculate the mass of hydrogen converted per second in the sun.

**CUMULATIVE PROBLEMS**

**62.** Examine the ratio of atomic mass to atomic number for the elements with *even* atomic number through calcium. This ratio is approximately the ratio of the average mass number to the atomic number.
(a) Which two elements stand out as different in this set of ten?
(b) What would be the ''expected'' atomic mass of argon, based on the correlation considered here?
(c) Show how the anomaly in the ordering of natural atomic masses of argon and potassium can be accounted for by the formation of ''extra'' $^{40}$Ar via decay of $^{40}$K atoms.

**63.** Hydrazine, $N_2H_4(\ell)$, reacts with oxygen in a rocket engine to form nitrogen and water vapor:

$$N_2H_4(\ell) + O_2(g) \longrightarrow N_2(g) + 2\,H_2O(g)$$

(a) Calculate $\Delta H°$ for this highly exothermic reaction at 25°C, using data from Appendix D.
(b) Calculate $\Delta E°$ of this reaction at 25°C.
(c) Calculate the total change in mass (in grams) during the reaction of 1.00 mol of hydrazine.

**64.** The long-lived isotope of radium, $^{226}$Ra, decays by alpha particle emission to its daughter radon, $^{222}$Rn, with a half-life of 1622 years. The energy of the alpha particle is 4.79 MeV. Suppose 1.00 g of $^{226}$Ra, freed of all its radioactive progeny, were placed in a calorimeter that contained 10.0 g of water, initially at 25°C. Neglecting the heat capacity of the calorimeter and heat loss to the surroundings, calculate the temperature the water would reach after 1.00 hour. Take the specific heat of water to be 4.18 J K$^{-1}$ g$^{-1}$.

**65.** The radioactive nuclide $^{232}_{90}$Th has a half-life of $1.39 \times 10^{10}$ years. It decays by a series of consecutive steps, the first two of which involve $^{228}_{88}$Ra (half-life 6.7 years) and $^{228}_{89}$Ac (half-life 6.13 hours).
(a) Write balanced equations for the first two steps in the decay of $^{232}$Th, indicating all decay products. Calculate the total kinetic energy carried off by the decay products.
(b) After a short initial time, the rate of formation of $^{228}$Ra becomes equal to its rate of decay. Express the number of $^{228}$Ra nuclei in terms of the number of $^{232}$Th nuclei, using the steady-state approximation from Section 11-4.

**66.** Zirconium is used in the fuel rods of most nuclear power plants. The following half-cell reduction potential applies to aqueous acidic solution:

$$ZrO_2(s) + 4\,H_3O^+(aq) + 4e^- \longrightarrow Zr(s) + 6\,H_2O(\ell)$$
$$\mathscr{E}° = -1.43 \text{ V}$$

(a) Predict whether zirconium can reduce water to hydrogen. Write a balanced equation for the overall reaction.
(b) Calculate $\Delta\mathscr{E}°$ and $K$ for the reaction in part (a).
(c) Can your answer to part (b) explain the release of hydrogen in the Three Mile Island accident and the much greater release of hydrogen (which subsequently exploded) at Chernobyl?

CHAPTER

# 13

*Calcium flame test.*

# Quantum Mechanics and Atomic Structure

Science can progress in different ways. On the one hand, the slow and steady accumulation of experimental results that support and refine existing models can lead to more satisfactory descriptions of natural phenomena. On the other hand, new experiments can give results that directly contradict previously accepted theories. In this case, a period of uncertainty ensues; it is resolved only through the eventual emergence of a new and more complete theory that is capable of explaining both the previously understood results and the new experiments. Such a process can be called a *scientific revolution*. In the first 25 years of this century, a revolution in physics led to the development of quantum theory. Since that time, chemistry has also been profoundly affected by the new concepts.

**480**

One of the fundamental assumptions of early science was that nature is continuous; that is, nature does not make jumps. On a macroscopic scale, this seems true enough. We can measure out an amount of graphite (carbon) of mass 9 kg, or 8.23 kg, or 6.4257 kg, and it appears that the mass can have any value provided our balance is sufficiently accurate. On an atomic scale, however, this apparently continuous behavior breaks down. An analogy may be useful here. A sand beach from a distance appears smooth and continuous, but a close look reveals that it is made up of individual grains of sand. This same "graininess" is found in matter observed on an atomic scale. The mass of carbon ($^{12}C$) comes in "packets," each of which weighs $1.99265 \times 10^{-26}$ kg. Although in principle two, three, or any integral number of such packets can be "weighed out," we cannot obtain $1\frac{1}{2}$ packets. Carbon is not a continuous material but comes in chunks, each containing the minimum measurable mass of carbon: that of an atom. Similarly, electric charge comes in packets of size $e$, as shown in Section 1-4, and we cannot obtain nonintegral numbers of packets in experiments involving chemical energies. The central idea of quantum theory is that energy also is not continuous but comes in discrete packets.

To understand the far-reaching nature of the quantum revolution, consider the state of physics at the end of the 19th century. The 200 years that followed the seminal work of Isaac Newton were the classical period in the study of mechanics, the branch of physics that predicts the motions of particles and the collections of particles that make up working mechanisms. By the end of that period, about 1900, physicists had achieved a deep understanding that successfully dealt with problems ranging from the motions of the planets in their orbits to the design of a bicycle. These attainments now make up the subject called *classical mechanics,* and they are eminently useful. In classical mechanics, a fundamental role is played by energy, which has two parts: kinetic energy, arising from the motion of particles, and potential energy, arising from the interactions of particles with one another or with an external field (such as gravity) and depending on the *positions* of the particles. Using classical mechanics, one can predict the future positions of a group of particles from their present positions and momenta and the forces between them. A major triumph of classical mechanics was the kinetic theory of gases outlined in Section 3-5.

At the end of the 19th century, it was naturally thought that the motion of elementary particles—such as the recently discovered electron—could be described by classical mechanics and that, once the correct laws of force were discovered, the properties of atoms and molecules could be predicted to any desired degree of accuracy by solving Newton's equations of motion. It was believed that all the fundamental laws of physics had been discovered. At the dedication of the Ryerson Physics Laboratory at the University of Chicago in 1894, A. A. Michelson said that "our future discoveries must be looked for in the sixth decimal place." Little did he imagine the far-reaching changes that would shake physics and chemistry during the next 30 years. Central to those changes was the discovery that all particles have wave-like properties, the effects of which are most pronounced for small, low-mass particles such as electrons in atoms. The incorporation of wave and particle aspects of matter into a single comprehensive theory is the achievement of *quantum mechanics,* the great 20th-century extension of classical mechanics. In this chapter we discuss the origins of the quantum theory, the theory's implications for the structure of atoms, and its connection to the periodic table. In subsequent chapters we shall consider the consequences of the quantum nature of matter for chemical bonds, coordination complexes, and the solid state.

| Table 13-1   Kinds of Waves | |
| --- | --- |
| **Wave** | **Oscillating Quantity** |
| Water | Height of water surface |
| Sound | Density of air |
| Light | Electric and magnetic fields |
| Chemical | Concentrations of chemical species |

## 13-1 WAVES AND LIGHT

Many kinds of waves are studied in physics and chemistry. A familiar example is water waves stirred up by the winds over the oceans, set off by a stone dropped into a quiet pool, or created by a laboratory water-wave machine. Sound waves, periodic compressions of the air that move from a source to a detector such as the human ear, are another example. Light waves, as we shall see, consist of oscillating electric and magnetic fields moving through space. Even some chemical reactions occur in such a way that waves of color pass through the sample as the reaction proceeds. Common to all these wave phenomena is the oscillatory variation of some property with time at a given fixed location in space (Table 13-1).

A snapshot of a water wave (Fig. 13-1) records the peaks and troughs present at some instant in time. The **amplitude** of the wave is the maximum displacement of the water surface over the undisturbed level of the water. The distance between two successive peaks (or troughs) is called the **wavelength** $\lambda$ (Greek lambda) of the wave, provided that this distance is reproducible from peak to peak. The **frequency** of a water wave can be measured by counting the number of peaks or troughs that pass a fixed point in space per second. The frequency, $\nu$ (Greek nu), has units of waves (or cycles) per second or simply $s^{-1}$. For example, if 12 water-wave peaks pass a certain point in 30 s, the frequency is

$$\text{frequency} = \nu = \frac{12}{30 \text{ s}} = 0.40 \text{ s}^{-1}$$

The wavelength and frequency of a wave determine its speed—the rate at which a particular wave crest moves through the medium. As Figure 13-1 shows, in a time interval $\Delta t = \nu^{-1}$ the wave moves through one wavelength, and so the speed (the distance traveled divided by the time elapsed) is

$$\text{speed} = \frac{\text{distance traveled}}{\text{time elapsed}} = \frac{\lambda}{\nu^{-1}} = \lambda \nu$$

The speed of a wave is the product of its wavelength and its frequency.

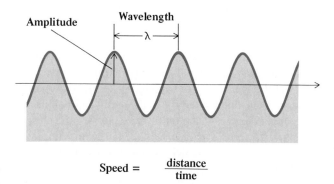

### Figure 13-1

As a water wave moves across an otherwise calm tank, its amplitude and wavelength can be determined. Its speed is found by dividing the travel distance of a particular wave crest by the time elapsed.

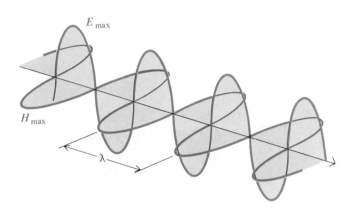

**Figure 13-2**

A laser emits a sharply focused beam of light with a very narrow range of wavelengths. The direction of motion of a laser beam can be manipulated by inserting mirrors in the beam's path. (Daniel Quat/ Phototake)

**Figure 13-3**

Light consists of waves of oscillating electric ($E$) and magnetic ($H$) fields that are perpendicular to each other and to the direction of propagation of the light.

## Electromagnetic Radiation

J. C. Maxwell had demonstrated by 1865 that light is **electromagnetic radiation.** A beam of light emitted by a laser (Fig. 13-2) consists of oscillating electric and magnetic fields perpendicular to the direction in which the light is propagating (Fig. 13-3). The speed $c$ of light passing through a vacuum is equal to the product $\lambda\nu$:

$$c = \lambda\nu = 2.9979 \times 10^8 \text{ m s}^{-1} \qquad \textbf{(13-1)}$$

The speed $c$ is a universal constant,[1] the same for all types of light, although the wavelength and frequency depend on the color of the light. Green light has a range of frequencies near $5.7 \times 10^{14} \text{ s}^{-1}$ and wavelengths near $5.3 \times 10^{-7}$ m (530 nm). Red light has a lower frequency and a longer wavelength than green, and violet light a higher frequency and shorter wavelength than green (Fig. 13-4). A laser such as that shown in Figure 13-2 emits nearly monochromatic light (light with a single frequency and wavelength), but white light contains the full range of visible wavelengths. White light can be resolved into its component wavelengths with a prism.

The light that is visible to the eye is only a very small part of the electromagnetic spectrum (Fig. 13-4). Light can have wavelengths longer than red light or shorter than violet. The warmth that radiates from a stone pulled out of a fire

---

[1] This constancy of $c$ has led to a new definition of the meter. One meter is now defined as the distance traveled by light in a vacuum during 1/299,792,458 second, and the speed of light is fixed at 299,792,458 m s$^{-1}$. The meter is no longer an independent unit but is defined in terms of the second.

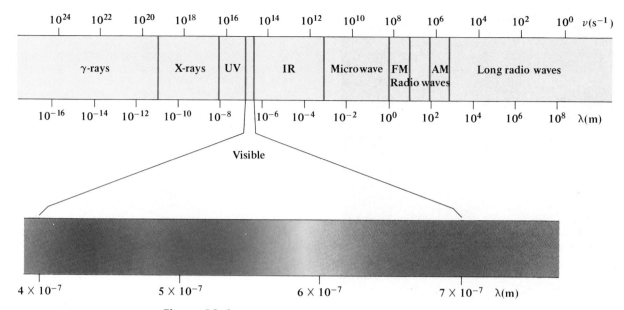

**Figure 13-4**

The electromagnetic spectrum. Note the small fraction that is visible to the human eye.

consists largely of infrared (IR) radiation, which has a wavelength longer than that of visible light. Microwave ovens employ radiation of greater than infrared wavelength, and still longer waves are employed in radio communication. The position of a radio station on the dial is given by its frequency in hertz (Hz), where 1 Hz is one cycle per second. Thus, FM stations typically broadcast at frequencies of tens to hundreds of megahertz (1 MHz = $10^6$ s$^{-1}$); AM stations have much lower broadcast frequencies, from hundreds to thousands of kilohertz (1 kHz = $10^3$ s$^{-1}$). Types of light with wavelengths shorter than those of visible light include ultraviolet (UV) light, x-rays, and gamma rays. The latter two kinds of radiation, with their high frequencies and short wavelengths, are more penetrating than visible or infrared radiation and so can be used to image bones or organs beneath the skin.

## Example 13-1

Almost all commercially available microwave ovens employ radiation with a frequency of $2.45 \times 10^9$ s$^{-1}$. Calculate the wavelength of this radiation.

### Solution

The wavelength is related to the frequency through

$$\lambda = \frac{c}{\nu} = \frac{3.00 \times 10^8 \text{ m s}^{-1}}{2.45 \times 10^9 \text{ s}^{-1}} = 0.122 \text{ m}$$

and so the wavelength is 12.2 cm.

*Related Problems:* 3, 4

## Blackbody Radiation

Around the turn of the century, results began to appear that could not be explained by classical physics. One of these arose from the study of **blackbody radiation.** As a body—for example, a bar of iron—is heated, it emits radiation and becomes first red, then orange, then yellow, then white as its temperature increases. The distribution of frequencies of radiated light changes with the temperature of the body. It is exactly this effect that gives stars of different temperatures their different colors. A perfectly absorbing material with a hollowed-out internal cavity is an idealized version of the iron bar. Only a small amount of radiation escapes from a small hole in the body and is detected. According to the classical theory of electromagnetism, the energy carried by a light wave is proportional to the square of its amplitude:

$$\text{energy} \propto (E_{\max})^2 \propto \text{intensity}$$

Figure 13-5 shows two typical experimental curves for the dependence of light intensity on wavelength. Also shown is the prediction of the classical theory of radiation,

$$\text{intensity} \propto kT/\lambda^4$$

which is in complete disagreement with experiment. This is referred to as an "ultraviolet catastrophe" because it predicts an infinite intensity at very short wavelengths.

   The blackbody radiation paradox was resolved by Max Planck in 1900. It had been postulated that radiation of a given frequency arises from a group of atoms on the surface of a solid body oscillating with that frequency. Because there are many more oscillators with high frequencies than with low, even a small amount of energy placed in them would shift the radiation to high frequencies, leading to the ultraviolet catastrophe. Planck made the daring hypothesis that it was *not* possible to put an arbitrarily small amount of energy into an oscillator of frequency $\nu$. Instead, he suggested that the energy must come in "packets," or *quanta,* of magnitude $h\nu$, and the minimum energy allowed is $h\nu$ itself. In this scheme, most of the high-frequency

**Figure 13-5**

The dependence of the intensity of blackbody radiation on wavelength, for two temperatures: 5000 K (red) and 7000 K (blue). The classical theory (dashed curves) disagrees with observation at shorter wavelengths. The sun has a blackbody temperature near 6000 K, and its light-intensity curve lies between the two shown.

oscillators can gain no energy because the packets of energy that are large enough to be accepted by them are scarce. Planck's hypothesis explains why the spectral intensity falls off at high frequencies and describes the experimental observations quite accurately. The constant $h$ in the relation

$$E = h\nu \tag{13-2}$$

has since been measured to very high precision. It is referred to as **Planck's constant** and has the value

$$h = 6.62608 \times 10^{-34}\,\text{J s}$$

## The Photoelectric Effect

A second paradox arose from the observation of the **photoelectric effect.** A beam of light shining onto a metal surface can eject electrons and cause an electric current (called a *photocurrent*) to flow. This fact in itself was not difficult to understand, because electromagnetic radiation carries energy. The problem came in explaining the dependence of the effect on the frequency of the light. It was found that for light frequencies less than a certain threshold frequency $\nu_0$, no electrons were ejected; once the frequency exceeded the threshold, the photocurrent increased rapidly. According to classical theory, the energy associated with electromagnetic radiation depends only on the intensity and not on the frequency. Why, then, could a very weak beam of blue light eject electrons from sodium when an intense red beam had no effect (Fig. 13-6a)? Further experiments measured the kinetic energy of the ejected electrons and revealed that the maximum kinetic energy varies linearly with frequency, as shown in Figure 13-6b. This behavior also was inexplicable in classical physics.

In 1905 Einstein used Planck's quantum hypothesis to explain the photoelectric effect. He suggested that light consists of a series of "packets" of energy called **photons,** each of which carries energy $E = h\nu$. If a low-frequency photon strikes an electron on a metal surface, it does not have enough energy to remove the electron. By contrast, a photon of high-frequency light carries sufficient energy to eject the electron. If it requires an energy of at least $h\nu_0$ to overcome the binding of the electrons to the atoms of the surface, then from the law of conservation of energy any excess energy $h\nu - h\nu_0$ will appear as kinetic energy of the outgoing electron:

$$\text{maximum kinetic energy} = \tfrac{1}{2}m_e v^2 = h(\nu - \nu_0) = h\nu - \Phi \tag{13-3}$$

That is what is observed experimentally (Fig. 13-6b). The constant $h$ obtained from this experiment is Planck's constant, identical to that from the blackbody radiation experiment; $h\nu_0$ is equal to the energy $\Phi$ binding the ejected electron to the surface and is called the **work function** of the material.

### Example 13-2

Light with a wavelength of 400 nm strikes the surface of cesium in a photocell, and the maximum kinetic energy of the electrons ejected is $1.54 \times 10^{-19}$ J. Calculate the work function of cesium and the longest wavelength of light that is capable of ejecting electrons from that metal.

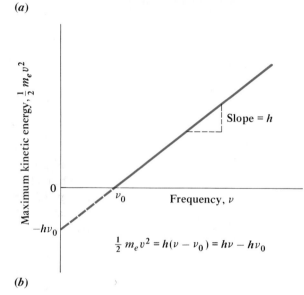

$$\tfrac{1}{2} m_e v^2 = h(\nu - \nu_0) = h\nu - h\nu_0$$

(a)

(b)

**Figure 13-6**

(a) The photoelectric effect. Blue light is effective in ejecting electrons from the surface of this metal, but red light is not. (b) The maximum kinetic energy of the ejected electrons varies linearly with the frequency of light used.

## Solution

The frequency of the light is

$$\nu = \frac{c}{\lambda} = \frac{3.00 \times 10^8 \text{ m s}^{-1}}{4.00 \times 10^{-7} \text{ m}} = 7.50 \times 10^{14} \text{ s}^{-1}$$

The binding energy $h\nu_0$ can be calculated from Einstein's formula:

$$(\tfrac{1}{2} m_e v^2)_{\text{max}} = h\nu - h\nu_0$$

$$1.54 \times 10^{-19} \text{ J} = (6.626 \times 10^{-34} \text{ J s})(7.50 \times 10^{14} \text{ s}^{-1}) - h\nu_0$$
$$= 4.97 \times 10^{-19} \text{ J} - h\nu_0$$

$$\Phi = h\nu_0 = (4.97 - 1.54) \times 10^{-19} \text{ J} = 3.43 \times 10^{-19} \text{ J}$$

The minimum frequency $\nu_0$ for the light to eject electrons is then

$$\nu_0 = \frac{3.43 \times 10^{-19} \text{ J}}{6.626 \times 10^{-34} \text{ J s}} = 5.18 \times 10^{14} \text{ s}^{-1}$$

From this, the maximum wavelength $\lambda_0$ is

$$\lambda_0 = \frac{c}{\nu_0} = \frac{3.00 \times 10^8 \text{ m s}^{-1}}{5.18 \times 10^{14} \text{ s}^{-1}} = 5.79 \times 10^{-7} \text{ m} = 579 \text{ nm}$$

*Related Problems:* 13, 14

Planck's and Einstein's results suggest that in some circumstances light displays wave-like properties, but in others it behaves like a stream of particles (photons). We shall see that this **wave–particle duality** is not a contradiction but rather part of the fundamental nature of light, and of matter as well.

## 13-2 ATOMIC SPECTROSCOPY AND THE BOHR MODEL

A third, and in some ways the most fundamental, paradox in the early 1900s was the stability of the atom. Rutherford's experiments (Section 1-4) had shown that the atom consists of a very small, massive nucleus surrounded by electrons. If the negatively charged electrons are stationary, what keeps them from falling into the nucleus in response to the attractive force of the positively charged nucleus? If the electrons move in orbits, why do they not spiral into the nucleus, emitting radiation continuously and with increasing frequency, as predicted by classical theory?

Closely related to this question of atomic stability were experimental observations of atomic spectra. Light that consists of a distribution of wavelengths (Fig. 13-7) can be resolved into its components by passing through a prism, because each wavelength is deviated through a different angle. One instrument used to separate light into its component wavelengths is called a **spectrograph** (Fig. 13-8); it consists of a narrow slit through which the light enters, the prism, and a photographic plate or other detector that records the position and intensity of slit images of the component wavelengths. The *spectrum* is an array of parallel lines corresponding to those slit images, if the light consists of discrete wavelengths, or a continuous band if the light contains all frequencies. Early experiments showed that, rather than emitting a continuous distribution of frequencies, atoms emit light at a discrete set of frequencies that are characteristic of that element (Fig. 13-9). In 1885 J. J. Balmer

**Figure 13-7**

When a gas is excited in an electrical discharge, it glows as it emits light. The colors of the light emitted by three gases—(*a*) neon, (*b*) argon, and (*c*) mercury—are shown. Each emission consists of several wavelengths of light, and the perceived color depends on which wavelength predominates. (Leon Lewandowski)

(*a*)

(*b*)

(*c*)

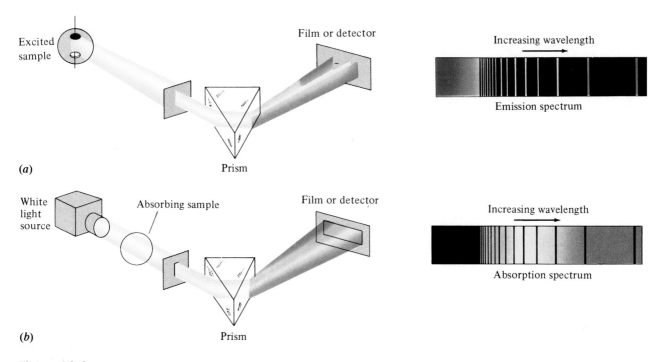

*(a)*                                    Prism

*(b)*                                    Prism

### Figure 13-8

*(a)* The emission spectrum of atoms or molecules is measured by passing the light emitted from an excited sample through a prism to separate it according to wavelength, then
recording the image on photographic film or with another detector. *(b)* In absorption
spectroscopy, white light from a source passes through the unexcited sample, which absorbs certain discrete wavelengths of light. Dark lines appear on a bright background.

discovered that when hydrogen atoms are excited by the passage of an electric
current through gaseous hydrogen, they emit a series of lines in the visible region,
with frequencies given by the simple formula

$$\nu = \left[ \frac{1}{4} - \frac{1}{n^2} \right] \times 3.29 \times 10^{15} \text{ s}^{-1} \qquad n = 3, 4, 5, \ldots$$

### Figure 13-9

Atoms of hydrogen, mercury, and
neon emit light at discrete wavelengths. The pattern seen is characteristic of the element under study.
$1 \text{ Å} = 10^{-10}$ m.

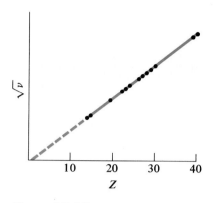

**Figure 13-10**

A graph of data from Moseley's experiments, showing how the frequency of the characteristic x-rays emitted by each element under a bombardment of electrons correlates with that element's atomic number.

In 1913 Henry Moseley measured the frequency of the characteristic x-rays most strongly emitted by different elements excited by bombardment and found the extremely simple relationship

$$\sqrt{\nu} = a(Z - b)$$

where $a$ and $b$ are constants and $Z$ is the atomic number (Fig. 13-10). The existence of discrete line spectra and various empirical equations that relate the frequencies of the lines challenged physicists in the early 1900s.

## The Bohr Atom

In 1913 Niels Bohr developed a model for the atom that accounted for the striking regularities seen in the spectrum of the hydrogen atom and one-electron ions such as $He^+$, $Li^{2+}$, and $Be^{3+}$. His model is sometimes referred to as the planetary model because in it an electron of mass $m_e$ is assumed to move in a circular orbit of radius $r$ about a fixed nucleus, as the earth orbits the sun. According to Coulomb's law, the potential energy of interaction of two charges, $Q_1$ and $Q_2$, separated by a distance $r$ is

$$\text{potential energy} = \frac{Q_1 Q_2}{4\pi\epsilon_0 r}$$

where $\epsilon_0$, called the *permittivity of the vacuum,* is a proportionality constant with numerical value $8.854 \times 10^{-12} \; C^2 \, J^{-1} \, m^{-1}$. For a nucleus (of charge $+Ze$) and an electron (of charge $-e$) this potential energy is

$$\text{potential energy} = -\frac{Ze^2}{4\pi\epsilon_0 r}$$

where the minus sign indicates a net attractive interaction, giving an algebraically lower energy at shorter distances. The total energy (kinetic and potential) of the electron in the atom is

$$E = \tfrac{1}{2}m_e v^2 - \frac{Ze^2}{4\pi\epsilon_0 r} \tag{13-4}$$

The Coulomb force of attraction on the electron, $F_{\text{coul}}$, is the derivative of the potential energy with respect to the separation $r$ and equals $Ze^2/4\pi\epsilon_0 r^2$. From Newton's law relating force and acceleration, $F_{\text{coul}} = m_e a$, and for uniform circular motion, the acceleration $a$ of the electron is $v^2/r$. Therefore,

$$F_{\text{coul}} = \frac{Ze^2}{4\pi\epsilon_0 r^2} = m_e a = m_e \frac{v^2}{r} \tag{13-5}$$

As mentioned earlier, classical physics predicts that a moving electron will emit light of continuously increasing frequency and will gradually spiral inward to the nucleus. Bohr avoided this by simply postulating that only certain discrete orbits (characterized by radius $r_n$ and energy $E_n$) are allowed and that light is emitted or absorbed only when the electron "jumps" from one stable orbit to another.

The *linear* momentum of an electron is the product of its mass and its velocity, $m_e v$. The *angular* momentum is a different quantity that describes turning motion about an axis. For the circular paths of the Bohr model, the angular momentum of the electron is the product of its mass, its velocity, and the radius of the orbit ($m_e v r$). Bohr assumed that the angular momentum is quantized and must be an integral multiple of $h/2\pi$, where $h$ is Planck's constant:

$$\text{angular momentum} = m_e v r = n\frac{h}{2\pi} \qquad n = 1, 2, 3, \ldots \qquad \text{(13-6)}$$

Once this quantization is assumed, the remainder of the calculation is straightforward. Equations 13-5 and 13-6 contain two unknowns, $v$ and $r$. Solving Equation 13-6 for $v$ ($=nh/2\pi m_e r$), inserting it into Equation 13-5, and solving for $r$ give

$$r_n = \frac{\epsilon_0 n^2 h^2}{\pi Z e^2 m_e} = \frac{n^2}{Z} a_0 \qquad \text{(13-7)}$$

where $a_0$, the **Bohr radius,** has the numerical value $5.29 \times 10^{-11}$ m $= 0.529$ Å. We have introduced a new unit of length, the *angstrom* (abbreviated Å), with $1$ Å $= 10^{-10}$ m. It is a convenient unit for expressing the sizes of atoms and molecules.

The velocity $v_n$ corresponding to the orbit with radius $r_n$ is

$$v_n = \frac{nh}{2\pi m_e r_n} = \frac{Ze^2}{2\epsilon_0 nh}$$

The results for $r_n$ and $v_n$ can be substituted into Equation 13-4 to give the energy:

$$E_n = \frac{-Z^2 e^4 m_e}{8\epsilon_0^2 n^2 h^2} = -(2.18 \times 10^{-18} \text{ J})\frac{Z^2}{n^2} \qquad n = 1, 2, 3, \ldots \quad \text{(13-8)}$$

As $n \to \infty$, the electron orbit is farther and farther from the nucleus ($r_n \to \infty$). The energy $E_n$ is 0 for the electron and proton at rest and at an infinite distance apart, and is negative for smaller distances. The state of lowest energy for the system of nucleus plus electron (the state with $n = 1$) is called the **ground state;** states with higher values of $n$ have higher energies and are called **excited states** (Fig. 13-11).

The **ionization energy** of an atom is the minimum energy needed to remove an electron from it. Removing an electron from a hydrogen atom in its ground state involves a change from the $n = 1$ state to the $n = \infty$ state, with the energy change

$$\Delta E = E_{\text{fin}} - E_{\text{init}} = 2.18 \times 10^{-18} \text{ J}$$

Multiplying this by Avogadro's number gives the ionization energy (*IE*) per mole of atoms:

$$IE = (6.022 \times 10^{23} \text{ mol}^{-1})(2.18 \times 10^{-18} \text{ J})$$
$$= 1.31 \times 10^6 \text{ J mol}^{-1} = 1310 \text{ kJ mol}^{-1}$$

This number agrees with the experimentally observed ionization energy of hydrogen atoms.

......................................................................................

## Example 13-3

Consider the $n = 2$ state of the $Li^{2+}$ ion. Using the Bohr model, calculate the radius of the electron orbit, the electron velocity, and the energy of the ion relative to the nucleus and electron separated by an infinite distance.

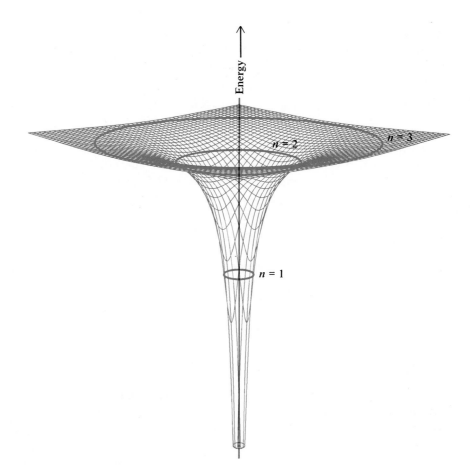

**Figure 13-11**

The interaction between the electron and nucleus has its lowest energy when the electron is closest to the nucleus. The electron moving away from the nucleus can be seen as moving up the sides of a steep potential energy well. In the Bohr theory, it can "catch" and stick on the sides only at certain allowed values of $r$, the radius, and $E$, the energy. The first three of these are shown by rings.

### Solution

Because $Z = 3$ for an $Li^{2+}$ ion (the nuclear charge is $+3e$) and $n = 2$, the radius is

$$r_2 = \frac{n^2}{Z}a_0 = \frac{4}{3}a_0 = \frac{4}{3}(0.529 \text{ Å}) = 0.705 \text{ Å}$$

The velocity is

$$v_2 = \frac{nh}{2\pi m_e r_2} = \frac{2(6.626 \times 10^{-34} \text{ J s})}{2\pi(9.11 \times 10^{-31} \text{ kg})(0.705 \times 10^{-10} \text{ m})} = 3.28 \times 10^6 \text{ m s}^{-1}$$

The energy is

$$E_2 = -\frac{(3)^2}{(2)^2}(2.18 \times 10^{-18} \text{ J}) = -4.90 \times 10^{-18} \text{ J}$$

*Related Problems:* 21, 22

## Atomic Spectra

When a one-electron atom or ion undergoes a transition from a state characterized by quantum number $n_i$ to a state lower in energy with quantum number $n_f$ ($n_i > n_f$), a photon is emitted to carry off the energy $h\nu$ lost by the atom. By conservation of energy, $E_i = E_f + h\nu$, so that

$$h\nu = \frac{Z^2 e^4 m_e}{8\epsilon_0^2 h^2}\left[\frac{1}{n_f^2} - \frac{1}{n_i^2}\right] \quad \text{(emission)}$$

As $n_i$ and $n_f$ take on a succession of integral values, lines are seen in the emission spectrum (Fig. 13-12) with frequencies

$$\nu = \frac{Z^2 e^4 m_e}{8\epsilon_0^2 h^3}\left(\frac{1}{n_f^2} - \frac{1}{n_i^2}\right) = (3.29 \times 10^{15}\ \text{s}^{-1})\, Z^2\left(\frac{1}{n_f^2} - \frac{1}{n_i^2}\right) \tag{13-9}$$

$$n_i > n_f = 1, 2, 3, \ldots \quad \text{(emission)}$$

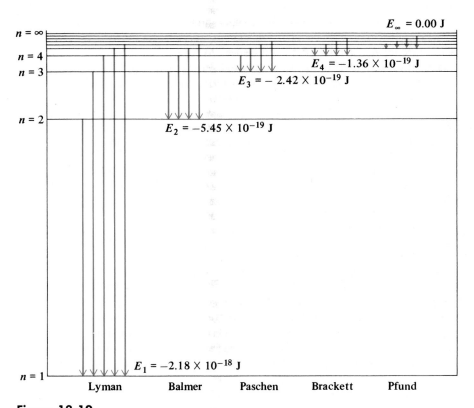

**Figure 13-12**

In the energy levels of the hydrogen atom, the separated electron and proton are arbitrarily assigned zero energy, and all other energies are more negative than that. Atoms emit light as they fall from higher to lower energy levels (blue arrows). Each series of related transitions is named after its discoverer.

On the other hand, an atom can *absorb* energy $h\nu$ from a photon as it undergoes a transition to a *higher* energy state ($n_f > n_i$). In this case, conservation of energy requires $E_i + h\nu = E_f$, and so the absorption spectrum shows a series of lines at frequencies

$$\nu = (3.29 \times 10^{15} \text{ s}^{-1}) Z^2 \left( \frac{1}{n_i^2} - \frac{1}{n_f^2} \right)$$

$$n_f > n_i = 1, 2, 3, \ldots \quad \text{(absorption)}$$

(13-10)

For hydrogen, which has $Z = 1$, the predicted emission spectrum with $n_f = 2$ corresponds to the series of lines in the visible region described by Balmer and mentioned at the beginning of this section. A series of lines at higher frequencies (in the ultraviolet region) is predicted for $n_f = 1$ (the Lyman series), and other series are predicted at lower frequencies (in the infrared region) for $n_f = 3, 4, \ldots$. In fact, the predicted and observed spectra of hydrogen and one-electron ions are in excellent agreement—a major triumph of the Bohr theory.[2]

The Bohr theory has a number of shortcomings, however. The most important is that it is incapable of predicting the energy levels and spectra of atoms and ions with more than one electron. The theory of quantum mechanics that was developed later retains certain features of the Bohr theory and replaces others. The concept of discrete, stationary states and transitions between them is retained, and so is the quantization of angular momentum, although the actual form is slightly different from that assumed by Bohr. On the other hand, the circular orbits of the Bohr theory do not appear in the current quantum mechanics.

## 13-3  WAVES, PARTICLES, AND THE SCHRÖDINGER EQUATION

In his theory for the photoelectric effect, Einstein showed that light has both wave-like and particle-like properties. In 1924 the French physicist Louis de Broglie posed the following question: If light can have particle-like properties, then why cannot matter have wave-like properties? He showed that Bohr's assumption about the quantization of angular momentum in the hydrogen atom could be rationalized by ascribing wave-like properties to the electron.

Up to now we have considered only one type of wave, a **traveling wave.** Electromagnetic radiation (light, x-rays, and gamma rays) is described by such a traveling wave, moving through space at speed $c$. Another type of wave is a **standing wave,** of which a simple example is a guitar string with fixed ends. A plucked string vibrates, but only certain oscillations of the string are possible. Because the ends are fixed, the only oscillations that can persist are those in which a whole number of half-wavelengths fits into the length of string $L$ (see Fig. 13-13). The condition on the allowed wavelengths is

$$n \frac{\lambda}{2} = L \qquad n = 1, 2, 3, \ldots$$

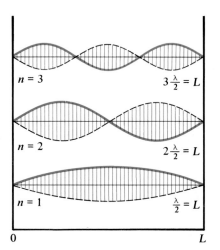

**Figure 13-13**

A string of length $L$ with fixed ends can vibrate in only a restricted set of ways. The positions of largest amplitude for the first three harmonics are shown here. In a standing wave such as this, the whole string is in motion except at its ends and at the nodes.

[2] Strictly speaking, the electron mass $m_e$ in the expressions for the energy levels of one-electron atoms should be replaced by the *reduced mass* $\mu$, equal to $m_e m_N/(m_e + m_N)$ where $m_N$ is the nuclear mass; $\mu$ differs from $m_e$ by less than 0.1%.

It is impossible to create a wave with any other value of $\lambda$ if the ends of the string are fixed. The oscillation with $n = 1$ is called the **fundamental** or first harmonic, and higher values of $n$ correspond to higher harmonics. At certain points within the standing wave the amplitude of oscillation is 0; these points are called **nodes.** (The fixed ends are not counted as nodes.) The higher the number of the harmonic $n$, the more numerous the nodes, the shorter the wavelength, the higher the frequency, and the higher the energy of the standing wave.

De Broglie realized that such standing waves are examples of quantization: only certain discrete states of vibration, characterized by the ''quantum number'' $n$, are allowed for the vibrating string. He suggested that the quantization of a one-electron atom might have the same origin, and that the electron might be associated with a standing wave, in this case a *circular* standing wave about the nucleus of the atom (see Fig. 13-14). For the wave's amplitude to be well defined (single-valued and smooth), a whole number of wavelengths must fit into the circumference of the circle $2\pi r$. The condition on the allowed wavelengths for standing circular waves is

$$n\lambda = 2\pi r \qquad n = 1, 2, 3, \ldots \qquad \text{(13-11)}$$

Bohr's assumption about quantization of the angular momentum of the electron was

$$m_e vr = n\frac{h}{2\pi}$$

which can be rewritten as

$$2\pi r = n\left[\frac{h}{m_e v}\right] \qquad \text{(13-12)}$$

Comparison of de Broglie's equation (13-11) with Bohr's (13-12) shows that the wavelength of the standing wave is related to the linear momentum $p$ of the electron by the simple formula

$$\lambda = \frac{h}{m_e v} = \frac{h}{p} \qquad \text{(13-13)}$$

De Broglie showed with the theory of relativity that exactly the same relationship holds between the wavelength and momentum of a *photon*. Although a photon has no rest mass, it does have momentum $p$, and its energy is $E = pc$. Setting this equal to $h\nu$ (following Einstein) gives $pc = h\nu = hc/\lambda$, or $\lambda = h/p$. De Broglie therefore proposed as a generalization that any particle moving with linear momentum $p$ has wave-like properties and a wavelength $\lambda = h/p$ associated with it.

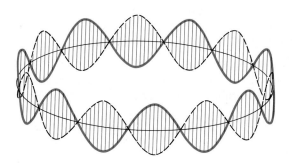

**Figure 13-14**

A circular standing wave on a closed loop. The state shown has $n = 7$, with seven full wavelengths around the circle.

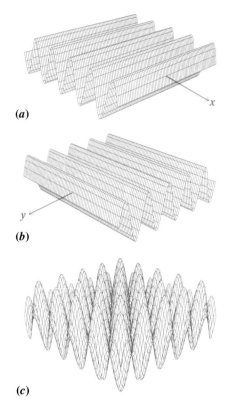

(a)

(b)

(c)

## Figure 13-15

When two waves (a and b) moving perpendicular to one another meet and overlap, their amplitudes add. Where two crests meet, constructive interference occurs and the resultant amplitude is greater than that of either wave alone (shown by the peaks in part c). Where a crest meets a trough, destructive interference gives an amplitude of zero.

Under what circumstances does the wave-like nature of particles become apparent? When waves from two sources pass through the same region of space, they *interfere* with each other. Consider water waves as an example. When two crests meet, *constructive* interference occurs and a higher crest (greater amplitude) appears; where a crest of one wave meets a trough of the other, *destructive* interference (smaller amplitude) occurs (Fig. 13-15). When x-rays are sent into a crystal, they are scattered by the atoms of the crystal (picture a water wave encountering a post embedded in the bottom of a lake). The x-rays scattered from each atom in the crystal interfere with one another to give a characteristic x-ray **diffraction pattern.** From de Broglie's generalization, such diffraction should also be observed when a beam of *electrons* with appropriate momentum (and therefore wavelength) is sent into a crystal. In 1927 C. Davisson and L. H. Germer showed that electrons are, in fact, diffracted by a crystal and that de Broglie's relationship correctly predicts their wavelength. The experiment provided a striking confirmation of de Broglie's hypothesis about the wave-like nature of matter. "Waves" and "particles" are idealized limits, and protons, electrons, and photons all possess both wave and particle aspects.

## Example 13-4

Calculate the de Broglie wavelengths of (a) an electron moving with velocity $1.0 \times 10^6$ m s$^{-1}$ and (b) a baseball of mass 0.145 kg, thrown with a velocity of 30 m s$^{-1}$.

## Solution

(a)
$$\lambda = \frac{h}{p} = \frac{h}{m_e v} = \frac{6.626 \times 10^{-34} \text{ J s}}{(9.11 \times 10^{-31} \text{ kg})(1.0 \times 10^6 \text{ m s}^{-1})}$$

$$= 7.3 \times 10^{-10} \text{ m} = 7.3 \text{ Å}$$

(b)
$$\lambda = \frac{h}{mv} = \frac{6.626 \times 10^{-34} \text{ J s}}{(0.145 \text{ kg})(30 \text{ m s}^{-1})}$$

$$= 1.5 \times 10^{-34} \text{ m} = 1.5 \times 10^{-24} \text{ Å}$$

The latter wavelength is far too small to be observed. For this reason we do not see the wave-like properties of baseballs or other macroscopic objects, but on a microscopic level, electrons moving in atoms show wave-like properties that are essential for atomic structure.

*Related Problems:* 27, 28

## The Heisenberg Indeterminacy Principle

In classical physics, particles have well-defined positions and momenta and follow precise trajectories. On an atomic scale, however, positions and momenta cannot be determined with precision because of the **indeterminacy principle**[3] stated by Werner Heisenberg in 1927.

---

[3] This is also known as the *uncertainty principle*; we use the term "indeterminacy" because it describes the true meaning of the principle more accurately.

The indeterminacy principle is directly connected with the process of measurement. How are the position and momentum of a macroscopic object such as a baseball in motion determined? The simplest way is to take a series of snapshots of it at different times, with each picture a record of the light (photons) scattered by the baseball. To observe an object, we must scatter something (such as a photon) from it (Fig. 13-16). Scattering photons from a baseball does not change the baseball's trajectory very much, but scattering photons from an electron can strongly affect its trajectory. To locate the *position* of the object accurately, we must use light with a wavelength comparable to or shorter than the size of the object. By de Broglie's relationship, however, the shorter the wavelength, the greater the momentum of the photon. We cannot then make an accurate measurement of the object's *momentum*, because it will be knocked away from its original trajectory by the photon. Thus, we cannot measure to arbitrary accuracy both the position and momentum of an object.

A rough estimate takes the minimum indeterminacy of the position measurement $\Delta x$ to be of order $\lambda$ and the indeterminacy of the momentum measurement $\Delta p$ to be of order $h/\lambda$ (because that is the momentum carried by the photon). Their product is then $(\Delta x)(\Delta p) \geq h$. A fuller analysis gives the result

$$(\Delta p)(\Delta x) \geq h/4\pi \qquad \textbf{(13-14)}$$

The indeterminacy principle places a fundamental limit on the accuracy to which the position and momentum of a particle can be known simultaneously. It reveals a fundamental defect in the Bohr model, which ascribes a definite position and momentum to the electron in its circular orbit around the nucleus.

Baseball trajectory

(a)

Electron trajectory

(b)

**Figure 13-16**
A photon, which has a negligible effect on the trajectory of a baseball (a), significantly perturbs the trajectory of the far less massive electron (b).

## Example 13-5

Suppose photons of green light (wavelength $5.3 \times 10^{-7}$ m) are used to locate the position of the baseball from Example 13-4 to an accuracy of one wavelength. Calculate the minimum indeterminacy in the *speed* of the baseball.

### Solution

The Heisenberg relation

$$(\Delta x)(\Delta p) \geq h/4\pi$$

gives

$$\Delta p \geq \frac{h}{4\pi\Delta x} = \frac{6.626 \times 10^{-34}\,\text{J s}}{4\pi\,(5.3 \times 10^{-7}\,\text{m})} = 9.9 \times 10^{-29}\,\text{kg m s}^{-1}$$

Because the momentum is just the mass (a constant) times the speed, the indeterminacy in the speed is

$$\Delta v = \frac{\Delta p}{m} \geq \frac{9.9 \times 10^{-29}\,\text{kg m s}^{-1}}{0.145\,\text{kg}} = 6.8 \times 10^{-28}\,\text{m s}^{-1}$$

This is such a tiny fraction of the speed of the baseball ($30\,\text{m s}^{-1}$) that the indeterminacy principle plays a negligible role in the measurement of the baseball's motion. Such is the case for all macroscopic objects.

*Related Problems: 29, 30*

## The Schrödinger Equation

De Broglie's work demonstrates that electrons in atoms have wave-like properties, and the indeterminacy principle shows that detailed trajectories of electrons cannot be defined; we must instead deal in terms of the *probability* of electrons having certain positions and momenta. These ideas are combined in the fundamental equation of quantum mechanics, the **Schrödinger equation,** discovered by the Austrian physicist Erwin Schrödinger in 1925. Schrödinger reasoned that if an electron (or any other particle) has wave-like properties, it should be described by a **wave function** that has a value at each position in space (just as it has for the height of a water wave or the amplitude of a classical electromagnetic wave). This wave function is symbolized by the Greek letter $\psi$ (psi), and $\psi(x, y, z)$ is the "height" of the wave at the point in space defined by the set of Cartesian coordinates $(x, y, z)$. Schrödinger wrote down the equation satisfied by $\psi$ for a given set of interactions between particles. Time-independent solutions of this equation—called **stationary states**— exist only for certain discrete values of the energy, and so energy quantization is a natural consequence of the Schrödinger equation.

A wave function, like an electromagnetic wave, can take positive values in some regions of space (it is said to have **positive phase** in these regions) and negative values in other regions (where it has **negative phase**). The points at which the wave function passes through 0 and changes sign are called nodes, just as in our guitar-string model for standing waves.

Solutions of the Schrödinger equation are possible for more than one value of the energy. The solution that corresponds to the lowest energy is called the **ground state** (just as in the Bohr model), and higher-energy solutions are called **excited states.** What is the physical meaning of the wave function $\psi$? We have no way of measuring $\psi$ directly, just as in classical wave optics we have no direct way of measuring the amplitudes of the electric and magnetic fields that constitute the light wave (Fig. 13-3). What *can* be measured in the latter case is the intensity of the light wave, which, as we have seen, is proportional to the square of the amplitude of the electric field:

$$\text{intensity} \propto (E_{\text{max}})^2$$

In turn, this intensity is proportional to the density of photons at that point in space or, in other words, to the *probability density* of finding a photon there.

By analogy, we propose that the square of the wave function $\psi^2$ for a particle is a probability density for that particle. In other words, $\psi^2(x, y, z)\Delta x \, \Delta y \, \Delta z$ is the probability that the particle will be found in a small volume $\Delta x \, \Delta y \, \Delta z$ about the point $(x, y, z)$. This probabilistic interpretation of the wave function, proposed by the German physicist Max Born, is now generally accepted because it provides a consistent picture of particle motion on a microscopic scale. We are obliged to acknowledge that our information about the location of a particle is statistical.

## 13-4 A Deeper Look . . . The Particle in a Box

The Schrödinger equation cannot be derived (just as Newton's laws cannot). Nonetheless, it is worthwhile to explore some properties of its mathematical form. We begin by considering a particle moving freely in one dimension with classical momentum $p$. Such a particle is associated with a wave of wavelength $\lambda = h/p$. Two "wave functions" with this wavelength are

$$\psi(x) = A \sin \frac{2\pi x}{\lambda} \quad \text{or} \quad B \cos \frac{2\pi x}{\lambda}$$

where $A$ and $B$ are constants. What equation do these trigonometric functions satisfy? Let us take the sine function as an example. From differential calculus, the derivative (or slope) of $\psi(x)$ is

$$\frac{d\psi(x)}{dx} = A \frac{2\pi}{\lambda} \cos \frac{2\pi x}{\lambda}$$

The slope of *this* is the second derivative of $\psi$, written $\frac{d^2\psi(x)}{dx^2}$ and equal to

$$\frac{d^2\psi(x)}{dx^2} = -A \left( \frac{2\pi}{\lambda} \right)^2 \sin \frac{2\pi x}{\lambda}$$

This is just a constant, $-(2\pi/\lambda)^2$, multiplied by the original wave function $\psi(x)$:

$$\frac{d^2\psi(x)}{dx^2} = -\left( \frac{2\pi}{\lambda} \right)^2 \psi(x)$$

This is an equation (called a *differential equation*) that is satisfied by the function $\psi(x) = A \sin (2\pi x/\lambda)$. It is easy to verify that this equation is also satisfied by the function $\psi(x) = B \cos (2\pi x/\lambda)$.

Let us now replace the wavelength $\lambda$ with the momentum $p$ from the de Broglie relation:

$$\frac{d^2\psi(x)}{dx^2} = -\left( \frac{2\pi}{h} p \right)^2 \psi(x)$$

We can rearrange this into a suggestive form by multiplying both sides by $-h^2/8\pi^2 m$, giving

$$-\frac{h^2}{8\pi^2 m} \frac{d^2\psi(x)}{dx^2} = \frac{p^2}{2m} \psi(x) = E_K \psi(x)$$

where $E_K = p^2/2m$ is the kinetic energy of the particle. This shows that there must be a relation between the second derivative of the wave function and the kinetic energy $E_K$. When one is large, the other is large as well.

If external forces are present, a potential energy term $V(x)$ (due to walls enclosing the particle or the presence of fixed charges, for example) must be included. Writing the total energy as $E = E_K + V(x)$ and substituting in the previous equation give

$$-\frac{h^2}{8\pi^2 m} \frac{d^2\psi(x)}{dx^2} + V(x)\psi(x) = E\psi(x) \quad \textbf{(13-15)}$$

This is the Schrödinger equation for a particle moving in one dimension. The calculation just given is not a derivation of this equation; rather, it indicates some useful intuitive connections to the de Broglie momentum–wavelength correspondence.

If $\psi^2(x)$ is to be a probability density, $\psi$ must have certain mathematical properties. It must be continuous, because the probability of finding a particle at every point in space must be defined and unique. Second, if all of space is subdivided into small intervals (in one dimension) or volumes (in three dimensions) and the probability of the particle's being in each volume is calculated, then the *sum* of those probabilities must be 1 because the particle is *somewhere*. Mathematically, this means

$$\int \psi^2(x) \, dx = 1 \quad \textbf{(13-16)}$$

When these conditions are imposed on $\psi(x)$, for reasonable choices of the potential $V(x)$ there are solutions $\psi_n(x)$ of the Schrödinger equation only for certain values of the energy. Quantization of the energy arises as a natural consequence of the Schrödinger equation, rather than being introduced in an ad hoc fashion as in the Bohr model.

The simplest problem for which the Schrödinger equation can be solved is the so-called particle in a box. This consists of a particle confined to a certain region of space (the "box"), which in one dimension becomes the interval between 0 and $L$, where $L$ is the length of the box (Fig. 13-17a). If the particle is truly to be confined, the potential energy $V(x)$ must rise abruptly to an infinite value at the two end walls, to prevent even fast-moving particles from escaping. On the other hand, inside the box the particle's motion is free and $V(x) = 0$. The solution of the Schrödinger equation for this case illustrates the general methods used for other potential-energy functions.

Wherever the potential energy $V$ is infinite, the probability of finding the particle must be 0. Hence, $\psi(x)$ and $\psi^2(x)$ must be 0 in these regions:

$$\psi(x) = 0 \quad \text{for} \quad x \leq 0 \quad \text{or} \quad x \geq L$$

Inside the box, where $V = 0$, the Schrödinger equation has the form

$$-\frac{h^2}{8\pi^2 m} \frac{d^2\psi(x)}{dx^2} = E\psi(x)$$

or, equivalently,

$$\frac{d^2\psi(x)}{dx^2} = -\frac{8\pi^2 mE}{h^2} \psi(x)$$

As already shown, the sine and cosine are two possible solutions to this equation, because their second derivative is a (negative) constant times the function itself. The cosine function can be eliminated in this case because, for the wave function to be continuous, $\psi(x)$ must be 0 at $x = 0$. This is true of the sine function but not of the cosine. Therefore,

$$\psi(x) = A \sin kx$$

If the wave function is also to be continuous at $x = L$, then

$$\psi(L) = 0$$

or

$$\psi(L) = A \sin kL = 0$$

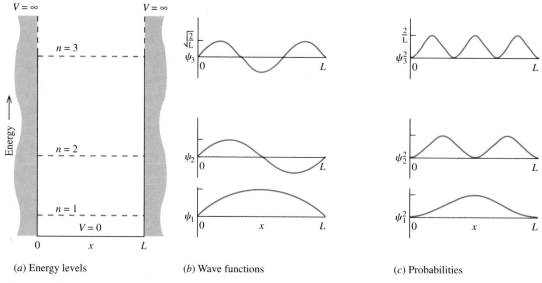

(a) Energy levels          (b) Wave functions          (c) Probabilities

**Figure 13-17**

(a) The potential energy for a particle in a box of length $L$, with the first three energy levels marked. (b) Wave functions, showing the ground state $\psi_1$ and the first two excited states. The more numerous the nodes, the higher the energy of the state. (c) The squares of the wave functions from part b, equal to the probability density for finding the particle at a particular point in the box.

This can be true only if

$$kL = n\pi \qquad n = 1, 2, 3, \ldots$$

because $\sin(n\pi) = 0$, so that

$$\psi(x) = A \sin\left(\frac{n\pi}{L}x\right) \qquad n = 1, 2, 3, \ldots$$

For the total probability of finding the particle *somewhere* in the box to be unity,

$$\int_0^L \psi^2(x)\, dx = 1$$

or

$$A^2 \int_0^L dx\, \sin^2\left(\frac{n\pi}{L}x\right) = 1$$

The integral has the value $L/2$, and so

$$A^2\left(\frac{L}{2}\right) = 1$$

$$A = \sqrt{\frac{2}{L}}$$

The wave function is then

$$\psi_n(x) = \sqrt{\frac{2}{L}} \sin\left(\frac{n\pi}{L}x\right) \qquad \text{(13-17)}$$

To find the energy $E_n$ for a particle with wave function $\psi_n$, calculate the second derivative:

$$\frac{d^2\psi_n(x)}{dx^2} = \frac{d^2}{dx^2}\left[\sqrt{\frac{2}{L}} \sin\left(\frac{n\pi}{L}x\right)\right]$$

$$= -\left(\frac{n\pi}{L}\right)^2\left[\sqrt{\frac{2}{L}} \sin\left(\frac{n\pi}{L}x\right)\right]$$

$$= -\left(\frac{n\pi}{L}\right)^2 \psi_n(x)$$

This must be equal to $-\dfrac{8\pi^2 m E_n}{h^2}\psi_n(x)$. Setting the coefficients equal gives

$$\frac{8\pi^2 m E_n}{h^2} = \frac{n^2\pi^2}{L^2}$$

$$E_n = \frac{h^2 n^2}{8mL^2} \qquad \text{(13-18)}$$

This solution of the Schrödinger equation demonstrates that the particle in the box is quantized. An energy $E_n$ and a wave function $\psi_n(x)$ correspond to each quantum number $n$ (a positive integer). These are the only allowed stationary states of the particle in a box. The spectrum of the particle in a box consists of a series of frequencies given by

$$h\nu = |E_n - E_n'|$$

where $n$ and $n'$ are positive integers and the $E_n$ are given by Equation 13-18. The wave functions $\psi_n(x)$ plotted in Figure 13-17b for the quantum states $n$ are exactly the standing waves of Figure 13-13. The guitar string and the particle in the box are physically analogous. Both systems are quantized by imposing the condition that the amplitude of the wave function $\psi$ be 0 at each end of the string or at each wall of the box.

From the variation of the square of the wave function of the particle with position (Fig. 13-17c), following Born's interpretation, we infer that in its ground state the particle has maximum probability of being in the middle of the box and lower probability of being near either wall. In its first excited state ($n = 2$), the probability is a maximum when it is near $L/4$ and $3L/4$ and zero near 0, $L/2$, and $L$. Wherever there is a node in the wave function, the probability is small that the particle will be near there. The wave function $\psi_n$ has $n - 1$ nodes, and the more numerous the nodes, the higher the energy of the particle.

Can a particle in a box have zero energy? If $n$ could have the value 0, the equation $E_n = h^2n^2/8mL^2$ would give $E_0 = 0$. But this is not possible. Setting $n$ equal to 0 in $\psi_n(x) = A \sin(n\pi x/L)$ makes $\psi_0(x)$ zero everywhere. In this case, $\psi_0^2(x)$ would be zero everywhere in the box, and therefore there could be no particle in the box. The same conclusion comes from the indeterminacy principle. If the energy of the lowest state could be 0, the momentum of the particle would also be 0. Moreover, the indeterminacy in the particle momentum, $\Delta p_x$, would be 0 and $\Delta x$ would be infinite, contradicting our condition that the particle be confined to a box of length $L$. Even at the absolute zero of temperature, where classical kinetic theory would suggest that all motion ceases, a finite quantity of energy remains for a bound system. It is required by the indeterminacy principle and is called the **zero-point energy.**

The Schrödinger equation is readily generalized to a particle in three dimensions. The wave function $\psi$ and potential $V$ now depend on three coordinates, $(x, y, z)$, and derivatives with respect to each coordinate appear. The allowed energies for a particle in a three-dimensional rectangular box are

$$E_{n_x n_y n_z} = \frac{h^2}{8m}\left[\frac{n_x^2}{L_x^2} + \frac{n_y^2}{L_y^2} + \frac{n_z^2}{L_z^2}\right] \quad \begin{cases} n_x = 1, 2, 3, \ldots \\ n_y = 1, 2, 3, \ldots \\ n_z = 1, 2, 3, \ldots \end{cases}$$

where $L_x$, $L_y$, and $L_z$ are the side lengths of the box. Here the state is designated by a set of three quantum numbers, $(n_x, n_y, n_z)$. Each independently ranges over the positive integers.

## Example 13-6

Consider the following two systems: (a) an electron in a one-dimensional box of length 1.0 Å and (b) a helium atom in a cube 30 cm on an edge. Calculate the energy difference between ground state and first excited state, expressing your answer in kJ mol$^{-1}$.

**Solution**

(a) For a one-dimensional box,

$$E_{\text{ground state}} = \frac{h^2}{8mL^2}(1^2)$$

$$E_{\text{first excited}} = \frac{h^2}{8mL^2}(2^2)$$

Then, for one electron in the box,

$$\Delta E = \frac{3h^2}{8mL^2}$$

$$= \frac{3(6.626 \times 10^{-34} \text{ J s})^2}{8(9.11 \times 10^{-31} \text{ kg})(1.0 \times 10^{-10} \text{ m})^2}$$

$$= 1.8 \times 10^{-17} \text{ J}$$

Multiplying this by $10^{-3}$ kJ J$^{-1}$ and by $N_0 = 6.022 \times 10^{23}$ mol$^{-1}$ gives

$$\Delta E = 11,000 \text{ kJ mol}^{-1}$$

(b) For a three-dimensional cube, $L_x = L_y = L_z = L$, and

$$E_{\text{ground state}} = \frac{h^2}{8mL^2}(1^2 + 1^2 + 1^2)$$

$$E_{\text{first excited}} = \frac{h^2}{8mL^2}(2^2 + 1^2 + 1^2)$$

In this case, the three states (2, 1, 1), (1, 2, 1), and (1, 1, 2) have the same energy.

$$\Delta E = \frac{3h^2}{8mL^2}$$

$$= \frac{3(6.626 \times 10^{-34} \text{ J s})^2}{8(4.0 \text{ u} \times 1.66 \times 10^{-27} \text{ kg u}^{-1})(0.30 \text{ m})^2}$$

$$= 2.8 \times 10^{-40} \text{ J}$$

$$= 1.7 \times 10^{-19} \text{ kJ mol}^{-1}$$

The energy levels are so close together in the latter case (due to the much greater dimensions of the box) that they appear continuous, and quantum effects play no role. The properties are almost those of a classical particle.

*Related Problems:* 31, 32

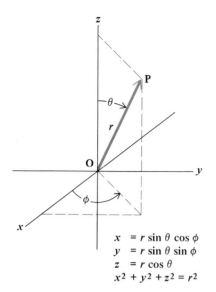

$$x = r \sin \theta \cos \phi$$
$$y = r \sin \theta \sin \phi$$
$$z = r \cos \theta$$
$$x^2 + y^2 + z^2 = r^2$$

**Figure 13-18**

The relationship of spherical coordinates $(r, \theta, \phi)$ to Cartesian coordinates $(x, y, z)$. Here $\theta$ is the angle with respect to the Cartesian $z$ axis, and $\phi$ the azimuthal angle (the angle between the $x$ axis and the projection onto the $xy$ plane of the arrow from the origin to P).

## 13-5 THE HYDROGEN ATOM

Exact solutions of the Schrödinger equation are possible for certain systems that are of interest in chemistry. One of the most important is a single electron interacting with a nucleus. The hydrogen atom is the simplest example of a one-electron atom or ion; other examples are $He^+$, $Li^{2+}$, and other ions in which all but one electron have been stripped off. They differ in the charge $Ze$ on the nucleus and therefore in the attractive force felt by the electron.

The Schrödinger equation for an isolated one-electron atom or ion can be solved most easily in spherical coordinates rather than in the more familiar Cartesian coordinates. Spherical coordinates are defined in Figure 13-18: $r$ is the distance of the electron at P from the nucleus at O, and the angles $\theta$ and $\phi$ are closely related to those used to locate points on the surface of the globe; $\theta$ is related to the latitude and $\phi$ to the longitude. Solutions of the Schrödinger equation for the one-electron atom exist only for particular values of the energy

$$E = E_n = -\frac{Z^2 e^4 m_e}{8\epsilon_0^2 n^2 h^2} \qquad n = 1, 2, \ldots$$

The energies are those found in the Bohr atom. Here, however, quantization comes about naturally from the solution of the Schrödinger equation rather than through an arbitrary assumption about the angular momentum.

There are $n^2$ independent solutions corresponding to energy $E_n$. These solutions (wave functions) can be written in the form

$$\psi_{n\ell m}(r, \theta, \phi) = R_{n\ell}(r)\, \chi_{\ell m}(\theta, \phi)$$

in which the total wave function is the product of a radial part $R_{n\ell}(r)$ and an angular part $\chi_{\ell m}(\theta, \phi)$, where $\chi$ is the Greek letter chi. Although the energy of a one-electron atom depends only on the **principal quantum number** $n$, two more quantum numbers are needed to describe the shape of the wave function. The **angular momentum quantum number** $\ell$ may take on any integral value from 0 to $n - 1$, and $m$, which may take on any integral value from $-\ell$ to $\ell$, is referred to as the **magnetic quantum number** because its value governs the behavior of the atom in an external magnetic field. For $n = 1$ (the ground state), the only allowed quantum numbers are $(\ell = 0, m = 0)$. For $n = 2$, there are $n^2 = 4$ allowed sets of quantum numbers, which are

$$(\ell = 0, m = 0), \quad (\ell = 1, m = 1), \quad (\ell = 1, m = 0), \quad (\ell = 1, m = -1)$$

The restrictions on $\ell$ and $m$ give rise to $n^2$ solutions of the Schrödinger equation for every value of $n$.

A wave function $\psi_{n\ell m}(r, \theta, \phi)$ that is a solution of the one-electron Schrödinger equation with quantum numbers $n$, $\ell$, and $m$ is called an **orbital**. This term recalls the circular orbits of the Bohr atom, but there is no real resemblance. An orbital is *not* the trajectory traced by an individual electron. Instead, it is a wave function $\psi$ that describes the standing wave corresponding to the electron, and the square of $\psi$ is the probability density of finding the electron at some point in space. The usual convention is to replace the angular momentum quantum number with a letter; we signify $\ell = 0$ with $s$, $\ell = 1$ with $p$, $\ell = 2$ with $d$, $\ell = 3$ with $f$, $\ell = 4$ with $g$, and on

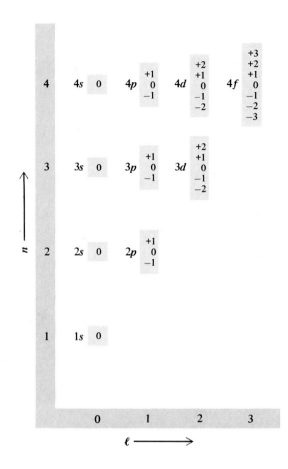

**Figure 13-19**

The allowed sets of quantum numbers for a one-electron atom or ion. The $m$ quantum numbers are shown to the right of each $(n, \ell)$ symbol.

through the alphabet. A wave function with $n = 1$ and $\ell = 0$ is thus called a $1s$ orbital, one with $n = 3$ and $\ell = 1$ is a $3p$ orbital, one with $n = 4$ and $\ell = 3$ is a $4f$ orbital, and so forth. The letters $s$, $p$, $d$, and $f$ come from early (pre–quantum mechanics) spectroscopy, in which certain spectral lines were referred to as sharp, principal, diffuse, and fundamental. These terms are not used in modern spectroscopy, but historical convention is maintained in the assignment of letters to the quantum number $\ell$. Figure 13-19 summarizes the allowed combinations of quantum numbers.

## Example 13-7

Give the names of all the orbitals with $n = 4$, and state how many $m$ values correspond to each type of orbital.

### Solution

The quantum number $\ell$ may range from 0 to $n - 1$, so its allowed values in this case are 0, 1, 2, and 3. The labels for the groups of orbitals are then

$$\ell = 0 \quad 4s \qquad \ell = 2 \quad 4d$$

$$\ell = 1 \quad 4p \qquad \ell = 3 \quad 4f$$

The quantum number $m$ ranges from $-\ell$ to $+\ell$, and so the number of $m$ values is $2\ell + 1$. This gives one $4s$ orbital, three $4p$ orbitals, five $4d$ orbitals, and seven $4f$ orbitals for a total of $16 = 4^2 = n^2$ orbitals with $n = 4$. They all have the same energy, but they differ in shape.

*Related Problems: 35, 36*

## Sizes and Shapes of Orbitals

The shapes and sizes of the hydrogen atom orbitals are important in chemistry. Table 13-2 gives the wave functions for the first few energy levels.

Begin with **s orbitals,** corresponding to $\psi_{n\ell m}$ with $\ell = 0$ (and therefore $m = 0$, as well). For all *s* orbitals, the angular part $\chi$ is a constant (Table 13-2). Because $\psi$

### Table 13-2   Angular and Radial Parts of Wave Functions for One-Electron Atoms

| Angular Part $\chi(\theta, \phi)$ | Radial Part $R_{n\ell}(r)$ |
|---|---|

$\ell = 0 \left\{ \chi(s) = \left(\dfrac{1}{4\pi}\right)^{1/2} \right.$

$R(1s) = 2\left(\dfrac{Z}{a_0}\right)^{3/2} \exp(-\sigma)$

$R(2s) = \dfrac{1}{2\sqrt{2}}\left(\dfrac{Z}{a_0}\right)^{3/2}(2-\sigma)\exp(-\sigma/2)$

$R(3s) = \dfrac{2}{81\sqrt{3}}\left(\dfrac{Z}{a_0}\right)^{3/2}(27-18\sigma+2\sigma^2)\exp(-\sigma/3)$

$\ell = 1 \begin{cases} \chi(p_x) = \left(\dfrac{3}{4\pi}\right)^{1/2}\sin\theta\cos\phi \\ \chi(p_y) = \left(\dfrac{3}{4\pi}\right)^{1/2}\sin\theta\sin\phi \\ \chi(p_z) = \left(\dfrac{3}{4\pi}\right)^{1/2}\cos\theta \end{cases}$

$R(2p) = \dfrac{1}{2\sqrt{6}}\left(\dfrac{Z}{a_0}\right)^{3/2}\sigma\exp(-\sigma/2)$

$R(3p) = \dfrac{4}{81\sqrt{6}}\left(\dfrac{Z}{a_0}\right)^{3/2}(6\sigma-\sigma^2)\exp(-\sigma/3)$

$\ell = 2 \begin{cases} \chi(d_{z^2}) = \left(\dfrac{5}{16\pi}\right)^{1/2}(3\cos^2\theta-1) \\ \chi(d_{xz}) = \left(\dfrac{15}{4\pi}\right)^{1/2}\sin\theta\cos\theta\cos\phi \\ \chi(d_{yz}) = \left(\dfrac{15}{4\pi}\right)^{1/2}\sin\theta\cos\theta\sin\phi \\ \chi(d_{xy}) = \left(\dfrac{15}{16\pi}\right)^{1/2}\sin^2\theta\sin 2\phi \\ \chi(d_{x^2-y^2}) = \left(\dfrac{15}{16\pi}\right)^{1/2}\sin^2\theta\cos 2\phi \end{cases}$

$R(3d) = \dfrac{4}{81\sqrt{30}}\left(\dfrac{Z}{a_0}\right)^{3/2}\sigma^2\exp(-\sigma/3)$

$$\sigma = \frac{Zr}{a_0} \qquad a_0 = \frac{\epsilon_0 h^2}{\pi e^2 m_e} = 0.529 \times 10^{-10}\text{ m}$$

depends on neither $\theta$ nor $\phi$, all $s$ orbitals are spherically symmetric about the nucleus. This means that the amplitude of an $s$ orbital (and therefore also the probability of finding the electron near some point in space) depends only on its distance $r$ from the nucleus and not on its direction in space. There are several ways to visualize the radial variation of the $ns$ orbitals with $n = 1, 2, 3, \ldots$. One way (Fig. 13-20$a$) is to make a "three-dimensional" picture in which the shading is heaviest where $\psi^2$ is largest and lighter where $\psi^2$ is smaller. A second way (Fig. 13-20$b$) is to plot the wave functions themselves: $\psi_{n00}(r) \propto R_{n0}(r)$. A third way (Fig. 13-20$c$) is to plot the **radial probability distribution** $r^2 R_{n0}^2(r)$. This is the probability density of finding the electron at any point in space at a distance $r$ from the nucleus. The factor $r^2$ in front accounts for the increasing volume of spherical shells at greater distances from the nucleus. The radial probability distribution is small near the nucleus, where the shell volumes (proportional to $r^2$) are small, and has its maximum at the distance where the electron is most likely to be found.

What is meant by the *size* of an orbital? Strictly speaking, the wave function of an electron in an atom stretches out to infinity, and so an atom has no boundary unless we set some arbitrary limit upon the extension of its wave function. We might define the size of an atom as that of the contour at which $\psi^2$ has fallen off to some particular numerical value, or as the extent of a "balloon skin" inside which the electron has a definite probability (for example, 90% or 99%) of being found. We

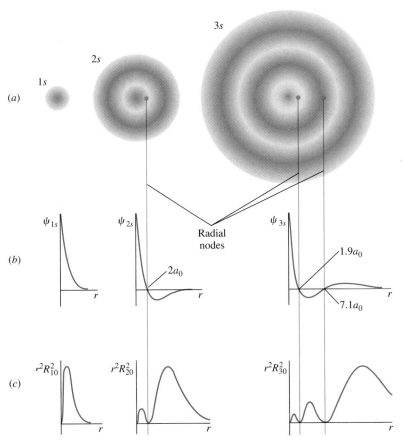

**Figure 13-20**

Three representations of hydrogen $s$ orbitals. (a) An electron-density representation of a hydrogen atom in its $1s$, $2s$, and $3s$ states. The spheres are cut off at a radius that encloses 90% of the total electron density. (b) The wave functions plotted against distance from the nucleus, $r$. (c) The radial probability distribution, equal to the square of the radial wave function multiplied by $r^2$.

$2p_x$

$yz$ nodal plane

$2p_y$

$xz$ nodal plane

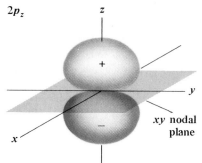

$2p_z$

$xy$ nodal plane

**Figure 13-21**

The distribution of electron density in the three $2p$ orbitals, with phases and nodal planes indicated.

have adopted the latter practice in pictures of orbitals for electrons. Figure 13-20a shows spheres that enclose 90% of the electron probability for the 1s, 2s, and 3s orbitals for hydrogen.

Note two features of s orbitals. First, the size of an orbital increases with increasing quantum number n, and so the electron is most likely to be found farther from the nucleus for larger n. A 3s orbital is larger than a 2s orbital, which in turn is larger than a 1s orbital. This is the quantum mechanical analog of the increase in radius of the Bohr orbits with n. Second, an ns orbital has $n - 1$ **radial nodes** (a radial node is a sphere about the nucleus on which $\psi$ and $\psi^2$ are 0). The more numerous the nodes, the higher the energy of the corresponding state.

It is more difficult to draw pictures of orbitals with angular quantum numbers $\ell$ different from 0, because these are not spherically symmetric. In addition, when two or more wave functions have the same energy, linear combinations of them also satisfy the Schrödinger equation. This is the case for the three wave functions with $\ell = 1$ (the **p orbitals**), for which the allowed m values are $-1, 0$, and $+1$. The usual convention is to take linear combinations that give three orbitals with the same shapes but different orientations, as shown (for $n = 2$) in Figure 13-21. The angular wave function for the $p_z$ orbital is proportional to $\cos \theta$ (or, in Cartesian coordinates, to z), and so such an orbital has maximum amplitude along the z axis (where $\theta = 0$ or $\pi$) and a node in the xy plane (where $\theta = \pi/2$, so that $\cos \theta = 0$). Therefore, it points along the z axis, with positive phase on one side of the xy plane and negative phase on the other side. An electron in a $p_z$ orbital has the highest probability of being found along the z axis and has zero probability of being found in the xy plane. This plane is a nodal plane or, more generally, an **angular node** across which the wave function changes sign. The $p_x$ orbital has the same shape but points along the x axis, and the $p_y$ orbital points along the y axis.

The radial parts of the np wave functions (represented by $R_{n1}$) are illustrated in Figure 13-22. The p orbitals, like the s orbitals, may also have radial nodes, in which the electron density vanishes at certain distances from the nucleus irrespective of direction. From Figure 13-22, the $R(2p)$ wave function possesses no nodes, the $R(3p)$ function has one node, and the $R(4p)$ function has two nodes. In general, the $R_{n1}$ radial wave functions have $n - 2$ nodes. Because the angular part of the np wave function always has a nodal plane, the total wave function has $n - 1$ nodes ($n - 2$ radial and 1 angular), which is the same number as an s orbital with the same principal quantum number. The $R_{21}(r)$ function [that is, $R(2p)$] in Table 13-2 contains the factor $\sigma$, which is proportional to r ($\sigma = Zr/a_0$) and causes it to vanish at the nucleus. This is true of all the radial wave functions except the ns functions, and it means that the probability density is zero for the electron to be at the nucleus for all wave functions with $\ell > 0$ (p, d, f, . . .).

The spatial distribution of the five d orbitals is somewhat more complicated (Fig. 13-23). The conventionally chosen linear combinations of these orbitals give four orbitals with the same shape but different orientations with respect to the Cartesian axes: $d_{xy}, d_{yz}, d_{xz}$, and $d_{x^2-y^2}$. A $d_{xy}$ orbital, for example, has four lobes, two with positive phase and two with negative phase; the maximum amplitude is at 45° to the x and y axes. The $d_{x^2-y^2}$ orbital has maximum amplitude along the x and y axes. The fifth orbital, $d_z^2$, has a different shape, with maximum amplitude along the z axis and a little "doughnut" in the xy plane. Each d orbital has two angular nodes (for example, for the $d_{xy}$ orbital the xz and yz planes are nodal surfaces). The radial

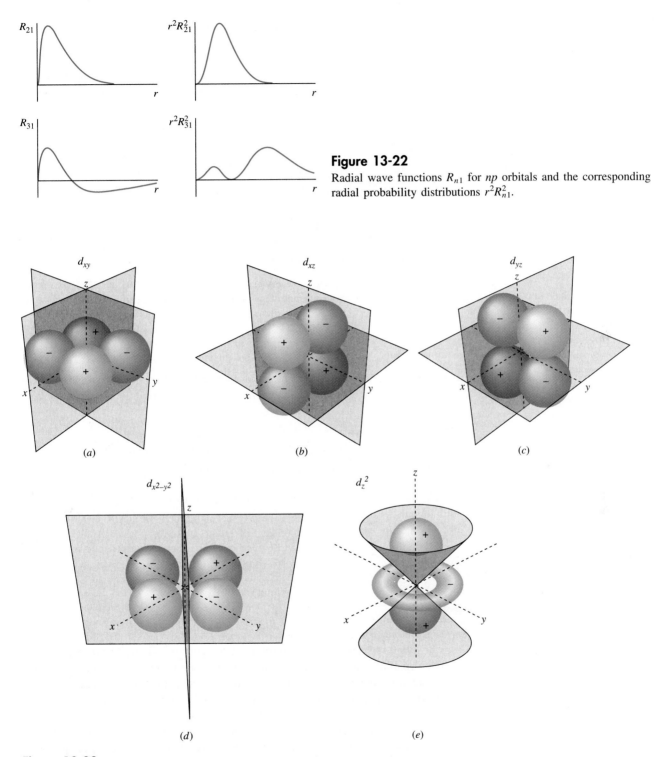

**Figure 13-22**

Radial wave functions $R_{n1}$ for $np$ orbitals and the corresponding radial probability distributions $r^2 R_{n1}^2$.

**Figure 13-23**

The shapes of the five $3d$ orbitals, with phases and nodal surfaces indicated.

functions $R_{n2}(r)$ have $n - 3$ radial nodes, giving once again $n - 1$ total nodes.

The wave functions for $f$ orbitals and orbitals of higher $\ell$ can be calculated, but they play a smaller role in chemistry than the $s$, $p$, and $d$ orbitals. We therefore stop at this point and summarize the important features of orbital shapes and sizes:

1. For a given value of $\ell$, an increase in $n$ leads to an increase in the average distance of the electron from the nucleus and therefore in the size of the orbital (Figs. 13-20 and 13-22).

2. An orbital with quantum numbers $n$ and $\ell$ has $\ell$ angular nodes and $n - \ell - 1$ radial nodes, giving a total of $n - 1$ nodes. For a one-electron atom or ion, the energy depends only on the number of nodes—that is, on $n$ but not on $\ell$ or $m$.

3. As $r$ approaches 0, $\psi(r, \theta, \phi)$ vanishes for all orbitals except $s$ orbitals; thus, only an electron in an $s$ orbital can ''penetrate to the nucleus,'' that is, have a finite probability of being found right at the nucleus.

These general statements will prove important in determining the electronic structure of many-electron atoms even though they are deduced from the one-electron case.

## Example 13-8

Compare the $3p$ and $4d$ orbitals of a hydrogen atom with respect to (a) number of radial and angular nodes and (b) energy of the corresponding atom.

### Solution

(a) The $3p$ orbital has a total of $n - 1 = 3 - 1 = 2$ nodes. Of these, one is angular ($\ell = 1$) and one is radial. The $4d$ orbital has $4 - 1 = 3$ nodes. Of these, two are angular ($\ell = 2$) and one is radial.

(b) The energy of a one-electron atom depends only on $n$. The energy of an atom with an electron in a $4d$ orbital is higher than that of an atom with an electron in a $3p$ orbital, because $4 > 3$.

*Related Problems:* 37, 38

## Electron Spin

Our discussion of one-electron atoms has so far omitted one important point. If a beam of hydrogen atoms in their ground state (with $n = 1$, $\ell = 0$, $m = 0$) is sent through a spatially varying magnetic field, it splits into two beams, each containing half of the atoms (Fig. 13-24). The result of this experiment is explained by introducing a *fourth* quantum number, $m_s$, which can take on two values, conventionally chosen to be $+\frac{1}{2}$ and $-\frac{1}{2}$. Because the behavior in the magnetic field is what would be predicted by classical theory if the electrons were balls of charge spinning about their own axes, the fourth quantum number is referred to as the **spin quantum number.** When $m_s = +\frac{1}{2}$ the electron spin is said to be ''up,'' and when $m_s = -\frac{1}{2}$ the spin is ''down.'' We should not take the classical image of a spinning ball of charge too literally, however. The spin quantum number comes from relativistic

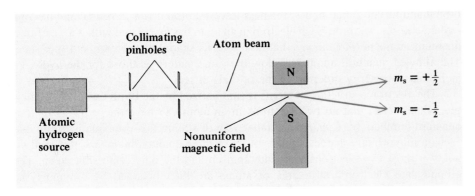

**Figure 13-24**
A beam of hydrogen atoms is split into two beams when it traverses a nonuniform magnetic field. Atoms with spin quantum number $m_s = +\frac{1}{2}$ follow one trajectory, and those with $m_s = -\frac{1}{2}$ follow another.

effects that are not included in the Schrödinger equation. For most practical purposes in chemistry, it is sufficient simply to solve the ordinary Schrödinger equation and then associate with each electron a spin quantum number $m_s = \pm\frac{1}{2}$, which does not affect the spatial probability density of the electron. The spin does affect the number of different quantum states with energy $E_n$, however, doubling it from $n^2$ to $2n^2$. This fact will assume considerable importance when we consider many-electron atoms in the next section.

## 13-6 MANY-ELECTRON ATOMS AND THE PERIODIC TABLE

As we move from one-electron to many-electron atoms, both the Schrödinger equation and its solutions become increasingly complicated. The simplest many-electron atom, helium, has two electrons and a nuclear charge of $+2e$. The positions of the two electrons in a helium atom can be described classically by two sets of Cartesian coordinates, $(x_1, y_1, z_1)$ and $(x_2, y_2, z_2)$. The wave function $\psi$ depends on all six of these variables: $\psi = \psi(x_1, y_1, z_1, x_2, y_2, z_2)$. Its square, $\psi^2(x_1, y_1, z_1, x_2, y_2, z_2)$, is the probability density of finding the first electron at point $(x_1, y_1, z_1)$ and, simultaneously, the second at $(x_2, y_2, z_2)$. The Schrödinger equation is now more complicated, and an explicit solution for helium is not possible. Nevertheless, modern computers have enabled us to solve this equation numerically to very high accuracy, and the predicted properties of helium are in excellent agreement with experiment.

Although these numerical calculations demonstrate conclusively the usefulness of the Schrödinger equation for predicting atomic properties, they suffer from two defects: first, they are somewhat difficult to interpret physically, and second, they become increasingly difficult to solve, even numerically, as the number of electrons increases. As a result, more approximate approaches to the many-electron Schrödinger equation have been developed. They rely on an **orbital approximation,** in which the electron charge-cloud density is assumed to be a superposition of charge clouds arising from the individual electrons. An electron in a many-electron atom dwells in an atomic orbital that has a characteristic size, shape, and energy and has a spin (up or down). These orbitals resemble the atomic orbitals of hydrogen in many ways. Their angular dependence is identical, and so it is straightforward to associate quantum numbers $\ell$ and $m$ with each atomic orbital. The radial dependence of the orbitals in many-electron atoms differs from that of one-electron orbitals, but a principal quantum number $n$ can still be defined. The lowest energy orbital is a $1s$

orbital and has no radial nodes, the next lowest $s$ orbital is a $2s$ orbital and has one radial node, and so forth. Each electron in an atom has associated with it a set of four quantum numbers ($n$, $\ell$, $m$, $m_s$) that describes its spatial distribution and spin state. The allowed quantum numbers follow the same pattern as those for the hydrogen atom, but the energy ordering of the orbitals differs.

The electronic structure of atoms is summarized by the **Pauli exclusion principle,** which states that no two electrons in an atom can have the same set of four quantum numbers ($n$, $\ell$, $m$, $m_s$). Another way of stating this principle is to say that each atomic orbital (characterized by a set of three quantum numbers, $n$, $\ell$, and $m$) holds at most two electrons, one with spin up and the other with spin down. The ground-state electronic structures of atoms are then obtained by arranging the atomic orbitals in order of increasing energy and filling them one electron at a time, starting with the lowest energy orbital and heeding the Pauli principle at all times. Starting at the energetic bottom and working up embodies what is called the **aufbau principle** (*Aufbau* is German for ''building up''). Often the atomic orbitals that are to be filled fall into groups that have closely similar or equal energies. A group of orbitals with the same energy (such as the three different $2p$ orbitals) form a **subshell,** and a group of subshells with similar energies make up a **shell.** The energy ordering of the orbitals is not absolutely fixed, however, but can vary from one atom to another or from an atom to its own ions.

## Building Up from Helium to Argon

Let us see how the aufbau principle works for the atoms from helium through neon. The lowest energy orbital is always the $1s$ orbital, and so helium has two electrons (with opposite spins) in that orbital. The ground-state **electron configuration** of the helium atom is symbolized as $1s^2$ and is conveniently illustrated by a ''pigeonhole'' diagram such as Figure 13-25. The $1s$ orbital in the helium atom is somewhat larger than the $1s$ orbital in the helium *ion* (He$^+$). In the ion the electron feels the full nuclear charge $+2e$, but in the atom each electron partially shields the other electron from the nuclear charge. The orbital in the helium atom can be described approximately in terms of an ''effective'' nuclear charge $Z_{\text{eff}}$ of 1.7, which lies between $+1$ (the value for complete shielding by the other electron) and $+2$ (no shielding).

An atom of lithium has three electrons. The third electron does not join the first two in the $1s$ orbital. It occupies a different orbital because, by the Pauli principle, at most two electrons may occupy any orbital. Will the third electron go into the $2s$ or the $2p$ orbital? In a one-electron atom these two orbitals have the same energy, because the energy depends only on the principal quantum number $n$. In a many-electron atom, the energies of orbitals depend on $\ell$ as well as $n$ (but not on $m$). An $s$ orbital penetrates more toward the nucleus than does a $p$ orbital, and so an electron in an $s$ orbital experiences (at least part of the time) an almost unshielded nuclear attraction. This lowers its energy relative to a $p$ orbital, which experiences the smaller attraction of a more highly shielded nucleus. The ground state of the lithium atom is therefore $1s^2 2s^1$, and $1s^2 2p^1$ is an excited state.

The next two elements present no difficulties. Beryllium has the ground-state configuration $1s^2 2s^2$. The $2s$ orbital is now filled, and boron, with five electrons, has the ground-state configuration $1s^2 2s^2 2p^1$. Because the three $2p$ orbitals of boron (for convenience, taken to be the $2p_x$, $2p_y$, and $2p_z$ orbitals) have the same energy, there is an equal chance for the electron to be in each one.

**Figure 13-25**
The ground-state electron configurations of first- and second-period atoms. The arrows represent electrons with spin quantum numbers $m_s = +\frac{1}{2}$ or $m_s = -\frac{1}{2}$; each circle represents an orbital.

With carbon, the sixth element, a new question arises. Will the sixth electron go into the same orbital as the fifth (for example, both into a $2p_x$ orbital with opposite spins), or will it go into the $2p_y$ orbital, which has equal energy? The answer is found in the observation that two electrons occupying the same atomic orbital experience stronger electron–electron repulsion than they would if they occupied orbitals in different regions of space. Thus, putting the last two electrons of carbon into two different $p$ orbitals, such as the $2p_x$ and $2p_y$, leads to lower energy than putting them into the same $p$ orbital. This is formalized in **Hund's rules:**

> When electrons are added to orbitals of equal energy, a single electron enters each orbital before a second one enters any orbital. In addition, the spins remain parallel if possible.

The electron configuration of carbon is then $1s^2 2s^2 2p_x^1 2p_y^1$ or, more simply, $1s^2 2s^2 2p^2$. It is shown in Figure 13-26.

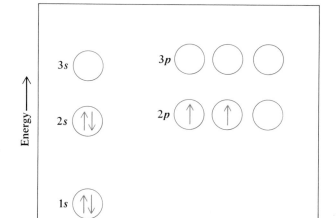

**Figure 13-26**

For a many-electron atom such as carbon, orbitals with different $\ell$ values (such as the 2s and 2p) have different energies. When two or more orbitals have the *same* energy (such as the three 2p orbitals here), electrons occupy different orbitals with parallel spins.

The presence of unpaired electrons in atoms of carbon affects their behavior in a magnetic field. A substance is **paramagnetic** if it is attracted into a magnetic field. Any substance that has one or more unpaired electrons in the atoms, molecules, or ions that compose it is paramagnetic; ground-state carbon atoms are an example. A substance in which all the electrons are paired is weakly **diamagnetic:** it is pushed *out* of a magnetic field, although the force it experiences is much smaller in magnitude than the force that pulls a typical paramagnetic substance into a magnetic field. Of the atoms examined so far, H, Li, B, and C are paramagnetic, but He and Be have all of their electron spins paired and so are diamagnetic.

The electron configurations from N through Ne follow from the stepwise filling of the 2p orbitals. The six elements from boron to neon are **p-block** elements because they appear during filling of p orbitals in the building-up process. The four elements that precede them (hydrogen through beryllium) are **s-block** elements.

The build-up of the third period, from sodium to argon, is an echo of what happened in the second; first the one 3s orbital is filled and then the three 3p orbitals. As the number of electrons in an atom reaches 15 or 20, it is frequently the practice to explicitly include only those electrons added in the building up beyond the last preceding noble-gas element. The configuration of that noble gas is then represented by its chemical symbol enclosed in brackets. The ground-state configuration of silicon, for example, is written $[Ne]3s^2 3p^2$ under this system.

### Example 13-9

Write the ground-state electron configurations for magnesium and sulfur. Are the gaseous atoms of these elements paramagnetic or diamagnetic?

### Solution

The noble-gas element preceding both elements is neon. Magnesium has two electrons beyond the neon core, which must be placed in the 3s orbital, the next higher in energy, to give the ground-state configuration $[Ne]3s^2$. A magnesium atom is diamagnetic because all its electrons are paired in orbitals.

Sulfur has six electrons beyond the neon core; the first two of these are in the $3s$ orbital, and the next four are in the $3p$ orbital. Sulfur's ground-state configuration is $[Ne]3s^23p^4$. When four electrons are put into three $p$ orbitals, two electrons must occupy one of the orbitals, and the other two occupy different orbitals to reduce electron–electron repulsion. By Hund's rules, the electrons' spins are parallel, and the sulfur atom is paramagnetic.

*Related Problems:* 41, 42, 43, 44

## Transition-Metal Elements and Beyond

After the three $3p$ orbitals are filled up with six electrons, the natural next step is a continuation of the building-up process using the $3d$ subshell. The aufbau principle requires that orbitals be filled in order of increasing energies, however. For neutral atoms at the beginning of the fourth period, it is an experimental fact that the $4s$ orbital ($n = 4$, $\ell = 0$) actually has a lower energy than the $3d$ orbital ($n = 3$, $\ell = 2$), although the two energies are very close; the $4s$ orbital is therefore filled first. The resulting ground-state configurations are $[Ar]4s^1$ for potassium and $[Ar]4s^2$ for calcium. The energy of the $3d$ orbitals does decline relative to the $4s$ orbital with increasing nuclear charge, however. In scandium, the very next element, the $3d$ orbital has a lower energy than the $4s$ orbital, and so an electron is removed more easily from the $4s$ orbital. The $Sc^+$ ion therefore has the ground-state configuration $[Ar]3d^14s^1$ rather than $[Ar]4s^2$, and $Sc^{2+}$ has the configuration $[Ar]3d^1$. The drop in energy of the $3d$ orbitals relative to the $4s$ continues throughout the transition-metal series from scandium to zinc, with the two $4s$ electrons removed more easily than the $3d$ electrons. The energy drop has important chemical consequences: the elements copper and zinc rarely have oxidation states higher than $+2$, and in the elements beyond zinc the $3d$ electrons are held so tightly that they play no chemical role. The ten elements from scandium to zinc are **d-block** elements because they appear with the filling of a $d$ orbital in the building-up process.

Experimental evidence shows that two of the transition-metal elements in the fourth period have ground-state electron configurations that are contrary to expectation based on the configurations that precede them. In its ground state, chromium has the configuration $[Ar]3d^54s^1$ rather than $[Ar]3d^44s^2$, and copper has the configuration $[Ar]3d^{10}4s^1$ rather than $[Ar]3d^94s^2$. Similar anomalies occur in the transition-metal atoms of the fifth period, and some other anomalies occur there as well, such as the ground-state configuration $[Kr]4d^75s^1$ that is observed for ruthenium in place of the expected $[Kr]4d^65s^2$.

In the sixth period, the filling of the $4f$ orbitals (and the generation of the **f-block** elements) begins as the rare-earth (lanthanide) elements from lanthanum to ytterbium are reached. The energies of the $4f$, $5d$, and $6s$ orbitals are comparable over much of the sixth period, and so their order of filling is erratic. (See Appendix F for the ground-state configurations of all the elements.) The sixth period ends with the radioactive noble gas radon. All of the elements in the seventh period, including the actinides, the second set of $f$-block elements, are radioactive.

The periodic table of Figure 13-27 groups elements in periods according to the subshell that is being filled as the atomic number increases. Exceptions to the "standard" order of filling are shown.

**1s**

| H | | | | | | | | | | | | | | | | | **1s** |
|---|---|---|---|---|---|---|---|---|---|---|---|---|---|---|---|---|---|
| | | | | | | | | | | | | | | | | | He |

**2s-filling** / **2p-filling**

| Li | Be | | B | C | N | O | F | Ne |
|---|---|---|---|---|---|---|---|---|

**3s-filling** / **3p-filling**

| Na | Mg | | Al | Si | P | S | Cl | Ar |
|---|---|---|---|---|---|---|---|---|

**4s-filling** — **3d-filling** — **4p-filling**

| K | Ca | Sc | Ti | V | Cr | Mn | Fe | Co | Ni | Cu | Zn | Ga | Ge | As | Se | Br | Kr |
|---|---|---|---|---|---|---|---|---|---|---|---|---|---|---|---|---|---|
| | | | | | $3d^5 4s^1$ | | | | | $3d^{10} 4s^1$ | | | | | | | |

**5s-filling** — **4d-filling** — **5p-filling**

| Rb | Sr | Y | Zr | Nb | Mo | Tc | Ru | Rh | Pd | Ag | Cd | In | Sn | Sb | Te | I | Xe |
|---|---|---|---|---|---|---|---|---|---|---|---|---|---|---|---|---|---|
| | | | | $4d^4 5s^1$ | $4d^5 5s^1$ | | $4d^7 5s^1$ | $4d^8 5s^1$ | $4d^{10}$ | $4d^{10} 5s^1$ | | | | | | | |

**6s-filling** — **5d-filling** — **6p-filling**

| Cs | Ba | Lu | Hf | Ta | W | Re | Os | Ir | Pt | Au | Hg | Tl | Pb | Bi | Po | At | Rn |
|---|---|---|---|---|---|---|---|---|---|---|---|---|---|---|---|---|---|
| | | | | | | | | | $5d^9 6s^1$ | $5d^{10} 6s^1$ | | | | | | | |

**7s-filling** — **6d-filling**

| Fr | Ra | Lw | Unq | Unp | Unh | Uns | Uno | Une | Uun | Uuu |
|---|---|---|---|---|---|---|---|---|---|---|

**4f-filling**

| La | Ce | Pr | Nd | Pm | Sm | Eu | Gd | Tb | Dy | Ho | Er | Tm | Yb |
|---|---|---|---|---|---|---|---|---|---|---|---|---|---|
| $5d^1 6s^2$ | $4f^1 5d^1 6s^2$ | | | | | | $4f^7 5d^1 6s^2$ | | | | | | |

**5f-filling**

| Ac | Th | Pa | U | Np | Pu | Am | Cm | Bk | Cf | Es | Fm | Md | No |
|---|---|---|---|---|---|---|---|---|---|---|---|---|---|
| $6d^1 7s^2$ | $6d^2 7s^2$ | $5f^2 6d^1 7s^2$ | $5f^3 6d^1 7s^2$ | $5f^4 6d^1 7s^2$ | | | $5f^7 6d^1 7s^2$ | | | | | | |

**Figure 13-27**

The filling of shells and the structure of the periodic table. Only the anomalous electron configurations are shown.

## Shells and the Periodic Table

A shell has been defined as a set of orbitals that have similar energies. This shell structure is also apparent in the electron densities of atoms, obtained from either experiment or calculation. The radial probability distribution for the electron density of argon (Fig. 13-28) shows clearly the different regions of space occupied by the first three shells, corresponding to the quantum numbers $n = 1$, 2, and 3. Of course there is some overlap of electrons between shells, but each peak can clearly be assigned to a shell.

**Valence electrons** are those that occupy the outermost shell, beyond the immediately preceding noble-gas configuration. Electrons in inner shells are closer to the nucleus, on the average, and lower in energy than valence electrons; they are called **core electrons.** A **closed-shell atom** is one whose outermost shell is filled with electrons so that any excitation (promotion of an electron to a higher energy orbital) requires a large amount of energy; the noble-gas atoms have closed shells, and the next higher unoccupied orbitals are separated from the occupied orbitals by a significant energy gap.

Two elements in the same group (column) of the periodic table have related valence electron configurations. An atom's bonding properties and therefore its

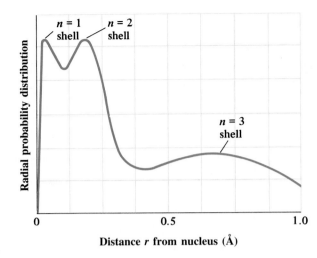

**Figure 13-28**

The shell structure of an argon atom is clearly visible in the three peaks of the radial probability distribution for the electrons in the atom.

chemical behavior flow from the configuration of its valence electrons. Sodium (configuration $[Ne]3s^1$) and potassium (configuration $[Ar]4s^1$), for example, each have a single valence electron in an $s$ orbital outside a closed shell; consequently, the two elements resemble each other closely in chemical properties. A major triumph of quantum mechanics is its ability to account for the periodic trends discovered by chemists many years earlier. Mendeleyev and others developed the periodic table of the elements purely on the basis of parallel chemical and physical properties (Section 2-1). Now we see that the structure of the periodic table can be derived from quantum mechanics. The ubiquitous octets in the Lewis electron-dot structures of second- and third-period atoms and ions arise from the eight available sites for electrons in the one $s$ orbital and three $p$ orbitals of the valence shell. The special properties of the transition-metal elements can be ascribed to the partial filling of their $d$ orbitals, a feature to which we shall return in Chapter 16.

### Sizes of Atoms and Ions

The sizes of atoms and ions influence how they interact in chemical compounds. These sizes can be estimated in several ways. If the electron density is known from theory or experiment, a contour surface of fixed electron density can be drawn, as demonstrated in Section 13-5 for one-electron atoms. Alternatively, if the atoms or ions in a crystal are assumed to be in contact with one another, a size can be defined from the measured distances between their centers (an approach to be explored in greater detail in Chapter 17). These and other measures of size are reasonably consistent with each other and allow us to tabulate sets of atomic and ionic radii, many of which are listed in Appendix F.

Certain systematic trends appear in these radii. For a series of elements or ions in the same group (column) of the periodic table, the radius usually increases with increasing atomic number. This occurs mainly because the Pauli exclusion principle effectively excludes added electrons from the region occupied by the core electrons and so forces an increase in size as more distant electron shells are occupied. On the other hand, Coulomb (electrostatic) forces cause the radii of atoms to *decrease* with

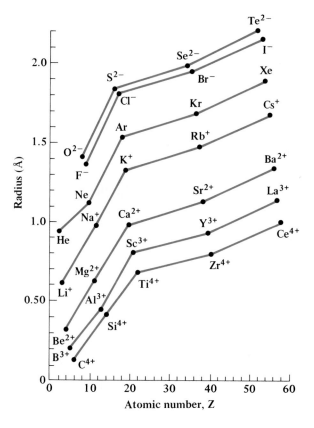

**Figure 13-29**

Ionic and atomic radii plotted versus atomic number. Each line connects a set of atoms or ions having the same charge; all species have noble-gas configurations.

increasing atomic number across a period. As the nuclear charge increases steadily, electrons join the same valence shell and are ineffective in shielding each other from its attraction.

Superimposed upon these broad trends are some subtler effects that have significant consequences in chemistry. One is shown dramatically in Figure 13-29. The radii of several sets of ions and atoms increase with atomic number in a given group, as already explained, but the *rate* of this increase changes considerably when the ions and atoms containing the same number of electrons as Ar are reached ($S^{2-}$, $Cl^-$, Ar, $K^+$, $Ca^{2+}$, $Sc^{3+}$, $Ti^{4+}$). The change in size from $Li^+$ to $Na^+$ to $K^+$ is quite great, for example, but the subsequent changes, to $Rb^+$ and $Cs^+$, are significantly smaller due to the filling of the *d* orbitals, which begins after $K^+$ is reached. Because atomic and ionic size decrease from left to right across a series of transition-metal elements (owing to increased nuclear charge), the radius of a main-group element when it is ultimately reached is smaller than it would have been had the transition series not intervened. A similar phenomenon occurs during the filling of the 4*f* orbitals in the lanthanide series and is called the *lanthanide contraction*. Its effect on the sizes of transition-metal atoms is discussed in Section 16-1.

A different measure of atomic size is the volume occupied by a mole of atoms of the element in the solid phase. Figure 13-30 shows the pronounced periodicity of the molar volume, with maxima occurring for the alkali metals. Two factors affect the experimentally measured molar volume: the ''size'' of the atoms and the geometry of the bonding that connects them. The large molar volumes of the alkali metals

**Figure 13-30**
The molar volumes (in $cm^3 mol^{-1}$) of some elements in their solid states. Note the high peaks corresponding to the alkali metals.

stem from both the large size of the atoms and the fact that they are organized in a rather open, loosely packed structure in the solid.

## Example 13-10

Predict which atom or ion in each of the following pairs should be larger: (a) Kr or Rb, (b) Y or Cd, (c) $F^-$ or $Br^-$.

### Solution

(a) Rb should be larger because it has an extra electron in a $5s$ orbital beyond the Kr closed shell.
(b) Y should be larger because nuclear charge increases through the transition series from Y to Cd.
(c) $Br^-$ should be larger because the extra outer electrons are excluded from the core.

*Related Problems:* 51, 52, 53, 54

## 13-7 EXPERIMENTAL MEASURES OF ORBITAL ENERGIES

The order of the energies of atomic orbitals determines the building up of electron configurations. How are these energies measured? An important experimental technique called **photoelectron spectroscopy** determines orbital energies directly. The technique is simply the photoelectric effect discussed in Section 13-1 applied not to metals but to free atoms or molecules. If radiation of sufficiently high frequency $\nu$ strikes an atom, an electron will be ejected with kinetic energy $\frac{1}{2}m_e v^2$. If the binding energy of that electron is $\Phi$, then, by the principle of conservation of energy,

$$h\nu = \Phi + \tfrac{1}{2}m_e v^2$$

as shown in Section 13-1. Experiments in photoelectron spectroscopy use high-energy radiation of a fixed frequency (in the ultraviolet or x-ray region of the spectrum), and the kinetic energies of the emitted electrons are determined. From these data the binding energies of the electrons in the atom are evaluated.

**Figure 13-31**

Sodium chloride consists of $Na^+$ ions (with the $[He]2s^22p^6$ configuration) and $Cl^-$ ions (with the $[Ne]3s^23p^6$ configuration). The portion of the photoelectron spectrum of NaCl that covers the binding energies of the outer electrons is shown here. Binding energy increases from right to left. The core electrons on the two ions, which are more tightly held than the outer electrons, would lie farther to the left.

Figure 13-31 shows an early experimental spectrum. The binding energy of the most weakly bound electron—the minimum amount of energy necessary to detach an electron from a ground-state atom—is called the **ionization energy.** Clearly, there can be no signal in the photoelectron spectrum for binding energies less than this. At greater binding energies, the spectrum consists of a series of peaks, each of which corresponds to the removal of an electron from a different orbital in the atom. Neon, for example, has a peak at the binding energy of the $2p$ electrons, then one for the $2s$ electrons, and finally, at the highest binding energy, one for the $1s$ electrons. Figure 13-32 shows measured orbital energies for neutral atoms.

### Periodic Trends in Ionization Energies

The ionization energy of an atom has just been defined as the minimum energy necessary to detach an electron from the neutral gaseous atom. It is the change in energy $\Delta E$ for the process

$$X(g) \longrightarrow X^+(g) + e^- \qquad\qquad \Delta E = IE_1$$

Ionization energies are always positive because ground-state atoms are stable systems. An input of energy is required to detach an electron from an atom. The *second*

**Figure 13-32**

The orbital energies of the first 97 elements, based on photoelectron spectroscopy experiments. Orbitals of the same principal quantum number $n$, such as the $2s$ and $2p$ orbitals, have similar energies and are well separated from orbitals of different $n$. This confirms the shell structure of many-electron atoms. Note, however, the $3d$ orbitals. In elements 21 through 30 (scandium through zinc), they lie well above the $3s$ and $3p$ orbitals and only slightly below the $4s$ orbitals. Their energy then drops rapidly as $Z$ increases further. The $4d$, $5d$, and $4f$ orbitals behave similarly. Note the logarithmic energy scale. One rydberg is $2.8 \times 10^{-18}$ J.

**Figure 13-33**

First and second ionization energies of atoms of the first three periods.

ionization energy, $IE_2$, is the minimum energy required to remove a second electron, or $\Delta E$ for the process

$$X^+(g) \longrightarrow X^{2+}(g) + e^- \qquad\qquad \Delta E = IE_2$$

The third, fourth, and higher ionization energies are defined in an analogous fashion. Ionization energies differ from the series of photoelectron energies. The photoelectron spectrum of an atomic gas measures the energies required to eject the different electrons from a neutral atom. In contrast, the ionization energies (except for $IE_1$) concern the removal of electrons from ions, not neutral atoms. Appendix F lists measured ionization energies of the elements, and Figure 13-33 shows the periodic trends in first and second ionization energies with atomic number.

The ionization energy influences how an atom interacts with other atoms to form chemical bonds. It is a measure of the stability of the outer electron configuration: the greater it is, the more energy is required to remove an outer electron, and therefore the more stable the electronic structure. Let us examine the periodic trends in the first ionization energy to learn about the stabilities of the various electron configurations. There is a large drop in $IE_1$ from helium to lithium, for two reasons: first, a $2s$ electron is much farther from the nucleus than a $1s$ electron; second, the $1s$ electrons screen the nucleus in lithium so effectively that the $2s$ electron "sees" a net positive charge close to $+1$ rather than the larger charge seen by the electrons in helium. Beryllium shows an increase in $IE_1$ because the nuclear charge has increased, but the electron being removed is still from a $2s$ orbital.

The $IE_1$ of boron is somewhat less than that of beryllium because the fifth electron is in a higher-energy (and therefore less stable) $2p$ orbital. In carbon and nitrogen, the additional electrons go into $2p$ orbitals as the nuclear charge increases to hold the outer electrons more tightly; hence, $IE_1$ increases. The nuclear charge is higher in oxygen than in nitrogen, which would give it a higher ionization energy; however, this is more than compensated for by the fact that oxygen must accommodate two electrons in the same $2p$ orbital, leading to greater electron–electron repulsion and diminished binding. Oxygen has a lower $IE_1$ than nitrogen. Fluorine and neon have successively higher first ionization energies because of increasing nuclear charge. The general trends of increasing ionization energy across a given period, as well as the dips that occur at certain points, can thus be understood through the orbital description of many-electron atoms.

The ionization energy tends to decrease down a group in the periodic table (for example, from lithium to sodium to potassium). As the principal quantum number increases, so does the distance of the outer electrons from the nucleus. There are exceptions to this trend, however, especially for the heavier elements in the middle of the periodic table. The first ionization energy of gold, for example, is higher than that of silver or of copper. This fact is crucial in making gold a ''noble metal,'' one that is resistant to attack by oxygen.

Similar trends are observed in the second ionization energy but are shifted higher in atomic number by one unit (see Fig. 13-33). Thus, $IE_2$ is very high for Li (because $Li^+$ has a filled $1s^2$ shell) but relatively low for Be (because $Be^+$ has a single electron in the outermost $2s$ orbital).

## Electron Affinity

The **electron affinity** $EA$ of an atom is the energy *released* when an electron is added to it. It is thus $-\Delta E$ for the reaction

$$X(g) + e^- \longrightarrow X^-(g)$$

The electron affinity of X is the same as the ionization energy of $X^-$—that is, $\Delta E$ for the process

$$X^-(g) \longrightarrow X(g) + e^-$$

An electron affinity is either positive or negative, depending on the element involved. A negative electron affinity means that the electron must be forced to ''stick'' and will spontaneously fly away from the negative ion if it is in free space. Appendix F lists the electron affinities of the elements.

The periodic trends in electron affinity (Fig. 13-34) parallel those in ionization energy for the most part, except that they are shifted one unit *lower* in atomic number. The reason is clear. Attaching an electron to F gives $F^-$, with the configuration $1s^2 2s^2 2p^6$, the same as that of Ne. Fluorine has a large affinity for electrons because the resulting closed-shell configuration is very stable. In contrast, the noble gases have negative electron affinities for the same reason that the alkali metals have small ionization energies: the last electron resides in a new shell far from the nucleus and is almost totally screened from the nuclear charge.

No gaseous atom has a positive electron affinity for a *second* electron, because a gaseous ion with a net charge of $-2e$ is always unstable with respect to ionization. Attaching a second electron means bringing it close to a species that is already negatively charged. The two repel each other, and the energy rises. In crystalline environments, however, doubly negative ions such as $O^{2-}$ can be stabilized by electrostatic interaction with neighboring positive ions.

**Figure 13-34**

The electron affinities (in kJ $mol^{-1}$) of gaseous atoms of the elements. Some of the atoms have negative electron affinities; this includes the noble gases, which are not visible at the extreme right.

## Example 13-11

Consider the elements selenium (Se) and bromine (Br). Which has the higher first ionization energy and which the higher electron affinity?

### Solution

These two atoms are adjacent to each other in the periodic table. Bromine has one more electron in the $4p$ subshell, and this electron should be more tightly bound than the $4p$ electrons in Se because of the extra unit of positive charge on the nucleus. Thus, $IE_1$ should be greater for Br.

Bromine has a greater electron affinity than selenium. Gaining the extra electron changes the Br atom into $Br^-$ ion, which has a particularly stable closed-shell electron configuration (the same as that of the noble gas atom Kr). No such configuration is created when an Se atom gains an additional electron.

*Related Problems:* 55, 56, 57, 58

## Interstellar Space

**PRACTICE EXERCISE**

The vast stretches of space between the stars are by no means empty. They contain both gases and dust particles at very low concentrations. Interstellar space extends so far that these low-density contents significantly affect the electromagnetic radiation arriving from distant stars and other sources, which is detected by telescopes. The gas in interstellar space consists primarily of hydrogen (either neutral or ionized) at a concentration on the order of one atom per cubic centimeter. The dust (thought to be mostly solid water, methane, or ammonia) is even less concentrated, with typically only a few dust particles (each $10^{-4}$ to $10^{-5}$ cm in radius) per cubic *kilometer.*

(a) The hydrogen in interstellar space near a star is largely ionized by the high-energy photons from the star. Such regions are called H II regions. Suppose a ground-state hydrogen atom absorbs a photon with a wavelength of 65 nm. Calcu-

Clouds of gas surround very hot stars in these galactic clusters. The red color arises from hydrogen radiation. (Anglo-Australian Telescope Board)

late the kinetic energy of the electron ejected. (*Note:* This is the gas-phase analog of the photoelectric effect for solids.)

(b) What is the de Broglie wavelength of the electron from part (a)?

(c) Free electrons in H II regions can be recaptured by hydrogen nuclei. In such an event, the atom emits a series of photons of increasing energy as the electrons cascade down through the quantum states of the hydrogen atom. The particle densities are so low that extremely high quantum states can be detected in interstellar space. In particular, the transition from the state $n = 110$ to $n = 109$ for the hydrogen atom has been detected. What is the Bohr radius of an electron in the state $n = 110$ for hydrogen?

(d) Calculate the wavelength of light emitted as an electron undergoes a transition from level $n = 110$ to $n = 109$. In what region of the electromagnetic spectrum does this lie?

(e) H II regions also contain ionized atoms heavier than hydrogen. Calculate the longest wavelength of light that will ionize a ground-state helium atom. Use data from Appendix F.

(f) The regions farther from stars are called H I regions. Here almost all of the hydrogen is neutral rather than ionized and is in its ground state. Will such hydrogen atoms absorb light in the Balmer series emitted by atoms in H II regions?

(g) We stated in Section 13-5 that the energy of the hydrogen atom depends only on the quantum number $n$. In fact, this is not quite true. The electron spin ($m_s$ quantum number) couples very weakly with the spin of the nucleus, making the ground state split into two states of almost equal energy. The radiation emitted in a transition from the upper to the lower of these levels has a wavelength of 21.2 cm and is of great importance in astronomy because it allows the H I regions to be studied. What is the energy difference between these two levels, both for a single atom and for a mole of atoms?

(h) The gas and dust particles between a star and the earth absorb the star's light more strongly in the blue region of the spectrum than in the red. As a result, stars appear slightly redder than they actually are. Will an estimate of the temperature of a star based on its apparent color give too high or too low a number?

**Answers**

(a) $8.8 \times 10^{-19}$ J

(b) 0.52 nm = 5.2 Å

(c) $6.40 \times 10^{-7}$ m = 6400 Å

(d) $5.98 \times 10^{-2}$ m = 5.98 cm, in the microwave region

(e) 50.4 nm

(f) No, because the lowest-energy absorption for a ground-state hydrogen atom is in the ultraviolet region of the spectrum.

(g) $9.37 \times 10^{-25}$ J; 0.564 J mol$^{-1}$

(h) Too low

## CONCEPTS & SKILLS

After studying this chapter and working the problems that follow, you should be able to

1. Relate the frequency, wavelength, and speed of light and other waves (Section 13-1, problems 1–8).

2. Describe blackbody radiation and the photoelectric effect, and discuss how

these paradoxes of classical physics were resolved by quantum mechanics (Section 13-1, problems 9–10).

3. Using the law of conservation of energy, relate the work function of a metal to the wavelength of light used to eject electrons in the photoelectric effect, and the kinetic energy of those electrons (Section 13-1, problems 11–14).

4. Use the Bohr model to calculate the energy levels of one-electron atoms and to find the frequencies and wavelengths of light emitted in transitions between energy levels (Section 13-2, problems 21–24).

5. Discuss the de Broglie relation and use it to calculate the wavelengths associated with particles in motion (Section 13-3, problems 27–28).

6. State the Heisenberg indeterminacy principle and use it to establish bounds within which the position and momentum of a particle can be known (Section 13-3, problems 29–30).

7. Determine the energy levels for particles in rigid rectangular boxes (Section 13-4, problems 31–32).

8. Give the quantum numbers that characterize one-electron atoms, and discuss the shapes, sizes, and nodal properties of the corresponding orbitals (Section 13-5, problems 33–38).

9. Use the aufbau principle to predict electron configurations of atoms and ions and to account for the structure of the periodic table (Section 13-6, problems 41–50).

10. Discuss the factors that lead to systematic variation of sizes of atoms and ion through the periodic table (Section 13-6, problems 51–54).

11. Describe the trends in ionization energy and electron affinity across the periodic table and relate them to the electronic structure of atoms (Section 13-7, problems 55–58).

## PROBLEMS

### Waves and Light

1. Some water waves reach the beach at a rate of one every 3.2 s, and the distance between their crests is 2.1 m. Calculate the speed of these waves.

2. The spacing between bands of color in a chemical wave from an oscillating reaction is measured to be 1.2 cm, and a new wave appears every 42 s. Calculate the speed of propagation of the chemical waves.

3. An FM radio station broadcasts at a frequency of $9.86 \times 10^7 \ s^{-1}$ (98.6 MHz). Calculate the wavelength of the radio waves.

4. The gamma rays emitted by $^{60}Co$ are used in radiation treatment of cancer. They have a frequency of $2.83 \times 10^{20} \ s^{-1}$. Calculate their wavelength, expressing your answer in meters and in angstroms.

5. Radio waves of wavelength $6.00 \times 10^2 \ m$ can be used to communicate with spacecraft over large distances.
   (a) Calculate the frequency of these radio waves.
   (b) Suppose a radio message is sent home by the astronauts in a spaceship approaching the planet Mars at a distance of $8.0 \times 10^{10} \ m$ from earth. How long (in minutes) will it take for the message to travel from the spaceship to earth?

6. An argon ion laser emits light of wavelength 488 nm.
   (a) Calculate the frequency of the light.
   (b) Suppose a pulse of light from this laser is sent from earth, is reflected from a mirror on the moon, and returns to its starting point. Calculate the time elapsed for the round trip, taking the distance from earth to moon to be $3.8 \times 10^5 \ km$.

7. The speed of sound in dry air at 20°C is $343.5 \ m \ s^{-1}$, and the frequency of the sound from the middle C note on a piano is $261.6 \ s^{-1}$ (according to the American standard pitch scale). Calculate the wavelength of this sound and the time it will take to travel 30.0 m across a concert hall.

8. Ultrasonic waves have frequencies too high to be detected by the human ear but can be produced and detected by vibrating crystals. Calculate the wavelength of an ultrasonic wave of frequency $5.0 \times 10^4 \ s^{-1}$ that is propagating through a sample of water at a speed of $1.5 \times 10^3 \ m \ s^{-1}$. Explain why ultra-

sound can be used to probe the size and position of the fetus inside its mother's abdomen. Could audible sound with a frequency of 8000 s$^{-1}$ be used for this purpose?

9. Both blue and green light eject electrons from the surface of potassium. In which case do the ejected electrons have the higher average kinetic energy?

10. When an intense beam of green light is directed onto a copper surface, no electrons are ejected. What will happen if the green light is replaced with red light?

11. Cesium is frequently employed in photocells because its work function ($3.43 \times 10^{-19}$ J) is the lowest of all the elements. Such photocells are efficient because the broadest range of wavelengths of light can eject electrons. What colors of light will eject electrons from cesium? What colors of light will eject electrons from selenium, which has a work function of $9.5 \times 10^{-19}$ J?

12. Alarm systems employ the photoelectric effect. A beam of light strikes a piece of metal in the photocell, ejecting electrons continuously and causing a small electric current to flow. When someone steps into the light beam, the current is interrupted and the alarm is triggered. What is the maximum wavelength of light that can be used in such an alarm system if the photocell metal is sodium, with a work function of $4.41 \times 10^{-19}$ J?

13. Light having a wavelength of $2.50 \times 10^{-7}$ m falls upon the surface of a piece of chromium in an evacuated glass tube. If the work function of the chromium is $7.21 \times 10^{-19}$ J, determine (a) the maximum kinetic energy of the emitted photoelectrons and (b) the speed of photoelectrons having this maximum kinetic energy.

14. Calculate the maximum wavelength of electromagnetic radiation if it is to cause detachment of electrons from the surface of metallic tungsten, which has a work function of $7.29 \times 10^{-19}$ J. If the maximum speed of the emitted photoelectrons is to be $2.00 \times 10^6$ m s$^{-1}$, what should the wavelength of the radiation be?

## Atomic Spectroscopy and the Bohr Model

15. Excited lithium atoms emit light strongly at a wavelength of 671 nm. This emission predominates when Li atoms are excited in a flame. Predict the color of the flame.

16. Excited mercury atoms emit light strongly at a wavelength of 454 nm. This emission predominates when Hg atoms are excited in a flame. Predict the color of the flame.

17. Barium atoms in a flame emit light as they undergo transitions from one energy level to another that is $3.6 \times 10^{-19}$ J lower in energy. Calculate the wavelength of light emitted and, by referring to Figure 13-4, predict the color visible in the flame.

18. Potassium atoms in a flame emit light as they undergo transitions from one energy level to another that is $4.9 \times 10^{-19}$ J lower in energy. Calculate the wavelength of light emitted

and, by referring to Figure 13-4, predict the color visible in the flame.

19. The sodium D-line is actually a pair of closely spaced spectroscopic lines seen in the emission spectrum of sodium atoms. The wavelengths are centered at 589.3 nm. The intensity of this emission makes it the major source of light (and causes the yellow color) in the sodium arc light.
(a) Calculate the energy change per sodium atom emitting a photon at the D-line wavelength.
(b) Calculate the energy change per mole of sodium atoms emitting photons at the D-line wavelength.
(c) If a sodium arc light is to produce 1.000 kilowatt (1000 J s$^{-1}$) of radiant energy at this wavelength, how many moles of sodium atoms must emit photons per second?

20. The power output of a laser is measured by its wattage, the number of joules of energy it radiates per second (1 W = 1 J s$^{-1}$). A 10-W laser produces a beam of green light with a wavelength of 520 nm ($5.2 \times 10^{-7}$ m).
(a) Calculate the energy carried by each photon.
(b) Calculate the number of photons emitted by the laser per second.

21. Use the Bohr model to calculate the radius and the energy of the B$^{4+}$ ion in the $n = 3$ state. How much energy would be required to remove the electrons from one mole of B$^{4+}$ ions in this state? What frequency and wavelength of light would be emitted in a transition from the $n = 3$ to the $n = 2$ state of this ion? Express all results in SI units.

22. He$^+$ ions are observed in stellar atmospheres. Use the Bohr model to calculate the radius and the energy of the He$^+$ ion in the $n = 5$ state. How much energy would be required to remove the electrons from one mole of He$^+$ in this state? What frequency and wavelength of light would be emitted in a transition from the $n = 5$ to the $n = 3$ state of this ion? Express all results in SI units.

23. The radiation emitted in the transition from $n = 3$ to $n = 2$ in a neutral hydrogen atom has a wavelength of 656.1 nm. What would be the wavelength of radiation emitted from a doubly ionized lithium atom (Li$^{2+}$) if a transition occurred from $n = 3$ to $n = 2$? In what region of the spectrum does this radiation lie?

24. The Be$^{3+}$ ion has a single electron. Calculate the frequencies and wavelengths of light in the ion's emission spectrum for the first three lines of each of the series that are analogous to the Lyman and the Balmer series of neutral hydrogen. In what region of the spectrum does this radiation lie?

## Waves, Particles, and the Schrödinger Equation

25. A guitar string with fixed ends has a length of 50 cm.
(a) Calculate the wavelengths of its fundamental mode of vibration (i.e., its first harmonic) and its third harmonic.
(b) How many nodes does the third harmonic have?

26. Suppose we picture an electron in a chemical bond as being a

wave with fixed ends. Take the length of the bond to be 1.0 Å.

(a) Calculate the wavelength of the electron wave in its ground state and in its first excited state.

(b) How many nodes does the first excited state have?

27. Calculate the de Broglie wavelength of the following:

(a) An electron moving at a speed of $1.00 \times 10^3$ m s$^{-1}$

(b) A proton moving at a speed of $1.00 \times 10^3$ m s$^{-1}$

(c) A baseball with a mass of 145 g, moving at a speed of 75 km hr$^{-1}$

28. Calculate the de Broglie wavelength of the following:

(a) Electrons that have been accelerated to a kinetic energy of $1.20 \times 10^7$ J mol$^{-1}$

(b) A helium atom moving at a speed of 353 m s$^{-1}$ (the root-mean-square speed of helium atoms at 20 K)

(c) A krypton atom moving at a speed of 299 m s$^{-1}$ (the root-mean-square speed of krypton atoms at 300 K)

29. (a) The position of an electron is known to within 10 Å $(1.0 \times 10^{-9}$ m). What is the minimum indeterminacy in its velocity?

(b) Repeat the calculation of part (a) for a helium atom.

30. No object can travel faster than the speed of light, so it would seem evident that the indeterminacy in the speed of any object is at most $3 \times 10^8$ m s$^{-1}$.

(a) What is the minimum indeterminacy in the position of an electron, given that we know nothing about its speed except that it is less than the speed of light?

(b) Repeat the calculation of part (a) for the position of a helium atom.

## The Particle in a Box

31. In Chapter 2 we introduced the concept of a double bond between carbon atoms, represented by C=C, with a length near 1.34 Å. The motion of an electron in such a bond can be treated very crudely as motion in a one-dimensional box. Calculate the energy of an electron in each of its three lowest allowed states if it is confined to move in a one-dimensional box of length 1.34 Å. Calculate the wavelength of light necessary to excite the electron from its ground state to the first excited state.

32. When metallic sodium is dissolved in liquid sodium chloride, electrons are released into the liquid. These dissolved electrons absorb light with a wavelength near 800 nm. Suppose we treat the positive ions surrounding an electron very crudely as defining a three-dimensional cubic box of edge $L$, and we assume that the absorbed light excites the electron from its ground state to the first excited state. Calculate the edge length $L$ within this simple model.

## The Hydrogen Atom

33. Which of the following combinations of quantum numbers are allowed for an electron in a one-electron atom, and which are not?

(a) $n = 2$, $\ell = 2$, $m = 1$, $m_s = \frac{1}{2}$

(b) $n = 3$, $\ell = 1$, $m = 0$, $m_s = -\frac{1}{2}$

(c) $n = 5$, $\ell = 1$, $m = 2$, $m_s = \frac{1}{2}$

(d) $n = 4$, $\ell = -1$, $m = 0$, $m_s = \frac{1}{2}$

34. Which of the following combinations of quantum numbers are allowed for an electron in a one-electron atom and which are not?

(a) $n = 3$, $\ell = 2$, $m = 1$, $m_s = 0$

(b) $n = 2$, $\ell = 0$, $m = 0$, $m_s = -\frac{1}{2}$

(c) $n = 7$, $\ell = 2$, $m = -2$, $m_s = \frac{1}{2}$

(d) $n = 3$, $\ell = -3$, $m = -2$, $m_s = -\frac{1}{2}$

35. Label the orbitals described by each of the following sets of quantum numbers.

(a) $n = 4$, $\ell = 1$

(b) $n = 2$, $\ell = 0$

(c) $n = 6$, $\ell = 3$

36. Label the orbitals described by each of the following sets of quantum numbers.

(a) $n = 3$, $\ell = 2$

(b) $n = 7$, $\ell = 4$

(c) $n = 5$, $\ell = 1$

37. How many radial nodes and how many angular nodes does each of the orbitals in problem 35 have?

38. How many radial nodes and how many angular nodes does each of the orbitals in problem 36 have?

39. Use the mathematical expression for the $2p_z$ wave function of a one-electron atom (see Table 13-2) to show that the probability of finding an electron in that orbital anywhere in the $xy$ plane is 0. What are the nodal planes for a $d_{xz}$ orbital and for a $d_{x^2-y^2}$ orbital?

40. (a) Use the radial wave function for the $3p$ orbital of a hydrogen atom (see Table 13-2) to calculate the value of $r$ for which a node exists.

(b) Find the values of $r$ for which nodes exist for the $3s$ wave function of the hydrogen atom.

## Many-Electron Atoms and the Periodic Table

41. Give the ground-state electron configurations of the following elements.

(a) C

(b) Se

(c) Fe

42. Give the ground-state electron configurations of the following elements.

(a) P

(b) Tc

(c) Ho

43. Write ground-state electron configurations for the ions Be$^+$, C$^-$, Ne$^{2+}$, Mg$^+$, P$^{2+}$, Cl$^-$, As$^+$, and I$^-$. Which do you expect will be paramagnetic due to the presence of unpaired electrons?

44. Write ground-state electron configurations for the ions Li$^-$, B$^+$, F$^-$, Al$^{3+}$, S$^-$, Ar$^+$, Br$^+$, and Te$^-$. Which do you expect

to be paramagnetic due to the presence of unpaired electrons?

45. Identify the atom or ion corresponding to each of the following descriptions:
    (a) An atom with ground-state electron configuration $[Kr]4d^{10}5s^25p^1$
    (b) An ion with charge $-2$ and ground-state electron configuration $[Ne]3s^23p^6$
    (c) An ion with charge $+4$ and ground-state electron configuration $[Ar]3d^3$

46. Identify the atom or ion corresponding to each of the following descriptions:
    (a) An atom with ground-state electron configuration $[Xe]4f^{14}5d^66s^2$
    (b) An ion with charge $-1$ and ground-state electron configuration $[He]2s^22p^6$
    (c) An ion with charge $+5$ and ground-state electron configuration $[Kr]4d^6$

47. Predict the atomic number of the (as yet undiscovered) element in the seventh period that is a halogen.

48. (a) Predict the atomic number of the (as yet undiscovered) alkali-metal element in the eighth period.
    (b) Suppose the eighth-period alkali-metal atom turned out to have atomic number 137. What explanation would you give for such a high atomic number (recall that the atomic number of francium is only 87)?

49. Suppose that the spin quantum number did not exist, so that only one electron could occupy each orbital of a many-electron atom. Give the atomic numbers of the first three noble-gas atoms in this case.

50. Suppose that the spin quantum number had three allowed values ($m_s = 0, +\frac{1}{2}, -\frac{1}{2}$). Give the atomic numbers of the first three noble-gas atoms in this case.

## Experimental Measures of Orbital Energies

51. For each of the following pairs of atoms or ions, state which you expect to have the larger radius.
    (a) Na or K
    (b) Cs or Cs$^+$
    (c) Rb$^+$ or Kr
    (d) K or Ca
    (e) Cl$^-$ or Ar

52. For each of the following pairs of atoms or ions, state which you expect to have the larger radius.
    (a) Sm or Sm$^{3+}$
    (b) Mg or Ca
    (c) I$^-$ or Xe
    (d) Ge or As
    (e) Sr$^+$ or Rb

53. Predict the larger ion in each of the following pairs. Give reasons for your answers.
    (a) O$^-$, S$^{2-}$
    (b) Co$^{2+}$, Ti$^{2+}$

(c) Mn$^{2+}$, Mn$^{4+}$
(d) Ca$^{2+}$, Sr$^{2+}$

54. Predict the larger ion in each of the following pairs. Give reasons for your answers.
    (a) S$^{2-}$, Cl$^-$
    (b) Tl$^+$, Tl$^{3+}$
    (c) Ce$^{3+}$, Dy$^{3+}$
    (d) S$^-$, I$^-$

55. For each of the following pairs of atoms, state which you expect to have the higher first ionization energy.
    (a) Rb or Sr
    (b) Po or Rn
    (c) Xe or Cs
    (d) Ba or Sr

56. For each of the following pairs of atoms, state which you expect to have the higher first ionization energy.
    (a) Bi or Xe
    (b) Se or Te
    (c) Rb or Y
    (d) K or Ne

57. For each of the following pairs of atoms, state which you expect to have the greater electron affinity.
    (a) Xe or Cs
    (b) Pm or F
    (c) Ca or K
    (d) Po or At

58. For each of the following pairs of atoms, state which you expect to have the higher electron affinity.
    (a) Rb or Sr
    (b) I or Rn
    (c) Ba or Te
    (d) Bi or Cl

59. The cesium atom has the lowest ionization energy of all the neutral atoms in the periodic table, 375.7 kJ mol$^{-1}$. What is the longest wavelength of light that could ionize a cesium atom? In which region of the electromagnetic spectrum does this light fall?

60. Until recently, it was thought that the Ca$^-$ ion was unstable, so that the calcium atom had a negative electron affinity. Some new experiments have now measured an electron affinity of $+2.0$ kJ mol$^{-1}$ for Ca. What is the longest wavelength of light that could remove an electron from a Ca$^-$ ion? In which region of the electromagnetic spectrum does this light fall?

## Additional Problems

61. A piano tuner uses a tuning fork that emits sound with a frequency of 440 s$^{-1}$. Calculate the wavelength of the sound from this tuning fork and the time the sound takes to travel 10.0 m across a large room. Take the speed of sound in air to be 343 m s$^{-1}$.

62. The distant galaxy called Cygnus A is one of the strongest

sources of radio waves reaching the earth. The distance of this galaxy from the earth is $3 \times 10^{24}$ m. How long (in years) does it take a radio wave of wavelength 10 m to reach the earth? What is the frequency of this radio wave?

63. Hot objects can emit blackbody radiation that appears red, orange, white, or bluish white but never green. Explain.

64. Compare the energy (in joules) carried by an x-ray photon (wavelength $\lambda = 0.20$ nm) with that carried by an AM radio wave photon ($\lambda = 200$ m). Calculate the energy of 1.00 mol of each type of photon. What effect do you expect each type of radiation to have in terms of inducing chemical reactions in the substances through which it passes?

65. When ultraviolet light of wavelength 131 nm strikes a polished nickel surface, the maximum kinetic energy of ejected electrons is measured to be $7.04 \times 10^{-19}$ J. Calculate the work function of nickel.

66. Express the velocity of the electron in the Bohr model in terms of fundamental constants ($m_e, e, h, \epsilon_0$), the nuclear charge $Z$, and the quantum number $n$. Evaluate the velocity of an electron in the ground states of an $He^+$ ion and a $U^{91+}$ ion. Compare these velocities with the speed of light $c$. As the velocity of an object approaches the speed of light, relativistic effects become important. In which kinds of atoms do you expect relativistic effects to be greatest?

67. Photons are emitted in the Lyman series, as hydrogen atoms undergo transitions from various excited states to the ground state. If ground-state $He^+$ ions are present in the same gas (near stars, for example), can they absorb these photons? Explain.

*68. Name a transition in the $C^{5+}$ ion that will lead to the absorption of green light.

69. The energies of macroscopic objects as well as those of microscopic ones are quantized, but the effects of the quantization are not seen because the difference in energy between adjacent states is so small. Apply Bohr's quantization of angular momentum to the revolution of the earth (mass $6.0 \times 10^{24}$ kg), which moves with a velocity of $3.0 \times 10^4$ m s$^{-1}$ in a circular orbit (radius $1.5 \times 10^{11}$ m) about the sun. The sun can be treated as fixed. Calculate the value of the quantum number $n$ for the present state of the earth–sun system. What would be the effect of an increase in $n$ by 1?

70. Sound waves, like light waves, can interfere with each other, giving maximum and minimum levels of sound. Suppose a listener standing directly between two loudspeakers hears the same tone being emitted from both. This listener observes that, when one of the speakers is moved 0.16 m farther away, the perceived intensity of the tone decreases from a maximum to a minimum.
(a) Calculate the wavelength of the sound.
(b) Calculate its frequency, using 343 m s$^{-1}$ as the speed of sound.

71. (a) If the kinetic energy of an electron is known to lie between $1.59 \times 10^{-19}$ J and $1.61 \times 10^{-19}$ J, what is the smallest distance within which it can be known to lie?

(b) Repeat the calculation of part (a) for a helium atom instead of an electron.

72. It is interesting to speculate on the properties of a universe with different values for the fundamental constants.
(a) In a universe in which Planck's constant had the value $h = 1$ J s, what would be the the de Broglie wavelength of a 145-g baseball moving at a speed of 20 m s$^{-1}$?
(b) Suppose the velocity of the ball from part (a) is known to lie between 19 and 21 m s$^{-1}$. What is the smallest distance within which it can be known to lie?
(c) Suppose that in this universe the mass of the electron is 1 g and the charge on the electron is 1 C. Calculate the Bohr radius of the hydrogen atom in this universe.

73. The normalized wave function for a particle in a one-dimensional box, in which the potential energy is zero, is $\psi(x) = \sqrt{2/L} \sin{(n\pi x/L)}$, where $L$ is the length of the box. What is the probability that the particle will lie between $x = 0$ and $x = L/4$ if the particle is in its $n = 2$ state?

74. A particle of mass $m$ is placed in a three-dimensional rectangular box with edge lengths $2L$, $L$, and $L$. Inside the box the potential energy is zero and outside it is infinite, and so the wave function goes smoothly to zero at the sides of the box. Calculate the energies, and give the quantum numbers of the ground state and the first five excited states (or sets of states of equal energy) for the particle in the box.

75. How does the $3d_{xy}$ orbital of an electron in an $O^{7+}$ ion resemble, and how does it differ from, the $3d_{xy}$ orbital of an electron in a hydrogen atom?

*76. The wave function of an electron in the lowest (i.e., ground) state of the hydrogen atom is

$$\psi(r) = \left(\frac{1}{\pi a_0^3}\right)^{1/2} \exp\left(-\frac{r}{a_0}\right)$$

$$a_0 = 0.529 \times 10^{-10} \text{ m}$$

(a) What is the probability of finding the electron inside a sphere of volume 1.0 pm$^3$, centered at the nucleus (1 pm = $10^{-12}$ m)?
(b) What is the probability of finding the electron in a volume of 1.0 pm$^3$ at a distance of 52.9 pm from the nucleus, in a fixed but arbitrary direction?
(c) What is the probability of finding the electron in a spherical shell of 1.0 pm thickness, at a distance of 52.9 pm from the nucleus?

77. An atom of sodium has the electron configuration [Ne]$6s^1$. Explain how this is possible.

*78. (a) The nitrogen atom has one electron in each of the $2p_x$, $2p_y$, and $2p_z$ orbitals. By using the form of the angular wave functions, show that the total electron density, $\psi^2(2p_x) + \psi^2(2p_y) + \psi^2(2p_z)$, is spherically symmetric (that is, it is independent of the angles $\theta$ and $\phi$). The neon atom, which has *two* electrons in each $2p$ orbital, is also spherically symmetric.
(b) The same result as in part (a) applies to $d$ orbitals, so that

a filled or half-filled subshell of $d$ orbitals is spherically symmetric. Identify the spherically symmetric atoms or ions among the following: $F^-$, Na, Si, $S^{2-}$, $Ar^+$, Ni, Cu, Mo, Rh, Sb, W, Au.

79. Chromium(IV) oxide is used in making magnetic recording tapes because it is paramagnetic. It can be described as a solid made up of $Cr^{4+}$ and $O^{2-}$ ions. Give the electron configuration of $Cr^{4+}$ in $CrO_2$, and determine the number of unpaired electrons on each chromium ion.

80. Use the data from Appendix F to graph the variation of atomic radius with atomic number for the rare-earth elements from lanthanum to lutetium.
    (a) What is the general trend in these radii? How do you account for it?
    (b) Which two elements in the series present exceptions to the trend?

81. Arrange the following six atoms or ions in order of size, from the smallest to the largest: K, $F^+$, Rb, $Co^{25+}$, Br, F, $Rb^-$.

82. Which is higher, the third ionization energy of lithium or the energy required to eject a $1s$ electron from a lithium atom in a photoelectron spectroscopy experiment? Explain.

83. The outermost electron in an alkali-metal atom is sometimes described as resembling an electron in the corresponding state of a one-electron atom. Compare the first ionization energy of Li with the binding energy of a $2s$ electron in a one-electron atom that has nuclear charge $Z_{eff}$, and determine the value of $Z_{eff}$ that is necessary for the two energies to agree. Repeat the calculation for the $3s$ electron of Na and the $4s$ electron of K.

*84. In two-photon ionization spectroscopy, the combined energies carried by two different photons are used to remove an electron from an atom or molecule. In such an experiment a potassium atom in the gas phase is to be ionized by two different light beams, one of which has wavelength 650 nm. What is the maximum wavelength for the second beam that will cause two-photon ionization?

*85. The x-ray emission lines observed by Moseley (Fig. 13-10) arise from transitions in which an electron drops from an occupied level with principal quantum number $n = 2$ to an empty level with $n = 1$ (the electron originally in that level has been ejected earlier). Assuming that the energy of the $n = 2$ level corresponds to that of a single-electron atom with effective nuclear charge $Z_{eff} = Z - \sigma_2$, and that of the $n = 1$ level with $Z_{eff} = Z - \sigma_1$, use the Bohr energy formula to estimate the value of the constant $a$ in the Moseley formula. (*Hint:* Assume that $Z$ is much larger than $\sigma_1$ and $\sigma_2$, and look at the coefficient of the $Z^2$ term in the expression for the frequency $\nu$.)

## CUMULATIVE PROBLEMS

86. As stated in Example 13-1, microwave ovens use radiation with a frequency of $2.45 \times 10^9 \ s^{-1}$. Calculate the number of such photons that must be absorbed to raise the temperature of 100.0 g of water from 15°C to its boiling point, 100°C. Take the specific heat capacity of water to be $4.2 \ J \ K^{-1} \ g^{-1}$.

*87. Because of its particle nature, light can exert pressure on a surface. Suppose a 1.0-W laser is focused on a circular area 0.10 mm in radius. The wavelength of the laser light is 550 nm. Assume that the photons travel perpendicular to the surface and are perfectly reflected, so that their momenta change sign after reflection from the wall. Using results from the kinetic theory of gases (Section 3-5), calculate the pressure exerted by the light on the surface. The momentum $p$ of a photon is related to its energy through $E = pc$.

*88. Chapter 12 stated that the nuclei of atoms contain protons and neutrons rather than protons and electrons.
    (a) Show by a calculation using the Heisenberg indeterminacy principle, $\Delta p \ \Delta x \geq h/4\pi$, that an electron cannot be confined within a nucleus. Assume the radius of a nucleus to be $\sim 1 \times 10^{-15}$ m.
    (b) Repeat the calculation for a proton or neutron inside the nucleus.

89. The standard energy change (at 25°C and atmospheric pressure) for the process

$$K(g) + Cl(g) \longrightarrow KCl(s)$$

has been measured to be $-653$ kJ. Use this fact, together with data on ionization energies and electron affinities from Appendix F, to calculate the lattice energy per mole of the KCl lattice, defined as the energy change required to separate the solid into its component ions:

$$KCl(s) \longrightarrow K^+(g) + Cl^-(g)$$

*Liquid oxygen is attracted to the region between the pole faces of a magnet because it is paramagnetic.*

# Chemical Bonding

**M**odern quantum mechanics embraces the disquieting notion that matter and energy have dual natures, part particle and part wave, and from this notion it correctly predicts the sizes, energy levels, and spectra of atoms and ions. This is a substantial achievement, but if it were all quantum mechanics could do, it would matter little to chemists, because chemistry addresses the atomic interactions that form molecules and extended solid structures with useful properties. In fact, quantum mechanics goes far beyond atoms, giving a deep underpinning to the simple ideas of bonding used in the Lewis electron-dot model.

In Chapter 2 we posed the question of why some combinations of atoms form molecules of great stability but others join together only for fleeting moments or not at all. Another important question regards the shapes of molecules: What determines the three-dimensional arrangement of atoms in space, and how does this shape affect molecular properties and reactivity? A third question, unanswered by the simple Lewis dot model, is why oxygen is paramagnetic and reactive when nitrogen is diamagnetic and relatively inert.

In this chapter we explore those questions and others by extending some of the ideas introduced in Chapter 13. *Molecular* orbitals emerge as a natural extension of atomic orbitals and help to explain the stability of molecules and the natures of their bonds. We develop the electron configurations of molecules and present an aufbau principle analogous to that for atoms. The rearrangement of electrons in polyatomic molecules leads to the VSEPR model for molecular geometry. Our overall goal is to understand how changes in the wave functions of electrons lead to the formation of chemical bonds.

## 14-1 PROPERTIES OF THE CHEMICAL BOND

The noble-gas atoms, with their filled valence shells, are quite unreactive because of the high cost in energy for removing or adding electrons to form a chemical bond. In contrast, atoms with partially filled valence shells (**open-shell atoms**) can be quite reactive, attaining stability by exchanging or sharing electrons with other atoms to form molecules or large-scale networks of interacting atoms. Before we consider the types of chemical bonds, let us examine two of the crucial properties that characterize bonds: length and energy. These two properties are related, in turn, to another property: the bond order.

### Bond Lengths

The first question to ask about a molecule is: What is its structure? For a diatomic molecule, the only relevant parameter is the bond length, the distance between the nuclei of the two atoms. For a polyatomic molecule, the structure is given by two or more bond lengths and one or more bond angles. Of course, molecules are not rigid structures with unchanging bond lengths and angles; their atoms can vibrate, usually with small amplitudes, about their average positions. Average bond lengths and bond angles can be measured by spectroscopic techniques (Section 15-1) and by x-ray diffraction (Section 17-1).

Table 14-1 lists the bond lengths of a number of diatomic molecules. Certain systematic trends are immediately obvious. Within a group in the periodic table, bond lengths usually increase with increasing atomic number. Thus, bond lengths increase in the series $F_2$ to $Cl_2$ to $Br_2$ to $I_2$. Similarly, ClF has a shorter bond length than ClBr. These trends are directly related to the corresponding trends in atomic sizes, which usually increase with increasing atomic number in a given group of the periodic table. This qualitative connection cannot be made quantitative, however, because bond lengths are not simply sums of atomic radii. Too much rearrangement of electron density occurs during bond formation to allow any such approximation.

**Table 14-1 Properties of Diatomic Molecules**

|  | Average Bond Length (Å) | Bond Enthalpy (kJ mol$^{-1}$) |
|---|---|---|
| $H_2$ | 0.751 | 436 |
| $N_2$ | 1.100 | 945 |
| $O_2$ | 1.211 | 498 |
| $F_2$ | 1.417 | 158 |
| $Cl_2$ | 1.991 | 243 |
| $Br_2$ | 2.286 | 193 |
| $I_2$ | 2.669 | 151 |
| HF | 0.926 | 568 |
| HCl | 1.284 | 432 |
| HBr | 1.424 | 366 |
| HI | 1.620 | 298 |
| ClF | 1.632 | 255 |
| BrF | 1.759 | 285 |
| BrCl | 2.139 | 219 |
| ICl | 2.324 | 211 |
| NO | 1.154 | 632 |
| CO | 1.131 | 1076 |

A significant result from experimental studies is that the length of the bond between a given pair of atoms changes very little from one molecule to another. A C—H bond has almost the same length in acetylsalicylic acid (aspirin, $C_9H_8O_4$) as in methane ($CH_4$). Table 14-2 shows that the lengths of O—H, C—C, and C—H bonds in numerous molecules are constant to within a few percent.

**Table 14-2 Reproducibility of Bond Lengths**

| Bond | Molecule | Bond Length (Å) |
|---|---|---|
| O—H | $H_2O$ | 0.958 |
|  | $H_2O_2$ | 0.960 |
|  | HCOOH | 0.95 |
|  | $CH_3OH$ | 0.956 |
| C—C | diamond | 1.5445 |
|  | $C_2H_6$ | 1.536 |
|  | $CH_3CHF_2$ | 1.540 |
|  | $CH_3CHO$ | 1.50 |
| C—H | $CH_4$ | 1.091 |
|  | $C_2H_6$ | 1.107 |
|  | $C_2H_4$ | 1.087 |
|  | $C_6H_6$ | 1.084 |
|  | $CH_3Cl$ | 1.11 |
|  | $CH_2O$ | 1.06 |

## Bond Enthalpies

Energetics is the crucial factor in the stability of molecules. A chemical bond forms because the involved atoms have lower energy when they are close to each other than when they are far apart. The molecular geometry that is observed is the one that gives the molecule the lowest energy. The energy absorbed when a bond is broken, $\Delta E_d$ (''d'' stands for *dissociation* here), reveals the strength of that bond. It is more common to tabulate the enthalpy change for breaking the bond, $\Delta H_d$ (that is, the heat absorbed at constant pressure), because the **bond enthalpy** $\Delta H_d$ is measured directly in calorimetric experiments at constant pressure. Because breaking the bond produces two fragments, $\Delta E_d$ and $\Delta H_d$ are related (Section 8-4) through

$$\Delta E_d = \Delta H_d - RT$$

per mole of bonds broken, where $R$ is the gas constant, $8.315 \text{ J K}^{-1} \text{ mol}^{-1}$.

Table 14-1 lists bond enthalpies for selected diatomic molecules. Again, certain systematic trends with changes in atomic number are evident. Bonds generally grow weaker with increasing atomic number, as shown by the decrease in the bond enthalpies of the hydrogen halides in the order HF > HCl > HBr > HI. Note, however, the unusual weakness of the bond in the fluorine molecule $F_2$ (its bond enthalpy is significantly *smaller* than that of $Cl_2$ and comparable to that of $I_2$). Bond strength decreases dramatically in the diatomic molecules from $N_2$ ($945 \text{ kJ mol}^{-1}$) to $O_2$ ($498 \text{ kJ mol}^{-1}$) to $F_2$ ($158 \text{ kJ mol}^{-1}$). What accounts for this behavior? A successful theory of bonding must explain both the general trends and the reasons for particular exceptions.

Bond enthalpies, like bond lengths, are fairly reproducible (within about 10%) from one compound to another. It is therefore possible to tabulate *average* bond enthalpies from measurements on a series of compounds (Table 8-3). The enthalpies of any given bond in different compounds will deviate somewhat from those shown, but in most cases the deviations are small.

## Bond Order

Sometimes the length and enthalpy of the bond between two specific kinds of atoms are *not* reproduced from one chemical compound to another but are sharply different. Table 14-3 shows the great differences in bond lengths and bond enthalpies of carbon–carbon bonds in ethane ($H_3CCH_3$), ethylene ($H_2CCH_2$), and acetylene (HCCH). Carbon–carbon bonds from many other molecules fit into one of the three classes given in the table (that is, some carbon–carbon bond lengths are close to 1.54 Å, others are close to 1.34 Å, and still others are close to 1.20 Å). This con-

### Table 14-3   Three Types of Carbon–Carbon Bonds

| Bond | Molecule | Bond Length (Å) | Bond Enthalpy (kJ mol$^{-1}$) |
|------|----------|-----------------|-------------------------------|
| C—C | $C_2H_6$ (or $H_3CCH_3$) | 1.536 | 348 |
| C=C | $C_2H_4$ (or $H_2CCH_2$) | 1.337 | 615 |
| C≡C | $C_2H_2$ (or HCCH) | 1.204 | 812 |

firms the existence of not one, but three, types of carbon–carbon bonds. The weakest and longest (as in ethane) is a single bond represented by C—C; that of intermediate strength (as in ethylene) is a double bond, C=C; and the strongest and shortest (as in acetylene) is a triple bond, C≡C. The **bond order** is the number of shared electron pairs for these bonds: 1, 2, and 3, respectively.

Even these three types do not cover all the carbon–carbon bonds found in nature, however. In benzene ($C_6H_6$) the experimental carbon–carbon bond length is 1.397 Å and the bond enthalpy is 505 kJ mol$^{-1}$. This bond is intermediate between a single bond and a double bond (it has bond order $1\frac{1}{2}$). In fact, we shall see in Section 14-5 that the bonding in compounds such as benzene differs from that in many other compounds. Although many bonds have properties that depend primarily on the two atoms forming the bond (and thus are similar from one compound to another), bonding in benzene and certain related molecules depends on the nature of the whole molecule. In sum, bond properties are often quite reproducible between compounds, but we must be alert for exceptions that may signal new types of chemical bonding.

## 14-2 IONIC AND COVALENT BONDS

Chemical bonds arise from the sharing or transfer of electrons between two or more atoms. When electron density is mainly shared between two atoms, the bond between them is **covalent.** When electron density is transferred from one atom to another, the resulting chemical bond is **ionic.** Most real bonds are neither fully ionic nor fully covalent but instead possess a mixture of ionic and covalent character. Bonds in which there is a partial transfer of charge are **polar covalent.** We present a classical picture of chemical bonding in this section and defer a quantum-mechanical description to Section 14-4. Let us start with a most useful concept for distinguishing between ionic and covalent bonds: electronegativity.

### Electronegativity

In 1932 Linus Pauling proposed a helpful way to quantify the extent of ionic bonding between a particular pair of atoms, A and B, in a molecule. Suppose the dissociation energy of an A—A bond is $\Delta E_{AA}$, and that of a B—B bond is $\Delta E_{BB}$; both bonds are covalent because the atoms in them are identical. Then an estimate of the *covalent* contribution to the dissociation energy of an A—B bond is the (geometric) mean of these two energies, $\sqrt{\Delta E_{AA}\Delta E_{BB}}$. However, the actual A—B bond includes ionic character because some charge transfer occurs between the atoms. The ionic character tends to strengthen the bond and give a greater value for $\Delta E_{AB}$. Pauling suggested that the difference between the actual and covalent bond energies,

$$\Delta \equiv \Delta E_{AB} - \sqrt{\Delta E_{AA}\Delta E_{BB}}$$

is a measure of the **electronegativity** difference between the two atoms A and B. Specifically, he defined

$$\chi_A - \chi_B = 0.102\Delta^{1/2}$$

where $\chi_A$ and $\chi_B$ (Greek chi) are the electronegativities of A and B and $\Delta$ is measured in kJ mol$^{-1}$. The atom that more readily accepts an electron (and thus tends to

**Figure 14-1**

Average electronegativities of atoms, computed with the method developed by Linus Pauling. Electronegativity values have no units.

carry a net negative charge) has the larger value of $\chi$. On the Pauling scale shown in Figure 14-1 and used in Appendix F, electronegativities range from 3.98 (for fluorine) to 0.79 (for cesium). These numerical values are useful for exploring periodic trends and making semiquantitative comparisons. They are not highly precise experimental results but average numbers based on measured bond energies from a large range of compounds.

Robert Mulliken proposed an alternative definition of electronegativity two years after Pauling's proposal. He observed that elements toward the lower left corner of the periodic table have low ionization energies and small electron affinities. This means that they give up electrons readily and do poorly at accepting electrons. They tend to act as electron *donors* in interactions with other elements. In contrast, elements in the upper right corner of the periodic table have high ionization energies and also (except for the noble gases) have large electron affinities. As a result, these elements accept electrons easily but give them up only reluctantly; such atoms act as electron *acceptors.*

Mulliken then defined electronegativity as proportional to the average of the ionization energy and the electron affinity.

$$\text{electronegativity (Mulliken)} \propto \tfrac{1}{2}(IE_1 + EA)$$

Electron acceptors (such as the halogens) have high ionization energies and electron affinities and are thus highly **electronegative.** Electron donors (such as the alkali metals) have low ionization energies and electron affinities and therefore low electronegativities; they are **electropositive.** The noble gases only rarely participate in chemical bonding. Their high ionization energies and negative electron affinities mean that they are reluctant either to give up or to accept electrons. Electronegativities are generally not assigned to the noble gases. Mulliken's scale turned out to give electronegativities approximately proportional to those on the Pauling Scale.

The periodic trends in electronegativity, as shown in Figure 14-1, are quite interesting. High electronegativity is favored by small size and a large effective charge felt by the outer electrons. Thus, electronegativity decreases down a group at the left or right side of the periodic table; that is, cesium is less electronegative than lithium and iodine is less electronegative than fluorine. In these groups, atomic size increases sufficiently rapidly with atomic number to make it the predominant factor and cause the electronegativity to decrease. For a group of late transition elements such as Cu, Ag, and Au, however, the size changes very slowly. The rise in effective nuclear charge then predominates so that electronegativity *increases* down the group's column in the table.

The absolute value of the difference in electronegativity of two bonded atoms tells the degree of polarity in their bond. A large difference (greater than about 2.0) means that the bond is ionic and electrons transfer completely or nearly completely to the more electronegative atom. A small difference (less than about 0.4) means that the bond is largely covalent, with electrons in the bond fairly evenly shared. Intermediate values of the difference signify a polar covalent bond with intermediate character.

## Example 14-1

Using Figure 14-1, arrange the following bonds in order of decreasing polarity: H—C, O—O, H—F, I—Cl, Cs—Au.

### Solution

The differences in electronegativity between the five pairs of atoms (without regard to sign) are 0.35, 0.00, 1.78, 0.50, and 1.75, respectively. The order of decreasing polarity is the order of decrease in this difference: H—F, Cs—Au, I—Cl, H—C, and O—O. The last bond in this listing is nonpolar.

*Related Problems:* 3, 4

## Ionic Bonding

Consider the energy change that accompanies the formation of an isolated ionic bond. Suppose a very electropositive element, such as potassium, reacts with a very electronegative element, such as fluorine. Being electropositive, potassium is a good electron donor; its ionization energy is

$$K \longrightarrow K^+ + e^- \qquad\qquad \Delta E = +419 \text{ kJ mol}^{-1}$$

Fluorine is a good electron acceptor; its electron affinity ($EA = -\Delta E$) is

$$F + e^- \longrightarrow F^- \qquad\qquad \Delta E = -328 \text{ kJ mol}^{-1}$$

Adding these two equations and their energy changes gives

$$K + F \longrightarrow K^+ + F^- \qquad \Delta E_\infty = 419 + (-328) = +91 \text{ kJ mol}^{-1}$$

The subscript "$\infty$" on $\Delta E$ indicates that the two reactant atoms are infinitely distant from each other, as are the product ions. Even in this favorable case (with very different electronegativities), there is a *cost* in energy to transfer an electron from a potassium atom to a fluorine atom. An examination of the ionization energies and electron affinities of the elements shows that this is always the case. The smallest ionization energy (for Cs, 376 kJ mol$^{-1}$) is greater in magnitude than the largest electron affinity (for Cl, 349 kJ mol$^{-1}$). For atoms separated by large distances in the gas phase, electron transfer to form ions is always energetically unfavorable.

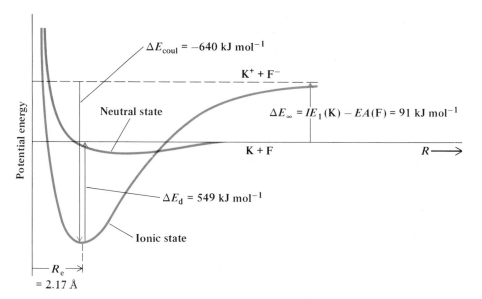

**Figure 14-2**

The potential energies of the ionic and neutral states of K + F as a function of separation $R$. Note that the two curves cross, so that the ionic state has a lower energy at short separations and the neutral state has a lower energy at large separations.

How, then, does an ionic bond form? As two ions approach each other, an attractive Coulomb force between them gives rise to an interaction energy:

$$\Delta E_{Coulomb} = \frac{Q_1 Q_2}{4\pi\epsilon_0 R}$$

where $R$ is the distance between the ion centers and $Q_1$ and $Q_2$ are the charges on the two ions (for $K^+$ and $F^-$, $Q_1$ is $+e$ and $Q_2$ is $-e$). Here $\epsilon_0$ is the proportionality constant introduced in Section 13-3, equal to $8.854 \times 10^{-12} \, C^2 \, J^{-1} \, m^{-1}$. Because $Q_1$ and $Q_2$ have opposite signs in this case, this Coulomb energy is negative and the total energy is *lowered* as the ions move toward each other. At short enough distances, the Coulomb attraction more than compensates for the energy required to transfer the electron, and an ionic bond forms. Figure 14-2 shows how the energy varies with separation $R$ for the ionic and neutral states. For large separations the neutral atoms are always more stable, but at shorter distances the ionic species may be favored. At very short distances the electron clouds of the two atoms or ions begin to repel each other and the energy rises steeply. The equilibrium bond length $R_e$ of an ionic bond is determined by a balance of attractive and repulsive forces.

For an ionic bond such as the one in KF, the energy of dissociation to neutral atoms can be approximated as

$$\Delta E_d \approx -\frac{Q_1 Q_2}{4\pi\epsilon_0 R_e} - \Delta E_\infty \qquad (14\text{-}1)$$

where $\Delta E_\infty = IE_1(K) - EA(F)$ in this case. In writing this equation we have assumed the bond to be purely ionic and have neglected the relatively small contribution of short-range repulsive interactions at the bond length $R_e$. The first term in this equation is the energy cost to separate the two *ions* (a positive number because $Q_2$ is negative), and the second term is the energy released when electrons are transferred between ions to give neutral atoms (a negative number including the minus sign).

## Example 14-2

Estimate the energy of dissociation to neutral atoms for KF, which has a bond length of 2.17 Å $= 2.17 \times 10^{-10}$ m.

### Solution

$$\Delta E_{\mathrm{d}} \approx -\frac{Q_1 Q_2}{4\pi\epsilon_0 R_{\mathrm{e}}} - \Delta E_{\infty}$$

$$= -\frac{-(1.602 \times 10^{-19}\,\mathrm{C})^2 (6.022 \times 10^{23}\,\mathrm{mol}^{-1})}{(4)(3.1416)(8.854 \times 10^{-12}\,\mathrm{C}^2\,\mathrm{J}^{-1}\,\mathrm{m}^{-1})(2.17 \times 10^{-10}\,\mathrm{m})}$$
$$\qquad - 9.1 \times 10^4\,\mathrm{J\,mol}^{-1}$$

$$= 64.0 \times 10^4\,\mathrm{J\,mol}^{-1} - 9.1 \times 10^4\,\mathrm{J\,mol}^{-1}$$

$$= 549\,\mathrm{kJ\,mol}^{-1}$$

This estimate compares fairly well with the experimentally measured dissociation energy of 498 kJ mol$^{-1}$. Note that we multiplied the Coulomb energy per molecule of KF by Avogadro's number to obtain the result in joules per mole of KF.

*Related Problems:* 7, 8

There are several reasons for the discrepancy between our simple calculations and experiment. First, our calculations omitted repulsive interactions; including them would tend to reduce $\Delta E_{\mathrm{d}}$. Second, we assumed the KF bond to be purely ionic, whereas in fact the bond has a small amount of covalent (nonionic) character. Third, we assumed the charge distributions of the ions to be point charges centered on the respective nuclei, ignoring the distortion in the charge distribution of the F$^-$ ion caused by the electric field of the nearby K$^+$ ion. This distortion, termed **polarization,** means that the charge distribution around each nucleus is not strictly symmetric.

Gaseous molecules of potassium fluoride exist only at high temperatures. At ordinary temperatures, a collection of such molecules would condense in a highly exothermic process to form an ionic solid. At room temperature, KF forms a three-dimensional lattice of great stability in which positive and negative ions occupy alternate sites.

## Covalent Bonding

Charge transfer followed by Coulomb attraction can account for the formation of ionic bonds. What happens when two atoms with similar or equal electronegativities interact? The $H_2$ molecule is quite stable, for example, with a bond dissociation energy of 432 kJ mol$^{-1}$, yet it consists of two identical atoms. There is no possibility of a net charge transfer from one to the other to form an ionic bond. The explanation of the stability of $H_2$ comes from a different type of bonding, *covalent* bonding, which arises from the sharing of electrons between atoms.

Consider the bonding of the $H_2^+$ molecular ion. This, the simplest of all polyatomic species, is stable but rather reactive. Experiment and theoretical calculations agree that the H—H bond length $R_{\mathrm{e}}$ is 1.06 Å and the dissociation energy $\Delta E_{\mathrm{d}}$ is

## Figure 14-3

Electrons attract hydrogen nuclei because the two have opposite charge. (*a*) If the electron lies between two hydrogen nuclei, the attraction creates a net force that pulls the two nuclei together. (*b*) If the electron lies outside the internuclear region, it exerts a much greater force on the nearer nucleus, giving a net force that pulls the two nuclei apart.

255.5 kJ mol$^{-1}$. To explain the existence of $H_2^+$, we use a simple classical model for chemical bonding. The electron is treated as a point negative charge interacting with the two hydrogen nuclei that are separated by a distance $R$ (Fig. 14-3). The electron exerts an attractive Coulomb force on the nuclei. If the electron lies *between* the two (Fig. 14-3*a*), the force tends to pull the nuclei together, *strengthening* the bond. If, on the other hand, it lies *outside* the region between the nuclei (Fig. 14-3*b*), the force tends to pull the two apart, *weakening* the bond. According to this simple picture, covalent bonding occurs when an electron spends most of its time in the region between nuclei so that it is shared between them. When, in contrast, an electron spends most of its time outside the region between the nuclei, the bond is weakened and the molecule tends to dissociate.

This picture of bonding in the $H_2^+$ molecular ion can be applied to other molecules. The shared valence-electron pairs of the Lewis model (Chapter 2) are positioned between two nuclei, where they experience net attractive interactions with each nucleus and contribute to the strengthening of the bond through the mechanism of Figure 14-3*a*. In Section 14-4 we will extend this simple classical model to a quantum-mechanical description in terms of electrons in molecular orbitals. The fundamental conclusion reached there will be the same: covalent bonding arises from electrons shared in the region between two nuclei.

### Dipole Moments and Percent Ionic Character

In a bond that is almost purely ionic, such as that of KF, almost complete transfer of an electron from the electropositive to the electronegative species takes place; KF can be written fairly accurately as $K^+F^-$ with charges $+e$ and $-e$ on the two ions. The charge distribution for a molecule such as HF, with significant covalent character, is more complex. If we wish to approximate it by two point charges, then it is best described as $H^{\delta+}F^{\delta-}$, where some fraction $\delta$ of the full charge $\pm e$ is on each nucleus. A useful measure of ionic character and of electronegativity differences, especially for diatomic molecules, is the **dipole moment** of the molecule. If two charges of equal magnitude and opposite sign, $+Q$ and $-Q$, are separated by a distance $R$, the dipole moment $\mu$ (Greek mu) of that charge distribution is

$$\mu = QR \tag{14-2}$$

In SI units, $\mu$ is measured in coulomb meters, an inconveniently large unit for discussing molecules. The unit most often used is the debye (D), which is related to SI units by[1]

$$1 \text{ debye} = 3.336 \times 10^{-30} \text{ coulomb meter}$$

---

[1]This definition appears unusual in SI units. If electrostatic units (esu) (which we do not use here) rather than coulombs are used to measure charge, then 1 debye = $10^{-18}$ esu cm. The debye is named after the Dutch-born Nobel laureate Peter Debye, who pioneered the measurement of dipole moments and the theory of polar molecules.

The debye can also be defined as the dipole moment of two charges $\pm e$ that are 0.2082 Å apart. If $\delta$ is the fraction of a unit charge on each atom in a diatomic molecule ($Q = e\delta$), then

$$\mu \text{ (debye)} = \frac{R(\text{Å})}{0.2082 \text{ Å D}^{-1}} \delta$$

Dipole moments are measured experimentally by electrical and spectroscopic methods and provide useful information about the nature of bonding. In HF, for example, the value of $\delta$ calculated from the dipole moment ($\mu = 1.82$ D) and bond length ($R = 0.917$ Å) is 0.41, substantially less than the value of 1 for a purely ionic bond. We convert $\delta$ to a "percent ionic character" by multiplying by 100% and say that the bond in HF is 41% ionic. Deviations from 100% ionic bonding occur for two reasons: first, covalent contributions lead to electron sharing between atoms; second, the electron charge distribution around one ion is distorted by the electric field of the other ion (polarization). When polarization is extreme, regarding the ions as point charges is no longer a good approximation, and a more accurate description of the distribution of electric charge is necessary.

Table 14-4 gives a scale of ionic character for diatomic molecules, based on the definition of $\delta$. The ionic character defined from the dipole moment is reasonably well correlated with the Pauling electronegativity difference (Fig. 14-4): high ionic character usually corresponds to large electronegativity difference, with the more electropositive atom carrying the charge $+\delta$. There are exceptions to this general trend, however. Carbon is less electronegative than oxygen, so one would predict a charge distribution of $C^{\delta+}O^{\delta-}$ in the CO molecule. In fact, the measured dipole moment is quite small in magnitude and in the other direction: $C^{\delta-}O^{\delta+}$, with $\delta = 0.02$. The discrepancy arises because of the lone-pair electron density on the carbon

**Table 14-4  Dipole Moments of Diatomic Molecules**

| Molecule | Average Bond Length (Å) | Dipole Moment (D) | Percent Ionic Character (100 $\delta$) |
|---|---|---|---|
| $H_2$ | 0.751 | 0 | 0 |
| CO | 1.131 | 0.112 | 2 |
| NO | 1.154 | 0.159 | 3 |
| HI | 1.620 | 0.448 | 6 |
| ClF | 1.632 | 0.888 | 11 |
| HBr | 1.424 | 0.828 | 12 |
| HCl | 1.284 | 1.109 | 18 |
| HF | 0.926 | 1.827 | 41 |
| CsF | 2.347 | 7.884 | 70 |
| LiCl | 2.027 | 7.129 | 73 |
| LiH | 1.604 | 5.882 | 76 |
| KBr | 2.824 | 10.628 | 78 |
| NaCl | 2.365 | 9.001 | 79 |
| KCl | 2.671 | 10.269 | 82 |
| KF | 2.176 | 8.593 | 82 |
| LiF | 1.570 | 6.327 | 84 |
| NaF | 1.931 | 8.156 | 88 |

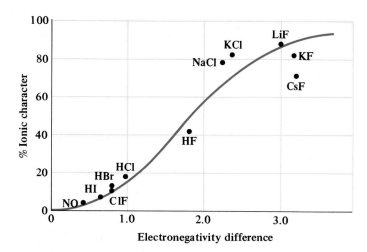

**Figure 14-4**

Two measures of ionic character for diatomic molecules are the electronegativity difference (from Fig. 14-1) and the percent ionic character $100\delta$ (calculated from the observed dipole moment and bond length). The curve in this graph shows that the two correlate approximately but that there are many exceptions.

atom (which, interestingly, is reflected in the formal charge of $-1$ carried by that atom, as discussed in Section 2-3).

## Electronegativity and Oxoacid Strength

One of the many chemical consequences of electronegativity and bond polarity is their effect on the strengths of oxoacids. These are acids in which the hydrogen ion or ions donated in aqueous solution are ones that previously were bonded to oxygen atoms. Examples include sulfuric acid ($H_2SO_4$), nitric acid ($HNO_3$), and phosphoric acid ($H_3PO_4$). If the central atom is designated X, then oxoacids have the structure

$$-X-O-H$$

where X can be bonded to additional —OH groups, to oxygen atoms, or to hydrogen atoms. How does the strength of the oxoacid change with the electronegativity of X? Consider first the extreme case in which X is a highly electropositive element, such as an alkali metal. Of course, NaOH is not an acid at all, but a base. The sodium atom in Na—O—H gives up a full electron to make $Na^+$ and $OH^-$ ions. Because the X—O bond here is almost completely ionic, the $OH^-$ group has a net negative charge that holds the $H^+$ tightly to the oxygen and prevents formation of $H^+$ ions. The less electropositive alkaline-earth elements behave similarly. They form hydroxides such as $Mg(OH)_2$ that are somewhat weaker bases than NaOH but do not in any way act as acids.

Now suppose the central atom X becomes more electronegative, reaching values between 2 and 3, as in the oxoacids of the elements B, C, P, As, S, Se, Br, and I. As X becomes more effective at withdrawing electron density from the oxygen atom, the X—O bond becomes more covalent. This leaves less negative charge on

$$X \underset{|}{\overset{|}{-}} O-H \longrightarrow X + O-H$$

(a)

Bond breaks

$$X-O \underset{|}{\overset{|}{-}} H \longrightarrow X-O + H$$

(b)

**Figure 14-5**

In part *a*, the atom X is electropositive, and so extra electron density (blue areas) accumulates on the OH group. The X—O bond then breaks easily, making the compound a base. In part *b*, X is electronegative, and so electron density is drawn from the H atom to the X—O bond. Now it is the O—H bond that breaks easily, and the compound is an acid.

the oxygen atom, and consequently the oxoacid releases $H^+$ more readily (Fig. 14-5). Other things being equal, acid strength should increase with increasing electronegativity of the central atom. This is, in fact, usually observed.

## 14-3 THE SHAPES OF MOLECULES

Molecular geometry is governed by energy; a molecule has the geometry that gives it the lowest potential energy. It is therefore reasonable to ask how changes in bond angles affect the energy of a molecule. Because a covalent bond is formed by the sharing of a pair of electrons between two atoms (as in the Lewis structure model of Section 2-3), changes in bond angles change the relative positions of the electron pairs around a given central atom. Electrons tend to repel each other for two reasons: electrostatic (Coulomb) repulsion between like charges, and Pauli repulsion. In Section 13-6 we quoted one version of the Pauli exclusion principle: two electrons in the same atom may not have the same set of four quantum numbers. A related statement is that the probability of finding two electrons of the same spin in the same region of space at the same time is zero. As a result of Pauli exclusion and Coulomb repulsion, it is desirable in terms of energy for electrons (especially those of the same spin) to avoid each other. This gives rise to a *correlation* of the instantaneous electron positions in a many-electron atom. If one electron is in one region of space, then a second electron is most likely to be in a different region. Electron repulsion is the basis for a simple and useful description of molecular geometries called the VSEPR theory.

### The VSEPR Theory

The **valence shell electron-pair repulsion (VSEPR) theory** starts with the fundamental idea that electron pairs in the valence shell of an atom repel each other. This includes both lone pairs, which are localized on the atom and not involved in bonding, and bonding pairs, which are covalently shared with other atoms. The electron pairs position themselves as far apart as possible to minimize their repulsions. The molecular geometry, which is defined by the positions of the *atoms,* is then traced from the relative locations of the electron pairs.

The arrangement that minimizes repulsions naturally depends on the number of electron pairs. Figure 14-6 shows the configuration of minimum energy for two to six electron pairs around a central atom. Two electron pairs place themselves on opposite sides of the atom in a linear arrangement, three pairs form a trigonal planar

2: Linear

3: Trigonal planar

4: Tetrahedral

5: Trigonal bipyramidal

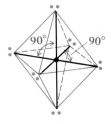

6: Octahedral

**Figure 14-6**

The positions of minimum energy for electron pairs on a sphere centered on an atom's nucleus. The angles between the electron pairs are indicated. For two, three, and four electron pairs they are 180°, 120°, and 109.47°, respectively.

structure, four arrange themselves at the corners of a tetrahedron, five define a trigonal bipyramid, and six an octahedron. To find which geometry applies, we determine the **steric number** $SN$ of the central atom, which is defined by

$$SN = \left(\begin{array}{c}\text{number of atoms}\\\text{bonded to central atom}\end{array}\right) + \left(\begin{array}{c}\text{number of lone pairs}\\\text{on central atom}\end{array}\right)$$

The steric number of an atom in a molecule can be determined by drawing the Lewis structure of the molecule and adding the number of atoms bonded to it and its number of lone pairs.

### Example 14-3

Calculate steric numbers for iodine in $IF_4^-$ and for bromine in $BrO_4^-$. These molecular ions have central $I^-$ or $Br^-$ surrounded by the other four atoms.

### Solution

The central $I^-$ has eight valence electrons (two in the $5s$ orbital and six in the three $5p$ orbitals). Each F atom has seven valence electrons of its own and needs to share one of the electrons from the $I^-$ to achieve a noble-gas configuration. Thus, four of the $I^-$ valence electrons take part in covalent bonds, leaving the remaining four to form two lone pairs. The steric number is given by

$$SN = 4 \text{ (bonded atoms)} + 2 \text{ (lone pairs)} = 6$$

In $BrO_4^-$, each oxygen atom needs to share two of the electrons from the $Br^-$ in order to achieve a noble-gas configuration. Because this accounts for all eight of the $Br^-$ valence electrons, there are no lone pairs on the central atom and

$$SN = 4 \text{ (bonded atoms)} + 0 \text{ (lone pairs)} = 4$$

Double-bonded or triple-bonded atoms count the same as single-bonded atoms in determining the steric number. In $CO_2$, for example, two double-bonded oxygen atoms are attached to the central carbon and there are no lone pairs on that atom, and so $SN = 2$.

The steric number is used to predict molecular geometries. In molecules $XY_n$ in which there are no lone pairs on the central atom X (the simplest case),

$$SN = \text{number of bonded atoms} = n$$

The $n$ bonding electron pairs (and therefore the outer atoms) position themselves as shown in Figure 14-6 to minimize electron-pair repulsion. Thus, $CO_2$ is predicted (and found experimentally) to be linear, $BF_3$ is trigonal planar, $CH_4$ tetrahedral, $PF_5$ trigonal bipyramidal, and $SF_6$ octahedral.

When lone pairs are present, the situation changes slightly. There can now be three different types of repulsions: (1) bonding pair against bonding pair, (2) bonding pair against lone pair, and (3) lone pair against lone pair. Consider the ammonia molecule ($NH_3$), for example, which has three bonding electron pairs and one lone pair (Fig. 14-7a). The steric number is four, and the electron pairs arrange themselves into an approximately tetrahedral structure. The lone pair is not identical to the three bonding pairs, however, and so there is no reason why the electron pair structure should be *exactly* tetrahedral. It is found that lone pairs tend to occupy more space than bonding pairs (because they are held closer to the central atom), and so the angles of bonds opposite them are reduced. The geometry of the *molecule,* as distinct from that of the electron pairs, is named for the sites occupied by actual atoms. The description of the molecular geometry makes no reference to lone pairs that may be present on the central atom, even though their presence affects that geometry. The structure of the ammonia molecule is thus predicted to be a trigonal pyramid in which the H—N—H bond angle is smaller than the tetrahedral angle of 109.5°. The observed structure has an H—N—H bond angle of 107.3°. The H—O—H bond angle in water, which has two lone pairs and two bonding pairs, is still smaller at 104.5° (Fig. 14-7b).

A similar distortion takes place when two types of outer atoms are present. In $CH_3Cl$, the bonding electron pair in the C—Cl bond is not the same as those in the C—H bonds, and so the structure is a distorted tetrahedron (Fig. 14-7c). Because Cl is more electronegative than H, it tends to attract electrons away from the central atom, reducing the electron-pair repulsion. This allows the Cl—C—H bond angles to become 108.5°, smaller than tetrahedral, while the H—C—H angles become 110.5°, larger than tetrahedral. In effect, electropositive substituents repel other substituents more strongly than do electronegative substituents.

The fluorides $PF_5$, $SF_4$, $ClF_3$, and $XeF_2$ all have steric number 5 but have different numbers of lone pairs (0, 1, 2, and 3, respectively). What shapes do their molecules have? We have already mentioned that $PF_5$ is trigonal bipyramidal. Two of the fluorine atoms occupy **axial** sites (Fig. 14-8a), and the other three occupy **equatorial** sites. Because the two kinds of sites are not equivalent, there is no reason for all of the P—F bond lengths to be equal. Experiment shows that the equatorial P—F bond length is 1.534 Å, shorter than the axial P—F lengths, which are 1.577 Å.

$SF_4$ has four bonded atoms and one lone pair. Does the lone pair occupy an axial or an equatorial site? In the VSEPR theory, electron pairs repel each other much more strongly when they form a 90° angle with respect to the central atom than when the angle is larger. A single lone pair therefore finds a position that minimizes the number of 90° repulsions it has with bonding electron pairs. It occupies an equatorial position with two 90° repulsions (Fig. 14-8b) rather than an axial position with three 90° repulsions (Fig. 14-8c). The axial S—F bonds are bent slightly away from the lone pair, and so the molecular structure of $SF_4$ is a distorted seesaw. A second lone pair (in $ClF_3$, for example) also takes an equatorial position, leading to a distorted T-shaped molecular structure (Fig. 14-8d). A third lone pair (in $XeF_2$ or $I_3^-$, for example) occupies the third equatorial position, and the molecular geometry is linear (Fig. 14-8e).

## Example 14-4

Predict the geometry of the following molecules and ions: (a) $ClO_3^+$, (b) $ClO_2^+$, (c) $SiH_4$, (d) $IF_5$.

### Solution

(a) The central Cl atom has all of its valence electrons (six, because the ion has a net positive charge) involved in bonds to the surrounding three oxygen atoms and has no lone pairs. Its steric number is 3. In the molecular ion the central Cl should be surrounded by the three O atoms in a trigonal planar structure.

(b) The central Cl atom in this ion also has a steric number of 3, composed of two

(a)

(b)

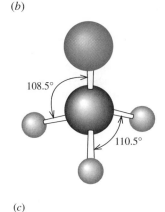

(c)

**Figure 14-7**

(a) Ammonia ($NH_3$) has a pyramidal structure in which the bond angles are smaller than 109.5°. (b) Water ($H_2O$) has a bent structure, with a bond angle smaller than 109.5°. (c) $CH_3Cl$ has a distorted tetrahedral structure.

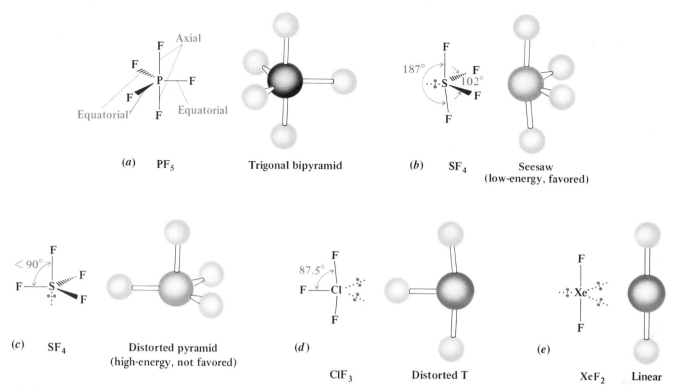

(a)  PF₅            Trigonal bipyramid        (b)  SF₄        Seesaw
                                                             (low-energy, favored)

(c)  SF₄     Distorted pyramid          (d)                         (e)
             (high-energy, not favored)       ClF₃     Distorted T       XeF₂    Linear

**Figure 14-8**

Molecules with steric number 5. Except for $PF_5$, the molecular geometry is named for the sites occupied by the *atoms,* not the underlying trigonal–bipyramidal structure of electron pairs.

bonded atoms and a single lone pair. The predicted molecular geometry is a bent molecule with an angle somewhat less than 120°.

(c) The central Si has a steric number of 4 and no lone pairs. The molecular geometry should consist of the Si atom surrounded by a regular tetrahedron of H atoms.

(d) Iodine has seven valence electrons of which five are shared in bonding pairs with F atoms. This leaves two electrons to form a lone pair, and so the steric number is 5 (bonded atoms) + 1 (lone pair) = 6. The structure will be based on the octahedron of electron pairs from Figure 14-6, with five F atoms and one lone pair. The lone pair can be placed on any one of the six equivalent sites and will cause the four F atoms to bend away from it toward the fifth F atom, giving the distorted structure shown in Figure 14-9.

*Related Problems:* 13, 14, 15, 16

**Figure 14-9**

The structure of IF₅. Note the distortions of F—I—F bond angles from 90° because of the lone pair at the bottom (not shown).

The VSEPR theory is a simple but remarkably powerful model for predicting the geometries and approximate bond angles of molecules that have a central atom. In fact, there are many cases in which it is more successful than theories that require extensive calculation. It has its limitations, however. VSEPR theory does not account for the fact that observed bond angles in the Group V and VI hydrides $H_2S$ (92°), $H_2Se$ (91°), $PH_3$ (93°), and $AsH_3$ (92°) are so far from tetrahedral (109.5°) and so close to right angles (90°).

*(a)*

*(b)*

*(c)*

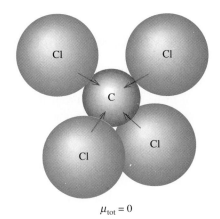

## Figure 14-10

The total dipole moment of a molecule is obtained by vector addition of its bond dipoles. This operation is performed by placing the arrows head to tail. (*a*) $CO_2$. (*b*) OCS. (*c*) $H_2O$. (*d*) $CCl_4$.

*(d)*

## Dipole Moments

Polyatomic molecules, like the diatomic molecules considered in the last section, may have dipole moments. Often, a good approximation is to assign a dipole moment (which is now a vector, shown as an arrow) to each bond and then obtain the total moment by carrying out a vector sum of the bond dipoles. In $CO_2$, for example, each C—O has a bond dipole (Fig. 14-10*a*). Because the dipoles are equal in magnitude and point in opposite directions, however, the total molecular dipole moment vanishes. In OCS, which is also linear with a central carbon atom, the C—O and C—S bond dipole moments have different magnitudes, leaving a net nonzero dipole moment (Fig. 14-10*b*). In the water molecule (Fig. 14-10*c*) the two bond dipoles add vectorially to give a net molecular dipole moment. The more symmetric molecule $CCl_4$, on the other hand, has no net dipole moment (Fig. 14-10*d*). Even though each of the four C—Cl bonds (pointing from the four corners of a tetrahedron) has a dipole moment, their vector sum is 0. Molecules such as $H_2O$ and OCS, with nonzero dipole moments, are **polar;** those such as $CO_2$ and $CCl_4$, with no net dipole moment, are **nonpolar,** even though they contain polar bonds. As discussed in Section 4-2, intermolecular forces differ between polar and nonpolar molecules, a fact that significantly affects their physical and chemical properties.

## Example 14-5

Predict whether the molecules $NH_3$ and $SF_6$ will have dipole moments.

## Solution

The $NH_3$ molecule has a dipole moment, because the three N—H bond dipoles add to give a net dipole pointing downward from the N atom to the base of the pyramid that defines the $NH_3$ structure. The $SF_6$ molecule has no dipole moment, because each S—F bond dipole is balanced by an equal one pointing in the opposite direction on the other side of the molecule.

*Related Problems:* 19, 20

## 14-4 MOLECULAR ORBITALS IN DIATOMIC MOLECULES

Section 14-2 pointed out that electron pairs between nuclei exert electrostatic forces that pull the nuclei together, but in Chapter 13 the quantum nature of electrons was seen to be important. From the Heisenberg indeterminacy principle, an electron cannot simply be "at" some point in space but must be described as occupying an orbital with a characteristic wave function whose square is the probability density for finding the electron at a point. In a molecule, such an orbital is spread out, or *delocalized,* over more than one atom and is called a **molecular orbital.** This section discusses the molecular orbitals of diatomic molecules, and the following section considers polyatomic molecules.

### Molecular Orbitals and Covalent Bonding

The simplest bonded species is the one-electron molecular ion $H_2^+$. The wave function of the electron is the solution of the Schrödinger equation for an electron bound in the electrostatic field of *two* nuclei (protons, here) separated by a distance $R$ (Fig. 14-3). A full calculation can be carried out numerically, but for many purposes a simpler description is sufficient and provides useful insight. When the electron is very close to nucleus A, it experiences a potential that is not very different from that in an isolated hydrogen atom. The ground-state wave function for the electron near A should therefore resemble a $1s$ atomic wave function $\psi_{1s}^A$. Near B, the wave function should resemble $\psi_{1s}^B$. A simple way to construct a molecular wave function with these properties is to take it to be a linear combination of both atomic wave functions, obtained by adding or subtracting the two with constant coefficients:

$$\psi = C_A \psi_{1s}^A + C_B \psi_{1s}^B$$

This wave function is an approximate molecular orbital for the electron in $H_2^+$. The constants $C_A$ and $C_B$ give the relative weights of the two atomic orbitals. If $C_A$ were greater in magnitude than $C_B$, the $\psi_{1s}^A$ orbital would be more heavily weighted and the electron would be more likely to be found near nucleus A, and vice versa. Because the two nuclei in the $H_2^+$ ion are identical, however, the electron should have the same probability density near one as near the other. Therefore, the magnitudes of $C_A$ and $C_B$ must be equal and either $C_A = C_B$ or $C_A = -C_B$. For these two choices $\psi^2$ is symmetric in the two nuclei.

In this way two molecular orbitals are constructed, designated $\sigma_{1s}$ and $\sigma_{1s}^*$, with the Greek letter $\sigma$ (sigma) indicating that the electron density is distributed symmetrically about the bond axis:

$$\sigma_{1s} = C_1[\psi_{1s}^A + \psi_{1s}^B]$$

$$\sigma_{1s}^* = C_2[\psi_{1s}^A - \psi_{1s}^B]$$

where $C_1$ and $C_2$ are proportionality constants chosen to ensure that the total probability of finding the electron *somewhere* is unity. The distributions of electron probability are obtained by squaring the wave functions:

$$[\sigma_{1s}]^2 = C_1^2[(\psi_{1s}^A)^2 + (\psi_{1s}^B)^2 + 2\psi_{1s}^A\psi_{1s}^B]$$

$$[\sigma_{1s}^*]^2 = C_2^2[(\psi_{1s}^A)^2 + (\psi_{1s}^B)^2 - 2\psi_{1s}^A\psi_{1s}^B]$$

These can be compared with the electron probability distribution for a noninteracting (n.i.) system (obtained by averaging the probabilities for $H_A + H_B^+$ and $H_A^+ + H_B$), which is

$$\psi^2(\text{n.i.}) = C_3^2[(\psi_{1s}^A)^2 + (\psi_{1s}^B)^2]$$

The values of $\psi^2$ along the internuclear axis are plotted in Figure 14-11 (right side). An electron in a $\sigma$ orbital has an enhanced probability of being found where the two atomic orbitals overlap (between the nuclei), and so $\sigma$ is a **bonding orbital**. In contrast, an electron in a $\sigma^*$ orbital has a *reduced* probability of being found between the nuclei; thus, $\sigma^*$ is an **antibonding orbital**. Note that the $\sigma^*$ orbital has a node between the two nuclei, and so the probability density of finding the electron at that point vanishes. Figure 14-12 shows the potential energy of the $H_2^+$ ion in its $\sigma_{1s}$ and $\sigma_{1s}^*$ states. The force between the nuclei in the antibonding state is everywhere repulsive, but in the bonding state the nuclei are attracted to each other and form a bound state at the distance corresponding to the lowest potential energy. The antibonding orbital has a higher potential energy because it has a node, and so in the ground state of $H_2^+$ the electron occupies the $\sigma_{1s}$ molecular orbital.

**Figure 14-11**

Antibonding and bonding molecular orbitals of $H_2^+$ along the internuclear axis. For comparison, the red lines show the electron probability distribution $\psi^2(\text{n.i.})$ for a noninteracting system. Using this reference system, the bonding orbital shows *increased* probability density between the nuclei, but the antibonding orbital shows *decreased* probability density in this region.

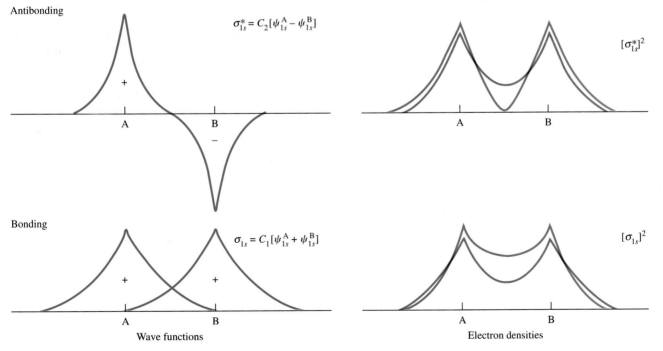

Antibonding

$$\sigma_{1s}^* = C_2[\psi_{1s}^A - \psi_{1s}^B]$$

$[\sigma_{1s}^*]^2$

Bonding

$$\sigma_{1s} = C_1[\psi_{1s}^A + \psi_{1s}^B]$$

$[\sigma_{1s}]^2$

Wave functions

Electron densities

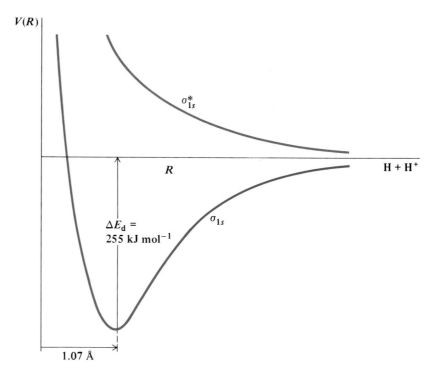

**Figure 14-12**

Potential energy of $H_2^+$ in a $\sigma_{1s}$ (bonding) and $\sigma_{1s}^*$ (antibonding) molecular orbital, shown as a function of internuclear separation $R$.

A molecular orbital, like an atomic orbital, can hold two electrons, one with spin up and the other with spin down. The ground-state $H_2$ molecule therefore accommodates two electrons with opposite spins in a $\sigma_{1s}$ bonding molecular orbital. In bonding molecular orbitals, covalent bonding arises from the sharing of electrons (most often electron pairs with opposite spins). The average electron density is greatest between the nuclei and tends to pull them together. Electron sharing alone is not sufficient for chemical bond formation. Electrons that are shared in an *antibonding* molecular orbital tend to force the nuclei apart, reducing the bond strength.

The energy difference between the two molecular orbitals is illustrated by a **correlation diagram** (Fig. 14-13), in which two $1s$ atomic orbitals are combined to give a lower-energy $\sigma_{1s}$ molecular orbital and a higher-energy $\sigma_{1s}^*$ molecular orbital. We can then apply an aufbau principle, analogous to that for atomic orbitals, in which electrons available from the two atoms are "fed" into the molecular orbitals, starting with the one of lowest energy. At most two electrons can occupy each

**Figure 14-13**

Correlation diagram for first-period diatomic molecules. The red arrows indicate the electron filling for the $H_2$ molecule.

molecular orbital. In the Lewis theory of chemical bonding, a shared pair of electrons corresponds to a single bond, two shared pairs to a double bond, and so forth. In the molecular orbital picture, the ground-state $H_2$ molecule, which has a pair of electrons in a bonding molecular orbital, has a single bond (i.e., a **bond order** of 1). Electrons in antibonding orbitals reduce the bond strength and therefore also the bond order. In summary, the bond order can be defined as follows:

$$\text{bond order} = \tfrac{1}{2}(\text{number of electrons in bonding molecular orbitals}$$
$$- \text{ number of electrons in antibonding molecular orbitals})$$

## Example 14-6

Give the ground-state electron configuration and the bond order of the $He_2^+$ molecular ion.

### Solution

The $He_2^+$ ion has three electrons, which are placed in molecular orbitals to give the ground-state configuration $(\sigma_{1s})^2(\sigma_{1s}^*)^1$, indicating that the ion has a doubly occupied $\sigma_{1s}$ orbital (bonding) and a singly occupied $\sigma_{1s}^*$ orbital (antibonding). The bond order is

$$\text{bond order} = \tfrac{1}{2}(2 \text{ electrons in } \sigma_{1s} - 1 \text{ electron in } \sigma_{1s}^*) = \tfrac{1}{2}$$

This should be a weaker bond than that in $H_2$.

## Homonuclear Diatomic Molecules

Table 14-5 lists the molecular orbital configurations of **homonuclear** diatomic molecules and molecular ions made from first-period elements. These configurations are simply the occupied molecular orbitals in order of increasing energy, together with the number of electrons in each orbital. Higher bond order corresponds to higher bond energies and shorter bond lengths. The species $He_2$ has bond order 0 and does not form a true chemical bond.

A general prescription for obtaining a molecular orbital description from atomic orbitals can now be stated:

1. Form linear combinations of atomic orbitals to give molecular orbitals. The total number of molecular orbitals formed in this way must equal the number of atomic orbitals used.

**Table 14-5   Configurations and Bond Orders for First-Row Diatomic Molecules**

| Species | Electron Configuration | Bond Order | Bond Energy ($kJ\ mol^{-1}$) | Bond Length (Å) |
|---------|------------------------|------------|------------------------------|-----------------|
| $H_2^+$ | $(\sigma_{1s})^1$ | $\tfrac{1}{2}$ | 255 | 1.06 |
| $H_2$ | $(\sigma_{1s})^2$ | 1 | 431 | 0.74 |
| $He_2^+$ | $(\sigma_{1s})^2(\sigma_{1s}^*)^1$ | $\tfrac{1}{2}$ | 251 | 1.08 |
| $He_2$ | $(\sigma_{1s})^2(\sigma_{1s}^*)^2$ | 0 | ~0 | Large |

**2.** Place the molecular orbitals in order from lowest to highest energy.
**3.** Put in electrons (at most two per molecular orbital), starting from the orbital of lowest energy. Apply Hund's rules where appropriate.

This procedure is readily applied to second-period homonuclear diatomic molecules. The $2s$ atomic orbitals of the two atoms are combined in the same fashion as $1s$ orbitals, giving a $\sigma_{2s}$ bonding orbital and a $\sigma_{2s}^*$ antibonding orbital. The $2p$ orbitals form different molecular orbitals depending on whether they are parallel or perpendicular to the internuclear (bond) axis. The $z$ axis is taken to lie along the bond. Then the $2p_z$ orbitals of the two atoms combine to give either a bonding $\sigma_{2p_z}$ or an antibonding $\sigma_{2p_z}^*$ molecular orbital (Fig. 14-14). The bonding orbital has increased electron density between the nuclei, whereas the antibonding orbital has a node. These orbitals are given the designation "$\sigma$" because, like the $\sigma_{1s}$ and $\sigma_{1s}^*$ orbitals, their electron density is symmetrically distributed about the internuclear axis.

The two $2p_x$ orbitals, which are perpendicular to the bond axis, also combine to form a bonding molecular orbital and an antibonding molecular orbital (Fig. 14-15). These orbitals have a nodal plane containing the internuclear axis (in this case, the $yz$ plane) and are designated by the letter $\pi$ rather than $\sigma$. The $\pi_{2p_x}$ orbital is bonding, and the $\pi_{2p_x}^*$ is antibonding. In the same way, $\pi_{2p_y}$ and $\pi_{2p_y}^*$ orbitals can be formed from the $2p_y$ atomic orbitals. Their lobes project above and below the $xz$ nodal plane, which is the plane of the page in Figure 14-15.

It is important to remember that the $+$ and $-$ signs in Figures 14-14 and 14-15 refer not to positive and negative *charges* but to positive and negative *phases* of the wave function. The relative phases of the two atomic orbitals are critical in deter-

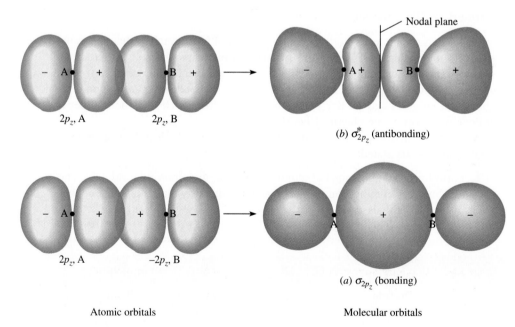

Atomic orbitals                    Molecular orbitals

**Figure 14-14**
Formation of (*a*) $\sigma_{2p_z}$ bonding and (*b*) $\sigma_{2p_z}^*$ antibonding molecular orbitals from $2p_z$ orbitals on atoms A and B. The signs refer to phases of the wave function.

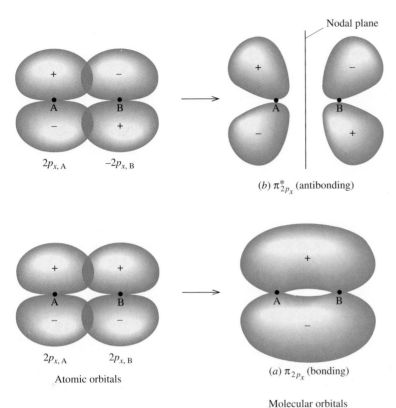

Figure 14-15

Formation of (*a*) $\pi_{2p_x}$ bonding and (*b*) $\pi^*_{2p_x}$ antibonding molecular orbitals from $2p_x$ orbitals on atoms A and B.

mining whether the resulting molecular orbital is bonding or antibonding. Bonding orbitals form from the overlap of wave functions with the same phase; antibonding orbitals form from the overlap of wave functions with opposite phase.

The next step is to determine the energy ordering of the molecular orbitals. In general, that would require a calculation beyond the scope of this book, but we can apply some simple qualitative rules:

1. The average energy of a bonding–antibonding pair of molecular orbitals lies approximately at the energy of the original atomic orbitals.
2. The energy difference between a bonding–antibonding pair becomes greater as the overlap of the atomic orbitals increases.

In second-period diatomic molecules, the $1s$ orbitals of the two atoms scarcely overlap; because the $\sigma_{1s}$ bonding and $\sigma^*_{1s}$ antibonding orbitals are both doubly occupied, they have little net effect on bonding properties and need not be considered. For this reason, chemistry is dominated by valence electron behavior. Figure 14-16 comprises correlation diagrams for the molecular orbitals formed from the $2s$ and $2p$ orbitals. There are two different energy orderings for diatomic molecules formed from second-period elements. The first (Fig. 14-16*a*) applies to the molecules with atoms Li through N (i.e., the first part of the period) and their positive and negative ions. The second (Fig. 14-16*b*) applies to the later elements, O, F, and Ne, and their positive and negative ions. Usually the splitting between $\sigma_{2p}$ and $\sigma^*_{2p}$ orbitals is larger than that between $\pi_{2p}$ and $\pi^*_{2p}$ because $2p$ orbitals directed along

**Figure 14-16**

Correlation diagrams for second-period diatomic molecules. In each case, the atomic orbitals at the sides combine to give the molecular orbitals in the center. The arrows represent the valence electrons in $B_2$ and $O_2$, both paramagnetic molecules.

the bond axis have greater overlap than those perpendicular to it, giving the result shown in Figure 14-16b. The different pattern of Figure 14-16a arises largely from electron–electron repulsions. The electron density from the filled $\sigma_{2s}$ and $\sigma_{2s}^*$ orbitals is concentrated along the bond axis. The $\pi_{2p}$ orbitals are filled before the $\sigma_{2p}$ orbitals because the electron density in a $\pi$ orbital is concentrated away from the bond axis, leading to reduced electron–electron repulsion. Toward the end of the second period (beginning with oxygen) the nuclear charge is great enough that the $\sigma_{2s}$ and $\sigma_{2s}^*$ electrons stay close to the nuclei and interact less strongly with electrons in a $\sigma_{2p_z}$ orbital.

The valence electrons are now placed into the molecular orbitals of Figure 14-16 to determine the electron configurations, and the bond orders can be calculated as before.

### Example 14-7

Determine the ground-state electron configuration and bond order of the $F_2$ molecule.

#### Solution

Each atom of fluorine has 7 valence electrons, and so 14 electrons are placed in the molecular orbitals to represent bonding in the $F_2$ molecule. The correlation diagram of Figure 14-16b gives the electron configuration

$$(\sigma_{2s})^2(\sigma_{2s}^*)^2(\sigma_{2p_z})^2(\pi_{2p})^4(\pi_{2p}^*)^4$$

Because there are eight valence electrons in bonding orbitals and six in antibonding orbitals, the bond order is

$$\text{bond order} = \tfrac{1}{2}(8 - 6) = 1$$

and the $F_2$ molecule has a single bond.

*Related Problems:* 25, 26, 27, 28

Table 14-6 summarizes the properties of second-period homonuclear diatomic molecules. Note the close relationship among bond order, bond length, and bond energy, and the fact that the bond orders calculated from the molecular orbitals agree completely with the results of the Lewis electron-dot model.

An important prediction comes from the two correlation diagrams in Figure 14-16. In the ground states of both $B_2$ and $O_2$, the last two electrons should follow Hund's rules and have parallel spins. Thus, the molecular orbital approach predicts paramagnetism for $B_2$ and $O_2$ because they contain unpaired electrons. This paramagnetism is exactly what is found experimentally (Fig. 14-17). In contrast, in the Lewis electron-dot structure for $O_2$,

$$:\overset{..}{O}::\overset{..}{O}:$$

all the electrons appear to be paired. Moreover, the high chemical reactivity of molecular oxygen can be rationalized as resulting from the readiness of the two $\pi^*$ electrons, unpaired and in different regions of space, to find additional bonding partners in other molecules.

## Heteronuclear Diatomic Molecules

Diatomic molecules such as CO and NO, formed from atoms of two different elements, are called **heteronuclear.** To construct molecular orbitals from the $2s$ atomic orbital of carbon and the $2s$ atomic orbital of oxygen, for example, the atomic wave

(a)

(b)

**Figure 14-17**

(a) Oxygen is paramagnetic; liquid oxygen poured between the pole faces of a magnet is attracted and held there. (b) When the experiment is repeated with liquid nitrogen, which is diamagnetic, the liquid pours straight through. (Leon Lewandowski)

**Table 14-6   Molecular Orbitals of Homonuclear Diatomic Molecules**

| Species | Number of Valence Electrons | Valence Electron Configuration | Bond Order | Bond Length (Å) | Bond Energy (kJ mol$^{-1}$) |
|---------|------|-------------------------------|-----------|-------------|--------------|
| $H_2$ | 2 | $(\sigma_{1s})^2$ | 1 | 0.74 | 431 |
| $He_2$ | 4 | $(\sigma_{1s})^2(\sigma_{1s}^*)^2$ | 0 | Not observed | |
| $Li_2$ | 2 | $(\sigma_{2s})^2$ | 1 | 2.67 | 105 |
| $Be_2$ | 4 | $(\sigma_{2s})^2(\sigma_{2s}^*)^2$ | 0 | 2.45 | 9 |
| $B_2$ | 6 | $(\sigma_{2s})^2(\sigma_{2s}^*)^2(\pi_{2p})^2$ | 1 | 1.59 | 289 |
| $C_2$ | 8 | $(\sigma_{2s})^2(\sigma_{2s}^*)^2(\pi_{2p})^4$ | 2 | 1.24 | 599 |
| $N_2$ | 10 | $(\sigma_{2s})^2(\sigma_{2s}^*)^2(\pi_{2p})^4(\sigma_{2p_z})^2$ | 3 | 1.10 | 942 |
| $O_2$ | 12 | $(\sigma_{2s})^2(\sigma_{2s}^*)^2(\sigma_{2p_z})^2(\pi_{2p})^4(\pi_{2p}^*)^2$ | 2 | 1.21 | 494 |
| $F_2$ | 14 | $(\sigma_{2s})^2(\sigma_{2s}^*)^2(\sigma_{2p_z})^2(\pi_{2p})^4(\pi_{2p}^*)^4$ | 1 | 1.41 | 154 |
| $Ne_2$ | 16 | $(\sigma_{2s})^2(\sigma_{2s}^*)^2(\sigma_{2p_z})^2(\pi_{2p})^4(\pi_{2p}^*)^4(\sigma_{2p_z}^*)^2$ | 0 | Not observed | |

functions combine linearly. The combinations produce a bonding molecular orbital (without a node)

$$\sigma_{2s} = C_A \psi_{2s}^A + C_B \psi_{2s}^B$$

and an antibonding molecular orbital (with a node)

$$\sigma_{2s}^* = C_A' \psi_{2s}^A - C_B' \psi_{2s}^B$$

where A and B refer to the two different atoms in the molecule. In the homonuclear case we argued that $C_A = C_B$ and $C_A' = C_B'$ because the electron should spend the same amount of time near one nucleus as near the other. If the two nuclei are different, however, such reasoning does not apply. If atom B is more electronegative than atom A, then the lower-energy $\sigma_{2s}$ orbital will have $C_B > C_A$ (so that the electron spends more time on the electronegative atom); the higher-energy $\sigma_{2s}^*$ orbital will have $C_A' > C_B'$ and more closely resemble a $2s_A$ atomic orbital.

Molecular orbital correlation diagrams show the energy levels of the more electronegative atom displaced *downward*, because that atom attracts valence electrons more strongly than does the less electronegative atom. Figure 14-18 shows the result for many heteronuclear diatomic molecules of second-period elements (those

### Figure 14-18

Correlation diagram for heteronuclear diatomic molecules, AB. The atomic orbitals for the more electronegative atom (B) are displaced downward because they have lower energies than those for A. The orbital filling shown is that for NO.

in which the electronegativity difference is not too great). This diagram can now be filled with the valence electrons to obtain the molecular orbital configuration. As an example, consider NO, which has 11 valence electrons (5 from N, 6 from O). Its ground-state configuration is

$$(\sigma_{2s})^2(\sigma_{2s}^*)^2(\pi_{2p})^4(\sigma_{2p_z})^2(\pi_{2p}^*)^1$$

From the fact that eight electrons are in bonding orbitals and three are in antibonding orbitals, NO has a bond order of $\frac{1}{2}(8-3) = 2\frac{1}{2}$ and is paramagnetic. Its bond energy is predicted to be (and is) smaller than the bond energy of CO, which has one fewer electron and a bond order of 3.

Atomic orbitals mix significantly to form molecular orbitals only if they are fairly close in energy and have appropriate symmetries. In the HF molecule, for example, the $1s$ orbital of the F atom is far too low in energy to mix with the hydrogen $1s$ orbital, and the same is true of the $2s$ F orbital (Fig. 14-19$a$). The net overlap of the $1s$ orbital of hydrogen with the $2p_x$ or $2p_y$ F orbital is zero (see Fig. 14-19$b$), because the positive and negative phases of the combined wave function sum to zero. This leaves only the $2p_z$ orbital, which combines with the hydrogen $1s$ orbital to give both $\sigma$ and $\sigma^*$ orbitals (Fig. 14-19$c,d$). Figure 14-20 is the correlation diagram for HF. The $2s$, $2p_x$, and $2p_y$ orbitals of fluorine do not mix with the $1s$ of hydrogen and so remain as atomic (nonbonding) states. Electrons in these orbitals

(a) Nonbonding
(negligible overlap)

(b) Nonbonding
(phases cancel)

(c) Bonding, $\sigma$
(significant overlap)

(d) Antibonding, $\sigma^*$

**Figure 14-19**
Overlap of atomic orbitals in HF.

**HF molecule**

$\sigma^*$ **(antibonding)**

**H atom**

$1s$

**F atom**

$2p$

$2p_x$   $2p_y$

(nonbonding)

$\sigma$ (bonding)

Energy

$2s$

$2s$ (nonbonding)

**Figure 14-20**
Correlation diagram for HF. The $2s$, $2p_x$, and $2p_y$ atomic orbitals of fluorine do not mix with the $1s$ atomic orbital of hydrogen and so remain nonbonding.

do not contribute significantly to the chemical bonding. Because fluorine is more electronegative than hydrogen, its $2p$ orbitals lie below the $1s$ hydrogen orbital in energy. The $\sigma$ orbital then contains more fluorine $2p_z$ character, and the $\sigma^*$ orbital more closely resembles a hydrogen $1s$ atomic orbital. When the eight valence electrons are put in for HF, the result is the molecular orbital configuration

$$(2s_F)^2(2p_F)^4(\sigma)^2$$

The net bond order is 1 because electrons in nonbonding atomic orbitals do not affect bond order. The electrons in the $\sigma$ orbital are more likely to be found near the fluorine atom than near the hydrogen, and so HF has the dipole moment $H^{\delta+}F^{\delta-}$.

If a more electropositive atom (such as Na or K) is substituted for H, the energy of its outermost $s$ orbital will be higher than that of the hydrogen atom, because its ionization energy is lower. In this case, the $\sigma$ orbital will resemble a fluorine $2p_z$ orbital even more (that is, the coefficient $C_F$ of the fluorine wave function will be close to 1, and $C_A$ for the alkali atom will be very small). In this limit the molecule can be described as having the valence electron configuration $(2s_F)^2(2p_F)^6$, which corresponds to the ionic species $Na^+F^-$ or $K^+F^-$. The magnitudes of the coefficients in the molecular orbital wave function are thus closely related to the ionic–covalent character of the bonding and to the dipole moment.

## 14-5 POLYATOMIC MOLECULES

There are two ways to describe bonding in polyatomic molecules. The first employs *delocalized* molecular orbitals, with electron wave functions that are spread out over the entire molecule. These molecular orbitals are formed as linear combinations of valence atomic orbitals, then placed in order of increasing energy, and finally filled with the available valence electrons. The electrons that lead to bonding are spread out over all the atoms in the molecule. One problem with this approach is of a fundamentally chemical nature. Bonds have fairly universal characteristics such as energy and length (Section 14-1). If electrons are described by molecular orbitals that are spread out over the entire molecule, then why should the properties of a bond be so weakly dependent on the particular molecule involved? Out of such chemical considerations a second approach was developed which employs molecular orbitals that are localized, either as bonding orbitals between particular pairs of atoms or as lone-pair orbitals on individual atoms. This approach is related to the qualitative Lewis theory, in which a similar localization of electrons is assumed.

In this section, localized orbitals are used for $\sigma$ bonds and, wherever possible, for $\pi$ bonds. In some molecules, however, delocalized molecular orbitals are useful to describe $\pi$ bonds correctly. This approach gives physical insight with relatively few mathematical complications. The following section describes more general approaches that are useful for quantitatively accurate calculations.

### Hybridization

The $BeH_2$ molecule is readily described by the localized orbital method. This molecule has four valence electrons (two from the Be and one from each H atom), all of which are involved in bonding in the Lewis model. From the VSEPR theory, the steric number is 2, and so the molecule is predicted to be linear, with one hydrogen

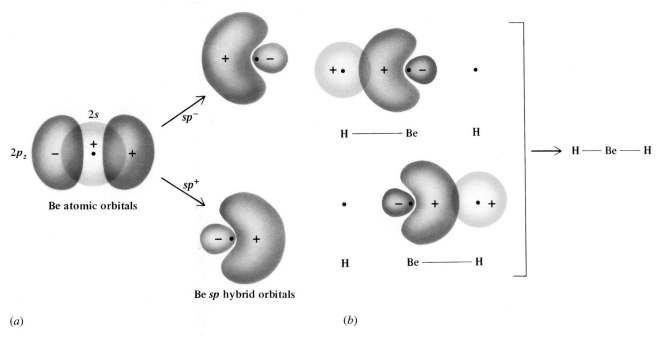

(a)

(b)

**Figure 14-21**

(a) The two $sp$ hybrid orbitals formed from the $2s$ and $2p_z$ orbitals on the Be atom. (b) The two $\sigma$ bonds that form from the overlap of the $sp$ hybrid orbitals with the hydrogen $1s$ orbitals, making two single bonds in the $BeH_2$ molecule.

atom on each side of the central beryllium atom. In the localized orbital method, atomic orbitals of the central atom are mixed to form new, singly occupied *atomic* orbitals, each of which can pair with a singly occupied orbital of another atom to form a $\sigma$ bond along the internuclear axis. In $BeH_2$, the $2s$ and $2p_z$ orbitals of Be are mixed to form two *sp* **hybrid atomic orbitals** on the central Be atom:

$$sp_{Be}^- = 2s - 2p_z$$

$$sp_{Be}^+ = 2s + 2p_z$$

Figure 14-21a shows the form of these orbitals. An electron in a hybrid $sp_{Be}^-$ orbital is localized in that it is much more likely to be found on the left side of the nucleus than on the right; the opposite is true of an electron in the $sp_{Be}^+$ orbital. Once the hybrid atomic orbitals of the central atom form, the two hydrogen atoms can be brought near, and each shares its electron with the corresponding hybrid orbital to form two single $\sigma$ bonds (Fig. 14-21b).

What happens in the hydrides of the other second-period elements? The boron atom has valence electron configuration $2s^2 2p$, and from the VSEPR theory $BH_3$ would have steric number 3, corresponding to a trigonal planar structure. Promotion of one of the $2s$ electrons creates the excited-state configuration $2s^1 2p_x^1 2p_y^1$, and these three atomic orbitals can be mixed to form the three equivalent *sp²* **hybrid atomic orbitals** shown in Figure 14-22. They lie in a plane with an angle of 120° between them. Experimentally, $BH_3$ molecules turn out to be unstable and react rapidly to form $B_2H_6$ or other higher compounds called "boranes." However, the closely related $BF_3$ molecule has the trigonal planar geometry characteristic of $sp^2$ hybridization.

In $CH_4$, VSEPR theory assigns a steric number of 4 to the central carbon atom. The $2s$ and three $2p$ orbitals of the central carbon atom can be combined to form

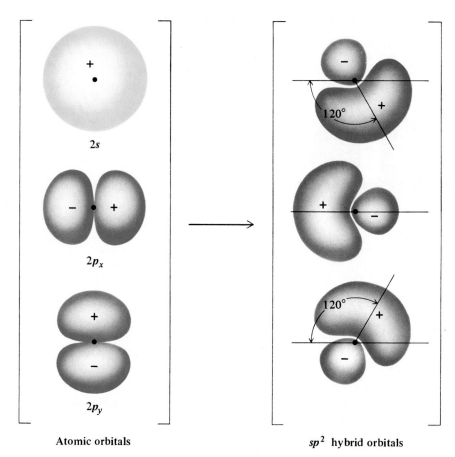

**Atomic orbitals**

$sp^2$ **hybrid orbitals**

**Figure 14-22**

Shapes and relative dispositions of the three $sp^2$ hybrid orbitals, formed from the $2s$, $2p_x$, and $2p_y$ atomic orbitals on a single atom.

**Figure 14-23**

Shapes and relative dispositions of the four $sp^3$ hybrid orbitals, which point to the corners of a tetrahedron.

four equivalent $sp^3$ **hybrid atomic orbitals,** which point toward the vertices of a tetrahedron (Fig. 14-23). Each can overlap a $1s$ orbital of one of the hydrogen atoms to give an overall tetrahedral structure for $CH_4$.

Lone-pair electrons can occupy hybrid orbitals. The nitrogen atom in $NH_3$ also has steric number 4, and its bonding can be described in terms of $sp^3$ hybridization. Of the eight valence electrons in $NH_3$, six are involved in $\sigma$ bonds between nitrogen

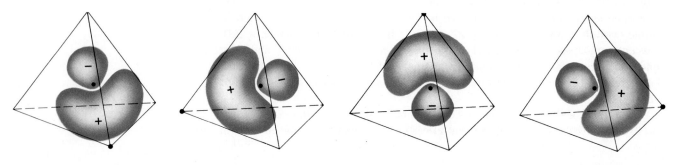

and hydrogen, and the other two occupy the fourth $sp^3$ hybrid orbital as a lone pair. Oxygen in $H_2O$ likewise has steric number 4 and can be described with $sp^3$ hybridization, with two lone pairs in $sp^3$ orbitals. This hybrid orbital description for $NH_3$ and $H_2O$ is consistent with bond angles close to the central angle of a tetrahedron (109.5°).

There is a close relationship between the VSEPR theory of Section 14-3 and the hybrid orbital approach, with steric numbers of 2, 3, and 4 corresponding to $sp$, $sp^2$, and $sp^3$ hybridization, respectively. Both theories are based on the energy savings resulting from the reduction of electron–electron repulsion. The physical importance of hybrid orbitals should not be overemphasized, though; they are really no more than a device to rationalize observed bond angles and equivalent bonds in some molecules.

### Example 14-8

Predict the structure of hydrazine ($H_2NNH_2$) by writing down its Lewis structure and using VSEPR theory. What is the hybridization of the two nitrogen atoms?

### Solution

The Lewis structure is

$$\begin{array}{c} \text{H} \\ \phantom{:}\overset{..}{\text{N}}\phantom{:}\overset{..}{\text{N}}\phantom{:} \\ \text{H}:\underset{..}{\text{N}}:\underset{..}{\text{N}}:\text{H} \\ \text{H} \end{array}$$

Both nitrogen atoms have steric number 4 and are $sp^3$ hybridized, with H—N—H and H—N—N angles of approximately 109.5°. The extent of rotation about the N—N bond cannot be predicted from the VSEPR theory or the hybrid orbital model. Figure 14-24 shows the full three-dimensional structure of hydrazine.

*Related Problems:* 39, 40, 41, 42

## Triatomic Nonhydrides

Many triatomic molecules and molecular ions can be formed from atoms of elements in the second and third periods. Some are linear (examples are $CO_2$, $N_2O$, OCS, and $NO_2^+$), and others are bent ($NO_2$, $O_3$, $NF_2$, and $NO_2^-$). We use localized orbitals for the $\sigma$ bonds and delocalized orbitals for the $\pi$ bonds of these molecules.

For linear molecules it is logical to use $sp$ hybridization to describe the central atom. This leaves two $p$ orbitals ($p_x$ and $p_y$) perpendicular to the bond axis. Hybridization of the outer atoms is not necessary to treat such molecules; the outer atoms are most simply described as having fully occupied $s$ orbitals not involved in bonding, a $p_z$ orbital that takes part in $\sigma$ bonding, and $p_x$ and $p_y$ orbitals that form delocalized $\pi$ bonds with the other atoms. The $\sigma$ bonds (which extend along the bond axes) can be obtained by combining the $p_z$ orbital of one of the outer atoms with the $sp$ hybrid orbital of the central atom that points toward it to form a localized bonding orbital, and the $p_z$ orbital of the other atom forms a bonding orbital with the other $sp$ hybrid. These two localized $\sigma$ bonds are filled by two pairs of electrons.

Linear combinations of the $p_x$ and $p_y$ orbitals of all three atoms make up the delocalized molecular orbitals involved in $\pi$ bonding. How do three $p_x$ orbitals

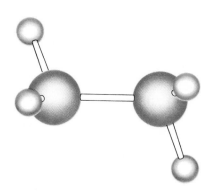

**Figure 14-24**

The structure of hydrazine, $N_2H_4$.

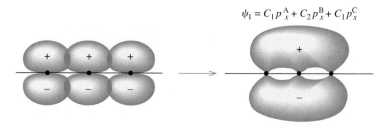

$$\psi_1 = C_1 p_x^A + C_2 p_x^B + C_1 p_x^C$$

(a) $\pi_x$ bonding

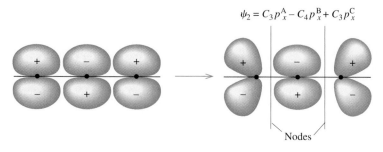

$$\psi_2 = C_3 p_x^A - C_4 p_x^B + C_3 p_x^C$$

Nodes

(b) $\pi_x^*$   antibonding

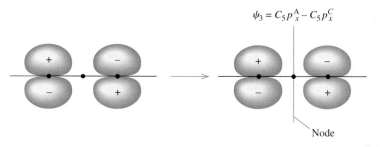

$$\psi_3 = C_5 p_x^A - C_5 p_x^C$$

Node

(c) $\pi_x^{nb}$ nonbonding

## Figure 14-25

Pi bonding in linear triatomic molecules. From three $p$ orbitals lying perpendicular to the bond axis can be constructed one bonding, one antibonding, and one nonbonding molecular orbital. A second group of molecular orbitals can be constructed from the $p$ orbitals that lie perpendicular to the plane of the page.

combine to form three molecular orbitals? Two of the combinations are obvious. One (Fig. 14-25a) is fully bonding, with no nodes, and another (Fig. 14-25b) is fully antibonding, with two nodes, one between each pair of atoms. The third linear combination is less obvious. Mathematical analysis shows that it has the form of Figure 14-25c, with a single node and with the coefficient of the wave function on the central atom being zero. This third orbital is called nonbonding ($\pi_x^{nb}$) because there is neither a node nor a region of increased electron density between the central atom and its neighbors. There is, however, weak antibonding between the outer atoms. The three $p_y$ orbitals can be combined to give the same types of molecular orbitals as the $p_x$. Figure 14-26 is the resulting correlation diagram for the $\pi$ levels

**Molecular orbitals**

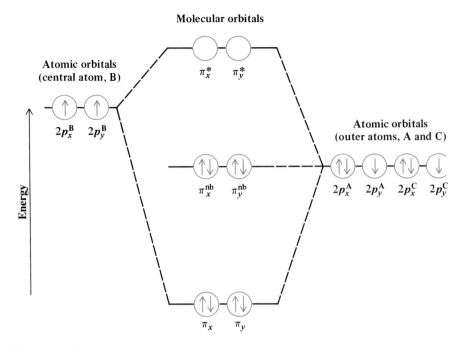

**Atomic orbitals
(central atom, B)**

$\pi_x^*$    $\pi_y^*$

$2p_x^B$    $2p_y^B$

**Atomic orbitals
(outer atoms, A and C)**

$\pi_x^{nb}$    $\pi_y^{nb}$

$2p_x^A$    $2p_y^A$    $2p_x^C$    $2p_y^C$

Energy

$\pi_x$    $\pi_y$

**Figure 14-26**

Correlation diagram for $\pi$ electrons in linear triatomic molecules. The orbital filling shown is that for $CO_2$.

only, with the nonbonding orbitals lying between the bonding and the antibonding orbitals in energy. As before, the energy increases with the number of nodes.

Consider the $CO_2$ molecule. It has 16 valence electrons (4 from the carbon and 6 from each oxygen). Two electrons on each oxygen are nonbonding electrons in $2s$ atomic orbitals. Of the remaining 12 electrons, 4 take part in localized $\sigma$ bonds, 2 between each oxygen and the carbon. This leaves 8 electrons for the $\pi$ system. Placing them in the correlation diagram of Figure 14-26 gives two electron pairs in $\pi$ orbitals and two pairs in $\pi^{nb}$ orbitals. The total bond order for the molecule is 4, because there are two pairs of electrons in $\sigma$ bonding orbitals (from the $sp$ hybrids) and two pairs in $\pi$ bonding orbitals (the nonbonding $\pi^{nb}$ and atomic $2s_{oxygen}$ orbitals neither increase nor decrease the bond order). Each bond is then of order 2, in agreement with the Lewis structure result for $CO_2$. The meaning is slightly different here, however, in that some of the bonding electrons (those in the $\pi$ orbital system) are delocalized over the full molecule rather than shared by only two of the atoms.

Nonlinear triatomic molecules can be described through $sp^2$ hybridization of the central atom. If the molecule lies in the $xy$ plane, then the $s$, $p_x$, and $p_y$ orbitals of the central atom can be combined to form three $sp^2$ hybrid orbitals with an angle close to 120° between each pair. One of these orbitals holds a lone pair of electrons, and the other two take part in $\sigma$ bonds with the outer atoms. The fourth orbital, a $p_z$ orbital, takes part in delocalized $\pi$ bonding (see Fig. 14-27a). On the outer atoms, the $p$ orbital pointing toward the central atom takes part in a localized $\sigma$ bond and the $p_z$ orbital takes part in $\pi$ bonding; the third $p$ orbital and the $s$ orbital are atomic orbitals that do not participate in bonding. The three $p_z$ atomic orbitals can be

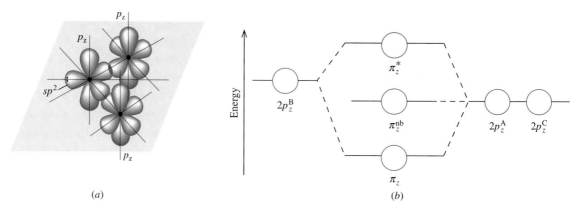

$(a)$                                                          $(b)$

**Figure 14-27**

$(a)$ Molecular orbitals for bent triatomic molecules. The central atom has three $sp^2$ hybrid orbitals lying in the plane of the molecule (green). From the three $p_z$ orbitals perpendicular to this plane (blue), three $\pi$ orbitals can be constructed. $(b)$ Correlation diagram for the $\pi$ orbitals.

combined into bonding, nonbonding, and antibonding $\pi$ orbitals much as in the linear molecule case (Fig. 14-27$b$). Here there is only one of each type of orbital ($\pi$, $\pi^{nb}$, $\pi^*$) rather than two as for linear molecules.

Consider a specific example, $NO_2^-$, with 18 electrons. Two electrons are placed in each oxygen $2s$ orbital and two more in each nonbonding oxygen $2p$ orbital, so that a total of eight electrons are localized on oxygen atoms. Two electrons also are placed as a lone pair in the third $sp^2$ orbital of the nitrogen atom. Of the remaining eight electrons, four are involved in the two $\sigma$ bonds between nitrogen and the oxygen atoms. The last four are placed into the $\pi$ electron system: two into the bonding $\pi$ orbital and two into the nonbonding $\pi^{nb}$ orbital. Because a total of six electrons are in bonding orbitals and none are in antibonding, the net bond order for the molecule is 3, or $1\frac{1}{2}$ per bond. In the Lewis picture, two resonance forms are needed to represent $NO_2^-$. The awkwardness of the resonance model is avoided by treating the electrons in the bonding $\pi$ molecular orbital as delocalized over the three atoms in the molecule.

Experiment shows that triatomic nonhydrides with 16 or fewer valence electrons generally are linear, whereas those with 17 to 20 valence electrons generally are bent. This is consistent with the VSEPR theory, because the former have steric number 2 (giving linear molecules), whereas the latter have steric number 3 or 4 (giving bent molecules with one or two lone-pair orbitals on the central atom). For molecules involving larger atoms later in the periodic table, both the molecular orbital and hybridization methods become more complicated and less useful for qualitative interpretation. The success of the simple VSEPR theory in predicting molecular geometries is all the more striking for these heavier atoms.

### Bonding in Organic Molecules

The structures of many organic molecules (compounds of carbon) can be described through the simple molecular orbital picture presented earlier in this section. A helpful first step in analyzing molecular bonding is to write the Lewis electron-dot

Low-energy structure

High-energy structure

**Figure 14-28**

As the 2-butene molecule is twisted about the C=C bond, the overlap of the two $p$ orbitals decreases, giving a higher energy.

structure. As an example, consider the molecule 2-butene ($CH_3CHCHCH_3$). Its Lewis structure is

$$\begin{array}{cccc} H & H & H & H \\ \cdot\cdot & \cdot\cdot & & \cdot\cdot & \cdot\cdot \\ H : C : C : & : C : C : H \\ \cdot\cdot & \cdot\cdot & & \cdot\cdot & \cdot\cdot \\ H & & & H \end{array}$$

From the VSEPR theory, the steric number of the two outer carbon atoms is 4 (and so they have $sp^3$ hybridization), and that of the two central carbon atoms is 3 (giving $sp^2$ hybridization). The bonding about the outer carbon atoms will be tetrahedral, and that about the central ones will be trigonal planar. Electron pairs are placed into each localized ($\sigma$) molecular orbital, giving a single bond between each pair of bonding atoms. In the case of 2-butene, these placements use 22 out of the 24 available valence electrons, forming a total of 11 single bonds.

Next, the remaining $p$ orbitals that were not involved in hybridization are combined to form $\pi$ molecular orbitals. The $p_z$ orbitals from the two central carbon atoms can be mixed to form a $\pi$ (bonding) molecular orbital and a $\pi^*$ (antibonding) molecular orbital. The remaining two valence electrons are placed into the $\pi$ orbital, giving a double bond between the central carbon atoms. If the $p_z$ orbital of one of these atoms is rotated about the central C—C axis, its overlap with the $p_z$ orbital of the other carbon atom changes (Fig. 14-28). The overlap is greatest, and the energy lowest, when the two $p_z$ orbitals are parallel to each other. In the most stable molecular geometry, the hydrogen atoms on the central carbon atoms lie in the same plane as the C—C—C—C carbon skeleton. This prediction is verified by experiment.

In 2-butene there are two possible placements for the outer $CH_3$ groups relative to the double bond (Fig. 14-29). In one, called *cis*-2-butene, the two methyl ($CH_3$) groups lie on the same side of the double bond; in the other, *trans*-2-butene, they lie on opposite sides. Converting one form to the other requires breaking the central $\pi$ bond (by rotating the two $p_z$ orbitals 180° with respect to each other, as in Fig.

*cis*-2-butene

*trans*-2-butene

**Figure 14-29**

The two *cis-trans* isomers of 2-butene.

14-28), then re-forming it in the other configuration. Because breaking a $\pi$ bond costs a significant amount of energy, both *cis* and *trans* forms are stable at room temperature, and interconversion between the two is very slow. Such molecules are called **geometrical isomers** when they have the same chemical formula but different structures.

---

### Example 14-9

The Lewis structure of propyne ($CH_3CCH$) is

$$\begin{array}{c} \text{H} \\ \overset{\cdot\cdot}{\text{H}:\text{C}}\text{—C}\equiv\text{C}:\text{H} \\ \underset{\cdot\cdot}{\text{H}} \end{array}$$

Discuss its bonding and predict its geometry.

### Solution

The leftmost carbon atom in the structure is $sp^3$ hybridized, and the other two carbon atoms are $sp$ hybridized. The atoms in the molecule are on a single straight line with the exception of the three hydrogen atoms on the leftmost carbon atom, which point outward toward three of the vertices of a tetrahedron. There is a $\sigma$ bond between each pair of bonded atoms. The $p_x$ and $p_y$ orbitals on carbon atoms 2 and 3 combine to form two $\pi$ orbitals and two $\pi^*$ orbitals; only the former are occupied in the ground-state electron configuration.

*Related Problems:* 49, 50

---

A second type of isomerism characteristic of organic molecules is **optical isomerism,** or **chirality.** A carbon atom that forms single bonds to four different atoms or groups of atoms can exist in two forms that are mirror images of each other but that cannot be interconverted without breaking and re-forming bonds (Fig. 14-30). If a mixture of the two forms is resolved into its optical isomers, the two forms rotate the plane of polarized light in different directions, and so such molecules are said to be ''optically active.'' Although paired optical isomers have identical physical properties, their chemical properties can differ when they interact with other optically active molecules. As we shall see in Section 23-4, proteins and other biomolecules are optically active. One goal of pharmaceutical research is to prepare particular

**Figure 14-30**

A molecule such as CHBrClF, which has four different atoms or groups of atoms bonded to a single carbon atom, exists in two mirror-image forms which cannot be superimposed by rotation. Such pairs of molecules are optical isomers; the carbon atom is called a chiral center.

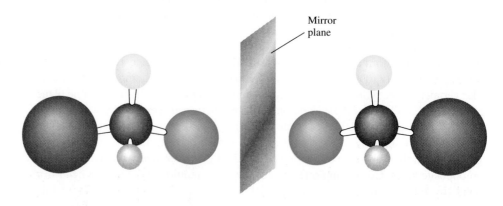

optical isomers of carbon compounds for medicinal use. In many cases one optical isomer is beneficial and the other is useless or even harmful.

## Conjugated Molecules

When two or more double or triple bonds occur close to each other in a molecule, a delocalized molecular orbital picture of the bonding should be used. One example is 1,3-butadiene ($CH_2CHCHCH_2$), which has the Lewis structure

$$\begin{array}{cccc} H & H & H & H \\ \cdot\cdot & \cdot\cdot & \cdot\cdot & \cdot\cdot \\ H:C: & :C:C: & :C:H \end{array}$$

All four carbon atoms have steric number 3, and so all are $sp^2$ hybridized. The remaining $p_z$ orbitals have maximum overlap when the four carbon atoms lie in the same plane, and so this molecule is predicted to be planar. From these four $p_z$ atomic orbitals, four molecular orbitals can be constructed by combining their phases as shown in Figure 14-31. The four electrons that remain after the $\sigma$ orbitals are filled

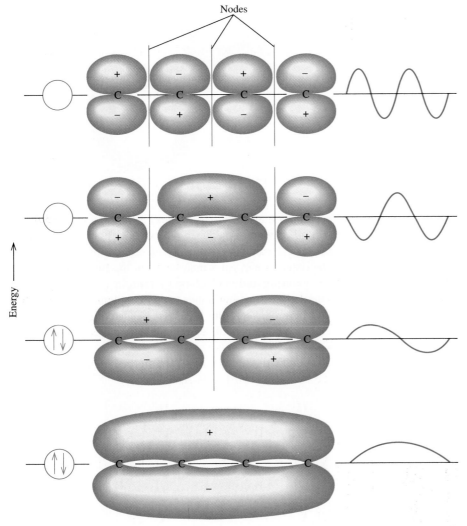

**Figure 14-31**

The four $\pi$ molecular orbitals formed from four $2p_z$ atomic orbitals in 1,3-butadiene, viewed from the side. Planar nodes are indicated by red lines. Note the similarity in nodal properties to the first four harmonics of a vibrating string (*right*). Only the two lowest-energy orbitals are occupied in the molecule in the ground state (*left*).

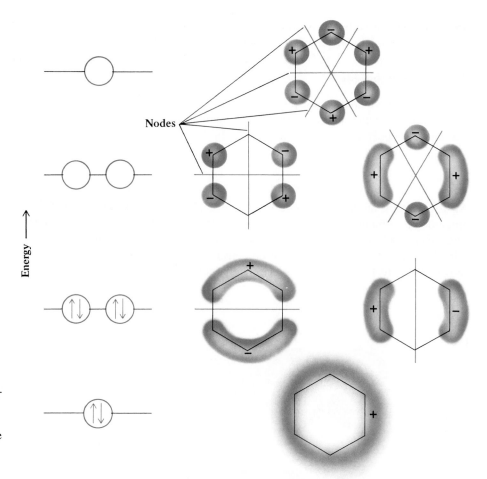

**Figure 14-32**
The six $\pi$ molecular orbitals for benzene, viewed from the top, formed from the six $2p_z$ atomic orbitals lying perpendicular to the plane of the molecule. Note the similarity in nodal properties to the standing waves on a loop shown in Figure 13-14. Only the three lowest-energy orbitals are occupied in molecules of benzene in the ground state.

are placed in the two lowest $\pi$ orbitals. The first of these is bonding among all four carbon atoms, and the second is bonding between the outer carbon atom pairs but antibonding between the central pair. Therefore, 1,3-butadiene has stronger and shorter bonds between the outer carbon pairs than between the two central carbon atoms. It is an example of a **conjugated $\pi$ electron system,** in which two or more double or triple bonds alternate with single bonds. Such conjugated systems have lower energies than would be predicted from localized bond models and are best described with delocalized molecular orbitals extending over the entire $\pi$ electron system.

Benzene is a ring molecule with the formula $C_6H_6$. In the language of Chapter 2, benzene is a resonance hybrid of two Lewis structures:

The modern view is that the molecule does not jump between two structures but rather has a single, time-independent electron distribution in which the $\pi$ bonding is described through delocalized molecular orbitals. Each carbon atom has $sp^2$ hybridization, and the remaining six $p_z$ orbitals combine to give six molecular orbitals delocalized over the entire molecule. Figure 14-32 shows the $\pi$ orbitals and their energy level diagram. The lowest-energy $\pi$ orbital has no nodes, the next two have two nodes, the next two have four nodes, and the highest-energy orbital has six nodes. The $C_6H_6$ molecule has 30 valence electrons, of which 24 occupy $sp^2$ hybrid orbitals and form $\sigma$ bonds. When the six remaining valence electrons are placed in the three lowest-energy $\pi$ orbitals, the resulting electron distribution is the same in all six carbon–carbon bonds. As a result, benzene has six carbon–carbon bonds of equal length with properties intermediate between those of single and double bonds.

## 14-6 A Deeper Look . . . Quantum Calculations for Molecules

The wide availability of high-speed computers has greatly expanded the range of calculations of molecular properties using the principles of quantum mechanics. The simple molecular orbital approaches described up to now are not only useful for gaining qualitative insight; they also provide a starting point for highly accurate predictions of the properties of small molecules.

### The Variational Principle

The total potential energy of a molecule includes three terms:

$$V = V_{el-el} + V_{el-nuc} + V_{nuc-nuc}$$

These are the repulsion of each electron by each other electron, the attraction of each electron by each nucleus, and the mutual repulsions of the nuclei. To proceed, let us take the nuclei to have fixed positions (which can be varied later), making the last of these terms an easily calculated constant. As discussed in Section 13-6, the wave function for an $N$-electron system depends on the positions of all $N$ electrons and can be written as $\psi(x_1, y_1, z_1, \ldots, x_N, y_N, z_N)$. The Schrödinger equation can be written in the symbolic form

$$\mathcal{H}\psi = E\psi$$

where $E$ is the energy of the molecule and $\mathcal{H}$ is an **operator** that acts on the wave function. This equation is a generalization of Equation 13-15 to many electrons in three spatial dimensions. The operator $\mathcal{H}$ involves second derivatives with respect to electron positions and multiplication by the total potential energy.

As stated in Chapter 13, exact solutions of the Schrödinger equation are not available for many-electron systems. Suppose, however, that an approximate solution $\psi_{app}$ is known. Then the corresponding approximate energy $E_{app}$ can be calculated according to the following prescription from quantum mechanics. Make the operator $\mathcal{H}$ act on $\psi_{app}$, multiply the result by $\psi_{app}$, and integrate over all the electron positions:

$$E_{app} = \int (\mathcal{H}\psi_{app}) \, \psi_{app} \, dx_1 \ldots dz_N$$

Mathematical analysis shows that this approximate energy is always greater than or equal to the exact energy:

$$E_{app} \geq E$$

with the equals sign applying only if the exact wave function is used in the integral.

This result is known as the **variational principle** in quantum mechanics and is of fundamental importance in calculating molecular properties. It enables us to systematically improve an approximate wave function by varying one or more of its parameters. The "best" wave function is the one that gives the lowest total energy, and that approximate energy lies above the exact energy of the system.

### The Self-Consistent Field Approximation

If each electron in a molecule could be treated separately, calculation would be easy. However, the electron–electron repulsion term $V_{el-el}$ in the total potential energy couples the electrons in a complicated fashion. In the **self-consistent field (SCF) approximation**, the total wave function is taken to be a product of single-electron wave functions. Each of these, in turn, is written as a linear combination of atomic orbitals; the coefficients are determined by minimizing the energy, using the variational principle. In the SCF theory, each electron is taken to interact with the *average* field of the other $N - 1$ electrons, calculated from their single-electron wave functions. The process is iterative, with a series of one-electron variational calculations that use the average field exerted by the remaining electrons. Each newly calculated wave function is reinserted and used to calculate an improved average field for the next electron. The process is continued until the wave functions stop changing; they are then said to be "self-consistent."

Large-scale SCF calculations (employing linear combinations of atomic orbitals) have been carried out for small molecules, with the bond lengths and angles further varied to find

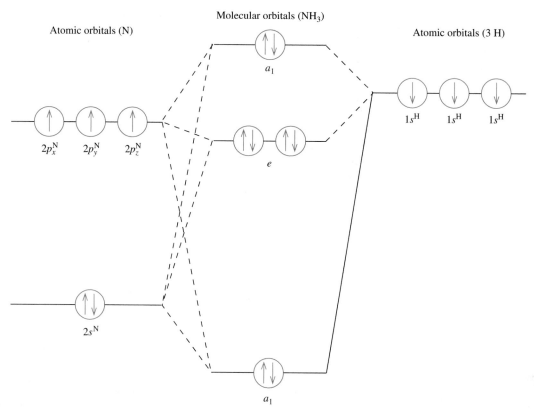

**Figure 14-33**

Molecular orbital correlation diagram for $NH_3$, calculated from the SCF model. The lower-energy $a_1$ orbital and the $e$ orbitals are bonding orbitals, while the upper $a_1$ is largely a nonbonding lone pair. Three higher-energy antibonding orbitals (unoccupied in $NH_3$) are not shown.

the stable (lowest-energy) structures. Reasonably accurate predictions of molecular properties usually result. One example of a quantitative SCF calculation is the one for ammonia, $NH_3$. The calculated H—N—H bond angle for this molecule is 107.2° (experiment: 106.7°), and the equilibrium N—H bond lengths are 1.00 Å (experiment: 1.012 Å).

*Orbital* shapes and energies can also be compared with those from simple bonding theories. Figure 14-33 is the orbital energy correlation diagram for ammonia. The lowest-energy ($a_1$) and $e$ orbitals are bonding molecular orbitals between the nitrogen atom and the three hydrogen atoms. They contain altogether six electrons, corresponding to three single bonds. The higher-energy $a_1$ orbital, also doubly occupied, is largely a nonbonding lone-pair orbital on the nitrogen. Notice that this description of orbital energies is quite different from the result of the hybrid orbital picture. In that case, one would expect $sp^3$ hybridization to give three identical bonding orbitals and one lone pair. The discrepancy illustrates one limitation of the hy-

brid orbital picture. The total bonding electron distribution in $NH_3$ is evenly distributed among the three N—H bonds, but it arises from three molecular orbitals of which only two have equal energies.

Calculations on molecules such as ammonia are not limited to the equilibrium geometry. Suppose the H atoms are moved upward until the N atom lies in a plane with them. The H atoms can then "pop through" to the other side, giving an inverted geometry in a process analogous to the inversion of an umbrella in a strong wind. The inverted molecule has the same symmetry, structure, and energy as the original molecule, but the intermediate planar state has a higher energy. With the SCF method, this energy is calculated to lie 21 kJ mol$^{-1}$ above the stable pyramidal state (the experimental value is 24 kJ mol$^{-1}$). SCF calculations give potential energy surfaces for chemical reactions (of the type used to study reaction kinetics in Section 11-6), but they become less accurate in cases in which bonds are made or broken.

### Electron Correlation

The SCF method is a useful approximation, but it ignores one important fact: electrons actually respond not only to the average charge density of other electrons but to the instantaneous positions of those electrons. In other words, a **correlation** exists between different electron positions, and it is not correct to take the wave function to be a simple product of one-electron wave functions. One of the effects of correlation is to produce a long-range attractive force (Section 3-7) between atoms and molecules, even when they are nonpolar. In a pair of argon atoms, for example, there is an instantaneous correlation between the positions of electrons on the two atoms, giving a fluctuating pair of dipole moments (Fig. 14-34) that interact with each other, lowering the energy of the pair and causing the two atoms to attract each other.

Electrons in molecules are also correlated. In the simple SCF picture that fact is ignored, and it is possible for both electrons in an electron-pair bond to be instantaneously located on one of the two bonded atoms. There are several ways to incorporate electron correlation into quantum-mechanical calculations. One is to work with wave functions that depend explicitly on the distance between a pair of electrons—a difficult method for all but two-electron atoms and molecules. A second way is to choose wave functions that never allow a pair of bonding electrons to occupy the same atom, which is the basis of the **valence bond method.** In its simplest form, this approach tends to exaggerate the importance of electron correlation, but it serves as a starting point for more accurate approaches. A third way to include electron correlation is through **configuration interaction.** In this approach, the SCF wave function is supplemented with many other wave functions that represent excited states in the one-electron picture. Variational calculations then lead to highly accurate predictions of molecular properties.

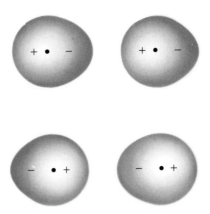

**Figure 14-34**
The fluctuating charge distribution in an atom can give rise to an instantaneous dipole moment. When two atoms are close to each other, their fluctuating dipole moments are correlated, leading to attraction and lowering the potential energy of the pair.

### Iodine

The shiny purple-black crystals of elemental iodine were first prepared in 1811 from the ashes of seaweed. Several species of seaweed concentrate the iodine that is present in very low proportions in seawater, and for many years seaweed was the major practical source of this element. Most iodine is now produced from natural brines via oxidation of iodide ion with chlorine.

(a) Iodine is an essential trace element in the human diet, and iodine deficiency causes goiter, the enlargement of the thyroid gland. To prevent goiter, much salt intended for human consumption is ''iodized'' by the addition of small quantities of sodium iodide. Calculate the electronegativity difference between sodium and iodine. Is sodium iodide an ionic or a covalent compound? What is its chemical formula?

(b) Iodine is an important reagent in synthetic organic chemistry because bonds between carbon and iodine form readily. Use electronegativities to determine whether the C—I bond is ionic, purely covalent, or polar covalent in character.

(c) Give the steric numbers for the iodine atom and identify the geometries of the following ions containing iodine and oxygen: $IO_3^-$, $IO_6^{5-}$, and $IO_4^-$.

(d) What is the ground-state electron configuration of the valence electrons of iodine molecules ($I_2$)? Is iodine paramagnetic or diamagnetic?

(e) What is the electron configuration of the $I_2^+$ molecular ion? Is its bond stronger or weaker than that in $I_2$? What is its bond order?

Iodine sublimes from the bottom of the beaker and condenses on the bottom of the round-bottom flask.

**Answers**

(a) The electronegativity difference is 1.73, and so the compound is largely ionic, with formula NaI.

(b) The electronegativity difference is 0.11, and so the C—I bond is largely covalent, with nearly equal sharing of electrons between the atoms.

(c) $IO_3^-$: $SN = 4$, structure is pyramidal; $IO_6^{5-}$: $SN = 6$, structure is octahedral; $IO_4^-$: $SN = 4$, structure is tetrahedral.

(d) $(\sigma_{5s})^2(\sigma_{5s}^*)^2(\sigma_{5p_z})^2(\pi_{5p})^4(\pi_{5p}^*)^4$; diamagnetic

(e) $(\sigma_{5s})^2(\sigma_{5s}^*)^2(\sigma_{5p_z})^2(\pi_{5p})^4(\pi_{5p}^*)^3$; stronger; bond order is $\frac{3}{2}$ versus 1.

## CONCEPTS & SKILLS

After studying this chapter and working the problems that follow, you should be able to

1. Describe the relationships between bond order, bond length, and bond enthalpy (Section 14-1, problems 1–2).
2. Calculate the energy of dissociation of gaseous diatomic ionic compounds into neutral atoms and ions (Section 14-2, problems 7–8).
3. Estimate the percent ionic character of a bond from its dipole moment (Section 14-2, problems 9–12).
4. Predict the geometries of molecules by use of the VSEPR model (Section 14-3, problems 13–18).
5. Determine whether a polyatomic molecule is polar or nonpolar (Section 14-3, problems 19–20).
6. Show how molecular orbitals can be constructed from the atomic orbitals of two atoms that form a chemical bond, and explain how the electron density between the atoms is related to the molecular orbital (Section 14-4).
7. Construct correlation diagrams for diatomic molecules formed from second- and third-period main-group elements. From these diagrams, give the electron configurations, work out the bond orders, and comment on their magnetic properties (Section 14-4, problems 25–36).
8. Describe the hybrid-atomic-orbital basis for representing the structures of molecules (Section 14-5, problems 39–46).
9. Discuss the delocalization of $\pi$ electrons in polyatomic molecules (Section 14-5, problems 47–48).
10. Describe the bonding in organic molecules and determine their geometries (Section 14-5, problems 49–50).
11. Discuss the SCF approach to calculating molecular properties, and comment on the role played by electron correlation (Section 14-6).

## PROBLEMS

### Properties of the Chemical Bond

1. The bond lengths of the X—H bonds in $NH_3$, $PH_3$, and $SbH_3$ are 1.02, 1.42, and 1.71 Å, respectively. Estimate the length of the As—H bond in $AsH_3$, the gaseous compound that decomposes on a heated glass surface in Marsh's test for arsenic. Which of these four hydrides has the weakest X—H bond?

2. Arrange the following covalent diatomic molecules in order of the lengths of the bonds: BrCl, ClF, IBr. Which of the three has the weakest bond (the smallest bond enthalpy)?

## Ionic and Covalent Bonds

3. Ionic compounds tend to have higher melting and boiling points and to be less volatile (that is, have lower vapor pressures) than covalent compounds. For each of the following pairs, use electronegativity differences to predict which compound has the higher vapor pressure at room temperature.
   (a) $CI_4$ or $KI$
   (b) $BaF_2$ or $OF_2$
   (c) $SiH_4$ or $NaH$

4. For each of the following pairs, use electronegativity differences to predict which compound has the higher boiling point.
   (a) $MgBr_2$ or $PBr_3$
   (b) $OsO_4$ or $SrO$
   (c) $Cl_2O$ or $Al_2O_3$

5. Pauling related bond dissociation energies to electronegativity differences between atoms. Convert the bond *enthalpies* for $H_2$, $Cl_2$, and $HCl$ (Table 14-1) to bond energies, and use them in the Pauling formula to estimate the electronegativity difference between H and Cl. Compare your estimate with the result from Figure 14-1.

6. Use the electronegativity difference between hydrogen and iodine (Fig. 14-1) and the bond enthalpies of $H_2$ and $I_2$ (Table 14-1) to estimate the bond enthalpy of $HI$. Compare the result with the measured value in Table 14-1. (*Note:* Convert from bond enthalpy to bond energy before using the Pauling formula.)

7. In a gaseous $KCl$ molecule, the internuclear distance is 2.67 Å. Using data from Appendix F and neglecting the small short-range repulsion between the ion cores of $K^+$ and $Cl^-$, estimate the dissociation energy of gaseous $KCl$ into K and Cl atoms (in kilojoules per mole).

8. In a gaseous $RbF$ molecule, the bond length is 2.274 Å.
   (a) Using data from Appendix D on $Rb(g)$ and $F(g)$, together with $\Delta H_f^\circ (RbF(g)) = -331.4$ kJ mol$^{-1}$, calculate the bond enthalpy and the bond dissociation energy of RbF.
   (b) Using part (a) and the electron affinity of F (328 kJ mol$^{-1}$), and neglecting the small short-range repulsion between the ion cores of $Rb^+$ and $F^-$ at the equilibrium bond length, estimate the first ionization energy of Rb. Compare your estimate with the directly measured experimental value, 403 kJ mol$^{-1}$.

9. Estimate the percent ionic character of the bond in each of the following diatomic molecules, based upon the dipole moment.

| | Average Bond Length (Å) | Dipole Moment (D) |
|---|---|---|
| ClO | 1.573 | 1.239 |
| KI | 3.051 | 10.82 |
| TlCl | 2.488 | 4.543 |
| InCl | 2.404 | 3.79 |

10. Estimate the percent ionic character of the bond in each of the following species. All the species are unstable or reactive under ordinary laboratory conditions, but they can be observed in interstellar space.

| | Average Bond Length (Å) | Dipole Moment (D) |
|---|---|---|
| OH | 0.980 | 1.66 |
| CH | 1.131 | 1.46 |
| CN | 1.175 | 1.45 |
| $C_2$ | 1.246 | 0 |

11. The percent ionic character of a bond can be approximated by the formula $16\Delta + 3.5\Delta^2$, where $\Delta$ is the magnitude of the difference in the electronegativities of the atoms (see Fig. 14-1). Calculate the percent ionic character of HF, HCl, HBr, HI, and CsF, and compare the results with those in Table 14-4.

12. The percent ionic character of the bonds in several interhalogen molecules (as estimated from their measured dipole moments and bond lengths) are ClF (11%), BrF (15%), BrCl (5.6%), ICl (5.8%), and IBr (10%). Estimate the percent ionic characters for each of these molecules, using the equation in the preceding problem, and compare them with the given values.

## The Shapes of Molecules

13. For each of the following molecules, give the steric number and sketch and name the approximate molecular geometry. In each case, the central atom is listed first and the other atoms are all bonded directly to it.
    (a) $CBr_4$
    (b) $SO_3$
    (c) $SeF_6$
    (d) $SOCl_2$
    (e) $ICl_3$

14. For each of the following molecules or molecular ions, give the steric number and sketch and name the approximate molecular geometry. In each case, the central atom is listed first and the other atoms are all bonded directly to it.
    (a) $PF_3$
    (b) $SO_2Cl_2$
    (c) $PF_6^-$
    (d) $ClO_2^-$
    (e) $GeH_4$

15. For each of the following molecules or molecular ions, give the steric number, sketch and name the approximate molecular geometry, and describe the directions of any *distortions* from the approximate geometry due to lone pairs. In each case, the central atom is listed first and the other atoms are all bonded directly to it.

(a) $ICl_4^-$

(b) $OF_2$

(c) $BrO_3^-$

(d) $CS_2$

16. For each of the following molecules or molecular ions, give the steric number, sketch and name the approximate molecular geometry, and describe the direction of any *distortions* from the approximate geometry due to lone pairs. In each case, the central atom is listed first and the other atoms are all bonded directly to it.

(a) $TeH_2$

(b) $AsF_3$

(c) $PCl_4^+$

(d) $XeF_5^+$

17. Give an example of a molecule or ion having a formula of each of the following types and structures.

(a) $AB_3$ (planar)

(b) $AB_3$ (pyramidal)

(c) $AB_2^-$ (bent)

(d) $AB_3^{2-}$ (planar)

18. Give an example of a molecule or ion having a formula of each of the following types and structures.

(a) $AB_4^-$ (tetrahedral)

(b) $AB_2$ (linear)

(c) $AB_6^-$ (octahedral)

(d) $AB_3^-$ (pyramidal)

19. For each of the answers in problem 13, state whether the species is polar or nonpolar.

20. For each of the answers in problem 14, state whether the species is polar or nonpolar.

21. The molecules of a certain compound contain one atom each of nitrogen, fluorine, and oxygen. Two possible structures are NOF (O as central atom) and ONF (N as central atom). Does the information that the molecule is bent limit the choice to one of these two possibilities? Explain.

22. Mixing $SbCl_3$ and $GaCl_3$ in a one-to-one molar ratio (using liquid sulfur dioxide as a solvent) gives a solid ionic compound of empirical formula $GaSbCl_6$. A controversy arises over whether this compound is $(SbCl_2^+)(GaCl_4^-)$ or $(GaCl_2^+)(SbCl_4^-)$.

(a) Predict the molecular structures of the two anions.

(b) It is learned that the cation in the compound has a bent structure. Based on this fact, which formulation is more likely to be correct?

23. (a) Use the VSEPR theory to predict the structure of the NNO molecule.

(b) The substance NNO has a small dipole moment. Which end of the molecule is more likely to be the positive end, based only on electronegativity?

24. Ozone ($O_3$) has a nonzero dipole moment. In the molecule of $O_3$, one of the oxygen atoms is directly bonded to the other two, which are not bonded to each other.

(a) Based on this information, state which of the following structures are possible for the ozone molecule: symmetric linear, unsymmetric linear (for example, different O—O bond lengths), and bent. (*Note:* Even an O—O bond can have a bond dipole if the two oxygen atoms are bonded to different atoms or if only one of the oxygen atoms is bonded to a third atom.)

(b) Use the VSEPR theory to predict which of the structures of part (a) is observed.

## Molecular Orbitals in Diatomic Molecules

25. If an electron is removed from a fluorine molecule, an $F_2^+$ molecular ion forms.

(a) Give the molecular electron configurations for $F_2$ and $F_2^+$.

(b) Give the bond order of each species.

(c) Predict which species should be paramagnetic.

(d) Predict which species has the greater bond dissociation energy.

26. When one electron is added to an oxygen molecule, a superoxide ion ($O_2^-$) is formed. The addition of two electrons gives a peroxide ion ($O_2^{2-}$). Removal of an electron from $O_2$ leads to $O_2^+$.

(a) Construct the correlation diagram for $O_2^-$.

(b) Give the molecular electron configuration for each of the following species: $O_2^+$, $O_2$, $O_2^-$, $O_2^{2-}$.

(c) Give the bond order of each species.

(d) Predict which species are paramagnetic.

(e) Predict the order of increasing bond dissociation energy among the species.

27. Predict the valence electron configuration and the total bond order for the molecule $S_2$, which forms in the gas phase when sulfur is heated to a high temperature. Will $S_2$ be paramagnetic or diamagnetic?

28. Predict the valence electron configuration and the total bond order for the molecule $I_2$. Will $I_2$ be paramagnetic or diamagnetic?

29. For each of the following valence electron configurations of a homonuclear diatomic molecule or molecular ion, identify the element X, Q, or Z, and determine the total bond order.

(a) $X_2$: $(\sigma_{2s})^2(\sigma_{2s}^*)^2(\sigma_{2p_z})^2(\pi_{2p})^4(\pi_{2p}^*)^4$

(b) $Q_2^+$: $(\sigma_{2s})^2(\sigma_{2s}^*)^2(\pi_{2p})^4(\sigma_{2p_z})^1$

(c) $Z_2^-$: $(\sigma_{2s})^2(\sigma_{2s}^*)^2(\sigma_{2p_z})^2(\pi_{2p})^4(\pi_{2p}^*)^3$

30. For each of the following valence electron configurations of a homonuclear diatomic molecule or molecular ion, identify the element X, Q, or Z, and determine the total bond order.

(a) $X_2$: $(\sigma_{2s})^2(\sigma_{2s}^*)^2(\sigma_{2p_z})^2(\pi_{2p})^4(\pi_{2p}^*)^2$

(b) $Q_2^-$: $(\sigma_{2s})^2(\sigma_{2s}^*)^2(\pi_{2p})^3$

(c) $Z_2^{2+}$: $(\sigma_{2s})^2(\sigma_{2s}^*)^2(\sigma_{2p_z})^2(\pi_{2p})^4(\pi_{2p}^*)^2$

31. For each of the electron configurations in problem 29, determine whether the molecule or molecular ion is paramagnetic or diamagnetic.

32. For each of the electron configurations in problem 30, determine whether the molecule or molecular ion is paramagnetic or diamagnetic.

33. Following the pattern of Figure 14-18, work out the correlation diagram for the CN molecule, showing the relative energy levels of the atoms and the bonding and antibonding orbitals of the molecule. Indicate the occupation of the molecular orbitals with arrows. State the order of the bond, and comment upon the magnetic properties of CN.

34. Following the pattern of Figure 14-18, work out the correlation diagram for the BeN molecule, showing the relative energy levels of the atoms and the bonding and antibonding orbitals of the molecule. Indicate the occupation of the molecular orbitals with arrows. State the order of the bond, and comment upon the magnetic properties of BeN.

35. The bond length of the transient diatomic molecule CF is 1.291 Å; that of the molecular ion $CF^+$ is 1.173 Å. Explain why the CF bond shortens with the loss of an electron. Refer to the proper molecular orbital correlation diagram.

36. The compound nitrogen oxide (NO) forms when the nitrogen and oxygen in air are heated. Predict whether the nitrosyl ion ($NO^+$) will have a shorter or a longer bond than the NO molecule. Will $NO^+$ be paramagnetic like NO, or diamagnetic?

37. What would be the electron configuration for an $HeH^-$ molecular ion? What bond order would you predict? How stable should such a species be?

38. The molecular ion $HeH^+$ has an equilibrium bond length of 0.774 Å. Draw an electron correlation diagram for this ion, indicating the occupied molecular orbitals. Is $HeH^+$ paramagnetic? When the $HeH^+$ ion dissociates, is a lower energy state reached by forming $He + H^+$ or $He^+ + H$?

## Polyatomic Molecules

39. Formulate a localized bond picture for the amide ion ($NH_2^-$). What hybridization do you expect the central nitrogen atom to have, and what geometry do you predict for the molecular ion?

40. Formulate a localized bond picture for the hydronium ion ($H_3O^+$). What hybridization do you expect the central oxygen atom to have, and what geometry do you predict for the molecular ion?

41. Write a Lewis electron-dot structure for each of the following molecules and ions. Formulate the hybridization for the central atom in each case, and give the molecular geometry.
    (a) $CCl_4$
    (b) $CO_2$
    (c) $OF_2$
    (d) $CH_3^-$
    (e) $BeH_2$

42. Write a Lewis electron-dot structure for each of the following molecules and ions. Formulate the hybridization for the central atom in each case, and give the molecular geometry.
    (a) $BF_3$
    (b) $BH_4^-$
    (c) $PH_3$

(d) $CS_2$
(e) $CH_3^+$

43. Describe the hybrid orbitals on the chlorine atom in the $ClO_3^+$ and $ClO_2^+$ molecular ions. Sketch the expected geometries of these ions.

44. Describe the hybrid orbitals on the chlorine atom in the $ClO_4^-$ and $ClO_3^-$ molecular ions. Sketch the expected geometries of these ions.

45. The sodium salt of the unfamiliar orthonitrate ion ($NO_4^{3-}$) has been prepared. What hybridization is expected on the N atom at the center of this ion? Predict the geometry of the $NO_4^{3-}$ ion.

46. Describe the hybrid orbitals used by the carbon atom in $N\equiv C—Cl$. Predict the geometry of the molecule.

47. The azide ion ($N_3^-$) is a weakly bound molecular ion. Formulate its molecular orbital structure in terms of localized $\sigma$ bonds and delocalized $\pi$ bonds. Do you expect $N_3$ and $N_3^+$ to be bound as well? Which of the three species do you expect to be paramagnetic?

48. Formulate the molecular orbital structure of $NO_2^+$ in terms of localized $\sigma$ bonds and delocalized $\pi$ bonds. Is it linear or nonlinear? Do you expect it to be paramagnetic? Repeat the analysis for $NO_2$ and for $NO_2^-$.

49. Acetic acid can be made by the oxidation of acetaldehyde ($CH_3CHO$). Molecules of acetaldehyde have a $—CH_3$ group, an oxygen atom, and a hydrogen atom attached to a carbon atom. Draw the Lewis structure for this molecule, give the hybridization of each carbon atom, and describe the $\pi$ orbitals and the number of electrons that occupy each one. Draw the three-dimensional structure of the molecule, showing all angles.

50. Acrylic fibers are polymers made from a starting material called acrylonitrile, $H_2C(CH)CN$. In acrylonitrile, a $—C\equiv N$ group replaces an H atom on ethylene. Draw the Lewis structure for this molecule, give the hybridization of each carbon atom, and describe the $\pi$ orbitals and the number of electrons that occupy each one. Draw the three-dimensional structure of the molecule, showing all angles.

51. Discuss the nature of the bonding in the nitrite ion ($NO_2^-$). Draw the possible Lewis resonance structures for this ion. Use the VSEPR theory to determine the steric number, the hybridization of the central N atom, and the geometry of the ion. Show how the use of resonance structures can be avoided by introducing a delocalized $\pi$ molecular orbital. What bond order does the molecular orbital model predict for the N—O bonds in the nitrite ion?

52. Discuss the nature of the bonding in the nitrate ion ($NO_3^-$). Draw the possible Lewis resonance structures for this ion. Use the VSEPR theory to determine the steric number, the hybridization of the central N atom, and the geometry of the ion. Show how the use of resonance structures can be avoided by introducing a delocalized $\pi$ molecular orbital. What bond order is predicted by the molecular orbital model for the N—O bonds in the nitrate ion?

## Additional Problems

53. The C—O bond length in free carbon monoxide is 1.13 Å. When four CO molecules bond to a central nickel atom to form $Ni(CO)_4$, the C—O bond length increases to 1.15 Å. (Recall from the Practice Exercise in Chapter 9 that this compound is used to purify nickel.) Is the C—O bond stronger or weaker in the complex than in a free CO molecule?

54. Ordinarily two metals, when mixed, form alloys that maintain their metallic character. If the two metals differ sufficiently in electronegativity, they can form compounds with significant ionic character. Compare the solid produced by mixing equal chemical amounts of Cs and Rb with that produced by mixing Cs and Au. Compute the electronegativity difference in each case, and determine whether either compound has significant ionic character. If either is ionic or partially ionic, which atom carries the net negative charge? Are there alkali halides with similar or smaller electronegativity differences?

55. Consult Figure 14-1, and compute the difference in electronegativity between the atoms in LiCl and those in HF. Based on their physical properties (see below), are the two similar or different in terms of bonding?

|               | **LiCl** | **HF**    |
|---------------|----------|-----------|
| Melting point | 605°C    | −83.1°C   |
| Boiling point | 1350°C   | 19.5°C    |

56. Use the data in Appendix F to compute the energy changes ($\Delta E$) of the following pairs of reactions.
    (a) $Na(g) + I(g) \rightarrow Na^+(g) + I^-(g)$   and
    $$Na(g) + I(g) \rightarrow Na^-(g) + I^+(g)$$
    (b) $K(g) + Cl(g) \rightarrow K^+(g) + Cl^-(g)$   and
    $$K(g) + Cl(g) \rightarrow K^-(g) + Cl^+(g)$$
    Explain why $Na^+I^-$ and $K^+Cl^-$ form in preference to $Na^-I^+$ and $K^-Cl^+$.

*57. At large interatomic separations, an alkali halide molecule MX has a lower energy as two neutral atoms, M + X; at short separations, the ionic form $(M^+)(X^-)$ has a lower energy. At a certain distance $R_c$ the energies of the two forms become equal, and it is near this distance that the electron will jump from the metal to the halogen atom during a collision. Because the forces between neutral atoms are weak at large distances, a reasonably good approximation can be made by ignoring any variation in potential $V(R)$ for the neutral atoms between $R_c$ and $R = \infty$. For the ions in this distance range, $V(R)$ is dominated by their Coulomb attraction.
    (a) Express $R_c$ in terms of the first ionization energy of the metal M and the electron affinity of the halogen X.
    (b) Calculate $R_c$ for LiF, KBr, and NaCl, using data from Appendix F.

58. The gaseous potassium chloride molecule has a measured dipole moment of 10.3 D, which indicates that it is a very polar molecule. The separation between the nuclei in this molecule is 2.67 Å. What would the dipole moment of a KCl

molecule be if there were opposite charges of one fundamental unit ($1.60 \times 10^{-19}$ C) at the nuclei?

59. (a) Predict the geometry of the $SbCl_5^{2-}$ ion, using the VSEPR method.
    (b) The ion $SbCl_6^{3-}$ is prepared from $SbCl_5^{2-}$ by treatment with $Cl^-$ ion. Determine the steric number of the central antimony atom in this ion, and discuss the extension of the VSEPR theory that would be needed for the prediction of its molecular geometry.

60. The element xenon (Xe) is by no means chemically inert; it forms a number of chemical compounds with electronegative elements such as fluorine and oxygen. The reaction of xenon with varying amounts of fluorine produces $XeF_2$ and $XeF_4$. Subsequent reaction of one or the other of these compounds with water produces (depending on conditions) $XeO_3$, $XeO_4$, and $H_4XeO_6$ as well as mixed compounds such as $XeOF_4$. Predict the structures of these six xenon compounds, using the VSEPR theory.

61. Predict the arrangement of the atoms about the sulfur atom in $F_4S{=}O$, assuming that double-bonded atoms require more space than single-bonded atoms.

62. Draw Lewis structures and predict the geometries of the following molecules. State which are polar and which nonpolar.
    (a) ONCl        (d) $SCl_4$
    (b) $O_2NCl$     (e) $CHF_3$
    (c) $XeF_2$

*63. Suppose that any given kind of bond, such as an O—H bond, has a characteristic electric dipole. In other words, suppose that electric dipole moments can be assigned to bonds just as bond enthalpies can be. Both are usefully accurate approximations. Consider the water molecule

$$H \overset{\text{O}}{\underset{\theta}{\diagup \diagdown}} H$$

Show that if $\mu_{OH}$ is the dipole moment of the OH bond, then the dipole moment of water is $\mu(H_2O) = 2\mu_{OH} \cos(\theta/2)$. What is the dipole moment $\mu_{OH}$ if $\mu(H_2O)$ is 1.86 D?

64. The carbon–carbon bond length in $C_2H_2$ is 1.20 Å; that in $C_2H_4$ is 1.34 Å, and that in $C_2H_6$ is 1.53 Å. Near which of these values would you predict the bond length of $C_2$ to lie? Is the experimentally observed value, 1.31 Å, consistent with your prediction?

65. (a) Sketch the occupied molecular orbitals of the valence shell for the $N_2$ molecule. Label the orbitals as $\sigma$ or $\pi$ orbitals, and specify which are bonding and which are antibonding.
    (b) If one electron is removed from the highest occupied orbital of $N_2$, will the equilibrium N—N distance become longer or shorter? Explain briefly.

66. Calcium carbide ($CaC_2$) is an intermediate in the manufacture of acetylene ($C_2H_2$). It is the calcium salt of the carbide (also called acetylide) ion ($C_2^{2-}$). What is the electron configuration of this molecular ion? What is its bond order?

67. Show how the fact that the $B_2$ molecule is paramagnetic indi-

cates that the energy ordering of the orbitals in this molecule is given by Figure 14-16a rather than 14-16b.

**68.** The $Be_2$ molecule has been detected experimentally. It has a bond length of 2.45 Å and a bond dissociation enthalpy of 9.46 kJ mol$^{-1}$. Write the ground-state electron configuration of $Be_2$ and predict its bond order, using the theory developed in the text. Compare the experimental bonding data on $Be_2$ with those recorded for $B_2$, $C_2$, $N_2$, and $O_2$ in Table 14-6. Is the prediction that stems from the simple theory seriously incorrect?

**\*69.** (a) The ionization energy of molecular hydrogen ($H_2$) is *greater* than that of atomic hydrogen (H), but that of molecular oxygen ($O_2$) is *lower* than that of atomic oxygen (O). Explain. (*Hint:* Think about the stability of the molecular ion that forms, in relation to bonding and antibonding electrons.) (b) What prediction would you make for the relative ionization energies of atomic and molecular fluorine (F and $F_2$)?

**70.** The molecular ion $HeH^+$ has an equilibrium bond length of 0.774 Å. Draw an electron correlation diagram for this molecule, indicating the occupied molecular orbitals. If the lowest-energy molecular orbital has the form $C_1\psi_{1s}^H + C_2\psi_{1s}^{He}$, do you expect $C_2$ to be larger or smaller than $C_1$?

**\*71.** The molecular orbital of the ground state of a heteronuclear diatomic molecule AB is

$$\psi = C_A\psi_A + C_B\psi_B$$

If a bonding electron spends 90% of its time in an orbital $\psi_A$ on atom A and 10% of its time in $\psi_B$ on atom B, what are the values of $C_A$ and $C_B$? (Neglect the overlap of the two orbitals.)

**72.** The stable molecular ion $H_3^+$ is triangular, with H—H distances of 0.87 Å. Sketch the molecule, indicating the region of greatest electron density of the lowest-energy molecular orbital.

**\*73.** According to recent spectroscopic results, nitramide

has a nonplanar molecule. It was previously thought to be planar.
(a) Predict the bond order of the N—N bond in the nonplanar structure.
(b) If the molecule really were planar after all, what would be the bond order of the N—N bond?

**74.** *Trans*-tetrazene ($N_4H_4$) consists of a chain of four nitrogen atoms with each of the two end atoms bonded to two hydrogen atoms. Use the concepts of steric number and hybridization to predict the overall geometry of the molecule. Give the expected structure of *cis*-tetrazene.

**75.** Compare the bonding in formic acid (HCOOH) with that in its conjugate base formate ion (HCOO$^-$). Each molecule has a central carbon atom bonded to the two oxygen atoms and to a hydrogen atom. Draw Lewis structures, determine the steric numbers and hybridization of the central carbon atom, and give the molecular geometries. How do the $\pi$ orbitals differ in formic acid and the formate molecular ion? The bond lengths of the C—O bonds in HCOOH are 1.23 Å (for the bond to the lone oxygen) and 1.36 Å (for the bond to the oxygen with a hydrogen atom attached). In what range of lengths do you predict the C—O bond length in the formate ion to lie?

**76.** In Section 14-5 we showed that the compound 2-butene exists in two isomeric forms, which can be interconverted only by breaking a bond (in that case, the central double bond). How many possible isomers correspond to each of the following chemical formulas? Remember that a simple rotation of an entire molecule does not give a different isomer. Each molecule contains a central C=C double bond.
(a) $C_2H_2Br_2$
(b) $C_2H_2BrCl$
(c) $C_2HBrClF$

**\*77.** The pyridine molecule ($C_5H_5N$) is obtained by replacing one C—H group in benzene with a nitrogen atom. Because nitrogen is more electronegative than the C—H group, orbitals with electron density on nitrogen are lower in energy. How do you expect the $\pi$ molecular orbitals and energy levels of pyridine to differ from those of benzene?

**\*78.** For each of the following molecules, construct the $\pi$ molecular orbitals from the $2p_z$ atomic orbitals perpendicular to the plane of the carbon atoms.

(a) Cyclobutadiene

(b) Allyl radical

Indicate which, if any, of these orbitals have identical energies on the basis of symmetry considerations. Show the number of electrons occupying each $\pi$ molecular orbital in the ground state, and indicate whether either or both of the molecules are paramagnetic. (*Hint:* Refer to Figures 14-31 and 14-32.)

## CUMULATIVE PROBLEMS

**\*79.** The electronegativities of the elements in Group IV of the periodic table are as follows: C, 2.55; Si, 1.90; Ge, 2.01; Sn, 1.88; Pb, 2.10. Explain the absence of a clear-cut trend. Is any trend seen in Group III? (*Hint:* Consider the relative influences of the size and the nuclear charge of the elements on the electronegativity.)

*80. A stable triatomic molecule can be formed that contains one atom each of nitrogen, sulfur, and fluorine. Three bonding structures are possible, depending on which is the central atom: NSF, SNF, and SFN.
(a) Write a Lewis structure for each of these molecules, indicating the formal charge on each atom.
(b) Frequently, the structure with the least separation of formal charge is the most stable. Is this statement consistent with the observed structure for this molecule—namely NSF, which has a central sulfur atom?
(c) Does consideration of the electronegativities of N, S, and F from Figure 14-1 help to rationalize this observed structure? Explain.

81. A compound is being tested for use as a rocket propellant. Analysis reveals that it contains 18.54% F, 34.61% Cl, and 46.85% O.
(a) Determine the empirical formula for this compound.
(b) Assuming that the molecular formula is the same as the empirical formula, draw a Lewis structure for this molecule. Review examples elsewhere in the chapter to decide which atom is most likely to lie at the center.

(c) Use the VSEPR theory to predict the structure of the molecule from part (b).

82. Use bond enthalpies from Table 8-3 to estimate the barrier to rotation about the central $C{=}C$ bond in 2-butene (Fig. 14-28). (*Hint:* Think about what bonds are broken in the process.) Do you expect that conversion of *cis*-2-butene to *trans*-2-butene will be faster or slower than the inversion of ammonia described in Section 14-6?

83. (a) Use thermodynamic data from Appendix D to calculate the enthalpy change when 1.00 mol of gaseous benzene is formed from carbon and hydrogen *atoms,* all in the gas phase.
(b) Compare this result with your calculated enthalpy change for forming one of the Lewis resonance structures for benzene. Use the bond enthalpies of $C{=}C$, $C{-}C$, and $C{-}H$ in Table 8-3.
(c) The additional lowering of the enthalpy found experimentally (the more negative value of $\Delta H$) is due to *resonance stabilization,* the delocalization of the electrons as they are free to move over the whole carbon ring. How great is the enthalpy change due to resonance stabilization in benzene?

*Bright sun rays shining through a gap between Douglas firs in Oregon. (Gary Braasch/Tony Stone Images)*

# The Interaction of Light with Molecules

In this chapter we consider the interaction between light (radiant energy) and molecules. Light is one of the most important tools for measuring the properties of molecules: their geometries, bond lengths, and bond energies. Light also plays crucial roles in changing molecules, by transferring the energy that causes bonds to break or rearrange, and by inducing chemical reactions. Light from the sun is the ultimate source for almost all the energy used on the earth, as it is harvested by plants through photosynthesis. Combustion then converts the energy stored in plants to other forms of energy. The atmosphere plays a crucial role in transporting some of the sun's radiation to the earth's surface and absorbing or reradiating other portions into space. Living species have transformed the composition of the atmosphere; over the last 200 years, industrialization has greatly increased the rate of change, raising serious questions about the sustainability of life on earth should present trends continue.

## 15-1 MOLECULAR SPECTROSCOPY

In Chapter 13 we discussed atomic spectroscopy and showed how the frequencies of the light absorbed or emitted by an atom are related to the energy differences between the atom's quantum states. In particular, the spectrum of the hydrogen atom

577

**Table 15-1  Spectroscopic Experiments**

| Spectral Region | Frequency ($s^{-1}$) | Energy Levels Involved | Information Obtained |
|---|---|---|---|
| Radio waves | $10^7 - 10^9$ | Nuclear spin states | Electronic structure near the nucleus |
| Microwave, far infrared | $10^9 - 10^{12}$ | Rotational | Bond lengths and bond angles |
| Near infrared | $10^{12} - 10^{14}$ | Vibrational | Stiffness of bonds |
| Visible, ultraviolet | $10^{14} - 10^{17}$ | Valence electrons | Electron configuration |
| X-ray | $10^{17} - 10^{19}$ | Core electrons | Core electron energies |

can be interpreted in terms of solutions of the Schrödinger equation for that atom. Spectroscopy also has been widely used to study the energy levels of molecules, and careful interpretation of the results leads to useful conclusions about molecular structure, bond lengths, and bond energies (Table 15-1). In fact, molecular spectroscopy is the main source of information about chemical bonding. In this section we consider the interaction between longer-wavelength (lower-energy) photons and *nuclear* motion in molecules. Section 15-2 will describe the electronic excitations that occur when molecules absorb visible light or radiation of still shorter wavelengths.

## Nuclear Magnetic Resonance

In **nuclear magnetic resonance (NMR) spectroscopy,** low-energy radio waves (photons carrying energies between 0.00002 and 0.00020 kJ mol$^{-1}$) ''tickle'' the nuclei in a molecule. Some nuclei have spins (just as electrons have spin), and the energies of different spin states are split apart by the magnetic field employed in a magnetic resonance spectrometer. Hydrogen ($^1$H) nuclei, for example, have ''spin-up'' and ''spin-down'' states that differ in energy in a magnetic field. These nuclei can change spin state by absorbing or emitting photons of FM radio frequencies. Nuclear spins are sensitive both to the chemical environment of a nucleus (and thus to the bonding of the atom) and to other nearby nuclear spins. Nuclear magnetic resonance spectroscopy accordingly offers a way to identify the bonding groups in a molecule and to measure distances between atoms.

An adaptation of NMR spectroscopy used in medical diagnosis is called magnetic resonance imaging (MRI). It relies on emission by the protons in the water contained by the body's organs. The patient is placed in the opening of a large magnet (Fig. 15-1), and a radio transmitter raises the proton spins in the relevant part of the body to their high-energy state. The radio-frequency photons subsequently emitted are then detected by a radio receiver coil. The amplitude of the signal indicates the density of water present. Thus, MRI can identify tumors by the excess water in their cells. The time delay until emission occurs is related to the type of tissue. Figure 15-2 shows an MRI brain scan.

## Rotations of Molecules

Three types of nuclear motion occur in gas-phase molecules: overall translational motion of a molecule through its container, rotational motion in which the molecule turns about one or more axes, and vibrational motion in which the nuclei move relative to each other as bond lengths or angles change. All three motions are subject

**Figure 15-1**

A magnetic resonance imaging (MRI) machine. The patient is placed on the platform and rolled forward into the magnet opening. (Peter Arnold)

to the laws of quantum mechanics, but in a gas the translational energy states are so close in energy (they correspond to the particle-in-a-box states of Section 13-4) that quantum effects are not apparent.

Rotational energies depend on the molecular **moments of inertia.** For a diatomic molecule, only a single moment of inertia $I$ enters, defined by

$$I = \mu R_e^2$$

where $\mu$ is the **reduced mass** of the molecule:

$$\mu = \frac{m_1 m_2}{m_1 + m_2}$$

with $m_1$ and $m_2$ being the masses of the two atoms and $R_e$ the bond length. Polyatomic molecules have up to three different moments of inertia, corresponding to rotations about three axes. Quantum mechanics demonstrates that rotational motion is quantized, and only certain discrete rotational energy levels are permitted. In a linear molecule, for example, the rotational energy can take on only the values

$$E_{\text{rot},J} = \frac{h^2}{8\pi^2 I} J(J + 1) \qquad J = 0, 1, 2, \ldots$$

where $J$ is the rotational quantum number. Typical rotational energy differences range from 0.001 to 1 kJ mol$^{-1}$.

Radiation in the far (long-wavelength) infrared and microwave regions of the electromagnetic spectrum excites rotational states of molecules. From gas-phase absorption spectra in these regions, the moments of inertia can be calculated and (because the atomic masses are known) bond lengths and angles can be determined. For a heteronuclear diatomic molecule, absorption of light is possible only between states that differ by 1 in rotational quantum number ($\Delta J = 1$). The allowed absorption frequencies are

$$\nu = \frac{\Delta E}{h} = \frac{h}{8\pi^2 I} [J_f(J_f + 1) - J_i(J_i + 1)]$$

$$= \frac{h}{8\pi^2 I} [(J_i + 1)(J_i + 2) - J_i(J_i + 1)]$$

$$= \frac{h}{4\pi^2 I} (J_i + 1) \qquad J_i = 0, 1, 2, \ldots$$

**Figure 15-2**

Computer-enhanced MRI scan of a normal human brain with the pituitary gland highlighted. (Scott Camazine/ Photo Researchers, Inc.)

on the basis of the fact that the quantum number of the final rotational state $J_f$ must be 1 greater than that of the initial rotational state $J_i$. The rotational absorption spectrum of such a molecule shows a series of equally spaced lines with a frequency separation of $h/4\pi^2 I$. The spectra for nonlinear polyatomic molecules are more complex, but their interpretation has enabled chemists to determine to high accuracy the molecular geometries for many small molecules.

## Vibrations of Molecules

Consider first the vibrational motion of a diatomic molecule. If the bond is stretched, the two atoms experience a restoring force that tends to bring them back to their original separation $R_e$, just as a restoring force acts on the ends of a spring that has been stretched. For a small change in bond length $R - R_e$, the force is proportional to that change:

$$F = -k(R - R_e)$$

where the **force constant** $k$ determines the stiffness of the bond. In SI units, $k$ is measured in joules per square meter, equivalent to kilograms per square second. According to classical mechanics, when a stretched bond is released, the atoms will oscillate back and forth about their average separation, much as two balls connected by a spring do. The oscillation frequency $\nu$ is given by

$$\nu = \frac{1}{2\pi}\sqrt{\frac{k}{\mu}}$$

where $\mu$ is again the reduced mass of the harmonic oscillator. In classical mechanics, any vibrational energy is permitted for the oscillator, but in quantum mechanics, the Schrödinger equation gives a discrete set of vibrational energy levels:

$$E_{\text{vib},v} = h\nu(v + \tfrac{1}{2}) \qquad v = 0, 1, 2, \ldots$$

where $h$ is Planck's constant and $v$ the vibrational quantum number. Note that even in the ground state ($v = 0$) the vibrational energy is not 0. The residual energy $\tfrac{1}{2}h\nu$ is the **zero-point energy**; it arises from the quantum nature of vibrational motion. Typical energy differences between the ground ($v = 0$) and first excited ($v = 1$) vibrational states range from 2 to 40 kJ mol$^{-1}$; they are greater than rotational energy differences but less than electronic energy differences (Table 15-1). These energy differences correspond to the absorption of radiation in the infrared region of the electromagnetic spectrum, with wavelengths longer than those of visible light.

In a polyatomic molecule, several types of vibrational motion are possible (Fig. 15-3). Each has a different frequency, and each gives rise to a series of allowed

## Figure 15-3

The three types of vibrational motion that are possible for a bent triatomic molecule. Arrows show the displacement of each atom during each type of vibration.

Symmetric stretch

Asymmetric stretch

Bend

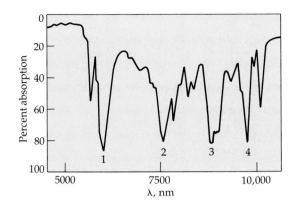

**Figure 15-4**

The infrared absorption spectrum of vitamin C (ascorbic acid, $C_6H_8O_6$). Strong absorptions at **1** (6100 nm), **2** (7600 nm), **3** (8800 nm), and **4** (9600 nm) are characteristic of particular features of the molecule. For example, band 1 is primarily due to a C=C bond stretch, and band 4, to a C—OH stretch.

quantum vibrational states. Infrared absorption spectra provide useful information about vibrational frequencies and force constants in molecules. For larger molecules, vibrational spectra are useful for identifying compounds, because each functional group (such as a C=O double bond or a C—OH bond) has its own characteristic frequency in the spectrum, largely independent of the mass, geometry, and other features of the molecule of which it is a part. The spectrum thus provides information about the functional groups in a given molecule (Fig. 15-4). Strong absorption is observed only for transitions between states that differ by 1 in vibrational quantum number ($\Delta v = 1$).

## Example 15-1

The spectrum of gaseous NaH (isotope: $^{23}Na^1H$) was determined via absorption spectroscopy. The photon wavelength required to excite the molecules rotationally from the state $J = 0$ to $J = 1$ was measured to be $1.02 \times 10^{-3}$ m, and the wavelength needed to excite them vibrationally from the state $v = 0$ to $v = 1$ was $8.53 \times 10^{-6}$ m. Calculate (a) the bond length and (b) the vibrational force constant of the NaH molecule. Use isotope atomic masses from Table 12-1.

### Solution

(a) The reduced mass is

$$\mu = \frac{m_{Na}m_H}{m_{Na} + m_H} = \frac{(22.9898 \text{ u})(1.0078 \text{ u})}{22.9898 + 1.0078 \text{ u}}$$

$$= 0.9655 \text{ u} = 1.603 \times 10^{-27} \text{ kg}$$

The energy change is

$$\Delta E = h\nu = \frac{hc}{\lambda}$$

which in this case is

$$\Delta E_{rot} = \frac{(6.626 \times 10^{-34} \text{ J s})(2.998 \times 10^8 \text{ m s}^{-1})}{1.02 \times 10^{-3} \text{ m}} = 1.95 \times 10^{-22} \text{ J}$$

$$= \frac{h^2}{8\pi^2 I}[(1)(2) - (0)(1)] = \frac{h^2}{4\pi^2 I}$$

Solving for the moment of inertia $I$ gives

$$I = \frac{h^2}{4\pi^2 \, \Delta E_{\text{rot}}} = \frac{(6.626 \times 10^{-34} \, \text{J s})^2}{4\pi^2 (1.95 \times 10^{-22} \, \text{J})} = 5.70 \times 10^{-47} \, \text{kg m}^2$$

The moment of inertia $I$ is related to the bond length $R_e$ by $I = \mu R_e^2$, and so

$$R_e^2 = \frac{I}{\mu} = \frac{5.70 \times 10^{-47} \, \text{kg m}^2}{1.603 \times 10^{-27} \, \text{kg}} = 3.56 \times 10^{-20} \, \text{m}^2$$

$$R_e = 1.89 \times 10^{-10} \, \text{m} = 1.89 \, \text{Å}$$

(b) The vibrational frequency is

$$\nu = \frac{c}{\lambda} = \frac{2.998 \times 10^8 \, \text{m s}^{-1}}{8.53 \times 10^{-6} \, \text{m}} = 3.515 \times 10^{13} \, \text{s}^{-1} = \frac{1}{2\pi}\sqrt{\frac{k}{\mu}}$$

Solving for the constant $k$ gives

$$k = \mu (2\pi\nu)^2 = (1.603 \times 10^{-27} \, \text{kg})(2\pi)^2 (3.515 \times 10^{13} \, \text{s}^{-1})^2$$

$$= 78.2 \, \text{kg s}^{-2} = 78.2 \, \text{J m}^{-2}$$

*Related Problems:* 3, 4, 5, 6, 7, 8

## Molecular Energy Levels and Their Thermal Occupation

Figure 15-5 schematically illustrates the energy levels of a typical diatomic molecule. Two electronic states (to be discussed further in the next section) as well as numerous rotational and vibrational states are shown. A polyatomic molecule would have an even more complicated set of energy levels, each characterized by several vibrational quantum numbers and by up to three rotational quantum numbers. Most spectral lines observed in absorption or emission correspond to changes in more than one kind of energy. For example, a line in the visible region of the emission spectrum may connect two energy levels that differ in rotational, vibrational, and electronic energy (see the red arrow in Figure 15-5).

As we saw in Section 3-5, molecules have a distribution of speeds, which is described by the Maxwell–Boltzmann distribution for thermal equilibrium at absolute temperature $T$. This corresponds to a distribution $P(E)$ of kinetic energies that falls off exponentially:

$$P(E) \propto \exp(-E/k_{\text{B}}T)$$

where $k_{\text{B}}$ is the Boltzmann constant, equal to $1.38 \times 10^{-23} \, \text{J K}^{-1}$. (If energies per *mole* are used, $k_{\text{B}}$ is replaced by the gas constant $R$.) This distribution gives rise to the Arrhenius dependence of gas-phase reaction rates on temperature (Section 11-5).

A similar **Boltzmann distribution** applies to the probability that a gas-phase molecule will be in a particular quantum state $i$ with energy $E_i$, except that now $i$ is a discrete label, and not all energies are allowed the molecule by quantum mechanics. The label $i$ stands for the set of rotational and vibrational quantum numbers, in addition to designating the electronic state under consideration. Then, at thermal equilibrium, the probability that a particular molecule has energy $E_i$ is

$$P_i \propto \exp(-E_i/k_{\text{B}}T)$$

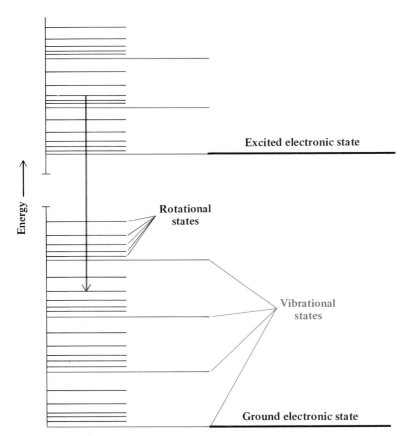

**Figure 15-5**

An energy-level diagram for a diatomic molecule, showing the electronic, vibrational, and rotational levels schematically. The arrow indicates one of the many possible transitions; in this case the rotational, vibrational, and electronic states of the molecule all change.

Equivalently, if a sample contains a great many molecules, the fraction in state $i$ will be $P_i$ at temperature $T$. This exponential falloff means that very few molecules have energies that are large compared to $k_B T$ at thermal equilibrium. The *ratio* of occupation numbers for two states, $i$ and $j$, is

$$P_i/P_j = \exp[-(E_i - E_j)/k_B T]$$

At room temperature, $k_B T$ has the value $4 \times 10^{-21}$ J (equivalent to 2.5 kJ mol$^{-1}$). This is large compared to the typical spacing of rotational levels, indicating that at room temperature many rotational states are occupied by a collection of molecules. This value is also a little smaller than typical vibrational level spacings, and so most small molecules at room temperature are in their ground vibrational states, but a measurable fraction occupy excited states. Finally, $k_B T$ is small compared to typical electronic energy spacings, and so only the ground electronic state is occupied at room temperature. When molecules are heated, the occupation of more highly excited levels increases.

## 15-2 EXCITED ELECTRONIC STATES

Excited states of atoms result from the promotion of one or more electrons into orbitals of higher energy. Excited electronic states in molecules arise similarly and can be described in the framework of molecular orbital theory. As an example,

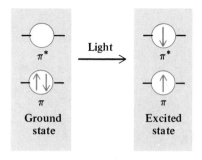

**Figure 15-6**

Occupancy of the $\pi$ molecular orbitals in the ground state and first excited state of ethylene, $C_2H_4$. Ethylene in the excited state is produced by irradiation of ground-state ethylene with ultraviolet light in the appropriate frequency range.

consider the ethylene molecule ($C_2H_4$). The bonding of the two carbon atoms in ethylene can be described through $sp^2$ hybridization, with the two remaining carbon $2p_z$ orbitals combined into a $\pi$ (bonding) molecular orbital and a $\pi^*$ (antibonding) molecular orbital (Fig. 15-6). The lowest-energy state of ethylene has two electrons in the $\pi$ orbital, but an excited state can be attained by removing one electron from the $\pi$ orbital and placing it in the $\pi^*$ orbital. An ethylene molecule in such an excited state has quite different properties from one in the ground state. The electron in the $\pi^*$ antibonding orbital cancels the effect of the electron in the $\pi$ bonding orbital, and the net bond order is reduced from 2 to 1 (the $\sigma$ bond). Consequently, the molecule in the excited state has a lower bond enthalpy and a longer C—C bond length. The observed wavelength for this $\pi \rightarrow \pi^*$ transition in ethylene is 162 nm, which lies in the ultraviolet region of the spectrum. Because ethylene does not absorb strongly in the visible region, visible light passes almost unaffected through samples of the substance, which is therefore colorless.

Now consider a molecule in which several double bonds alternate with single bonds in a conjugated $\pi$ bonding system such as that discussed for butadiene or benzene in Section 14-5. Observe that the lowest-energy $\pi$ orbital in Figure 14-31 extends over the full conjugated $\pi$ system and resembles a standing wave with a wavelength proportional to the length of the molecule. As more alternating single and double bonds are added, the characteristic wavelengths become longer and the frequencies and energies become lower (Table 15-2). For a sufficiently long chain, the first absorption is shifted into the visible region of the spectrum, and the substance takes on color. The perceived color is related to the absorption spectrum of the material, but indirectly. What we see is the light transmitted *through* the material, not the light absorbed. In other words, the color perceived is **complementary** to the color most strongly absorbed by the sample (Fig. 15-7a). Examples are the dyes indigo and beta-carotene. Figure 15-7b plots the strength of the absorption of these

*(a)*

*(b)*

**Figure 15-7**

*(a)* The color wheel, showing complementary colors opposite each other.
*(b)* The absorption spectra of the dyes beta-carotene and indigo differ in the visible region and hence have different colors. The letters stand for the colors of the light at particular wavelengths (violet, blue, green, yellow, orange, and red).

Table 15-2 **Absorption of Light by Molecules with Conjugated $\pi$ Electron Systems**

| Molecule | Number of C=C Bonds | Wavelength of Maximum Absorption (nm) |
|---|---|---|
| $C_2H_4$ | 1 | 162 |
| $C_4H_6$ | 2 | 217 |
| $C_6H_8$ | 3 | 251 |
| $C_8H_{10}$ | 4 | 304 |

two dyes at different wavelengths. Indigo absorbs in the yellow-orange region of the spectrum, and so to an observer it appears violet-blue (the colors complementary to yellow and orange). Beta-carotene absorbs at shorter wavelengths in the blue and violet regions and appears orange or yellow. Indigo is used to dye blue jeans; beta-carotene is responsible for the orange color of carrots, the colors of some processed foods, and the yellow and orange in certain bird feathers. Most bird colors, however, arise from diffraction, not absorption, of light.

## Example 15-2

Suppose you set out to design a new green dye. Over what range of wavelengths would you want your trial compound to absorb light?

### Solution

A good green dye must transmit green light and absorb other colors of light. This can be achieved by a molecule that absorbs in the violet-blue *and* orange-red regions of the spectrum. Thus, you would want a dye with strong absorptions in both these regions.

The naturally occurring substance chlorophyll, which is responsible for the green colors of grass and leaves, absorbs light over just these wavelength ranges, converting solar energy to chemical energy for the growth of the plant. It is used commercially as a green dye.

*Related Problems:* 13, 14

## Fullerenes

Among the most interesting conjugated $\pi$ electron systems is a molecule discovered in 1985: buckminsterfullerene, $C_{60}$. Previously, only two forms of carbon (diamond and graphite) were known. In 1985, Harold Kroto, Robert Curl, and Richard Smalley were studying certain long-chain carbon molecules that had been discovered spectroscopically by radioastronomers in the vicinity of red giant stars. They sought to duplicate the conditions near those stars by vaporizing a graphite target with a laser beam. Analysis of the products by mass spectrometry revealed not only the hoped-for molecules but a large proportion of molecules of molar mass 720 g $mol^{-1}$, which corresponds to the molecular formula $C_{60}$. Although the amounts of $C_{60}$ present were far too small to isolate for direct determination of molecular

(a)

(b)

**Figure 15-8**
(a) The structure of $C_{60}$, buckminsterfullerene. Note the pattern of hexagons and pentagons. (b) The design on the surface of a soccer ball has the same pattern as the structure of $C_{60}$. (T. J. S. Dennis, University of Sussex, UK)

structure, Kroto, Curl, and Smalley correctly suggested the cage structure shown in Figure 15-8a and named the molecule buckminsterfullerene after the architect Buckminster Fuller, the inventor of the geodesic dome—which the molecular structure of $C_{60}$ resembles.

Molecules of $C_{60}$ have a highly symmetric structure: 60 carbon atoms arranged in a closed net with 20 hexagonal faces and 12 pentagonal faces. The pattern is exactly the design on the surface of a soccer ball (Fig. 15-8b). Every C atom has a steric number of 3; all 60 atoms are accordingly $sp^2$ hybridized, although one of the three bond angles at each carbon atom must be distorted from the usual 120° $sp^2$ bond angle down to 108°. As in benzene, the $\pi$ electrons of the double bonds are delocalized: the 60 p orbitals (one from each carbon atom) mix to give 60 molecular orbitals spread over both sides of the entire closed surface of the molecule. The lowest 30 of these molecular orbitals are occupied by the 60 $\pi$ electrons.

In 1990 scientists succeeded in synthesizing $C_{60}$ in gram quantities by striking an electric arc between two carbon rods held under an inert atmosphere. The carbon vapors condensed to a soot, which was extracted with an organic solvent. Using chromatography, the scientists could separate the $C_{60}$—as a solution of a delicate magenta hue (Fig. 15-9) and finally as a crystalline solid (Fig. 2-4c)—from various impurities. Since 1990, $C_{60}$ has been found to form in sooting flames when hydrocarbons are burned. Thus, the newest form of carbon has been (in the words of one of its discoverers) "under our noses since time immemorial." In 1994 the first buckminsterfullerene molecules were brought back from outer space in the form of an impact crater from a tiny meteorite on an orbiting spacecraft.

Buckminsterfullerene is not the only new form of carbon to emerge from the chaos of carbon vapor condensing at high temperature. Synthesis of $C_{60}$ simultaneously produces a whole family of closed-cage carbon molecules called **fullerenes.** All the fullerenes have even numbers of atoms, with formulas ranging up to $C_{400}$ and higher. These materials offer exciting prospects for technical applications. For example, because $C_{60}$ readily accepts and donates electrons to or from its $\pi$ molecular orbitals, it has possible applications in batteries. It forms compounds (such as $Rb_3C_{60}$) that superconduct (have zero resistance to the passage of an electric current) up to 30 K. Fullerenes also can encapsulate foreign atoms that are present during synthesis. If a graphite disk is soaked with a solution of $LaCl_3$, dried, and used as a laser target, the substance $La@C_{60}$ forms, where the symbol @ means that the La atom is trapped within the 60-atom carbon cage.

## Photochemistry

What happens to molecules after they absorb radiation? Some reemit a photon almost immediately and return to the ground electronic state, although they usually end up in rotational and vibrational states different from the original ones. This process is called **fluorescence** and typically occurs within about $10^{-9}$ s. A second type of light emission occurs much more slowly, taking seconds (or in some cases hours), and is called **phosphorescence.** To understand phosphorescence, it is necessary to consider the spins of the electrons in an excited state. The absorption of light (shown in Fig. 15-6) does not change the electron spin and so gives an excited state (called a *singlet*) with one spin up and one down. Another possible excited electronic state (called a *triplet*) has both spins up (Fig. 15-10) and, by Hund's rule, has a lower energy than the singlet. Although the triplet state cannot be reached directly by absorption of light, there is a small probability that an excited molecule will cross over to the triplet state as it evolves in time or undergoes collisions. Once there, it remains for a long time because emission from a triplet state to a ground singlet state is very slow.

Figure 15-11 schematically illustrates these two pathways for loss of energy by an electronically excited molecule, along with two other possible outcomes. One is that the molecule crosses over to the ground electronic state and gradually loses its energy as heat to the surroundings as it cascades down to lower vibrational and rotational levels. The final possibility is that the energy of the molecule is high enough that bonds break, producing a chemical reaction. Molecules in excited states have enhanced reactivity and can often fragment or rearrange to give new molecules. **Photochemistry,** the study of the chemical reactions that follow the excitation of molecules to higher electronic states through absorption of photons, is an active area of research.

Some compounds are so photochemically sensitive that they must be stored in the dark, because they react rapidly to form products when exposed to light. An example is anhydrous hydrogen peroxide, which reacts explosively in light to give water and oxygen:

$$H_2O_2 \longrightarrow 2\,OH \longrightarrow H_2O + \tfrac{1}{2}O_2$$

**Figure 15-9**
(*Left*) A solution of $C_{60}$ in an organic solvent has a delicate magenta color. (*Right*) The related fullerene $C_{70}$ is distinctly orange in the same solvent. (Kurt Nassau/Phototake)

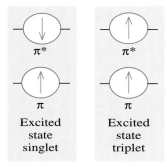

**Figure 15-10**
The first excited singlet and triplet states have the same orbital occupancy but different spin states. The triplet is generally lower in energy.

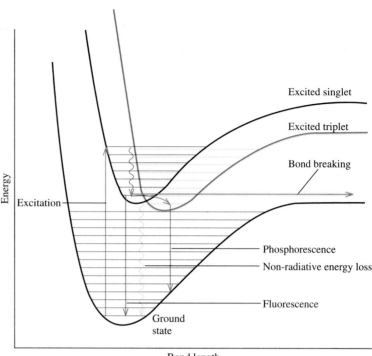

**Figure 15-11**

A schematic picture of the loss of energy by a molecule in an excited electronic state. After the molecule absorbs a photon (purple arrow), it cascades down through a series of levels to the bottom of the excited singlet state (wavy purple arrow). Then four outcomes are possible. The molecule can return to the ground state by fluorescence (green arrow), it can cross over to the triplet state and return from there to the ground state by phosphorescence (red arrow), it can cross over to the ground state and cascade down without radiating (wavy yellow arrow), or a bond can break (orange arrow).

In a process called **photodissociation,** when the energy of a visible photon is added to the molecule, the rather weak O—O bond in $H_2O_2$ breaks apart. Another type of photochemical reaction involves changes in molecular structure without the complete breaking of bonds. Molecules such as those of *trans*-2-butene (Fig. 14-29) absorb ultraviolet light by the excitation of an electron from a $\pi$ to a $\pi^*$ molecular orbital, just as ethylene does. In the excited electronic state of *trans*-2-butene, the carbon–carbon double bond is effectively reduced to a single bond, and the one $CH_3$ group can rotate relative to the other to form *cis*-2-butene. The ultraviolet light has led to the formation of a geometric isomer of the original molecule.

### New Developments in Spectroscopy

Modern molecular spectroscopy aims not only to determine energy levels (and, from them, information about molecular structure and bonding) but to give us an understanding of the dynamics of excited molecules—a detailed series of events that may include nuclear motion, light emission, and electron transfer. Such an understanding requires **time-resolved spectroscopy,** in which short light pulses are used to excite molecules and the emitted photons are identified not only by wavelength but also by the time delay until each appears. Light pulses with durations of picoseconds ($10^{-12}$ s) and even femtoseconds ($10^{-15}$ s) can be made with modern lasers, and the very rapid evolution of excited molecules is monitored by measuring the absorption of a second light pulse, which arrives at a variable time after the initial pulse, or by counting photons emitted with time (Fig. 15-12).

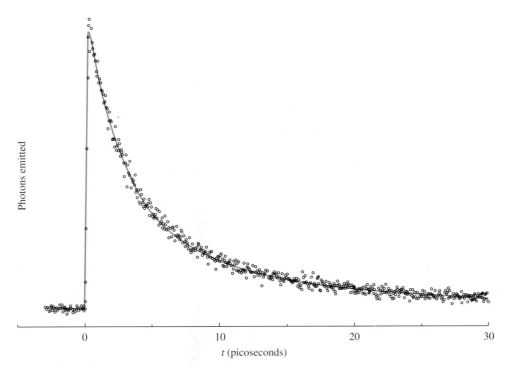

Photons emitted

*t* (picoseconds)

## Figure 15-12

The initial chemical step in photosynthesis, as carried out by the purple bacterium
*R. sphaeroides,* is the transfer of an electron from the excited state of a pair of chlorophyll molecules to a pheophytin molecule located 17 Å away. (Section 15-3 describes the
structure of the reaction center in greater detail.) This electron transfer occurs very rapidly and with essentially 100% efficiency. Ultrafast spectroscopy monitors the electron
transfer by detecting the disappearance of the excited state of the pair of chlorophyll
molecules through the decay of its fluorescence. The data shown here reveal that the
electron moves to the pheophytin with a half-life of 2.8 picoseconds ($2.8 \times 10^{-12}$ s).
(Graham Fleming)

A promising approach to studying the spectra of complex molecules involves
cooling them in supersonic jets. A typical, moderate-size molecule in solution has a
rather broad and featureless spectrum in the visible and ultraviolet regions (Fig.
15-13, red curve) because there are many vibrational and rotational levels, and
interactions with solvent molecules at different distances shift their energies, giving
a blurred overall picture. (The same blurring of rotational levels within a particular
vibration state produced the broad bands in Fig. 15-4.) If a molecule can be vaporized into the gas state, the effects of solvent are removed and additional structure
appears (Fig. 15-13, blue curve); however, molecular collisions and the numerous
thermally occupied energy levels still give bands that are too broad to be interpreted
quantitatively. To simplify the spectrum, it would be desirable to cool the molecule
to low temperatures so that very few levels were occupied. In general, however, the
gas would then condense to a liquid or solid and defeat the purpose of isolating the
molecules. In a supersonic jet, molecules, along with a carrier gas such as helium,

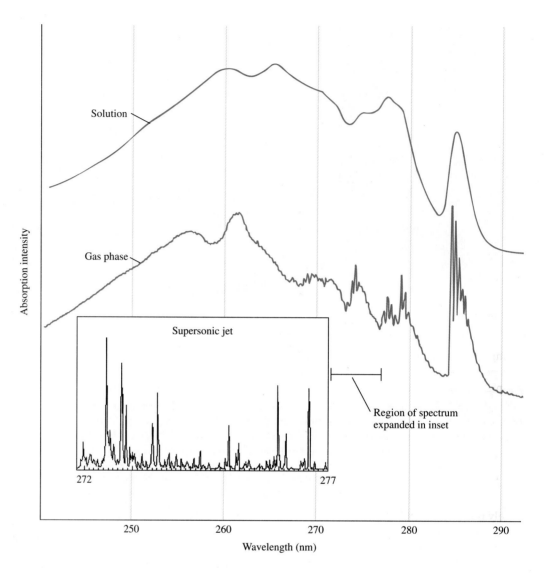

**Figure 15-13**

A portion of the absorption spectrum of indole, $C_8H_7N$, in the ultraviolet region. In heptane solution (red curve) the spectrum has broad and feature-less bands, while in the gas phase (blue curve) additional structure appears. Cooling indole molecules in a supersonic jet reveals a large number of sharp absorption lines (see inset, which covers a very small range of wavelengths). (D. Levy, H. Strickland)

are forced through a nozzle at supersonic speeds. The gas expands adiabatically, which cools the molecules (Section 8-5). Collisions redistribute molecules into lower energy levels, until downstream in the jet the effective temperatures are quite low and much more structure appears in the spectra (Fig. 15-13, inset).

Surprising as it may seem, another way to simplify spectra is to examine one molecule at a time. Chemists have studied the spectra of molecules that are present in low concentrations in a low-temperature solid. Working with thin samples and focused laser beams, they can probe a region of the sample with a relatively small number of those molecules. Using extraordinarily sensitive detectors, they measure signals from one molecule at a time (Fig. 15-14). The chemists can even see how a signal changes with time; as the molecules near the probed molecule move, the frequency of the probed molecule's transition shifts as well. Thus, chemists not only can ''see'' single atoms with scanning tunneling microscopes (Fig. 1-7), they can also probe the spectra of single molecules.

**Figure 15-14**
Images of the fluorescence of single molecules. The height of each peak gives the intensity of light emitted by that molecule. The axis into the page is spatial position (scanned over a distance of $4 \times 10^{-5}$ m), and the horizontal axis scans a narrow range of wavelengths for the exciting light. (W. E. Moerner, IBM)

## 15-3 PHOTOSYNTHESIS

Living species harvest light and store its energy by carrying out chemical reactions with positive free energy changes. This process of **photosynthesis** has transformed our planet and permitted the evolution of human life. Before we consider green plants (which provide most of the photosynthetic energy storage today), let us examine the reaction mechanisms in existent photosynthetic bacteria that are thought to have evolved earlier, and are considerably simpler, than plants. These bacteria have recently yielded up some important secrets in experimental investigations.

The purple photosynthetic bacteria include the species *Rhodobacter sphaeroides* and *Rhodopseudomonas viridis*. Although these two differ in the details of their chemical makeup, the broad features of their mechanisms appear to be closely related. Their purple color reflects their absorption of light predominantly in the red region of the spectrum, with wavelengths up to 925 nm. *R. sphaeroides* and *R. viridis* use a variety of pigment molecules, including derivatives of chlorophyll (Fig. 15-15b), as "antennas" to absorb photons that reach their surface. The antenna

**Figure 15-15**
Structures of (*a*) porphine, (*b*) bacteriochlorophyll, and (*c*) heme. Carbon atoms are understood to lie at intersections of bonds where no other atom is indicated. In bacteriopheophytin, the central $Mg^{2+}$ ion shown in part *b* is replaced by two $H^+$ ions, producing a ring like that in part *a*. Cytochromes are proteins containing heme groups at their centers.

(*a*) **Porphine**

(*b*) **Bacteriochlorophyll**

(*c*) **Heme**

**Figure 15-16**

Computer-generated diagrams of the reaction center of the purple bacterium *R. sphaeroides,* based on x-ray diffraction data. (*a*) The reaction center has been sliced through the middle to show the donor-acceptor molecules surrounded by protein (dark green, red, and blue portions). (*b*) The protein has been removed entirely. The four bacteriochlorophyll molecules are shown in white (for the special pair) and purple (for the other two); bacteriopheophytin molecules are light green; ubiquinone molecules are yellow. An iron(II) ion is shown between the ubiquinone molecules. The A branch is on the right, and the B branch on the left. (© J. Norris)

molecules store the light energy through promotion of electrons to excited singlet states (Fig. 15-6), then transfer this energy rapidly to other molecules nearby. In such a transfer, the initially excited molecule returns to its ground state while a neighboring molecule reaches an excited singlet state. Within about $10^{-10}$ s (100 picoseconds), the energy migrates to the **reaction center,** where it is trapped and is ready to be used to carry out chemical reactions.

The initial trapping site in the reaction center consists of a **special pair** of bacteriochlorophyll molecules. The existence of this special pair was postulated in the 1960s by scientists who used magnetic resonance and absorption spectroscopy to probe the dynamics of the reaction center. Dramatic confirmation came in 1984 when three German chemists (Hartmut Michel, Johann Deisenhofer, and Robert Huber) succeeded in preparing crystals of the membrane-bound reaction-center protein in *R. viridis* and determining its chemical structure by x-ray diffraction (Section 17-1). Not only did they locate the bacteriochlorophyll molecules that form the special pair, they also identified the positions of the molecules involved in the subsequent steps in photosynthesis.

Figure 15-16 shows the reaction center of the related bacterium *R. sphaeroides*; its structure is much more evident when the surrounding protein is stripped away. This complex consists of four bacteriochlorophyll molecules (the special pair and two others), two bacteriopheophytin molecules (which are bacteriochlorophyll molecules in which the central Mg ion is replaced by two hydrogen ions), two ubiquinone molecules (Fig. 15-17), and an iron(II) ion. Interestingly, these molecules are

(*b*)

(*a*)

**Figure 15-17**

The structures of (a) ubiquinone (UQ) and (b) its reduced form, UQH$_2$. Note the tail that contains a C$_5$H$_8$ group repeated ten times.

arranged in an almost perfectly symmetric fashion, with two branches (called A and B). However, the surrounding protein breaks the symmetry of the branches and causes the energy flow to pass almost entirely through the A branch (on the right in Fig. 15-16). Any energy passing through the B branch is too little to detect.

Time-resolved spectroscopy (Figure 15-12) has shown that within about 2.8 picoseconds ($2.8 \times 10^{-12}$ s) an electron is transferred from the special pair to the bacteriopheophytin molecule in the A branch, creating an anion on that site and leaving a cation on the special pair. Within 200 picoseconds, the electron reaches the ubiquinone molecule in the A branch and then is transferred across to the other ubiquinone molecule, in the B branch. The special pair picks up an electron from the iron atom in the heme group (Fig. 15-15c) on a cytochrome protein outside the membrane and is reduced back to its original state. It then absorbs a second photon and transfers a second electron to the same B-branch ubiquinone molecule. This doubly reduced ubiquinone is less tightly bound to the reaction-center protein and moves away, being replaced by an unreduced ubiquinone molecule. The reduced ubiquinone gives up its electrons and hydrogen ions to another protein, which releases H$^+$ ions outside the cell membrane. The resulting hydrogen ion (proton) gradient across the cell wall stores energy (it is equivalent to the concentration cell considered in Chapter 10) and can carry out chemical reactions needed by the cell, ultimately transferring hydrogen ions back into the cell to close the cycle.

The chemical reactions at the reaction center can be summarized by the following six steps:

**1.** $(BChl)_2 + UQ \xrightarrow{\text{light}} (BChl)_2^+ + UQ^-$

**2.** $(BChl)_2^+ + Cyt \longrightarrow (BChl)_2 + Cyt^+$

**3.** $(BChl)_2 + UQ^- \xrightarrow{\text{light}} (BChl)_2^+ + UQ^{2-}$

**4.** $(BChl)_2^+ + Cyt \longrightarrow (BChl)_2 + Cyt^+$

**5.** $UQ^{2-} + 2\,H_{in}^+ \longrightarrow UQH_2$

**6.** $UQH_2 + 2\,H_{in}^+ + 2\,Cyt^+ \longrightarrow UQ + 4\,H_{out}^+ + 2\,Cyt$

Overall: $4\,H_{in}^+ \longrightarrow 4\,H_{out}^+$

Here $(BChl)_2$ stands for the special pair of bacteriochlorophyll molecules, UQ for ubiquinone, and Cyt for the cytochrome protein. Steps 1 and 3 involve excitation of bacteriochlorophyll and transfer of a pair of electrons to a ubiquinone molecule; steps 2 and 4 restore the special pair to its initial state; steps 5 and 6 transfer hydrogen ions outside the membrane wall and restore the cytochrome to its reduced form. The net reaction is the light-driven movement of hydrogen ions from inside to outside the cell.

Purple bacteria have only a single type of reaction center, but green plants have two types, referred to as photosystem I (PS I) and photosystem II (PS II). These absorb light at somewhat different wavelengths, allowing plants to use light energy from the sun more efficiently than purple bacteria do. Moreover, the two photosystems are linked to one another, so that the excitation of PS II is followed by the excitation of PS I with a photon of longer wavelength. The cumulative effect of the two photons absorbed in series is the storage of a larger amount of energy than is available to purple bacteria. Whereas the bacteria use $H_2S$ and other reduced compounds as sources of hydrogen, green plants are able to split abundant water molecules and in the process give off oxygen, which is, of course, vital to the existence and survival of animal species on the planet.

## 15-4  ATMOSPHERIC PHOTOCHEMISTRY

The photochemical reactions in green plants and photosynthetic bacteria are not the only reactions induced by sunlight. The ability of light to cause chemical reactions is also apparent in the earth's atmosphere. In fact, photosynthesis and atmospheric chemistry are intimately connected. Reactions in the atmosphere determine the intensities and wavelengths of the light that reach the earth's surface to be harvested by living species. At the same time, oxygen, the gas produced by green-plant photosynthesis, has transformed the atmosphere; before photosynthesis began, there was almost no free oxygen at the earth's surface.

Although the chemical composition summarized in Table 3-1 describes the average makeup of the portion of the atmosphere closest to the earth's surface, it does not do justice to the variation in chemical properties with height, to the dramatic role of local fluctuations in trace gases, and to the dynamics underlying observed average concentrations. The atmosphere is a complex chemical system that is far from equilibrium. Its properties are determined by an intricate combination of thermodynamic and kinetic factors. It is a multilayered structure, bathed in radiation from the sun and interacting at the bottom with the oceans and land masses. At least four layers are identifiable, each with a characteristic variation of temperature (Fig. 15-18). In the outer two layers (the **thermosphere** and the **mesosphere**) atmospheric density is low, and the intense radiation from the sun causes extensive ionization of the particles that are present. The third layer, the **stratosphere,** is the region from 12 to 50 km (approximately) above the earth's surface. The **troposphere** is the lowest region, extending 12 km out from the earth's surface. In the troposphere, warmer air lies beneath cooler air. This is a dynamically unstable situation because warm air is less dense and tends to rise, and so convection takes place, mixing the gases in the troposphere and determining the weather. In the stratosphere, the temperature increases with height and there is little vertical mixing from convection. Mixing across the borders between the layers is also slow, and so most chemical processes in each layer can be described separately.

Atmospheric chemistry dates back to the 18th century. Cavendish, Priestley, Lavoisier, and Ramsay were the first scientists to study the composition of the atmosphere. In recent years, atmospheric chemistry has developed in two different but related directions. First, the sensitivity of chemical analysis has greatly improved, and analyses for substances at concentrations below the part-per-billion

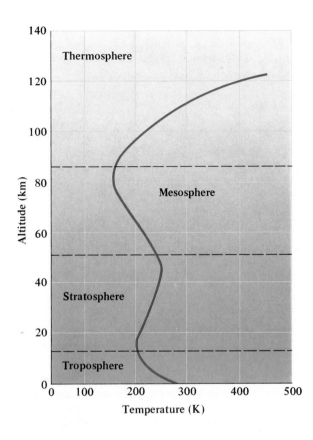

**Figure 15-18**

The variation of the temperature of the atmosphere with altitude, showing the layered structure of the atmosphere.

(ppb) level are now carried out routinely. Airplanes and satellites enable scientists to map the global distributions of trace substances. Second, advances in gas-phase chemical kinetics have led to a better quantitative understanding of the ways in which substances in the atmosphere react with one another and with light. Much of the impetus for these studies of atmospheric chemistry comes from concern about the effect of air pollution on life.

## Stratospheric Chemistry

As Figure 13-5 shows, the sun emits light over a broad range of wavelengths, with the highest intensity at about 500 nm, in the visible region of the spectrum. The intensities at wavelengths down to 100 nm in the ultraviolet region are quite substantial, and because the energy $h\nu$ carried by a photon is inversely proportional to the wavelength $\lambda$ ($h\nu = hc/\lambda$), the ultraviolet photons carry much more energy than do photons of visible light. If these photons could penetrate to the earth's surface in substantial numbers, they could do a great deal of damage to living organisms. The outer portions of the atmosphere (especially the thermosphere) play a crucial role in preventing this penetration through the photodissociation of oxygen molecules:

$$O_2 + h\nu \longrightarrow 2\,O$$

where "$h\nu$" symbolizes a photon. This reaction reduces the number of high-energy photons reaching the lower parts of the atmosphere, especially those with wavelengths less than 200 nm.

## Example 15-3

The bond dissociation energy of $O_2$ is 496 kJ mol$^{-1}$. Calculate the maximum wavelength of light that can photodissociate an oxygen molecule.

### Solution

Because 496 kJ dissociates 1.00 mol of $O_2$ molecules, the energy to dissociate one molecule is found by dividing by Avogadro's number:

$$\frac{496 \times 10^3 \text{ J mol}^{-1}}{6.022 \times 10^{23} \text{ mol}^{-1}} = 8.24 \times 10^{-19} \text{ J}$$

A photon carrying this energy has a wavelength $\lambda$, given by

$$8.24 \times 10^{-19} \text{ J} = h\nu = \frac{hc}{\lambda}$$

so that

$$\lambda = \frac{hc}{8.24 \times 10^{-19} \text{ J}} = \frac{(6.626 \times 10^{-34} \text{ J s})(2.998 \times 10^8 \text{ m s}^{-1})}{8.24 \times 10^{-19} \text{ J}}$$

$$= 2.41 \times 10^{-7} \text{ m} = 241 \text{ nm}$$

Any photons with wavelengths *shorter* than this are energetic enough to dissociate oxygen molecules. Those with wavelengths shorter than 200 nm are the most efficient in causing photodissociation.

*Related Problems:* 39, 40

---

Rather few photons with wavelengths less than 200 nm can penetrate to the stratosphere, but those that do establish a small concentration of oxygen atoms in that layer. These atoms can collide with the much more prevalent oxygen molecules to form highly excited molecules of ozone, symbolized by $O_3^*$:

$$O + O_2 \rightleftharpoons O_3^*$$

The excited-state $O_3^*$ can dissociate in a unimolecular reaction back to O and $O_2$, as indicated by the reverse arrow in this equilibrium. Alternatively, if another atom or molecule collides with it soon enough, some of its excess energy can be transferred to that atom or molecule, a process represented by

$$O_3^* + M \longrightarrow O_3 + M$$

where M stands for the atom or molecule with which the $O_3^*$ collides (the most likely are oxygen and nitrogen molecules, because they are the most abundant species in the atmosphere). The net effect of these two reactions is to produce a small concentration of ozone in the stratosphere.

Under laboratory conditions, ozone is an unstable compound. Its conversion to oxygen is thermodynamically spontaneous

$$O_3(g) \longrightarrow \tfrac{3}{2} O_2(g) \qquad\qquad \Delta G° = -163 \text{ kJ}$$

but takes place quite slowly in the absence of light. In the stratosphere, ozone photodissociates readily to $O_2$ and O in a reaction that requires about 106 kJ per mole of ozone, much less than the dissociation of $O_2$:

$$O_3 + h\nu \longrightarrow O_2 + O$$

This process occurs most efficiently for wavelengths between 200 and 350 nm. The energy of light in this range of wavelengths is too small to be absorbed by molecular oxygen but quite large enough to damage organisms at the earth's surface. The ozone layer shields the earth's surface from 200- to 350-nm ultraviolet radiation coming from the sun. The balance between formation and photodissociation leads to a steady-state concentration of more than $10^{15}$ molecules of ozone per liter in the stratosphere.

It is important to know whether molecules being released in the lower atmosphere can reach the stratosphere and affect its amount of ozone. Certain types of air pollution give rise to *radicals* that catalyze ozone depletion. A radical is a chemical species that contains an odd (unpaired) electron, and it is usually formed by the breaking of a covalent bond to form a pair of neutral species. One pressing concern involves chlorofluorocarbons (CFCs)—compounds of chlorine, fluorine, and carbon that are used as refrigerants and as propellants in some aerosol sprays. CFCs are nonreactive at sea level but can photodissociate in the stratosphere:

$$CCl_2F_2 + h\nu \longrightarrow CClF_2 + Cl$$

The atomic chlorine thus released soon reacts with O atoms or $O_3$ to give ClO:

$$Cl + O \longrightarrow ClO$$

$$Cl + O_3 \longrightarrow ClO + O_2$$

The ClO radical is the immediate culprit in the destruction of stratospheric ozone. Local increases in ClO concentration are directly correlated with decreases in $O_3$ concentration. ClO catalyzes the destruction of ozone, probably by the mechanism

$$2\,ClO + M \longrightarrow ClOOCl + M$$

$$ClOOCl + h\nu \longrightarrow ClOO + Cl$$

$$ClOO + M \longrightarrow Cl + O_2 + M$$

$$\underline{2 \times (Cl + O_3 \longrightarrow ClO + O_2)}$$

net reaction:   $2\,O_3 \longrightarrow 3\,O_2$

where M stands for $N_2$ and $O_2$ molecules. This catalytic cycle is ordinarily disrupted rather quickly as Cl reacts with other stratospheric species to form less reactive "reservoir molecules" such as HCl and $ClONO_2$. The lack of mixing in the stratosphere keeps the reservoir molecules around for long periods, however.

It is believed that atomic Cl escapes its reservoirs through heterogeneous reactions on the polar stratospheric clouds that form during the intense Antarctic winter. The escaped Cl gives rise to the annual "ozone holes" above the Antarctic. During these episodes, more than 70% of the total column ozone is depleted before the values rise again (Fig. 15-19). Researchers have demonstrated that the extent of depletion above the Antarctic is increasing, and measurements have shown smaller but still serious depletions over other parts of the globe. International agreement (the Montreal Protocol) has led to major reductions in the production of chlorofluorocar-

METEOR 3/TOMS October 2, 1994

**Figure 15-19**

This color image shows total ozone amounts in the Southern Hemisphere for Oct. 2, 1994. The Dobson unit measures the stratospheric concentration of ozone. Preliminary minimum ozone values for this day were below 100 Dobson Units, while the ozone hole area was at its maximum extent in a region near the South Pole. The total ozone values were measured by the Total Ozone Mapping Spectrometer (TOMS) instrument aboard the Russian Meteor-3 satellite, launched in August 1991. The color scale at the bottom of the figure shows the total ozone values. The Antarctic ozone hole is located in the center of the figure and is indicated by all colors contained within the blue boundary. Prior to 1975, values of total ozone between 250 and 300 Dobson units (blue to green) were normal, indicating a decrease of about 150 Dobson Units. (NASA) bons, with a complete phase-out planned by 1996. Even if this goal is met, stratospheric ozone concentrations will continue to decrease for some time to come as already-released chlorofluorocarbons mix upward into the stratosphere.

### Tropospheric Chemistry

The troposphere is the part of the atmosphere in contact with the earth's surface. It is therefore most directly and immediately influenced by human activities, especially by the gases or small particles put into the air by automobiles, power plants, and factories. Some pollutants have long lifetimes and are spread fairly evenly over the earth's surface; others attain large concentrations only around particular cities or industrial areas.

The oxides of nitrogen are major air pollutants. Their persistence in the atmosphere demonstrates the importance of kinetics, as opposed to thermodynamics, in the chemistry of the atmosphere. All of the oxides of nitrogen are thermodynamically unstable with respect to the elements at 25°C, as shown by their positive free energies of formation. They form whenever air is heated to high enough temperatures, either in an industrial process or in the engine of a car. They accumulate to concentrations much higher than equilibrium because they decompose slowly. The interconversion among $NO$, $NO_2$, and $N_2O_4$ is rapid and strongly temperature-dependent, and so they are generally grouped together as "$NO_x$" in pollution reports. Photochemical smog is formed by the action of light on nitrogen dioxide, followed by reaction to produce ozone:

$$NO_2 + h\nu \longrightarrow NO + O$$

$$O + O_2 + M \longrightarrow O_3 + M \qquad (M = N_2 \text{ or } O_2)$$

It may seem paradoxical that concern about ozone in the stratosphere involves its potential *depletion*, whereas concern about the troposphere involves the *production* of ozone. Although ozone is beneficial in preventing radiation from penetrating to the earth's surface, it is quite harmful in direct contact with organisms because of its strong oxidizing power; levels of 10 to 15 ppm are sufficient to kill small mammals, and a concentration as low as 3 ppm is enough to trigger an "ozone alert." In addition, ozone reacts with incompletely oxidized organic compounds from gasoline and with nitrogen oxides in the air to produce harmful irritants such as methyl nitrate ($CH_3NO_3$).

The oxides of sulfur create global pollution problems because they have longer lifetimes in the atmosphere than the oxides of nitrogen. Some of the $SO_2$ and $SO_3$ in the air originates from biological processes and from volcanoes, but much comes from the oxidation of sulfur in petroleum and in coal that is burned for fuel. If the sulfur is not removed from the fuel or the exhaust gas, $SO_2$ enters the atmosphere as a stable but reactive pollutant. Further oxidation by radicals leads to sulfur trioxide:

$$SO_2 + OH \longrightarrow SO_2OH$$

$$SO_2OH + O_2 \longrightarrow SO_3 + OOH$$

$$OOH \longrightarrow O + OH$$

The atomic oxygen thus produced can react with molecular oxygen to form ozone. **Acid rain** results from the reaction of $NO_2$ and $SO_3$ with hydroxyl radical and water vapor in the air to form nitric acid ($HNO_3$) and sulfuric acid ($H_2SO_4$). The acids dissolve in water and return to the earth in the rain.

## The Greenhouse Effect

As we have seen, most of the short-wavelength photons from the sun are absorbed by the outer atmosphere and do not reach the earth's surface. The radiation that does reach the earth is in balance with reradiation into space and helps to maintain a livable temperature. Certain gases in the troposphere play a crucial role in this balance, because they absorb infrared radiation emitted by the warm surface of the earth rather than letting it pass out to space. In winter, the temperature does not fall as low on cloudy nights as it does on clear nights because the water vapor in the clouds provides a thermal blanket, absorbing outgoing radiation with wavelengths near 20,000 nm and reradiating it to the earth's surface.

Two other gases that absorb infrared radiation to significant extents are carbon dioxide and methane. Both are uniformly distributed in fairly low concentrations through the troposphere, but their concentrations have increased steadily over the 200 to 300 years since the beginning of the industrial revolution (Fig. 15-20). The tremendous increase in the burning of fossil fuels for energy is a major contributor to this trend. It is estimated that by the year 2050 the concentration of $CO_2$ in the atmosphere will be double the premodern value. Chlorofluorocarbons also absorb infrared radiation, especially over some key wavelength ranges left open by $CO_2$ and water vapor.

Such changes are viewed with alarm because of a phenomenon called the **greenhouse effect**—a global increase in average surface temperature that occurs if heat given off by the earth's surface is prevented from escaping by higher concentrations of gases that absorb and reradiate infrared radiation. An increase in average surface temperature of 2 to 5°C over the next century would have major repercussions, including the melting of some polar ice, a consequent increase in the level of the oceans, and the conversion of arable land to desert. To prevent these undesirable changes, it will be necessary to develop new energy sources not based on the burning of fossil fuels by early in the 21st century.

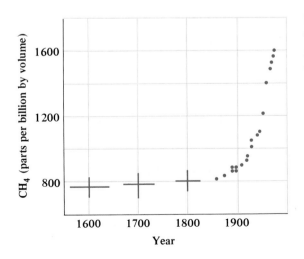

**Figure 15-20**

The concentration of methane in the lower atmosphere has doubled since 1600. Note the dramatic increase in the last 100 years.

**PRACTICE EXERCISE**

A chemical plant uses bromine to manufacture brominated flame retardants for plastics and fibers. (Ethyl Corporation)

## Bromine

Elemental bromine is a brownish red liquid that was first isolated in 1826. It is currently produced by oxidation of bromide ion in natural brines with elemental chlorine.

(a) Bromine compounds have been known and used for centuries. The deep purple color that symbolized imperial power in ancient Rome originated with the compound dibromoindigo, which was extracted in tiny quantities from purple snails (about 8000 snails per gram of compound). What color and maximum wavelength of *absorbed* light would give a deep purple (violet) color?

(b) What excited electronic state is responsible for the brownish red color of bromine? Refer to Figures 14-16 and 15-6.

(c) The two naturally occurring isotopes of bromine are $^{79}$Br and $^{81}$Br, with masses of 78.918 and 80.916 u, respectively. The wavelength of the $J = 0$ to $J = 1$ rotational transition in $^{79}$Br$^{81}$Br is measured to be 6.18 cm. Use this information to calculate the bond length in the $Br_2$ molecule, and compare the result with that listed in Table 14-1.

(d) The wavelength of the vibrational transition in the $^{79}$Br$^{81}$Br molecule is $3.09 \times 10^{-5}$ m. Calculate the force constant for the bond in this molecule.

(e) The action of light on bromine compounds released into the air (such as by leaded gasoline) causes the formation of the BrO radical. Give the bond order of this species by comparing it to the related radical OF.

(f) There is concern that synthetic bromine-containing compounds, in addition to chlorofluorocarbons, are helping to destroy ozone in the stratosphere. The BrO (see part (e) can take part with ClO in the following catalytic cycle:

$$Cl + O_3 \longrightarrow ClO + O_2$$

$$Br + O_3 \longrightarrow BrO + O_2$$

$$ClO + BrO \longrightarrow Cl + Br + O_2$$

Write the overall equation for this cycle.

### Answers

(a) Yellow light, near 530 nm

(b) The lowest-energy excited state, which arises from the excitation of an electron from the filled $\pi_{4p}^*$ orbital to the unfilled $\sigma_{4p_z}^*$ orbital

(c) 2.28 Å (from Table 14-1: 2.286 Å)

(d) 247 J m$^{-2}$

(e) $\frac{3}{2}$ order

(f) Overall: $2\,O_3 \longrightarrow 3\,O_2$

**CONCEPTS & SKILLS**

After studying this chapter and working the problems that follow, you should be able to

1. Relate the moments of inertia, bond lengths, and vibrational force constants of diatomic molecules to their rotational and vibrational spectra (Section 15-1, problems 1–8).

**2.** Discuss the Boltzmann distribution of population among possible molecular quantum states (Section 15-1, problems 9–10).

**3.** Describe the preparation of excited electronic states by absorption of radiation, and the subsequent flow of energy (Section 15-2, problems 11–16).

**4.** Discuss several new techniques in molecular spectroscopy and the purposes they serve (Section 15-2).

**5.** Explain the mechanism by which photosynthetic bacteria and green plants convert light energy to chemical energy (Section 15-3, problems 21–22).

**6.** Indicate processes in the atmosphere that are beneficial and those that are potentially damaging to the ecosystem (Section 15-4, problems 23–26).

## Molecular Spectroscopy

**1.** Use data from Tables 12-1 and 14-1 to predict the energy spacing between the ground state and the first excited rotational state of the $^{14}N^{16}O$ molecule.

**2.** Use data from Tables 12-1 and 14-1 to predict the energy spacing between the ground state and the first excited rotational state of the $^{1}H^{19}F$ molecule.

**3.** The first three absorption lines in the pure rotational spectrum of gaseous $^{12}C^{16}O$ are found to have the frequencies $1.15 \times 10^{11}$, $2.30 \times 10^{11}$, and $3.46 \times 10^{11}$ s$^{-1}$. Calculate:
(a) The moment of inertia $I$ of CO (in kg m$^2$)
(b) The energies of the $J = 1$, $J = 2$, and $J = 3$ rotational levels of CO, measured from the $J = 0$ state (in joules)
(c) The C—O bond length (in angstroms)

**4.** Four consecutive absorption lines in the pure rotational spectrum of gaseous $^{1}H^{35}Cl$ are found to have the frequencies $2.50 \times 10^{12}$, $3.12 \times 10^{12}$, $3.74 \times 10^{12}$, and $4.37 \times 10^{12}$ s$^{-1}$. Calculate:
(a) The moment of inertia $I$ of HCl (in kg m$^2$)
(b) The energies of the $J = 1$, $J = 2$, and $J = 3$ rotational levels of HCl, measured from the $J = 0$ state (in joules)
(c) The H—Cl bond length (in angstroms)
(d) The initial and final $J$ states for the observed absorption lines

**5.** The Li$_2$ molecule ($^7$Li isotope) shows a very weak infrared line in its vibrational spectrum at a wavelength of $2.85 \times 10^{-5}$ m. Calculate the force constant for the Li$_2$ molecule.

**6.** The Na$_2$ molecule ($^{23}$Na isotope) shows a very weak infrared line in its vibrational spectrum at a wavelength of $6.28 \times 10^{-5}$ m. Calculate the force constant for the Na$_2$ molecule, and compare your result with that of problem 5. Give a reason for any difference.

**7.** The "signature" infrared absorption that indicates the presence of a C—H stretching motion in a molecule occurs at wavelengths near $3.4 \times 10^{-6}$ m. Use this information to estimate the force constant of the C—H stretch. Take the reduced mass in this motion to be approximately equal to the mass of

the hydrogen atom (a good approximation when the H atom is attached to a heavy group).

**8.** Repeat the calculation of the preceding problem for the N—H stretch, where absorption occurs near $2.9 \times 10^{-6}$ m. Which bond is stiffer: N—H or C—H?

**9.** Estimate the ratio of the number of molecules in the first excited vibrational state of the molecule N$_2$ to the number in the ground state, at a temperature of 450 K. The vibrational frequency of N$_2$ is $7.07 \times 10^{13}$ s$^{-1}$.

**10.** The vibrational frequency of the ICl molecule is $1.15 \times 10^{13}$ s$^{-1}$. For every million ($1.00 \times 10^6$) molecules in the ground vibrational state, how many will be in the first excited vibrational state at a temperature of 300 K?

## Excited Electronic States

**11.** Suppose that an ethylene molecule gains an additional electron to give the C$_2$H$_4^-$ ion. Will the bond order of the carbon–carbon bond increase or decrease? Explain.

**12.** Suppose that an ethylene molecule is ionized by a photon to give the C$_2$H$_4^+$ ion. Will the bond order of the carbon–carbon bond increase or decrease? Explain.

**13.** The color of the dye "indanthrene brilliant orange" is evident from its name. In what wavelength range would you expect to find the maximum in the absorption spectrum of this molecule? Refer to the color spectrum in Figure 13-4.

**14.** In what wavelength range would you expect to find the maximum in the absorption spectrum of the dye "crystal violet"?

**15.** The structure of the molecule cyclohexene is

Does the absorption of ultraviolet light by cyclohexene occur at longer or at shorter wavelengths than the absorption by benzene? Explain.

16. The naphthalene molecule has a structure that corresponds to two benzene molecules fused together:

The $\pi$ electrons in this molecule are delocalized over the entire molecule. The wavelength of maximum absorption in benzene is 255 nm. Will the corresponding wavelength in naphthalene be shorter or longer than 255 nm?

17. To satisfy the octet rule, fullerene must have double bonds. How many? Give a simple rule for one way of placing them in the structure shown in Figure 15-8$a$.

18. It has been suggested that a compound of formula $C_{12}B_{24}N_{24}$ might exist and have a structure like that of $C_{60}$ (buckminsterfullerene).
    (a) Explain the logic of this suggestion by comparing the number of valence electrons in $C_{60}$ and $C_{12}B_{24}N_{24}$.
    (b) Propose the most symmetric pattern of C, B, and N atoms in $C_{12}B_{24}N_{24}$ to occupy the 60 atom sites in the buckminsterfullerene structure. Where could the double bonds be placed in such a structure?

19. Use data from Table 14-1 to give an upper bound on the wavelengths of light that are capable of dissociating a molecule of ClF. (*Hint:* Remember to convert from bond enthalpy to bond energy.)

20. Use data from Table 14-1 to give an upper bound on the wavelengths of light that are capable of dissociating a molecule of ICl. (*Hint:* Remember to convert from bond enthalpy to bond energy.)

## Photosynthesis

21. One way in which photosynthetic bacteria store chemical energy is through the conversion of a compound called adenosine diphosphate (ADP), together with hydrogen phosphate ion, to adenosine triphosphate (ATP):

$$ADP^{3-} + HPO_4^{2-} + H_3O^+ \longrightarrow ATP^{4-} + 2\,H_2O$$
$$\Delta G = +34.5 \text{ kJ (pH 7)}$$

Suppose some chlorophyll molecules absorb 1.00 mol of photons of blue light with wavelength 430 nm. If *all* this energy could be used to convert ADP to ATP at room conditions and pH 7, how many molecules of ATP would be produced per photon absorbed? (The actual number is smaller because the conversion is not 100% efficient.)

22. Repeat the calculation of the preceding problem for red light with wavelength 700 nm.

## Atmospheric Photochemistry

23. The bond dissociation energy of a typical C—F bond in a chlorofluorocarbon is approximately 440 kJ $mol^{-1}$. Calculate the maximum wavelength of light that can photodissociate a molecule of $CCl_2F_2$, breaking such a C—F bond.

24. The bond dissociation energy of a typical C—Cl bond in a chlorofluorocarbon is approximately 330 kJ $mol^{-1}$. Calculate the maximum wavelength of light that can photodissociate a molecule of $CCl_2F_2$, breaking such a C—Cl bond.

25. Draw a Lewis structure(s) for the ozone molecule ($O_3$). Determine the steric number and hybridization of the central oxygen atom, and identify the molecular geometry. Describe the nature of the $\pi$ bonds and give the bond order of the O—O bonds in ozone.

26. The compounds carbon dioxide ($CO_2$) and sulfur dioxide ($SO_2$) are formed by the burning of coal. Their apparently similar formulas mask underlying differences in molecular structure. Determine the shapes of these two types of molecules, identify the hybridization at the central atom of each, and compare the natures of their $\pi$ bonds.

## Additional Problems

27. What are the moments of inertia of $^1H^{19}F$ and $^1H^{81}Br$, expressed in kg $m^2$? Compute the spacings $\nu = \Delta E/h$ of the rotational states, in $s^{-1}$, between $J = 0$ and 1 and between $J = 1$ and 2. Explain, in one sentence, why the large change in mass from 19 to 81 causes only a small change in rotational energy differences.

28. The average bond length of a molecule can change slightly with vibrational state. In $^{23}Na^{35}Cl$, the frequency of light absorbed in a change from the $J = 1$ to the $J = 2$ rotational state in the ground vibrational state ($v = 0$) was measured to be $\nu = 2.60511 \times 10^{10}$ $s^{-1}$, and that for a change from $J = 1$ to $J = 2$ in the first excited vibrational state ($v = 1$) was $\nu = 2.58576 \times 10^{10}$ $s^{-1}$. Calculate the average bond lengths of NaCl in these two vibrational states, taking the relative atomic mass of $^{23}Na$ to be 22.9898 and that of $^{35}Cl$ to be 34.9689.

29. The vibrational frequencies of $^{23}Na^1H$, $^{23}Na^{35}Cl$, and $^{23}Na^{127}I$ are $3.51 \times 10^{13}$ $s^{-1}$, $1.10 \times 10^{13}$ $s^{-1}$, and $0.773 \times 10^{13}$ $s^{-1}$, respectively. Their bond lengths are 1.89, 2.36, and 2.71 Å. What are their reduced masses? What are their force constants? If NaH and NaD have the same force constant, what is the vibrational frequency of NaD? D is $^2H$.

30. Recall that nuclear spin states in nuclear magnetic resonance are typically separated by energies of $2 \times 10^{-5}$ to $2 \times 10^{-4}$ kJ $mol^{-1}$. What are the ratios of occupation probability between a pair of such levels at thermal equilibrium and a temperature of 25°C?

31. The vibrational temperature of a molecule prepared in a supersonic jet can be estimated from the observed populations of its vibrational levels, assuming a Boltzmann distribution.

The vibrational frequency of HgBr is $5.58 \times 10^{12}$ s$^{-1}$, and the ratio of the number of molecules in the $v = 1$ state to the number in the $v = 0$ state is 0.127. Estimate the vibrational temperature under these conditions.

**32.** An electron in the $\pi$ orbital of ethylene ($C_2H_4$) is excited by a photon to the $\pi^*$ orbital. Do you expect the equilibrium bond length in the excited ethylene molecule to be greater or less than that in ground-state ethylene? Will the vibrational frequency in the excited state be higher or lower than in the ground state? Explain your reasoning.

**33.** One isomer of retinal is converted to a second isomer by the absorption of a photon:

This process is a key step in the chemistry of vision. Although free retinal (in the form shown to the left of the arrow) has an absorption maximum at 376 nm, in the ultraviolet region of the spectrum this absorption shifts into the visible range when the retinal is bound in a protein, as it is in the eye.
(a) How many of the C=C double bonds are *cis* and how many are *trans* in each of the preceding structures? (When assigning labels, consider the relative positions of the two largest groups attached at each double bond.) Describe the motion that takes place upon absorption of a photon.
(b) If the ring and the —CHO group in retinal were replaced by —CH$_3$ groups, would the absorption maximum in the molecule shift to longer or shorter wavelengths?

**\*34.** The ground-state electron configuration of the $H_2^+$ molecular ion is $(\sigma_{1s})^1$.
(a) A molecule of $H_2^+$ absorbs a photon and is excited to the $\sigma_{1s}^*$ molecular orbital. Predict what happens to the molecule.
(b) Another molecule of $H_2^+$ absorbs even more energy in an interaction with a photon and is excited to the $\sigma_{2s}$ molecular orbital. Predict what happens to this molecule.

**\*35.** (a) Write a Lewis structure for formaldehyde ($H_2CO$), and decide the hybridization of the central carbon atom.
(b) Formulate the molecular orbitals for the molecule.
(c) A strong absorption is observed in the ultraviolet region of the spectrum and is attributed to a $\pi \rightarrow \pi^*$ transition. Another, weaker transition is observed at lower frequencies. What electronic excitation causes the weaker transition?

**36.** Do you expect the energy of the special pair of bacteriochlorophyll molecules to be higher or lower than the energy of an isolated bacteriochlorophyll? (*Hint:* Think about the analogy between the mixing of atomic orbitals to make molecular orbitals and the mixing of molecular orbitals on two nearby molecules.)

**37.** Draw a schematic diagram of the steps in bacterial photosynthesis, numbering them in sequence and showing the approximate spatial relations of the involved molecules.

**38.** Write balanced chemical equations that describe the formation of nitric acid and sulfuric acid in rain, starting with the sulfur in coal and the oxygen, nitrogen, and water vapor in the atmosphere.

**39.** Compare and contrast the roles of ozone ($O_3$) and nitrogen dioxide ($NO_2$) in the stratosphere and in the troposphere.

**40.** Describe the greenhouse effect and its mechanism of operation. Give three examples of energy sources that contribute to increased $CO_2$ in the atmosphere and three that do not.

## CUMULATIVE PROBLEMS

**41.** At thermal equilibrium, is the rate at which a molecule is excited from $v = 0$ to the $v = 1$ level greater than or less than the rate for the reverse process? What is the ratio of the rate constants? (*Hint:* Think of the analogy with the chemical equilibrium between two species.)

**42.** The standard enthalpy of combustion of $C_{60}(s)$ has been found to be $-25,891$ kJ mol$^{-1}$. Compute the standard enthalpy of formation of $C_{60}(s)$.

**43.** The hydroxyl radical has been referred to as the "chief cleanup agent in the troposphere." Its concentration is approximately zero at night and becomes as high as $1 \times 10^7$ molecules per cm$^3$ in highly polluted air.
(a) Calculate the maximum mole fraction and partial pressure of OH in polluted air at 25°C and atmospheric pressure.
(b) Write an equation for the reaction of HO with $NO_2$ in the atmosphere. How does the oxidation state of nitrogen change in this reaction? What is the ultimate fate of the product of this reaction?

**44.** In unpolluted air at 300 K, the hydroxyl radical OH reacts with CO with a bimolecular rate constant of $1.6 \times 10^{11}$ L mol$^{-1}$ s$^{-1}$ and with $CH_4$ with a rate constant of $3.8 \times 10^9$ L mol$^{-1}$ s$^{-1}$. Take the partial pressure of CO in air to be constant at $1.0 \times 10^{-7}$ atm and that of $CH_4$ to be $1.7 \times 10^{-6}$ atm, and assume that these are the primary mechanisms by which OH is consumed in the atmosphere. Calculate the half-life of OH under these conditions.

# 16

*The purple color of this amethyst results from the absorption of light by $[FeO_4]^{4-}$ color center defects in quartz.*

# Transition Metals and Coordination Complexes

The partially filled *d* electron shells of the transition-metal elements manifest a range of physical properties and chemical reactions that contrast with those of the main-group elements. The presence of unpaired electrons in the transition-metal elements and their compounds, the availability of low-lying unoccupied orbitals, and the facility with which transition-metal oxidation states change all give rise to a fascinating and rich chemistry. Of particular interest in this chapter are the magnetic properties and colors of transition-metal compounds, which are variable and striking, and the wide variety of geometrical arrangements in which ions and small molecules position themselves around metal atoms. We describe how some of these properties vary through the transition-metal series and present bonding models that seek to explain the observed properties. We also describe some organometallic compounds, which contain bonds between metal and carbon atoms. Organometallic substances play an important role in the catalysis of reactions in organic chemistry and biochemistry.

# 16-1 CHEMISTRY OF THE TRANSITION METALS

Let us begin by surveying some key physical and chemical properties of the transition-metal elements, with particular emphasis on the fourth-period elements (those
from scandium through zinc in which the $3d$ shell is progressively filled), and by
exploring connections to the quantum theory of atomic structure discussed in Section 13-6.

## Physical Properties

Table 16-1 lists some key physical properties of the fourth-period transition elements, mostly taken from Appendix F. As the nuclear charge increases across the
period, the first and second ionization energies tend to increase, although this process is by no means smooth. The energies of the $4s$ and $3d$ orbitals are so close that
details of the neutral atom and ion electron configurations are not easy to predict
from a qualitative model of atomic structure.

The increasing nuclear charge also causes electrons to be held closer and the
atomic radius to decrease across each period, as Table 16-1 and Figure 16-1 show.
Near the end of each period, however, the increased electron–electron repulsion in a
nearly filled $d$ shell outweighs the effect of nuclear charge and forces the atom to
expand. Atoms of the fifth-period elements (from yttrium to cadmium) are larger
than those of the fourth-period elements, but almost no further increase occurs from
the fifth to the sixth period (Fig. 16-1) because of the lanthanide contraction: the
filling of the $4f$ shell prior to the sixth-period transition elements makes the outer $d$
electrons feel a greater effective charge and causes the atom to be smaller than it
would be otherwise. (A parallel effect, shown in Figure 13-29 for main-group atoms
and ions, arose from the filling of the $3d$ shell.) The atomic and ionic radii of
hafnium, in the sixth period, are essentially the same as those of zirconium, in the
fifth period. Because these elements are similar in both valence configuration *and*
size, they have quite similar properties and are difficult to separate from each other.

The binding strength of atoms in an elemental metal has a maximum in the
middle of each transition series and falls off on both sides. This pattern shows up in

## Table 16-1 Properties of the Fourth-Period Transition Elements

| Element | Sc | Ti | V | Cr | Mn | Fe | Co | Ni | Cu | Zn |
|---|---|---|---|---|---|---|---|---|---|---|
| $IE_1$ (kJ mol$^{-1}$) | 631 | 658 | 650 | 653 | 717 | 759 | 758 | 737 | 745 | 906 |
| $IE_2$ (kJ mol$^{-1}$) | 1235 | 1310 | 1414 | 1592 | 1509 | 1562 | 1648 | 1753 | 1958 | 1733 |
| Radius (Å) | 1.61 | 1.45 | 1.31 | 1.25 | 1.37 | 1.24 | 1.25 | 1.25 | 1.28 | 1.34 |
| $\Delta H_f^\circ(M(g))$ (kJ mol$^{-1}$) | 378 | 470 | 514 | 397 | 281 | 416 | 425 | 430 | 338 | 131 |
| Boiling point (°C) | 2831 | 3287 | 3380 | 2672 | 1962 | 2750 | 2870 | 2732 | 2567 | 907 |
| Melting point (°C) | 1541 | 1660 | 1890 | 1857 | 1244 | 1535 | 1495 | 1453 | 1083 | 420 |
| $M^{2+}$ configuration | $d^1$ | $d^2$ | $d^3$ | $d^4$ | $d^5$ | $d^6$ | $d^7$ | $d^8$ | $d^9$ | $d^{10}$ |
| $\Delta H_{hyd}(M^{2+})^\dagger$ (kJ mol$^{-1}$) | | | | −2799 | −2740 | −2839 | −2902 | −2985 | −2989 | −2937 |
| $\mathscr{E}°(M^{2+}|M)$ (V) | | −1.63 | −1.2 | −0.56 | −1.03 | −0.41 | −0.28 | −0.23 | +0.34 | −0.76 |

$^\dagger$ Defined as $\Delta H_f^\circ(M^{2+}(aq)) - \Delta H_f^\circ(M^{2+}(g))$. Because aqueous species are defined relative to $\Delta H_f^\circ(H^+(aq)) = 0$, this is not an absolute enthalpy of
hydration.

**Figure 16-1**

Variation of atomic radius through the fourth-, fifth-, and sixth-period transition elements. Symbols shown are for the fourth-period elements.

the variation of the enthalpy of formation of the gaseous atoms $\Delta H_f^\circ(M(g))$ and in the boiling points of the fourth-period elements (Table 16-1). It is also evident in the melting points of the transition metals (Fig. 16-2). All three properties correlate roughly with the number of unpaired electrons in the involved elements, which reaches a maximum in the middle of the corresponding series. Another approach is to think in terms of the formation of delocalized orbitals in the solid from outer $d$

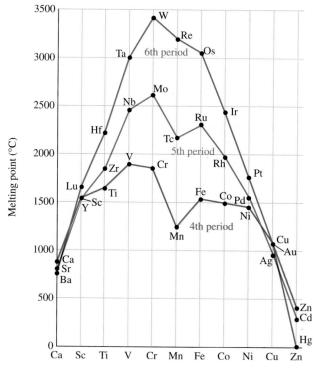

**Figure 16-2**

Variation of melting point through the three periods of transition elements.

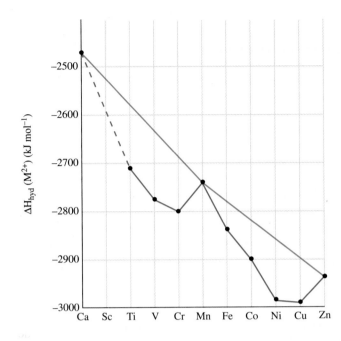

**Figure 16-3**

Enthalpies of hydration of $M^{2+}$ ions, defined as $\Delta H_f^\circ(M^{2+}(aq)) - \Delta H_f^\circ(M^{2+}(g))$. The crystal field stabilization energy (discussed in Section 16-4) stabilizes certain ions, lowering $\Delta H_{hyd}$ from the red line to the blue line.

orbitals on the atoms. Lower-energy orbitals in the solid are primarily bonding (and are progressively filled through the first half of each transition series), and higher-energy antibonding orbitals become filled after that, giving a lower bond strength. Tungsten, near the middle of the sixth period, has a very high melting point (3410°C), which makes it useful in light-bulb filaments; mercury, at the end of the same period, has a melting point well below room temperature (−39°C).

Table 16-1 and Figure 16-3 illustrate the variation of the enthalpy change of hydration for transition-metal ions $M^{2+}$. This variation is connected to the heat released in the process

$$M^{2+}(g) \longrightarrow M^{2+}(aq)$$

or (recognizing the existence of hydrated complex ions of the transition-metal ions)

$$M^{2+}(g) + 6\ H_2O(\ell) \longrightarrow [M(H_2O)_6]^{2+}(aq)$$

The enthalpy of hydration becomes more negative later in the fourth period, but superimposed on this general trend is a secondary trend that gives a maximum near $Mn^{2+}$, a $d^5$ species. Similar anomalies near Mn appear in Figures 16-1 and 16-2 and are related to the fact that Mn contains a half-filled $d$ shell. Section 16-4 will discuss the origin of the trend shown in Figure 16-3.

## Redox Chemistry of the Transition Elements

Figure 16-4 shows the characteristic oxidation states of compounds of the transition metals and should be compared with Figure 2-13 for main-group elements. The early members of each period have maximum oxidation numbers that correspond to the participation of all the outer $s$ and $d$ electrons in ionic or covalent bonds. Thus, elements in the scandium group have only the +3 oxidation state, and manganese has a maximum oxidation state of +7, as do the other elements in its group, forming

**Figure 16-4**

Some of the oxidation states found in compounds of the transition-metal elements.

compounds such as $HReO_4$ and the dark red liquid $Mn_2O_7$. An interesting trend in Figure 16-4 is the increasing tendency toward higher oxidation states among the heavier transition elements, which is the opposite of the trend among main-group elements. Thus, the chemistry of iron is dominated by the +2 and +3 oxidation states, as in the common oxides FeO and $Fe_2O_3$, but the +8 state, which is nonexistent for iron, is important for the later members of the iron group, ruthenium and osmium. The oxide $OsO_4$, for example, is a volatile yellow solid that melts at 41°C and boils at 131°C. Its selective reaction with C=C double bonds makes it useful in organic synthesis and as a biological stain. The chemistry of nickel is almost entirely that of the +2 oxidation state, but the later elements in the nickel group, palladium and platinum, have chemistries increasingly dominated by the +4 state. For example, $NiF_2$ is the only stable fluoride of nickel, but both $PdF_2$ and $PdF_4$ exist; $PtF_2$ is not found, but both $PtF_4$ and $PtF_6$ have been prepared.

Compounds in which the transition-metal element has a high oxidation number tend to be relatively covalent, whereas those with lower oxidation numbers are more ionic. Oxides are an example: $Mn_2O_7$ is a covalent compound that is a liquid at room temperature (crystallizing only at 6°C), but $Mn_3O_4$ is an ionic compound, containing both Mn(II) and Mn(III), that melts at 1564°C. Covalent oxides tend to be acid anhydrides, whereas ionic oxides are basic, just as happens in the main-group elements (compare Fig. 2-21). Table 16-2 shows that the oxides of two transition elements, chromium and manganese, demonstrate these tendencies.

The transition elements show such a range of oxidation states because their partially filled $d$ orbitals can accept or donate electrons in chemical reactions. This same property makes many of their compounds effective catalysts, both homogeneous and heterogeneous. Because an element such as iron can exist as either $Fe^{2+}$ or $Fe^{3+}$ in solution, it can catalyze electron-transfer reactions by shuttling back and forth between oxidized and reduced forms without ever leaving solution, unlike an ion such as $K^+$ that has only one oxidation state in solution. Iron oxides are also used as heterogeneous catalysts for the production of ammonia from the elements and solid $V_2O_5$ catalyzes the production of $SO_3$ from $SO_2$. (Chapter 21 will discuss

**Table 16-2  Acid–Base Character of Oxides of Chromium and Manganese**

| Oxidation State | Chromium | | | Manganese | | |
|---|---|---|---|---|---|---|
| | Oxide | Hydroxide | Acidic/Basic | Oxide | Hydroxide | Acidic/Basic |
| +2 | CrO Black | $Cr(OH)_2$ | Basic | MnO Green | $Mn(OH)_2$ | Basic |
| +3 | $Cr_2O_3$ Green | $Cr(OH)_3$ | Amphoteric | $Mn_2O_3$ Black | $Mn(OH)_3$ | Weakly basic |
| +4 | | | | $MnO_2$ Black | $H_2MnO_3$ | Amphoteric |
| +6 | $CrO_3$ Dark red | $H_2CrO_4$ | Weakly acidic | $MnO_3$ Reddish | $H_2MnO_4$ | Acidic |
| | | $H_2Cr_2O_7$ | Acidic | | | |
| +7 | | | | $Mn_2O_7$ Dark red | $HMnO_4$ | Strongly acidic |

both of these processes.) Many enzymes (biological catalysts) contain transition-metal ions that undergo redox reactions at their active sites.

Table 16-1 lists the standard reduction potentials to the metallic states of $M^{2+}$ transition elements. All but copper have negative reduction potentials, indicating that the metals should be oxidized by 1 M acidic solutions; several metals (Ti, V, and Mn) are strong enough reducing agents that they should even react with water. In practice, however, each of these metals in bulk rapidly gains an oxide coating that protects (passivates) it against further reaction. The positive reduction potential of copper ions means that copper is oxidized only with difficulty. The stability of elemental copper arises at least in part from the high second ionization energy of its atoms relative to the other transition metals (Table 16-1). Silver and gold, the fifth- and sixth-period metals in the copper group, have ions with even higher reduction potentials and are still more resistant to oxidation. Consequently, copper, silver, and gold have served as coinage metals throughout history.

Several transition elements in higher oxidation states are good oxidizing agents, as their large positive reduction potentials indicate. Examples are acidic solutions of permanganate ion (containing Mn(VII))

$$MnO_4^- + 8\,H_3O^+ + 5\,e^- \longrightarrow Mn^{2+} + 12\,H_2O \qquad \mathscr{E}° = +1.51\text{ V}$$

and of dichromate ion (containing Cr(VI)):

$$Cr_2O_7^{2-} + 14\,H_3O^+ + 6\,e^- \longrightarrow 2\,Cr^{3+} + 21\,H_2O \qquad \mathscr{E}° = +1.33\text{ V}$$

Dichromate ion is isoelectronic with manganese(VII) oxide, the explosive red liquid. The chromium-containing species that is isoelectronic with $MnO_4^-$ is the chromate ion, $CrO_4^{2-}$. Chromate and dichromate ion take part in an equilibrium that depends on pH:

$$2\,CrO_4^{2-} + 2\,H_3O^+ \rightleftharpoons Cr_2O_7^{2-} + 3\,H_2O \qquad K = 4.2 \times 10^{14}$$

Low pH (acidic conditions) favors formation of the orange dichromate ion, and high pH (basic conditions) favors the yellow chromate ion.

## 16-2 THE FORMATION OF COORDINATION COMPLEXES

One of the key features of the transition metals is their ability to form complexes with small molecules and ions. For example, solid copper(II) sulfate is made by reacting copper and hot concentrated sulfuric acid ("oil of vitriol"). Its common name, "blue vitriol," recalls this origin and reports the color that is its most obvious property. There is more to this compound than copper and sulfate, however; it contains water as well. The water in blue vitriol is important, because when it is driven away by strong heat, the blue color vanishes, leaving greenish white anhydrous copper(II) sulfate (Fig. 16-5). The blue of blue vitriol comes from a **coordination complex** in which $H_2O$ molecules bond directly to $Cu^{2+}$ ions to form composite ions with the formula $[Cu(H_2O)_4]^{2+}$. As a Lewis acid, the $Cu^{2+}$ ion *coordinates* four water molecules into a group by accepting electron density from a lone pair on each. By acting as electron-pair donors and sharing electron density with the $Cu^{2+}$ ion, the four water molecules, which in this interaction are called **ligands,** come into the **coordination sphere** of the ion. Blue vitriol has the chemical formula $[Cu(H_2O)_4]SO_4 \cdot H_2O$; the fifth water molecule is not coordinated directly to copper.

The positive ions of every metal in the periodic table accept electron density to some degree and can therefore coordinate surrounding electron donors, even if only weakly. The solvation of the $K^+$ ion by $H_2O$ molecules in aqueous solution (Fig. 2-15) is an example of weak coordination. The ability to make fairly strong, *directional* bonds by accepting electron pairs from neighboring molecules or ions is characteristic of the transition-metal elements. Coordination occupies a middle

**Figure 16-5**

Hydrated copper(II) sulfate, $CuSO_4 \cdot 5H_2O$, is blue (*left*), but the anhydrous compound, $CuSO_4$, is greenish white (*right*). A structural study of the solid compound reveals that four of the water molecules are closely associated with the copper and the fifth is not. Thus, a better representation of the hydrated compound is $[Cu(H_2O)_4]SO_4 \cdot H_2O$. (Leon Lewandowski)

Carbonate ion, $CO_3^{2-}$

Oxalate ion, $C_2O_4^{2-}$

Ethylenediamine, $NH_2CH_2CH_2NH_2$

**Figure 16-6**

These three ligands are bidentate; each is capable of donating two pairs of electrons.

place energetically between the weak intermolecular attractions in solids (see Chapter 17) and the stronger covalent and ionic bonding (Chapter 14). Thus, heating blue vitriol disrupts the $Cu-H_2O$ bonds at temperatures well below those required to break the covalent bonds in the $SO_4^{2-}$ group. The enthalpy change in breaking a $+2$ transition-metal ion away from a coordinated water molecule falls in the range of 170 to 210 kJ mol$^{-1}$. This is far less than the enthalpy changes for the strongest chemical bonds, but it is by no means small. Metal ions with $+3$ charges have still stronger coordinate bonds with water.

The total number of metal-to-ligand bonds in a complex (usually two to six) is the **coordination number** of the metal. Some common ligands are the halide ions ($F^-$, $Cl^-$, $Br^-$, $I^-$), ammonia ($NH_3$), carbon monoxide ($CO$), and water (Table 16-3). Each of these is capable of forming only a single bond to a central metal atom. They are called *monodentate* (from Latin *mono*, meaning "one," plus *dens*, meaning "tooth," indicating that they bind at only one point). Other ligands can form two or more such bonds and are referred to as *bidentate, tridentate,* and so forth. Ethylenediamine ($NH_2CH_2CH_2NH_2$), in which two $NH_2$ groups are held together by a carbon backbone, is a particularly important bidentate ligand. Both nitrogen atoms in ethylenediamine have lone electron pairs to share. If all the nitrogen donors of three ethylenediamine molecules bind to a single $Co^{3+}$ ion, then that ion has a coordination number of 6, and the formula of the resulting complex is $[Co(en)_3]^{3+}$ (where "en" is the abbreviation for ethylenediamine). Complexes in which a ligand coordinates via two or more donors to the same central atom are called **chelates** (from Greek *chele*, meaning "claw," because the ligand grabs onto the central atom like a pincers). Figure 16-6 shows the structures of some important chelating ligands.

Brackets are used in chemical formulas of coordination complexes to group together the symbols of the central atom and its coordinated ligands. In the formula $[Pt(NH_3)_6]Cl_4$, the portion in brackets represents a positively charged coordination complex in which Pt coordinates six $NH_3$ ligands. The brackets emphasize that a complex is a distinct chemical entity with its own properties. Within the brackets, the symbol of the central atom comes first. The electric charge on a coordination complex is the sum of the oxidation number of the metal ion and the charges of the ligands that surround it. Thus, the complex of copper(II) ($Cu^{2+}$) with four $Br^-$ ions is an anion with a $-2$ charge, $[CuBr_4]^{2-}$.

**Table 16-3  Common Ligands and Their Names**

| Ligand | Name |
|--------|------|
| $^\dagger :NO_2^-$ | Nitro |
| $:OCO_2^{2-}$ | Carbonato |
| $:ONO^-$ | Nitrito |
| $:CN^-$ | Cyano |
| $:SCN^-$ | Thiocyanato |
| $:NCS^-$ | Isothiocyanato |
| $:OH^-$ | Hydroxo |
| $:OH_2$ | Aqua |
| $:NH_3$ | Ammine |
| $:CO$ | Carbonyl |
| $:NO^+$ | Nitrosyl |

$^\dagger$The ligating atom is indicated by a pair of dots ( : ) to show a lone pair of electrons. In the $CO_3^{2-}$ ligand, either one or two of the oxygen atoms can donate a lone pair to a metal.

## Example 16-1

Determine the oxidation state of the coordinated metal atom in each of the following compounds:
(a) $K[Co(NH_3)_2(CN)_4]$, (b) $[Os(CO)_5]$, (c) $Na[Co(H_2O)_3(OH)_3]$.

### Solution

(a) The oxidation state of K is known to be $+1$, and so the complex in brackets is an anion with a $-1$ charge, $[Co(NH_3)_2(CN)_4]^-$. The charge on the two $NH_3$ ligands is 0, and the charge on each of the four $CN^-$ ligands is $-1$. The oxidation state of the Co must then be $+3$, because $4 \times -1$ (for the $CN^-$) $+ 2 \times 0$ (for the $NH_3$) $+ 3$ (for Co) equals the required $-1$.

(b) The ligand CO has zero charge, and the complex has zero charge as well. Therefore, the oxidation state of the osmium is 0.

(c) There are three neutral ligands (the water molecules) and three ligands with $-1$ charges (the hydroxide ions). The $Na^+$ ion contributes only $+1$, and so the oxidation state of the cobalt must be $+2$.

*Related Problems:* 9, 10

Coordination modifies the chemical and physical properties of both the central atom and the ligands. Consider the chemistry of aqueous cyanide $(CN^-)$ and iron(II) $(Fe^{2+})$ ions. The first reacts immediately with acid to generate gaseous HCN, a deadly poison, and the second, when mixed with aqueous base, instantly precipitates a gelatinous hydroxide. Reaction between the two ions gives the complex ion $[Fe(CN)_6]^{4-}(aq)$, which undergoes neither these two reactions nor any others considered diagnostic of simple $CN^-$ and $Fe^{2+}$. Because coordination changes a ligand's chemical behavior, a given ligand may be present in multiple forms in the same compound. The two $Cl^-$ ions in $[Pt(NH_3)_3Cl]Cl$ differ chemically, because one is coordinated and one is not. Treatment of an aqueous solution of this substance with $Ag^+$ ion immediately precipitates the uncoordinated $Cl^-$ as $AgCl(s)$, but *not* the coordinated $Cl^-$.

Ionic coordination complexes of opposite charges can combine with each other—just as any positive ion can combine with a negative ion—to form a salt. For example, the cation $[Pt(NH_3)_4]^{2+}$ and the anion $[PtCl_4]^{2-}$ form a doubly complex ionic compound of formula $[Pt(NH_3)_4][PtCl_4]$. This compound and the four compounds

$$[Pt(NH_3)_2Cl_2] \qquad\qquad [Pt(NH_3)_3Cl][Pt(NH_3)Cl_3]$$
$$[Pt(NH_3)_3Cl]_2[PtCl_4] \qquad [Pt(NH_3)_4][Pt(NH_3)Cl_3]_2$$

all contain Pt, $NH_3$, and Cl in the ratio of 1 to 2 to 2; that is, they have the same percentage composition. Two pairs even have the same molar mass. Yet the five compounds differ in structure and in physical and chemical properties. The concept of coordination organizes an immense number of chemical compositions as combinations of ligands linked in varied ratios with central metal atoms or ions. An otherwise bewildering collection of information on chemical reactivity is rationalized in terms of one ligand substituting for another.

### Naming Coordination Compounds

Up to now, we have used only the chemical symbols for coordination compounds, but for many purposes names are also useful. Some substances have names that were given to them before their structures were known. Thus, $K_3[Fe(CN)_6]$ was called potassium ferricyanide, and $K_4[Fe(CN)_6]$ was potassium ferrocyanide (these are complexes of $Fe^{3+}$ (ferric) and $Fe^{2+}$ (ferrous) ions, respectively). The older names still find some use but are being replaced by systematic names based on the following set of rules:

1. The name of a coordination complex is written as a single word built from the names of the ligands, a prefix for each to indicate how many are present, and a name for the central metal.

2. If a compound contains a coordination complex, it is named as a simple ionic compound is named: the positive ion is named first, followed (after a space) by the name of the negative ion, regardless of which is the complex ion.

3. The name of an anionic ligand is obtained by replacing the usual ending with the suffix -o. The names of neutral ligands are unchanged. Exceptions to the latter rule are *aqua* (for water), *ammine* (for $NH_3$), and *carbonyl* (for CO). See Table 16-3.

4. Greek prefixes (*di-, tri-, tetra-, penta-, hexa-*) are used to indicate the number of ligands of a given type attached to the central ion, if there is more than one. The prefix *mono-* (for one) is not used. If the name of the ligand itself contains a term such as *mono-* or *di-* (as in ethylene*di*amine), then the name of the ligand is placed in parentheses and the prefixes *bis-, tris-,* and *tetrakis-* are used instead of *di-, tri-,* and *tetra-*.

5. The ligands are listed in alphabetical order, without regard for the prefixes that tell how often each type of ligand occurs in the coordination sphere.

6. A Roman numeral enclosed in parentheses immediately following the name of the metal gives the oxidation state of the central metal atom. If the complex ion has a net negative charge, the ending -*ate* is added to the stem of the name of the metal.

The following list of complexes illustrates systematic naming.

| Complex | Systematic Name |
|---|---|
| $K_3[Fe(CN)_6]$ | Potassium hexacyanoferrate(III) |
| $K_4[Fe(CN)_6]$ | Potassium hexacyanoferrate(II) |
| $Fe(CO)_5$ | Pentacarbonyliron(0) |
| $[Co(NH_3)_5CO_3]Cl$ | Pentaamminecarbonatocobalt(III) chloride |
| $K_3[Co(NO_2)_6]$ | Potassium hexanitrocobaltate(III) |
| $[Cr(H_2O)_4Cl_2]Cl$ | Tetraaquadichlorochromium(III) chloride |
| $[Pt(NH_2CH_2CH_2NH_2)_3]Br_4$ | Tris(ethylenediamine)platinum(IV) bromide |
| $K_2[CuCl_4]$ | Potassium tetrachlorocuprate(II) |

## Example 16-2

Interpret the names and write the formulas of these coordination compounds: (a) sodium tricarbonatocobaltate(III), (b) diamminediaquadichloroplatinum(IV) bromide, (c) sodium tetranitratoborate(III).

### Solution

(a) In the anion, three carbonate ligands (with $-2$ charges) are coordinated to a cobalt atom in the $+3$ oxidation state. Because the complex ion thus has an overall charge of $-3$, three sodium cations are required, and the correct formula is $Na_3[Co(CO_3)_3]$.

(b) The ligands coordinated to a Pt(IV) include two ammonia molecules, two water molecules, and two chloride ions. Ammonia and water are electrically neutral, but the two chloride ions contribute a total charge of $2 \times (-1) = -2$ that adds with the $+4$ of the platinum and gives the complex ion a $+2$ charge. Two bromide anions are required to balance this, and so the formula is $[Pt(NH_3)_2(H_2O)_2Cl_2]Br_2$.

(c) The complex anion has four $-1$ nitrate ligands coordinated to a central boron(III). This gives a net charge of $-1$ on the complex ion and requires one sodium ion in the formula $Na[B(NO_3)_4]$.

*Related Problems:* 11, 12

## Ligand Substitution Reactions

If the yellow crystalline solid nickel(II) sulfate is exposed to moist air at room temperature, it takes up six water molecules per formula unit. These water molecules coordinate the nickel ions to form a bright green complex:

$$NiSO_4(s) + 6\,H_2O(g) \longrightarrow [Ni(H_2O)_6]SO_4(s)$$

Yellow        Colorless                    Green

Heating the green hexaaquanickel(II) sulfate to a temperature well above the boiling point of water drives off the water and regenerates the yellow $NiSO_4$ in a reversal of the reaction just shown. A similar coordination reaction generates a product with completely different color when yellow $NiSO_4(s)$ is exposed to gaseous ammonia, $NH_3(g)$. This time, the product is a blue-violet complex:

$$NiSO_4(s) + 6\,NH_3(g) \longrightarrow [Ni(NH_3)_6]SO_4(s)$$

Yellow        Colorless                    Blue-violet

Heating the blue-violet product drives off ammonia, and the color of the solid then changes back to yellow. Given these facts, it is not hard to explain the observation that a green $[Ni(H_2O)_6]^{2+}(aq)$ solution turns blue-violet when treated with $NH_3(aq)$ (Fig. 16-7). The $NH_3$ must be displacing the $H_2O$ from the coordination sphere.

$$[Ni(H_2O)_6]^{2+} + 6\,NH_3(aq) \longrightarrow [Ni(NH_3)_6]^{2+} + 6\,H_2O$$

Green        Colorless                    Blue-violet        Colorless

Complexes that undergo rapid substitution of one ligand for another are **labile.** Those in which substitution proceeds slowly or not at all are **inert.** In an inert complex, a high activation energy for ligand substitution prevents rapid reaction even though there may be a thermodynamic tendency to proceed. In the substitution reaction

$$[Co(NH_3)_6]^{3+}(aq) + 6\,H_3O^+(aq) \longrightarrow [Co(H_2O)_6]^{3+} + 6\,NH_4^+(aq)$$

the products are favored thermodynamically by an enormous amount (the equilibrium constant is about $10^{64}$). Yet the inert $[Co(NH_3)_6]^{3+}$ complex ion lasts for weeks in acidic solution because no low-energy path for the reaction exists. The $[Co(NH_3)_6]^{3+}$ ion is thermodynamically unstable relative to $[Co(H_2O)_6]^{3+}$ yet kinetically stable (that is, inert). The closely related cobalt(II) complex, $[Co(NH_3)_6]^{2+}$, undergoes a similar substitution reaction in a few seconds:

$$[Co(NH_3)_6]^{2+}(aq) + 6\,H_3O^+(aq) \longrightarrow [Co(H_2O)_6]^{2+} + 6\,NH_4^+(aq)$$

The hexaaminecobalt(II) complex is thermodynamically unstable and also labile.

**Figure 16-7**
When ammonia is added to the green solution of nickel(II) sulfate on the left (which contains $[Ni(H_2O)_6]^{2+}$ ions), ligand substitution occurs to give the blue-violet solution on the right (which contains $[Ni(NH_3)_6]^{2+}$ ions). (Leon Lewandowski)

Substitution of one ligand for another can proceed in stages in the coordination sphere of a metal. By controlling the reaction conditions, substitution can usually be stopped at intermediate stages. For example, all of the possible four-coordinate compositions of Pt(II) with the two ligands $NH_3$ and $Cl^-$

$$[Pt(NH_3)_4]^{2+} \quad [Pt(NH_3)_3Cl]^+ \quad [Pt(NH_3)_2Cl_2] \quad [Pt(NH_3)Cl_3]^- \quad [PtCl_4]^{2-}$$

can be prepared. Such mixed-ligand complexes add greatly to the variety and richness of coordination chemistry.

## 16-3 STRUCTURES OF COORDINATION COMPLEXES

The Alsatian-Swiss chemist Alfred Werner pioneered the field of coordination chemistry in the late 19th century. At that time, a number of compounds of cobalt (III) chloride with ammonia were known. They had the following chemical formulas and colors:

| Compound 1: | $CoCl_3 \cdot 6NH_3$ | Orange-yellow |
| Compound 2: | $CoCl_3 \cdot 5NH_3$ | Purple |
| Compound 3: | $CoCl_3 \cdot 4NH_3$ | Green |
| Compound 4: | $CoCl_3 \cdot 3NH_3$ | Green |

The fact that treatment of these compounds with aqueous hydrochloric acid did not remove the ammonia suggested that the ammonia was somehow closely bound with the cobalt ions. Treatment with aqueous silver nitrate at 0°C, on the other hand, gave interesting results. With compound 1, all of the chloride present precipitated as solid AgCl. With compound 2, only two thirds of the chloride precipitated, and with compound 3, only one third. Compound 4 did not react at all with the silver nitrate. Werner accounted for these facts by positing the existence of coordination complexes with six ligands (either chloride ions or ammonia molecules) attached to each $Co^{3+}$ ion. Specifically, he wrote the formulas for compounds 1 through 4 as

Compound 1:  $[Co(NH_3)_6]^{3+} (Cl^-)_3$
Compound 2:  $[Co(NH_3)_5Cl]^{2+} (Cl^-)_2$
Compound 3:  $[Co(NH_3)_4Cl_2]^+ (Cl^-)$
Compound 4:  $[Co(NH_3)_3Cl_3]$

Only those chloride ions that were *not* ligands attached directly to cobalt were precipitated upon addition of cold aqueous silver nitrate.

Werner realized that with his proposed complexes came predictions about the electrical conductivity of aqueous solutions of the salts of these complex ions. Compound 1, for example, should have a molar conductivity close to that of $Al(NO_3)_3$, which also forms one $3+$ ion and three $1-$ ions per formula unit dissolved in water. His experiments confirmed this resemblance and showed that compound 2 resembled $Mg(NO_3)_2$ and compound 3 resembled $NaNO_3$ in conductivity. Compound 4 behaved as a nonelectrolyte, with a very low electrical conductivity, as expected from the absence of ions in its formula.

Werner and other chemists studied a variety of other coordination complexes, using both physical and chemical techniques. Their research has shown that the most common coordination number by far is 6, as in the cobalt complexes just discussed. Coordination numbers ranging from 2 to 12 have been observed, however, including coordination numbers 2 (as in $[Ag(NH_3)_2]^+$), 4 (as in $[PtCl_4]^{2-}$), and 5 (as in $[Ni(CN)_5]^{3-}$).

## Three-Dimensional Structures

If the central cobalt ion in $[Co(NH_3)_6]^{3+}$ is bonded to six ammonia ligands, what is the geometrical structure of the complex? This question naturally occurred to Werner, who suggested that the arrangement should be the simplest and most symmetric possible, with ligands at the six vertices of a regular octahedron (Fig. 16-8). Modern methods of x-ray diffraction (Section 17-1) enable us to make very precise determinations of atomic positions in crystals and completely confirm Werner's hypothesis of octahedral coordination for this complex. Such techniques were not available a century ago, however, and so Werner turned to a study of the properties of substituted complexes to test his hypothesis.

If one ammonia ligand is replaced with a chloride ion, the resulting complex has the formula $[Co(NH_3)_5Cl]^{2+}$, with one vertex of the octahedron occupied by $Cl^-$ and the other five by $NH_3$. Only one structure of this type is possible, because all six vertices of a regular octahedron are equivalent and the various singly substituted complexes $[MA_5B]$ (where $A = NH_3$, $B = Cl^-$, $M = Co^{3+}$, for example) can be superimposed upon one another. Now suppose a second $NH_3$ ligand is replaced with

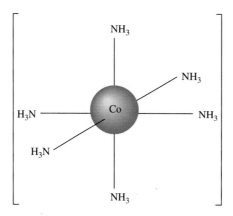

**Figure 16-8**

The octahedral structure of the $Co(NH_3)_6^{3+}$ ion. All six corners of the octahedron are equivalent.

(a) cis-[Co(NH$_3$)$_4$Cl$_2$]$^+$          (b) trans-[Co(NH$_3$)$_4$Cl$_2$]$^+$

**Figure 16-9**
The cis-[Co(NH$_3$)$_4$Cl$_2$]$^+$ and trans-[Co(NH$_3$)$_4$Cl$_2$]$^+$ ions. The cis complex is purple in solution, but the trans is green.

Cl$^-$. The second Cl$^-$ can either lie in one of the four positions closest to the first Cl$^-$ (in the plane perpendicular to the Co–Cl$^-$ line; see Fig. 16-9a) or lie in the sixth position, on the opposite side of the central metal atom (Fig. 16-9b). The former arrangement, in which the two Cl$^-$ ligands are closer to each other, is cis-[Co(NH$_3$)$_4$Cl$_2$]$^+$, and the latter, with the two Cl$^-$ ligands farther apart, is trans-[Co(NH$_3$)$_4$Cl$_2$]$^+$. The octahedral structure model predicts that exactly two different ions with the chemical formula [Co(NH$_3$)$_4$Cl$_2$]$^+$ exist. Such species are called **geometrical isomers.** When Werner began his work, only the green trans form was known, but by 1907 he had prepared the cis isomer and shown that it differed from the trans isomer in color (it was violet rather than green) and other physical properties. The isolation of two, and only two, geometrical isomers of this ion was good (although not conclusive) evidence that the octahedral structure was correct. Similar isomerism is displayed by the complex ion [CoCl$_2$(en)$_2$]$^+$, which also has a purple cis form and a green trans form (Fig. 16-10).

### Example 16-3

How many geometrical isomers exist for the octahedral coordination compound [Co(NH$_3$)$_3$Cl$_3$]?

### Solution

We begin with the two isomers of [Co(NH$_3$)$_4$Cl$_2$]$^+$ and see how many different structures can be made by replacing one more ammonia molecule with Cl$^-$.

Starting with the trans form (Fig. 16-9b), it is clear that replacement of any of the four NH$_3$ ligands at site 2, 3, 4, or 5 gives an equivalent structure that can be superimposed by rotation. Figure 16-11a shows this isomer. What isomers can be made from the cis form of Fig. 16-9a? If either the ammonia ligand at site 4 or that at site 6 (i.e., one of the two that are trans to existing Cl$^-$ ligands) is replaced, the result is simply a rotated version of Figure 16-11a, and so these replacements do *not* give another isomer. Replacement of the ligand at site 3 or 5, however, gives a different structure, shown in Figure 16-11b. (Note that the 1, 2, 3 replacements give the same structure as the 1, 2, 5 replacements.)

We conclude that there are two, and only two, possible isomers of the octahedral complex structure MA$_3$B$_3$. In fact, only one form of [Co(NH$_3$)$_3$Cl$_3$] has

**Figure 16-10**
The complex ion [CoCl$_2$(en)$_2$]$^+$ is an octahedral complex that has cis and trans isomers, according to the relative positions of the two Cl$^-$ ligands. Salts of the cis isomers are purple, and salts of the trans isomers are green. (Charles Steele)

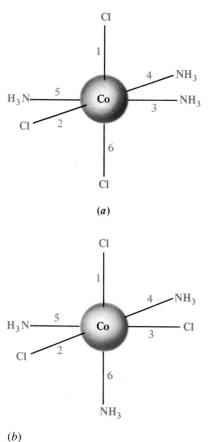

*(a)*

*(b)*

**Figure 16-11**

Two structural isomers of the coordination compound $Co(NH_3)_3Cl_3$.

been prepared to date, presumably because the two isomers interconvert rapidly, but two isomers are known for the closely related coordination complex $[Cr(NH_3)_3(NO_2)_3]$.

*Related Problems:* 21, 22

## Square-Planar, Tetrahedral, and Linear Structures

Complexes with coordination numbers of 4 typically are either tetrahedral or square planar. The tetrahedral geometry (Fig. 16-12*a*) predominates for four-coordinate complexes of the early transition metals (those toward the left side of the *d* block of elements in the periodic table). Geometrical isomerism is not possible for tetrahedral complexes of the general form $MA_2B_2$, because all such structures are superimposable.

The square-planar geometry (Fig. 16-12*b,c*) is common for four-coordinate complexes of $Au^{3+}$, $Ir^+$, and $Rh^+$ and, most especially, ions with the $d^8$ valence electron configurations: $Ni^{2+}$, $Pd^{2+}$, and $Pt^{2+}$. The $Ni^{2+}$ ion forms a few tetrahedral structures, but four-coordinate $Pd^{2+}$ and $Pt^{2+}$ are nearly exclusively square planar. Square-planar complexes of the type $MA_2B_2$ can have isomers, as Figures 16-12*b* and 16-12*c* illustrate for *cis-* and *trans-*$[Pt(NH_3)_2Cl_2]$. The *cis* form of this compound is a potent and widely used anticancer drug called cisplatin, but the *trans* form has no therapeutic properties.

Finally, linear complexes with coordination numbers of 2 exist, especially for ions with $d^{10}$ configurations such as $Cu^+$, $Ag^+$, $Au^+$, and $Hg^{2+}$. The central silver atom in a complex such as $[Ag(NH_3)_2]^+$ in aqueous solution strongly attracts several water molecules as well, however, and so its actual coordination number under these circumstances may be greater than 2.

## Chiral Structures

The complex ion $[Pt(en)_3]^{4+}$ displays a type of isomerism that differs from the geometrical isomerism discussed to this point. The two structures shown in Figure 16-13 are mirror-image pairs that cannot be superimposed via rotation. Section 14-5

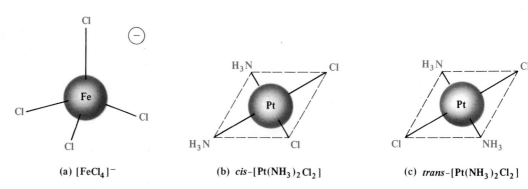

(a) $[FeCl_4]^-$          (b) *cis-*$[Pt(NH_3)_2Cl_2]$          (c) *trans-*$[Pt(NH_3)_2Cl_2]$

**Figure 16-12**

Four-coordinate complexes. (*a*) Tetrahedral, $[FeCl_4]^-$. (*b, c*) Square planar, illustrating (*b*) the *cis* and (*c*) the *trans* forms of $[Pt(NH_3)_2Cl_2]$.

discussed such optical isomers in connection with tetrahedrally bonded carbon atoms (recall Fig. 14-30); like such carbon atoms, the central atom in an optical isomer is called a chiral center.

## Example 16-4

Suppose that the complex ion $[Co(NH_3)_2(H_2O)_2Cl_2]^+$ is synthesized with the two ammine ligands *cis* to each other, the two aqua ligands *cis* to each other, and the two chloro ligands *cis* to each other (Fig. 16-14*a*). Is this complex optically active?

### Solution

We indicate a mirror with a dashed line and create the mirror image by making each point in it lie the same distance from the dashed line as does its generating point in the original structure (Fig. 16-14). Comparison of the original and the mirror image (Fig. 16-14*b*) shows that *cis, cis*-$[Co(NH_3)_2(H_2O)_2Cl_2]^+$ is chiral, because the two structures cannot be superimposed even after they are turned. As this example proves, nonchelates can be chiral.

*Related Problems:* 23, 24

The hexadentate ligand EDTA (ethylenediaminetetraacetate ion) forms chiral complexes. Figure 16-15 shows the structure of this chelating ligand, as coordinated to a $Co^{3+}$ ion. The central metal ion is literally "enveloped" by the ligand as it coordinates simultaneously at the six corners of the coordination octahedron. Such a chelating agent has a very strong affinity for certain metal ions and can **sequester** them effectively in solution. EDTA solubilizes the scummy precipitates that $Ca^{2+}$ ion forms with anionic constituents of soap by forming a stable complex with $Ca^{2+}$.

Mirror plane

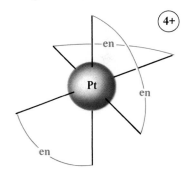

en = $H_2\ddot{N}-CH_2CH_2-\ddot{N}H_2$

## Figure 16-13

Enantiomers of the $[Pt(en)_3]^{4+}$ ion. Reflection through the mirror plane transforms one enantiomer into the other. The two cannot be superimposed by simple rotation.

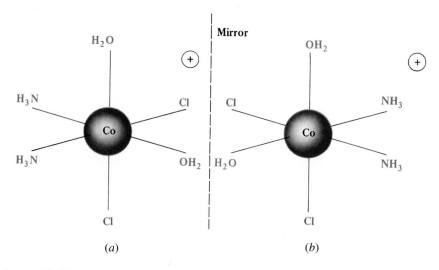

## Figure 16-14

The structure of the all-*cis* $[Co(NH_3)_2(H_2O)_2Cl_2]^+$ complex ion, together with its mirror image.

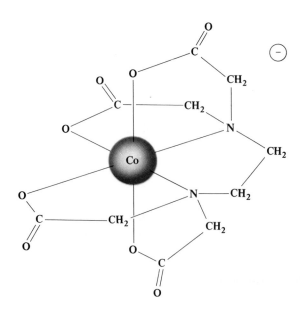

**Figure 16-15**

The chelation complex of EDTA with cobalt(III). Each EDTA ion has six donor sites at which it can bind (by donating lone pair electrons) to the central metal ion.

**EDTA anion**

Thus, it breaks up the main contributor to bathtub rings and is a "miracle ingredient" in some bathtub cleaners. EDTA is used to recover trace contaminants from water (some metal ions, especially heavy ones, are toxic). It has been used as an antidote for lead poisoning because of its great affinity for $Pb^{2+}$ ions. Complexes of EDTA with iron in plant foods permit a slow release of iron to the plant. EDTA also sequesters copper and nickel ions in edible fats and oils. Because these metal ions catalyze the oxidation reactions that turn oils rancid, EDTA preserves freshness.

## 16-4 CRYSTAL FIELD THEORY AND MAGNETIC PROPERTIES

What is the nature of the bonding in coordination complexes of the transition metals that leads to their special properties? Why does Pt(IV) form only octahedral complexes while Pt(II) forms square-planar ones, and under what circumstances does Ni(II) form octahedral, square-planar, and tetrahedral complexes? Can trends in the length and strength of metal to ligand bonds be understood? To answer these questions, we need a theoretical description of bonding in coordination complexes.

### Crystal Field Theory

A simple and useful model for the energies of coordination complexes is **crystal field theory,** which is based on an ionic description of the metal–ligand bonds. It treats the complex as a central metal ion perturbed by the approach of negatively charged ligands. In an octahedral complex, negative charges from the lone pairs of six ligands are brought up along the $\pm x$, $\pm y$, and $\pm z$ coordinate axes toward a metal atom or ion at the origin. In an atom or ion of a transition metal in free space, the energy of an electron is the same in each of the five $d$ orbitals, but when the external charges are brought up (Fig. 16-16), the energies of the $d$ orbitals change by different amounts because of differing amounts of repulsion between the electrons on the ligands and those in the $d$ orbitals. An electron in a $d_{x^2-y^2}$ or $d_{z^2}$ orbital on the central

**Large repulsions**

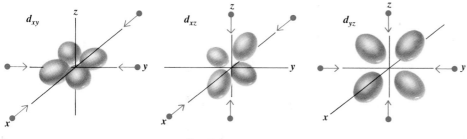

**Smaller repulsions**

**Figure 16-16**
Octahedral crystal field splitting of $3d$ orbital energies by ligands. As the external charges approach the five $3d$ orbitals, the largest repulsions arise in the $d_{z^2}$ and $d_{x^2-y^2}$ orbitals, which point directly at two or four of the approaching charges.

metal atom is most likely to be found along the coordinate axes, where it comes into close contact with electrons from the ligand. This proximity raises the energy of the orbital. By contrast, an electron in a $d_{xy}$, $d_{yz}$, or $d_{xz}$ orbital of the metal has greatest probability density *between* the coordinate axes, and so it experiences less repulsion from an octahedral array of approaching charges and has a lower energy than the $d_{x^2-y^2}$ and $d_{z^2}$ orbitals. In the octahedral field of the ligand lone pairs, the $d$ orbital energy levels split into two groups (Fig. 16-17). The three lower-energy orbitals, called $t_{2g}$ levels, correspond to the $d_{xy}$, $d_{yz}$, and $d_{xz}$ orbitals of the transition-metal atom; the two higher-energy orbitals, called $e_g$ levels, correspond to the $d_{x^2-y^2}$ and $d_{z^2}$ orbitals. The energy difference between the two sets of levels is $\Delta_o$, the **crystal field splitting energy** for the octahedral complex that has formed.

Consider now the electron configuration for a transition-metal ion experiencing such a crystal field. The $Cr^{3+}$ ion, for example, has three $d$ electrons. According to Hund's rules (see Section 13-6), one electron will enter each of the three $t_{2g}$ levels and the three electrons will have parallel spins, so that complexes of $Cr^{3+}$ will be paramagnetic. In an ion such as $Mn^{3+}$, which has a fourth $d$ electron, two ground-state electron configurations are possible. The fourth electron can be placed either in a $t_{2g}$ level, with its spin opposite that of the electron already in that level (Fig. 16-17a), or in an $e_g$ level with spin parallel to those of the three $t_{2g}$ electrons (Fig. 16-17b). The former happens when the splitting $\Delta_o$ is large, because in that case the energy cost of placing an electron in an $e_g$ orbital is prohibitively high. If the splitting is small, however, the $e_g$ orbital will be occupied to avoid the electron–electron repulsion that would occur if two electrons were to occupy one of the $t_{2g}$ orbitals.

In central metal atoms or ions with four to seven $d$ electrons, two types of electron configurations are possible. When $\Delta_o$ is large, **low-spin complexes** form in

(a) $[Mn(CN)_6]^{3-}$ (low spin)

**Figure 16-17**

An octahedral field increases the energies of all five $d$ orbitals, but the increase is greater for the $d_{z^2}$ and $d_{x^2-y^2}$ orbitals. As a result, the orbitals are split into two sets that differ by the energy $\Delta_o$. The orbital occupancy shown is for (a) the low-spin (large $\Delta_o$) and (b) the high-spin (small $\Delta_o$) complexes of Mn(III).

(b) $[Mn(H_2O)_6]^{3+}$ (high spin)

which electrons are paired in the lower-energy $t_{2g}$ orbital, and the $e_g$ orbitals are not occupied until the $t_{2g}$ levels are filled. When $\Delta_o$ is small, **high-spin complexes** occur in which electrons are placed singly, with parallel spins to the fullest extent possible, in both $t_{2g}$ and $e_g$ orbitals.

Table 16-4 summarizes the electron configurations discussed so far and provides the **crystal field stabilization energies (CFSE)** of the complexes. The CFSE is the energy change of an octahedral complex relative to that from a hypothetical spherical crystal field. In an octahedral field, the energy of the three $t_{2g}$ orbitals is lowered by $\frac{2}{5}\Delta_o$ and that of the two $e_g$ orbitals is raised by $\frac{3}{5}\Delta_o$ relative to the spherical field (Fig. 16-17). If these five orbitals are half or fully occupied (as in a $d^{10}$ or a high-spin $d^5$ complex), then the energy is the same as in a spherical field: the CFSE is zero. If the lower-energy levels are preferentially occupied, however, the configuration is stabilized. For example, in a low-spin $d^4$ complex, the four electrons in the $t_{2g}$ orbitals each lower the energy by $\frac{2}{5}\Delta_o$, giving a total CFSE of $-\frac{8}{5}\Delta_o$.

The crystal field stabilization energy helps to explain the shape of the curve shown in Figure 16-3 for the enthalpies of hydration of transition-metal ions. If each measured value (blue curve) is adjusted by correcting for the CFSE of that complex ion, results quite close to the straight red lines are obtained. The relatively small magnitude of the enthalpy of hydration for $Mn^{2+}$ arises from the high-spin $d^5$

**Table 16-4  Electron Configurations and Crystal Field Stabilization Energies for High- and Low-Spin Octahedral Complexes**

| Configuration | | $d^1$ | $d^2$ | $d^3$ | $d^4$ | $d^5$ | $d^6$ | $d^7$ | $d^8$ | $d^9$ | $d^{10}$ |
|---|---|---|---|---|---|---|---|---|---|---|---|
| Examples | | $Ti^{3+}$ | $Ti^{2+}, V^{3+}$ | $V^{2+}, Cr^{3+}$ | $Mn^{3+}, Re^{3+}$ | $Mn^{2+}, Fe^{3+}$ | $Fe^{2+}, Pd^{4+}$ | $Co^{2+}, Rh^{2+}$ | $Ni^{2+}, Pt^{2+}$ | $Cu^{2+}$ | $Zn^{2+}, Ag^+$ |
| **HIGH SPIN** | $e_g$ | — — | — — | — — | ↑ — | ↑ ↑ | ↑ ↑ | ↑ ↑ | ↑ ↑ | ↑⇅ ↑ | ⇅ ⇅ |
| | $t_{2g}$ | ↑ _ _ | ↑ ↑ _ | ↑ ↑ ↑ | ↑ ↑ ↑ | ↑ ↑ ↑ | ⇅ ↑ ↑ | ⇅ ⇅ ↑ | ⇅ ⇅ ⇅ | ⇅ ⇅ ⇅ | ⇅ ⇅ ⇅ |
| | CFSE | $-\frac{2}{5}\Delta_o$ | $-\frac{4}{5}\Delta_o$ | $-\frac{6}{5}\Delta_o$ | $-\frac{3}{5}\Delta_o$ | 0 | $-\frac{2}{5}\Delta_o$ | $-\frac{4}{5}\Delta_o$ | $-\frac{6}{5}\Delta_o$ | $-\frac{3}{5}\Delta_o$ | 0 |
| **LOW SPIN** | $e_g$ | | | | — — | — — | — — | ↑ — | | | |
| | $t_{2g}$ | Same as high spin | | | ⇅ ↑ ↑ | ⇅ ⇅ ↑ | ⇅ ⇅ ⇅ | ⇅ ⇅ ⇅ | Same as high spin | | |
| | CFSE | | | | $-\frac{8}{5}\Delta_o$ | $-\frac{10}{5}\Delta_o$ | $-\frac{12}{5}\Delta_o$ | $-\frac{9}{5}\Delta_o$ | | | |

configuration of $Mn^{2+}$, which results in a CFSE of zero. On either side of this ion, the negative CFSE lowers the enthalpy of hydration.

## Magnetic Properties

The existence of high-spin and low-spin configurations accounts for the magnetic properties of many different coordination compounds. As discussed in Section 14-4, substances can be classified as paramagnetic or diamagnetic according to whether they are attracted into a magnetic field. Figure 16-18 depicts an experiment to demonstrate the universal susceptibility of substances to the influence of magnetic fields. A cylindrical sample is suspended so that its bottom is between the poles of a powerful magnet but its top extends out of the field. It is weighed very accurately

Magnet

(a) Paramagnetic          (b) Diamagnetic

**Figure 16-18**

(a) If a sample of a paramagnetic substance is dipped into a magnetic field, it is drawn down into the field and weighs more than it does in the absence of the field. (b) If a diamagnetic substance is dipped into the field in the same way, it is buoyed up and weighs less than it would in the absence of the field.

and then reweighed when the magnet is moved away. The net force on the sample is found to change in the presence of the magnetic field. Substances that are repelled by a nonuniform magnetic field weigh less when dipped into one and are *diamagnetic*; substances that are attracted by a magnetic field weigh more and are *paramagnetic*. The weighings just described give numerical values for the **magnetic susceptibility** of a substance, its tendency to interact with a magnetic field. The susceptibility of a diamagnet is negative and small, while that of a paramagnet is positive and can be quite large.

As Section 14-4 explained, paramagnetism is associated with atoms, ions, or molecules that contain one or more electrons with unpaired spins. Diamagnetic substances have the spins of all of their electrons paired. Thus, measurements of magnetic susceptibility reveal which substances have unpaired electron spins and which have completely paired electron spins. The number of unpaired electrons per molecule in a paramagnet can even be counted on the basis of the magnitude of the magnetic susceptibility of the sample. On a molar basis, a substance with two unpaired electrons per molecule is pulled into a magnetic field more strongly than a substance with only one unpaired electron per molecule.

These facts emerge in connection with coordination complexes because paramagnetism is prevalent among transition-metal complexes, whereas the great majority of other chemical substances are diamagnetic. Among complexes of a given metal ion, the number of unpaired electrons, as observed by magnetic susceptibility, varies with the identities of the ligands. Both $[Co(NH_3)_6]^{3+}$ and $[CoF_6]^{3-}$ have six ligands surrounding a central $Co^{3+}$ ion, yet the former is diamagnetic (because it is a strong-field, low-spin complex) and the latter is paramagnetic to the extent of four unpaired electrons (because it is a weak-field, high-spin complex). Similarly, $[Fe(CN)_6]^{4-}$ is diamagnetic, but $[Fe(H_2O)_6]^{2+}$ has four unpaired electrons; these complexes also correspond to the two $d^6$ configurations shown in Table 16-4.

### Example 16-5

The octahedral complex ions $[FeCl_6]^{3-}$ and $[Fe(CN)_6]^{3-}$ are both paramagnetic, but the former is high-spin and the latter is low-spin. Identify the $d$ electron configurations in these two octahedral complex ions. In which is the octahedral field splitting greater? How does the crystal field stabilization energy differ between the complexes?

### Solution

The $Fe^{3+}$ ion has five $d$ electrons. A high-spin complex such as $[FeCl_6]^{3-}$ has five unpaired spins $(t_{2g}^3 e_g^2)$; a low-spin complex such as $[Fe(CN)_6]^{3-}$ has one unpaired spin $(t_{2g}^5)$. The splitting $\Delta_o$ must be greater for cyanide than for chloride ion ligands. The CFSE for the $[FeCl_6]^{3-}$ complex is zero, while that for the $[Fe(CN)_6]^{3-}$ complex is $-2\Delta_o$.

*Related Problems:* 27, 28

### Square-Planar and Tetrahedral Complexes

Crystal field theory applies to square-planar and tetrahedral complexes as well as to octahedral complexes. Let us begin with a square-planar complex in which four ligands are brought up toward the metal ion along the $\pm x$ and $\pm y$ axes. The $d_{x^2-y^2}$

(*a*) **Square planar**

(*b*) **Tetrahedral**

**Figure 16-19**

Energy-level structures of the 3*d* orbitals in square-planar and tetrahedral crystal fields.

orbital on the metal has a higher energy than a $d_{z^2}$ orbital in such a field, because an electron in the $d_{x^2-y^2}$ orbital has maximum probability density along the $x$ and $y$ axes, where it is most strongly repelled by ligand electrons. In the same way, the energy of the $d_{xy}$ orbital lies above that of the $d_{xz}$ and $d_{yz}$ in a square-planar field because the former lies in the plane of the ligands, whereas each of the latter wave functions has a node in the $xy$ plane. Figure 16-19*a* shows the resulting level structure.

Another way to think of the square-planar crystal field splittings is to consider what happens to the octahedral energy levels from Figure 16-17*a* as the two ligands on the $\pm z$ axis move away. The two $e_g$ orbitals then no longer have the same energy, and the energy of the $d_{z^2}$ orbital falls. Meanwhile, the three $t_{2g}$ orbitals split, with the $d_{xz}$ and $d_{yz}$ orbitals decreasing in energy. Some coordination complexes can be described as having distorted octahedral structures, with the two $z$-axis ligands moved outward but not removed completely. The level splittings in these cases are intermediate between the octahedral and the square-planar patterns.

Tetrahedral complexes result from bringing up ligands to four of the eight corners of an imaginary cube with the metal ion at its center. The $d_{x^2-y^2}$ and $d_{z^2}$ orbitals point toward the centers of the cube faces, but the other three orbitals point toward the centers of the cube edges, which are closer to corners where the ligands are and therefore are more strongly repelled. Figure 16-19*b* depicts the result; it is the reverse of the energy-level ordering found for octahedral complexes. The magnitude of the splitting $\Delta_t$ is significantly smaller than that of $\Delta_o$.

Octahedral complexes are most common because the formation of six bonds to ligands, rather than four, confers greater stability. Square-planar complexes are important primarily for low-spin $d^8$ electron configurations with large splittings. Formation of an octahedral complex in such cases forces the last two electrons into a high-energy $e_g$ orbital; in the square-planar complex, they are in a lower-energy $d_{xy}$ orbital. Square-planar complexes predominate for the $d^8$ ions Pt(II) and Pd(II), which have large splittings, but $Ni^{2+}$, which has significantly smaller ligand field splittings, more often forms octahedral complexes. Tetrahedral complexes are less common because the coordination number is lower, the lowering of ligand field energy is smaller, and electrons must be placed into the upper level at an earlier stage in the filling process.

## 16-5 THE SPECTROCHEMICAL SERIES AND LIGAND FIELD THEORY

The most striking properties of the complexes of the transition metals are their colors (Fig. 16-20). Complexes of Co(III), for example, have these colors:

| | |
|---|---|
| $[Co(NH_3)_6]^{3+}$ | Orange |
| $[Co(NH_3)_4Cl_2]^+$ | A green form and a violet form |
| $[Co(NH_3)_5(H_2O)]^{3+}$ | Purple |

Colors arise because coordination complexes often absorb light in portions of the visible spectrum. As Section 15-3 explained, the color perceived in a sample is the color *complementary* to that which is most strongly absorbed. When white light enters an aqueous solution containing $[Co(NH_3)_5Cl]^{2+}$, the ion absorbs strongly near a wavelength of 530 nm, in the yellow-green region of the spectrum. Only the blue and red components of the white light are transmitted by the solution, which appears purple. A material that absorbs across all visible wavelengths appears gray or black, and one that absorbs weakly or not at all in the visible range is colorless.

The colors of many octahedral transition-metal compounds arise from the excitation of electrons from occupied $t_{2g}$ levels to empty $e_g$ levels. The frequency $\nu$ of

**Figure 16-20**
Several colored coordination compounds. Clockwise from top left, they are $Cr(CO)_6$ (white), $K_3[Fe(C_2O_4)_3]$ (green), $[Co(en)_3]I_3$ (orange), $[Co(NH_3)_5(H_2O)]Cl_3$ (red), and $K_3[Fe(CN)_6]$ (red-orange).

**Figure 16-21**
The colors of the hexaaqua complexes of metal ions (from left) $Mn^{2+}$, $Fe^{3+}$, $Co^{2+}$, $Ni^{2+}$, $Cu^{2+}$, and $Zn^{2+}$, prepared from their nitrate salts. Note that the $d^{10}$ $Zn^{2+}$ complex is colorless. The green color of the $Ni^{2+}$ is due to absorption of both red and blue light that passes through the solution. The yellow color of the solution containing $Fe(H_2O)_6^{3+}$ is caused by hydrolysis of that ion to form $Fe(OH)(H_2O)_5^{2+}$; if this reaction is suppressed, the solution is pale violet. (Leon Lewandowski)

light that is capable of inducing such a transition is related to the energy difference between the two states, $\Delta_o$, by

$$h\nu = \Delta_o$$

The larger the crystal field splitting, the higher the frequency of light absorbed most strongly and the shorter its wavelength. In $[Co(NH_3)_6]^{3+}$, an orange compound that absorbs most strongly in the violet region of the spectrum, the crystal field splitting $\Delta_o$ is larger than in $[Co(NH_3)_5Cl]^{2+}$, a violet compound that absorbs most strongly at lower frequencies (longer wavelengths) in the yellow-green region of the spectrum. A $d^{10}$ complex (such as those of $Zn^{2+}$ or $Cd^{2+}$) is colorless because all the $d$ levels (both $t_{2g}$ and $e_g$) are filled; consequently, a $t_{2g} \rightarrow e_g$ transition cannot take place, and the absorption in the visible region of the spectrum is very slight. High-spin $d^5$ complexes such as $[Mn(H_2O)_6]^{2+}$ and $[Fe(H_2O)_6]^{3+}$ also show only very weak absorption because processes in which a spin is reversed on excitation by light (required by the Pauli principle in these cases) occur only rarely. For this reason, the $Mn^{2+}$ complex ion with six water ligands has a very pale color (Fig. 16-21).

From measurements of the magnetic properties and absorption spectra of transition-metal complexes, it is possible to rank ligands from those that interact most weakly with the metal ion (and therefore give the smallest crystal field splitting) to those that interact most strongly and give the largest splitting. Although such an ordering is not invariably followed for every ligand with every central metal atom, it is nonetheless useful. The resulting order for a selection of ligands from weakest to strongest is called the **spectrochemical series:**

| | | |
|---|---|---|
| $I^- < Br^- < Cl^-$ | $< F^-, OH^- < H_2O <$ | $:NCS^- < NH_3 < en < CO, CN^-$ |
| Weak-field ligands (high spin) | Intermediate-field ligands | Strong-field ligands (low spin) |

Transition metals with $4d$ and $5d$ valence electrons have larger crystal field splittings than those with $3d$ valence electrons; even weak-field ligands yield low-spin complexes in these cases.

### Example 16-6

Predict which of the following octahedral complexes should have the shortest wavelength of absorption in the visible spectrum: $[FeF_6]^{3-}$, $[Fe(CN)_6]^{3-}$, $[Fe(H_2O)_6]^{3+}$.

## Solution

The $[Fe(CN)_6]^{3-}$ has the strongest-field ligands of the three complexes, and so its energy levels are split by the greatest amount. The frequency of light absorbed should be greatest, and the absorption wavelength should be the shortest, for this ion.

The observed color of $[Fe(CN)_6]^{3-}$ solutions is red, indicating absorption of blue and violet light. Solutions of $[Fe(H_2O)_3]^{3+}$ are a very pale violet, due to weak absorption of red light, and $[FeF_6]^{3-}$ solutions are colorless, indicating that the absorption lies beyond the long wavelength limit of the visible spectrum.

*Related Problems:* 37, 38, 39, 40

## Ligand Field Theory

The ordering of the spectrochemical series cannot be explained within the crystal field model and indeed points up an unsatisfactory feature of that model. The bonding in coordination compounds *cannot* be fully ionic, as assumed in the crystal field description, because neutral ligands such as $NH_3$ and $CO$ give *larger* splittings than those induced by a small negative ion such as $F^-$. The crystal field theory has been modified to include covalent as well as ionic aspects of coordination. The resulting **ligand field theory** gives a powerful description of transition-metal complexes.

Ligand field theory begins by constructing molecular orbitals from the valence $d$, $s$, and $p$ orbitals of the central metal atom and from the six ligand orbitals that point along the metal–ligand bonds in an octahedral complex. The ligand orbital would be the $sp^3$ orbital containing a lone pair in $NH_3$, or the $p_z$ orbital from a $Cl^-$ ion lying on the $z$ axis. Because the phases of the wave function alternate in its different lobes, there is no net overlap between the $d_{xy}$, $d_{yz}$, or $d_{xz}$ orbital of the metal atom and a ligand orbital directed along the bond axis (this is analogous to the absence of net overlap between a $p$ orbital perpendicular to the bond axis and an $s$ orbital on the second atom, shown in Fig. 14-19$b$). As a result, these three $d$ orbitals are nonbonded orbitals on the metal atom, and their energy shifts neither up nor down. The other two metal $d$ orbitals ($d_{x^2-y^2}$ and $d_{z^2}$), together with the metal $s$ and $p$ valence orbitals (a total of six atomic orbitals), *can* be combined with the six ligand orbitals to form a total of twelve molecular orbitals, six of which are $\sigma$ bonding orbitals and six $\sigma^*$ antibonding orbitals. Figure 16-22 is the resulting orbital correlation diagram.

The molecular orbitals, beginning with those of lowest energy, are then filled with electrons. Twelve electrons are available from the bonding electron pairs on the six ligands, and these fill the six $\sigma$ bonding orbitals. The metal $d$ electrons then are placed in the nonbonding $d_{xy}$, $d_{xz}$, and $d_{yz}$ orbitals ($t_{2g}$) and the pair of antibonding $\sigma^*$ $d_{x^2-y^2}$ and $\sigma^*$ $d_{z^2}$ orbitals that lie above them ($e_g$). These five orbitals have exactly the same level structure as in the crystal field theory (Fig. 16-17), so that all of the results from that theory can be carried over to this one. The magnitude of the splitting $\Delta_o$ arises here from the increase in energy of the antibonding $\sigma^*$ molecular orbitals relative to the nonbonding metal-atom $d$ orbitals. The ligand field picture allows for partial covalent and partial ionic character in the bonding of the complex. The lower the energy of the original ligand orbitals relative to the metal $d$ orbitals, the more closely the $\sigma$ bonding orbitals resemble the pure ligand orbitals. In this limit the bonding becomes purely ionic. The situation closely parallels the discusshown for $PF_5$ in Figure 14-8. In organometallic compounds, special stability occurs when the central metal atom is surrounded by 18 valence electrons, which give it a closed shell (10 $d$, 6 $p$, and 2 $s$ electrons). The iron atom in $Fe(CO)_5$ is surrounded by 8 of its own valence electrons plus 10 from the five CO ligand molecules, giving a total of 18 electrons. Other novel structures (Fig. 16-25) include metal–metal bonds and three-center bridge bonds formed by carbonyl groups. In molecules of $Mn_2(CO)_{10}$, each Mn atom is surrounded by 7 of its own valence electrons, 10 electrons from five CO molecules, and 1 from the electron-pair bond formed with the second Mn atom, for a total of 18. Some compounds do not fit the "rule of 18," however; molecules of the blue-green crystalline compound $V(CO)_6$ have a central vanadium atom that shares only 17 valence electrons. The odd number of electrons makes this compound paramagnetic.

Other organometallic compounds are "sandwich compounds," in which a transition-metal atom is situated between two planar hydrocarbon rings with delocalized $\pi$ electrons. One of the most stable sandwich compounds is ferrocene, $Fe(C_5H_5)_2$, in which an $Fe^{2+}$ ion is coordinated to two cyclopentadienyl anions ($C_5H_5^-$), giving the structure shown in Figure 16-26. The $C_5H_5^-$ anion itself is a resonance hybrid of five Lewis structures, two of which are

(a) $Mn_2(CO)_{10}$

(b) $Co_2(CO)_8$

**Figure 16-25**
The structures of two binuclear carbonyls. In part $b$, note the bridge bonding by CO molecules.

Molecules of ferrocene have 18 electrons about the iron ion: 6 $d$ electrons from $Fe^{2+}$ and 6 $\pi$ electrons from each $C_5H_5^-$ ion. The compound is remarkably stable, forming orange crystals that melt at 173°C to an orange liquid that boils at 249°C. The compound is thermally stable to about 500°C. A related, more weakly bound 18-electron compound has molecules with a chromium atom sandwiched between two benzene rings, $Cr(C_6H_6)_2$.

One reason to take particular interest in metal–carbon bonds is the catalytic role transition metals often play in making organic compounds. An example is the Monsanto process, which is now the predominant industrial route to acetic acid. The overall reaction is

$$CH_3OH + CO \longrightarrow CH_3COOH$$

with both starting materials (methanol and carbon monoxide) produced economically through the partial oxidation of methane from natural gas. An organometallic complex of rhodium(II) and iodine catalyzes the synthesis. In the presence of this compound, the production of acetic acid proceeds smoothly under mild conditions (a pressure of 1 atm and a temperature of 175°C). The mechanism of the reaction is believed to be that shown in Figure 16-27. Methanol reacts with hydrogen iodide (formed from the iodine) to give methyl iodide and water (see central box of figure):

$$CH_3OH + HI \longrightarrow CH_3I + H_2O$$

The methyl iodide adds to a four-coordinated rhodium complex to give a six-coordinated complex with an Rh—$CH_3$ bond (step $a$). In the crucial step of the mechanism, one of the carbonyl groups already attached to the Rh atom swings up and

**Figure 16-26**
The structure of ferrocene, $Fe(C_5H_5)_2$, in the gaseous state. Each flat pentagon represents a $C_5H_5$ ring.

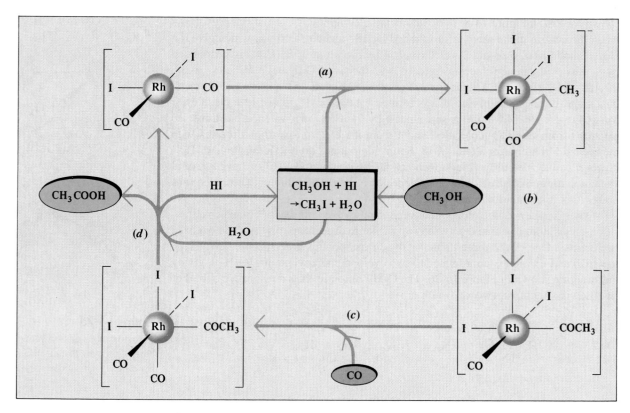

**Figure 16-27**

The Monsanto process for making acetic acid (shown in green) from methanol and carbon monoxide (red). This cyclic process proceeds clockwise around the diagram. The rhodium coordination complex (*upper left*) returns to its initial state after each cycle. The crucial step is *b*, in which a CO ligand swings up (colored arrow) and inserts itself into the Rh—CH₃ bond.

inserts itself into the Rh—CH₃ bond (step *b*). Another CO molecule then coordinates to the Rh atom to replace the CO lost to this insertion (step *c*). Upon addition of water, the —COCH₃ group detaches from the rhodium to give acetic acid (step *d*). The iodine atom leaves at the same time, regenerating hydrogen iodide, and the rhodium complex is returned to its original state, ready to start another catalytic cycle.

## Coordination Complexes in Biology

Coordination complexes, particularly chelates, play fundamental roles in the biochemistry of both plants and animals. Trace amounts of at least nine transition elements are essential to life—vanadium, chromium, manganese, iron, cobalt, nickel, copper, zinc, and molybdenum.

Several of the most important complexes are based on *porphine,* a compound of carbon, nitrogen, and hydrogen whose structure was shown in Figure 15-15*a*. Re-

**Figure 16-28**

The structure of vitamin $B_{12}$. Carbon atoms are understood to lie at the intersections of bonds where no other atoms are indicated.

moval of two central hydrogen ions leaves four nitrogen atoms that can bind to a metal ion $M^{2+}$ and form a tetradentate chelate structure. Such a structure, modified by the addition of several side groups, gives a complex with $Fe^{2+}$ ions called *heme* (Fig. 15-15c). Recall from Section 15-3 that photosynthesis involves tetradentate chelate structures in which the four nitrogen atoms bind to $Mg^{2+}$ (in chlorophyll) or $Fe^{2+}$ (in the heme group in the cytochrome protein). *Hemoglobin,* the compound that transports oxygen in the blood, contains four such heme groups. The fifth coordination site of each iron(II) ion binds *globin* (a high-molar-mass protein) via nitrogen atoms on that protein, and the sixth site is occupied by water or molecular oxygen. Oxygen is a strong-field ligand that makes the $d^6$ $Fe^{2+}$ diamagnetic and gives the short-wavelength absorption that is responsible for the red color of blood in the arteries. After the oxygen is delivered to the cells, it is replaced by the weaker-field ligand water, and a paramagnetic complex forms that absorbs at longer wavelengths, producing the bluish tint of blood in the veins. In cases of carbon monoxide poisoning, CO molecules occupy the sixth coordination site and block the binding and transport of oxygen; the equilibrium constant for CO binding is 200 times that for $O_2$ binding.

Vitamin $B_{12}$ is useful in the treatment of pernicious anemia and other diseases. Its structure (Fig. 16-28) has certain similarities to that of heme. Again, a metal ion is coordinated by a planar tetradentate ligand, with two other donors completing the coordination octahedron by occupying *trans* positions. In vitamin $B_{12}$, the metal ion is cobalt and the planar ring is *corrin,* which is somewhat similar to porphine. In the human body, enzymes derived from vitamin $B_{12}$ accelerate a range of important reactions, including those involved in producing red blood cells.

## PRACTICE EXERCISE

Crystals of potassium hexachloro-
platinate(IV) ($K_2PtCl_6$). (ASTRA)

### Platinum

The precious metal platinum was first used by South American Indians, who found impure, native samples of it in the gold mines of what is now Ecuador and used the samples to make small items of jewelry. Platinum's high melting point (1772°C) makes it harder to work than gold (m.p. 1064°C) and silver (m.p. 962°C), but this same property and a high resistance to chemical attack make platinum suitable as a material for high-temperature crucibles. Although platinum is a noble metal, in the +4 and +2 oxidation states it forms a wide variety of compounds, many of which are coordination complexes. Its coordinating abilities make it an important catalyst for organic and inorganic reactions.

(a) The anticancer drug cisplatin, $cis$-[Pt(NH$_3$)$_2$Cl$_2$] (Fig. 16-12$b$), can be prepared from $K_2PtCl_6$ via reduction with $N_2H_4$ (hydrazine), giving $K_2PtCl_4$, followed by replacement of two chloride ion ligands with ammonia. Give systematic names to the three platinum complexes referred to in this statement.

(b) The coordination compound diamminetetracyanoplatinum(IV) has been prepared, but salts of the hexacyanoplatinate(IV) ion have not. Write the chemical formulas of these two species.

(c) Platinum forms organometallic complexes quite readily. In one of the simplest of these, Pt(II) is coordinated to two chloride ions and two molecules of ethylene ($C_2H_4$) to give an unstable yellow crystalline solid. Can this complex have more than one isomer? If so, describe the possible isomers.

(d) Platinum(IV) is readily complexed by ethylenediamine. Draw the structures of both enantiomers of the complex ion $cis$-[Pt(Cl)$_2$(en)$_2$]$^{2+}$. In this compound the Cl$^-$ ligands are $cis$ to one another.

(e) In platinum(IV) complexes, the octahedral crystal field splitting $\Delta_o$ is relatively large. Is $K_2PtCl_6$ diamagnetic or paramagnetic? What is its $d$ electron configuration?

(f) Is cisplatin diamagnetic or paramagnetic?

(g) The salt $K_2$[PtCl$_4$] is red, but [Pt(NH$_3$)$_4$]Cl$_2 \cdot$H$_2$O is colorless. In what regions of the spectrum do the dominant absorptions for these compounds lie?

(h) When the two salts from part (g) are dissolved in water and the solutions mixed, a green precipitate called Magnus's green salt forms. Propose a chemical formula for this salt, and assign the corresponding systematic name.

### Answers

(a) $cis$-diamminedichloroplatinum(II), potassium hexachloroplatinate(IV), and potassium tetrachloroplatinate(II)

(b) [Pt(NH$_3$)$_2$(CN)$_4$] and [Pt(CN)$_6$]$^{2-}$

(c) Pt(II) forms square-planar complexes. There are two possible forms, arising from $cis$ and $trans$ placement of the ethylene molecules. These are analogous to the two isomers shown in Figure 16-12$b$ and $c$.

(e) The six $d$ electrons in Pt(IV) are all in the lower $t_{2g}$ level in a low-spin, large-$\Delta_o$ complex. All are paired, and so the compound is diamagnetic.

(f) Diamagnetic, with the bottom four levels in the square-planar configuration (all but $d_{x^2-y^2}$) occupied

(g) Red transmission corresponds to absorption of green light by $K_2$[PtCl$_4$]. A colorless solution has absorptions at either higher or lower frequency than visible. Because Cl$^-$ is a weaker field ligand than NH$_3$, the absorption frequency should be higher for the $Pt(NH_3)_4^{2+}$ complex, putting it in the ultraviolet region of the spectrum.

(h) $[Pt(NH_3)_4][PtCl_4]$, tetraammineplatinum(II) tetrachloroplatinate(II)

After studying this chapter and working the problems that follow, you should be able to

1. Discuss the systematic variation of physical and chemical properties of the transition elements through the periodic table (Section 16-1, problems 1–6).
2. Define the terms "coordination compound," "coordination number," "ligand," and "chelation" (Section 16-2).
3. Name coordination compounds, given their molecular formulas (Section 16-2, problems 9–14).
4. Draw geometrical isomers of octahedral, tetrahedral, and square-planar complexes (Section 16-3, problems 21–24).
5. Use crystal field theory to interpret the magnetic properties of coordination compounds in terms of the electron configurations of their central ions (Section 16-4, problems 25–28).
6. Relate the colors of coordination compounds to their crystal field splitting energies and crystal field stabilization energies (Section 16-5, problems 33–40).
7. Use ligand field theory to order the energy levels in coordination compounds and to account for the spectrochemical series (Section 16-5).
8. Describe several organometallic compounds, including compounds that have roles in biology (Section 16-6, problems 41–44).

### Chemistry of the Transition Metals

1. Of the compounds $PtF_4$ and $PtF_6$, predict (a) which is more soluble in water and (b) which is more volatile.
2. The melting point of $TiCl_4$ ($-24°C$) lies below those of $TiF_4$ ($284°C$) and $TiBr_4$ ($38°C$). Explain why by considering the covalent–ionic nature of these compounds and the intermolecular forces in each case.
3. The decavanadate ion is a complex species with chemical formula $V_{10}O_{28}^{6-}$. It reacts with an excess of acid to form the dioxovanadium ion, $VO_2^+$, and water. Write a balanced chemical equation for this reaction. In what oxidation state is vanadium before and after this reaction? What vanadium oxide has the same oxidation state?
4. What is the chemical formula of vanadium(III) oxide? Do you predict this compound to be more basic or more acidic than the vanadium oxide of problem 3? Write a balanced chemical equation for the reaction of vanadium(III) oxide with a strong acid.

5. Titanium(III) oxide is prepared by reaction of titanium(IV) oxide with hydrogen at high temperature. Write a balanced chemical equation for this reaction. Which oxide do you expect to have more strongly basic properties?
6. Treatment of cobalt(II) oxide with oxygen at high temperatures gives $Co_3O_4$. Write a balanced chemical equation for this reaction. What is the oxidation state of cobalt in $Co_3O_4$?

### The Formation of Coordination Complexes

7. Will methylamine ($CH_3NH_2$) be a monodentate or a bidentate ligand? With which of its atoms will it bind to a metal ion?
8. Show how the glycinate ion ($H_2N$—$CH_2$—$COO^-$) can act as a bidentate ligand. (Draw a Lewis structure if necessary.) Which atoms in the glycinate ion will bind to a metal ion?
9. Determine the oxidation state of the metal in each of the following coordination complexes: $[V(NH_3)_4Cl_2]$, $[Mo_2Cl_8]^{4-}$, $[Co(H_2O)_2(NH_3)Cl_3]^-$, $[Ni(CO)_4]$.

10. Determine the oxidation state of the metal in each of the following coordination complexes: $Mn_2(CO)_{10}$, $[Re_3Br_{12}]^{3-}$, $[Fe(H_2O)_4(OH)_2]^+$, $[Co(NH_3)_4Cl_2]^+$.

11. Give the chemical formula that corresponds to each of the following compounds.
    (a) Sodium tetrahydroxozincate(II)
    (b) Dichlorobis(ethylenediamine)cobalt(III) nitrate
    (c) Triaquabromoplatinum(II) chloride
    (d) Tetraamminedinitroplatinum(IV) bromide

12. Give the chemical formula of each of the following compounds.
    (a) Silver hexacyanoferrate(II)
    (b) Potassium tetraisothiocyanatocobaltate(II)
    (c) Sodium hexafluorovanadate(III)
    (d) Potassium trioxalatochromate(III)

13. Assign a systematic name to each of the following chemical compounds.
    (a) $NH_4[Cr(NH_3)_2(NCS)_4]$    (c) $K[Mn(CN)_5]$
    (b) $[Tc(CO)_5]I$    (d) $[Co(NH_3)_4(H_2O)Cl]Br_2$

14. Give the systematic name for each of the following chemical compounds.
    (a) $[Ni(H_2O)_4(OH)_2]$    (c) $K_4[Os(CN)_6]$
    (b) $[HgClI]$    (d) $[FeBrCl(en)_2]Cl$

15. Describe the behavior of a coordination complex that is both thermodynamically stable and labile.

16. Describe the behavior of a coordination complex that is both thermodynamically unstable and inert.

17. Calculate the standard free energy change for the dissociation of tetraamminecopper(II) ion to $Cu^{2+}(aq)$ and $NH_3(aq)$ in basic aqueous solution, using data from Appendix D. Is this complex thermodynamically stable with respect to this reaction under standard conditions? Why does it dissociate under acidic conditions? Write a balanced equation for the dissociation reaction at pH = 0, and calculate the standard free energy change for that reaction.

18. The $\Delta G_f^\circ$ of the tetracyanonickelate(II) anion is +472.1 kJ mol$^{-1}$ in aqueous solution. Use this information, together with data from Appendix D, to calculate the equilibrium constant at 25°C for the dissociation

$$[Ni(CN)_4]^{2-}(aq) \longrightarrow Ni^{2+}(aq) + 4\,CN^-(aq)$$

Isotopically labeled $Na^{13}CN$ is added to a solution of $K_2Ni(CN)_4$. Subsequent examination of the $Ni(CN)_4^{2-}$ present reveals extensive incorporation of $^{13}CN^-$ into the complex ion. Comment on these observations, and discuss the difference between lability and thermodynamic stability as applied to complexes of nickel(II) with the cyanide ion.

## Structures of Coordination Complexes

19. Suppose 0.010 mol of each of the following compounds is dissolved (separately) in 1.0 L of water: $KNO_3$, $[Co(NH_3)_6]Cl_3$, $Na_2[PtCl_6]$, $[Cu(NH_3)_2Cl_2]$. Rank the re-

sulting four solutions in order of conductivity, from lowest to highest.

20. Suppose 0.010 mol of each of the following compounds is dissolved (separately) in 1.0 L of water: $BaCl_2$, $K_4[Fe(CN)_6]$, $[Cr(NH_3)_4Cl_2]Cl$, $[Fe(NH_3)_3Cl_3]$. Rank the resulting four solutions in order of conductivity, from lowest to highest.

21. Draw the structures of all possible isomers for the following complexes. Indicate which isomers are enantiomer pairs.
    (a) Diamminebromochloroplatinum(II) (square-planar)
    (b) Diaquachlorotricyanocobaltate(III) ion (octahedral)
    (c) Trioxalatovanadate(III) ion (octahedral)

22. Draw the structures of all possible isomers for the following complexes. Indicate which isomers are enantiomer pairs.
    (a) Bromochloro(ethylenediamine)platinum(II) (square-planar)
    (b) Tetraamminedichloroiron(III) ion (octahedral)
    (c) Amminechlorobis(ethylenediamine)iron(III) ion (octahedral)

23. Iron(III) forms octahedral complexes. Sketch the structures of all the distinct isomers of $[Fe(en)_2Cl_2]^+$, indicating which pairs of structures are mirror images of each other.

24. Platinum(IV) forms octahedral complexes. Sketch the structures of all the distinct isomers of $[Pt(NH_3)_2Cl_2F_2]$, indicating which pairs of structures are mirror images of each other.

## Crystal Field Theory and Magnetic Properties

25. For each of the following ions, draw diagrams like those in Figure 16-17 to show orbital occupancies in both weak and strong octahedral fields. Indicate the total number of unpaired electrons in each case.
    (a) $Mn^{2+}$
    (b) $Zn^{2+}$
    (c) $Cr^{3+}$
    (d) $Mn^{3+}$
    (e) $Fe^{2+}$

26. Repeat the work of the preceding problem for the following ions.
    (a) $Cr^{2+}$
    (b) $V^{3+}$
    (c) $Ni^{2+}$
    (d) $Pt^{4+}$
    (e) $Co^{2+}$

27. Experiments can measure not only whether a compound is paramagnetic but also the number of unpaired electrons. It is found that the octahedral complex ion $[Fe(CN)_6]^{3-}$ has fewer unpaired electrons than the octahedral complex ion $[Fe(H_2O)_6]^{3+}$. How many unpaired electrons are present in each species? Explain. In each case, express the crystal field stabilization energy in terms of $\Delta_o$.

28. The octahedral complex ion $[MnCl_6]^{3-}$ has more unpaired spins than the octahedral complex ion $[Mn(CN)_6]^{3-}$. How many unpaired electrons are present in each species? Explain.

In each case, express the crystal field stabilization energy in terms of $\Delta_o$.

29. Explain why octahedral coordination complexes with three and eight $d$ electrons on the central metal atom are particularly stable. Under what circumstances would you expect complexes with five or six $d$ electrons on the central metal atom to be particularly stable?

30. The following reduction potentials have been measured in acidic aqueous solution:

$$Mn^{3+} + e^- \longrightarrow Mn^{2+} \qquad \mathscr{E}° = 1.51 \text{ V}$$

$$Fe^{3+} + e^- \longrightarrow Fe^{2+} \qquad \mathscr{E}° = 0.77 \text{ V}$$

$$Co^{3+} + e^- \longrightarrow Co^{2+} \qquad \mathscr{E}° = 1.84 \text{ V}$$

Explain why the reduction potential for $Fe^{3+}$ lies below those for the +3 oxidation states of the elements on both sides of it in the periodic table.

## The Spectrochemical Series and Ligand Field Theory

31. An aqueous solution of zinc nitrate contains the $[Zn(H_2O)_6]^{2+}$ ion and is colorless. What conclusions can be drawn about the absorption spectrum of the $[Zn(H_2O)_6]^{2+}$ complex ion?

32. An aqueous solution of sodium hexaiodoplatinate(IV) is black. What conclusions can be drawn about the absorption spectrum of the $[PtI_6]^{2-}$ complex ion?

33. Estimate the wavelength of maximum absorption for the octahedral ion hexacyanoferrate(III) from the fact that light transmitted by a solution of it is red. Estimate the crystal field splitting energy $\Delta_o$ (in kJ mol$^{-1}$).

34. Estimate the wavelength of maximum absorption for the octahedral ion hexaaquanickel(II) from the fact that its solutions are colored green by transmitted light. Estimate the crystal field splitting energy $\Delta_o$ (in kJ mol$^{-1}$).

35. Estimate the crystal field stabilization energy for the complex in problem 33. (*Note:* This is a high-field (low-spin) complex.)

36. Estimate the crystal field stabilization energy for the complex in problem 34.

37. The chromium(III) ion in aqueous solution has a blue-violet color.
(a) What is the complementary color to blue-violet?
(b) Estimate the wavelength of maximum absorption for a $Cr(NO_3)_3$ solution.
(c) Will the wavelength of maximum absorption increase or decrease if cyano ligands are substituted for the coordinated water? Explain.

38. An aqueous solution containing the hexaamminecobalt(III) ion has a yellow color.
(a) What is the complementary color to yellow?
(b) Estimate this solution's wavelength of maximum absorption in the visible spectrum.

39. (a) An aqueous solution of $Fe(NO_3)_3$ has only a pale color, but an aqueous solution of $K_3[Fe(CN)_6]$ is bright red. Do you expect a solution of $K_3[FeF_6]$ to be brightly colored or pale? Explain your reasoning.
(b) Would you predict a solution of $K_2[HgI_4]$ to be colored or colorless? Explain.

40. (a) An aqueous solution of $Mn(NO_3)_2$ has a very pale pink color, but an aqueous solution of $K_4[Mn(CN)_6]$ is deep blue. Explain why the two differ so much in the intensities of their colors.
(b) Predict which of the following compounds would be colorless in aqueous solution: $K_2[Co(NCS)_4]$, $Zn(NO_3)_2$, $[Cu(NH_3)_4]Cl_2$, $CdSO_4$, $AgClO_3$, $Cr(NO_3)_2$.

## Organometallic Compounds and Catalysis

41. Give the number of valence electrons surrounding the central transition-metal ion or atom in each of the following: $[Cr(CO)_4]$, $[Os(CO)_5]$, $[H_2Fe(CO)_4]$, $K_3[Fe(CN)_5CO]$. The hydrogen atoms are covalently bonded to the central iron atom in the third case.

42. Give the number of valence electrons surrounding the central transition-metal ion or atom in each of the following: $[HMn(CO)_5]$, $[V(CO)_6]$, $Na[Co(CO)_4]$, $[HTc(CO)_5]$. The hydrogen atoms are covalently bonded to the central metal atom in each case.

43. Vitamin $B_{12s}$ is 4.43% cobalt by mass. Determine the molar mass of vitamin $B_{12s}$ if each molecule of it contains one atom of cobalt.

44. Hemoglobin, the compound that transports oxygen in the bloodstream, has a molar mass of $6.8 \times 10^4$ g mol$^{-1}$. It is also 0.33% iron by mass. Determine how many iron atoms are in a single hemoglobin molecule.

## Additional Problems

45. Of the ten fourth-period transition metal elements in Table 16-1, which one has particularly low melting and boiling points? How can you explain this in terms of the electronic configuration of this element?

*46. Although copper lies between nickel and zinc in the periodic table, the reduction potential of $Cu^{2+}$ is above that for both $Ni^{2+}$ and $Zn^{2+}$ (see Table 16-1). Use other data from Table 16-1 to account for this observation. (*Hint:* Think of a multi-step process to convert a metal atom in the solid to a metal ion in solution. For each step, compare the relevant energy or enthalpy changes for Cu with those for Ni or Zn.)

47. A reference book lists five different values for the electronegativity of molybdenum, a different value for each oxidation state from +2 through +6. Predict which electronegativity is highest and which is lowest.

48. If *trans*-$[Cr(en)_2(NCS)_2]SCN$ is heated, it forms gaseous ethylenediamine and solid $[Cr(en)_2(NCS)_2][Cr(en)(NCS)_4]$.

Write a balanced chemical equation for this reaction. What are the oxidation states of the chromium ions in the reactant and in the two complex ions in the product?

49. A coordination complex has the molecular formula $[Ru_2(NH_3)_6Br_3](ClO_4)_2$. Determine the oxidation state of ruthenium in this complex.

50. Heating 2.0 mol of a coordination compound gives 1.0 mol $NH_3$, 2.0 mol $H_2O$, 1.0 mol $HCl$, and 1.0 mol $(NH_4)_3[Ir_2Cl_9]$. Write the formula of the original (six-coordinate) coordination compound and name it.

51. Explain why ligands are usually negative or neutral in charge, and only rarely positive.

52. Match each compound in the group on the left with the compound on the right most likely to have the same electrical conductivity per mole in aqueous solution.

    (a) $[Fe(H_2O)_5Cl]CO_3$      HCN
    (b) $[Mn(H_2O)_6]Cl_3$        $Fe_2(SO_4)_3$
    (c) $[Zn(H_2O)_3(OH)]Cl$      NaCl
    (d) $[Fe(NH_3)_6]_2(CO_3)_3$  $MgSO_4$
    (e) $[Cr(NH_3)_3Br_3]$        $Na_3PO_4$
    (f) $K_3[Fe(CN)_6]$           $GaCl_3$

53. Three different compounds are known to have the empirical formula $CrCl_3 \cdot 6H_2O$. When exposed to a dehydrating agent, compound 1 (which is dark green) loses 2 mol of water per mole of compound, compound 2 (light green) loses 1 mol of water, and compound 3 (violet) loses no water. What are the probable structures of these compounds? If an excess of silver nitrate solution is added to 100.0 g of each of these compounds, what mass of silver chloride will precipitate in each case?

54. The octahedral structure is not the only possible six-coordinate structure. Other possibilities include a planar hexagonal structure and a triangular prism structure. In the latter, the ligands are arranged in two parallel triangles, one lying above the metal atom and the other below the metal atom with its corners directly in line with the corners of the first triangle. Show that the existence of two and only two isomers of $[Co(NH_3)_4Cl_2]^+$ is evidence against both of these possible structures.

55. Cobalt(II) forms more tetrahedral complexes than any other ion except zinc(II). Draw the structure(s) of the tetrahedral complex $[CoCl_2(en)]$. Could this complex exhibit geometrical or optical isomerism? If one of the $Cl^-$ ligands is replaced by $Br^-$, what kinds of isomerism, if any, are possible in the resulting compound?

56. Is the coordination compound $[Co(NH_3)_6]Cl_2$ diamagnetic or paramagnetic?

57. A coordination compound has the empirical formula $PtBr(en)(SCN)_2$ and is diamagnetic.
    (a) Examine the $d$ electron configurations on the metal atoms, and explain why the formulation $[Pt(en)_2(SCN)_2]$ $[PtBr_2(SCN)_2]$ is preferred for this substance.
    (b) Name this compound.

58. We used crystal field theory to order the energy-level splittings induced in the five $d$ orbitals. The same procedure could be applied to $p$ orbitals. Predict the level splittings (if any) induced in the three $p$ orbitals by octahedral and square-planar crystal fields.

59. The three complex ions $[Mn(CN)_6]^{5-}$, $[Mn(CN)_6]^{4-}$, and $[Mn(CN)_6]^{3-}$ have all been synthesized, and all are low-spin octahedral complexes. For each complex, determine the oxidation number of Mn, the configuration of the $d$ electrons (how many $t_{2g}$ and how many $e_g$), and the number of unpaired electrons present.

60. Is the $Mn^{2+}(aq)$ ion a strong reducing agent? Is it a strong oxidizing agent? (Use Appendix E.) What can you conclude about the stability of $Mn^{2+}(aq)$?

61. The following ionic radii (in angstroms) are estimated for the $+2$ ions of selected elements of the first transition-metal series, based on the structures of their oxides: $Ca^{2+}(0.99)$, $Ti^{2+}(0.71)$, $V^{2+}(0.64)$, $Mn^{2+}(0.80)$, $Fe^{2+}(0.75)$, $Co^{2+}(0.72)$, $Ni^{2+}(0.69)$, $Cu^{2+}(0.71)$, $Zn^{2+}(0.74)$. Draw a graph of ionic radius versus atomic number in this series, and account for its shape. The oxides take the rock salt structure. Are these solids better described as high-spin or low-spin transition-metal complexes?

62. The coordination geometries of $[Mn(NCS)_4]^{2-}$ and $[Mn(NCS)_6]^{4-}$ are tetrahedral and octahedral, respectively. Explain why the two have the same room-temperature molar magnetic susceptibility.

63. The complex ion $CoCl_4^{2-}$ has a tetrahedral structure. How many $d$ electrons are on the Co? What is its electronic configuration? Why is the tetrahedral structure stable in this case?

64. In the coordination compound $(NH_4)_2[Fe(H_2O)F_5]$, the Fe is octahedrally coordinated.
    (a) Based on the fact that $F^-$ is a weak-field ligand, predict whether this compound is diamagnetic or paramagnetic. If it is paramagnetic, tell how many unpaired electrons it has.
    (b) By comparison with other complexes discussed in the chapter, discuss the likely color of this compound.
    (c) Determine the $d$ electron configuration of the iron in this compound.
    (d) Name this compound.

65. The compound $Cs_2[CuF_6]$ is bright orange and paramagnetic. Determine the oxidation number of copper in this compound, the most likely geometry of the coordination around the copper, and the possible configurations of the $d$ electrons of the copper.

66. In what directions do you expect the bond length and vibrational frequency of a free CO molecule to change when it becomes a CO ligand in an $Ni(CO)_4$ molecule? Explain your reasoning.

67. Give the number of valence electrons surrounding the central transition-metal ion in each of the following known organometallic compounds or complex ions: $[Co(C_5H_5)_2]^+$, $[Fe(C_5H_5)(CO)_2Cl]$, $[Mo(C_5H_5)_2Cl_2]$, $[Mn(C_5H_5)(C_6H_6)]$.

**68.** Molecular nitrogen ($N_2$) can act as a ligand in certain coordination complexes. Predict the structure of $[V(N_2)_6]$, which is isolated by condensing V with $N_2$ at 25 K. Is this compound diamagnetic or paramagnetic? What is the formula of the carbonyl compound of vanadium that has the same number of electrons?

**69.** How is the rule of 18, discussed in Section 16-6, related to the ligand field theory of Section 16-5? What levels are occupied in a complex such as hexacarbonylchromium(0)? Are any electrons placed into antibonding orbitals that are derived from the chromium $d$ orbitals?

**\*70.** The compound $WH_2(C_5H_5)_2$ acts as a base, but $TaH_3(C_5H_5)_2$ does not. Explain.

**71.** Discuss the role of transition-metal complexes in biology. Consider such aspects as their absorption of light, the existence of many different structures, and the possibility of multiple oxidation states.

## CUMULATIVE PROBLEMS

**72.** Predict the volume of hydrogen generated at 1.00 atm and 25°C by the reaction of 4.53 g of scandium with excess aqueous hydrochloric acid.

**\*73.** Mendeleyev's early periodic table (Fig. 2-5) placed manganese and chlorine in the same group. Discuss the chemical evidence for these placements, focusing on the oxides of the two elements and their acid–base and redox properties. Is there a connection between the electronic structures of their atoms? In what ways are the elements different?

**74.** A 0.10 M solution of $Na_2Cr_2O_7$ is prepared, and the pH is adjusted to 7.00. Calculate the equilibrium concentration of $CrO_4^{2-}$ that results.

**\*75.** Methylamine ($NH_2CH_3$) is a monodentate ligand, whereas ethylenediamine ($H_2NCH_2CH_2NH_2$) is bidentate. For the two reactions

$$Cd^{2+}(aq) + 4\,NH_2CH_3(aq) \longrightarrow Cd(NH_2CH_3)_4^{2+}(aq)$$

$$Cd^{2+} + 2\,H_2NCH_2CH_2NH_2(aq) \longrightarrow Cd(en)_2^{2+}(aq)$$

the energy and enthalpy changes are quite similar, because the bonds formed in the complexes are much the same. Nevertheless, the formation constants for the two complexes differ by more than four orders of magnitude. Explain these facts by predicting which reaction will have the less negative entropy change. Which complex is more stable? This effect of chelation on the entropy change is sometimes referred to as the *chelate effect*.

**76.** An orange-yellow osmium carbonyl compound is heated to release CO and leave elemental osmium behind. Treatment of 6.79 g of the compound releases 1.18 L of $CO(g)$ at 25°C and 2.00 atm pressure. What is the empirical formula of this compound? Propose a possible structure for it by comparing with the metal carbonyls shown in Figure 16-25.

*The structure of diamond consists of a network of interlocking tetrahedral carbon atoms.*

# Structure and Bonding in Solids

**A**lthough the symmetry and beauty of crystals have always excited curiosity and wonder, the science of **crystallography** did not begin until the latter part of the 18th century. In those closing days of the Age of Enlightenment, when Lavoisier led the modern approach to chemistry, another brilliant French thinker established the fundamental laws of crystallography. René Just Haüy was struck by the fact that when he accidentally dropped a crystal of calcite (a form of calcium carbonate), it fractured into smaller crystals that had the same interfacial angles between their plane surfaces as did the original crystal. The constancy of the interfacial angles observed when crystals are cleaved is known as Haüy's law. Haüy drew the conclusion that the outward symmetry of crystals (Fig. 17-1) implies an internal regularity and the existence of a smallest crystal unit. His inferences were correct; what distinguishes the crystalline state from the gaseous and liquid states is the nearly perfect positional order of the atoms, ions, or molecules in crystals.

This chapter begins with a look at the microscopic structure of a perfect crystal. We then examine the sources of cohesion in matter, identifying the forces that hold together different kinds of solids. More disordered condensed phases of matter— defective crystals and amorphous solids—are then considered. The chapter closes with a look at liquid crystals, states of matter that are intermediate between solids and liquids in their physical properties.

**Figure 17-1**

Microcrystals of realgar, $As_4S_4$. (Julius Weber)

## 17-1 CRYSTAL SYMMETRY AND THE UNIT CELL

Crystals can be classified according to the natures of the bonds between the atoms, ions, or molecules that compose them, and such classification will later prove useful (Section 17-3). A more fundamental classification of crystals employs the numbers and kinds of their **symmetry elements.** When the result of rotation, reflection, or inversion of an object can be exactly superimposed on (matched point for point to) the original object, the structure is said to contain the corresponding symmetry element—an axis of rotation, a plane of reflection (mirror plane), or a central point (inversion center), as shown in Figure 17-2. These symmetry operations can be applied to geometrical shapes, to physical objects, and to molecular structures.

Consider a cube as an example. Suppose the center of the cube is placed at the origin of its coordinate system and the symmetry operations that transform it into identity with itself are counted (Fig. 17-3). The $x$, $y$, and $z$ coordinate axes are 4-fold axes of rotational symmetry, denoted by $C_4$, because a cube that is rotated through a multiple of 90° ($= 360°/4$) about any one of these axes is indistinguishable from the original cube. Similarly, a cube has four 3-fold axes of rotational symmetry, designated $C_3$, that are the body diagonals of the cube, connecting opposite vertices. In addition, a cube has six 2-fold rotational axes of symmetry, defined by the six axes that pass through the centers of edges and through the coordinate origin. Next, the

**Figure 17-2**

Three types of symmetry operation: rotation about an $n$-fold axis, reflection in a plane, and inversion through a point.

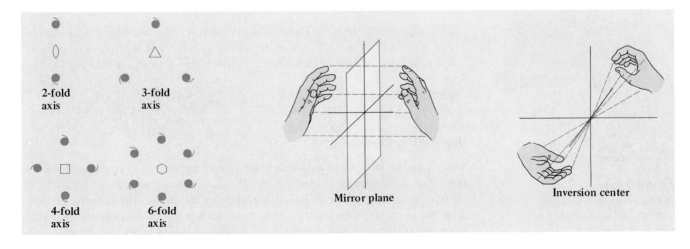

2-fold axis

3-fold axis

4-fold axis

6-fold axis

Mirror plane

Inversion center

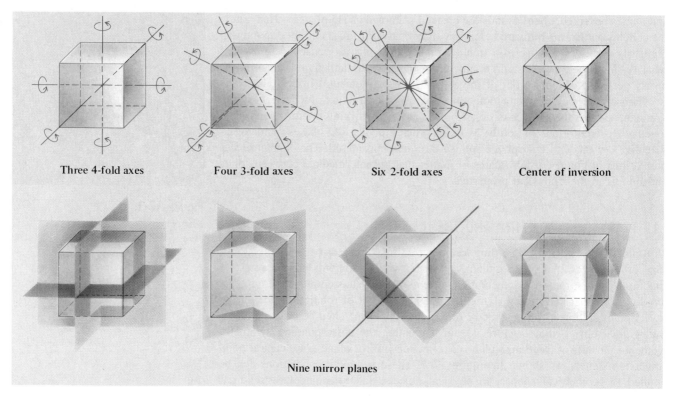

Three 4-fold axes     Four 3-fold axes     Six 2-fold axes     Center of inversion

Nine mirror planes

**Figure 17-3**

Symmetry operations acting on a cube.

cube has nine mirror planes of symmetry (designated by the symbol *m*), which reflect any point in one half of the cube into an equivalent point in the other half. Finally, a cube has a center of inversion (reflection through a point, designated *i*).

### Example 17-1

Identify the symmetry elements of the ammonia molecule ($NH_3$).

### Solution

If the ammonia molecule is drawn as a pyramid with the nitrogen atom at the top (Fig. 17-4), then the only axis of rotational symmetry is a 3-fold axis passing downward through the N atom. Three mirror planes intersect at this 3-fold axis.

*Related Problems:* 3, 4

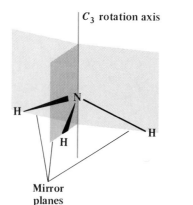

**Figure 17-4**

An ammonia molecule has a 3-fold axis of rotation and three mirror planes of symmetry.

### Unit Cells in Crystals

The symmetry operations just described can be applied to crystals as well as to individual molecules or shapes. Identical sites within a crystal recur regularly because of long-range order in the organization of the atoms. The three-dimensional array made up of all the points within a crystal that have the same environment in the same orientation is a **crystal lattice.** Such a lattice is an abstraction "lifted

**Figure 17-5**

Unit cells always have three pairs of mutually parallel faces. Only six pieces of information are required to construct a scale model of a unit cell: the three cell edges ($a$, $b$, and $c$) and the three angles between the edges ($\alpha$, $\beta$, and $\gamma$). By convention, $\gamma$ is the angle between edges $a$ and $b$, $\alpha$ the angle between $b$ and $c$, and $\beta$ the angle between $a$ and $c$.

away'' from a real crystal, embodying the scheme of repetition at work in that crystal. The lattice of highest possible symmetry is that of the **cubic system.** This lattice is obtained by filling space with a series of identical cubes, which are the **unit cells** in the system. A single unit cell contains all structural information about its crystal, because in principle the crystal could be constructed by making a great many copies of a single original unit cell and stacking them. Unit cells fill space.

Other crystal systems can be achieved with other unit cells, but constraints of symmetry permit only seven types of three-dimensional lattices. Each has a unit cell with the shape of a parallelepiped (Fig. 17-5), whose size and shape are fully described by three edge lengths ($a$, $b$, and $c$) and the three angles between those edges ($\alpha$, $\beta$, and $\gamma$). These lengths and angles are the **cell constants.** The symmetry that defines each of the seven crystal systems imposes conditions on the shape of the unit cell (Table 17-1 and Fig. 17-6). The unit cell chosen is the smallest unit that has all the symmetry elements of the crystal lattice; there is no benefit in using large cells when small ones will do. A unit cell of the minimum size is **primitive** and shares each of the eight **lattice points** at its corners with seven other unit cells, giving one lattice point per unit cell.

Other cells with the same volume (an infinite number, in fact) could be constructed, and each could generate the macroscopic crystal by repeated elementary translations, but only those shown in Figure 17-6 possess the symmetry elements of their crystal systems. Figure 17-7 illustrates a few of the infinite number of cells that can be constructed for a two-dimensional rectangular lattice. Only the rectangular cell B in the figure has three 2-fold rotation axes and two mirror planes. Although the other cells all have the same area, each of them has only one 2-fold axis and no mirror planes; they are therefore not acceptable unit cells.

| Table 17-1   The Seven Crystal Systems | | |
|---|---|---|
| **Crystal System** | **Minimum Essential Symmetry** | **Conditions on Unit-Cell Edges and Angles** |
| Hexagonal | One 6-fold rotation | $a = b$; $\alpha = \beta = 90°$, $\gamma = 120°$ |
| Cubic | Four independent 3-fold rotations[†] | $a = b = c$; $\alpha = \beta = \gamma = 90°$ |
| Tetragonal | One 4-fold rotation | $a = b$; $\alpha = \beta = \gamma = 90°$ |
| Trigonal | One 3-fold rotation | $a = b = c$; $\alpha = \beta = \gamma \neq 90°$ |
| Orthorhombic | Three mutually perpendicular 2-fold rotations | $\alpha = \beta = \gamma = 90°$ |
| Monoclinic | One 2-fold rotation | $\alpha = \gamma = 90°$ |
| Triclinic | No symmetry required | None |

[†]Each of these axes makes 70.53° angles with the other three.

$a = b = c$
$\alpha = \beta = \gamma = 90°$

**CUBIC**

$a = b \neq c$
$\alpha = \beta = \gamma = 90°$

**TETRAGONAL**

$a \neq b \neq c$
$\alpha = \beta = \gamma = 90°$

**ORTHORHOMBIC**

**TRIGONAL**
$a = b = c$
$\alpha = \beta = \gamma \neq 90°$

$a = b \neq c$
$\alpha = \beta = 90°$
$\gamma = 120°$

**HEXAGONAL**

$a \neq b \neq c$
$\alpha = \gamma = 90°$
$\beta \neq 90°$

**MONOCLINIC**

$a \neq b \neq c$
$\alpha \neq \beta \neq \gamma \neq 90°$

**TRICLINIC**

**Figure 17-6**

Shapes of the unit cells in the seven crystal systems. The more symmetric crystal systems have more symmetric cells.

Sometimes the smallest, or primitive, unit cell does not have the full symmetry of the crystal lattice. If so, a larger *nonprimitive* unit cell that does have the characteristic symmetry is deliberately chosen (Fig. 17-8). Only three types of nonprimitive cells are commonly used in the description of crystals: **body-centered, face-centered,** and **side-centered.** They are shown in Figure 17-9.

**Figure 17-7**

Cells with and without mirror planes in a two-dimensional lattice. Each cell has the same area, but only cell B possesses mirror planes in addition to the 2-fold rotation axes required by the lattice.

644

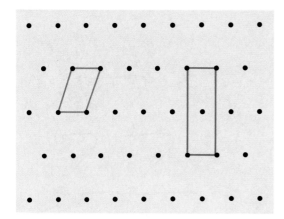

**Figure 17-8**

In this two-dimensional lattice, every lattice point is at the intersection of a horizontal mirror line and a vertical mirror line. It is possible to draw a primitive unit cell (red), but the larger, centered unit cell (blue) is preferred because it also has two mirror lines, the full symmetry of the lattice. A similar argument applies to the choice of unit cells on three-dimensional lattices.

## Scattering of X-Rays by Crystals

In the 19th century, crystallographers could classify crystals into the seven crystal systems only on the basis of their external symmetries. They could not measure the dimensions of unit cells or the positions of atoms within them. Wilhelm Roentgen's discovery of x-rays in 1895 provided a tool of enormous power for determining the structures of crystals. Max von Laue made the original suggestion that crystals might serve as three-dimensional gratings for the diffraction of electromagnetic radiation with a wavelength comparable to the distance between planes of atoms. Friedrich and Knipping demonstrated experimentally in 1912 that this was indeed the case, and von Laue was awarded the Nobel prize in physics in 1914 for his theory of the diffraction of x-rays by crystals. At about the same time, W. H. Bragg and W. L. Bragg (father and son) at Cambridge University also demonstrated the diffraction of x-rays by crystals and shared the Nobel prize in physics the following year. (W. L. Bragg was 22 and still a student at Cambridge when he discovered the diffraction law.) The formulation of the diffraction law proposed by the Braggs is

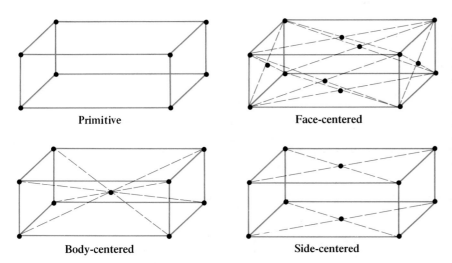

Primitive

Face-centered

Body-centered

Side-centered

**Figure 17-9**

Centered lattices, like all lattices, have lattice points at the eight corners of the unit cell. A body-centered lattice has an additional lattice point at the center of the cell, a face-centered lattice has additional points at the centers of the six faces, and a side-centered lattice has points at the centers of two parallel sides of the unit cell. (*Note:* The dots in these diagrams are not atoms but lattice points.)

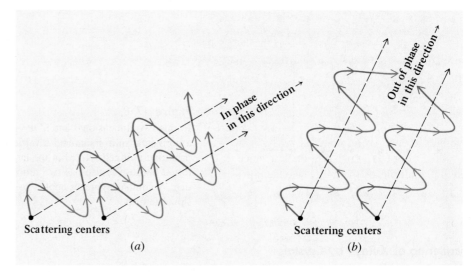

**Figure 17-10**

A beam of x-rays (not shown) is striking two scattering centers, which emit scattered radiation. The difference in the lengths of the paths followed by the scattered waves determines whether they interfere (*a*) constructively or (*b*) destructively. This path difference depends on both the distance between the centers and the direction in which the scattered waves are moving.

equivalent to von Laue's suggestion and somewhat simpler to visualize. We shall follow an approach similar to theirs.

When electromagnetic radiation passes through matter, it interacts with the electrons in atoms, and some of it is scattered in all directions. Suppose that x-radiation strikes two neighboring scattering centers. The expanding spheres of scattered waves soon encounter each other and interfere. In some directions, the waves are in phase and reinforce each other, or interfere *constructively* (Fig. 17-10*a*); in others they are out of phase and cancel each other out, or interfere *destructively* (Fig. 17-10*b*). Constructive interference occurs when the paths traversed by two waves differ in length by a whole number of wavelengths. The amplitudes of waves that interfere constructively add to one another, and the intensity of the scattered radiation in that direction is proportional to the square of the total amplitude.

Figure 17-11 illustrates the constructive interference of x-rays scattered by the electrons in atoms in equally spaced planes separated by the distance *d*. A parallel

**Figure 17-11**

Constructive interference of x-rays scattered by atoms in lattice planes. Three beams of x-rays, scattered by atoms in three successive layers of a simple cubic crystal, are shown. Note that the phases of the waves are the same along the line CH, indicating constructive interference at this scattering angle $2\theta$.

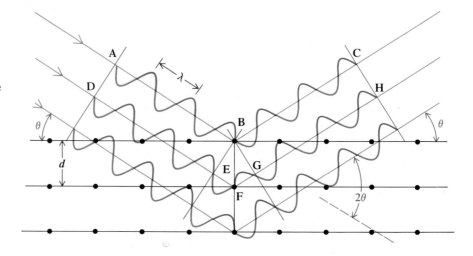

bundle of coherent x-rays of a single known wavelength is allowed to fall on the surface of a crystal, making an angle $\theta$ with a set of parallel planes of atoms in the crystal. The scattering angle $2\theta$ is then varied by rotating the crystal about an axis perpendicular to the plane of the figure. Line AD in Figure 17-11 represents a wave front of waves that are in phase as they approach the crystal. The wave that is scattered at B follows the path ABC, and the one that is scattered at F follows the path DFH. The second wave travels a greater distance than the first, and the difference in path length is the sum of the two segments EF and FG. To achieve constructive interference in the scattered waves (that is, for the phases to be the same along the wave front CH), this additional distance traveled by the second wave must be an integral multiple of the x-ray wavelength $\lambda$:

$$EF + FG = n\lambda \qquad n = 1, 2, 3, \ldots$$

From trigonometry, the lengths of these two segments are equal to each other and to $d \sin \theta$, where $d$ is the interplanar spacing. Therefore, constructive interference occurs only when

$$n\lambda = 2d \sin \theta \qquad n = 1, 2, 3, \ldots \qquad \textbf{(17-1)}$$

It is easy to verify that for angles that meet this condition, the waves scattered from the third and subsequent planes are also in phase with the waves scattered from the first two planes.

The preceding condition on allowed wavelengths is referred to as the **Bragg law,** and the corresponding angles are called **Bragg angles** for that particular set of parallel planes of atoms. It appears as though the beam of x-rays has been reflected symmetrically from those crystal planes, and we often speak colloquially of the ''Bragg reflection'' of x-rays. The x-rays have not been reflected, however, but have undergone constructive interference, more commonly called diffraction. The case $n = 1$ is called first-order Bragg diffraction, $n = 2$ is second-order, and so forth.

We now possess a tool of immense value for determining the interplanar spacings of crystals. If a crystal is turned through different directions, other parallel sets of planes with different separations are brought into the Bragg condition. The symmetry of the resulting diffraction pattern identifies the crystal system, and the Bragg angles determine the cell constants. Moreover, the *intensities* of the diffracted beams permit the locations of the atoms in the unit cell to be determined.

Analogous scattering techniques use beams of neutrons. Here the scattering interaction is between the magnetic moments of neutrons and nuclei, but the principles are the same as for x-ray diffraction. Of course, it is the wave character of neutrons (in particular, their de Broglie wavelength; see Chapter 13) that is relevant for neutron diffraction. Recall that the de Broglie wavelength and neutron momentum are related by $\lambda = h/p$.

## Example 17-2

A diffraction pattern of aluminum is obtained by using x-rays with wavelength $\lambda = 0.709$ Å. The second-order Bragg diffraction from the parallel faces of the cubic unit cells is observed at the angle $2\theta = 20.2°$. Calculate the lattice parameter $a$.

**Solution**

From the Bragg condition for $n = 2$,

$$2\lambda = 2d \sin \theta$$

the spacing between planes, which is the lattice parameter, is

$$d = \frac{\lambda}{\sin \theta} = \frac{0.709 \text{ Å}}{\sin (10.1°)} = 4.04 \text{ Å} = a$$

*Related Problems:* 5, 6

## 17-2 CRYSTAL STRUCTURE

Certain elements crystallize in particularly simple solid structures, in which an atom is situated at each point of the lattice. Polonium is the only element known to crystallize in the **simple cubic lattice,** with its atoms at the intersections of three sets of equally spaced planes that meet at right angles. Each unit cell contains one Po atom, separated from each of its six nearest neighbors by 3.35 Å.

The alkali metals crystallize in the **body-centered cubic (b.c.c.) structure** at atmospheric pressure (Fig. 17-12). A unit cell of this structure contains two lattice points, one at the center of the cube and the other at any one of the eight corners. A single alkali-metal atom is associated with each lattice point. An alternative way to visualize this is to realize that each of the eight atoms that lie at the corners of a b.c.c. unit cell is shared by the eight unit cells that meet at those corners. The contribution of the atoms to one unit cell is therefore $8 \times \frac{1}{8} = 1$ atom, to which is added the atom that lies wholly within that cell at its center.

The metals aluminum, nickel, copper, and silver, among others, crystallize in the **face-centered cubic (f.c.c.) structure** shown in Figure 17-13. This unit cell contains four lattice points, with a single atom associated with each point. No atom lies wholly within the unit cell; there are atoms at the centers of its six faces, each of which is shared with another cell (contributing $6 \times \frac{1}{2} = 3$ atoms), and an atom at each corner of the cell (contributing $8 \times \frac{1}{8} = 1$ atom), for a total of four atoms per unit cell.

The volume of a unit cell is given by the formula

$$V_c = abc\sqrt{1 - \cos^2\alpha - \cos^2\beta - \cos^2\gamma + 2 \cos \alpha \cos \beta \cos \gamma} \quad \text{(17-2)}$$

When the angles are all 90° (so that their cosines are 0), this formula reduces to the simple result $V = abc$ for the volume of a rectangular box. If the mass of the unit cell contents is known, the theoretical cell density can be computed. This density must come close to the measured density of the crystal, a quantity that can be determined by entirely independent experiments. For an *element* whose crystal contains $n_c$ atoms per unit cell, the calculated cell density is

$$\text{density} = \rho = \frac{\text{mass}}{\text{volume}} = \frac{n_c \dfrac{\mathcal{M}}{N_0}}{V_c} = \frac{n_c \mathcal{M}}{N_0 V_c}$$

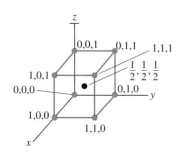

**Figure 17-12**

The b.c.c. structure. An atom is located at the center of each cubic cell (yellow) as well as at each corner of the cube (blue). The atoms are reduced slightly in size to make their positions clear.

where we have used the fact that the molar mass $\mathcal{M}$ of the element divided by Avogadro's number $N_0$ is the mass of a single atom. This equation can also be used to calculate Avogadro's number from the measured density and cell constants, as the following example illustrates.

## Example 17-3

Sodium has a density of $\rho = 0.9700$ g cm$^{-3}$ at 20°C, and its lattice parameter is $a = 4.2856$ Å. What is the value of Avogadro's number, given that the molar mass of sodium is 22.9898 g mol$^{-1}$?

### Solution

Sodium has two atoms per unit cell because its structure is b.c.c., and the volume of the unit cell is $a^3$. Solve the foregoing equation for Avogadro's number:

$$N_0 = \frac{n_c \mathcal{M}}{\rho a^3} = \frac{(2)(22.9898 \text{ g mol}^{-1})}{(0.9700 \text{ g cm}^{-3})(4.2856 \times 10^{-8} \text{ cm})^3} = 6.022 \times 10^{23} \text{ mol}^{-1}$$

*Related Problems: 15, 16*

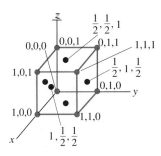

**Figure 17-13**

The f.c.c. structure. Atoms are located at the centers of the faces (yellow) as well as at the corners of the cube (blue). The atoms are reduced slightly in size to make their positions clear.

It is useful to define the locations of atoms in the unit cell with a set of three numbers measured in units of the lattice parameter(s). For this purpose, one corner of the unit cell is taken to be at the origin of the coordinate axes appropriate to the crystal system, and an atom at that lattice point has the coordinates $(0, 0, 0)$. Equivalent atoms at the seven remaining corners of the cell then have the coordinates $(1, 0, 0)$, $(0, 1, 0)$, $(0, 0, 1)$, $(1, 0, 1)$, $(1, 1, 0)$, $(0, 1, 1)$, and $(1, 1, 1)$. These atom positions are generated from an atom at $(0, 0, 0)$ by successive translations through unit distances along the three axes. An atom in a body-centered site has coordinates $(\frac{1}{2}, \frac{1}{2}, \frac{1}{2})$, and the prescription for locating it is to proceed from the coordinate origin at $(0, 0, 0)$ a distance $a/2$ along $a$, then a distance $b/2$ along $b$, and finally $c/2$ along $c$. In the same way, an atom in a face-centered site has coordinates such as $(\frac{1}{2}, \frac{1}{2}, 0)$, $(\frac{1}{2}, 0, \frac{1}{2})$, or $(0, \frac{1}{2}, \frac{1}{2})$.

So far, we have considered only cubic metals in which one atom corresponds to each lattice point. More complicated structures also occur, even for the elements, in which atoms occupy positions that are not lattice points. Diamond, for example, has an f.c.c. structure with eight atoms (not four) per unit cell. Boron has a tetragonal structure with a very complex unit cell that contains 50 atoms. In molecular crystals, the number of atoms per unit cell can be still greater; for example, in a protein with a molecular mass of $10^5$, there may be tens of thousands of atoms per unit cell. The protein shown in Figure 15-16 is an example of a complex structure revealed by modern x-ray crystallography.

## Atomic Packing in Crystals

We have already discussed two measures of the "size" of an atom or molecule. In Section 3-7, the van der Waals parameter $b$ was related to the volume excluded per mole of molecules, so that $b/N_0$ is one measure of molecular size. In Section 13-6, an approximate radius of an atom was obtained by using the distance at which the

**Table 17-2   Structural Properties of Cubic Lattices**

|  | Simple Cubic | Body-Centered Cubic | Face-Centered Cubic |
|---|---|---|---|
| Lattice points per cell | 1 | 2 | 4 |
| Number of nearest neighbors | 6 | 8 | 12 |
| Nearest-neighbor distance | $a$ | $a\sqrt{3}/2 = 0.866a$ | $a\sqrt{2}/2 = 0.707a$ |
| Atomic radius | $a/2$ | $a\sqrt{3}/4 = 0.433a$ | $a\sqrt{2}/4 = 0.354a$ |
| Packing fraction | $\dfrac{\pi}{6} = 0.524$ | $\dfrac{\sqrt{3}\pi}{8} = 0.680$ | $\dfrac{\sqrt{2}\pi}{6} = 0.740$ |

electron density had fallen off to a particular value, or the distance at which a certain fraction of the total electron density lay within a sphere of that radius. A third, and related, measure of atomic size is based on the interatomic separations in a crystal. The radius of a noble-gas or metallic atom, for example, can be approximated as half the distance between the center of an atom and the center of its nearest neighbor in the crystal. Crystal structures are pictured as resulting from the packing of spheres in which nearest neighbors are in contact.

The nearest-neighbor separation in a simple cubic crystal is equal to the lattice parameter $a$, and so the atomic radius in that case is $a/2$. For the b.c.c. lattice, the central atom in the unit cell "touches" each of the eight atoms at the corners of the cube, but those at the corners do not touch one another, as Figure 17-12 shows. The nearest-neighbor distance is calculated from the Cartesian coordinates of the atom at the origin $(0, 0, 0)$ and that at the cell center $(a/2, a/2, a/2)$. By the Pythagorean theorem, the distance between these points is $\sqrt{(a/2)^2 + (a/2)^2 + (a/2)^2} = a\sqrt{3}/2$, so that the atomic radius is $a\sqrt{3}/4$. Figure 17-13 shows that in an f.c.c. crystal the atom at the center of a face [such as at $(0, a/2, a/2)$] touches each of the neighboring corner atoms [such as at $(0, 0, 0)$], so that the nearest-neighbor distance is $\sqrt{(a/2)^2 + (a/2)^2} = a\sqrt{2}/2$ and the atomic radius is $a\sqrt{2}/4$. Table 17-2 summarizes the results for cubic lattices.

What is the densest crystal packing that can be achieved? To answer this question, construct a crystal by first putting down a plane of atoms with the highest possible density, shown in Figure 17-14a. Each sphere is in contact with six other

**Figure 17-14**

Close-packing of spheres. (a) One layer, with atoms centered on sites labeled a. (b) Two layers, with the atoms of the second layer centered on sites labeled b. The third layer can be placed on the sites labeled c (giving cubic close-packing) or over those marked a (giving hexagonal close-packing).

(a)

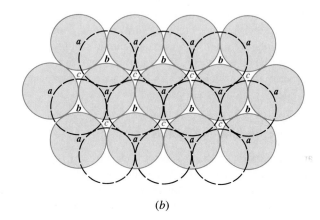

(b)

spheres in the plane. Then put down a second close-packed plane on top of the first one (Fig. 17-14b) in such a way that each sphere in the second plane is in contact with three spheres in the plane below it; that is, each sphere in the second plane forms a tetrahedron with three spheres beneath. This can be accomplished by placing the second plane of spheres above the sites marked b in Figure 17-13a. When the third plane is laid down, there are two possibilities. The spheres can be placed in the depressions marked c or in those marked a, which lie directly above the spheres in the first layer.

Clearly, there are two choices for placing each plane, and an infinite number of crystal structures can be generated that have the same atomic packing density. The two simplest such structures correspond to the periodic layer sequences *abcabcabc* . . . and *abababab*. . . . The first of these is the face-centered cubic structure already discussed, and the second is a close-packed structure in the hexagonal crystal system termed **hexagonal close-packed (h.c.p.).** In each of these simple structures, atoms occupy 74.0% of the unit cell volume, as the following example shows. (Atoms that crystallize in the b.c.c. structure occupy only 68.0% of the crystal volume, and the packing fraction for a simple cubic array is only 52.4%.)

## Example 17-4

Calculate (a) the atomic radius of an aluminum atom and (b) the fraction of the volume of aluminum that is occupied by its atoms.

### Solution

(a) Aluminum crystallizes in the f.c.c. crystal system, and its unit cell therefore contains four atoms. Because a face-centered atom touches each of the atoms at the corners of its face, the atomic radius $r_1$ (see the triangle in Fig. 17-15) can be expressed by

$$4r_1 = a\sqrt{2}$$

Using the value $a = 4.04$ Å derived from x-ray diffraction (Example 17-2) and solving for $r_1$ give

$$r_1 = 1.43 \text{ Å}$$

(b) The ratio of the volume occupied by four atoms to the volume of a unit cell is

$$f = \frac{4\left[\frac{4}{3}\pi r_1^3\right]}{a^3} = \frac{4\frac{4}{3}\pi\left[\frac{a\sqrt{2}}{4}\right]^3}{a^3} = 0.740$$

*Related Problem: 23*

### Interstitial Sites

The ways in which the empty volume is distributed in a crystal are both interesting and important. For the close-packed f.c.c. structure, two types of **interstitial sites,** upon which the free volume in the unit cell is centered, are identifiable. An **octahedral site** is surrounded at equal distances by six nearest-neighbor atoms. Figure 17-15 shows that such sites lie at the midpoints of the edges of the f.c.c. unit cell. A

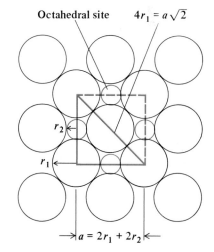

**Figure 17-15**

Octahedral sites in an f.c.c. lattice. The geometric procedure for relating the site radius $r_2$ to the atom radius $r_1$ is shown.

cell has 12 edges, each of which is shared by four unit cells, so that the edges contribute three octahedral interstitial sites per cell. In addition, the site at the center of the unit cell is also octahedral, and so the total number of octahedral sites per f.c.c. unit cell is four, the same as the number of atoms in the unit cell.

With a bit of simple geometry we can calculate the size of an octahedral site in an f.c.c. structure or, more precisely, the radius $r_2$ of a smaller atom that would fit in the site without overlapping its neighboring atoms. Figure 17-15 represents a cell face in which the length of the diagonal is $4r_1$ and the length of the cell edge is $2r_1 + 2r_2$, where $r_1$ is the radius of the host atoms and $r_2$ is the radius of the octahedral site. From the figure,

$$r_1 = a\frac{\sqrt{2}}{4}$$

$$a = 2r_2 + 2r_1 = 2r_2 + 2a\frac{\sqrt{2}}{4}$$

$$r_2 = \frac{a}{2} - a\frac{\sqrt{2}}{4} = 0.146a$$

and the ratio of the octahedral-site radius to the host-atom radius is

$$\frac{r_2}{r_1} = \frac{0.146a}{a\sqrt{2}/4} = 0.414$$

The second type of interstitial site, known as a **tetrahedral site,** lies at the center of the space defined by four touching spheres. In an f.c.c. cell, a tetrahedral site occurs in the volume between a corner atom and the three face-centered atoms nearest to it. Geometrical reasoning like that used for the octahedral site gives the following ratio of the radius of a tetrahedral site to that of a host atom:

$$\frac{r_2}{r_1} = 0.225$$

The f.c.c. unit cell contains eight tetrahedral sites, twice the number of atoms in the cell.

Interstitial sites are important when a crystal contains atoms of several kinds with considerably different radii. We shall return to this shortly when we consider the structures of ionic crystals.

## 17-3 COHESION IN SOLIDS

At the beginning of this chapter, we stated that the nature of the bonding forces between atoms might provide a useful way to classify solids. Such a classification does indeed lead to an understanding of the remarkable differences in the chemical and physical properties of different materials. For this purpose, we now consider crystals with bonding interactions classified as primarily molecular, ionic, metallic, or covalent.

### Molecular Crystals

Molecular crystals include the noble gases; oxygen; nitrogen; the halogens; compounds such as carbon dioxide; metal halides of low ionicity such as $Al_2Cl_6$, $FeCl_3$, and $BiCl_3$; and the vast majority of organic compounds. All these molecules are held

**Figure 17-16**

The van der Waals radii of the carbon and nitrogen atoms superimposed on an outline of the molecular structure of cyanuric triazide, $C_3N_{12}$, to show the volume of space from which each molecule excludes the others. Van der Waals forces in the molecular crystal hold the molecules in contact in a pattern that minimizes empty space. The thin black lines emphasize the 3-fold symmetry of the pattern.

in their lattice sites by the intermolecular forces discussed in Sections 3-7 and 4-2. The trade-off between attractive and repulsive forces among even small molecules in a molecular crystal is quite complex because so many atoms are involved. A useful simplification is to picture a molecule as a set of fused spheres centered at each nucleus. The radius of each sphere is the van der Waals radius of the element involved. In molecular crystals, such shapes pack together so that no molecules overlap but empty space is minimized. Figure 17-16 depicts such a "space-filling model" of cyanuric triazide ($C_3N_{12}$), showing how nature solves the problem of efficiently packing many copies of the rather complicated molecular shape of $C_3N_{12}$ in a single layer. In the three-dimensional molecular crystal of $C_3N_{12}$, many such layers stack up with a slight offset that minimizes unfilled space between layers.

Van der Waals forces are much weaker than the forces that operate in ionic, metallic, and covalent crystals. Consequently, molecular crystals typically have low melting points and are soft and easily deformed. Although at atmospheric pressure the noble-gas elements crystallize in the highly symmetric face-centered cubic lattice shown in Figure 17-13, molecules (especially those with complex geometries) more often form crystals of low symmetry in the monoclinic or triclinic systems. Few display the striking collective physical properties of electrical conductivity and magnetism that are characteristic of metals. The interesting properties of these crystals are primarily the properties of their molecular units.

Molecular crystals are of great scientific value. If proteins and other macromolecules are obtained in the crystalline state, their structures can be determined by x-ray diffraction. Knowing the three-dimensional structures of biological molecules is the starting point for understanding how they work.

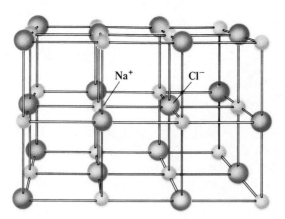

**Figure 17-17**

The sodium chloride, or rock-salt, structure. The crystal lattice is face-centered cubic, as can be seen from the fact that translating each atom by half a unit cell in the direction of the face diagonal superimposes identical atoms. There are eight ions (4 Na⁺ and 4 Cl⁻) per unit cell.

## Ionic Crystals

Compounds formed by atoms with significantly different electronegativities are largely ionic, and to a first approximation the ions can be treated as hard, charged spheres that occupy positions on the crystal lattice (see the ionic radii in Appendix F). All the elements of Groups I and II of the periodic table react with Group VI and VII elements to form ionic compounds, the great majority of which crystallize in the cubic system. Specifically, the alkali-metal halides (except for the cesium halides), the ammonium halides, and the oxides and sulfides of the alkaline-earth metals all crystallize in the **rock-salt,** or **sodium chloride, structure** shown in Figure 17-17. It may be viewed as an f.c.c. lattice of anions whose octahedral sites are all occupied by cations or, equivalently, as an f.c.c. lattice of cations whose octahedral sites are all occupied by anions. Either way, each ion is surrounded by six equidistant ions of the opposite charge. The rock-salt structure is a stable crystal structure when the cation–anion radius ratio lies between 0.414 and 0.732, if cations and anions are assumed to behave as incompressible charged spheres.

When the hard-sphere cation–anion radius ratio exceeds 0.732, as it does for the cesium halides, a different crystal structure called the **cesium chloride structure,** is more stable. It may be looked on as two interpenetrating simple cubic lattices, one of anions and the other of cations, as shown in Figure 17-18. When the cation–anion radius is less than 0.414, the **zinc blende,** or **sphalerite, structure** (named after the structure of ZnS) results. This crystal consists of an f.c.c. lattice of $S^{2-}$ ions, with $Zn^{2+}$ ions occupying half of the available tetrahedral sites in alternation, as Figure 17-19a illustrates. **Fluorite** ($CaF_2$) has yet another structure; the unit cell is based on an f.c.c. lattice of $Ca^{2+}$ ions. The $F^-$ ions occupy all eight of the tetrahedral sites, and so the unit cell contains four $Ca^{2+}$ and eight $F^-$ ions (Fig. 17-19b). The radius ratios (0.414 and 0.732) at which crossovers from one type of crystal to another occur are not accidental numbers. Recall from Section 17-2 that 0.414 is the ratio of the octahedral-site radius to the host-atom radius for an f.c.c. lattice; only when this size ratio is exceeded does the ion inserted into that site come into contact with ions of the opposite sign in the rock-salt structure. The number 0.732 comes from a corresponding calculation of the radius ratio of the interstitial site at the center of a simple cubic unit cell (see problem 25). It is important to realize that the radius-ratio criterion for the stability limits of the structures of binary ionic compounds depends on the ions' being incompressible and the wave functions' not overlapping. The criterion fails when these approximations are not met.

**Figure 17-18**

The cesium chloride structure. This crystal lattice is simple cubic, with one Cs⁺ and one Cl⁻ ion per unit cell.

The strength and range of the electrostatic attractions make ionic crystals hard, high-melting, brittle solids that are electrical insulators. Melting an ionic crystal, however, disrupts the lattice and sets the ions free to move, and so ionic liquids are good electrical conductors.

## Metallic Crystals

The characteristic property of metals is their good ability to conduct electricity and heat. Both phenomena are due to the ease with which valence electrons move; electrical conduction is a result of the flow of electrons from regions of high potential energy to those of low potential energy, and heat conduction is a result of the flow of electrons from high-temperature regions (where their kinetic energies are high) to low-temperature regions (where their kinetic energies are low). Why are electrons so mobile in a metal but so tightly bound to atoms in an insulating solid, such as diamond or sodium chloride?

The answer lies in the type of bonding found in metals, which is quite different from that in other crystals. The valence electrons in a metal are delocalized in huge molecular orbitals that extend over the entire crystal. To understand the origin of these molecular orbitals, suppose just two sodium atoms are brought together, each in its electronic ground state with the configuration $1s^2 2s^2 2p^6 3s^1$. As the atoms approach each other, the wave functions of their $3s$ electrons combine to form two molecular orbitals—one in which their phases are symmetric ($\sigma_{3s}$) and another in which they are antisymmetric ($\sigma_{3s}^*$). Solving the Schrödinger equation yields two energy states, one above and the other below the energy of the atomic $3s$ levels. (This is analogous to the formation of a hydrogen molecule from two hydrogen atoms, considered in Chapter 14.) If both valence electrons are put into the level of lower energy with spins opposed, the result is an $Na_2$ molecule. If a third sodium atom is added, the $3s$ atomic levels of the atoms split into three sublevels (Fig. 17-20). Two electrons occupy the lowest level with their spins opposed, and the third electron occupies the middle level. The three energy levels and the three electrons belong collectively to the three sodium atoms. A fourth sodium atom could be added so that there would be four closely spaced energy sublevels, and this process could be carried on without limit.

The foregoing is not a mere "thought experiment." Sodium vapor contains about 17% $Na_2$ molecules at its normal boiling point. Larger sodium clusters have been produced in molecular beam experiments, and mass spectrometry shows that such clusters can contain any desired number of atoms. For each added Na atom, another energy sublevel is added. Because the sublevels are so very closely spaced in a solid (with, say, $10^{23}$ atoms), the manifold of sublevels can be regarded as a **band.** Figure 17-20 depicts the generation of bands of sublevels that broaden as the spacing between the nuclear centers decreases. The electrons that belong to the $1s$, $2s$, and $2p$ atomic levels of sodium are only very slightly broadened at the equilibrium internuclear separation of the crystal, but even they would broaden into bands if the spacing were diminished further.

Now suppose a small electric potential difference is applied across a sodium crystal. The spin-paired electrons that lie deep in the band cannot be accelerated by a weak electric field, because there are occupied levels just above them. They have no place to go (recall the Pauli principle, which states that an energy level contains at most two electrons). At the top of the "sea" of occupied levels, however, there is an uppermost electron-occupied or half-occupied level called the **Fermi level.** Elec-

Sphalerite (ZnS)

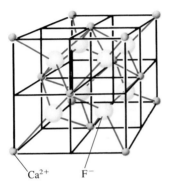

Fluorite ($CaF_2$)

**Figure 17-19**

Two ionic lattices in the f.c.c. system. A single (non-primitive) cubic unit cell of each is shown.

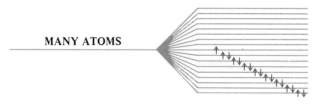

**Figure 17-20**

As sodium atoms are brought together, the molecular orbitals formed from their 3s atomic orbitals spread out into a band of levels, half occupied by electrons.

trons that lie near that level have the highest kinetic energy of all the valence electrons in the crystal and can be accelerated by the electric field to occupy the levels above. They are free to migrate in response to the electric field so that they conduct an electric current. These same electrons at the Fermi level are responsible for the high thermal conductivities of metals. They are also the electrons freed by the photoelectric effect, when a photon gives them sufficient kinetic energy to escape from the metal (Section 13-2).

A natural question arises: Why are the alkaline earth elements metals, given the argument just presented? A metal such as magnesium contains two 3s electrons, and one might expect the band derived from the broadened 3s level to be completely filled. The answer to this question is that the energies of the 3s orbitals and 3p orbitals for magnesium are not greatly different. When the internuclear separation becomes small enough, the 3p band overlaps the 3s band, and many unoccupied sublevels are then available above the highest filled level.

Most metals have crystal structures of high symmetry and crystallize in b.c.c., f.c.c., or h.c.p. lattices (Fig. 17-21). Relatively few metals (Ga, In, Sn, Sb, Bi, and Hg) have more complex crystal structures. Many metals undergo phase transitions to other structures when the temperature or pressure is changed. Liquids as well as crystals can be metals; in fact, the conductivity usually drops by only a small amount when a metal melts. The electron sea provides very strong binding in most

| Li 1.52 | Be 1.13 | | | | | | | | | | | | | | |
| Na 1.86 | Mg 1.60 | | | | | | | | | | | Al 1.43 | | | |
| K 2.27 | Ca 1.97 | Sc 1.61 | Ti 1.45 | V 1.31 | Cr 1.25 | Mn 1.37 | Fe 1.24 | Co 1.25 | Ni 1.25 | Cu 1.28 | Zn 1.34 | Ga 1.22 | | As 1.21 | |
| Rb 2.47 | Sr 2.15 | Y 1.78 | Zr 1.59 | Nb 1.43 | Mo 1.36 | Tc 1.35 | Ru 1.32 | Rh 1.34 | Pd 1.38 | Ag 1.44 | Cd 1.49 | In 1.63 | Sn 1.40 | Sb 1.41 | |
| Cs 2.65 | Ba 2.17 | * Lu 1.72 | Hf 1.56 | Ta 1.43 | W 1.37 | Re 1.34 | Os 1.34 | Ir 1.36 | Pt 1.37 | Au 1.44 | Hg 1.50 | Tl 1.70 | Pb 1.75 | Bi 1.55 | Po 1.67 |
| Fr 2.7 | Ra 2.23 | ** Lr | Unq | Unp | Unh | Uns | Uno | Une | Uun | Uuu | | | | | |

**Po is simple cubic**

| * | La 1.87 | Ce 1.82 | Pr 1.82 | Nd 1.81 | Pm 1.81 | Sm 1.80 | Eu 2.00 | Gd 1.79 | Tb 1.76 | Dy 1.75 | Ho 1.74 | Er 1.73 | Tm 1.72 | Yb 1.94 |
| ** | Ac 1.88 | Th 1.80 | Pa 1.61 | U 1.38 | Np 1.30 | Pu 1.51 | Am 1.84 | Cm | Bk | Cf | Es | Fm | Md | No |

b.c.c.   f.c.c.   h.c.p.   Tetragonal   Orthorhombic   Trigonal

**Figure 17-21**

Crystal structures of the metallic elements at 25°C and 1 atm pressure. Atomic radii (in angstroms) are calculated from the hard sphere contact model, as in Example 17-4.

metals, as shown by their high boiling points. Metals also have a very large range of melting points. Gallium melts at 29.78°C (Fig. 17-22), and mercury stays liquid at temperatures that freeze water. On the other hand, many transition metals require temperatures in excess of 1000°C to melt, and tungsten, the highest-melting elemental metal, melts at 3410°C (Section 16-1).

Chapter 19 will consider the production and uses of metals in more detail.

## Covalent Crystals

We turn finally to a class of crystalline solids whose atoms are linked by covalent bonds rather than by the electrostatic attractions of ions or the valence-electron "glue" in a metal. The archetype of the covalent crystal is diamond, which belongs to the cubic system. The ground-state electron configuration of a carbon atom is $1s^2 2s^2 2p^2$, and, as shown in Section 14-5, its bonding can be described by four hybrid $sp^3$ orbitals that are directed to the four corners of a regular tetrahedron. Each of the equivalent hybrid orbitals contains one electron that can spin-pair with the electron in one of the $sp^3$ orbitals of another carbon atom. Each carbon atom can

**Figure 17-22**

Molten gallium has a low melting point, low enough to melt from the heat of the body. (Leon Lewandowski)

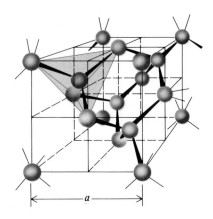

**Figure 17-23**

The structure of diamond. Each carbon atom has four nearest neighbors surrounding it at the corners of a tetrahedron.

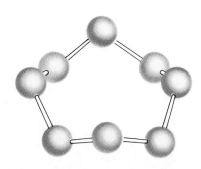

**Figure 17-24**

The structure of the sulfur molecule. The orthorhombic unit cell of rhombic sulfur, the most stable form of elemental sulfur at room temperature, is large and contains 16 of these $S_8$ molecules for a total of 128 atoms of sulfur.

thus link covalently to four others to yield the space-filling network shown in Figure 17-23. Covalent crystals are also called "network crystals," for obvious reasons. In a sense, every atom in a covalent crystal is part of one giant molecule that is the crystal itself. These crystals have very high melting points because of the strong attractions between covalently bound atoms. They are hard and brittle. Chapter 20 will discuss the mechanical and electrical properties of covalent crystals.

## Structures of the Elements

The elements provide examples of three of the four classes of crystalline solids described in this section. Only ionic solids are excluded, because a single element cannot have the two types of atoms of different electronegativities needed to form an ionic material. We have already discussed some of the structures formed by metallic elements, which are sufficiently electropositive that their atoms readily give up electrons to form the electron sea of metallic bonding. The nonmetallic elements are more complex in their structures, reflecting a competition between inter- and intramolecular bonding and producing molecular or covalent solids with varied properties.

Each halogen atom has seven valence electrons and can react with one other halogen atom to form a diatomic molecule. Once this single bond forms, there is no further bonding capacity; the halogen diatomic molecules interact with one another only through relatively weak van der Waals forces and form molecular solids with low melting and boiling points.

The Group VI elements oxygen, sulfur, and selenium display dissimilar structures in the solid state. Each oxygen atom (with six valence electrons) can form one double or two single bonds. Except in ozone, its high-free-energy form, oxygen uses up all its bonding capacity with an intramolecular double bond, forming a molecular liquid and a molecular solid that are only weakly bound. In contrast, diatomic sulfur molecules (S=S) are relatively rare, being encountered only in high-temperature vapors. The favored forms of sulfur involve the bonding of every atom to two other sulfur atoms. This leads to either rings or chains, and both are observed. The stable form of sulfur at room temperature consists of $S_8$ molecules, with eight sulfur atoms arranged in a puckered ring (Fig. 17-24). The weak interactions between $S_8$ molecules make elemental sulfur a rather soft molecular solid. Above 160°C, the rings in molten sulfur break open and relink to form long, tangled chains, producing a highly viscous liquid. An unstable ring form of selenium ($Se_8$) is known, but the thermodynamically stable form of this element is a gray crystal of metallic appearance that consists of very long spiral chains with weak interchain interaction. Crystalline tellurium has a similar structure. The Group VI elements thus show a trend (moving down the periodic table) away from the formation of multiple bonds and toward the chains and rings characteristic of atoms that each form two single bonds.

A similar trend is evident in Group V. Only nitrogen forms diatomic molecules with triple bonds, in which all the bonding capacity is used between pairs of atoms. Elemental phosphorus exists in three forms, in all of which each phosphorus atom forms three single bonds rather than one triple bond. White phosphorus (Fig. 17-25a) consists of tetrahedral $P_4$ molecules, which interact with each other through weak van der Waals forces. Black phosphorus and red phosphorus (Fig. 17-25b,c) are higher-melting network solids in which the three bonds formed by each atom connect it directly or indirectly with all the other atoms in the sample. Unstable solid (*a*) White phosphorus       (*b*) Black phosphorus

(*c*) Red phosphorus

**Figure 17-25**
Structures of elemental phosphorus.

forms of arsenic and antimony that consist of $As_4$ or $Sb_4$ tetrahedra like those in white phosphorus can be prepared by rapid cooling of the vapor. The stable forms of these elements have structures related to that of black phosphorus.

The elements considered so far lie on the border between covalent and molecular solids. Other elements, those of intermediate electronegativity, exist as solids on the border between metallic and covalent and are called **metalloids.** Antimony has a metallic luster, for example, but is a rather poor conductor of electricity and heat. Silicon and germanium are **semiconductors,** with electrical conductivities far lower than those of metals but still significantly higher than those of true insulators such as diamond. Section 20-4 will examine the special properties of these materials more closely.

Some elements of intermediate electronegativity exist in two crystalline forms with very different properties. White tin has a tetragonal crystal structure and is a metallic conductor. Below 13°C it crumbles slowly to form a powder of gray tin (with the diamond structure) that is a poor conductor. Its formation at low temperature is known as the ''tin disease'' and can be prevented by the addition of small amounts of bismuth or antimony. The thermodynamically stable form of carbon at room conditions is not the insulator diamond, but graphite. Graphite consists of sheets of fused hexagonal rings with only rather weak interactions between layers (Fig. 17-26). Each carbon atom shows $sp^2$ hybridization, with its remaining $p$ orbital (perpendicular to the graphite layers) taking part in extended $\pi$-bonding interactions over the whole plane. Graphite can be pictured as a series of interlocked benzene rings, with $\pi$-electron delocalization contributing significantly to its stability. The delocalized electrons give graphite a conductivity in the planes of fused hexagons approaching that of the metallic elements. The conductivity and relative chemical inertness of graphite make it useful for electrodes in electrochemistry.

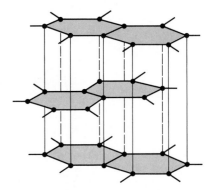

**Figure 17-26**
The structure of graphite.

## 17-4 A Deeper Look . . . Lattice Energies of Crystals

The **lattice energy** of a crystal is the energy required to separate the crystal into its component atoms, molecules, or ions at 0 K. In this section we examine the calculation and measurement of lattice energies for molecular and ionic crystals.

### Lattice Energy of a Molecular Crystal

The lattice energy of a molecular crystal can be estimated by using the simple Lennard-Jones potential of Section 3-7:

$$V_{LJ}(R) = 4\epsilon \left[ \left( \frac{\sigma}{R} \right)^{12} - \left( \frac{\sigma}{R} \right)^{6} \right]$$

Table 3-4 lists the values of $\epsilon$ and $\sigma$ for various atoms and molecules. To obtain the total potential energy for 1 mol, sum over all pairs of atoms or molecules:

$$V_{tot} = \frac{1}{2} \sum_{i=1}^{N_0} \sum_{j=1}^{N_0} V_{LJ}(R_{ij})$$

where $R_{ij}$ is the distance between atom $i$ and atom $j$. The factor $\frac{1}{2}$ arises because each interaction between a pair of atoms should be counted only once, not twice. For a crystal of macroscopic size, this can be rewritten as

$$V_{tot} = \frac{N_0}{2} \sum_{j=1}^{N_0} V_{LJ}(R_{ij})$$

where $i$ is taken to be some atom in the middle of the crystal. If the nearest-neighbor distance is $R_0$, then we define a ratio of distances $p_{ij} = R_{ij}/R_0$ and rewrite $V_{tot}$ for the Lennard–Jones potential as

$$V_{tot} = \frac{N_0}{2}(4\epsilon) \left[ \sum_{j} \left( \frac{\sigma}{p_{ij} R_0} \right)^{12} - \sum_{j} \left( \frac{\sigma}{p_{ij} R_0} \right)^{6} \right]$$

$$= 2\epsilon N_0 \left[ \left( \frac{\sigma}{R_0} \right)^{12} \sum_{j} (p_{ij})^{-12} - \left( \frac{\sigma}{R_0} \right)^{6} \sum_{j} (p_{ij})^{-6} \right]$$

The two summations are dimensionless properties of the lattice structure, and accurate values can be obtained by summing over the first few sets of nearest neighbors (Table 17-3). The resulting total energy for the f.c.c. lattice is

$$V_{tot} = 2\epsilon N_0 \left[ 12.132 \left( \frac{\sigma}{R_0} \right)^{12} - 14.454 \left( \frac{\sigma}{R_0} \right)^{6} \right]$$

The equilibrium atomic spacing at $T = 0$ K should be close to the one that gives a minimum in $V_{tot}$, which can be calculated by differentiating the preceding expression with respect to $R_0$ and setting the derivative to 0. The result is

$$R_0 \approx 1.09\sigma$$

and the value of $V_{tot}$ at this value of $R_0$ is

$$V_{tot} = -8.61\epsilon N_0$$

The corresponding potential energy when the atoms or molecules are completely separated from one another is zero. The lattice energy is the difference between these quantities and is a positive number:

$$\text{lattice energy} = -V_{tot} = 8.61\epsilon N_0$$

This overestimates the true lattice energy because of the quantum-mechanical effect of zero-point energy (Section 15-1). When a quantum correction is applied, the binding energy is reduced by 28%, 10%, 6%, and 4% for Ne, Ar, Kr, and Xe, respectively. Table 17-4 shows the resulting predictions for crystal lattice energies and nearest-neighbor distances. The agreement with experiment is quite reasonable, considering the approximations inherent in the use of a Lennard–Jones potential derived entirely from gas-phase data. For helium the amplitude of zero-point motion is so great that if a crystal did form, it would immediately melt. Consequently, helium remains liquid down to absolute zero at atmospheric pressure.

| Table 17-3 | Lattice Sums for Molecular Crystals (f.c.c. Structure) | | | |
|---|---|---|---|---|
| | **Number, $n$** | $p_{ij}$ | $n(p_{ij})^{-12}$ | $n(p_{ij})^{-6}$ |
| Nearest neighbors | 12 | 1 | 12 | 12 |
| Second nearest neighbors | 6 | $\sqrt{2}$ | 0.0938 | 0.750 |
| Third nearest neighbors | 24 | $\sqrt{3}$ | 0.0329 | 0.889 |
| Fourth nearest neighbors | 12 | 2 | 0.0029 | 0.188 |
| Fifth nearest neighbors | 24 | $\sqrt{5}$ | 0.0015 | 0.192 |
| | ⋮ | ⋮ | ⋮ | ⋮ |
| Total | | | 12.132 | 14.454 |

**Table 17-4    Properties of Noble-Gas Crystals[†]**

|  | $R_0$ (Å) | | Lattice Energy (kJ mol$^{-1}$) | |
| --- | --- | --- | --- | --- |
|  | Predicted | Observed | Predicted | Observed |
| Ne | 3.00 | 3.13 | 1.83 | 1.88 |
| Ar | 3.71 | 3.76 | 7.72 | 7.74 |
| Kr | 3.92 | 4.01 | 11.5 | 11.2 |
| Xe | 4.47 | 4.35 | 15.2 | 16.0 |

[†] All data are extrapolated to 0 K and zero pressure.

### Lattice Energy of an Ionic Crystal

In Section 14-2 we calculated the potential energy of a gaseous diatomic ionic molecule relative to the separated ions by means of Coulomb's law:

$$V = \frac{Q_1 Q_2}{4\pi\epsilon_0 R_0}$$

where $R_0$ is the equilibrium internuclear separation. With a few modifications, the same considerations apply to a calculation of the lattice energies of ionic compounds in the crystalline state.

For simplicity, consider a hypothetical one-dimensional crystal (Fig. 17-27), in which ions of charge $+e$ and $-e$ alternate with an internuclear separation of $R_0$. One ion, selected to occupy an arbitrary origin, will interact attractively with all ions of opposite sign to make the following contribution to the crystal energy:

$$-\frac{e^2}{4\pi\epsilon_0 R_0}[2(1) + 2(\tfrac{1}{3}) + 2(\tfrac{1}{5}) + \cdots]$$

Here the factors of 2 come from the fact that there are *two* ions of opposite sign at a distance $R_0$ from a given ion, two at a distance $3R_0$, two at $5R_0$, and so forth. The negative sign occurs because the ions that occupy odd-numbered sites have a charge opposite that of the ion at the origin, and their interaction with the ion at the origin is attractive. The ion at the origin also interacts repulsively with all ions of the same sign to make the following contribution to the crystal energy:

$$+\frac{e^2}{4\pi\epsilon_0 R_0}[2(\tfrac{1}{2}) + 2(\tfrac{1}{4}) + 2(\tfrac{1}{6}) + \cdots]$$

**Figure 17-27**
Lattice energy for a one-dimensional ionic crystal.

The net interaction of $N_0$ such ions of each sign with one another is:

$$-\frac{N_0 e^2}{4\pi\epsilon_0 R_0}\left[2 - \frac{2}{2} + \frac{2}{3} - \frac{2}{4} + \frac{2}{5} - \frac{2}{6} + \cdots\right]$$

We must be very careful with factors of 2. To obtain the potential energy for the interaction of $N_0$ positive ions with $N_0$ negative ions, it is necessary to multiply the total potential energy of a given ion due to all others by $2N_0$ and then divide by 2 to avoid counting the interaction of a given pair of ions twice. This gives the preceding result.

If such a calculation is carried out for a real three-dimensional crystal, the result is a series (such as that just given in brackets) whose value sums to a dimensionless number that depends upon the crystal structure. That number is called the **Madelung constant**, $M$, and its value is independent of the unit-cell dimensions. Table 17-5 lists the values of the Madelung constant for several crystal structures. The lattice energy is again the opposite of the total potential energy. Expressed in terms of the Madelung constant, it is

$$\text{lattice energy} = \frac{N_0 e^2}{4\pi\epsilon_0 R_0} M$$

### Example 17-5

Calculate the electrostatic part of the lattice energy of sodium chloride, given that the internuclear separation between $Na^+$ and $Cl^-$ ions is 2.82 Å.

**Table 17-5    Madelung Constants**

| Lattice | $M$ |
| --- | --- |
| Rock salt | 1.7476 |
| CsCl | 1.7627 |
| Zinc blende | 1.6381 |
| Fluorite | 2.5194 |

## Solution

Using the Madelung constant of 1.7476 for this structure gives

$$\frac{(6.02 \times 10^{23}\ \text{mol}^{-1})(1.602 \times 10^{-19}\ \text{C})^2(1.7476)}{(4\pi)(8.854 \times 10^{-12}\ \text{C}^2\ \text{J}^{-1}\ \text{m}^{-1})(2.82 \times 10^{-10}\ \text{m})}$$

$$= 8.61 \times 10^5\ \text{J mol}^{-1} = 861\ \text{kJ mol}^{-1}$$

*Related Problems:* 31, 32

Ionic lattice energies can be measured experimentally by means of a thermodynamic cycle developed by Max Born and Fritz Haber. The **Born–Haber cycle** is an application of Hess's law (the first law of thermodynamics). It is illustrated by a calculation of the lattice energy of sodium chloride, which is $\Delta E$ for the reaction

$$\text{NaCl}(s) \longrightarrow \text{Na}^+(g) + \text{Cl}^-(g) \qquad \Delta E = ?$$

This reaction can be represented as a series of steps, each with a measurable energy or enthalpy change. In the first step, the ionic solid is converted to the elements in their standard states:

$$\text{NaCl}(s) \longrightarrow \text{Na}(s) + \tfrac{1}{2}\text{Cl}_2(g)$$
$$\Delta E_1 \approx \Delta H = -\Delta H_f^\circ(\text{NaCl}(s)) = +411.2\ \text{kJ}$$

In the second step, the elements are transformed into gas-phase atoms:

$$\text{Na}(s) \longrightarrow \text{Na}(g) \qquad \Delta E \approx \Delta H = \Delta H_f^\circ(\text{Na}(g))$$
$$= +107.3\ \text{kJ}$$
$$\tfrac{1}{2}\text{Cl}_2(g) \longrightarrow \text{Cl}(g) \qquad \Delta E \approx \Delta H = \Delta H_f^\circ(\text{Cl}(g))$$
$$= +121.7\ \text{kJ}$$
$$\text{Na}(s) + \tfrac{1}{2}\text{Cl}_2(g) \longrightarrow \text{Na}(g) + \text{Cl}(g) \qquad \Delta E_2 = +229.0\ \text{kJ}$$

Finally, in the third step, electrons are transferred from the sodium atoms to the chlorine atoms to give ions:

$$\text{Na}(g) \longrightarrow \text{Na}^+(g) + e^- \qquad \Delta E = IE_1(\text{Na}) = 496\ \text{kJ}$$
$$\text{Cl}(g) + e^- \longrightarrow \text{Cl}^-(g) \qquad \Delta E = -EA(\text{Cl}) = -349\ \text{kJ}$$
$$\text{Na}(g) + \text{Cl}(g) \longrightarrow \text{Na}^+(g) + \text{Cl}^-(g) \qquad \Delta E_3 = +147\ \text{kJ}$$

Here $EA(\text{Cl})$ is the electron affinity of Cl, and $IE_1(\text{Na})$ is the first ionization energy of Na. The total energy change is

$$\Delta E = \Delta E_1 + \Delta E_2 + \Delta E_3 = 411 + 229 + 147 = +787\ \text{kJ}$$

The small differences between $\Delta E$ and $\Delta H$ were neglected in this calculation. If their difference is taken into account (using $\Delta H = \Delta E + RT\,\Delta n_g$, where $\Delta n_g$ is the change in the number of moles of gas molecules in each step of the reaction), then $\Delta E_2$ is decreased by $\tfrac{3}{2}RT$ and $\Delta E_1$ by $\tfrac{1}{2}RT$, giving a net decrease of $2RT$ and changing $\Delta E$ to $+782\ \text{kJ mol}^{-1}$. Comparing this experimental lattice energy with the energy calculated in Example 17-5 shows that the latter is approximately 10% greater, presumably because short-range repulsive interactions and zero-point energy were not taken into account in the lattice energy calculation.

## 17-5 DEFECTS AND AMORPHOUS SOLIDS

Although real crystals display beautiful symmetries to the eye, they are not perfect. As a practical matter, it is impossible to rid a crystal of all impurities or to ensure that it contains perfect periodic ordering. Amorphous solids are states of matter in which so many defects are present that crystalline order is destroyed.

### Point Defects

Point defects in a pure crystalline substance include **vacancies,** in which atoms are missing from lattice sites, and **interstitials,** in which atoms are inserted in sites different from their normal sites. In real crystals, a small fraction of the normal atom sites remain unoccupied. Such vacancies are called **Schottky defects,** and their concentration depends on temperature:

$$N = N_s \exp(-\Delta G/RT)$$

where $N$ is the number of lattice vacancies per unit volume, $N_s$ is the number of atom sites per unit volume, and $\Delta G$ is the molar free energy of formation of vacancies. Figure 17-28a illustrates Schottky defects in the crystal structure of a metal or noble gas. Schottky defects also occur in ionic crystals but with the restriction that the imperfect crystals remain electrically neutral. Thus, in sodium chloride, for every missing $Na^+$ ion there must also be a missing $Cl^-$ ion (see Fig. 17-28b).

In certain kinds of crystals, atoms or ions are displaced from their regular lattice sites to interstitial sites, and the crystal defect consists of the lattice vacancy plus the interstitial atom or ion. Figure 17-28c illustrates this type of lattice imperfection, known as a **Frenkel defect.** The silver halides (AgCl, AgBr, AgI) are examples of crystals in which Frenkel disorder is extreme. The crystal structures of these compounds are established primarily by the anion lattice, and the silver ions occupy highly disordered, almost random, sites. The rate of diffusion of silver ions in these solids is exceptionally high, as studies employing radioactive isotopes of silver have shown. Both Frenkel and Schottky defects in crystals are mobile, jumping from one lattice site to a neighboring site with frequencies that depend on the temperature and the strengths of the atomic forces. Diffusion in crystalline solids is due largely to the presence and mobility of point defects; it is a thermally activated process, just like the rates of chemical reactions considered in Chapter 11. The coefficient of self-diffusion has the form

$$D = D_0 \exp[-E_a/RT]$$

where $E_a$ is the activation energy. The rates of diffusive motion in crystalline solids vary enormously from one substance to another. In a crystal of a low-melting metal such as sodium, an average atom undergoes about $10^8$ diffusive jumps per second at 50°C, whereas in a metal such as tungsten that melts at 3410°C, an average atom jumps to another lattice site less than once per year at 1000°C!

If an alkali halide crystal such as NaCl is irradiated with x-rays, ultraviolet radiation, or high-energy electrons, some $Cl^-$ ions may lose an electron:

$$Cl^- + h\nu \longrightarrow Cl + e^-$$

The resulting Cl atom, being uncharged and much smaller than a $Cl^-$ ion, is no longer strongly bound in the crystal and can diffuse to the surface and escape. The electron can migrate through the crystal quite freely until it encounters an anion vacancy and is trapped in the Coulomb field of the surrounding cations (Fig. 17-29). This crystal defect is called an **F-center** (from the German word *Farbenzentrum,*

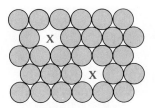

(a) Schottky defects in a metal

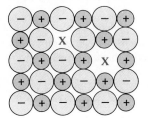

(b) Schottky defect in an ionic crystal

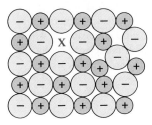

(c) Frenkel defect in an ionic crystal

**Figure 17-28**

Point imperfections in a lattice. The red X's denote vacancies.

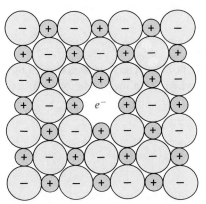

**Figure 17-29**

An F-center in a crystal.

**Figure 17-30**

Pure calcium fluoride, $CaF_2$, is white, but the natural sample of calcium fluoride (fluorite) shown here is a rich purple because F-centers are present. These are lattice sites at which the $F^-$ anion is replaced by an electron only. (Leon Lewandowski)

meaning **color center**). It is the simplest of a family of electronic crystal defects. As the name suggests, it imparts a color to ionic crystals (Fig. 17-30).

## Nonstoichiometric Compounds

As Chapter 1 emphasized, the law of definite proportions was one of the principal pieces of evidence that led to the acceptance of Dalton's atomic theory. It is now recognized that a great many solid-state binary compounds do *not* have fixed and unvarying compositions but exist over a range of compositions in a single phase. Thus, FeO (wüstite) has the composition range $Fe_{0.85}O_{1.00}$ to $Fe_{0.95}O_{1.00}$ and is never found with its nominal 1:1 composition. The compounds NiO and $Cu_2S$ also deviate considerably from their nominal stoichiometries.

The explanation depends on the existence of more than one oxidation state for the metal. In wüstite, iron can exist in either the +2 or the +3 oxidation state. Suppose a solid were to begin at the hypothetical composition of $Fe_{1.00}O_{1.00}$, with iron entirely in the +2 oxidation state. For every two $Fe^{3+}$ ions introduced, three $Fe^{2+}$ ions must be removed to maintain overall charge neutrality. The total number of moles of iron is then less than that in the ideal FeO stoichiometry. The departure from the nominal stoichiometry can be far more extreme than that found in wüstite. The composition of "TiO" ranges from $Ti_{0.75}O$ to $Ti_{1.45}O$. Nickel oxide varies only from $Ni_{0.97}O$ to NiO in composition, but the variation is accompanied by a dramatic change in properties. When the compound is prepared in the 1:1 composition, it is pale green and is an electrical insulator. When it is prepared in an excess of oxygen, it is black and conducts electricity fairly well. In the black material, a small fraction of $Ni^{2+}$ ions are replaced by $Ni^{3+}$ ions, and compensating vacancies occur at some nickel atom sites in the crystal.

## Example 17-6

The composition of a sample of wüstite is $Fe_{0.930}O_{1.00}$. What percentage of the iron is in the form of iron(III)?

### Solution

For every 1.00 mol of oxygen atoms in this sample, there is 0.930 mol of iron atoms. Suppose $y$ mol of the iron is in the +3 oxidation state and $0.930 - y$ is in the +2 oxidation state. Then the total positive charge from the iron (in moles of electron charge) is

$$+3y + 2(0.930 - y)$$

This positive charge must exactly balance the 2 mol of negative charge carried by the mole of oxygen atoms (recall that each oxygen atom has oxidation number $-2$). We conclude that

$$3y + 2(0.930 - y) = +2$$

Solving this equation for $y$ gives

$$y = 0.140$$

The percentage of iron in the form of $Fe^{3+}$ is then the ratio of this to the total number of moles of iron, 0.930, multiplied by 100%:

$$\% \text{ iron in form of } Fe^{3+} = \frac{0.140}{0.930} \times 100\% = 15.1\%$$

*Related Problems:* 37, 38

## Alloys

The nonstoichiometric compounds just described are ionic materials with compositional disorder. A related type of disorder is exhibited by an **alloy,** a mixture of elements that displays metallic properties.

There are two types of alloys. In a **substitutional alloy,** some of the metal atoms in a crystal lattice are replaced by other atoms (usually of comparable size). Examples are brass, in which approximately one third of the atoms in a copper crystal are replaced by zinc atoms, and pewter, an alloy of tin that contains 7% copper, 6% bismuth, and 2% antimony. In an **interstitial alloy,** atoms of one or more additional elements enter the interstitial sites of the host metal lattice. An example is steel, in which carbon atoms occupy interstitial sites of an iron crystal, making the material stronger and harder than pure iron. Mild steel contains less than 0.2% C and is used for nails, whereas high-carbon steels can contain up to 1.5% C and are used in specialty applications such as tools and springs. *Alloy steels* are both substitutional and interstitial, as atoms from metals such as chromium and vanadium substitute for iron atoms, with carbon remaining in interstitial sites. Alloy steels have a variety of specialized purposes, ranging from cutlery to bicycle frames.

## Amorphous Solids and Glasses

The atoms, ions, or molecules in crystalline solids exhibit high degrees of spatial order. Now let us briefly consider solids that lack this characteristic. **Amorphous solids,** commonly called **glasses,** resemble crystalline solids in many respects; they may have chemical compositions, mechanical properties such as hardness and elasticity, and electrical and magnetic properties that are similar to those of crystals. Like crystals, glasses may have molecular, ionic, covalent, or metallic bonding. On an atomic scale, however, amorphous solids lack the regular periodic structure of crystals. They are states of matter in which so many defects are present that crystalline order is destroyed.

Some substances have a strong tendency to solidify as glasses. The best example is the material used in common window panes, with the approximate chemical formula $Na_2O \cdot CaO \cdot (SiO_2)_6$. This is a partly ionic, partly covalent material with $Na^+$ and $Ca^{2+}$ ions distributed through a covalently bonded Si—O network. Glass-forming ability is not restricted to a few special materials, however. If a substance can be liquefied, it can almost certainly be prepared in an amorphous state. Even metals, which are known primarily in the crystalline state, have been made into amorphous solids. The trick is to bypass crystallization by cooling molten material very fast. One technique involves shooting a jet of liquid metal at a rapidly rotating cold cylinder, which produces a continuous ribbon of amorphous metal at a rate up to 2 km per minute.

On the molecular level, a strong tendency to form glasses is associated with the presence of long or irregularly shaped molecules that can easily become tangled and

**Figure 17-31**
Glass fibers of specialized composition can conduct a beam of light for miles without substantial loss. (Comstock)

disordered. Even slowly cooling a liquid assembly of such molecules may not afford enough time for them to organize into a crystalline lattice before solidification. Instead of a sharp liquid-to-crystal transition, such glass-formers transform continuously, over a range of temperature, into amorphous solids. They lend themselves to fabrication into articles of every conceivable shape, because the flow properties of the work piece can be managed by controlling its temperature. This plasticity is the reason that glass has played an indispensable role in science, industry, and the arts.

One of the most exciting new uses of a glass is the transmission of voice messages, television images, and data as light pulses. Tens of thousands of audio messages can be transmitted simultaneously through glass fibers no greater in diameter than a human hair by encoding the audio signal into electronic impulses that modulate light from a laser source. The light passes down the glass fiber as though it were a tube. Chemical control of the glass composition reduces light loss and permits messages to travel many kilometers without amplification (Fig. 17-31).

## 17-6 LIQUID CRYSTALS

**Liquid crystals** constitute an interesting state of matter with properties intermediate between those of true liquids and those of crystals. Unlike glasses, liquid-crystal states are thermodynamically stable. Many organic materials do not show a single solid-to-liquid transition but rather a cascade of transitions involving new intermediate phases. In recent years, liquid crystals have been used in a wide variety of practical applications, ranging from temperature sensors to displays on calculators and other electronic devices.

## The Structure of Liquid Crystals

Substances that form liquid crystals are usually characterized by molecules with elongated, rod-like shapes. An example is terephthal-bis-(4-*n*-butylaniline), called TBBA, whose molecular structure can be represented as

$$H_9C_4 - \hspace{-0.5em}\bigcirc\hspace{-0.5em} - N{=}C - \hspace{-0.5em}\bigcirc\hspace{-0.5em} - C{=}N - \hspace{-0.5em}\bigcirc\hspace{-0.5em} - C_4H_9$$
$$\qquad\qquad\quad \overset{\displaystyle H}{|}\qquad\quad\ \overset{\displaystyle H}{|}$$

(a) Liquid

(b) Nematic phase

(c) Smectic A phase

(d) Molecular crystal

with hydrocarbon groups at the ends separated by a relatively rigid backbone of benzene rings and N=C double bonds. Such rod-like molecules tend to line up even in the liquid phase, as Figure 17-32a shows. Ordering in this phase persists only over small distances, however, and on average a given molecule is equally likely to take any orientation.

The simplest type of liquid-crystal phase is the **nematic phase** (Fig. 17-32b). In a nematic liquid crystal, the molecules display a preferred orientation in a particular direction, but their centers are distributed at random, as they would be in an ordinary liquid. Although liquid-crystal phases are characterized by a net orientation of molecules over large distances, not all the molecules point in exactly the same direction. There are fluctuations in the orientation of each molecule, and only *on average* do the molecules have a greater probability of pointing in a particular direction.

Some liquid crystals form one or more **smectic phases.** These display a variety of microscopic structures that are indicated by the letters A, B, C, and so forth. Figure 17-32c shows one of them, the smectic A structure; the molecules continue to display net orientational ordering, but now, unlike in the nematic phase, the centers of the molecules also tend to lie in layers. Within each layer, however, these centers are distributed at random as in an ordinary liquid.

At low enough temperatures, a liquid crystal freezes into a crystalline solid (Fig. 17-31d) in which the molecules' orientations are ordered and their centers lie on a regular three-dimensional lattice. The progression of structures in Figure 17-32 illustrates the meaning of the term "liquid crystal." Liquid crystals are solid-like in showing orientational ordering but liquid-like in the random distribution of the centers of their molecules.

In taking a particular orientation, a liquid crystal is very sensitive both to the positions and natures of the surfaces with which it is in contact and to small electric or magnetic fields. This sensitivity is the basis for the use of nematic liquid crystals in electronic display devices such as digital watches and calculators (Fig. 17-33) as well as in large-screen liquid-crystal displays.

## Figure 17-32

Different states of structural order for rod-shaped molecules. The figure is only schematic; in a real sample, the lining up of the molecules would not be so nearly perfect.

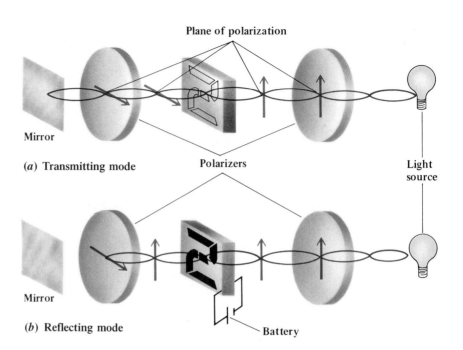

Plane of polarization

Mirror

(a) Transmitting mode          Polarizers

Light source

Mirror

(b) Reflecting mode          Battery

**Figure 17-33**
The mode of operation of a liquid-crystal display device. (a) The light has a polarization that permits it to pass through the second polarizing filter and strike the mirror, giving a bright display. (b) Imposition of a potential difference across some portion of the display causes the liquid-crystal molecules to rotate, creating a different polarization of light. Because the "rotated" light is blocked by the second filter, it does not reach the mirror, and that part of the display appears black.

## PRACTICE EXERCISE

Two forms of elemental phosphorus: white and red.

## Phosphorus

Solid elemental phosphorus appears in a rich variety of forms, with crystals in all seven crystal systems reported under various conditions of temperature, pressure, and sample preparation.

(a) The thermodynamically stable form of phosphorus under room conditions is black phosphorus. Its unit cell is orthorhombic with edges of lengths 3.314, 4.376, and 10.48 Å. Calculate the volume of one unit cell, and determine the number of phosphorus atoms per unit cell, if the density of this form of phosphorus is 2.69 g cm$^{-3}$.

(b) The form of phosphorus that is easiest to prepare from the liquid or gaseous state is white phosphorus, which consists of $P_4$ molecules in a cubic lattice. When x-rays of wavelength 2.29 Å are scattered from the parallel faces of its unit cells, the first-order Bragg diffraction is observed at an angle $2\theta$ of 7.10°. Calculate the length of the unit-cell edge for white phosphorus. At what angle will third-order Bragg diffraction be seen?

(c) Amorphous red phosphorus has been reported to convert to monoclinic, triclinic, tetragonal, and cubic red forms with different heat treatments. Identify the changes in the shape of the unit cell as a cubic lattice is converted first to tetragonal, then monoclinic, then triclinic.

(d) A monoclinic form of red phosphorus has been studied that has cell edge lengths 9.21, 9.15, and 22.60 Å, with an angle $\beta$ of 106.1°. Each unit cell contains 84 atoms of phosphorus. Estimate the density of this form of phosphorus.

(e) Phosphorus forms many compounds with other elements. Describe the nature of the bonding in the solids white elemental phosphorus ($P_4$), black elemental phosphorus, sodium phosphate ($Na_3PO_4$), and phosphorus trichloride ($PCl_3$).

**Answers**

(a) Volume is 152.0 Å³; eight atoms per unit cell

(b) 18.5 Å; angle $2\theta = 21.4°$

(c) Cubic to tetragonal: one cell edge is stretched or shrunk. Tetragonal to monoclinic: a second cell edge is stretched or shrunk and the angles between two adjacent faces (and their opposite faces) are changed from 90°. Monoclinic to triclinic: the remaining two angles between faces are deformed from 90°.

(d) Density is 2.36 g cm⁻³

(e) $P_4$(white) and $PCl_3$ are molecular solids; P(black) is covalent; $Na_3PO_4$ is ionic.

**CONCEPTS & SKILLS**

After studying this chapter and working the problems that follow, you should be able to

1. Identify the symmetry elements of different crystal systems (Section 17-1, problems 1–4).
2. Explain how x-rays and neutrons are diffracted by crystals, and use information from such experiments to calculate lattice spacings (Section 17-1, problems 5–10).
3. Describe the packing of atoms in simple crystal lattices (Section 17-2, problems 11–24).
4. Compare the natures of the forces that hold atoms or molecules in their lattice sites in molecular, ionic, metallic, and covalent crystals (Section 17-3, problems 27–28).
5. Calculate lattice energies of molecular and ionic crystals (Section 17-4, problems 31–34).
6. Describe the kinds of equilibrium defects that are present in crystalline solids, and the properties of amorphous solids (Section 17-5).
7. Determine the oxidation states present in nonstoichiometric solids (Section 17-5, problems 37–38).
8. Explain what a liquid crystal is, and state how nematic and smectic phases differ from ordinary liquids and crystalline solids (Section 17-6, problems 39–40).

**PROBLEMS**

### Crystal Symmetry and the Unit Cell

1. Which of the following has 3-fold rotational symmetry? Explain.
   (a) An isosceles triangle   (c) A tetrahedron
   (b) An equilateral triangle   (d) A cube
2. Which of the following has 4-fold rotational symmetry? Explain.
   (a) A cereal box (exclusive of the writing on the sides)
   (b) A stop sign (not counting the writing)
   (c) A tetrahedron
   (d) A cube
3. Identify the symmetry elements of the $CCl_2F_2$ molecule (Fig. 8-15).

4. Identify the symmetry elements of the $PF_5$ molecule (Fig. 14-8a).
5. The second-order Bragg diffraction of x-rays with $\lambda = 1.660$ Å from a set of parallel planes in copper occurs at an angle $2\theta = 54.70°$. Calculate the distance between the scattering planes in the crystal.
6. The second-order Bragg diffraction of x-rays with $\lambda = 1.237$ Å from a set of parallel planes in aluminum occurs at an angle $2\theta = 35.58°$. Calculate the distance between the scattering planes in the crystal.
7. The distance between members of a set of equally spaced planes of atoms in crystalline lead is 4.950 Å. If x-rays with $\lambda = 1.936$ Å are diffracted by this set of parallel planes, cal-

culate the angle $2\theta$ at which fourth-order Bragg diffraction will be observed.

8. The distance between members of a set of equally spaced planes of atoms in crystalline sodium is 4.28 Å. If x-rays with $\lambda = 1.539$ Å are diffracted by this set of parallel planes, calculate the angle $2\theta$ at which second-order Bragg diffraction will be observed.

9. The members of a series of equally spaced parallel planes of ions in crystalline LiCl are separated by 2.570 Å. Calculate all the angles $2\theta$ at which diffracted beams of various orders may be seen, if the x-ray wavelength used is 2.167 Å.

10. The members of a series of equally spaced parallel planes in crystalline vitamin $B_{12}$ are separated by 16.02 Å. Calculate all the angles $2\theta$ at which diffracted beams of various orders may be seen, if the x-ray wavelength used is 2.294 Å.

11. A crucial protein at the photosynthetic reaction center of the purple bacterium *Rhodopseudomonas viridis* (Section 15-3) has been separated from the organism, crystallized, and studied by x-ray diffraction. This substance crystallizes with a primitive unit cell in the tetragonal system. The cell dimensions are $a = b = 223.5$ Å and $c = 113.6$ Å.
(a) Determine the volume, in cubic angstroms, of this cell.
(b) One of the crystals in this experiment was box-shaped, with dimensions $1 \times 1 \times 3$ mm. Compute the number of unit cells in this crystal.

12. Compute the volume (in cubic angstroms) of the unit cell of potassium hexacyanoferrate(III) ($K_3Fe(CN)_6$), a substance that crystallizes in the monoclinic system with $a = 8.40$ Å, $b = 10.44$ Å, and $c = 7.04$ Å and with $\beta = 107.5°$.

13. The compound $Pb_4In_3B_{17}S_{18}$ crystallizes in the monoclinic system with a unit cell having $a = 21.021$ Å, $b = 4.014$ Å, $c = 18.898$ Å, and the only non-90° angle equal to 97.07°. There are two molecules in every unit cell. Compute the density of this substance.

14. Strontium chloride hexahydrate ($SrCl_2 \cdot 6H_2O$) crystallizes in the trigonal system in a unit cell with $a = 8.9649$ Å and $\alpha = 100.576°$. The unit cell contains three formula units. Compute the density of this substance.

15. At room temperature, the edge length of the cubic unit cell in elemental silicon is 5.431 Å, and the density of silicon at the same temperature is 2.328 g cm$^{-3}$. Each cubic unit cell contains eight silicon atoms. Using only these facts, perform the following operations.
(a) Calculate the volume (in cubic centimeters) of one unit cell.
(b) Calculate the mass (in grams) of silicon present in a unit cell.
(c) Calculate the mass (in grams) of an atom of silicon.
(d) The mass of an atom of silicon is 28.0855 u. Estimate Avogadro's number to four significant figures.

16. One form of crystalline iron has a body-centered cubic lattice with an iron atom at every lattice point. Its density at 25°C is 7.86 g cm$^{-3}$. The length of the edge of the cubic

unit cell is 2.87 Å. Use these facts to estimate Avogadro's number.

17. Sodium sulfate ($Na_2SO_4$) crystallizes in the orthorhombic system in a unit cell with $a = 5.863$ Å, $b = 12.304$ Å, and $c = 9.821$ Å. The density of these crystals is 2.663 g cm$^{-3}$. Determine how many $Na_2SO_4$ formula units are present in the unit cell.

18. The density of turquoise, $CuAl_6(PO_4)_4(OH)_8(H_2O)_4$, is 2.927 g cm$^{-3}$. This gemstone crystallizes in the triclinic system with cell constants $a = 7.424$ Å, $b = 7.629$ Å, $c = 9.910$ Å, $\alpha = 68.61°$, $\beta = 69.71°$, and $\gamma = 65.08°$. Calculate the volume of the unit cell, and determine how many copper atoms are present in each unit cell of turquoise.

19. An oxide of rhenium has a structure with an Re atom at each corner of the cubic unit cell and an O atom at the center of each edge of the cell. What is the chemical formula of this compound?

20. The mineral perovskite has a calcium atom at each corner of the unit cell, a titanium atom at the center of the unit cell, and an oxygen atom at the center of each face. What is the chemical formula of this compound?

21. Iron has a body-centered cubic structure with a density of 7.86 g cm$^{-3}$.
(a) Calculate the nearest-neighbor distance in crystalline iron.
(b) What is the lattice parameter for the cubic unit cell of iron?
(c) What is the atomic radius of iron?

22. The structure of aluminum is face-centered cubic and its density is $\rho = 2.70$ g cm$^{-3}$.
(a) How many Al atoms belong to a unit cell?
(b) Calculate $a$, the lattice parameter, and $d$, the nearest-neighbor distance.

23. Sodium has the body-centered cubic structure, and its lattice parameter is 4.28 Å.
(a) How many Na atoms does a unit cell contain?
(b) What fraction of the volume of the unit cell is occupied by Na atoms, if they are represented by spheres in contact with one another?

24. Nickel has a face-centered cubic structure with a density of 8.90 g cm$^{-3}$.
(a) Calculate the nearest-neighbor distance in crystalline nickel.
(b) What is the atomic radius of nickel?
(c) What is the radius of the largest atom that could fit into the interstices of a nickel lattice, approximating the atoms as spheres?

25. Calculate the ratio of the maximum radius of an interstitial atom at the center of a simple cubic unit cell to the radius of the host atom.

26. Calculate the ratio of the maximum radius of an interstitial atom at the center of each face of a body-centered cubic unit cell to the radius of the host atom.

## Cohesion in Solids and Liquids

27. Classify each of the following solids as molecular, ionic, metallic, or covalent.
    (a) $BaCl_2$
    (b) SiC
    (c) CO
    (d) Co

28. Classify each of the following solids as molecular, ionic, metallic, or covalent.
    (a) Rb
    (b) $C_5H_{12}$
    (c) B
    (d) $Na_2HPO_4$

29. By examining Figure 17-18, determine the number of nearest neighbors, second nearest neighbors, and third nearest neighbors of a $Cs^+$ ion in crystalline CsCl. The nearest neighbors of the $Cs^+$ ion are $Cl^-$ ions, and the second nearest neighbors are $Cs^+$ ions.

30. Repeat the determinations of the preceding problem for the NaCl crystal, referring to Figure 17-17.

## Lattice Energies of Crystals

31. Calculate the energy needed to dissociate 1.00 mol of crystalline RbCl into its gaseous ions if the Madelung constant for its structure is 1.7476 and the radii of $Rb^+$ and $Cl^-$ are 1.48 Å and 1.81 Å, respectively. Assume that the repulsive energy reduces the lattice energy by 10% from the pure Coulomb energy.

32. Repeat the calculation of problem 31 for CsCl, taking the Madelung constant from Table 17-5 and taking the radii of $Cs^+$ and $Cl^-$ to be 1.67 Å and 1.81 Å.

33. (a) Use the Born-Haber cycle, with data from Appendices D and F, to calculate the lattice energy of LiF.
    (b) Compare the result of part (a) with the Coulomb energy calculated by using an Li—F separation of 2.014 Å in the LiF crystal, which has the rock-salt structure.

34. Repeat the calculations of problem 33 for crystalline KBr, which has the rock-salt structure with a K—Br separation of 3.298 Å.

## Defects and Amorphous Solids

35. Will the presence of Frenkel defects change the measured density of a crystal?

36. What effect will the (unavoidable) presence of Schottky defects have on the determination of Avogadro's number via the method described in problems 15 and 16?

37. Iron(II) oxide is nonstoichiometric. A particular sample was found to contain 76.55% iron and 23.45% oxygen by mass.
    (a) Calculate the empirical formula of the compound (four significant figures).

(b) What percentage of the iron in this sample is in the +3 oxidation state?

38. A sample of nickel oxide contains 78.23% Ni by mass.
    (a) What is the empirical formula of the nickel oxide to four significant figures?
    (b) What fraction of the nickel in this sample is in the +3 oxidation state?

## Liquid Crystals

39. Compare the natures and extents of order in the smectic liquid-crystal and isotropic liquid phases of a substance. Which has the higher entropy? Which has the higher enthalpy?

40. Nematic liquid crystals form when a liquid of long rod-like molecules is cooled. What *additional* types of intermolecular interactions would you expect to favor the formation of a *smectic* phase?

## Additional Problems

41. Some water waves with a wavelength of 3.0 m are diffracted by an array of evenly spaced posts in the water. If the rows of posts are separated by 5.0 m, calculate the angle $2\theta$ at which the first-order ''Bragg diffraction'' of these water waves will be seen.

42. A crystal scatters x-rays of wavelength $\lambda = 1.54$ Å at an angle $2\theta$ of 32.15°. Calculate the wavelength of the x-rays in another experiment if this same diffracted beam from the same crystal is observed at an angle $2\theta$ of 34.46°.

43. The number of beams diffracted by a single crystal depends on the wavelength $\lambda$ of the x-rays used and on the volume associated with one lattice point in the crystal—that is, on the volume $V_p$ of a primitive unit cell. An approximate formula is

$$\text{number of diffracted beams} = \frac{4}{3}\pi\left(\frac{2}{\lambda}\right)^3 V_p$$

(a) Compute the volume of the conventional unit cell of crystalline sodium chloride. This cell is cubic and has an edge length of 5.6402 Å.
    (b) The NaCl unit cell contains four lattice points. Compute the volume of a primitive unit cell for NaCl.
    (c) Use the formula given in this problem to estimate the number of diffracted rays that will be observed if NaCl is irradiated with x-rays of wavelength 2.2896 Å.
    (d) Use the formula to estimate the number of diffracted rays that will be observed if NaCl is irradiated with x-rays having the shorter wavelength 0.7093 Å.

44. If the wavelength $\lambda$ of the x-rays is too large relative to the spacing of planes in the crystal, no Bragg diffraction will be seen because $\sin \theta$ would be larger than 1 in the Bragg equation, even for $n = 1$. Calculate the longest wavelength of x-rays that can give Bragg diffraction from a set of planes separated by 4.20 Å.

**45.** The crystal structure of diamond is face-centered cubic, and the atom coordinates in the unit cell are $(0, 0, 0)$, $(\frac{1}{2}, \frac{1}{2}, 0)$, $(\frac{1}{2}, 0, \frac{1}{2})$, $(0, \frac{1}{2}, \frac{1}{2})$, $(\frac{1}{4}, \frac{1}{4}, \frac{1}{4})$, $(\frac{3}{4}, \frac{1}{4}, \frac{3}{4})$, $(\frac{3}{4}, \frac{3}{4}, \frac{1}{4})$, and $(\frac{1}{4}, \frac{3}{4}, \frac{3}{4})$. The lattice parameter is $a = 3.57$ Å. What is the C—C bond distance in diamond?

**46.** Polonium is the only element known to crystallize in the simple cubic lattice.
(a) What is the distance between nearest-neighbor polonium atoms if the first-order diffraction of x-rays with $\lambda = 1.785$ Å from the parallel faces of its unit cells appears at an angle of $2\theta = 30.96°$ from these planes?
(b) What is the density of polonium in this crystal (in g cm$^{-3}$)?

**47.** At room temperature, monoclinic sulfur has the unit-cell dimensions $a = 11.04$ Å, $b = 10.98$ Å, $c = 10.92$ Å, and $\beta = 96.73°$. Each cell contains 48 atoms of sulfur.
(a) Explain why it is not necessary to give the values of the angles $\alpha$ and $\gamma$ in this cell.
(b) Compute the density of monoclinic sulfur in units of g cm$^{-3}$.

**48.** A compound contains three elements: sodium, oxygen, and chlorine. It crystallizes in a cubic lattice. The oxygen atoms are at the corners of the unit cells, the chlorine atoms are at the centers of the unit cells, and the sodium atoms are at the centers of the faces of the unit cells. What is the formula of the compound?

**\*49.** Show that the radius of the largest sphere that can be placed in a tetrahedral interstitial site in an f.c.c. lattice is $0.225r_1$, where $r_1$ is the radius of the atoms making up the lattice. (*Hint:* Consider a cube with the centers of four spheres placed at alternate corners, and visualize the tetrahedral site at the center of the cube. What is the relationship between $r_1$ and the length of a diagonal of a face? The length of a body diagonal?)

**50.** What is the closest packing arrangement possible for a set of thin circular discs lying in a plane? What fraction of the area of the plane is occupied by the discs? Show how the same reasoning can be applied to the packing of infinitely long, straight cylindrical fibers.

**51.** Name two elements that form molecular crystals, two that form metallic crystals, and two that form covalent crystals. What generalizations can you make about the portions of the periodic table where each type is found?

**52.** The nearest-neighbor distance in crystalline LiCl (rock-salt structure) is 2.570 Å; the bond length in a gaseous LiCl molecule is significantly shorter, 2.027 Å. Explain.

**53.** (a) Using the data of Table 3-4, estimate the lattice energy and intermolecular separation of nitrogen in its solid state, assuming an f.c.c. structure for the solid lattice.
(b) The density of cubic nitrogen is 1.026 g cm$^{-3}$. Calculate the lattice parameter $a$ and the nearest-neighbor distance. Compare your answer with that from part (a).

**\*54.** Solid CuI$_2$ is unstable relative to CuI at room temperature, but CuBr$_2$, CuCl$_2$, and CuF$_2$ are all stable relative to the copper(I) halides. Explain by considering the steps in the Born-Haber cycle for these compounds.

**55.** A crystal of sodium chloride has a density of 2.165 g cm$^{-3}$ in the absence of defects. Suppose a crystal of NaCl is grown in which 0.15% of the sodium ions and 0.15% of the chloride ions are missing. What is the density in this case?

**56.** The activation energy for the diffusion of sodium atoms in the crystalline state is 42.22 kJ mol$^{-1}$, and $D_0 = 0.145$ cm$^2$ s$^{-1}$.
(a) Calculate the diffusion constant $D = D_0 \exp(-E_a/RT)$ of sodium in the solid at its melting point (97.8°C).
(b) What is the root-mean-square displacement of an average sodium atom from an arbitrary origin after the lapse of 1.0 hour at $t = 97.8°C$? (*Hint:* Use Equation 3-17 in Chapter 3.)

**57.** A compound of titanium and oxygen contains 28.31% oxygen by mass.
(a) If the compound's empirical formula is Ti$_x$O, calculate $x$ to four significant figures.
(b) The nonstoichiometric compounds Ti$_x$O can be described as having a Ti$^{2+}$—O$^{2-}$ lattice in which certain Ti$^{2+}$ ions are missing or are replaced by Ti$^{3+}$ ions. Calculate the fraction of Ti$^{2+}$ sites in the nonstoichiometric compound that are vacant and the fraction that are occupied by Ti$^{3+}$ ions.

**58.** Classify the bonding in the following amorphous solids as molecular, ionic, metallic, or covalent.
(a) Amorphous silicon, used in photocells to collect light energy from the sun
(b) Polyvinyl chloride, a plastic of long-chain molecules composed of —CH$_2$CHCl— repeating units, and used in pipes and siding
(c) Soda–lime–silica glass, used in windows
(d) Copper–zirconium glass, an alloy of the two elements with approximate formula Cu$_3$Zr$_2$, used for its high strength and good conductivity

## CUMULATIVE PROBLEMS

**59.** Sodium hydride (NaH) crystallizes in the rock-salt structure, with four formula units of NaH per cubic unit cell. A beam of monoenergetic neutrons, selected to have a velocity of $2.639 \times 10^3$ m s$^{-1}$, is scattered in second order through an angle of $2\theta = 36.26°$ by the parallel faces of the unit cells of a sodium hydride crystal.

(a) Calculate the wavelength of the neutrons.

(b) Calculate the edge length of the cubic unit cell.

(c) Calculate the distance from the center of an $Na^+$ ion to the center of a neighboring $H^-$ ion.

(d) If the radius of an $Na^+$ ion is 0.98 Å, what is the radius of an $H^-$ ion, assuming the two ions are in contact?

60. Chromium(III) oxide has a structure in which chromium ions occupy two thirds of the octahedral interstitial sites in a hexagonal close-packed lattice of oxygen ions. What is the $d$-electron configuration on the chromium ion?

61. A useful rule of thumb is that in crystalline compounds every nonhydrogen atom occupies 18 $Å^3$, and the volume occupied by hydrogen atoms can be neglected. Using this rule, estimate the density of ice (in g $cm^{-3}$). Explain why the answer is so different from the observed density of ice.

62. Estimate, for the F-centers in $CaF_2$, the wavelength of maximum absorption in the visible region of the spectrum that will give rise to the color shown in Figure 17-30.

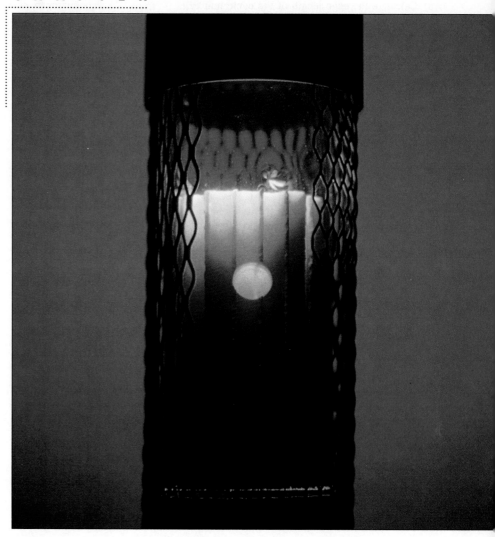

*A catalytic converter, used to allow
wood stoves to burn more efficiently.*

# Chemical Processes

The first 17 chapters of this book introduced the underlying principles of chemistry. Chapters 1 through 11 emphasized macroscopic aspects of chemistry: mass relationships, the periodic table, the states of matter, the nature of chemical equilibrium and the thermodynamic laws that underlie it, and the factors that determine the rates of chemical reactions. Chapters 12 through 17 took a microscopic approach, showing how electrons and nuclei combine to form atoms, which join together through chemical bonds to make molecules and complex ions. The further interactions between molecules give rise to the observed properties of bulk matter: gases, liquids, and solids.

So far we have looked at individual pieces: the chemical equilibrium in a particular reaction, the nature of bonding in a particular solid compound. Our understanding of chemistry cannot stop there, however. In the remaining chapters of the book, we begin to combine the pieces to see how reactions fit together in an overall chemical process that transforms a set of starting materials into a set of products. We introduce this **process-based approach** to chemistry by first examining the nature of the chemical industry and its major products. We then turn to geology and biology to consider the similarities and differences between the processes that occur in the natural world and those designed by chemists, and to see their points of contact in aspects such as the availability of raw materials and the effect of pollutants on the environment. Finally, we take a close look at the production and uses of the lightest chemical element, hydrogen, as an illustration of the approach followed in the remaining chapters of the book.

## 18-1 THE CHEMICAL INDUSTRY

A firm grasp of chemical principles is essential to the design of clean and efficient manufacturing processes for chemical products. These processes are continually developed and changed to improve product yield, reduce waste, and minimize costs. Chemical plants are usually expensive to construct, and a new process invariably passes through several **pilot-plant stages** after its development in the research laboratory. In these scaling-up operations, the process is optimized with respect to such variables as temperature, pressure, and time of reaction. Almost all laboratory-scale chemical processes are **batch processes,** in which the reactants are mixed in a container and allowed to react, and the products are removed in a sequence of operations. Many industrial-scale processes, in contrast, are **continuous processes,** designed so that reactants are fed in and products are removed continuously. It is rarely a simple matter to convert a process from batch to continuous operation, but the latter type is generally more economical.

We can represent processes in the chemical industry schematically (Fig. 18-1) as

starting materials  &longrightarrow;  desired products + byproducts

Energy, often a great deal of it, plays a part in such transformations. Sometimes energy must be put in to carry out the desired reaction; in other cases it is produced, usually as heat, and must be recycled or safely discarded. Let us examine each of the terms in this equation in light of basic principles of chemistry.

In the ideal case, the starting materials for a chemical process are easily obtainable and easily transportable to the site where they will be used. Starting materials tend to be in thermodynamic states of rather low free energy, simply because compounds with high free energy tend to react spontaneously to give products of lower free energy. Over the long history of the earth, compounds of high free energy have mostly disappeared. Thus, aluminum is very abundant in the earth's crust but is never found in elemental form. Rather, it is tied up in minerals of much lower free energy such as kaolinite, a clay with the chemical formula $Al_2Si_2O_5(OH)_4$ and a large negative free energy of formation ($\Delta G_f^\circ = -3800 \text{ kJ mol}^{-1}$). Some compounds of high free energy remain in nature because kinetics is involved as well as thermodynamics; states of high free energy can persist over long periods if the rate

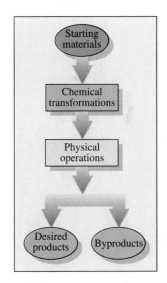

**Figure 18-1**

Schematic diagram of a chemical process, employing the color conventions that were used in Figure 16-27 and that will be followed in subsequent chapters. Starting materials (red) are converted to products (green) via a series of chemical reactions (blue) and physical operations (orange). Filtering and evaporation are examples of physical operations.

of reaction is sufficiently low. The compounds of high free energy that do occur in nature are particularly valuable resources for chemistry. Hydrocarbons such as octane occur in petroleum deposits and have high free energies relative to their combustion products:

$$C_8H_{18}(g) + \tfrac{25}{2} O_2(g) \longrightarrow 9 H_2O(\ell) + 8 CO_2(g) \qquad \Delta G° = -5272 \text{ kJ}$$

The list of uses for hydrocarbons is long. It is unfortunate that so much of the limited supply of hydrocarbons is being burned as fuel rather than conserved for future use as chemical starting materials for polymers, fibers, detergents, and other goods.

In many cases, a nonspontaneous reaction (one with $\Delta G > 0$) can be caused to proceed by linking it to another reaction for which $\Delta G < 0$—for instance, the combustion of a fossil fuel such as coal, oil, or natural gas. In many chemical processes, energy is supplied as electricity. The powerful oxidizing (bleaching) agent sodium chlorate, for example, is made by the overall reaction

$$NaCl(aq) + 3 H_2O(\ell) \longrightarrow NaClO_3(aq) + 3 H_2(g) \qquad \Delta G° = +835 \text{ kJ}$$

This reaction becomes possible only with an external source of electrical work. The electricity itself is generated by burning fossil fuels or harnessing the energy of a nuclear reaction so that, again, coupling to a spontaneous reaction effects a nonspontaneous reaction. Many chemical transformations are practical only if inexpensive energy sources are available. It is very desirable to use the energy from the sun directly to carry out chemical reactions. Nature, of course, does just this in photosynthesis (Section 15-3), the process in which light is harvested by chlorophyll and used to convert water and carbon dioxide (low-free-energy states) to glucose and other sugars (high-free-energy states). Much current research is aimed at understanding photosynthesis and copying its principles to carry out other chemical transformations.

In an ideal chemical process, the desired product would be obtained (1) in 100% yield at room temperature and pressure, (2) at a convenient and controllable rate, and (3) already pure or in a condition that allows easy separation and purification. In real processes, compromises are necessary. The yield of a chemical reaction is fundamentally limited by thermodynamics, and so conditions of pressure and temperature must be adjusted to optimize it, often by expensive equipment. The rate of a reaction depends on pressure and temperature but is also sensitive to the presence of catalysts. The control of reaction rates is so important in the chemical industry that researchers constantly search for catalysts that will speed the formation of desired products without leading to unwanted byproducts. Finally, the importance of the last point—separability of the desired product—cannot be overemphasized. A process that yields a product contaminated with large amounts of impurities (or with dangerous impurities) may be abandoned even if it gives good yields at a convenient rate.

This brings us to the last term in the schematic equation for industrial processes: the byproducts. In some cases the byproducts are relatively harmless, but the question of what to do with them remains. Each year, a small mountain of $CaSO_4$ is produced in the synthesis of phosphate fertilizers; although $CaSO_4$ is harmless in itself (it is used in gypsum wallboard), large volumes of it create awkward and expensive disposal problems. Other byproducts are toxic to plant and animal life, ranging from slightly irritating to deadly. In the past, many of these toxic wastes were buried in landfills. This practice carries the risk of contamination of water sources and direct exposure of nearby residents and is now thoroughly regulated.

Important social, economic, and legal incentives to eliminate or reduce worthless and hazardous byproducts already exist. Parts of the chemical industry have responded by modifying their processes to minimize the generation of wastes. Where this has not been possible, other measures have been taken to diminish the effects of production on the environment. Recycling of materials such as glass and paper and incineration of flammable wastes are the methods of choice, with adequate precautions taken to ensure complete combustion and with the use of alkaline scrubbers to remove acidic compounds from the effluent gas stream. In this way, even very hazardous organic compounds are converted to carbon dioxide and water vapor. Wastes that are not combustible can generally be made nonhazardous through chemical treatment. The relatively small fraction of wastes that cannot be incinerated or treated chemically can be put in a secure landfill.

The ideal way to dispose of byproducts is to find a use for them. The cost of waste disposal has encouraged research into ways to recycle the wastes or employ them differently. One example comes from the uranium fuel processing industry, in which scrap uranium contaminated with dirt and grease is oxidized by nitric acid to uranyl nitrate, $UO_2(NO_3)_2$, which is soluble in water and can therefore be purified. The oxidation reaction gives off the nitrogen oxides NO and $NO_2$, which are pollutants. Vented as waste, they would contribute to the formation of acid rain. Instead, these gases are caused to react first with oxygen and then with water to give nitric acid:

$$NO(g) + \tfrac{1}{2} O_2(g) \longrightarrow NO_2(g)$$

$$3\,NO_2(g) + H_2O(\ell) \longrightarrow 2\,HNO_3(aq) + NO(g)$$

The use of these reactions offers two advantages: reduction of the amount of nitrogen oxide released to the atmosphere and recycling of the nitric acid product for use in the oxidation reaction. A cost savings also results.

The picture of chemical processing we have painted so far has been a simplified description of raw materials being converted directly to final products. In fact, such conversions are usually multistep processes, with the involvement of many chemicals that do not appear in the finished product. Figure 18-2 shows how aspirin is made from a set of raw materials that can be traced all the way back to air, water, salt, sulfur, petroleum, and natural gas. Numerous chemicals appear only as inter-

Crystals of aspirin under polarized light. (Phillip A. Harrington/Fran Heyl Associates)

CH₃ — Group derived from acetic acid

Group derived from formic acid

Benzene ring

The structure of acetylsalicylic acid, the active component of aspirin. Note the substructures, which are those of three simpler compounds: benzene, formic acid, and acetic acid.

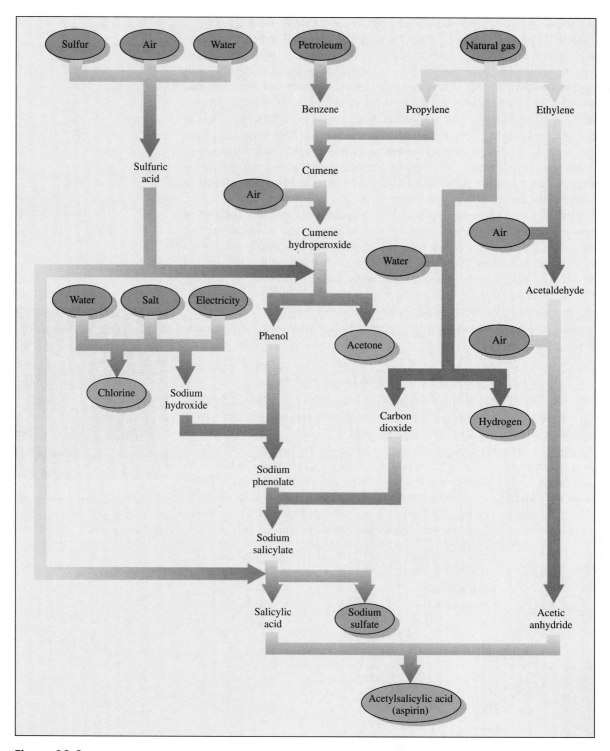

**Figure 18-2**

Process flow diagram for the synthesis of aspirin. The ultimate precursors of aspirin are air, water, salt, sulfur, petroleum, and natural gas. Many of the intermediates are organic chemicals that will appear in Chapter 23.

## Table 18-1  Chemicals Produced in Largest Volume in the United States (1994)[†]

| Rank | Name | Formula | U.S. Production (billions of kg) | Principal End Use |
|------|------|---------|----------------------------------|-------------------|
| 1 | Sulfuric acid | $H_2SO_4$ | 40.5 | Fertilizers, chemicals, processing |
| 2 | Nitrogen | $N_2$ | 30.6 | Fertilizers |
| 3 | Oxygen | $O_2$ | 22.5 | Steel, welding |
| 4 | Ethylene | $C_2H_4$ | 22.0 | Plastics, antifreeze |
| 5 | Lime | $CaO$ | 17.4 | Paper, chemicals, cement |
| 6 | Ammonia | $NH_3$ | 17.2 | Fertilizers |
| 7 | Propylene | $C_3H_6$ | 13.1 | Gasoline, plastics |
| 8 | Sodium hydroxide | $NaOH$ | 11.7 | Chemical processing, aluminum production, soap |
| 9 | Phosphoric acid | $H_3PO_4$ | 11.5 | Fertilizers |
| 10 | Chlorine | $Cl_2$ | 11.0 | Bleaches, plastics, water purification |
| 11 | Sodium carbonate | $Na_2CO_3$ | 9.3 | Glass |
| 12 | Ethylene dichloride | $C_2H_4Cl_2$ | 8.5 | Plastics, drycleaning |
| 13 | Nitric acid | $HNO_3$ | 8.0 | Fertilizers, explosives |
| 14 | Ammonium nitrate | $NH_4NO_3$ | 8.0 | Fertilizers, mining |
| 15 | Urea | $(NH_2)_2CO$ | 7.3 | Fertilizers |
| 16 | Vinyl chloride | $C_2H_3Cl$ | 6.7 | Plastics |

[†]Not included as "chemicals" are steel and concrete. Adapted from *Chemical and Engineering News,* April 10, 1995, p. 17.

mediates; there is no sulfuric acid in the aspirin that is eventually produced, but sulfuric acid plays a key role in several steps of the manufacturing process. The figure also shows the interdependency built into the chemical industry. An aspirin manufacturer does not purchase salt, sulfur, and the other primary materials but rather buys intermediate chemicals that may have applications in a wide variety of other fields. Instead of manufacturing sulfuric acid in small quantities, the aspirin company buys it from a firm that makes far larger quantities—for use by steel mills, for example.

Table 18-1 shows the amounts of the chemicals produced in the greatest volumes in the United States during 1994. Relatively few are used *directly* as consumer products. Household ammonia, for example, is a negligible fraction of the total produced. Most ammonia undergoes further chemical transformations for use as fertilizer in forms that include ammonium nitrate, urea, and ammonium phosphate. The raw materials of the heavy-chemicals industry are relatively few in number— chiefly air, water, limestone, salt, coal, natural gas and petroleum, sulfur, and phosphate rock—simply because these are abundantly available at low cost. The following chapters will discuss some of the methods by which the high-volume chemicals are produced. Some chemicals that are not made in huge quantities are nevertheless important. Pharmaceuticals, for example, are vital to health care even though they are produced in relatively small amounts.

## 18-2 GEOCHEMICAL AND BIOCHEMICAL PROCESSES

Geochemistry and biochemistry are subjects to which entire books cannot do justice, let alone a short section of a single chapter. Our goal here is only to point out some of the similarities and differences between chemical processes in the natural world

| Table 18-2   Sources of Major Raw Materials for Chemistry | |
|---|---|
| **Atmosphere** | **Hydrosphere** |
| Oxygen ($O_2$) | Water ($H_2O$) |
| Nitrogen ($N_2$) | Sodium chloride (NaCl) |
| Noble gases | Sodium bromide (NaBr) |
|  | Sodium iodide (NaI) |
|  | Magnesium chloride ($MgCl_2$) |
| **Lithosphere** | **Biosphere** |
| Silica ($SiO_2$) | Coal (C) |
| Limestone ($CaCO_3$) | Petroleum |
| Sodium chloride (NaCl) | Natural gas ($CH_4$) |
| Potassium chloride (KCl) | Sulfur (S) |
| Sodium carbonate ($Na_2CO_3$) | Phosphate rock ($Ca_5(PO_4)_3F$) |
| Sodium sulfate ($Na_2SO_4$) | Organic natural products |
| Metal ores |  |

The atmosphere is the envelope of gases that surrounds the earth. The lithosphere includes the outer parts of the solid earth. The hydrosphere is the earth's water in its many forms. The biosphere comprises living organisms, their immediate surroundings, and their products.

and those in chemical industry, and to discuss the sources for the important raw materials of chemistry (see Table 18-2).

Geochemical transformations are slow. Time scales range from years (for the weathering of the softest rocks) to millions and even billions of years, as Section 19-2 will discuss further. One reason for this slowness is that geochemical processes are frequently solid-state reactions, and the diffusion of atoms through solids is many orders of magnitude slower than that through liquids and gases. One might suppose that because of this slowness, geochemical processes would not be very useful models for industrial chemistry. In fact, they offer important lessons. The manufacture of synthetic diamonds employs very high pressures to carry out in a short time the same transformations that occur more slowly, but also under pressure, beneath the surface of the earth. Chemists also synthesize materials that resemble minerals formed by geochemical processes. Examples are modern ceramic materials (Section 20-3), which are related to ceramics made from natural clays, and zeolite molecular sieves (Section 19-1), which exist in both synthetic and mineral versions and are used to catalyze reactions and to separate mixtures.

If geochemical processes are characterized by their slowness, biochemical processes are notable for their speed and selectivity. The key to these features is the presence of enzyme catalysts (Section 11-7). Enzymes can greatly accelerate the rate of a reaction and are quite specific both in the reactants upon which they act and in the products formed. The study of biochemical processes has influenced the design of industrial catalysts, which are becoming increasingly selective as processes are improved. There is still a long way to go before scientists will be able to duplicate the remarkable catalytic actions of enzymes, however.

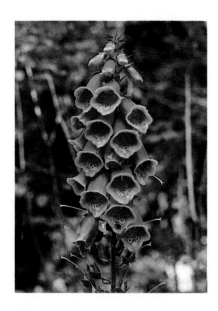

The foxglove plant produces digitalis, an important drug in the treatment of heart disease. (Pamela Harper)

## Raw Materials for Chemistry

Natural products obtained from plants and animals play a large role in chemistry; they are used both directly and as precursors for chemical syntheses of other substances. So many natural products are in common use that it is impossible to list even a small fraction of them. Penicillin, for example, was first produced from the accidental contamination of a bacterial culture by mold. The terpenes are a class of compounds found in almost all plants; they are a major component of turpentine, which is obtained from the resin of pine trees. Many terpenes are used directly as artificial scents in the perfume industry; others serve as precursors in the manufacture of vitamin A (Fig. 18-3).

(a) β-ionone

(b) Vitamin A

### Figure 18-3

(a) The compound β-ionone, $C_{13}H_{20}O$, is a terpene that is used extensively in perfumes. (b) Its more significant use is as a precursor in the synthesis of vitamin A. Note the close connection between the structure of vitamin A and that of retinal, a molecule crucial for vision (problem 33 in Chapter 15).

The distinction between chemical synthesis and biological synthesis has blurred with time. On the one hand, there is increasing use of living agents such as specialized bacteria to carry out the synthesis of complex products that are difficult to produce by conventional chemical means. On the other hand, certain natural products are more easily obtained by synthesizing them from other starting materials than by extracting them from existing sources. Methanol, for example, used to be obtained by the destructive distillation of wood—hence the name "wood alcohol"— but it is now produced mainly from synthesis gas (discussed in the next section).

The fossil fuels are important chemical raw materials produced by biological activity. The decay of plant matter through the action of bacterial reducing agents creates first peat, then lignite, and finally bituminous coal. Crude oil (petroleum) is thought to have formed from the decomposition of the remains of small marine organisms under reducing conditions. Hydrocarbons of low molar mass that formed in this way are gases rather than liquids. Pockets of natural gas (primarily methane, with smaller amounts of hydrogen, ethane, and propane) are found in association with deposits of petroleum. Although all of these fossil fuels have biological origins, geological transformations have brought them to the forms in which they are found today.

A chemical raw material with a somewhat more unexpected biological origin is sulfur. Although sulfur is widely distributed in compounds, its most convenient commercial source is the deposits of elemental sulfur often associated with salt domes under the surface of the earth. Sulfur is thought to have precipitated in these settings as calcium sulfate, in contact with natural gas and carbon dioxide. Reaction of these substances to give elemental sulfur and calcium carbonate was then catalyzed by enzymes secreted by certain bacteria.

$$4\,CaSO_4(s) + 3\,CH_4(g) + CO_2(g) \longrightarrow 4\,CaCO_3(s) + 4\,S(s) + 6\,H_2O(\ell)$$

The origins of phosphate rock, which has the approximate chemical formula $Ca_5(PO_4)_3F$, are less clear. The rock is found in large deposits in certain parts of the world, notably Florida and North Africa. Phosphate ions are present in rather low concentration in seawater, and the solubility of calcium phosphate is comparable to that of calcium carbonate. Therefore, it is surprising that the phosphate is found in such concentration rather than scattered through the world's limestone deposits. The concentrated deposits of phosphate rock may have resulted from extensive dissolution and reprecipitation of phosphate–carbonate mixtures, perhaps in a small but critical pH range in which the phosphate precipitated and the carbonate remained in solution. It is more likely, however, that the phosphate rock resulted from accumulation by marine organisms. Phosphate rock is the source of phosphorus and phosphate-based fertilizers. As such, it is the raw material used in greatest volume in the chemical industry, as Chapter 21 will discuss further.

Other raw materials of importance in chemistry come from the hydrosphere and lithosphere. The classifications of Table 18-2 are not absolute, because many materials move between hydrosphere and lithosphere as they dissolve and reprecipitate, or they are shaped by both biological and geochemical processes. Sodium chloride, for example, is extracted from salt brines (concentrated aqueous solutions of mixed salts) in addition to occurring in mineral form. Calcium carbonate, which is found most extensively in limestone deposits, also forms the shells of many marine animals and is obtained economically from them. It is the source of lime (CaO), the cheapest base available and a chemical of major importance (Fig. 18-4).

**Figure 18-4**

Limestone being quarried as a source of lime. (Ireco)

## 18-3 HYDROGEN

Hydrogen is the most abundant atomic species in the universe; at the earth's surface it ranks third in abundance, behind only oxygen and silicon. This section discusses how elemental hydrogen is produced and used.

### Production of Hydrogen

Hydrogen constitutes 11.2% of the mass of water and is therefore abundantly available nearly everywhere on the earth's surface. From the point of view of easy access to starting materials, the ideal method for producing elemental hydrogen would be to split water:

$$H_2O(\ell) \longrightarrow H_2(g) + \tfrac{1}{2} O_2(g)$$

This reaction can be carried out in an electrolytic cell, as Section 10-7 discusses (see also Fig. 1-6). Commercial cells use solutions of sodium hydroxide, rather than pure water, to permit the passage of large currents. Although electrolysis is useful when very pure hydrogen is needed, it is not a major source of hydrogen today because of the prohibitive cost of the electricity. Considerable research is being devoted to the possibility of using sunlight directly to split water in the presence of a catalyst, but trial processes of this type are far from being practical for the large amounts of hydrogen needed.

The reaction of a strong acid with iron produces gaseous hydrogen:

$$H_2SO_4(aq) + Fe(s) \longrightarrow H_2(g) + FeSO_4(aq)$$

The French scientist Jacques Charles used this reaction in 1783 to generate hydrogen for the first human flight in a hydrogen balloon (Fig. 18-5). The similar reaction of zinc with acid is a useful laboratory-scale reaction for the production of hydrogen. Neither of these reactions is useful for making hydrogen on a large scale, however.

Currently, the major source for hydrogen is the methane in natural gas, and Figure 18-6 shows the method of production. There are two crucial steps in this process. The first is the **reforming reaction:**

$$CH_4(g) + H_2O(g) \longrightarrow CO(g) + 3\,H_2(g) \qquad \Delta H° = +206 \text{ kJ}$$

**Figure 18-5**
The French chemist Jacques Charles took part in the first hydrogen balloon ascent from Paris on December 1, 1783. (The Granger Collection)

**Figure 18-6**
The reforming and shift reactions produce hydrogen and carbon dioxide from methane and steam.

This reaction is endothermic and proceeds with an increase in the chemical amount of gas from 2 mol to 4. Accordingly, it is carried out at high temperatures (750 to 1000°C) and low total pressures to increase yield. An excess of steam is added to drive the reaction toward completion, and a nickel catalyst is used. The gas mixture produced, called **synthesis gas,** is used directly to make methanol ($CH_3OH$) and other chemicals. If a feedstock other than methane is used, synthesis gas is obtained with different ratios of carbon monoxide to hydrogen. With propane, for example, the reforming reaction is

$$C_3H_8(g) + 3\ H_2O(g) \longrightarrow 3\ CO(g) + 7\ H_2(g)$$

The second step in the production of hydrogen is called the **shift reaction.** In it, additional steam reacts with the carbon monoxide from the first step:

$$CO(g) + H_2O(g) \longrightarrow CO_2(g) + H_2(g) \qquad \Delta H° = -41\ \text{kJ}$$

This reaction is exothermic and so occurs to a greater extent at low temperatures. As is frequently the case with exothermic processes, however, the temperature cannot be lowered too far because the reaction then becomes too slow. The best compromise comes at a temperature of about 350°C, with an iron oxide catalyst. Two things are accomplished with the shift reaction: (1) additional hydrogen beyond that from the reforming reaction is obtained, and (2) carbon dioxide is obtained as a byproduct. The shift reaction is the major commercial source of carbon dioxide, which has many uses. Carbon dioxide is frozen and sold for refrigeration (Dry Ice). In liquid form it is used to fill fire extinguishers and to make carbonated water for sparkling beverages. Gaseous $CO_2$ is used in chemical synthesis, such as in the production of salicylic acid for aspirin (Fig. 18-2).

At present, natural gas is the primary raw material for making hydrogen and carbon dioxide. As reserves of natural gas are consumed, other sources must be developed. One older process that is attracting new attention is the reaction of coal with steam:

$$C(s) + H_2O(g) \longrightarrow CO(g) + H_2(g) \qquad \Delta H° = +131\ \text{kJ}$$

This mixture of product gases is called **water gas** and burns with a blue flame because of its carbon monoxide content. It has less hydrogen proportionately than the synthesis gas produced from methane or higher hydrocarbons. Water gas can react further, in the shift reaction, to give additional hydrogen and carbon dioxide. Once a mixture of CO and $H_2$ is prepared with the appropriate proportions the reforming reaction just described can be reversed to make methane for use as a fuel; the overall process is referred to as the gasification of coal (Fig. 18-7).

## The Uses of Hydrogen

Hydrogen does not appear on the list of the top 16 chemicals in Table 18-1, or even on a list of the top 50 in a given year. One reason is that it is the lightest element, and so it ranks low on a mass basis. A second reason is that the list refers to the sales or shipments of chemicals and does not include amounts that are **captive**—that is, used directly after production to make other chemical end-products. This is very much the case with hydrogen, which is a far more important chemical than its commercial ranking indicates. Its major use is for reaction with nitrogen to make ammonia for fertilizers, a process Section 21-3 considers in greater detail. Chemical plants that produce ammonia use natural gas, steam, and nitrogen as starting materials, and

**Figure 18-7**
Reserves of coal are much more extensive than those of petroleum or natural gas. In this demonstration plant in Texas, researchers are studying the production of gas from coal for use as fuel and in chemical synthesis. (Shell Oil Company)

hydrogen plays only the role of an intermediate. Another captive use for hydrogen is in the creation of methanol. As already mentioned, this synthesis is carried out with the product from the reforming reaction, synthesis gas:

$$CO(g) + 2\,H_2(g) \longrightarrow CH_3OH(g) \qquad\qquad \Delta H° = -90\ \text{kJ}$$

Low temperatures (300°C) and high pressures (250–300 atm) are used. In most cases, the same plant makes ammonia, methanol, carbon dioxide, and hydrogen.

One final use of hydrogen is its addition to carbon–carbon double bonds (which are present in what are called **unsaturated hydrocarbons** because they are capable of taking up more hydrogen) to convert them to single bonds (**saturated compounds).** The simplest example is the reduction of ethylene to ethane:

$$H_2C{=}CH_2 + H_2 \longrightarrow H_3C{-}CH_3$$

Hydrogen is used extensively in petroleum refining to eliminate carbon–carbon double bonds. It is also employed in food processing to convert liquid vegetable oils to solids. For example, a large proportion of sunflower oil is an oily liquid composed of molecules with the structural formula

$$
\begin{aligned}
&\qquad\qquad\qquad\qquad\qquad\qquad\qquad\qquad\ \overset{\displaystyle O}{\overset{\|}{}}\\
&CH_3(CH_2)_4CH{=}CHCH_2CH{=}CH(CH_2)_7C{-}OCH_2\\
&CH_3(CH_2)_4CH{=}CHCH_2CH{=}CH(CH_2)_7COOCH\\
&CH_3(CH_2)_4CH{=}CHCH_2CH{=}CH(CH_2)_7C{-}OCH_2\\
&\qquad\qquad\qquad\qquad\qquad\qquad\qquad\qquad\ \underset{\displaystyle O}{\underset{\|}{}}
\end{aligned}
$$

This is a **polyunsaturated oil** because there are six C=C double bonds per molecule. **Hydrogenation** with 6 mol of $H_2$ in the presence of a catalyst saturates this oil, and the product has a high enough melting point to make it a solid at room conditions. The use of solid fats (or solidified oils) offers advantages in food processing and preservation, so that "hydrogenated vegetable oil" is an ingredient in many foodstuffs. Heavy consumption of saturated fats has been linked to diseases of the heart and circulatory system, however, and the presence of *un*hydrogenated (polyunsaturated) oils in foods is now heavily advertised.

## CONCEPTS & SKILLS

After studying this chapter and working the problems that follow, you should be able to

1. Describe how the chemical industry operates, and the variety of considerations that must be taken into account in the design of a new chemical process (Section 18-1, problems 1–2).
2. Compare geochemical and biochemical processes to processes in the chemical industry (Section 18-2).
3. Outline the sources of important raw materials for chemistry (Section 18-2, problems 3–4).
4. Describe methods for the production of hydrogen and the uses of this element (Section 18-3, problems 5–12).

## PROBLEMS

### Chemical Processes

1. Define the terms "batch process" and "continuous process." How do the designs of the two types of processes differ?
2. Define the terms "byproducts" and "waste products." Is there a distinction between the two?
3. Identify a useful source of each of the following chemical elements as specifically as you can, based on information in the chapter.
   (a) Sulfur        (c) Phosphorus
   (b) Carbon        (d) Calcium
4. In what ways are biochemical processes analogous to and different from processes in chemical industry? Draw a process diagram for the production of sulfur, beginning with calcium sulfate, carbon dioxide, and marine bacteria.

### Hydrogen

5. You need a small supply of hydrogen for a chemical reaction. Describe a good way to prepare it in the laboratory, using a balanced chemical equation.
6. When water is electrolyzed on the laboratory scale to produce hydrogen and oxygen, a small amount of sulfuric acid is added to the water.
   (a) Why is the acid added?
   (b) How could you prove that the hydrogen and oxygen come primarily from the water and not from the acid?
7. The production of synthesis gas from methane in natural gas and steam

$$CH_4(g) + H_2O(g) \longrightarrow CO(g) + 3\,H_2(g)$$

is an endothermic process, and so it is favored at sufficiently high temperatures. Estimate the temperature at which the equilibrium constant for the reaction is equal to 1 (that is, at which $\Delta G = 0$).

8. The shift reaction

$$CO(g) + H_2O(g) \longrightarrow CO_2(g) + H_2(g)$$

is an exothermic process, and so it is favored at sufficiently low temperatures. Estimate the temperature at which the equilibrium constant for the reaction is 1 (that is, at which $\Delta G = 0$).

9. One procedure for generating gaseous hydrogen involves the action of steam on iron to give iron oxide ($Fe_3O_4$).
   (a) Write a balanced chemical equation for this reaction.
   (b) Calculate the enthalpy change $\Delta H°$ for this reaction, and determine whether the equilibrium yield would be enhanced or reduced by an increase in temperature.

10. Iron oxide produced as described in the preceding problem is subsequently reduced to iron by the carbon monoxide present in synthesis gas.
   (a) Write a balanced chemical equation for this step.
   (b) Calculate the enthalpy change $\Delta H°$ for this second step. What will be the effect of an increase in temperature on the equilibrium yield of this step? An increase in pressure?

11. Calculate the volume of hydrogen at 0°C and 1.00 atm that is required to convert 500.0 g of linoleic acid ($C_{18}H_{32}O_2$) to stearic acid ($C_{18}H_{36}O_2$).

12. A chemist determines that 4.20 L of hydrogen at 298 K and a pressure of 1.00 atm is required to completely hydrogenate 48.5 g of the unsaturated compound oleic acid to stearic acid ($C_{18}H_{36}O_2$). How many units of unsaturation (where a unit of unsaturation is one double bond) are in a molecule of oleic acid?

### Additional Problems

13. Identify the state of lower free energy in each of the following pairs.
   (a) Aspirin versus: coal, hydrogen, air

(b) Seawater versus: pure water, sodium chloride, magnesium carbonate

(c) Vitamin A, oxygen versus: carbon dioxide, water

**14.** Calcium hydride ($CaH_2$) is used as a convenient source of gaseous hydrogen in the laboratory. Calculate the volume of hydrogen produced at 25°C and a pressure of 1.00 atm when 12.21 g of $CaH_2$ is added to a large excess of water.

**15.** Explain why reacting iron or zinc with sulfuric acid is not a practical means for the large-scale generation of hydrogen.

**16.** As the technical chief in a soft-drink factory, you buy a tank that contains 900 lb of liquid carbon dioxide and set up a machine that uses the carbon dioxide to carbonate the company's product. After 800 lb of $CO_2$ has been taken out of the tank, the pressure gauge on the cylinder still reads 64 atm as it did on the day it arrived. The boss concludes that the gauge must be broken and tells you to buy a costly replacement gauge. Use Figure 4-6 to explain to the boss why the gauge is almost certainly *not* broken. What is a reliable way to measure the amount of $CO_2$ that remains in the cylinder?

**17.** In a recent year, the total consumption of carbonated beverages in the United States corresponded to 63 *billion* 12-oz cans or bottles. If a 12-oz can has a volume of 355 mL, and the concentration of dissolved $CO_2$ is 0.15 M, calculate the total mass of $CO_2$ produced and used for making carbonated beverages during that year.

**18.** Compute the oxidation number of carbon in ethylene and in ethane. Is hydrogen oxidized or reduced when it is used to saturate the multiple bonds in a hydrocarbon?

**19.** Classify the following as redox, dissolution, precipitation, or acid–base reactions.

(a) $4\,CaSO_4(s) + 3\,CH_4(g) + CO_2(g) \rightarrow$
$$4\,CaCO_3(s) + 4\,S(s) + 6\,H_2O(\ell)$$

(b) $CO(g) + 2\,H_2(g) \rightarrow CH_3OH(g)$

(c) $H_2C{=}CH_2 + H_2 \rightarrow H_3C{-}CH_3$

(d) $H_2SO_4(aq) + Fe(s) \rightarrow H_2(g) + FeSO_4(aq)$

19

*Azurite is a basic copper carbonate with chemical formula $Cu_3(CO_3)_2(OH)_2$.*

# The Lithosphere

The study of the solid earth involves techniques from chemistry, physics, and geology. Seismic data (shock-wave propagation velocities from earthquakes and man-made explosions) show that the earth consists of three distinct parts: the **core,** the **mantle,** and the **crust** (Table 19-1). The core of the earth is believed to consist primarily of iron, probably with smaller amounts of nickel and iron sulfide (FeS). This would be consistent with the composition of meteorites, which are thought to resemble the material from which the earth formed. The outer part of the core is probably liquid, and there may be a solid region at the center. Outside the core is the mantle, whose seismic properties are consistent with a composition largely of magnesium-iron silicates.

The crust of the earth has an average thickness of only 17 km, and so it contributes relatively little to the total mass. The crust shows greater variation in rock types than does the mantle. In contrast with the core and the mantle, the crust of the earth (also called the **lithosphere**) is in a state of continual transformation. Heat and matter flow from its hot interior; it is subjected to the chemical action of its atmosphere and hydrosphere; and it is modulated by its interaction with the biosphere (living matter on the surface of the earth). Contact of the lithosphere with these zones outside it has a profound effect on the distribution of the elements within it.

This chapter begins with a look at the types of silicate compounds that make up much of the lithosphere and continues with a discussion of the chemical transformations they undergo at and beneath the surface of the earth. The final two sections then focus on the winning of metals from their ore sources in the earth.

**Table 19-1  The Shell Structure of the Earth**

| | Thickness (km) | Volume ($10^{20}$ m$^3$) | Mean Density (g cm$^{-3}$) | Mass ($10^{24}$ kg) | Mass (percent) |
|---|---|---|---|---|---|
| Atmosphere | — | — | — | 0.000005 | 0.00009 |
| Hydrosphere | 3.8 | 0.0137 | 1.03 | 0.00141 | 0.024 |
| Crust | 17 | 0.08 | 2.8 | 0.024 | 0.4 |
| Mantle | 2883 | 8.99 | 4.5 | 4.016 | 67.2 |
| Core | 3471$^\dagger$ | 1.75 | 11.0 | 1.936 | 32.4 |
| Whole earth | 6371$^\dagger$ | 10.83 | 5.52 | 5.976 | 100 |

$^\dagger$Radius.

## 19-1  SILICATES

Silicon and oxygen make up most of the earth's crust, with oxygen accounting for 47% and silicon for 28% of its mass. The silicon–oxygen bond is strong and partially ionic. It forms the basis for a class of minerals called **silicates,** which make up the bulk of the rocks, clays, sand, and soils in the earth's crust. From time immemorial, silicates have provided the ingredients for building materials such as bricks, cement, concrete, ceramics, and glass.

The structure-building properties of silicates (Table 19-2) originate in the tetrahedral orthosilicate anion ($SiO_4^{4-}$) in which the negative charge of the silicate ion is balanced by the compensating charge of one or more cations. The simplest silicates consist of individual $SiO_4^{4-}$ anions (Fig. 19-1$a$), with cations arranged around them on a regular crystalline lattice. Such silicates are properly called **orthosilicates.** Examples are forsterite ($Mg_2SiO_4$) and fayalite ($Fe_2SiO_4$), which are the extreme members of a class of minerals called **olivines,** [Mg,Fe]$_2$SiO$_4$. There is a continuous range of proportions of magnesium and iron in the olivines.

Other silicate structures form when two or more $SiO_4^{4-}$ tetrahedra link and share oxygen vertices. The simplest such minerals are the **disilicates**—such as thortveitite, $Sc_2(Si_2O_7)$—in which two $SiO_4^{4-}$ tetrahedra are linked (Fig. 19-1$b$). Additional linkages of tetrahedra create the ring, chain, sheet, and network struc-

**Table 19-2  Silicate Structures**

| Structure | Figure | Corners Shared at Each Si | Repeat Unit | Si:O Ratio | Example |
|---|---|---|---|---|---|
| Tetrahedra | 19-1$a$ | 0 | $SiO_4^{4-}$ | 1:4 | Olivines |
| Pairs of tetrahedra | 19-1$b$ | 1 | $Si_2O_7^{6-}$ | 1:3$\frac{1}{2}$ | Thortveitite |
| Closed rings | 19-1$c$ | 2 | $SiO_3^{2-}$ | 1:3 | Beryl |
| Infinite single chains | 19-1$d$ | 2 | $SiO_3^{2-}$ | 1:3 | Pyroxenes |
| Infinite double chains | 19-1$e$ | 2$\frac{1}{2}$ | $Si_4O_{11}^{6-}$ | 1:2$\frac{3}{4}$ | Amphiboles |
| Infinite sheets | 19-1$f$ | 3 | $Si_2O_5^{2-}$ | 1:2$\frac{1}{2}$ | Talc |
| Infinite network | 19-1$g$ | 4 | $SiO_2$ | 1:2 | Quartz |

(a) Tetrahedral unit    (b) Disilicate   (c) Cyclosilicate   (d) Infinite single chain

(e) Infinite double chain

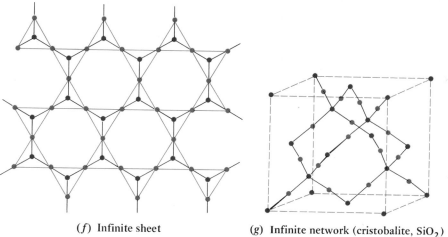

(f) Infinite sheet              (g) Infinite network (cristobalite, $SiO_2$)

**Figure 19-1**

The classes of silicate structures. Part *a* shows the symbol used for the $SiO_4^{4-}$ tetrahedron. Bear in mind that all of these structures are actually three-dimensional. Planar projections are used here for convenience of representation.

tures shown in Figure 19-1 and listed in Table 19-2. In each of these the fundamental tetrahedron is readily identified, but the Si:O ratio is no longer 4 because oxygen atoms are shared at the linkages.

### Example 19-1

By referring to Table 19-2, predict the structural class in which the mineral Egyptian blue ($CaCuSi_4O_{10}$) belongs. Give the oxidation state of each of its atoms.

### Solution

Because the Si:O ratio is 4:10, or $1:2\frac{1}{2}$, this mineral should have an infinite sheet structure with the repeating unit $Si_2O_5^{2-}$. The oxidation states of Si and O are +4 and −2, as usual, and that of Ca is +2. In order that the total oxidation number per formula unit sum to 0, the oxidation state of Cu must be +2.

*Related Problems:* 3, 4

The physical properties of the silicates correlate closely with their structures. An example of an infinite layered structure (Fig. 19-1*f*) is talc, $Mg_3(Si_4O_{10})(OH)_2$. In talc, all of the bonding interactions among the atoms occur in a single layer. Layers of talc sheets are attracted to one another only by van der Waals interactions, which (being weak) permit one layer to slip easily across another. This accounts for

**Figure 19-2**

The mineral quartz is one form of silica, $SiO_2$. (Leon Lewandowski)

the slippery feel of talc (called talcum powder). When all four vertices of each tetrahedron are linked to other tetrahedra (Fig. 19-1*g*), three-dimensional network structures such as quartz result (Fig. 19-2). Note that the quartz network carries no charge; consequently, there are no cations in its structure. Three-dimensional network silicates such as quartz are much stiffer and harder than the linear and layered silicates, and they resist deformation well.

**Asbestos** is a generic term for a group of naturally occurring hydrated silicates that can be processed mechanically into long fibers (Fig. 19-3). Some of these silicates, such as tremolite, $Ca_2Mg_5(Si_4O_{11})_2(OH)_2$, show the infinite double-chain structure of Figure 19-1*e*. Another kind of asbestos mineral is chrysotile, $Mg_3(Si_2O_5)(OH)_4$. As the formula indicates, this mineral has a sheet structure (Fig. 19-1*f*), but the sheets are rolled into long tubes. Asbestos minerals are fibrous because the bonds along the strand-like tubes are stronger than those that hold different tubes together. Asbestos is an excellent thermal insulator that does not burn, resists acids, and is very strong. For many years, it was used in cement for pipes and ducts and woven into fabric to make fire-resistant roofing paper and floor tiles. Its use has decreased significantly in recent years because inhalation of its small fibers can cause the lung disease asbestosis. The risk comes with breathing asbestos dust that is raised during mining and manufacturing processes or is released in buildings in which asbestos-containing materials are fraying, crumbling, or being removed.

**Figure 19-3**
The fibrous structure of asbestos is apparent in this sample. (Leon Lewandowski)

## Aluminosilicates

An important class of minerals called **aluminosilicates** results from the replacement of some of the silicon atoms in silicates with aluminum atoms. Aluminum is the third most abundant element in the earth's crust (8% by mass), where it occurs largely in the form of aluminosilicates. Aluminum in minerals can be a simple cation ($Al^{3+}$) or it can replace silicon in tetrahedral coordination. When it replaces silicon, it contributes only three electrons to the bonding framework in place of the four electrons of silicon atoms. The additional required electron is supplied by the ionization of a metal atom such as sodium or potassium; the resulting alkali-metal ions occupy nearby sites in the aluminosilicate structure.

The most abundant and important of the aluminosilicate minerals in the earth's surface are the **feldspars,** which result from the substitution of aluminum for silicon in three-dimensional silicate networks such as quartz. The aluminum ions must be accompanied by other cations such as sodium, potassium, or calcium to maintain overall charge neutrality. Albite is a feldspar with chemical formula $NaAlSi_3O_8$. In the high-temperature form of this mineral, the aluminum and silicon atoms are distributed at random (in 1 : 3 proportion) over the tetrahedral sites available to them. At lower temperatures, other crystal structures become thermodynamically stable, with partial ordering of the Al and Si sites.

If one of the four silicon atoms in the structural unit of talc, $Mg_3(Si_4O_{10})(OH)_2$, is replaced by an aluminum atom, and a potassium atom is furnished to supply the fourth electron needed for bonding in the tetrahedral silicate framework, the result is the composition $KMg_3(AlSi_3O_{10})(OH)_2$, which belongs to the family of **micas** (Fig. 19-4). Mica is harder than talc, and its layers slide less readily over one another, although the crystals still cleave easily into sheets. The cations occupy sites between the infinite sheets, and the van der Waals bonding that holds adjacent sheets together

**Figure 19-4**
The mechanical properties of crystals of mica are quite anisotropic. Thin sheets can be peeled off a crystal of mica by hand, but the sheets resist stresses in other directions well. Transparent, thin sheets of mica, sometimes called isinglass, have been used for heat-resistant windows in stoves or in place of window glass. (Julius Weber)

in talc is augmented by an ionic contribution. The further replacement of the three $Mg^{2+}$ ions in $KMg_3(AlSi_3O_{10})(OH)_2$ with two $Al^{3+}$ ions gives the mineral muscovite, $KAl_2(AlSi_3O_{10})(OH)_2$. Writing its formula in this way indicates that there are aluminum atoms in two kinds of sites in the structure: one Al atom per formula unit occupies a tetrahedral site, substituting for an Si atom, and the other two Al atoms are between the two adjacent layers. The formulas that mineralogists and crystallographers use convey more information than the usual empirical chemical formula of a compound.

## Clay Minerals

**Clays** are minerals produced by the weathering action of water and heat on primary minerals. Their compositions can vary widely as a result of the replacement of one element with another. Invariably they are microcrystalline or powdered in form and are usually hydrated. Often they are used as supports for catalysts, as fillers in paint, and as ion-exchange vehicles. The clays that readily absorb water and swell are used as lubricants and bore-hole sealers in the drilling of oil wells.

The derivation of clays from talcs and micas provides a direct way to understand the clays' structures. The infinite-sheet mica pyrophyllite, $Al_2(Si_4O_{10})(OH)_2$, serves as an example. If one out of six $Al^{3+}$ ions in the pyrophyllite structure is replaced by one $Mg^{2+}$ ion and one $Na^+$ ion (which together carry the same charge), a type of clay called montmorillonite, $MgNaAl_5(Si_4O_{10})_3(OH)_6$, results. This clay readily absorbs water, which infiltrates between the infinite sheets and hydrates the $Mg^{2+}$ and $Na^+$ ions there, causing the montmorillonite to swell (Fig. 19-5).

### Figure 19-5

The structure of the clay mineral montmorillonite. Insertion of variable amounts of water causes the distance between layers to swell from 9.6 Å to more than 20 Å.

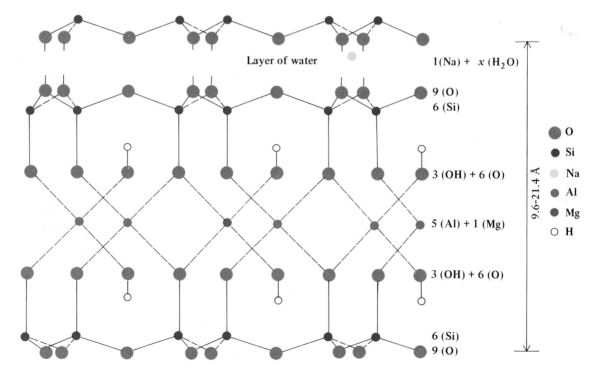

Layer of water     1(Na) + $x$ (H₂O)

9 (O)
6 (Si)

3 (OH) + 6 (O)

5 (Al) + 1 (Mg)

3 (OH) + 6 (O)

6 (Si)
9 (O)

● O
● Si
● Na
● Al
● Mg
○ H

9.6–21.4 Å

A different clay derives from the layered mineral talc, $Mg_3(Si_4O_{10})(OH)_2$. If iron(II) and aluminum replace magnesium and silicon in varying proportions, and if water molecules are allowed to take up positions between the layers, the swelling clay vermiculite results. When heated, vermiculite pops like popcorn, as the steam generated by the vaporization of water between the layers puffs the flakes up into a light, fluffy material with air inclusions. Because of its porous structure, vermiculite is used for thermal insulation or as an additive to soils to loosen them.

## Zeolites

**Zeolites** are a class of three-dimensional aluminosilicates. Like the feldspars, they carry a negative charge on the aluminosilicate framework that is compensated by neighboring cations such as the alkali-metal and alkaline-earth ions. Zeolites differ from feldspars in having much more open structures consisting of polyhedral cavities connected by tunnels (Fig. 19-6). Many zeolites are found in nature, but they can also be synthesized under conditions controlled to favor cavities of very uniform size and shape. Most zeolites accommodate water molecules in their cavities, where they provide a mobile phase for the migration of the charge-compensating cations. This enables zeolites to serve as ion-exchange materials (in which one kind of positive ion can be readily exchanged for another) and is the key to their ability to soften water. Water "hardness" arises from soluble calcium and magnesium salts such as $Ca(HCO_3)_2$ and $Mg(HCO_3)_2$. Such salts are converted to insoluble carbonates (boiler scale) when the water is heated and form objectionable precipitates (bathtub ring) with soaps. When hard water is passed through a column packed with a zeolite that has sodium ions in its structure, the calcium and magnesium ions exchange with the sodium ions and are removed from the water phase:

$$2\,NaZ(s) + Ca^{2+}(aq) \rightleftharpoons CaZ_2(s) + 2\,Na^+(aq)$$

When the ion-exchange capacity of the zeolite is exhausted, this reaction can be reversed by passing a concentrated solution of sodium chloride through the zeolite to regenerate it in the sodium form. Ion exchange also works to separate chemically similar cations of the rare-earth elements or the actinides from one another on the basis of their slightly different tendencies to associate with zeolites used as the stationary phase in column chromatography (Section 5-6).

A second use of zeolites derives from the ease with which they adsorb small molecules. Their sponge-like affinity for water makes them useful as drying agents; they are put between the panes of double-pane glass windows to prevent moisture from condensing on the inner surfaces. The pore size of zeolites can be selected to allow molecules that are smaller than a certain size to pass through but hold back larger molecules. Such zeolites serve as "molecular sieves"; they have been used to capture nitrogen molecules in a gas stream while permitting oxygen molecules to pass through.

Perhaps the most exciting use of zeolites is as catalysts. Molecules of varying sizes and shapes have different rates of diffusion through a zeolite; this feature enables chemists to enhance the rates and yields of desired reactions and suppress unwanted reactions. The most extensive applications of zeolites at present are in the catalytic cracking of crude oil, a process that involves breaking down long-chain hydrocarbons and re-forming them into branched-chain molecules of lower molecular mass for use in high-octane unleaded gasoline. A relatively new process employs

**Figure 19-6**

The structure of the synthetic zeolite ZSM-5, showing a benzene molecule passing through one of the channels. ZSM-5 consists of a network of aluminum, silicon, and oxygen atoms, with cations to balance the overall charge. It is used as a catalyst in the conversion of crude oil to petroleum products. (Mobil Corporation) "shape-selective" zeolite catalysts to convert methanol ($CH_3OH$) to high-quality gasoline. Plants have been built to make gasoline by this process, using methanol derived from coal or from natural gas.

## 19-2 GEOCHEMISTRY

**Geochemistry** is the study of the chemistry of the earth. It attempts to explain the distribution of elements and compounds throughout the earth and the migrations and chemical transformations they undergo. This section presents a brief introduction to this fascinating and very useful subject, based on the types of rocks found in the crust: igneous, sedimentary, and metamorphic. These rock types differ in both origin and chemical composition, and their study requires the application of almost every principle discussed in the first 17 chapters of this book, from acid–base equilibrium to thermodynamics to the structures of solids.

### Igneous Rocks

**Igneous rocks** make up more than 95% of the earth's crust and contain a wide variety of minerals with the silicate structures described in Table 19-2. Such rocks form during the crystallization of a **magma,** a hot, molten silicate fluid. Although there is some variation in the compositions of magmas, the primary elements present are oxygen, silicon, and aluminum, with smaller amounts of iron, magnesium, calcium, sodium, and potassium. High-temperature melts of the resultant oxides have been studied in the laboratory, and the observed order of crystallization of minerals from these melts can be correlated with properties of igneous rocks.

The crystallization of a magma takes place in two different reaction series, as Figure 19-7 illustrates. In the continuous series, feldspars of compositions varying

**Figure 19-7**

The reaction series of a magma. As the magma cools, minerals form along the two reaction series from top to bottom.

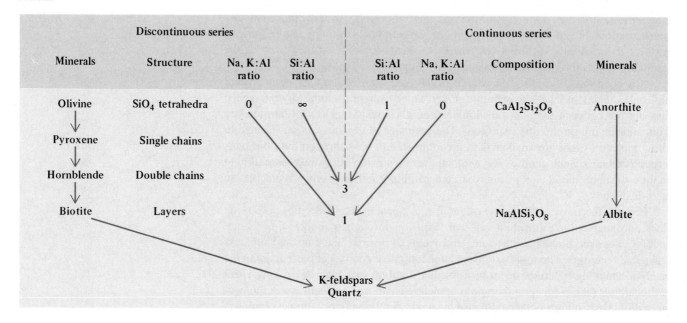

continuously from $CaAl_2Si_2O_8$ (anorthite) to $NaAlSi_3O_8$ (albite) crystallize successively. These two minerals form a nearly ideal solid solution, and the graph of temperature versus composition for the liquid–solid equilibrium is closely analogous to the graph for the vaporization of an ideal liquid mixture (Fig. 4-21). The second reaction series is a discontinuous one that passes from silicates with low connectivities (olivines) through chain and layer structures to network crystals. This series is discontinuous because only a finite number of discrete structure types are formed, in contrast with the continuous composition change in the feldspar series.

As a magma cools, the first solids that crystallize are olivines from the discontinuous series and calcium-rich feldspars from the continuous series. At this stage, two possibilities exist. If these solids stay in contact with the remaining melt, they can react with it as the temperature falls, giving pyroxene and feldspars somewhat richer in sodium, the next minerals in the reaction series. On the other hand, if the solids are separated from the melt (a process called **fractionation**), the remaining liquid is deficient in silicon relative to aluminum and deficient in calcium relative to sodium; it goes on to crystallize into later members of the two series, such as biotite and albite. The greater the degree of fractionation, the farther the reaction series proceeds, leading eventually to a melt of silica and potassium feldspars, which crystallize at the lowest temperature. In a quantitative treatment of magma crystallization, both the effect of pressure on the equilibria and the catalytic effect of small amounts of water vapor must be taken into account. The latter effect is particularly dramatic, with a relatively small mass fraction of water depressing the freezing point of $SiO_2$ by 600°C and accelerating reaction rates by many orders of magnitude.

Besides the major elements in magmas discussed so far, other elements are present. They often enter the silicate structures in place of the more common ions. If a minor element has approximately the same charge and ionic radius as a major element, it is said to be "camouflaged" in the crystal containing the major element, and the ratio of abundances of the two elements is roughly constant throughout a whole rock series. Gallium ($Ga^{3+}$) is camouflaged in aluminum minerals, and cobalt ($Co^{2+}$) is camouflaged in magnesium compounds. A minor element ion with similar radius but higher charge than a major element ion, or of the same charge but a smaller radius, is said to be "captured" and generally appears in the early stages of crystallization of the corresponding major element, because it is held to the corresponding lattice site more tightly by electrostatic forces. Barium can replace potassium in potassium feldspars because of its similar ionic radius and higher charge. Ions whose sizes and charges do not permit them to substitute for major ions remain in the residual melt, a silica-rich liquid containing water and other volatile compounds such as $CO_2$, $H_2S$, HCl, and $N_2$. The pressure of the volatiles forces the residual liquid into cracks in the surrounding rock. In the cracks, it eventually crystallizes as **pegmatites,** which are rich in the less abundant elements, such as beryllium, molybdenum, tin, uranium, and the lanthanides. A knowledge of how elements' ions fit into crystal lattices therefore helps in the search for elements that are economically important.

## Sedimentary Rocks

**Sedimentary rocks** form from the products of weathering of other rocks. Water (especially moving water in streams) helps to break rocks into smaller pieces and to partially dissolve them. Carbon dioxide (which forms carbonic acid when dissolved)

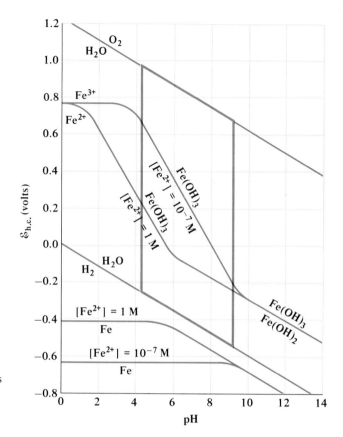

**Figure 19-8**

A potential-pH diagram for the ions and hydroxides of iron. The blue lines illustrate the dependence of half-cell voltage on pH for the redox couples indicated.

causes further dissolution and chemical reaction. Subsequent precipitation of solids leads to a variety of minerals, many of which show strong **differentiation,** with elemental abundances very different from those of the original material.

Resistance to weathering increases along the reaction series of Figure 19-7. Olivines and calcium feldspars erode and dissolve quite readily, but quartz (a network solid) is very resistant and is therefore left behind in the weathering process, often in the form of sandstone that contains as much as 99% $SiO_2$. Layered aluminosilicates such as biotite (a mica) absorb water readily to become clays. The compositions of clays depend strongly on the pH of the waters from which they form. Because aluminum ions are amphoteric, $Al_2O_3$ is more soluble in mildly acidic solutions ($pH < 4$) than is $SiO_2$, and so clays with high aluminum content tend to form at lower pH. Shales, which are rocks derived from layers of clay, are the most abundant sedimentary rocks. In some cases they contain extensive deposits of crude oil; as a result, they are of great interest to petroleum geologists.

Elements that do not remain in the silica residue and that are not incorporated into clays remain in aqueous solution. Certain ions are subject to oxidation or reduction, and the transformations they undergo can depend strongly on pH. These relationships are most clearly seen in a reduction potential–pH diagram, such as Figure 19-8 for the oxidation states of iron. The standard reduction potential for the half-reaction

$$Fe^{3+} + e^- \longrightarrow Fe^{2+}$$

is 0.770 V; it is shown as the corresponding intercept on the left side of the figure. Because $H_3O^+$ and $OH^-$ do not appear explicitly in this half-reaction, the half-cell potential does not change with a moderate increase of pH. As the pH increases further, however, hydroxides begin to form, and the half-cell reaction in basic solution is

$$Fe(OH)_3(s) + e^- \longrightarrow Fe(OH)_2(s) + OH^-(aq)$$

According to the Nernst equation, its reduction potential is

$$\mathcal{E} = \mathcal{E}°(Fe(OH)_3 | Fe(OH)_2) - 0.0592 \log_{10} [OH^-]$$
$$= -0.56 \text{ V} + (0.0592 \text{ V})(14 - \text{pH})$$

**Figure 19-9**
This striking rosette is one naturally occurring form of hematite, $Fe_2O_3$. (Julius Weber)

which changes linearly with pH at high pH. The region of intermediate pH is somewhat more complicated, because the complex ions $FeOH^{2+}$, $FeOH^+$, and $Fe(OH)_2^+$ form. A second curve in Figure 19-8 shows the pH dependence of the $Fe^{2+} | Fe$ half-cell equilibrium. In a strongly oxidizing environment, iron(III) is the stable oxidation state (top of diagram), but in a strongly reducing environment iron(0) would be stable (bottom of diagram). Between the two is the stability region of iron(II). Potential-pH diagrams help in analyzing what minerals will form under given conditions. Thus, hematite ($Fe_2O_3$) (Fig. 19-9) contains iron(III) and tends to form at high pH and under oxidizing conditions, whereas siderite ($FeCO_3$) contains iron(II) and forms at low pH and under reducing conditions.

Ions that do *not* react to form solids remain in solution until enough water evaporates that the volume decreases below the limit of their solubility. Ocean water is saturated with respect to calcium carbonate so that, when evaporation begins, $CaCO_3$ is the first solid to precipitate, usually in the form of limestone. Once the solution has been reduced to about one fifth of its original volume, calcium sulfate precipitates as gypsum ($CaSO_4 \cdot 2H_2O$). After further evaporation of seawater (a reduction to about one-tenth the original volume), halite (NaCl) precipitates in great salt "domes," which are mined as the starting material for the chlor-alkali industry (Chapter 22). The brine that remains behind at this stage is rich in potassium, magnesium, and sulfate ions.

The formation of sedimentary deposits involves an interplay of oxidation–reduction, acid–base, and solubility equilibria. The result is substantial differentiation of elements that began rather uniformly distributed in igneous rocks.

## Metamorphic Rocks

The minerals that make up the earth's crust continually undergo transformations in structure and composition. As we have seen, igneous rocks form through the crystallization of fluid magmas, and sedimentary rocks arise from weathering and subsequent hydration and ion exchange. A third type of change, called **metamorphism,** is induced by heat and pressure below the earth's surface and leads to the formation of metamorphic rocks. The conversion of limestone to crystalline marble with the same composition is one example of a change in texture caused by metamorphism. Metamorphism involves much deeper burial (higher pressures) and higher temperatures than typical interconversions of sedimentary structures, but not such high temperatures as to cause melting and magma formation.

Purely structural changes can occur relatively rapidly when crystals are subjected to changes in temperature or pressure. Chemical reactions, on the other hand, lead to changes in composition and require the transport of material through a rock, either by solid-state diffusion over geological time scales or under the action of volatile substances such as water, which can catalyze solid–solid reactions by dissolving and reprecipitating the reacting solids. The latter mechanism can be important, as demonstrated by laboratory experiments on reactions such as

$$2 \, MgO(s) + SiO_2(s) \longrightarrow Mg_2SiO_4(s)$$

This reaction occurs slowly under dry conditions, even at 1300°C, but much more rapidly in the presence of water vapor at only 600°C. The kinetics and mechanisms of such fluid-catalyzed reactions deserve further study because they can be a major contributor to metamorphism.

Because metamorphic rocks form below the earth's surface, pressure has an important effect, which can be analyzed in qualitative terms through Le Châtelier's principle: an increase in pressure at constant temperature displaces an equilibrium toward substances that have smaller volumes and greater densities. The silicates that are most characteristic of metamorphic rocks have chain and layer structures and are denser than those with the more open network structures, such as quartz and the feldspars. The denser minerals are stabilized at higher pressures.

A simple example of the effect of pressure on solid-solid equilibrium is a substance such as $CaCO_3$ that exists in two crystalline forms (calcite and aragonite) with different molar volumes. In a pressure–temperature phase diagram (Section 4-1) the line of phase coexistence has a slope given by the **Clapeyron equation:**

$$\left(\frac{dp}{dT}\right)_{coex} = \frac{\Delta S}{\Delta V} = \frac{\Delta H}{T \, \Delta V} \tag{19-1}$$

where $\Delta S$ is the entropy change for the transition between the two forms and $\Delta V$ is the change in volume (see problem 38). Because $\Delta G = 0$ along the coexistence curve, it must also be true that at each point $\Delta H = T \, \Delta S$, which gives the second equality in Eq. 19-1. The denser (lower-volume) form of the solid (which in the case of calcium carbonate is aragonite) lies on the high-pressure side of the coexistence curve.

Now consider a chemical reaction between solid compounds. At atmospheric pressure, the temperature at which $\Delta G = 0$ for such a reaction is $T = \Delta H°/\Delta S°$ (recall problem 33 in Chapter 9). At higher pressures, the temperature that divides the range of stable products from stable reactants shifts according to an equation closely analogous to Eq. 19-1:

$$\left(\frac{dp}{dT}\right)_{\Delta G = 0} = \frac{\Delta S}{\Delta V}$$

where $\Delta S$ and $\Delta V$ are now the enthalpy change and volume change of reaction. If these depend weakly on pressure (as they do for most reactions involving only solids), the result is a straight line for the equilibrium curve between reactants and products ($\Delta G = 0$), as Figure 19-10a shows.

If gases are either reactants or products in a reaction, the effect of pressure on the equilibrium can be much greater. An example is the reaction

$$\underset{\text{Dolomite}}{CaMg(CO_3)_2(s)} + \underset{\text{Quartz}}{2 \, SiO_2(s)} \longrightarrow \underset{\text{Diopside}}{CaMgSi_2O_6(s)} + \underset{\text{Carbon dioxide}}{2 \, CO_2(g)}$$

**Figure 19-10**

Equilibrium curves in metamorphism, showing the locus of states for which $\Delta G = 0$ for the transformations indicated.

Because the volume of the carbon dioxide is much greater than that of the solids taking part, an increase in pressure shifts the reaction to the left, decreasing the amounts of diopside and $CO_2$. The equilibrium curve for this reaction with $\Delta G = 0$ can be found by noting that $\Delta H$ is approximately independent of pressure, and $\Delta S$ depends on pressure primarily through the resulting compression of the $CO_2(g)$. If the carbon dioxide is at a pressure $P$, then

$$\Delta S \approx \Delta S^\circ - (\Delta n_g)R \ln P$$

where $\Delta n_g$ is the change in the number of moles of gas in the chemical equation, and we have used the result from Section 9-1 that the entropy change on isothermal compression of $\Delta n_g$ mol of an ideal gas from 1 atm to $P$ atm is $-(\Delta n_g)R \ln P$. Setting $\Delta G$ to 0 gives

$$T = \frac{\Delta H}{\Delta S} = \frac{\Delta H^\circ}{\Delta S^\circ - (\Delta n_g)R \ln P} \qquad \textbf{(19-2)}$$

with the characteristic nonlinear dependence of $T$ on $P$ shown in Figure 19-10$b$.

## Example 19-2

The standard enthalpies of formation of dolomite and diopside are $-2326.3$ and $-3206.2$ kJ mol$^{-1}$, respectively, and their standard entropies are 155.18 and 142.93 J K$^{-1}$ mol$^{-1}$. Use this information, together with data from Appendix D, to calculate the temperature at which $\Delta G = 0$ for the reaction of dolomite with quartz (a) at atmospheric pressure and (b) at $P = 1000$ atm.

## Solution

(a) $\Delta H^\circ = 2\,\Delta H_f^\circ(CO_2) + \Delta H_f^\circ(CaMgSi_2O_6) - \Delta H_f^\circ(CaMg(CO_3)_2) - 2\,\Delta H_f^\circ(SiO_2)$
$\quad = +155$ kJ

$\Delta S^\circ = 2\,S^\circ(CO_2) + S^\circ(CaMgSi_2O_6) - S^\circ(CaMg(CO_3)_2) - 2\,S^\circ(SiO_2)$
$\quad = +331$ J K$^{-1}$

$$T = \frac{\Delta H^\circ}{\Delta S^\circ} = \frac{155{,}000 \text{ J}}{331 \text{ J K}^{-1}} = 468 \text{ K} = 195°C$$

At this temperature, $K = 1$ for the reaction.

(b) Using $\Delta n_g = 2$ mol in Equation 19-2 gives

$$T = \frac{155,000 \text{ J}}{331 \text{ J K}^{-1} - (2 \text{ mol})(8.315 \text{ J K}^{-1} \text{ mol}^{-1}) \ln 1000} = 717 \text{ K} = 444°\text{C}$$

This result is only approximate because we neglected pressure effects on the enthalpy and entropy of the solids and assumed ideal gas behavior even at very high pressure.

*Related Problems:* 9, 10

As metamorphism proceeds, the elemental composition of the involved rocks becomes more uniform. High-grade metamorphism eventually merges (at high temperatures) with magma formation. In fact, a full geochemical cycle can be described in which a primary magma crystallizes to give igneous rocks, weathering yields sedimentary rocks, and finally metamorphism leads to remelting and magma formation once again (Fig. 19-11). The composition tends to become differentiated (less uniform) as the magma is converted to igneous and then sedimentary rocks and then, at later stages in the cycle, becomes more uniform through metamorphism.

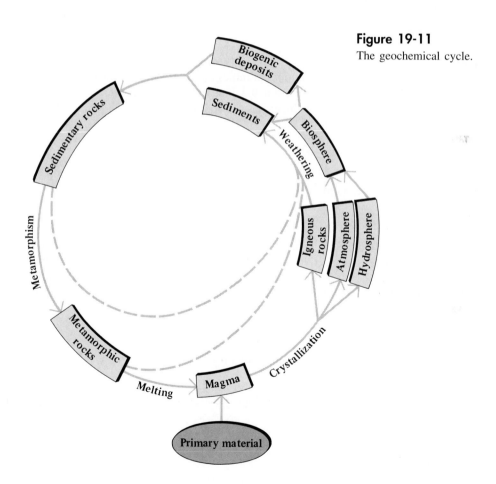

**Figure 19-11**

The geochemical cycle.

# 19-3 EXTRACTIVE METALLURGY

The recovery of metals from their sources in the earth is the science of **extractive metallurgy,** a discipline that draws on chemistry, physics, and engineering for its methods. As a science it is a comparatively recent subject, but its beginnings, which were evidently in the Near East about 6000 years ago, marked the emergence of humanity from the Stone Age. The earliest known metals were undoubtedly gold, silver, and copper, because they could be found in their native (elemental) states (Fig. 19-12). Gold and silver were valued for their ornamental uses, but they are too soft to have been made into tools. Iron was also found in elemental form (although rarely) in meteorites.

Most metals are combined with other elements, such as oxygen and sulfur, in ores, and chemical processes are required to free them. As Table 19-3 shows, the free energies of formation of most metal oxides are negative, indicating that the reverse reactions, which yield the free metal and oxygen, have positive free energy changes. Scientists can carry out such a reverse reaction only by coupling it with a second, spontaneous chemical reaction. The greater the cost in free energy, the more difficult the production of the free metal. Thus, silver and gold (at the bottom of Table 19-3) exist in nature as elements, and mercury can be released from its oxide or sulfide ore (cinnabar) merely by moderate heating (see Fig. 1-8). Winning copper, zinc, and iron requires more stringent conditions; the ores of these metals are reduced in chemical reactions at high temperatures, collectively called **pyrometallurgy.** These reactions are carried out in huge furnaces, in which a fuel such as coke (coal from which volatile components have been expelled) serves both as the reducing agent and as the source of heat to maintain the required high temperatures. The

**Figure 19-12**

A specimen of native copper. (Leon Lewandowski)

## Table 19-3    Metal Oxides Arranged According to Ease of Reduction

| Metal Oxide | Metal | $n^†$ | $\Delta G_f^\circ / n^†$ (kJ mol$^{-1}$) | A Method of Production of the Metal |
|---|---|---|---|---|
| MgO | Mg | 2 | −285 | Electrolysis of MgCl$_2$ |
| Al$_2$O$_3$ | Al | 6 | −264 | Electrolysis |
| TiO$_2$ | Ti | 4 | −222 | Reaction with Mg |
| Na$_2$O | Na | 2 | −188 | Electrolysis of NaCl |
| Cr$_2$O$_3$ | Cr | 6 | −176 | Electrolysis, reduction by Al |
| ZnO | Zn | 2 | −159 | Smelting of ZnS |
| SnO$_2$ | Sn | 4 | −130 | Smelting |
| Fe$_2$O$_3$ | Fe | 6 | −124 | Smelting |
| NiO | Ni | 2 | −106 | Smelting of nickel sulfides |
| PbO | Pb | 2 | −94 | Smelting of PbS |
| CuO | Cu | 2 | −65 | Smelting of CuFeS$_2$ |
| HgO | Hg | 2 | −29 | Moderate heating of HgS |
| Ag$_2$O | Ag | 2 | −6 | Found in elemental form |
| Au$_2$O$_3$ | Au | 6 | >0 | Found in elemental form |

$^†$ The standard free energies of formation of the metal oxides in kJ mol$^{-1}$ are adjusted for fair comparison by dividing them by $n$, the total decrease in oxidation state required to reduce the metal atoms contained in the oxide to oxidation states of 0. Thus, the reduction of Cr$_2$O$_3$ involves a change in the oxidation state of two chromium atoms from +3 to 0, so that $n = 2 \times 3 = 6$.

process, called **smelting,** involves both chemical change and melting. The combustion of carbon

$$C(s) + O_2(g) \longrightarrow CO_2(g) \qquad\qquad \Delta G^\circ = -394 \text{ kJ mol}^{-1}$$

provides the driving force for the overall reaction. Even smelting is not sufficient for the metals at the top of Table 19-3, which have the most negative free energies of formation for their oxides (and for their sulfides as well). Electrometallurgical methods, to be considered in Section 19-4, are needed.

When the oxide of a metal is reduced to elemental form, the molar free energies of formation in Table 19-3 are divided by the total change in the oxidation number. This allows a fair comparison of the relative ease of reduction of two compounds that have the metals in different oxidation states. Reducing 1 mol of $Fe_2O_3$ to elemental iron, for example, requires oxidizing three times as much carbon (to $CO_2$) as reducing 1 mol of ZnO, and so its free energy of formation is divided by 6 in the table; the free energy of formation of ZnO is divided by only 2. Rank in the table tells which metals can reduce which metal oxides. The standard free energy change for the reduction

$$Cr_2O_3(s) + 3\,Mg(s) \longrightarrow 2\,Cr(s) + 3\,MgO(s)$$

is

$$\Delta G^\circ = 3\,\Delta G_f^\circ(MgO) - \Delta G_f^\circ(Cr_2O_3)$$

$$= 6 \times (\tfrac{1}{2}\,\Delta G_f^\circ(MgO) - \tfrac{1}{6}\,\Delta G_f^\circ(Cr_2O_3))$$

Because MgO lies above $Cr_2O_3$ in the table, the difference enclosed in parentheses is negative, and this reaction is spontaneous. A similar computation for any pair of entries in the table confirms that a metal will spontaneously reduce the oxide of any element below it in the table. Thus, magnesium is used commercially to produce titanium from its ore, rutile ($TiO_2$). Once the reaction is initiated, aluminum reacts spectacularly with iron(III) oxide in the thermite reaction (Fig. 8-11) to give iron and $Al_2O_3$. The metals at the top of the table are the most difficult to produce in terms of thermodynamics and, once made, are the most reactive. It is no accident that a correlation exists between the reactivity of a metal and its date of first isolation in elemental form.

The remainder of this section is devoted to the production and uses of two of the most important metals produced by chemical reduction, copper and iron. Section 19-4 discusses two other important metals, aluminum and magnesium, which require electrochemical processing. Four other metals have been introduced through the Practice Exercises earlier in the book: titanium (Chapter 1), nickel (Chapter 9), manganese (Chapter 10), and platinum (Chapter 16).

## Copper and Its Alloys

Copper occurs in a variety of minerals in the form of veins that are frequently exposed as outcrops at the earth's surface. Chalcopyrite ($CuFeS_2$), chalcocite ($Cu_2S$), and bornite ($Cu_5FeS_4$) are chief among the copper-bearing ores. On exposure to the weathering action of the atmosphere, they are oxidized to beautiful green or blue basic copper carbonates such as malachite, $Cu_2CO_3(OH)_2$, and azurite, $Cu_3(CO_3)_2(OH)_2$, both of which are semiprecious stones (Fig. 19-13). We speculate that metallurgy began when a potter introduced an ore such as malachite into his

**Figure 19-13**

Several gems from copper minerals. Clockwise from top: malachite with azurite, malachite, azurite with chalcopyrite, and turquoise. (Leon Lewandowski)

kiln, perhaps as a colored glaze, and found a globule of a red metal after firing the ore to high temperatures in the reducing atmosphere generated by the burning of carbon (charcoal):

$$Cu_2CO_3(OH)_2(s) + C(s) \longrightarrow 2\,Cu(s) + 2\,CO_2(g) + H_2O(g)$$

This would not have been pure copper, of course. Pure copper is not hard enough to make serviceable cutting tools, although to some extent it can be "work-hardened" by hammering. Implements such as daggers and axes that date to the third millennium B.C. are found to contain a few percent of arsenic, which gives copper its hardness. We cannot be sure whether tin originated by design or from the accidental use of ores that contained minor concentrations of arsenic as an impurity. But its appearance, at first in trace amounts and increasing to about 10% as the centuries passed, surely marks the birth of a remarkably sophisticated industry. Bronze is an alloy that contains about 10% tin and 90% copper (Fig. 19-14). It is hard without being brittle, can be cast in a mold, and melts at a temperature (950°C) substantially below the melting point of pure copper (1084°C).

Brass is a second important alloy of copper, containing up to 42% zinc along with copper. Brass was first discovered by the Romans, and to this day it is prized as a material for precision instruments because it is hard, easy to machine, and resistant to corrosion.

The modern high demand for copper stems largely from its use in wires for conducting electricity and has led to worldwide exploitation of copper ores, of which the most abundant is chalcopyrite ($CuFeS_2$). Open-pit mining operations on a mammoth scale (Fig. 19-15) remove and crush tons of rock that may contain less

**Figure 19-14**

A bronze axe-head from Luristan (Persia) that dates from about 1200 B.C. (Leon Lewandowski)

**Figure 19-15**

An open-pit copper mine in Utah. The ore contains less than 1% copper, but the metal can nevertheless be extracted economically, depending on fluctuations in the world market price. (Don Green/Phototake)

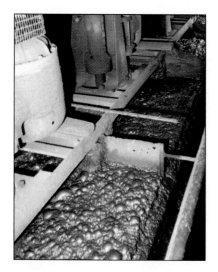

**Figure 19-16**

Froth flotation for the enrichment of ore. Frothing agents are added to a water suspension of the pulverized ore. As compressed air is bubbled in, the desired mineral ($CuFeS_2$, in the case of copper production) is concentrated at the surface of a foam and borne upward while the unwanted rocky material settles to the bottom.

than 1% copper. **Froth flotation** is used to concentrate the copper ores (Fig. 19-16). In this process, crushed low-grade ore is mixed with water, a special oil, and a detergent. A stream of compressed air produces a foamy, frothy mixture in which oil-wetted particles of ore are buoyed up by clinging to air bubbles, while water-wetted earth and rock sink to the bottom. The copper-rich froth is skimmed off, the oil is separated for reuse, and (after drying) a solid mass enriched in $CuFeS_2$ (typically to 25–30% copper) remains.

The next step is the **roasting** of this enriched ore—reaction with air at high temperatures to convert the iron to its oxide while leaving the copper as a sulfide:

$$6\,CuFeS_2(s) + 13\,O_2(g) \longrightarrow 3\,Cu_2S(s) + 2\,Fe_3O_4(s) + 9\,SO_2(g)$$

In the next stage, the mixture of $Cu_2S$ and $Fe_3O_4$ is put in a furnace, in which the flame is reflected from the roof to play down on the material being heated. The temperature is raised to 1100°C, and limestone ($CaCO_3$) and silica ($SiO_2$) are added as a flux to remove the iron:

$$CaCO_3(s) + Fe_3O_4(s) + 2\,SiO_2(s) \longrightarrow CO_2(g) + CaO{\cdot}FeO{\cdot}Fe_2O_3{\cdot}(SiO_2)_2(\ell)$$
(Slag, composition variable)

Because $Cu_2S(\ell)$ is not soluble in the calcium–iron slag and exceeds it in density, it settles to the bottom, and the molten slag layer can be poured off. The $Cu_2S(\ell)$ is run into another furnace (the converter) through which additional air is blown to cause the redox reaction

$$Cu_2S(\ell) + O_2(g) \longrightarrow 2\,Cu(\ell) + SO_2(g)$$

The liquid copper is then cooled and cast into large slabs of "blister copper" for further purification. The final refinement routinely employs an electrochemical method (Section 19-4) to recover the small amounts of gold and silver that have accompanied the copper to this stage.

Three major problems afflict the smelting of copper: the pollution that can be caused by the $SO_2$ byproduct, the disposal of the calcium–iron–silicate slag that can contain toxic heavy-metal impurities such as cadmium, and the energy costs associated with operating high-temperature furnaces. Emission control regulations have required new processes in which nearly all of the $SO_2$ is captured and converted to commercially important sulfuric acid (see Chapter 21). Careful attention to recycling water in newer smelters has greatly reduced the amount of heavy metals leached out from the slag. Finally, energy costs are reduced by new **hydrometallurgical** methods for separating the copper and iron in the ore. When an aqueous mixture of copper(II) chloride and iron(III) chloride is added to the ore, the copper-bearing minerals are reduced in the reactions

$$CuFeS_2(s) + 3\,CuCl_2(aq) \longrightarrow 4\,CuCl(s) + FeCl_2(aq) + 2\,S(s)$$

$$CuFeS_2(s) + 3\,FeCl_3(aq) \longrightarrow CuCl(s) + 4\,FeCl_2(aq) + 2\,S(s)$$

Excess aqueous sodium chloride is added to the solids that form to convert the $CuCl(s)$ to the soluble complex ion $[CuCl_2]^-(aq)$, which is separated from the sulfur and converted to copper and $CuCl_2$ in a disproportionation reaction:

$$2\,CuCl_2^-(aq) \longrightarrow Cu(s) + CuCl_2(aq) + 2\,Cl^-(aq)$$

The copper(II) chloride is recycled to reduce more ore.

## Iron and Steel

The chief sources of iron are the minerals hematite ($Fe_2O_3$), magnetite ($Fe_3O_4$), goethite (FeO(OH)), limonite (FeO(OH)$\cdot n$H$_2$O), and siderite ($FeCO_3$). The metallurgy of iron requires considerably higher temperatures than that of copper, and even though forced drafts were used to fan the fires, the molten iron produced in early times merely formed globules interspersed in a semiliquid slag. By repeatedly heating and hammering the red-hot mass, the more fluid slag could be squeezed out to yield a product known as wrought iron. Such iron was fairly soft and usually inferior to bronze, until artisans learned that prolonged heating of iron in contact with charcoal caused it to absorb carbon. When iron containing several percent of dissolved carbon was quenched in water, it produced an extremely hard metal, a form of steel. This technology, developed by the Hittites around 1200 B.C., was further refined in Greco-Roman times and then was little changed until the invention of the **blast furnace** in the late 14th and early 15th centuries.

Figure 19-17 illustrates the construction of a blast furnace for smelting iron ore. It consists of a cylindrical steel shell some 10 m in diameter at its widest part and about 30 m high, lined with a refractory "fire brick." Designed for continuous operation, it is run for about five years, after which its lining must be torn out and replaced. A charge of iron ore, coke, and limestone ($CaCO_3$) or dolomite ($CaMg(CO_3)_2$) enters the top of the stack through a hopper at intervals and works its way downward as the reaction proceeds. Preheated air enters the stack through nozzles called **tuyeres** above the **hearth,** where the molten iron collects.

High-grade iron ore is not pure $Fe_2O_3$ (70% Fe) but contains up to about 27% silica, alumina, clay, and minor impurities that must be converted to a fluid slag for the process to be continuous. That is why limestone or dolomite is added with the charge; each converts silica and alumina (melting points 1700° and 2050°C, respectively) to lower-melting calcium aluminum silicates (m.p. approximately 1550°C):

$$CaCO_3(s) \longrightarrow CaO(s) + CO_2(g)$$

$$CaO(s) + Al_2O_3(s) + 2\,SiO_2(s) \longrightarrow CaO\cdot Al_2O_3\cdot(SiO_2)_2(\ell)$$
(Composition variable)

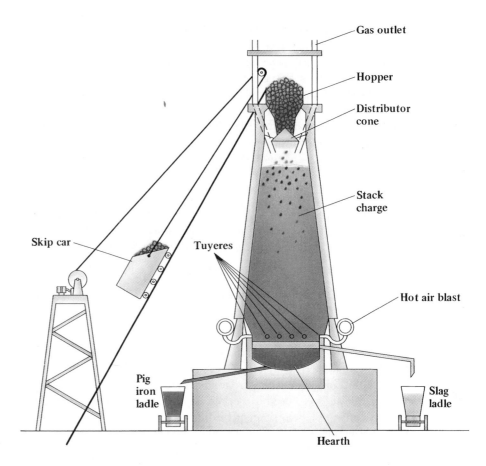

**Figure 19-17**

A blast furnace for smelting iron ore.

Hematite is reduced not by direct reaction with carbon itself but by carbon monoxide in the hot gases as they pass up through the charge under forced draft:

$$Fe_2O_3(s) + 3\,CO(g) \longrightarrow 2\,Fe(s) + 3\,CO_2(g)$$

The carbon monoxide is produced in reactions between hot coke and the preheated air that enters through the tuyeres. At this level in the blast furnace, the temperature is a maximum (about 1875°C) because of the highly exothermic reaction

$$C(s) + O_2(g) \longrightarrow CO_2(g)$$

At this high temperature, carbon dioxide is in equilibrium with carbon monoxide in the reaction

$$C(s) + CO_2(g) \rightleftharpoons 2\,CO(g)$$

The equilibrium constant of this reaction at 1875°C is readily calculated from data in Appendix D if we assume temperature-independent enthalpies and entropies; it is large, about $1 \times 10^5$. The reduction of hematite to iron occurs as a heterogeneous gas–solid reaction fairly high in the stack, where the temperature is still well below the melting point of pure iron (1535°C). The result is a spongy form of iron interspersed with the other components of the charge. As the mass slowly descends through the stack, it encounters increasingly higher temperatures until, just above the tuyeres, both the iron and the slag components liquefy and trickle down into the hearth of the furnace. Molten calcium aluminum silicate (slag) is less dense than molten iron and floats on it. Periodically, the two liquids are separately drawn off into ladle cars, the iron to be further refined and the slag to be solidified into a clinker (fused ash), which is ground up for use in portland cement or crushed for use as road ballast. The iron produced in the blast furnace, called **pig iron,** contains a total of 5 to 10% carbon, silicon, phosphorus, and other impurities, which are removed in a subsequent process. A modern blast furnace produces about 1500 metric tons of pig iron and 750 metric tons of slag per day while consuming about 1200 metric tons of coke.

Steel is iron mixed with a small amount of carbon (up to about 1.7% by mass). The properties of a steel depend on the exact percentage of carbon and on the identities and levels of other alloying elements. Steel is made from pig iron by removing silicon, phosphorus, manganese, sulfur, and other impurities and lowering the carbon content to a target value that depends on the intended end use. Modern methods of conversion rely on the oxidation of the unwanted elements in the molten pig iron. The oxides escape as gases or react with a flux, which is added for this purpose, to form a slag phase that separates from the hot metal. The greatest quantities of steel are produced in **basic oxygen converters**—barrel-shaped furnaces open at one end and lined with refractory bricks. A converter stands 10 meters high, has a diameter of $5\frac{1}{2}$ meters, and can be tilted to allow the starting materials to be tipped in or product to be poured out. Such a converter is charged with some 150 to 250 tons of molten pig iron, which is transferred in insulated ladles from a blast furnace (Fig. 19-18). Cold scrap steel is also often added. The converter is then turned upright under a hood (to collect gases) and a high-velocity jet of pure oxygen is blown down onto the charge through nozzles in a water-cooled retractable lance positioned above

**Figure 19-18**

Molten iron being charged into a basic oxygen furnace. The basic oxygen process for steelmaking is an *oxygen* process because it employs a blast of oxygen, rather than air, to oxidize impurities. It is a *basic* process because the flux contains a base, lime, to capture the acidic oxides (such as $SO_2$ and $P_4O_{10}$) of those impurities. (Bethlehem Steel)

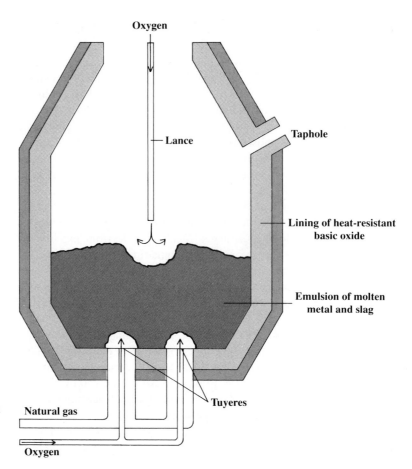

**Figure 19-19**

The operation of a basic oxygen converter. Blasts of pure oxygen from above (the lance) and below (the tuyeres) refine pig iron into steel. After processing, the converter is tipped on its side and the molten steel is tapped off through the taphole. The impurity-containing slag floats atop the steel and is easily separated.

the surface of the molten charge (Fig. 19-19). More oxygen is blown in through bottom-mounted tuyeres. Each tuyere is protected from burning by injection of a small amount of propane gas through an annular space at its end. The propane forms a sheath around the tuyeres. Its endothermic decomposition

$$C_3H_8 \longrightarrow 3\,C + 4\,H_2$$

cools and protects the tuyeres.

Within seconds after blowing starts, strong oxidation commences in the pig iron. The flux, which is powdered burnt lime (CaO) or burnt dolomite (CaO·MgO), is blown into the mixture along with the oxygen. Some iron is oxidized to FeO and is rapidly distributed throughout the charge:

$$2\,Fe + O_2 \longrightarrow 2\,FeO$$

Silicon burns directly to silicon dioxide, $SiO_2$, or reacts with FeO to give the same product:

$$Si + O_2 \longrightarrow SiO_2$$

$$2\,FeO + Si \longrightarrow 2\,Fe + SiO_2$$

The acidic $SiO_2$ then reacts with the basic flux and forms a product that separates from the hot metal as slag:

$$SiO_2 + CaO \xrightarrow[\text{In slag}]{} CaO \cdot SiO_2(\ell)$$

Oxidation of the phosphorus and its removal into the slag proceed similarly, the overall equation being

$$5\,FeO + 2\,P + 3\,CaO \xrightarrow[\text{In slag}]{} 5\,Fe + (CaO)_3 \cdot P_2O_5(\ell)$$

The slag also takes up MnO (as it forms by oxidation of manganese in the pig iron) and some FeO; both of these metal oxides are acidic. Carbon, the major impurity in the pig iron, is oxidized to gaseous carbon monoxide:

$$C + FeO \longrightarrow Fe + CO$$

$$C + \tfrac{1}{2}O_2 \longrightarrow CO$$

Under the conditions in the converter, essentially no carbon dioxide ($CO_2(g)$) is produced. The escape of $CO(g)$ from the melt causes a boiling action that mixes the slag and metal phases thoroughly and hastens the transfer of oxides into the slag. All of the oxidation reactions are exothermic. The rising temperature in the converter melts the scrap steel and keeps the slag molten. Intermittently throughout the blow, a probe is plunged into the melt to measure the temperature (a good indicator of carbon content) and oxygen content of the steel. The data are used under automatic computer control to adjust the blowing rate, the distance of the lance to the surface of the melt, the rate of addition of flux, and other operating conditions to reach the desired carbon content and temperature simultaneously. The process is complete in 25 to 30 minutes.

Oxygen, hydrogen, and nitrogen are soluble in molten steel and must be removed before the steel can be cast. Otherwise, the escape of the dissolved gases would cause large void spaces (blowholes) in the casting. Dissolved oxygen also can react with alloying elements to form particles of oxides that remain in the finished steel as internal and surface defects. The next stage of steel refining is therefore **vacuum degassing.** Large vacuum pumps reduce the pressure above the steel as argon gas, which is not soluble in molten steel, is blown through the melt from the bottom. These actions circulate the liquid and speed the escape of dissolved gases. Any desired alloying elements are also added during degassing. Additional oxygen is blown in at the start of degassing to oxidize the remaining carbon. During degassing, the low pressure above the melt shifts the equilibrium

$$C(\textit{dissolved}) + \tfrac{1}{2}O_2(\textit{dissolved}) \rightleftharpoons CO(g)$$

to the right and favors low levels of both carbon and oxygen in the steel. Toward the end of the degassing, aluminum is added. This further reduces the level of oxygen in the steel by the reactions

$$2\,Al + \tfrac{3}{2}O_2 \longrightarrow Al_2O_3$$

$$2\,Al + 3\,FeO \longrightarrow Al_2O_3 + 3\,Fe$$

The insoluble aluminum oxide forms an easily separated layer.

The carbon content of fully refined steel is far less than that of the original pig iron. For example, in steel intended for automobile body sheets, carbon is reduced to ultralow levels (0.002% C by mass). Ultralow-carbon steels have better press-forming characteristics than ordinary low-carbon or mild steel (0.1 to 0.4% C by mass). The basic oxygen process allows the production, at short intervals, of large batches of steel of controlled chemical make-up. It has major advantages over other ways of making steel, particularly when the steel is taken directly for the continuous casting of sheets. In recent years, the basic oxygen process has entirely supplanted the formerly dominant open-hearth process, in which pig iron was cooked for several hours in oxygen-enriched air with a basic flux.

## 19-4 ELECTROMETALLURGY

The methods of pyrometallurgy and hydrometallurgy considered in Section 19-3 are unsuitable for production of metals such as aluminum and the alkali and alkaline-earth elements from their ore sources. These metals have particularly high free energies relative to their compounds in readily available ores. **Electrometallurgy,** or electrolytic production, provides the best ways of winning such elements from their ores. Electrochemical cells are also used to purify the metals produced by the techniques of pyrometallurgy.

### Aluminum

Aluminum is the third most abundant element in the earth's crust (after oxygen and silicon), accounting for 8.2% of the total mass. It occurs most commonly in association with silicon in the aluminosilicates of feldspars and micas and in clays, the products of weathering of these rocks. The most important ore for aluminum production is bauxite, a hydrated aluminum oxide that contains 50 to 60% $Al_2O_3$; 1 to 20% $Fe_2O_3$; 1 to 10% silica; minor concentrations of titanium, zirconium, vanadium, and other transition-metal oxides; and the balance (20 to 30%) water. Bauxite is purified via the **Bayer process,** which takes advantage of the fact that the amphoteric oxide alumina is soluble in strong bases but iron(III) oxide is not. Crude bauxite is dissolved in sodium hydroxide

$$Al_2O_3(s) + 2\,OH^-(aq) + 3\,H_2O(\ell) \longrightarrow 2\,Al(OH)_4^-(aq)$$

and separated from hydrated iron oxide and other insoluble impurities by filtration. Pure hydrated aluminum oxide precipitates when the solution is cooled to supersaturation and seeded with crystals of the product:

$$2\,Al(OH)_4^-(aq) \longrightarrow Al_2O_3{\cdot}3H_2O(s) + 2\,OH^-(aq)$$

The water of hydration is removed by calcining at high temperature (1200°C).

Compared with copper, iron, gold, and lead, which were known in antiquity, aluminum is a relative newcomer. Sir Humphry Davy obtained it as an alloy of iron and proved its metallic nature in 1809. It was first prepared in relatively pure form by H. C. Oersted in 1825, by reduction of aluminum chloride with an amalgam of potassium dissolved in mercury,

$$AlCl_3(s) + 3\,K(Hg)_x(\ell) \longrightarrow 3\,KCl(s) + Al(Hg)_{3x}(\ell)$$

**Figure 19-20**

An electrolytic cell used in the Hall–Héroult process for production of aluminum.

after which the mercury was removed by distillation. Aluminum remained largely a laboratory curiosity until 1886, when Charles Hall in the United States (then a 21-year-old graduate of Oberlin College) and Paul Héroult (a Frenchman of the same age) independently invented an efficient process for its production. In 1991 the worldwide production of aluminum by the **Hall–Héroult process** was approximately $1.5 \times 10^7$ metric tons.

The Hall–Héroult process involves the cathodic deposition of aluminum, from molten cryolite ($Na_3AlF_6$) containing dissolved $Al_2O_3$, in electrolysis cells (Fig. 19-20). Each cell consists of a rectangular steel box some 6 m long, 2 m wide, and 1 m high, which serves as the cathode, and massive graphite anodes that extend through the roof of the cell into the molten cryolite bath. Enormous currents (50,000 to 100,000 A) are passed through the cell, and as many as 100 such cells may be connected in series.

Molten cryolite, which is completely dissociated into $Na^+$ and $AlF_6^{3-}$ ions, is an excellent solvent for aluminum oxide, giving rise to an equilibrium distribution of ions such as $Al^{3+}$, $AlF^{2+}$, $AlF_2^+$, ..., $AlF_6^{3-}$, and $O^{2-}$ in the electrolyte. Cryolite melts at 1000°C, but its melting point is lowered by dissolved aluminum oxide, so that the operating temperature of the cell is about 950°C. Compared with the melting point of pure $Al_2O_3$ (2050°C), this is a low temperature, and it is the reason the Hall-Héroult process has succeeded. Molten aluminum is somewhat denser than the melt at 950°C and so collects at the bottom of the cell, from which it is tapped periodically (Fig. 19-21). Oxygen is the primary anode product, but it reacts with the graphite electrode to produce carbon dioxide. The overall cell reaction is

$$2\ Al_2O_3 + 3\ C \longrightarrow 4\ Al + 3\ CO_2$$

Aluminum and its alloys have a tremendous variety of applications. Many of these make use of aluminum's low density (Table 19-4), an advantage over iron or steel when weight savings are desirable—such as in the transportation industry, which uses aluminum in vehicles from automobiles to satellites. Aluminum's high electrical conductivity and low density make it useful for electrical transmission lines. For structural and building applications, its resistance to corrosion is an important feature, as is the fact that it becomes stronger at subzero temperatures. (Steel

| Table 19-4 | Densities of Selected Metals |
|---|---|
| **Metal** | **Density (g cm$^{-3}$) at Room Conditions** |
| Li | 0.534 |
| Na | 0.971 |
| Mg | 1.738 |
| Al | 2.702 |
| Ti | 4.54 |
| Zn | 7.133 |
| Fe | 7.874 |
| Ni | 8.902 |
| Cu | 8.96 |
| Ag | 10.50 |
| Pb | 11.35 |
| U | 18.95 |
| Au | 19.32 |
| Pt | 21.45 |

**Figure 19-21**

Molten aluminum being tapped from an electrolytic cell. (Alcoa)

and iron sometimes become brittle under these circumstances.) Household products that contain aluminum include foil, soft drink cans, and cooking utensils.

## Magnesium

Like aluminum, magnesium is a very abundant element on the surface of the earth, but it is not easy to prepare in elemental form. Although ores such as dolomite ($CaMg(CO_3)_2$) and carnallite ($KCl \cdot MgCl_2 \cdot 6H_2O$) exist, the major commercial source of magnesium and its compounds is seawater. Magnesium forms the second most abundant positive ion in the sea, and scientists separate $Mg^{2+}$ from the other cations in seawater ($Na^+$, $Ca^{2+}$, and $K^+$, in particular) by taking advantage of the fact that magnesium hydroxide is the least soluble hydroxide of the group. Economical recovery of magnesium requires a low-cost base to treat large volumes of seawater and efficient methods for separating the $Mg(OH)_2(s)$ that precipitates from the solution. One base that is used in this way is calcined dolomite, prepared by heating dolomite to high temperatures to drive off carbon dioxide:

$$CaMg(CO_3)_2(s) \longrightarrow CaO \cdot MgO(s) + 2\,CO_2(g)$$

The great solubility of calcium hydroxide ($K_{sp} = 5.5 \times 10^{-6}$) relative to magnesium hydroxide ($K_{sp} = 1.2 \times 10^{-11}$) leads to the reaction

$$CaO \cdot MgO(s) + Mg^{2+}(aq) + 2\,H_2O(\ell) \longrightarrow 2\,Mg(OH)_2(s) + Ca^{2+}(aq)$$

The magnesium hydroxide that is produced in this process includes not only the magnesium from the seawater but also that from the dolomite.

An interesting alternative to dolomite as a base for magnesium production is a process used off the coast of Texas (Fig. 19-22). Oyster shells (composed largely of $CaCO_3$) are calcined to give lime (CaO), which is added to the seawater to yield

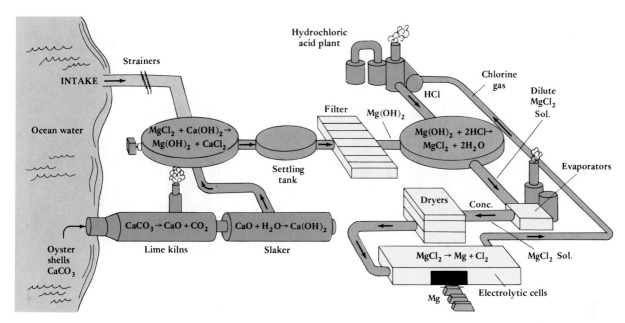

**Figure 19-22**

The production of magnesium hydroxide starts with the addition of lime (CaO) to seawater. Reaction of the magnesium hydroxide with hydrochloric acid produces magnesium chloride, which, after drying, is electrolyzed to give magnesium.

magnesium hydroxide. The Mg(OH)$_2$ slurry (a suspension in water) is washed and filtered in huge nylon filters.

After purification, Mg(OH)$_2$ can be used as "milk of magnesia." Alternatively, reaction with carbon dioxide gives magnesium carbonate, used for coating sodium chloride in table salt to prevent caking and for antacid remedies. Reaction with sulfuric acid gives magnesium sulfate heptahydrate (MgSO$_4$·7H$_2$O), which is known as Epsom salts and is used as a sizing for fabrics. A final alternative is to add hydrochloric acid to the magnesium hydroxide to neutralize it and yield hydrated magnesium chloride:

$$Mg(OH)_2(s) + 2\,HCl(aq) \longrightarrow MgCl_2(aq) + 2\,H_2O(\ell)$$

After the water is evaporated, the solid magnesium chloride is melted (m.p. 708°C) in a large steel electrolysis cell that holds as much as 10 tons of the molten salt. The steel in the cell acts as the cathode during electrolysis, with graphite anodes suspended from the top. The cell reaction is

$$MgCl_2(\ell) \longrightarrow Mg(\ell) + Cl_2(g)$$

The molten magnesium liberated at the cathode floats to the surface and is dipped out periodically, while the chlorine released at the anodes is collected and reacted with steam at high temperatures to produce hydrochloric acid. This is recycled for further reaction with magnesium hydroxide.

Until 1918, elemental magnesium was used mainly in fireworks and flashbulbs, which took advantage of its great reactivity with the oxygen in air and the bright light given off in that reaction (Fig. 2-17). Since that time, many further uses for the metal and its alloys have been developed. Magnesium is even less dense than aluminum and is used in alloys with aluminum to lower its density and improve its resistance to corrosion under basic conditions. As discussed in Section 10-6, magnesium is used as a sacrificial anode to prevent the oxidation of another metal with

**Figure 19-23**

In the electrolytic refining of copper, many slabs of impure copper, which serve as anodes, alternate with thin sheets of pure copper (the cathodes). Both are dipped into a dilute acidic solution of copper. As the copper is oxidized from the impure anodes, it enters the solution and migrates to the cathodes, where it plates out in purer form. (ASARCO)

Cathode
−

Anode
+

Thin sheet of
pure copper

Solution of
$CuSO_4$
and $H_2SO_4$

Slab of
impure copper

which it is in contact. It is also used as a reducing agent to produce other metals such as titanium, uranium, and beryllium from their compounds.

### Electrorefining and Electroplating

Metals such as copper, silver, nickel, and tin that have been produced by pyrometallurgical methods are too impure for many purposes, and **electrorefining** is used to purify them further. Crude metallic copper, for example, is cast into slabs, which are used as anodes in electrolysis cells that contain a solution of $CuSO_4$ in aqueous $H_2SO_4$. Thin sheets of pure copper serve as cathodes, and the copper that dissolves at the anodes is deposited in purer form on the cathodes (Fig. 19-23). Impurities that are more easily oxidized than copper, such as nickel, dissolve along with the copper but remain in solution; elements such as silver and gold that are less easily oxidized do not dissolve but fall away from the anode as a metallic slime. Periodically the anode slime and the solution are removed and further processed for recovery of the elements they contain.

A related process is **electroplating,** in which electrolysis is used to plate out a thin layer of a metal on another material, frequently a second metal. In chrome plating, the piece of metal to be plated is placed in a hot bath of sulfuric acid and chromic acid ($H_2CrO_4$) and is made the cathode in an electrolytic cell. As current passes through the cell, chromium is reduced from the +6 oxidation state in chromic acid to elemental chromium and plates out on the cathode. A decorative chromium layer can be as thin as $2.5 \times 10^{-5}$ cm (corresponding to 2 g of Cr per square meter of surface). Thicker layers ranging up to $10^{-2}$ cm are found in hard chromium plate, prized for its resistance to wear and used in automobile trim. Steel can be plated with cadmium to improve its resistance to corrosion in marine environments. Gold and silver are used both for decorative plating and (because they are good conductors of electricity) on electronic devices.

### Example 19-3

Suppose a layer of chromium $3.0 \times 10^{-3}$ cm thick is to be plated onto an automobile bumper with a surface area of $2.0 \times 10^3$ cm². If a current of 250 A is used, how long must current be passed through the cell to achieve the desired thickness? The density of chromium is $7.2 \text{ g cm}^{-3}$.

## Solution

The volume of the Cr is the product of the thickness of the layer and the surface area:

$$\text{volume} = (3.0 \times 10^{-3} \text{ cm})(2.0 \times 10^3 \text{ cm}^2) = 6.0 \text{ cm}^3$$

The mass of chromium is the product of this volume and the density:

$$\text{mass Cr} = (6.0 \text{ cm}^3)(7.2 \text{ g cm}^{-3}) = 43.2 \text{ g}$$

From this, the chemical amount of Cr that must be reduced is

$$\frac{43.2 \text{ g}}{52.00 \text{ g mol}^{-1}} = 0.831 \text{ mol Cr}$$

Because Cr is being reduced from oxidation state +6 in $H_2CrO_4$ to 0 in the elemental form, six electrons are required for each atom of Cr deposited. The number of moles of electrons is then

$$0.831 \text{ mol Cr} \times \left(\frac{6 \text{ mol } e^-}{1 \text{ mol Cr}}\right) = 4.98 \text{ mol } e^-$$

$$\text{total charge} = (4.98 \text{ mol})(96{,}485 \text{ C mol}^{-1}) = 4.81 \times 10^5 \text{ C}$$

The required electrolysis time is the total charge divided by the current (in amperes):

$$\text{time} = \frac{4.81 \times 10^5 \text{ C}}{250 \text{ C s}^{-1}} = 1.9 \times 10^3 \text{ s} = 32 \text{ min}$$

*Related Problems:* 29, 30

After studying this chapter and working the problems that follow, you should be able to

**CONCEPTS & SKILLS**

1. Show how the fundamental silicate tetrahedral unit ($SiO_4^{4-}$) links to other silicate tetrahedra to form rings, chains, double chains, sheets, and space-filling crystalline networks (Section 19-1, problems 1–4).
2. Describe the chemical compositions and structures of aluminosilicates, clays, and zeolites (Section 19-1, problems 5–6).
3. Outline the crystallization sequences of silicate minerals from magma and their subsequent weathering to form sedimentary rocks (Section 19-2).
4. Apply the Clapeyron relationship to calculate the effects of pressure and temperature on the phase equilibria of minerals (Section 19-2, problems 9–12).
5. Outline the chemistry involved in the reduction of copper and iron ores in smelters and blast furnaces (Section 19-3, problems 19–22).
6. Describe the Hall–Héroult process for the production of aluminum (Section 19-4, problem 25).

7. Summarize the methods used to recover magnesium compounds from seawater and the compounds' subsequent conversion to elemental magnesium (Section 19-4, problem 26).
8. Use Faraday's laws to relate current to metal deposited in electrorefining and electroplating operations (Section 19-4, problems 29–30).

## PROBLEMS

### Silicates

1. Draw a Lewis electron-dot structure for the disilicate ion $(Si_2O_7^{6-})$. What changes in this structure would be necessary to produce the structure of the pyrophosphate ion $(P_2O_7^{4-})$ and the pyrosulfate ion $(S_2O_7^{2-})$? What is the analogous compound of chlorine?

2. Draw a Lewis electron-dot structure for the cyclosilicate ion $Si_6O_{18}^{12-}$, which forms part of the structures of beryl and emerald.

3. By referring to Table 19-2, predict the structure of each of the following silicate minerals (network, sheets, double chains, and so forth). Give the oxidation state of each atom.
   (a) Andradite, $Ca_3Fe_2(SiO_4)_3$
   (b) Vlasovite, $Na_2ZrSi_4O_{10}$
   (c) Hardystonite, $Ca_2ZnSi_2O_7$
   (d) Chrysotile, $Mg_3Si_2O_5(OH)_4$

4. By referring to Table 19-2, predict the structure of each of the following silicate minerals (network, sheets, double chains, and so forth). Give the oxidation state of each atom.
   (a) Tremolite, $Ca_2Mg_5(Si_4O_{11})_2(OH)_2$
   (b) Gillespite, $BaFeSi_4O_{10}$
   (c) Uvarovite, $Ca_3Cr_2(SiO_4)_3$
   (d) Barysilate, $MnPb_8(Si_2O_7)_3$

5. By referring to Table 19-2, predict the structure of each of the following aluminosilicate minerals (network, sheets, double chains, and so forth). In each case, the aluminum atoms grouped with the silicon and oxygen in the formula substitute for Si in tetrahedral sites. Give the oxidation state of each atom.
   (a) Keatite, $Li(AlSi_2O_6)$
   (b) Muscovite, $KAl_2(AlSi_3O_{10})(OH)_2$
   (c) Cordierite, $Al_3Mg_2(AlSi_5O_{18})$

6. By referring to Table 19-2, predict the structure of each of the following aluminosilicate minerals (network, sheets, double chains, and so forth). In each case, the aluminum atoms grouped with the silicon and oxygen in the formula substitute for Si in tetrahedral sites. Give the oxidation state of each atom.
   (a) Amesite, $Mg_2Al(AlSiO_5)(OH)_4$
   (b) Phlogopite, $KMg_3(AlSi_3O_{10})(OH)_2$
   (c) Thomsonite, $NaCa_2(Al_5Si_5O_{20})\cdot6H_2O$

### Geochemistry

7. Sketch the dependence of cell potential on pH for the half-reactions corresponding to the reduction of $Co^{2+}$ at 1 M concentration (or $Co(OH)_2$) to $Co(s)$. Use data from Appendix E.

8. Sketch the dependence of cell potential on pH for the half-reactions corresponding to the reduction of $Pb^{2+}$ at 1 M concentration (or PbO) to $Pb(s)$. Use data from Appendix E.

9. Calcite ($CaCO_3$) reacts with quartz to form wollastonite ($CaSiO_3$) and carbon dioxide.
   (a) Write a balanced equation for this reaction.
   (b) Use the data in Appendix D to calculate $\Delta H°$ and $\Delta S°$ for this reaction.
   (c) Calculate the temperature at which $\Delta G = 0$ ($K = 1$) for this reaction at atmospheric pressure.
   (d) Calculate the temperature at which $\Delta G = 0$ for this reaction at $P = 500$ atm, assuming ideal gas behavior for the carbon dioxide.

10. Tremolite, $Ca_2Mg_5Si_8O_{22}(OH)_2$, reacts with calcite ($CaCO_3$) and quartz to give diopside ($CaMgSi_2O_6$), water vapor, and carbon dioxide.
    (a) Write a balanced chemical equation for this reaction.
    (b) The standard enthalpies of formation of tremolite and diopside are $-12360$ and $-3206.2$ kJ mol$^{-1}$, respectively, and their standard entropies are 548.9 and 142.93 J K$^{-1}$ mol$^{-1}$. Use this information, together with data from Appendix D, to calculate $\Delta H°$ and $\Delta S°$ for this reaction.
    (c) Calculate the temperature at which $\Delta G = 0$ ($K = 1$) for this reaction at atmospheric pressure.
    (d) Calculate the temperature at which $\Delta G = 0$ for this reaction at $P = 600$ atm, assuming ideal gas behavior for the water vapor and carbon dioxide.

11. Calcium carbonate exists in two stable forms, calcite (molar volume 36.94 cm$^3$ mol$^{-1}$) and aragonite (molar volume 34.16 cm$^3$ mol$^{-1}$).
    (a) Use data from Appendix D to calculate $\Delta H°$, $\Delta S°$, and $\Delta G°$ for the conversion of calcite to aragonite at 25°C and 1 atm pressure.
    (b) Which form is thermodynamically stable under these conditions?
    (c) Which form will be stable at very low temperature?
    (d) Which form will be stable at high pressures?

(e) Use the Clapeyron equation to calculate the slope of the coexistence curve for this reaction at $P = 1$ atm.

**12.** Silica exists in several stable forms, including quartz (molar volume 22.69 cm$^3$ mol$^{-1}$) and cristobalite (molar volume 25.74 cm$^3$ mol$^{-1}$).
(a) Use data from Appendix D to calculate $\Delta H°$, $\Delta S°$, and $\Delta G°$ for the conversion of quartz to cristobalite at 25°C and 1 atm pressure.
(b) Which form is thermodynamically stable under these conditions?
(c) Which form will be stable at very high temperatures?
(d) Which form will be stable at high pressures?
(e) Use the Clapeyron equation to calculate the slope of the coexistence curve for this reaction at $P = 1$ atm.

## Extractive Metallurgy

**13.** The principal ore of manganese is pyrolusite ($MnO_2$), which can be reduced to elemental manganese with aluminum. Write a balanced equation for the reaction, and calculate the standard enthalpy and free energy changes per mole of manganese produced from pyrolusite, using data in Appendix D.

**14.** Uranium can be prepared in elemental form by reducing $U_3O_8$ with calcium. Write a balanced equation for the reaction, and calculate the standard enthalpy and free energy changes per mole of uranium produced from $U_3O_8$, using the following thermodynamic data:

| | $\Delta H_f°$ (kJ mol$^{-1}$) | $S°$ (J K$^{-1}$ mol$^{-1}$) |
|---|---|---|
| $U_3O_8(s)$ | −3575 | 282.6 |
| $U(s)$ | 0 | 50.2 |
| $CaO(s)$ | −635 | 39.8 |
| $Ca(s)$ | 0 | 41.4 |

**15.** As mentioned in the text, mercury(II) oxide (HgO) decomposes to mercury and oxygen when heated. Use data from Appendix D to estimate the temperature at which this change becomes spontaneous—that is, at which the free energy change ($\Delta G$) becomes negative.

**16.** Use data from Appendix D to predict the temperature above which CuO decomposes spontaneously to $Cu(s)$ and $\frac{1}{2} O_2(g)$ at atmospheric pressure.

**17.** Explain why it is possible to reduce tungsten(VI) oxide ($WO_3$) to metal at elevated temperature with hydrogen. At what temperature does the equilibrium constant become unity? Use data from Appendix D.

**18.** $WO_3$ can be reduced to tungsten if it is heated with carbon in an electric furnace:

$$2\,WO_3(s) + 3\,C(s) \longrightarrow 2\,W(s) + 3\,CO_2(g)$$

(a) Calculate the standard free energy change $\Delta G°$ for the reduction of $WO_3$ to W, and comment on the feasibility of the process.

(b) What must be done to make the process thermodynamically feasible, assuming that $\Delta H$ and $\Delta S$ do not vary much with temperature?

**19.** Analysis of a copper-bearing seam of rock shows it to contain three copper-containing substances: 1.1% chalcopyrite ($CuFeS_2$), 0.42% covellite (CuS), and 0.51% bornite ($Cu_5FeS_4$) by mass. Calculate the total mass of copper present in 1.0 metric ton ($1.0 \times 10^3$ kg) of this copper-bearing ore.

**20.** Calculate (a) the mass of copper and (b) the volume of sulfur dioxide at 0°C and 1 atm that are produced in the smelting of 1.0 metric ton ($1.0 \times 10^3$ kg) of chalcopyrite ($CuFeS_2$).

**21.** (a) Write balanced equations for the reduction with carbon monoxide of hematite ($Fe_2O_3$), magnetite ($Fe_3O_4$), and siderite ($FeCO_3$) to iron.
(b) Calculate the standard free energy change $\Delta G°$ per mole of iron produced from each of the three reactions in part (a). From your results, predict which iron compound should be easiest to reduce from the standpoint of thermodynamics.

**22.** Calculate $\Delta G°$ for each of the following reactions:

$$Fe_2O_3(s) + \tfrac{3}{2}\,C(s) \longrightarrow 2\,Fe(s) + \tfrac{3}{2}\,CO_2(g)$$

$$Fe_2O_3(s) + 3\,C(s) \longrightarrow 2\,Fe(s) + 3\,CO(g)$$

$$Fe_2O_3(s) + 3\,CO(g) \longrightarrow 2\,Fe(s) + 3\,CO_2(g)$$

From your results, decide which is the better reducing agent—C (being oxidized to CO or $CO_2$) or CO.

## Electrometallurgy

**23.** In the Downs process, molten sodium chloride is electrolyzed to produce sodium. A valuable byproduct is chlorine. Write equations representing the processes taking place at the anode and at the cathode in the Downs process.

**24.** The first element to be prepared by electrolysis was potassium. In 1807, Humphry Davy, then 29 years old, passed an electric current through molten potassium hydroxide (KOH), obtaining liquid potassium at one electrode and water and oxygen at the other. Write equations to represent the processes taking place at the anode and at the cathode.

**25.** A current of 55,000 A is passed through a series of 100 Hall-Héroult cells for a period of 24 hours. Calculate the maximum theoretical mass of aluminum that can be recovered.

**26.** A current of 75,000 A is passed through an electrolysis cell containing molten $MgCl_2$ for a period of 7.0 days. Calculate the maximum theoretical mass of magnesium that can be recovered.

**27.** An important use for magnesium is to make titanium. In the Kroll process, magnesium reduces titanium(IV) chloride to elemental titanium in a sealed vessel at 800°C. Write a balanced chemical equation for this reaction. What mass of magnesium is needed, in theory, to produce 100 kg of titanium from titanium(IV) chloride?

**28.** Calcium is used to reduce vanadium(V) oxide to elemental vanadium in a sealed steel vessel. Vanadium is used in vanadium steel alloys for jet engines, high-quality knives, and tools. Write a balanced chemical equation for this process. What mass of calcium is needed, in theory, to produce 20.0 kg of vanadium from vanadium(V) oxide?

**29.** Galvanized steel consists of steel with a thin coating of zinc to reduce corrosion. The zinc can be deposited electrolytically, by making the steel object the cathode and a block of zinc the anode in an electrochemical cell containing a dissolved zinc salt. Suppose a steel garbage can is to be galvanized and requires that a total mass of 7.32 g of zinc be coated to the required thickness. For how long should a current of 8.50 A be passed through the cell to achieve this?

**30.** In the electroplating of a silver spoon, the spoon acts as the cathode and a piece of pure silver as the anode. Both dip into a solution of silver cyanide (AgCN). Suppose that a current of 1.5 A is passed through such a cell for a period of 22 min and that the spoon has a surface area of 16 cm$^2$. Calculate the average thickness of the silver layer deposited on the spoon, taking the density of silver to be 10.5 g cm$^{-3}$.

## Additional Problems

**31.** Predict the structure of each of the following silicate minerals (network, sheets, double chains, and so forth). Give the oxidation state of each atom.
(a) Apophyllite, $KCa_4(Si_8O_{20})F \cdot 8H_2O$
(b) Rhodonite, $CaMn_4(Si_5O_{15})$
(c) Margarite, $CaAl_2(Al_2Si_2O_{10})(OH)_2$

**32.** By referring to Table 19-2, predict the kind of structure formed by manganpyrosmalite, a silicate mineral with chemical formula $Mn_{12}FeMg_3(Si_{12}O_{30})(OH)_{10}Cl_{10}$. Give the oxidation state of each atom in this formula unit.

**33.** A reference book lists the chemical formula of one form of vermiculite as

$$[(Mg_{2.36}Fe_{0.48}Al_{0.16})(Si_{2.72}Al_{1.28})O_{10}(OH)_2][Mg_{0.32}(H_2O)_{4.32}]$$

Determine the oxidation state of the iron in this mineral.

**34.** The most common feldspars are those containing potassium, sodium, and calcium cations. They are called, respectively, orthoclase ($KAlSi_3O_8$), albite ($NaAlSi_3O_8$), and anorthite ($CaAl_2Si_2O_8$). The solid solubility of orthoclase in albite is limited, and its solubility in anorthite is almost negligible. Albite and anorthite, however, are completely miscible at high temperatures and show complete solid solution. Offer an explanation for these observations, based on the tabulated radii of the $K^+$, $Na^+$, and $Ca^{2+}$ ions from Appendix F.

**35.** The clay mineral kaolinite ($Al_2Si_2O_5(OH)_4$) is formed by the weathering action of water containing dissolved carbon dioxide on the feldspar mineral anorthite ($CaAl_2Si_2O_8$). Write a balanced chemical equation for the reaction that occurs. The $CO_2$ forms $H_2CO_3$ as it dissolves. As the pH is lowered, will the weathering occur to a greater or a lesser extent?

**36.** Certain kinds of zeolite have the general formula $M_2O \cdot Al_2O_3 \cdot ySiO_2 \cdot wH_2O$ where M is an alkali metal such as Na or K, $y$ is 2 or more, and $w$ is any integer. Compute the mass percentage of Al in a zeolite that has M = K, $y = 4$, and $w = 6$.

**37.** (a) Use data from Tables 6-2 and 7-2 to calculate the solubility of $CaCO_3$ in water at pH 7.
(b) Will the solubility increase or decrease if the pH is lowered and the water becomes more acidic?
(c) Calculate the maximum amount of limestone (primarily calcium carbonate) that could dissolve per year in a river at pH 7 with an average flow rate of $1.0 \times 10^6$ m$^3$ per hour.

**\*38.** Show how the Clapeyron equation follows from the second law of thermodynamics, expressed as

$$dG = V\,dP - S\,dT$$

**39.** Talc, $Mg_3Si_4O_{10}(OH)_2$, reacts with forsterite ($Mg_2SiO_4$) to form enstatite ($MgSiO_3$) and water vapor.
(a) Write a balanced chemical equation for this reaction.
(b) If the water pressure is equal to the total pressure, will formation of products be favored or disfavored with increasing total pressure?
(c) The entropy change for this reaction is positive. Will the slope of the coexistence curve (pressure plotted against temperature) be positive or negative?

**40.** (a) On a single graph, plot $\Delta G$ as a function of temperature from 0°C to 1500°C for the reactions

$$2\,C(s, gr) + O_2(g) \longrightarrow 2\,CO(g)$$
$$2\,Fe(s) + O_2(g) \longrightarrow 2\,FeO(s)$$
$$2\,Ni(s) + O_2(g) \longrightarrow 2\,NiO(s)$$
$$\tfrac{4}{3}\,Al(s) + O_2(g) \longrightarrow \tfrac{2}{3}\,Al_2O_3(s)$$

Assume that $\Delta H$ and $\Delta S$ for these reactions are independent of temperature, and approximate FeO by the nonstoichiometric compound wüstite.
(b) Use your graph to predict the temperature range over which coke can be used to reduce NiO and FeO to the elements.
(c) Can $Al_2O_3$ be reduced to elemental aluminum by coke over this temperature range?

**41.** Compute the equilibrium constant of the reaction

$$CO_2(g) + C(s) \rightleftharpoons 2\,CO(g)$$

at room temperature (25°C) and at 1875°C.

**42.** The principal ore of chromium is the mineral chromite ($FeCr_2O_4$), which has a standard free energy of formation of ($\Delta G_f^\circ$) of $-1344$ kJ mol$^{-1}$. Because the main use of chromium is in chromium-containing steels, the ore is generally reduced directly to the alloy ferrochrome by the reaction

$$FeCr_2O_4(s) + 4\,C(s) \longrightarrow Fe(s) + 2\,Cr(s) + 4\,CO(g)$$

Calculate the standard free energy change for the production of ferrochrome according to this equation, using Appendix D for other information you may need. (*Note:* The free energy change in mixing iron and chromium in the alloy is small and can be neglected.)

**43.** Compare and contrast the smelting of copper with that of iron. Include in your discussion the differences in starting materials, in temperatures and reducing agents employed, and in byproducts and their disposal.

**44.** The radius of the earth is 6370 km. Estimate the amount of aluminum (in kilograms) in the top 1 km of the earth's crust, assuming the average density of the crust to be approximately 2.8 g cm$^{-3}$.

**45.** Compare and contrast the production of aluminum with that of magnesium. Include in your discussion the differences in starting materials, in reactions preparatory to the electrolysis, and in reaction byproducts.

**46.** Use data from Appendix D to predict whether elemental magnesium should react with water on purely thermodynamic grounds. What is observed in practice? Explain.

**47.** If pure chromium is to be obtained from its chromite ore, $FeCr_2O_4(s)$, the process commonly used is to fuse it with sodium carbonate. The unbalanced equation for this reaction is

$$FeCr_2O_4(s) + Na_2CO_3(\ell) + O_2(g) \longrightarrow$$
$$Fe_2O_3(s) + Na_2CrO_4(\ell) + CO_2(g)$$

(a) Balance the preceding chemical equation.

(b) Calculate the standard enthalpy change of the reaction, assuming that the enthalpies of fusion of $Na_2CO_3$ and $Na_2CrO_4$ are approximately equal. The standard molar enthalpy of formation of $FeCr_2O_4(s)$ is $-1445$ kJ mol$^{-1}$, and that of $Na_2CrO_4(s)$ is $-1342$ kJ mol$^{-1}$. Use Appendix D for the additional data you require.

(c) The sodium chromate produced in the reaction is soluble in water and can be leached from the solidified melt. Suggest a procedure for converting it to elemental chromium.

**48.** A 55.5-kg slab of crude copper from a smelter has a copper content of 98.3%. Estimate the time required to purify it electrochemically, if it is used as the anode in a cell that has acidic copper(II) sulfate as its electrolyte and a current of $2.00 \times 10^3$ A is passed through the cell.

**49.** Sheet iron can be galvanized by passing a direct current through a cell containing a solution of zinc sulfate between a graphite anode and the iron sheet. Zinc plates out on the iron. The process can be made continuous if the iron sheet is a coil that unwinds as it passes through the electrolysis cell and coils up again after it emerges from a rinse bath. Calculate the cost of the electricity required to deposit a 0.250-mm-thick layer of zinc on both sides of an iron sheet that is 1.00 m wide and 100 m long, if a current of 25 A at a voltage of 3.5 V is used and the energy efficiency of the process is 90%. The cost of electricity is 10 cents per kilowatt-hour (1 kWh = 3.6 MJ). Consult Appendix F for data on zinc.

# 20

# Ceramics and Semiconductors

Throughout history, discoveries of new materials from which to fashion the structures of everyday life have set off great changes in human affairs. Manufactured materials fall into three broad classes: metals, plastics, and ceramics. Metals were introduced in Chapter 17 and discussed further in Chapter 19, and organic plastics are among the polymers to be considered in Chapter 23. This chapter focuses on ceramics, the oldest of the three types of materials. New discoveries in ceramics are occurring at a startling rate and will certainly generate major change.

Ceramics have value both as structural materials and for their wide range of electronic properties. Traditionally, ceramics have served as electrical insulators, but in recent years their contributions as semiconductors in integrated circuits and as high-temperature superconductors have gained importance.

## 20-1  THE PROPERTIES OF CERAMICS

The term **ceramics** encompasses synthetic materials that have as their essential components inorganic, nonmetallic materials. This broad definition includes cement, concrete, and glass in addition to the more traditional fired clay products such as bricks, roof tiles, pottery, and porcelain. The use of ceramics predates recorded history; fragments of pottery chronicle the emergence of civilization from a primitive state. No one knows when human hands first shaped small vessels from moist clay and left them to harden in the heat of the sun. Such containers held nuts, grains, and berries well but transformed into formless mud when water was poured into them. Then someone discovered (perhaps by accident) that if clay was placed in the glowing embers of a fire, it became as hard as rock and withstood water well. Molded figures made 24,000 years ago and found in what is now Czechoslovakia are the earliest fired ceramic objects discovered so far, and fired clay vessels from the Near East date from 8000 B.C. With the action of fire on clay, the art and science of ceramics began.

Ceramics offer stiffness, hardness, resistance to wear, and resistance to corrosion (particularly by oxygen and water), even at high temperatures. They are less dense than most metals, which makes them desirable metal substitutes when weight is a factor. Most are good electrical insulators at ordinary temperatures, a property that is exploited in electronics and power transmission (Fig. 20-1). Ceramics retain their strength well at high temperatures. Several important structural metals soften or melt at temperatures a thousand degrees below the melting points of their chemical compounds in ceramics. Aluminum, for example, melts at 660°C, whereas aluminum oxide ($Al_2O_3$), an important compound in many ceramics, does not melt until it reaches a temperature of 2051°C.

Against these advantages must be listed some serious disadvantages. Ceramics are generally brittle and low in tensile strength. They tend to have high thermal expansion but low thermal conductivity, making them subject to **thermal shock,** in

**Figure 20-1**
This insulating ceramic, comprising several interlocking pieces, transmits no electric current despite a potential difference of hundreds of thousands of volts between one end and the other. (Niagara Mohawk Power Company)

which a sudden local temperature change causes cracking or shattering. Whereas metals and plastics dent or deform under stress, ceramics cannot absorb stress in that way; they break instead. A major drawback of ceramics as structural materials is their tendency to fail unpredictably and catastrophically during use. Moreover, some ceramics lose mechanical strength as they age. This tendency is an insidious and serious problem in many applications.

## Composition and Structure of Ceramics

Ceramics employ a wide variety of chemical compounds, and useful ceramic bodies are nearly always mixtures of several compounds. **Silicate ceramics,** of which common pots, dishes, and bricks are made, consist of aluminosilicate clay minerals. All contain the tetrahedral $SiO_4$ grouping discussed in Section 19-1. In **oxide ceramics,** a number of metals combine with oxygen to give a compound such as alumina ($Al_2O_3$), magnesia (MgO), or yttria ($Y_2O_3$); silicon is a minor or nonexistent component. **Nonoxide ceramics** contain compounds that are free of oxygen as a principal component. Some important examples are silicon nitride ($Si_3N_4$), silicon carbide (SiC), and boron carbide (approximate composition $B_4C$).

One important property of ceramics is their porosity. Porous ceramics have small openings into which fluids (typically air or water) can infiltrate. Fully dense ceramics have no channels of this sort. Two ceramic pieces can have the same chemical composition but quite different densities if the first is porous and the second is not.

A **ceramic phase** is any portion of the whole body that is physically homogeneous and bounded by a surface that separates it from other parts. Distinct phases are visible at a glance in coarse-grained ceramic pieces; in a fine-grained piece, phases can be seen with a microscope. When examined on a still finer scale, most ceramics, like metals, are microcrystalline, consisting of small crystalline grains cemented together (Fig. 20-2). The **microstructure** of a ceramic object includes the sizes and shapes of the grains, the sizes and distribution of voids (openings between grains) and cracks, the identity and distribution of impurity grains, and the stresses within the structure. Microstructural variations have enormous importance in ceramics because slight changes at this level strongly influence the properties of individual ceramic pieces. This is less true for plastic and metallic objects.

### Figure 20-2

Microstructures of aluminosilicate ceramics, revealed by the different colors of light emitted after bombardment by electrons. (*a*) Forsterite (red), spinel (green), and periclase (dark brown) grains. (*b*) Periclase (blue) and oldhamite (yellow) grains. (Musa Karakus and Richard Hagni)

(*a*)

(*b*)

The microstructure of a ceramic body depends markedly on the details of its fabrication. The techniques of forming and firing a ceramic piece are as important as its chemical composition in determining ultimate behavior, because they confer a unique microstructure. This fact calls attention to the biggest problem with ceramics as structural materials: inconsistent quality. Ceramic engineers can produce parts that are stronger than steel, but not *reliably* because of the difficulties of monitoring and controlling microstructure. Gas-turbine engines fabricated of silicon nitride (Fig. 20-3), for example, run well at 1370°C, which is hot enough to soften or melt most metals. The high operating temperature increases engine efficiency, and the ceramic turbines weigh less than metal ones, further boosting fuel economy. Despite these advantages, ceramic gas turbines are not made commercially. Acceptable ceramic turbines have to be built from selected, pretested components. The testing costs and rejection rates are so high that economical mass production has been impossible so far.

## Making Ceramics

The manufacture of most ceramics involves four steps: (1) preparation of the raw material; (2) forming into the desired shape, often by mixing a powder with water or other binder and molding the resulting plastic mass; (3) drying and firing the piece, also called **densification** because pores (voids) in the dried ceramic fill in; and (4) finishing the piece by sawing, grooving, grinding, or polishing.

The raw materials for traditional ceramics are natural clays that come from the earth as powders or thick pastes and become plastic enough after adjustment of their water content to be formed freehand or on a potter's wheel. In contrast, the modern special ceramics (both oxide and nonoxide) require chemically pure raw materials that have been produced synthetically. Close control of the purity of the starting materials for these ceramics is essential for production of finished pieces with the desired properties. In addition to being formed by hand or in open molds, special ceramic pieces can be shaped by squeezing (compacting) the dry or semidry powders in a strong, closed mold of the desired shape at either ordinary or elevated temperatures (the latter is called hot pressing).

Firing a ceramic causes **sintering,** in which the fine particles of the ceramic start to merge by diffusion at high temperatures. Atoms migrate out of their individual grains, and **necks** grow at the points of contact to connect adjoining particles. The density of the material increases as the voids between grains partially fill. Sintering occurs below the melting point of the material and shrinks the ceramic body. In addition to merging the grains of the ceramic, firing causes partial melting, chemical reactions among different phases, reactions with gases in the atmosphere of the firing chamber, and recrystallization of compounds with an accompanying growth in crystal size. All of these changes influence the microstructure of a piece and must be understood and controlled. Firing accelerates physical and chemical changes, of course, but only rarely is thermodynamic equilibrium reached in a fired ceramic piece. Kinetic factors, including the rate of heating, the length of time each temperature is held, and the rate of cooling, influence microstructure. As a result, the use of microwave radiation (as in microwave ovens) rather than kilns to fire ceramics is under development in ceramic factories because it promises more exact control of the heating rate and thereby more reliable quality.

**Figure 20-3**

A part for a gas-turbine engine, freshly fabricated from silicon nitride. (GTE Labs)

## 20-2 SILICATE CERAMICS

The silicate ceramics include materials with a wide variety of compositions, structures, and uses, ranging from simple earthenware bricks and pottery to cement, fine porcelain, and glass. The structural strength of these materials is based on the same linking of silicate ion tetrahedra that gives structure to silicate minerals in nature.

### Pottery and Structural Clay Products

Section 19-2 noted that aluminosilicate clays are the products of the weathering of primary minerals. When water is added to such a clay in a moderate amount, the result is a thick paste that is easily molded into a variety of shapes. Clays expand as water invades the spaces between adjacent aluminosilicate sheets of the mineral, but they release most of this water to a dry atmosphere and thus shrink. A small fraction of the water or hydroxide ions remains rather tightly bound by ion–dipole forces to cations between the aluminosilicate sheets and is lost only when the clay is heated to a high temperature. The firing of aluminosilicate clays simultaneously causes irreversible chemical changes. The clay kaolinite ($Al_2Si_2O_5(OH)_4$) undergoes the following reaction:

$$3 Al_2Si_2O_5(OH)_4 \longrightarrow Al_6Si_2O_{13}(s) + 4 SiO_2(s) + 6 H_2O(g)$$
$$\text{(Kaolinite)} \qquad\qquad \text{(Mullite)} \qquad \text{(Silica)} \qquad \text{(Water)}$$

The fired ceramic body is a mixture of two phases, mullite and silica. Mullite, a rare mineral in nature, takes the form of needle-like crystals that interpenetrate and confer strength on the ceramic. When the temperature is above 1470°C, the silica phase forms as minute grains of cristobalite, a crystalline form of $SiO_2$.

If chemically pure kaolinite is fired, the finished ceramic object is white. Such purified clay minerals are the raw materials for fine china. Clays in nature contain impurities, such as transition-metal oxides, that affect the colors of both the unfired material and the fired ceramic object. The colors of the metal oxides arise from their absorption of light at visible wavelengths (see Section 16-5). Common colors for natural ceramics are yellow or greenish yellow, brown, and red. Red bricks, for instance, have been made from a clay with a high iron content.

Before a clay is fired in a kiln, it must first be freed of moisture by slow heating to about 500°C. If a clay body dried at room temperature were placed directly into a hot kiln, it would literally explode from the sudden, uncontrolled expulsion of water. A fired ceramic shrinks somewhat as it cools, tending to crack. Cracks limit the strength of the fired object and are undesirable. Their occurrence can be minimized by coating the surface of a partially fired clay object with a thin **glaze** that holds the ceramic in a state of tension as it cools. Glazes, as their name implies, are glasses that have no sharp melting temperatures, but rather harden and develop resistance to shear stresses increasingly as the temperature of the high-fired clay object gradually falls. Glazes are usually aluminosilicates that have high aluminum content to raise their viscosity and thereby reduce the tendency to run off the clay surface during firing. Glazes also provide the means of coloring the surfaces of fired clays and imparting decorative designs to them. Transition-metal oxides (particularly those of Ti, V, Cr, Mn, Fe, Co, Ni, and Cu) are responsible for the colors. The oxidation state of the transition metal in the glaze is critical in determining the color produced and is controlled by regulating the composition of the atmosphere in the kiln; atmospheres rich in oxygen give high oxidation states, and those low or lacking in oxygen give low oxidation states.

## Glass

Glassmaking probably originated in the Near East about 3500 years ago. It is one of the oldest domestic arts, but its beginnings, like those of metallurgy, are obscure. To initiate and contain the necessary chemical reactions, both glassmaking and metallurgy required high-temperature charcoal-fueled ovens and vessels made of materials that did not melt easily. In early glassmaking, artisans fabricated the desired shapes by sculpting them from solid chunks of glass. Later, molten glass was poured in successive layers over a core of sand. The invention of glassblowing, probably in the first century B.C., was a great advance. The glassblower dipped a long iron tube into molten glass and caused a rough ball of viscous material to accumulate on its end by rotating the tube. Blowing into the iron tube forced the soft glass to take the form of a hollow ball (Fig. 20-4) that could be further shaped into a vessel and severed from the blowing tube with a blade. Artisans in the Roman Empire developed glassblowing to a high degree, but with the decline of that civilization, glassmaking in Europe deteriorated until the Venetians redeveloped the lost techniques a thousand years later.

**Figure 20-4**
Glassblowing. (Leon Lewandowski)

Glasses are amorphous solids of widely varying composition whose physical properties were discussed in Section 17-5. In this chapter, the term "glass" is used in the restricted and familiar sense to refer to materials formed from silica, usually in combination with metal oxides. The absence of long-range order in glasses makes them **isotropic;** that is, their physical properties are the same in all directions. This has certain advantages in technology, including the fact that glasses expand uniformly in all directions with an increase in temperature. The mechanical strength of glass is intrinsically very high, exceeding the tensile strength of steel, provided the surface is free of scratches and other imperfections. Flaws in the surface provide sites where fractures can start when the glass is stressed. When a glass object of intricate shape and nonuniform thickness is suddenly cooled, internal stresses are locked in and may be relieved catastrophically when the object is heated or struck (Fig. 20-5). **Annealing,** or slowly heating a strained object to a temperature somewhat below its softening point and holding it there for a while before allowing it to cool, provides a chance for short-range diffusion of atoms to occur and eliminate internal stresses.

The softening and annealing temperatures and other properties, such as density, of a glass depend on its chemical composition (Table 20-1). Silica itself ($SiO_2$) forms a glass if it is heated above its melting point and then cooled rapidly to avoid crystallization. The resulting vitreous (glassy) silica has limited uses because the high temperatures required to shape it make it quite expensive. Sodium silicate glasses are formed in the high-temperature reaction of silica sand with anhydrous sodium carbonate (soda ash, $Na_2CO_3$):

$$Na_2CO_3(s) + nSiO_2(s) \longrightarrow Na_2O \cdot (SiO_2)_n(s) + CO_2(g)$$

The melting point of the nonvolatile product is about 900°C, and if cooling through that temperature is rapid, the glassy state results. The product, called "water glass," is water-soluble and thus unsuitable for making vessels. Its aqueous solutions, however, are used in some detergents and as adhesives for sealing cardboard boxes.

**Figure 20-5**
The strains associated with internal stresses in rapidly cooled glass can be made visible by polarized light. They appear as brightly colored regions. (David Guizak/Phototake)

**Table 20-1   Composition and Properties of Various Glasses**

| Silica Glass | Soda-Lime Glass | Borosilicate Glass | Aluminosilicate Glass | Leaded Glass |
|---|---|---|---|---|
| *Composition* | | | | |
| $SiO_2$, 99.9% | $SiO_2$, 73% | $SiO_2$, 81% | $SiO_2$, 63% | $SiO_2$, 56% |
| $H_2O$, 0.1% | $Na_2O$, 17% | $B_2O_3$, 13% | $Al_2O_3$, 17% | PbO, 29% |
| | CaO, 5% | $Na_2O$, 4% | CaO, 8% | $K_2O$, 9% |
| | MgO, 4% | $Al_2O_3$, 2% | MgO, 7% | $Na_2O$, 4% |
| | $Al_2O_3$, 1% | | $B_2O_3$, 5% | $Al_2O_3$, 2% |
| *Coefficient of Linear Thermal Expansion* ($°C^{-1} \times 10^7$)[†] | | | | |
| 5.5 | 93 | 33 | 42 | 89 |
| *Softening Point* (°C) | | | | |
| 1580 | 695 | 820 | 915 | 630 |
| *Annealing Point* (°C) | | | | |
| 1050 | 510 | 565 | 715 | 435 |
| *Density* (g cm$^{-3}$) | | | | |
| 2.20 | 2.47 | 2.23 | 2.52 | 3.05 |
| *Refractive Index*[‡] *at* $\lambda = 589$ *nm* | | | | |
| 1.459 | 1.512 | 1.474 | 1.530 | 1.560 |

[†] The coefficient of linear thermal expansion is defined as the fractional increase in the length of a body when its temperature is increased by 1°C.
[‡] The refractive index is a vital property of glass for optical applications. It is defined by $n = \sin \theta_i / \sin \theta_r$ where $\theta_i$ is the angle of incidence of a ray of light on the surface of the glass and $\theta_r$ is the angle of refraction of the ray of light in the glass.

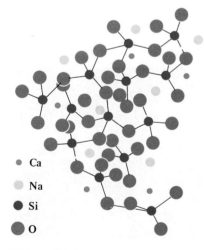

● Ca

○ Na

● Si

● O

**Figure 20-6**

The structure of a soda-lime glass. Note the tetrahedral coordination of oxygen atoms around each silicon atom.

An insoluble glass with useful structural properties results if lime (CaO) is added to the sodium carbonate–silica starting materials. **Soda-lime glass** is the product, with the approximate composition $Na_2O \cdot CaO \cdot (SiO_2)_6$. Soda-lime glass is easy to melt and shape and is used in applications ranging from bottles to window glass. More than 90% of all the glass manufactured today is soda-lime glass. Figure 20-6 shows the structure of this ionic glass schematically. It is a three-dimensional network of the type discussed in Section 19-1 but with random coordination of the silicate tetrahedra. The network is a giant "polyanion," with $Na^+$ and $Ca^{2+}$ ions distributed in the void spaces to compensate for the negative charge on the network.

Replacing lime and some of the silica in a glass with other oxides ($Al_2O_3$, $B_2O_3$, $K_2O$, or PbO) modifies its properties noticeably. For example, the thermal conductivity of ordinary (soda-lime) glass is quite low, and its coefficient of thermal expansion is high. This means that local subjection of its surface to extreme heat or cold sets up internal stresses, and the glass may shatter. The coefficient of thermal expansion is appreciably lower in certain borosilicate glasses, in which boron occupies many of the silicon sites. Pyrex, the most familiar of these glasses, has a coefficient of linear expansion about one-third that of ordinary soda-lime glass and is the material of choice for laboratory glassware and household ovenware. A glass called Vycor, which has an even smaller coefficient of expansion (approximating that of fused silica), is made by chemical treatment of a borosilicate glass to leach out its sodium; this leaves a porous structure that is densified by raising the temperature and shrinking the glass to its final volume.

## Cements

Hydraulic cement was first developed by the ancient Romans, who found that a mixture of lime (CaO) and dry volcanic ash reacted slowly with water, even at low temperatures, to form a durable solid. They used this knowledge to build the Pantheon, a circular building whose concrete dome, spanning 143 ft without internal support, still stands more than 2000 years after its construction! Knowledge of cement making was lost for centuries after the fall of the Roman Empire. An English bricklayer, Joseph Aspdin, rediscovered it in 1824 and patented a process for calcining a mixture of limestone and clay. He called the product **portland cement** because, when mixed with water, it hardened to a material that resembled a kind of limestone found on the Isle of Portland. Portland cement is now manufactured in every major country on earth, and worldwide production exceeds that of all other materials. Portland cement opened up a new age in the methods of constructing highways and buildings. Now rock could be crushed, then molded in cement, rather than shaped with cutting tools.

Portland cement is a finely ground (powdered) mixture of compounds produced by the high-temperature reaction of lime, silica, alumina, and iron oxide. The lime (CaO) may come from limestone or chalk deposits, and the silica ($SiO_2$) and alumina ($Al_2O_3$) are often obtained in clays or slags. The blast furnaces of steel mills are a common source of slag, which is a byproduct of the smelting of iron ore (Section 19-3). The composition of slag varies, but it can be represented as a calcium aluminum silicate of approximate formula $CaO \cdot Al_2O_3 \cdot (SiO_2)_2$. When quenched in water, molten slag solidifies into ''blast furnace clinkers.'' This material is crushed and ground to a fine powder, blended with lime in the correct proportion, and burned again in a horizontal rotary kiln at temperatures up to 1500°C to produce ''cement clinker.'' A final stage of grinding and the addition of about 5% gypsum ($CaSO_4 \cdot 2H_2O$) to lengthen the setting time complete the manufacture. Table 20-2 gives the composition of a typical portland cement in the form of percentages of the separate oxides. These simple materials, which are the ''elements'' of cement making, combine in the cement in more complex compounds such as tricalcium silicate, $(CaO)_3 \cdot SiO_2$, and tricalcium aluminate, $(CaO)_3 \cdot Al_2O_3$.

Cement sets when the semiliquid slurry first formed by the addition of water to the powder becomes a solid of low strength. Subsequently, it gains strength in a

### Table 20-2  Composition of Portland Cement

| Oxide | Percentage by Mass |
| --- | --- |
| Lime (CaO) | 61–69 |
| Silica ($SiO_2$) | 18–24 |
| Alumina ($Al_2O_3$) | 4–8 |
| Iron(III) oxide ($Fe_2O_3$) | 1–8 |
| Minor oxides (MgO, $Na_2O$, $K_2O$, $SO_3$) | 2–4 |

slower hardening process. Setting and hardening involve a complex group of exothermic reactions in which several hydrated compounds form. Portland cement is a **hydraulic cement,** because it hardens not by loss of water but by chemical reactions that incorporate water into the final body. The main reaction during setting is the hydration of the tricalcium aluminate, which can be approximated by the equation

$$(CaO)_3 \cdot Al_2O_3(s) + 3\ (CaSO_4 \cdot 2H_2O)(s) + 26\ H_2O(\ell) \longrightarrow$$
$$(CaO)_3 \cdot Al_2O_3 \cdot (CaSO_4)_3 \cdot 32H_2O(s)$$

After 5 or 6 hours, the product forms as a microscopic forest of long crystalline needles that lock together to solidify the cement. Later, the calcium silicates react with water to harden the cement. An example is

$$6\ (CaO)_3 \cdot SiO_2(s) + 18\ H_2O(\ell) \longrightarrow (CaO)_5 \cdot (SiO_2)_6 \cdot 5H_2O(s) + 13\ Ca(OH)_2(s)$$

The hydrated calcium silicates develop as strong tendrils that coat and enclose unreacted grains of cement, one another, and other particles that may be present, binding them in a robust network. Most of the strength of cement comes from these entangled networks, which, in turn, ultimately depend for their strength on chains of O—Si—O—Si silicate bonds. Hardening is slower than setting; it may take as long as a year for a cement to attain its final strength.

Portland cement is rarely used alone. Generally, it is combined with sand, water, and lime to make **mortar,** which is applied with a trowel to bond bricks or stone to one another and assemble a structure. When portland cement is mixed with sand and aggregate (crushed stone or pebbles) in proportions of 1:3.75:5 by volume, the mixture is called **concrete.** Concrete has outstanding resistance to compressive forces and therefore is the material normally used for the foundations of buildings and the construction of dams, on which the compressive loads are enormous. The stiffness (resistance to bending) of concrete is high, but its fracture toughness (resistance to impact) is substantially lower, and its tensile strength is relatively poor. To compensate, the concrete used in structural elements that are subject to transverse or tensile stresses, such as beams, is usually reinforced with steel rods.

As excess water evaporates from cement during hardening, pores form that typically constitute 25 to 30% of the volume of the solid. This porosity weakens concrete, and recent research has shown that the fracture strength of concrete is related inversely to the size of the largest pores in the cement. A new material called "macro-defect-free" (MDF) cement has pores only a few micrometers across (rather than about a millimeter as in conventional cement). The smaller pores result from the addition of water-soluble polymers that make a dough-like "liquid" ce-

Architectural forms executed in concrete in a California shopping mall. (Coco McCoy/Rainbow)

ment. Liquid cement can be molded with the use of far less water. Unset MDF cement is mechanically kneaded and extruded into the desired shape. The final product possesses substantially increased bending resistance and fracture toughness. MDF cement can even be molded into springs and shaped on a conventional lathe. Reinforcement with organic fibers further increases its toughness. The advent of MDF cement is a good illustration of the way in which chemistry and engineering collaborate to furnish new materials in the emerging discipline of **materials science.**

## 20-3 NONSILICATE CERAMICS

Many useful ceramics exist that are *not* based on the Si—O bond and the $SiO_4$ tetrahedron. They have important uses in electronics, optics, and the chemical industry. Some are oxides, but others contain neither silicon nor oxygen.

### Oxide Ceramics

Oxide ceramics are materials that contain oxygen in combination with any of a number of metals. These materials are named by adding an *-ia* ending to the stem of the name of the metallic element. Thus, if the main chemical component of an oxide ceramic is $Be_2O_3$, it is a *beryllia* ceramic; if the main component is $Y_2O_3$, it is an *yttria* ceramic; and if it is MgO, it is a *magnesia* ceramic. As Table 20-3 shows, the melting points of these and other oxides are substantially higher than the melting points of the elements themselves. Such high temperatures are hard to achieve and maintain, and the molten oxides corrode most container materials. Oxide ceramic bodies are therefore not shaped by melting the appropriate oxide and pouring it into a mold. Instead, these ceramics are fabricated by sintering, like the silicate ceramics.

   **Alumina** ($Al_2O_3$) is the most important nonsilicate ceramic material. It melts at 2051°C and retains strength even at temperatures of 1500 to 1700°C. Alumina has a high electrical resistivity and withstands thermal shock and corrosion well. These properties make it a good material for spark plug insulators.

   High-density alumina is fabricated in such a way that open pores between the grains are nearly completely eliminated; the grains are small, having an average diameter as low as 1.5 $\mu$m. Unlike most other ceramics, high-density alumina has good mechanical strength against impact, which has led to its use in armor plating. The ceramic absorbs the energy of an impacting projectile by breaking, and so penetration does not occur. High-density alumina is also used in high-speed cutting tools for machining metals. Such a use is perhaps surprising, but the temperature

**Table 20-3  Melting Points of Some Metals and Their Oxides**

| Metal | Melting Point (°C) | Oxide | Melting Point (°C) |
|---|---|---|---|
| Be | 1287 | BeO | 2570 |
| Mg | 651 | MgO | 2800 |
| Al | 660 | $Al_2O_3$ | 2051 |
| Si | 1410 | $SiO_2$ | 1723 |
| Ca | 865 | CaO | 2572 |
| Y | 1852 | $Y_2O_3$ | 2690 |

**Figure 20-7**
The femoral head of this artificial hip is made of high-density alumina. (Nelson–Bohart/Tom Stack and Associates)

resistance of the ceramic allows much higher cutting speeds, and a ceramic cutting edge has no tendency to weld to the metallic workpiece as metallic tools do. These properties make alumina cutting tools superior to metallic tools as long as they do not break too easily. High-density alumina is also used in artificial joints (Fig. 20-7).

If $Al_2O_3$ doped with a small percentage of MgO is fired in a vacuum or a hydrogen atmosphere (instead of air) at a temperature of 1800 to 1900°C, even very small pores, which scatter light and make the material white, are removed. The resulting ceramic is translucent. This material is used to contain the sodium in high-intensity sodium discharge lamps. With envelopes of high-density alumina (Fig. 20-8), such lamps can be operated at 1500°C to give a whiter and more intense light. Old-style sodium-vapor lamps with glass envelopes were limited to a temperature of 600°C because the sodium vapor reacted with the glass at higher temperatures. At low temperature, the light from a sodium-vapor lamp has an undesirable yellow color.

**Magnesia** (MgO) is mainly used as a **refractory**—a ceramic material that withstands a temperature of more than 1500°C without melting (MgO melts at 2800°C). A major use of magnesia is as insulation in electrical heating devices, because it combines high thermal conductivity with excellent electrical resistance. Magnesia is prepared from magnesite ores, which consist of $MgCO_3$ and a variety of impurities. When purified magnesite is heated to 800–900°C, carbon dioxide is driven off to form MgO($s$) in fine grains:

$$MgCO_3(s) \longrightarrow MgO(s) + CO_2(g)$$

After cooling, fine-grained MgO reacts avidly with water to form magnesium hydroxide:

$$MgO(s) + H_2O(\ell) \longrightarrow Mg(OH)_2(s)$$

Heating fine-grained MgO($s$) to 1700°C causes the MgO grains to sinter. Fully sintered MgO consists of large crystals and does not react with water. The $\Delta G°$ of the reaction between MgO and $H_2O$ does not change when magnesia is heated. The altered microstructure (larger crystals versus small) makes the reaction with water exceedingly slow, however.

## Superconducting Ceramics

Each of the oxide ceramics discussed so far has consisted of a single chemical compound, except for minor additives. A natural idea for new ceramics is to make materials that contain two (or more) oxides in equal or nearly equal molar amounts. Thus, if $BaCO_3$ and $TiO_2$ are mixed and heated to a high temperature, they react to give the ceramic barium titanate:

$$BaCO_3(s) + TiO_2(s) \longrightarrow BaTiO_3(s) + CO_2(g)$$

Barium titanate, which has many novel properties, is a **mixed oxide ceramic** (writing the formula as $BaO \cdot TiO_2$ shows this). It has the same structure as the mineral **perovskite,** $CaTiO_3$ (see Fig. 20-9) except, of course, that Ba replaces Ca. Perovskites typically have two metal atoms for every three oxygen atoms, taking the general formula $ABO_3$ where A stands for a metal atom at the center of the unit cube and B for an atom of a different metal at the cube corners.

Recently, interest in perovskite ceramic compositions has grown explosively following the discovery that some of them become **superconducting** at relatively high temperatures. A superconducting material offers no resistance to the flow of an

**Figure 20-8**
The white inner envelope of high-density alumina in this high-intensity sodium vapor light is visible through the outer envelope of glass. (Leon Lewandowski)

electric current. The Dutch physicist Heike Kamerlingh-Onnes discovered super-conductivity in 1911 when he cooled mercury below its superconducting transition temperature, 4 K. Such very low temperatures are hard to achieve and maintain, but if practical superconductors could be made to work at higher temperatures (or even at room temperature!), then power transmission, electronics, transportation, medicine, and many other aspects of human life would be transformed. More than 60 years of research with metallic systems culminated in 1973 in the discovery of a niobium–tin alloy with a world-record superconducting transition temperature of 23.3 K. Progress toward higher-temperature superconductors then stalled until 1986, when K. Alex Müller and J. Georg Bednorz had the inspired notion to look for higher transition temperatures among perovskite ceramics and they found a Ba—La—Cu—O perovskite phase with transition temperature of 35 K. This result motivated other scientists, who soon discovered another rare-earth-containing perovskite ceramic that became a superconductor at 90 K. This result was particularly exciting because 90 K exceeds the boiling point of liquid nitrogen (77 K), a relatively cheap refrigerant (Fig. 20-10). This 1-2-3 compound (so called because its formula $YBa_2Cu_3O_{(9-x)}$ has one yttrium, two barium, and three copper atoms per formula unit) is not an ideal perovskite because it has fewer than nine oxygen atoms in combination with its six metal atoms. The deficiency makes $x$ in the formula somewhat greater than 2, depending on the exact method of preparation. Figure 20-11 illustrates the structure of this nonstoichiometric solid. More recently, the maximum superconducting transition temperature has risen above 135 K in another class of ceramics containing mercury. Still higher transition temperatures have been achieved by applying high pressures.

A crucial objective in the application of superconducting ceramics is to devise ways to fabricate the new materials in desired forms, such as wires. This is quite a challenge since these superconductors are ceramics and have the brittleness and fragility typical of ceramic materials. Superconducting wires and tapes as long as 100 m are now commercially available.

**Figure 20-9**

The structure of perovskite ($CaTiO_3$). A stack of three unit cells is shown, with some additional O atoms from neighboring cells. Each Ti (blue) is surrounded by six O's; each Ca (green) has eight O's as nearest neighbors.

## Nonoxide Ceramics

In nonoxide ceramics, nitrogen or carbon takes the place of oxygen in combination with silicon or boron. Examples include boron nitride (BN), boron carbide ($B_4C$), the silicon borides ($SiB_4$ and $SiB_6$), silicon nitride ($Si_3N_4$), and silicon carbide

**Figure 20-10**

The levitation of a small magnet above a disk of superconducting material. A superconducting substance cannot be penetrated by an external magnetic field. At room temperature, the magnet rests on the ceramic disk. When the disk is cooled with liquid nitrogen, it becomes superconducting and excludes the magnet's field, forcing the magnet into the air. Here the magnet floats above a disk of chilled 1-2-3 ceramic. (Yoav Levy/Phototake)

**Figure 20-11**

The structure of $YBa_2Cu_3O_{9-x}$ is a layered perovskite. The layers differ because one third of them contain yttrium (orange) and two thirds of them contain barium (green). Compare this structure with that of perovskite in Figure 20-9. Every third Ba in the hypothetical "$BaCuO_3$" is replaced by a Y, and the O's in the layer occupied by the Y are removed. Note the puckering in the Cu—O layers, as well. The elliptical atom "shapes" come from movement of the atoms, as revealed by x-ray diffraction.

| Table 20-4 | Bond Enthalpies in Nonoxide Ceramics | |
|---|---|
| **Bond** | **Bond Enthalpy (kJ mol⁻¹)** |
| B—N | 389 |
| B—C | 448 |
| Si—N | 439 |
| Si—C | 435 |
| B—Si | 289 |
| C—C | 350 |

(SiC). All of these compounds possess strong, short covalent bonds. They are hard and strong but brittle. Table 20-4 lists the enthalpies of the chemical bonds in these compounds.

Much research has aimed at making gas-turbine and other engines from ceramics. Heat engines run more efficiently at higher temperatures, and above 1000°C some ceramics have better properties (strength, corrosion resistance, density) for engines than do metals in all respects except their intrinsic brittleness. Of the oxide ceramics, only alumina and zirconia ($ZrO_2$) are strong enough for use in engines, but both have insufficient resistance to thermal shock. Attention has therefore turned to the nonoxide **silicon nitride** ($Si_3N_4$). In this network solid (Fig. 20-12), every silicon atom bonds to four nitrogen atoms that surround it at the corners of a tetrahedron; the tetrahedra link into a three-dimensional network by sharing corners. The Si—N bond is covalent and strong (the bond enthalpy is 439 kJ mol⁻¹). The similarity to the joining of $SiO_4$ units in silicate minerals (Section 19-1) is clear.

At first, silicon nitride seems chemically unpromising as a high-temperature structural material. It is unstable in contact with water because the reaction

$$Si_3N_4(s) + 6\, H_2O(\ell) \longrightarrow 3\, SiO_2(s) + 4\, NH_3(g)$$

has a negative $\Delta G°$. This reaction causes finely ground $Si_3N_4$ powder to give off an odor of ammonia in moist air at room temperature. Silicon nitride is also thermodynamically unstable in air, reacting spontaneously with oxygen:

$$Si_3N_4(s) + 3\, O_2(g) \longrightarrow 3\, SiO_3(s) + 2\, N_2(g)$$

**Figure 20-12**

The structure of silicon nitride ($Si_3N_4$). Each Si atom is bonded to four nitrogen atoms, and each N atom is bonded to three Si atoms. The result is a strong network.

The reaction has a $\Delta G°$ of $-1927$ kJ mol$^{-1}$. In practice, neither reaction occurs at a perceptible rate when $Si_3N_4$ is in bulk form. Initial contact of oxygen or water with $Si_3N_4(s)$ forms a surface film of $SiO_2(s)$ that protects the bulk of the $Si_3N_4(s)$ from further attack. When *strongly* heated in air—to about 1900°C—$Si_3N_4$ does decompose, violently, but below that temperature it resists attack.

Fully dense silicon nitride parts are stronger than metallic alloys at high temperatures. Ball bearings made of dense silicon nitride work well without lubrication at temperatures up to 700°C, for example, lasting longer than steel ball bearings. Because the strength of silicon nitride increases with the density attained in the production process, the trick is to form a dense piece of silicon nitride into the desired shape. Reaction bonding is one method for making useful shapes of silicon nitride. Powdered silicon is compacted in molds, removed, and then fired under an atmosphere of nitrogen at 1250 to 1450°C. The reaction

$$3\,Si(s) + 2\,N_2(g) \longrightarrow Si_3N_4(s)$$

forms the ceramic. The parts neither swell nor shrink significantly during the chemical conversion from Si to $Si_3N_4$, making it possible to fabricate complex shapes reliably. Unfortunately, reaction-bonded $Si_3N_4$ is still somewhat porous and is not strong enough for many applications. In the "hot-pressed" forming process, $Si_3N_4$ powder is prepared in the form of exceedingly small particles by reaction of silicon tetrachloride with ammonia:

$$3\,SiCl_4(g) + 4\,NH_3(g) \longrightarrow Si_3N_4(s) + 12\,HCl(g)$$

The solid $Si_3N_4$ forms as a smoke. It is captured as a powder, mixed with a carefully controlled amount of MgO additive, placed in an enclosed mold, and sintered at 1850°C under a pressure of 230 atm. The resulting ceramic shrinks to nearly full density (no pores). Because the material does not flow well (to fill a complex mold completely), only simple shapes are possible. Hot-pressed silicon nitride is impressively tough and can be machined only with great difficulty, with diamond tools.

In the silicate minerals discussed in Chapter 19, $AlO_4$ units routinely substitute for $SiO_4$ tetrahedra as long as positive ions of some type are present to balance the electric charge. This fact suggests that in silicon nitride some $Si^{4+}$ ions, which lie at the centers of tetrahedra of nitrogen atoms, could be replaced by $Al^{3+}$ if a compensating replacement of $O^{2-}$ for $N^{3-}$ were made simultaneously. Experiments show that ceramic alloying of this type works well, producing many new ceramics with great potential, called **sialons** (named for the four elements Si—Al—O—N). Sialons exemplify the way in which a structural theme in a naturally occurring material can guide the search for new materials.

Boron has one fewer valence electron than carbon, and nitrogen has one more valence electron. **Boron nitride** (BN) is therefore isoelectronic with carbon, and it is not surprising that it has two structural modifications that resemble the structures of graphite and diamond. In hexagonal boron nitride, the boron and nitrogen atoms take alternate places in an extended ''chicken-wire'' sheet in which the B—N distance is 1.45 Å. The sheets stack in such a way that each boron atom has a nitrogen atom directly above it and directly below it, and vice versa. The cubic form of boron nitride has the diamond structure, is comparable in hardness to diamond, and resists oxidation better. Cubic boron nitride is often prepared by **chemical vapor deposition,** a method used to fabricate several other ceramics as well. In this method, a controlled chemical reaction of gases on a contoured, heated surface gives a solid product of the desired shape. If a cup made of BN is needed, a cup-shaped mold is heated to a temperature exceeding 1000°C, and a mixture of $BCl_3(g)$ and $NH_3(g)$ is passed over its surface. The reaction

$$BCl_3(g) + NH_3(g) \longrightarrow BN(s) + 3\,HCl(g)$$

deposits a cup-shaped layer of BN(s). Boron nitride cups and tubes are used to contain and evaporate molten metals.

**Silicon carbide** (SiC) is diamond in which half of the C atoms are replaced by Si atoms. Also known by its trade name, Carborundum, silicon carbide was originally developed as an abrasive but is now used primarily as a refractory and as an additive in steel manufacture. It is formed and densified by methods similar to those used with silicon nitride. Silicon carbide is often produced in the form of small plates or whiskers to reinforce other ceramics. Fired silicon carbide whiskers are quite small (0.5 $\mu$m in diameter and 50 $\mu$m long) but very strong. They are mixed with a second ceramic material before that material is shaped. Firing then yields a **composite ceramic.** A composite is stronger and tougher than an unreinforced body of the same primary material. Whiskers serve to reinforce the main material by stopping cracks; they either deflect advancing cracks or soak up their energy, and a widening crack must dislodge them to proceed (Fig. 20-13).

### Figure 20-13

The microstructure of a composite ceramic. The light-colored areas are $TiB_2$, the gray areas are SiC, and the dark areas are voids. The $TiB_2$ acts to toughen the SiC matrix. Note that the crack shown passing through from left to right is forced to deflect around the $TiB_2$ particles. (Carborundum Company)

## 20-4 SILICON AND SEMICONDUCTORS

As a Group IV element, silicon resembles carbon chemically. Both silicon and carbon exhibit valences of 4 and the capacity to form extended molecular structures. As elements, both crystallize in the cubic system and possess the same structure. The bonding in solid silicon can be described by $sp^3$ hybrid orbitals directed to the four corners of a regular tetrahedron. Each atom is covalently linked to four others in a space-filling "giant molecule" network just like that of diamond (Fig. 17-22). There are just enough valence electrons to provide a single bond between a silicon atom and each of its four neighbors in the network.

An alternative to the localized orbital explanation of the bonding in silicon is to consider the electrons as occupying "molecular" orbitals that are spread out over the entire silicon crystal. Figure 17-19 employed an analogous procedure to construct a band of half-occupied levels in metallic sodium. In the case of silicon, the $4N_0$ valence atomic orbitals from 1 mol ($N_0$ atoms) of silicon split into *two* bands in the silicon crystal, each containing $2N_0$ very closely spaced levels (Fig. 20-14). The lower band is called the **valence band** and the upper one the **conduction band.** Between the top of the valence band and the bottom of the conduction band is an energy region that is off limits to electrons. The magnitude of the separation in energy between the valence band and the lowest level of the conduction band is called the **band gap,** $E_g$, which for pure silicon is $1.94 \times 10^{-19}$ J. This is the amount of energy an electron must gain to be excited from the top of the valence band to the bottom of the conduction band. For 1 mol of electrons to be excited in this way, the energy increases by a factor of Avogadro's number $N_0$, giving

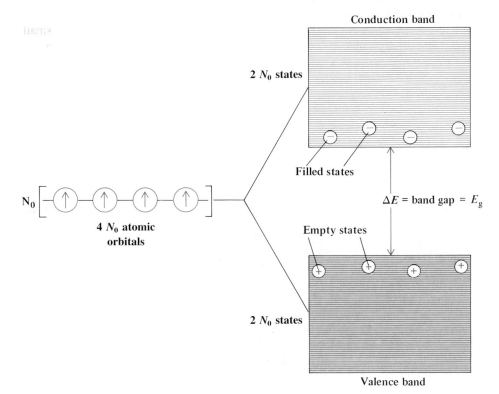

**Figure 20-14**

The valence orbitals of the silicon atom combine in crystalline silicon to give two bands of very closely spaced levels. The valence band is almost completely filled, and the conduction band is almost empty.

117 kJ mol$^{-1}$. (Another unit used for band gaps is the *electron volt*, defined in Section 12-2 as $1.60218 \times 10^{-19}$ J. In this unit, the band gap in Si is 1.21 eV.)

Each silicon atom in the crystal contributes four valence electrons to the bands of orbitals in Figure 20-14, for a total of $4N_0$ per mole. This is a sufficient number for two electrons to be placed in each level of the valence band (with opposing spins), with the conduction band left empty. There are no low-lying energy levels for the electrons at the top of the valence band to enter if their energy is raised by a small increment. The band gap means that an electron at the top of the filled valence band must acquire an energy of at least $1.94 \times 10^{-19}$ J, equivalent to 117 kJ mol$^{-1}$, to jump to the lowest empty level of the conduction band. This is a large amount of energy. If a thermal source were to supply it, the temperature of the source would have to be on the order of

$$ T = \frac{\Delta E}{R} = \frac{117,000 \text{ J mol}^{-1}}{8.315 \text{ J K}^{-1} \text{ mol}^{-1}} = 14,000 \text{ K} $$

At room temperature, only a few electrons per mole in the extreme tail of the Boltzmann distribution have enough energy to jump the gap, and so the conduction band in pure silicon is very sparsely populated with electrons. The result is that silicon is not a good conductor of electricity; good electrical conductivity requires a net motion of many electrons under the impetus of a small electric potential difference. Silicon has an electrical conductivity that is 11 orders of magnitude smaller than that of a typical metal (copper) at room temperature. Silicon is called a **semiconductor** because its electrical conductivity, although less than that of a metal, is far greater than that of an **insulator** such as diamond, which has a larger band gap. The conductivity of a semiconductor increases with temperature as more electrons are excited into levels in the conduction band. Another way to increase the conductivity of a semiconductor is to irradiate it with a beam of electromagnetic radiation that has a frequency high enough to excite electrons from the valence band to the conduction band. This process resembles the photoelectric effect described in Chapter 13, with the difference that now the electrons are not removed from the material but only moved into the conduction band, ready to conduct a current if a potential difference is imposed.

## Example 20-1

Calculate the longest wavelength of light that can excite electrons from the valence band to the conduction band in silicon. In what region of the spectrum does this wavelength fall?

### Solution

The energy carried by a photon is $h\nu = hc/\lambda$, where $h$ is Planck's constant, $\nu$ the photon frequency, $c$ the speed of light, and $\lambda$ the photon wavelength. For a photon to just excite an electron across the band gap, this energy must equal the energy of that band gap, $E_g = 1.94 \times 10^{-19}$ J.

$$ \frac{hc}{\lambda} = E_g $$

Solving for the wavelength gives

$$\lambda = \frac{hc}{E_g} = \frac{(6.626 \times 10^{-34} \text{ J s}) \times (2.998 \times 10^8 \text{ m s}^{-1})}{1.94 \times 10^{-19} \text{ J}}$$

$$= 1.02 \times 10^{-6} \text{ m} = 1020 \text{ nm}$$

This wavelength falls in the infrared region of the spectrum. Photons with shorter wavelengths (for example, visible light) carry more than enough energy to excite electrons to the conduction band in silicon.

*Related Problems:* 15, 16

## Doped Semiconductors

When certain other elements are added in a process called **doping,** silicon acquires interesting electronic properties. For example, if atoms of a Group V element such as arsenic or antimony are diffused into silicon, they substitute for silicon atoms in the network. Such an atom has five valence electrons, and so each one introduces one more electron into the silicon crystal than is needed for bonding. The extra electrons occupy energy levels just below the lowest level of the conduction band. Very little energy is required to promote electrons from such a **donor impurity** level into the conduction band, and the electrical conductivity of the silicon crystal is increased without a rise in temperature. Silicon doped with atoms of a Group V element is called an ***n*-type semiconductor** to indicate that the charge carrier is negative.

On the other hand, if a Group III element such as gallium is used as a dopant, the valence band contains one fewer electron per dopant atom because Group III elements have only three valence electrons, not four. This situation corresponds to creating one **hole** in the valence band, with an effective charge of +1, for each Group III atom added. If a voltage difference is impressed across a crystal that is doped in this way, it causes the positively charged holes to move toward the negative source of potential. Equivalently, a valence-band electron next to the (positive) hole moves in the *opposite* direction (that is, toward the positive source of potential). Whether we think of holes or electrons as the mobile charge carrier in the valence band, the result is the same; we are simply using different words to describe one physical phenomenon (Fig. 20-15). Silicon that has been doped with a Group III element is called a ***p*-type semiconductor** to indicate that the carrier has an effective positive charge.

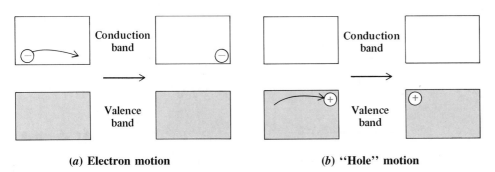

(*a*) **Electron motion**          (*b*) **"Hole" motion**

**Figure 20-15**

(*a*) In an *n*-type semiconductor, a small number of electrons occupy levels in the conduction band. When an electric field is imposed, each electron moves (black arrows) into one of the numerous nearby vacant energy levels in the conduction band. (*b*) In a *p*-type semiconductor, a small number of levels in the valence band are unoccupied. Conduction occurs as electrons in the numerous occupied levels of the valence band jump into the sparsely distributed unoccupied levels. The arrow shows the motion of an electron to occupy a previously empty level. The process can also be described as the motion of positively charged holes in the opposite direction.

A different class of semiconductors is based not on silicon but on equimolar compounds of Group III and Group V elements. Gallium arsenide, for example, is isoelectronic to the Group IV semiconductor germanium. Doping GaAs with the Group VI element tellurium produces an *n*-type semiconductor; doping with zinc, which has one *fewer* valence electron than gallium, gives a *p*-type semiconductor. Other III–V combinations have different band gaps and are useful in particular applications. Indium antimonide (InSb), for example, has a small enough band gap that absorption of infrared radiation causes electrons to be excited from the valence to the conduction band, and an electric current then flows when a small potential difference is applied. This compound is therefore used as a detector of infrared radiation. Still other compounds of the zinc group (zinc, cadmium, and mercury) and Group VI elements such as sulfur have the same average number of valence electrons per atom as silicon and make useful semiconductors.

Semiconductors can fulfill many electronic functions, the simplest being **rectification,** the conversion of alternating current to direct current. Suppose thin crystals of *n*-type and *p*-type silicon are placed in contact with each other and connected to a battery. In Figure 20-16*a* the positive pole of the battery is connected to the *n*-type silicon, and the negative pole to the *p*-type silicon. Only a very small transient current can flow through the circuit in this case, because when the electrons in the

(*a*) **Current blocked**

(*b*) **Current flows**

(*c*) **Half-wave rectification of alternating current**

**Figure 20-16**

Current rectification by an *n*–*p* transistor.

conduction band of the $n$-type silicon have flowed out to the positive pole of the battery, there are none to take their place and current ceases to flow. However, if the negative pole of the battery is connected to the $n$-type silicon and the positive pole to the $p$-type silicon (Fig. 20-16$b$), a steady current flows because electrons and holes move in opposite directions and recombine at the $n$–$p$ junction. In effect, electrons flow toward the $n$–$p$ junction in the $n$-type material, holes flow toward the $n$–$p$ junction in the $p$-type material, and the junction is a "sink" where electrons fill the holes in the valence band and neutralize one another. If, instead of a galvanic cell, an alternating current source were connected to the $n$–$p$ rectifier, current would flow in one direction only, creating pulsed direct current (Fig. 20-16$c$).

Semiconductors perform a wide range of electronic functions that formerly required the use of vacuum tubes. Vacuum tubes occupy much more space, generate great amounts of heat, and require considerably more energy to operate than **transistors,** their semiconductor counterparts. Semiconductors can also be built into integrated circuits (Fig. 20-17) that store and process information at great speeds. Solar cells based on silicon or gallium arsenide convert the radiant energy of the sun directly to electrical work by a technology that is virtually nonpolluting (Fig. 20-18). Currently, the high capital costs of solar cells make them uncompetitive with conventional fossil-fuel sources of energy, but as reserves of fossil fuels dwindle, solar energy will become an important option.

**Figure 20-17**

An integrated circuit consists of a series of connected electrical circuits printed onto a tiny silicon wafer. The actual size of this portion of circuit is approximately 6 mm on a side. (W. Marin/Phototake)

## Pigments and Phosphors

The band gap of an insulator or semiconductor has a significant effect on its color. Pure diamond has a large band gap, and so even blue light does not have enough energy to excite electrons from the valence band to the conduction band. As a result, light passes through diamond without being absorbed, and diamond is colorless.

**Figure 20-18**

This California photovoltaic power plant uses silicon solar collectors on a large scale to harvest energy from the sun. (Comstock)

**Figure 20-19**
Mixed crystals of two semiconductors with different band gaps, CdS (yellow) and CdSe (black), show a range of colors. The band-gap energy decreases as the composition of the mixture becomes richer in Se. (Kurt Nassau/Phototake)

Cadmium sulfide (CdS) (Fig. 20-19) has a band gap of $4.2 \times 10^{-19}$ J, which corresponds to a wavelength of 470 nm, in the visible region of the spectrum. It therefore absorbs violet and blue light but strongly transmits yellow and has a deep yellow color. Cadmium sulfide is, in fact, the pigment called cadmium yellow. Cinnabar (HgS) (Fig. 20-20) has a smaller band gap, $3.2 \times 10^{-19}$ J, and absorbs all light except red. It has a deep red color and is the pigment vermillion. Semiconductors with band gaps of less than $2.8 \times 10^{-19}$ J absorb all wavelengths of visible light and appear black. They include silicon (Example 20-1), germanium, and gallium arsenide.

Doping silicon brings donor levels close enough to the conduction band, or acceptor levels close enough to the valence band, that thermal excitation causes electrons to move into a conducting state. The corresponding doping of *insulators,* or wide-band-gap semiconductors, brings donor or acceptor states into positions where *visible light* can be absorbed or emitted. This changes the colors and optical properties of the materials. Nitrogen doped into diamond produces a donor impurity level in the band gap. Transitions to this level can absorb some blue light, giving the diamond an undesirable yellowish color. On the other hand, boron doped into diamond gives an acceptor level that absorbs red light most strongly and gives the highly prized and rare "blue diamond."

**Phosphors** are wide-band-gap materials with dopants selected to create new levels such that particular colors of light are emitted. Electrons in these materials are excited by light of other wavelengths or by electrons hitting their surfaces, and light is then emitted as they return to lower-energy states. A fluorescent lamp, for example, is a mercury-vapor lamp in which the inside of the tube has been coated with phosphors. The phosphors absorb the violet and ultraviolet light emitted by mercury vapor and emit at lower energies and longer wavelengths, giving a nearly white light that is more desirable than the bluish light that comes from a mercury-vapor lamp without the phosphors.

Phosphors are also used in television screens. The picture is formed by scanning a beam of electrons (from an electron gun) over the screen. The electrons strike the phosphors coating the screen, exciting their electrons and causing them to emit light. In a black-and-white television tube, the phosphors are a mixture of silver doped

**Figure 20-20**
Crystalline cinnabar (HgS). (Julius Weber)

into ZnS, which gives blue light, and silver doped into $Zn_xCd_{1-x}S$, which gives yellow light. The combination of the two provides a reasonable approximation of white. A color television uses three different electron guns with three corresponding types of phosphor on the screen. Silver doped into ZnS gives blue; manganese doped into $Zn_2SiO_4$ is used for green; and europium doped into $YVO_4$ gives red light. Masks are used to ensure that each electron beam encounters only the phosphors that correspond to the desired color.

**CONCEPTS & SKILLS**

After studying this chapter and working the problems that follow, you should be able to

1. Describe the structure of ceramic materials and the ways in which they are formed (Section 20-1).
2. Outline the properties of pottery, glass, and cement and the chemical reactions that give them structural strength (Section 20-2, problems 1–8).
3. List several important oxide and mixed oxide ceramics, and give some of their uses (Section 20-3, problems 9–10).
4. Discuss the special properties of nonoxide ceramics and the kinetic and thermodynamic factors that make them useful (Section 20-3, problems 11–14).
5. Describe the mechanism of action of intrinsic, *n*-type, and *p*-type semiconductors (Section 20-4, problems 17–20).
6. Relate the band gap of a semiconductor or phosphor to the frequencies of electromagnetic radiation absorbed or emitted when electrons make transitions between the valence and conduction bands (Section 20-4, problems 21–24).

**PROBLEMS**

### Silicate Ceramics

1. A ceramic much used by artisans for small carved figurines is based on the mineral steatite (commonly known as soapstone). Steatite is a hydrated magnesium silicate that has the composition $Mg_3Si_4O_{10}(OH)_2$. It is remarkably soft—a fingernail can scratch it. When heated in a furnace to about 1000°C, chemical reaction transforms it into a hard, two-phase composite of magnesium silicate ($MgSiO_3$) and quartz in much the same way that clay minerals are converted to mullite ($Al_6Si_2O_{13}$) and cristobalite ($SiO_2$) on firing. Write a balanced chemical equation for this reaction.

2. A clay mineral that is frequently used together with or in place of kaolinite is pyrophyllite ($Al_2Si_4O_{10}(OH)_2$). Write a balanced chemical equation for the production of mullite and cristobalite on the firing of pyrophyllite.

3. Calculate the volume of carbon dioxide produced at 0°C and 1 atm pressure when a 2.50-kg sheet of ordinary glass is made from its starting materials—sodium carbonate, calcium carbonate, and silica. Take the composition of the glass to be $Na_2O \cdot CaO \cdot (SiO_2)_6$.

4. Calculate the volume of steam produced when a 4.0-kg brick made from pure kaolinite is completely dehydrated at 600°C and a pressure of 1.00 atm.

5. A sample of soda-lime glass contains the following percentages by mass of oxides: $SiO_2$, 72.4%; $Na_2O$, 18.1%; CaO, 8.1%; $Al_2O_3$, 1.0%; MgO, 0.2%; BaO, 0.2%. (The elements are not actually present as binary oxides, but this is the way compositions are usually given.) Calculate the chemical amounts of silicon, sodium, calcium, aluminum, magnesium, and barium atoms per mole of oxygen atoms in this sample.

6. A sample of portland cement contains the following percentages by mass of oxides: CaO, 64.3%; $SiO_2$, 21.2%; $Al_2O_3$, 5.9%; $Fe_2O_3$, 2.9%; MgO, 2.5%; $SO_3$, 1.8%; $Na_2O$, 1.4%. Calculate the chemical amounts of calcium, silicon, aluminum, iron, magnesium, sulfur, and sodium atoms per mole of oxygen atoms in this sample.

7. The most important contributor to the strength of hardened portland cement is tricalcium silicate, $(CaO)_3 \cdot SiO_2$, for which the measured standard enthalpy of formation is $-2929.2 \text{ kJ mol}^{-1}$. Calculate the standard enthalpy change

for the production of 1.00 mol of tricalcium silicate from quartz and lime.

**8.** One of the simplest heat-generating reactions that takes place when water is added to cement is the production of calcium hydroxide (slaked lime) from lime. Write a balanced chemical equation for this reaction, and use data from Appendix D to calculate the amount of heat generated by the reaction of 1.00 kg of lime with water under room conditions.

## Nonsilicate Ceramics

**9.** Calculate the average oxidation number of the copper in $YBa_2Cu_3O_{9-x}$ if $x = 2$. Assume that the rare-earth element yttrium is in its usual $+3$ oxidation state.

**10.** The mixed oxide ceramic $Tl_2Ca_2Ba_2Cu_3O_{10+x}$ has zero electrical resistance at 125 K. Calculate the average oxidation number of the copper in this compound if $x = 0.50$ and thallium is in the $+3$ oxidation state.

**11.** Silicon carbide (SiC) is made by the high-temperature reaction of silica sand (quartz) with coke; the byproduct is carbon monoxide.
(a) Write a balanced chemical equation for this reaction.
(b) Calculate the standard enthalpy change per mole of SiC produced.
(c) Predict (qualitatively) the conductivity, melting point, and hardness of silicon carbide.

**12.** Boron nitride (BN) is made by the reaction of boron trichloride with ammonia.
(a) Write a balanced chemical equation for this reaction.
(b) Calculate the standard enthalpy change per mole of BN produced, given that the standard molar enthalpy of formation of BN(s) is $-254.4$ kJ mol$^{-1}$.
(c) Predict (qualitatively) the conductivity, melting point, and hardness of boron nitride.

**13.** The standard free energy of formation of cubic silicon carbide (SiC) is $-62.8$ kJ mol$^{-1}$. Determine the standard free energy change when 1.00 mol of SiC reacts with oxygen to form $SiO_2$ (s, quartz) and $CO_2(g)$. Is silicon carbide thermodynamically stable in air under room conditions?

**14.** The standard free energy of formation of boron carbide ($B_4C$) is $-71$ kJ mol$^{-1}$. Determine the standard free energy change when 1.00 mol of $B_4C$ reacts with oxygen to form $B_2O_3(s)$ and $CO_2(g)$. Is boron carbide thermodynamically stable in air under room conditions?

## Silicon and Semiconductors

**15.** Electrons in a semiconductor are excited from the valence band to the conduction band by absorbing photons with energies exceeding the band gap. At room temperature, indium phosphide (InP) is a semiconductor that absorbs light only at wavelengths less than 920 nm. Calculate the band gap in InP.

**16.** Both GaAs and CdS are semiconductors that are under study for possible use in solar cells to generate electric current from sunlight. Their band gaps are $2.29 \times 10^{-19}$ J and $3.88 \times 10^{-19}$ J, respectively, at room temperature. Calculate the longest wavelength of light that is capable of exciting electrons across the band gap in each of these substances. In which region of the electromagnetic spectrum do these wavelengths fall? Use this result to explain why CdS-based sensors are employed in some cameras to estimate the proper exposure conditions.

**17.** The number of electrons excited to the conduction band per cubic centimeter in a semiconductor can be estimated from the equation

$$n_e = (4.8 \times 10^{15} \text{ cm}^{-3} \text{ K}^{-3/2}) \, T^{3/2} \, e^{-E_g/(2\,RT)}$$

where $T$ is the temperature in kelvins and $E_g$ the band gap in joules *per mole*. The band gap of diamond at 300 K is $8.7 \times 10^{-19}$ J. How many electrons are thermally excited to the conduction band at this temperature in a 1.00-cm$^3$ diamond crystal?

**18.** The band gap of pure crystalline germanium is $1.1 \times 10^{-19}$ J at 300 K. How many electrons are excited from the valence band to the conduction band in a 1.00-cm$^3$ crystal of germanium at 300 K? Use the equation in the preceding problem.

**19.** Describe the nature of electrical conduction in (a) Si doped with P and (b) InSb doped with Zn.

**20.** Describe the nature of electrical conduction in (a) Ge doped with In and (b) CdS doped with As.

**21.** In a light-emitting diode (LED), which is used in displays on electronic equipment, watches, and clocks, a voltage is imposed across an $n$–$p$ semiconductor junction. The electrons on the $n$ side combine with the holes on the $p$ side and emit light at the frequency of the band gap. This process can also be described as the emission of light as electrons fall from levels in the conduction band to empty levels in the valence band. It is the reverse of the production of electric current by illumination of a semiconductor.

Many LEDs are made from semiconductors having the general composition $GaAs_{1-x}P_x$. When $x$ is varied between 0 and 1, the band gap changes and, with it, the color of light emitted by the diode. When $x = 0.4$, the band gap is $2.9 \times 10^{-19}$ J. Determine the wavelength and color of the light emitted by this LED.

**22.** When the LED described in the preceding problem has the composition $GaAs_{0.14}P_{0.86}$ (that is, $x = 0.86$), the band gap has increased to $3.4 \times 10^{-19}$ J. Determine the wavelength and color of the light emitted by this LED.

**23.** The pigment zinc white (ZnO) turns bright yellow when heated, but the white color returns when the sample is cooled. Does the band gap increase or decrease when the sample is heated?

**24.** Mercury(II) sulfide (HgS) exists in two different crystalline forms. In cinnabar, the band gap is $3.2 \times 10^{-19}$ J; in metacinnabar, it is $2.6 \times 10^{-19}$ J. In some old paintings with improperly formulated paints, the pigment vermillion (cinnabar) has transformed into metacinnabar after prolonged exposure to light. Describe the color change that results.

## Additional Problems

25. Iron oxides are red when the average oxidation state of iron is high and black when it is low. To impart each of these colors to a pot made from clay that contains iron oxides, would you employ an air-rich or a smoky atmosphere in the kiln? Explain.

26. In what ways does soda-lime glass resemble, and in what ways does it differ from, a pot made from the firing of kaolinite? Include composition, structure, physical properties, and method of preparation in your discussion.

27. Leaded glass is used for fine crystal in place of soda-lime-silica glass. Table 20-1 shows that it is "heavier" (denser) and has a higher refractive index. The latter property gives the leaded glass its attractive sparkle. What is the ratio of the number of lead atoms to silicon atoms for leaded glass of the composition given in Table 20-1?

28. Refractories can be classified as acidic or basic, depending on the properties of the oxides in question. A basic refractory must not be used in contact with acid, and an acidic refractory must not be used in contact with a base. Classify magnesia and silica as acidic or basic refractories.

29. Dolomite bricks are used in the linings of furnaces in the cement and steel industries. Pure dolomite contains 45.7%

$MgCO_3$ and 54.3% $CaCO_3$ by mass. Determine the empirical formula of dolomite.

30. A book states correctly that beryllia (BeO) ceramics have some use but possess only poor resistance to strong acids and bases. Write likely chemical equations for the reaction of BeO with a strong acid and with a strong base.

31. Silicon nitride resists all acids except hydrofluoric, with which it reacts to give silicon tetrafluoride and ammonia. Write a balanced chemical equation for this reaction.

32. Compare alumina and magnesia oxide ceramics that have significant ionic character, with covalently bonded nonoxide ceramics such as SiC and $B_4C$ (problems 13 and 14) with respect to thermodynamic stability at ordinary conditions.

33. Compare the hybridization of silicon atoms in Si(s) with that of carbon atoms in graphite (see Fig. 17-25). If silicon were to adopt the graphite structure, would its electrical conductivity be high or low?

34. Describe how the band gap varies from a metal to a semiconductor to an insulator.

35. Suppose some people are sitting in a row at a movie theater, with a single empty seat on the left end of the row. Every five minutes, a person moves into a seat on his or her left if it is empty. In what direction and with what speed does the empty seat "move" along the row? Comment on the connection of this scenario with hole motion in $p$-type semiconductors.

36. A sample of silicon doped with antimony is an $n$-type semiconductor. Suppose a small amount of gallium is added to such a semiconductor. Describe how the conduction properties of the solid vary with the amount of gallium added.

# CHAPTER

# 21

*Aerial view of sulfur piled in the harbor in Vancouver, British Columbia, Canada. (© 1990 Stuart McCall)*

# Sulfur, Phosphorus, and Nitrogen

S ulfur, phosphorus, and nitrogen are three elements of intermediate electronegativity that are of fundamental importance in industrial and biological processes. The "workhorse" acids of chemical industry—sulfuric, phosphoric, and nitric, together with hydrochloric acid—are produced in quantities of tens of billions of kilograms each year and are used directly or indirectly in virtually every chemical manufacturing process. These three elements also play a central role in the production of chemical fertilizers.

## 21-1 SULFURIC ACID AND ITS USES

In the 20th century, the consumption of sulfuric acid has become a barometer of the health not only of a country's chemical industry but of its entire manufacturing base. This is because sulfuric acid, the chemical produced in greatest amount in the world,

744

is used in so many ways, ranging from metal treatment to the production of pharma-ceuticals. How did this strongly acidic, extremely reactive chemical come to have such an important place in chemistry and in modern industry?

The role of sulfuric acid was established between 1750 and 1900. In 1750 sulfuric acid was produced and used on a very modest scale in metal assaying and treatment and by some doctors for cures that had no scientific basis. New uses for the acid stimulated the search for new ways to produce it; as the price went down, further uses were discovered and exploited. From 1750 to 1900, the price of sulfuric acid declined steadily, and the amount produced grew explosively. This mutual stimulation between new uses and new processes is the hallmark of the chemical industry, and the story of sulfuric acid is a good way to illustrate it.

Sulfuric acid was probably first generated by alchemists who heated crystalline green vitriol, or iron(II) sulfate heptahydrate, in a retort:

$$FeSO_4 \cdot 7H_2O(s) \longrightarrow H_2SO_4(\ell) + FeO(s) + 6\,H_2O(g)$$

By the 17th century it was produced on a limited commercial scale from iron pyrite ores. The $FeS_2$ in the ores was converted to $FeSO_4$, then oxidized to solid iron(III) sulfate, $Fe_2(SO_4)_3$. This solid was broken up and strongly heated in small clay vessels, a treatment that decomposed it into iron(III) oxide and sulfur trioxide:

$$Fe_2(SO_4)_3(s) \longrightarrow Fe_2O_3(s) + 3\,SO_3(g)$$

The gas produced was then absorbed in water to make sulfuric acid:

$$SO_3(g) + H_2O(\ell) \longrightarrow H_2SO_4(\ell)$$

Production of sulfuric acid was stimulated by the discovery in 1744 that the valued dyestuff indigo, used for many years to dye cotton, could also be used as a wool dye after treatment with concentrated sulfuric acid.

## The Lead-Chamber Process

The lead-chamber process, which was developed in the second half of the 18th century, probably also had its origins in the laboratories of alchemists, who burned sulfur in earthenware vessels. The small amounts of $SO_3$ that were produced (along-side the $SO_2$ that was the main product) were condensed and absorbed into water to make sulfuric acid. An accidental discovery revealed that the addition of sodium nitrate or potassium nitrate improves the yield of $SO_3$. These salts decompose to give nitrogen dioxide, which reacts with $SO_2$ to give $SO_3$:

$$SO_2(g) + NO_2(g) \longrightarrow SO_3(g) + NO(g)$$

In 1736 Joshua Ward took the next important step by replacing the earthenware vessels in which the sulfur was burned with large glass bottles arranged in series, to speed up the process.

The development of the room-size **lead chamber,** first employed by John Roe-buck in 1746, dramatically expanded the manufacture of sulfuric acid. The output of one of the older earthenware vessels was several ounces, and Ward's glass bottles had outputs on the order of several pounds. By contrast, the lead chamber allowed sulfuric acid to be produced in amounts of hundreds of pounds to tons, reducing the price of the product through economy of scale and lower labor costs. In the lead-chamber process, a mixture of sulfur and potassium nitrate was placed in a ladle and

**Figure 21-1**

The Glover–Gay-Lussac process for making sulfuric acid. The portion of the diagram to the left of the yellow line shows the original lead-chamber process, which produced a weaker acid and consumed and emitted more nitrogen oxides than the full process as it developed later.

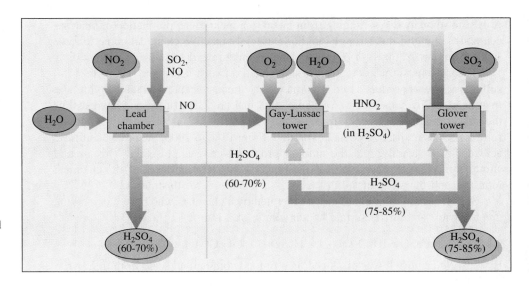

ignited inside a large chamber lined with lead, the floor of which was covered with water. The gases condensed on the walls and were absorbed in the water. After the process was repeated several times, the dilute sulfuric acid was removed and boiled down to concentrate it further. Later developments included blowing in steam to speed up the reaction with water and disperse the gases, and separating the burning chamber from the absorption chamber.

Joseph Gay-Lussac took a significant step forward in 1835 when he built a tower to recover the NO that had previously been vented and to convert it back to $NO_2$ by reaction with oxygen. More precisely, in a Gay-Lussac tower, NO was converted to nitrous acid ($HNO_2$) dissolved in aqueous sulfuric acid:

$$2\,NO(g) + \tfrac{1}{2}\,O_2(g) + H_2O(\ell) \longrightarrow 2\,HNO_2(aq)$$

The nitrous acid then reacted in a second tower—named after its developer, John Glover—to oxidize sulfur dioxide:

$$2\,HNO_2(aq) + SO_2(g) \longrightarrow H_2SO_4(aq) + 2\,NO(g)$$

The overall reaction from these steps is simply

$$SO_2(g) + \tfrac{1}{2}\,O_2(g) \longrightarrow SO_3(g)$$

Recycling the oxides of nitrogen greatly reduced the consumption of sodium nitrate or potassium nitrate, which was now needed only to make up for losses in the process. In addition, the Glover tower produced more concentrated sulfuric acid—75 to 85% $H_2SO_4$ by mass compared with 60 to 70% obtained by earlier methods. Figure 21-1 is a flow diagram for the full process, after the development of the Glover tower in 1859.

### The Contact Process

As early as 1831, the Englishman Peregrine Phillips observed that platinum could catalyze the conversion of $SO_2$ to $SO_3$, which lies at the heart of the production of sulfuric acid. Little attention was paid to this discovery until the 1870s, when the growth of the German dye industry stimulated a search for a method of producing sulfuric acid more concentrated than that from the Glover tower. Platinum catalysts were rediscovered and patented but initially had limited usefulness because they were poisoned and rendered ineffective by impurities in the sulfur dioxide feed. As a result, the early applications of this method used, somewhat paradoxically, sulfuric acid from the Glover tower as a raw material. The sulfuric acid was decomposed by heating,

$$H_2SO_4(aq) \longrightarrow SO_2(g) + H_2O(\ell) + \tfrac{1}{2}O_2(g)$$

and the relatively pure $SO_2$ was converted to $SO_3$ over the catalyst and then back to $H_2SO_4$. In this way, a highly concentrated acid was obtained but at great expense. Over the next 30 years, researchers recognized the role of arsenic and other impurities in poisoning the catalyst, and took steps to remove the impurities from the $SO_2$ feed, allowing direct production of concentrated sulfuric acid without the intermediate lead-chamber step. Different catalysts were studied as well, leading eventually to the selection of an oxide of vanadium ($V_2O_5$) that is the primary catalyst in use today.

The catalytic production of sulfuric acid from $SO_2$ is called the **contact process** and is outlined in Figure 21-2. In the reaction of $SO_2$ with oxygen,

$$SO_2(g) + \tfrac{1}{2}O_2(g) \longrightarrow SO_3(g)$$

the number of moles of gas decreases. This suggests running the reaction under high total pressures to increase the yield of the product. The small advantage gained by using high pressures in this reaction is more than offset by the greater cost of the necessary equipment, however, and so the reaction is carried out at atmospheric pressure. A thermodynamic analysis gives

$$\Delta H° = -98.9\ \text{kJ} \qquad \Delta S° = -94.0\ \text{J K}^{-1}$$

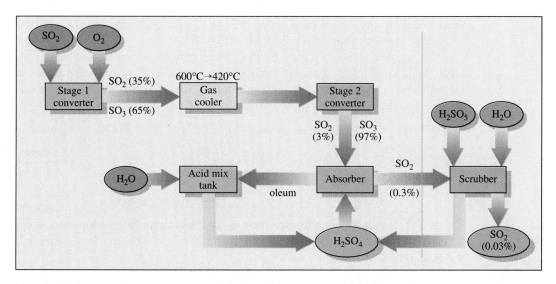

**Figure 21-2**
Flow chart of the contact process for making sulfuric acid. The portion to the right of the yellow line represents a process used in some modern plants to reduce air pollution from residual $SO_2$.

Because the reaction is exothermic, the lower the temperature, the higher the equilibrium constant and the greater the degree of conversion to products at equilibrium. The change in free energy is 0 (and the equilibrium constant 1) when

$$\Delta G = \Delta H - T\,\Delta S \approx \Delta H° - T\,\Delta S° = 0$$

or

$$T = \frac{\Delta H°}{\Delta S°} = \frac{-98{,}900\ \text{J}}{-94.0\ \text{J K}^{-1}} = 1050\ \text{K}$$

or approximately 780°C. The temperature must be maintained substantially below this point to achieve significant yield of product.

The problem is that at lower temperature the reaction becomes slower, although the catalyst does help to speed it up. A two- to four-stage process is typically used. The entering $SO_2(g)$ reaches the first catalyst at a temperature of 420°C. As the reaction begins, heat is evolved and the temperature of the reacting gas mixture rises. After a few seconds, the mixture has reached equilibrium at about 600°C with conversion of 60 to 70% of the $SO_2$. The gases are then cooled back to 420°C and allowed to react one or two more times over the catalyst, with lower temperatures and longer exposure periods used. The result is conversion of about 97% of the $SO_2$ to $SO_3$. For even greater conversion, the gases are then passed into a tower where $SO_3$ dissolves in sulfuric acid. This removes the reaction product, so that the reaction again shifts to the right when the unreacted $SO_2$ is passed over the catalyst for a last time. The $SO_3$ from this last stage is then absorbed, giving an overall yield of about 99.7% of the $SO_2$ introduced originally. Careful consideration of thermodynamics and kinetics has led to a highly efficient process. Almost all sulfuric acid manufactured today is made by the contact process.

If $SO_3$ were absorbed in water rather than in sulfuric acid, the product would be more dilute and less $SO_3$ would be absorbed. In addition, the direct reaction of $SO_3$ with water produces a fine acidic mist that is difficult to condense. Absorption of $SO_3$ into sulfuric acid gives fuming sulfuric acid, or **oleum,** that can be used directly or diluted with water to the desired strength. An equimolar amount of $SO_3$ dissolved in $H_2SO_4$ gives disulfuric acid ($H_2S_2O_7$) (Fig. 21-3).

The amount of $SO_2$ released by the contact process to pollute the air is very small, but further removal of objectionable $SO_2$ from the tail gases can be achieved (at additional expense) in another step. Some $H_2SO_4$, which is available in abundance, of course, is electrolytically oxidized to peroxydisulfuric acid ($H_2S_2O_8$):

$$2\ H_2SO_4(aq) + 2\ H_2O(\ell) \longrightarrow H_2S_2O_8(aq) + 2\ H_3O^+(aq) + 2\ e^- \qquad \text{(anode)}$$

$$2\ H_3O^+(aq) + 2\ e^- \longrightarrow H_2(g) + 2\ H_2O(\ell) \qquad \text{(cathode)}$$

The $H_2S_2O_8$ quickly reacts:

$$H_2O(\ell) + H_2S_2O_8(aq) \longrightarrow H_2SO_4(aq) + H_2SO_5(aq)$$

The $H_2SO_5$ is called peroxymonosulfuric acid. The exiting gases are passed into a scrubber to mingle with a solution of this powerful oxidizing agent. In the scrubber, the reaction

$$SO_2(g) + H_2SO_5(aq) + H_2O(\ell) \longrightarrow 2\ H_2SO_4(aq)$$

converts $SO_2(g)$ to sulfuric acid. More than 90% of the already small amount of residual $SO_2(g)$ is removed in this way. The product, dilute sulfuric acid, is cycled back into the main process.

**Figure 21-3**
Disulfuric acid, $H_2S_2O_7$.

## Sources for Sulfur and Sulfur Dioxide

The primary raw material for making sulfuric acid is sulfur (Fig. 21-4) or sulfur dioxide. Sources for these chemicals have changed over the years, driven by considerations of price and the desirability of reducing atmospheric pollution. Centuries ago, the sulfur mines of Sicily were the major source of elemental sulfur. Increases in price, especially after a cartel was established to exploit the mines, led to the search for other sources. As the use of metals increased in the 19th century, sulfur was obtained in the form of $SO_2$ as a byproduct of the roasting of sulfide ores of zinc, iron, or copper, through reactions such as

$$ZnS(s) + \tfrac{3}{2} O_2(g) \longrightarrow ZnO(s) + SO_2(g)$$

Capture of sulfur dioxide and its conversion to sulfuric acid created a cheap starting material that reduced air pollution from $SO_2$.

In the late 1890s, a new development shifted interest from sulfide ores back to elemental sulfur as a starting material, at least in the United States. This was the discovery by Herman Frasch of a new method to extract sulfur from underground deposits. The salt domes of Louisiana, Texas, and Mexico contained substantial reserves of sulfur in porous limestones, but extraction of the sulfur had proved difficult. The **Frasch process** (Fig. 21-5) is an ingenious mining method in which sulfur is liquefied by steam injected into the deposit and pumped to the surface. The sulfur is usually shipped as a liquid in heated vessels to sulfuric acid plants, where it is used to make sulfuric acid.

Since the 1970s, the Frasch process has declined in importance as a source of sulfur, owing to a competing process that recovers sulfur impurities in natural gas

**Figure 21-4**

A sulfur crystal. (Leon Lewandowski)

**Figure 21-5**

The Frasch process uses three concentric pipes driven into the ground. Superheated water (at a temperature of 160°C) is pumped under pressure through the outermost pipe into the sulfur-bearing rock formation. This heats the rock above the melting point of sulfur, 119°C. The molten sulfur is heavier than water and collects in a pool, where heated, compressed air pumped through the innermost pipe works it into a froth that rises to the surface through the third pipe.

Compressed air

Hot sulfur froth

Hot water

Soil and rock

Sulfur-bearing rock

**Figure 21-6**
Sulfuric acid should always be added to water, rather than the reverse. The dissolution reaction is quite exothermic; water can boil and splatter acid out if water is added to acid, because the heat is generated in the region where the less dense water touches the dense acid. Adding acid to water mixes the two much more effectively. Here, the indicator in the water, methyl orange, is changing color from yellow to red as acid is added. (Leon Lewandowski)

and petroleum. The latter process was developed to reduce sulfur oxide pollution from the burning of oil and gas. Hydrogen sulfide and other sulfide impurities are acidic and so are preferentially extracted into solutions of basic potassium carbonate:

$$H_2S(g) + CO_3^{2-}(aq) \longrightarrow HS^-(aq) + HCO_3^-(aq)$$

After removal from this solution, a portion of the $H_2S$ is then burned in air at 1000°C over an aluminum oxide catalyst to give sulfur dioxide:

$$H_2S(g) + \tfrac{3}{2} O_2(g) \longrightarrow SO_2(g) + H_2O(g)$$

This can be used directly to make sulfuric acid or converted back to elemental sulfur by reaction with additional $H_2S$ in the **Claus process,**

$$SO_2(g) + 2 H_2S(g) \longrightarrow 3 S(\ell) + 2 H_2O(g)$$

which uses an $Fe_2O_3$ catalyst. The Claus process greatly reduces pollution from the burning of natural gas and petroleum. New processes to reduce emission of sulfur oxides from the burning of coal are greatly needed now.

## Uses of Sulfuric Acid and Its Derivatives

Pure sulfuric acid is a colorless, viscous liquid that freezes at 10.4°C and boils at 279.6°C. It reacts avidly with water (Fig. 21-6) and with organic compounds. It may be mixed with water in any proportion, with provisions for the large amounts of heat that are liberated. Despite its corrosive power, sulfuric acid is easily handled and transported in steel drums. This fact, together with its acid strength and low cost, has given it a tremendous range of uses. In metals processing, sulfuric acid is used to leach copper, uranium, and vanadium from their ores and to ''pickle,'' or descale, steel. In pickling, layers of oxides on the surfaces of the metal are dissolved away by reaction with the acid. Much sulfuric acid is used as a **dehydrating agent** (Fig. 21-7) in the synthesis of organic chemicals and in the processing of petrochemicals.

**Figure 21-7**
Concentrated sulfuric acid is an excellent dehydrating agent. It can remove water from a carbohydrate (here, sucrose, $C_{12}H_{22}O_{11}$), leaving mostly carbon.

Lesser but still important amounts are used in making hydrochloric and hydrofluoric acids and $TiO_2$ pigments for paint. As will be discussed, the largest single use of sulfuric acid is in the fertilizer industry. All of these uses are *indirect,* however, in the sense that the sulfur from the sulfuric acid does not become incorporated as an ingredient in products. Rather, it ends up either as sulfate wastes or as **spent acid,** acid that has been diluted and contaminated by its use in processing. The *direct* uses of sulfuric acid are surprisingly few.

An important chemical made from sulfuric acid is sodium sulfate ($Na_2SO_4$). It is produced in the **Hargreaves process,** in which the immediate reactant is $SO_2$ rather than $H_2SO_4$:

$$4\, NaCl + 2\, SO_2 + 2\, H_2O + O_2 \longrightarrow 2\, Na_2SO_4 + 4\, HCl$$

The $SO_2$ for this reaction often comes from the decomposition of spent sulfuric acid. Sodium sulfate is also a byproduct of the manufacture of rayon (Section 23-3). In addition, there are major natural salt sources for sodium sulfate in California and Texas. Sodium sulfate is used as a phosphate substitute in detergents and in the manufacture of paper products.

## Wood Pulp and Paper Processing

One of the most important applications of sulfur chemistry is in processing wood to wood pulp for use in paper and cardboard. Wood has three major components: cellulose, lignin, and oils and resins. Cellulose is a fibrous polymeric compound of carbon, oxygen, and hydrogen that provides a supporting structure. Lignin, which is also a complex polymeric material, acts as a matrix to bind the fibers of cellulose together. To produce wood pulp, the cellulose must be separated from the lignin. The other materials in the wood are removed by solvent extraction and in many cases have important uses (for example, as turpentine, pine oil, rosin, soaps, and tannin).

There are four steps in making paper from trees:

1. Logging and milling operations to collect and debark trees and convert the logs to wood chips
2. Digestion, the chemical process by which lignin is separated from cellulose
3. Bleaching to lighten the color of the cellulose
4. Production of paper or cardboard from the cellulose fibers

The first and last steps are mechanical operations and will not concern us here. The next chapter will discuss bleaching. We discuss only the second step, digestion, here. The chemical reaction involved is a hydrolysis, whereby the addition of water under proper conditions breaks down the lignin polymer into alcohols and organic acids containing long chains of carbon atoms. The hydrolysis proceeds only in a slow digestive process in the presence of other chemicals. The two primary methods used are the sulfate (Kraft) process and the sulfite process.

The **sulfate process** (Fig. 21-8) is the more widely used of these two methods because it works for almost all kinds of wood and produces the strongest paper (*Kraft* is German for "strength"). The digestion liquor is an aqueous solution of NaOH and $Na_2S$, the active agents in the hydrolysis of lignin. After digestion for several hours at 175°C, the pulp (now primarily cellulose) is separated by filtration, and the liquor is treated to extract any desired organic compounds. The next and

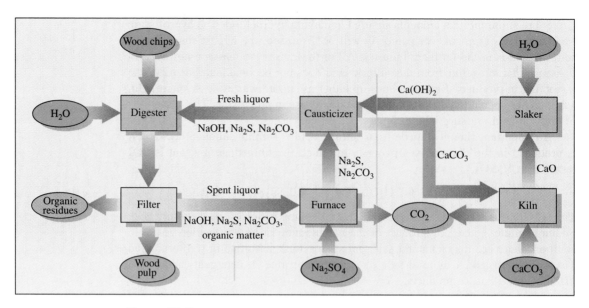

**Figure 21-8**

The sulfate process for pulping wood. The overall reaction is: wood chips + water →
wood pulp + organic residues. The entire part of the flow diagram to the right of the
yellow line serves only to make up losses in the process.

crucial step is the removal and recycling of the inorganic components from the spent
liquor. First, sodium sulfate is added to make up for the inevitable losses in the
process. Then the liquor is concentrated and its organic component is burned, generating heat to run the paper mill. During the burning, the sodium sulfate is reduced by
the organic matter in the liquor, a process indicated schematically as

$$Na_2SO_4(s) + 2\,[C] \longrightarrow Na_2S(s) + 2\,CO_2(g)$$

The sodium hydroxide in the liquor reacts with the carbon dioxide produced by the
oxidation of the organic matter, giving sodium carbonate:

$$2\,NaOH(s) + CO_2(g) \longrightarrow Na_2CO_3(s) + H_2O(g)$$

To regenerate the sodium hydroxide for a new cycle of digestion, the solid residue
from the burning is dissolved in water. Limestone ($CaCO_3$) is decomposed to lime
(CaO) and carbon dioxide in a separate lime kiln, and the lime is hydrated with
water to produce slaked lime, $Ca(OH)_2$. The addition of this base to the dissolved
residue precipitates the carbonate ion that was produced in the preceding step and
generates hydroxide ion:

$$CO_3^{2-}(aq) + Ca(OH)_2(s) \longrightarrow 2\,OH^-(aq) + CaCO_3(s)$$

The sodium hydroxide is thus restored to the digestion liquor. The calcium carbonate is itself separated and recycled to the kiln to produce additional lime. In this
process the material that is consumed chemically is water; sodium sulfate and limestone are added only as needed to make up for losses from what would ideally be a
closed cycle. The major problem with the Kraft process is the occurrence of side
reactions that lead to the formation of $H_2S$ and other sulfur compounds, which smell
vile and cause serious air and water pollution if they are not captured.

The sulfate process involves a basic digestion medium, but the **sulfite process** uses a mildly acidic medium, with pH near 4.5. It is limited to certain soft woods such as spruce and hemlock, and it produces a lighter pulp that requires less bleaching and is suitable for making high-quality writing paper. In the digestion liquor, gaseous sulfur dioxide is added to a solution containing hydroxide ion (generally prepared from MgO) to produce hydrogen sulfite ions:

$$SO_2(g) + OH^-(aq) \longrightarrow HSO_3^-(aq)$$

The action of the hydrogen sulfite ion at 60°C over 6 to 12 hours hydrolyzes the cellulose–lignin complex and dissolves the lignin, under conditions milder than those required for the sulfate process. After the process is complete, the pulp is recovered by filtration and the liquor is evaporated. The organic residues are again burned to recover heat; during this procedure the $Mg(HSO_3)_2$ decomposes according to the equation

$$Mg(HSO_3)_2(s) \longrightarrow MgO(s) + H_2O(g) + 2\ SO_2(g)$$

Both the $SO_2$ and MgO are recovered and recycled through the process.

## 21-2 PHOSPHORUS CHEMISTRY

Until rather recently, no single use of sulfuric acid dominated the others. The situation has changed with the growing production of phosphate fertilizers (Fig. 21-9), which are processed with sulfuric acid and which now account for well over half of the annual consumption of sulfuric acid in the United States. Phosphorus occurs in nature primarily as the $PO_4^{3-}$ ion in phosphate rock, the main component of which is

**Figure 21-9**
The world's largest linear disc reclaimer, reclaiming phosphate shale at an elemental phosphorus plant in Idaho. (Manley Prim/FMC Corporation)

**Figure 21-10**
Fluorapatite, $Ca_5(PO_4)_3F$, is one of several apatite minerals that are the commercial sources of phosphorus. Pure $Ca_5(PO_4)_3F$ is white; the purple in this sample results from manganese impurities. (Julius Weber)

fluorapatite, $Ca_5(PO_4)_3F$ (Fig. 21-10). Originally, the rock was simply ground up and applied to fields to make them more fertile. A more active fertilizer is made by treating a slurry of ground-up apatite with sulfuric acid to produce **superphosphate,** a mixture of calcium dihydrogen phosphate and gypsum:

$$2\,Ca_5(PO_4)_3F(s) + 7\,H_2SO_4(aq) + 17\,H_2O(\ell) \longrightarrow$$
$$3\,(Ca(H_2PO_4)_2\cdot2H_2O(s)) + 7\,(CaSO_4\cdot2H_2O(s)) + 2\,HF(g)$$

Superphosphate is more effective because the essential phosphorus nutrient is in a more soluble form. This mixture provides the important secondary nutrients calcium and sulfur, as well as phosphorus, to growing crops.

A variant of the superphosphate process, the **wet-acid process,** uses an excess of sulfuric acid to produce phosphoric acid. The chemical reaction in this case is

$$Ca_5(PO_4)_3F(s) + 5\,H_2SO_4(aq) + 10\,H_2O(\ell) \longrightarrow$$
$$3\,H_3PO_4(aq) + 5\,(CaSO_4\cdot2H_2O(s)) + HF(g)$$

The hydrogen fluoride that is liberated is carried to an absorption tower containing $SiF_4$. In this tower it reacts to give $H_2SiF_6$, which is used in aqueous solution to fluoridate drinking water. The solid gypsum ($CaSO_4\cdot2H_2O$) is filtered out, and the dilute solution of phosphoric acid ($H_3PO_4$) is concentrated by evaporation. The phosphoric acid made by this process is not pure because of residual gypsum and impurities from the phosphate rock. The major use for wet-process phosphoric acid is again fertilizer manufacture, where it is substituted for sulfuric acid to produce **triple superphosphate** fertilizer, which does not contain gypsum.

$$Ca_5(PO_4)_3F + 7\,H_3PO_4 + 5\,H_2O \longrightarrow 5\,Ca(H_2PO_4)_2\cdot2H_2O + HF(g)$$

This fertilizer contains a greater percentage of phosphorus, which reduces transportation costs.

The use of triple superphosphate fertilizer has fallen in recent years. The major phosphorus-containing fertilizer now is ammonium phosphate, which is obtained by the reaction of phosphoric acid with ammonia:

$$H_3PO_4(aq) + 3\ NH_3(aq) \longrightarrow (NH_4)_3PO_4(aq)$$

This supplies nitrogen, a necessary nutrient, as well as phosphorus. The next section will discuss the manufacture of ammonia and other nitrogen-based fertilizers.

An alternative to the wet-acid process, used to make purer phosphoric acid, is called the **furnace process.** Phosphate rock, $SiO_2$ (in the form of sand), and elemental carbon (in the form of coke) are fed continuously into an electric-arc furnace that operates at 2000°C. The overall reaction is approximately

$$12\ Ca_5(PO_4)_3F(\ell) + 43\ SiO_2(\ell) + 90\ C(s) \longrightarrow$$
$$90\ CO(g) + 20\ (3CaO \cdot 2SiO_2)(\ell) + 3\ SiF_4(g) + 9\ P_4(g)$$

The elemental white phosphorus is burned in air to tetraphosphorus decaoxide ($P_4O_{10}$) (Fig. 21-11):

$$P_4(g) + 5\ O_2(g) \longrightarrow P_4O_{10}(s)$$

This is a powerful drying agent that dehydrates $HNO_3$ to $N_2O_5$ and even $H_2SO_4$ to $SO_3$. The ultimate product of its reaction with water is phosphoric acid:

$$P_4O_{10}(s) + 6\ H_2O(\ell) \longrightarrow 4\ H_3PO_4(aq)$$

● Phosphorus

● Oxygen

**Figure 21-11**
The structure of $P_4O_{10}$ is closely related to the tetrahedral $P_4$ structure in white phosphorus.

## Derivatives of Phosphoric Acid

The major nonfertilizer use of wet-process phosphoric acid is in the manufacture of detergents and cleaning agents that are sodium salts of phosphoric acid and its derivatives. Phosphoric acid in excess reacts with sodium carbonate to give sodium dihydrogen phosphate, a weak acid:

$$2\ H_3PO_4(aq) + Na_2CO_3(aq) \longrightarrow 2\ NaH_2PO_4(aq) + CO_2(aq) + H_2O(\ell)$$

When the sodium carbonate is in excess, the product is sodium hydrogen phosphate, a weak base:

$$H_3PO_4(aq) + 2\ Na_2CO_3(aq) \longrightarrow 2\ NaHCO_3(aq) + Na_2HPO_4(aq)$$

Sodium carbonate is too weak a base to remove the last hydrogen ion from $HPO_4^{2-}$, and so the stronger (and more expensive) base NaOH must be added:

$$NaOH(aq) + Na_2HPO_4(aq) \longrightarrow Na_3PO_4(aq) + H_2O(\ell)$$

Sodium phosphate ($Na_3PO_4$), also called trisodium phosphate, or TSP, gives strongly basic solutions that are excellent for cleaning.

Heating solid sodium hydrogen phosphate ($Na_2HPO_4$) expels one molecule of water for every two formula units:

$$\overset{\ominus}{:\!O\!:} \qquad \overset{\ominus}{:\!O\!:} \qquad \overset{\ominus}{:\!O\!:} \quad \overset{\ominus}{:\!O\!:}$$

$$O\!=\!P\!-\!O\!-\!H + H\!-\!O\!-\!P\!=\!O \longrightarrow O\!=\!P\!-\!O\!-\!P\!=\!O + H_2O$$

$$\underset{\ominus}{:\!O\!:} \qquad \underset{\ominus}{:\!O\!:} \qquad \underset{\ominus}{:\!O\!:} \quad \underset{\ominus}{:\!O\!:}$$

**Figure 21-12**

The structure of sodium tripoly-phosphate.

The product, sodium diphosphate ($Na_4P_2O_7$), is the sodium salt of diphosphoric acid ($H_4P_2O_7$). This is the phosphorus analog of the disulfuric acid ($H_2S_2O_7$) mentioned in Section 21-1. In diphosphates, each phosphorus atom is tetrahedrally coordinated to four oxygen atoms (as in the original $PO_4^{3-}$ ion), but two tetrahedra now share a single corner. A third tetrahedron can be linked to the other two through the high-temperature reaction

$$2 \, Na_2HPO_4(s) + NaH_2PO_4(s) \longrightarrow Na_5P_3O_{10}(s) + 2 \, H_2O(g)$$

with one molecule of water expelled for each linkage that forms. The product sodium tripolyphosphate ($Na_5P_3O_{10}$), or STPP (Fig. 21-12), is used as a **builder** in detergents. It helps to sequester $Ca^{2+}$ and $Mg^{2+}$ ions, to maintain the pH in an appropriate range, to keep dirt in suspension, and to increase the efficiency of the detergent itself. The use of detergents that contain phosphates has diminished in recent years because phosphates released into the groundwater fertilize algae in lakes. Excessive algal growth can destroy a lake through **eutrophication** (the word comes from the Greek for "well nourished"). When the algae die, their bodies fall to the bottom of the lake. As the lake becomes shallower, bottom-rooted plants can take hold; eventually the lake becomes a marsh and, finally, a meadow. This phenomenon has inspired legislation to limit the use of phosphates in detergents.

The purer phosphoric acid from the furnace process is used widely in the food industry. Its acidity gives a pleasing sour taste to soft drinks (called "phosphates" when they were first introduced). Calcium dihydrogen phosphate monohydrate, $Ca(H_2PO_4)_2 \cdot H_2O$, is mixed with $NaHCO_3$ in some baking powders. When water is added, these ionic compounds dissolve, and the acidic dihydrogen phosphate ions react with the basic hydrogen carbonate ions to generate carbon dioxide:

$$H_2PO_4^-(aq) + HCO_3^-(aq) \longrightarrow HPO_4^{2-}(aq) + CO_2(g) + H_2O(\ell)$$

The liberated gas makes baked goods rise. Another phosphoric acid derivative used in the food industry is sodium hydrogen phosphate ($Na_2HPO_4$), which is used as an emulsifier to distribute butterfat uniformly through pasteurized processed cheese.

## 21-3 NITROGEN FIXATION

Nitrogen at the earth's surface exists almost entirely (99.9%) as gaseous diatomic molecules ($N_2$), which make up 78% by volume of the atmosphere. The N—N bond is a triple bond, and its enthalpy of dissociation is 945 kJ mol$^{-1}$, making it the largest of any diatomic molecule except carbon monoxide. The nitrogen molecule is remarkably unreactive in comparison with other triple-bonded molecules. This is apparent in a simple thermochemical comparison. In each of the two parallel reactions,

$$N\!\equiv\!N(g) + 3\,H_2(g) \longrightarrow 2\,NH_3(g) \qquad\qquad \Delta H° = -92.2\,\text{kJ}$$

$$HC\!\equiv\!CH(g) + 3\,H_2(g) \longrightarrow 2\,CH_4(g) \qquad\qquad \Delta H° = -376.4\,\text{kJ}$$

a triple bond and three H—H bonds are broken and six X—H single bonds are formed (X is N or C). The enthalpy changes in the formation of an N—H bond and a C—H bond are roughly equal, and so the large difference in the $\Delta H°$'s of the reactions must mean that the triple bond in $N_2$ is substantially stronger than the triple bond in $C_2H_2$. The great stability of the $N_2$ molecule diminishes its reactivity and means that an $N_2$ atmosphere can be used to prevent air oxidation in metallurgical, chemical, and food-processing applications. For most biological and industrial purposes, however, what is needed is not molecular nitrogen but compounds of nitrogen with other elements. The formation of such compounds is referred to as **nitrogen fixation.**

## Natural Sources of Fixed Nitrogen

Figure 21-13 illustrates how nitrogen moves through a worldwide cycle involving biological, geological, and atmospheric processes. Herbivores and omnivores use nitrogen from the plants and vegetable matter they have consumed in a variety of

**Figure 21-13**

The nitrogen cycle.

essential biological functions, including the synthesis of proteins. Eventually, the nitrogen is excreted and returned to the soil in forms such as urea, $(NH_2)_2CO$. A portion is then reused by other plants as they grow, but some is converted to molecular nitrogen by bacteria and returned to the atmosphere. To continue the cycle, a way is needed to fix the nitrogen again.

Nature has at least two ways of fixing atmospheric nitrogen. One is the action of lightning on air, in which the high temperature of the electrical discharge causes the reaction of nitrogen with oxygen to form nitrogen oxide (NO). The nitrogen oxide is subsequently oxidized to $NO_2$, which reacts with hydroxyl radicals (OH) in the atmosphere to form dilute nitric acid. The $HNO_3$ falls to the earth in rain and provides fixed nitrogen (in the form of the nitrate ion) for plant growth. Nitrogen fixed in this way is thought to amount to at least 10% of the annual total. A second way in which nitrogen is fixed is by the action of certain bacteria that live on the roots of clover and alfalfa plants.

By the late 18th century, the population of Europe had grown sufficiently that methods of increasing crop yields were needed. The first step was to reuse animal wastes efficiently. Sailing ships went as far as the coastal regions of South America to bring back guano, the excrement dropped by seagulls, because it was rich in nitrogen and easily mined. Mineral deposits of sodium nitrate in Chile were exploited for fertilizer. By the end of the 19th century some fixed nitrogen was being recovered from the waste liquors of natural gas plants in the form of ammonia. It was generally converted to ammonium sulfate, $(NH_4)_2SO_4$, by reaction with sulfuric acid, and it was used in this form as a fertilizer.

## Early Industrial Processes

In 1898, in a presidential address to the British Association, Sir William Crookes said,

> England and all civilised nations stand in deadly peril of not having enough to eat. As mouths multiply, food resources dwindle. . . . It is the chemist who must come to the rescue of the threatened communities. It is through the laboratory that starvation may ultimately be turned to plenty.[1]

He concluded by adding that "there is a gleam of light amid this darkness and despondency . . . the fixation of atmospheric nitrogen" which would be "one of the great discoveries awaiting the ingenuity of chemists." Within ten years of his prediction, three methods had been developed for fixing atmospheric nitrogen.

The first was the **electric-arc process,** which was based on the observation by Cavendish in the 1780s that a spark causes the combination of nitrogen and oxygen in air. A century later, electrical technology was sufficiently advanced to create electric arcs in which the temperature was raised to 2000 to 3000°C. The NO that formed was cooled, oxidized to $NO_2$, and passed into an absorption tower where it reacted with water to make dilute nitric acid ($HNO_3$). This process closely resembled the fixing of nitrogen by lightning discharges. The energy costs for running the electric arc were high, and it was difficult to continuously obtain large amounts of fixed nitrogen. As a result, the electric-arc process was never a significant factor in nitrogen fixation.

[1]Quoted in L. F. Haber, *The Chemical Industry, 1900–1930* (Clarendon Press, Oxford, 1971), p. 84.

More promising was the **cyanamide process,** also called the **Frank–Caro process** after its developers. This method uses calcium carbide ($CaC_2$) as a starting material. The discovery of this compound has an interesting history. In 1892 T. L. Willson in North Carolina was trying to make elemental calcium from lime (CaO) and tar in an electric-arc furnace. He obtained a product that was clearly not calcium and threw it into a stream, where, to his surprise, it reacted with water to liberate large amounts of a combustible gas. The compound was calcium carbide, and its reaction with water gives acetylene ($C_2H_2$):

$$CaC_2(s) + 2\,H_2O(\ell) \longrightarrow C_2H_2(g) + Ca(OH)_2(aq)$$

Calcium carbide is now made from lime and coke in an electric-arc furnace at 2000 to 2200°C:

$$CaO(s) + 3\,C(s) \longrightarrow CaC_2(s) + CO(g) \qquad \Delta H° = +465 \text{ kJ}$$

In the cyanamide process, calcium carbide then reacts (at 1100°C in an electric furnace) with nitrogen, obtained from liquefying and distilling air, to give calcium cyanamide:

$$CaC_2(s) + N_2(g) \longrightarrow CaCN_2(s) + C(s) \qquad \Delta H° = -291 \text{ kJ}$$

Steam is added to convert this to ammonia:

$$CaCN_2(s) + 4\,H_2O(g) \longrightarrow Ca(OH)_2(s) + CO_2(g) + 2\,NH_3(g)$$

Between 1900 and 1930, calcium cyanamide was itself used as a fertilizer because it furnishes the soil with ammonia and lime. The major current use of $CaCN_2$ is as a starting material for making plastics.

## The Haber–Bosch Process

Both the electric-arc and the cyanamide processes have been superseded by the **Haber–Bosch process,** which is the predominant source of industrial fixed nitrogen today. The German chemists Fritz Haber and Walther Nernst carried out studies from 1900 to 1910 on the effects of pressure and temperature on the equilibrium between ammonia and its component elements:

$$N_2(g) + 3\,H_2(g) \rightleftharpoons 2\,NH_3(g)$$

The practical ammonia synthesis Haber developed with chemical engineer Kurt Bosch required both deep chemical insight and equipment capable of operating under conditions of high temperature and pressure. At 298 K, the standard free energy and standard enthalpy of formation of 2 mol of ammonia from its elements are

$$\Delta G° = -33.0 \text{ kJ} \qquad \text{and} \qquad \Delta H° = -92.2 \text{ kJ}$$

and the value for the equilibrium constant is

$$\ln K_{298} = \frac{-\Delta G°}{RT} = \frac{33{,}000 \text{ J mol}^{-1}}{(8.315 \text{ J K}^{-1}\text{ mol}^{-1})(298 \text{ K})} = 13.32$$

$$K_{298} = e^{13.32} = 6 \times 10^5$$

which is large relative to 1; the formation of ammonia is favored on thermodynamic grounds. Unfortunately, the reaction is very slow at 298 K, and the temperature must be raised to speed it. Because the process is exothermic, increased temperature

**Figure 21-14**

A modern chemical plant in Louisiana that produces ammonia via the Haber–Bosch process. (Tom Carroll)

**Figure 21-15**

The active ingredients in the fire retardants being dropped by this airplane are ammonium sulfate and ammonium phosphate. After the fire, these compounds act as soil fertilizers, speeding the regrowth of vegetation in burned areas. The red color comes from a dye that turns to a neutral earth tone after exposure to sunlight for a few days. (Phototake)

lowers the equilibrium constant and decreases product yield. Thus, competition occurs between thermodynamic factors, which favor a high yield at low temperature, and kinetic factors, which favor the use of a high temperature.

The compromise adopted in industrial production is a temperature of 700 to 900 K. At 800 K, for example, the equilibrium constant can be estimated from the van't Hoff equation (Section 9-5):

$$\ln \frac{K_{800}}{K_{298}} = -\frac{\Delta H^\circ}{R}\left(\frac{1}{T_2} - \frac{1}{T_1}\right)$$

$$= \frac{92{,}200 \text{ J mol}^{-1}}{8.315 \text{ J K}^{-1} \text{ mol}^{-1}}\left(\frac{1}{800 \text{ K}} - \frac{1}{298 \text{ K}}\right) = -23.35$$

Solving this for the equilibrium constant at 800 K gives

$$K_{800} = (7 \times 10^{-11})\, K_{298} = 4 \times 10^{-5}$$

This value is only approximate because $\Delta H$ changes with temperature between 298 and 800 K. The qualitative result, however, is that the product yield is greatly reduced at the higher temperature. To increase the yield, high total pressures are used, which favor the product in this process because 2 mol of gaseous product is formed from 4 mol of gaseous reactants (recall Fig. 5-7). At a total pressure of 200 atm, the reaction yields 15 to 30% of the amount of $NH_3$ that would be produced by complete consumption of the reactants.

A catalyst increases the rate of this reaction. Haber used osmium, a rare and expensive metal, in his original work. Later he developed a less expensive catalyst based on partially oxidized iron containing small amounts of aluminum, which has been used ever since. The ammonia produced is liquefied to separate it from the unreacted nitrogen and hydrogen, which are recycled.

Although the principles of the Haber–Bosch process were developed before World War I, its implementation in Germany was greatly speeded by the need for fixed nitrogen to make military explosives when the supplies of nitrates from Chile were cut off. It is one of the ironies of history that the commercialization of this process, which more than any other has expanded the food-producing capacity of the world, began as a military project that probably prolonged World War I by at least a year.

In spite of the significant role the Haber–Bosch process has played since that time, its use to fix nitrogen has some problems. Fertilizer factories (Fig. 21-14) consume large amounts of energy and require expensive structural materials to operate at high pressure and high temperature. Dependence on hydrogen as a reactant is another problem. Because most hydrogen is currently obtained from petroleum and natural gas, both nonrenewable resources, an alternative hydrogen production process that uses coal or, ultimately, water will eventually have to be found. Scientists are also trying to design chemical catalysts that imitate the action of nitrogen-fixing bacteria as an economical way of converting atmospheric nitrogen to ammonia or to some other fixed form.

## 21-4 CHEMICALS FROM AMMONIA

Once molecular nitrogen has been fixed in the form of ammonia, further reactions to produce many other compounds are possible. Reactions with sulfuric, phosphoric, and nitric acid give ammonium sulfate, ammonium phosphate, and ammonium nitrate, respectively (Fig. 21-15). All three have been used as chemical fertilizers, with the latter two being particularly popular in recent years. Some ammonia is also converted to urea, $(NH_2)_2CO$. Modern methods for producing urea use the reaction of $CO_2$ with ammonia in hot aqueous solution (180 to 200°C) under pressure to form the intermediate ammonium carbamate ($NH_2COONH_4$), which is then decomposed to urea by heat:

$$2 NH_3(aq) + CO_2(aq) \longrightarrow NH_2COONH_4(aq)$$

$$NH_2COONH_4(aq) \longrightarrow (NH_2)_2CO(s) + H_2O(g)$$

Urea is a widely used solid-nitrogen fertilizer and is an ingredient in other products ranging from glues to skin creams to disinfectants. Urea has historical significance as well, having been first synthesized by Friedrich Wöhler in 1828, from ammonia and cyanic acid (HCNO). Wöhler's work was important in demonstrating that an "organic" compound, formed in human and animal metabolism (and excreted in urine), could be synthesized from strictly inorganic starting materials.

Urea, $(NH_2)_2CO$, is an important solid nitrogen fertilizer.

## Nitric Acid

One of the most important products made from ammonia is nitric acid, which is generated by the overall reaction

$$NH_3 + 2 O_2 \longrightarrow HNO_3 + H_2O$$

This reaction is deceptively simple as written, because direct combination of ammonia with oxygen will *not* produce nitric acid. The most commonly used industrial process, the **Ostwald process** (Fig. 21-16), involves three steps that, added together, give the preceding overall reaction. Let us examine these steps.

The first step is the partial oxidation of ammonia in air (Fig. 21-17):

$$4 NH_3(g) + 5 O_2(g) \longrightarrow 4 NO(g) + 6 H_2O(g) \qquad \Delta H° = -905.5 \text{ kJ}$$

This is a strongly exothermic reaction, but it occurs very slowly at room temperature. In addition, once some NO is produced, a competing reaction occurs,

$$4 NH_3(g) + 6 NO(g) \longrightarrow 5 N_2(g) + 6 H_2O(g)$$

which is undesirable because it returns nitrogen to its elemental state. To obtain practical yields of NO, the reaction is carried out at elevated temperatures (800–950°C) and a catalyst, consisting of a fine gauze of noble metals such as platinum

### Figure 21-16

Flow diagram for the synthesis of nitric acid from ammonia. Both the second step (oxidation of NO to $NO_2$) and the third step (absorption into water) described in the text take place in the absorber.

**Figure 21-17**
When a heated platinum wire is placed just above a solution of concentrated aqueous ammonia, it glows for as long as half an hour as it catalyzes the continued exothermic oxidation of $NH_3$ to NO by oxygen from the air. (Leon Lewandowski)

**Figure 21-18**
This gold–palladium gauze catalyst is used to make nitric acid from ammonia. (Johnson Matthey)

and rhodium or gold and palladium, is used (Fig. 21-18). Such a catalyst promotes the desired reaction in preference to the competing one, and this first step occurs with extraordinary speed: only about $3 \times 10^{-4}$ s of contact time between gases and the gauze is needed to effect excellent conversion.

The second, and rate-determining, step is the reaction of nitrogen oxide with oxygen to produce nitrogen dioxide:

$$2\,NO(g) + O_2(g) \longrightarrow 2\,NO_2(g) \qquad \Delta H° = -114.1 \text{ kJ}$$

Example 11-3 and Section 11-5 discussed the kinetics of this reaction, which requires no catalyst, and showed that the rate law has the form

$$\text{rate} = k[NO]^2[O_2]$$

and the rate constant $k$ *decreases* with increasing temperature. Because the reaction occurs faster at low temperatures, this step is carried out at the lowest temperature conveniently reached by the available cooling water, 10 to 40°C.

The third step is the absorption of the $NO_2(g)$ into water:

$$3\,NO_2(g) + 3\,H_2O(\ell) \longrightarrow 2\,H_3O^+(aq) + 2\,NO_3^-(aq) + NO(g)$$
$$\Delta H° = -113.5 \text{ kJ}$$

The NO(g) produced reacts again in the second step to give more $NO_2(g)$. The nitric acid has a concentration in the range of 50 to 65% by mass. Distilling off the water does not increase the concentration beyond 69% $HNO_3$, which is the "concentrated nitric acid" in normal laboratory use. Addition of a powerful dehydrating agent (concentrated sulfuric acid) and distillation separate more water to give a solution that contains 95 to 98% nitric acid (fuming nitric acid).

We have analyzed the production of nitric acid in some detail not only because nitric acid is an important chemical but also because the Ostwald process illustrates some of the problems involved in practical chemistry. The original overall reaction is quite simple and is strongly favored thermodynamically:

$$\frac{[H_3O^+][NO_3^-]}{P_{NH_3}(P_{O_2})^2} = K_{298} = 10^{52}$$

Nevertheless, kinetic effects and the presence of competing equilibria require the careful design of a multistep process to obtain practical yields.

## Hydrazine

Another important product made from ammonia is hydrazine ($N_2H_4$). In principle, many reactions are capable of producing hydrazine from ammonia, such as

$$2\,NH_3 + \tfrac{1}{2}O_2 \longrightarrow N_2H_4 + H_2O$$

This and other direct pathways fail as practical methods, however. Competing reactions such as

$$2\,NH_3 + \tfrac{3}{2}O_2 \longrightarrow N_2 + 3\,H_2O$$

are thermodynamically favored because the $N_2$ molecule is so stable. The method used to make hydrazine commercially is the reaction of ammonia with hypochlorite ion ($OCl^-$) in aqueous solution, known as the **Raschig synthesis:**

$$2\,NH_3(aq) + OCl^-(aq) \longrightarrow N_2H_4(aq) + H_2O(\ell) + Cl^-(aq)$$

Pure liquid hydrazine resembles hydrogen peroxide in many of its physical properties, boiling at 114°C and freezing at 2°C ($H_2O_2$ boils at 152°C and freezes at −1.7°C). Hydrazine has had some use as a rocket fuel because its reaction with oxygen is extremely exothermic:

$$N_2H_4(\ell) + O_2(g) \longrightarrow N_2(g) + 2\,H_2O(g) \quad \Delta H° = -534\ \text{kJ}$$

A small mass of liquid hydrazine produces a tremendous thrust as the hot gaseous products rush out at high speed from the rocket engine.

Nitrogen in hydrazine has an oxidation state of −2, intermediate between those of molecular nitrogen (0) and nitrogen in ammonia (−3). As a result, hydrazine (like hydrogen peroxide) can act either as a reducing agent or as an oxidizing agent in aqueous solution. In basic solution (at pH 14), the standard reduction potential for the half-reaction

$$N_2(g) + 4\,H_2O(\ell) + 4\,e^- \longrightarrow N_2H_4(aq) + 4\,OH^-(aq)$$

is $\mathscr{E}° = -1.16$ V. This means that hydrazine tends to be easily oxidized to nitrogen under these conditions and so should act as a good reducing agent. In a strongly acidic medium, hydrazine is almost completely converted to its conjugate acid, the hydrazinium ion $N_2H_5^+$. The standard reduction potential $\mathscr{E}° = 1.275$ V for the half-reaction

$$N_2H_5^+(aq) + 3\,H_3O^+(aq) + 2\,e^- \longrightarrow 2\,NH_4^+(aq) + 3\,H_2O(\ell)$$

at pH 0 suggests that $N_2H_5^+$ should be easily reduced to $NH_4^+$ and therefore should serve as an excellent oxidizing agent under these conditions. This reaction is slow, however, and hydrazine is mainly known as a reducing agent.

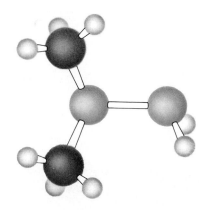

The compound dimethyl hydrazine, $(CH_3)_2NNH_2$, fueled the lunar excursion modules in their departures from the surface of the moon. The oxidant was the nitrogen oxide $N_2O_4$. Ignition was a certainty because the two burst into flame upon contact.

Hydrazine is used to control boiler corrosion through the reaction

$$6\,Fe_2O_3(s) + N_2H_4(aq) \longrightarrow 4\,Fe_3O_4(s) + N_2(g) + 2\,H_2O(\ell)$$

The red $Fe_2O_3$ oxide (ordinary rust) is reduced to $Fe_3O_4$ that forms a protective black layer to hinder further rusting. Hydrazine is also used to remove certain metal ions from the wastewaters of chemical plants. Chromate ion ($CrO_4^{2-}$), for example, is reduced to $Cr^{3+}$ in the reaction

$$20\,H_3O^+(aq) + 4\,CrO_4^{2-}(aq) + 3\,N_2H_4(aq) \longrightarrow$$
$$4\,Cr^{3+}(aq) + 3\,N_2(g) + 36\,H_2O(\ell)$$

The $Cr^{3+}$ is then precipitated by adding base:

$$Cr^{3+}(aq) + 3\,OH^-(aq) \longrightarrow Cr(OH)_3(s)$$

Hydrazine also removes halogens ($X_2$, X = F, Cl, Br, I) from industrial wastewater by the reaction

$$N_2H_4(aq) + 2\,X_2(aq) \longrightarrow N_2(g) + 4\,HX(aq)$$

## Explosives

An **explosive** is a material that, when subjected to shock, decomposes rapidly and exothermically to produce a large volume of gas. Useful explosives need to be thermodynamically unstable but kinetically stable so that they can be stored safely for long periods before they are detonated. Table 21-1 lists some explosives and their decomposition products. It is noteworthy that all of them contain nitrogen. The stability of gaseous nitrogen, a product of the detonation in all of these cases, contributes to the exothermicity of the reactions. The evolution of heat alone, however, is not as important as the *rate* at which the heat is released. The combustion of gasoline, for example, releases about 48 kJ per gram, far more than the heat released by the explosives listed in the table. The difference is that the detonation of an explosive is a chain reaction, in which the release of heat takes place at a tremendous rate.

**Table 21-1  Explosives**

| Name | Formula | Products | $\Delta H$ of Explosion (kJ g$^{-1}$) |
|---|---|---|---|
| Gunpowder | $2\,KNO_3 + 3\,C + S$ | $N_2 + 3\,CO_2 + K_2S$ | $-2.1$ |
| Nitrocellulose | $C_{24}H_{29}O_9(NO_3)_{11}$ | $20.5\,CO + 3.5\,CO_2 + 14.5\,H_2O + 5.5\,N_2$ | $-4.5$ |
| Nitroglycerin | $C_3H_5(NO_3)_3$ | $3\,CO_2 + 2.5\,H_2O + 1.5\,N_2 + 0.25\,O_2$ | $-6.4$ |
| Ammonium nitrate | $NH_4NO_3$ | $2\,H_2O + N_2 + 0.5\,O_2$ | $-1.6$ |
| TNT | $C_6H_2CH_3(NO_2)_3$ | $3.5\,CO + 3.5\,C + 2.5\,H_2O + 1.5\,N_2$ | $-4.4$ |
| Picric acid | $C_6H_2(OH)(NO_2)_3$ | $5.5\,CO + 1.5\,H_2O + 0.5\,C + 1.5\,N_2$ | $-4.4$ |
| Ammonium picrate | $C_6H_2(NO_2)_3ONH_4$ | $4\,CO + 2\,C + 3\,H_2O + 2\,N_2$ | $-2.6$ |
| Tetryl | $C_7H_5N_5O_8$ | $5.5\,CO + 2.5\,H_2O + 1.5\,C + 2.5\,N_2$ | $-4.7$ |
| Mercury fulminate | $Hg(ONC)_2$ | $Hg + 2\,CO + N_2$ | $-1.5$ |
| Lead azide | $PbN_6$ | $Pb + 3\,N_2$ | $-1.5$ |

**Figure 21-19**
When it is dry, nitrogen triiodide ($NI_3$) is so sensitive to shock that it will explode at the gentlest touch.

Explosives can be characterized as **primary** (or *initiating*) and **secondary.** Primary explosives such as mercury fulminate and lead azide are very sensitive and explode when shocked or heated (Fig. 21-19). Sodium azide, $NaN_3$, is used in automobile air bags; if the car decelerates rapidly, a steel ball causes the $NaN_3$ to explode, and the gaseous nitrogen generated rapidly fills the bag. Primary explosives are also used in small amounts as detonators to start the explosion of larger amounts of secondary explosives. The latter are less sensitive to shock but explode when detonated by a primary explosive.

The earliest explosive was gunpowder, developed before A.D. 1000 in China. Early gunpowder (black powder) was a mixture of potassium nitrate, sulfur, and charcoal (carbon) that was moistened and ground up. The dried product was granulated and loaded into cartridges or bombs. This gunpowder was used in both military and civilian applications until the middle of the 19th century. In 1845 nitrocellulose, or guncotton, was discovered as the product of the treatment of cotton with nitric acid. It is far more powerful than the early gunpowder and is the major constituent of modern gunpowder, although it is unsuitable for commercial purposes because it is difficult to handle. Trinitrotoluene, or TNT, is used for large-scale military purposes. It is manufactured by reaction of toluene ($C_6H_5CH_3$) with a mixture of nitric acid and sulfuric acid, resulting in the substitution of three nitro groups ($NO_2^-$) on the benzene ring:

$$C_6H_5CH_3 + 3\,HNO_3\ (\text{in } H_2SO_4) \longrightarrow C_6H_2CH_3(NO_2)_3 + 3\,H_2O$$

Civilian uses of explosives include mining and road construction. The development of dynamite in the second half of the 19th century advanced these fields tremendously. The major constituent of dynamite is nitroglycerine, made by adding glycerin to a mixture of concentrated nitric and sulfuric acids:

$$C_3H_5(OH)_3 + 3\,HNO_3\ (\text{in } H_2SO_4) \longrightarrow C_3H_5(ONO_2)_3 + 3\,H_2O$$

The molecular structure of TNT, $C_6H_2(CH_3)(NO_2)_3$.

**Figure 21-20**

Ammonium nitrate ($NH_4NO_3$) decomposes explosively in the presence of powdered zinc and a catalyst containing the chloride ion.

The Swedish industrialist Alfred Nobel discovered that nitroglycerine could be exploded by shock, and he developed a detonator for it. Nitroglycerine exploded too easily for routine use, however, until Nobel found that absorbing it onto diatomaceous earth (shells of tiny marine organisms) stabilized it greatly. This mixture was the first dynamite, and Nobel made a vast fortune selling it throughout the world. His will endowed the annual Nobel prizes in medicine, chemistry, physics, literature, and peace. Modern dynamite still employs nitroglycerine, but it is mixed with other substances such as ammonium nitrate or sodium nitrate.

In recent years, ammonium nitrate mixed with small amounts of fuel oil has increasingly supplanted dynamite as an industrial explosive. Recall that the major use of ammonium nitrate is as fertilizer; although it is thermodynamically unstable, it is quite safe to handle in small amounts. The advantage of this formulation is its inherent safety. If a misfire occurs and the charge is not set off, the volatile fuel oil evaporates and the remaining ammonium nitrate is once again safe. Also, the fuel oil can be added only after the ammonium nitrate has been placed in the bore hole, meaning that the risks of transporting explosive materials can be avoided. It is unfortunate that ammonium nitrate and fuel oil are readily obtained by terrorists. A mixture of these was used in the 1995 bombing in Oklahoma City.

Ammonium nitrate decomposes nonexplosively above 200°C,

$$NH_4NO_3(s) \longrightarrow N_2O(g) + 2\,H_2O(g) \qquad\qquad \Delta G^\circ = -169\ \text{kJ}$$

providing a commercial source of dinitrogen oxide (nitrous oxide), which is used as an anesthetic ("laughing gas") and an aerosol propellant. When the mass of ammonium nitrate exceeds a critical value (many tons) or when the ammonium nitrate contains chloride impurities, it may detonate (Fig. 21-20):

$$NH_4NO_3(s) \longrightarrow N_2(g) + 2\,H_2O(g) + \tfrac{1}{2}O_2(g) \qquad\qquad \Delta G^\circ = -273\ \text{kJ}$$

After a shipload of ammonium nitrate blew up in the harbor of Texas City, Texas, in 1947, devastating the town, the U.S. government set limits on the quantity of ammonium nitrate that may be stored in one place.

## 21-5 OXOACIDS OF SULFUR, PHOSPHORUS, AND NITROGEN

The production of the oxoacids of sulfur, phosphorus, and nitrogen was discussed earlier in this chapter. Here we compare two facets of their aqueous chemistry: their ability to act as oxidizing (or reducing) agents and their acid strengths.

Figure 21-21 shows the reduction potentials for these oxoacids and their conjugate bases. The species listed are those that predominate at the pH given (0 or 14). In acidic solution, nitrous acid exists largely as $HNO_2$ molecules, for example, but in basic solution almost all these molecules have ionized to form $NO_2^-$ ions. Nitric acid, on the other hand, is a strong acid and is almost fully dissociated to nitrate and hydronium ions, even at pH 0. Because hydronium ion or hydroxide ion participates in the reduction half-reactions, the half-cell potentials depend strongly on pH, as is seen by comparing the left half of the figure with the right half.

The reduction potentials of nitrogen, sulfur, and phosphorus oxoacids on the left side of the table show that the phosphorus oxoacids are the most difficult to reduce and therefore are the least effective oxidizing agents. The oxoacids of sulfur are intermediate in this regard, and nitric acid is the strongest oxidizing acid listed.

$$\mathscr{E}° \text{(V)}$$

$$HNO_2 + H_3O^+ + e^- \rightarrow NO + 2\,H_2O$$ 0.99

$$NO_3^- + 3\,H_3O^+ + 2\,e^- \rightarrow HNO_2 + 4\,H_2O$$ 0.94

$$2\,SO_2 + 2\,H_3O^+ + 4\,e^- \rightarrow S_2O_3^{2-} + 3\,H_2O$$ 0.40

$$HSO_4^- + 3\,H_3O^+ + 2\,e^- \rightarrow SO_2 + 5\,H_2O$$ 0.17

$$H_3PO_4 + 2\,H_3O^+ + 2\,e^- \rightarrow H_3PO_3 + 3\,H_2O$$ −0.28

$$H_3PO_3 + 2\,H_3O^+ + 2\,e^- \rightarrow H_3PO_2 + 3\,H_2O$$ −0.50

**Acidic solution pH 0**

**Basic solution pH 14**

$$\mathscr{E}° \text{(V)}$$

$$NO_3^- + H_2O + 2\,e^- \rightarrow NO_2^- + 2\,OH^-$$ 0.01

$$NO_2^- + H_2O + e^- \rightarrow NO + 2\,OH^-$$ −0.46

$$2\,SO_3^{2-} + 3\,H_2O + 4\,e^- \rightarrow S_2O_3^{2-} + 6\,OH^-$$ −0.58

$$SO_4^{2-} + H_2O + 2\,e^- \rightarrow SO_3^{2-} + 2\,OH^-$$ −0.98

$$PO_4^{3-} + 2\,H_2O + 2\,e^- \rightarrow HPO_3^{2-} + 3\,OH^-$$ −1.12

$$HPO_3^{2-} + 2\,H_2O + 2\,e^- \rightarrow H_2PO_2^- + 3\,OH^-$$ −1.57

**Figure 21-21**

The standard reduction potentials for oxides of nitrogen, sulfur, and phosphorus are higher in acidic solution at pH 0 (*left*) than in basic solution at pH 14 (*right*).

In concentrated form, it will attack and dissolve many metals, such as copper (Fig. 3-8), that phosphoric acid will not attack. In applications needing a strong acid that is *not* a strong oxidizing agent, phosphoric acid is preferable to nitric or sulfuric acid (Fig. 21-22).

The right side of Figure 21-21 shows that the oxidizing strengths of all of these oxoacids are drastically curtailed when they are replaced by their basic forms at pH 14. The highest reduction potential achieved (that for nitrate ion) is only +0.01 V at pH 14, and so basic conditions should be avoided if oxidation is desired. On the other hand, the lower oxidation states of the sulfur and phosphorus compounds can be excellent *reducing* agents at pH 14. Both sulfite ($SO_3^{2-}$) and thiosulfate ($S_2O_3^{2-}$) ions are fairly strong reducing agents in basic solution, as shown by their appearance on the right sides of half-equations with negative reduction potentials (−0.98 and −0.58 V, respectively). The bases of the lower oxoacids of phosphorus ($H_2PO_2^-$ and $HPO_3^{2-}$) are even stronger reducing agents.

From Figure 21-21 we can identify which species of intermediate oxidation state are unstable with respect to disproportionation under the conditions given.

**Figure 21-22**

Concentrated nitric acid reacts with sugar after some time (*left*), whereas sulfuric acid (*center*) reacts immediately. By contrast, sugar and phosphoric acid (*right*) form a viscous solution, but no reaction is seen. (Leon Lewandowski)

**Figure 21-23**

(a) The simplest Lewis diagram that can be drawn for $H_3PO_3$ gives an incorrect structure. This acid would be triprotic, like $H_3PO_4$. (b) The observed structure of $H_3PO_3$, which requires assigning formal charge to the P atom and the lone O atom. The hydrogen atom attached to the P is not released into acidic solution, and so the acid is diprotic.

Nitrous acid, for example, disproportionates spontaneously in acidic solution according to

$$3\,HNO_2 \longrightarrow NO_3^- + 2\,NO + H_3O^+ \qquad \Delta\mathscr{E}° = 0.99 - 0.94 = +0.05\ V$$

In basic solution, however, the nitrite ion is stable against disproportionation to NO and $NO_3^-$ because the hypothetical reaction

$$3\,NO_2^- + H_2O \longrightarrow NO_3^- + 2\,NO + 2\,OH^-$$

has $\Delta\mathscr{E}° = -0.46 - 0.01 = -0.47\ V < 0$, and so the equilibrium lies strongly to the left. The $NO_2^-$ ion is still thermodynamically unstable with respect to $NO_3^-$ and $N_2O$ or $N_2$, but these reactions occur so slowly that $NO_2^-$ can be kept in basic solution indefinitely.

One might guess at first that having an element in a very high oxidation state would make a compound a strong oxidizing agent—as is the case with permanganate solutions, which contain Mn(VII) and certainly do have great oxidizing power. Figure 21-21 shows that this is not true in general, however. Nitrous acid (N oxidation state = +3) would be a stronger oxidizing agent than nitric acid (+5) at pH 0, were it stable against disproportionation. Aqueous sulfur dioxide (S oxidation state = +4) is a stronger oxidizing agent than hydrogen sulfate ion (+6). The important factor is not the oxidation number itself but rather the relative stabilities of the redox pair of chemical species. Nitrous acid is a strong oxidizing agent precisely because it is much less stable than its reduced form, nitrogen oxide. Thermodynamic stability changes with oxidation state in an irregular fashion.

Oxidation state also affects acid strength for the oxoacids of S, P, and N as well as those of other nonmetals. Higher oxidation states should correspond to more covalent X—O bonds and thus higher acid ionization constants (recall Fig. 14-5). An even better correlation than oxidation state, however, is a related quantity: the number of lone oxygen atoms attached to the central atom. The rule is that the strength of oxoacids with a given central element increases with the number of lone oxygen atoms attached to the central atom. If the formula of these acids is written as $XO_n(OH)_m$, the corresponding acid strengths fall into distinct classes according to the value of $n$, the number of lone oxygen atoms (see Table 21-2). Each increase of 1 in $n$ increases the acid ionization constant $K_a$ by a factor of about $10^5$. Another

**Table 21-2   Acid Ionization Constants for Oxoacids of the Nonmetals**

| $X(OH)_m$, Very Weak | $K_a$ | $XO(OH)_m$, Weak | $K_a$ | $XO_2(OH)_m$, Strong | $K_a$ | $XO_3(OH)_m$, Very Strong | $K_a$ |
|---|---|---|---|---|---|---|---|
| $Cl(OH)$ | $3 \times 10^{-8}$ | $H_2PO(OH)$ | $8 \times 10^{-2}$ | $SeO_2(OH)_2$ | $10^3$ | $ClO_3(OH)$ | $2 \times 10^7$ |
| $Te(OH)_6$ | $2 \times 10^{-8}$ | $IO(OH)_5$ | $2 \times 10^{-2}$ | $ClO_2(OH)$ | $5 \times 10^2$ | | |
| $Br(OH)$ | $2 \times 10^{-9}$ | $SO(OH)_2$ | $2 \times 10^{-2}$ | $SO_2(OH)_2$ | $1 \times 10^2$ | | |
| $As(OH)_3$ | $6 \times 10^{-10}$ | $ClO(OH)$ | $1 \times 10^{-2}$ | $NO_2(OH)$ | $2 \times 10^1$ | | |
| $B(OH)_3$ | $6 \times 10^{-10}$ | $HPO(OH)_2$ | $1 \times 10^{-2}$ | $IO_2(OH)$ | $1.6 \times 10^{-1}$ | | |
| $Ge(OH)_4$ | $4 \times 10^{-10}$ | $PO(OH)_3$ | $8 \times 10^{-3}$ | | | | |
| $Si(OH)_4$ | $2 \times 10^{-10}$ | $AsO(OH)_3$ | $5 \times 10^{-3}$ | | | | |
| $I(OH)$ | $4 \times 10^{-11}$ | $SeO(OH)_2$ | $3 \times 10^{-3}$ | | | | |
| | | $TeO(OH)_2$ | $3 \times 10^{-3}$ | | | | |
| | | $NO(OH)$ | $5 \times 10^{-4}$ | | | | |

way to describe this effect is to focus on the stability of the conjugate base, $XO_{n+1}(OH)_{m-1}^-$, of the oxoacid. The greater the number of lone oxygen atoms attached to the central atom, the more easily the net negative charge can be spread out over the ion, and therefore the more stable the base. This leads to a larger $K_a$.

An unusual and interesting structural result can be obtained from Table 21-2. Figure 21-23$a$ shows the simplest Lewis structure for phosphorous acid ($H_3PO_3$), in which each atom achieves an octet configuration. Such a structure could also be written $P(OH)_3$ and would be analogous to $As(OH)_3$, which has no lone oxygen atoms bonded to the central atom ($n = 0$). On the basis of this analogy, the expected value of $K_a$ for $P(OH)_3$ is on the order of $10^{-9}$ (a very weak acid). In fact, however, $H_3PO_3$ is only a moderately weak acid ($K_a = 1 \times 10^{-2}$) and fits better into the class of acids with one lone oxygen atom bonded to the central atom. X-ray diffraction measurements support this analysis based on chemical properties. The true structure of $H_3PO_3$ is best represented by Figure 21-23$b$ and corresponds either to a Lewis structure with more than eight electrons around the central phosphorus atom or to one with formal charges on the central phosphorus and lone oxygen atoms. The formula of this acid is written as $HPO(OH)_2$ in Table 21-2. Unlike phosphoric acid ($H_3PO_4$), which is a triprotic acid, $H_3PO_3$ is a diprotic acid. The third hydrogen atom, the one directly bonded to the phosphorus atom, does not ionize even in strongly basic aqueous solution.

After studying this chapter and working the problems that follow, you should be able to

1. Compare the lead-chamber and contact processes for the manufacture of sulfuric acid (Section 21-1, problems 1–6).
2. Outline the role of sulfur in the manufacture of paper pulp from wood (Section 21-1, problem 7).
3. Describe methods for making phosphoric acid and its uses in fertilizer manufacture and consumer chemistry (Section 21-2, problems 9–12).
4. Give thermodynamic analyses of different ways of fixing nitrogen in the form of ammonia (Section 21-3, problems 13–18).
5. Describe the stages in the industrial synthesis of nitric acid and the uses of this chemical (Section 21-4, problems 19–20).
6. Discuss the structures and chemistries of nitrogen compounds with oxidation states intermediate between ammonia and elemental nitrogen (Section 21-4, problems 23–29).
7. List several nitrogen-containing explosives and describe their mechanisms of action (Section 21-4, problems 33–34).
8. Compare the redox and acid–base chemistries of the aqueous oxoacids of S, P, and N (Section 21-5, problems 35–36).

### Sulfuric Acid and Its Uses

1. Propose several ways in which predictions from Le Châtelier's principle could be used to improve the yield of $SO_3(g)$ from $SO_2(g)$ and air.

2. Give a thermodynamic analysis of the Claus process; that is, determine the $\Delta H°$ and $\Delta S°$ of the main reaction and predict conditions that will favor a high yield of sulfur. Take S($s$) as the physical form of sulfur.

3. Residual $SO_2$ can be removed from the effluent gases of sulfuric acid plants by oxidation with $H_2SO_5$, as explained in the text. Suppose that the electrolysis system that supplies the $H_2SO_5$ failed. Would it be thermodynamically feasible to substitute an aqueous solution of $H_2O_2$ as the oxidant for the $SO_2$? Consult Appendix D for necessary data.

4. Amounts of sulfur in a sample can be measured by burning the sample in oxygen and then collecting the oxides of sulfur in a dilute solution of hydrogen peroxide. What is the product of this reaction? Write a balanced equation. Suggest a good way to determine quantitatively the amount of product that is formed.

5. Write a balanced overall chemical equation for the production of sulfuric acid from zinc sulfide ore.

6. Write a balanced overall chemical equation for the production of sulfuric acid from hydrogen sulfide in natural gas.

7. Describe (via chemical equations) how sodium sulfide for use in the Kraft process can be prepared starting from sulfuric acid and other easily accessible materials.

8. Predict the product(s) of the reaction of disulfuric acid ($H_2S_2O_7$) with water.

## Phosphorus Chemistry

9. Hot concentrated sulfuric acid reacts with phosphorus to give sulfur dioxide, phosphoric acid, and water. Write a balanced chemical equation for this reaction.

10. Describe (via chemical equations) how sodium tripolyphosphate for detergents can be prepared starting from phosphoric acid and other easily accessible materials.

11. Phosphoric acid ($H_3PO_4$) can be kept pure only in the crystalline state. Above its melting point of 42.4°C, pairs of $H_3PO_4$ units gradually lose water. What is the other product of this dehydration reaction? What is its structure?

12. The P—O—P linkages in diphosphoric acid are extended by adding other phosphate groups (through reaction with $H_3PO_4$ and elimination of water). The resulting long-chain phosphate is called "metaphosphoric acid." Determine its empirical formula, assuming that the chain is very long.

## Nitrogen Fixation

13. A minor source of fixed nitrogen is emission from the inside of the earth in volcanic eruptions. One estimate of the total amount of fixed nitrogen per year from this source is $2 \times 10^8$ kg. How many atoms of fixed nitrogen are emitted per second, on the average, by volcanoes?

14. Observations of lightning suggest that the average lightning stroke produces $10^{27}$ molecules of NO. The rate of lightning strokes worldwide is about 100 per second. According to these data, how much nitrogen (in metric tons) is fixed annually by lightning strokes?

15. Write an overall equation to represent the production of $NH_3$ by the cyanamide process as described in the text. Assume

that the starting materials are limestone ($CaCO_3$), coke (C), nitrogen ($N_2$), and water. Determine the $\Delta H°$ of the process represented by your equation, per mole of $NH_3$ produced.

16. Suppose that, in the preceding problem, the calcium hydroxide and carbon dioxide produced were recycled to give the calcium carbonate starting material. Write the overall equation in this case, and calculate the $\Delta H°$ per mole of $NH_3$ produced.

17. Some chemical plants manufacture ammonia via the Haber–Bosch process using pressures as high as 700 atm rather than the 200 atm cited in the text. What are the advantages and disadvantages of working at these higher pressures? For a given yield of product, should a higher or lower temperature be used if the pressure is raised? Will the reaction then be faster or slower?

18. Suppose a catalyst were developed that allowed the rapid production of ammonia from the elements at 600 K. Would this be useful in the Haber-Bosch process? Estimate the equilibrium constant for the formation of 2 mol of $NH_3$ at this temperature, using the van't Hoff equation.

## Chemicals from Ammonia

19. For each of the three steps in the Ostwald process for making nitric acid from ammonia, state whether a high yield of product will be favored by low or high temperature.

20. For each of the three steps in the Ostwald process for making nitric acid from ammonia, state whether a high yield of product will be favored by low or high total pressure.

21. Write balanced equations to represent the reaction of concentrated nitric acid with Cr($s$) and with Fe($s$). Assume that the nitric acid is reduced to $NO_2(g)$ and that Cr and Fe are oxidized to the +3 oxidation state.

22. Write balanced equations to represent the reaction of concentrated nitric acid with Ni($s$) and with Tl($s$), assuming that the nitric acid is reduced to $NO(g)$ and that Ni and Tl are oxidized to the +2 and +3 oxidation states, respectively.

23. *Trans*-tetrazene has the formula $N_4H_4$. It consists of a chain of four nitrogen atoms with the two terminal nitrogen atoms bonded to two hydrogen atoms each. Draw a Lewis structure for this compound.

24. Diimine ($N_2H_2$) has not been isolated, but there is evidence that it exists in solution.
   (a) Draw a Lewis structure for this compound.
   (b) Would the conversion of hydrazine to diimine in solution require an oxidizing agent or a reducing agent?

25. Two reduction half-reactions involving hydrazine ($N_2H_4$) are

$$N_2H_4 + 2\,H_2O + 2\,e^- \longrightarrow 2\,NH_3 + 2\,OH^-$$
$$\mathscr{E}° = 0.1 \text{ V}$$

$$N_2 + 4\,H_2O + 4\,e^- \longrightarrow N_2H_4 + 4\,OH^-$$
$$\mathscr{E}° = -1.16 \text{ V}$$

Determine whether hydrazine is thermodynamically stable against disproportionation to $N_2$ and $NH_3$ at pH 14.

**26.** In our discussion of conversions among $N_2$, $N_2H_4$, and $NH_3$, we did not mention a compound of nitrogen named hydroxylamine ($NH_2OH$).
(a) What is the oxidation number of nitrogen in $NH_2OH$?
(b) Two reduction half-reactions involving $NH_2OH$ are

$$2\,NH_2OH + 2\,e^- \longrightarrow N_2H_4 + 2\,OH^- \qquad \mathscr{E}° = -0.73 \text{ V}$$

$$N_2 + 4\,H_2O + 2\,e^- \longrightarrow 2\,NH_2OH + 2\,OH^-$$
$$\mathscr{E}° = -1.59 \text{ V}$$

Determine whether hydroxylamine is thermodynamically stable against disproportionation to $N_2$ and $N_2H_4$ at pH 14.

**27.** The suggestion is made that the reaction between liquid hydrazine and liquid nitric acid be used to boost a specialized type of rocket because the products, $N_2(g)$ and $H_2O(\ell)$, are completely innocuous in the atmosphere. Write a balanced equation, and compute $\Delta H°$ for this reaction.

**28.** Write a balanced chemical equation for a reaction between hydrazine and dinitrogen tetraoxide that gives only nitrogen and water. Compute $\Delta H°$ for this reaction.

**29.** Hydrazine ($N_2H_4$) is less basic than ammonia, and $NF_3$ is not basic at all. Formulate a general principle about the effect of substituents attached to nitrogen on the nitrogen's ability to accept hydrogen ions (or, equivalently, to donate electrons). Do you expect hydroxylamine ($NH_2OH$) to be more basic or less basic than ammonia?

**30.** Suppose the hydrogen atoms in phosphine ($PH_3$) are replaced by chlorine atoms to give $PCl_3$. Do you expect $PCl_3$ to be a stronger or a weaker Lewis base than $PH_3$? Explain your reasoning.

**31.** Describe a practical reaction sequence by which $N_2O_4(s)$ could be obtained from the elements.

**32.** Describe a practical reaction sequence by which $N_2O(g)$ could be obtained from the elements.

**33.** During World War II, the Allies used to grind up downed Nazi war planes, which contained aluminum, and mix them in with ammonium nitrate to make bombs, which were then dropped over Germany. Explain the reason for putting in the aluminum from the airplanes.

**34.** As stated in Section 21-4, ammonium nitrate mixed with fuel oil is an inexpensive and widely used explosive. The fuel oil consists of long-chain hydrocarbons, $(CH_2)_n$, with $n = 8$ to 15, and it reacts with the oxygen produced in the decomposition of ammonium nitrate to give $H_2O(g)$ and $CO_2(g)$. This increases both the exothermicity of the reaction and the gas volume generated, both of which lead to greater explosive power. Calculate the optimal *mass fraction* of fuel oil that should be mixed with ammonium nitrate to achieve this goal.

**35.** Figure 21-21 can be extended to include species in which the oxidation states of sulfur, phosphorus, and nitrogen are 0 or negative. In basic solution, for example, the following addition reduction potentials are measured at pH 14:

$$P_4(s) + 12\,H_2O(\ell) + 12\,e^- \longrightarrow 4\,PH_3(g) + 12\,OH^-$$
$$\Delta\mathscr{E}° = -0.89 \text{ V}$$

$$4\,H_2PO_2^- + 4\,e^- \longrightarrow P_4(s) + 8\,OH^- \qquad \Delta\mathscr{E}° = -2.05 \text{ V}$$

(a) Will $P_4$ disproportionate spontaneously in basic solution?
(b) Which is the stronger reducing agent at pH 14—$P_4$ or $PH_3$?

**36.** Extending Figure 21-21, the following additional reduction potentials are measured at pH 14:

$$S(s) + H_2O + 2\,e^- \longrightarrow HS^- + OH^-$$
$$\Delta\mathscr{E}° = -0.51 \text{ V}$$

$$S_2O_3^{2-} + 3\,H_2O(\ell) + 4\,e^- \longrightarrow 2\,S(s) + 6\,OH^-$$
$$\Delta\mathscr{E}° = -0.74 \text{ V}$$

(a) Will sulfur disproportionate spontaneously under basic conditions?
(b) Which is the stronger reducing agent—S or $HS^-$?

## Additional Problems

**37.** Use thermodynamics to compare oxidation of $SO_2$ in the absence of $NO_2$

$$SO_2(g) + \tfrac{1}{2}O_2(g) \longrightarrow SO_3(g)$$

with oxidation of $SO_2$ by $NO_2$:

$$SO_2(g) + NO_2(g) \longrightarrow SO_3(g) + NO(g)$$

Which has the greater thermodynamic driving force $(-\Delta G)$ at 25°C? At low temperatures, does the $NO_2$ act by increasing the yield or by speeding up the reaction?

**38.** In the lead-chamber process, the sulfuric acid was contained in a lead container. Using the information in Table 7-1, explain the great resistance of lead to corrosion by sulfuric acid.

**39.** The standard reduction potential for the half-reaction

$$S_2O_8^{2-} + 2\,e^- \longrightarrow 2\,SO_4^{2-}$$

is 2.0 V. Use the Nernst equation to compute the minimum voltage required to operate an electrolysis unit that converts 1.0 M $SO_4^{2-}$ to 0.50 M $S_2O_8^{2-}$ and hydrogen at a pressure of 0.10 atm if the pH is 0.

**40.** The world production of pure $H_2SO_4$ in 1991 was $5.51 \times 10^{10}$ kg. Suppose that all locations converted sulfur to $H_2SO_4$ at 99.97% efficiency but lost the rest of the sulfur as $SO_2$ to the atmosphere. What, then, was the total mass of emitted sulfur dioxide (in kilograms)?

**41.** Write a chemical equation for (a) a precipitation reaction, (b) an acid–base reaction, and (c) an oxidation–reduction reaction, in each of which aqueous sulfuric acid is a reactant.

**42.** The $Na_2S$ for the Kraft process can be made by heating $Na_2SO_4$ with coke (C). Write a balanced chemical equation for this reaction. Can this reaction be used to supply heat for the digestion of the wood, or does it consume heat?

**43.** Most sulfuric acid (other than that used in making fertilizer) does not form part of the final product. It lingers inconveniently as spent sulfuric acid that is both diluted and contaminated. Spent acid must not be dumped in the environment.

You propose to regenerate $H_2SO_4$ by heating spent acid to high temperatures (about 1000 K) in the presence of natural gas ($CH_4$) to make $SO_2(g)$ by the reaction

$$4\,SO_3(g) + CH_4(g) \longrightarrow CO_2(g) + 4\,SO_2(g) + 2\,H_2O(g)$$

and then recycling the $SO_2(g)$ to make more $H_2SO_4$.
(a) Estimate $\Delta G$ for your proposed reaction.
(b) A critic asserts that the reaction will probably proceed as follows:

$$SO_3(g) + CH_4(g) \longrightarrow CO_2(g) + H_2O(g) + H_2S(g)$$

Estimate $\Delta G$ at 1000 K for the critic's proposed reaction.
(c) How can you tell which reaction will occur? Explain.

44. Suppose a fertilizer plant produces phosphoric acid through the wet-acid process and uses it to make triple superphosphate fertilizer ($Ca(H_2PO_4)_2 \cdot H_2O$). Write a balanced equation for the overall reaction. How many moles of gypsum are generated per mole of $Ca(H_2PO_4)_2 \cdot H_2O$?

45. In fertilizer manufacture, it is conventional to measure relative amounts of phosphorus in different fertilizers by arbitrarily assuming that the phosphorus is present as $P_2O_5$, dividing the mass of $P_2O_5$ that would be present by the total mass, and expressing that mass as a percentage. For example, the formula of triple superphosphate is rewritten as

$$Ca(H_2PO_4)_2 \cdot H_2O = (P_2O_5)(CaO)(H_2O)_3$$

The ratio of the mass of $P_2O_5$ to the total mass is then the ratio of the molar mass of $P_2O_5$ to the molar mass of the formula unit, or $141.94/252.07 = 0.563$, and so this fertilizer is 56.3% $P_2O_5$. Carry out the corresponding calculation for ordinary superphosphate fertilizer, which contains 7 mol of gypsum for every 3 mol of $Ca(H_2PO_4)_2 \cdot H_2O$. Account for the origin of the word "triple" in "triple superphosphate."

46. The disposal of byproduct gypsum ($CaSO_4 \cdot 2H_2O(s)$) is a significant problem for fertilizer manufacturers.
(a) A typical fertilizer plant may produce 100 metric tons of phosphoric acid per day by the wet-acid process (1 metric ton = $10^3$ kg). What mass of gypsum is produced per *year*?
(b) The density of crystalline gypsum is 2.32 g cm$^{-3}$. What volume of gypsum is generated per year? (In fact, the volume is several times *larger* because of the porosity of the gypsum produced.)

47. During the early 1970s, when the price of petroleum and natural gas rose rapidly, the price of nitrogen-based fertilizers increased rapidly as well. Explain why the two commodities are so closely linked.

48. A piece of hot platinum foil is inserted into a container filled with a mixture of ammonia and oxygen. Brown fumes appear near the surface of the foil. Write balanced equations for the reactions taking place.

49. Nitrogen oxides ($NO_x$) from power stations, steam generators, and other fossil-fuel-burning sources are serious air pollutants. One promising technology (called NOxOUT) for

removing $NO_x$ from stack emissions injects an aqueous solution of urea, $(NH_2)_2CO$, into various points of the combustion system. Urea reacts with the nitrogen oxides to give $N_2$, $CO_2$, and $H_2O$.
(a) Write a balanced chemical equation to represent the reaction of urea with $NO_2$.
(b) Write a second balanced chemical equation to represent the reaction of urea with NO.

50. The concentration of $NO_3^-(aq)$ in a solution can be determined by adding sulfuric acid and mercury and then collecting the gaseous NO that the reaction generates:

$$2\,NO_3^-(aq) + 8\,H_3O^+(aq) + 3\,SO_4^{2-}(aq) + 6\,Hg(\ell) \longrightarrow$$
$$3\,Hg_2SO_4(s) + 12\,H_2O(\ell) + 2\,NO(g)$$

A 357-mL sample of a solution is treated in this way and generates 11.46 mL of NO($g$), measured at a pressure of 0.965 atm and a temperature of 18.8°C. Compute the concentration of nitrate ion in the solution.

51. The formation of hydrazine in the Raschig synthesis is thought to be a two-step process that involves $NH_2Cl$ as an intermediate. Write balanced equations for the two steps in the Raschig synthesis.

52. Molecules of hydrazine ($N_2H_4$) have a sizable dipole moment. Sketch a structure that this information definitely *eliminates* as the structure of the hydrazine molecule.

53. (a) Compute the $\Delta G°$ of the reaction

$$6\,Fe_2O_3(s) + N_2H_4(aq) \longrightarrow$$
$$4\,Fe_3O_4(s) + N_2(g) + 2\,H_2O(\ell)$$

(b) As explained in the text, this reaction is used to prevent further rusting of the walls of boilers. A boilermaker asks you about a proposal to improve this procedure by making the solution of hydrazine strongly acidic (to "burn off the worst of the rust as the hydrazine simultaneously seals off the rest"). Evaluate this idea.

54. Determine how many grams of each ingredient you would need to include in a mixture to make 100 g of gunpowder. Use the information in Table 21-1.

55. The first acid ionization constant of the oxoacid $H_3PO_2$ is $8 \times 10^{-2}$. What molecular structure do you predict for $H_3PO_2$? Will this acid be monoprotic, diprotic, or triprotic in aqueous solution?

56. Oxoacids can be formed that involve several central atoms of the same chemical element. (Sodium salts of these polyphosphoric acids are used as "builders" in detergents to improve their cleaning power.) An example is $H_3P_3O_9$, which can be written $P_3O_6(OH)_3$. In such a case, we would expect acid strength to correlate approximately with the *ratio* of the number of lone oxygen atoms to the number of central atoms (this ratio is 6:3 for $H_3P_3O_9$, for example). Rank the following in order of increasing acid strength: $H_3PO_4$, $H_3P_3O_9$, $H_4P_2O_6$, $H_4P_2O_7$, $H_5P_3O_{10}$. Assume that no hydrogen atoms are directly bonded to phosphorus in these compounds.

*Halogens.* Left, *Iodine (I₂) is a grayish black solid that sublimes to form violet vapor.* Center, *Bromine (Br₂) is a reddish brown liquid that forms vapor of the same color.* Right, Chlorine (Cl₁₂) is a greenish yellow gas at room temperature. (Richard Megna/ Fundamental Photographs, New York)

# The Halogen Family and the Noble Gases

The elements that make up the last two main groups of the periodic table are a study in contrasts. The halogens are among the most reactive of all the elements and are found terrestrially only in combination with other elements. Except for oxygen itself, they are the only elements that are classified as oxidizing agents. In contrast, the noble gases are so unreactive that they are found in nature only in elemental form.

This chapter begins with a historical survey of the industrial-scale preparation of chlorine and its linkage with the two important bases sodium carbonate and sodium hydroxide. After a comparative look at the chemistries of chlorine, bromine, and iodine, the special properties of fluorine are introduced. This element and some of its compounds are highly reactive and toxic, but other compounds are at the opposite extreme and are deliberately chosen for uses demanding inert, nontoxic materials. Under proper conditions, fluorine forms compounds with nearly all of the other elements, the only exceptions being helium, neon, and argon.

## 22-1 CHEMICALS FROM SALT

By the second half of the 18th century, several important industries in Europe and North America faced a growing crisis in the supply of raw materials. The glass industry relied on soda ash (sodium carbonate, $Na_2CO_3$) or potash (potassium carbonate, $K_2CO_3$) as a flux to lower the viscosity of the molten glass. As their common names imply, the two carbonates were derived from the leachings of ashes: soda ash from the ashes of sea plants, and potash from the ashes of forest trees and shrubs. By 1750, demand for glass was exceeding the supplies from traditional sources, which were of variable quality in any case.

Soapmakers faced a supply problem of their own, because they needed a cheap base to react with vegetable oils and animal fats. One base, lime, was readily available. It was produced by high-temperature decomposition of the limestone that was abundantly available throughout the world:

$$CaCO_3(s) \xrightarrow{850°C} CaO(s) + CO_2(g)$$

In the 18th century, lime kilns were among the largest industrial structures in Europe. Quicklime (CaO) could be **slaked** by adding water to give a sparingly soluble hydroxide:

$$CaO(s) + H_2O(\ell) \longrightarrow Ca(OH)_2(s)$$

In this form, it served well to neutralize acids but had no use in the soap industry because the calcium salts that formed from its reaction with oils were insoluble in water. A soap must be water-soluble.

An expanding demand for fine clothing greatly increased the need to bleach cotton and linen in the mid-18th century. In early bleaching methods, the cloth or fibers were repeatedly exposed to lactic acid from sour milk, bases from wood ashes, and the bright light of the sun. Cotton took up to three months to bleach, and linen up to six, and the entire process was very dependent on the weather and the supply of milk. In the closing years of the 18th century, the lead-chamber process (Section 21-1) began to supply sulfuric acid that could replace the sour milk and greatly hasten that part of the process. The need for a cheap base and a bleaching agent more effective than sunlight created major bottlenecks.

This section describes how salt (sodium chloride, NaCl) provided the answer to these problems. Processes were developed to convert salt to sodium carbonate and sodium hydroxide, and the byproducts of that conversion included compounds of chlorine that could be made into effective bleaches. Several competing processes rose and fell, with utilization of waste materials for economy playing a central role. These processes, together with the production of sulfuric acid (examined in Chapter 21), dominated the 19th-century chemical industry and continue to play a very important role today.

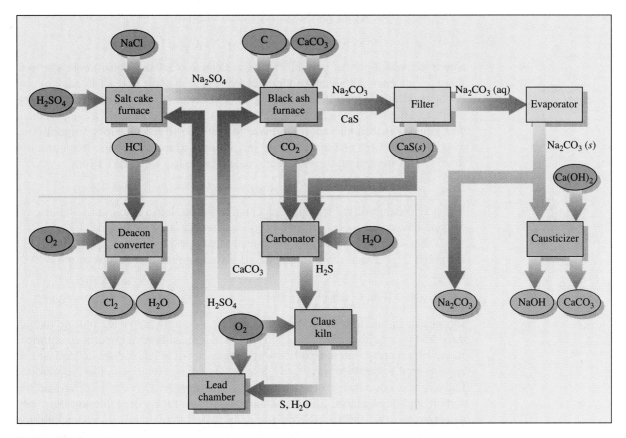

**Figure 22-1**

In the original version of the Leblanc process, HCl, CO$_2$, and CaS were all waste products generated along with the desired Na$_2$CO$_3$. Later versions of the process (below the yellow line) converted HCl to useful Cl$_2$ and recovered sulfur from the CaS to make sulfuric acid.

## The Leblanc Process

In 1775 the French Academy of Sciences offered a prize for a satisfactory process to make sodium carbonate from sodium chloride. A French physician and amateur chemist, Nicolas Leblanc, solved the problem during the next 15 years, although he did not receive credit for the discovery within his lifetime. Figure 22-1 outlines the process named after him.

The first step in the **Leblanc process** is the production of salt cake, or sodium sulfate (Na$_2$SO$_4$). The process Leblanc employed was the standard one at the time and involved the addition of sulfuric acid to sodium chloride at temperatures of 800 to 900°C:

$$2\,NaCl(s) + H_2SO_4(aq) \longrightarrow Na_2SO_4(aq) + 2\,HCl(g)$$

The sodium sulfate was ground up, mixed with limestone and charcoal or powdered coal, and put in trays in a furnace. The reactions that took place can be summarized in two equations:

$$Na_2SO_4(s) + 2\,C(s) \longrightarrow Na_2S(s) + 2\,CO_2(g)$$

$$Na_2S(s) + CaCO_3(s) \longrightarrow Na_2CO_3(s) + CaS(s)$$

The result was a porous gray-black mixture called black ash, which consisted primarily of calcium sulfide and sodium carbonate, with some unreacted calcium carbonate and other impurities. Filtering systems were developed to leach out the soluble sodium carbonate, leaving behind the solid calcium sulfide as a byproduct. The liquid was then burned in a furnace to remove sulfur and carbon impurities as $SO_2(g)$ and $CO_2(g)$ and leave behind the desired sodium carbonate. If sodium hydroxide was needed, the sodium carbonate was boiled with slaked lime.

$$Na_2CO_3(aq) + Ca(OH)_2(aq) \longrightarrow CaCO_3(s) + 2\,NaOH(aq)$$

The precipitation of the insoluble calcium carbonate left a solution of sodium hydroxide that could be concentrated further or evaporated to dryness.

The use of the Leblanc process spread rapidly, especially in Great Britain, and increased the demand for sulfuric acid as a starting material. A look at the overall equation for the process

$$2\,NaCl + H_2SO_4 + 2\,C + CaCO_3 \longrightarrow Na_2CO_3 + 2\,CO_2 + CaS + 2\,HCl$$

shows that in addition to the primary product, sodium carbonate, and the carbon dioxide, two significant byproducts result—calcium sulfide and hydrogen chloride. In the 19th century, calcium sulfide was dumped at sea or left in piles on land, where it weathered to produce noxious $H_2S$ and $SO_2$ gases. Hydrogen chloride was initially vented as an acidic gas, which laid waste to vegetation for miles around the sodium carbonate plants. Chimneys hundreds of feet high were built, but they only dispersed the HCl farther. The next remedy passed the gaseous hydrogen chloride into towers in which water trickled down over a packing of coke. The hydrogen chloride dissolved to create a dilute solution of hydrochloric acid that was run off into streams and rivers, substituting water pollution for air pollution. In 1863 the Alkali Act prohibited such pollution in Great Britain. Although some condemned the law as excessive interference by the government in private business, it stimulated the search for ways of using the hydrochloric acid. Efficient processes were developed to convert it to bleaching powder, making a noxious waste into a useful product.

Before we consider hydrochloric acid conversion processes, let us return to a somewhat earlier period to examine the role of chlorine and its compounds as bleaching agents.

## Chlorine and Bleaches

The Swedish chemist Karl Wilhelm Scheele discovered chlorine in 1774 and prepared it in its elemental form through the reaction of hydrochloric acid with pyrolusite, a mineral containing $MnO_2$:

$$4\,HCl(aq) + MnO_2(s) \longrightarrow Cl_2(g) + MnCl_2(aq) + 2\,H_2O(\ell)$$

Scheele did not realize that the greenish yellow gas he had produced was an element, and it remained for Humphry Davy to identify it as such in 1811 and to name it (from the Greek word *chloros,* meaning "green"). In the meantime, Berthollet and de Saussure had already described chlorine's bleaching properties in 1786. The

Addition of aqueous HBr from the funnel to solid $MnO_2$ produces gaseous bromine; a similar reaction generates chlorine from HCl.

chlorine was unsatisfactory in several ways, however: it would disintegrate the cloth unless monitored carefully, it was difficult to handle, and at the time it could not be transported. The Scottish chemist Charles Tennant made a significant advance in 1799 when he patented a material that he called **bleaching powder,** which was made by saturating slaked lime with chlorine:

$$Ca(OH)_2(s) + Cl_2(g) \longrightarrow CaCl(OCl)(s) + H_2O(\ell)$$

Together with sulfuric acid, bleaching powder greatly reduced the time needed to bleach cotton and linen, so that by the 1830s one week (rather than several months) was sufficient to bleach cotton goods.

Bleaching powder acts in aqueous solution by releasing hypochlorite ion ($OCl^-$),

$$CaCl(OCl)(s) \longrightarrow Ca^{2+}(aq) + Cl^-(aq) + OCl^-(aq)$$

which in turn can carry out the bleaching. One disadvantage of bleaching powder is that, on standing, it decomposes to calcium chloride and oxygen:

$$CaCl(OCl)(s) \longrightarrow CaCl_2(s) + \tfrac{1}{2}O_2(g)$$

This can be avoided by preparing calcium hypochlorite, $Ca(OCl)_2$, which also produces twice as much hypochlorite ion in solution, making it more effective on the basis of mass. Modern household bleach uses the sodium salt of hypochlorite ion ($NaOCl$) rather than the calcium salt. Sodium chlorite ($NaClO_2$) is an even stronger bleach that is used in industrial applications such as paper bleaching.

For almost a century after chlorine's discovery in 1774, the major method of making chlorine for bleach was the original reaction used by Scheele. This is an extremely wasteful method, because all of the manganese and much of the chlorine is lost as $MnCl_2$. By the mid-19th century, hydrochloric acid (the noxious byproduct of the Leblanc process) was in extensive use in bleach manufacture, and a less wasteful method of oxidizing it was needed.

Between 1868 and 1874, the British chemist and industrialist Henry Deacon developed a process to convert gaseous hydrogen chloride into chlorine over a copper chloride catalyst:

$$2\,HCl(g) + \tfrac{1}{2}O_2(g) \longrightarrow Cl_2(g) + H_2O(g)$$

The **Deacon process** was widely adopted, and hydrochloric acid from the Leblanc process became the major source of chlorine for bleaches. As time went on, in fact, the sodium carbonate from the Leblanc process became the less important product economically and was sold at a loss. The Leblanc process yielded profits only because bleaching powder was coproduced from hydrogen chloride, the former waste product. This turnabout illustrates the way in which the chemical industry adapts to changing circumstances.

## The Solvay Process

In the mid-19th century, a new process for making sodium carbonate was being developed in Belgium. Figure 22-2 is a flow diagram for the **Solvay process.** The first and most important reaction is

$$2\,NaCl(aq) + 2\,NH_3(aq) + 2\,CO_2(g) + 2\,H_2O(\ell) \longrightarrow$$
$$2\,NaHCO_3(s) + 2\,NH_4Cl(aq)$$

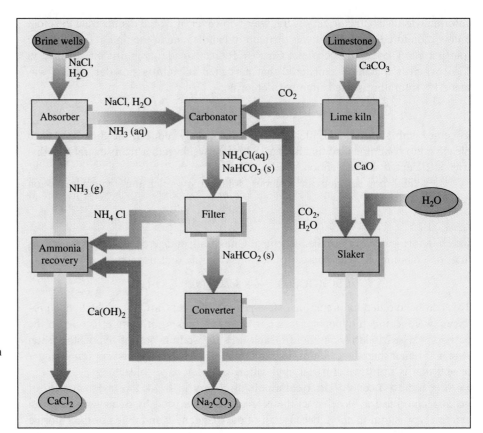

**Figure 22-2**
The net reaction in the Solvay process is the production of sodium carbonate and calcium chloride from salt and calcium carbonate. Ammonia plays a crucial role in making the process work but is not consumed (except for inevitable small losses).

This reaction is carried out in a carbonating tower, a tall reactor in which brine (concentrated aqueous NaCl) is saturated with ammonia and carbon dioxide. The sodium hydrogen carbonate ($NaHCO_3$) that forms is relatively insoluble and is removed by filtration on huge rotating drum filters. It is then decomposed thermally in a rotary dryer above 270°C to give sodium carbonate as the product:

$$2\,NaHCO_3(s) \longrightarrow Na_2CO_3(s) + H_2O(g) + CO_2(g)$$

The carbon dioxide freed in this reaction is returned to the carbonating tower, supplemented by $CO_2$ generated in a lime kiln:

$$CaCO_3(s) \longrightarrow CaO(s) + CO_2(g)$$

The lime is slaked to $Ca(OH)_2(aq)$,

$$CaO(s) + H_2O(\ell) \longrightarrow Ca(OH)_2(aq)$$

which is then used in a final step to recover ammonia for reuse in the carbonating tower:

$$2\,NH_4Cl(aq) + Ca(OH)_2(aq) \longrightarrow 2\,NH_3(aq) + CaCl_2(aq) + 2\,H_2O(\ell)$$

The sum of these five steps gives the remarkably simple overall reaction for the Solvay process:

$$2\,NaCl(aq) + CaCO_3(s) \longrightarrow Na_2CO_3(s) + CaCl_2(aq)$$

Although this "ammonia–soda" process had been described and studied by many chemists beginning in 1800, it had languished because it was uneconomical. For many years, the stumbling block was the inefficient recovery of the ammonia. The high cost of this material meant that losses had to be kept below about 5% per cycle to make the process feasible. In the 1860s, the Belgian chemist and industrialist Ernest Solvay developed and patented improved carbonators and stills to recover ammonia, and the process grew rapidly in importance, taking over markets from the Leblanc process. Raw material and fuel costs are lower in the Solvay process than in the Leblanc process (lower temperatures are required, and sulfuric acid is not a reactant). The Solvay process also produces a much purer grade of sodium carbonate in higher yield. It has the final advantage of being a continuous rather than a batch process, and it was the first such continuous process to achieve widespread commercial success.

Sodium carbonate is used primarily in the manufacture of glass, especially the soda-lime glass discussed in Section 20-2. Another use for sodium carbonate is in the production of sodium hydrogen carbonate ($NaHCO_3$), which has the common name bicarbonate of soda. Although $NaHCO_3$ appears as an intermediate in the Solvay process, it is not taken out for direct use because it contains too many impurities, especially ammonia, which gives it an odor. Instead, a saturated aqueous solution of sodium carbonate is prepared and carbonated:

$$Na_2CO_3(aq) + CO_2(g) + H_2O(\ell) \longrightarrow 2\,NaHCO_3(s)$$

The suspension of sodium hydrogen carbonate is filtered, washed, and dried, leaving a product of about 99.9% purity. It is used in the kitchen as baking soda because it reacts with acids, such as the lactic acid in sour milk or potassium hydrogen tartrate, to generate carbon dioxide, which makes dough rise. It is also used in dry chemical fire extinguishers (recall problem 54 in Chapter 5).

In the 20th century, even the Solvay process has yielded to other sources of bases. One reason for the decline of the Solvay process is its byproduct, calcium chloride. This salt is used for dust control in summer and to melt ice on streets in winter (a saturated $CaCl_2$ solution freezes at $-50°C$, as compared with $-20°C$ for a saturated NaCl solution). Beyond these uses, it has little commercial value and poses a difficult disposal problem. Although the Solvay process remains important worldwide, it has been almost completely supplanted in the United States by the mining of a double salt of sodium carbonate and sodium hydrogen carbonate called **trona,** with the approximate composition $2Na_2CO_3 \cdot NaHCO_3 \cdot 2H_2O$. Heating this ore gives sodium carbonate.

## Electrolysis and the Chlor-Alkali Industry

In the 1890s, a new process for the production of bases and chlorine was developed and grew rapidly in importance: the electrolytic generation of sodium hydroxide and chlorine from salt brines. The dominant technique for electrolysis in the United States is the **diaphragm cell** (Fig. 22-3). If an electric current is passed through a sufficiently concentrated solution of sodium chloride, the anode reaction is

$$2\,Cl^-(aq) \longrightarrow Cl_2(g) + 2\,e^-$$

and the cathode reaction is

$$2\,H_2O(\ell) + 2\,e^- \longrightarrow H_2(g) + 2\,OH^-(aq)$$

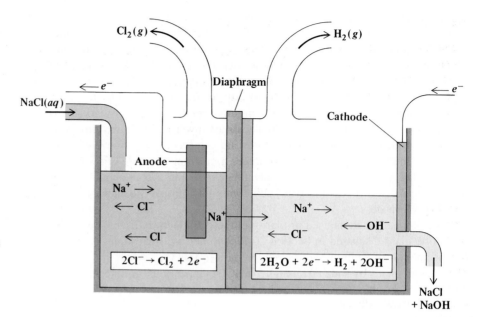

**Figure 22-3**

In a diaphragm cell, the cathode is a perforated steel box. Note that the level of the solution is deliberately made higher in the anode compartment than in the cathode to minimize migration of OH⁻ through the diaphragm to the anode, where it could react with the chlorine being generated.

The overall reaction is the sum of these half-reactions:

$$2\,H_2O(\ell) + 2\,Cl^-(aq) \longrightarrow 2\,OH^-(aq) + H_2(g) + Cl_2(g) \qquad \Delta G° = +422\ kJ$$

The gaseous hydrogen and chlorine bubble out and are collected and dried, while the hydroxide ion forms part of a mixed solution of sodium hydroxide and sodium chloride. The less soluble sodium chloride precipitates upon evaporation and can be removed by filtration, leaving a solution containing 50% by mass of sodium hydroxide in water. This solution may be used directly or evaporated further to make solid sodium hydroxide.

Early diaphragm cells used graphite anodes, but these had to be replaced frequently and have been superseded by anodes made of titanium coated with platinum, ruthenium, or iridium. The cathodes are steel boxes with perforated sides. The diaphragm prevents the mixing of the $H_2$ and $Cl_2$ gases but allows ions to pass. Asbestos has been the most widely used material for these diaphragms, but new diaphragms are being developed based on fluorinated polymers. Another approach is to replace the diaphragm with an ion-exchange membrane, which permits only sodium ions to pass to the cathode (Fig. 22-4). By preventing the passage of chloride ions, this method generates pure aqueous sodium hydroxide at the cathode.

The electrolytic production of sodium hydroxide and chlorine was proposed in the mid-19th century, but the only source of electricity at the time was the battery. This made the process so expensive that it was little more than a laboratory curiosity. By the 1890s, however, the development of the dynamo (generator) changed the situation dramatically. The electrolysis industry grew up rapidly around sources of cheap electricity from water power—in Norway and near Niagara Falls in the United States. This industry's first effect was to doom the Leblanc process, which had survived so long only from the profits in the coproduction of chlorine with sodium carbonate. With competing chlorine produced in greater purity electrolytically, the Leblanc process quickly disappeared. Over a longer period, the electrolytic chlor-alkali process has reduced the role of the Solvay process as well, by yielding a base (NaOH) that competes effectively with sodium carbonate in many applications.

The laws of chemistry require that each mole of chlorine produced by the electrolysis of brine be accompanied by two moles of sodium hydroxide. Because the demand for these two products is rarely in ideal balance, the prices of both commodities fluctuate. A byproduct of the reaction is hydrogen. It can react directly with chlorine to give gaseous hydrogen chloride of high purity, although there are less expensive sources for this chemical. Alternatively, the chlor-alkali process can be coupled with a fertilizer plant, and the hydrogen can be reacted with nitrogen to make ammonia (Section 21-3).

## 22-2 THE CHEMISTRY OF CHLORINE, BROMINE, AND IODINE

Chlorine, bromine, and iodine resemble one another much more closely than they do fluorine, whose special properties are discussed in Section 22-3. Section 22-1 described the production of chlorine and its role in making bleaches. That role is now overshadowed by the importance of chlorine in the manufacture of chlorinated hydrocarbons for use as solvents, pesticides, and plastics such as polyvinyl chloride (Sec. 23-3). Growing concern about the environmental effects and toxic properties of chlorinated hydrocarbons (see Fig. 22-5) has slowed the growth of the industry. Another major use for chlorine is the purification of titanium dioxide. This substance occurs in impure form; it reacts above 900°C with chlorine and coke to give titanium tetrachloride:

$$TiO_2(s) + 2\,C(s) + 2\,Cl_2(g) \longrightarrow TiCl_4(\ell) + 2\,CO(g)$$

Because $TiCl_4$ is a volatile liquid (b.p. 136°C), it can easily be purified by distillation and then oxidized to pure titanium dioxide, regenerating chlorine:

$$TiCl_4(\ell) + O_2(g) \longrightarrow TiO_2(s) + 2\,Cl_2(g)$$

Titanium dioxide is the major white pigment used in paint. It is more opaque than other pigments, and so paints containing it have high covering power.

Bromine is obtained largely from naturally occurring brines, where it occurs as aqueous bromide ion ($Br^-$). Treatment of these brines with chlorine (Fig. 22-6) results in the reaction

$$2\,Br^-(aq) + Cl_2(aq) \longrightarrow 2\,Cl^-(aq) + Br_2(aq)$$

**Figure 22-4**

State-of-the-art membrane technology installed in a Washington State chlor-alkali plant that produces chlorine and sodium hydroxide. (Occidental Petroleum)

**Figure 22-5**

One of the many polychlorinated biphenyls (PCBs), which were once used as insulating fluids in electrical transformers. These compounds are being phased out because of their toxic effect on marine life.

**Figure 22-6**

Bromide ion (*left*) is colorless. Oxidation by chlorine yields bromine (*center*), which is colored in aqueous solution. The color is even more intense in an organic solvent (upper layer on *right*).

This equilibrium lies to the right because chlorine is a stronger oxidizing agent than bromine. The bromine liberated is swept out with steam, distilled, and purified. Until recently, the most important use for bromine was in the production of dibromoethane ($C_2H_4Br_2$), which was used in gasoline to scavenge the lead deposited from the breakdown of tetraethyl lead; the use of this "antiknock" additive is decreasing, however, because of concerns about contamination of the environment with lead. A major current use for bromine is in the production of flame-retardant organic compounds. Aqueous solutions of $CaBr_2$ and $ZnBr_2$ are also employed as high-density fluids to recover oil from deep wells.

Iodine occurs in seawater to the extent of only $6 \times 10^{-7}\%$, but it is concentrated in certain species of kelp, from whose ash its recovery is commercially feasible. Iodine is present in the growth-regulating hormone thyroxin, produced by the thyroid gland. Much of the table salt sold has 0.01% NaI added to prevent goiter, the enlargement of the thyroid gland. Silver iodide is used in high-speed photographic film, whereas silver bromide and silver chloride are used in slower-speed film and in photographic printing paper.

The hydrogen halides HCl, HBr, and HI are all gases at room conditions, and all three dissolve in water to form strong acids. Hydrogen chloride is usually prepared by the direct combination of hydrogen and chlorine over a platinum catalyst or as a byproduct of organic chemicals processing. Its aqueous solution, hydrochloric acid, is a major industrial acid, used extensively for the cleaning of metal surfaces. Hydrogen bromide and hydrogen iodide have fewer uses. Their aqueous solutions are prepared on a laboratory scale by the hydrolysis of phosphorus tribromide or phosphorus triiodide:

$$PBr_3(\ell) + 3 H_2O(\ell) \longrightarrow 3 HBr(aq) + H_3PO_3(aq)$$

### Solution Chemistry of the Halogens

When chlorine, bromine, or iodine is added to water, the halogen disproportionates in part to form its halo- and hypohalous acids according to the equilibrium

$$X_2(aq) + 2 H_2O(\ell) \rightleftharpoons HOX(aq) + X^-(aq) + H_3O^+(aq)$$

The equilibrium constant for this reaction is very small, but the reaction can be shifted to the right in basic solution, where HOX is neutralized and the overall reaction is

$$X_2(aq) + 2 OH^-(aq) \rightleftharpoons OX^-(aq) + X^-(aq) + H_2O(\ell)$$

No complications arise with chlorine, and stable solutions of chloride and hypochlorite ions result at 25°C. In this way, a solution of sodium hypochlorite (NaOCl)— used as a laundry bleach—can be prepared. The hypochlorite ion is thermodynamically unstable with respect to disproportionation into chloride and chlorate ions,

$$3 ClO^-(aq) \longrightarrow 2 Cl^-(aq) + ClO_3^- \qquad \Delta G° = -90.7 \text{ kJ}$$

but at ordinary temperatures this reaction is very slow. When the temperature is increased to about 75°C, however, the rate of disproportionation is rapid.

The equilibrium constant for the disproportionation of bromine in water to bromide and hypobromite ions is even smaller than that for the disproportionation of chlorine; if the concentration of base is increased, the further disproportionation of hypobromite to bromide and bromate ions occurs rapidly even at 25°C. The same is true of iodine, but to an even greater extent. The solubility of iodine in water at 25°C is only $1.34 \times 10^{-3}$ M, but in the presence of iodide ion, much more dissolves as triiodide ion forms:

$$I^-(aq) + I_2(s) \rightleftharpoons I_3^-(aq) \qquad\qquad K\,(25°C) = 714$$

The $I_3^-$ ion has a linear structure that is symmetric in solution and in certain crystalline salts but unsymmetric (unequal I—I distances) in others.

Among the possible acids of formula $HXO_3$, only iodic acid exists in anhydrous form, a crystalline solid that is stable up to its melting point. The alkali chlorates, bromates, and iodates have the generic formula $MXO_3$, and all are well-characterized salts that can be crystallized from aqueous solution. Acidic solutions of the salts are powerful oxidizing agents that react quantitatively with their corresponding halide ions to yield the halogen:

$$IO_3^-(aq) + 5\,I^-(aq) + 6\,H_3O^+(aq) \longrightarrow 3\,I_2(aq) + 9\,H_2O(\ell)$$

When potassium chlorate is carefully heated in the absence of catalysts, it is converted to potassium perchlorate:

$$4\,KClO_3(s) \longrightarrow 3\,KClO_4(s) + KCl(s)$$

Perchloric acid can be formed by heating potassium perchlorate with concentrated sulfuric acid, but this is a somewhat hazardous reaction that is prone to explosions. A safer procedure is to oxidize an aqueous solution of an alkali chlorate at the anode of an electrolytic cell:

$$ClO_3^-(aq) + 3\,H_2O(\ell) \longrightarrow ClO_4^-(aq) + 2\,H_3O^+(aq) + 2\,e^-$$

Perchloric acid is a colorless liquid that boils at 82°C. In its anhydrous form it is a powerful oxidizing agent, but in aqueous solution at moderate concentration it does not display oxidizing ability. Aqueous perbromic acid, on the other hand, is a powerful oxidizing agent that is also formed electrolytically.

Aqueous periodic acid is a more useful oxidizing agent than either perbromic or perchloric acid, because its behavior is more controllable. Periodic acid exists in several forms, the simplest of which is metaperiodic acid ($HIO_4$). Like iodic acid, metaperiodic acid is a crystalline solid at room temperature. It is possible to expand the coordination sphere of iodine(VII) from four to six ligands by the addition of two water molecules, and the resulting orthoperiodic acid has the formula $H_5IO_6$. This substance is a crystalline solid that decomposes at 140°C.

## Halogen Oxides

Chlorine, bromine, and iodine form several oxides, most of which are rather unstable. Dichlorine monoxide is prepared by reaction of chlorine with mercury(II) oxide:

$$2\,Cl_2(g) + 2\,HgO(s) \longrightarrow Cl_2O(g) + HgCl_2 \cdot HgO(s)$$

In basic solution it forms hypochlorites, as it is the anhydride of hypochlorous acid:

$$Cl_2O(g) + 2\,OH^-(aq) \longrightarrow 2\,OCl^-(aq) + H_2O(\ell)$$

Chlorine dioxide is a very unstable gas that is prepared by reaction of chlorine with silver chlorate:

$$Cl_2(g) + 2\,AgClO_3(s) \longrightarrow 2\,ClO_2(g) + 2\,AgCl(s) + O_2(g)$$

**Figure 22-7**

Iodine reacts with starch to form a dark-colored complex. (Leon Lewandowski)

It is a very powerful oxidizing agent that is used to bleach wheat flour. Like nitrogen dioxide, it consists of "odd-electron" molecules, but unlike that molecule, it shows little tendency to dimerize. In basic solution it disproportionates to form chlorite and chlorate ions, and can be regarded as a mixed anhydride of chlorous and chloric acids:

$$2\,ClO_2(g) + 2\,OH^-(aq) \longrightarrow ClO_2^-(aq) + ClO_3^-(aq) + H_2O(\ell)$$

The most stable of the chlorine oxides is dichlorine heptaoxide ($Cl_2O_7$), a colorless liquid that boils without decomposing at 82°C. It is prepared by dehydration of perchloric acid with $P_4O_{10}$:

$$12\,HClO_4(\ell) + P_4O_{10}(s) \longrightarrow 6\,Cl_2O_7(\ell) + 4\,H_3PO_4(\ell)$$

The oxides of bromine, $Br_2O$ and $BrO_2$, are prepared by methods that parallel those for the corresponding oxides of chlorine. The most important of the iodine oxides is diiodine pentaoxide, a white crystalline solid that is prepared by thermally dehydrating iodic acid:

$$2\,HIO_3(s) \longrightarrow I_2O_5(s) + H_2O(g)$$

Diiodine pentaoxide is useful in analytical chemistry for the detection and quantitative determination of carbon monoxide, which it readily oxidizes:

$$5\,CO(g) + I_2O_5(s) \longrightarrow I_2(s) + 5\,CO_2(g)$$

The iodine released in this reaction is detected visually by means of the deep blue-black colored complex it forms with starch (Fig. 22-7).

## 22-3 FLUORINE AND ITS COMPOUNDS

Fluorine ranks 13th in abundance among the elements in the rocks of the earth's crust, and its compounds were used as early as 1670 for the decorative etching of glass. Yet heroic efforts to isolate it as a free element failed for many years, until

1886. The difficulty of preparing elemental fluorine results from the fact that is is among the strongest oxidizing agents. The other halogens are moderate to very good at accepting electrons in aqueous media, as shown by their standard reduction potentials, 0.535 V (for $I_2$), 1.065 V (for $Br_2$), and 1.358 V (for $Cl_2$). Fluorine is *avid* for electrons:

$$F_2(g) + 2\,e^- \longrightarrow 2\,F^-(aq) \qquad\qquad \mathscr{E}° = 2.87\text{ V}$$

Because fluor*ine* is so good at gaining electrons, fluor*ides* strongly resist oxidation.

The major source of fluorine is the ore fluorspar, which contains fluorite ($CaF_2$) (Fig. 17-29). Figure 17-18*b* showed the cubic structure of $CaF_2$. Another source of fluorine is fluorapatite, $Ca_5(PO_4)_3F$, which is mined in great quantity to make phosphate fertilizer (Section 21-2). At one time, fluorine-containing byproducts (HF, $SiF_4$, and $H_2SiF_6$) from fertilizer plants were valueless and created environmental pollution comparable to the hydrogen chloride pollution caused by the Leblanc process (Section 22-1). Now the fluorine from fluorapatite has economic value in making fluorine compounds.

## Hydrogen Fluoride

Almost half of the fluorspar that is mined is used in steelmaking to increase the fluidity of the slag. Most of the remainder is reacted with concentrated sulfuric acid to make hydrogen fluoride:

$$CaF_2(s) + H_2SO_4(aq) \longrightarrow 2\,HF(g) + CaSO_4(s)$$

The resulting gaseous HF is condensed and redistilled to free it from sulfur oxides. Anhydrous hydrogen fluoride is a colorless liquid that freezes at $-83.37°C$ and boils at $19.54°C$, near room temperature. The boiling point far exceeds what the trend among the other hydrogen halides predicts and is strong evidence for association of the HF molecules (by hydrogen bonds) in the liquid. In this and other physical properties, HF is closer to water and ammonia than to HCl, HBr, or HI. As pure liquids, both water and hydrogen fluoride contain polar molecules but few ions; both are poor electrical conductors. Liquid hydrogen fluoride undergoes autoionization much as water does:

$$2\,HF(\ell) \rightleftharpoons H^+ + FHF^-$$

The reaction has an equilibrium constant (at 19.5°C) of $2.0 \times 10^{-14}$, a value close to the room-temperature autoionization constant of water ($1.0 \times 10^{-14}$).

Hydrogen fluoride in aqueous solution is a weak acid ($K_a = 6.6 \times 10^{-4}$ at 25°C), quite unlike HCl, HBr, and HI, which are strong acids. Its weakness as an acid is due both to the strength of the H—F bond and to the extensive hydrogen bonding in its aqueous solutions. Fluoride ions interact so strongly with hydronium ions, $H_3O^+$, that they remain bound in $F^- \cdots H_3O^+$ complexes rather than separating as chloride, bromide, and iodide ions do.

Anhydrous hydrogen fluoride is quite poisonous, causes serious burns, and corrodes silicate glass by the reaction

$$4\,HF(\ell) + SiO_2(s) \longrightarrow SiF_4(g) + 2\,H_2O(\ell)$$

This reaction underlies one long-standing use of aqueous hydrogen fluoride: etching decorative patterns in glass (Fig. 22-8). Hydrogen fluoride does not attack Teflon

**Figure 22-8**
Glass can be etched in intricate patterns with aqueous hydrogen fluoride. This English plate depicting Venus and Cupid dates from 1890. The border and background were formed by protecting portions of the glass with wax and dipping the piece in HF($aq$); the cameo figures were carved. (George Woodall)

and certain other fluorine-containing plastics. Their availability in recent years has made it easier to handle hydrogen fluoride safely. Another major use of HF($g$) is in the aluminum industry. It reacts with NaAlO$_2$($aq$), itself prepared from bauxite ore (Al$_2$O$_3$($s$)) and NaOH($aq$), to make synthetic cryolite according to the equation

$$\text{NaAlO}_2(aq) + 2\,\text{NaOH}(aq) + 6\,\text{HF}(g) \longrightarrow \text{Na}_3\text{AlF}_6(aq) + 4\,\text{H}_2\text{O}(\ell)$$

The very important Hall–Héroult process to produce aluminum (Fig. 19-20) depends on the electrolysis of solutions of Al$_2$O$_3$ in molten cryolite.

**The Preparation of Fluorine**

Elemental fluorine is produced from hydrogen fluoride by a method similar to that first used by Moissan—the electrolysis of molten potassium hydrogen fluoride (KF·2HF), a solution of KF in liquid HF. In the solution, F$^-$ ion associates strongly with HF to form FHF$^-$ (the hydrogen difluoride ion), the species that actually loses the electrons:

$$\text{FHF}^- \longrightarrow \text{F}_2(g) + \text{H}^+ + 2\,e^- \qquad \text{(anode, oxidation)}$$

$$\underline{\phantom{2\text{ HF}}\quad 2\,\text{H}^+ + 2\,e^- \longrightarrow \text{H}_2(g) \qquad\qquad \text{(cathode, reduction)}}$$

$$2\,\text{HF} \longrightarrow \text{H}^+ + \text{FHF}^- \longrightarrow \text{F}_2(g) + \text{H}_2(g) \qquad \text{(overall)}$$

This process resembles the electrolytic production of Cl$_2$($g$) (Section 22-1). As fluorine and hydrogen evolve, HF($g$) is fed in continuously to keep the electrolyte composition constant at KF·2HF. A commercial fluorine cell uses a steel cathode and carbon anode. The cells are carefully engineered to prevent contact between the product gases, which react at catastrophic speed.

Fluorine is a very pale yellow gas. It condenses to a canary-yellow liquid at −188.14°C and solidifies at −219.62°C. It is an exceedingly toxic substance. Concentrations exceeding 25 ppm in air quickly damage the eyes, nose, lungs, and skin. Fluorine's pungent, irritating odor makes it first detectable in air at a concentration of about 3 ppm, well below the level of acute poisoning but above the recommended safe working level of 0.1 ppm. The ferocious reactivity of fluorine demands the utmost care in its use, and handling procedures that minimize the hazards have been developed. Certain metals (nickel, copper, steel, and the nickel–copper alloy Monel metal) resist attack by fluorine by forming a layer of a fluoride salt on their surfaces. Reactions can be carried out at room temperature in vessels made of these metals. The pure element is regularly packaged and shipped as a compressed gas in special cylinders, but it is also often generated at the point of use.

The role of elemental fluorine in the chemical industry has grown to major proportions since 1945. Before then, elemental fluorine was mostly a laboratory curiosity. The motive for finding ways to prepare, purify, and handle it was the need to synthesize a volatile uranium compound that could separate the $^{235}U$ and $^{238}U$ isotopes by gaseous diffusion, as described in Section 3-6. Uranium hexafluoride met that requirement, and gaseous diffusion is still used for preparing $^{235}U$-enriched reactor fuel. At present, more than half of the world's production of $F_2(g)$ is used to make $UF_6$ from uranium(IV) oxide by these reactions:

$$UO_2(g) + 4\,HF(g) \longrightarrow UF_4(s) + 2\,H_2O(g)$$

$$UF_4(s) + F_2(g) \longrightarrow UF_6(g)$$

Most of the remaining fluorine production goes to make sulfur hexafluoride by the reaction

$$S(g) + 3\,F_2(g) \longrightarrow SF_6(g)$$

This unreactive gas is a good electrical insulator. Its presence within heavy-duty electrical switches minimizes arcing or sparking when the switches are thrown. Electrical discharges that do occur in an $SF_6$ atmosphere decompose the gas to an extent (into sulfur and fluorine), but the dissociation products rapidly recombine. Some triple-pane windows are manufactured with $SF_6$ sealed in the "dead-air" spaces between the panes because it cuts heat transmission and muffles sound better than air.

## Fluorine and the Other Halogens

The differences between fluorine and the other halogens are extreme enough to place fluorine in a class by itself. It is much more reactive than chlorine, bromine, and iodine; indeed, it is by far the most reactive of all the elements. In forming a compound, a fluorine atom either shares electrons in a single covalent bond or gains one electron to form the fluoride ion ($F^-$). In either case, it always attains a noble-gas electron configuration ($1s^2 2s^2 2p^6$). The other halogens tend to behave similarly but tolerate numerous exceptions; fluorine does not. The rules for assigning oxidation numbers (Section 2-4) recognize the special character of fluorine by setting its oxidation number in compounds always equal to $-1$. Chlorine, bromine, and iodine may have positive oxidation numbers.

The chemical and physical properties of fluorine (and its compounds) deviate from the values that would be predicted by extrapolation of the trends among the other halogens. Figure 22-9 presents some data from Appendix F that show this deviation. The bond dissociation energy of $F_2$ is *smaller* than would be expected from the trend among the other halogens, but the first ionization energy of F is *greater*. Despite the position of F as the most electronegative of the elements, electron-affinity data show that $F(g)$ accepts another electron (to form the gaseous anion) *less* readily than $Cl(g)$ and only slightly more readily than $Br(g)$. The unex-

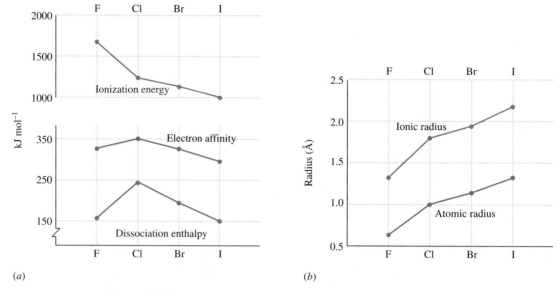

**Figure 22-9**

The measured physical properties of fluorine often do not fall on the trend lines defined by the behavior of the other halogens. (*a*) Here the ionization energy of fluorine is unexpectedly high, and its electron affinity and bond dissociation enthalpy are unexpectedly low. (*b*) The sizes of the fluorine atom and the fluoride ion are both smaller than predicted by comparison with the other halogens.

pectedly low electron affinity of a fluorine atom is related to the contracted size of the atom. As Figure 22-9*b* shows, F(*g*) and F⁻(*g*) are smaller than an extrapolation of the trends among the other halogens would predict, causing crowding and greater mutual repulsion among the electrons in their $n = 2$ quantum levels.

Strong bonds, both covalent and ionic, characterize the interaction of fluorine with most other elements. Figure 22-10*a* shows that the H—F, C—F, and Na—F bond dissociation enthalpies exceed those of H, C, and Na combined with the other halogens. Fluorine's strong bonds with other elements and weak bond with itself (in the $F_2$ molecule the bond dissociation enthalpy is only 158 kJ mol⁻¹) together explain its terrific reactivity. Both factors are traceable to the small size of the fluorine atom. The two atoms in the fluorine molecule are only 1.417 Å apart. This puts their lone-pair electrons closer to each other than those in the other halogens: repulsion among them lowers the bond dissociation energy. Small fluorine makes short bonds to nonfluorine atoms, as well. Lone-pair repulsion of the type just described becomes important only with elements that, like fluorine itself, are small and electronegative. Hence, O—F and N—F bonds are, like the F—F bond, weaker than would otherwise be expected (Fig. 22-10*b*).

Because fluorine atoms are small, many of them fit around a given central atom before crowding each other. Therefore, atoms often attain higher coordination numbers with fluorine than with other elements. In many fluorides, such as $K_2[NiF_6]$, $Cs[AuF_6]$, and $PtF_6$, the central element has an unexpectedly high oxidation number. The other halogens offer no comparable compounds.

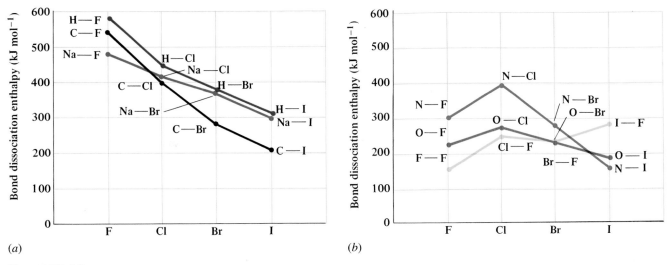

**Figure 22-10**

(*a*) Fluorine forms strong bonds with more electropositive elements such as C, H, and Na. (*b*) In contrast, fluorine forms unexpectedly weak bonds with the electronegative elements N, O, and F in comparison to the bonds formed by the other halogens.

## Reactions of Fluorine

In general, fluorine reacts with other elements and compounds by oxidizing them. It displaces the other, less reactive, halogens from compounds, just as chlorine displaces bromine and iodine, the elements located below it in the periodic table, and bromine displaces iodine. An example is the reaction

$$2\,NaCl(s) + F_2(g) \longrightarrow 2\,NaF(s) + Cl_2(g)$$

Fluorine sometimes breaks this pattern by going much further. Thus, if $F_2(g)$ is heated with potassium chloride, the reaction is

$$KCl(s) + 2\,F_2(g) \longrightarrow KClF_4(s)$$

Here the chlorine is not displaced from the compound but oxidized to the $+3$ state and surrounded by fluorine atoms in the "tetrafluorochlorate(III)" ion. Further heating of $KClF_4(s)$ with $F_2$ eventually does displace chlorine from potassium but in an unexpected form, $ClF_5(g)$:

$$KClF_4(s) + F_2(g) \longrightarrow KF(s) + ClF_5(g)$$

In a variation, fluorine oxidizes iodine in potassium iodide and also displaces it:

$$KI(s) + F_2(s) \longrightarrow KF(s) + IF(g)$$

Under different conditions with the same reactants, the theme is carried to an extreme:

$$KI(s) + 4\,F_2(g) \longrightarrow KF(s) + IF_7(g)$$

Fluorine also can displace oxygen from its compounds:

$$2\,CaO(s) + 2\,F_2(g) \longrightarrow 2\,CaF_2(s) + O_2(g)$$

**Figure 22-11**

In SOF$_2$, oxygen and fluorine both bond to the central sulfur atom. The VSEPR model correctly predicts this geometry.

With other oxides, only a portion of the oxygen is displaced, as is the case with the reaction of SO$_2(g)$ and F$_2(g)$ to give thionyl fluoride (Fig. 22-11):

$$2\,SO_2(g) + 2\,F_2(g) \longrightarrow 2\,SOF_2(g) + O_2(g)$$

The most important oxide of all is water. Water vapor at 100°C *burns* in F$_2$ with a weak, luminous flame. The reaction is

$$2\,F_2(g) + 2\,H_2O(g) \longrightarrow 4\,HF(g) + O_2(g)$$

As is often the case, different conditions favor different products. An excess of fluorine in contact with liquid water or ice gives hydrogen fluoride and highly unstable hydrogen oxyfluoride:

$$F_2(g) + H_2O(s) \longrightarrow HF(\ell) + HOF(\ell)$$

Oxygen difluoride also forms,

$$2\,F_2(g) + H_2O(s) \longrightarrow 2\,HF(\ell) + OF_2(g)$$

and the reaction that gives O$_2$ and HF still takes place to a limited extent. In all three of these reactions fluorine disrupts stable H$_2$O, taking oxygen from the $-2$ to the 0 or $+2$ state. Few materials other than fluorine can oxidize water in this way.

### Metal Fluorides

The alkali and alkaline-earth metals form ionic salts with fluorine of the expected stoichiometry. Except for lithium fluoride, the alkali fluorides are all quite soluble in water. The differences between the alkaline-earth fluorides (BeF$_2$, MgF$_2$, CaF$_2$, SrF$_2$, and BaF$_2$) and the corresponding chlorides, bromides, and iodides are striking: BeF$_2$ is ionic and soluble in water, but BeCl$_2$ is covalent enough to sublime; CaF$_2$ is insoluble in water, but the other calcium halides are quite soluble in water. All these fluorides are prepared most easily by reaction of the corresponding hydroxide or carbonate with aqueous hydrofluoric acid. In their lower oxidation states, the transition metals and Group III and IV metals also form fluoride salts with a range of solubilities; ZnF$_2$, SnF$_2$, and PbF$_2$ are examples.

Low concentrations (about 1 mg L$^{-1}$) of fluoride ion in drinking water help to prevent tooth decay. Fluoride ion substitutes for hydroxide ion in tooth enamel, changing some of the hydroxyapatite, Ca$_5$(PO$_4$)$_3$OH, in the enamel to fluorapatite, Ca$_5$(PO$_4$)$_3$F. The replacement works because the F$^-$ and OH$^-$ ions have the same charge ($-1$) and similar radii (1.33 Å for F$^-$ versus approximately 1.2 Å for OH$^-$). Fluorapatite is less soluble than hydroxyapatite in the acidic oral environment fostered by the consumption of sweets. The deliberate fluoridation of civic water supplies to cut down tooth decay is a common public health measure in the United States. In addition, fluoride salts such as SnF$_2$ or NaF are often added to toothpaste.

Fluorine forms salts with most metals in their lower oxidation states but is capable of oxidizing many metals to exceptionally high oxidation states as well. The chemistry of cobalt centers on the $+2$ oxidation state, for example, except in complex ions, where the $+3$ state is not uncommon. Cobalt(III) fluoride, obtained by direct fluorination of the metal, is one of the very few simple ionic salts of that element in which the cation has an oxidation number of $+3$. Platinum is directly oxidized to platinum(IV) by fluorine, and (at more extreme temperatures) to platinum(VI) fluoride. The latter is a dark red solid with a low melting point (56.7°C), suggesting its largely covalent nature. Other heavy metals that form volatile hexafluorides include osmium, iridium, molybdenum, tungsten, and uranium. Fluorides of lower oxidation state are also known for all of these metals.

Normally, one thinks of silver as having only an oxidation state of $+1$, and in the great majority of its compounds this is the case. Silver in the $+2$ oxidation state, however, can be formed by heating silver(I) fluoride in a stream of fluorine:

$$2\,AgF(s) + F_2(g) \longrightarrow 2\,AgF_2(s)$$

Even the $+3$ state of silver can be formed in a ternary compound when potassium fluoride is mixed with silver fluoride and heated in fluorine:

$$KF(s) + AgF(s) + F_2(g) \longrightarrow KAgF_4(s)$$

In these reactions the $4d$ subshell of the silver ion must be opened, at a cost of 2073 and 3360 kJ mol$^{-1}$, respectively, to form silver in the $+2$ and $+3$ oxidation states.

## Covalent Fluorides

Fluorine forms compounds with all the elements in Groups III through VII. The Group III elements (boron, aluminum, gallium, indium, and thallium) all form trifluorides, and thallium forms TlF as well. Boron trifluoride is one of the most powerful Lewis acids, readily forming the complex ion $[BF_4]^-$ in solutions that contain $F^-$. Boron trifluoride is a gas at room temperature, but the trifluorides of aluminum and the higher members of Group III are solids that melt near 1000°C. They are intermediate between ionic and covalent in their bonding, and in the crystalline state their structures consist of extended arrays of metal atoms in sixfold coordination with fluorine atoms.

The transitional character of bonding is even more evident in the fluorides of Group IV. At room temperature, carbon tetrafluoride and silicon tetrafluoride are gases whose molecules are covalent; germanium tetrafluoride is a molecular liquid, but the tetrafluorides of tin and lead are solids with considerable ionic character. As already mentioned, the difluorides of tin and lead are ionic salts. This fact illustrates a general rule: the halides of the heavier transition and posttransition metals are more covalent when the oxidation number of the metal is high, and more ionic when it is low.

The fluorides of the Group V elements are just as varied in composition and behavior. Nitrogen forms fluorides with the compositions $N_2F_2$ (dinitrogen difluoride), $N_2F_4$ (dinitrogen tetrafluoride), and $NF_3$ (nitrogen trifluoride). All are gases at room temperature. Nitrogen trifluoride is rather inert and exhibits no electron-donor properties; like $CF_4$, it is not hydrolyzed by water. It fluorinates certain metals when heated:

$$2\,NF_3(g) + Cu(s) \longrightarrow N_2F_4(g) + CuF_2(s)$$

Dinitrogen tetrafluoride (Fig. 22-12) is much more reactive than nitrogen trifluoride and undergoes hydrolysis in water. The trifluoride and pentafluoride of phosphorus are also both gases at room temperature. Unlike $NF_3$, both react readily with water:

$$PF_3(g) + 3\,H_2O(\ell) \longrightarrow H_3PO_3(aq) + 3\,HF(aq)$$

$$PF_5(g) + 4\,H_2O(\ell) \longrightarrow H_3PO_4(aq) + 5\,HF(aq)$$

The PF$_3$ molecule has a trigonal pyramidal structure like NH$_3$, as the VSEPR model predicts. Unlike ammonia, however, it does not donate its lone electron pair readily,

*trans*

*gauche*

**Figure 22-12**

Molecules of $N_2F_4$ exist in two isomeric forms.

**Figure 22-13**

The molecular structures of $SF_4$ and $SF_6$.

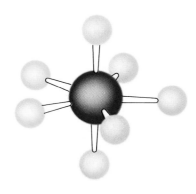

**Figure 22-14**

The molecular structure of $IF_7$ is based on a pentagonal bipyramid.

presumably because the highly electronegative fluorine atoms withdraw electron density from the central atom. Like $BF_3$, $PF_5$ is a very powerful Lewis acid and is useful as a catalyst for the polymerization of hydrocarbons. The remaining Group V elements (arsenic, antimony, and bismuth) resemble phosphorus in forming both trifluorides and pentafluorides that readily hydrolyze in aqueous solution. They are powerful fluorinating agents, capable of introducing fluorine into other molecules by substitution reactions.

The highly toxic gas oxygen difluoride ($OF_2$) can be made by passing fluorine through aqueous alkaline solutions:

$$2 F_2(g) + 2 OH^-(aq) \longrightarrow OF_2(g) + 2 F^-(aq) + H_2O(\ell)$$

It is worthwhile to compare $OF_2$ with the analogous compound formed between chlorine and oxygen. Both $Cl_2O$ and $OF_2$ are gases under room conditions, and both have oxygen atoms bonded to two halogens. In both compounds the steric number of the oxygen atom is 4, with two bonds, two lone pairs on the central oxygen atom, and a bent shape. Despite these similarities, $Cl_2O$ and $OF_2$ differ profoundly in reactivity. Oxygen difluoride contains ''positive oxygen'' (note the $+2$ oxidation number for oxygen), because fluorine is highly electronegative. It reacts with water by oxidizing it to liberate $O_2$:

$$OF_2(g) + H_2O(\ell) \longrightarrow O_2(g) + 2 HF(aq)$$

Dichlorine oxide contains ordinary ''negative oxygen'' and behaves quite differently with water. It reacts to give hypochlorous acid without oxidation or reduction:

$$Cl_2O(g) + H_2O(\ell) \longrightarrow 2 HOCl(aq)$$

The chemistries of the compounds diverge along the same lines in many other reactions.

The later Group VI elements also form fluorides. The two most important fluorides of sulfur are quite dissimilar in their properties. Both sulfur tetrafluoride and sulfur hexafluoride (Fig. 22-13) are gases at room temperature; the former is highly reactive, and the latter is extremely inert. Both selenium and tellurium form a tetrafluoride and a hexafluoride with fluorine. These compounds, like the fluorides of their Group V neighbors arsenic and antimony, are excellent fluorinating agents. Their strikingly enhanced reactivity compared with sulfur hexafluoride is not easy to explain, except by noting that the larger central atoms are not as well shielded by their fluorine atoms.

Fluorine reacts with the other halogens to form several **interhalogen** compounds; they are liquids at room temperature except for the chlorofluorides, which are gases. In these compounds, fluorine atoms, as always in the $-1$ oxidation state, surround a central chlorine, bromine, or iodine atom as $XF$, $XF_3$, $XF_5$, where X is Cl, Br, or I. The only $XF_7$ compound occurs with iodine. The VSEPR model provides accurate predictions of the structures of these compounds. In $IF_7$ iodine has a steric number of 7, and the structure is based on the pentagonal bipyramid (Fig. 22-14). As liquids, the interhalogens have low electrical conductivities, reflecting their largely covalent character (see, however, problem 27).

### Fluorinated Hydrocarbons and Chlorofluorocarbons

Fluorine can replace hydrogen in hydrocarbons, giving rise to an important class of compounds called fluorinated hydrocarbons. The C—F bond is short and strong, an average of 1.2 times stronger than the C—H bond and 1.4 to 2.2 times stronger than the bonds between carbon and the other halogens; it is the strongest single bond. Fluorination of hydrocarbons with no double or triple bonds therefore increases thermal and chemical stability. The reverse is true in the reactions of the other halogens with hydrocarbons. When all the hydrogen atoms in a hydrocarbon are replaced by fluorine atoms, compounds called **fluorocarbons** (or **perfluorocarbons**) result. The electron density on the fluorine atoms in fluorocarbons is tightly held and is not easily distorted by approaching atoms or ions. This reduces the reactivity of these compounds and weakens their intermolecular forces. The compounds have low melting and boiling points and low enthalpies of vaporization.

One provocative use of fluorine compounds is in the preparation of blood substitutes. Many fluorocarbons are excellent solvents for oxygen. A milliliter of liquid $C_7F_{16}$, for example, dissolves 1900 times more oxygen at room conditions than a milliliter of water. Small laboratory animals can survive total immersion in oxygenated fluorocarbon liquids, as their lungs successfully extract the oxygen they need from solution. Fluorocarbons are insoluble in water. For use in the bloodstream, they are prepared as fine-grained emulsions in water. Rats have survived for days without apparent harm after their blood was completely replaced with a concentrated emulsion of $C_8F_{17}Br$ in water. Over time, the composition of the rats' blood reverted to normal as the blood substitute was excreted and new red blood cells grew. Practical fluorocarbon blood substitutes would need no matching of blood types, could be stored for emergency use, and would not transmit disease.

If an excess of elemental fluorine is mixed with a hydrocarbon under ordinary conditions, the result is usually a fire or explosion to give HF and $CF_4$. Controlled cryogenic (very cold) conditions can avoid such outcomes, and the treatment of hydrocarbons with elemental fluorine can give reasonable yields of partially and completely fluorinated products. The best-known perfluorocarbon is the solid polytetrafluoroethylene (Teflon), formulated as $-(CF_2-CF_2-)_n$, where $n$ is a large number. Chemically, this compound is nearly completely inert, resisting attack from boiling sulfuric acid, molten potassium hydroxide, gaseous fluorine, and other aggressive chemicals (Fig. 22-15). Physically, it has excellent heat stability (a working temperature up to 260°C), is a very good electrical insulator, and has a low coefficient of friction that makes it useful for bearing surfaces in machines as well as coating frying pans (''no-stick'' Teflon). In Teflon the carbon atoms lie in a long chain that is encased by tightly bound fluorine atoms (Fig. 22-16). Even reactants with a strong innate ability to disrupt C—C bonds (such as fluorine itself) fail to attack Teflon at observable rates because there is no route of attack past the surrounding fluorine atoms and their tightly held electrons.

**Chlorofluorocarbons** (often called CFCs) contain one or more carbon atoms with both fluorine and chlorine atoms attached as side groups. Chlorofluorocarbons are chemically inert, nontoxic gases or low-boiling liquids (Fig. 22-17). They are relatively cheap to make and have boiling points conveniently near room temperature. These properties, together with their lack of reactivity, fit them to serve as

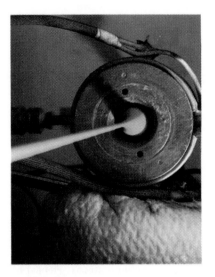

**Figure 22-15**

Teflon being extruded over copper wire for electrical insulation. Such coated wires are used in telephone and computer cables. (DuPont)

**Figure 22-16**

A model of the molecular structure of Teflon shows that the fluorine atoms shield the carbon chain very effectively. Note that the chain must twist to accommodate the fluorine atoms.

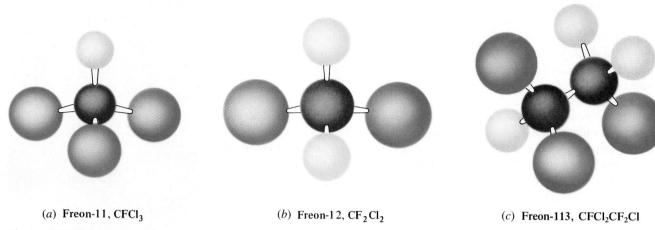

(a) **Freon-11, CFCl₃**          (b) **Freon-12, CF₂Cl₂**          (c) **Freon-113, CFCl₂CF₂Cl**

**Figure 22-17**

Some chlorofluorocarbons (Freons) of principal commercial importance.

propellants in aerosol sprays and as working fluids in air-conditioning machinery. Soon after their discovery, chlorofluorocarbons replaced $SO_2$ and $NH_3$ in the latter application. The inertness of chlorofluorocarbons has drawbacks as well, because the molecules that escape into the air persist unchanged. In time they rise into the stratosphere, where ultraviolet radiation from the sun finally breaks them down. Chlorine atoms generated in this decomposition catalyze the destruction of ozone ($O_3$) in the upper atmosphere, as Section 15-4 discussed in detail.

## 22-4  COMPOUNDS OF FLUORINE AND THE NOBLE GASES

For many years the noble gases were believed to be chemically inert. They failed to react with the strongest oxidizing agents available, and an early (1895) attempt by Moissan to make argon react with fluorine also failed. In 1933 Linus Pauling suggested on theoretical grounds that xenon and fluorine should form compounds. Soon thereafter, chemists subjected a mixture of the two gases to an electric discharge but saw no evidence of reaction. This result, other negative results, and the dangers of working with elemental fluorine discouraged further experiments. It seemed clear that if fluorine could not force a noble gas into combination, then nothing could. Interest in the subject was further stifled because generally accepted simple bonding theory appeared to ratify the complete chemical inertness of the noble gases. Then, in 1962, Neil Bartlett treated gaseous xenon with the powerful fluorinating agent $PtF_6$, and a yellow-orange solid compound of platinum, fluorine, and xenon resulted. In 1965 an irony emerged: the same fluorine–xenon mixtures that did not react in the 1933 electric discharge experiment *did* give $XeF_2(s)$ simply upon exposure to sunlight!

Bartlett's discovery spurred research on the reactions of the noble gases with fluorine, and soon afterward the synthesis of xenon tetrafluoride by direct combination of the elements at high pressure was reported:

$$Xe(g) + 2\,F_2(g) \longrightarrow XeF_4(s)$$

Varying the relative amounts of xenon and fluorine permitted synthesis of $XeF_2(s)$ and $XeF_6(s)$, as well. All three xenon fluorides are colorless crystalline solids at room temperature (Fig. 22-18). The VSEPR theory correctly predicts the linear structure of $XeF_2$ molecules and the square-planar structure of $XeF_4$ molecules. Because the steric number of the central Xe is 7 and not 6, $XeF_6$ molecules are not octahedral; however, the VSEPR theory does not correctly predict their complex structure.

The three xenon fluorides are moderate to powerful fluorinating agents but are thermodynamically stable with respect to decomposition into xenon and fluorine at room temperature. Their behavior toward water is quite diverse. Xenon difluoride hydrolyzes only very slowly in water, and solutions more concentrated than 0.1 M can be prepared at low temperatures. On the other hand, $XeF_4$ and $XeF_6$ react vigorously with water to form xenon trioxide ($XeO_3$):

$$3\,XeF_4(s) + 6\,H_2O(\ell) \longrightarrow XeO_3(aq) + 2\,Xe(g) + 12\,HF(aq) + \tfrac{3}{2}O_2(g)$$

$$XeF_6(s) + 3\,H_2O(\ell) \longrightarrow XeO_3(aq) + 6\,HF(aq)$$

Xenon trioxide dissolves in aqueous solution without ionizing, but in basic solution it acts as a Lewis acid, accepting $OH^-$:

$$XeO_3(aq) + OH^- \longrightarrow HXeO_4^-(aq)$$

The hydrogen xenate ion then disproportionates to form xenon and the perxenate ion, in which the oxidation number of xenon is $+8$:

$$2\,HXeO_4^-(aq) + 2\,OH^- \longrightarrow XeO_6^{4-}(aq) + Xe(g) + O_2(g) + 2\,H_2O(\ell)$$

Only one simple krypton halide has been prepared. Irradiating a mixture of krypton and fluorine at low temperature with electrons forms krypton difluoride, a thermodynamically unstable substance. It is a volatile, colorless, crystalline solid. Experiment confirms the linear structure predicted by the VSEPR theory. Krypton difluoride forms salt-like compounds with strong fluoride ion acceptors such as $AsF_5$ and $SbF_5$:

$$KrF_2 + AsF_5 \longrightarrow [KrF]^+[AsF_6]^-$$

Krypton difluoride is a stronger oxidizing agent than fluorine itself. It has a total bond enthalpy (for *both* Kr—F bonds) of only 96 kJ mol$^{-1}$; this is substantially less than the bond enthalpy of $F_2(g)$, 158 kJ mol$^{-1}$. Krypton difluoride oxidizes $Xe(g)$ to $XeF_6(s)$ at room temperature and oxidizes $I_2$ directly to $IF_7$. A solution of $KrF_2$ in liquid HF even attacks gold:

$$7\,KrF_2(g) + 2\,Au(s) \longrightarrow 2\,KrF[AuF_6] + 5\,Kr$$

Heating the $KrF[AuF_6]$ to 60°C generates orange $AuF_5$, in which gold has a $+5$ oxidation number. By way of comparison, $F_2(g)$ itself requires a temperature of 250°C to act on powdered gold and then generates only $AuF_3$.

**Figure 22-18**
Crystals of xenon tetrafluoride, formed in the creation of the first binary noble gas compound. (Argonne National Laboratory)

---

After studying this chapter and working the problems that follow, you should be able to

**CONCEPTS & SKILLS**

1. Outline the historical development of the chlor-alkali industry, including the linkage between production of chlorine and the bases sodium carbonate and sodium hydroxide (Section 22-1, problems 1–6).

2. Describe the production of and uses for important compounds of chlorine, bromine, and iodine (Section 22-2, problems 7–12).
3. Compare and contrast the chemical properties of the oxoacids and oxides of chlorine, bromine, and iodine (Section 22-2, problems 13–16).
4. Discuss the preparation of elemental fluorine and the uses of this element and its compounds (Section 22-3).
5. Compare the properties of fluorine and its compounds with those of the other halogens (Section 22-3, problems 17–20).
6. Discuss the physical and chemical properties of fluorides of the elements through the periodic table (Section 22-3, problems 21–32).
7. Describe the chemical reactions undergone by the heavier noble-gas elements (Section 22-4, problems 33–38).

## PROBLEMS

### Chemicals from Salt

1. Calculate the equilibrium constant at 25°C for the production of sodium hypochlorite bleach from dichlorine oxide and sodium hydroxide:

$$Cl_2O(g) + 2\,OH^-(aq) \rightleftharpoons 2\,OCl^-(aq) + H_2O(\ell)$$

Relevant data can be found in Appendix D.

2. Calculate the equilibrium constant at 25°C for the preparation of chlorine by the reaction of hydrochloric acid with pyrolusite, used by Scheele in 1774:

$$2\,Cl^-(aq) + 4\,H_3O^+(aq) + MnO_2(s) \rightleftharpoons$$
$$Cl_2(g) + Mn^{2+}(aq) + 6\,H_2O(\ell)$$

Relevant data can be found in Appendix D.

3. Calculate the standard enthalpy change per mole of $Na_2CO_3$ produced by the Leblanc process, using data from Appendix D. From your calculation, what do you conclude about the overall heat requirements of that process?

4. Calculate the standard enthalpy change per mole of $Na_2CO_3$ produced by the Solvay process, using data from Appendix D. From your calculation, what do you conclude about the overall heat requirements of that process?

5. A recent magazine article about the chlor-alkali process in the United States reported that "production of chlorine in 1987 . . . total[ed] about 10.9 million tons. . . . Caustic soda production as usual [was] 4 to 5 percent higher than chlorine production." Explain the use of the phase "as usual" by computing the theoretical ratio of the yields of these two chemicals in the process.

6. Calculate the minimum theoretical external cell voltage to drive a chlor-alkali cell in which all reactants and products are in their standard states.

### The Chemistry of Chlorine, Bromine, and Iodine

7. At 25°C, 33.6 g of bromine will dissolve in 1 L of water. Use this fact to estimate $\Delta G°$ for the reaction

$$Br_2(\ell) \longrightarrow Br_2(aq)$$

Compare your estimate with the result calculated from data in Appendix D.

8. At 25°C, the solubility of $I_2$ in water is only 0.33 g L$^{-1}$. Estimate $\Delta G°$ for the reaction

$$I_2(s) \longrightarrow I_2(aq)$$

Compare your estimate with the result calculated from data in Appendix D.

9. Calculate the equilibrium constant at 25°C for the reaction

$$2\,Br^-(aq) + Cl_2(g) \rightleftharpoons Br_2(g) + 2\,Cl^-(aq)$$

Use your result to comment on the feasibility of using cheap chlorine gas to extract bromine from seawater.

10. Hypobromite ion in aqueous solution disproportionates into bromide ion and bromate ion according to the equation

$$3\,BrO^-(aq) \rightleftharpoons 2\,Br^-(aq) + BrO_3^-(aq)$$

Calculate the equilibrium constant at 25°C for this reaction, given the following thermodynamic data:

|  | $\Delta H_f°$ (kJ mol$^{-1}$) | $S°$ (J K$^{-1}$ mol$^{-1}$) |
|---|---|---|
| $Br^-(aq)$ | −122 | 82 |
| $BrO^-(aq)$ | −94 | 42 |
| $BrO_3^-(aq)$ | −67 | 162 |

11. Gaseous hydrogen bromide can be produced by the action of a nonoxidizing acid such as phosphoric acid on a solution of sodium bromide.

(a) Write a balanced chemical equation for this reaction.

(b) What volume of hydrogen bromide at 0°C and 1.00 atm will be produced when an excess of phosphoric acid is added to 100 mL of a solution with an NaBr concentration of 0.050 M?

12. When chlorine is bubbled through a basic solution containing iodide ions, chloride ions and iodate ions are produced.

(a) Write a balanced chemical equation for this reaction.

(b) Suppose 0.216 L of chlorine at 0°C and 1.00 atm is needed to react completely with 50.0 mL of the iodide solution in part (a). Determine the initial concentration of iodide in that solution.

13. Given the reduction potentials,

$$ClO_2 + e^- \longrightarrow ClO_2^- \qquad \mathscr{E}° = 0.954 \text{ V}$$

$$ClO_3^- + H_2O + e^- \longrightarrow ClO_2 + 2\,OH^- \qquad \mathscr{E}° = -0.25 \text{ V}$$

decide whether $ClO_2$ is stable with respect to disproportionation in basic aqueous solution.

14. Given the following reduction potentials

$$ClO^- + H_2O + 2\,e^- \longrightarrow Cl^- + 2\,OH^- \qquad \mathscr{E}° = 0.90 \text{ V}$$

$$ClO_3^- + 2\,H_2O + 4\,e^- \longrightarrow ClO^- + 4\,OH^- \qquad \mathscr{E}° = 0.48 \text{ V}$$

decide whether hypochlorite ion ($ClO^-$) is stable with respect to disproportionation in basic aqueous solution under standard conditions.

15. The reduction potential diagram for iodine in aqueous solution at pH 14 and 25°C has the form

$$IO_3^- \xrightarrow{0.14\text{ V}} IO^- \xrightarrow{0.45\text{ V}} I_2 \xrightarrow{0.535\text{ V}} I^-$$
$$\underset{0.49\text{ V}}{\underline{\qquad\qquad\qquad}}$$

(a) Calculate the stable products that result when $I_2$ is dissolved in water at pH 14 and 25°C. Explain your answer.

(b) Calculate the standard half-cell potential at pH 14 for the half-reaction

$$IO_3^-(aq) + 3\,H_2O(\ell) + 6\,e^- \longrightarrow I^-(aq) + 6\,OH^-(aq)$$

16. The reduction potential diagram for bromine in aqueous solution at pH 14 and 25°C has the form

$$BrO_3^- \xrightarrow{0.54\text{ V}} BrO^- \xrightarrow{0.45\text{ V}} Br_2 \xrightarrow{1.06\text{ V}} Br^-$$
$$\underset{0.76\text{ V}}{\underline{\qquad\qquad\qquad}}$$

(a) Calculate the stable products that result when $Br_2$ is dissolved in water at pH 14 and 25°C. Explain your answer.

(b) Calculate the standard half-cell potential at pH 14 for the half-reaction

$$BrO_3^-(aq) + 3\,H_2O(\ell) + 6\,e^- \longrightarrow Br^-(aq) + 6\,OH^-(aq)$$

## Fluorine and Its Compounds

17. Write balanced equations for the reaction of elemental fluorine with the following:
(a) $SrO(s)$
(b) $O_2(g)$
(c) $UF_4(s)$

18. Write balanced equations for the reaction of elemental fluorine with the following:
(a) $NaCl(s)$
(b) $MgO(s)$
(c) $Na(s)$

19. The alkali metals form one or more acid fluorides, with formulas $MF\cdot HF$, $MF\cdot 2HF$, and $MF\cdot 3HF$ (where M stands for the alkali metal). In these compounds, additional HF molecules link with $F^-$ ions by means of hydrogen bonds, in a manner similar to the incorporation of water of hydration into crystalline hydrates. Nothing similar is observed with the other alkali-metal halides.

Suppose a compound is found to contain 2.48 g of fluorine for every gram of sodium. What is the empirical formula of this compound? Assume that any hydrogen present is not detected in this analysis.

20. There are ways to combine carbon and fluorine beyond those mentioned in the text. When carbon in the form of graphite is treated with elemental fluorine, F atoms occupy interstitial spaces between the layers of C atoms, and compounds of empirical formula $C_4F$, $C_2F$, and $CF$ result. The bonds between carbon and fluorine in all three of these materials are covalent. Graphite fluorides are used as electrodes in advanced primary batteries and as lubricants.

Suppose a graphite fluoride is analyzed to contain 2.53 g of carbon for every gram of fluorine. What is the empirical formula of this compound?

21. A gaseous binary compound of chlorine and fluorine has a density at 0°C and 1.00 atm of $4.13 \text{ g L}^{-1}$ and is 38.35% chlorine by mass. Determine its molecular formula.

22. A compound has the molecular formula $ClO_3F$. Estimate the volume occupied by 225 g of this compound at 0°C and 1.00 atm.

23. Predict the structures of the following fluorine-containing molecules.
(a) $OF_2$  (d) $BrF_5$
(b) $BF_3$  (e) $IF_7$
(c) $BrF_3$  (f) $SeF_6$

24. Predict the structures of the following fluorine-containing molecular ions.
(a) $SiF_6^{2-}$  (d) $AsF_6^-$
(b) $ClF_2^+$  (e) $BF_4^-$
(c) $IF_4^+$  (f) $BrF_6^+$

25. Thionyl and selenyl difluoride have the compositions $SOF_2$ and $SeOF_2$, with the oxygen and fluorine atoms directly linked to the central sulfur or selenium atom. They can be

prepared by reaction of a strong fluorinating agent, such as $PF_5$, with sulfur dioxide or selenium dioxide:

$$SO_2(g) + PF_5(g) \longrightarrow SOF_2(g) + POF_3(g)$$

(a) Use the VSEPR theory to predict the structures of $SOF_2$ and $SeOF_2$. Draw the structures.
(b) Thionyl difluoride can coordinate with $BF_3$ through its lone pair. Does it act as a Lewis acid or as a Lewis base in this reaction?

**26.** When two oxygen atoms take the place of four fluorine atoms in $SF_6$, the compound has the composition $SO_2F_2$ and is called "sulfuryl difluoride." Like $SF_6$, $SO_2F_2$ is comparatively unreactive and hydrolyzes with difficulty.
(a) Use the VSEPR theory to predict the structure of $SO_2F_2$. Draw this structure.
(b) In the conversion of thionyl difluoride ($SOF_2$) to sulfuryl difluoride,

$$2 SOF_2 + O_2 \longrightarrow 2 SO_2F_2$$

does the thionyl difluoride act as a Lewis acid or as a Lewis base?

**27.** Although $BrF_3$ is a covalent compound, $BrF_3(\ell)$ is slightly conductive due to the autoionization of $BrF_3$ molecules. This is analogous to the small conductivity of water arising from autoionization:

$$2 H_2O(\ell) \rightleftharpoons H_3O^+(aq) + OH^-(aq)$$

Write a balanced chemical equation for the autoionization of $BrF_3$. Can this reaction be described as a Lewis acid–base reaction?

**28.** The molecules in liquid $BrF_3$ are linked through "fluorine bonds" that are analogous to the hydrogen bonds in water. Would you expect the entropy of vaporization of $BrF_3(\ell)$ to be larger or smaller than the Trouton's rule value, $88 J K^{-1} mol^{-1}$?

**29.** Rank the following compounds in order of normal boiling point, from lowest to highest: $CaF_2$, $PtF_6$, $PtF_4$.

**30.** For each of the following pairs of fluorine compounds, predict which will have the higher melting point.
(a) NaF or $PF_3$
(b) $AsF_3$ or $AsF_5$
(c) $CF_4$ or $C_5F_{12}$

**31.** One of the hazards in making Teflon is that the starting material, tetrafluoroethylene, $C_2F_4(g)$, can explode, giving C(s, graphite) and $CF_4(g)$. The standard enthalpy of formation of $C_2F_4(g)$ is $-651 kJ mol^{-1}$. Use this fact, together with data from Appendix D, to estimate the amount of heat that would be released if a tank containing 1.00 kg of $C_2F_4$ were to explode.

**32.** Compute the $\Delta G°$ of the reaction

$$CF_4(g) + 2 H_2O(\ell) \longrightarrow CO_2(aq) + 4 HF(aq)$$

using data from Appendix D. Explain how $CF_4$ spontane-

ously decomposes according to the preceding equation, yet in practice is stable up to 500°C.

## Compounds of Fluorine and the Noble Gases

**33.** The standard enthalpy change $\Delta H°$ for the reaction

$$XeF_6(g) + 3 H_2(g) \longrightarrow Xe(g) + 6 HF(g)$$

is $-1282 kJ$. Using this fact and information from Appendix D, calculate the following:
(a) The standard molar enthalpy of formation $\Delta H_f°$ of $XeF_6(g)$
(b) The average enthalpy of an Xe—F bond in $XeF_6(g)$

**34.** The standard enthalpy change $\Delta H°$ for the reaction

$$XeF_4(g) + 2 H_2(g) \longrightarrow Xe(g) + 4 HF(g)$$

is $-887 kJ$. Using this fact and information from Appendix D, calculate the following:
(a) The standard molar enthalpy of formation $\Delta H_f°$ of $XeF_4(g)$
(b) The average enthalpy of an Xe—F bond in $XeF_4(g)$

**35.** Xenon oxotetrafluoride ($XeOF_4$) is a liquid at room temperature. It can be prepared by the controlled hydrolysis of xenon hexafluoride:

$$XeF_6(g) + H_2O(\ell) \longrightarrow XeOF_4(\ell) + 2 HF(g)$$

The standard enthalpies of formation of the noble-gas compounds in this reaction are

$$\Delta H_f°(XeF_6(g)) = -298 kJ mol^{-1}$$
$$\Delta H_f°(XeOF_4(\ell)) = +148 kJ mol^{-1}$$

Calculate $\Delta H°$ for the hydrolysis reaction. Use Appendix D for additional data.

**36.** The energy of explosion of $XeO_3(s)$ into its gaseous elements was measured in a constant-volume calorimeter.

$$XeO_3(s) \longrightarrow Xe(g) + \tfrac{3}{2} O_2(g)$$

(a) A $2.763 \times 10^{-4}$ mol sample released 112 J of heat when it was exploded. Calculate the standard energy of formation ($\Delta E_f°$) of $XeO_3(s)$.
(b) Will the standard *enthalpy* of formation of $XeO_3(s)$ be larger or smaller than your answer to part (a)?

**37.** The reduction of perxenate ion ($XeO_6^{4-}$) to xenon is so favored thermodynamically in acidic solution that it will even oxidize $Mn^{2+}$ to $MnO_4^-$, one of the strongest oxidizing agents ordinarily encountered, with evolution of oxygen. Write balanced half-equations and an equation for the overall reaction in this process.

**38.** The following standard reduction potentials have been measured for the oxides of xenon in acidic aqueous solution:

$$H_4XeO_6(aq) + 2 H_3O^+(aq) + 2 e^- \longrightarrow$$
$$XeO_3(aq) + 5 H_2O(\ell) \quad \mathscr{E}° = +2.36 V$$

$$XeO_3(aq) + 6\,H_3O^+(aq) + 6\,e^- \longrightarrow$$
$$Xe(g) + 9\,H_2O(\ell) \qquad \mathscr{E}° = +2.12\text{ V}$$

(a) Would you classify perxenic acid ($H_4XeO_6$) as an oxidizing agent or as a reducing agent?

(b) In the preceding, $XeO_3$ acts as an oxidizing agent in one half-reaction and as a reducing agent in the other. At pH 0, is $XeO_3$ stronger as an oxidizing agent or as a reducing agent?

(c) Is $XeO_3$ stable with respect to disproportionation to $Xe(g)$ and $H_4XeO_6(aq)$ in acidic aqueous solution?

## Additional Problems

**39.** Chlorine-containing bleaches act by oxidizing. By referring to Appendix E, predict whether bromine will be a stronger or a weaker bleach than chlorine.

**40.** In the Deacon process, an equilibrium is reached among chlorine, steam, hydrogen chloride, and oxygen at 450°C.

(a) Use the $\Delta H_f°$ and $\Delta S_f°$ values from Appendix D to estimate an equilibrium constant at 450°C for the reaction

$$2\,HCl(g) + \tfrac{1}{2}O_2(g) \rightleftharpoons Cl_2(g) + H_2O(g)$$

(b) Suppose that bromine replaces chlorine in the preceding reaction. After referring to the appropriate bond enthalpies in Table 14-1, state whether the equilibrium constant at 450°C will be larger or smaller. Explain.

**41.** Suppose a chlor-alkali plant has 250 cells, through each of which a current of 100,000 A passes continuously.

(a) Calculate the mass of chlorine that this plant can produce per 24-hour day.

(b) If the cell voltage is 3.5 V, calculate the total electrical energy consumed by the plant in one 24-hour day, and express it both in joules and in kilowatt-hours.

(c) If electricity costs $0.05 per kilowatt-hour, calculate the cost for the needed electricity needed to run the plant for one 24-hour day.

**42.** The solubility of iodine in water at 25°C is 0.033 g per 100 g. Calculate the solubility of iodine in 1.0 M KI solution at this temperature, taking into account the equilibrium

$$I^-(aq) + I_2(aq) \rightleftharpoons I_3^-(aq)$$

Relevant data can be found in Appendix D.

**43.** Astatine is a little-studied element whose longest-lived isotope, $^{210}At$, has a half-life of only 8.3 hours.

(a) Predict whether HAt will be a stronger or a weaker acid than HI.

(b) The $At^-$ ion is produced when elemental astatine reacts in aqueous solution with zinc, but no reaction is seen with 1 M $Fe^{2+}(aq)$. What range of reduction potentials for conversion of $At_2$ to $At^-$ is consistent with these observations?

(c) If $Cl_2(g)$ is bubbled through a solution containing $At^-(aq)$, what reaction will result?

(d) Write a balanced equation for the reaction that should occur when $At_2(s)$ is dissolved in basic aqueous solution.

**44.** Solid iodine has a small electrical conductivity at room temperature, which increases with temperature. By considering the neighbors of iodine in the periodic table, explain how this may be true. Do you predict that the conductivity of solid bromine at −20°C will be greater than or less than that of solid iodine at the same temperature? Why?

**45.** The reaction

$$K_2[NiF_6] + TiF_4(s) \longrightarrow K_2[TiF_6](s) + NiF_2(s) + F_2(g)$$

generates elemental fluorine. Does $TiF_4(s)$ play the role of a Lewis acid or a Lewis base in this reaction? Explain.

**46.** A commercial fluorine cell generates 3.3 kg of fluorine per hour by electrolysis of $KF\cdot2HF$. Compute the average current passing through this cell.

**47.** Suppose an aqueous solution of NaF is electrolyzed between platinum electrodes. Will $F_2(g)$ be produced? Why or why not?

**48.** (a) Predict the vapor density of $HF(g)$ at 1.00 atm pressure and its normal boiling point, 20°C, assuming ideal gas behavior.

(b) The observed vapor density under the conditions in part (a) is 3.11 g $L^{-1}$. What is the explanation for the large discrepancy with the calculated result of part (a)?

**49.** The reaction

$$4\,HF(\ell) + SiO_2(s) \longrightarrow SiF_4(g) + 2\,H_2O(\ell)$$

can be used to release gold that is distributed in certain quartz veins of hydrothermal origin. Let us say the quartz contains $1.0 \times 10^{-3}\%$ gold by mass and the gold has a market value of $350 per troy ounce. Will the process be economically feasible if commercial (50% by mass) aqueous hydrogen fluoride (density 1.17 g $cm^{-3}$) costs $0.25 per liter? (1 troy ounce = 31.3 g.)

**50.** Ingestion of 5 to 10 g of NaF is lethal to a 70-kg man. Smaller doses can be treated with calcium therapy, which restores $Ca^{2+}$ levels in the body. Write a chemical equation for the effect of NaF on the body.

**51.** Dioxygen difluoride, $O_2F_2$, is a particularly potent fluorinating agent, made by irradiating a mixture of $O_2$ and $F_2$ at the temperature of liquid nitrogen. The O-to-O distance in $O_2F_2$ is nearly as short as the distance in $O_2$, and the O-to-F distances are quite long. Draw the Lewis structure for this molecule, determine the bond order of all bonds, and describe its geometry. What is the name of the analogous compound of oxygen and hydrogen?

**52.** The fluorinating agent dioxygen difluoride, $O_2F_2$ (see the preceding problem), is important because it converts the plutonium in almost any plutonium-containing material to $PuF_6$ under mild conditions. The volatility of $PuF_6$ then allows the separation of radioactive Pu from various impurities. Write balanced chemical equations for the reaction of $PuO_2$ with $O_2F_2$ and the reaction of Pu with $O_2F_2$.

**53.** When $SF_4(g)$ reacts with $CsF(s)$, the $SF_5^-$ ion is formed. Use the VSEPR theory to predict its geometry.

**54.** The compound $S_2F_{10}$ was not mentioned in the chapter.
(a) Taking the known bond-forming tendencies of fluorine into account, suggest a probable Lewis structure for this compound.
(b) Use the VSEPR model to predict the molecular geometry of $S_2F_{10}$.

**55.** The result of the accidental first preparation of Teflon was immediately subjected to elemental analysis for both chlorine and fluorine. The report on the new, mysterious white powder was that it contained no chlorine and 48.4% fluorine by mass. Compare these first values to the actual percentages of fluorine and chlorine in Teflon.

**56.** The ionization energy of $O_2$ is 1180 kJ mol$^{-1}$. In 1962 Neil Bartlett reported that it reacts with $PtF_6(g)$ to form the solid ionic compound $O_2^+PtF_6^-$. By referring to data in Appendix F, show why it occurred to Bartlett that xenon might also form a compound with $PtF_6$.

**57.** Calculate the equilibrium constant for the reaction

$$Xe(g) + 2\,F_2(g) \rightleftharpoons XeF_4(s)$$

at 25°C. The standard molar free energy of formation of $XeF_4(s)$ is $-134$ kJ mol$^{-1}$.

**58.** Xenon difluoride reacts with arsenic pentafluoride to give an ionic compound:

$$2\,XeF_2 + AsF_5 \longrightarrow [Xe_2F_3]^+[AsF_6]^-$$

Identify which species are Lewis acids in this reaction and which are Lewis bases.

# 23

*A sugar crystal, viewed under polarized light.*

# Organic Chemistry and Polymers

C arbon is unique among the elements in the large number of compounds it forms and in the variety of their structures. In combination with hydrogen, it forms molecules with single, double, and triple bonds; chains; rings; branched structures; and cages (Fig. 23-1). The thousands of stable hydrocarbons are in sharp contrast with the mere two stable compounds between oxygen and hydrogen (water and hydrogen peroxide). Even the rather versatile elements nitrogen and oxygen form only six nitrogen oxides.

The unique properties of carbon relate to its position in the periodic table. As a second-period element, carbon has relatively small atoms, which allow it to easily form the double and triple bonds that are rare in the compounds of related elements, such as silicon. As a Group IV element, carbon can form four bonds, more than the other second-period elements—which give it wide scope for structural elaboration. Finally, as an element of intermediate electronegativity, carbon forms covalent compounds both with relatively electronegative elements, such as oxygen, nitrogen, and the halogens, and with relatively electropositive elements, such as hydrogen and the heavy metals mercury and lead.

**Figure 23-1**

One simple and unusual hydrocarbon is cubane ($C_8H_8$), in which the eight carbon atoms are arranged at the corners of a cube.

The study of the compounds of carbon is the discipline traditionally called **organic chemistry,** although the chemistry of carbon is intimately bound up with that of inorganic elements and with biochemistry. This chapter begins with a look at the composition, refining, and chemical processing of petroleum, the primary starting material for the production of hydrocarbons and their derivatives. It turns to the types of compounds that result when elements such as chlorine, oxygen, and nitrogen combine with carbon and hydrogen. Sections 23-3 and 23-4 then examine the synthetic and natural polymers that result from the chemical combination of small-molecule building blocks into long-chain polymers.

## 23-1  PETROLEUM REFINING AND THE HYDROCARBONS

When the first oil well was drilled in 1859 near Titusville, Pennsylvania, the future effects of the exploitation of petroleum on everyday life could not have been anticipated. Today, the petroleum and petrochemical industries span the world and touch the most minute details of our daily lives. In the early years of this century, the development of the automobile, fueled by low-cost gasoline derived from petroleum, dramatically changed many people's lifestyles. The subsequent use of gasoline and oil to power trains and planes, tractors and harvesters, pumps and coolers, transformed travel, agriculture, and industry. Natural gas and heating oil warm the vast majority of homes in the United States. Finally, the spectacular growth of the petrochemical industry since 1945 has led to the introduction of innumerable new products ranging from pharmaceuticals to plastics and synthetic fibers. More than half of the chemical compounds produced in greatest volumes stem directly or indirectly from petroleum.

As we look forward to the next century, the prospects for the continued enjoyment of cheap petroleum and petrochemicals are clouded. Many wells have been drained, and the remaining petroleum is relatively difficult and costly to extract. Petroleum is not easy to make. It originated from the deposition and decay of organic matter (of animal or vegetable origin) in oxygen-poor marine sediments. Subsequently, petroleum migrated to the porous sandstone rocks from which it is extracted today. In the last 100 years, we have consumed a significant fraction of the petroleum accumulated in the earth over many millions of years. The imperative for the future is to save the remaining reserves for uses for which few substitutes are available (such as the manufacture of specialty chemicals) while finding other sources of heat and energy.

### Petroleum Distillation and the Normal Alkanes

Although crude petroleum contains small amounts of oxygen, nitrogen, and sulfur, its major constituents are **hydrocarbons,** compounds of carbon and hydrogen. The most prevalent hydrocarbons in petroleum are the **straight-chain alkanes** (also called normal alkanes, or *n*-alkanes), which consist of chains of carbon atoms bonded to one another by single bonds, with enough hydrogen atoms on each carbon atom to bring it to the maximum bonding capacity of four. These alkanes have the generic formula $C_nH_{2n+2}$; Table 23-1 lists the names and formulas of the first few. The ends of the molecules are methyl ($-CH_3$) groups, with methylene ($-CH_2-$)

**Table 23-1    Straight-Chain Alkanes**

| Name | Formula | Name | Formula |
|------|---------|------|---------|
| Methane | $CH_4$ | Decane | $C_{10}H_{22}$ |
| Ethane | $C_2H_6$ | Undecane | $C_{11}H_{24}$ |
| Propane | $C_3H_8$ | Dodecane | $C_{12}H_{26}$ |
| Butane | $C_4H_{10}$ | Tridecane | $C_{13}H_{28}$ |
| Pentane | $C_5H_{12}$ | Tetradecane | $C_{14}H_{30}$ |
| Hexane | $C_6H_{14}$ | Pentadecane | $C_{15}H_{32}$ |
| Heptane | $C_7H_{16}$ | ⋮ | ⋮ |
| Octane | $C_8H_{18}$ | Triacontane | $C_{30}H_{62}$ |
| Nonane | $C_9H_{20}$ | | |

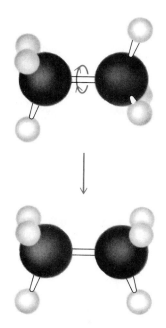

**Figure 23-2**

The two —$CH_3$ groups in ethane rotate easily about the bond that joins them.

groups between. We could write pentane ($C_5H_{12}$) as $CH_3CH_2CH_2CH_2CH_3$ to indicate the structure more explicitly or, in abbreviated fashion, as $CH_3(CH_2)_3CH_3$.

Each carbon atom in a straight-chain alkane forms four single bonds, which point to the corners of a tetrahedron (exhibiting $sp^3$ hybridization). Although bond *lengths* do not change much through vibration, rotation about a C—C single bond occurs quite easily (Fig. 23-2), and so a given hydrocarbon molecule in a gas or liquid constantly changes its conformation. The term "straight chain" refers only to the bonding pattern in which each carbon atom is bonded to the next one in a sequence; it does not mean that the carbon atoms are positioned along a straight line. An alkane molecule with 10 to 20 carbon atoms looks quite different from its curled-up form when its bonds are extended to give a "stretched" molecule (Fig. 23-3). These two extreme conformations and many others interconvert rapidly at room temperature.

**Figure 23-3**

Two of the very many possible conformations of the alkane $C_{17}H_{36}$. The carbon atoms are not shown explicitly, but they lie at every intersection of four lines.

**Figure 23-4**

The melting and boiling points of the straight-chain alkanes increase with chain length *n*. Note the alternation in the melting points: alkanes with *n* odd tend to have lower melting points because they are more difficult to pack into a crystal lattice.

Figure 23-4 shows the melting and boiling points of the straight-chain alkanes; both increase with the number of carbon atoms and thus with molecular mass. This is a consequence of the increasing strength of dispersion forces between heavier molecules, as discussed in Section 3-7. Methane, ethane, propane, and butane are all gases at room temperature, but the hydrocarbons that follow them in the alkane series are liquids. Alkanes beyond about $C_{17}H_{36}$ are waxy solids at 20°C, with melting points that increase with the number of carbon atoms present. Paraffin wax, a low-melting solid, is a mixture of alkanes that have 20 to 30 carbon atoms per molecule. Petrolatum (petroleum jelly, or Vaseline) is a different mixture that is semisolid at room temperature.

A mixture of hydrocarbons such as petroleum does not boil at a single, sharply defined temperature. Instead, as such a mixture is heated, the compounds with lower boiling points (the most volatile) boil off first and, as the temperature rises, more and more of the material vaporizes. The existence of a boiling-point range permits components of a mixture to be separated by distillation, as Section 4-6 discussed. The earliest petroleum distillation was a simple batch process: the crude oil was heated in a still, the volatile fractions were removed at the top and condensed to

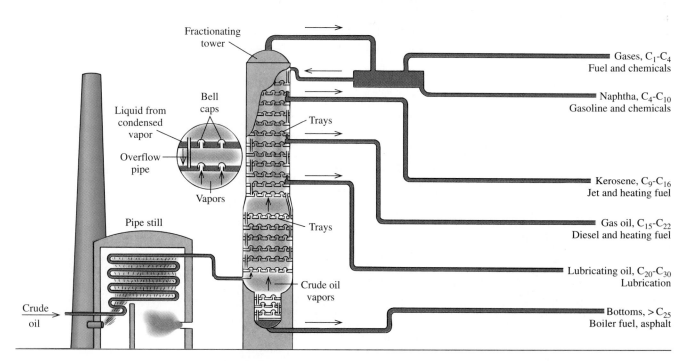

**Figure 23-5**

In the distillation of petroleum, the lighter, more volatile hydrocarbon fractions are removed from higher up the column, and the heavier fractions from lower down.

gasoline, and the still was cleaned for another batch. Modern petroleum refineries use much more sophisticated and efficient distillation methods, in which crude oil is added continuously and fractions of different volatility are tapped off at various points up and down the distillation column (Figs. 23-5 and 23-6). To save on energy costs, heat exchangers are employed to capture the heat from condensation of the liquid products.

Distillation allows hydrocarbons to be separated by boiling point and thus by molecular mass. The gases that emerge from the top of the column resemble the natural gas that collects in rock cavities above petroleum deposits. The gas mixtures can be separated further by redissolving the ethane, propane, and butane in a liquid solvent such as hexane. The methane-rich mixture of gases that remains is used for chemical synthesis or is shipped by pipeline to heat homes. Redistilling the hexane and its dissolved gases allows their separation and their use as chemical starting materials. Propane and butane are also bottled under pressure as liquefied petroleum gas (LPG), which is used for fuel in rural areas. After the gases, the next fraction to emerge from the petroleum distillation column is **naphtha,** which is used primarily in the manufacture of gasoline. Subsequent fractions of successively higher molecular mass are employed for jet and diesel fuel, heating oil, and machine lubricating oil. The heavy, involatile sludge that remains at the bottom of the distillation unit is pitch or asphalt, which is used for roofing and paving.

## Cyclic and Branched-Chain Alkanes

The straight-chain alkanes are not the only hydrocarbons in petroleum. Two other important classes of compounds, the cyclic and branched-chain alkanes, are also represented. A **cycloalkane** consists of at least one chain of carbon atoms attached

**Figure 23-6**
Distillation towers in a modern oil refinery. (Robin Smith and Tony Stone/Worldwide)

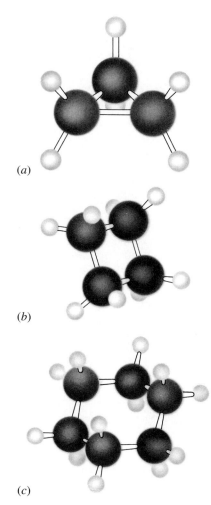

**Figure 23-7**
Three cyclic hydrocarbons. (*a*) Cyclopropane, $C_3H_6$. (*b*) Cyclobutane, $C_4H_8$. (*c*) Cyclohexane, $C_6H_{12}$.

at the ends to form a closed loop. In the formation of this additional C—C bond, two hydrogen atoms must be eliminated, and so the general formula for cycloalkanes with one ring is $C_nH_{2n}$ (Fig. 23-7). The smallest cycloalkanes, cyclopropane and cyclobutane, are very strained compounds because the C—C—C bond angle is 60° (in $C_3H_6$) or 90° (in $C_4H_8$), far less than the normal tetrahedral angle of 109.5°. As a result, these compounds are more reactive than the heavier cycloalkanes or their straight-chain analogs, propane and butane.

**Branched-chain alkanes** are hydrocarbons that contain only C—C and C—H single bonds but in which the carbon atoms are no longer arranged in a straight chain. One or more carbon atoms in each molecule are bonded to three or four other carbon atoms rather than to only one or two as in the normal alkanes or cycloalkanes. The simplest branched-chain molecule (Fig. 23-8*b*) is that of 2-methylpropane, sometimes referred to as isobutane. This molecule has the same molecular formula as butane ($C_4H_{10}$) but a different bonding structure, in which the central carbon atom is bonded to three —$CH_3$ groups and only one hydrogen atom. The compounds butane and 2-methylpropane are **isomers.** Their molecules have the same formula but different three-dimensional structures that can be interconverted only by breaking and re-forming chemical bonds.

The number of possible isomers increases rapidly with increasing numbers of carbon atoms in the hydrocarbon molecule. Thus, butane and 2-methylpropane are the only two isomers of chemical formula $C_4H_{10}$, but there are three isomers of $C_5H_{12}$, five of $C_6H_{14}$, nine of $C_7H_{16}$, and millions of $C_{30}H_{62}$. A systematic procedure for naming these isomers has been codified by the International Union of Pure and Applied Chemistry (IUPAC). The following set of rules is a part of that procedure.

**1.** Find the longest continuous chain of carbon atoms in the molecule. The molecule is named as a derivative of this alkane. In Figure 23-8*b* a chain of three carbon atoms can be found, and so the molecule is a derivative of propane.

(*a*)                                    (*b*)

**Figure 23-8**
Two isomeric hydrocarbons with the molecular formula $C_4H_{10}$. (*a*) Butane. (*b*) 2-methylpropane.

**2.** The hydrocarbon groups attached to the chain are called **alkyl groups.** Their names are obtained by dropping the ending *-ane* from the corresponding alkane and replacing it with *-yl* (Table 23-2). The methyl group, —$CH_3$, is derived from methane ($CH_4$), for example. Note also the isopropyl group, which attaches by its middle carbon atom.

**3.** Number the carbon atoms along the chain identified in rule 1. Identify each alkyl group by the number of the carbon atom at which it is attached to the chain. The methyl group in the molecule in Figure 23-8*b* is attached to the second of the three carbon atoms in the propane chain, and so the molecule is called 2-methylpropane. The carbon chain is numbered from the end that gives the lowest number for the position of the first attached group.

**4.** If more than one alkyl group of the same type is attached to the chain, use the prefixes *di-* (two), *tri-* (three), *tetra-* (four), *penta-* (five), and so forth to specify the total number of such attached groups in the molecule. Thus, 2,2,3-trimethylbutane has two methyl groups attached to the second carbon atom and one to the third carbon atom of the four-atom butane chain. It is an isomer of heptane ($C_7H_{16}$).

**5.** If several types of alkyl groups appear, name them in alphabetical order. Ethyl is listed before methyl, which appears before propyl.

| Table 23-2 | Alkyl Side Groups |
|---|---|
| **Name** | **Formula** |
| Methyl | —$CH_3$ |
| Ethyl | —$CH_2CH_3$ |
| Propyl | —$CH_2CH_2CH_3$ |
| Isopropyl | —$CH(CH_2)_2$ |
| Butyl | —$CH_2CH_2CH_2CH_3$ |

## Example 23-1

Name the following branched-chain alkane:

$$CH_3 \diagdown \quad CH_2 \diagdown \quad CH_2{-}CH_3$$
$$\quad \quad C \quad \quad \quad C$$
$$CH_3 \diagup \; H \; \; CH_3 \diagup \; \diagdown CH_2{-}CH_3$$

### Solution

The longest continuous chain of carbon atoms is six, and so this is a derivative of hexane. Number the carbon atoms starting from the left.

$$CH_3 \diagdown \quad {}^{3}CH_2 \diagdown \quad {}^{5}CH_2{-}{}^{6}CH_3$$
$$\quad \; {}^{2}C \quad \quad \; {}^{4}C$$
$$\; {}^{1}CH_3 \diagup \; H \; \; CH_3 \diagup \; \diagdown CH_2{-}CH_3$$

Methyl groups are attached to carbon atoms 2 and 4, and an ethyl group is attached to atom 4. The name is thus 4-ethyl-2,4-dimethylhexane. Note that if we had started numbering from the right, the higher number 3 would have appeared for the position of the first methyl group, and so the numbering from the left is preferred.

*Related Problems:* 9, 10

The fraction of branched-chain alkanes in a gasoline affects how it burns in an engine. Gasoline composed entirely of straight-chain alkanes burns very unevenly, causing "knocking" that can damage the engine. Blends that are richer in branched-chain and cyclic alkanes burn with much less knocking. Smoothness of combustion is rated quantitatively via the **octane number** of the gasoline, which was defined in 1927 by selecting two compounds that lie at extremes in the knocking they cause.

Pure 2,2,4-trimethylpentane (commonly known as isooctane) burns very smoothly and was assigned an octane number of 100. Of the compounds examined at the time, pure heptane caused the most knocking and was assigned octane number 0. Mixtures of heptane and isooctane cause intermediate amounts of knocking. Standard mixtures of these two compounds define a scale for evaluating the knocking caused by real gasolines, which are complex mixtures of branched-chain and straight-chain hydrocarbons. If a gasoline sample produces the same amount of knocking in a test engine as a mixture of 90% (by volume) 2,2,4-trimethylpentane and 10% heptane, it is assigned the octane number 90.

Certain additives increase the octane rating of gasoline. The least expensive of these is tetraethyllead, $Pb(C_2H_5)_4$, a compound that has very weak bonds between the central lead atom and the ethyl carbon atoms. It readily releases ethyl radicals ($\cdot C_2H_5$) into the gasoline during combustion; these reactive species speed and smooth the combustion process, reducing knocking and giving better fuel performance. The lead is released into the atmosphere and causes serious long-term health hazards, however. As a result, the use of lead in gasoline is being phased out, and other low-cost additives are being sought to increase octane number. Chemical processing to make branched-chain compounds from straight-chain compounds represents another solution to the problem, although only a partial one.

## Alkenes and Alkynes

The hydrocarbons discussed so far are referred to as **saturated,** because all the carbon–carbon bonds are single bonds. Hydrocarbons that have double and triple carbon–carbon bonds are referred to as **unsaturated** (Fig. 23-9). Ethylene ($C_2H_4$) has a double bond between its carbon atoms and is called an **alkene.** The simplest **alkyne** is acetylene ($C_2H_2$), which has a triple bond between its carbon atoms. In naming these compounds, the *-ane* ending of the corresponding alkane is replaced by *-ene* when a double bond is present and by *-yne* when a triple bond is present. Ethene is thus the systematic name for ethylene, and ethyne for acetylene, although we shall continue to use their more common names. For any compound with a carbon backbone of four or more carbon atoms, it is necessary to specify the location of the double or triple bond. This is done by numbering the carbon–carbon bonds and putting the number of the lower-numbered carbon involved in the multi-

**Figure 23-9**

One way to distinguish alkanes from alkenes is by their reactions with aqueous $KMnO_4$. This strong oxidizing agent undergoes no reaction with hexane and retains its purple color (*left*). When $KMnO_4$ is placed in contact with 1-hexene, however, a redox reaction takes place in which the brown solid $MnO_2$ forms (*right*) and —OH groups are added to both sides of the double bond in the 1-hexene, giving a compound with the formula $CH_3(CH_2)_3CH(OH)CH_2OH$. (Charles Steele)

ple bond before the name of the alkene or alkyne. Thus, the two different isomeric alkynes with the formula $C_4H_6$ are

$$HC\equiv C-CH_2-CH_3 \qquad \text{1-butyne}$$

$$CH_3-C\equiv C-CH_3 \qquad \text{2-butyne}$$

There is yet another subtlety in alkenes. We recall from Section 14-5 that rotation does not take place readily about a carbon–carbon double bond as it does about a single bond. Many alkenes therefore exist in contrasting isomeric forms, depending on whether bonding groups are on the same side or on opposite sides of the double bond. There is only a single isomer of 1-butene but two of 2-butene:

These compounds differ in melting and boiling points, density, and other physical and chemical properties. Compounds with two double bonds are called dienes, those with three are trienes, and so forth. The compound 1,3-pentadiene, for example, is a derivative of pentane with two double bonds:

$$CH_2{=}CH-CH{=}CH-CH_3$$

In such **polyenes**, each double bond may lead to *cis* and *trans* conformations, depending on its neighboring groups, and so there may be several isomers with the same bonding patterns but different molecular geometries and physical properties.

Alkenes are not present to a significant extent in crude petroleum. They are key compounds for the synthesis of organic chemicals and polymers, however, and so their production from alkanes is of great importance. One way to achieve this production is by the **cracking** of the petroleum by heat or with catalysts. In **catalytic cracking,** the heavier fractions from the distillation column (compounds of $C_{12}$ or higher) are passed over a silica–alumina catalyst at temperatures of 450 to 550°C. Reactions such as

$$CH_3(CH_2)_{12}CH_3 \longrightarrow CH_3(CH_2)_4{-}CH{=}CH_2 + C_7H_{16}$$

occur to break the long chain into fragments. This type of reaction accomplishes two purposes. First, the shorter-chain hydrocarbons have lower boiling points and can be added to gasoline. Second, the alkenes that result have higher octane numbers than the corresponding alkanes and perform better in the engine. Moreover, these alkenes can react with alkanes to give the more highly branched alkanes that are desirable in gasoline. **Thermal cracking** uses higher temperatures of 850 to 900°C and no catalyst. It produces shorter-chain alkenes, such as ethylene and propylene, through reactions such as

$$CH_3(CH_2)_{10}CH_3 \longrightarrow CH_3(CH_2)_8CH_3 + CH_2{=}CH_2$$

The short-chain alkenes are too volatile to be good components of gasoline but are among the most important starting materials for chemical synthesis.

## Aromatic Hydrocarbons

One last group of hydrocarbons in crude petroleum is the **aromatic hydrocarbons,** of which benzene is the simplest example. The benzene molecule consists of a six-carbon ring in which delocalized $\pi$ electrons significantly increase the molecular stability (Section 14-5). Benzene is sometimes represented by its chemical formula $C_6H_6$, and sometimes (to show structure) by a hexagon with a circle inside it:

The six points of the hexagon represent the six carbon atoms, with the hydrogen atoms omitted for simplicity. The circle represents the delocalized $\pi$ electrons, which are spread out evenly over the ring. The molecules of other aromatic compounds contain benzene rings with various side groups or two (or more) benzene rings linked by alkyl chains or fused side by side, as in naphthalene ($C_{10}H_8$):

Besides benzene, the most prevalent aromatic compounds in petroleum are toluene, in which one hydrogen atom on the benzene ring is replaced by a methyl group, and the xylenes, in which two such replacements are made:

Toluene          o-Xylene          m-Xylene          p-Xylene

This set of compounds is referred to as BTX (for *b*enzene-*t*oluene-*x*ylene). The BTX in petroleum is very important to polymer synthesis, as we shall see in Section 23-3. These components also significantly increase octane number and are used to make high-performance fuels with octane numbers above 100, as are required in modern aviation.

A major advance in petroleum refining has been the development of **reforming reactions,** which produce BTX aromatics from straight-chain alkanes containing the same numbers of carbon atoms. A fairly narrow distillation fraction containing only $C_6$ to $C_8$ alkanes is taken as the starting material. The reactions employ high temperatures and transition-metal catalysts such as platinum or rhenium on alumina supports, and their detailed mechanisms are not fully understood. Apparently a normal alkane such as hexane is cyclized and dehydrogenated to give benzene as the primary product (Fig. 23-10). Heptane yields mostly toluene, and octane yields a mixture of xylenes. Toluene is of increasing importance as a solvent because tests on laboratory animals show it to be far less carcinogenic than benzene. Benzene is

$$CH_3(CH_2)_4CH_3 \rightleftharpoons CH_3(CH_2)_3CH=CH_2$$

n-hexane        1-hexene

**Figure 23-10**

The reforming reaction that produces benzene from hexane uses a catalyst of platinum on alumina. The platinum facilitates the removal of hydrogen, and the alumina ($Al_2O_3$) facilitates the opening and closing of rings. One typical multistep sequence is shown; other intermediates can form, as well.

more important for chemical synthesis, however, and so a large fraction of the toluene produced is converted to benzene by **hydrodealkylation:**

This reaction is carried out at high temperatures (550–650°C) and pressures of 40 to 80 atm.

## 23-2 FUNCTIONAL GROUPS AND ORGANIC SYNTHESIS

The first section of this chapter described the diverse compounds formed from the two elements carbon and hydrogen and the winning of these compounds from petroleum. We now consider the effect of the attachment on or insertion into a hydrocarbon of additional chemical elements, such as oxygen, nitrogen, or a halogen. In doing so, we shift our attention from structures of entire molecules to the properties of **functional groups,** which consist of the noncarbon atoms plus the portions of the molecule immediately adjacent to them. Functional groups tend to be the reactive sites in organic molecules, and their chemical properties depend only rather weakly on the natures of the hydrocarbons to which they are attached. This fact permits us

**Table 23-3   Common Functional Groups**

| Functional Group[†] | Type of Compound | Examples |
|---|---|---|
| R—F, —Cl, —Br, —I | Alkyl or aryl halide | $CH_3CH_2Br$ (bromoethane) |
| R—OH | Alcohol | $CH_3CH_2OH$ (ethanol) |
| | Phenol | ![phenol structure] —OH <br> (phenol) |
| R—O—R′ | Ether | $CH_3$—O—$CH_3$  (dimethyl ether) |
| R—C(=O)H (aldehyde structure) | Aldehyde | $CH_3CH_2CH_2$—C(=O)—H <br> (butyraldehyde, or butanal) |
| R, R′ C=O (ketone structure) | Ketone | $CH_3$—C(=O)—$CH_3$ <br> (propanone, or acetone) |
| R—C(=O)OH (carboxylic acid structure) | Carboxylic acid | $CH_3COOH$ <br> (acetic acid, or ethanoic acid) |
| R—C(=O)O—R′ (ester structure) | Ester | $CH_3$—C(=O)—O—$CH_3$ <br> (methyl acetate) |
| R—$NH_2$ | Amine | $CH_3NH_2$ (methylamine) |
| R—C(=O) with N(R′)(R″) (amide structure) | Amide | $CH_3$—C(=O)—$NH_2$   (acetamide) |

[†]The symbols R, R′, and R″ stand for hydrocarbon radicals. In some cases they may represent H.

to regard an organic molecule as a hydrocarbon frame, which mainly governs size and shape, to which are attached functional groups that mainly determine the molecule's chemistry. Table 23-3 shows some of the most important functional groups.

A small number of hydrocarbon building blocks from petroleum and natural gas (methane, ethylene, propylene, benzene, and xylene) are the starting points for the synthesis of most of the high-volume organic chemicals produced today. How are these chemicals synthesized?

## Halides

One of the simplest functional groups consists of a single halogen atom, which we take to be chlorine for illustrative purposes. **Alkyl halides** form when mixtures of alkanes and halogens (except iodine) are heated or exposed to light.

$$CH_4 + Cl_2 \xrightarrow{250\text{--}400°C \text{ or light}} CH_3Cl + HCl$$

Methane    Chlorine                                Chloromethane    Hydrogen chloride

The mechanism is a chain reaction, as described in Section 11-4. Ultraviolet light initiates the reaction by dissociating a small number of chlorine molecules into highly reactive atoms. These take part in linked reactions of the form

$$Cl\cdot + CH_4 \longrightarrow HCl + \cdot CH_3 \quad \text{(propagation)}$$

$$\cdot CH_3 + Cl_2 \longrightarrow CH_3Cl + Cl\cdot \quad \text{(propagation)}$$

Chloromethane (also called methyl chloride) is used in synthesis to add methyl groups to organic molecules. If sufficient chlorine is present, more highly chlorinated methanes form. This phenomenon provides a method for the industrial synthesis of dichloromethane ($CH_2Cl_2$, also called methylene chloride), trichloromethane ($CHCl_3$, chloroform), and tetrachloromethane ($CCl_4$, carbon tetrachloride). All three are used as solvents and have vapors with anesthetic or narcotic effects.

A more important industrial route to alkyl halides than the free-radical reactions just described is the addition of chlorine to C=C double bonds. Billions of kilograms of 1,2-dichloroethane (commonly called ethylene dichloride) are manufactured each year, making this compound the largest-volume derived organic chemical. It is made by adding chlorine to ethylene over an iron(III) oxide catalyst at moderate temperatures (40–50°C), either in the vapor phase or in a solution of 1,2-dibromoethane:

$$CH_2{=}CH_2 + Cl_2 \longrightarrow ClCH_2CH_2Cl$$

Almost all the 1,2-dichloroethane produced is used to make chloroethylene (vinyl chloride, $CH_2{=}CHCl$). This is accomplished by heating the 1,2-dichloroethane to 500°C over a charcoal catalyst to abstract HCl:

$$ClCH_2CH_2Cl \longrightarrow CH_2{=}CHCl + HCl$$

The HCl can be recovered and converted to $Cl_2$ (Section 22-1) for further production of 1,2-dichloroethane from ethylene. Vinyl chloride has a much lower boiling point than 1,2-dichloroethane ($-13°C$ compared with 84°C), and so the two are easily separated by fractional distillation. Vinyl chloride is used in the production of polyvinyl chloride plastic (Section 23-3).

## Alcohols and Phenols

**Alcohols** have the —OH functional group attached to an alkyl chain. The simplest alcohol is methanol ($CH_3OH$), which is made from synthesis gas, as described in Section 18-3. The next higher alcohol, ethanol ($CH_3CH_2OH$), can be produced from the fermentation of sugars. Although fermentation is the major source of ethanol for alcoholic beverages and for "gasohol" (automobile fuel made up of 90% gasoline and 10% ethanol), it is not significant for industrial production, which uses the direct hydration of ethylene:

$$CH_2{=}CH_2 + H_2O \longrightarrow CH_3CH_2OH$$

Temperatures of 300 to 400°C and pressures of 60 to 70 atm are employed, with a phosphoric acid catalyst. Both methanol and ethanol are widely used as solvents and as intermediates for further chemical synthesis.

Two three-carbon alcohols exist, depending on whether the —OH group is attached to a terminal carbon atom or the central carbon atom. They are 1-propanol and 2-propanol:

$$CH_3CH_2CH_2OH \qquad\qquad CH_3\underset{\displaystyle \underset{OH}{|}}{C}HCH_3$$

1-Propanol          2-Propanol

The two are frequently referred to as propyl alcohol and isopropyl alcohol, respectively. The systematic names of alcohols are obtained by replacing the *-ane* ending of the corresponding alkane with *-anol* and using a numerical prefix, where necessary, to identify the carbon atom to which the —OH group is attached.[1] Isopropyl alcohol is made from propylene by means of an interesting hydration reaction that is catalyzed by sulfuric acid. The first step is addition of $H^+$ to the double bond,

$$CH_3—CH{=}CH_2 + H^+ \longrightarrow CH_3—\overset{\oplus}{C}H—CH_3$$

producing a transient charged species in which the positive charge is centered on the central carbon atom. Attack by negative $HSO_4^-$ ions occurs at this positive site and leads to a neutral intermediate that can be isolated:

$$CH_3—\underset{\displaystyle \underset{OSO_3H}{|}}{C}H—CH_3$$

Further reaction with water then causes the replacement of the —$OSO_3H$ group with an —OH group and the regeneration of the sulfuric acid:

$$H_2O + CH_3—\underset{\displaystyle \underset{OSO_3H}{|}}{C}H—CH_3 \longrightarrow CH_3—\underset{\displaystyle \underset{OH}{|}}{C}H—CH_3 + H_2SO_4$$

This mechanism explains why only 2-propanol, and no 1-propanol, forms.

The compound 1-propanol is a **primary alcohol:** the carbon atom to which the —OH group is bonded has exactly one other carbon atom attached to it. The isomeric compound 2-propanol is a **secondary alcohol** because the carbon atom to which the —OH group is attached has two carbon atoms (in the two methyl groups) attached to it. The simplest **tertiary alcohol** (in which the carbon atom attached to the —OH group is also bonded to three other carbon atoms) is 2-methyl-2-propanol:

$$CH_3—\overset{\displaystyle \overset{OH}{|}}{\underset{\displaystyle \underset{CH_3}{|}}{C}}—CH_3$$

Primary, secondary, and tertiary alcohols differ in chemical properties.

**Phenols** are compounds in which an —OH group is attached directly to an aromatic ring. The simplest example is phenol itself ($C_6H_5OH$). Phenols differ from alcohols in both physical and chemical properties. One of the most important differences is acidity. Phenol (also called carbolic acid) has an acid ionization constant of $1 \times 10^{-10}$, much greater than the $K_a$'s of typical alcohols, which range from $10^{-16}$ to $10^{-18}$. The reason for this difference is the greater stability of the conjugate base

---

[1] Contrast the names of alcohols with the corresponding names of alkyl halides. If the —OH group were replaced by a chlorine atom, the names of these compounds would be 1-chloropropane and 2-chloropropane.

(the phenoxide ion, $C_6H_5O^-$) due to the spreading out of the negative charge over the aromatic ring. Phenol, although not a strong acid, does react readily with sodium hydroxide to form the salt sodium phenoxide:

$$C_6H_5OH + NaOH \longrightarrow C_6H_5O^-Na^+ + H_2O$$

The corresponding reaction between NaOH and alcohols does not occur to a significant extent, although sodium ethoxide can be prepared by reaction of sodium metal with anhydrous ethanol:

$$Na + C_2H_5OH \longrightarrow C_2H_5ONa + \tfrac{1}{2}H_2$$

The manufacture of phenols uses quite different types of reactions from those employed to make alcohols. One method, introduced in 1924 and still employed to a small extent today, involves the chlorination of the benzene ring followed by reaction with sodium hydroxide:

$$C_6H_6 + Cl_2 \longrightarrow HCl + C_6H_5Cl$$

$$C_6H_5Cl + 2\,NaOH \longrightarrow C_6H_5O^-\,Na^+ + NaCl + H_2O$$

This approach illustrates a characteristic difference between the reactions of aromatics and alkenes. When chlorine reacts with an alkene, it *adds* across the double bond (as just shown in the production of 1,2-dichloroethane). When an aromatic ring is involved, substitution of chlorine for hydrogen occurs instead, and the aromatic $\pi$-bonding structure is preserved.

The method used to make almost all phenol today is a different one, however. It involves, first, the acid-catalyzed reaction of benzene with propylene to give cumene, or isopropyl benzene:

The structure of phenol, $C_6H_5OH$.

$$C_6H_6 + \underset{\substack{| \\ CH_3 \quad H}}{\overset{\substack{CH_2 \\ \parallel \\ C}}{}} \xrightarrow{H^+} C_6H_5-\underset{CH_3}{\overset{CH_3}{\underset{|}{\overset{|}{C}}}}-H$$

Cumene

As in the production of 2-propanol (see p. 814), the first step is the addition of $H^+$ to propylene to give $CH_3-CH^+-CH_3$. This ion then attaches to the benzene ring through its central carbon atom to give the cumene and regenerate the $H^+$ ion. Subsequent reaction of cumene with oxygen (Fig. 23-11) gives phenol and acetone,

**Figure 23-11**

The production of phenol and acetone from cumene is a two-step process involving insertion of $O_2$ to make a peroxide, followed by acid-catalyzed migration of the —OH group to form the products.

cumene      cumene hydroperoxide      phenol      acetone

an important compound to be discussed later in this section. The ultimate products of the manufacture of phenol are mostly polymers and aspirin (Fig. 18-2).

## Ethers

**Ethers** are characterized by the —O— functional group, in which an oxygen atom provides a link between two separate alkyl or aromatic groups. One important ether is diethyl ether, often called simply ether, in which two ethyl groups are linked to the same oxygen atom:

$$C_2H_5—O—C_2H_5$$

This is a useful solvent for organic reactions and was formerly used as an anesthetic. It can be produced by a **condensation reaction** (a reaction in which a small molecule such as water is split out) between two molecules of ethanol in the presence of concentrated sulfuric acid as a dehydrating agent:

$$CH_3CH_2\boxed{OH + H}OCH_2CH_3 \xrightarrow{H_2SO_4} CH_3CH_2—O—CH_2CH_3 + H_2O$$

Another ether of growing importance is methyl *t*-butyl ether (MTBE):

$$\begin{array}{c} CH_3 \\ | \\ CH_3—O—C—CH_3 \\ | \\ CH_3 \end{array}$$

This compound is increasingly replacing tetraethyllead as an additive to gasolines to increase their octane ratings. The bond between the oxygen and the *t*-butyl group is weak, and it breaks to form radicals that assist the smooth combustion of gasoline.

In a cyclic ether, oxygen forms part of a ring with carbon atoms, as in the common solvent tetrahydrofuran (Fig. 23-12). The smallest such ring has two carbon atoms bonded to each other and to the oxygen atom; it occurs in ethylene oxide,

$$\begin{array}{c} O \\ / \backslash \\ CH_2—CH_2 \end{array}$$

which is made by direct oxidation of ethylene over a silver catalyst:

$$CH_2{=}CH_2 + \tfrac{1}{2}O_2 \xrightarrow{Ag} \begin{array}{c} O \\ / \backslash \\ CH_2—CH_2 \end{array}$$

Such ethers with three-membered rings are called **epoxides.** The major use of ethylene oxide is in the preparation of ethylene glycol:

$$\begin{array}{c} O \\ / \backslash \\ CH_2—CH_2 \end{array} + H_2O \longrightarrow HO—CH_2—CH_2—OH$$

This reaction is carried out either at 195°C under pressure or at lower temperatures (50–70°C) with sulfuric acid as a catalyst. Ethylene glycol is a dialcohol, or **diol,** in which two —OH groups are attached to adjacent carbon atoms. Its primary use is as antifreeze: when it is added to water, it substantially lowers the freezing point. It is commonly combined with water in automobile radiators for this reason.

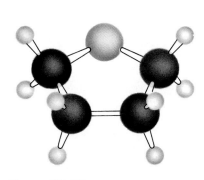

**Figure 23-12**
The structure of tetrahydrofuran, $C_4H_8O$.

## Aldehydes and Ketones

An **aldehyde** contains the characteristic $-\overset{\displaystyle O}{\overset{\|}{C}}-H$ functional group in its molecules and can be described as the result of the dehydrogenation of a primary alcohol. Formaldehyde, for example, results from the dehydrogenation of methanol at high temperatures with an iron oxide–molybdenum oxide catalyst:

$$CH_3OH \longrightarrow \overset{H}{\underset{H}{}}C=O + H_2$$

**Figure 23-13**
The structure of acetaldehyde, $CH_3CHO$.

Another reaction that gives the same primary product is the oxidation reaction

$$CH_3OH + \tfrac{1}{2} O_2 \longrightarrow \overset{H}{\underset{H}{}}C=O + H_2O$$

Formaldehyde is readily soluble in water, and a 40% aqueous solution called formalin is used to preserve biological specimens. It is a component of wood smoke and helps to preserve smoked meat and fish, probably by reacting with nitrogen-containing groups in the proteins of attacking bacteria. Its major use is in making polymer adhesives and insulating foam.

The next aldehyde in the series is acetaldehyde, the dehydrogenation product of ethanol (Fig. 23-13). Industrially produced acetaldehyde is made not from ethanol but by the oxidation of ethylene, using a $PdCl_2$ catalyst.

**Ketones** have the $\diagdown C=O$ functional group in which a carbon atom forms a double bond to an oxygen atom and single bonds to two separate alkyl or aromatic groups. Such compounds can be described as the products of dehydrogenation or oxidation of secondary alcohols, just as aldehydes come from primary alcohols. The simplest ketone, acetone, is in fact made by the dehydrogenation of 2-propanol over a copper oxide or zinc oxide catalyst at 500°C:

$$CH_3-\underset{\underset{OH}{|}}{CH}-CH_3 \longrightarrow CH_3-\underset{\underset{O}{\|}}{C}-CH_3 + H_2$$

The structure of acetone, $CH_3COCH_3$.

Acetone is also produced (in greater volume) as the coproduct with phenol of the oxidation of cumene, as discussed earlier. It is a widely used solvent and is the starting material for the synthesis of a number of polymers.

## Carboxylic Acids and Esters

**Carboxylic acids** contain the $-\overset{\displaystyle O}{\overset{\|}{C}}-OH$ functional group (also written as $-COOH$). They are the products of the oxidation of aldehydes, just as aldehydes are the products of the oxidation of primary alcohols. (The turning of wine to vinegar is a two-step oxidation leading from ethanol through acetaldehyde to acetic acid.) Industrially, acetic acid can be produced by the air oxidation of acetaldehyde over a manganese acetate catalyst at 55 to 80°C:

$$CH_3C\overset{\displaystyle O}{\underset{\displaystyle H}{\diagup}} + \tfrac{1}{2} O_2 \xrightarrow{Mn(CH_3COO)_2} CH_3C\overset{\displaystyle O}{\underset{\displaystyle OH}{\diagup}}$$

The reaction now preferred for acetic acid production, for reasons of economy, is the combination of methanol with carbon monoxide (both derived from natural gas) over a catalyst containing rhodium and iodine. The overall reaction is

$$CH_3OH(g) + CO(g) \xrightarrow{\text{Rh,I}_2} CH_3COOH(g)$$

and can be described as a **carbonylation,** or the insertion of CO into the methanol C—O bond. Figure 16-27 showed the mechanism of the reaction.

Acetic acid is a member of a series of carboxylic acids with formulas H—$(CH_2)_n$—COOH. Before acetic acid (with $n = 1$) comes the very simplest of these carboxylic acids, formic acid (HCOOH), in which $n = 0$. This compound was first isolated from extracts of the crushed bodies of ants, and its name stems from the Latin word *formica,* meaning "ant." Formic acid is the strongest acid of the series, and acid strength decreases with increasing length of the hydrocarbon chain. The longer-chain carboxylic acids are called fatty acids. Sodium stearate, the sodium salt of stearic acid, $CH_3(CH_2)_{16}COOH$, is a typical component of soap. It cuts grime by simultaneously interacting with grease particles at its hydrocarbon tail and with water at its carboxylate ion end group. This makes the grease soluble in water.

Carboxylic acids react with alcohols or phenols to give **esters,** forming water as the coproduct. An example is the condensation of acetic acid with methanol to give methyl acetate:

$$CH_3C \overset{O}{\diagup} \boxed{OH + H}OCH_3 \longrightarrow CH_3C \overset{O}{\diagup} OCH_3 + H_2O$$

Esters are named by stating the name of the alkyl group of the alcohol (the methyl group in this case), followed by the name of the carboxylic acid with the ending *-ate* (acetate). One of the most important esters in commercial production is vinyl acetate, with the structure

$$CH_3 - C \overset{O}{\diagdown} \underset{O-CH=CH_2}{}$$

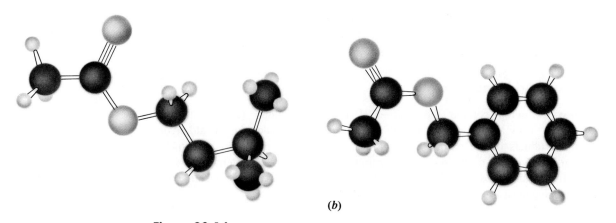

(a)                                                    (b)

**Figure 23-14**
The structures of (a) isoamyl acetate, $CH_3COO(CH_2)_2CH(CH_3)_2$, and (b) benzyl acetate, $CH_3COOCH_2C_6H_5$.

It is prepared by the reaction of acetic acid not with an alcohol but with ethylene and oxygen over a catalyst such as $CuCl_2$ and $PdCl_2$:

$$CH_3\overset{O}{\overset{\|}{C}}-OH + CH_2=CH_2 + \tfrac{1}{2}O_2 \xrightarrow{CuCl_2} CH_3\overset{O}{\overset{\|}{C}}-O-CH=CH_2 + H_2O$$

Esters are colorless, volatile liquids that frequently have pleasant odors. Many occur naturally in flowers and fruits. Isoamyl acetate (Fig. 23-14a) is generated in apples as they ripen and contributes to the flavor and odor of the fruit. Benzyl acetate, the ester formed from acetic acid and benzyl alcohol (Fig. 23-14b), is a major component of oil of jasmine and is used in the preparation of perfumes.

## Amines and Amides

The **amines** are derivatives of ammonia with the general formula $R_3N$, where R can represent a hydrocarbon group or hydrogen. If only one hydrogen atom of ammonia is replaced by a hydrocarbon group, the result is a **primary amine.** Examples are ethylamine and aniline:

Ethylamine    Aniline    Dimethylamine    Trimethylamine

If two hydrocarbon groups replace hydrogen atoms in the ammonia molecule, the compound is a **secondary amine** (such as dimethylamine), and three replacements make a **tertiary amine** (trimethylamine). Amines are bases because the lone electron pair on the nitrogen can accept a hydrogen ion in the same way that the lone pair on the nitrogen in ammonia does.

A primary or secondary amine (or ammonia itself) can react with a carboxylic acid to form an **amide.** This is another condensation reaction and is analogous to the formation of an ester from reaction of an alcohol with a carboxylic acid. An example of amide formation is

$$CH_3\overset{O}{\overset{\|}{C}}-\boxed{OH + H}-N(CH_3)_2 \longrightarrow CH_3\overset{O}{\overset{\|}{C}}-N(CH_3)_2 + H_2O$$

If ammonia is the reactant, an $-NH_2$ group replaces the $-OH$ group in the carboxylic acid as the amide is formed:

$$CH_3\overset{O}{\overset{\|}{C}}-OH + NH_3 \longrightarrow CH_3\overset{O}{\overset{\|}{C}}-NH_2 + H_2O$$
Acetamide

Amide linkages are present in the backbone of every protein molecule and therefore have great importance in biochemistry (Section 23-4).

## 23-3 SYNTHETIC POLYMERS

The organic compounds discussed so far have had relatively small molecules, ranging from those with four or five atoms (such as methane or formaldehyde) to long-chain hydrocarbons, with around 20 carbon atoms and relative molecular masses of

When sulfur is heated to above 150°C, the $S_8$ rings break open and join together to form long polymeric chains. The resulting red material is a highly viscous liquid.

several hundred. In addition to these smaller molecules, carbon atoms string together in stable chains of essentially unlimited length. Such chains provide the backbones of truly huge molecules that may contain hundreds of thousands or even millions of atoms. Such compounds, called **polymers,** are formed by the linking of numerous separate small **monomer units** in strands and webs.

Although many synthetic polymers are based on the ability of carbon to form stable long-chain molecules with various functional groups attached, carbon is not unique in this ability. Recall from Chapter 19 the chains, sheets, and networks found in natural silicates, in which the elements silicon and oxygen join together to form extended structures. This section, however, focuses on organic polymers.

### Polymerization Reactions

To construct a polymer, very many monomers must add to a growing polymer molecule, and the reaction must not falter after the first few molecules have reacted. This is achieved by having the polymer molecule retain highly reactive functional groups at all times during its synthesis. The two major types of polymer growth are addition polymerization and condensation polymerization.

In **addition polymerization,** monomers react to form a polymer chain without net loss of atoms. The most common type of addition polymerization involves the free-radical chain reaction of molecules that have C=C double bonds. As in the chain reactions considered in Section 11-4, the overall process consists of three steps: initiation, propagation (repeated many times to build up a long chain), and termination. As an example, consider the polymerization of vinyl chloride (chloroethene, $CH_2$=CHCl) to polyvinyl chloride (Fig. 23-15). This process can be initi-

(a)

(b)

### Figure 23-15

(a) This chemical plant in Texas produces several billion kilograms of polyvinyl chloride each year. The spherical tanks store gaseous raw material. (b) A pipefitting of polyvinyl chloride. (a, Occidental Petroleum; b, Leon Lewandowski) ated by a small concentration of molecules that have bonds weak enough to be broken by the action of light or heat, giving radicals. An example of such an **initiator** is a peroxide, which can be represented as R—O—O—R′, where R and R′ represent alkyl groups. The weak O—O bonds break

$$R—\overset{..}{\underset{..}{O}}—\overset{..}{\underset{..}{O}}—R' \longrightarrow R—\overset{..}{\underset{..}{O}}\cdot + \cdot \overset{..}{\underset{..}{O}}—R' \quad \text{(initiation)}$$

to give radicals, whose oxygen valence shells are incomplete. The radicals remedy this by reacting avidly with vinyl chloride, accepting electrons from the C=C double bonds to reestablish a closed-shell electron configuration on the oxygen atoms:

$$R—\overset{..}{\underset{..}{O}}\cdot + CH_2{=}CHCl \longrightarrow R—\overset{..}{\underset{..}{O}}—CH_2—\overset{\overset{\textstyle H}{|}}{\underset{\underset{\textstyle Cl}{|}}{C}}\cdot \quad \text{(propagation)}$$

One of the two $\pi$ electrons in the vinyl chloride double bond has been used to form a single bond with the R—O· radical. The other remains on the second carbon atom, leaving it as a seven-valence-electron atom that will react with another vinyl chloride molecule:

$$R—O—CH_2—\overset{\overset{\textstyle H}{|}}{\underset{\underset{\textstyle Cl}{|}}{C}}\cdot + CH_2{=}\overset{\overset{\textstyle H}{|}}{\underset{\underset{\textstyle Cl}{|}}{C}} \longrightarrow R—O—CH_2—\overset{\overset{\textstyle H}{|}}{\underset{\underset{\textstyle Cl}{|}}{C}}—CH_2—\overset{\overset{\textstyle H}{|}}{\underset{\underset{\textstyle Cl}{|}}{C}}\cdot \quad \text{(propagation)}$$

At each stage, the end group of the lengthening chain is one electron short of a valence octet and remains quite reactive. The reaction can continue, building up long-chain molecules of high molecular mass. The vinyl chloride monomers always attach to the growing chain with their $CH_2$ group, because the odd electron is more stable on a CHCl end group. This gives the polymer a regular alternation of —$CH_2$— and —CHCl— groups. Its chemical formula is —$(CH_2CHCl)_n$.

Termination occurs when the radical end groups on two different chains encounter each other and the two chains couple to give a longer chain:

$$R—O{-}(CH_2{-}CHCl)_{\overline{m}}CH_2—\overset{\overset{\textstyle H}{|}}{\underset{\underset{\textstyle Cl}{|}}{C}}\cdot + \cdot\overset{\overset{\textstyle H}{|}}{\underset{\underset{\textstyle Cl}{|}}{C}}—CH_2{-}(CHCl{-}CH_2)_{\overline{n}}O—R' \longrightarrow$$

$$R—O{-}(CH_2{-}CHCl)_{\overline{m}}CH_2—\overset{\overset{\textstyle H}{|}}{\underset{\underset{\textstyle Cl}{|}}{C}}—\overset{\overset{\textstyle H}{|}}{\underset{\underset{\textstyle Cl}{|}}{C}}—CH_2{-}(CHCl{-}CH_2)_{\overline{n}}O—R'$$

Alternatively, a hydrogen atom may transfer from one end group to the other:

$$R—O{-}(CH_2{-}CHCl)_{\overline{m}}\overset{\overset{\textstyle \textcircled{H}\,\, H}{|\quad|}}{\underset{\underset{\textstyle H\,\,\, Cl}{|\quad\,\,|}}{C—C}}\cdot + \cdot\overset{\overset{\textstyle H}{|}}{\underset{\underset{\textstyle Cl}{|}}{C}}—CH_2{-}(CHCl{-}CH_2)_{\overline{n}}O—R' \longrightarrow$$

$$R—O{-}(CH_2{-}CHCl)_{\overline{m}}CH{=}CHCl + CH_2Cl{-}CH_2{-}(CHCl{-}CH_2)_{\overline{n}}O—R'$$
$$\text{(termination)}$$

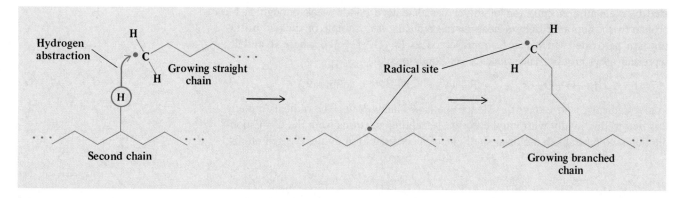

**Figure 23-16**

A growing branched chain is created during free-radical polymerization when a radical at one end of a growing straight chain extracts a hydrogen atom from the middle of a second chain.

The latter termination step leaves a double bond on one chain end and a $-CH_2Cl$ group on the other. When the polymer molecules are long, the exact natures of the end groups have little effect on the physical and chemical properties of the material. A different type of hydrogen transfer step often has a much greater effect on the properties of the resulting polymer. Suppose that a hydrogen atom transfers not from the monomer unit on the *end* of a second chain but from a monomer unit in the middle of that chain (Fig. 23-16). Then the first chain stops growing, but the radical site moves to the middle of the second chain, and growth resumes from that point, forming a *branched* polymeric chain with very different properties.

Addition polymerization can be initiated by ions as well as by free radicals. An example is the polymerization of acrylonitrile:

$$n\ CH_2{=}CH \longrightarrow \left[CH_2CH\right]_n$$
$$\qquad\quad | \qquad\qquad\qquad |$$
$$\qquad\quad C{\equiv}N \qquad\qquad\quad C{\equiv}N$$

A suitable initiator for this process is butyl lithium, $(CH_3CH_2CH_2CH_2)^-Li^+$. The butyl anion (abbreviated $Bu^-$) reacts with the end carbon atom in a molecule of acrylonitrile to give a new anion:

$$Bu^{\ominus}Li^{\oplus} + CH_2{=}CH \longrightarrow Bu{-}CH_2{-}CH^{\ominus}Li^{\oplus} \qquad \text{(initiation)}$$
$$\qquad\qquad\qquad | \qquad\qquad\qquad\qquad\qquad |$$
$$\qquad\qquad\qquad C{\equiv}N \qquad\qquad\qquad\qquad\quad C{\equiv}N$$

The new anion then reacts with an additional molecule of acrylonitrile:

$$Bu{-}CH_2{-}CH^{\ominus}Li^{\oplus} + CH_2{=}CH \longrightarrow$$
$$\qquad\quad | \qquad\qquad\qquad\qquad |$$
$$\qquad\quad C{\equiv}N \qquad\qquad\qquad C{\equiv}N$$

$$Bu{-}CH_2{-}CH{-}CH_2{-}CH^{\ominus}Li^{\oplus} \qquad \text{(propagation)}$$
$$\qquad\qquad\quad | \qquad\qquad\quad |$$
$$\qquad\qquad\quad C{\equiv}N \qquad\qquad C{\equiv}N$$

The process continues, building up a long-chain polymer.

Ionic polymerization differs from free-radical polymerization because the negatively charged end groups repel one another, ruling out termination by the coupling of two chains. The ionic group at the end of the growing polymer is stable at each stage. Once the supply of monomer has been used up, the polymer can exist indefinitely with its ionic end group, in contrast with the free-radical case, in which some reaction must take place to terminate the process. Ion-initiated polymers are called "living" polymers because, when additional monomer is added (even months later), they resume growth and increase in molecular mass. Termination can be achieved by adding water to replace the $Li^+$ with a hydrogen ion:

$$+(CH_2-CH)_n CH_2-CH^\ominus Li^\oplus + H_2O \longrightarrow$$
$$\quad\quad\quad\; | \quad\quad\quad | $$
$$\quad\quad\quad C\equiv N \quad\quad C\equiv N$$

$$+(CH_2-CH)_n CH_2-CH_2 + Li^\oplus + OH^\ominus \quad\quad \text{(termination)}$$
$$\quad\quad\quad\quad | \quad\quad\quad\quad |$$
$$\quad\quad\quad\quad C\equiv N \quad\quad C\equiv N$$

A second important mechanism of polymerization is **condensation polymerization,** in which a small molecule (frequently water) is split off as each monomer unit is attached to the growing polymer.[2] An example is the polymerization of 6-aminohexanoic acid. The first two molecules react upon heating according to

An amide linkage and water form from the reaction of an amine with a carboxylic acid. The new molecule still has an amine group on one end and a carboxylic acid group on the other, and so it can react with two more molecules of 6-aminohexanoic acid. The process repeats to build up a long-chain molecule. For each monomer unit added, one molecule of water is split off. The final polymer in this case is called nylon 6 and is used in fiber-belted radial tires and in carpets.

Both addition and condensation polymerization can be carried out with mixtures of two or more types of monomers present in the reaction mixture. The result is a **random copolymer** that incorporates both types of monomers in an irregular sequence along the chain. For example, a 1:6 molar ratio of styrene to butadiene monomers is used to make styrene–butadiene rubber (SBR) for automobile tires, and a 2:1 ratio gives a copolymer that is an ingredient in latex paints.

## Cross-Linking

If every monomer forming a polymer has only two reactive sites, then only chains and rings can be made. The 6-aminohexanoic acid used in making nylon 6, for example, has one amine group and one carboxylic acid group per molecule. When both functional groups react, one link is forged in the polymer chain, but that link cannot react further. If some or all of the monomers in a polymer have three or more reactive sites, however, then cross-linking to form sheets or networks is possible.

---

[2]Condensation reactions have appeared several times outside the context of polymer synthesis. For example, two molecules of $H_2SO_4$ condense to form disulfuric acid ($H_2S_2O_7$) (Fig. 21-3), and a carboxylic acid condenses with an alcohol to form an ester (Section 23-2).

**Figure 23-17**
When a mixture of phenol ($C_6H_5OH$) and formaldehyde ($CH_2O$) dissolved in acetic acid is treated with concentrated hydrochloric acid, a phenol-formaldehyde polymer grows.

One important example of cross-linking involves phenol-formaldehyde copolymers (Fig. 23-17). When these two compounds are mixed (with the phenol in excess in the presence of an acid catalyst), straight-chain polymers form. The first step is the addition of formaldehyde to phenol to give methylolphenol:

Molecules of methylolphenol then undergo condensation reactions (releasing water) to form a linear polymer called novalac:

If, on the other hand, the reaction is carried out with an excess of formaldehyde, di- and trimethylolphenols form:

Each of these monomers has more than two reactive sites and can react with up to three others to form a cross-linked polymer that is much stronger and more impact-resistant than the linear polymer. The very first synthetic plastic, Bakelite, was made in 1907 from cross-linked phenol and formaldehyde. Modern phenol-formaldehyde polymers are used as adhesives for plywood; more than a billion kilograms are produced per year in the United States.

Cross-linking is often desirable because it leads to a stronger material. Sometimes cross-linking agents are added deliberately to form additional bonds between polymer chains. Polybutadiene, for example, contains double bonds that can be linked upon addition of appropriate oxidizing agents. One especially important kind of cross-linking occurs through sulfur chains in rubber, as we shall see.

## Uses for Polymers

The three largest uses for polymers are in fibers, plastics, and elastomers (rubbers). We can distinguish between these three types of materials on the basis of their physical properties, especially their resistance to stretching. A typical fiber strongly resists stretching and can be elongated by less than 10% before breaking. Plastics are intermediate in their resistance to stretching and elongate 20 to 100% before breaking. Finally, elastomers stretch readily, with elongations of 100 to 1000% (that is, some types of rubber can be stretched by a factor of 10 without breaking). The remainder of this section examines the major kinds of polymers and their uses.

## Fibers

Many important fibers, including cotton and wool, are naturally occurring polymers. The first commercially successful synthetic polymers were made not by polymerization reactions but through the chemical regeneration of the natural polymer cellulose, a condensation polymer of the sugar glucose that is made by plants:

In the viscose rayon process, still employed today, cellulose is digested in a concentrated solution of NaOH to convert the —OH groups to $—O^- Na^+$ ionic groups. Reaction with $CS_2$ leads to the formation of about one "xanthate" group for every two glucose monomer units:

Xanthate

Such substitutions reduce the hydrogen-bond forces holding polymer chains together. In the ripening step, some of these xanthate groups are removed with regen-

**Figure 23-18**

Filter paper (cellulose) will dissolve in a concentrated ammonia solution containing $[Cu(NH_3)_4]^{2+}$ ions. When the solution is extruded into aqueous sulfuric acid, a dark blue thread of rayon (regenerated cellulose) precipitates. (Leon Lewandowski)

eration of $CS_2$, and others migrate to the —$CH_2OH$ groups from the ring —OH groups. Afterward, sulfuric acid is added to neutralize the NaOH and to remove the remaining xanthate groups. At the same time, the viscose rayon is spun out to form fibers (Fig. 23-18) while new hydrogen bonds form.

Rayon is a "semisynthetic" fiber because it is prepared from a natural polymeric starting material. The first truly synthetic polymeric fiber was nylon, developed in the 1930s by Wallace Carothers at Du Pont Company. He knew of the condensation of an amine with a carboxylic acid to form an amide linkage (Section 23-2) and noted that, if each molecule had *two* amine or carboxylic acid functional groups, long-chain polymers could form. The specific starting materials upon which Carothers settled, after numerous attempts, were adipic acid and hexamethylenediamine:

$$\underset{\text{Adipic acid}}{HO-\overset{\overset{\displaystyle O}{\|}}{C}-(CH_2)_4-\overset{\overset{\displaystyle O}{\|}}{C}-OH} \qquad \underset{\text{Hexamethylenediamine}}{H_2N-(CH_2)_6-NH_2}$$

The two react with loss of water, according to the equation

$$HO-\overset{\overset{\displaystyle O}{\|}}{C}-(CH_2)_4-\overset{\overset{\displaystyle O}{\|}}{C}-\boxed{OH+H}-\underset{\underset{\displaystyle H}{|}}{N}-(CH_2)_6-NH_2 \longrightarrow$$

$$HO-\overset{\overset{\displaystyle O}{\|}}{C}-(CH_2)_4-\overset{\overset{\displaystyle O}{\|}}{C}-\underset{\underset{\displaystyle H}{|}}{N}-(CH_2)_6-NH_2 + H_2O$$

The resulting molecule has a carboxylic acid group on one end (which can react with another molecule of hexamethylenediamine) and an amine group on the other end (which can react with another molecule of adipic acid). The process can continue indefinitely, leading to a polymer with the formula

$$\left[\overset{\overset{\displaystyle O}{\|}}{C}-(CH_2)_4-\overset{\overset{\displaystyle O}{\|}}{C}-\underset{\underset{\displaystyle H}{|}}{N}-(CH_2)_6-\underset{\underset{\displaystyle H}{|}}{N}\right]_n$$

called nylon 66 (Fig. 23-19). The nylon is extruded as a thread or spun as a fiber from the melt. The combination of well-aligned polymer molecules and N—H$\cdots$O hydrogen bonds between chains makes nylon one of the strongest materials known. The designation "66" indicates that this nylon has six carbon atoms on the starting carboxylic acid and six on the diamine. Other nylons can be made with different numbers of carbon atoms.

Just as a carboxylic acid reacts with an amine to give an amide, so it reacts with an alcohol to give an ester. This suggests the possible reaction of a dicarboxylic acid and a glycol (dialcohol) to form a polymer. The polymer produced most extensively in this way is polyethylene terephthalate, which is built up from terephthalic acid (a

**Table 23-4   Fibers**

| Name | Structural Units | Properties | Sample Uses |
|------|------------------|------------|-------------|
| Rayon | Regenerated cellulose | Absorbent, soft, easy to dye, poor wash and wear | Dresses, suits, coats, curtains, blankets |
| Acetate | Acetylated cellulose | Fast drying, supple, shrink-resistant | Dresses, shirts, draperies, upholstery |
| Nylon | Polyamide | Strong, lustrous, easy to wash, smooth, resilient | Carpeting, upholstery, tents, sails, hosiery, stretch fabrics, rope |
| Dacron | Polyester | Strong, easy to dye, shrink-resistant | Permanent-press fabrics, rope, sails, thread |
| Acrylic (Orlon) | $-(CH_2-CH)_n-$ with $C\equiv N$ | Warm, lightweight, resilient, quick-drying | Carpeting, sweaters, baby clothes, socks |

Adapted from P. J. Chenier, *Survey of Industrial Chemistry* (John Wiley & Sons, New York, 1986), Table 18.4.

benzene ring with —COOH groups on both ends) and ethylene glycol. The first two molecules react according to

**Figure 23-19**

Hexamethylenediamine is dissolved in water (lower layer), and adipyl chloride, a derivative of adipic acid, is dissolved in hexane (upper layer). At the interface between the layers, nylon forms and is drawn out onto the stirring bar.

Further reaction then builds up the polymer, which is called polyester and sold under trade names such as Dacron. The planar benzene rings in this polymer make it stiffer than nylon, which has no aromatic groups in its backbone, and help make polyester fabrics crush-resistant. The same polymer formed in a thin sheet rather than a fiber becomes Mylar, a very strong film used for audio and video tapes.

Table 23-4 summarizes the structures, properties, and uses of some important fibers.

## Plastics

**Plastics** are loosely defined as polymeric materials that can be molded or extruded into desired shapes and that harden upon cooling or solvent evaporation. Rather than being spun into threads in which their molecules are aligned, as in fibers, plastics are cast into three-dimensional forms or spread into films for packaging applications. Although celluloid articles were fabricated by plastic processing by the late 1800s, the first important synthetic plastic was Bakelite, the phenol-formaldehyde resin

**Table 23-5   Plastics**

| Name | Structural Units | Properties | Sample Uses |
|---|---|---|---|
| Polyethylene | $-(CH_2-CH_2)_n$ | High density: hard, strong, stiff<br>Low density: soft, flexible, clear | Molded containers, lids, toys, pipe<br>Packaging, trash bags, squeeze bottles |
| Polypropylene | $-(CH_2-CH)_n$<br>$\quad\quad\;$ $CH_3$ | Stiffer, harder than high-density polyethylene, higher melting point | Containers, lids, carpeting, luggage, rope |
| Polyvinyl chloride | $-(CH_2-CH)_n$<br>$\quad\quad\;$ $Cl$ | Nonflammable, resistant to chemicals | Water pipes, roofing, credit cards, records |
| Polystyrene | $-(CH_2-CH)_n$<br> (phenyl ring) | Brittle, flammable, not resistant to chemicals, easy to process and dye | Furniture, toys, refrigerator linings, insulation |
| Phenolics | Phenol-formaldehyde copolymer | Resistant to heat, water, chemicals | Plywood adhesive, Fiberglas binder, circuit boards |

Adapted from P. J. Chenier, *Survey of Industrial Chemistry* (John Wiley & Sons, New York, 1986), pp. 252–264.

whose cross-linking was discussed earlier in this section. Table 23-5 lists some of the most important plastics and their properties.

Ethylene ($CH_2=CH_2$) is the simplest monomer that will polymerize. Through free-radical-initiated addition polymerization at high pressures (1000–3000 atm) and temperatures (300–500°C), it forms polyethylene:

$$n\,CH_2=CH_2 \longrightarrow \;-[CH_2-CH_2]_n$$

The polyethylene formed in this way is not the perfect linear chain implied by this simple equation. Free radicals frequently abstract hydrogen from the middles of chains in this synthesis, and so the polyethylene is heavily branched with hydrocarbon side chains of varying length. It is **low-density polyethylene** (LDPE) because the difficulty of packing the irregular side chains gives it a lower density ($<0.94\ g\ cm^{-3}$) than that of perfectly linear polyethylene. This irregularity also makes it relatively soft, and so its primary uses are in coatings, plastic packaging, trash bags, and squeeze bottles in which softness is an advantage, not a drawback.

A major breakthrough occurred in 1954, when the German chemist Karl Ziegler showed that ethylene could also be polymerized with a catalyst consisting of $TiCl_4$ and an organoaluminum compound (for example, $Al(C_2H_5)_3$). The addition of ethylene takes place at each stage within the coordination sphere of the titanium atom, so that monomers can add only at the end of the growing chain. The result is linear polyethylene, also called **high-density polyethylene** (HDPE) because of its density

(a)

(b)

(c)

**Figure 23-20**
The structures of (a) isotactic, (b) syndiotactic, and (c) atactic polypropylene. In these structures the red spheres represent —CH₃ (methyl) side groups on the long chain.

(0.96 g cm$^{-3}$). Because its linear chains are regular, HDPE contains large crystalline regions, which make it much harder than LDPE and thus suitable for molding into plastic bowls, lids, and toys.

A third kind of polyethylene introduced in the late 1970s is called **linear low-density polyethylene** (LLDPE). It is made by the same metal-catalyzed reactions as HDPE, but it is a deliberate copolymer with other 1-alkenes such as 1-butene. It has some side groups (which reduce the crystallinity and density), but they are of a controlled short length as opposed to the irregular, long side branches in LDPE. LLDPE is stronger and more rigid than LDPE; it is also less expensive because lower pressures and temperatures are used in its manufacture.

If one of the hydrogen atoms of the ethylene monomer unit is replaced with a different type of atom or functional group, the plastics that form upon polymerization have different properties. Substitution of a methyl group (that is, the use of propylene as monomer) leads to polypropylene:

$$\left[ CH_2-CH \atop \qquad\quad CH_3 \right]_n$$

This reaction cannot be carried out successfully by free-radical polymerization, however. It was first achieved in 1953–1954 by Ziegler and the Italian chemist Giulio Natta, who used the Ziegler catalyst later employed in making HDPE. In polypropylene, the methyl groups attached to the carbon backbone can be arranged in different conformations (Fig. 23-20). In the **isotactic form,** all the methyl groups are arranged on the same side, whereas in the **syndiotactic form** they alternate in a regular fashion. The **atactic form** shows a random positioning of methyl groups. Natta showed that the Ziegler catalyst led to isotactic polypropylene, and he developed another catalyst, using $VCl_4$, that gave the syndiotactic form. Polypropylene plastic is stiffer and harder than HDPE and has a higher melting point, and so it is particularly useful in applications requiring high temperatures (such as the sterilization of medical instruments).

In polystyrene, a benzene ring replaces one hydrogen atom of each ethylene monomer unit. Because such a ring is bulky, atactic polystyrene does not crystallize to any significant extent. The most familiar application of this polymer is in the polystyrene foam used in disposable containers for food and drinks and as insulation. A volatile liquid or a compound that dissociates to gaseous products on heating is added to the molten polystyrene. It forms bubbles that remain as the polymer is cooled and molded. The gas-filled pockets in the final product make it a good thermal insulator.

## Rubber

An **elastomer** is a polymer that can be deformed to a great extent and still recover its original form when the deforming stress is removed. The term ''rubber'' was introduced by Joseph Priestley, who observed that such materials can be used to rub out pencil marks. Natural rubber is a polymer of isoprene (2-methylbutadiene). The isoprene molecule contains two double bonds of which polymerization removes only one, and so natural rubber is unsaturated, containing one double bond per isoprene unit. In polymeric isoprene, the geometry at each double bond can be either *cis* or *trans* (Fig. 23-21). Natural rubber is all *cis*-polyisoprene. The all-*trans* form also occurs in nature in the sap of certain trees and is called gutta-percha. This material is used to cover golf balls because it is particularly tough. Isoprene can be polymerized by free-radical addition polymerization, but the resulting polymer contains a mixture of *cis* and *trans* double bonds and is useless as an elastomer.

Even pure natural rubber is of limited usefulness because it melts, is soft, and does not fully spring back to its original form after being stretched. In 1839 the American inventor Charles Goodyear discovered that if sulfur is added to rubber and the mixture is heated, the rubber hardens, becomes more resilient, and does not melt. This process is referred to as **vulcanization** and involves the formation of sulfur bridges between the methyl side groups on different chains. Small amounts of sulfur

**Figure 23-21**

In the polymerization of isoprene, a *cis* or *trans* configuration can form at each double bond in the polymer. The blue arrows show the redistribution of the electrons upon bond formation.

(<5%) yield an elastic material in which sulfur links between chains remain after stretching and enable the rubber to regain its original form when the external force is removed. Large amounts of sulfur give the very hard, nonelastic material ebonite.

Research on synthetic substitutes for natural rubber began in the United States and Europe before World War II. Attention focused on copolymers of butadiene with styrene (now called SBR rubber) and with acrylonitrile (NBR rubber). The Japanese occupation of the rubber-producing countries of Southeast Asia sharply curtailed the supply of natural rubber to the Allied nations, and rapid steps were taken to increase production of synthetic rubber. The initial production goal was 40,000 tons per year of SBR. By 1945, U.S. production had reached an incredible total of more than 600,000 tons per year. During those few years, many advances were made in production techniques, quantitative analysis, and basic understanding of rubber elasticity. Styrene–butadiene rubber production continued after the war, and in 1950 SBR exceeded natural rubber in overall production volume for the first time. More recently, several factors have favored natural rubber: the increasing cost of the hydrocarbon feed stock for synthetic rubber, gains in productivity of natural rubber, and the growing preference for belted radial tires, which use more natural rubber.

The development of the Ziegler–Natta catalysts has affected rubber production as well. First, it facilitated the synthesis of all-*cis* polyisoprene and the demonstration that its properties were nearly identical to those of natural rubber. (A small amount of ''synthetic natural rubber'' is produced today.) Second, a new kind of synthetic rubber was developed: all-*cis* polybutadiene. It now ranks second in production after styrene–butadiene rubber.

## 23-4 NATURAL POLYMERS

All the products of human ingenuity in the design of polymers pale beside the products of nature. Plants and animals employ a tremendous variety of long-chain molecules with different functions: some for structural strength, others to act as catalysts, and still others to provide instructions for the synthesis of vital components of the cell. In this section we discuss these three important classes of natural polymers: polysaccharides, proteins, and nucleic acids.

### Carbohydrates and Polysaccharides

**Carbohydrates** form a class of compounds of carbon with hydrogen and oxygen. The name comes from the chemical formulas of these compounds, which can be written $C_n(H_2O)_m$, suggesting a ''hydrate'' of carbon. Simple **sugars,** or **monosaccharides,** are carbohydrates with the chemical formula $C_nH_{2n}O_n$. Sugars with three, four, five, and six carbon atoms are called trioses, tetroses, pentoses, and hexoses, respectively.

Glucose is a hexose sugar that exists in several forms in solution (Fig. 23-22). There is a rapid equilibrium between a straight-chain form (a six-carbon molecule with five —OH groups and one aldehyde —CHO group) and a cyclic form, in which the ring is composed of five carbon atoms and one oxygen, with four —OH side groups and one —CH$_2$OH side group. In the straight-chain form, four of the carbon

**Figure 23-22**

D-glucose exists in two ring forms in solution (*a* and *c*), which interconvert via an open-chain form (*b*). The two rings differ in the placement of the —OH and —H groups on carbon atom 1.

atoms (those numbered 2 through 5) are chiral centers, with four different groups bonded to them. As discussed in Section 14-5 (see Fig. 14-30), each such carbon atom can exist in two configurations, giving rise to $2^4 = 16$ distinct hexose sugars. The glucose formed in plant photosynthesis always has the chirality shown in Figure 23-22*b*. Of the 15 other straight-chain hexose sugars, the only ones found in nature are D-galactose (in the milk sugar lactose) and D-mannose (a plant sugar).

Figure 23-22 shows that glucose actually has two different ring forms, depending on whether the —OH group created from the aldehyde by the closing of the ring lies above or below the plane of the ring. Another way to see this is to note that closing the ring creates a fifth chiral carbon atom. The two ring forms of D-glucose are called $\alpha$-D-glucose (Fig. 23-22*a*) and $\beta$-D-glucose (Fig. 23-22*c*). In aqueous solution, these two forms interconvert rapidly via the open-chain glucose form and cannot be separated. They can be isolated separately in crystalline form, however. D-fructose, a common sugar found in fruit and honey, has the same molecular formula as D-glucose but is a member of a class of hexose sugars that are ketones rather than aldehydes. In their straight-chain forms, these sugars have the C=O double bond at carbon atom 2 rather than carbon atom 1 (Fig. 23-23).

Many plant cells do not stop the synthesis process with simple sugars such as glucose, but continue by linking sugars together to form more complex carbohydrates. **Disaccharides** are composed of two simple sugars linked together by a condensation reaction with the elimination of water. Examples shown in Figure 23-24 are the milk sugar lactose and the plant sugar sucrose (ordinary table sugar, extracted from sugarcane and sugar beets). Further linkages of sugar units lead to polymers called **polysaccharides.** The position of the oxygen atom linking the monomer units has a fundamental effect on the properties and functions of the

(a) Five-membered ring     (b) Open-chain form     (c) Six-membered ring

## Figure 23-23

In aqueous solutions of the sugar D-fructose, an equilibrium exists among a five-atom ring, an open chain, and a six-atom ring. In addition to the $\beta$ isomers shown here, both ring forms have $\alpha$ isomers, in which the —$CH_2OH$ and —OH on carbon 2 are exchanged.

## Figure 23-24

Two disaccharides. Their derivations from monosaccharide building blocks are shown.

Lactose

Sucrose

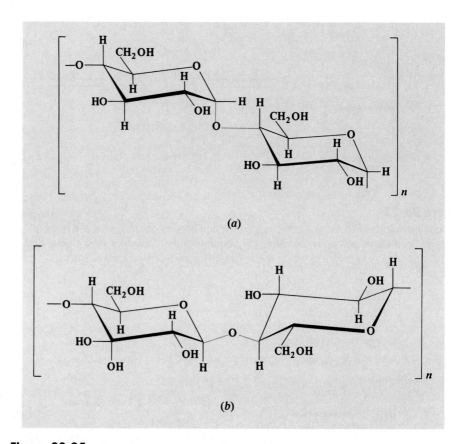

**Figure 23-25**

Both starch (*a*) and cellulose (*b*) are polymers of glucose. In starch, all the cyclic glucose units are α-D-glucose. In cellulose, all the monomer units are β-D-glucose.

**Figure 23-26**

The structure of glycine. On the left side is the carboxylic acid group (—COOH), and on the right is the amine group (—NH$_2$).

polymers that result. Starch (Fig. 23-25*a*) is a polymer of α-D-glucose and is metabolized by humans and animals. Cellulose (Fig. 23-25*b*), a polymer of β-D-glucose, cannot be digested except by certain bacteria that live in the digestive tracts of goats, cows, and other ruminants and in some insects, such as termites. It forms the structural fiber of trees and plants and is present in linen, cotton, and paper. It is the most abundant organic compound on earth.

## Protein Building Blocks: Amino Acids

The monomeric building blocks of the biopolymers called proteins are the α-amino acids. The simplest amino acid is glycine, which has the molecular structure shown in Figure 23-26. An amino acid, as indicated by the name, must contain an amine group (—NH$_2$) and a carboxylic acid group (—COOH). In α-amino acids, the two groups are bonded to the same carbon atom. In acidic aqueous solution, the amine group is protonated to form —NH$_3^+$; in basic solution, the carboxylic acid group loses a proton to form —COO$^-$. At intermediate pH, both reactions occur. The net result is that the simple amino acid form shown in Figure 23-26 is almost never present in aqueous solution (see problem 55).

Two glycine molecules can condense with loss of water to form an amide:

$$
\underset{\text{HO}}{\overset{\text{O}}{\diagdown}}\!\!C\!-\!CH_2\!-\!\underset{\boxed{H \quad HO}}{\overset{H}{N}} \quad + \quad \underset{\text{HO}}{\overset{\text{O}}{\diagdown}}\!\!C\!-\!CH_2\!-\!NH_2 \quad \longrightarrow
$$

$$
\underset{\text{HO}}{\overset{\text{O}}{\diagdown}}\!\!C\!-\!CH_2\!-\!\underset{\underset{H}{|}}{N}\!-\!\overset{\overset{\text{O}}{\|}}{C}\!-\!CH_2\!-\!NH_2 + H_2O
$$

The amide functional group connecting two amino acids is referred to as a **peptide linkage,** and the resulting molecule is a *dipeptide*—in this case, diglycine. Because the two ends of the molecule still have carboxylic acid and amine groups, further condensation reactions to form a **polypeptide,** a polymer composed of many amino acid groups, are possible. If glycine were the only amino acid available, the result would be polyglycine, a rather uninteresting protein. There is a close similarity between this naturally occurring condensation polymer and the synthetic polyamide nylon. Polyglycine could be called ''nylon 2,'' a simple polyamide in which each repeating unit contains two carbon atoms.

Nature does not stop with glycine as a monomer unit. Instead, any of 20 different $\alpha$-amino acids are found in most natural polypeptides. In each of these, one of the hydrogen atoms on the central carbon atom of glycine is replaced by another side group. Alanine is the next simplest $\alpha$-amino acid after glycine; it has a $-CH_3$ group in place of an $-H$ atom. This substitution has a profound consequence. In alanine, four different groups are attached to a central carbon: $-COOH$, $-NH_2$, $-CH_3$, and $-H$. There are two ways in which four different groups can be arranged in a tetrahedral structure about a central atom (Fig. 14-30). The two optical isomers of alanine are designated by the prefixes L- and D- for *levo* and *dextro* (Latin for ''left'' and ''right,'' respectively).

If a mixture of L- and D-alanine were caused to polymerize, nearly all the polymer molecules would have different structures, because their sequences of D-alanine and L-alanine monomer units would differ. To create polymers with definite structures for particular roles, there is only one recourse: to build all polypeptides from one of the optical isomers so that the properties will be reproducible from molecule to molecule. Nearly all naturally occurring $\alpha$-amino acids are of the L form, and most earthly organisms have no use for D-$\alpha$-amino acids in making polypeptides. Terrestrial life could presumably have begun equally well using mainly D-amino acids (all biomolecules would be mirror images of their present forms). The mechanism by which the established preference was initially selected is not known.

The $-H$ group of glycine and $-CH_3$ group of alanine give just the first two amino acid building blocks. Table 23-6 shows all 20 important $\alpha$-amino acids, arranged by side group. Note the variety in their chemical and physical properties. Some side groups contain basic groups; others are acidic. Some are compact; others are bulky. Some can take part in hydrogen bonds; others can complex readily with metal ions to form coordination complexes. This variety in properties leads to even more variety in the polymers derived from the $\alpha$-amino acids.

## Table 23-6  α-Amino Acid Side Groups

| | Symbol | Structure of Side Group |
|---|---|---|
| **Hydrogen "Side Group"** | | |
| Glycine | Gly | —H |
| **Alkyl Side Groups** | | |
| Alanine | Ala | —CH$_3$ |
| Valine | Val | —CH—CH$_3$ <br>    &#124; <br>    CH$_3$ |
| Leucine | Leu | —CH$_2$—CH—CH$_3$ <br>         &#124; <br>         CH$_3$ |
| Isoleucine | Ile | —CH—CH$_2$—CH$_3$ <br>    &#124; <br>    CH$_3$ |
| Proline | Pro (structure of entire amino acid) | |
| **Aromatic Side Groups** | | |
| Phenylalanine | Phe | |
| Tyrosine | Tyr | |
| Tryptophan | Trp | |

*(Continued)*

## Proteins

A **protein** is a long polypeptide made up of α-amino acids joined by peptide linkages. The term is usually applied to polymers with more than about 50 amino acid groups; large proteins may contain many thousand such groups. Given the fact that any one of 20 α-amino acids may appear at each point in the chain, the number of possible sequences of amino acids in even small proteins is staggering. Moreover, the amino acid sequence describes only one aspect of the molecular structure of a protein. It contains no information about the three-dimensional conformation adopted by the protein. The carbonyl group and the amine group in each amino acid along the protein chain are potential sites for hydrogen bonds, which may also involve functional groups on the amino acid side chains. Also, the cysteine side

| Table 23-6   (cont.) | | |
|---|---|---|
| | **Symbol** | **Structure of Side Group** |
| **Alcohol-Containing Side Groups** | | |
| Serine | Ser | $-CH_2OH$ |
| Threonine | Thr | $\underset{\displaystyle CH_3}{\overset{\displaystyle OH}{-CH}}$ |
| **Basic Side Groups** | | |
| Lysine | Lys | $-CH_2CH_2CH_2CH_2NH_2$ |
| Arginine | Arg | $-CH_2CH_2CH_2NH-C\overset{\displaystyle NH}{\underset{\displaystyle NH_2}{}}$ |
| Histidine | His | $-CH_2-C=CH$ <br> $HN \quad N$ <br> $C$ <br> $H$ |
| **Acidic Side Groups** | | |
| Aspartic acid | Asp | $-CH_2COOH$ |
| Glutamic acid | Glu | $-CH_2CH_2COOH$ |
| **Amide-Containing Side Groups** | | |
| Asparagine | Asn | $-CH_2\overset{\displaystyle O}{\overset{\displaystyle \|}{C}}-NH_2$ |
| Glutamine | Gln | $-CH_2CH_2\overset{\displaystyle O}{\overset{\displaystyle \|}{C}}-NH_2$ |
| **Sulfur-Containing Side Groups** | | |
| Cysteine | Cys | $-CH_2-SH$ |
| Methionine | Met | $-CH_2CH_2-S-CH_3$ |

groups ($-CH_2-SH$) can react with one another, with loss of hydrogen, to form $-CH_2-S-S-CH_2-$ disulfide bridges between different cysteine groups in a single chain or between neighboring chains (the same kind of cross-linking by sulfur occurs in the vulcanization of rubber). As a result of these strong intrachain interactions, the molecules of a given protein have a rather well-defined conformation even in solution, as compared with the much more varied range of conformations available to a simple alkane chain (Fig. 23-3). The three-dimensional structures of many proteins have been determined by x-ray diffraction.

There are two primary categories of proteins: fibrous and globular. **Fibrous proteins** are usually structural materials and consist of polymer chains linked in sheets or twisted in long fibers. Silk is a fibrous protein in which the monomer units

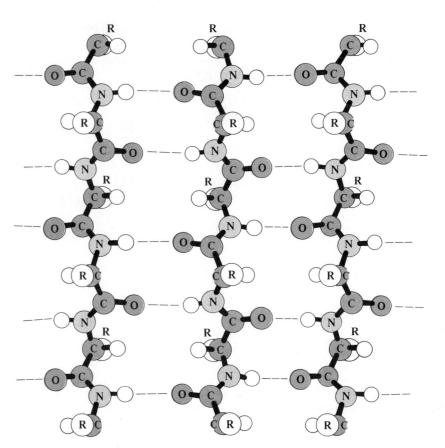

**Figure 23-27**

The structure of silk. The amino acid side groups (shown as R) must be small to fit into a sheet-like structure. Glycine (R = H) and alanine (R = CH₃) predominate.

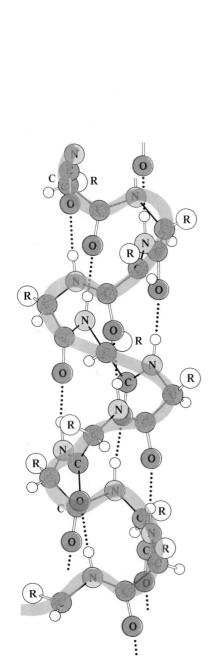

**Figure 23-28**

The structure of an α-helix. The superimposed yellow line highlights the helical structure, which is maintained by hydrogen bonds (dotted lines).

are primarily glycine and alanine, with smaller amounts of serine and tyrosine. The protein chains are cross-linked by hydrogen bonds to form sheet-like structures (Fig. 23-27) that are arranged so that the nonhydrogen side groups all lie on one side of the sheet; the sheets then stack in layers. The relatively weak forces between sheets give silk its characteristic smooth feel. The amino acids in wool and hair have side chains that are larger, bulkier, and less regularly distributed than those in silk, and so sheet structures do not form. Instead, the protein molecules twist into a right-handed coil called an **α-helix** (Fig. 23-28). In this structure, each carbonyl group is hydrogen-bonded to the amine group of the fourth amino acid farther along the chain; the bulky side groups jut out from the helix and do not interfere with one another.

The second type of protein is the **globular protein.** Globular proteins include the carriers of oxygen in the blood (hemoglobin) and in cells (myoglobin). They have irregular folded structures (Fig. 23-29) and typically consist of 100 to 1000 amino acid groups in one or more chains. Globular proteins frequently have parts of their structures in α-helices and sheets, with other portions in more disordered forms. Hydrocarbon side groups tend to cluster in regions that exclude water, whereas charged and polar side groups tend to remain in close contact with water.

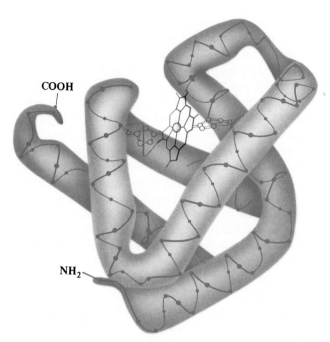

**Figure 23-29**

A computer-generated model of the structure of myoglobin. The central heme group (red) is shown in greater detail in Figure 15-15c; two histidine groups (green) extend toward the central iron atom. Much of the protein (blue tube) is coiled in α-helices.

The sequences of amino acid units for many such proteins have been worked out by cleaving them into smaller pieces and analyzing the structure of the fragments. It took Frederick Sanger ten years to complete the first such determination of sequence for the 51 amino acids in bovine insulin (Fig. 23-30). His persistence earned him the

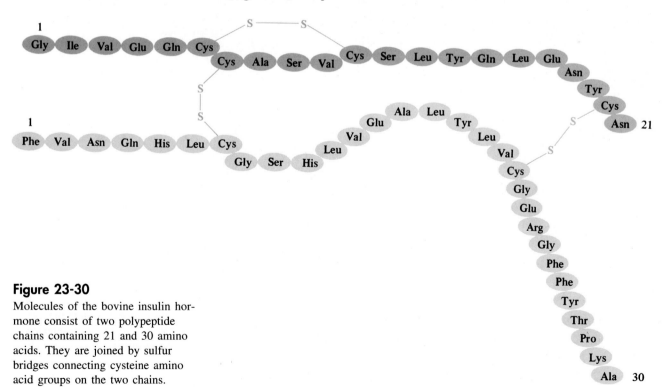

**Figure 23-30**

Molecules of the bovine insulin hormone consist of two polypeptide chains containing 21 and 30 amino acids. They are joined by sulfur bridges connecting cysteine amino acid groups on the two chains.

Nobel prize in chemistry in 1958. Now, automated procedures enable scientists to rapidly determine amino acid sequences in much longer protein molecules.

**Enzymes** constitute a very important class of globular proteins. They catalyze particular reactions in the cell, such as the synthesis and breakdown of proteins, the transport of substances across cell walls, and the recognition and resistance of foreign bodies. Enzymes act by lowering the activation barrier for a reaction, and they must be selective so as to act only on a restricted group of substrates.

Let us examine the enzyme carboxypeptidase A, whose structure has been determined by x-ray diffraction. It removes amino acids one at a time from the carboxylic acid end of a polypeptide. Figure 23-31 shows the structure of the active site (with a peptide chain in place, ready to be cleaved). A special feature of this enzyme is the role played by the zinc ion, which is coordinated to two histidine residues in the enzyme and to a carboxylate group on a nearby glutamic acid residue. The zinc ion helps to remove electrons from the carbonyl group of the peptide linkage, making it more positive and thereby more susceptible to attack by water or by the carboxylate group of a second glutamic acid residue. The side chain on the outer amino acid of the peptide being cleaved is positioned in a hydrophobic cavity, which favors large aromatic or branched side chains (such as that in tyrosine) over smaller hydrophilic side chains (such as that in aspartic acid). Carboxypeptidase A is thus selective in the rates with which it cleaves peptide chains.

The molecular "engineering" that lies behind nature's design of carboxypeptidase A and other enzymes is truly remarkable. The amino acid residues that form the

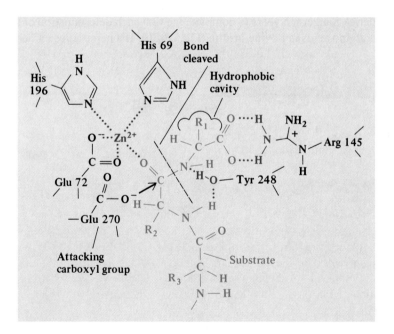

**Figure 23-31**

The active site of carboxypeptidase A. Shown in red is a substrate polypeptide that is being cleaved by the enzyme. Green is used to show the role of $Zn^{2+}$ as a complexing ion, and blue, to show the hydrogen bonds that maintain the geometry.

active site and determine its catalytic properties are *not* adjacent to one another in the protein chain. As indicated by the numbers after the residues in Figure 23-31, the two glutamic acid residues are the 72nd and 270th amino acids along the chain. The enzyme adopts a conformation in which the key residues, distant from one another in terms of chain position, are nonetheless quite close in three-dimensional space, allowing the enzyme to carry out its specialized function.

## Nucleic Acids

We have seen that proteins are copolymers made up typically of 20 types of monomer units. Simply mixing the amino acids and letting them dehydrate to form polymer chains at random would never lead to the particular structures needed by living cells. How does the cell preserve information about the amino acid sequences that make up its proteins, and how does it transmit this information to daughter cells through the reproductive process? These questions lie in the field of molecular genetics, an area in which chemistry plays the central role.

The primary genetic material is deoxyribonucleic acid (DNA). This biopolymer is made up of four types of monomer units called **nucleotides.** Each nucleotide is composed of three parts:

1. One molecule of a pyrimidine or purine base. The four bases are thymine, cytosine, adenine, and guanine (Fig. 23-32a).
2. One molecule of the sugar D-deoxyribose ($C_5H_{10}O_4$). D-ribose is a pentose sugar with a five-membered ring.
3. One molecule of phosphoric acid ($H_3PO_4$).

The cyclic sugar molecule links the base to the phosphate group, undergoing two condensation reactions, with loss of water, to form the nucleotide (Fig. 23-32b). The first key to discovering the structure of DNA was the observation that, although the proportions of the four bases in DNA from different organisms were quite variable, the chemical amount of cytosine (C) was always approximately equal to that of guanine (G), and the chemical amount of adenine (A) was always approximately equal to that of thymine (T). This suggested some type of base pairing in DNA that could lead to association of C with G and A with T. The second crucial observation was an x-ray diffraction study by Rosalind Franklin and Maurice Wilkins that suggested the presence of helical structures of more than one chain in DNA.

James Watson and Francis Crick put together these two pieces of information in their famous 1953 proposal of a double-helix structure for DNA. They concluded that DNA consists of two interacting helical strands of nucleic acid polymer (Fig. 23-33), with each cytosine on one strand linked through hydrogen bonds to a guanine on the other, and each adenine to a thymine. This accounted for the observed molar ratios of the bases, and it also provided a model for the replication of the molecule, which is crucial for passing on information during the reproductive process. One DNA strand serves as a template upon which a second DNA strand is synthesized. A DNA molecule reproduces by starting to unwind at one end. As it does so, new nucleotides are guided into position opposite the proper bases on each of the two strands. If the nucleotide does not fit the template, it cannot link to the polymeric strand under construction. The result of the polymer synthesis is two

**Figure 23-32**

(*a*) The structures of the purine and pyrimidine bases. Black dots indicate hydrogen bonding between pairs of bases. (*b*) The structure of the nucleotide adenosine monophosphate (AMP).

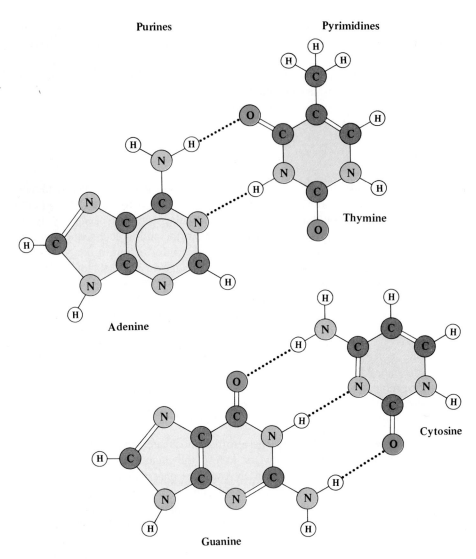

(*a*)  Structures of the pyrimidine and purine bases

(*b*)  Adenosine monophosphate

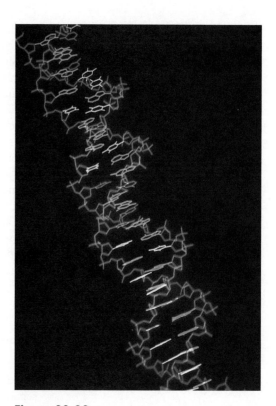

**Figure 23-33**
The double-helix structure of DNA. (*Right*, Dan McCoy/R. Langridge/Rainbow)

double-helix molecules, each containing one strand from the original and one new strand that is identical to the original in every respect.

Information is encoded in DNA in the sequence of the base pairs. Subsequent research has broken this genetic code and established the connection between the base sequence in a segment of DNA and the amino acid sequence of the protein synthesized according to the directions in that segment. The code in a nucleic acid is read as consecutive, nonoverlapping triplets of bases, with each triplet standing for a particular amino acid. Thus, a nucleic acid strand consisting of pure cytosine gives a polypeptide of pure proline, meaning that the triplet CCC codes for proline. The nucleic acid strand AGAGAGAG . . . is read as the alternating triplets AGA and GAG and gives a polypeptide consisting of alternating arginine (coded by AGA) and glutamic acid (coded by GAG) monomer units. There are 64 ($4^3$) possible triplets, and so typically more than one code exists for a particular amino acid. Some triplets serve as signals to terminate a polypeptide chain. Remarkably, the genetic code appears to be universal, independent of the species of plant or animal, a finding that suggests a common origin for all terrestrial life.

## CONCEPTS & SKILLS

After studying this chapter and working the problems that follow, you should be able to

1. Identify important hydrocarbons in crude petroleum and describe how they are separated (Section 23-1, problems 1–4).
2. Write names and structural formulas for hydrocarbons (Section 23-1, problems 5–10).
3. Identify important functional groups and outline chemical processes by which important chemical compounds are synthesized (Section 23-2, problems 11–18).
4. Contrast the methods of addition and condensation polymerization (Section 23-3, problems 19–24).
5. Give several examples of fibers, plastics, and rubbers and describe how they are made and used (Section 23-3, problems 25–28).
6. Describe the formation of polysaccharides from sugars, proteins from amino acids, and DNA from nucleotides and the roles of these biopolymers in living cells (Section 23-4, problems 29–34).

## PROBLEMS

### Petroleum Refining and the Hydrocarbons

1. (a) Use average bond enthalpies from Table 8-3 to estimate the standard enthalpy of formation of cyclopropane ($C_3H_6$).
(b) The measured enthalpy change for the combustion of cyclopropane

$$C_3H_6(g) + \tfrac{9}{2} O_2(g) \longrightarrow 3\,CO_2(g) + 3\,H_2O(g)$$

is $\Delta H° = -1959$ kJ. Use this, together with data from Appendix D, to calculate the standard enthalpy of formation of cyclopropane.

(c) By comparing your answers from (a) and (b), estimate the "strain enthalpy" associated with forming a ring of three carbon atoms (with C—C—C bond angles of only 60°) in cyclopropane.

2. (a) Use average bond enthalpies from Table 8-3 to estimate the standard enthalpy of formation of cyclobutane ($C_4H_8$).
(b) The measured enthalpy change for the combustion of cyclobutane

$$C_4H_8(g) + 6\,O_2(g) \longrightarrow 4\,CO_2(g) + 4\,H_2O(g)$$

is $\Delta H° = -2568$ kJ. Use this, together with data from Appendix D, to calculate the standard enthalpy of formation of cyclobutane.

(c) By comparing your answers from (a) and (b), estimate the "strain enthalpy" associated with forming a ring of four carbon atoms (with C—C—C bond angles of only 90°) in cyclobutane.

3. (a) Write a chemical equation involving structural formulas for the catalytic cracking of decane into an alkane and an alkene that contain equal numbers of carbon atoms. Assume that both products have straight chains of carbon atoms.

   (b) Draw and name one other isomer of the alkene.

4. (a) Write an equation involving structural formulas for the catalytic cracking of 2,2,3,4,5,5-hexamethylhexane. Assume that the cracking occurs between carbon atoms 3 and 4.

   (b) Draw and name one other isomer of the alkene.

5. Write structural formulas for the following:

   (a) 2,3-dimethylpentane
   (b) 3-ethyl-2-pentene
   (c) Methylcyclopropane
   (d) 2,2-dimethylbutane
   (e) 3-propyl-2-hexene
   (f) 3-methyl-1-hexene
   (g) 4-ethyl-2-methylheptane
   (h) 4-ethyl-2-heptyne

6. Write structural formulas for the following:

   (a) 2,3-dimethyl-1-cyclobutene
   (b) 2-methyl-2-butene
   (c) 2-methyl-1,3-butadiene
   (d) 2,3-dimethyl-3-ethylhexane
   (e) 4,5-diethyloctane
   (f) Cyclooctene
   (g) Propadiene
   (h) 2-pentyne

7. Write structural formulas for *trans*-3-heptene and *cis*-3-heptene.

8. Write structural formulas for *cis*-4-octene and *trans*-4-octene.

9. Name the following hydrocarbons.

   (a)
$$CH_2=C=C-CH_2-CH_2-CH_3$$
with H above the third C

   (b)
$$H_2C=C-C=C-CH_2$$
with H H above and H H below

   (c)
$$H_2C=C-CH_2-CH_2-CH_2-CH_3$$
with CH₃ above the second C

   (d) $CH_3-CH_2-C\equiv C-CH_2-CH_3$

10. Name the following hydrocarbons.

   (a)
$$CH_2=C-C=CH_2$$
with CH₃ and CH₃ below

   (b)
$$CH_3-C=C-C=C-CH_3$$
with H H H H below

   (c)
$$CH_3-C-CH_2-CH_3$$
with CH₃ above and CH₃ below

   (d)
$$CH_3-C-CH_3$$
with CH₂ below (double bond)

11. State the hybridization of each of the carbon atoms in the hydrocarbon structures in problem 9.

12. State the hybridization of each of the carbon atoms in the hydrocarbon structures in problem 10.

## Functional Groups and Organic Synthesis

13. Write balanced equations for the following reactions. Use structural formulas to represent the organic compounds.

   (a) The production of butyl acetate from butanol and acetic acid
   (b) The conversion of ammonium acetate to acetamide and water
   (c) The dehydrogenation of 1-propanol
   (d) The complete combustion (to $CO_2$ and $H_2O$) of heptane

14. Write balanced equations for the following reactions. Use structural formulas to represent the organic compounds.

   (a) The complete combustion (to $CO_2$ and $H_2O$) of cyclopropanol
   (b) The reaction of isopropyl acetate with water to give acetic acid and isopropanol
   (c) The dehydration of ethanol to give ethylene
   (d) The reaction of 1-butyl iodide with water to give 1-butanol

15. Outline, using chemical equations, the synthesis of the following from easily available petrochemicals and inorganic starting materials.

   (a) Vinyl bromide ($CH_2=CHBr$)
   (b) 2-butanol
   (c) Acetone ($CH_3COCH_3$)

16. Outline, using chemical equations, the synthesis of the following from easily available petrochemicals and inorganic starting materials.

   (a) Vinyl acetate ($CH_3COOCH=CH_2$)
   (b) Formamide ($HCONH_2$)
   (c) 1,2-difluoroethane

17. Write a general equation (using R to represent a general alkyl group) for the formation of an ester by the condensation of a tertiary alcohol with a carboxylic acid.

18. Explain why it is impossible to form an amide by the condensation of a tertiary amine with a carboxylic acid.

19. In a recent year, the United States produced $6.26 \times 10^9$ kg of ethylene dichloride (1,2-dichloroethane) and $15.87 \times 10^9$ kg

of ethylene. Assuming that all significant quantities of ethylene dichloride were produced from ethylene, what fraction of the ethylene production went into making ethylene dichloride? What mass of chlorine was required for this conversion?

20. In a recent year, the United States produced $6.26 \times 10^9$ kg of ethylene dichloride (1,2-dichloroethane) and $3.73 \times 10^9$ kg of vinyl chloride. Assuming that all significant quantities of vinyl chloride were produced from ethylene dichloride, what fraction of the ethylene dichloride production went into making vinyl chloride? What mass of hydrogen chloride was generated as a byproduct?

## Synthetic Polymers

21. Write a balanced chemical equation to represent the addition polymerization of 1,1-dichloroethylene. The product of this reaction is Saran, used as a plastic wrap.

22. Write a balanced chemical equation to represent the addition polymerization of tetrafluoroethylene. The product of this reaction is Teflon.

23. A polymer produced by addition polymerization consists of $-(CH_2-O)-$ groups joined in a long chain. What was the starting monomer?

24. The polymer polymethyl methacrylate is used to make Plexiglas. It has the formula

Draw the structural formula of the starting monomer.

25. The monomer glycine ($NH_2-CH_2-COOH$) can undergo condensation polymerization to form polyglycine, in which the structural units are joined by amide linkages.
(a) What molecule is split off in the formation of polyglycine?
(b) Draw the structure of the repeat unit in polyglycine.

26. The polymer $-(NH-CH(CH_3)-\overset{\overset{\displaystyle O}{\|}}{C})_n$ forms upon condensation polymerization with loss of water. Draw the structure of the starting monomer.

27. Determine the mass of adipic acid and the mass of hexamethylenediamine needed to make $1.00 \times 10^3$ kg of nylon 66 fiber.

28. Determine the mass of terephthalic acid and the mass of ethylene glycol needed to make 10.0 kg of polyester fiber.

29. In a recent year, 4.37 billion kilograms of low-density polyethylene was produced in the United States. What volume of gaseous ethylene at 0°C and 1.00 atm would give this amount?

30. In a recent year, 2.84 billion kilograms of polystyrene was produced in the United States. Polystyrene is the addition polymer formed from the styrene monomer, $C_6H_5CH=CH_2$. How many styrene monomer units were incorporated in that 2.84 billion kilograms of polymer?

## Natural Polymers

31. By referring to Figure 23-24a, draw the structure of the ring form of $\beta$-D-galactose. How many asymmetric carbon atoms (chiral centers) are there in the molecule?

32. By referring to Figure 23-32b, draw the structure of the ring form of D-ribose. How many asymmetric carbon atoms (chiral centers) are there in the molecule?

33. How many tripeptides can be synthesized using just three different species of $\alpha$-amino acids?

34. How many different polypeptides, each containing ten amino acids, can be made from the amino acids listed in Table 23-6? How many different polypeptides, each containing 100 amino acids, can be made?

35. Draw the structure of the pentapeptide alanine–leucine–phenylalanine–glycine–isoleucine. Assume that the free $-NH_2$ group is at the alanine end of the peptide chain. Would this compound be more likely to dissolve in water or in octane? Explain.

36. Draw the structure of the pentapeptide aspartic acid–serine–lysine–glutamic acid–tyrosine. Assume that the free $-NH_2$ group is at the aspartic acid end of the peptide chain. Would this compound be more likely to dissolve in water or in octane? Explain.

37. Suppose a long-chain polypeptide is constructed entirely from phenylalanine monomer units. What is its empirical formula? How many amino acids does it contain if its molar mass is 17,500 g mol$^{-1}$?

38. A typical bacterial DNA has a molar mass of $4 \times 10^9$ g mol$^{-1}$. Approximately how many nucleotides does it contain?

## Additional Problems

39. The standard enthalpy of formation of heptane is $-187.82$ kJ mol$^{-1}$, and the standard enthalpy of formation of isooctane (2,2,4-trimethylpentane) is $-224.13$ kJ mol$^{-1}$.
(a) Compute the standard enthalpy of combustion of 1.00 mol of each of the two if they are burned to $CO_2(g)$ and $H_2O(g)$ in an internal-combustion engine.
(b) Which compound delivers more heat per gallon? (Isooctane has a density of 5.77 lb per U.S. gallon, and heptane has a density of 5.71 lb per U.S. gallon.)

40. Ethylene ($C_2H_4$) is an organic compound of great economic significance that is often transported by pipeline. It has a critical temperature of 282.65 K and a critical pressure of 50.096 atm.

(a) Express the critical temperature and pressure of ethylene in degrees Fahrenheit and in pounds per square inch.

(b) A pipeline contains ethylene at a pressure of 54 atm and a temperature of 15°C. Discuss the physical state of the ethylene within the pipe.

(c) The pipeline is 100 km long and has an interior diameter of 250 mm. Estimate the mass of ethylene in the pipeline, assuming it is an ideal gas.

(d) The density of the ethylene within the pipe from part (c) is 0.20 g cm$^{-3}$. Compute the actual mass of ethylene in the pipe.

(e) If the temperature of the entire pipeline and its contents falls to 0°C on a cold day, suggest what the operators could do to make sure that the ethylene in the pipe does not change its physical state.

**41.** *Trans*-cyclodecene boils at 193°C, but *cis*-cyclodecene boils at 195.6°C. Write structural formulas for these two compounds.

**42.** Consider the following proposed structures for benzene, each of which is consistent with the molecular formula C$_6$H$_6$.

(i)

(ii)

(iii)

(iv) CH$_3$—C≡C—C≡C—CH$_3$

(v) CH$_2$=CH—C≡C—CH=CH$_2$

(a) When benzene reacts with chlorine to give C$_6$H$_5$Cl, only one isomer of that compound forms. Which of the five proposed structures for benzene are consistent with this observation?

(b) When C$_6$H$_5$Cl reacts further with chlorine to give C$_6$H$_4$Cl$_2$, exactly three isomers of the latter compound form. Which of the five proposed structures for benzene are consistent with this observation?

**43.** How does *dehydration* differ from *dehydrogenation*? Give an example of each.

**44.** When an ester forms from an alcohol and a carboxylic acid,

an oxygen atom links the two parts of each ester molecule. This atom could have come originally from the alcohol, from the carboxylic acid, or randomly from either. Propose an experiment using isotopes to determine which is the case.

**45.** Hydrogen can be added to a certain unsaturated hydrocarbon in the presence of a platinum catalyst to form hexane. When the same hydrocarbon is oxidized with KMnO$_4$, it yields acetic acid and butanoic acid. Identify the hydrocarbon and write balanced chemical equations for the reactions.

**46.** (a) It is reported that ethylene is released when pure ethanol is passed over alumina (Al$_2$O$_3$) that is heated to 400°C, but diethyl ether is obtained at a temperature of 230°C. Write balanced equations for both of these dehydration reactions.

(b) If the temperature is raised well above 400°C, an aldehyde forms. Write a chemical equation for this reaction.

**47.** In the addition polymerization of acrylonitrile, a very small amount of butyl lithium causes a reaction that can consume hundreds of pounds of the monomer; however, the butyl lithium is called an initiator, not a catalyst. Explain why.

**48.** Based on the fact that the free-radical polymerization of ethylene is spontaneous, and the fact that polymer molecules are less disorganized than the starting monomers, decide whether the polymerization reaction is exothermic or endothermic. Explain.

**49.** Approximately 950 million lb of ethylene dichloride was exported from the United States in a recent year, according to a trade journal. The article states that ''between 500 million and 550 million pounds of PVC could have been made from that ethylene dichloride.'' Compute the range of percentage yields of PVC from ethylene dichloride that is implied by these figures.

**50.** The complete hydrogenation of natural rubber (the addition of H$_2$ to all double bonds) gives a product that is indistinguishable from the product of the complete hydrogenation of gutta-percha. Explain how this strengthens the conclusion that these two substances are isomers of each other.

**51.** A reducing solution breaks S—S bonds in proteins, whereas an oxidizing solution allows them to re-form. Discuss how such solutions might be used to carry out the permanent waving of hair.

**52.** L-sucrose tastes sweet, but it is not metabolized. It has been suggested as a potential nonnutritive sweetener. Draw the molecular structure of L-sucrose, using Figure 23-24*b* as a starting point.

**53.** Polypeptides are synthesized from a 50:50 mixture of L-alanine and D-alanine. How many different isomeric molecules containing 22 monomer units are possible?

**54.** An osmotic pressure measurement taken on a solution containing hemoglobin shows that the molar mass of that protein is approximately 65,000 g mol$^{-1}$. A chemical analysis shows it to contain 0.344% of iron by mass. How many iron atoms does each hemoglobin molecule contain?

**\*55.** At very low pH, alanine is a diprotic acid that can be represented as $H_3N^+$—$CH(CH_3)$—COOH. The p$K_a$ of the carboxyl group is 2.3, and the p$K_a$ of the —$NH_3^+$ group is 9.7. (a) At pH 7, what fraction of the amino acid molecules dissolved in an aqueous solution will have the form $H_3N^+$—$CH(CH_3)$—COO$^-$?

(b) What fraction of the molecules at this pH will have the form $H_2N$—$CH(CH_3)$—COOH?

56. The sequence of bases in one strand of DNA reads ACTTGACCG. Write the sequence of bases in the complementary strand.

*A crystal of elemental bismuth.*

# APPENDICES

# APPENDIX A

# Scientific Notation and Experimental Error

## A-1 SCIENTIFIC NOTATION

Very large and very small numbers are common in chemistry. Repeatedly writing such numbers in the ordinary way (for example, the important number 602,213,700,000,000,000,000,000) would be tedious and would engender errors. **Scientific notation** offers a better way. A number in scientific notation is expressed as a number from 1 to 10 multiplied by 10 raised to some power. Any number can be represented in this way, as the following examples show.

$$643.8 = 6.438 \times 10^2$$

$$-19,000,000 = -1.9 \times 10^7$$

$$0.0236 = 2.36 \times 10^{-2}$$

$$602,213,700,000,000,000,000,000 = 6.022137 \times 10^{23}$$

A simple rule of thumb is that the power to which 10 is raised is $n$ if the decimal point is moved $n$ places to the left and is $-n$ if the decimal is moved $n$ places to the right.

When two or more numbers written in scientific notation are to be added or subtracted, they should first be expressed as multiples of the *same* power of 10:

$$
\begin{array}{rcl}
6.431 \times 10^4 & \longrightarrow & 6.431 \times 10^4 \\
+\ 2.1\ \ \times 10^2 & \longrightarrow & +\ 0.021 \times 10^4 \\
+\ 3.67\ \ \times 10^3 & \longrightarrow & +\ 0.367 \times 10^4 \\
\hline
? & & 6.819 \times 10^4
\end{array}
$$

When two numbers in scientific notation are multiplied, the coefficients are multiplied and then the powers of 10 are multiplied (by adding the exponents):

$$1.38 \times 10^{-16} \times 8.80 \times 10^3 = (1.38 \times 8.80) \times 10^{(-16+3)}$$
$$= 12.1 \times 10^{-13} = 1.21 \times 10^{-12}$$

We divide one number by a second by dividing the coefficients and then multiplying by 10 raised to the first exponent minus the second:

$$\frac{6.63 \times 10^{-27}}{2.34 \times 10^{-16}} = \frac{6.63}{2.34} \times \frac{10^{-27}}{10^{-16}}$$
$$= 2.83 \times 10^{[-27-(-16)]} = 2.83 \times 10^{-11}$$

Any calculator or computer equipped to perform scientific and engineering calculations can accept and display numbers in scientific notation. It cannot determine whether the input has an error, however, or whether the answer makes sense.

That is your responsibility! Develop the habit of mentally estimating the order of magnitude of the answer as a rough check on your calculator's result.

## A-2 EXPERIMENTAL ERROR

Chemistry is an experimental science in which every quantitative measurement is subject to some degree of error. We can seek to reduce error by carrying out additional measurements or by changing our experimental apparatus, but we can never eliminate error altogether. It is important, therefore, to be able to assess the results of an experiment quantitatively to establish the limits of the experiment's validity. Errors are of two types: random (lack of precision) and systematic (lack of accuracy).

### Precision and Random Errors

**Precision** refers to the degree of agreement in a collection of experimental results and is estimated by repeating the measurement under conditions as nearly identical as possible. If the conditions are truly identical, then differences among the trials are due to random error. As a specific example, consider some actual results of an early, important experiment by American physicist Robert Millikan in 1909, to measure the charge $e$ on the electron. The experiment (discussed in greater detail in Chapter 12) involved a study of the motion of charged oil drops suspended in air in an electric field. Millikan made hundreds of measurements on many different oil drops, but we shall consider only a set of results for $e$ found for one particular drop (Table A-1). The values he found ranged from 4.894 to $4.941 \times 10^{-10}$ esu. What do we choose to report as the best estimate for $e$? The proper procedure is to first examine the data to see whether any of the results are especially far from the rest (a value above $5 \times 10^{-10}$ esu would fall into this category). Such values are likely to result from some mistake in carrying out or reporting that particular measurement and therefore are excluded from further consideration (although there have been cases in science where just such exceptional results have led to significant breakthroughs). In Millikan's data, no such points should be excluded. To obtain our best estimate for $e$, we calculate the **mean,** or **average value,** by adding up the values found and dividing by the number of measurements. We can write the average value of any property after a series of $N$ measurements $x_1, x_2, \ldots, x_N$ as

$$\bar{x} = \frac{1}{N}(x_1 + x_2 + \cdots + x_N) = \frac{1}{N}\sum_{i=1}^{N} x_i$$

| Table A-1 | | | | | | | | | | | | |
|---|---|---|---|---|---|---|---|---|---|---|---|---|
| | | | | | **Measurement Number** $e$ $(10^{-10}$ esu) | | | | | | | |
| 1 | 2 | 3 | 4 | 5 | 6 | 7 | 8 | 9 | 10 | 11 | 12 | 13 |
| 4.915 | 4.920 | 4.937 | 4.923 | 4.931 | 4.936 | 4.941 | 4.902 | 4.927 | 4.900 | 4.904 | 4.897 | 4.894 |

From R. A. Millikan, *Phys. Rev.* 32:349, 1911. [1 esu $= 3.3356 \times 10^{-10}$ C] where a capital Greek sigma ($\Sigma$) is introduced to indicate a summation of $x_i$ over values of $i$ from 1 to $N$. In the present case, this gives an average for $e$ of $4.917 \times 10^{-10}$ esu.

This average by itself does not convey any estimate of uncertainty. If all of the measurements had given results between 4.91 and $4.92 \times 10^{-10}$ esu, the uncertainty would be less than if the results had ranged from $4 \times 10^{-10}$ to $6 \times 10^{-10}$ esu. Furthermore, an average of 100 measurements should have less uncertainty than an average of 5. How are these ideas made quantitative? A statistical measure of the spread of data, called the **standard deviation** $\sigma$, is useful in this regard. It is given by the formula

$$\sigma = \sqrt{\frac{(x_1 - \bar{x})^2 + (x_2 - \bar{x})^2 + \cdots + (x_N - \bar{x})^2}{N - 1}}$$

$$= \sqrt{\frac{1}{N-1} \sum_{i=1}^{N} (x_i - \bar{x})^2}$$

The standard deviation is found by adding up the squares of the deviations of the individual data points from the average value $\bar{x}$, dividing by $N - 1$, and taking the square root. Table A-2 shows how $\sigma$ is used quantitatively. A **confidence limit** is defined as

$$\text{confidence limit} = \pm \frac{t\sigma}{\sqrt{N}}$$

The table gives the factor $t$ for various numbers of measurements, $N$, and for various levels of confidence.

**Table A-2**

| N (Number of Observations) | Factor $t$ for Confidence Interval of | | | |
|---|---|---|---|---|
| | 80% | 90% | 95% | 99% |
| 2 | 3.08 | 6.31 | 12.7 | 63.7 |
| 3 | 1.89 | 2.92 | 4.30 | 9.92 |
| 4 | 1.64 | 2.35 | 3.18 | 5.84 |
| 5 | 1.53 | 2.13 | 2.78 | 4.60 |
| 6 | 1.48 | 2.02 | 2.57 | 4.03 |
| 7 | 1.44 | 1.94 | 2.45 | 3.71 |
| 8 | 1.42 | 1.90 | 2.36 | 3.50 |
| 9 | 1.40 | 1.86 | 2.31 | 3.36 |
| 10 | 1.38 | 1.83 | 2.26 | 3.25 |
| 11 | 1.37 | 1.81 | 2.23 | 3.17 |
| 12 | 1.36 | 1.80 | 2.20 | 3.11 |
| 13 | 1.36 | 1.78 | 2.18 | 3.06 |
| 14 | 1.35 | 1.77 | 2.16 | 3.01 |
| 15 | 1.34 | 1.76 | 2.14 | 2.98 |
| $\infty$ | 1.29 | 1.64 | 1.96 | 2.58 |

For Millikan's data, $N = 13$ and $\sigma = 0.017 \times 10^{-10}$. For 95% confidence with 13 measurements, the table shows $t = 2.18$ and the confidence limit is

$$\text{confidence limit} = \pm \frac{(2.18)(0.017 \times 10^{-10})}{\sqrt{13}} = \pm 0.010 \times 10^{-10} \text{ esu}$$

Thus, a 95% probability exists that the *true* average (obtained by repeating the experiment under the same conditions an infinite number of times) will lie within $\pm 0.010 \times 10^{-10}$ esu of the average $4.917 \times 10^{-10}$ esu. Within this 95% confidence level, our best estimate for $e$ is written as

$$(4.917 \pm 0.010) \times 10^{-10} \text{ esu}$$

For other confidence levels and other numbers of measurements, the factor $t$ and therefore the confidence limit change.

## Accuracy and Systematic Errors

The charge $e$ on the electron has been measured by several different techniques since Millikan's day. The current best estimate for $e$ is

$$e = (4.8032068 \pm 0.0000015) \times 10^{-10} \text{ esu}$$
$$= (1.60217733 \pm 0.00000049) \times 10^{-19} \text{ C}$$

This value lies outside the range of uncertainty we estimated from Millikan's original data. In fact, it lies well below the smallest of the 13 measurements of $e$. Why?

To understand this discrepancy, we need to remember that there is a second source of error in any experiment: *systematic* error that causes a shift in the measured values from the true value and reduces the **accuracy** of the result. By making more measurements, we can reduce the uncertainty due to *random* errors and improve the *precision* of our result, but if systematic errors are present, the average value will continue to deviate from the true value. Such systematic errors may result from a miscalibration of the experimental apparatus or from a fundamental inadequacy in the technique for measuring a property. In the case of Millikan's experiment, the then-accepted value for the viscosity of air (used in calculating the charge $e$) was subsequently found to be wrong. This caused his results to be systematically too high.

Error thus arises from two sources. Lack of precision (random errors) can be estimated by a statistical analysis of a series of measurements. Lack of accuracy (systematic errors) is much more problematical. If a systematic error is known to be present, we should do our best to correct for it before reporting the result. (For example, if our apparatus has not been calibrated correctly, it should be recalibrated.) The problem is that systematic errors of which we have no knowledge may be present. In this case, the experiment should be repeated with different apparatus to eliminate the systematic error caused by a particular piece of equipment; better still, a different and independent way to measure the property might be devised. Only after enough independent experimental data are available can we be convinced of the accuracy of a result—that is, how closely it approximates the true result.

## A-3 SIGNIFICANT FIGURES

The number of **significant figures** is the number of digits used to express a measured or calculated quantity, excluding zeros that may precede the first nonzero digit. Suppose the mass of a sample of sodium chloride is measured to be 8.241 g and the uncertainty is estimated to be ±0.001 g. The mass is said to be given to four significant figures because we are confident of the first three digits (8, 2, 4) and the uncertainty appears in the fourth (1), which nevertheless is still significant. Writing additional digits beyond the 1 would not be justified, however, unless the accuracy of the weighing could be improved. When we record a volume as 22.4 L, we imply that the uncertainty in the measurement is in the last digit written ($V = 22.4 \pm 0.3$ L, for example). A volume written as 22.43 L, on the other hand, implies that the uncertainty is far less and appears only in the fourth significant figure.

In the same way, writing 20.000 m is quite different from writing 20.0 m. The second measurement (with three significant figures) could easily be made with a common meterstick. The first (with five significant figures) would require a more precise method. We should avoid reporting results such as ''700 m,'' however, because the two trailing zeros may or may not be significant. The uncertainty in the measurement could be of order ±1 m or ±10 m or perhaps ±100 m; it is impossible to tell which without further information. To avoid this ambiguity, we can write such measurements using the scientific notation described in Section A-1. The measurement ''700 m'' translates into any of the following:

$$7.00 \times 10^2 \text{ m} \qquad \text{Three significant figures}$$
$$7.0 \times 10^2 \text{ m} \qquad \text{Two significant figures}$$
$$7 \times 10^2 \text{ m} \qquad \text{One significant figure}$$

Frequently it is necessary to combine several different experimental measurements to obtain a final result. Some operations involve addition or subtraction, and others entail multiplication or division. These operations affect the number of significant figures that should be retained in the calculated result. Suppose, for example, that a weighed sample of 8.241 g of sodium chloride is dissolved in 160.1 g of water. What will be the mass of the solution that results? It is tempting to simply write $160.1 + 8.241 = 168.341$ g, but this is *not* correct. In saying that the mass of water is 160.1 g, we imply that there is some uncertainty about the number of tenths of a gram measured. This uncertainty must also apply to the sum of the masses, and so the last two digits in the sum are not significant and should be **rounded off,** leaving 168.3 as the final result.

> Following addition or subtraction, round off the result to the leftmost decimal place that contained an uncertain digit in the original numbers.

Rounding off is a straightforward operation. It consists of first discarding the digits that are not significant and then adjusting the last remaining digit. If the first discarded digit is less than 5, the remaining digits are left as they are (for example, 168.341 is rounded down to 168.3 because the first discarded digit, 4, is less than 5).

If the first discarded digit is greater than 5, or if it is equal to 5 and is followed by one or more nonzero digits, then the last digit is increased by 1 (for example, 168.364 and 168.3503 both become 168.4 when rounded off to four digits). Finally, if the first digit discarded is 5 and all subsequent digits are zeros, the last digit remaining is rounded to the nearest even digit (for example, both 168.35 and 168.45 would be rounded to 168.4). This last rule is chosen so that, on the average, as many numbers are rounded up as down. Other conventions are sometimes used.

In multiplication or division it is not the number of decimal places that matters (as in addition or subtraction) but the number of significant figures in the least precisely known quantity. Suppose, for example, the measured volume of a sample is 4.34 cm$^3$ and its mass is 8.241 g. The density, found by dividing the mass by the volume on a calculator, for example, is

$$\frac{8.241 \text{ g}}{4.34 \text{ cm}^3} = 1.89884 \ldots \text{g cm}^{-3}$$

How many significant figures should we report? Because the volume is the less precisely known quantity (three significant figures as opposed to four for the mass), it controls the precision that may properly be reported in the answer. Only three significant figures are justified, and so the result is rounded to 1.90 g cm$^{-3}$.

> The number of significant figures in the result of a multiplication or division is the smallest of the numbers of significant figures used as input.

It is best to carry out the arithmetical operations and *then* round the final answer to the correct number of significant figures, rather than round off the input data first. The difference is usually small, but this recommendation is nevertheless worth following. For example, the correct way to add the three distances 15 m, 6.6 m, and 12.6 m is

$$
\begin{array}{r}
15 \;\;\text{m} \\
+\;\; 6.6 \text{ m} \\
+\; 12.6 \text{ m} \\
\hline
34.2 \text{ m} \longrightarrow 34 \text{ m}
\end{array}
\qquad \text{rather than} \qquad
\begin{array}{r}
15 \;\;\text{m} \longrightarrow 15 \text{ m} \\
6.6 \text{ m} \longrightarrow \;\;7 \text{ m} \\
12.6 \text{ m} \longrightarrow 13 \text{ m} \\
\hline
35 \text{ m}
\end{array}
$$

For the same reason, we frequently carry extra digits through the intermediate steps of a worked example and round off only for the final answer. If a calculation is done entirely on a scientific calculator or a computer, several extra digits are usually carried along automatically. Before the final answer is reported, however, it is important to round off to the proper number of significant figures.

Sometimes pure constants appear in expressions. In this case, the accuracy of the result is determined by the accuracy of the other factors. The uncertainty in the volume of a sphere, $\frac{4}{3}\pi r^3$, depends only on the uncertainty in the radius $r$; 4 and 3 are pure constants (4.000 . . . and 3.000 . . ., respectively), and $\pi$ can be given to as many significant figures (3.14159265 . . .) as are warranted by the radius.

## PROBLEMS

### Scientific Notation

1. Express the following in scientific notation.
   (a) 0.0000582
   (b) 1402
   (c) 7.93
   (d) −6593.00
   (e) 0.002530
   (f) 1.47
2. Express the following in scientific notation.
   (a) 4579
   (b) −0.05020
   (c) 2134.560
   (d) 3.825
   (e) 0.0000450
   (f) 9.814
3. Convert the following from scientific notation to decimal form.
   (a) $5.37 \times 10^{-4}$
   (b) $9.390 \times 10^6$
   (c) $-2.47 \times 10^{-3}$
   (d) $6.020 \times 10^{-3}$
   (e) $2 \times 10^4$
4. Convert the following from scientific notation to decimal form.
   (a) $3.333 \times 10^{-3}$
   (b) $-1.20 \times 10^7$
   (c) $2.79 \times 10^{-5}$
   (d) $3 \times 10^1$
   (e) $6.700 \times 10^{-2}$
5. A certain chemical plant produces $7.46 \times 10^8$ kg of polyethylene in one year. Express this amount in decimal form.
6. A microorganism contains 0.0000046 g of vanadium. Express this amount in scientific notation.

### Experimental Error

7. A group of students took turns using a laboratory balance to weigh the water contained in a beaker. The results they reported were 111.42 g, 111.67 g, 111.21 g, 135.64 g, 111.02 g, 111.29 g, and 111.42 g.
   (a) Should any of the data be excluded before the average is calculated?
   (b) From the remaining measurements, calculate the average value of the mass of the water in the beaker.
   (c) Calculate the standard deviation $\sigma$ and, from it, the 95% confidence limit.
8. By measuring the sides of a small box, a group of students made the following estimates for its volume: 544 cm³, 590 cm³, 523 cm³, 560 cm³, 519 cm³, 570 cm³, and 578 cm³.

(a) Should any of the data be excluded before the average is calculated?
(b) Calculate the average value of the volume of the box from the remaining measurements.
(c) Calculate the standard deviation $\sigma$ and, from it, the 90% confidence limit.
9. Of the measurements in problems 7 and 8, which is more precise?
10. A more accurate determination of the mass in problem 7 (using a better balance) gives the value 104.67 g, and an accurate determination of the volume in problem 8 gives the value 553 cm³. Which of the two measurements in problems 7 and 8 is more accurate, in the sense of having the smaller systematic error relative to the actual value?

### Significant Figures

11. State the number of significant figures in each of the following measurements.
    (a) 13.604 L
    (b) −0.00345°C
    (c) 340 lb
    (d) $3.40 \times 10^2$ miles
    (e) $6.248 \times 10^{-27}$ J
12. State the number of significant figures in each of the following measurements.
    (a) −0.0025 in
    (b) 7000 g
    (c) 143.7902 s
    (d) $2.670 \times 10^7$ Pa
    (e) $2.05 \times 10^{-19}$ J
13. Round off each of the measurements in problem 11 to two significant figures.
14. Round off each of the measurements in problem 12 to two significant figures.
15. Round off the measured number 2,997,215.548 to nine significant digits.
16. Round off the measured number in problem 15 to eight, seven, six, five, four, three, two, and one significant digits.
17. Express the results of the following additions and subtractions to the proper number of significant figures. All of the numbers are measured quantities.
    (a) 67.314 + 8.63 − 243.198 =
    (b) 4.31 + 64 + 7.19 =
    (c) $3.2156 \times 10^{15} - 4.631 \times 10^{13} =$
    (d) $2.41 \times 10^{-26} - 7.83 \times 10^{-25} =$
18. Express the results of the following additions and subtractions to the proper number of significant figures. All of the numbers are measured quantities.
    (a) 245.876 + 4.65 + 0.3678 =

(b) $798.36 - 1005.7 + 129.652 =$

(c) $7.98 \times 10^{17} + 6.472 \times 10^{19} =$

(d) $(4.32 \times 10^{-15}) + (6.257 \times 10^{-14}) - (2.136 \times 10^{-13}) =$

19. Express the results of the following multiplications and divisions to the proper number of significant figures. All of the numbers are measured quantities.

(a) $\dfrac{-72.415}{8.62} =$

(b) $52.814 \times 0.00279 =$

(c) $(7.023 \times 10^{14}) \times (4.62 \times 10^{-27}) =$

(d) $\dfrac{4.3 \times 10^{-12}}{9.632 \times 10^{-26}} =$

20. Express the results of the following multiplications and divisions to the proper number of significant figures. All of the numbers are measured quantities.

(a) $129.587 \times 32.33 =$

(b) $\dfrac{4.7791}{3.21 \times 5.793} =$

(c) $\dfrac{10566.9}{3.584 \times 10^{29}} =$

(d) $(5.247 \times 10^{13}) \times (1.3 \times 10^{-17}) =$

21. Compute the area of a triangle (according to the formula $A = \frac{1}{2}ba$) if its base and altitude are measured to equal 42.07 cm and 16.0 cm, respectively. Justify the number of significant figures in the answer.

22. An inch is defined as exactly 2.54 cm. The length of a table is measured as 505.16 cm. Compute the length of the table in inches. Justify the number of significant figures in the answer.

# APPENDIX B

# SI Units and Unit Conversions

**M**easurements in chemistry and physics require the use of units—reference standards to which experimentally observed quantities can be related. Over the course of history, different countries evolved different sets of units to express length, mass, and many other physical dimensions. Gradually, those diverse sets of units are being replaced by international standards that facilitate easy comparison of measurements in different localities and avoid complications and confusion. The unified system of units recommended by international agreement is called SI, standing for "Système International d'Unités," or the International System of Units. In this appendix we outline the use of SI units and discuss interconversions with other systems of units.

The SI uses seven **base units,** which are listed in Table B-1. All other units can be written as combinations of the base units. In writing the units for a measurement, we abbreviate them (Table B-1) and use exponential notation to denote the power to which a unit is raised; a minus sign appears in the exponent when the unit is in the denominator. Velocity, for example, is a quantity with dimensions of length divided by time, and so in SI it is expressed in meters per second, or $m\ s^{-1}$. Some derived units that are used frequently have special names. Energy, for example, is the product of mass and the square of the velocity. Therefore, it is measured in units of kilogram square meters per square seconds ($kg\ m^2\ s^{-2}$), and $1\ kg\ m^2\ s^{-2}$ is called a *joule.* Other derived units, such as the pascal for measuring pressure, appear in Table B-2. Although these names provide a useful shorthand, it is important to remember their meanings in terms of the base units.

Because scientists work on scales ranging from the microscopic to the astronomical, there is a tremendous range in the magnitudes of measured quantities. Consequently, a set of **prefixes** has been incorporated into the International System

### Table B-1  Base Units in the International System of Units

| Quantity | Unit | Symbol |
|---|---|---|
| Length | meter | m |
| Mass | kilogram | kg |
| Time | second | s |
| Temperature | kelvin | K |
| Chemical amount (of substance) | mole | mol |
| Electric current | ampere | A |
| Luminous intensity | candela | cd |

| Table B-2 | Derived Units in SI | | |
|---|---|---|---|
| **Quantity** | **Unit** | **Symbol** | **Definition** |
| Energy | joule | J | $kg\ m^2\ s^{-2}$ |
| Force | newton | N | $kg\ m\ s^{-2} = J\ m^{-1}$ |
| Power | watt | W | $kg\ m^2\ s^{-3} = J\ s^{-1}$ |
| Pressure | pascal | Pa | $kg\ m^{-1}\ s^{-2} = N\ m^{-2}$ |
| Electric charge | coulomb | C | $A\ s$ |
| Electric potential difference | volt | V | $kg\ m^2\ s^{-3}\ A^{-1} = J\ C^{-1}$ |

of Units to simplify the description of small and large quantities (Table B-3). The prefixes specify various powers of 10 times the base and derived units. Some of them are quite familiar in everyday use: the kilometer, for example, is $10^3$ m. Others may sound less familiar—for instance, the femtosecond (1 fs = $10^{-15}$ s) or the gigapascal (1 GPa = $10^9$ Pa).

In addition to base and derived SI units, several units that are not officially sanctioned are in this book. The first is the liter (abbreviated L), a very convenient size for volume measurements in chemistry; a liter is $10^{-3}$ m$^3$ or 1 cubic decimeter (dm$^3$):

$$1\ L = 1\ dm^3 = 10^{-3}\ m^3 = 10^3\ cm^3$$

Second, we use the *angstrom* (abbreviated Å) as a unit of length for atoms and molecules:

$$1\ \text{Å} = 10^{-10}\ m = 100\ pm = 0.1\ nm$$

This unit is used simply because most atomic sizes and chemical bond lengths fall in the range of one to several angstroms, and the use of either picometers or nanometers is slightly awkward. Next, we use the *atmosphere* (abbreviated atm) as a unit of pressure. It is not a simple power of 10 times the SI unit of the pascal, but rather is defined by

$$1\ atm = 101,325\ Pa = 0.101325\ MPa$$

| Table B-3 | Prefixes in SI | | | | |
|---|---|---|---|---|---|
| **Fraction** | **Prefix** | **Symbol** | **Factor** | **Prefix** | **Symbol** |
| $10^{-1}$ | *deci-* | d | 10 | *deca-* | da |
| $10^{-2}$ | *centi-* | c | $10^2$ | *hecto-* | h |
| $10^{-3}$ | *milli-* | m | $10^3$ | *kilo-* | k |
| $10^{-6}$ | *micro-* | $\mu$ | $10^6$ | *mega-* | M |
| $10^{-9}$ | *nano-* | n | $10^9$ | *giga-* | G |
| $10^{-12}$ | *pico-* | p | $10^{12}$ | *tera-* | T |
| $10^{-15}$ | *femto-* | f | | | |
| $10^{-18}$ | *atto-* | a | | | |

This unit is used because most chemistry is carried out at pressures near atmospheric pressure, for which the pascal is too small a unit to be convenient. In addition, expressions for equilibrium constants (Chapter 5) are simplified when pressures are expressed in atmospheres.

The non-SI temperature units require special mention. The two most important temperature scales in the United States are the Fahrenheit scale and the Celsius scale, which employ the Fahrenheit degree (°F) and the Celsius degree (°C), respectively. The *size* of the Celsius degree is the same as that of the SI temperature unit, the kelvin, but the two scales are shifted by 273.15°C:

$$T_K = \frac{1 \text{ K}}{1°\text{C}}(t_{°C}) + 273.15 \text{ K}$$

The size of the degree Fahrenheit is 5/9 the size of the Celsius degree. The two scales are related by

$$t_{°C} = \frac{5°\text{C}}{9°\text{F}}(t_{°F} - 32°\text{F}) \qquad \text{or} \qquad t_{°F} = \frac{9°\text{F}}{5°\text{C}}(t_{°C}) + 32°\text{F}$$

The advantage of a unified system of units is that if all the quantities in a calculation are expressed in SI units, the final result must come out in SI units. Nevertheless, it is important to become familiar with the ways in which units are interconverted, because units other than SI base units often appear in calculations. The **unit conversion method** provides a systematic approach to this problem.

As a simple example, suppose the mass of a sample is measured to be 64.3 g. If this is to be used in a formula involving SI units, it should be converted to kilograms (the SI base unit of mass). To do this, we use the fact that 1 kg = 1000 g and write

$$\frac{64.3 \text{ g}}{1000 \text{ g kg}^{-1}} = 0.0643 \text{ kg}$$

Note that this is, in effect, division by 1; because 1000 g = 1 kg, 1000 g kg$^{-1}$ = 1, and we cancel units between numerator and denominator to obtain the final result. This unit conversion could also be written as

$$64.3 \text{ g}\left(\frac{1 \text{ kg}}{1000 \text{ g}}\right) = 0.0643 \text{ kg}$$

In this book we use the more compact first version of the unit conversion. Instead of *dividing* by 1000 g kg$^{-1}$, we can equally well *multiply* by $1 = 10^{-3}$ kg g$^{-1}$.

$$(64.3 \text{ g})(10^{-3} \text{ kg g}^{-1}) = 0.0643 \text{ kg}$$

Other unit conversions may involve more than just powers of 10, but they are equally easy to carry out. For example, to express 16.4 inches in meters, we use the fact that 1 inch = 0.0254 m (or 1 = 0.0254 m inch$^{-1}$) so that

$$(16.4 \text{ inches})(0.0254 \text{ m inch}^{-1}) = 0.417 \text{ m}$$

More complicated combinations are possible. For example, to convert from liter-atmospheres to joules (the SI unit of energy), two separate unit conversions are used:

$$(1 \text{ L atm})(10^{-3} \text{ m}^3 \text{ L}^{-1})(101{,}325 \text{ Pa atm}^{-1}) = 101.325 \text{ kg m}^2 \text{ s}^{-2}$$
$$= 101.325 \text{ Pa m}^3 = 101.325 \text{ J}$$

When doing chemical calculations, it is very important to write out units explicitly and cancel units in intermediate steps in order to obtain the correct units for the final result. This practice is a way of checking to make sure units have not been incorrectly mixed without unit conversions or an incorrect formula has not been used. If a result that is supposed to be a temperature comes out with units of $m^3 \, s^{-2}$, a mistake has been made!

## PROBLEMS

1. Rewrite the following in scientific notation, using only the base units of Table B-1, without prefixes.
    (a) 65.2 nanograms
    (b) 88 picoseconds
    (c) 5.4 terawatts
    (d) 17 kilovolts

2. Rewrite the following in scientific notation, using only the base units of Table B-1, without prefixes.
    (a) 66 $\mu$K
    (b) 15.9 MJ
    (c) 0.13 mg
    (d) 62 GPa

3. Express the following temperatures in degrees Celsius.
    (a) 9001°F
    (b) 98.6°F (the normal body temperature of human beings)
    (c) 20°F above the boiling point of water at 1 atm pressure
    (d) −40°F

4. Express the following temperatures in degrees Fahrenheit.
    (a) 5000°C
    (b) 40.0°C
    (c) 212°C
    (d) −40°C

5. Express the temperatures given in problem 3 in kelvins.

6. Express the temperatures given in problem 4 in kelvins.

7. Express the following in SI units, either base or derived.
    (a) 55.0 miles per hour (1 mile = 1609.344 m)
    (b) 1.15 g cm$^{-3}$
    (c) $1.6 \times 10^{-19}$ C Å
    (d) 0.15 mol L$^{-1}$
    (e) $5.7 \times 10^3$ L atm day$^{-1}$

8. Express the following in SI units, either base or derived.
    (a) 67.3 atm
    (b) $1.0 \times 10^4$ V cm$^{-1}$
    (c) 7.4 Å year$^{-1}$
    (d) 22.4 L mol$^{-1}$
    (e) 14.7 lb inch$^{-2}$ (1 inch = 2.54 cm; 1 lb = 453.59 g)

9. The kilowatt-hour (kWh) is a common unit in measurements of the consumption of electricity. What is the conversion factor between the kilowatt-hour and the joule? Express 15.3 kWh in joules.

10. A car's rate of fuel consumption is often measured in miles per gallon (mpg). Determine the conversion factor between miles per gallon and the SI unit of meters per cubic decimeter (1 gallon = 3.785 dm$^3$, and 1 mile = 1609.344 m). Express 30.0 mpg in SI units.

11. A certain V-8 engine has a displacement of 404 in$^3$. Express this volume in cubic centimeters (cm$^3$) and in liters.

12. Light travels in a vacuum at a speed of $3.00 \times 10^8$ m s$^{-1}$.
    (a) Convert this speed to miles per second.
    (b) Express this speed in furlongs per fortnight, a little-used unit of speed. (A furlong, a distance used in horse racing, is 660 ft; a fortnight is exactly 2 weeks.)

# Mathematics for General Chemistry

Mathematics is an essential tool in chemistry. This appendix reviews some of the most important mathematical techniques for general chemistry.

## C-1  USING GRAPHS

In many situations in science, we are interested in how one quantity (measured or predicted) depends on another quantity. The position of a moving car depends on the time, for example, or the pressure of a gas depends on the volume of the gas at a given temperature. A very useful way to show such a relation is through a **graph,** in which one quantity is plotted against another.

The usual convention in drawing graphs is to use the horizontal axis for the variable over which we have control and use the vertical axis for the measured or calculated quantity. After a series of measurements, the points on the graph frequently lie along a recognizable curve, and that curve can be drawn through the points. Because any experimental measurement involves some degree of uncertainty, there will be some scatter in the points, so there is no purpose in drawing a curve that passes precisely through every measured point. If there is a relation between the quantities measured, however, the points will display a systematic trend, and a curve can be drawn that represents that trend.

The curves plotted on graphs can have many shapes. The simplest and most important shape is a straight line. A straight line is a graph of a relation such as

$$y = 4x + 7$$

or, more generally,

$$y = mx + b$$

where the variable $y$ is plotted along the vertical axis and the variable $x$ along the horizontal axis (Fig. C-1). The quantity $m$ is the **slope** of the line that is plotted, and $b$ is the **intercept**—the point at which the line crosses the $y$ axis. This can be demonstrated by setting $x$ equal to 0 and noting that $y$ is then equal to $b$. The slope is a measure of the steepness of the line; the greater the value of $m$, the steeper the line. If the line goes up and to the right, the slope is positive; if it goes down and to the right, the slope is negative.

The slope of a line can be determined from the coordinates of two points on the line. Suppose, for example, that when $x = 3$, $y = 5$, and when $x = 4$, $y = 7$. These two points can be written in shorthand notation as (3, 5) and (4 ,7). The slope of the

Graph of the equation $y = 2x - 1$

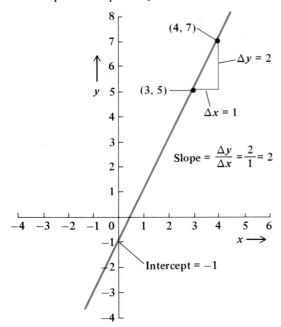

**Figure C-1**

A straight-line, or linear, relationship between two experimental quantities is a very desirable result because it is easy to graph and easy to represent mathematically. The equation of this straight line ($y = 2x - 1$) fits the general form $y = mx + b$. The line's $y$ intercept is $-1$ and its slope is 2.

line is then defined as the "rise over the run"—the change in the $y$ coordinate divided by the change in the $x$ coordinate:

$$\text{slope} = m = \frac{\Delta y}{\Delta x} = \frac{7 - 5}{4 - 3} = \frac{2}{1} = 2$$

The symbol $\Delta$ (capital Greek delta) indicates the change in a quantity—the final value minus the initial value. In chemistry, if the two quantities being graphed have dimensions, the slope has dimensions as well. If a graph of distance traveled (in meters) against time (in seconds) is a straight line, its slope has dimensions of meters per second (m s$^{-1}$).

## C-2 SOLUTION OF ALGEBRAIC EQUATIONS

Frequently in chemistry it is necessary to solve an algebraic equation for an unknown quantity, such as a concentration or a partial pressure in an equilibrium–constant expression. Let us represent the unknown quantity with the symbol $x$. If the equation is linear, the method of solution is straightforward:

$$5x + 9 = 0$$

$$5x = -9$$

$$x = -\frac{9}{5}$$

or, more generally, if $ax + b = 0$, then $x = -b/a$.

Nonlinear equations are of many kinds. One of the most common in chemistry is the quadratic equation, which can always be rearranged into the form

$$ax^2 + bx + c = 0$$

where each of the constants ($a$, $b$, and $c$) may be positive, negative, or 0. The two solutions to a quadratic equation are given by the quadratic formula:

$$x = \frac{-b \pm \sqrt{b^2 - 4ac}}{2a}$$

As an example, suppose that the equation

$$x = 3 + \frac{7}{x}$$

arises in a chemistry problem. Multiplying by $x$ and rearranging the terms gives

$$x^2 - 3x - 7 = 0$$

Inserting $a = 1$, $b = -3$, and $c = -7$ into the quadratic formula gives

$$x = \frac{-(-3) \pm \sqrt{(-3)^2 - 4(1)(-7)}}{2} = \frac{3 \pm \sqrt{9 + 28}}{2} = \frac{3 \pm \sqrt{37}}{2}$$

The two roots of the equation are

$$x = 4.5414 \qquad \text{and} \qquad x = -1.5414$$

In a chemistry problem, the choice of the proper root can frequently be made on physical grounds. If $x$ corresponds to a concentration, for example, the negative root is unphysical and can be discarded.

The solution of cubic or higher-order algebraic equations (or more complicated equations involving sines, cosines, logarithms, or exponentials) becomes more difficult, and approximate or numerical methods must be used. As an illustration, consider the equation

$$x^2 \left( \frac{2.00 + x}{3.00 - x} \right) = 1.00 \times 10^{-6}$$

If $x$ is assumed to be small relative to both 2.00 and 3.00, we obtain the simpler approximate equation

$$x^2 \left( \frac{2.00}{3.00} \right) \approx 1.00 \times 10^{-6}$$

which leads to the roots $x = \pm 0.00122$. We immediately confirm that the solutions obtained in this way *are* small compared with 2.00 (and 3.00) and that our approximation was a good one. When a quantity ($x$ in this case) is added to or subtracted from a larger quantity in a complicated equation, it is usually worthwhile, in solving the equation, to simply neglect the occurrence of the small quantity. Note that this tactic works only for addition and subtraction, never for multiplication or division.

Suppose now that the equation is changed to

$$x^2 \left( \frac{2.00 + x}{3.00 - x} \right) = 1.00 \times 10^{-2}$$

In this case, the equation that comes from neglecting $x$ compared with 2.00 and 3.00 is

$$x^2\left(\frac{2.00}{3.00}\right) \approx 1.00 \times 10^{-2}$$

which is solved by $x \approx \pm 0.122$. The number 0.122 is smaller than 2.00 and 3.00, but not so small that it can be ignored. In this case, more precise results can be obtained by **iteration.** Let us simply add and subtract the approximate positive root $x \approx +0.122$ as specified inside the parentheses and solve again for $x$:

$$x^2\left(\frac{2.00 + 0.122}{3.00 - 0.122}\right) = 1.00 \times 10^{-2}$$

$$x = 0.116$$

This new value can again be inserted into the original equation and the process repeated:

$$x^2\left(\frac{2.00 + 0.116}{3.00 - 0.116}\right) = 1.00 \times 10^{-2}$$

$$x = 0.117$$

Once the successive values of $x$ agree to within the desired accuracy, the iteration can be stopped. Another root of the equation is obtained in a similar fashion if the iterative procedure starts with $x = -0.122$.

In some cases, iteration fails. Suppose, for example, we have the equation

$$x^2\left(\frac{2.00 + x}{3.00 - x}\right) = 10.0$$

There is no particular reason to believe that $x$ should be small compared with 2.00 or 3.00, but if we nevertheless assume that it is, we find $x = \pm\sqrt{15} = \pm 3.873$. Putting $x = +3.873$ back in for an iterative cycle leads to the equation

$$x^2 = -1.49$$

which has no real solutions. Starting with $x = -3.873$ succeeds no better; it gives

$$x^2 = -36.7$$

One way to overcome these difficulties is to solve the original equation graphically. We plot the left side of the equation against $x$ and see at which values it becomes equal to 10 (the right side). We might initially calculate

| $x$ | $x^2\left(\dfrac{2.00 + x}{3.00 - x}\right)$ |
|-----|-----|
| 0 | 0 |
| 1 | 1.5 |
| 2 | 16 |

We observe that for $x = 1$ the left side is less than 10, whereas for $x = 2$ it is greater than 10. Somewhere in between there must be an $x$ for which the left side is *equal to*

10. We can pinpoint it by selecting values of $x$ between 1 and 2; if the left side is less than 10, $x$ should be increased, and if it is greater than 10, $x$ should be decreased.

| $x$ | $x^2\left(\dfrac{2.00 + x}{3.00 - x}\right)$ |
|---|---|
| 1.5 | 5.25 |
| Increase to  1.8 | 10.26 |
| Decrease to  1.75 | 9.19 |
| Increase to  1.79 | 10.04 |

Thus, 1.79 is quite close to a solution of the equation. Improved values are easily obtained by further adjustments of this type.

## C-3 POWERS AND LOGARITHMS

Raising a number to a power and the inverse operation, taking a logarithm, are important in many chemical problems. Although the ready availability of electronic calculators makes the mechanical execution of these operations quite routine, it remains important to understand what is involved in such "special functions."

The mathematical expression $10^4$ implies multiplying 10 by itself 4 times to give 10,000. Ten, or any other number, when raised to the power 0 always gives 1:

$$10^0 = 1$$

Negative powers of 10 give numbers less than 1 and are equivalent to raising 10 to the corresponding *positive* power and then taking the reciprocal:

$$10^{-3} = 1/10^3 = 0.001$$

We can extend the idea of raising to a power to include powers that are not whole numbers. For example, raising to the power $\frac{1}{2} = 0.5$ is the same as taking the square root:

$$10^{0.5} = \sqrt{10} = 3.1623\ldots$$

Scientific calculators have a $10^x$ (or INV LOG) key that can be used for calculating powers of 10 in cases where the power is not a whole number.

Numbers other than 10 can be raised to powers, as well; these numbers are referred to as **bases.** Many calculators have a $y^x$ key that lets any positive number $y$ be raised to any power $x$. One of the most important bases in scientific problems is the transcendental number called $e$ (2.7182818 . . .). The $e^x$ (or INV LN) key on a calculator is used to raise $e$ to any power $x$. The quantity $e^x$ is called the **exponential** of $x$. A key property of powers is that a base raised to the sum of two powers is equivalent to the product of the base raised separately to these powers. Thus, we can write

$$10^{21+6} = 10^{21} \times 10^6 = 10^{27}$$

The same type of relationship holds for any base, including $e$.

Logarithms also occur frequently in chemistry problems. The logarithm of a number is the exponent to which some base has to be raised to obtain the number.

The base is almost always either 10 or the transcendental number $e$. Thus,

$$B^a = n \quad \text{and} \quad \log_B n = a$$

where $a$ is the logarithm, $B$ is the base, and $n$ is the number.

**Common logarithms** are base-10 logarithms; that is, they are powers to which 10 has to be raised in order to give the number. For example, $10^3 = 1000$, and so $\log_{10} 1000 = 3$. We shall frequently omit the 10 when showing common logarithms and write this equation as $\log 1000 = 3$. Only the logarithms of 1, 10, 100, 1000, and so on are whole numbers; the logarithms of other numbers are decimal fractions. The decimal point in a logarithm divides it into two parts. To the left of the decimal point is the *characteristic*; to the right is the *mantissa*. Thus, the logarithm in the equation

$$\log (7.310 \times 10^3) = 3.8639$$

has a characteristic of 3 and a mantissa of 0.8639. As may be verified with a calculator, the base-10 logarithm of the much larger (but closely related) number $7.310 \times 10^{23}$ is 23.8639. As this case illustrates, the characteristic is determined solely by the location of the decimal point in the number and not by the number's precision, so it is *not* included when counting significant figures. The mantissa should be written with as many significant figures as the original number.

A logarithm is truly an exponent and as such follows the same rules of multiplication and division as other exponents. In multiplication and division we have

$$\log (n \times m) = \log n + \log m$$

$$\log \left(\frac{n}{m}\right) = \log n - \log m$$

Furthermore,

$$\log n^m = m \log n$$

and so the logarithm of $3^5 = 243$ is

$$\log 3^5 = 5 \log 3 = 5 \times 0.47712 = 2.3856$$

There is no such thing as the logarithm of a negative number, because there is no power to which 10 (or any other base) can be raised to give a negative number.

A frequently used base for logarithms is the number $e$ ($e = 2.7182818\ldots$). The logarithm to the base $e$ is called the **natural logarithm** and is indicated by $\log_e$ or ln. Base-$e$ logarithms are related to base-10 logarithms by the formula

$$\ln n = 2.3025851 \log n$$

As already stated, calculations of logarithms and powers are inverse operations. Thus, if we want to find the number for which 3.8639 is the common logarithm, we simply calculate

$$10^{3.8639} = 7.310 \times 10^3$$

If we need the number for which the natural logarithm is 2.108, we calculate

$$e^{2.108} = 8.23$$

As before, the number of significant digits in the answer should correspond to the number of digits in the *mantissa* of the logarithm.

## C-4  SLOPES OF CURVES AND DERIVATIVES

Very frequently in science, one measured quantity depends on a second one. If the property $y$ depends on a second property $x$, we can write $y = f(x)$, where $f$ is a function that expresses the dependence of $y$ on $x$. Often we are interested in the effect of a small change $\Delta x$ on the dependent variable $y$. If $x$ changes to $x + \Delta x$, then $y$ will change to $y + \Delta y$. How is $\Delta y$ related to $\Delta x$? Suppose we have the simple linear relation between $y$ and $x$

$$y = mx + b$$

where $a$ and $b$ are constants. If we substitute $y + \Delta y$ and $x + \Delta x$, we find

$$y + \Delta y = m(x + \Delta x) + b$$

Subtracting the first equation from the second leaves

$$\Delta y = m\,\Delta x$$

or

$$\frac{\Delta y}{\Delta x} = m$$

The change in $y$ is proportional to the change in $x$, with a proportionality constant equal to the slope of the line in the graph of $y$ against $x$.

Suppose now we have a slightly more complicated relationship such as

$$y = ax^2$$

If we substitute $y = y + \Delta y$ and $x = x + \Delta x$ here, we find

$$y + \Delta y = a(x + \Delta x)^2$$
$$= ax^2 + 2ax\,\Delta x + (\Delta x)^2$$

Subtracting as before leaves

$$\Delta y = 2ax\,\Delta x + (\Delta x)^2$$

Here we have a more complicated, nonlinear relationship between $\Delta y$ and $\Delta x$. If $\Delta x$ is small enough, however, the term $(\Delta x)^2$ will be small relative to the term proportional to $\Delta x$, and we may write

$$\Delta y \approx 2ax\,\Delta x \qquad (\Delta x \text{ small})$$

$$\frac{\Delta y}{\Delta x} \approx 2ax \qquad (\Delta x \text{ small})$$

How do we indicate this graphically? The graph of $y$ against $x$ is no longer a straight line, and so we need to generalize the concept of slope. We define a **tangent line** at the point $x_0$ as the line that touches the graph of $f(x)$ at $x = x_0$ without crossing it.[1] If $\Delta y$ and $\Delta x$ are small (see Fig. C-2), the slope of the tangent line is

$$\frac{\Delta y}{\Delta x} = 2ax_0 \qquad \text{at} \qquad x = x_0$$

---

[1] This is an intuitive, rather than a rigorously mathematical, definition of a tangent line.

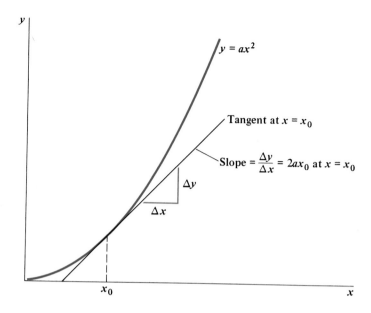

**Figure C-2**
The slope of a curve at a point.

for the example just discussed. Clearly the slope is not constant but changes with $x_0$. If we define the slope at each point on the curve by the slope of the corresponding tangent line, we obtain a *new* function that gives the slope of the curve $f(x)$ at each point $x$. We call this new function the **derivative** of $f(x)$ and represent it with $df/dx$. We have already calculated the derivatives of two functions:

$$f(x) = mx + b \implies \frac{df}{dx} = m$$

$$f(x) = ax^2 \implies \frac{df}{dx} = 2ax$$

Table C-1 shows derivatives of several other functions that are important in basic chemistry.

| Table C-1   Derivatives of Simple Functions | |
| --- | --- |
| **Function $f(x)$** | **Derivative $\dfrac{df}{dx}$** |
| $mx + b$ | $m$ |
| $ax^2$ | $2ax$ |
| $\dfrac{a}{x} = ax^{-1}$ | $-\dfrac{a}{x^2} = -ax^{-2}$ |
| $ax^n$ | $nax^{n-1}$ |
| $e^{ax}$ | $ae^{ax}$ |
| $\ln ax$ | $\dfrac{1}{x}$ |
| $\sin ax$ | $a \cos ax$ |
| $\cos ax$ | $-a \sin ax$ |

The constants $a$, $b$, $m$, and $n$ may be positive or negative.

The derivative gives the slope of the tangent line to $f(x)$ at each point $x$ and can be used to approximate the response to a small perturbation $\Delta x$ in the independent variable $x$:

$$\Delta y \approx \frac{df}{dx}\Delta x$$

As an example, consider 1.00 mol of an ideal gas at 0°C that obeys the law

$$P\text{ (atm)} = \frac{22.414 \text{ L atm}}{V \text{ (L)}}$$

Suppose $V$ is 20.00 L. If the volume $V$ is changed by a small amount $\Delta V$ at fixed $T$ and number of moles, what will be the corresponding change $\Delta P$ in the pressure? We write

$$P = f(V) = \frac{22.414}{V} = \frac{a}{V}$$

$$\frac{dP}{dV} = \frac{df}{dV} = -\frac{a}{V^2} = -\frac{22.414}{V^2} \qquad \text{(from Table C-1)}$$

$$\Delta P \approx \frac{dP}{dV}\Delta V = -\frac{22.414}{V^2}(\Delta V)$$

If $V = 20.00$ L, then

$$\Delta P \approx -\frac{22.414 \text{ L atm}}{(20.00 \text{ L})^2}\Delta V = -(0.0560 \text{ atm L}^{-1})\,\Delta V$$

## C-5 AREAS UNDER CURVES AND INTEGRALS

Another mathematical operation that arises frequently in science is the calculation of the area under a curve. Some areas are those of simple geometric shapes and are easy to calculate. If the function $f(x)$ is a constant,

$$f(x) = a$$

then the area under a graph of $f(x)$ between the two points $x_1$ and $x_2$ is that of a rectangle (Fig. C-3a) and is easily calculated as

$$\text{area} = \text{height} \times \text{base} = a(x_2 - x_1) = [ax_2] - [ax_1]$$

If $f(x)$ is a straight line that is neither horizontal nor vertical,

$$f(x) = mx + b$$

then the area is that of a trapezoid (Fig. C-3b) and is equal to

$$\text{area} = \text{average height} \times \text{base}$$
$$= \tfrac{1}{2}(mx_1 + b + mx_2 + b)(x_2 - x_1)$$
$$= [\tfrac{1}{2}mx_2^2 + bx_2] - [\tfrac{1}{2}mx_1^2 + bx_1]$$

Suppose now that $f(x)$ is a more complicated function, such as that shown in Figure C-3c. We can estimate the area under this graph by approximating it with a series of small rectangles of width $\Delta x$ and varying heights. If the height of the $i$th

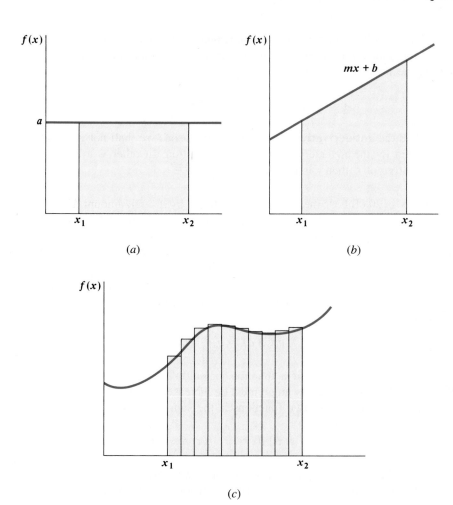

**Figure C-3**
The integral as the area under a curve.

rectangle is $y_i = f(x_i)$, then we have, approximately,

$$\text{area} \approx f(x_1)\,\Delta x + f(x_2)\,\Delta x + \cdots$$

$$= \sum_i f(x_i)\,\Delta x$$

As the widths of the rectangles $\Delta x$ become small, this becomes a better approximation. We *define* the area under a curve $f(x)$ between two points as the limiting value of this sum as $\Delta x$ approaches 0, called the **definite integral:**

$$\text{area} = \int_{x_1}^{x_2} f(x)\,dx$$

We have already worked out two examples of such integrals:

$$f(x) = a \implies \int_{x_1}^{x_2} a\,dx = [ax_2] - [ax_1]$$

$$f(x) = ax + b \implies \int_{x_1}^{x_2} (ax + b)\,dx = [\tfrac{1}{2}ax_2^2 + bx_2] - [\tfrac{1}{2}ax_1^2 + bx_1]$$

These integrals have an interesting form, which can be generalized to other cases as well: some function $F$ evaluated at the upper limit $(x_2)$ minus that function $F$ evaluated at the lower limit $(x_1)$.

$$\int_{x_1}^{x_2} f(x)\, dx = F(x_2) - F(x_1)$$

$F$ is called the **antiderivative** of $f$ because, for reasons we shall not go into here, it is obtained by the inverse of a derivative operation. In other words, if $F$ is the antiderivative of $f$, then $f$ is the derivative of $F$:

$$\frac{dF}{dx} = f(x)$$

To calculate integrals, therefore, we invert the results of Table C-1, because we need to find those functions $F(x)$ whose derivatives are equal to $f(x)$. Table C-2 lists several of the most important integrals for basic chemistry.

Several additional mathematical properties of integrals are important at this point. If a function is multiplied by a constant $c$, then the integral is multiplied by the same constant:

$$\int_{x_1}^{x_2} cf(x)\, dx = c \int_{x_1}^{x_2} f(x)\, dx$$

The reason is self-evident: if a function is multiplied everywhere by a constant factor, then the area under its graph must be increased by the same factor. Second, the integral of the sum of two functions is the sum of the separate integrals:

$$\int_{x_1}^{x_2} [\,f(x) + g(x)]\, dx = \int_{x_1}^{x_2} f(x)\, dx + \int_{x_1}^{x_2} g(x)\, dx$$

Finally, if the upper and lower limits of integration are reversed, the sign of the integral changes as well. This is easily seen from the antiderivative form:

$$\int_{x_2}^{x_1} f(x)\, dx = F(x_1) - F(x_2) = -[F(x_2) - F(x_1)] = -\int_{x_1}^{x_2} f(x)\, dx$$

### Table C-2   Integrals of Simple Functions

| Function $f(x)$ | Antiderivative $F(x)$ | Integral $\displaystyle\int_{x_1}^{x_2} f(x)\, dx$ |
|---|---|---|
| $a$ | $ax$ | $ax_2 - ax_1$ |
| $mx + b$ | $\frac{1}{2} mx^2 + bx$ | $[\frac{1}{2} mx_2^2 + bx_2] - [\frac{1}{2} mx_1^2 - bx_1]$ |
| $x^n \ (n \neq -1)$ | $\dfrac{1}{n+1} x^{n+1}$ | $\dfrac{1}{n+1}(x_2^{n+1} - x_1^{n+1})$ |
| $\dfrac{1}{x} = x^{-1} \ (x > 0)$ | $\ln x$ | $(\ln x_2 - \ln x_1) = \ln \dfrac{x_2}{x_1}$ |
| $\dfrac{1}{x^2} = x^{-2}$ | $-\dfrac{1}{x}$ | $-\left(\dfrac{1}{x_2} - \dfrac{1}{x_1}\right)$ |
| $e^{ax}$ | $\dfrac{e^{ax}}{a}$ | $\dfrac{1}{a}(e^{ax_2} - e^{ax_1})$ |

The constants $a$, $b$, $m$, and $n$ may be positive or negative.

## Using Graphs

1. A particular plot of distance traveled against time elapsed is found to be a straight line. After an elapsed time of 1.5 hours, the distance traveled was 75 miles, and after an elapsed time of 3.0 hours, the distance traveled was 150 miles. Calculate the slope of the plot of distance against time, and give its units.

2. The pressure of a gas in a rigid container is measured at several different temperatures, and it is found that a plot of pressure against temperature is a straight line. At 20.0°C the pressure is 4.30 atm, and at 100.0°C the pressure is 5.47 atm. Calculate the slope of the plot of pressure against temperature and give its units.

3. Rewrite each of the following linear equations in the form $y = mx + b$, and give the slope and intercept of the corresponding plot. Then draw the graph.
   (a) $y = 4x - 7$
   (b) $7x - 2y = 5$
   (c) $3y + 6x - 4 = 0$

4. Rewrite each of the following linear equations in the form $y = mx + b$, and give the slope and intercept of the corresponding plot. Then draw the graph.
   (a) $y = -2x - 8$
   (b) $-3x + 4y = 7$
   (c) $7y - 16x + 53 = 0$

5. Graph the relation
$$y = 2x^3 - 3x^2 + 6x - 5$$
   from $y = -2$ to $y = +2$. Is the plotted curve a straight line?

6. Graph the relation
$$y = \frac{8 - 10x - 3x^2}{2 - 3x}$$
   from $x = -3$ to $x = +3$. Is the plotted curve a straight line?

## Solution of Algebraic Equations

7. Solve the following linear equations for $x$.
   (a) $7x + 5 = 0$
   (b) $-4x + 3 = 0$
   (c) $-3x = -2$

8. Solve the following linear equations for $x$.
   (a) $6 - 8x = 0$
   (b) $-2x - 5 = 0$
   (c) $4x = -8$

9. Solve the following quadratic equations for $x$.
   (a) $4x^2 + 7x - 5 = 0$
   (b) $2x^2 = -3 - 6x$
   (c) $2x + \dfrac{3}{x} = 6$

10. Solve the following quadratic equations for $x$.
    (a) $6x^2 + 15x + 2 = 0$
    (b) $4x = 5x^2 - 3$
    (c) $\dfrac{1}{2 - x} + 3x = 4$

11. Solve each of the following equations for $x$ using the approximation of small $x$, iteration, or graphical solution, as appropriate.
    (a) $x(2.00 + x)^2 = 2.6 \times 10^{-6}$
    (b) $x(3.00 - 7x)(2.00 + 2x) = 0.230$
    (c) $2x^3 + 3x^2 + 12x = -16$

12. Solve each of the following equations for $x$ using the approximation of small $x$, iteration, or graphical solution, as appropriate.
    (a) $x(2.00 + x)(3.00 - x)(5.00 + 2x) = 1.58 \times 10^{-15}$
    (b) $x\dfrac{(3.00 + x)(1.00 - x)}{2.00 - x} = 0.122$
    (c) $12x^3 - 4x^2 + 35x = 10$

## Powers and Logarithms

13. Calculate each of the following expressions, giving your answers the proper numbers of significant figures.
    (a) $\log (3.56 \times 10^4)$
    (b) $e^{-15.69}$
    (c) $10^{8.41}$
    (d) $\ln (6.893 \times 10^{-22})$

14. Calculate each of the following expressions, giving your answers the proper numbers of significant figures.
    (a) $10^{-16.528}$
    (b) $\ln (4.30 \times 10^{13})$
    (c) $e^{14.21}$
    (d) $\log (4.983 \times 10^{-11})$

15. What number has a common logarithm of 0.4793?

16. What number has a natural logarithm of $-15.824$?

17. Determine the common logarithm of $3.00 \times 10^{121}$. It is quite likely that your calculator will *not* give a correct answer. Explain why.

18. Compute the value of $10^{-107.8}$. It is quite likely that your calculator will *not* give the correct answer. Explain why.

19. The common logarithm of 5.64 is 0.751. Without using a calculator, determine the common logarithm of $5.64 \times 10^7$ and of $5.64 \times 10^{-3}$.

20. The common logarithm of 2.68 is 0.428. Without using a calculator, determine the common logarithm of $2.68 \times 10^{192}$ and of $2.68 \times 10^{-289}$.

21. Use the graphical method and a calculator to solve the equation
$$\log \ln x = -x$$
   for $x$. Give $x$ to four figures.

**22.** Use a calculator to find a number that is equal to the reciprocal of its own natural logarithm. Report the answer to four digits.

## Slopes of Curves and Derivatives

**23.** Calculate the derivatives of the following functions.
(a) $y = 4x^2 + 4$
(b) $y = \sin 3x + 4 \cos 2x$
(c) $y = 3x + 2$
(d) $y = \ln 7x$

**24.** Calculate the derivatives of the following functions.
(a) $y = 6x^{19}$
(b) $y = 7x^2 + 6x + 2$
(c) $y = e^{-6x}$
(d) $y = \cos 2x + \dfrac{7}{x}$

## Areas Under Curves and Integrals

**25.** Calculate the following integrals.
(a) $\int_2^4 (3x + 1)\, dx$
(b) $\int_0^5 x^6\, dx$
(c) $\int_1^4 e^{-2x}\, dx$

**26.** Calculate the following integrals.
(a) $\int_{-1}^3 4\, dx$
(b) $\int_2^{100} \dfrac{1}{x}\, dx$
(c) $\int_2^4 \dfrac{5}{x^2}\, dx$

# Standard Chemical
# Thermodynamic Properties

T his table lists standard enthalpies of formation $\Delta H_f^\circ$, standard third-law entropies $S^\circ$, standard free energies of formation $\Delta G_f^\circ$, and molar heat capacities at constant pressure, $C_P$, for a variety of substances, all at 25°C (298.15 K) and 1 atm. The table proceeds from the left side to the right side of the periodic table. Binary compounds are listed under the element that occurs to the left in the periodic table, except that binary oxides and hydrides are listed with the other element. Thus, KCl is listed with potassium and its compounds, but $ClO_2$ is listed with chlorine and its compounds.

Note that the *solution-phase* entropies are not absolute entropies but are measured relative to the arbitrary standard $S^\circ(H^+(aq)) = 0$. Consequently, some of them (as well as some of the heat capacities) are negative.

Most of the thermodynamic data in these tables were taken from the *NBS Tables of Chemical Thermodynamic Properties* (1982) and changed, where necessary, from a standard pressure of 0.1 MPa to 1 atm. The data for organic compounds $C_nH_m(n > 2)$ were taken from the *Handbook of Chemistry and Physics* (1981).

| Substance | $\Delta H_f^\circ$ (25°C) kJ mol$^{-1}$ | $S^\circ$ (25°C) J K$^{-1}$ mol$^{-1}$ | $\Delta G_f^\circ$ (25°C) kJ mol$^{-1}$ | $C_P$ (25°C) J K$^{-1}$ mol$^{-1}$ |
|---|---|---|---|---|
| H($g$) | 217.96 | 114.60 | 203.26 | 20.78 |
| H$_2$($g$) | 0 | 130.57 | 0 | 28.82 |
| H$^+$($aq$) | 0 | 0 | 0 | 0 |
| H$_3$O$^+$($aq$) | −285.83 | 69.91 | −237.18 | 75.29 |
| Li($s$) | 0 | 29.12 | 0 | 24.77 |
| Li($g$) | 159.37 | 138.66 | 126.69 | 20.79 |
| Li$^+$($aq$) | −278.49 | 13.4 | −293.31 | 68.6 |
| LiH($s$) | −90.54 | 20.01 | −68.37 | 27.87 |
| Li$_2$O($s$) | −597.94 | 37.57 | −561.20 | 54.10 |
| LiF($s$) | −615.97 | 35.65 | −587.73 | 41.59 |
| LiCl($s$) | −408.61 | 59.33 | −384.39 | 47.99 |
| LiBr($s$) | −351.21 | 74.27 | −342.00 | — |
| LiI($s$) | −270.41 | 86.78 | −270.29 | 51.04 |
| Na($s$) | 0 | 51.21 | 0 | 28.24 |
| Na($g$) | 107.32 | 153.60 | 76.79 | 20.79 |
| Na$^+$($aq$) | −240.12 | 59.0 | −261.90 | 46.4 |
| Na$_2$O($s$) | −414.22 | 75.06 | −375.48 | 69.12 |
| NaOH($s$) | −425.61 | 64.46 | −379.53 | 59.54 |
| NaF($s$) | −573.65 | 51.46 | −543.51 | 46.86 |

*(Continued)*

| Substance | $\Delta H_f^\circ$ (25°C) kJ mol$^{-1}$ | $S^\circ$ (25°C) J K$^{-1}$ mol$^{-1}$ | $\Delta G_f^\circ$ (25°C) kJ mol$^{-1}$ | $C_P$ (25°C) J K$^{-1}$ mol$^{-1}$ |
|---|---|---|---|---|
| NaCl(s) | −411.15 | 72.13 | −384.15 | 50.50 |
| NaBr(s) | −361.06 | 86.82 | −348.98 | 51.38 |
| NaI(s) | −287.78 | 98.53 | −286.06 | 52.09 |
| NaNO$_3$(s) | −467.85 | 116.52 | −367.07 | 92.88 |
| Na$_2$S(s) | −364.8 | 83.7 | −349.8 | — |
| Na$_2$SO$_4$(s) | −1387.08 | 149.58 | −1270.23 | 128.20 |
| NaHSO$_4$(s) | −1125.5 | 113.0 | −992.9 | — |
| Na$_2$CO$_3$(s) | −1130.68 | 134.98 | −1044.49 | 112.30 |
| NaHCO$_3$(s) | −950.81 | 101.7 | −851.1 | 87.61 |
| K(s) | 0 | 64.18 | 0 | 29.58 |
| K(g) | 89.24 | 160.23 | 60.62 | 20.79 |
| K$^+$(aq) | −252.38 | 102.5 | −283.27 | 21.8 |
| KO$_2$(s) | −284.93 | 116.7 | −239.4 | 77.53 |
| K$_2$O$_2$(s) | −494.1 | 102.1 | −425.1 | — |
| KOH(s) | −424.76 | 78.9 | −379.11 | 64.9 |
| KF(s) | −567.27 | 66.57 | −537.77 | 49.04 |
| KCl(s) | −436.75 | 82.59 | −409.16 | 51.30 |
| KClO$_3$(s) | −397.73 | 143.1 | −296.25 | 100.25 |
| KBr(s) | −393.80 | 95.90 | −380.66 | 52.30 |
| KI(s) | −327.90 | 106.32 | −324.89 | 52.93 |
| KMnO$_4$(s) | −837.2 | 171.71 | −737.7 | 117.57 |
| K$_2$CrO$_4$(s) | −1403.7 | 200.12 | −1295.8 | 145.98 |
| K$_2$Cr$_2$O$_7$(s) | −2061.5 | 291.2 | −1881.9 | 219.24 |
| Rb(s) | 0 | 76.78 | 0 | 31.06 |
| Rb(g) | 80.88 | 169.98 | 53.09 | 20.79 |
| Rb$^+$(aq) | −251.17 | 121.50 | −283.98 | — |
| RbCl(s) | −435.35 | 95.90 | −407.82 | 52.38 |
| RbBr(s) | −394.59 | 109.96 | −381.79 | 52.84 |
| RbI(s) | −333.80 | 118.41 | −328.86 | 53.18 |
| Cs(s) | 0 | 85.23 | 0 | 32.17 |
| Cs(g) | 76.06 | 175.49 | 49.15 | 70.79 |
| Cs$^+$(aq) | −258.28 | 133.05 | −292.02 | −10.5 |
| CsF(s) | −553.5 | 92.80 | −525.5 | 51.09 |
| CsCl(s) | −443.04 | 101.17 | −414.55 | 52.47 |
| CsBr(s) | −405.81 | 113.05 | −391.41 | 52.93 |
| CsI(s) | −346.60 | 123.05 | −340.58 | 52.80 |
| II    Be(s) | 0 | 9.50 | 0 | 16.44 |
| Be(g) | 324.3 | 136.16 | 286.6 | 20.79 |
| BeO(s) | −609.6 | 14.14 | −580.3 | 25.52 |
| Mg(s) | 0 | 32.68 | 0 | 24.89 |
| Mg(g) | 147.70 | 148.54 | 113.13 | 20.79 |
| Mg$^{2+}$(aq) | −466.85 | −138.1 | −454.8 | — |
| MgO(s) | −601.70 | 26.94 | −569.45 | 37.15 |
| MgCl$_2$(s) | −641.32 | 89.62 | −591.82 | 71.38 |
| MgSO$_4$(s) | −1284.9 | 91.6 | −1170.7 | 96.48 |
| Ca(s) | 0 | 41.42 | 0 | 25.31 |
| Ca(g) | 178.2 | 154.77 | 144.33 | 20.79 |
| Ca$^{2+}$(aq) | −542.83 | −53.1 | −553.58 | — |

*(Continued)*

| Substance | $\Delta H_f^\circ$ (25°C) kJ mol$^{-1}$ | $S^\circ$ (25°C) J K$^{-1}$ mol$^{-1}$ | $\Delta G_f^\circ$ (25°C) kJ mol$^{-1}$ | $C_P$ (25°C) J K$^{-1}$ mol$^{-1}$ |
|---|---|---|---|---|
| CaH$_2$(s) | −186.2 | 42 | −147.2 | — |
| CaO(s) | −635.09 | 39.75 | −604.05 | 42.80 |
| CaS(s) | −482.4 | 56.5 | −477.4 | 47.40 |
| Ca(OH)$_2$(s) | −986.09 | 83.39 | −898.56 | 87.49 |
| CaF$_2$(s) | −1219.6 | 68.87 | −1167.3 | 67.03 |
| CaCl$_2$(s) | −795.8 | 104.6 | −748.1 | 72.59 |
| CaBr$_2$(s) | −682.8 | 130 | −663.6 | — |
| CaI$_2$(s) | −533.5 | 142 | −528.9 | — |
| Ca(NO$_3$)$_2$(s) | −938.39 | 193.3 | −743.20 | 149.37 |
| CaC$_2$(s) | −59.8 | 69.96 | −64.9 | 62.72 |
| CaCO$_3$(s, calcite) | −1206.92 | 92.9 | −1128.84 | 81.88 |
| CaCO$_3$(s, aragonite) | −1207.13 | 88.7 | −1127.80 | 81.25 |
| CaSO$_4$(s) | −1434.11 | 106.9 | −1321.86 | 99.66 |
| CaSiO$_3$(s) | −1634.94 | 81.92 | −1549.66 | 85.27 |
| CaMg(CO$_3$)$_2$(s, dolomite) | −2326.3 | 155.18 | −2163.4 | 157.53 |
| Sr(s) | 0 | 52.3 | 0 | 26.4 |
| Sr(g) | 164.4 | 164.51 | 130.9 | 20.79 |
| Sr$^{2+}$(aq) | −545.80 | −32.6 | −559.48 | — |
| SrCl$_2$(s) | −828.9 | 114.85 | −781.1 | 75.60 |
| SrCO$_3$(s) | −1220.0 | 97.1 | −1140.1 | 81.42 |
| Ba(s) | 0 | 62.8 | 0 | 28.07 |
| Ba(g) | 180 | 170.24 | 146 | 20.79 |
| Ba$^{2+}$(aq) | −537.64 | 9.6 | −560.77 | — |
| BaCl$_2$(s) | −858.6 | 123.68 | −810.4 | 75.14 |
| BaCO$_3$(s) | −1216.3 | 112.1 | −1137.6 | 85.35 |
| BaSO$_4$(s) | −1473.2 | 132.2 | −1362.3 | 101.75 |
| Sc(s) | 0 | 34.64 | 0 | 25.52 |
| Sc(g) | 377.8 | 174.68 | 336.06 | 22.09 |
| Sc$^{3+}$(aq) | −614.2 | −255 | −586.6 | — |
| Ti(s) | 0 | 30.63 | 0 | 25.02 |
| Ti(g) | 469.9 | 180.19 | 425.1 | 24.43 |
| TiO$_2$(s, rutile) | −944.7 | 50.33 | −889.5 | 55.02 |
| TiCl$_4$($\ell$) | −804.2 | 252.3 | −737.2 | 145.18 |
| TiCl$_4$(g) | −763.2 | 354.8 | −726.8 | 95.4 |
| Cr(s) | 0 | 23.77 | 0 | 23.35 |
| Cr(g) | 396.6 | 174.39 | 351.8 | 20.79 |
| Cr$_2$O$_3$(s) | −1139.7 | 81.2 | −1058.1 | 118.74 |
| CrO$_4^{2-}$(aq) | −881.15 | 50.21 | −727.75 | — |
| Cr$_2$O$_7^{2-}$(aq) | −1490.3 | 261.9 | −1301.1 | — |
| W(s) | 0 | 32.64 | 0 | 24.27 |
| W(g) | 849.4 | 173.84 | 807.1 | 21.31 |
| WO$_2$(s) | −589.69 | 50.54 | −533.92 | 56.11 |
| WO$_3$(s) | −842.87 | 75.90 | −764.08 | 73.76 |
| Mn(s) | 0 | 32.01 | 0 | 26.32 |
| Mn(g) | 280.7 | 238.5 | 173.59 | 20.79 |
| Mn$^{2+}$(aq) | −220.75 | −73.6 | −228.1 | 50 |

*(Continued)*

| Substance | $\Delta H_f^\circ$ (25°C) kJ mol$^{-1}$ | $S^\circ$ (25°C) J K$^{-1}$ mol$^{-1}$ | $\Delta G_f^\circ$ (25°C) kJ mol$^{-1}$ | $C_P$ (25°C) J K$^{-1}$ mol$^{-1}$ |
|---|---|---|---|---|
| MnO($s$) | −385.22 | 59.71 | −362.92 | 45.44 |
| MnO$_2$($s$) | −520.03 | 53.05 | −465.17 | 54.14 |
| MnO$_4^-$($aq$) | −541.4 | 191.2 | −447.2 | −82.0 |
| Fe($s$) | 0 | 27.28 | 0 | 25.10 |
| Fe($g$) | 416.3 | 180.38 | 370.7 | 25.68 |
| Fe$^{2+}$($aq$) | −89.1 | −137.7 | −78.9 | — |
| Fe$^{3+}$($aq$) | −48.5 | −315.9 | −4.7 | — |
| Fe$_{0.947}$O($s$, wüstite) | −266.27 | 57.49 | −245.12 | 48.12 |
| Fe$_2$O$_3$($s$, hematite) | −824.2 | 87.40 | −742.2 | 103.85 |
| Fe$_3$O$_4$($s$, magnetite) | −1118.4 | 146.4 | −1015.5 | 143.43 |
| Fe(OH)$_3$($s$) | −823.0 | 106.7 | −696.5 | — |
| FeS($s$) | −100.0 | 60.29 | −100.4 | 50.54 |
| FeCO$_3$($s$) | −740.57 | 93.1 | −666.72 | 82.13 |
| Fe(CN)$_6^{3-}$($aq$) | 561.9 | 270.3 | 729.4 | — |
| Fe(CN)$_6^{4-}$($aq$) | 455.6 | 95.0 | 695.1 | — |
| Co($s$) | 0 | 30.04 | 0 | 24.81 |
| Co($g$) | 424.7 | 179.41 | 380.3 | 23.02 |
| Co$^{2+}$($aq$) | −58.2 | −113 | −54.4 | — |
| Co$^{3+}$($aq$) | 92 | −305 | 134 | — |
| CoO($s$) | −237.94 | 52.97 | −214.22 | 55.23 |
| CoCl$_2$($s$) | −312.5 | 109.16 | −269.8 | 78.49 |
| Ni($s$) | 0 | 29.87 | 0 | 26.07 |
| Ni($g$) | 429.7 | 182.08 | 384.5 | 25.36 |
| Ni$^{2+}$($aq$) | −54.0 | −128.9 | −45.6 | — |
| NiO($s$) | −239.7 | 37.99 | −211.7 | 44.31 |
| Pt($s$) | 0 | 41.63 | 0 | 25.86 |
| Pt($g$) | 565.3 | 192.30 | 520.5 | 25.53 |
| PtCl$_6^{2-}$($aq$) | −668.2 | 219.7 | −482.7 | — |
| Cu($s$) | 0 | 33.15 | 0 | 24.44 |
| Cu($g$) | 338.32 | 166.27 | 298.61 | 20.79 |
| Cu$^+$($aq$) | 71.67 | 40.6 | 49.98 | — |
| Cu$^{2+}$($aq$) | 64.77 | −99.6 | 65.49 | — |
| CuO($s$) | −157.3 | 42.63 | −129.7 | 42.30 |
| Cu$_2$O($s$) | −168.6 | 93.14 | −146.0 | 63.64 |
| CuCl($s$) | −137.2 | 86.2 | −119.88 | 48.5 |
| CuCl$_2$($s$) | −220.1 | 108.07 | −175.7 | 71.88 |
| CuSO$_4$($s$) | −771.36 | 109 | −661.9 | 100.0 |
| Cu(NH$_3$)$_4^{2+}$($aq$) | −348.5 | 273.6 | −111.07 | — |
| Ag($s$) | 0 | 42.55 | 0 | 25.35 |
| Ag($g$) | 284.55 | 172.89 | 245.68 | 20.79 |
| Ag$^+$($aq$) | 105.58 | 72.68 | 77.11 | 21.8 |
| AgCl($s$) | −127.07 | 96.2 | −109.81 | 50.79 |
| AgNO$_3$($s$) | −124.39 | 140.92 | −33.48 | 93.05 |
| Ag(NH$_3$)$_2^+$($aq$) | −111.29 | 245.2 | −17.12 | — |
| Au($s$) | 0 | 47.40 | 0 | 25.42 |
| Au($g$) | 366.1 | 180.39 | 326.3 | 20.79 |
| Zn($s$) | 0 | 41.63 | 0 | 25.40 |

*(Continued)*

| Substance | $\Delta H_f^\circ$ (25°C) kJ mol$^{-1}$ | $S^\circ$ (25°C) J K$^{-1}$ mol$^{-1}$ | $\Delta G_f^\circ$ (25°C) kJ mol$^{-1}$ | $C_P$ (25°C) J K$^{-1}$ mol$^{-1}$ |
|---|---|---|---|---|
| $Zn(g)$ | 130.73 | 160.87 | 95.18 | 20.79 |
| $Zn^{2+}(aq)$ | −153.89 | −112.1 | −147.06 | 46 |
| $ZnO(s)$ | −348.28 | 43.64 | −318.32 | 40.25 |
| $ZnS(s,\ sphalerite)$ | −205.98 | 57.7 | −201.29 | 46.0 |
| $ZnCl_2(s)$ | −415.05 | 111.46 | −369.43 | 71.34 |
| $ZnSO_4(s)$ | −982.8 | 110.5 | −871.5 | 99.2 |
| $Zn(NH_3)_4^{2+}(aq)$ | −533.5 | 301 | −301.9 | — |
| $Hg(\ell)$ | 0 | 76.02 | 0 | 27.98 |
| $Hg(g)$ | 61.32 | 174.85 | 31.85 | 20.79 |
| $HgO(s)$ | −90.83 | 70.29 | −58.56 | 44.06 |
| $HgCl_2(s)$ | −224.3 | 146.0 | −178.6 | — |
| $Hg_2Cl_2(s)$ | −265.22 | 192.5 | −210.78 | — |

| | Substance | $\Delta H_f^\circ$ (25°C) kJ mol$^{-1}$ | $S^\circ$ (25°C) J K$^{-1}$ mol$^{-1}$ | $\Delta G_f^\circ$ (25°C) kJ mol$^{-1}$ | $C_P$ (25°C) J K$^{-1}$ mol$^{-1}$ |
|---|---|---|---|---|---|
| III | $B(s)$ | 0 | 5.86 | 0 | 11.09 |
| | $B(g)$ | 562.7 | 153.34 | 518.8 | 20.80 |
| | $B_2H_6(g)$ | 35.6 | 232.00 | 86.6 | 56.90 |
| | $B_5H_9(g)$ | 73.2 | 275.81 | 174.9 | 96.78 |
| | $B_2O_3(s)$ | −1272.77 | 53.97 | −1193.70 | 62.93 |
| | $H_3BO_3(s)$ | −1094.33 | 88.83 | −969.02 | 81.38 |
| | $BF_3(g)$ | −1137.00 | 254.01 | −1120.35 | 50.46 |
| | $BF_4^-(aq)$ | −1574.9 | 180 | −1486.9 | — |
| | $BCl_3(g)$ | −403.76 | 289.99 | −388.74 | 62.72 |
| | $BBr_3(g)$ | −205.64 | 324.13 | −232.47 | 67.78 |
| | $Al(s)$ | 0 | 28.33 | 0 | 24.35 |
| | $Al(g)$ | 326.4 | 164.43 | 285.7 | 21.38 |
| | $Al^{3+}(aq)$ | −531 | −321.7 | −485 | — |
| | $Al_2O_3(s)$ | −1675.7 | 50.92 | −1582.3 | 79.04 |
| | $AlCl_3(s)$ | −704.2 | 110.67 | −628.8 | 91.84 |
| | $Ga(s)$ | 0 | 40.88 | 0 | 25.86 |
| | $Ga(g)$ | 277.0 | 168.95 | 238.9 | 25.36 |
| | $Tl(s)$ | 0 | 64.18 | 0 | 26.32 |
| | $Tl(g)$ | 182.21 | 180.85 | 147.44 | 20.79 |

| | Substance | $\Delta H_f^\circ$ (25°C) kJ mol$^{-1}$ | $S^\circ$ (25°C) J K$^{-1}$ mol$^{-1}$ | $\Delta G_f^\circ$ (25°C) kJ mol$^{-1}$ | $C_P$ (25°C) J K$^{-1}$ mol$^{-1}$ |
|---|---|---|---|---|---|
| IV | $C(s,\ graphite)$ | 0 | 5.74 | 0 | 8.53 |
| | $C(s,\ diamond)$ | 1.895 | 2.377 | 2.900 | 6.11 |
| | $C(g)$ | 716.682 | 157.99 | 671.29 | 20.84 |
| | $CH_4(g)$ | −74.81 | 186.15 | −50.75 | 35.31 |
| | $C_2H_2(g)$ | 226.73 | 200.83 | 209.20 | 43.93 |
| | $C_2H_4(g)$ | 52.26 | 219.45 | 68.12 | 43.56 |
| | $C_2H_6(g)$ | −84.68 | 229.49 | −32.89 | 52.63 |
| | $C_3H_8(g)$ | −103.85 | 269.91 | −23.49 | 73.0 |
| | $n\text{-}C_4H_{10}(g)$ | −124.73 | 310.03 | −15.71 | 97.5 |
| | $C_4H_{10}(g,\ isobutane)$ | −131.60 | 294.64 | −17.97 | 96.8 |
| | $n\text{-}C_5H_{12}(g)$ | −146.44 | 348.40 | −8.20 | 120 |
| | $C_6H_6(g)$ | 82.93 | 269.2 | 129.66 | 81.6 |
| | $C_6H_6(\ell)$ | 49.03 | 172.8 | 124.50 | 136 |
| | $CO(g)$ | −110.52 | 197.56 | −137.15 | 29.14 |
| | $CO_2(g)$ | −393.51 | 213.63 | −394.36 | 37.11 |
| | $CO_2(aq)$ | −413.80 | 117.6 | −385.98 | — |

*(Continued)*

| Substance | $\Delta H_f^\circ$ (25°C) kJ mol$^{-1}$ | $S^\circ$ (25°C) J K$^{-1}$ mol$^{-1}$ | $\Delta G_f^\circ$ (25°C) kJ mol$^{-1}$ | $C_P$ (25°C) J K$^{-1}$ mol$^{-1}$ |
|---|---|---|---|---|
| $CS_2(\ell)$ | 89.70 | 151.34 | 65.27 | 75.7 |
| $CS_2(g)$ | 117.36 | 237.73 | 67.15 | 45.40 |
| $H_2CO_3(aq)$ | −699.65 | 187.4 | −623.08 | — |
| $HCO_3^-(aq)$ | −691.99 | 91.2 | −586.77 | — |
| $CO_3^{2-}(aq)$ | −677.14 | −56.9 | −527.81 | — |
| $HCOOH(\ell)$ | −424.72 | 128.95 | −361.42 | 99.04 |
| $HCOOH(aq)$ | −425.43 | 163 | −372.3 | — |
| $COOH^-(aq)$ | −425.55 | 92 | −351.0 | −87.9 |
| $CH_2O(g)$ | −108.57 | 218.66 | −102.55 | 35.40 |
| $CH_3OH(\ell)$ | −238.66 | 126.8 | −166.35 | 81.6 |
| $CH_3OH(g)$ | −200.66 | 239.70 | −162.01 | 43.89 |
| $CH_3OH(aq)$ | −245.93 | 133.1 | −175.31 | — |
| $H_2C_2O_4(s)$ | −827.2 | — | — | 117 |
| $HC_2O_4^-(aq)$ | −818.4 | 149.4 | −698.34 | — |
| $C_2O_4^{2-}(aq)$ | −825.1 | 45.6 | −673.9 | — |
| $CH_3COOH(\ell)$ | −484.5 | 159.8 | −390.0 | 124.3 |
| $CH_3COOH(g)$ | −432.25 | 282.4 | −374.1 | 66.5 |
| $CH_3COOH(aq)$ | −485.76 | 178.7 | −396.46 | — |
| $CH_3COO^-(aq)$ | −486.01 | 86.6 | −369.31 | −6.3 |
| $CH_3CHO(\ell)$ | −192.30 | 160.2 | −128.12 | — |
| $C_2H_5OH(\ell)$ | −277.69 | 160.7 | −174.89 | 111.46 |
| $C_2H_5OH(g)$ | −235.10 | 282.59 | −168.57 | 65.44 |
| $C_2H_5OH(aq)$ | −288.3 | 148.5 | −181.64 | — |
| $CH_3OCH_3(g)$ | −184.05 | 266.27 | −112.67 | 64.39 |
| $CF_4(g)$ | −925 | 261.50 | −879 | 61.09 |
| $CCl_4(\ell)$ | −135.44 | 216.40 | −65.28 | 131.75 |
| $CCl_4(g)$ | −102.9 | 309.74 | −60.62 | 83.30 |
| $CHCl_3(g)$ | −103.14 | 295.60 | −70.37 | 65.69 |
| $COCl_2(g)$ | −218.8 | 283.53 | −204.6 | 57.66 |
| $CH_2Cl_2(g)$ | −92.47 | 270.12 | −65.90 | 50.96 |
| $CH_3Cl(g)$ | −80.83 | 234.47 | −57.40 | 40.75 |
| $CBr_4(s)$ | 18.8 | 212.5 | 47.7 | 144.3 |
| $CH_3I(\ell)$ | −15.5 | 163.2 | 13.4 | 126 |
| $HCN(g)$ | 135.1 | 201.67 | 124.7 | 35.86 |
| $HCN(aq)$ | 107.1 | 124.7 | 119.7 | — |
| $CN^-(aq)$ | 150.6 | 94.1 | 172.4 | — |
| $CH_3NH_2(g)$ | −22.97 | 243.30 | 32.09 | 53.1 |
| $CO(NH_2)_2(s)$ | −333.51 | 104.49 | −197.44 | 93.14 |
| $Si(s)$ | 0 | 18.83 | 0 | 20.00 |
| $Si(g)$ | 455.6 | 167.86 | 411.3 | 22.25 |
| $SiC(s)$ | −65.3 | 16.61 | −62.8 | 26.86 |
| $SiO_2(s, \text{quartz})$ | −910.94 | 41.84 | −856.67 | 44.43 |
| $SiO_2(s, \text{cristobalite})$ | −909.48 | 42.68 | −855.43 | 44.18 |
| $Ge(s)$ | 0 | 31.09 | 0 | 23.35 |
| $Ge(g)$ | 376.6 | 335.9 | 167.79 | 30.73 |
| $Sn(s, \text{white})$ | 0 | 51.55 | 0 | 26.99 |
| $Sn(s, \text{gray})$ | −2.09 | 44.14 | 0.13 | 25.77 |
| $Sn(g)$ | 302.1 | 168.38 | 267.3 | 21.26 |
| $SnO(s)$ | −285.8 | 56.5 | −256.9 | 44.31 |

*(Continued)*

| Substance | $\Delta H_f^\circ$ (25°C) kJ mol$^{-1}$ | $S^\circ$ (25°C) J K$^{-1}$ mol$^{-1}$ | $\Delta G_f^\circ$ (25°C) kJ mol$^{-1}$ | $C_P$ (25°C) J K$^{-1}$ mol$^{-1}$ |
|---|---|---|---|---|
| $SnO_2(s)$ | −580.7 | 52.3 | −519.6 | 52.59 |
| $Sn(OH)_2(s)$ | −561.1 | 155 | −491.7 | — |
| $Pb(s)$ | 0 | 64.81 | 0 | 26.44 |
| $Pb(g)$ | 195.0 | 161.9 | 175.26 | 20.79 |
| $Pb^{2+}(aq)$ | −1.7 | 10.5 | −24.43 | — |
| $PbO(s,$ yellow$)$ | −217.32 | 68.70 | −187.91 | 45.77 |
| $PbO(s,$ red$)$ | −218.99 | 66.5 | −188.95 | 45.81 |
| $PbO_2(s)$ | −277.4 | 68.6 | −217.36 | 64.64 |
| $PbS(s)$ | −100.4 | 91.2 | −98.7 | 49.50 |
| $PbI_2(s)$ | −175.48 | 174.85 | −173.64 | 77.36 |
| $PbSO_4(s)$ | −919.94 | 148.57 | −813.21 | 103.21 |
| $N_2(g)$ | 0 | 191.50 | 0 | 29.12 |
| $N(g)$ | 472.70 | 153.19 | 455.58 | 20.79 |
| $NH_3(g)$ | −46.11 | 192.34 | −16.48 | 35.06 |
| $NH_3(aq)$ | −80.29 | 111.3 | −26.50 | — |
| $NH_4^+(aq)$ | −132.51 | 113.4 | −79.31 | 79.9 |
| $N_2H_4(\ell)$ | 50.63 | 121.21 | 149.24 | 98.87 |
| $N_2H_4(aq)$ | 34.31 | 138 | 128.1 | — |
| $NO(g)$ | 90.25 | 210.65 | 86.55 | 29.84 |
| $NO_2(g)$ | 33.18 | 239.95 | 51.29 | 37.20 |
| $NO_2^-(aq)$ | −104.6 | 123.0 | −32.2 | −97.5 |
| $NO_3^-(aq)$ | −205.0 | 146.4 | −108.74 | −86.6 |
| $N_2O(g)$ | 82.05 | 219.74 | 104.18 | 38.45 |
| $N_2O_4(g)$ | 9.16 | 304.18 | 97.82 | 77.28 |
| $N_2O_5(s)$ | −43.1 | 178.2 | 113.8 | 143.1 |
| $HNO_2(g)$ | −79.5 | 254.0 | −46.0 | 45.6 |
| $HNO_3(\ell)$ | −174.10 | 155.49 | −80.76 | 109.87 |
| $NH_4NO_3(s)$ | −365.56 | 151.08 | −184.02 | 139.3 |
| $NH_4Cl(s)$ | −314.43 | 94.6 | −202.97 | 84.1 |
| $(NH_4)_2SO_4(s)$ | −1180.85 | 220.1 | −901.90 | 187.49 |
| $P(s,$ white$)$ | 0 | 41.09 | 0 | 23.84 |
| $P(s,$ red$)$ | −17.6 | 22.80 | −12.1 | 21.21 |
| $P(g)$ | 314.64 | 163.08 | 278.28 | 20.79 |
| $P_2(g)$ | 144.3 | 218.02 | 103.7 | 32.05 |
| $P_4(g)$ | 58.91 | 279.87 | 24.47 | 67.15 |
| $PH_3(g)$ | 5.4 | 210.12 | 13.4 | 37.11 |
| $H_3PO_4(s)$ | −1279.0 | 110.50 | −1119.2 | 106.06 |
| $H_3PO_4(aq)$ | −1288.34 | 158.2 | −1142.54 | — |
| $H_2PO_4^-(aq)$ | −1296.29 | 90.4 | −1130.28 | — |
| $HPO_4^{2-}(aq)$ | −1292.14 | −33.5 | −1089.15 | — |
| $PO_4^{3-}(aq)$ | −1277.4 | −222 | −1018.7 | — |
| $PCl_3(g)$ | −287.0 | 311.67 | −267.8 | 71.84 |
| $PCl_5(g)$ | −374.9 | 364.47 | −305.0 | 112.80 |
| $As(s,$ gray$)$ | 0 | 35.1 | 0 | 24.64 |
| $As(g)$ | 302.5 | 174.10 | 261.0 | 20.79 |
| $As_2(g)$ | 222.2 | 239.3 | 171.9 | 35.00 |
| $As_4(g)$ | 143.9 | 314 | 92.4 | — |
| $AsH_3(g)$ | 66.44 | 222.67 | 68.91 | 38.07 |

*(Continued)*

| Substance | $\Delta H_f^\circ$ (25°C) kJ mol$^{-1}$ | $S^\circ$ (25°C) J K$^{-1}$ mol$^{-1}$ | $\Delta G_f^\circ$ (25°C) kJ mol$^{-1}$ | $C_P$ (25°C) J K$^{-1}$ mol$^{-1}$ |
|---|---|---|---|---|
| $As_4O_6(s)$ | −1313.94 | 214.2 | −1152.53 | 191.29 |
| $Sb(s)$ | 0 | 45.69 | 0 | 25.33 |
| $Sb(g)$ | 262.3 | 180.16 | 222.1 | 20.79 |
| $Bi(s)$ | 0 | 56.74 | 0 | 25.52 |
| $Bi(g)$ | 207.1 | 186.90 | 168.2 | 20.79 |

| VI | Substance | $\Delta H_f^\circ$ | $S^\circ$ | $\Delta G_f^\circ$ | $C_P$ |
|---|---|---|---|---|---|
| VI | $O_2(g)$ | 0 | 205.03 | 0 | 29.36 |
| | $O(g)$ | 249.17 | 160.95 | 231.76 | 21.91 |
| | $O_3(g)$ | 142.7 | 238.82 | 163.2 | 39.20 |
| | $OH^-(aq)$ | −229.99 | −10.75 | −157.24 | −148.5 |
| | $H_2O(\ell)$ | −285.83 | 69.91 | −237.18 | 75.29 |
| | $H_2O(g)$ | −241.82 | 188.72 | −228.59 | 35.58 |
| | $H_2O_2(\ell)$ | −187.78 | 109.6 | −120.42 | 89.1 |
| | $H_2O_2(aq)$ | −191.17 | 143.9 | −134.03 | — |
| | $S(s, \text{rhombic})$ | 0 | 31.80 | 0 | 22.64 |
| | $S(s, \text{monoclinic})$ | 0.30 | 32.6 | 0.096 | — |
| | $S(g)$ | 278.80 | 167.71 | 238.28 | 23.67 |
| | $S_8(g)$ | 102.30 | 430.87 | 49.66 | 156.44 |
| | $H_2S(g)$ | −20.63 | 205.68 | −33.56 | 34.23 |
| | $H_2S(aq)$ | −39.7 | 121 | −27.83 | — |
| | $HS^-(aq)$ | −17.6 | 62.8 | 12.08 | — |
| | $SO(g)$ | 6.26 | 221.84 | −19.87 | 30.17 |
| | $SO_2(g)$ | −296.83 | 248.11 | −300.19 | 39.87 |
| | $SO_3(g)$ | −395.72 | 256.65 | −371.08 | 50.67 |
| | $H_2SO_3(aq)$ | −608.81 | 232.2 | −537.81 | — |
| | $HSO_3^-(aq)$ | −626.22 | 139.7 | −527.73 | — |
| | $SO_3^{2-}(aq)$ | −635.5 | −29 | −486.5 | — |
| | $H_2SO_4(\ell)$ | −813.99 | 156.90 | −690.10 | 138.91 |
| | $HSO_4^-(aq)$ | −887.34 | 131.8 | −755.91 | −84 |
| | $SO_4^{2-}(aq)$ | −909.27 | 20.1 | −744.53 | −293 |
| | $SF_6(g)$ | −1209 | 291.71 | −1105.4 | 97.28 |
| | $Se(s, \text{black})$ | 0 | 42.44 | 0 | 25.36 |
| | $Se(g)$ | 227.07 | 176.61 | 187.06 | 20.82 |

| VII | Substance | $\Delta H_f^\circ$ | $S^\circ$ | $\Delta G_f^\circ$ | $C_P$ |
|---|---|---|---|---|---|
| VII | $F_2(g)$ | 0 | 202.67 | 0 | 31.30 |
| | $F(g)$ | 78.99 | 158.64 | 61.94 | 22.74 |
| | $F^-(aq)$ | −332.63 | −13.8 | −278.79 | −106.7 |
| | $HF(g)$ | −271.1 | 173.67 | −273.2 | 29.13 |
| | $HF(aq)$ | −320.08 | 88.7 | −296.82 | — |
| | $XeF_4(s)$ | −261.5 | — | — | — |
| | $Cl_2(g)$ | 0 | 222.96 | 0 | 33.91 |
| | $Cl(g)$ | 121.68 | 165.09 | 105.71 | 21.84 |
| | $Cl^-(aq)$ | −167.16 | 56.5 | −131.23 | −136.4 |
| | $HCl(g)$ | −92.31 | 186.80 | −95.30 | 29.12 |
| | $ClO^-(aq)$ | −107.1 | 42 | −36.8 | — |
| | $ClO_2(g)$ | 102.5 | 256.73 | 120.5 | 41.97 |
| | $ClO_2^-(aq)$ | −66.5 | 101.3 | 17.2 | — |
| | $ClO_3^-(aq)$ | −103.97 | 162.3 | −7.95 | — |
| | $ClO_4^-(aq)$ | −129.33 | 182.0 | −8.52 | — |

*(Continued)*

| Substance | $\Delta H_f^\circ$ (25°C) kJ mol$^{-1}$ | $S^\circ$ (25°C) J K$^{-1}$ mol$^{-1}$ | $\Delta G_f^\circ$ (25°C) kJ mol$^{-1}$ | $C_P$ (25°C) J K$^{-1}$ mol$^{-1}$ |
|---|---|---|---|---|
| $Cl_2O(g)$ | 80.3 | 266.10 | 97.9 | 45.40 |
| $HClO(aq)$ | −120.9 | 142 | −79.9 | — |
| $ClF_3(g)$ | −163.2 | 281.50 | −123.0 | 63.85 |
| $Br_2(\ell)$ | 0 | 152.23 | 0 | 75.69 |
| $Br_2(g)$ | 30.91 | 245.35 | 3.14 | 36.02 |
| $Br_2(aq)$ | −2.59 | 130.5 | 3.93 | — |
| $Br(g)$ | 111.88 | 174.91 | 82.41 | 20.79 |
| $Br^-(aq)$ | −121.55 | 82.4 | −103.96 | −141.8 |
| $HBr(g)$ | −36.40 | 198.59 | −53.43 | 29.14 |
| $BrO_3^-(aq)$ | −67.07 | 161.71 | 18.60 | — |
| $I_2(s)$ | 0 | 116.14 | 0 | 54.44 |
| $I_2(g)$ | 62.44 | 260.58 | 19.36 | 36.90 |
| $I_2(aq)$ | 22.6 | 137.2 | 16.40 | — |
| $I(g)$ | 106.84 | 180.68 | 70.28 | 20.79 |
| $I^-(aq)$ | −55.19 | 111.3 | −51.57 | −142.3 |
| $I_3^-(aq)$ | −51.5 | 239.3 | −51.4 | — |
| $HI(g)$ | 26.48 | 206.48 | 1.72 | 29.16 |
| $ICl(g)$ | 17.78 | 247.44 | −5.44 | 35.56 |
| $IBr(g)$ | 40.84 | 258.66 | 3.71 | 36.44 |
| VIII   $He(g)$ | 0 | 126.04 | 0 | 20.79 |
| $Ne(g)$ | 0 | 146.22 | 0 | 20.79 |
| $Ar(g)$ | 0 | 154.73 | 0 | 20.79 |
| $Kr(g)$ | 0 | 163.97 | 0 | 20.79 |
| $Xe(g)$ | 0 | 169.57 | 0 | 20.79 |

# A P P E N D I X   E

# Standard Reduction Potentials at 25°C

| Half-Reaction | $\mathscr{E}°$ (volts) |
|---|---|
| $F_2(g) + 2\,e^- \longrightarrow 2\,F^-$ | 2.87 |
| $H_2O_2 + 2\,H_3O^+ + 2\,e^- \longrightarrow 4\,H_2O$ | 1.776 |
| $PbO_2(s) + SO_4^{2-} + 4\,H_3O^+ + 2\,e^- \longrightarrow PbSO_4(s) + 6\,H_2O$ | 1.685 |
| $Au^+ + e^- \longrightarrow Au(s)$ | 1.68 |
| $MnO_4^- + 4\,H_3O^+ + 3\,e^- \longrightarrow MnO_2(s) + 6\,H_2O$ | 1.679 |
| $HClO_2 + 2\,H_3O^+ + 2\,e^- \longrightarrow HClO + 3\,H_2O$ | 1.64 |
| $HClO + H_3O^+ + e^- \longrightarrow \frac{1}{2}\,Cl_2(g) + 2\,H_2O$ | 1.63 |
| $Ce^{4+} + e^- \longrightarrow Ce^{3+}$ (1 M $HNO_3$ solution) | 1.61 |
| $2\,NO(g) + 2\,H_3O^+ + 2\,e^- \longrightarrow N_2O(g) + 3\,H_2O$ | 1.59 |
| $BrO_3^- + 6\,H_3O^+ + 5\,e^- \longrightarrow \frac{1}{2}\,Br_2(\ell) + 9\,H_2O$ | 1.52 |
| $Mn^{3+} + e^- \longrightarrow Mn^{2+}$ | 1.51 |
| $MnO_4^- + 8\,H_3O^+ + 5\,e^- \longrightarrow Mn^{2+} + 12\,H_2O$ | 1.491 |
| $ClO_3^- + 6\,H_3O^+ + 5\,e^- \longrightarrow \frac{1}{2}\,Cl_2(g) + 9\,H_2O$ | 1.47 |
| $PbO_2(s) + 4\,H_3O^+ + 2\,e^- \longrightarrow Pb^{2+} + 6\,H_2O$ | 1.46 |
| $Au^{3+} + 3\,e^- \longrightarrow Au(s)$ | 1.42 |
| $Cl_2(g) + 2\,e^- \longrightarrow 2\,Cl^-$ | 1.3583 |
| $Cr_2O_7^{2-} + 14\,H_3O^+ + 6\,e^- \longrightarrow 2\,Cr^{3+} + 21\,H_2O$ | 1.33 |
| $O_3(g) + H_2O + 2\,e^- \longrightarrow O_2 + 2\,OH^-$ | 1.24 |
| $O_2(g) + 4\,H_3O^+ + 4\,e^- \longrightarrow 6\,H_2O$ | 1.229 |
| $MnO_2(s) + 4\,H_3O^+ + 2\,e^- \longrightarrow Mn^{2+} + 6\,H_2O$ | 1.208 |
| $ClO_4^- + 2\,H_3O^+ + 2\,e^- \longrightarrow ClO_3 + 3\,H_2O$ | 1.19 |
| $Br_2(\ell) + 2\,e^- \longrightarrow 2\,Br^-$ | 1.065 |
| $NO_3^- + 4\,H_3O^+ + 3\,e^- \longrightarrow NO(g) + 6\,H_2O$ | 0.96 |
| $2\,Hg^{2+} + 2\,e^- \longrightarrow Hg_2^{2+}$ | 0.905 |
| $Ag^+ + e^- \longrightarrow Ag(s)$ | 0.7996 |
| $Hg_2^{2+} + 2\,e^- \longrightarrow 2\,Hg(\ell)$ | 0.7961 |
| $Fe^{3+} + e^- \longrightarrow Fe^{2+}$ | 0.770 |
| $O_2(g) + 2\,H_3O^+ + 2\,e^- \longrightarrow H_2O_2 + 2\,H_2O$ | 0.682 |
| $BrO_3^- + 3\,H_2O + 6\,e^- \longrightarrow Br^- + 6\,OH^-$ | 0.61 |
| $MnO_4^- + 2\,H_2O + 3\,e^- \longrightarrow MnO_2(s) + 4\,OH^-$ | 0.588 |
| $I_2(s) + 2\,e^- \longrightarrow 2\,I^-$ | 0.535 |
| $Cu^+ + e^- \longrightarrow Cu(s)$ | 0.522 |
| $O_2(g) + 2\,H_2O + 4\,e^- \longrightarrow 4\,OH^-$ | 0.401 |
| $Cu^{2+} + 2\,e^- \longrightarrow Cu(s)$ | 0.3402 |
| $PbO_2(s) + H_2O + 2\,e^- \longrightarrow PbO(s) + 2\,OH^-$ | 0.28 |
| $Hg_2Cl_2(s) + 2\,e^- \longrightarrow 2\,Hg(\ell) + 2\,Cl^-$ | 0.2682 |
| $AgCl(s) + e^- \longrightarrow Ag(s) + Cl^-$ | 0.2223 |
| $SO_4^{2-} + 4\,H_3O^+ + 2\,e^- \longrightarrow H_2SO_3 + 5\,H_2O$ | 0.20 |
| $Cu^{2+} + e^- \longrightarrow Cu^+$ | 0.158 |

| Half-Reaction | $\mathscr{E}°$ (volts) |
|---|---|
| $S_4O_6^{2-} + 2\,e^- \longrightarrow 2\,S_2O_3^{2-}$ | 0.0895 |
| $NO_3^- + H_2O + 2\,e^- \longrightarrow NO_2^- + 2\,OH^-$ | 0.01 |
| $2\,H_3O^+ + 2\,e^- \longrightarrow H_2(g) + 2\,H_2O(\ell)$ | 0.000 exactly |
| $Pb^{2+} + 2\,e^- \longrightarrow Pb(s)$ | −0.1263 |
| $Sn^{2+} + 2\,e^- \longrightarrow Sn(s)$ | −0.1364 |
| $Ni^{2+} + 2\,e^- \longrightarrow Ni(s)$ | −0.23 |
| $Co^{2+} + 2\,e^- \longrightarrow Co(s)$ | −0.28 |
| $PbSO_4(s) + 2\,e^- \longrightarrow Pb(s) + SO_4^{2-}$ | −0.356 |
| $Mn(OH)_3(s) + e^- \longrightarrow Mn(OH)_2(s) + OH^-$ | −0.40 |
| $Cd^{2+} + 2\,e^- \longrightarrow Cd(s)$ | −0.4026 |
| $Fe^{2+} + 2\,e^- \longrightarrow Fe(s)$ | −0.409 |
| $Cr^{3+} + e^- \longrightarrow Cr^{2+}$ | −0.424 |
| $Fe(OH)_3(s) + e^- \longrightarrow Fe(OH)_2(s) + OH^-$ | −0.56 |
| $PbO(s) + H_2O + 2\,e^- \longrightarrow Pb(s) + 2\,OH^-$ | −0.576 |
| $2\,SO_3^{2-} + 3\,H_2O + 4\,e^- \longrightarrow S_2O_3^{2-} + 6\,OH^-$ | −0.58 |
| $Ni(OH)_2(s) + 2\,e^- \longrightarrow Ni(s) + 2\,OH^-$ | −0.66 |
| $Co(OH)_2(s) + 2\,e^- \longrightarrow Co(s) + 2\,OH^-$ | −0.73 |
| $Cr^{3+} + 3\,e^- \longrightarrow Cr(s)$ | −0.74 |
| $Zn^{2+} + 2\,e^- \longrightarrow Zn(s)$ | −0.7628 |
| $2\,H_2O + 2\,e^- \longrightarrow H_2(g) + 2\,OH^-$ | −0.8277 |
| $Cr^{2+} + 2\,e^- \longrightarrow Cr(s)$ | −0.905 |
| $SO_4^{2-} + H_2O + 2\,e^- \longrightarrow SO_3^{2-} + 2\,OH^-$ | −0.92 |
| $Mn^{2+} + 2\,e^- \longrightarrow Mn(s)$ | −1.029 |
| $Mn(OH)_2(s) + 2\,e^- \longrightarrow Mn(s) + 2\,OH^-$ | −1.47 |
| $Al^{3+} + 3\,e^- \longrightarrow Al(s)$ | −1.706 |
| $Sc^{3+} + 3\,e^- \longrightarrow Sc(s)$ | −2.08 |
| $Ce^{3+} + 3\,e^- \longrightarrow Ce(s)$ | −2.335 |
| $La^{3+} + 3\,e^- \longrightarrow La(s)$ | −2.37 |
| $Mg^{2+} + 2\,e^- \longrightarrow Mg(s)$ | −2.375 |
| $Mg(OH)_2(s) + 2\,e^- \longrightarrow Mg(s) + 2\,OH^-$ | −2.69 |
| $Na^+ + e^- \longrightarrow Na(s)$ | −2.7109 |
| $Ca^{2+} + 2\,e^- \longrightarrow Ca(s)$ | −2.76 |
| $Ba^{2+} + 2\,e^- \longrightarrow Ba(s)$ | −2.90 |
| $K^+ + e^- \longrightarrow K(s)$ | −2.925 |
| $Li^+ + e^- \longrightarrow Li(s)$ | −3.045 |

All voltages are standard reduction potentials (relative to the standard hydrogen electrode) at 25°C and 1 atm pressure. All species are in aqueous solution unless otherwise indicated.

# Physical Properties of the Elements

## Hydrogen and the Alkali Metals (Group I Elements)

|  | Hydrogen | Lithium | Sodium | Potassium | Rubidium | Cesium | Francium |
|---|---|---|---|---|---|---|---|
| Atomic number | 1 | 3 | 11 | 19 | 37 | 55 | 87 |
| Atomic mass | 1.00794 | 6.941 | 22.98977 | 39.0983 | 85.4678 | 132.9054 | (223) |
| Melting point (°C) | −259.14 | 180.54 | 97.81 | 63.65 | 38.89 | 28.40 | 25 |
| Boiling point (°C) | −252.87 | 1347 | 903.8 | 774 | 688 | 678.4 | 677 |
| Density at 25°C (g cm$^{-3}$) | 0.070 (−253°C) | 0.534 | 0.971 | 0.862 | 1.532 | 1.878 | |
| Color | Colorless | Silver | Silver | Silver | Silver | Silver | |
| Ground-state electron configuration | $1s^1$ | $[He]2s^1$ | $[Ne]3s^1$ | $[Ar]4s^1$ | $[Kr]5s^1$ | $[Xe]6s^1$ | $[Rn]7s^1$ |
| Ionization energy[†] | 1312.0 | 520.2 | 495.8 | 418.8 | 403.0 | 375.7 | ≈400 |
| Electron affinity[†] | 72.770 | 59.63 | 52.867 | 48.384 | 46.884 | 45.505 | est. 44 |
| Electronegativity | 2.20 | 0.98 | 0.93 | 0.82 | 0.82 | 0.79 | 0.70 |
| Ionic radius (Å) | 1.46(H$^-$) | 0.68 | 0.98 | 1.33 | 1.48 | 1.67 | ≈1.8 |
| Atomic radius (Å) | 0.37 | 1.52 | 1.86 | 2.27 | 2.47 | 2.65 | ≈2.7 |
| Enthalpy of fusion[†] | 0.1172 | 3.000 | 2.602 | 2.335 | 2.351 | 2.09 | |
| Enthalpy of vaporization[†] | 0.4522 | 147.1 | 97.42 | 89.6 | 76.9 | 67.8 | |
| Bond enthalpy of M$_2$[†] | 436 | 102.8 | 72.6 | 54.8 | 51.0 | 44.8 | |
| Standard reduction potential (volts) | 0 H$^+$/H$_2$ | −3.045 Li$^+$/Li | −2.7109 Na$^+$/Na | −2.924 K$^+$/K | −2.925 Rb$^+$/Rb | −2.923 Cs$^+$/Cs | ≈2.9 Fr$^+$/Fr |

[†]In kilojoules per mole.

## The Alkaline-Earth Metals (Group II Elements)

| | Beryllium | Magnesium | Calcium | Strontium | Barium | Radium |
|---|---|---|---|---|---|---|
| Atomic number | 4 | 12 | 20 | 38 | 56 | 88 |
| Atomic mass | 9.01218 | 24.3050 | 40.078 | 87.62 | 137.327 | (226) |
| Melting point (°C) | 1283 | 648.8 | 839 | 769 | 725 | 700 |
| Boiling point (°C) | 2484 | 1105 | 1484 | 1384 | 1640 | |
| Density at 25°C (g cm$^{-3}$) | 1.848 | 1.738 | 1.55 | 2.54 | 3.51 | 5 |
| Color | Gray | Silver | Silver | Silver | Silver-yellow | Silver |
| Ground-state electron configuration | [He]$2s^2$ | [Ne]$3s^2$ | [Ar]$4s^2$ | [Kr]$5s^2$ | [Xe]$6s^2$ | [Rn]$7s^2$ |
| Ionization energy[†] | 899.4 | 737.7 | 589.8 | 549.5 | 502.9 | 509.3 |
| Electron affinity[†] | <0 | <0 | 2.0 | 4.6 | 13.95 | >0 |
| Electronegativity | 1.57 | 1.31 | 1.00 | 0.95 | 0.89 | 0.90 |
| Ionic radius (Å) | 0.31 | 0.66 | 0.99 | 1.13 | 1.35 | 1.43 |
| Atomic radius (Å) | 1.13 | 1.60 | 1.97 | 2.15 | 2.17 | 2.23 |
| Enthalpy of fusion[†] | 11.6 | 8.95 | 8.95 | 9.62 | 7.66 | 7.15 |
| Enthalpy of vaporization[†] | 297.6 | 127.6 | 154.7 | 154.4 | 150.9 | 136.7 |
| Bond enthalpy of $M_2$[†] | 9.46 | | | | | |
| Standard reduction potential (volts) | −1.70 $Be^{2+}/Be$ | −2.375 $Mg^{2+}/Mg$ | −2.76 $Ca^{2+}/Ca$ | −2.89 $Sr^{2+}/Sr$ | −2.90 $Ba^{2+}/Ba$ | −2.916 $Ra^{2+}/Ra$ |

## Group III Elements

| | Boron | Aluminum | Gallium | Indium | Thallium |
|---|---|---|---|---|---|
| Atomic number | 5 | 13 | 31 | 49 | 81 |
| Atomic mass | 10.811 | 26.98154 | 69.723 | 114.82 | 204.3833 |
| Melting point (°C) | 2300 | 660.37 | 29.78 | 156.61 | 303.5 |
| Boiling point (°C) | 3658 | 2467 | 2403 | 2080 | 1457 |
| Density at 25°C (g cm$^{-3}$) | 2.34 | 2.702 | 5.904 | 7.30 | 11.85 |
| Color | Yellow | Silver | Silver | Silver | Blue-white |
| Ground-state electron configuration | [He]$2s^2 2p^1$ | [Ne]$3s^2 3p^1$ | [Ar]$3d^{10}4s^2 4p^1$ | [Kr]$4d^{10}5s^2 5p^1$ | [Xe]$4f^{14}5d^{10}6s^2 6p^1$ |
| Ionization energy[†] | 800.6 | 577.6 | 578.8 | 558.3 | 589.3 |
| Electron affinity[†] | 26.7 | 42.6 | 29 | 29 | ≈20 |
| Electronegativity | 2.04 | 1.61 | 1.81 | 1.78 | 1.83 |
| Ionic radius (Å) | 0.23 (+3) | 0.51 (+3) | 0.62 (+3) 0.81 (+1) | 0.81 (+3) | 0.95 (+3) 1.47 (+1) |
| Atomic radius (Å) | 0.88 | 1.43 | 1.22 | 1.63 | 1.70 |
| Enthalpy of fusion[†] | 22.6 | 10.75 | 5.59 | 3.26 | 4.08 |
| Enthalpy of vaporization[†] | 508 | 291 | 272 | 243 | 182 |
| Bond enthalpy of $M_2$[†] | 295 | 167 | 116 | 106 | ≈63 |
| Standard reduction potential (volts) | −0.890 $B(OH)_3/B$ | −1.706 $Al^{3+}/Al$ | −0.560 $Ga^{3+}/Ga$ | −0.338 $In^{3+}/In$ | 0.719 $Tl^{3+}/Tl$ |

[†] In kilojoules per mole.

## Group IV Elements

|                                      | Carbon | Silicon | Germanium | Tin | Lead |
|--------------------------------------|--------|---------|-----------|-----|------|
| Atomic number                        | 6 | 14 | 32 | 50 | 82 |
| Atomic mass                          | 12.011 | 28.0855 | 72.61 | 118.710 | 207.2 |
| Melting point (°C)                   | 3550 | 1410 | 937.4 | 231.9681 | 327.502 |
| Boiling point (°C)                   | 4827 | 2355 | 2830 | 2270 | 1740 |
| Density at 25°C (g cm$^{-3}$)        | 2.25 (gr) 3.51 (dia) | 2.33 | 5.323 | 5.75 (gray) 7.31 (white) | 11.35 |
| Color                                | Black (gr) Colorless (dia) | Gray | Gray-white | Silver | Blue-white |
| Ground-state electron configuration  | [He]$2s^2 2p^2$ | [Ne]$3s^2 3p^2$ | [Ar]$3d^{10}4s^2 4p^2$ | [Kr]$4d^{10}5s^2 5p^2$ | [Xe]$4f^{14}5d^{10}6s^2 6p^2$ |
| Ionization energy[†]                 | 1086.4 | 786.4 | 762.2 | 708.6 | 715.5 |
| Electron affinity[†]                 | 121.85 | 133.6 | ≈120 | ≈120 | 35.1 |
| Electronegativity                    | 2.55 | 1.90 | 2.01 | 1.88 | 2.10 |
| Ionic radius (Å)                     | 0.15 (+4) 2.60 (−4) | 0.42 (+4) 2.71 (−4) | 0.53 (+4) 0.73 (+2) 2.72 (−4) | 0.71 (+4) 0.93 (+2) | 0.84 (+4) 1.20 (+2) |
| Atomic radius (Å)                    | 0.77 | 1.17 | 1.22 | 1.40 | 1.75 |
| Enthalpy of fusion[†]                | 105.0 | 50.2 | 34.7 | 6.99 | 4.774 |
| Enthalpy of vaporization[†]          | 718.9 | 359 | 328 | 302 | 195.6 |
| Bond enthalpy of M$_2$[†]            | 178 | 317 | 280 | 192 | 61 |
| Standard reduction potential (volts) |  |  | −0.13 H$_2$GeO$_3$,H$^+$/Ge | −0.1364 Sn$^{2+}$/Sn | −0.1263 Pb$^{2+}$/Pb |

## Group V Elements

|                                      | Nitrogen | Phosphorus | Arsenic | Antimony | Bismuth |
|--------------------------------------|----------|------------|---------|----------|---------|
| Atomic number                        | 7 | 15 | 33 | 51 | 83 |
| Atomic mass                          | 14.00674 | 30.97376 | 74.92159 | 121.75 | 208.9804 |
| Melting point (°C)                   | −209.86 | 44.1 | 817 (28 atm.) | 630.74 | 271.3 |
| Boiling point (°C)                   | −195.8 | 280 | 613 (subl.) | 1750 | 1560 |
| Density at 25°C (g cm$^{-3}$)        | 0.808 (−196°C) | 1.82 (white) 2.20 (red) 2.69 (black) | 5.727 | 6.691 | 9.747 |
| Color                                | Colorless |  | Gray | Blue-white | White |
| Ground-state electron configuration  | [He]$2s^2 2p^3$ | [Ne]$3s^2 3p^3$ | [Ar]$3d^{10}4s^2 4p^3$ | [Kr]$4d^{10}5s^2 5p^3$ | [Xe]$4f^{14}5d^{10}6s^2 6p^3$ |
| Ionization energy[†]                 | 1402.3 | 1011.7 | 947 | 833.7 | 703.3 |
| Electron affinity[†]                 | −7 | 72.03 | ≈80 | 103 | 91.3 |
| Electronegativity                    | 3.04 | 2.19 | 2.18 | 2.05 | 2.02 |
| Ionic radius (Å)                     | 1.71 (−3) | 0.44 (+3) 2.12 (−3) | 0.46 (+5) 0.58 (+3) 2.22 (−3) | 0.62 (+5) 0.76 (+3) 2.45 (−3) | 0.96 (+3) |
| Atomic radius (Å)                    | 0.70 | 1.10 | 1.21 | 1.41 | 1.55 |
| Enthalpy of fusion[†]                | 0.720 | 6.587 | 27.72 | 20.91 | 10.88 |
| Enthalpy of vaporization[†]          | 5.608 | 59.03 | 334 | 262.5 | 184.6 |
| Bond enthalpy of M$_2$[†]            | 945 | 485 | 383 | 289 | 194 |
| Standard reduction potential (volts) | 0.96 NO$_3^-$,H$^+$/NO | −0.276 H$_3$PO$_4$/H$_3$PO$_3$ | 0.234 As$_2$O$_3$,H$^+$/As | 0.1445 Sb$_2$O$_3$,H$^+$/Sb | −0.46 Bi$_2$O$_3$,OH$^-$/Bi |

[†] In kilojoules per mole.

## The Chalcogens (Group VI Elements)

| | Oxygen | Sulfur | Selenium | Tellurium | Polonium |
|---|---|---|---|---|---|
| Atomic number | 8 | 16 | 34 | 52 | 84 |
| Atomic mass | 15.9994 | 32.066 | 78.96 | 127.60 | (209) |
| Melting point (°C) | −218.4 | 119.0 (mon) 112.8 (rhom) | 217 | 449.5 | 254 |
| Boiling point (°C) | −182.962 | 444.674 | 684.9 | 989.8 | 962 |
| Density at 25°C (g cm$^{-3}$) | 1.14 (−183°C) | 1.957 (mon) 2.07 (rhom) | 4.79 | 6.24 | 9.32 |
| Color | Pale blue ($\ell$) | Yellow | Gray | Silver | Silver-gray |
| Ground-state electron configuration | [He]$2s^2 2p^4$ | [Ne]$3s^2 3p^4$ | [Ar]$3d^{10}4s^2 4p^4$ | [Kr]$4d^{10}5s^2 5p^4$ | [Xe]$4f^{14}5d^{10}6s^2 6p^4$ |
| Ionization energy[†] | 1313.9 | 999.6 | 940.9 | 869.3 | 812 |
| Electron affinity[†] | 140.97676 | 200.4116 | 194.967 | 190.15 | ≈180 |
| Electronegativity | 3.44 | 2.58 | 2.55 | 2.10 | 2.00 |
| Ionic radius (Å) | 1.40 (−2) | 0.29 (+6) 1.84 (−2) | 0.42 (+6) 1.98 (−2) | 0.56 (+6) 2.21 (−2) | 0.67 (+6) 2.30 (−2) |
| Atomic radius (Å) | 0.66 | 1.04 | 1.17 | 1.43 | 1.67 |
| Enthalpy of fusion[†] | 0.4187 | 1.411 | 5.443 | 17.50 | 10 |
| Enthalpy of vaporization[†] | 6.819 | 238 | 207 | 195 | 90 |
| Bond enthalpy of M$_2$[†] | 498 | 429 | 308 | 225 | |
| Standard reduction potential (volts) | 1.229 $O_2, H^+/H_2O$ | −0.508 $S/S^{2-}$ | −0.78 $Se/Se^{2-}$ | −0.92 $Te/Te^{2-}$ | ≈−1.4 $Po/Po^{2-}$ |

## The Halogens (Group VII Elements)

| | Fluorine | Chlorine | Bromine | Iodine | Astatine |
|---|---|---|---|---|---|
| Atomic number | 9 | 17 | 35 | 53 | 85 |
| Atomic mass | 18.998403 | 35.4527 | 79.904 | 126.90447 | (210) |
| Melting point (°C) | −219.62 | −100.98 | −7.25 | 113.5 | 302 |
| Boiling point (°C) | −188.14 | −34.6 | 58.78 | 184.35 | 337 |
| Density at 25°C (g cm$^{-3}$) | 1.108 (−189°C) | 1.367 (−34.6°C) | 3.119 | 4.93 | |
| Color | Yellow | Yellow-green | Deep red | Violet-black | |
| Ground-state electron configuration | [He]$2s^2 2p^5$ | [Ne]$3s^2 3p^5$ | [Ar]$3d^{10}4s^2 4p^5$ | [Kr]$4d^{10}5s^2 5p^5$ | [Xe]$4f^{14}5d^{10}6s^2 6p^5$ |
| Ionization energy[†] | 1681.0 | 1251.1 | 1139.9 | 1008.4 | ≈930 |
| Electron affinity[†] | 328.0 | 349.0 | 324.7 | 295.2 | ≈270 |
| Electronegativity | 3.98 | 3.16 | 2.96 | 2.66 | 2.20 |
| Ionic radius (Å) | 1.33 | 1.81 | 1.96 | 2.20 | ≈2.27 |
| Atomic radius (Å) | 0.64 | 0.99 | 1.14 | 1.33 | 1.40 |
| Enthalpy of fusion[†] | 0.511 | 6.410 | 10.55 | 15.78 | 23.9 |
| Enthalpy of vaporization[†] | 6.531 | 20.347 | 29.56 | 41.950 | |
| Bond enthalpy of M$_2$[†] | 158 | 243 | 193 | 151 | 110 |
| Standard reduction potential (volts) | 2.87 $F_2/F^-$ | 1.358 $Cl_2/Cl^-$ | 1.065 $Br_2/Br^-$ | 0.535 $I_2/I^-$ | ≈0.2 $At_2/At^-$ |

[†] In kilojoules per mole.

## The Noble Gases (Group VIII Elements)

|  | Helium | Neon | Argon | Krypton | Xenon | Radon |
|---|---|---|---|---|---|---|
| Atomic number | 2 | 10 | 18 | 36 | 54 | 86 |
| Atomic mass | 4.00260 | 20.1797 | 39.948 | 83.80 | 131.29 | (222) |
| Melting point (°C) | −272.2 (26 atm) | −248.67 | −189.2 | −156.6 | −111.9 | −71 |
| Boiling point (°C) | −268.934 | −246.048 | −185.7 | −152.30 | −107.1 | −61.8 |
| Density at 25°C (g cm$^{-3}$) | 0.147 (−270.8°C) | 1.207 (−246.1°C) | 1.40 (−186°C) | 2.155 (−152.9°C) | 3.52 (−109°C) | 4.4 (−52°C) |
| Color | Colorless | Colorless | Colorless | Colorless | Colorless | Colorless |
| Ground-state electron configuration | $1s^2$ | $[He]2s^22p^6$ | $[Ne]3s^23p^6$ | $[Ar]3d^{10}4s^24p^6$ | $[Kr]4d^{10}5s^25p^6$ | $[Xe]4f^{14}5d^{10}6s^26p^6$ |
| Ionization energy[†] | 2372.3 | 2080.6 | 1520.5 | 1350.7 | 1170.4 | 1037.0 |
| Electron affinity[†] | <0 | <0 | <0 | <0 | <0 | <0 |
| Atomic radius (Å) | 0.32 | 0.69 | 0.97 | 1.10 | 1.30 | 1.45 |
| Enthalpy of fusion[†] | 0.02093 | 0.3345 | 1.176 | 1.637 | 2.299 | 2.9 |
| Enthalpy of vaporization[†] | 0.1005 | 1.741 | 6.288 | 9.187 | 12.643 | 18.4 |

## The Transition Elements

|  | Scandium | Yttrium | Lutetium | Titanium | Zirconium | Hafnium |
|---|---|---|---|---|---|---|
| Atomic number | 21 | 39 | 71 | 22 | 40 | 72 |
| Atomic mass | 44.95591 | 88.90585 | 174.967 | 47.88 | 91.224 | 178.49 |
| Melting point (°C) | 1541 | 1522 | 1656 | 1660 | 1852 | 2227 |
| Boiling point (°C) | 2831 | 3338 | 3315 | 3287 | 4504 | 4602 |
| Density at 25°C (g cm$^{-3}$) | 2.989 | 4.469 | 9.840 | 4.54 | 6.506 | 13.31 |
| Color | Silver | Silver | Silver | Silver | Gray-white | Silver |
| Ground-state electron configuration | $[Ar]3d^14s^2$ | $[Kr]4d^15s^2$ | $[Xe]4f^{14}5d^16s^2$ | $[Ar]3d^24s^2$ | $[Kr]4d^25s^2$ | $[Xe]4f^{14}5d^26s^2$ |
| Ionization energy[†] | 631 | 616 | 523.5 | 658 | 660 | 654 |
| Electron affinity[†] | 18.1 | 29.6 | ≈50 | 7.6 | 41.1 | ≈0 |
| Electronegativity | 1.36 | 1.22 | 1.27 | 1.54 | 1.33 | 1.30 |
| Ionic radius (Å) | 0.81 | 0.93 | 0.848 (+3) | 0.68 | 0.80 | 0.78 |
| Atomic radius (Å) | 1.61 | 1.78 | 1.72 | 1.45 | 1.59 | 1.56 |
| Enthalpy of fusion[†] | 11.4 | 11.4 | 19.2 | 18.62 | 20.9 | 25.5 |
| Enthalpy of vaporization[†] | 328 | 425 | 247 | 426 | 590 | 571 |
| Standard reduction potential (volts) | −2.08 $Sc^{3+}/Sc$ | −2.37 $Y^{3+}/Y$ | −2.30 $Lu^{3+}/Lu$ | −0.86 $TiO_2,H^+/Ti$ | −1.43 $ZrO_2,H^+/Zr$ | −1.57 $HfO_2,H^+/Hf$ |

[†] In kilojoules per mole.

## The Transition Elements (cont.)

| | **Vanadium** | **Niobium** | **Tantalum** | **Chromium** | **Molybdenum** | **Tungsten** |
|---|---|---|---|---|---|---|
| Atomic number | 23 | 41 | 73 | 24 | 42 | 74 |
| Atomic mass | 50.9415 | 92.90638 | 180.9479 | 51.9961 | 95.94 | 183.85 |
| Melting point (°C) | 1890 | 2468 | 2996 | 1857 | 2617 | 3410 |
| Boiling point (°C) | 3380 | 4742 | 5425 | 2672 | 4612 | 5660 |
| Density at 25°C (g cm$^{-3}$) | 6.11 | 8.57 | 16.654 | 7.18 | 10.22 | 19.3 |
| Color | Silver-white | Gray-white | Steel gray | Silver | Silver | Steel gray |
| Ground-state electron configuration | [Ar]$3d^3 4s^2$ | [Kr]$4d^4 5s^1$ | [Xe]$4f^{14}5d^3 6s^2$ | [Ar]$3d^5 4s^1$ | [Kr]$4d^5 5s^1$ | [Xe]$4f^{14}5d^4 6s^2$ |
| Ionization energy[†] | 650 | 664 | 761 | 652.8 | 684.9 | 770 |
| Electron affinity[†] | 50.7 | 86.2 | 31.1 | 64.3 | 72.0 | 78.6 |
| Electronegativity | 1.63 | 1.60 | 1.50 | 1.66 | 2.16 | 2.36 |
| Ionic radius (Å) | 0.59 (+5) | 0.69 (+5) | 0.68 (+5) | | 0.62 (+6) | 0.62 (+6) |
| | 0.63 (+4) | 0.74 (+4) | | 0.63 (+3) | 0.70 (+4) | 0.70 (+4) |
| | 0.74 (+3) | | | 0.89 (+2) | | |
| | 0.88 (+2) | | | | | |
| Atomic radius (Å) | 1.31 | 1.43 | 1.43 | 1.25 | 1.36 | 1.37 |
| Enthalpy of fusion[†] | 21.1 | 26.4 | | 20.9 | 27.8 | 35.4 |
| Enthalpy of vaporization[†] | 512 | 722 | 781 | 394.7 | 589.2 | 819.3 |
| Standard reduction potential (volts) | −1.2 $V^{2+}/V$ | −0.62 $Nb_2O_5,H^+/Nb$ | −0.71 $Ta_2O_5,H^+/Ta$ | −0.74 $Cr^{3+}/Cr$ | 0.0 $H_2MoO_4,H^+/Mo$ | −0.09 $WO_3,H^+/W$ |

| | **Manganese** | **Technetium** | **Rhenium** | **Iron** | **Ruthenium** | **Osmium** |
|---|---|---|---|---|---|---|
| Atomic number | 25 | 43 | 75 | 26 | 44 | 76 |
| Atomic mass | 54.93805 | (98) | 186.207 | 55.847 | 101.07 | 190.2 |
| Melting point (°C) | 1244 | 2172 | 3180 | 1535 | 2310 | 3045 |
| Boiling point (°C) | 1962 | 4877 | 5627 | 2750 | 3900 | 5027 |
| Density at 25°C (g cm$^{-3}$) | 7.21 | 11.50 | 21.02 | 7.874 | 12.41 | 22.57 |
| Color | Gray-white | Silver-gray | Silver | Gray | White | Blue-white |
| Ground-state electron configuration | [Ar]$3d^5 4s^2$ | [Kr]$4d^5 5s^2$ | [Xe]$4f^{14}5d^5 6s^2$ | [Ar]$3d^6 4s^2$ | [Kr]$4d^7 5s^1$ | [Xe]$4f^{14}5d^6 6s^2$ |
| Ionization energy[†] | 717.4 | 702 | 760 | 759.3 | 711 | 840 |
| Electron affinity[†] | <0 | ≈53 | ≈14 | 15.7 | ≈100 | ≈106 |
| Electronegativity | 1.55 | 1.90 | 1.90 | 1.90 | 2.2 | 2.20 |
| Ionic radius (Å) | 0.80 (+2) | | 0.56 (+7) | 0.60 (+3) | 0.67 (+4) | 0.69 (+6) |
| | | | 0.72 (+4) | 0.72 (+2) | | 0.88 (+4) |
| Atomic radius (Å) | 1.37 | 1.35 | 1.34 | 1.24 | 1.32 | 1.34 |
| Enthalpy of fusion[†] | 14.6 | 23.8 | 33.1 | 15.19 | 26.0 | 31.8 |
| Enthalpy of vaporization[†] | 279 | 585 | 778 | 414 | 649 | 678 |
| Standard reduction potential (volts) | −0.183 $Mn^{3+}/Mn$ | 0.738 $TcO_4^-,H^+/TcO_2$ | 0.3 $Re^{3+}/Re$ | −0.036 $Fe^{3+}/Fe$ | 0.49 $Ru^{4+}/Ru^{3+}$ | 0.85 $OsO_4,H^+/Os$ |

[†]In kilojoules per mole.

## The Transition Elements (cont.)

| | Cobalt | Rhodium | Iridium | Nickel | Palladium | Platinum |
|---|---|---|---|---|---|---|
| Atomic number | 27 | 45 | 77 | 28 | 46 | 78 |
| Atomic mass | 58.93320 | 102.90550 | 192.22 | 58.69 | 106.42 | 195.08 |
| Melting point (°C) | 1495 | 1966 | 2410 | 1453 | 1552 | 1772 |
| Boiling point (°C) | 2870 | 3727 | 4130 | 2732 | 3140 | 3827 |
| Density at 25°C (g cm$^{-3}$) | 8.9 | 12.41 | 22.42 | 8.902 | 12.02 | 21.45 |
| Color | Steel gray | Silver | Silver | Silver | Steel white | Silver |
| Ground-state electron configuration | [Ar]$3d^7 4s^2$ | [Kr]$4d^8 5s^1$ | [Xe]$4f^{14} 5d^7 6s^2$ | [Ar]$3d^8 4s^2$ | [Kr]$4d^{10}$ | [Xe]$4f^{14} 5d^9 6s^1$ |
| Ionization energy[†] | 758 | 720 | 880 | 736.7 | 805 | 868 |
| Electron affinity[†] | 63.8 | 110 | 151 | 111.5 | 51.8 | 205.1 |
| Electronegativity | 1.88 | 2.28 | 2.20 | 1.91 | 2.20 | 2.28 |
| Ionic radius (Å) | 0.63 (+3) 0.72 (+2) | 0.68 (+3) | 0.68 (+4) | 0.69 (+2) | 0.65 (+4) 0.80 (+2) | 0.65 (+4) 0.80 (+2) |
| Atomic radius (Å) | 1.25 | 1.34 | 1.36 | 1.25 | 1.38 | 1.37 |
| Enthalpy of fusion[†] | 16.2 | 21.5 | 26.4 | 17.6 | 17.6 | 19.7 |
| Enthalpy of vaporization[†] | 373 | 557 | 669 | 428 | 353 | 564 |
| Standard reduction potential (volts) | −0.28 $Co^{2+}/Co$ | 1.43 $Rh^{4+}/Rh^{3+}$ | 0.1 $Ir_2O_3/Ir,OH^-$ | −0.23 $Ni^{2+}/Ni$ | 0.83 $Pd^{2+}/Pd$ | 1.2 $Pt^{2+}/Pt$ |

| | Copper | Silver | Gold | Zinc | Cadmium | Mercury |
|---|---|---|---|---|---|---|
| Atomic number | 29 | 47 | 79 | 30 | 48 | 80 |
| Atomic mass | 63.546 | 107.8682 | 196.96654 | 65.39 | 112.411 | 200.59 |
| Melting point (°C) | 1083.4 | 961.93 | 1064.43 | 419.58 | 320.9 | −38.87 |
| Boiling point (°C) | 2567 | 2212 | 2807 | 907 | 765 | 356.58 |
| Density at 25°C (g cm$^{-3}$) | 8.96 | 10.50 | 19.32 | 7.133 | 8.65 | 13.546 |
| Color | Red | Silver | Yellow | Blue-white | Blue-white | Silver |
| Ground-state electron configuration | [Ar]$3d^{10} 4s^1$ | [Kr]$4d^{10} 5s^1$ | [Xe]$4f^{14} 5d^{10} 6s^1$ | [Ar]$3d^{10} 4s^2$ | [Kr]$4d^{10} 5s^2$ | [Xe]$4f^{14} 5d^{10} 6s^2$ |
| Ionization energy[†] | 745.4 | 731.0 | 890.1 | 906.4 | 867.7 | 1007.0 |
| Electron affinity[†] | 118.5 | 125.6 | 222.749 | <0 | <0 | <0 |
| Electronegativity | 1.90 | 1.93 | 2.54 | 1.65 | 1.69 | 2.00 |
| Ionic radius (Å) | 0.72 (+2) 0.96 (+1) | 0.89 (+2) 1.26 (+1) | 0.85 (+2) 1.37 (+1) | 0.74 (+2) | 0.97 (+2) 1.14 (+1) | 1.10 (+2) 1.27 (+1) |
| Atomic radius (Å) | 1.28 | 1.44 | 1.44 | 1.34 | 1.49 | 1.50 |
| Enthalpy of fusion[†] | 13.3 | 11.95 | 12.36 | 7.39 | 6.11 | 2.300 |
| Enthalpy of vaporization[†] | 304 | 285 | 365 | 131 | 112 | 59.1 |
| Standard reduction potential (volts) | 0.340 $Cu^{2+}/Cu$ | 0.800 $Ag^+/Ag$ | 1.42 $Au^{3+}/Au$ | −0.763 $Zn^{2+}/Zn$ | −0.403 $Cd^{2+}/Cd$ | 0.796 $Hg_2^{2+}/Hg$ |

[†] In kilojoules per mole.

## The Lanthanide Elements

| | Lanthanum | Cerium | Praseodymium | Neodymium | Promethium | Samarium | Europium |
|---|---|---|---|---|---|---|---|
| Atomic number | 57 | 58 | 59 | 60 | 61 | 62 | 63 |
| Atomic mass | 138.9055 | 140.115 | 140.90765 | 144.24 | (145) | 150.36 | 151.965 |
| Melting point (°C) | 921 | 798 | 931 | 1010 | ≈1080 | 1072 | 822 |
| Boiling point (°C) | 3457 | 3257 | 3212 | 3127 | ≈2400 | 1778 | 1597 |
| Density at 25°C (g cm$^{-3}$) | 6.145 | 6.657 | 6.773 | 6.80 | 7.22 | 7.520 | 5.243 |
| Color | Silver | Gray | Silver | Silver | | Silver | Silver |
| Ground-state electron configuration | [Xe]$5d^16s^2$ | [Xe]$4f^15d^16s^2$ | [Xe]$4f^36s^2$ | [Xe]$4f^46s^2$ | [Xe]$4f^56s^2$ | [Xe]$4f^66s^2$ | [Xe]$4f^76s^2$ |
| Ionization energy[†] | 538.1 | 528 | 523 | 530 | 536 | 543 | 547 |
| Electron affinity[†] | 50 | est. 50 | | | | | |
| Electronegativity | 1.10 | 1.12 | 1.13 | 1.14 | | 1.17 | |
| Ionic radius (Å) | 1.15 | 0.92 (+4) 1.034 (+3) | 0.90 (+4) 1.013 (+3) | 0.995 (+3) | 0.979 (+3) | 0.964 (+3) | 0.950 (+3) 1.09 (+2) |
| Atomic radius (Å) | 1.87 | 1.82 | 1.82 | 1.81 | 1.81 | 1.80 | 2.00 |
| Enthalpy of fusion[†] | 5.40 | 5.18 | 6.18 | 7.13 | 12.6 | 8.91 | (10.5) |
| Enthalpy of vaporization[†] | 419 | 389 | 329 | 324 | | 207 | 172 |
| Standard reduction potential (volts) | −2.37 La$^{3+}$/La | −2.335 Ce$^{3+}$/Ce | −2.35 Pr$^{3+}$/Pr | −2.32 Nd$^{3+}$/Nd | −2.29 Pm$^{3+}$/Pm | −2.30 Sm$^{3+}$/Sm | −1.99 Eu$^{3+}$/Eu |

| | Gadolinium | Terbium | Dysprosium | Holmium | Erbium | Thulium | Ytterbium |
|---|---|---|---|---|---|---|---|
| Atomic number | 64 | 65 | 66 | 67 | 68 | 69 | 70 |
| Atomic mass | 157.25 | 158.92534 | 162.50 | 164.93032 | 167.26 | 168.93421 | 173.04 |
| Melting point (°C) | 1311 | 1360 | 1409 | 1470 | 1522 | 1545 | 824 |
| Boiling point (°C) | 3233 | 3041 | 2335 | 2720 | 2510 | 1727 | 1193 |
| Density at 25°C (g cm$^{-3}$) | 7.900 | 8.229 | 8.550 | 8.795 | 9.066 | 9.321 | 6.965 |
| Color | Silver | Silver-gray | Silver | Silver | Silver | Silver | Silver |
| Ground-state electron configuration | [Xe]$4f^75d^16s^2$ | [Xe]$4f^96s^2$ | [Xe]$4f^{10}6s^2$ | [Xe]$4f^{11}6s^2$ | [Xe]$4f^{12}6s^2$ | [Xe]$4f^{13}6s^2$ | [Xe]$4f^{14}6s^2$ |
| Ionization energy[†] | 592 | 564 | 572 | 581 | 589 | 596.7 | 603.4 |
| Electron affinity[†] | | | | est. 50 | | | |
| Electronegativity | 1.20 | | 1.22 | 1.23 | 1.24 | 1.25 | |
| Ionic radius (Å) | 0.938 (+3) | 0.84  (+4) 0.923 (+3) | 0.908 (+3) | 0.894 (+3) | 0.881 (+3) | 0.869 (+3) | 0.858 (+3) 0.93  (+2) |
| Atomic radius (Å) | 1.79 | 1.76 | 1.75 | 1.74 | 1.73 | 1.72 | 1.94 |
| Enthalpy of fusion[†] | 15.5 | 16.3 | 17.2 | 17.2 | 17.2 | 18.2 | 9.2 |
| Enthalpy of vaporization[†] | 301 | 293 | 165 | 285 | 280 | 240 | 165 |
| Standard reduction potential (volts) | −2.28 Gd$^{3+}$/Gd | −2.31 Tb$^{3+}$/Tb | −2.29 Dy$^{3+}$/Dy | −2.33 Ho$^{3+}$/Ho | −2.32 Er$^{3+}$/Er | −2.32 Tm$^{3+}$/Tm | −2.22 Yb$^{3+}$/Yb |

[†] In kilojoule per mole.

## The Actinide Elements

| | Actinium | Thorium | Protactinium | Uranium | Neptunium | Plutonium | Americium |
|---|---|---|---|---|---|---|---|
| Atomic number | 89 | 90 | 91 | 92 | 93 | 94 | 95 |
| Atomic mass | (227) | 232.0381 | 231.03588 | 238.0289 | (237) | (244) | (243) |
| Melting point (°C) | 1050 | 1750 | 1600 | 1132.3 | 640 | 624 | 994 |
| Boiling point (°C) | 3200 | 4790 | | 3818 | 2732 | 3232 | 2607 |
| Density at 25°C (g cm$^{-3}$) | 10.07 | 11.72 | 15.37 | 18.95 | 20.25 | 19.84 | 13.67 |
| Color | Silver | Silver | Silver | Silver | Silver | Silver | Silver |
| Ground-state electron configuration | [Rn]$6d^17s^2$ | [Rn]$6d^27s^2$ | [Rn]$5f^26d^17s^2$ | [Rn]$5f^36d^17s^2$ | [Rn]$5f^46d^17s^2$ | [Rn]$5f^67s^2$ | [Rn]$5f^77s^2$ |
| Ionization energy[†] | 499 | 587 | 568 | 587 | 597 | 585 | 578 |
| Electronegativity | 1.1 | 1.3 | 1.5 | 1.38 | 1.36 | 1.28 | 1.3 |
| Ionic radius (Å) | 1.11 (+3) | 0.99 (+4) | 0.89 (+5) | 0.80 (+6) | 0.71 (+7) | 0.90 (+4) | 0.89 (+4) |
| | | | 0.96 (+4) | 0.93 (+4) | 0.92 (+4) | 1.00 (+3) | 0.99 (+3) |
| | | | 1.05 (+3) | 1.03 (+3) | 1.01 (+3) | | |
| Atomic radius (Å) | 1.88 | 1.80 | 1.61 | 1.38 | 1.30 | 1.51 | 1.84 |
| Enthalpy of fusion[†] | 14.2 | 18.8 | 16.7 | 12.9 | 9.46 | 3.93 | 14.4 |
| Enthalpy of vaporization[†] | 293 | 575 | 481 | 536 | 337 | 348 | 238 |
| Standard reduction potential (volts) | $-2.6$ Ac$^{3+}$/Ac | $-1.90$ Th$^{4+}$/Th | $-1.0$ PaO$_2^+$,H$^+$/Pa | $-1.8$ U$^{3+}$/U | $-1.9$ Np$^{3+}$/U | $-2.03$ Pu$^{3+}$/Pu | $-2.32$ Am$^{3+}$/Am |

| | Curium | Berkelium | Californium | Einsteinium | Fermium | Mendelevium | Nobelium |
|---|---|---|---|---|---|---|---|
| Atomic number | 96 | 97 | 98 | 99 | 100 | 101 | 102 |
| Atomic mass | (247) | (247) | (251) | (252) | (257) | (258) | (259) |
| Melting point (°C) | 1340 | | | | | | |
| Boiling point (°C) | | | | | | | |
| Density at 25°C (g cm$^{-3}$) | 13.51 | 14 | | | | | |
| Color | Silver | Silver | Silver | Silver | | | |
| Ground-state electron configuration | [Rn]$5f^76d^17s^2$ | [Rn]$5f^97s^2$ | [Rn]$5f^{10}7s^2$ | [Rn]$5f^{11}7s^2$ | [Rn]$5f^{12}7s^2$ | [Rn]$5f^{13}7s^2$ | [Rn]$5f^{14}7s^2$ |
| Ionization energy[†] | 581 | 601 | 608 | 619 | 627 | 635 | 642 |
| Electronegativity | 1.3 | 1.3 | 1.3 | 1.3 | 1.3 | 1.3 | 1.3 |
| Ionic radius (Å) | 0.88 (+4) | 0.87 (+4) | 0.86 (+4) | 0.85 (+4) | 0.84 (+4) | 0.84 (+4) | 0.83 (+4) |
| | 1.01 (+3) | 1.00 (+3) | 0.99 (+3) | 0.98 (+3) | 0.97 (+3) | 0.96 (+3) | 0.95 (+3) |
| | 1.19 (+2) | 1.18 (+2) | 1.17 (+2) | 1.16 (+2) | 1.15 (+2) | 1.14 (+2) | 1.13 (+2) |
| Standard reduction potential (volts) | $-2.06$ Cm$^{3+}$/Cm | $-1.05$ Bk$^{3+}$/Bk | $-1.93$ Cf$^{3+}$/Cf | $-2.0$ Es$^{3+}$/Es | $-1.96$ Fm$^{3+}$/Fm | $-1.7$ Md$^{3+}$/Md | $-1.2$ No$^{3+}$/No |

[†] In kilojoules per mole.

# A P P E N D I X   G

# Solutions to Selected
# Odd-Numbered Problems

## Chapter 1

1. Mercury is an element; water and sodium chloride are compounds. The other materials are mixtures: seawater and air are homogeneous; table salt and wood are heterogeneous. Mayonnaise appears homogeneous to the naked eye, but under magnification reveals itself as water droplets suspended in oil.

3. Substances

5. 16.9 g

7. (a) 2.005, and 1.504 g   (b) 2.005/1.504 = 1.333 = 4/3. SiN (or a multiple).

9. 2, 3, 4, 5

11. HO (or any multiple such as $H_2O_2$)

13. 2.0 L $N_2O$, 3.0 L $O_2$

15. (a) 145/94 = 1.54   (b) 94 electrons

17. 95 protons, 146 neutrons, 95 electrons

19. 28.086

21. 11.01

23. $2.107300 \times 10^{-22}$ g

25. (a) 283.89   (b) 115.36   (c) 164.09   (d) 158.03   (e) 132.13

27. The total count of $2.52 \times 10^9$ atoms of gold has a mass of only $8.3 \times 10^{-13}$ g, far too small to detect with a balance.

29. 1.041 mol

31. 2540 $cm^3$ = 2.54 L

33. $7.03 \times 10^{23}$ atoms

35. Pt: 47.06%, F: 36.67%, Cl: 8.553%, O: 7.720%

37. $N_4H_6$, $H_2O$, LiH, $C_{12}H_{26}$

39. 0.225%

41. $Zn_3P_2O_8$

43. $Fe_3Si_7$

45. BaN, $Ba_3N_2$

47. (a) 0.923 g C, 0.077 g H   (b) No   (c) 92.3% C, 7.7% H   (d) CH

49. $C_4F_8$

51. (a) 62.1   (b) 6   (c) 56   (d) Si (atomic mass 28.1), N (atomic mass 14.0)
    (e) $Si_2H_6$

53. (a) $3\,H_2 + N_2 \rightarrow 2\,NH_3$
    (b) $2\,K + O_2 \rightarrow K_2O_2$
    (c) $PbO_2 + Pb + 2\,H_2SO_4 \rightarrow 2\,PbSO_4 + 2\,H_2O$
    (d) $2\,BF_3 + 3\,H_2O \rightarrow B_2O_3 + 6\,HF$
    (e) $2\,KClO_3 \rightarrow 2\,KCl + 3\,O_2$
    (f) $CH_3COOH + 2\,O_2 \rightarrow 2\,CO_2 + 2\,H_2O$
    (g) $2\,K_2O_2 + 2\,H_2O \rightarrow 4\,KOH + O_2$
    (h) $3\,PCl_5 + 5\,AsF_3 \rightarrow 3\,PF_5 + 5\,AsCl_3$

55. (a) 12.06 g   (b) 1.258 g   (c) 4.692 g

**57.** 7.83 g $K_2Zn_3[Fe(CN)_6]_2$
**59.** 0.134 g $SiO_2$
**61.** $1.18 \times 10^3$ g
**63.** 418 g KCl; 199 g $Cl_2$
**65.** (a) 58.8   (b) Probably Ni
**67.** 42.49% NaCl, 57.51% KCl
**69.** 14.7 g $NH_4Cl$, 5.3 g $NH_3$
**71.** 303.0 g Fe; 83.93%


## Chapter 2

**1.** Melting point 1250°C (obs. 1541°C), boiling point 2386°C (obs. 2831°C), density 3.02 g cm$^{-3}$ (obs. 2.99)

**3.** $SbH_3$, HBr, $SnH_4$, $H_2Se$

**5.** (a) $\ddot{Rn}\!:$ $86e^-$ (8 valence, 78 core)

  (b) Sr $\cdot$ $^+$ $37e^-$ (1 valence, 36 core)

  (c) $:\!\ddot{Se}\!:^{2-}$ $36e^-$ (8 valence, 28 core)

  (d) $\cdot\ddot{Sb}\cdot$ $^-$ $52e^-$ (6 valence, 46 core)

**7.** (a) CsCl, cesium chloride Cs $\cdot$ + $:\!\ddot{Cl}\cdot$ → $(Cs^+)(:\!\ddot{Cl}\!:^-)$

  (b) $CaAt_2$, calcium astatide Ca $\cdot$ + 2 $:\!\ddot{At}\cdot$ → $(Ca^{2+})(:\!\ddot{At}\!:^-)_2$

  (c) $Al_2S_3$, aluminum sulfide 2 $\cdot\!\dot{Al}\cdot$ + 3 $\cdot\ddot{S}\cdot$ → $(Al^{3+})_2(:\!\ddot{S}\!:^{2-})_3$

  (d) $K_2Te$, potassium telluride 2 K $\cdot$ + $\cdot\ddot{Te}\cdot$ → $(K^+)_2(:\!\ddot{Te}\!:^{2-})$

**9.** (a) Aluminum oxide   (b) Rubidium selenide   (c) Ammonium sulfide
  (d) Calcium nitrate   (e) Cesium sulfate   (f) Potassium hydrogen carbonate

**11.** (a) AgCN   (b) $Ca(OCl)_2$   (c) $K_2CrO_4$   (d) $Ga_2O_3$   (e) $KO_2$   (f) $Ba(HCO_3)_2$

**13.** $Na_3PO_4$, sodium phosphate

**15.** (a)                    (b)

  (c)                    (d)

**17.**

The first is favored.

**19.** (a) Group IV, $CO_2$   (b) Group VII, $Cl_2O_7$   (c) Group V, $NO_2^-$
  (d) Group VI, $HSO_4^-$

**21.** (a) H—$\ddot{As}$—H   (b) H—$\ddot{O}$—$\ddot{Cl}\!:$
        |
        H

  (c) $:\!\ddot{Xe}\!:\!\ddot{F}\!:$   (d)

**23.**

```
        : O :
         ‖
  H—N—C—N—H
    |       |
    H       H
```

Bond lengths: N—H $1.01 \times 10^{-10}$ m, N—C $1.47 \times 10^{-10}$ m, C=O $1.20 \times 10^{-10}$ m

**25.**

```
  : S—S—S :
   : S :     : S :
  : S—S—S :
```

**27. (a)**

```
    H   : F :
         |  (-1)
  H—N—B—F :
   (+1)|   |
    H   : F :
```

**(b)**

```
     H        O                H        O (-1)
     |       //                |       /
  H—C—C            ↔        H—C—C
     |       \                 |       \\
     H        O (-1)           H        O
```

**(c)**

```
         O                      O (-1)
        //                     /
  H—O—C           ↔       H—O—C
        \                      \\
         O                      O
         (-1)
```

**29.**

```
  (-1)                  (-1)
  : O—N=O  ↔  O=N—O :
```

Between $1.18 \times 10^{-10}$ m and $1.43 \times 10^{-10}$ m

**31.**

```
    H                    H   (+1) (-1)            H   (-1)      (+1)
    |                    |                        |
  H—C—N=C=O  ↔  H—C—N≡C—O :  ↔  H—C—N—C≡O :
    |                    |                        |
    H                    H                        H
```

**33. (a)**

```
     : F :  : F :
        \   /
         P
        / \
     : F :  : F :
        : F :
```

**(b)**

```
       : F :   : F :
          \   /
    : F—S—F :
```

**(c)**

```
    : O :   : F :
       \\   /
        Xe
       //   \
    : O :   : F :
```

**35.** $\overset{+2\ -1}{\text{SrBr}_2}$   $\overset{+2\ -2\ +1}{\text{Zn(OH)}_4^{2-}}$   $\overset{-4\ +1}{\text{SiH}_4}$   $\overset{+2\ +4\ -2}{\text{CaSiO}_3}$

$\overset{+6\ -2}{\text{Cr}_2\text{O}_7^{2-}}$   $\overset{+2\ +5\text{-}2}{\text{Ca}_5(\text{PO}_4)_3}\overset{-1}{\text{F}}$   $\overset{+1\ -\frac{1}{2}}{\text{KO}_2}$   $\overset{+1\ -1}{\text{CsH}}$

**37. (a)** $SiO_2$   **(b)** $(NH_4)_2CO_3$   **(c)** $PbO_2$   **(d)** $P_2O_5$   **(e)** $CaI_2$   **(f)** $Fe(NO_3)_3$

**39. (a)** Copper(I) sulfide and copper(II) sulfide   **(b)** Sodium sulfate
**(c)** Tetraarsenic hexaoxide   **(d)** Zirconium(IV) chloride   **(e)** Dichlorine heptaoxide
**(f)** Gallium(I) oxide

**41. (a)** $Ag^+(aq) + Cl^-(aq) \rightarrow AgCl(s)$
**(b)** $K_2CO_3(s) + 2\,H^+(aq) \rightarrow 2\,K^+(aq) + CO_2(g) + H_2O(\ell)$
**(c)** $2\,Cs(s) + 2\,H_2O(\ell) \rightarrow 2\,Cs^+(aq) + 2\,OH^-(aq) + H_2(g)$
**(d)** $2\,MnO_4^-(aq) + 16\,H^+(aq) + 10\,Cl^-(aq) \rightarrow 5\,Cl_2(g) + 2\,Mn^{2+}(aq) + 8\,H_2O(\ell)$

**43. (a)** $Ca(OH)_2(aq) + 2\,HF(aq) \rightarrow CaF_2(aq) + 2\,H_2O(\ell)$
Hydrofluoric acid, calcium hydroxide, calcium fluoride

**(b)** $2\,RbOH(aq) + H_2SO_4(aq) \rightarrow Rb_2SO_4(aq) + 2\,H_2O(\ell)$
Sulfuric acid, rubidium hydroxide, rubidium sulfate
**(c)** $Zn(OH)_2(s) + 2\,HNO_3(aq) \rightarrow Zn(NO_3)_2(aq) + 2\,H_2O(\ell)$
Nitric acid, zinc hydroxide, zinc nitrate
**(d)** $KOH(aq) + CH_3COOH(aq) \rightarrow KCH_3COO(aq) + H_2O(\ell)$
Acetic acid, potassium hydroxide, potassium acetate
**45.** Sodium sulfide
**47. (a)** $CaO(s) + H_2O(\ell) \rightarrow Ca(OH)_2(s)$
**(b)** The CaO acts as a Lewis base, donating a pair of electrons (located on the oxide ion) to one of the hydrogen ions (the Lewis acid) on the water molecule.
**49. (a)** Fluoride acceptor
**(b)** Acids: $BF_3$, $TiF_4$; bases: $ClF_3O_2$, $KF$

**51. (a)** $2\,\overset{+3}{P}F_2I(\ell) + 2\,\overset{0}{Hg}(\ell) \rightarrow \overset{+2}{P_2}F_4(g) + \overset{+1}{Hg_2}I_2(s)$

**(b)** $2\,K\overset{+5}{Cl}\overset{-2}{O_3}(s) \rightarrow 2\,K\overset{-1}{Cl}(s) + 3\,\overset{0}{O_2}(g)$

**(c)** $4\,\overset{-3}{N}H_3(g) + 5\,\overset{0}{O_2}(g) \rightarrow 4\,\overset{+2}{N}\overset{-2}{O}(g) + 6\,H_2\overset{-2}{O}(g)$

**(d)** $2\,\overset{0}{As}(s) + 6\,Na\overset{+1}{O}H(\ell) \rightarrow 2\,Na_3\overset{+3}{As}O_3(s) + 3\,\overset{0}{H_2}(g)$

**53.** $2\,\overset{0}{Au}(s) + 6\,H_2\overset{+6}{Se}O_4(aq) \rightarrow \overset{+3}{Au_2}(\overset{+6}{Se}O_4)_3(aq) + 3\,H_2\overset{+4}{Se}O_3(aq) + 3\,H_2O(\ell)$
Gold is oxidized and selenium (half of it) is reduced.
**55.** $Ba(s) + H_2(g) \rightarrow BaH_2(s)$
$BaH_2(s) + 2\,H_2O(\ell) \rightarrow Ba^{2+}(aq) + 2\,OH^-(aq) + 2\,H_2(g)$
$BaH_2(s) + 2\,HCl(aq) \rightarrow Ba^{2+}(aq) + 2\,Cl^-(aq) + 2\,H_2(g)$
$BaH_2(s) + 2\,H_2O(\ell) + Zn^{2+}(aq) + SO_4^{2-}(aq) \rightarrow BaSO_4(s) + Zn(OH)_2(s) + 2\,H_2(g)$
**57. (a)** Base, $Mg(OH)_2$ **(b)** Acid, $HOCl$ **(c)** Acid, $H_2SO_4$ **(d)** Base, $CsOH$
**59.** $SnO(s) + 2\,HCl(aq) \rightarrow Sn^{2+}(aq) + 2\,Cl^-(aq) + H_2O(\ell)$
$SnO(s) + NaOH(aq) + H_2O(\ell) \rightarrow Sn(OH)_3^-(aq) + Na^+(aq)$
**61. (a)** $H_2Te(aq) + H_2O(\ell) \rightarrow H_3O^+(aq) + HTe^-(aq)$; acid–base
**(b)** $SrO(s) + CO_2(g) \rightarrow SrCO_3(s)$; Lewis acid–base
**(c)** $2\,HI(aq) + CaCO_3(s) \rightarrow Ca^{2+}(aq) + 2\,I^-(aq) + CO_2(aq) + H_2O(\ell)$; acid–base
**(d)** $Na_2O(s) + 2\,NH_4Br(s) \rightarrow 2\,Na^+(aq) + 2\,Br^-(aq) + 2\,NH_3(aq) + H_2O(\ell)$; acid–base

## Chapter 3

**1.** $NH_4HS(s) \rightarrow NH_3(g) + H_2S(g)$
**3.** $NH_4Br(s) + NaOH(aq) \rightarrow NH_3(g) + H_2O(\ell) + NaBr(aq)$
**5.** 10.3 m
**7.** $1.40 \times 10^4$ ft
**9.** 1697.5 atm, $1.7200 \times 10^3$ bar
**11.** 0.857 atm
**13.** 8.00 L
**15.** 14.3 gills
**17.** 134 L
**19.** 35.2 psi
**21. (a)** 19.8 atm **(b)** 23.0 atm
**23.** The mass of a gas in a given volume changes proportionately to the absolute temperature. This statement will only be true at $-19.8°C$.
**25. (a)** $2\,Na(s) + 2\,HCl(g) \rightarrow 2\,NaCl(s) + H_2(g)$ **(b)** 4.23 L

**27.** $3.0 \times 10^6$ L $H_2 = 3.0 \times 10^3$ m$^3$

**29.** 24.2 L

**31.** **(a)** 932 L $H_2S$   **(b)** 1.33 kg, 466 L $SO_2$

**33.** $X_{SO_3} = 0.135$; $P_{SO_3} = 0.128$ atm

**35.** $X_{N_2} = 0.027$; $P_{N_2} = 1.6 \times 10^{-4}$ atm

**37.** **(a)** $X_{CO} = 0.444$   **(b)** $X_{CO} = 0.33$

**39.** **(a)** $5.8 \times 10^{17}$   **(b)** 520 L

**41.** **(a)** $1.93 \times 10^3$ m s$^{-1}$ = 1.93 km s$^{-1}$   **(b)** 226 m s$^{-1}$

**43.** 6100 m s$^{-1}$ (6000 K), 790 m s$^{-1}$ (100 K)

**45.** Greater. As $T$ increases, the Maxwell–Boltzmann distribution shifts to higher speeds.

**47.** $6.1 \times 10^{-9}$ m

**49.** 92.6 g mol$^{-1}$

**51.** 1840 stages

**53.** $7.4 \times 10^{-7}$ atm; 20 m$^2$ s$^{-1}$

**55.** 162 atm = $2.39 \times 10^3$ psi

**57.** **(a)** 27.8 atm   **(b)** 24.6 atm; attractive forces dominate.


Chapter 4

**1.** 0.9344 g L$^{-1}$

**3.** 2.92 g $CaCO_3$

**5.** 0.69 atm; a fraction of 0.31 (31%) lies below.

**7.** No phase change will occur.

**11.** **(a)** Above. If gas and solid coexist at $-84.0°C$, their coexistence must extend upward in temperature to the triple point.
    **(b)** The solid will sublime at some temperature below $-84°C$.

**13.** Heavier gases have stronger attractive forces, favoring the liquid and solid states.

**15.** Ne < NO < $NH_3$ < RbCl; nonpolar < polar < hydrogen bonded < ionic

**17.** An eight-membered ring of alternating H's and O's is reasonable. The ring is probably not planar.

**19.** The two have comparable molar masses, but the hydrogen bonding in hydrazine should give it a higher boiling point.

**21.** **(a)** $5.53 \times 10^{-3}$ M   **(b)** $5.5 \times 10^{-3}$ molal   **(c)** 3.79 L

**23.** Molarity = 12.39 M in HCl; mole fraction of HCl = 0.2324; molality = 16.81 molal in HCl

**25.** 8.9665 molal

**27.** 0.00643 g $H_2O$

**29.** **(a)** 1.33 g mL$^{-1}$   **(b)** 164 mL

**31.** 1.06 M in NaOH

**33.** 16.8 mL of 7.91 M $HNO_3$

**35.** $6.74 \times 10^3$ L $CO_2(g)$

**37.** **(a)** $PF_3 + 3 H_2O \rightarrow H_3PO_3 + 3 HF$   **(b)** $[H_3PO_3] = 0.0882$ M; [HF] = 0.265 M

**39.** $7.175 \times 10^{-3}$ M $Fe^{2+}(aq)$

**41.** 0.04841 M $HNO_3$

**43.** 0.2985 atm

**45.** 3.34 K kg mol$^{-1}$

**47.** 340 g mol$^{-1}$

**49.** $1.7 \times 10^2$ g mol$^{-1}$

**51.** $-2.8°C$. As the solution becomes more concentrated, its freezing point will decrease further.

**53.** 2.70 particles (complete dissociation of $Na_2SO_4$ gives 3 particles)

**55.** $1.88 \times 10^4$ g mol$^{-1}$

**57.** $7.46 \times 10^3$ g mol$^{-1}$

**59. (a)** 0.17 mol $CO_2$   **(b)** Because the partial pressure of $CO_2$ in the atmosphere is much less than 1 atm, the excess $CO_2$ will bubble out from the solution and escape when the cap is removed.

**61.** $4.13 \times 10^2$ atm

**63.** 0.774

**65. (a)** 0.491   **(b)** 0.250 atm   **(c)** 0.575

## Chapter 5

**1. (a)** $\dfrac{P_{H_2O}^2}{P_{H_2}^2 P_{O_2}} = K$   **(b)** $\dfrac{P_{XeF_6}}{P_{Xe} P_{F_2}^3} = K$   **(c)** $\dfrac{P_{CO_2}^{12} P_{H_2O}^6}{P_{C_6H_6}^2 P_{O_2}^{15}} = K$

**3.** $P_4(g) + 6\,Cl_2(g) + 2\,O_2(g) \rightleftarrows 4\,POCl_3(g)$   $\dfrac{P_{POCl_3}^4}{P_{P_4} P_{Cl_2}^6 P_{O_2}^2} = K$

**5. (a)** $\dfrac{P_{CO_2} P_{H_2}}{P_{CO} P_{H_2O}} = K$   **(b)** 0.056 atm

**7.** $1.04 \times 10^{-4}$

**9.** 14.6

**11. (a)** $P_{SO_2} = P_{Cl_2} = 0.58$ atm; $P_{SO_2Cl_2} = 0.14$ atm   **(b)** 2.4

**13.** $K_1 = (K_2)^3$

**15.** $K_2/K_1$

**17. (a)** 0.180 atm   **(b)** 0.756

**19.** $P_{PCl_5} = 0.078$ atm; $P_{PCl_3} = P_{Cl_2} = 0.409$ atm

**21.** $P_{Br_2} = 0.0116$ atm; $P_{I_2} = 0.0016$ atm; $P_{IBr} = 0.0768$ atm

**23.** $P_{N_2} = 0.52$ atm; $P_{O_2} = 0.70$ atm; $P_{NO} = 3.9 \times 10^{-16}$ atm

**25.** $P_{N_2} = 0.0148$ atm; $P_{H_2} = 0.0445$ atm; $P_{NH_3} = 0.941$ atm

**27.** $5.6 \times 10^{-5}$ mol $L^{-1}$

**29. (a)** $9.83 \times 10^{-4}$   **(b)** Net consumption

**31.** $K > 5.1$

**33. (a)** 0.80, left   **(b)** $P_{P_4} = 5.12$ atm, $P_{P_2} = 1.77$ atm   **(c)** Dissociation

**35. (a)** Shifts left   **(b)** Shifts right   **(c)** Shifts right   **(d)** The volume must have been increased to keep the total pressure constant; shifts left   **(e)** No effect

**37. (a)** Exothermic   **(b)** Decrease

**39.** Run the first step at low temperature and high pressure, and the second step at high temperature and low pressure.

**41. (a)** $\dfrac{(P_{H_2S})^8}{(P_{H_2})^8} = K$   **(b)** $\dfrac{(P_{COCl_2})(P_{H_2})}{P_{Cl_2}} = K$   **(c)** $P_{CO_2} = K$   **(d)** $\dfrac{1}{(P_{C_2H_2})^3} = K$

**43. (a)** $\dfrac{[Zn^{2+}]}{[Ag^+]^2} = K$   **(b)** $\dfrac{[VO_3(OH)^{2-}][OH^-]}{[VO_4^{3-}]} = K$   **(c)** $\dfrac{[HCO_3^-]^6}{[As(OH)_6^{3-}]^2 P_{CO_2}^6} = K$

**45. (a)** The graph is a straight line passing through the origin.   **(b)** The experimental $K$'s range from $3.42 \times 10^{-2}$ to $4.24 \times 10^{-2}$ with a mean of $3.84 \times 10^{-2}$.

**47.** $2.0 \times 10^{-2}$

**49. (a)** 0.31 atm   **(b)** 1.65 atm, 0.15 atm

**51. (a)** $Q = 2.05$; reaction shifts to right.   **(b)** $Q = 3.27$; reaction shifts to left.

**53. (a)** $8.46 \times 10^{-5}$   **(b)** 0.00336 atm

**55.** 76

**57. (a)** $K_1 = 1.64 \times 10^{-2}$; $K_2 = 5.4$   **(b)** 330

## Chapter 6

**1. (a)** $Cl^-$ cannot act as a Brønsted-Lowry acid.   **(b)** $SO_4^{2-}$   **(c)** $NH_3$   **(d)** $NH_2^-$   **(e)** $OH^-$

3. $HCO_3^-(aq)$ serves as the base.

5. 3.70

7. $3 \times 10^{-7}$ M $< [H_3O^+] < 3 \times 10^{-6}$ M; $3 \times 10^{-9}$ M $< [OH^-] < 3 \times 10^{-8}$ M

9. $[H_3O^+] = 1.0 \times 10^{-8}$ M; $[OH^-] = 1.7 \times 10^{-6}$ M

11. The first reaction is more likely to be correct. In the second case, the reactant $H_3O^+$ would be present at very low concentration ($10^{-7}$ M) and would give neither a fast nor a vigorous reaction.

13. (a) $C_{10}H_{15}ON(aq) + H_2O(\ell) \rightleftarrows C_{10}H_{15}ONH^+(aq) + OH^-(aq)$   (b) $7.1 \times 10^{-11}$
    (c) stronger

15. $K = 24$; $HClO_2$ is the stronger acid and $NO_2^-$ the stronger base.

17. (a) methyl orange   (b) 3.8 to 4.4

19. 2.35

21. (a) 2.45   (b) 0.72 mol

23. 1.16

25. $1.2 \times 10^{-6}$

27. 10.4

29. 0.083 M

31. The reaction gives a base of moderate strength, the acetate ion, in solution, so the pH $> 7$.

33. $HCl < NH_4Br < KI < NaCH_3COO < NaOH$

35. 8.08

37. (a) 4.36   (b) 4.63

39. *m*-chlorobenzoic acid

41. 639 mL

43. 13.88, 11.24, 7.00, 2.77

45. 2.86, 4.72, 8.71, 11.00

47. 11.89, 11.52, 10.79, 8.20, 6.05, 3.90, 1.95

49. 0.97 g

51. 0.0872 g, pH = 6.24, bromothymol blue

53. $4 \times 10^{-7}$

55. $[H_3AsO_4] = 8.0 \times 10^{-2}$ M; $[H_2AsO_4^-] = [H_3O^+] = 2.0 \times 10^{-2}$ M;
    $[HAsO_4^{2-}] = 9.3 \times 10^{-8}$ M; $[AsO_4^{3-}] = 1.4 \times 10^{-17}$ M

57. $[PO_4^{3-}] = 0.020$ M; $[HPO_4^{2-}] = [OH^-] = 0.030$ M; $[H_2PO_4^-] = 1.61 \times 10^{-7}$ M;
    $[H_3PO_4] = 7.1 \times 10^{-18}$ M

59. $[H_2CO_3] = 8.5 \times 10^{-6}$ M; $[HCO_3^-] = 1.5 \times 10^{-6}$ M; $[CO_3^{2-}] = 2.8 \times 10^{-11}$ M

61. 6.87

63. 1.51, 1.61, 2.07, 4.01, 6.07, 8.77, 9.29, 11.51

## Chapter 7

1. 4.65 L

3. About 48°C

5. $Fe_2(SO_4)_3(s) \rightleftarrows 2\, Fe^{3+}(aq) + 3\, SO_4^{2-}(aq)$
   $[Fe^{3+}]^2[SO_4^{2-}]^3 = K_{sp}$

7. 0.0665 g per 100 mL water

9. 14.3 g $L^{-1}$

11. $[I^-] = 6.2 \times 10^{-10}$ M, $[Hg_2^{2+}] = 3.1 \times 10^{-10}$ M

13. $1.9 \times 10^{-12}$

15. $1.6 \times 10^{-8}$

17. Yes. The initial reaction quotient is $6.2 \times 10^{-10} > K_{sp}$.

19. No

21. $[Pb^{2+}] = 2.3 \times 10^{-10}$ M; $[IO_3^-] = 0.033$ M

23. $[Ag^+] = 1.8 \times 10^{-2}$ M; $[CrO_4^{2-}] = 6.2 \times 10^{-9}$ M

**25.** $2.4 \times 10^{-8}$ mol $L^{-1}$

**27.** (a) $3.4 \times 10^{-6}$ M  (b) $1.6 \times 10^{-14}$ M

**29.** $9.1 \times 10^{-3}$ M

**31.** In pure water: $1.2 \times 10^{-4}$ M; at pH 7: 0.15 M

**33.** (a) Unchanged  (b) Increase  (c) Increase

**35.** (a) $8.6 \times 10^{-4}$ M, $Pb^{2+}$ in solid  (b) $3.1 \times 10^{-7}$

**37.** $[I^-] = 5.3 \times 10^{-4}$ M

**39.** $2 \times 10^{-13}$ M

**41.** pH = 2.4; $[Pb^{2+}] = 6 \times 10^{-11}$ M

**43.** $[Cu(NH_3)_4^{2+}] = 0.10$ M; $[Cu^{2+}] = 6 \times 10^{-14}$ M

**45.** $[K^+] = 0.0051$ M and $[Na^+] = 0.0076$ M

**47.** More will dissolve in 1 M NaCl ($3 \times 10^{-5}$ versus $1.3 \times 10^{-5}$ mol $L^{-1}$). Less will dissolve in 0.100 M NaCl ($3 \times 10^{-6}$ versus $1.3 \times 10^{-5}$ mol $L^{-1}$).

**49.** $Cu^{2+}(aq) + 2 H_2O(\ell) \rightleftarrows CuOH^+(aq) + H_3O^+(aq)$  or
$Cu(H_2O)_4^{2+}(aq) + H_2O(\ell) \rightleftarrows Cu(H_2O)_3OH^+(aq) + H_3O^+(aq)$

**51.** 5.3

**53.** $K_a = 9.6 \times 10^{-10}$

**55.** In the first case, solid $Pb(OH)_2$ will precipitate; $[Pb^{2+}] = 4.2 \times 10^{-13}$ M, $[Pb(OH)_3^-] = 0.17$ M. In the second case, solid $Pb(OH)_2$ will not precipitate; $[Pb^{2+}] = 1 \times 10^{-13}$ M, $[Pb(OH)_3^-] = 0.050$ M.

## Chapter 8

**1.** $-2.16 \times 10^4$ L atm $= -2.19 \times 10^6$ J

**3.** 86.6 m

**5.** 24.8, 28.3, 29.6, 31.0, 32.2 J $K^{-1}$ $mol^{-1}$. Extrapolating gives about 33.5 J $K^{-1}$ $mol^{-1}$ for Fr.

**7.** 26.1, 25.4, 25.0, 24.3, 25.4, 27.6 J $K^{-1}$ $mol^{-1}$

**9.** 0.468 J $K^{-1}$ $g^{-1}$

**11.** $q_1 = Mc_{s_1} \Delta T_1 = -q_2 = -Mc_{s_2} \Delta T_2 \qquad \dfrac{C_{s_1}}{C_{s_2}} = -\dfrac{\Delta T_2}{\Delta T_1}$

**13.** 314 J $g^{-1}$ (modern value is 333 J $g^{-1}$)

**15.** $w = -323$ J; $\Delta E = -394$ J; $q = -71$ J

**17.** (a) 38.3 L  (b) $w = -1.94 \times 10^3$ J; $q = 0$; $\Delta E = -1.94 \times 10^3$ J  (c) 272 K

**19.** $\Delta E = +11.2$ kJ; $q = 0$; $w = +11.2$ kJ

**21.** (a) $w$ is zero, $q$ is positive, and $\Delta E$ is positive.
(b) $w$ is zero, $q$ is negative, and $\Delta E$ is negative.
(c) $(w_1 + w_2)$ is zero. $(\Delta E_1 + \Delta E_2) = (q_1 + q_2)$. The latter two sums could be any of three possibilities: both positive, both negative, or both zero.

**23.** (a) $-6.68$ kJ  (b) $+7.49$ kJ  (c) $+0.594$ kJ

**25.** 41.3 kJ

**27.** $+513$ J

**29.** 10.4°C

**31.** $-623.5$ kJ

**33.** Diamond. This is recommended as a source of heat only as a last resort, however!

**35.** $-555.93$ kJ

**37.** (a) $-878.26$ kJ  (b) $-1.35 \times 10^7$ kJ of heat absorbed

**39.** (a) $-81.4$ kJ  (b) 55.1°C

**41.** $\Delta H_f^\circ = -152.3$ kJ $mol^{-1}$

**43.** (a) $C_{10}H_8(s) + 12 O_2(g) \rightarrow 10 CO_2(g) + 4 H_2O(\ell)$
(b) $-5157$ kJ  (c) $-5162$ kJ  (d) $+84$ kJ $mol^{-1}$

**45.** $-264$ kJ $mol^{-1}$

**47.** $-1.58 \times 10^3$ kJ $mol^{-1}$

**49.** Each side of the equation has three B—Br and three B—Cl bonds.
**51.** $w = -6.87$ kJ; $q = +6.87$ kJ; $\Delta E = \Delta H = 0$
**53.** $T_f = 144$ K; $w = \Delta E = -3.89$ kJ; $\Delta H = -6.48$ kJ
**55.** **(a)** 0.333  **(b)** $q = -1000$ J  **(c)** $w = -500$ J
**57.** $w = 91$ J; $q = 891$ J

## Chapter 9

**1.** $9.61$ J $K^{-1}$ mol$^{-1}$
**3.** $29$ kJ mol$^{-1}$
**5.** $\Delta E = \Delta H = 0$; $w = -1.22 \times 10^4$ J; $q = +1.22 \times 10^4$ J; $\Delta S = +30.5$ J $K^{-1}$
**7.** $\Delta S_{sys} = +30.2$ J $K^{-1}$; $\Delta S_{surr} = -30.2$ J $K^{-1}$; $\Delta S_{univ} = 0$
**9.** $\Delta S_{Fe} = -8.24$ J $K^{-1}$; $\Delta S_{H_2O} = +9.49$ J $K^{-1}$; $\Delta S_{tot} = +1.25$ J $K^{-1}$
**11.** **(a)** $6 \times 6 = 36$  **(b)** 1 in 36
**13.** The tendency for entropy to increase
**15.** $10^{-3.62 \times 10^{23}}$ (i.e., one part in $10^{3.62 \times 10^{23}}$)
**17.** **(a)** $\Delta S > 0$  **(b)** $\Delta S > 0$  **(c)** $\Delta S < 0$
**19.** **(a)** $-116.58$ J $K^{-1}$  **(b)** Lower (more negative)
**21.** $\Delta S° = -162.54, -181.12, -186.14, -184.72, -191.08$ J $K^{-1}$. The entropy changes in these reactions become increasingly negative with increasing atomic mass, except that the rubidium reaction is out of line.
**23.** $\Delta S_{surr}$ must be positive, and greater in magnitude than $44.7$ J $K^{-1}$.
**25.** $\Delta S_{sys} > 0$ because the gas produced (oxygen) has many possible microstates.
**27.** **(a)** $+740$ J  **(b)** $2.65$ kJ  **(c)** No  **(d)** 196 K
**29.** $q = +38.7$ kJ; $w = -2.9$ kJ; $\Delta E = +35.8$ kJ; $\Delta S_{sys} = +110$ J $K^{-1}$; $\Delta G = 0$
**31.** Overall reaction: $2$ Fe$_2$O$_3$(s) + 3 C(s) → 4 Fe(s) + 3 CO$_2$(g)
$\Delta G = +840 + 3(-400) = -360 < 0$
**33.** **(a)** $0 < T < 3000$ K  **(b)** $0 < K < 1050$ K  **(c)** Spontaneous at all temperatures
**35.** WO$_3$(s) + 3 H$_2$(g) → W(s) + 3 H$_2$O(g)
$\Delta H° = 117.41$ kJ; $\Delta S° = 131.19$ J $K^{-1}$. $\Delta G < 0$ for $T > \Delta H°/\Delta S° = 895$ K.
**37.** The $\Delta S$ of the process is evidently negative.
**39.** $\Delta G° = -550.23$ kJ; $K = 2.5 \times 10^{96}$
**41.** **(a)** $2.6 \times 10^{12}$  $\dfrac{P_{SO_3}}{(P_{SO_2})(P_{O_2})^{1/2}} = K$
**(b)** $5.4 \times 10^{-35}$  $(P_{O_2})^{1/2} = K$
**(c)** $5.3 \times 10^3$  $[Cu^{2+}][Cl^-]^2 = K$
**43.** Low temperature and high pressure
**45.** $-58$ kJ
**47.** **(a)** $-56.9$ kJ  **(b)** $\Delta H° = -55.6$ kJ, $\Delta S° = +4.2$ J $K^{-1}$
**49.** $4.3 \times 10^{-3}$
**51.** **(a)** $23.8$ kJ mol$^{-1}$  **(b)** $T_b = 240$ K
**53.** $80.4°C$

## Chapter 10

**1.** **(a)** $2$ VO$_2^+$(aq) + SO$_2$(g) → 2 VO$^{2+}$(aq) + SO$_4^{2-}$(aq)
**(b)** Br$_2$(ℓ) + SO$_2$(g) + 6 H$_2$O(ℓ) → 2 Br$^-$(aq) + SO$_4^{2-}$(aq) + 4 H$_3$O$^+$(aq)
**(c)** Cr$_2$O$_7^{2-}$(aq) + 3 Np$^{4+}$(aq) + 2 H$_3$O$^+$(aq) → 2 Cr$^{3+}$(aq) + 3 NpO$_2^+$(aq) + 3 H$_2$O(ℓ)
**(d)** 5 HCOOH(aq) + 2 MnO$_4^-$(aq) + 6 H$_3$O$^+$ aq) → 5 CO$_2$(g) + 2 Mn$^{2+}$(aq) + 14 H$_2$O(ℓ)
**(e)** 3 Hg$_2$HPO$_4$(s) + 2 Au(s) + 8 Cl$^-$(aq) + 3 H$_3$O$^+$(aq) →
6 Hg(ℓ) + 3 H$_2$PO$_4^-$(aq) + 2 AuCl$_4^-$(aq) + 3 H$_2$O(ℓ)
**3.** **(a)** 2 Cr(OH)$_3$(s) + 3 Br$_2$(aq) + 10 OH$^-$(aq) → 2 CrO$_4^{2-}$(aq) + 6 Br$^-$(aq) + 8 H$_2$O(ℓ)

     **(b)** $ZrO(OH)_2(s) + 2\,SO_3^{2-}(aq) \rightarrow Zr(s) + 2\,SO_4^{2-}(aq) + H_2O(\ell)$

     **(c)** $7\,HPbO_2^-(aq) + 2\,Re(s) \rightarrow 7\,Pb(s) + 2\,ReO_4^-(aq) + H_2O(\ell) + 5\,OH^-(aq)$

     **(d)** $4\,HXeO_4^-(aq) + 8\,OH^-(aq) \rightarrow 3\,XeO_6^{4-}(aq) + Xe(g) + 6\,H_2O(\ell)$

     **(e)** $N_2H_4(aq) + 2\,CO_3^{2-}(aq) \rightarrow N_2(g) + 2\,CO(g) + 4\,OH^-(aq)$

**5. (a)** $Fe^{2+}(aq) \rightarrow Fe^{3+}(aq) + e^-$      oxidation

     $H_2O_2(aq) + 2\,H_3O^+(aq) + 2\,e^- \rightarrow 4\,H_2O(\ell)$      reduction

     **(b)** $5\,H_2O(\ell) + SO_2(aq) \rightarrow HSO_4^-(aq) + 3\,H_3O^+(aq) + 2\,e^-$      oxidation

     $MnO_4^-(aq) + 8\,H_3O^+(aq) + 5\,e^- \rightarrow Mn^{2+}(aq) + 12\,H_2O(\ell)$      reduction

     **(c)** $ClO_2^-(aq) \rightarrow ClO_2(g) + e^-$      oxidation

     $ClO_2^-(aq) + 4\,H_3O^+(aq) + 4\,e^- \rightarrow Cl^-(aq) + 6\,H_2O(\ell)$      reduction

**7.** $3\,HNO_2(aq) \rightarrow NO_3^-(aq) + 2\,NO(g) + H_3O^+(aq)$

**11.** 0.180 mol

**13. (a)** $Zn(s) + Cl_2(g) \rightarrow Zn^{2+}(aq) + 2\,Cl^-(aq)$

     **(b)** $1.20 \times 10^3$ C; $1.24 \times 10^{-2}$ mol $e^-$

     **(c)** Decreases by 0.407 g

     **(d)** 0.152 L $Cl_2$ consumed

**15.** +2

**17. (a)** Anode: $Cl^- \rightarrow \frac{1}{2}\,Cl_2(g) + e^-$; cathode: $K^+ + e^- \rightarrow K(\ell)$;

     total: $Cl^- + K^+ \rightarrow K(\ell) + \frac{1}{2}\,Cl_2(g)$

     **(b)** Mass K = 14.6 g; mass $Cl_2$ = 13.2 g

**19.** $\Delta G° = -921$ J; $w_{max} = +921$ J

**21. (a)** Anode: $Co \rightarrow Co^{2+} + 2\,e^-$; cathode: $Br_2 + 2\,e^- \rightarrow 2\,Br^-$;

     total: $Co + Br_2 \rightarrow Co^{2+} + 2\,Br^-$

     **(b)** 1.34 V

**23. (a)** Anode: $Zn \rightarrow Zn^{2+} + 2\,e^-$; cathode: $In^{3+} + 3\,e^- \rightarrow In$

     **(b)** $-0.338$ V

**25.** Reducing agent

**27.** $Br_2$ less effective than $Cl_2$

**29. (a)** $BrO_3^-$   **(b)** Cr   **(c)** Co

**31. (a)** $\mathscr{E}° = -0.183$ V   **(b)** It will not disproportionate.

**33. (a)** No   **(b)** $Br^-$

**35.** 0.37 V

**37.** $-0.31$ V

**39.** 5.17

**41. (a)** 0.31 V   **(b)** $1 \times 10^{-8}$ M

**43.** $K = 3 \times 10^{31}$, orange

**45.** $3 \times 10^6$

**47.** pH = 2.53; $K_a$ = 0.0029

**49. (a)** 1.065 V   **(b)** $1.3 \times 10^{-11}$ M   **(c)** $7.6 \times 10^{-13}$

**51.** 2.041 V; 12.25 V

**53. (a)** $9.3 \times 10^6$ C   **(b)** $1.1 \times 10^8$ J

**55.** No, because $H_2SO_4(aq)$ is not the only substance whose amount changes during discharge. The accumulated $PbSO_4$ must also be removed and replaced by Pb and $PbO_2$.

**57.** 7900 J g$^{-1}$

**59.** $\Delta\mathscr{E}° = -0.419$ V, not spontaneous under standard conditions (pH = 14). If $[OH^-]$ is small enough, the equilibrium will shift to the right and the reaction will become spontaneous.

**61.** According to its reduction potential, yes. In practice, however, the sodium would react instantly and explosively with the water and would therefore be useless for this purpose.

**63. (a)** Ni   **(b)** 21.9 g (provided the electrolyte volume is very large)   **(c)** $H_2$

## Chapter 11

1. $5.3 \times 10^{-5}$ mol L$^{-1}$ s$^{-1}$

3. rate $= -\dfrac{\Delta[N_2]}{\Delta t} = -\dfrac{1}{3}\dfrac{\Delta[H_2]}{\Delta t} = \dfrac{1}{2}\dfrac{\Delta[NH_3]}{\Delta t}$

5. **(a)** Rate $= k[NO]^2[H_2]$; $k$ has the units L$^2$ mol$^{-2}$ s$^{-1}$
   **(b)** An increase by a factor of 18

7. Rate $= k[C_5H_5N][CH_3I]$   **(b)** $k = 75$ L mol$^{-1}$ s$^{-1}$   **(c)** $7.5 \times 10^{-8}$ mol L$^{-1}$ s$^{-1}$

9. $3.2 \times 10^4$ s

11. 0.0019 atm

13. $5.3 \times 10^{-3}$ s$^{-1}$

15. $2.34 \times 10^{-5}$ M

17. $2.9 \times 10^{-6}$ s

19. **(a)** Bimolecular, rate $= k[HCO][O_2]$
    **(b)** Termolecular, rate $= k[CH_3][O_2][N_2]$
    **(c)** Unimolecular, rate $= k[HO_2NO_2]$

21. **(a)** The first step is unimolecular; the others are bimolecular.
    **(b)** $H_2O_2 + O_3 \rightarrow H_2O + 2 O_2$
    **(c)** O, ClO, CF$_2$Cl, Cl

23. 0.26 L mol$^{-1}$ s$^{-1}$

25. **(a)** $A + B + E \rightarrow D + F$     rate $= \dfrac{k_1 k_2}{k_{-1}}\dfrac{[A][B][E]}{[D]}$

    **(b)** $A + D \rightarrow B + F$     rate $= \dfrac{k_1 k_2 k_3}{k_{-1} k_{-2}}\dfrac{[A][D]}{[B]}$

27. Only mechanism (b)

29. Only mechanism (a)

31. **(a)** rate $= \dfrac{k_2 k_1[A][B][E]}{k_2[E] + k_{-1}[D]}$
    **(b)** When $k_2[E] \ll k_{-1}[D]$

33. $\dfrac{d[Cl_2]}{dt} = \dfrac{k_1 k_2 [NO_2Cl]^2}{k_{-1}[NO_2] + k_2[NO_2Cl]}$

35. **(a)** $4.25 \times 10^5$ J   **(b)** $1.54 \times 10^{11}$ L mol$^{-1}$ s$^{-1}$

37. **(a)** $1.49 \times 10^{-3}$ L mol$^{-1}$ s$^{-1}$   **(b)** $1.30 \times 10^3$ s

39. **(a)** $4.3 \times 10^{13}$ s$^{-1}$   **(b)** $1.7 \times 10^5$ s$^{-1}$

41. 70.3 kJ mol$^{-1}$

43. $6.0 \times 10^9$ L mol$^{-1}$ s$^{-1}$

45. **(a)** $1.2 \times 10^{-3}$ mol L$^{-1}$ s$^{-1}$   **(b)** $5 \times 10^{-5}$ M

## Chapter 12

1. **(a)** $2\ {}^{12}_{6}C \rightarrow {}^{23}_{12}Mg + {}^{1}_{0}n$
   **(b)** ${}^{15}_{7}N + {}^{1}_{1}H \rightarrow {}^{12}_{6}C + {}^{4}_{2}He$
   **(c)** $2\ {}^{3}_{2}He \rightarrow {}^{4}_{2}He + 2\ {}^{1}_{1}H$

3. **(a)** $3.300 \times 10^{10}$ kJ mol$^{-1}$ = 342.1 MeV total; 8.551 MeV per nucleon
   **(b)** $7.313 \times 10^{10}$ kJ mol$^{-1}$ = 757.9 MeV total; 8.711 MeV per nucleon
   **(c)** $1.738 \times 10^{11}$ kJ mol$^{-1}$ = 1801.7 MeV total; 7.570 MeV per nucleon

5. The two ${}^4$He atoms are more stable, with a mass lower than that of one ${}^8$Be atom by $9.86 \times 10^{-5}$ u.

7. 16.96 MeV

9. **(a)** ${}^{39}_{17}Cl \rightarrow {}^{39}_{18}Ar + {}^{0}_{-1}e^- + \tilde{\nu}$
   **(b)** ${}^{22}_{11}Na \rightarrow {}^{22}_{10}Ne + {}^{0}_{1}e^+ + \nu$
   **(c)** ${}^{224}_{88}Ra \rightarrow {}^{220}_{86}Rn + {}^{4}_{2}He$
   **(d)** ${}^{82}_{38}Sr + {}^{0}_{-1}e^- \rightarrow {}^{82}_{37}Rb + \nu$

11. $^{19}$Ne decays to $^{19}$F by positron emission; $^{23}$Ne decays to $^{23}$Na by beta decay.

13. An electron, 0.78 MeV

15. $^{30}_{14}\text{Si} + ^{1}_{0}n \rightarrow ^{31}_{14}\text{Si} \rightarrow ^{31}_{15}\text{P} + ^{0}_{-1}e^{-} + \tilde{\nu}$

17. $^{210}_{84}\text{Po} \rightarrow ^{206}_{82}\text{Pb} + ^{4}_{2}\text{He}$
$^{4}_{2}\text{He} + ^{9}_{4}\text{Be} \rightarrow ^{12}_{6}\text{C} + ^{1}_{0}n$

19. $3.7 \times 10^{10}$ min$^{-1}$

21. (a) $1.0 \times 10^{6}$   (b) $5.9 \times 10^{4}$

23. $1.6 \times 10^{18}$ disintegrations s$^{-1}$

25. 4200 years

27. $3.0 \times 10^{9}$ years

29. $6.0 \times 10^{9}$ years

31. $^{11}_{6}\text{C} \rightarrow ^{11}_{5}\text{B} + ^{0}_{1}e^{+} + \nu$; $^{15}_{8}\text{O} \rightarrow ^{15}_{7}\text{N} + ^{0}_{1}e^{+} + \nu$

33. Exposure from $^{15}$O is greater by a factor of $1.72/0.99 = 1.74$.

35. (a) $2.3 \times 10^{10}$ Bq   (b) 2.4 mrad   (c) Yes. A dose of 500 rad has a 50% chance of being lethal; this dose is greater than 850 rad in the first 8 days.

37. (a) $^{90}_{38}\text{Sr} \rightarrow ^{90}_{40}\text{Zr} + 2\ ^{0}_{-1}e^{-} + 2\ \tilde{\nu}$
(b) 2.8 MeV   (c) $5.23 \times 10^{12}$ disintegrations s$^{-1}$.
(d) $4.44 \times 10^{11}$ disintegrations s$^{-1}$.

39. Increase (assuming no light isotopes of U are products of the decay)

41. $7.59 \times 10^{7}$ kJ g$^{-1}$

## Chapter 13

1. 0.66 m s$^{-1}$

3. 3.04 m

5. (a) $5.00 \times 10^{5}$ s$^{-1}$   (b) 4.4 min

7. $\lambda = 1.313$ m, time $= 0.0873$ s

9. Blue

11. Part of the yellow light, together with green and blue light, will eject electrons from Cs. No visible light will eject electrons from Se; ultraviolet light is required.

13. (a) $7.4 \times 10^{-20}$ J   (b) $4.0 \times 10^{5}$ m s$^{-1}$

15. Red

17. 550 nm, green

19. (a) $3.371 \times 10^{-19}$ J   (b) 203.0 kJ mol$^{-1}$   (c) $4.926 \times 10^{-3}$ mol s$^{-1}$

21. $r_3 = 0.0952$ nm, $E_3 = -6.06 \times 10^{-18}$ J, energy per mole $= 3.65 \times 10^{3}$ kJ mol$^{-1}$,
$\nu = 1.14 \times 10^{16}$ s$^{-1}$, $\lambda = 2.63 \times 10^{-8}$ m $= 26.3$ nm

23. 72.90 nm, ultraviolet

25. (a) 100 cm, 33 cm   (b) 2 nodes

27. (a) $7.27 \times 10^{-7}$ m   (b) $3.96 \times 10^{-10}$ m   (c) $2.2 \times 10^{-34}$ m

29. (a) $5.8 \times 10^{4}$ m s$^{-1}$   (b) 7.9 m s$^{-1}$

31. $E_1 = 3.36 \times 10^{-18}$ J; $E_2 = 1.34 \times 10^{-17}$ J; $E_3 = 3.02 \times 10^{-17}$ J; $\lambda = 1.97 \times 10^{-8}$ m

33. Only (b) is allowed.

35. (a) $4p$   (b) $2s$   (c) $6f$

37. (a) 2 radial, 1 angular   (b) 1 radial, 0 angular   (c) 2 radial, 3 angular

39. $\chi^2(p_z) \propto \cos^2 \theta = 0$ for $\theta = \pi/2$; $d_{xz}$ nodal planes are the $yz$ and $xy$ planes; $d_{x^2-y^2}$ nodal planes are two planes containing the $z$ axis at 45° from the $x$ and $y$ axes.

41. (a) $1s^2 2s^2 2p^2$   (b) $[\text{Ar}]3d^{10}4s^2 4p^4$   (c) $[\text{Ar}]3d^6 4s^2$

43. Be$^+$: $1s^2 2s^1$; C$^-$: $1s^2 2s^2 2p^3$; Ne$^{2+}$: $1s^2 2s^2 2p^4$; Mg$^+$: $[\text{Ne}]3s^1$; P$^{2+}$: $[\text{Ne}]3s^2 3p^1$; Cl$^-$: $[\text{Ne}]3s^2 3p^6$; As$^+$: $[\text{Ar}]3d^{10}4s^2 4p^2$; I$^-$: $[\text{Kr}]4d^{10}5s^2 5p^6$. All except Cl$^-$ and I$^-$ are paramagnetic.

45. (a) In   (b) S$^{2-}$   (c) Mn$^{4+}$

47. 117

49. First "noble gases" at $Z = 1, 5, 9$
51. **(a)** K   **(b)** Cs   **(c)** Kr   **(d)** K   **(e)** $Cl^-$
53. **(a)** $S^{2-}$   **(b)** $Ti^{2+}$   **(c)** $Mn^{2+}$   **(d)** $Sr^{2+}$
55. **(a)** Sr   **(b)** Rn   **(c)** Xe   **(d)** Sr
57. **(a)** Cs   **(b)** F   **(c)** K   **(d)** At
59. 318.4 nm, near ultraviolet

## Chapter 14

1. The As—H bond length will lie between 1.42 and 1.71 Å (observed: 1.52 Å). $SbH_3$ will have the weakest bond.
3. **(a)** $CI_4$   **(b)** $OF_2$   **(c)** $SiH_4$
5. Estimated: 1.05; from figure 14-1: 0.96.
7. $450 \text{ kJ mol}^{-1}$
9. ClO, 16%; KI, 74%; TlCl, 38%; InCl, 33%
11. HF, 40%; HCl, 19%; HBr, 14%; HI, 8%; CsF, 87%
13. **(a)** SN = 4, tetrahedral   **(b)** SN = 3, trigonal planar   **(c)** SN = 6, octahedral
    **(d)** SN = 4, pyramidal   **(e)** SN = 5, distorted T-shape
15. **(a)** SN = 6, square planar   **(b)** SN = 4, bent, angle < 109.5°
    **(c)** SN = 4, pyramidal, angle < 109.5°   **(d)** SN = 2, linear
17. **(a)** $SO_3$   **(b)** $NF_3$   **(c)** $NO_2^-$   **(d)** $CO_3^{2-}$
19. Only (d) and (e) are polar.
21. No, because VSEPR theory predicts a steric number of 3 and a bent molecule in both cases.
23. **(a)** Linear   **(b)** The N end
25. **(a)** $F_2$: $(\sigma_{2s})^2(\sigma_{2s}^*)^2(\sigma_{2p})^2(\pi_{2p})^4(\pi_{2p}^*)^4$
    $F_2^+$: $(\sigma_{2s})^2(\sigma_{2s}^*)^2(\sigma_{2p})^2(\pi_{2p})^4(\pi_{2p}^*)^3$
    **(b)** $F_2$: 1, $F_2^+$: $\frac{3}{2}$   **(c)** $F_2^+$ should be paramagnetic.   **(d)** $F_2 < F_2^+$
27. $(\sigma_{3s})^2(\sigma_{3s}^*)^2(\sigma_{3p})^2(\pi_{3p})^4(\pi_{3p}^*)^2$; bond order = 2; paramagnetic
29. **(a)** F, 1   **(b)** N, $2\frac{1}{2}$   **(c)** O, $1\frac{1}{2}$
31. (a) is diamagnetic; (b) and (c) are paramagnetic.
33. Bond order $2\frac{1}{2}$; paramagnetic
35. The outermost electron in CF is in a $\pi_{2p}^*$ molecular orbital. Removing it gives a stronger bond.
37. $(\sigma_{1s})^2(\sigma_{1s}^*)^2$, bond order 0, should be unstable.
39. $sp^3$ hybridization of central $N^-$, bent molecular ion
41. **(a)** $sp^3$ on C, tetrahedral   **(b)** $sp$ on C, linear   **(c)** $sp^3$ on O, bent
    **(d)** $sp^3$ on C, pyramidal   **(e)** $sp$ on Be, linear
43. $sp^2$ hybrid orbitals. $ClO_3^+$ is trigonal planar, $ClO_2^+$ is bent.
45. $sp^3$ hybrid orbitals, tetrahedral
47. Sixteen electrons, so molecule is linear; $sp$ hybridization of central N gives two $\sigma$ bonds with $2p_z$ orbitals on outer N atoms (four electrons).
    Lone pairs on both $2s$ orbitals on outer N atoms (four atoms). $\pi$ system as in Figure 14-26: $(\pi)^4(\pi^{nb})^4$ with eight electrons. Total bond order = 4; bond order 2 per N—N bond. $N_3$ and $N_3^+$ should be bound. $N_3$ and $N_3^+$ are paramagnetic.
49. —$CH_3$ carbon $sp^3$, other carbon $sp^2$
    A $\pi$ orbital with two electrons bonds the second C atom with the O atom.
    The three groups around the second carbon form an approximately trigonal planar structure, with bond angles near 120°. The geometry about the first carbon atom is approximately tetrahedral, with angles near 109.5°.

**51.** $\{O{=}N{-}O: \leftrightarrow :O{-}N{=}O\}$

$SN = 3$, $sp^2$ hybridization, bent molecule. The $2p_z$ orbitals perpendicular to the plane of the molecule can be combined into a $\pi$ molecular orbital containing one pair of electrons. This orbital adds bond order $\frac{1}{2}$ to each N—O bond, for a total bond order of $\frac{3}{2}$ per bond.

## Chapter 15

**1.** $6.736 \times 10^{-23}$ J = 40.56 J mol$^{-1}$

**3.** (a) $1.45 \times 10^{-46}$ kg m$^2$   (b) $E_1 = 7.65 \times 10^{-23}$ J; $E_2 = 2.30 \times 10^{-22}$ J; $E_3 = 4.59 \times 10^{-22}$ J   (c) 1.13 Å

**5.** 25.4 kg s$^{-2}$

**7.** $5.1 \times 10^2$ kg s$^{-2}$

**9.** $5.3 \times 10^{-4}$

**11.** Decrease, because the electron will enter a $\pi^*$ antibonding orbital

**13.** In the blue range, around 450 nm

**15.** Shorter; the $\pi$ bonding is localized in cyclohexene.

**17.** 30 double bonds; on the bonds shared by two hexagonal faces

**19.** 474 nm

**21.** 8 molecules

**23.** 270 nm

**25.** $\{O{=}O{-}O: \leftrightarrow :O{-}O{=}O\}$

$SN = 3$, $sp^2$ hybridization, bent molecule.

Two electrons in a $\pi$ orbital formed from the three $2p_z$ orbitals perpendicular to the molecular plane, total bond order of $\frac{3}{2}$ for each O—O bond.

## Chapter 16

**1.** (a) $PtF_4$   (b) $PtF_6$

**3.** $V_{10}O_{28}^{6-} + 16\,H_3O^+ \rightarrow 10\,VO_2^+ + 24\,H_2O$   + 5 state, $V_2O_5$

**5.** $2\,TiO_2 + H_2 \rightarrow Ti_2O_3 + H_2O$   Titanium(III) oxide

**7.** Monodentate, at the N-atom lone pair

**9.** +2, +2, +2, 0

**11.** (a) $Na_2[Zn(OH)_4]$   (b) $[CoCl_2(en)_2]NO_3$   (c) $[Pt(H_2O)_3Br]Cl$   (d) $[Pt(NH_3)_4(NO_2)_2]Br_2$

**13.** (a) Ammonium diamminediisothiocyanatochromate(III)
   (b) Pentacarbonyltechnetium(I) iodide
   (c) Potassium pentacyanomanganate(IV)
   (d) Tetraammineaquachlorocobalt(III) bromide

**15.** Such a complex will persist in solution but will exchange ligands with other complexes, as can be verified by isotopic labeling experiments.

**17.** $\Delta G° = +70.56$ kJ, stable against dissociation. In acid,
   $[Cu(NH_3)_4]^{2+}(aq) + 4\,H_3O^+(aq) \rightarrow Cu^{2+}(aq) + 4\,NH_4^+(aq) + 4\,H_2O(\ell)$
   $\Delta G° = -140.68$ kJ, spontaneous

**19.** $[Cu(NH_3)_2Cl_2] < KNO_3 < Na_2[PtCl_6] < [Co(NH_3)_6]Cl_3$

**25.** (a) Strong: 1, weak: 5   (b) Strong or weak: 0   (c) Strong or weak: 3
   (d) Strong: 2, weak: 4   (e) Strong: 0, weak: 4

**27.** $[Fe(CN)_6]^{3-}$: 1 unpaired, CFSE = $-2\,\Delta_o$; $[Fe(H_2O)_6]^{3+}$: 5 unpaired, CFSE = 0

**29.** $d^3$ gives half-filled and $d^8$ filled and half-filled shells for metal ions in an octahedral field. $d^5$ will be half filled and stable for high spin (small $\Delta_o$), and $d^6$ will be a filled subshell and stable for low spin (large $\Delta_o$).

**31.** This ion does not absorb any significant amount of light in the visible range.

**33.** $\lambda \approx 480$ nm; $\Delta_o \approx 250$ kJ mol$^{-1}$

**35.** $-500$ kJ mol$^{-1}$

**37.** **(a)** Orange-yellow  **(b)** Approximately 600 nm (actual: 575 nm)  **(c)** Decrease, because $CN^-$ is a strong-field ligand and will increase $\Delta_o$.

**39.** **(a)** Pale, because $F^-$ is an even weaker field ligand than $H_2O$, so it should be high-spin $d^5$. This is a half-filled shell and the complex will absorb only weakly.
**(b)** Colorless, because Hg(II) is a $d^{10}$ filled-subshell species

**41.** 14, 18, 18, 18

**43.** 1330 g mol$^{-1}$

## Chapter 17

**1.** (b), (c), and (d)

**3.** Two mirror planes, one 2-fold rotation axis

**5.** 3.613 Å

**7.** 102.9°

**9.** 49.87° and 115.0°

**11.** **(a)** $5.675 \times 10^6$ Å$^3$  **(b)** $5 \times 10^{14}$

**13.** 4.059 g cm$^{-3}$

**15.** **(a)** $1.602 \times 10^{-22}$ cm$^3$  **(b)** $3.729 \times 10^{-22}$ g  **(c)** $4.662 \times 10^{-23}$ g
**(d)** $6.025 \times 10^{23}$ (in error by 0.05%)

**17.** 1

**19.** $ReO_3$

**21.** **(a)** 2.48 Å  **(b)** 2.87 Å  **(c)** 1.24 Å

**23.** **(a)** 2  **(b)** 0.680

**25.** 0.732

**27.** **(a)** Ionic  **(b)** Covalent  **(c)** Molecular  **(d)** Metallic

**29.** 8, 6, 24

**31.** 664 kJ mol$^{-1}$

**33.** **(a)** 1041 kJ mol$^{-1}$  **(b)** 1206 kJ mol$^{-1}$

**35.** Not by a significant amount, because each vacancy is accompanied by an interstitial. In large numbers, such defects could cause a small bulging of the crystal and a decrease in its density.

**37.** **(a)** $Fe_{0.9352}O$  **(b)** 13.86% of the iron

**39.** The isotropic phase has higher entropy and higher enthalpy.

## Chapter 18

**1.** A batch process is carried out by mixing reactants, letting them react, and then extracting products in a series of separate steps. In a continuous process, all the steps run simultaneously. A batch process can take place in a single container with one opening, but a continuous process will either need to flow through several containers, or at least require different places into which reactants can be introduced and from which products are removed.

**3.** **(a)** In salt domes, produced by bacteria from $CaSO_4$, $CH_4$, and $CO_2$
**(b)** In coal, produced from plant matter via bacterial reduction, heat, and pressure
**(c)** In phosphate rock, $Ca_5(PO_4)_3F$
**(d)** In limestone (calcium carbonate, $CaCO_3$)

**5.** $Zn(s) + H_2SO_4(aq) \rightarrow Zn^{2+}(aq) + SO_4^{2-}(aq) + H_2(g)$

**7.** 960 K

**9.** **(a)** $3\,Fe(s) + 4\,H_2O(g) \rightarrow Fe_3O_4(s) + 4\,H_2(g)$
**(b)** $\Delta H° = -151.1$ kJ. An increase in $T$ will *reduce* the thermodynamic yield.

**11.** 79.92 L

## Chapter 19

**1.**

The structures of $P_2O_7^{4-}$ and $S_2O_7^{2-}$ have the same number of electrons, and are found by writing "P" or "S" in place of "Si" and adjusting the formal charges. The chlorine compound is $Cl_2O_7$, dichlorine heptaoxide.

**3. (a)** Tetrahedra; Ca: +2, Fe: +3, Si: +4, O: −2.

  **(b)** Infinite sheets; Na: +1, Zr: +2, Si: +4, O: −2

  **(c)** Pairs of tetrahedra; Ca: +2, Zn: +2, Si: +4, O: −2

  **(d)** Infinite sheets; Mg: +2, Si: +4, O: −2, H: +1

**5. (a)** Infinite network; Li: +1, Al: +3, Si: +4, O: −2

  **(b)** Infinite sheets; K: +1, Al: +3, Si: +4, O: −2, H: +1

  **(c)** Closed rings or infinite single chains; Al: +3, Mg: +2, Si: +4, O: −2

**9. (a)** $CaCO_3(s) + SiO_2(s) \rightarrow CaSiO_3(s) + CO_2(g)$

  **(b)** $\Delta H° = +89.41$ kJ; $\Delta S° = +160.81$ J K$^{-1}$  **(c)** 556 K  **(d)** 819 K

**11. (a)** $\Delta H° = -0.21$ kJ; $\Delta S° = -4.2$ J K$^{-1}$; $\Delta G° = +1.04$ kJ  **(b)** Calcite

  **(c)** Aragonite  **(d)** Aragonite  **(e)** $1.5 \times 10^6$ Pa K$^{-1}$ = 15 atm K$^{-1}$

**13.** $MnO_2(s) + \frac{4}{3} Al(s) \rightarrow Mn(s) + \frac{2}{3} Al_2O_3(s)$; $\Delta H° = -597.1$ kJ, $\Delta G° = -589.7$ kJ

**15.** 839 K = 566°C

**17.** $\Delta S°$ is positive, so $\Delta G$ will become negative at sufficiently high temperature (above 895 K).

**19.** 9.8 kg

**21. (a)** $Fe_2O_3(s) + 3 CO(g) \rightarrow 2 Fe(s) + 3 CO_2(g)$

  $Fe_3O_4(s) + 4 CO(g) \rightarrow 3 Fe(s) + 4 CO_2(g)$

  $FeCO_3(s) + CO(g) \rightarrow Fe(s) + 2 CO_2(g)$

  **(b)** From $Fe_2O_3$, −14.7 kJ mol$^{-1}$ of Fe; from $Fe_3O_4$, −4.4 kJ mol$^{-1}$ of Fe; from $FeCO_3$, +15.1 kJ mol$^{-1}$ of Fe. $Fe_2O_3$ is thermodynamically easiest to reduce per mole of Fe produced.

**23.** $2 Cl^- \rightarrow Cl_2(g) + 2 e^-$       anode

  $Na^+ + e^- \rightarrow Na(\ell)$       cathode

**25.** $4.4 \times 10^4$ kg

**27.** $2 Mg + TiCl_4 \rightarrow Ti + 2 MgCl_2$; 102 kg

**29.** 42.4 min

## Chapter 20

**1.** $Mg_3Si_4O_{10}(OH)_2(s) \rightarrow 3 MgSiO_3(s) + SiO_2(s) + H_2O(g)$

**3.** 234 L

**5.** 0.418 mol Si, 0.203 mol Na, 0.050 mol Ca, 0.0068 mol Al, 0.002 mol Mg, 0.0005 mol Ba

**7.** −113.0 kJ

**9.** $\frac{7}{3}$

**11. (a)** $SiO_2(s) + 3 C(s) \rightarrow SiC(s) + 2 CO(g)$

  **(b)** +624.6 kJ mol$^{-1}$

  **(c)** Low conductivity, high melting point, very hard

**13.** −1188.2 kJ, unstable

**15.** $2.16 \times 10^{-19}$ J

**17.** $6.1 \times 10^{-27}$ electrons; that is, no electrons

**19. (a)** Electron movement (*n*-type)   **(b)** Hole movement (*p*-type)
**21.** 680 nm, red
**23.** Decrease

## Chapter 21

**1.** Operate at high total pressures and low temperatures; continuously remove $SO_3$ as it is produced.

**3.** Yes. The reaction that would occur

$$H_2O_2(aq) + SO_2(g) \longrightarrow H_2SO_4(aq)$$

has $\Delta G° = -321.7$ kJ $< 0$, assuming $H_2SO_4(aq)$ is entirely $HSO_4^-(aq)$ and $H^+(aq)$. Thus, the process is thermodynamically feasible.

**5.** $ZnS(s) + 2 O_2(g) + H_2O(\ell) \rightarrow ZnO(s) + H_2SO_4(\ell)$

**7.** $H_2SO_4(\ell) + 2 NaOH(s) \rightarrow Na_2SO_4(s) + 2 H_2O(\ell)$
$Na_2SO_4(s) + 2 C(s) \rightarrow Na_2S(s) + 2 CO_2(g)$

**9.** $5 H_2SO_4(aq) + 2 P(s) \rightarrow 5 SO_2(g) + 2 H_3PO_4(aq) + 2 H_2O(\ell)$

**11.** $H_4P_2O_7$, pyrophosphoric acid. Two $PO_3$ groups are joined by an —O— linkage, with H atoms on four of the outer O atoms.

**13.** $3 \times 10^{26}$ s$^{-1}$

**15.** $CaCO_3(s) + 2 C(s) + N_2(g) + 4 H_2O(\ell) \rightarrow 2 NH_3(g) + 2 CO_2(g) + CO(g) + Ca(OH)_2(s)$;
$\Delta H° = +374.4$ kJ. Dividing by 2 gives $+187.2$ kJ mol$^{-1}$ of $NH_3(g)$ produced.

**17.** Higher pressures mean higher yields, but also more expensive equipment. At fixed yield, a higher temperature can be used at high pressure, giving a faster reaction.

**19.** High yield will be favored by low temperature for all three steps.

**21.** $3 HNO_3(aq) + Cr(s) + 3 H_3O^+(aq) \rightarrow 3 NO_2(g) + Cr^{3+}(aq) + 6 H_2O(\ell)$
$3 HNO_3(aq) + Fe(s) + 3 H_3O^+(aq) \rightarrow 3 NO_2(g) + Fe^{3+}(aq) + 6 H_2O(\ell)$

**23.**

**25.** Unstable

**27.** $5 N_2H_4(\ell) + 4 HNO_3(\ell) \rightarrow 7 N_2(g) + 12 H_2O(\ell)$; $\Delta H° = -2986.71$ kJ

**29.** It appears that electronegative atoms reduce the basicity of the N atom to which they are attached: the lone electron pair is drawn closer to the N atom by the inductive effect of electronegative atoms. $NH_2OH$ should be less basic than $NH_3$ ($K_b = 1.07 \times 10^{-8}$ for hydroxylamine versus $1.76 \times 10^{-5}$ for $NH_3$).

**31.** $N_2(g) + 3 H_2(g) \rightarrow 2 NH_3(g)$ (Haber-Bosch)
$4 NH_3(g) + 5 O_2(g) \rightarrow 4 NO(g) + 6 H_2O(g)$ (Pt catalyst, first step of Ostwald process)
$2 NO(g) + O_2(g) \rightarrow 2 NO_2(g)$ (second step of Ostwald process)
$2 NO_2(g) \rightarrow N_2O_4(s)$ (low temperature)

**33.** The decomposition of $NH_4NO_3$ gives $N_2 + 2 H_2O + \frac{1}{2} O_2$. The oxygen produced can react with the aluminum, releasing additional heat.

**35. (a)** Yes   **(b)** $P_4$

## Chapter 22

**1.** $3.2 \times 10^{16}$

**3.** $\Delta H° = +258.5$ kJ for the reaction
$2 NaCl(s) + H_2SO_4(\ell) + 2 C(s) + CaCO_3(s) \rightarrow Na_2CO_3(s) + 2 CO_2(g) + CaS(s) + 2 HCl(g)$
Endothermic, so heat must be supplied to make it run.

5. Two moles of NaOH should form for each mole of $Cl_2$. The ratio of masses is 1.13, corresponding to 13% more caustic soda than chlorine. Presumably losses or side reactions give the different yield ratio obtained in practice.

7. +3.87 kJ (tabulated value +3.93 kJ)

9. $K = 1.0 \times 10^9$. Because $K$ is so large, the process is feasible. For example, suppose $[Br^-] = 10^{-6}$ M, $[Cl^-] = 0.1$ M. Then from $K$ we calculate $P_{Br_2} = 0.1$ atm.

11. $H_3PO_4(aq) + Br^-(aq) \rightarrow HBr(g) + H_2PO_4^-(aq)$   (b) 0.11 L

13. Unstable

15. (a) We find $I_2(s) + 2\,OH^-(aq) \rightarrow I^-(aq) + IO^-(aq) + H_2O(\ell)$
$\Delta\mathscr{E}° = +0.085$ V, so this is a feasible process. However,
$3\,IO^-(aq) \rightarrow IO_3^-(aq) + 2\,I^-(aq)$
has $\Delta\mathscr{E}° = +0.35$ V. This process is also feasible, and because it consumes $IO^-(aq)$ it is the ultimate reaction. The stable products are $IO_3^-(aq)$ and $I^-(aq)$.
    (b) 0.26 V

17. (a) $2\,SrO(s) + 2\,F_2(g) \rightarrow 2\,SrF_2(s) + O_2(g)$
    (b) $O_2(g) + 2\,F_2(g) \rightarrow 2\,OF_2(g)$
    (c) $UF_4(s) + F_2(g) \rightarrow UF_6(g)$

19. $NaF \cdot 2HF$

21. $ClF_3$

23. (a) Bent   (b) Trigonal planar   (c) Distorted T   (d) Square pyramidal
    (e) Pentagonal bipyramid   (f) Octahedral

25. (a) Pyramidal   (b) Lewis base

27. $2\,BrF_3 \rightleftharpoons BrF_4^- + BrF_2^+$
    Here, $BrF_2^+$ and $BrF_3$ are both Lewis acids (electron-pair acceptors), and $F^-$ is the Lewis base (electron-pair donor).

29. $PtF_6 < PtF_4 < CaF_2$

31. 2740 kJ

33. (a) $-345$ kJ mol$^{-1}$   (b) 136 kJ mol$^{-1}$

35. +190 kJ mol$^{-1}$

37. $XeO_6^{4-}(aq) + 12\,H_3O^+(aq) + 8\,e^- \rightarrow Xe(g) + 18\,H_2O(\ell)$
$Mn^{2+}(aq) + 12\,H_2O(\ell) \rightarrow MnO_4^-(aq) + 8\,H_3O^+(aq) + 5\,e^-$
$5\,XeO_6^{4-}(aq) + 8\,Mn^{2+}(aq) + 6\,H_2O(\ell) \rightarrow 5\,Xe(g) + 8\,MnO_4^-(aq) + 4\,H_3O^+(aq)$

## Chapter 23

1. (a) $-64$ kJ mol$^{-1}$   (b) $+53$ kJ mol$^{-1}$   (c) $+117$ kJ mol$^{-1}$

3. (a) $C_{10}H_{22} \rightarrow C_5H_{10} + C_5H_{12}$
    (b) Two: 1-pentene and 2-pentene

5. (a)
    (b)
    (c)

**(d)**

$$CH_3 \quad H$$
$$H_3C-\underset{\underset{CH_3}{|}}{\overset{\overset{CH_3}{|}}{C}}-\underset{\underset{H}{|}}{\overset{\overset{H}{|}}{C}}-CH_3$$

**(e)**

$$\underset{H_3C}{\overset{H}{\diagdown}}C=C\underset{CH_2-CH_2-CH_3}{\overset{CH_2-CH_2-CH_3}{\diagup}}$$

**(f)**

$$\underset{H}{\overset{H}{\diagdown}}C=C\underset{\underset{H}{|}}{\overset{CH_3}{\diagup}}$$
$$\underset{CH_2-CH_2-CH_3}{\overset{C}{}}$$

**(g)**

$$CH_3 \qquad CH_2-CH_3$$
$$H_3C-\underset{\underset{H}{|}}{\overset{\overset{CH_3}{|}}{C}}-CH_2-\underset{\underset{H}{|}}{\overset{\overset{CH_2-CH_3}{|}}{C}}-CH_2-CH_2-CH_3$$

**(h)**

$$CH_2-CH_3$$
$$H_3C-C\equiv C-\underset{\underset{H}{|}}{\overset{\overset{CH_2-CH_3}{|}}{C}}-CH_2-CH_2-CH_3$$

**7.**

$$\underset{H_3C-CH_2}{\overset{H}{\diagdown}}C=C\underset{H}{\overset{CH_2-CH_2-CH_3}{\diagup}}$$

$$\underset{H}{\overset{H_3C-CH_2}{\diagdown}}C=C\underset{H}{\overset{CH_2-CH_2-CH_3}{\diagup}}$$

**9. (a)** 1,2-hexadiene  **(b)** 1,3,5-hexatriene  **(c)** 2-methyl-1-hexene  **(d)** 3-hexyne

**11. (a)** $sp^2$, $sp$, $sp^2$, $sp^3$, $sp^3$, $sp^3$
   **(b)** All $sp^2$
   **(c)** $sp^2$, $sp^2$; all others $sp^3$
   **(d)** $sp^3$, $sp^3$, $sp$, $sp$, $sp^3$, $sp^3$

**13. (a)** $CH_3CH_2CH_2CH_2OH + CH_3COOH \rightarrow CH_3COOCH_2CH_2CH_2CH_3 + H_2O$
   **(b)** $NH_4CH_3COO \rightarrow CH_3CONH_2 + H_2O$
   **(c)** $CH_3CH_2CH_2OH \rightarrow CH_3CH_2CHO$ (propionaldehyde) $+ H_2$
   **(d)** $CH_3(CH_2)_5CH_3 + 11\ O_2 \rightarrow 7\ CO_2 + 8\ H_2O$

**15. (a)** $CH_2{=}CH_2 + Br_2 \rightarrow CH_2BrCH_2Br$; $CH_2BrCH_2Br \rightarrow CH_2{=}CHBr + HBr$
   **(b)** $CH_3CH_2CH{=}CH_2 + H_2O \rightarrow CH_3CH_2CH(OH)CH_3$ (using $H_2SO_4$)
   **(c)** $CH_3CH{=}CH_2 + H_2O \rightarrow CH_3CH(OH)CH_3$ (using $H_2SO_4$);
   $CH_3CH(OH)CH_3 \rightarrow CH_3COCH_3 + H_2$ (copper or zinc oxide catalyst)

**17.** $R-\overset{\overset{O}{\|}}{C}-OH + (R')_3C-OH \rightarrow R-\overset{\overset{O}{\|}}{C}-O-C(R')_3 + H_2O$

**19.** 11.2%; $4.49 \times 10^9$ kg

**21.** $n\ CCl_2{=}CH_2 \rightarrow {+}(CCl_2-CH_2{)_n}$

**23.** Formaldehyde, $CH_2{=}O$

**25. (a)** $H_2O$  **(b)** ${-}(NH-CH_2-CO){-}$

**27.** 646 kg adipic acid and 513 kg hexamethylenediamine

**29.** $3.49 \times 10^{12}$ L $= 3.49 \times 10^9$ m$^3$ $= 3.49$ km$^3$

**31.** 5 chiral centers

**33.** 27

**35.** In octane, because the side groups are all hydrocarbon

**37.** $C_9H_9NO$; 119 units

## Appendix A

1. **(a)** $5.82 \times 10^{-5}$ **(b)** $1.402 \times 10^3$ **(c)** 7.93 **(d)** $-6.59300 \times 10^3$
   **(e)** $2.530 \times 10^{-3}$ **(f)** 1.47
3. **(a)** 0.000537 **(b)** 9,390,000 **(c)** −0.00247 **(d)** 0.006020 **(e)** 20,000
5. 746,000,000 kg
7. **(a)** The value 135.6 **(b)** 111.34 g **(c)** 0.22 g, 0.23 g
9. That of the mass in problem 7.
11. **(a)** 5 **(b)** 3 **(c)** Either two or three **(d)** 3 **(e)** 4
13. **(a)** 14 L **(b)** −0.0034°C **(c)** 340 lb = $3.4 \times 10^2$ lb **(d)** $3.4 \times 10^2$ miles
   **(e)** $6.2 \times 10^{-27}$ J
15. 2,997,215.55
17. **(a)** −167.25 **(b)** 76 **(c)** $3.1693 \times 10^{15}$ **(d)** $-7.59 \times 10^{-25}$
19. **(a)** −8.40 **(b)** 0.147 **(c)** $3.24 \times 10^{-12}$ **(d)** $4.5 \times 10^{13}$
21. $A = 337$ cm$^2$

## Appendix B

1. **(a)** $6.52 \times 10^{-11}$ kg **(b)** $8.8 \times 10^{-11}$ s **(c)** $5.4 \times 10^{12}$ kg m$^2$ s$^{-3}$
   **(d)** $1.7 \times 10^4$ kg m$^2$ s$^{-3}$ A$^{-1}$
3. **(a)** 4983°C **(b)** 37.0°C **(c)** 111°C **(d)** −40°C
5. **(a)** 5256 K **(b)** 310.2 K **(c)** 384 K **(d)** 233 K
7. **(a)** 24.6 m s$^{-1}$ **(b)** $1.51 \times 10^3$ kg m$^{-3}$ **(c)** $1.6 \times 10^{-29}$ A s m
   **(d)** $1.5 \times 10^2$ mol m$^{-3}$ **(e)** 6.7 kg m$^2$ s$^{-3}$ = 6.7 W
9. 1 kW-hr = $3.6 \times 10^6$ J; 15.3 kW-hr = $5.51 \times 10^7$ J
11. 6620 cm$^3$ or 6.62 L

## Appendix C

1. 50 miles per hour
3. **(a)** Slope = 4, intercept = −7
   **(b)** Slope = $\frac{7}{2}$ = 3.5, intercept = $-\frac{5}{2}$ = −2.5
   **(c)** Slope = −2, intercept = $\frac{4}{3}$ = 1.3333 . . .
5. The graph is not a straight line.
7. **(a)** $-\frac{5}{7}$ = −0.7142857 . . . **(b)** $\frac{3}{4}$ = 0.75 **(c)** $\frac{2}{3}$ = 0.6666 . . .
9. **(a)** 0.5447 and −2.295 **(b)** −0.6340 and −2.366 **(c)** 2.366 and 0.6340
11. **(a)** $6.5 \times 10^{-7}$ (also two complex roots) **(b)** $4.07 \times 10^{-2}$ (also 0.399 and −1.011)
   **(c)** −1.3732 (also two complex roots)
13. **(a)** 4.551 **(b)** $1.53 \times 10^{-7}$ **(c)** $2.6 \times 10^8$ **(d)** −48.7264
15. 3.015
17. 121.477
19. 7.751 and −2.249
21. $x = 1.086$
23. **(a)** $8x$ **(b)** $3 \cos 3x - 8 \sin 2x$ **(c)** 3 **(d)** $\dfrac{1}{x}$
25. **(a)** 20 **(b)** $\frac{78125}{7}$ **(c)** 0.0675

ammonia (*continued*)
  correlation diagram for, 568
  equilibrium law for, 177
  in coordination complexes, 267–69
  in Solvay process, 777–79
  production of, 191, 759–60
  structure of, 57i, 543i
ammonium chloride
  dissociation of, 90
  hydrolysis of, 224i
ammonium ion, 55
ammonium nitrate, decomposition of, 766, 771
ammonium perchlorate, 120
ammonium phosphate, 755, 760i
Amontons's law, 125
**amorphous solid**   A solid that lacks any
  long-range order in the arrangement of its
  atoms, 665–66
AMP. *See* adenosine monophosphate
amphoteric equilibrium, 242–43
**amphoterism**   The phenomenon of a single
  species serving both as an acid and as a
  base, 74, 210
  of complex ions, 272–73
**amplitude**   The distance from center to crest
  or center to trough of a standing or traveling wave, 482
**analysis**   Literally, ''taking apart''; the
  determination of identity and composition of compounds and mixtures, 4
angstrom, 491, A-11
angular momentum, 490–91
**angular momentum quantum number**   A
  quantum number for a one-electron atom
  that governs the shape of the orbital;
  takes integer values from $l = 0$ to $l =
  n - 1$, where $n$ is the principal quantum
  number, 502
**angular node**   A surface (often a plane) in a
  wave function at which the electron density equals zero and across which the
  wave function changes sign, 506
anhydride, 76–77
**anion**   a negatively charged ion, 53
  naming of, 55–56, 55t
**annealing**   The slow cooling of a piece of
  material to allow the release of internal
  stresses by short-range diffusion of
  atoms, 725
**annihilation radiation**   The radiation given
  off when a particle and its antiparticle
  encounter and annihilate each other, 454
**anode**   The site at which oxidation occurs in
  an electrochemical cell, 366
  sacrificial, 391
**antibonding orbital**   A molecular orbital
  whose occupation by electrons results in
  a reduction of bond strength, 547
antineutrino, 454

**antiparticle**   A body that annihilates its conjugate particle to give high-energy photons, 454
aragonite, 46i, 274i, 698
argon, 51, 89
**aromatic hydrocarbon**   A compound of hydrogen and carbon whose molecules contain one or more rings with conjugated
  $\pi$-electron systems, 810–11
Arrhenius, Svante, 70, 427
Arrhenius equation, 427
**asbestos**   A fibrous material, formerly used
  for thermal insulation, that is composed
  of double-chain or rolled-up layered silicates, 691
ascorbic acid, 37
aspartame, 43
Aspdin, Joseph, 727
asphalt, 805
aspirin
  production of, 678i
  structure of, 677i
astatine, 476, 799
Aston, Francis William, 19
atactic, 829i, 830
atmosphere
  composition of, 89
  photochemistry in, 594–99
  temperature of, 595i
  unit of pressure, 93, A-11
**atom**   The smallest particle of an element, 7,
  10
  composition of, 15–19
  size of, 25, 515–17, 606i, 657i
  spectra of, 488–90
  stability of, 488
**atomic mass, relative.**   *See* relative atomic
  mass
**atomic mass unit**   Unit of mass defined as
  exactly $\frac{1}{12}$ of the mass of a single atom
  of $^{12}C$, 449
**atomic number**   The number of electrons in
  the neutral atom; also, the number of
  protons in the nucleus of the atom, 18
**atomic theory of matter**   The postulate that
  matter is composed of indestructible
  atoms that combine in whole-number ratios to form compounds, 10
**aufbau principle**   The systematic filling of
  the atomic orbitals of an atom with electrons, beginning with the orbital of lowest energy and working up, a maximum
  of two electrons being placed in each
  orbital, 510, 548
**autoionization**   The spontaneous reaction of
  molecules of a substance to give pairs of
  ions, such as the reaction of water to
  give $H_3O^+$ plus $OH^-$ ions, 210–11

**average reaction rate**   The change in the
  concentration of a product of a reaction
  divided by the time interval over which
  the change occurs and also divided by
  the coefficient of the product in the
  chemical equation for the reaction, 405–6
average value, A-3
Avogadro, Amedeo, 14
**Avogadro's hypothesis**   The hypothesis that
  equal volumes of different gases held at
  the same temperature and pressure contain equal numbers of particles, 14–15,
  100
**Avogadro's number**   The number of atoms
  in exactly 12 g of $^{12}C$ (equal to
  $6.022137 \times 10^{23}$ mol$^{-1}$), 22
  and Brownian motion, 165
  and moles, 23
  determination of, 648–49
**axial**   Lying along the axis of symmetry perpendicular to the plane that contains the
  equator; used to describe a type of site
  for atom or group substitution in a molecule, 543
axis (of rotation), 641
**azeotrope**   A liquid solution of two volatile
  components whose vapor has the same
  composition, 162, 163i
azide ion, 59


back-bonding, 629–30
bacteriochlorophyll, 591i, 592
Bakelite, 827–28
baking soda, 250
balancing chemical equations
  by inspection, 29–31
  using half-reactions, 360–65
balancing nuclear equations, 450
balloon
  hot-air, 100i
  hydrogen, 122–23, 683
Balmer, Johann Jakob, 488
**band**   A very closely spaced manifold of
  energy levels in a solid, 655, 656i
  in semiconductors, 735–37
**band gap**   The energy difference between the
  top of one band and the bottom of the
  next higher band, 735–36
bar, unit of pressure, 93t
barium sulfate, 278
**barometer**   A device to measure the pressure
  of the atmosphere, 91, 92i
Bartlett, Neil, 694
**base**   A substance that, when dissolved in
  water, increases the amount of hydroxide
  ion over that present in pure water, 71
  Arrhenius, 71

**Brownian motion** (*continued*)
  motion of particles as they undergo collisions with surrounding solvent molecules, 165
buckminsterfullerene, 585
**buffer solution** A solution that maintains a nearly constant pH despite the addition of small amounts of acid or base; usually made by mixing a weak acid with its conjugate weak base, 225–31
  acidic and basic, 228
  capacity of, 230–31, 230i
  designing, 229–30
  in blood, 226
**builder** An ingredient in detergents used to sequester ions that hinder cleansing action and to maintain the pH, 756
buret, 25i, 146
butadiene, molecular orbitals for, 565–66
butane, 805
byproducts, 676

cadmium sulfide, 740
calcite, 46i, 274i, 698
calcium carbide, 759
calcium carbonate
  crystal forms of, 46i
  dissociation of, 90, 192, 193i
  geological formation of, 682, 697
  in seawater, 121
  reaction of, with acid, 90i
  solubility of, in acid, 263
calcium cyanamide, 759
calcium hydride, 74i
calcium hypochlorite, 31
calcium ion, analysis for, 147
calorie, 289
**calorimeter** An instrument used to measure amounts of heat flowing into or out of a system, 287
  bomb, 295i
  combustion, 295i
  ice, 287i
  Styrofoam cup, 289i
**canal rays** Positively charged particles (ions) in cathode ray tubes, 18
Cannizzaro, Stanislao, 15
**carbohydrate** A compound of general formula $C_n(H_2O)_m$, 831
carbolic acid. *See* phenol
carbon
  elemental, 48i
  isotopes of, 21
carbon-14 dating, 462–63
carbon dioxide
  and the greenhouse effect, 599
  correlation diagram for, 561
  dissolved in water, 237fn

in beverages, 160
manufacture of, 684
phase diagram of, 134i
production of, from carbonates, 90
structure of, 90i
carbon monoxide
  analysis for, 83–84
  dipole moment of, 539–40
  in coordination complexes, 629–30
  structure of, 58
carbon tetrachloride, 813
carbonate equilibria, 239–40, 273–74
carbonic acid, 237–40
**carbonylation** A reaction in which the C=O group is inserted into a previously existing molecule, 818
carborundum. *See* silicon carbide
**carboxylic acid** An organic compound whose molecules contain the —COOH functional group, 817–19
carboxypeptidase A, 840–41
Carnot, Sadi, 310
**Carnot cycle** A cyclic process in which a system reversibly traverses two isothermal and two adiabatic paths to return to its original state, 310–12
Carothers, Wallace, 826
Cartesian coordinate axes, 104i
catalysis
  enzyme, 436–38
  heterogeneous, 434–35
  homogeneous, 433–34
  zeolite, 693–94
**catalyst** A substance that takes part in a chemical reaction and speeds it up, but itself undergoes no permanent chemical change, 433–36
**catalytic converter** An assembly of catalysts used in automobiles to oxidize hydrocarbons and CO to carbon dioxide and water and to reduce oxides of nitrogen to their elements, 435
**catalytic cracking** The breaking down of longer chain hydrocarbons in crude petroleum into shorter chains by using catalysts, 809
**cathode** The site at which reduction occurs in an electrochemical cell, 366
**cathode rays** Electrons; so called when they emanate from the negatively charged electrode to carry the current in a low-pressure gas-discharge tube, 16
**cation** A positively charged ion, 53
  naming of, 54–55
Cavendish, Henry, 51, 594, 758
cell
  electrochemical, 365–68
  electrolytic, 367
  galvanic, 366i, 367

cell constants, 643
**cell voltage** The electrical potential difference across the electrodes of an electrochemical cell; a measure of the tendency of the cell redox reaction to occur spontaneously, 367
  standard, 372
cellulose, 825, 834
Celsius temperature scale, 96, A-12
cement, 727–29
  macro-defect-free, 728
ceramic phase, 722
**ceramics** Synthetic materials that have as their essential components inorganic, nonmetallic materials, 721–23
  nonoxide, 731–34
  oxide, 729–31
  silicate, 724–29
cesium chloride structure, 654i
CFC. *See* chlorofluorocarbon
CFSE. *See* crystal field stabilization energy
Chadwick, James, 467
**chain reaction** A reaction that proceeds through a series of linked elementary reactions, some of which are repeated many times, 425–26
  branching, 426
  nuclear, 468, 469i
**chalcogen** An element of Group VI in the periodic table, 47
chalcopyrite, 703
charge, of electron, 17–18
charge-to-mass ratio of electron, 16–17
Charles, Jacques, 96, 122–23, 683
**Charles's law** The statement that at constant pressure, the volume of a sample of gas is a linear function of its temperature, 96–98
**chelate** A coordination complex in which a ligand coordinates through two or more donors to the same central atom, 611, 620i
**chemical amount** Amount of a substance expressed in a unit that defines the number of chemically reactive entities present (the accepted unit is the mole), 23
chemical bond. *See* bond
**chemical equation** A representation of a chemical reaction that uses chemical formulas and conforms to the laws of mass and charge conservation, 26
  balancing, 29–31, 360–65
chemical equilibrium. *See* equilibrium
**chemical formula** A representation of the chemical composition of a substance; refers either to empirical formula or to molecular formula, 10, 26
**chemical vapor deposition** A method used in the fabrication of ceramics, 734

Chernobyl, 471

**chiral center** A central atom in a chiral structure; for tetrahedrally bonded carbon, the four attached groups must be different for the carbon atom to be a chiral center, 564, 618–19, 832

chlor-alkali industry, 779–81

chlorine

  chemistry of, 781–84

  discovery of, 776

  isotopes of, 20

  oxides of, 783–84

  production of, 777, 779–81

**chlorofluorocarbon** A compound in which chlorine and fluorine appear as side groups on a chain of carbon atoms, 304, 597, 793–94

  photodissociation of, 597

chloroform, 813

chlorophyll

  absorption of light by, 585

  role of, in photosynthesis, 591–94

cholesterol, 168

**chromatography** A method of separating substances that exploits the difference in the partition ratios of two or more solute species between two phases, such as a gas and a liquid absorbed on a porous support, 196

  column, 197i

  gas–liquid, 198–99

  ion exchange, 198t

  paper, 196i

  thin layer, 198t

chrome plating, 714–15

chromium

  oxides of, 609–10

  production of, 718–19

cinnabar, 740i

*cis* isomer, 563

  in coordination complexes, 617

cisplatin, 618, 634

citric acid, 43

**Clapeyron equation** An equation that relates the pressure and temperature changes at the coexistence of two phases to the enthalpy and volume changes, 698

classical mechanics, 481

**Claus process** The reaction of $SO_2$ with $H_2S$ to give sulfur and water, 750

Clausius, Rudolf, 104, 314

**clay** A hydrated aluminosilicate produced by the weathering of primary minerals, 692–93

  in ceramic manufacture, 724–25

**closed-shell species** An atom or ion whose valence shell contains the maximum possible number of electrons, 514

**closed system** A system whose boundaries

are impermeable to the flow of matter, 281

coal

  formation of, 682

  gasification of, 684

cobalt chloride, 26

cohesion, in solids, 652–58

**colligative properties** Physical properties of a solution that depend only on the concentration of solute species and not on their nature, 148–58

  with dissociating solutes, 152

collision rate

  and chemical kinetics, 431

  with walls, 110

  with other molecules, 112–14

**colloid** A mixture of two or more substances in which one is suspended in the second as a large number of small particles that nonetheless exceed molecular size, 164–65

color, of coordination compounds, 626–28

color, of light, 484

color center. *See* F-center

color wheel, 584i

column chromatography, 197

**combining volumes, law of** The statement that the volumes of reacting gases at constant temperature and pressure stand in the ratio of simple integers, 13

combustion train, 28i

**common-ion effect** The observation that if a solution and a solid salt to be dissolved in it have an ion in common, then the solubility of the salt is depressed, 260–62

**common logarithm** A logarithm for which the base is 10, A-19

**complementary color** The color opposite a given color on the color wheel, 584

**complex ion** An ion in which a central metal ion is bound to one or more ligand molecules or ions that are usually capable of independent existence in solution (*see also* coordination complexes), 267

  amphoterism in, 272–73

  equilibria, 267–73

  hydrolysis in, 272

**composite ceramic** A material in which one ceramic is reinforced by the admixture of whiskers of another ceramic before firing, 734

**compound** A substance containing two or more elements, 6

  binary, 6

  covalent, 56, 67–68

  ionic, 53–56

  nonstoichiometric, 9

  quaternary, 6

  ternary, 6

**compressibility factor** The ratio $PV/nRT$, which differs from 1 for a real gas, 115

**concentration** The amount of a solute present in a given amount of solvent or solution, 141

  in gas-phase equilibria, 186–87

concrete, 728

**condensation** The formation of a liquid or solid from a gas, 130

**condensation polymerization** Polymerization via condensation reactions, 823

**condensation reaction** The joining together of two molecules as a third, usually small, molecule is split out, 816, 819

**conduction band** A partially filled band of energy levels in a crystal through which the electrons can move easily under the influence of an electric field, 735

conductivity

  of coordination complexes in solution, 616

  of electrolyte solutions, 68, 69i

  of metals, 655–56

  of salts, 54

  of the elements, 51

confidence limit, A-4

configuration interaction, 569

**conjugate acid–base pair** A Brønsted–Lowry acid and the base formed when it donates a hydrogen ion, or a Brønsted–Lowry base and the acid formed when it accepts a hydrogen ion, 210

**conjugated $\pi$ electron system** A molecule or portion of a molecule in which double or triple bonds alternate with carbon–carbon single bonds, and electrons are delocalized across several atoms, 565–67

  absorption of light by, 584–85

conservation of energy, 291

**conservation of mass, law of** The statement that in every chemical operation an equal quantity of matter exists before and after the operation, 9

**contact process** A means for the production of sulfuric acid in which sulfur dioxide is oxidized to sulfur trioxide by oxygen on a vanadium oxide catalyst, 746–48

**continuous process** A chemical process in which reactants are fed in and products removed continuously, 675

  Solvay process as first, 779

**coordinate covalent bond** A covalent bond in which the shared electrons are both supplied by one of the bonded atoms; results from the interaction of a Lewis base and a Lewis acid, 71

**coordination complex** A compound in which metal atoms or ions are bonded via coordinate covalent bonds to a small

Lewis structure of, 62
molecular orbitals for, 554
nitrogen trifluoride, 791
nitrogen triiodide, 765i
nitroglycerin
decomposition of, 90
use in dynamite, 765–66
nitrous oxide. *See* dinitrogen oxide
Nobel, Alfred, 766
**noble gas** An element of Group VIII of the periodic table, 51
compounds of, with fluorine, 85, 794–95
in air, 89
properties of crystals of, 661t
stability of, 54
**node** A region of no vibration (of zero amplitude) of a standing wave, 495
nonideal gas. *See* real gas
**nonideal solution** A solution in which the vapor pressures of the components deviate from the predictions of Raoult's law, 148, 152
vapor pressures for, 159i
**nonmetal** A substance lacking a metallic luster, having poor ability to conduct electricity, and lacking the malleability of metals, 45
nonoxide ceramics, 722, 731–34
**nonpolar** Having no dipole moment, 545
liquids, 137
**nonstoichiometric compound** A solid compound in which the proportions of the constituent elements are continuously variable over some range, 9, 664–65, 731
**normal alkane** A straight-chain alkane, 802–5
melting and boiling points of, 804i
names of, 803t
**normal boiling point** The temperature at which the vapor pressure of a liquid equals 1 atm, 132
effect of intermolecular forces on, 136–38
novalac, 824
nuclear decay processes, 454–58, 460t
nuclear equation, 450
**nuclear fission** The exothermic splitting of massive nuclides into fragments of roughly equal mass, 458, 466–69
**nuclear fusion** The exothermic union of two light nuclei to form a heavier nuclide, 472–74
nuclear magnetic resonance spectroscopy, 578
nuclear power reactor, 470–71
nuclear stability, 449–54
**nucleosynthesis** The production of heavier nuclides from lighter ones, especially in the interiors of stars, 472–73
**nucleotide** A component of the genetic material DNA composed of a pyrimidine or

purine base, a sugar, and a phosphate group, 841, 842i
**nucleus** The small, positively charged kernel at the center of an atom, composed of protons and neutrons, 18
nuclide, 19
mass, 450t
stability of, 451i
nylon, 823, 826, 827t, 835

octahedral complexes, 616–18
crystal field splittings for, 620–22
octahedral site, 651–52
**octane number** A measure of the smoothness of combustion of gasoline in an automobile engine, 807–8
**octet** Eight electrons in the valence shell of an atom, 54
and orbitals, 515
octet deficient molecules, 62, 71–72
**octet rule** The statement that main group elements lose, gain, or share electrons in chemical bonding in such a way as to attain a valence octet, 57
breakdown of, 62–64
oil. *See* petroleum
oil-drop experiment, 17–18
Oklo, 469
oleum, 748
olivine, 689
opal, 164i
**open-shell atom** An atom with a partially filled valence shell, 530
**open system** A system whose boundaries permit the flow of matter into it or out of it, 282
orange tornado, 271i
operator, 567
optical atomic trap, 124, 357
optical isomers, 564–65
in coordination complexes, 618–19
**orbital** A wave function that is a solution of the one-electron Schrödinger equation, 502
antibonding, 547
approximation, 509
bonding, 547
energy, 518i
molecular, 546
size and shape of, 504–8
order (of a bond), 532–33
**order (of a reaction)** The power to which the concentration of a species is raised in the rate law for a reaction, 408
determination of, 408–10
overall, 411
organic chemistry, 801–19
organic molecules, bonding in, 562–67

**organometallic compound** A compound that contains metal-carbon bonds, 630–32
orpiment, 11i
orthosilicate, 689
osmium tetraoxide, 608
**osmotic pressure** The excess pressure that must be applied to a solution to prevent the net diffusion of pure solvent through a semipermeable membrane into the solution, 156–58
and thermodynamics, 350–51
**Ostwald process** The controlled oxidation of ammonia in three steps to produce nitric acid, 761–63
oxalic acid, structure of, 147i
**oxidation** A loss of electrons; an increase in oxidation number, 72
**oxidation number** A number assigned to an atom in a molecule that reflects its state of oxidation
assigning of, 65–67
common, for main group elements, 67i
common, for transition-series elements, 608i
in naming compounds, 67–68
relation to charge on atom, 66–67
**oxidation–reduction reactions** Reactions in which electrons are transferred; reactions in which the oxidation number of at least one element changes, 72
balancing equations for, 360–65
oxidation state. *See* oxidation number
oxide ceramics, 722, 729–31
oxides
acid–base character of, 77i
chemistry of, 75–78
of manganese, 66i
of the halogens, 783–84
**oxidizing agent** A substance that can cause the oxidation of another substance, while it is itself reduced, 375
halogens as, 783
oxoacids as, 768
**oxoacid** An acid containing hydrogen ion in combination with an oxoanion
as oxidizing agents, 766–68
strengths of, 540–41, 768–69
**oxoanion** A negatively charged molecular ion containing oxygen in combination with some other element, 55
oxygen
isotopic abundances, 21
paramagnetism of, 553
phase diagram of, 171
separation of, from air, 162, 163i
oxygen difluoride, 86, 792
ozone
as oxidizing agent, 375
determination of, 172
in the stratosphere, 596–98

substitution, of ligands, 614–15

substitutional alloy, 665

**substrate** A reactant molecule that is capable of binding to the active site of an enzyme, 436

sucrose, 832, 833i

sugar, 831–34

sulfate ion, structure of, 270i

**sulfate process** A process used in paper production in which the pulp is digested in a mixture of sodium hydroxide and sodium sulfide, 751–52

sulfides, solubility of, 264–67

**sulfite process** A process used in paper production in which the pulp is digested in a mildly acidic digestion liquor with a high concentration of hydrogen sulfite ions, 752

sulfur
  biological origin of, 682
  in nature, 6
  polymerization of, 820i
  reaction of, with oxygen, 76
  sources of, 749–50
  structure of, 658
  use of, in rubber, 830

sulfur dioxide
  and acid rain, 243–44, 438, 598–99
  condensation of, 128–29
  sources for, 749
  structure of, 90i

sulfur hexafluoride
  structure of, 63i, 792
  uses for, 787

sulfur trioxide
  and acid rain, 243–44, 598–99
  structure of, 91i

sulfuric acid
  as oxidizing agent, 766–67
  ionization of, 236
  production of, 199, 744–48
  raw materials for, 749–50
  safe use of, 750
  structure of, 71
  uses of, 750–51

sulfuryl chloride, 201

**superconductor** A material offering zero resistance to the passage of an electric current, 587, 730–31, 732i

**supercritical fluid** Term applied to a substance held at a temperature and pressure that exceed both its critical temperature and its critical pressure, 135

superheating, of liquids, 133

supernova, 472

**superoxide** A compound containing oxygen in the oxidation state $-\frac{1}{2}$, the form of the superoxide ion, $O_2^-$, 75–76, 78–79

superphosphate fertilizer, 754

**supersaturated solution** A solution that temporarily contains more of a solute than the equilibrium quantity, 252–53

supersonic nozzle, 323, 589–90

**surroundings** The part of the universe that lies outside the system, 281

symmetry element, 641

syndiotactic, 829–30

**synthesis** The combination of elements or simple compounds to make more complex compounds, 4

**synthesis gas** A mixture of hydrogen and carbon monoxide, produced by the reaction of water vapor and a hydrocarbon (or coke), that is used as the starting point in the synthesis of methanol and other organic chemicals, 684

syrup, maple, 165–66

**system** A real or imagined portion of the universe that is confined by boundaries or mathematical constraints, 281
  closed, 281
  open, 282

talc, 690–91

tangent line, A-20, A-21i

technetium, 465i, 472, 476

Teflon, 793

television screens, 740–41

**temperature** The degree of hotness or coldness of an object as measured on some defined scale, 96–98
  absolute, 98
  Boyle, 118
  effect on equilibrium constants, 345–47
  effect on reaction rates, 426–30
  effect on vapor pressure, 347–48
  meaning of, in kinetic theory, 110
  thermodynamic scale of, 314

Tennant, Charles, 777

**termination** The final stage in a chain reaction, 425, 821

**termolecular reaction** An elementary reaction that involves the simultaneous collision of three reactants, 417

terpene, 681

**tertiary** Used to designate an organic compound in which the substituent group occupies a site on a carbon atom that has no attached hydrogen atoms, or to derivatives of ammonia that have no hydrogen atoms attached to the nitrogen atom
  alcohol, 814
  amine, 819

tetrahedral complexes, 618
  crystal field splittings for, 625–26

tetrahedral site, 652

tetrahydrofuran, 816i

**theoretical yield** The amount of a product of a reaction predicted from stoichiometry assuming that the reaction proceeds without loss or side-reaction, 34

**thermal cracking** The breaking down of long-chain hydrocarbon molecules in petroleum through heat, 809

**thermal equilibrium** A macroscopic condition of a system characterized by the constancy of its pressure and temperature, 110

**thermal shock** The failure of a ceramic due to large, localized thermal expansion, 721–22

thermite reaction, 297i, 319

**thermochemistry** The measurement and prediction of heat effects in chemistry, 294–305

**thermodynamic efficiency** The ideal efficiency an engine would have if it could be operated in a purely reversible fashion, 312

**thermodynamic process** A process that leads to a change in the thermodynamic state of a system, 282
  irreversible, 282i, 283
  reversible, 283

**thermodynamic state** A macroscopic, time-independent condition of a system at thermal and mechanical equilibrium, characterized by a well-defined temperature and pressure, 282

thermometer, 96–98

thermonuclear reaction, 472

**thermosphere** The portion of the earth's atmosphere extending out from 80 km, 594

thin layer chromatography, 198t

thionyl fluoride, structure of, 790

thiosulfate ion, 270i, 767

**third law of thermodynamics** The statement that the entropy of any substance approaches zero at the absolute zero of temperature, 335–37

Thompson, Benjamin, 288

Thompson, William, 310

Thomson, J. J., 16, 19

Three Mile Island, 471

thymine, 842i

time-resolved spectroscopy, 588–89

tin
  alloys of, with copper, 703
  solid forms of, 136, 354, 659

titanium, 36
  production of, 400, 717

titanium dioxide, 78, 781

## Locations of Some Important Tables of Data

## A Note on the Names of Elements 104–111

The names from "unnilquadium" through "unununium" (and the symbols Unq through Uuu) are temporary, pending decisions on assigning credit for the discovery of these elements and final choices of names. A controversy has arisen between the groups involved in the discovery of several of these elements and a special committee of the International Union of Pure and Applied Chemistry (IUPAC), which was charged with deciding between rival claims for names. Below are listed the various names proposed by the different groups:

| Atomic Number | Temporary Symbol | Proposed Names (by Discovery Groups) | IUPAC Name | IUPAC Symbol |
|---|---|---|---|---|
| 104 | Unq | Rutherfordium (American) Kurchatovium (Russian) | Dubnium | Db |
| 105 | Unp | Hahnium (American) Nielsbohrium (Russian) | Joliotium | Jl |
| 106 | Unh | Seaborgium (American) | Rutherfordium | Rf |
| 107 | Uns | Nielsbohrium (German) | Bohrium | Bh |
| 108 | Uno | Hassium (German) | Hahnium | Hn |
| 109 | Une | Meitnerium (German) | Meitnerium | Mt |